DATE DUE

STEEL DESIGNERS' MANUAL

STEEL DESIGNERS' MANUAL

FOURTH EDITION

Prepared for the
CONSTRUCTIONAL STEEL RESEARCH
AND DEVELOPMENT ORGANISATION

A HALSTED PRESS BOOK

JOHN WILEY & SONS   NEW YORK

Granada Publishing Limited

First published in Great Britain 1955 by Crosby Lockwood & Son Ltd

Copyright © Fourth edition 1972 Constructional Steel Research and Development Organisation

All rights reserved. No part of this publication may be
reproduced, stored in a retrieval system, or transmitted,
in any form or by any means, electronic, mechanical,
photocopying, recording or otherwise, without the prior
permission of the publishers.

Library of Congress Cataloging in Publication Data
Main entry under title:
Steel designers' manual

"A Halsted Press book."
First and 2d editions by C. S. Gray and others.
Includes bibliographies and index.

   1. Building, Iron and steel—Handbooks, manuals, etc. 2. Steel, Structural—Handbooks, manuals, etc. I. Gray, Charles Stuart. Steel designers' manual. II. Constructional Steel Research and Development Organisation.

TA685.G67 1975   691'.7   75–19073
ISBN 0–470–16865–X

Printed in Great Britain

FOREWORD TO THE FOURTH EDITION

The Steel Designers' Manual was first published in 1955, at the instigation of the British Steel Producers' Conference and the British Iron & Steel Federation, in order to bridge the gap between the field of structural analysis and the practical application of steelwork in constructional engineering.

This new edition is in essence a metric version of the Third Edition, first published in 1966. The contents have been revised to take into account B.S. 4360: Part 2: 1969, Weldable structural steels, and B.S. 449: Part 2: 1969, The use of structural steel in building, both of which are in metric units. Where co-ordinated metric dimensions are not in operation, as in the case of steel sections, metric equivalents of the Imperial units have been employed. It should, however, be noted that as metric angles will be available with effect from January 1973, details are included of both the Imperial angles and the metric angles which will replace them.

It is hoped that this edition will be as acceptable to students and practising engineers in the metric environment as its predecessors have been in the Imperial system.

DIRECTOR
CONSTRADO

May, 1972

List of contributors of new chapters and those who have undertaken the revision of existing chapters in the Third and Fourth Editions.

B. O. ALLWOOD, F.I.STRUCT.E.
PROFESSOR SIR JOHN F. BAKER, O.B.E., F.R.S., M.A., SC.D., D.SC., F.I.C.E., F.I.STRUCT.E.
W. BATES, F.I.STRUCT.E.
J. H. CROSS, M.I.STRUCT.E.
E. P. GALLAGHER, M.ENG., M.I.C.E.
G. B. GODFREY, M.I.C.E., F.I.STRUCT.E.
C. S. GRAY, O.B.E., B.SC. (ENG.), M.I.C.E., M.I.STRUCT.E.
PROFESSOR J. HEYMAN, M.A., PH.D., F.I.C.E.
H. V. HILL, M.SC., M.I.C.E., M.I.STRUCT.E.
L. G. JOHNSON, M.A., F.I.STRUCT.E.
F. W. LAMBERT, M.I.C.E., M.I.STRUCT.E.
S. J. McMINN, M.ENG., PH.D., M.I.C.E., M.I.STRUCT.E.
W. BASIL SCOTT, F.I.STRUCT.E.
SIR FREDERICK S. SNOW, C.B.E., F.I.C.E., F.I.MECH.E., P P.I.STRUCT.E.
R. W. TURNER, M.I.C.E., M.I.STRUCT.E.
F. E. S. WEST, M.M., M.PHIL., M.I.C.E., M.I.STRUCT.E.
D. T. WILLIAMS, F.I.STRUCT.E.

CONTENTS

		PAGE
	FOREWORD TO THE FOURTH EDITION	v
	LIST OF CONTRIBUTORS	vi
	INTRODUCTION	ix
	THE WORK OF THE STEEL STRUCTURES RESEARCH COMMITTEE	x
1	BENDING AND AXIAL STRESSES	1
2	SIMPLY SUPPORTED BEAMS	17
3	FIXED, BUILT-IN OR ENCASTRE BEAMS	39
4	CONTINUOUS BEAMS	51
5	CANTILEVER AND SUSPENDED SPAN CONSTRUCTION	61
6	PROPPED CANTILEVERS	67
7	THE DEFLECTION OF BEAMS	81
8	THE DEFLECTION OF COMPOUND GIRDERS	103
9	BEAMS IN TORSION	107
10	FORCES IN PLANE FRAMES	119
11	THE DEFLECTION OF FRAMED STRUCTURES	135
12	INFLUENCE LINES	155
13	METHODS OF STRUCTURAL ANALYSIS	195
	(a) The Area-Moment Method	195
	(b) Moment Distribution	225
	(c) The Slope Deflection Method	276
14	SLOPES AND DEFLECTIONS IN RIGID FRAMES	289
15	FORMULAE FOR RIGID FRAMES	299
16	RIGID FRAME CHARTS	345
17	VIERENDEEL GIRDERS	413
18	FRAMING FOR SINGLE-STOREY SHEDS	429
19	SPACE FRAMES	441
20	DESIGN OF ANGLE STRUTS	487
21	ENGINEERING WORKSHOP DESIGN	497
22	PLASTIC THEORY AND DESIGN	531
23	COMPOSITE CONSTRUCTION	585

		PAGE
24	FOUNDATIONS	613
25	STEEL PILING	639
26	GIRDERS	645
27	MOMENTS OF INERTIA OF PLATE GIRDERS	681
28	CONNECTIONS	701
29	SURFACE PREPARATION OF STRUCTURAL STEELWORK	799
30	DESIGN OF MULTI-STOREY STANCHIONS	809
31	WIND ON MULTI-STOREY BUILDINGS	847
32	FLOORS	869
33	FLOOR PLATES	879
34	WELDING PRACTICE	883
35	STEEL SHEET ROOFING AND CLADDING	887
36	STEEL WINDOWS AND PATENT GLAZING	903
37	WALLS	911
38	USE OF COMPUTERS IN STRUCTURAL DESIGN	921
39	FIRE-RESISTING CONSTRUCTION	935
40	MISCELLANEOUS TABLES	973
	BRITISH STANDARDS	1077
	INDEX	1079

INTRODUCTION

The First Edition and the subsequently revised Second Edition of this Manual were prepared under the joint authorship of:

Charles S. Gray, O.B.E., B.Sc.(Eng.), M.I.C.E., M.I.Struct.E.

Lewis E. Kent, O.B.E., B.Sc.(Eng.), M.I.C.E., P.P.I.Struct.E.

W. E. Mitchell, PP.I.Struct.E.

G. Bernard Godfrey, M.I.C.E., F.I.Struct.E.

Several chapters by other contributors were also included.

This principle was extended in the Third Edition and twelve entirely new chapters were contributed by authors having special knowledge and experience of the particular subjects.

This Fourth Edition has been mainly based on the metrication of the Third Edition with the co-operation of the original authors where revision of content was required in the light of recent developments. The material has been largely prepared by a small working party consisting of W. Bates and W. Basil Scott under the chairmanship of D. T. Williams. A complete list of the names of the authors and of those responsible for the revision of existing chapters for both the Third and Fourth editions is to be found on page vi.

As in previous editions, the aim has been to present up-to-date practice in the design of steel-framed buildings in as concise a form as possible. Although not intended to be a textbook on structural theory, basic principles have been included where appropriate in order to provide a background for the worked examples. Some of the chapters dealing with more theoretical subjects already treated in numerous textbooks have been omitted to allow greater coverage of less commonly treated subjects.

A number of examples illustrate the use of more than one method of solution in order that their relative merits may be compared.

By kind permission of the British Standards Institution references are made to British Standards throughout the Manual and all material is in accordance with current Specifications and Codes of Practice at the time of going to print. It must be emphasised, however, that these are revised from time to time and it is important to ensure that the most recent publication is used. Copies of Standards may be obtained from the BSI Sales Department, Newton House, 101 Pentonville Road, London N1 9ND.

THE WORK OF THE STEEL STRUCTURES RESEARCH COMMITTEE

by
Sir John Baker
Formerly Technical Officer to the Committee

Most of the chapters in this Manual deal with elastic theory and the "elastic" design methods but Chapter 22 gives an outline of the plastic theory of structures, the study of which began 30 years ago, and of the plastic or "collapse" method of design which is coming more and more into prominence for certain types of structure in place of the long established elastic method.

These two methods of design, elastic and plastic, are fundamentally different but it would be surprising, in the practical world of engineering, if there had not been a link connecting them. This link was provided in the years from 1929 to 1936 when the Steel Structures Research Committee was attempting to produce for the British Steel Industry, an improved method of structural design.

All the methods of steelwork design now available are based on, or have been profoundly influenced by the work of this Committee. It is frequently alleged that whilst Britain is pre-eminent in science it is backward in the application of science. However, since the Committee began its work, not only has this country been in the forefront of steel structural research, but it has led the world in the derivation and the application of the most advanced methods of steelwork design and construction.

Following the realisation that building codes were based on too restrictive assumptions, the Steel Structures Research Committee embarked in 1929 on a comprehensive investigation of the application of modern theory to the design of all forms of steel structures although, in the event, its work was confined almost entirely to the multi-storey steel building frame. At the time the Committee began its work the construction of steel frames in this country was regulated generally by the London County Council (General Powers) Act, 1909, although some cities had their own regulations, a study of which showed curious discrepancies, particularly in regard to the live loads for which buildings of various usages had to be designed.

The "Code of Practice for the use of Structural Steel in Building" published in the Committee's First Report (1) in 1931 was accepted practically unchanged by the London County Council and became the first British Standard No. 449 and so controlled the design of steel structures in this country until the revision of the "simple method" of design as set out in the present edition.

While this Code of Practice was based mainly on the data in the hands of engineers at that time it incorporated the results of some new investigations. For instance a careful survey was made of the live loads in certain London office buildings (4) and Sir Thomas Stanton provided new data on wind loads. The most important investigation was into the strength of struts. As a result, the Perry formula was adopted, as it has been since for every British Standard dealing with steel compression members, with the values of the constants representing yield stress and imperfections recommended by Andrew Robertson. This formula gave the safe working load on a pin-ended strut. The Code allowed an effective length less than the storey height to be assumed for an intermediate length of a continuous stanchion. An attempt was made to deduce satisfactory values for effective length in any practical frame (5); later investigation of the behaviour of actual buildings showed that this concept of "effective length", though it stubbornly persists in building regulations, is really untenable.

The review of existing regulations left no doubts in the minds of the Committee that the method of designing steel-framed buildings then in common use, and still surviving as the "simple method" of B.S. 449, had no firm rational basis. No advance leading to economies could, therefore, be made until the real behaviour under load of the steel-framed building was

THE STEEL STRUCTURES RESEARCH COMMITTEE

understood. To this end the Committee embarked on a comprehensive research programme in which the first successful tests of actual buildings were made. The barrier to success in the past had, of course, been the lack of an instrument for measuring strains in steelwork accurate enough and yet robust enough to stand up to the rigorous conditions experienced on actual buildings in the field, or rather in the city street. The first move therefore was the design of such a strain-gauge (6). It was used with moderate success in the first tests on a full scale frame, the New Geological Museum, South Kensington, but it was too expensive for wide use and full success did not come until the discovery of the remarkable acoustic gauge, the Maihak of German design. With this accurate and robust instrument the Committee's experimental work went ahead with great speed. Exhaustive tests were made on a special two bay, three storey Experimental Frame constructed throughout of 8 in. x 4 in. x 18 lb. joists, on three representative buildings then under construction, the Cumberland Hotel, the Euston office block and a residential flats building, Latymer Court. On these last three, tests were made on the bare steelwork, after floors had been laid, stanchions cased and finally after walls and partitions had been constructed, so that the effect of cladding on the stresses in the steelwork could be determined. This work is fully described in the Committee's Second (2) and Final (3) Reports.

These tests showed quite clearly that the behaviour of an actual building was radically different from that assumed in the design methods then in common use. Though in every case the members were joined together by the usual bolted or riveted cleated connections, assumed in design to be hinges, the behaviour of each structure approximated closely to that of a rigidly jointed frame. Thus, when a vertical load was applied to a beam, appreciable restraining moments were developed at the ends, with corresponding bending moments in the stanchion lengths. In one of the Experimental Frames where the 8 in. x 4 in. I beams framed into the web of a stanchion, the connection consisting of a $3\frac{1}{2}$ in. x 3 in. x $\frac{5}{16}$ in. angle bracket 4 in. long with a similar top cleat, $\frac{1}{2}$ in. diameter black bolts being used throughout, the "equivalent eccentricity", that is the distance from the stanchion axis of the end of a similarly loaded simply supported beam which would have given rise to the same stanchion stresses, was as much as 10 in., five times what would have been assumed by the most conservative designer at that time. In the bare frame of the hotel building the equivalent eccentricity was as much as 44 in., in the office building 34 in. and in the flats building 13 in.; these eccentricities were increased, in general, in the finished clothed buildings. They clearly gave rise to large restraining moments at the ends of the beams which reduced the maximum stresses in the beams by 17 to 25 per cent but they also were responsible for large bending stresses in the stanchions, many times those assumed in design. Furthermore these bending stresses were appreciable not only in the stanchion lengths into which the beam framed but in those further removed thus making nonsense of the usual assumption of pin-ended struts and, as has already been mentioned, of the concept of effective length.

The problem facing the Committee was to deduce a practical design method for what was a highly redundant structure, by virtue of the near rigidity of the connections. The behaviour of the connections was clearly a critical factor and Dr. C. Batho, Professor of Civil Engineering at Birmingham University, concentrated on its elucidation. Accounts of his work on bolted and riveted connections of all kinds will be found in all three of the Committee's Reports. He was able to produce data on the restraining moments which could be relied upon from any practical cleated connection, so making economical beam design a simple matter.

These data covered not only the use of rivets, common in those days, but high tensile black bolts with controlled torque. In fact the Steel Structures Research Committee solved all the problems more than thirty years ago, designed and used two different forms of torque-control spanner and even drew up regulations for the substitution of high tensile steel bolts for rivets. Yet British industry made no use of this pioneer work for which it had paid, but waited twenty years until it had to pay again to the American licensors who had, in the meantime, patented the devices.

While the design of the beams was simple and produced economies, stanchion design was a much tougher proposition. However, for the first time the stress distribution in a continuous stanchion was understood and so the most rigorous conditions, which did not, of course, occur when all floors were loaded, could be picked out. A design method was derived (7) which was rational; it did not depend on unjustifiable assumptions of pin-ends or call for wild guesses of an "effective length". Inevitably it was more time-consuming in the design office than the "simple method" of B.S. 449; more serious, however, it did not lead to compensating economies because, in spite of the rigorous worst loading conditions inherent in the method, a timid Committee insisted in addition on an unnecessarily high load factor of 2. The result has been

that though the S.S.R.C. method is permitted under the clause "Semi-rigid design" in B.S. 449: 1969, it has been very little used. Disappointing though this is, the engineers responsible for the method of stanchion design are likely to have the compensation of seeing their work widely used after a lapse of thirty years since it forms the basis of the method of design advocated by the Joint Committee of The Institution of Structural Engineers and The Institute of Welding in their Report "Fully rigid-multi-storey welded steel frames" (December 1964).

It is hoped that this brief account has drawn attention to the wealth of information, the only information available, of the true behaviour and stress distribution in multi-storey steel framed buildings, contained in the three Reports of the Steel Structures Research Committee. Every ambitious steelwork designer should study the Reports. They are probably to be found in many works' libraries and in some university and public libraries but they are not otherwise widely available. This is because early in World War II Her Majesty's Stationery Office store containing the only supply of the Reports was bombed and destroyed. For this reason a summary of the Committee's work was published in 1954 (8) and is still in print.

Finally it can be said that the Committee's work had possibly its greatest influence in that out of it grew the plastic method of design. Whatever can be claimed for the Committee's final Recommendations for Design there is no doubt that the elastic behaviour of redundant structures, on which they depended, is too complicated to form the basis of a really comprehensive and satisfactory method of design. This was very clear to those who, for seven years, had carried out the investigations for the Committee. So when these were completed in 1936, the investigators turned their attention to the behaviour of structures carried out of the the elastic into the plastic range and so to collapse. Out of this grew the successful plastic method of design described in Chapter 22.

REFERENCES

1. First Report of the Steel Structures Research Committee. H.M.S.O. (1931).
2. Second Report of the Steel Structures Research Committee. H.M.S.O. (1934).
3. Final Report of the Steel Structures Research Committee. H.M.S.O. (1936).
4. WHITE, C. M. *Survey of Live Loads*, First Report S.S.R.C. H.M.S.O. (1931).
5. BAKER, J. F. *A Note on the Effective Length of a Pillar*, Second Report S.S.R.C. H.M.S.O. (1934).
6. BAKER, J. F. *Examination of Building in Course of Erection*, First Report S.S.R.C. H.M.S.O. (1931).
7. BAKER, J. F. & LEADER WILLIAMS, E. *The Design of Stanchions in Building Frames*, Final Report S.S.R.C. H.M.S.O. (1936).
8. BAKER, J. F. *The Steel Skeleton*, Vol. 1. Cambridge U.P. (1954).

STEEL DESIGNERS' MANUAL

1. BENDING AND AXIAL STRESSES

Ordinary Beam Theory

THE theory involved in the derivation of the formula for the moment of resistance (M.R.) of a beam section is based on the following assumptions:

(i) The beam section must have one axis of symmetry, say, a vertical axis, in the plane of which a bending moment is applied.
(ii) Simple or circular bending is assumed. This is the type produced by equal and opposite pure couples and in which shear is absent, as depicted by the portion AB of the beam shown in Fig. 1.
(iii) Sections of the beam which are plane before bending remain plane after bending.
(iv) The stress in any fibre is proportional to its strain.
(v) Young's Modulus of Elasticity (E) is constant.

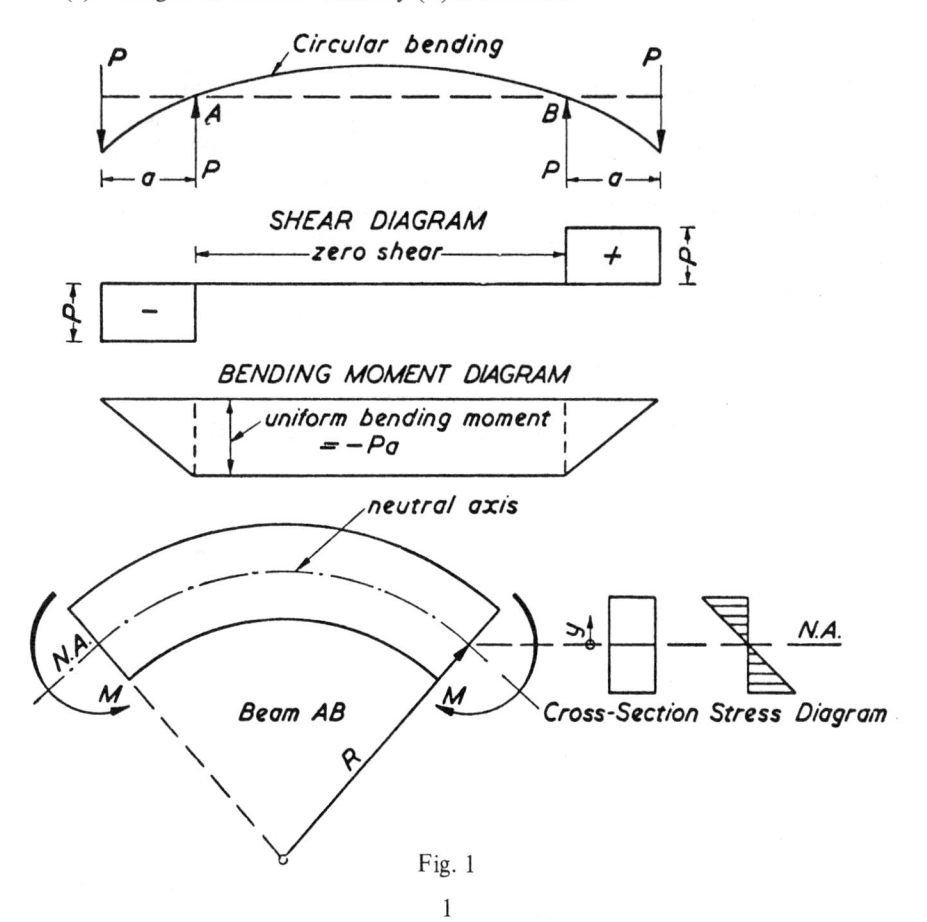

Fig. 1

BENDING AND AXIAL STRESSES

The results of the beam theory can be expressed as follows:

$$\frac{M}{I} = \frac{f}{y} = \frac{E}{R}$$

where M = the bending moment,
I = the moment of inertia,
f = the stress at any point,
y = the vertical distance from the neutral axis to the point under consideration
and R = the radius of curvature of the beam.

In using this expression f will normally represent a maximum stress, so that y, in that case, will be the distance from the neutral axis to an extreme fibre, top or bottom of the section as the case may be.

The expression

$$\frac{M}{I} = \frac{f}{y}$$

can be written

$$M = f\frac{I}{y}$$

But $\frac{I}{y} = Z$, the modulus of section. Hence:

$$M = fZ \quad \ldots \ldots \ldots \ldots \ldots \ldots \quad (1)$$

The foregoing remarks apply, in theory, only to beams subjected to circular bending. When shear is present there are shear strains, but as such strains are very small compared with those due to bending, they are normally neglected in practice. Consequently, the bending stresses in any beam with one axis of symmetry and subjected to bending only may be found from the standard formula, $M = fZ$.

If, however, a member is considered which is not symmetrical about its xx axis, say, a 'T'-bar, then two values of Z will exist and the maximum compressive stress will be different from the maximum tensile stress.

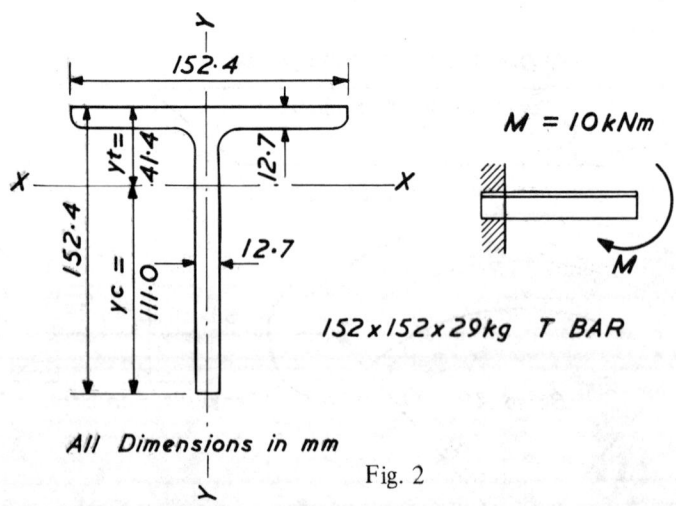

Fig. 2

Example 1. The 'T'-bar shown in Fig. 2 is subjected to a B.M. of 10 kNm. Calculate the maximum compressive and tensile stresses.

Z_t, the section modulus in terms of tension,

$$= \frac{I_{xx}}{y_t} = \frac{792.5}{4.14} = 191.4 \text{ cm}^3$$

Z_c, the section modulus in terms of compression

$$= \frac{I_{xx}}{y_c} = \frac{792.5}{11.10} = 71.4 \text{ cm}^3$$

Now $f = M/Z$.
Therefore, the maximum tensile stress,

$$f_t = \frac{M}{Z_t} = \frac{10 \text{ kNm}}{191.4 \text{ cm}^3} = 52.2 \text{ N/mm}^2$$

while the maximum compressive stress,

$$f_c = \frac{M}{Z_c} = \frac{10 \text{ kNm}}{71.4 \text{ cm}^3} = 14.0 \text{ N/mm}^2$$

Unsymmetrical Bending

Fig. 3 shows the section of a rolled-steel beam which is subjected to an oblique force P acting through the C.G. of the beam.

The resulting B.M. M may be resolved into two components along the *principal axes xx* and *yy*, so that

$$\left. \begin{array}{l} M_{xx} = M \cdot \sin \theta \\ M_{yy} = M \cdot \cos \theta \end{array} \right\} \quad \ldots \ldots (2)$$

Consequently, the stress at any fibre in the beam may be found from the expression

$$f = M \left(\pm \frac{y}{I_{xx}} \sin \theta \pm \frac{x}{I_{yy}} \cos \theta \right) \ldots (3)$$

where the positive values represent compressive stresses and the negative, tensile. In another form, the expression becomes

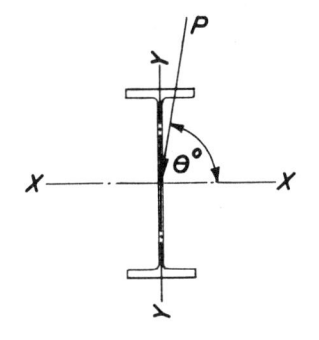

Fig. 3

$$f = \pm \frac{M_{xx}}{Z_{xx}} \pm \frac{M_{yy}}{Z_{yy}} \quad \ldots \ldots (4)$$

The same expressions may be employed for other symmetrical sections provided that the force is assumed to act at the C.G. of the section.

It should be noted that for most universal beams the resultant stress f increases rapidly as the angle θ (Fig. 3) departs from 90°. However, this effect is not so marked with universal columns.

Z-polygons

In the expression

$$f = M\left(\frac{y}{I_{xx}} \sin\theta + \frac{x}{I_{yy}} \cos\theta\right),$$

the section modulus

$$Z = \frac{I_{xx} \cdot I_{yy}}{I_{yy} \cdot y \cdot \sin\theta + I_{xx} \cdot x \cdot \cos\theta}$$

Hence, for any given direction of loading and for any given point in a section Z is a constant. Thus the value Z is a measure of the strength of a section for bending in any direction.

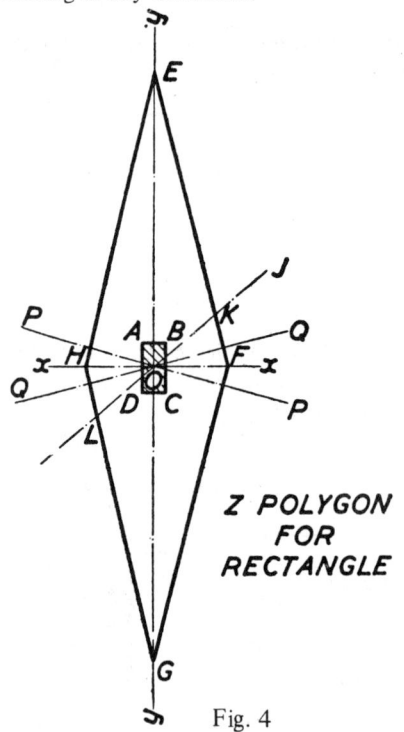

Equation (5) is a 'straight line' equation, so that the variation in Z for the critical points in any section may be easily represented on a figure known as a Z-polygon.

Consider Fig. 4, which shows a rectangle $ABCD$ with its xx and yy axes produced in both directions.

With O, the centroid of the section, as origin, set off to some scale OE and OG equal to the maximum value of Z_{xx}, i.e. with y relative to AB or CD. Similarly, set off OF and OH equal to the maximum value of Z_{yy}, i.e. with x relative to BC or DA. Then EF is the Z-line for the point B, FG the Z-line for C, and so on, while the full figure $EFGH$ is known as the Z-polygon for the rectangle.

Now, if a moment M is applied to the rectangle in the plane JO, then KO, which is intercepted by the Z-polygon, is the value of Z appropriate to the particular plane of the moment. If this value = Z_{KO}, then the maximum stress in the section may be found from the expression.

$$f = \frac{M}{Z_{KO}}.$$

Fig. 4 — Z POLYGON FOR RECTANGLE

It will be appreciated that the value of the intercept LO could equally well be used, as LO is equal to KO.

The Z-polygon provides several sources of information. For example, it is obvious that the plane of maximum strength is along the axis yy. There are two planes of minimum strength, i.e. PP and QQ, along which the distance from the origin to the polygon is least.

Fig. 5 shows Z-polygons for a 305 mm x 127 mm x 48 kg universal beam, and a 305 mm x 305 mm x 97 kg universal column.

BENDING AND AXIAL STRESSES

Z POLYGONS FOR UNIVERSAL SECTIONS

Fig. 5

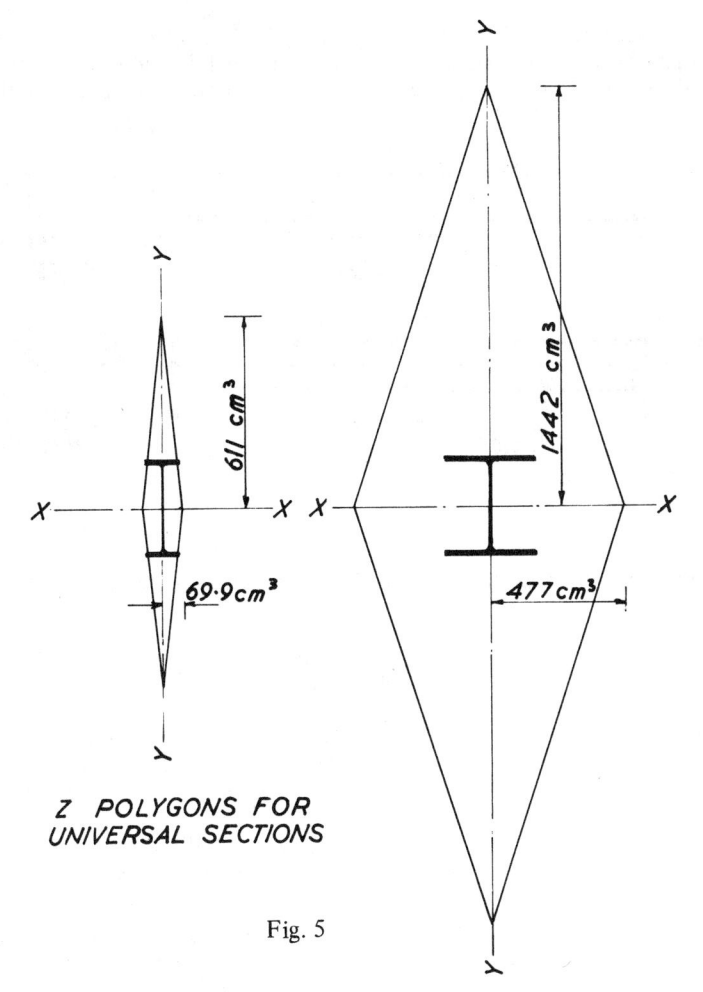

Pure Tension or Compression

When a tensile force P is applied axially to a member the resultant stress in any fibre is uniform and is given by the expression

$$f = -\frac{P}{A} \quad \quad \quad \quad \quad (6A)$$

where A = the cross-sectional area of the member.

Similarly, when a compressive force is applied the resultant stress in any fibre,

$$f = +\frac{P}{A} \quad \quad \quad \quad \quad (6B)$$

Bending and Axial Stresses

When a member is subjected to bending and axial forces, as shown in Fig. 6, the resulting maximum fibre stresses

$$f_{max} = \pm\frac{P}{A} \pm \frac{M_{xx}}{Z_{xx}} \quad \quad \quad \quad \quad (7)$$

A similar effect may be obtained in an eccentrically loaded column. Consider Fig. 7, which shows the cross-section of a column which is symmetrical about both the xx and yy axes. If a compressive force P is applied on the yy axis at a distance

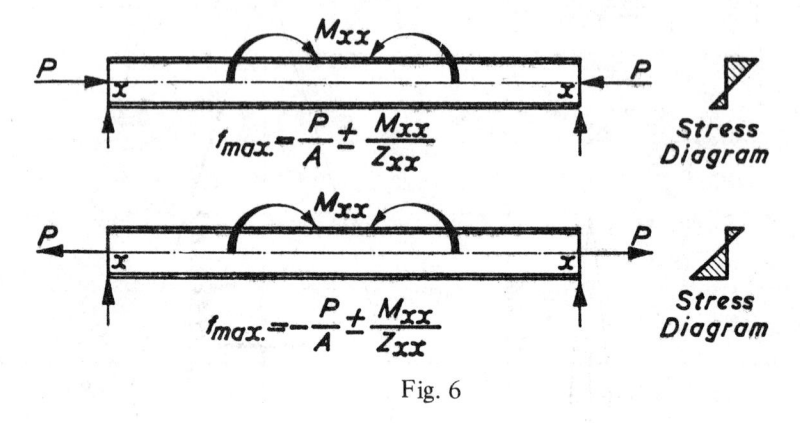

Fig. 6

$\pm e_x$ from the xx axis then the resultant fibre stress at any distance y from the xx axis

$$f = +\frac{P}{A} \pm \frac{P \cdot e_x \times y}{I_{xx}} \quad \ldots \ldots \ldots \ldots (8)$$

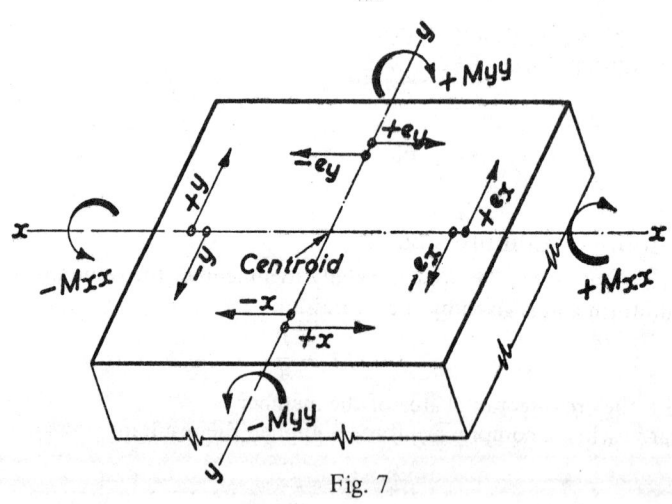

$$= +\frac{P}{A} \pm \frac{M_{xx} \cdot y}{I_{xx}}$$

Fig. 7

If the force P is applied on the xx axis at a distance $\pm e_y$ from the yy axis then the resultant stress,

$$f = \frac{P}{A} \pm \frac{M_{yy} \cdot x}{I_{yy}}$$

GENERAL EXPRESSION FOR STRESS

When the force may be compressive or tensile and eccentric about both axes, then the fibre stress at any point

$$f = \pm \frac{P}{A} \pm \frac{M_{xx} \cdot y}{I_{xx}} \pm \frac{M_{yy} \cdot x}{I_{yy}} \quad \ldots \ldots \ldots \ldots (9)$$

When y and x are maxima, equation (9) becomes

$$f_{max} = \pm \frac{P}{A} \pm \frac{M_{xx}}{Z_{xx}} \pm \frac{M_{yy}}{Z_{yy}} \quad \ldots \ldots \ldots \ldots (10)$$

Having given the formulae for a symmetrical section, it is convenient to consider the effect of any type of force acting eccentrically on any cross-section.

General Expression for Stress

Consider the cross-section of a compression member shown in Fig. 8, xx and yy being any rectangular axes through the centroid.

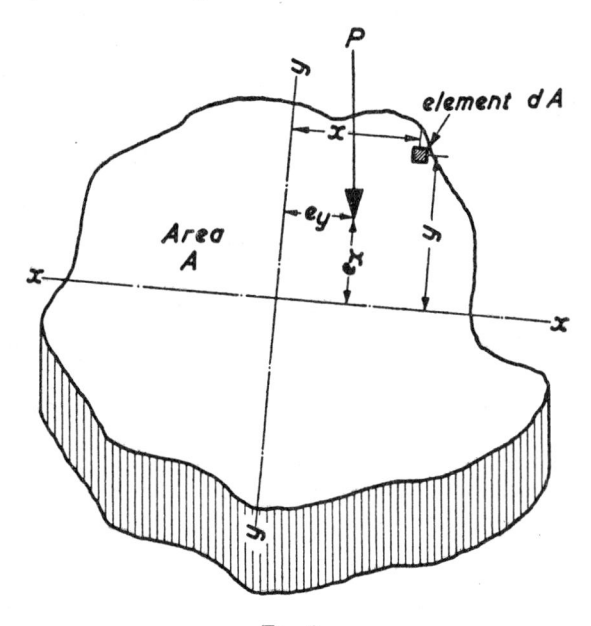

Fig. 8

Suppose that a force P is applied longitudinally to the member, its eccentricity about the xx and yy axes being e_x and e_y respectively. Then the effect upon the member is the same as if P is applied at the centroid and moments M_{xx} and M_{yy}, equal to $P \cdot e_x$ and $P \cdot e_y$ respectively, are applied about the xx and yy axes respectively.

If it is assumed that the member is not stressed beyond the elastic limit, then the unit stress f at any point $= a + bx + cy$, where x and y are the co-ordinates of the point.

Consider an element of area dA, having a total stress of $f \cdot dA$.

Then $\qquad P = \int f \cdot dA = a\int dA + b\int x \cdot dA + c\int y \cdot dA$

Since xx and yy pass through the centroid

$$\int x \, . \, dA = 0 \text{ and } \int y \, . \, dA = 0.$$

Hence
$$P = a\int dA = aA$$

or
$$a = \frac{P}{A} \quad \ldots \ldots \ldots \ldots \ldots \ldots \ldots \text{(a)}$$

Similarly,

$$M_{xx} = \int f \, . \, dA \, . \, y = a\int y \, . \, dA + b\int xy \, . \, dA + c\int y^2 \, . \, dA$$
$$= bI_{xy} + cI_{xx} \quad \ldots \ldots \ldots \ldots \ldots \ldots \text{(b)}$$
$$M_{yy} = \int f \, . \, dA \, . \, x = a\int x \, . \, dA + b\int x^2 \, . \, dA + c\int xy \, . \, dA$$
$$= cI_{xy} + bI_{yy} \quad \ldots \ldots \ldots \ldots \ldots \ldots \text{(c)}$$

Now,
$$M_{yy}I_{xx} = bI_{xx}I_{yy} + cI_{xx}I_{xy} \quad \ldots \ldots \ldots (c \times I_{xx})$$
$$M_{xx}I_{xy} = bI_{xy}^2 + cI_{xx}I_{xy} \quad \ldots \ldots \ldots (b \times I_{xy})$$

Subtracting,
$$M_{yy}I_{xx} - M_{xx}I_{xy} = b(I_{xx}I_{yy} - I_{xy}^2)$$

whence
$$b = \frac{M_{yy}I_{xx} - M_{xx}I_{xy}}{I_{xx}I_{yy} - I_{xy}^2} \quad \ldots \ldots \ldots \ldots \text{(d)}$$

Similarly;
$$c = \frac{M_{xx}I_{yy} - M_{yy}I_{xy}}{I_{xx}I_{yy} - I_{xy}^2} \quad \ldots \ldots \ldots \ldots \text{(e)}$$

Employing equations (a), (d) and (e),
$$f = a + bx + cy$$
$$= \frac{P}{A} + \frac{M_{yy}I_{xx} - M_{xx}I_{xy}}{I_{xx}I_{yy} - I_{xy}^2} \, . \, x + \frac{M_{xx}I_{yy} - M_{yy}I_{xy}}{I_{xx}I_{yy} - I_{xy}^2} \, . \, y \quad \ldots \ldots \text{(11A)}$$

If the numerator and denominator of the coefficient of x are divided by I_{xx} and those for y are divided by I_{yy}, the expression becomes

$$f = \frac{P}{A} + \frac{M_{yy} - M_{xx}(I_{xy}/I_{xx})}{I_{yy} - (I_{xy}^2/I_{xx})} \, . \, x + \frac{M_{xx} - M_{yy}(I_{xy}/I_{yy})}{I_{xx} - (I_{xy}^2/I_{yy})} \, . \, y \quad \ldots \text{(11B)}$$

Either formula may be used but the latter avoids the differences of large quantities which in slide-rule calculations may lead to appreciable errors.

Bending about Principal Axes

It will be noted that equations (11A) and (11B) are related to any pair of rectangular axes xx and yy which are chosen for convenience of calculation. Some designers prefer to work from principal axes, especially if the values of I are known for these axes, but normally there is very little to choose between the amount of calculation for each method.

Suppose that uu and vv are the appropriate principal axes, then equation (9) becomes

$$f = \pm \frac{P}{A} \pm \frac{M_{uu} \, . \, v}{I_{uu}} \pm \frac{M_{vv} \, . \, u}{I_{vv}} \quad \ldots \ldots \ldots \ldots \text{(12)}$$

where v and u are measured normal to the uu and vv axes respectively.

BENDING ABOUT PRINCIPAL AXES 9

Example 2. The 229 mm × 102 mm × 22.1 mm angle shown in Fig. 9 is subjected to a B.M. of 25 kNm in the plane of the *yy* axis and it is free to deflect both

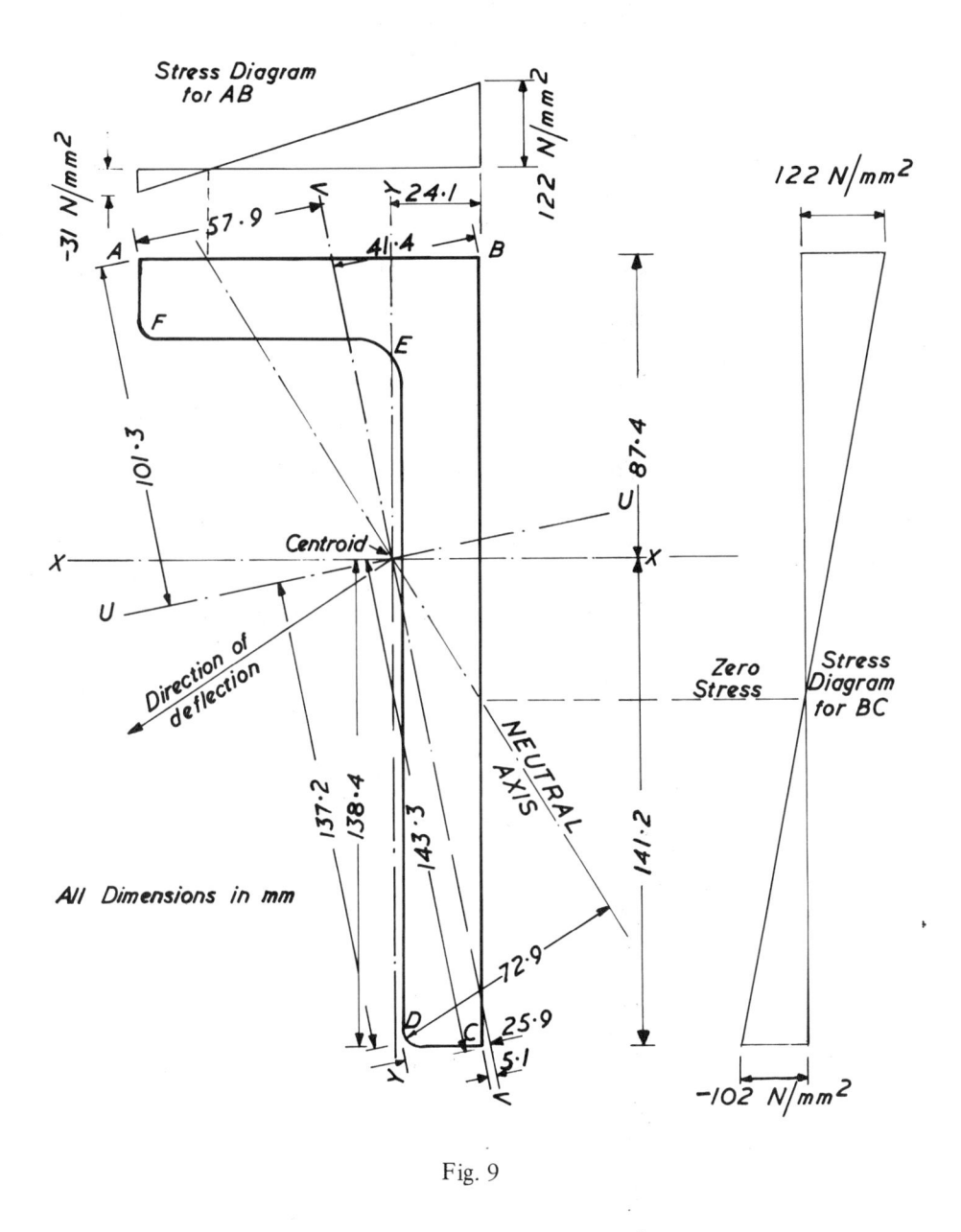

Fig. 9

downwards and laterally although the legs are restrained so that they remain parallel to their original position. Calculate the maximum stress.

BENDING AND AXIAL STRESSES

The properties of the section given in B.S.4: Part 1 : 1970 are as follows:

$$I_{xx} = 3\,606 \text{ cm}^4 \qquad I_{yy} = 447 \text{ cm}^4$$
$$I_{uu} = 3\,747 \text{ cm}^4 \qquad I_{vv} = 306 \text{ cm}^4$$
$$c_x = 8.73 \text{ cm} \qquad c_y = 2.41 \text{ cm}$$
$$\text{Tan } \theta = 0.21, \quad \text{i.e. } \theta = 11°52'$$

Consequently,
$$\sin \theta = 0.2056$$
$$\cos \theta = 0.9786$$
$$M_{uu} = M_{xx} \cos 11°\, 52' = 24.47 \text{ kNm}$$
$$M_{vv} = M_{xx} \sin 11°\, 52' = 5.14 \text{ kNm}$$

The principal problem is to find the point of maximum stress. Now the stress in any fibre of a member subjected to bending is directly proportional to the distance of that fibre from the neutral axis. Therefore, if the neutral axis is located the most highly stressed fibre is that most remote from it.

Consider point A, the co-ordinates of which, with respect to the uu and vv axes, are v and u respectively.

$$f_A = \frac{M_{uu} \cdot v}{I_{uu}} + \frac{M_{vv} \cdot u}{I_{vv}}$$

$$= \frac{24.47 \text{ kNm} \times 101.3 \text{ mm}}{3\,747 \text{ cm}^4} - \frac{5.14 \text{ kNm} \times 57.9 \text{ mm}}{306 \text{ cm}^4}$$

$$= -31.12 \text{ N/mm}^2$$

Considering point B,

$$f_B = \frac{24.47 \text{ kNm} \times 80.8 \text{ mm}}{3\,747 \text{ cm}^4} + \frac{5.14 \text{ kNm} \times 41.4 \text{ mm}}{306 \text{ cm}^4}$$

$$= +122.31 \text{ N/mm}^2$$

Considering point C,

$$f_C = -\frac{24.47 \text{ kNm} \times 143.3 \text{ mm}}{3\,747 \text{ cm}^4} - \frac{5.14 \text{ kNm} \times 5.1 \text{ mm}}{306 \text{ cm}^4}$$

$$= -102.11 \text{ N/mm}^2$$

Now there is a constant variation of stress from A to B and from B to C. Consequently, the neutral axis is located by joining the points of zero stress in AB and BC. It is evident that the point most remote from the neutral axis and therefore most highly stressed, is D.

$$f_D = -\frac{24.47 \text{ kNm} \times 137.2 \text{ mm}}{3\,747 \text{ cm}^4} - \frac{5.14 \text{ kNm} \times 25.9 \text{ mm}}{306 \text{ cm}^4}$$

$$= -133.10 \text{ N/mm}^2$$

i.e. the stress is tensile.

LOCATION OF NEUTRAL AXIS

It is worth noting that although the plane of the applied B.M. is in the YY axis, the plane of bending is at right angles to the neutral axis. Further, it will be observed that had the neutral axis been assumed to be located at right angles to the plane of bending, as in the case of symmetrical sections, the resulting stresses at A, B and C would have been incorrect numerically and, in addition, the stress at A would have been given as compressive whereas it is actually tensile.

Nevertheless, it must be repeated that the stresses have been computed on the assumption that the section is free to deflect in any manner but not to twist. If, however, lateral deflection is prevented, owing to restraint offered by the load or any other medium, and vertical deflection only is permitted, then the stresses will be correctly given by using the applied moment in conjunction with the appropriate section moduli for the XX axis.

Example 3. As an alternative solution the angle section in Example 2 will be analysed using equation (11B).

First calculate the value of I_{xy}.

$$\tan 2\theta = -\frac{2I_{xy}}{I_{xx} - I_{yy}}$$

$$= \frac{2I_{xy}}{I_{yy} - I_{xx}}.$$

Therefore
$$2I_{xy} = \tan 2\theta (I_{yy} - I_{xx})$$
$$= +0.439\,7(447 - 3\,606) \text{ cm}^4$$

and
$$I_{xy} = -694.5 \text{ cm}^4$$

M acts in the plane of the yy axis.

Hence $M_{xx} = 25$ kNm. $M_{yy} = 0$.

For point D, the co-ordinates are: $x = 2.5$ mm, $y = -138.4$ mm
Therefore,

$$f = -\frac{M_{xx}(I_{xy}/I_{xx})}{I_{yy} - (I_{xy}^2/I_{xx})} \cdot x + \frac{M_{xx}}{I_{xx} - (I_{xy}^2/I_{yy})} \cdot y$$

$$= \frac{-25 \text{ kNm} \times \left(\dfrac{-694.5 \text{ cm}^4}{3\,606 \text{ cm}^4}\right) \times 2.5 \text{ mm}}{447 \text{ cm}^4 - \left[\dfrac{(-694.5 \text{ cm}^4)^2}{3\,606 \text{ cm}^4}\right]} + \frac{25 \text{ kNm} \times -138.4 \text{ mm}}{3\,606 \text{ cm}^4 - \left[\dfrac{(-694.5 \text{ cm}^4)^2}{447 \text{ cm}^4}\right]}$$

$$= +3.84 \text{ N/mm}^2 - 136.9 \text{ N/mm}^2$$

$$= -133.06 \text{ N/mm}^2$$

which agrees very closely with the previous result.

Location of Neutral Axis

In Example 2 a method of locating the neutral axis was given. Now the neutral axis is the line running through points of zero stress. In sections subjected to bending only, i.e. with no longitudinal load P, one point of zero stress is the

BENDING AND AXIAL STRESSES

centroid of the section. If equation (11A) or (11B) is equated to zero, omitting the term P/A, the ratio y/x is obtained which gives the tangent of the angle ϕ which the neutral axis makes with the xx axis of the section.

Hence, if equation (11A) is used,

$$\tan \phi = -\frac{M_{yy}I_{xx} - M_{xx}I_{xy}}{M_{xx}I_{yy} - M_{yy}I_{xy}} \qquad (13)$$

Now the stress in any fibre of a member subjected to bending is directly proportional to the distance of that fibre from the neutral axis. Therefore, provided a member is subjected to bending only, the stress at any fibre,

$$f = \frac{M_{NA}}{I_{NA}} \cdot y_{NA} \qquad (14)$$

where M and I have the usual significance and y_{NA} is the lever arm about the neutral axis NA.

When, in addition to bending, a member is subjected to an axial force P, the stress at any fibre, including the centroid, due to $P = P/A$. Hence, the neutral axis shifts parallel to the position it occupies when there is only bending, the perpendicular distance travelled being

$$\frac{P}{A} \times \frac{I_{NA}}{M_{NA}}.$$

The neutral axis may also be located with reference to the principal axis uu by employing equation (12) without the term P/A. If the angle between the neutral and principal axes is θ, then

$$\tan \theta = \frac{M_{vv} \cdot I_{uu}}{M_{uu} \cdot I_{vv}}.$$

Example 4. Calculate the maximum stress in the angle section of Examples 2 and 3 using equations (13) and (14). See Fig. 9.

Now
$$\tan \phi = -\frac{M_{yy}I_{xx} - M_{xx}I_{xy}}{M_{xx}I_{yy} - M_{yy}I_{xy}}$$

$$= -\frac{0 - (25 \text{ kNm} \times -694.5 \text{ cm}^4)}{(25 \text{ kNm} \times 447 \text{ cm}^4) - 0}$$

$$= -\frac{694.5}{447} = -1.554$$

Therefore
$$\phi = -57° \ 12'$$

$$\sin \phi = -0.840 \ 6$$

$$\sin 2\phi = -0.910 \ 7$$

$$\cos \phi = +0.541 \ 7$$

and
$$M_{NA} = M \cdot \cos \phi$$

$$= 25 \text{ kNm} \times 0.541 \ 7$$

$$= 13.54 \text{ kNm}$$

y_{NA} for this point is found to be -72.9 mm, either by calculations or scaling from Fig. 9.

It may be shown that

$I_{NA} = \cos^2\phi I_{xx} + \sin^2\phi I_{yy} - \sin 2\phi I_{xy}$

$= (0.541\ 7)^2 \times 3\ 606\ cm^4 + (-0.840\ 6)^2 \times 447\ cm^4 + 0.910\ 7 \times (-694.5\ cm^4)$

$= 1\ 058 + 316 - 633$

$= 741\ cm^4$

Then $\quad f = \dfrac{M_{NA}}{I_{NA}} \cdot y_{NA} = \dfrac{13.54\ kNm \times -72.9\ mm}{741\ cm^4}$

$\qquad\qquad\qquad\quad = -133.2\ N/mm^2$

The Circle of Inertia

Unsymmetrical sections subjected to bending may be analysed graphically by using a construction known as the Circle of Inertia. The method of analysis is best described by taking a practical example.

Example 5. The member shown in Fig. 10, which consists of two 229 mm × 89 mm × 33 kg B.S. Channels, is subjected to a B.M. of 23 kNm in the plane of the YY axis and 34 kNm in the plane of the XX axis.

The properties of each channel section, given in B.S. 4:Part 1, are as follows:

$\qquad I_{xx} = 3\ 387\ cm^4 \qquad I_{yy} = 285\ cm^4$

$\qquad c_x = 0 \qquad\qquad\quad c_y = 2.53\ cm$

Cross-sectional area $A = 41.73\ cm^2$

The centroid of the cross-section of the member could be calculated in the usual way, but, obviously, it lies midway between the centroids of the individual channel sections.

Therefore, for the whole member, mentioning the left-hand channel first:

$I_{XX} = 3\ 387\ cm^4 + (41.73\ cm^2 \times 4.46\ cm \times 4.46\ cm) +$

$\qquad\qquad\qquad + 285\ cm^4 + (41.73\ cm^2 \times 4.46\ cm \times 4.46\ cm)$

$\quad\ = 5\ 332\ cm^4$

$I_{YY} = 285 + (41.73\ cm^2 \times 6.99\ cm \times 6.99\ cm) + 3\ 387 +$

$\quad\ = 7\ 708\ cm^4 \qquad\qquad\qquad + (41.73\ cm^2 \times 6.99\ cm \times 6.99\ cm)$

$I_{XY} = (A \times -x \times -y) + (A \times x \times y)$

$\quad\ = 2 \times 41.73\ cm^2 \times 6.99\ cm \times 4.46\ cm$

$\quad\ = 2\ 602\ cm^4$

Note that a symmetrical section such as a channel has no value I_{xy} about its own centroid, as the xx and yy axes are principal axes.

Now the Circle of Inertia may be drawn.

From the centroid of the member, O, set off vertically, i.e. along the YY axis, Oa and ab to some scale to represent I_{XX} and I_{YY} respectively.

14   BENDING AND AXIAL STRESSES

Let Ob be a diameter of the circle of inertia, c being the centre. Then any diameter can represent any pair of rectangular axes in the plane of the cross-section of the member. From a, and scaling horizontally, set off ad equal to I_{XY}, scaling to the right as the product of inertia is positive.

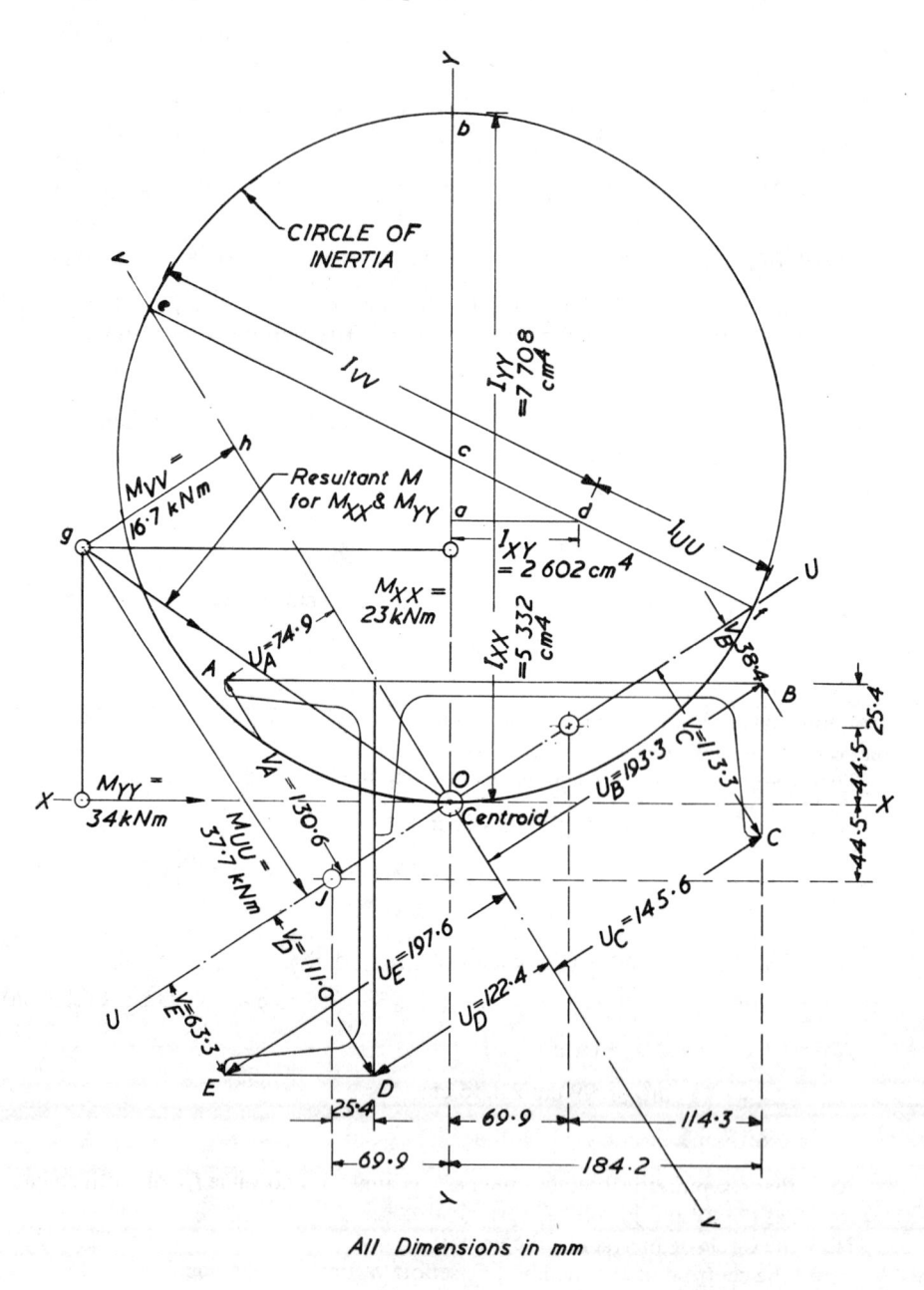

Fig. 10

Draw the diameter *ecdf* through *d*. Then *fd* and *de* represent the values of the moments of inertia I_{UU} and I_{VV} about the principal axes. When scaled I_{uu} = 3 660 cm⁴ and I_{vv} = 9 370 cm⁴.

Draw lines through *f* and *O* and through *e* and *O*. Then these lines are respectively the principal axes *UU* and *VV*.

Now set off the values of M_{XX} and M_{YY} to some scale on the *YY* and *XX* axes respectively. Then *gO* is the resultant of these moments and also the resultant of M_{vv} and M_{uu} which are represented by *gh* and *gj* respectively.

Therefore,
$$M_{VV} = 16.7 \text{ kNm}$$
and
$$M_{UU} = 37.7 \text{ kNm}$$

When the values of *v* and *u* have been scaled for each corner of the member, the fibre stresses may be calculated.

Now
$$f = \pm \frac{M_{UU} \cdot v}{I_{UU}} \pm \frac{M_{VV} \cdot u}{I_{VV}}$$

Hence,
$$f_A = + \frac{37.7 \text{ kNm} \times 130.6 \text{ mm}}{3\,660 \text{ cm}^4} + \frac{16.7 \text{ kNm} \times 74.9 \text{ mm}}{9\,370 \text{ cm}^4}$$
$$= + 149.9 \text{ N/mm}^2$$

$$f_B = - \frac{37.7 \times 38.4}{3\,660} + \frac{16.7 \times 193.3}{9\,370}$$
$$= - 74.0 \text{ N/mm}^2$$

$$f_C = - \frac{37.7 \times 113.3}{3\,660} - \frac{16.7 \times 145.6}{9\,370}$$
$$= - 142.7 \text{ N/mm}^2$$

$$f_D = - \frac{37.7 \times 111.0}{3\,660} + \frac{16.7 \times 122.4}{9\,370}$$
$$= - 92.6 \text{ N/mm}^2$$

$$f_E = - \frac{37.7 \times 63.3}{3\,660} + \frac{16.7 \times 197.6}{9\,370}$$
$$= - 30.0 \text{ N/mm}^2$$

Deflection

The value and direction of the deflection of members with unsymmetrical sections may be obtained by calculating the resultant of the deflections about the principal axes. As an alternative, a direct calculation, normal to the neutral axis, may be made when the neutral axis has been located. However, the position of the neutral axis varies with the plane of loading whereas the principal axes are functions of the section, and therefore are constant for any loading.

BENDING AND AXIAL STRESSES

Example 6. Calculate the deflection in the 229 mm x 102 mm x 22.1 mm angle employed for Examples 2 to 4 assuming that the moment M_{xx} of 25 kNm is caused by a U.D.L. of 80 kN over a span of 2.5 m.

Now for a U.D.L.

$$d = \frac{5WL^3}{384EI} = \frac{WL}{8} \cdot \frac{5L^2}{48EI}$$

$$= M \cdot \frac{5L^2}{48EI}$$

Assume that $E = 2.1 \times 10^5$ N/mm²

Considering the *uu* axis,

$$d_{uu} = M_{uu} \left(\frac{5 \times (2.5 \text{ m})^2}{48 \times 2.1 \times 10^5 \text{ N/mm}^2 \times I_{uu}} \right)$$

$$= 24.47 \text{ kNm} \left(\frac{5 \times (2.5 \text{ m})^2}{48 \times 2.1 \times 10^5 \text{ N/mm}^2 \times 3\,747 \text{ cm}^4} \right)$$

$$= 2.02 \text{ mm}$$

Considering the *vv* axis,

$$d_{vv} = M_{vv} \left(\frac{5 \times (2.5 \text{ m})^2}{48 \times 2.1 \times 10^5 \text{ N/mm}^2 \times I_{vv}} \right)$$

$$= 5.14 \text{ kNm} \left(\frac{5 \times (2.5 \text{ m})^2}{48 \times 2.1 \times 10^5 \text{ N/mm}^2 \times 306 \text{ cm}^4} \right)$$

$$= 5.19 \text{ mm}$$

The resultant deflection,

$$d_{NA} = \sqrt{(d_{uu}^2 + d_{vv}^2)}$$

$$= 5.57 \text{ mm}$$

2. SIMPLY SUPPORTED BEAMS

ALTHOUGH it is assumed that every reader of this book has an elementary knowledge of the theory of structures, it is advisable to consider the calculations required in obtaining the shear forces (S.F.s) and bending moments (B.M.s) in simply supported beams, not so much for their own sake but because these quantities, obtainable by simple statics, form the basis of many other calculations required for the analysis of built-in beams, continuous beams and other indeterminate structures.

It should be noted that appropriate formulae for simple beams and cantilevers under various types of loads are presented in tabular form on pp. 29–38.

In the case of simple beams, it is necessary to calculate the support reactions before the bending moments can be evaluated. This procedure is reversed for built-in or continuous beams. The following rules relate to the S.F. and B.M. diagrams for beams, viz.:

1. The shear force at any section is the algebraic sum of normal forces acting to one side of the section.
2. Shear is considered positive when the shear force calculated as above is upwards to the left of the section.
3. The B.M. at any section is the algebraic summation of the moments about that section of all forces to one side of the section.
4. Moments are considered positive when the middle of a beam sags with respect to its ends or when tension occurs in the lower fibres of the beam.
5. Under point loads only, the S.F. diagram will consist of a series of horizontal and vertical lines, whilst the B.M. diagram will consist of sloping straight lines, changes of slope occurring only at the loads.
6. For uniformly distributed loads (U.D.L.s) the S.F. diagram will consist of sloping straight lines, whilst the B.M. diagram will consist of second-degree parabolas.
7. The maximum B.M. occurs at the point of zero shear, where such exists, or at the point where the shear-force curve crosses the base line.

Loading, Shear Force and Bending Moment

The relationship between loading, S.F. and B.M. may be derived graphically by considering Fig. 1.

If *ABC* represents the intensity of loading upon a beam simply supported at *A* and *B*, then the total area enclosed by the curve and the base *AB* will represent the total load on the span.

If a second curve *DEF* is constructed on a base line *GD* such that the ordinate *X* at any point represents the area of the load intensity curve to, say, the right of the point, then it follows that the curve *DEF* represents the total load to any point and the ordinate *FG* will equal the total load on the beam. If this curve is then amended by the addition of the reactions as shown, then by the definition given earlier the curve *DEFHJ* will be the shear force diagram for the given loading, since the ordinate at any point will be the algebraic sum to the right of the section of the downward load and the upward reaction.

18 SIMPLY SUPPORTED BEAMS

If then the total load curve *DEFG* is treated in a similar manner and a third curve *KLM* is constructed on a base line *KN* such that the ordinate *Y* at any point represents the area of the total load curve to the right of the point, then the curve *KLM* represents the total moment about *B* of the load to the right of the point and

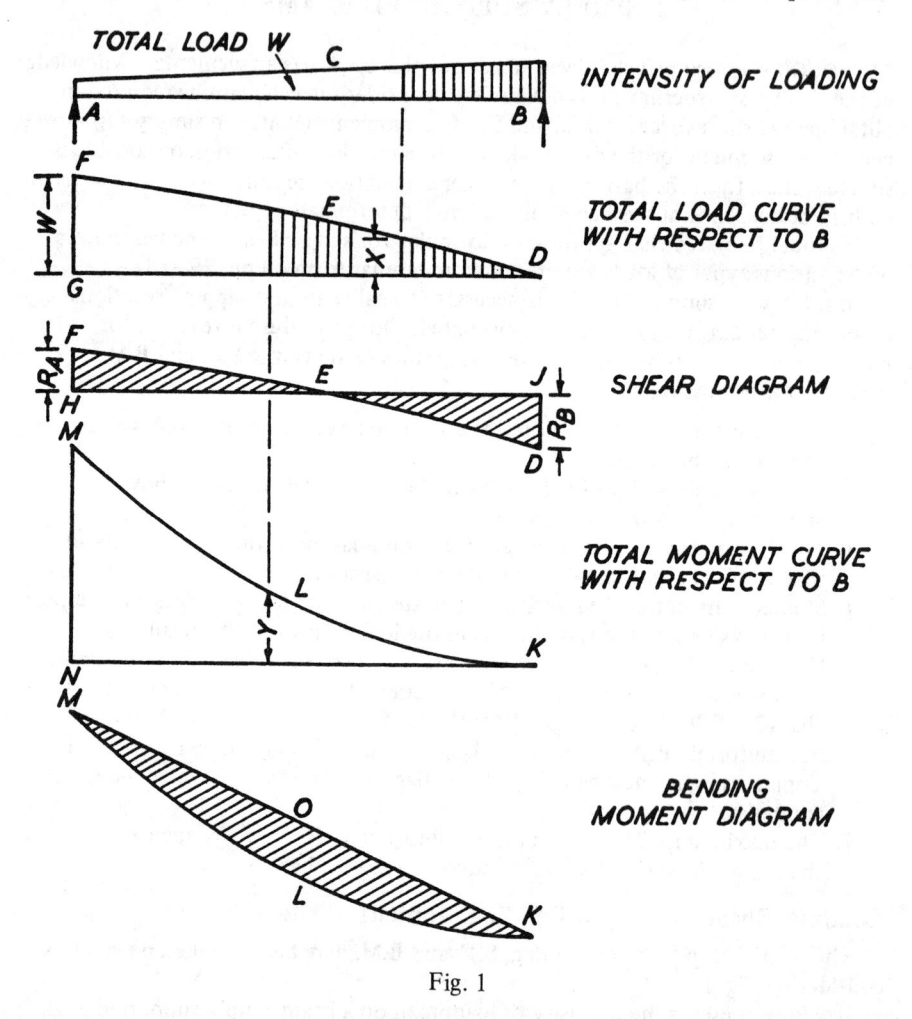

Fig. 1

the ordinate *MN* represents the moment of *W* about *B*. If this curve is then corrected by the line *KM* representing the effect of the moment of R_A about *B*, then the curve *KLMO* will be the bending moment diagram for the beam.

If the load intensity curve be called a primitive curve, then the shear curve is known as the sum curve of the load intensity curve and the bending moment curve will be the sum curve of the shear curve. The proof of this is as follows.

Consider any section *XX* and a portion of load *AB* distant *x* from the section as shown in Fig. 2.

The shear at *XX* due to this portion of the load will be the hatched area *ABCD*, and hence the total load up to *XX* will be represented by the total area of the load

LOADING, SHEAR FORCE AND BENDING MOMENT 19

intensity curve to that point. Hence the sum curve of the load intensity curve will be, by definition, the total shear curve.

Consider the bending moment at *XX*. The B.M. at this point due to the portion of load AB = the given portion of load x x. Then if E and F are the corresponding

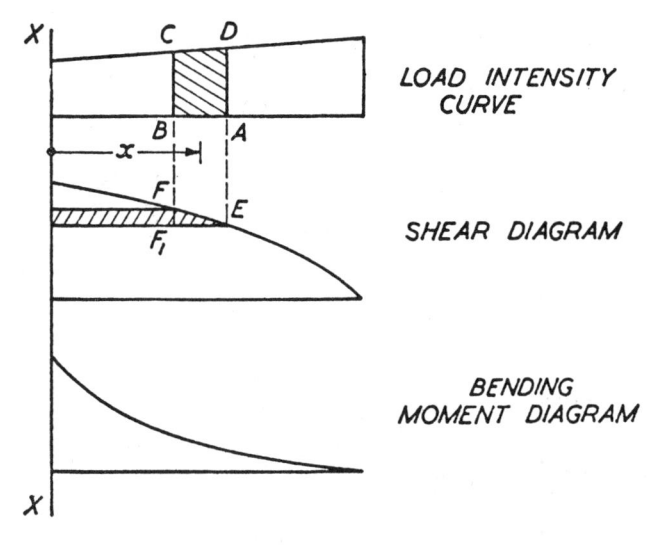

Fig. 2

points on the shear curve, the difference of the ordinates at E and F represents the load on the portion AB.

Therefore $\qquad$ load on $AB = FF_1$,

and $\qquad$ B.M. at XX due to this load $= FF_1 \times x$.

That is to say, the shaded portion of the shear curve represents the B.M. at *XX* due to the portion AB of the load. Hence the total B.M. at *XX* due to the load to the right equals the total area of the shear diagram to *XX*.

The sum curve of a horizontal straight line is a sloping straight line and the sum curve of a sloping straight line is a parabola of the second degree, whilst the sum curve of this will be a parabola of the third degree, etc.

The sum curve is obtained graphically as shown in Fig. 3.

Divide the primitive curve into a number of small parts 1, 2, 3, ... Find the mid-points of these divisions and project horizontally to a vertical line AB passing through the origin. From any point P on the base line CA produced draw a corresponding ray diagram 1, 2, 3, ... For the shear diagram draw parallels to the respective rays across the relevant spaces, to give a connected line as shown, which is the approximate first sum curve of the primitive curve. The true sum curve is a line joining the mid-points of each of the divisions of the sum curve illustrated.

The procedure is then repeated to produce the moment diagram which is the second sum curve.

The scales of the respective load, shear and moment diagrams are related as follows. Let 25 mm = W kN/m be the scale of the load intensity diagram, and let

p_1, the polar distance used for the construction of the shear curve, be measured on the space scale. Then the shear force scale is 25 mm = $p_1 W$ kN. Similarly, if p_2 is the polar distance used for the construction of the moment curve, again measured on the space scale, the B.M. scale is 25 mm = $p_2 p_1 W$ kNm.

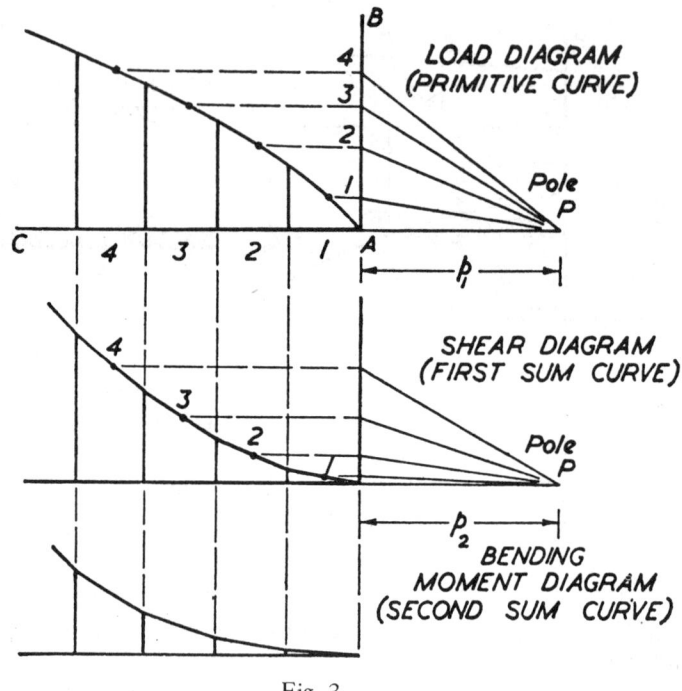

Fig. 3

Point of Maximum B.M.

At the point where the B.M. is a mathematical maximum (or minimum) the tangent to the B.M. curve must be horizontal and therefore the ray corresponding to this must also be horizontal. This ray, however, is related to the ordinate of the shear diagram, and since the pole is taken on the base line produced it follows that the shear force ordinate at this point must be zero for the line to the pole to be horizontal. Thus we have the rule that the maximum B.M. occurs where the shear is zero.

Mathematical Approach

The foregoing relationship between loading, shear force and bending moment, derived graphically, may be expressed mathematically as follows:

Let the intensity of load at a distance x from the origin be $F(x)$. Then:

$$\text{the shear at the point} = \int F(x)\,dx + C_1$$

and

$$\text{the B.M.} = \int F(x)x \cdot dx + C_1 x + C_2.$$

The integration constants C_1 and C_2 represent the effects of the base lines JH and KOM in Fig. 1.

Areas and Centres of Gravity

It has been noted earlier that the area of the load intensity diagram represents the total load on the beam. The area and the position of its centre of gravity may be found by link and vector polygons or by Simpson's Rule. Using this information, the reactions may be calculated.

To apply Simpson's Rule it is necessary to divide the loaded length into an even number of equal parts and the required area is approximately (to within about 0.1 per cent of the true value as given by integration) equal to the sum of the extreme ordinates plus four times the sum of the odd ordinates plus twice the sum of the remaining even ordinates all multiplied by one-third of the common distance between them.

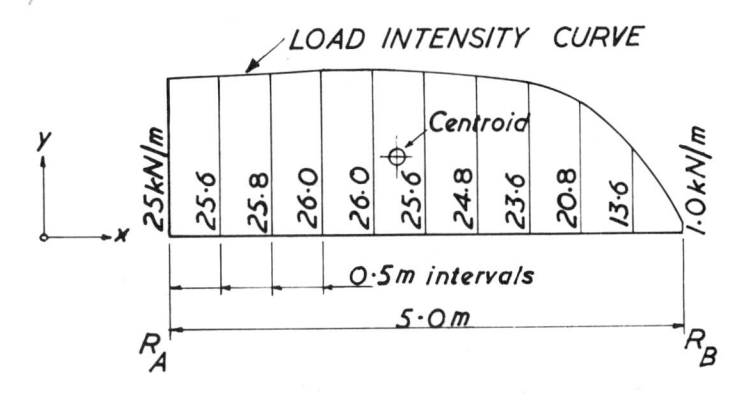

Fig. 4

Thus if the base is divided into n equal parts of length D, and the extreme ordinates are y_0 and y_n we have:

$$\text{Area} = \tfrac{1}{3}D\{[y_0 + y_n] + 4[y_1 + y_3 + y_5 + \ldots + y_{(n-1)}] + 2[y_2 + y_4 + y_6 + \ldots + y_{(n-2)}]\}.$$

Further, the distance of the centre of gravity of the load from one end is given by:

$$\bar{x} = \frac{\Sigma(Ax)}{\Sigma(A)}$$

where $\Sigma(Ax)$ is the summation of moments about that end of the individual parts of the loading diagram and $\Sigma(A)$ is the total area of the diagram.

Simpson's Rule will be applied to the beam shown in Fig. 4.
Tabulating:

$x =$ 0   0.5   1.0   1.5   2.0   2.5   3.0   3.5   4.0   4.5   5.0
$y =$ 25.0 25.6 25.8 26.0 26.0 25.6 24.8 23.6 20.8 13.6 1.0
$xy =$ 0   12.8 25.8 39.0 52.0 64.0 74.4 82.6 83.2 61.2 5.0

22 SIMPLY SUPPORTED BEAMS

$\Sigma(Ax) = \Sigma(xy)$

$= \frac{1}{3} \times 0.5\{(0 + 5) + 4(12.8 + \ldots + 61.2) + 2(25.8 + \ldots + 83.2)\}$

$= 252.4$

$\Sigma(A) = \Sigma(y)$

$= \frac{1}{3} \times 0.5\{(25 + 1) + 4(25.6 + \ldots + 13.6) + 2(25.8 + \ldots + 20.8)\}$

$= 113.0.$

Hence the distance of the centroid from R_A

$$= \frac{252.4}{113.0} = 2.23 \text{ m}$$

Taking moments about the left support A,

$$R_B \times 5 = \Sigma Ax$$

Whence $\qquad R_B = \dfrac{252.4}{5} = 50.48$ kN

and $\qquad R_A = 113.0 - 50.48 = 62.52$ kN

S.F. and B.M. Diagrams

Having stated the basic principles relating to simply supported beams, two examples will be given, the first utilising the arithmetical approach and the second the algebraic approach to analysis.

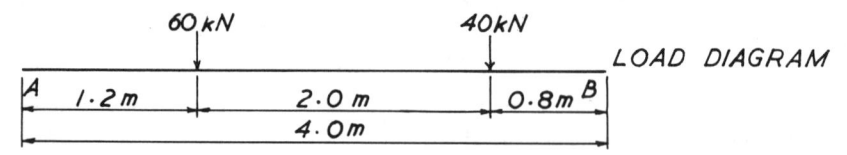

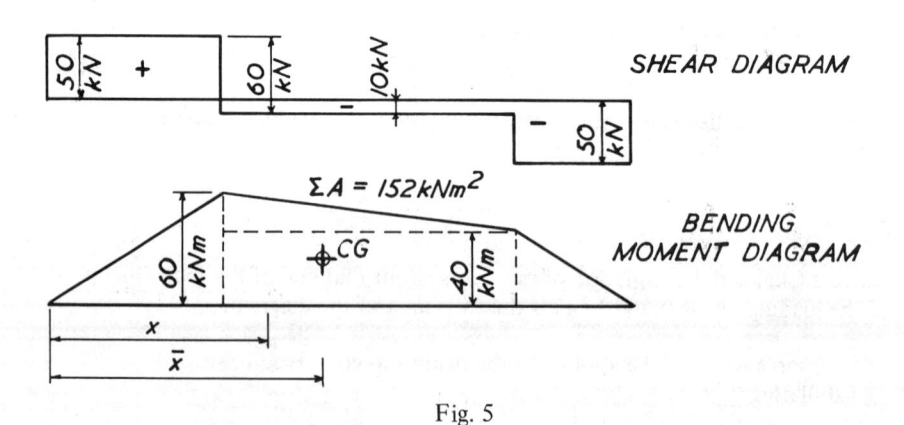

Fig. 5

Example 1. It is required to draw the S.F. and B.M. diagrams for the beam shown in Fig. 5 and to calculate the area and the position of the centre of gravity of the B.M. diagram.

S.F. AND B.M. DIAGRAMS

Taking moments about B,
$$R_A \times 4 = 40 \times 0.8 + 60 \times 2.8$$
and
$$R_A = \frac{200}{4} = 50 \text{ kN}$$

The total load = 100 kN. Hence:
$$R_B = 100 - 50 = 50 \text{ kN}$$

The S.F. diagram can now be drawn as shown. It will be observed that the curve crosses the base line under the 60 kN point load, i.e. at C, and consequently this will be the point at which the B.M. has its maximum value.

The
$$\text{B.M. at } C = M_C = R_A \times 1.2 = 50 \times 1.2$$
$$= 60 \text{ kNm}$$

Similarly,
$$M_D = R_A \times 3.2 - 60 \times 2$$
$$= R_B \times 0.8$$
$$= 40 \text{ kNm}$$

It is useful to note that the B.M. at any point is numerically equal to the area of the S.F. diagram on either side of that point. Considering the S.F. diagram to the left of D,
$$M_D = 50 \times 1.2 + (-10 \times 2.0)$$
$$= 40 \text{ kNm as found before.}$$

The area of the B.M. diagram may be found by dividing it into three triangles and a rectangle as shown.

Thus
$$\Sigma A = \frac{60 \times 1.2}{2} + 40 \times 2 + \frac{20 \times 2}{2} + \frac{40 \times 0.8}{2}$$
$$= 36 + 80 + 20 + 16$$
$$= 152 \text{ kNm}^2$$

The moment of the area of the B.M. diagram about the left support A
$$= \Sigma Ax = \left(36 \times \frac{2 \times 1.2}{3}\right) + \left[80 \times \left(1.2 + \frac{2}{2}\right)\right] + \left[20 \times \left(1.2 + \frac{2}{3}\right)\right] + \left[16 \times \left(1.2 + 2.0 + \frac{0.8}{3}\right)\right]$$
$$= 28.8 + 176 + 37.4 + 55.5$$
$$= 297.7 \text{ kNm}^3$$

Therefore, the distance to the C.G. of the B.M. diagram from the left support A:
$$\bar{x} = \frac{\Sigma Ax}{\Sigma A} = \frac{297.7}{152} = 1.96 \text{ m}$$

Example 2. It is required to draw the S.F. and B.M. diagrams for the beam shown in Fig. 6 and to calculate the area and the position of the C.G. of the B.M. diagram.

The beam carries a load of 100 kN varying at a constant rate from 10 kN/m at A to 40 kN/m at B. Hence the load may be considered as being of two components, one a uniformly distributed load (U.D.L.) of 10 kN/m and the other a triangular-shaped load varying from zero at A to 30 kN/m at B, the total weight of each of the

components being respectively 40 kN and 60 kN. The C.G. of the U.D.L. is in the middle of the beam while that of the triangular load occurs one-third of the distance from B to A.

Taking moments about B,

$$R_A \times 4 = 40 \times \frac{4}{2} + 60 \times \frac{4}{3}$$

and

$$R_A = 40 \text{ kN}$$

$$R_B = 100 - 40$$

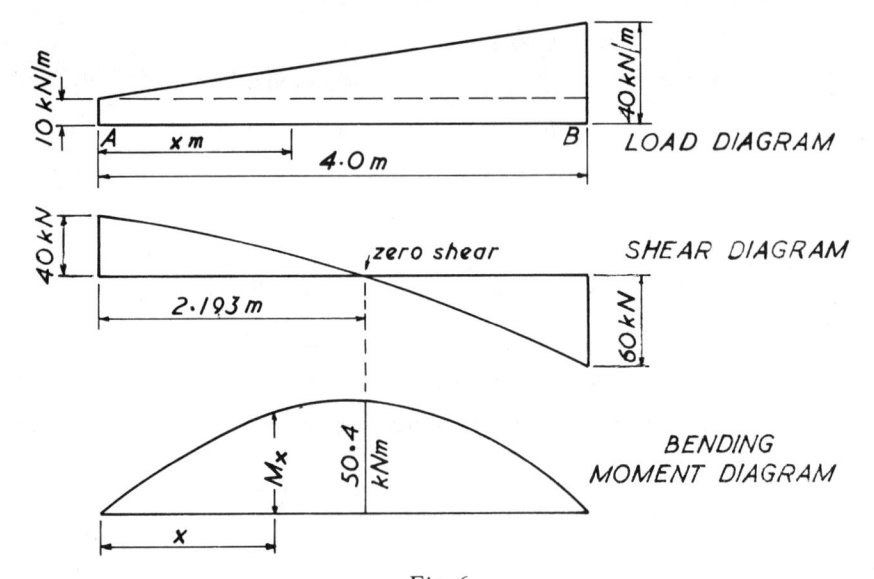

$$= 60 \text{ kN}$$

Fig. 6

Now the load to any point x ft. from A towards B

$$= 10 \times x + \frac{30}{L} \times x \times \frac{x}{2}$$

$$= 10x + 3.75x^2$$

Then the point of zero S.F. and maximum B.M. occurs where

$$R_A - (10x + 3.75x^2) = 0$$

or where

$$10x + 3.75x^2 = 40$$

i.e. at 2.193 m from A.

Alternatively, elementary differentiation may be used to find the point of maximum B.M.:

$$M_x = R_A \times x - 10 \times x \times \frac{x}{2} - \frac{30}{L} \times x \times \frac{x}{2} \times \frac{x}{3}$$

$$= 40x - 5x^2 - 1.25x^3$$

Maximum B.M. occurs when $\dfrac{dM_x}{dx} = 0$,

i.e. when $\quad 40 - 10x - 3.75x^2 = 0$

or $\quad 10x + 3.75x^2 = 40$

Whence $x = 2.193$ m, as before.
The S.F. diagram may now be drawn.

The maximum B.M. $= 40x - 5x^2 - 1.25x^3$

$\qquad = 40 \times 2.193 - 5 \times 2.193^2 - 1.25 \times 2.193^3$

$\qquad = 87.7 - 24.1 - 13.2$

$\qquad = 50.4$ kNm

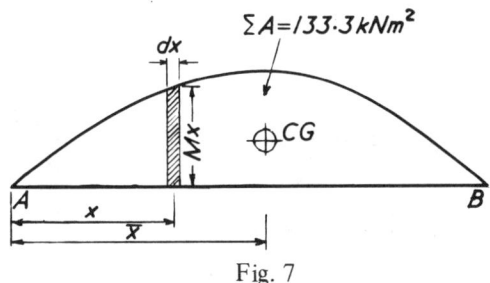

Fig. 7

Consider Fig. 7.
The area of the B.M. diagram

$$= \Sigma A = \int M \, dx$$

$$= \int_0^4 (40x - 5x^2 - 1.25x^3) \, dx$$

$$= \left[\dfrac{40x^2}{2} - \dfrac{5x^3}{3} - \dfrac{1.25x^4}{4} \right]_0^4$$

$$= 320 - 106.7 - 80$$

$$= 133.3 \text{ kNm}^2$$

The moment of the area of the B.M. diagram about the support A

$$= \Sigma Ax = \int Mx \, dx$$

$$= \int_0^4 (40x^2 - 5x^3 - 1.25x^4) \, dx$$

$$= \left[\dfrac{40x^3}{3} - \dfrac{5x^4}{4} - \dfrac{1.25x^5}{5} \right]_0^4$$

$$= 853.3 - 320 - 256$$

$$= 277.3 \text{ kNm}^3$$

Hence, the distance from the support A to the C.G. of the B.M. diagram.

$$= \bar{x} = \frac{\Sigma Ax}{\Sigma A} = \frac{277.3}{133.3} = 2.08 \text{ m}$$

It should be noted that the B.M. curve is a composite one resulting from the addition of the second-degree parabola associated with a U.D.L. and the third-degree parabola associated with a triangular load. The area under a second-degree parabola or under the third-degree parabola for this particular triangular load = $WL^2/12$. The total W in the example = 100 kN. Hence, the area under the curve should be 100 × 4 × 4/12 = 133.3 kNm², which checks the result calculated above.

Examples of Sum Curve

The area relationships between the load intensity, shear and moment curves derived on pages 17–20 are useful in the case of a beam continuously supported, such as a stiff foundation carrying imposed point or partially distributed loads, as the examples shown in Figs. 8 and 9 indicate.

Example 1.
It is necessary to find the net load intensity diagram as shown in Fig. 8.

Then:

Shear at B = area of load curve to left of B

= 60 kN/m × 2 m = 120 kN

Shear at C = area of load curve to left of C

= 120 kN − (90 kN/m × 2 m) = 120 − 180 = −60 kN

etc.

Since the load intensity diagram is composed of horizontal straight lines, the shear diagram will consist of sloping straight lines and is drawn as shown.

Points of zero shear occur at G, H and J and at each of these the tangent to the B.M. curve will be horizontal and the B.M. value a maximum or minimum.

Then:

B.M. at B = area of shear curve to left of B

$$= \frac{120 \text{ kN} \times 2 \text{ m}}{2} = 120 \text{ kNm}$$

B.M. at G = area of shear curve to left of G

$$= \frac{120 \text{ kN} \times 3.33 \text{ m}}{2} = 200 \text{ kNm}$$

B.M. at C = 200 kNm − $\frac{60 \times 0.67}{2}$ = 180 kNm

B.M. at H = 180 kNm − $\frac{60 \times 1}{2}$ = 150 kNm

The principle of symmetry operates, and the values for the remaining points D, J and E will be the same as those calculated for C, G and B respectively.

Since the shear diagram is composed of sloping straight lines, the B.M. diagram will be composed of second-degree parabolas, and is drawn as shown. Note that the slopes of the parabolas change with the reversal of slopes in the shear diagram. The B.M. diagram will be a smooth curve throughout.

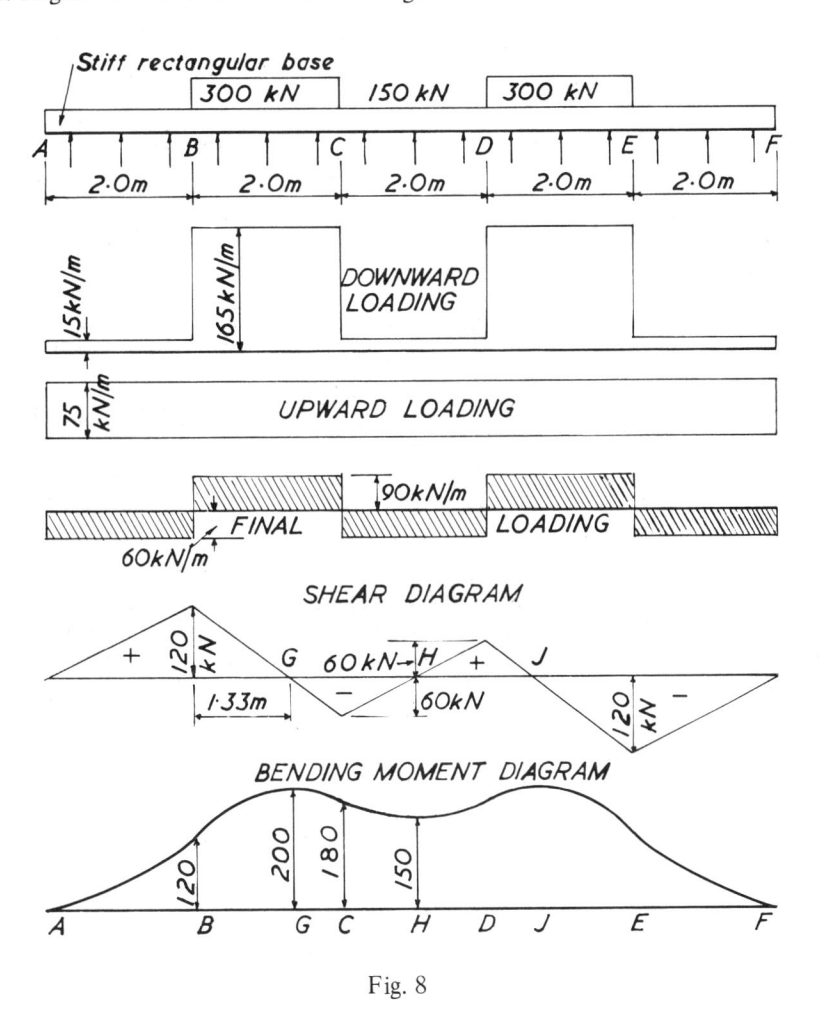

Fig. 8

Example 2.

Consider as an alternative an example with point loads of 300 kN in lieu of the loads distributed over the two lengths of 6 ft. The downward loads are thus two point loads of 300 kN each, and the upward load is 600 kN uniformly distributed. Using the summation of areas as before for the shear due to the uniform load and remembering that there will be a step in the shear diagram at each of the point loads, the S.F. diagram is as shown in Fig. 9.

The B.M. diagram is based upon the areas of the shear diagram and it should be noted that steps in the latter are accompanied by cusps in the former.

SIMPLY SUPPORTED BEAMS

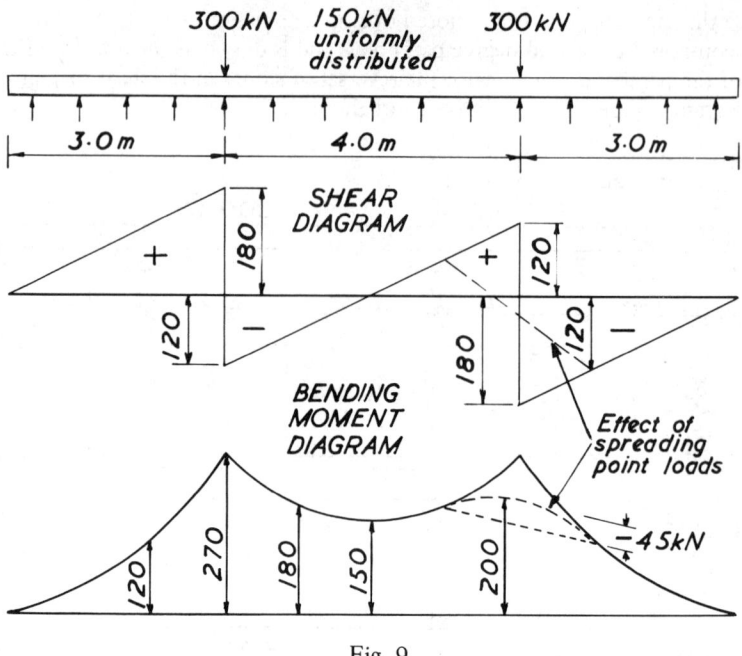

Fig. 9

It will be observed that the B.M. diagram in Fig. 8 could be obtained from that in Fig. 9 by calculating the moments at D and E and on the line joining these ordinates constructing a parabola of height $WL/8$ (−45 kNm), where W = 180 kN, as shown on the shear diagram, and L = 2 m.

CANTILEVERS

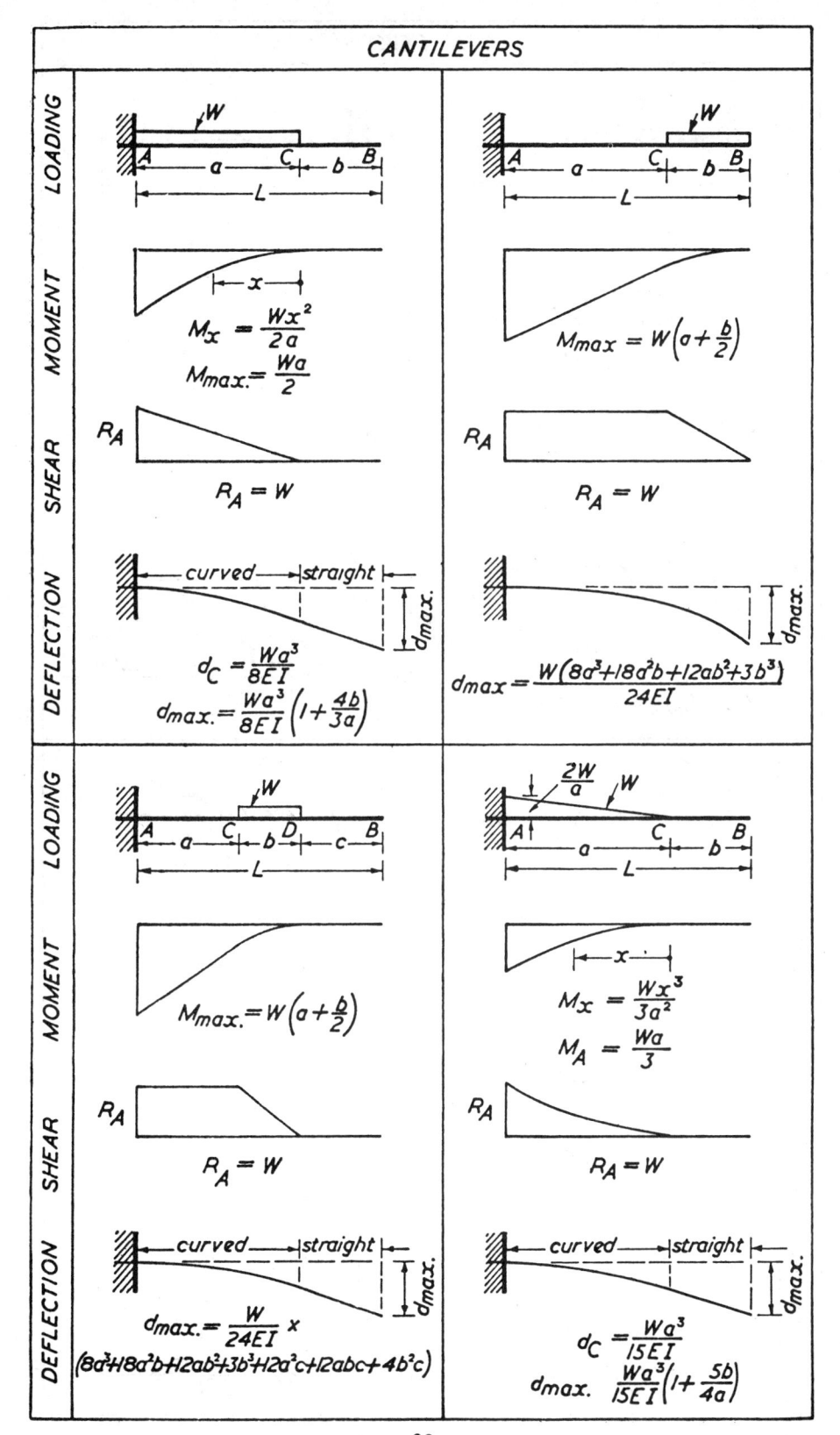

Case 1 (top left)

LOADING: Uniform load W over length a from A to C; b from C to B; total length L.

MOMENT:
$$M_x = \frac{Wx^2}{2a}$$
$$M_{max} = \frac{Wa}{2}$$

SHEAR: $R_A = W$

DEFLECTION: curved — straight
$$d_C = \frac{Wa^3}{8EI}$$
$$d_{max} = \frac{Wa^3}{8EI}\left(1 + \frac{4b}{3a}\right)$$

Case 2 (top right)

LOADING: Uniform load W over length b from C to B; a from A to C; total length L.

MOMENT:
$$M_{max} = W\left(a + \frac{b}{2}\right)$$

SHEAR: $R_A = W$

DEFLECTION:
$$d_{max} = \frac{W(8a^3 + 18a^2 b + 12ab^2 + 3b^3)}{24EI}$$

Case 3 (bottom left)

LOADING: Uniform load W over length b from C to D; a from A to C; c from D to B; total length L.

MOMENT:
$$M_{max} = W\left(a + \frac{b}{2}\right)$$

SHEAR: $R_A = W$

DEFLECTION: curved — straight
$$d_{max} = \frac{W}{24EI} \times (8a^3 + 18a^2 b + 12ab^2 + 3b^3 + 12a^2 c + 12abc + 4b^2 c)$$

Case 4 (bottom right)

LOADING: Triangular load, peak $\frac{2W}{a}$ at A, over length a from A to C; b from C to B; total length L.

MOMENT:
$$M_x = \frac{Wx^3}{3a^2}$$
$$M_A = \frac{Wa}{3}$$

SHEAR: $R_A = W$

DEFLECTION: curved — straight
$$d_C = \frac{Wa^3}{15EI}$$
$$d_{max} = \frac{Wa^3}{15EI}\left(1 + \frac{5b}{4a}\right)$$

CANTILEVERS

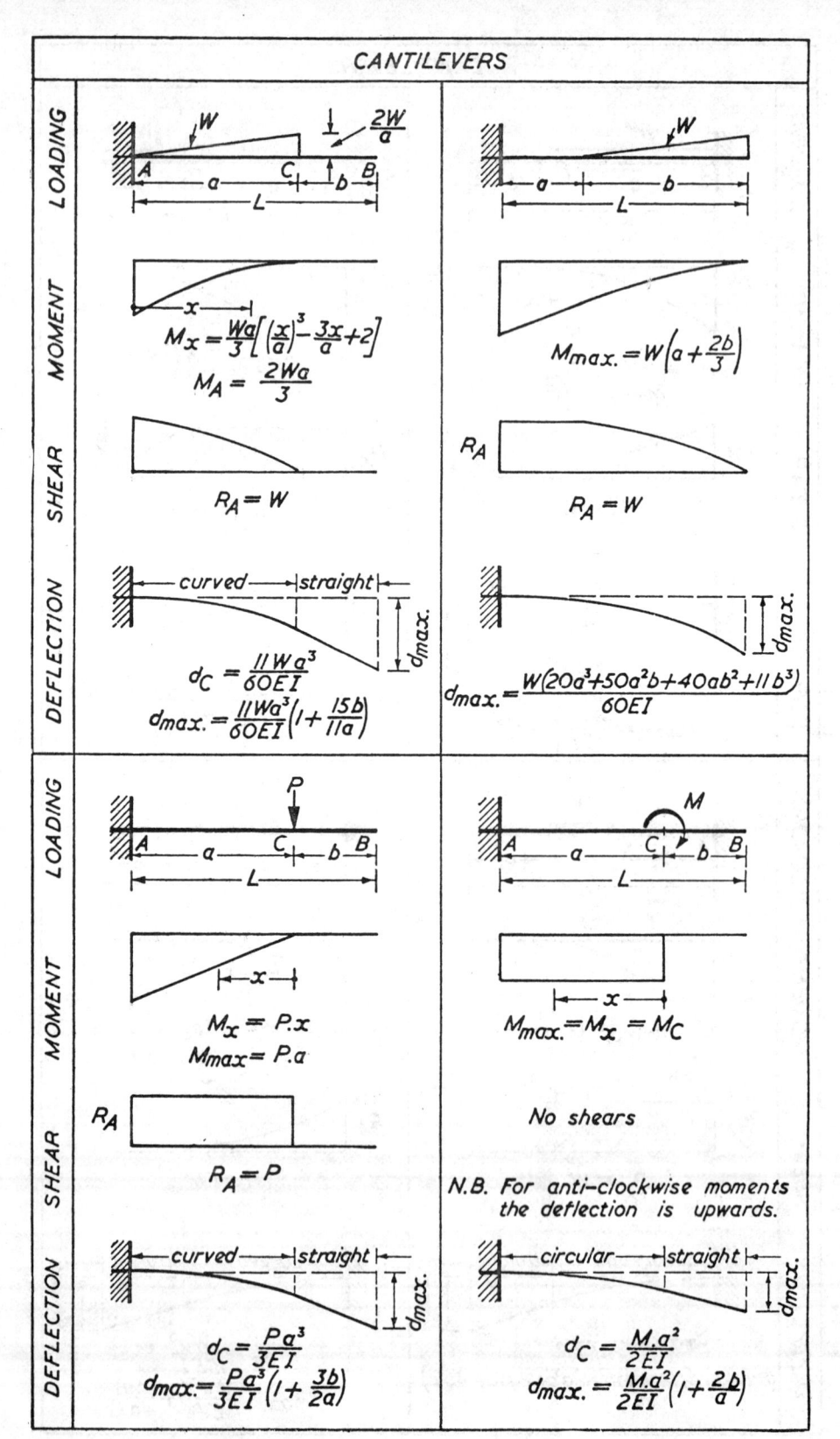

Case 1: Triangular load (increasing toward C), partial span

LOADING: Load W distributed triangularly with maximum $\frac{2W}{a}$ at C; spans a (A to C) and b (C to B); total length L.

MOMENT:
$$M_x = \frac{Wa}{3}\left[\left(\frac{x}{a}\right)^3 - \frac{3x}{a} + 2\right]$$
$$M_A = \frac{2Wa}{3}$$

SHEAR: $R_A = W$

DEFLECTION: curved — straight, d_{max}
$$d_C = \frac{11Wa^3}{60EI}$$
$$d_{max} = \frac{11Wa^3}{60EI}\left(1 + \frac{15b}{11a}\right)$$

Case 2: Triangular load over full span

LOADING: Load W triangularly distributed over full length with spans a and b; total L.

MOMENT:
$$M_{max} = W\left(a + \frac{2b}{3}\right)$$

SHEAR: $R_A = W$

DEFLECTION:
$$d_{max} = \frac{W(20a^3 + 50a^2b + 40ab^2 + 11b^3)}{60EI}$$

Case 3: Point load P at C

LOADING: Point load P at C; spans a and b; total L.

MOMENT:
$$M_x = P.x$$
$$M_{max} = P.a$$

SHEAR: $R_A = P$

DEFLECTION: curved — straight
$$d_C = \frac{Pa^3}{3EI}$$
$$d_{max} = \frac{Pa^3}{3EI}\left(1 + \frac{3b}{2a}\right)$$

Case 4: Applied moment M at C

LOADING: Moment M applied at C; spans a and b; total L.

MOMENT:
$$M_{max} = M_x = M_C$$

SHEAR: No shears

N.B. For anti-clockwise moments the deflection is upwards.

DEFLECTION: circular — straight
$$d_C = \frac{M.a^2}{2EI}$$
$$d_{max} = \frac{M.a^2}{2EI}\left(1 + \frac{2b}{a}\right)$$

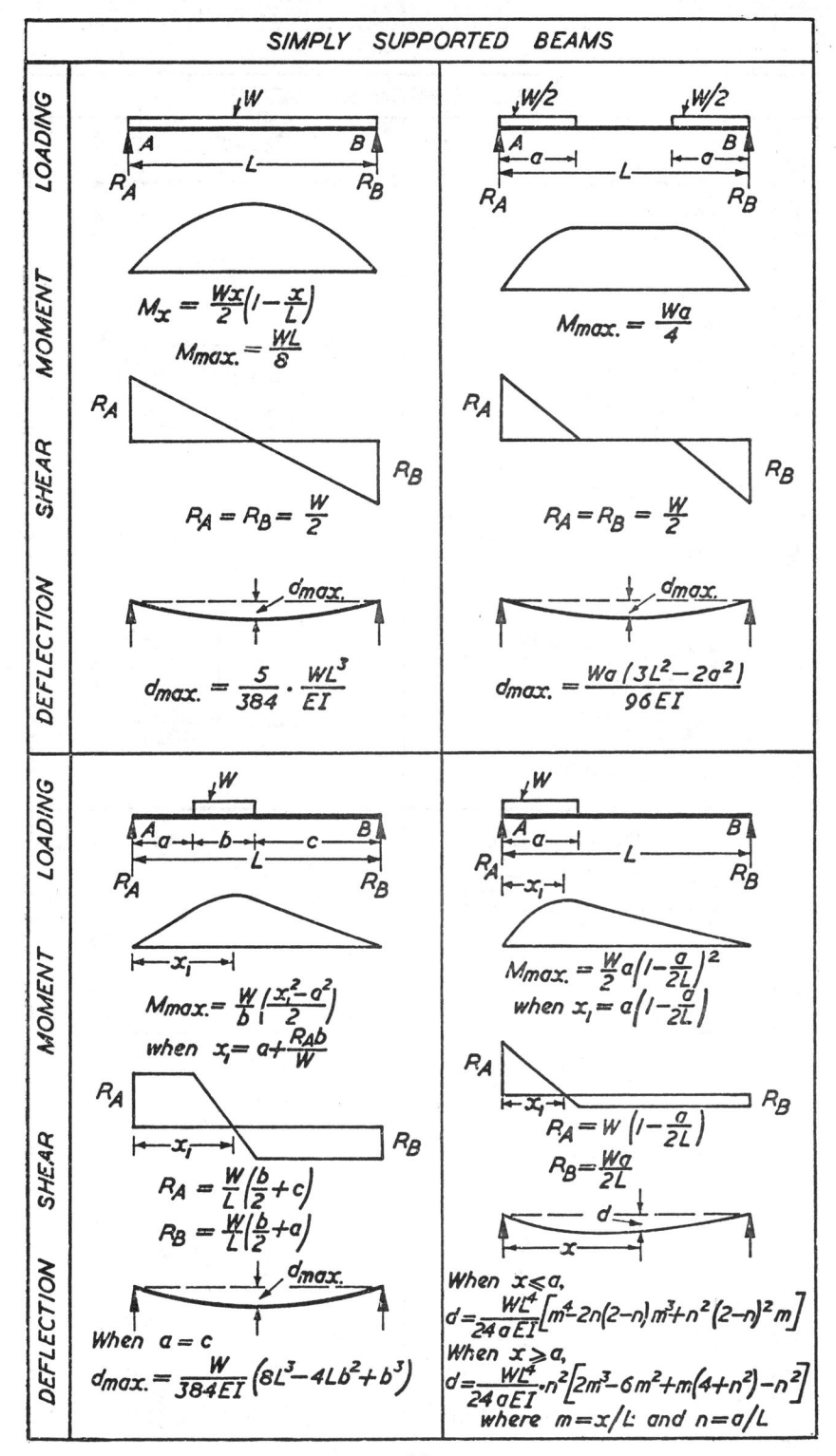

SIMPLY SUPPORTED BEAMS

Case 1: Uniform load W over span L

LOADING: Simply supported beam A–B, length L, uniform load W, reactions R_A, R_B.

MOMENT:
$$M_x = \frac{Wx}{2}\left(1 - \frac{x}{L}\right)$$
$$M_{max} = \frac{WL}{8}$$

SHEAR:
$$R_A = R_B = \frac{W}{2}$$

DEFLECTION:
$$d_{max} = \frac{5}{384} \cdot \frac{WL^3}{EI}$$

Case 2: Two equal loads W/2 at distance a from each support

LOADING: Simply supported beam A–B, length L, loads W/2 each at distance a from A and B.

MOMENT:
$$M_{max} = \frac{Wa}{4}$$

SHEAR:
$$R_A = R_B = \frac{W}{2}$$

DEFLECTION:
$$d_{max} = \frac{Wa(3L^2 - 2a^2)}{96\,EI}$$

Case 3: Partial uniform load W over segment b (with a on left, c on right)

LOADING: Simply supported beam, load W on length b between distances a and c.

MOMENT:
$$M_{max} = \frac{W}{b}\left(\frac{x_1^2 - a^2}{2}\right)$$
$$\text{when } x_1 = a + \frac{R_A b}{W}$$

SHEAR:
$$R_A = \frac{W}{L}\left(\frac{b}{2} + c\right)$$
$$R_B = \frac{W}{L}\left(\frac{b}{2} + a\right)$$

DEFLECTION: When $a = c$,
$$d_{max} = \frac{W}{384\,EI}\left(8L^3 - 4Lb^2 + b^3\right)$$

Case 4: Concentrated load W at distance a from A

LOADING: Simply supported beam, point load W at distance a from A.

MOMENT:
$$M_{max} = \frac{W}{2}\,a\left(1 - \frac{a}{2L}\right)^2$$
$$\text{when } x_1 = a\left(1 - \frac{a}{2L}\right)$$

SHEAR:
$$R_A = W\left(1 - \frac{a}{2L}\right)$$
$$R_B = \frac{Wa}{2L}$$

DEFLECTION:
When $x \leqslant a$,
$$d = \frac{WL^4}{24\,aEI}\left[m^4 - 2n(2-n)\,m^3 + n^2(2-n)^2\,m\right]$$
When $x \geqslant a$,
$$d = \frac{WL^4}{24\,aEI}\cdot n^2\left[2m^3 - 6m^2 + m(4+n^2) - n^2\right]$$
where $m = x/L$ and $n = a/L$

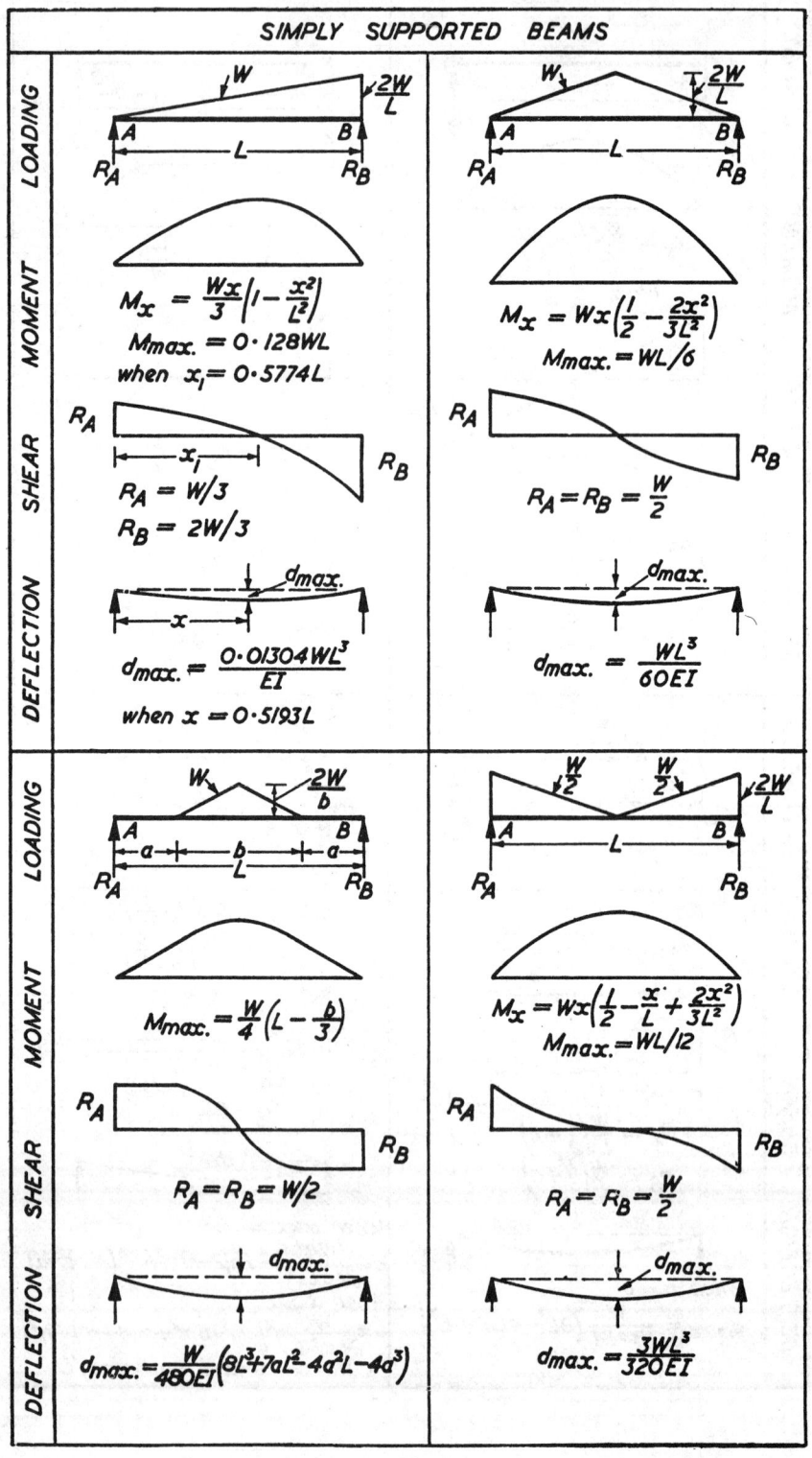

SIMPLY SUPPORTED BEAMS

Loading (top left): Triangular load increasing from A to B, peak $2W/L$ at B, total load W, span L, reactions R_A, R_B.

Moment:
$$M_x = \frac{Wx}{3}\left(1 - \frac{x^2}{L^2}\right)$$
$M_{max.} = 0.128WL$ when $x_1 = 0.5774L$

Shear:
$R_A = W/3$
$R_B = 2W/3$

Deflection:
$$d_{max.} = \frac{0.01304\,WL^3}{EI}$$
when $x = 0.5193L$

Loading (top right): Triangular load peaking at centre, max $2W/L$, total W, span L.

Moment:
$$M_x = Wx\left(\frac{1}{2} - \frac{2x^2}{3L^2}\right)$$
$M_{max.} = WL/6$

Shear:
$R_A = R_B = \dfrac{W}{2}$

Deflection:
$$d_{max.} = \frac{WL^3}{60EI}$$

Loading (bottom left): Triangular load peaking at distance a from A, peak $2W/b$, span $L = a + b + a$.

Moment:
$$M_{max.} = \frac{W}{4}\left(L - \frac{b}{3}\right)$$

Shear:
$R_A = R_B = W/2$

Deflection:
$$d_{max.} = \frac{W}{480EI}\left(8L^3 + 7aL^2 - 4a^2L - 4a^3\right)$$

Loading (bottom right): Triangular loads with $W/2$ at ends decreasing to zero at centre, peak $2W/L$, span L.

Moment:
$$M_x = Wx\left(\frac{1}{2} - \frac{x}{L} + \frac{2x^2}{3L^2}\right)$$
$M_{max.} = WL/12$

Shear:
$R_A = R_B = \dfrac{W}{2}$

Deflection:
$$d_{max.} = \frac{3WL^3}{320EI}$$

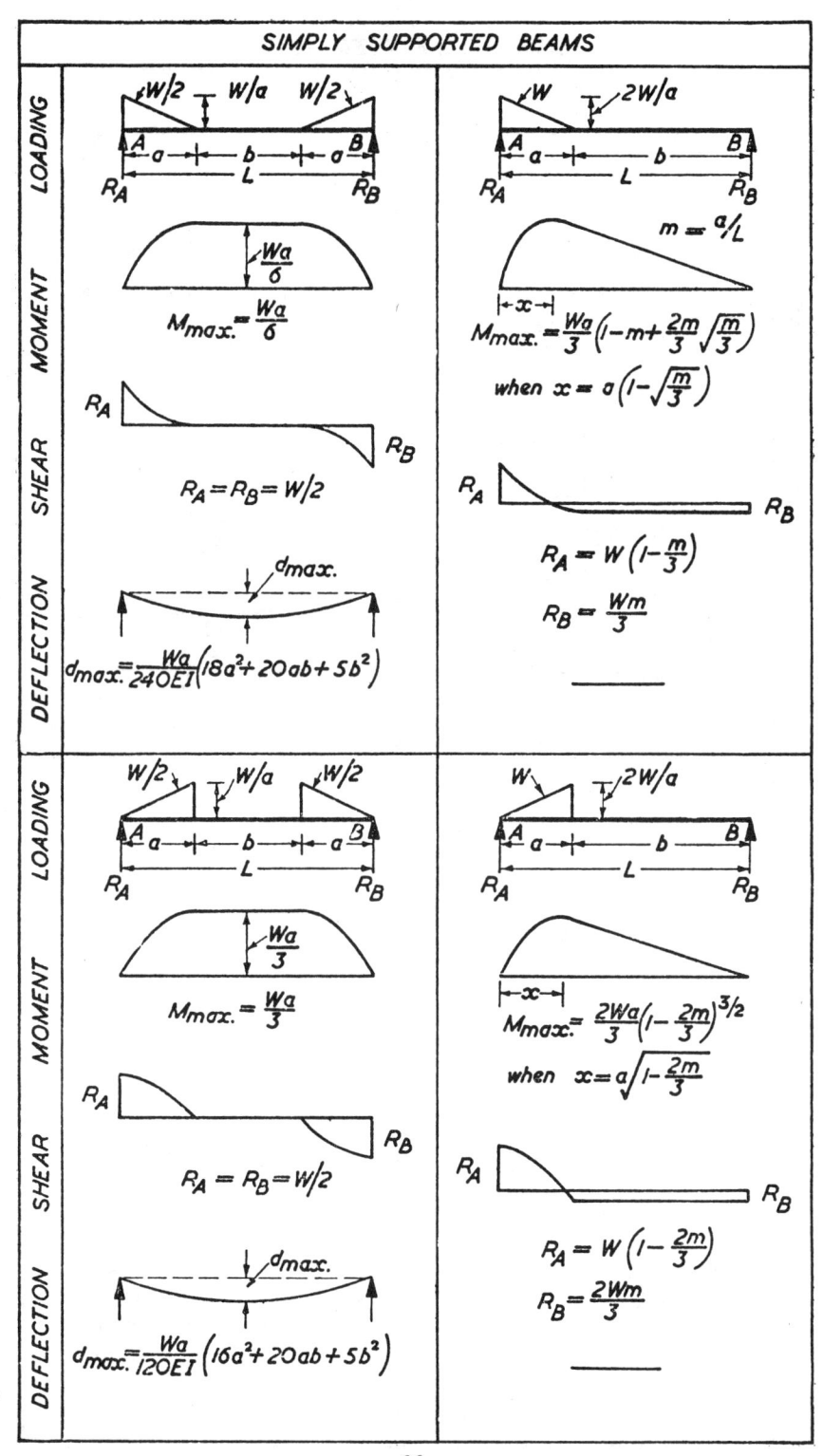

SIMPLY SUPPORTED BEAMS

LOADING (top-left): Symmetric trapezoidal load, $w/2$ at ends rising to w/a over length a, uniform w/a over middle length b, spans $a + b + a = L$.

MOMENT:
$$M_{max.} = \frac{Wa}{6}$$

SHEAR:
$$R_A = R_B = W/2$$

DEFLECTION:
$$d_{max.} = \frac{Wa}{240EI}(18a^2 + 20ab + 5b^2)$$

LOADING (top-right): Triangular load rising from 0 at A to $2W/a$, with $m = a/L$, spans $a + b = L$.

MOMENT:
$$M_{max.} = \frac{Wa}{3}\left(1 - m + \frac{2m}{3}\sqrt{\frac{m}{3}}\right)$$
when $x = a\left(1 - \sqrt{\frac{m}{3}}\right)$

SHEAR:
$$R_A = W\left(1 - \frac{m}{3}\right)$$
$$R_B = \frac{Wm}{3}$$

LOADING (bottom-left): Symmetric triangular peaks $w/2$ to w/a at inner edges of a, with middle length b.

MOMENT:
$$M_{max.} = \frac{Wa}{3}$$

SHEAR:
$$R_A = R_B = W/2$$

DEFLECTION:
$$d_{max.} = \frac{Wa}{120EI}(16a^2 + 20ab + 5b^2)$$

LOADING (bottom-right): Triangular load from W at A down to $2W/a$, spans $a + b = L$.

MOMENT:
$$M_{max.} = \frac{2Wa}{3}\left(1 - \frac{2m}{3}\right)^{3/2}$$
when $x = a\sqrt{1 - \frac{2m}{3}}$

SHEAR:
$$R_A = W\left(1 - \frac{2m}{3}\right)$$
$$R_B = \frac{2Wm}{3}$$

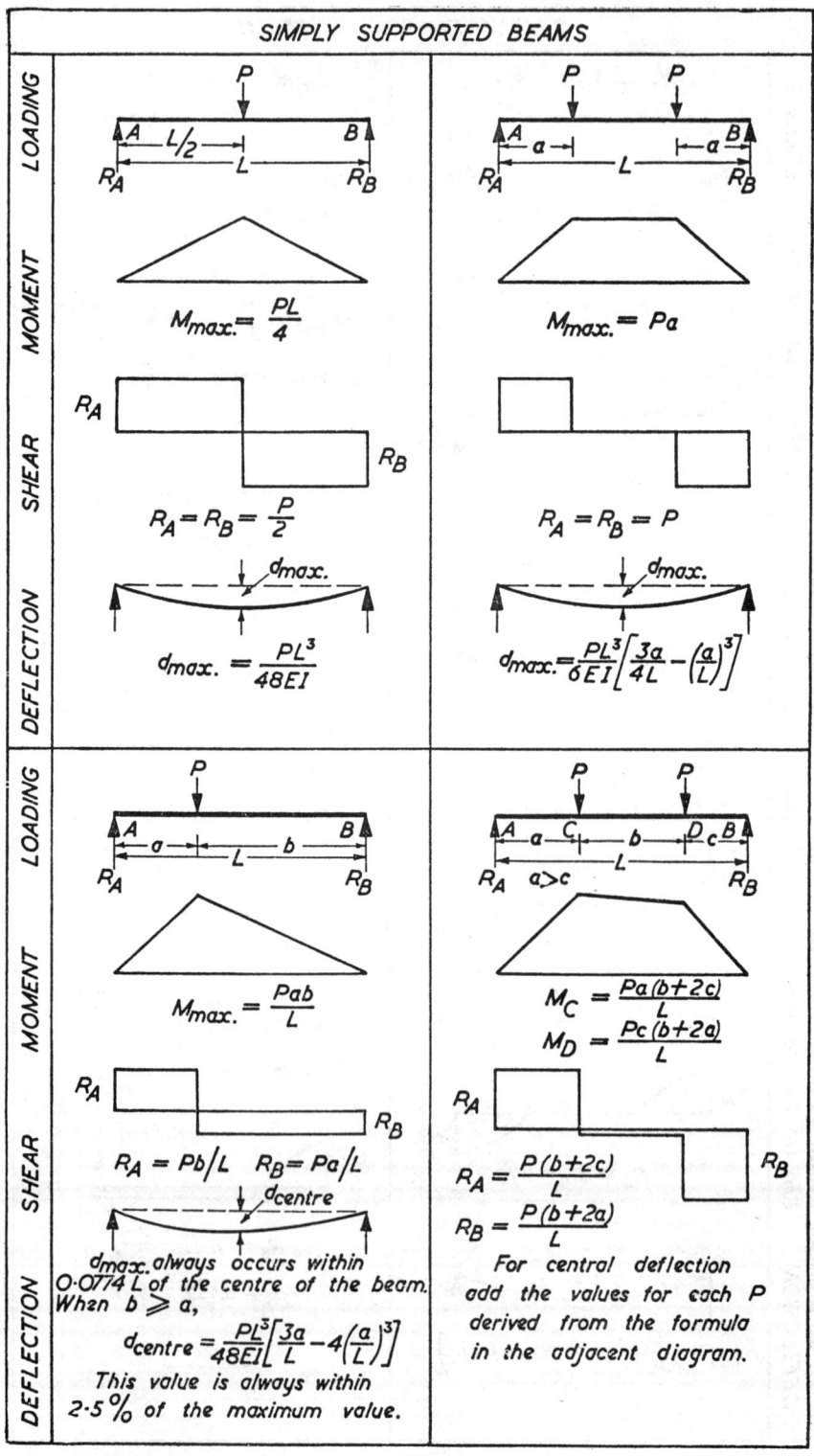

SIMPLY SUPPORTED BEAMS

LOADING — Single central point load P at $L/2$, supports R_A, R_B.

MOMENT: $M_{max.} = \dfrac{PL}{4}$

SHEAR: $R_A = R_B = \dfrac{P}{2}$

DEFLECTION: $d_{max.} = \dfrac{PL^3}{48EI}$

LOADING — Two equal point loads P at distance a from each support.

MOMENT: $M_{max.} = Pa$

SHEAR: $R_A = R_B = P$

DEFLECTION: $d_{max.} = \dfrac{PL^3}{6EI}\left[\dfrac{3a}{4L} - \left(\dfrac{a}{L}\right)^3\right]$

LOADING — Single point load P at distance a from A, distance b from B.

MOMENT: $M_{max.} = \dfrac{Pab}{L}$

SHEAR: $R_A = Pb/L \qquad R_B = Pa/L$

DEFLECTION: $d_{max.}$ always occurs within $0.0774\,L$ of the centre of the beam. When $b \geqslant a$,
$$d_{centre} = \dfrac{PL^3}{48EI}\left[\dfrac{3a}{L} - 4\left(\dfrac{a}{L}\right)^3\right]$$
This value is always within 2.5% of the maximum value.

LOADING — Two unequal point loads P at C (distance a from A) and D (distance c from B), with $a > c$.

MOMENT:
$M_C = \dfrac{Pa(b+2c)}{L}$
$M_D = \dfrac{Pc(b+2a)}{L}$

SHEAR:
$R_A = \dfrac{P(b+2c)}{L}$
$R_B = \dfrac{P(b+2a)}{L}$

DEFLECTION: For central deflection add the values for each P derived from the formula in the adjacent diagram.

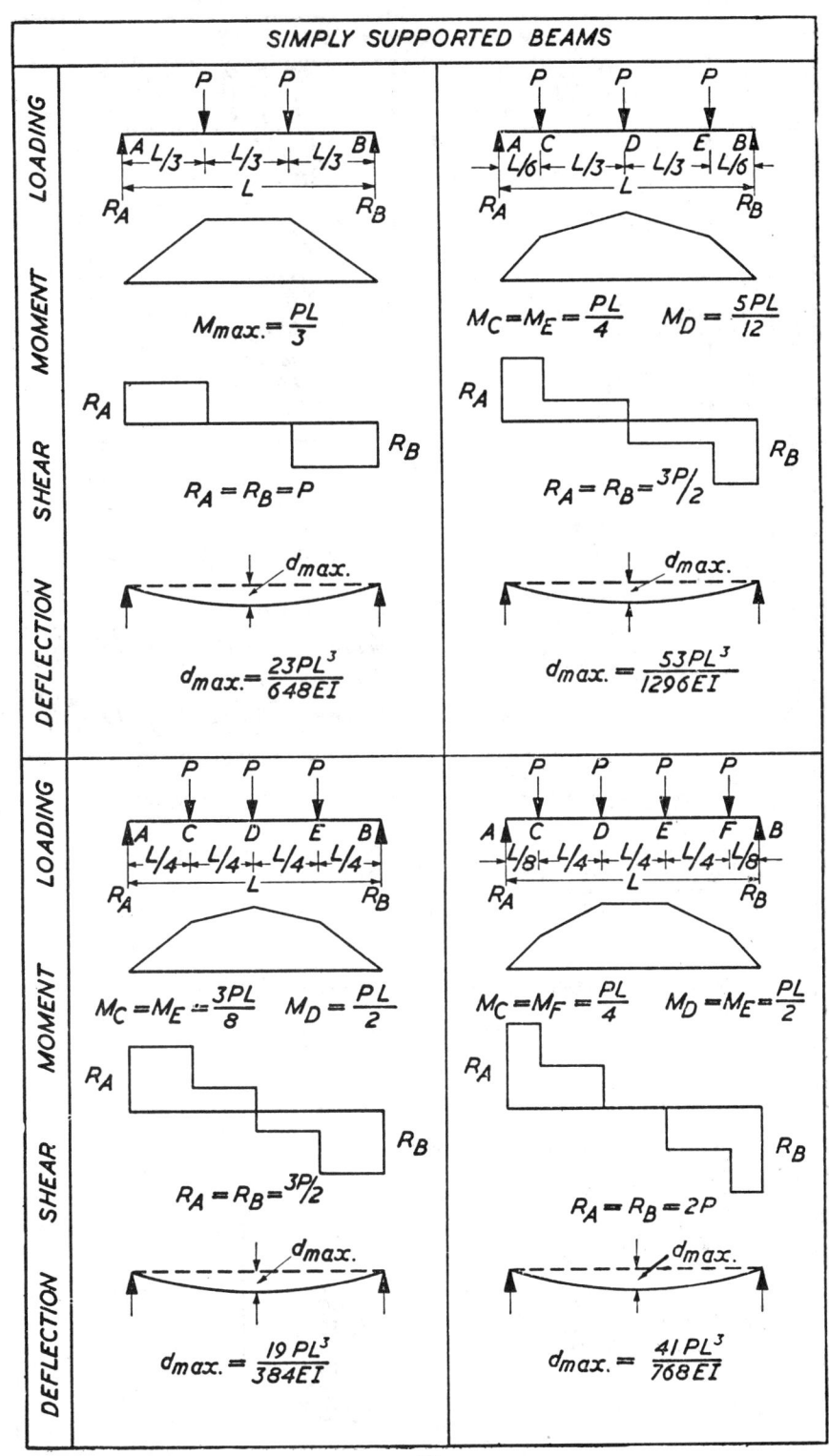

SIMPLY SUPPORTED BEAMS

Loading: Two loads P at $L/3$ from each end (points at $L/3$, $L/3$, $L/3$).

$$M_{max.} = \frac{PL}{3}$$

$$R_A = R_B = P$$

$$d_{max.} = \frac{23 PL^3}{648 EI}$$

Loading: Three loads P at C, D, E spaced $L/6$, $L/3$, $L/3$, $L/6$.

$$M_C = M_E = \frac{PL}{4} \qquad M_D = \frac{5PL}{12}$$

$$R_A = R_B = \frac{3P}{2}$$

$$d_{max.} = \frac{53 PL^3}{1296 EI}$$

Loading: Three loads P at C, D, E spaced $L/4$ each.

$$M_C = M_E = \frac{3PL}{8} \qquad M_D = \frac{PL}{2}$$

$$R_A = R_B = \frac{3P}{2}$$

$$d_{max.} = \frac{19 PL^3}{384 EI}$$

Loading: Four loads P at C, D, E, F spaced $L/8$, $L/4$, $L/4$, $L/4$, $L/8$.

$$M_C = M_F = \frac{PL}{4} \qquad M_D = M_E = \frac{PL}{2}$$

$$R_A = R_B = 2P$$

$$d_{max.} = \frac{41 PL^3}{768 EI}$$

SIMPLY SUPPORTED BEAMS

LOADING / MOMENT / SHEAR / DEFLECTION

Left case (5 equal segments, 4 point loads P at C, D, E, F):

$$M_C = M_F = \frac{2PL}{5} \qquad M_D = M_E = \frac{3PL}{5}$$

$$R_A = R_B = 2P$$

$$d_{max.} = \frac{63\,PL^3}{1000\,EI}$$

Right case ((n−1) forces P spaced at L/n):

When n is odd,
$$M_{max.} = \frac{(n^2-1)\,PL}{8n}$$

When n is even,
$$M_{max.} = n \cdot PL/8$$

$$R_A = R_B = (n-1)\,P/2$$

When n is odd
$$d_{max.} = \frac{PL^3}{192\,EI}\left[n - \frac{1}{n}\right]\left[3 - \frac{1}{2}\left(1 - \frac{1}{n^2}\right)\right]$$

When n is even
$$d_{max.} = \frac{PL^3}{192\,EI} \cdot n\left[3 - \frac{1}{2}\left(1 + \frac{4}{n^2}\right)\right]$$

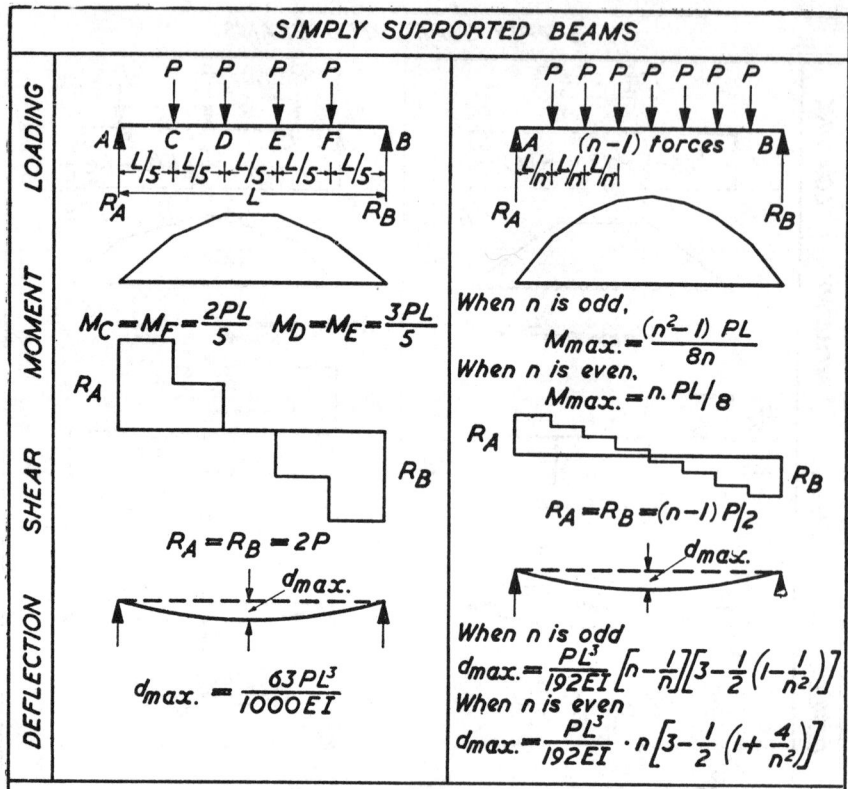

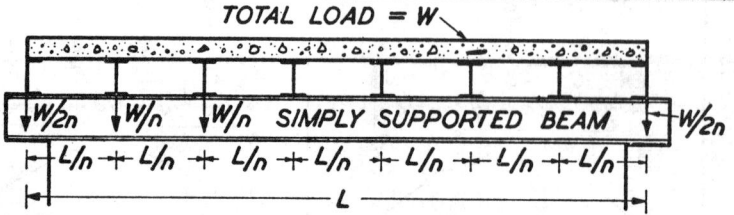

TOTAL LOAD = W — SIMPLY SUPPORTED BEAM
End loads W/2n, interior loads W/n, spacing L/n.

When n > 10, consider the load uniformly distributed

The reaction at the supports = W/2, but the maximum S.F.
at the ends of the beam = $\dfrac{W(n-1)}{2n} = A.W$

The value of the maximum bending moment = C.WL

The value of the deflection at the centre of the span = $k \cdot \dfrac{WL^3}{EI}$

Value of n	A	C	k
2	0·2500	0·1250	0·0105
3	0·3333	0·1111	0·0118
4	0·3750	0·1250	0·0124
5	0·4000	0·1200	0·0126
6	0·4167	0·1250	0·0127
7	0·4286	0·1224	0·0128
8	0·4375	0·1250	0·0128
9	0·4444	0·1236	0·0129
10	0·4500	0·1250	0·0129

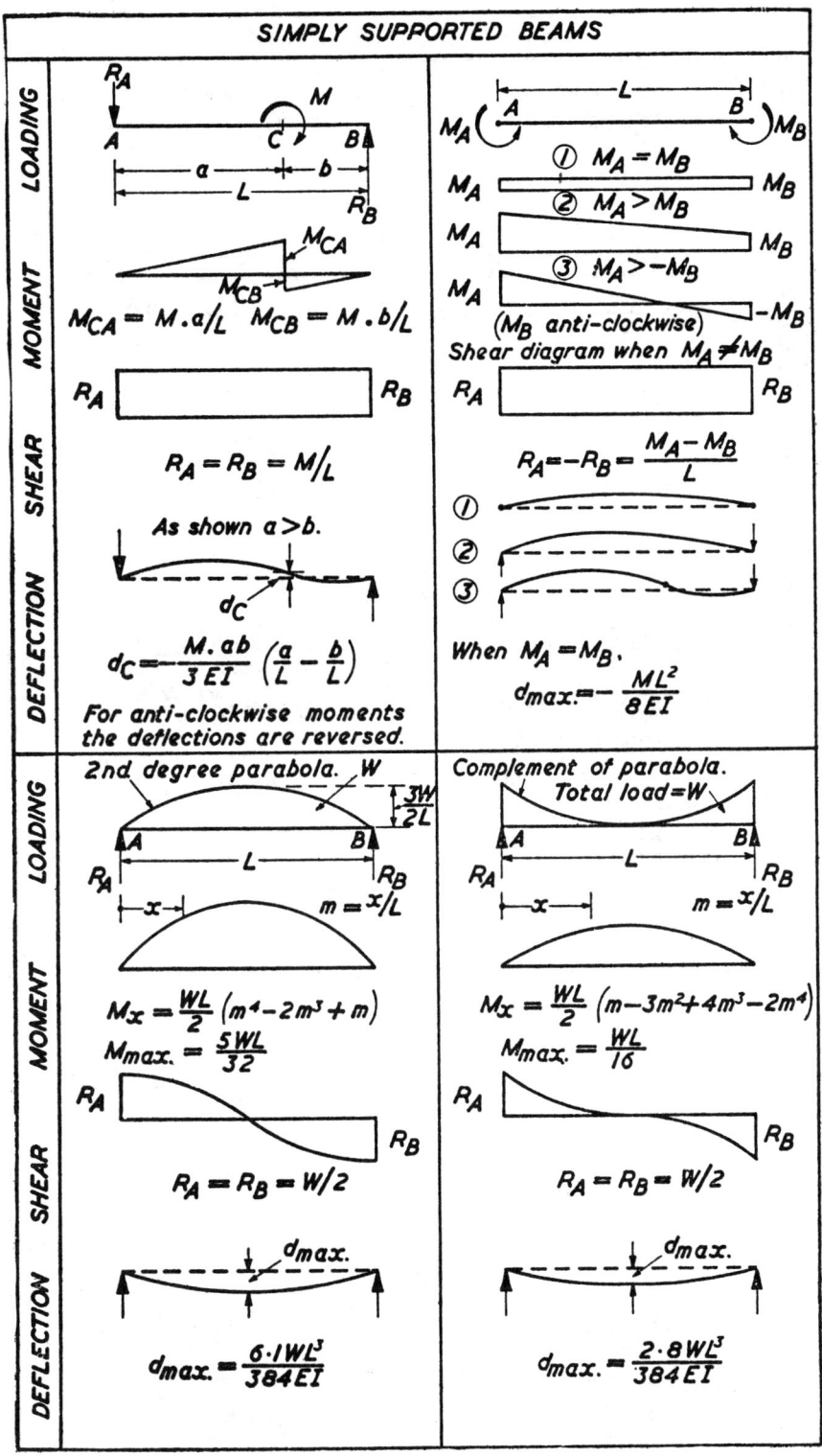

SIMPLY SUPPORTED BEAMS

Panel 1 (top-left): Moment M applied at point C

LOADING: Beam AB with moment M applied at C, distance a from A, distance b from B, total length L. Reactions R_A, R_B.

MOMENT:
$$M_{CA} = M \cdot a/L \qquad M_{CB} = M \cdot b/L$$

SHEAR:
$$R_A = R_B = M/L$$

DEFLECTION: As shown $a > b$.
$$d_C = -\frac{M \cdot ab}{3EI}\left(\frac{a}{L} - \frac{b}{L}\right)$$

For anti-clockwise moments the deflections are reversed.

Panel 2 (top-right): End moments M_A and M_B

LOADING: Beam with end moments M_A and M_B, length L.

MOMENT diagrams:
① $M_A = M_B$
② $M_A > M_B$
③ $M_A > -M_B$ (M_B anti-clockwise)

Shear diagram when $M_A \ne M_B$

SHEAR:
$$R_A = -R_B = \frac{M_A - M_B}{L}$$

DEFLECTION: When $M_A = M_B$,
$$d_{max.} = -\frac{ML^2}{8EI}$$

Panel 3 (bottom-left): 2nd degree parabola load

LOADING: 2nd degree parabola. Total load W, peak $\frac{3W}{2L}$, length L, $m = x/L$.

MOMENT:
$$M_x = \frac{WL}{2}(m^4 - 2m^3 + m)$$
$$M_{max.} = \frac{5WL}{32}$$

SHEAR:
$$R_A = R_B = W/2$$

DEFLECTION:
$$d_{max.} = \frac{6 \cdot 1 WL^3}{384 EI}$$

Panel 4 (bottom-right): Complement of parabola

LOADING: Complement of parabola. Total load $= W$, length L, $m = x/L$.

MOMENT:
$$M_x = \frac{WL}{2}(m - 3m^2 + 4m^3 - 2m^4)$$
$$M_{max.} = \frac{WL}{16}$$

SHEAR:
$$R_A = R_B = W/2$$

DEFLECTION:
$$d_{max.} = \frac{2 \cdot 8 WL^3}{384 EI}$$

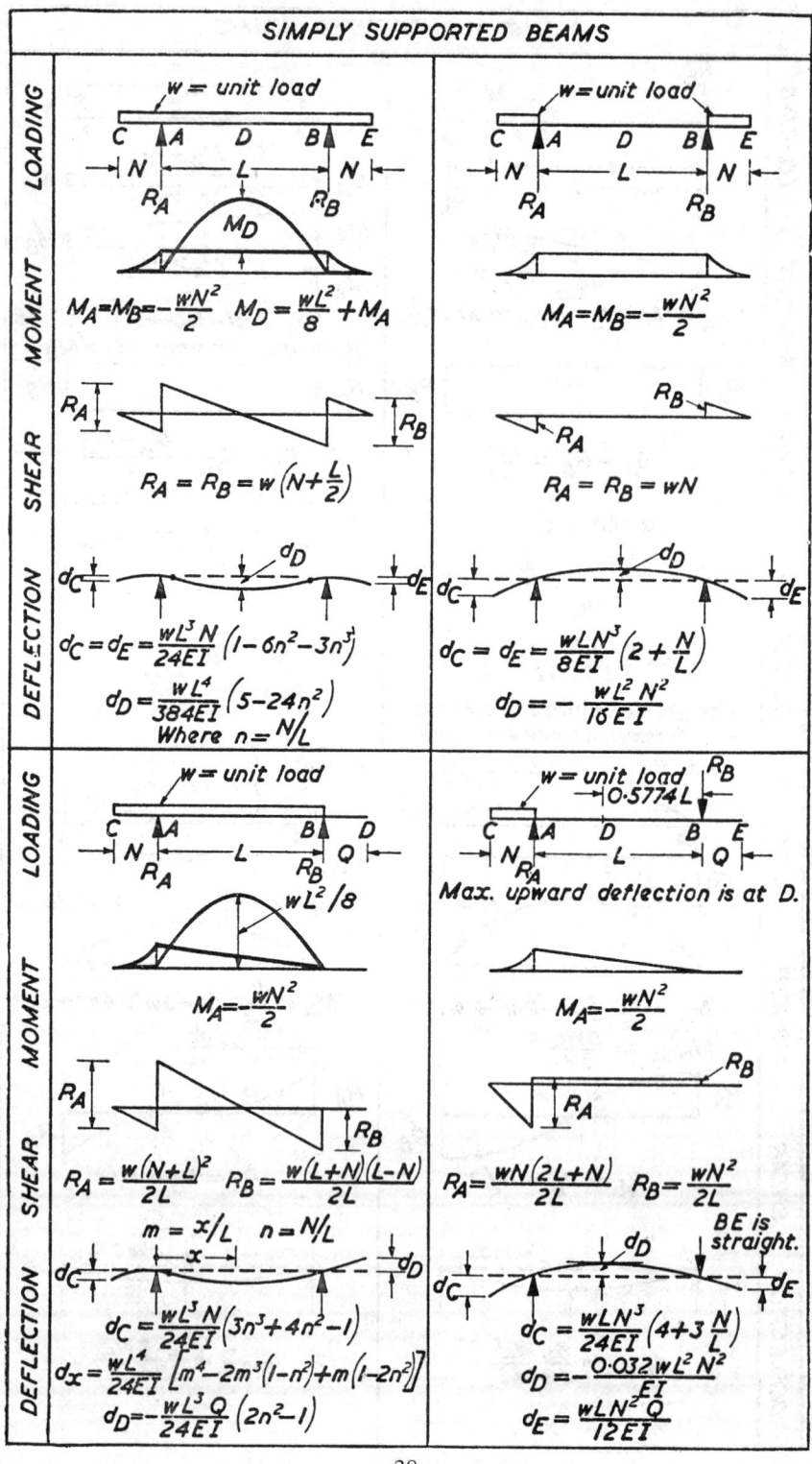

SIMPLY SUPPORTED BEAMS

Case 1: Uniform load with overhangs on both sides

LOADING: w = unit load, beam with points C, A, D, B, E; overhangs N on each side, span L between supports R_A and R_B.

MOMENT:
$$M_A = M_B = -\frac{wN^2}{2} \qquad M_D = \frac{wL^2}{8} + M_A$$

SHEAR:
$$R_A = R_B = w\left(N + \frac{L}{2}\right)$$

DEFLECTION:
$$d_C = d_E = \frac{wL^3 N}{24EI}\left(1 - 6n^2 - 3n^3\right)$$
$$d_D = \frac{wL^4}{384EI}\left(5 - 24n^2\right)$$
Where $n = N/L$

Case 2: Load only on overhangs

LOADING: w = unit load on overhangs only.

MOMENT:
$$M_A = M_B = -\frac{wN^2}{2}$$

SHEAR:
$$R_A = R_B = wN$$

DEFLECTION:
$$d_C = d_E = \frac{wLN^3}{8EI}\left(2 + \frac{N}{L}\right)$$
$$d_D = -\frac{wL^2 N^2}{16EI}$$

Case 3: Uniform load with overhang on one side

LOADING: w = unit load, overhang N at left, span L, free end Q at right.

MOMENT: $wL^2/8$
$$M_A = -\frac{wN^2}{2}$$

SHEAR:
$$R_A = \frac{w(N+L)^2}{2L} \qquad R_B = \frac{w(L+N)(L-N)}{2L}$$

DEFLECTION:
$m = x/L \qquad n = N/L$
$$d_C = \frac{wL^3 N}{24EI}\left(3n^3 + 4n^2 - 1\right)$$
$$d_x = \frac{wL^4}{24EI}\left[m^4 - 2m^3(1-n^2) + m(1-2n^2)\right]$$
$$d_D = -\frac{wL^3 Q}{24EI}\left(2n^2 - 1\right)$$

Case 4: Partial uniform load

LOADING: w = unit load over $0.5774L$, R_B at right.
Max. upward deflection is at D.

MOMENT:
$$M_A = -\frac{wN^2}{2}$$

SHEAR:
$$R_A = \frac{wN(2L+N)}{2L} \qquad R_B = \frac{wN^2}{2L}$$

DEFLECTION: BE is straight.
$$d_C = \frac{wLN^3}{24EI}\left(4 + 3\frac{N}{L}\right)$$
$$d_D = -\frac{0.032\, wL^2 N^2}{EI}$$
$$d_E = \frac{wLN^2 Q}{12EI}$$

3. FIXED, BUILT-IN OR ENCASTRE BEAMS

WHEN the ends of a beam are firmly held so that they cannot rotate under the action of the superimposed loads, the beam is known as a fixed, built-in or encastré beam. The B.M. diagram for such a beam is in two parts, viz.: the free or positive B.M. diagram, which would have resulted had the ends been simply supported, i.e. free to rotate, and the fixing or negative B.M. diagram which results from the restraints imposed upon the ends of the beam.

Normally, the supports for built-in beams are on the same level and the ends of the beams are horizontal. This type will be considered first.

Beams with Supports at the same Level

The conditions for solution, derived by Mohr, are two, viz.:

(i) The area of the fixing or negative B.M. diagram is equal to that of the free or positive B.M. diagram.
(ii) The centres of gravity of the two diagrams lie in the same vertical line, i.e. are equidistant from a given end of the beam.

Fig. 1

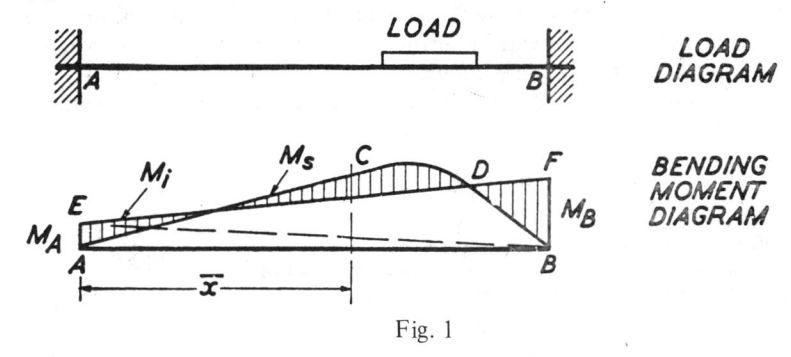

Fig. 1 shows a typical B.M. diagram for a built-in beam.

$ACDB$ is the diagram of the free moment M_s and the trapezium $AEFB$ is the diagram of the fixing moment M_i, the portions shaded representing the final diagram.

Let A_s = the area of the free B.M. diagram
and A_i = the area of the fixing moment diagram.

Then from condition (i) above, $A_s = A_i$, while from condition (ii) their centres of gravity lie in the same vertical line, say, distance $\bar{x}$ from the left-hand support A.

Now AE = the fixing moment M_A
and BF = the fixing moment M_B.

Therefore $$\frac{M_A + M_B}{2} \times L = A_i$$

and $$M_A + M_B = \frac{2A_i}{L} \quad \dots \dots \dots \dots \text{(a)}$$

FIXED, BUILT-IN OR ENCASTRE BEAMS

Divide the trapezium $AEFB$ by drawing the diagonal EB and take area moments about the support A.

Then
$$A_i \cdot \bar{x} = \left(\frac{M_A \times L}{2} \times \frac{L}{3}\right) + \left(\frac{M_B \times L}{2} \times \frac{2L}{3}\right)$$

$$= \frac{L^2}{6}(M_A + 2M_B)$$

$$M_A + 2M_B = \frac{6A_i\bar{x}}{L^2} \quad \ldots \ldots \ldots \ldots \ldots \ldots \text{(b)}$$

But
$$M_A + M_B = \frac{2A_i}{L} \quad \ldots \ldots \ldots \ldots \ldots \text{(a)}$$

Also,
$$A_s = A_i$$

Subtracting (a) from (b) and substituting A_s for A_i:

$$M_B = \frac{6A_s\bar{x}}{L^2} - \frac{2A_s}{L}.$$

Similarly
$$M_A = \frac{4A_s}{L} - \frac{6A_s\bar{x}}{L^2}.$$

It will be seen, therefore, that the fixing moments for any built-in beam on level supports can be calculated provided that the area of the free B.M. diagram and the position of its centre of gravity are known.

For point loads, however, the principle of reciprocal moments provides the simplest solution.

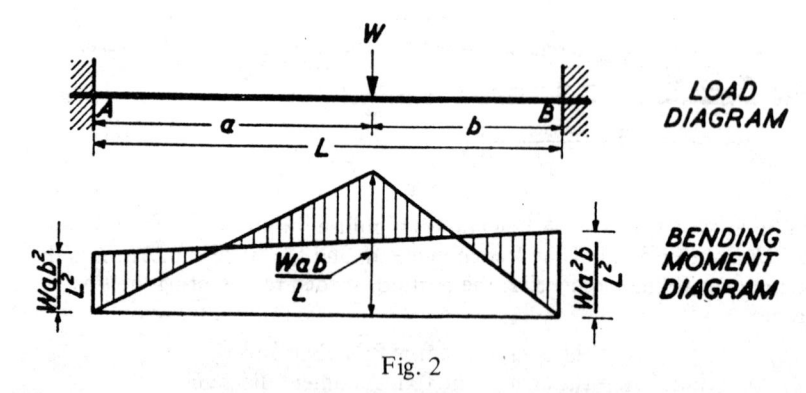

Fig. 2

With reference to Fig. 2,

$$M_A = \frac{Wab}{L} \times \frac{b}{L} = \frac{Wab^2}{L^2}$$

$$M_B = \frac{Wab}{L} \times \frac{a}{L} = \frac{Wa^2b}{L^2},$$

i.e the fixing moments are in reciprocal proportion to the distances of the ends of the beam from the point load.

SHEAR FORCES IN FIXED BEAMS

In the case of several isolated loads, this principle is applied to each load in turn and the results summated.

It should be noted that appropriate formulae for built-in beams are given on pp. 43 to 49.

Beams with Supports at Different Levels

The ends are assumed, as before, to be horizontal.

The bent form of the unloaded beam as shown in Fig. 3 is similar to the bent form of two simple cantilevers which can be achieved by cutting the beam at the

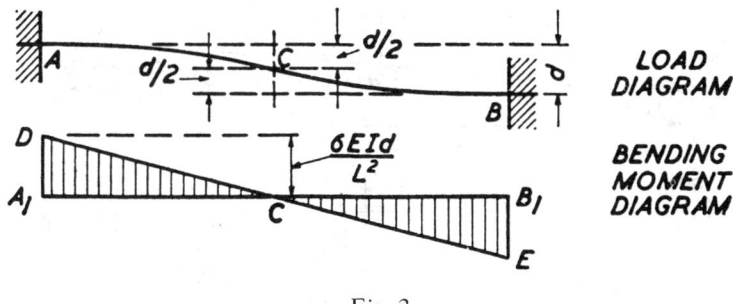

Fig. 3

centre C, and placing downward and upward loads at the free ends of the cantilevers such that the deflection at the end of each cantilever is $d/2$.

Therefore $\dfrac{d}{2} = \dfrac{P(L/2)^3}{3EI}$, (being the standard deflection formula)

or $$P = \dfrac{12EId}{L^3}.$$

This load would cause a B.M. at A or B equal to

$$P \times \dfrac{L}{2} = \dfrac{12EId}{L^3} \times \dfrac{L}{2} = \dfrac{6EId}{L^2}.$$

The solution in any given case consists of adding to the ordinary diagram of B.M.s, the B.M. diagram $A_1 DCEB_1$.

Shear Forces in Fixed Beams

It must be noted that in the case of fixed beams, it is necessary to evaluate the B.M.s before the S.F.s can be determined. This is the converse of the procedure in the case of simply supported beams.

The S.F. at the ends of a beam is found in the following manner:

$$\text{S.F.}_A = \text{the simple support reaction at } A + \dfrac{M_A - M_B}{L}$$

$$\text{S.F.}_B = \text{the simple support reaction at } B + \dfrac{M_B - M_A}{L},$$

FIXED, BUILT-IN OR ENCASTRE BEAMS

where M_A and M_B are the numerical values of the moments at the ends of the beam.

These formulae must be followed exactly with respect to the signs shown since if M_A is smaller than M_B the signs will adjust themselves.

It will be seen that for symmetrical loads where $M_A = M_B$, the reactions will be the same as for simply supported beams.

As an example, consider Fig. 4

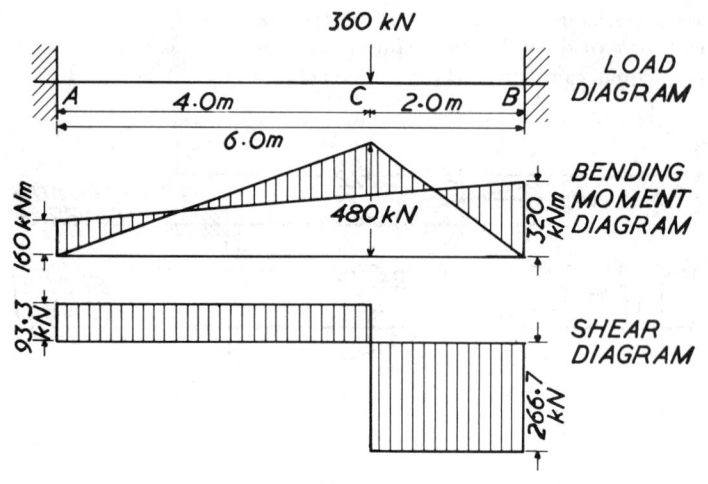

Fig. 4

If AB were simply supported, then the maximum M, at C, would be:

$$\frac{Wab}{L} = \frac{360 \times 4 \times 2}{6} = 480 \text{ kNm}$$

while R_A would be 120 kN and R_B 240 kN.

By the principle of reciprocal moments (p. 26)

$$M_A = 160 \text{ kNm} \quad \text{and} \quad M_B = 320 \text{ kNm}$$

Now M_A could be considered as being caused by a downward force at B equal to $M_A/L = 26.67$ kN, which would have to be balanced at A, for the equilibrium of vertical loads, by an upward force of 26.67 kN. Similarly M_B could be caused by a downward force of 53.33 kN at A, with an upward balancing force of 53.33 kN at B.

Hence, the final shear force at A

$$= \text{the simple support reaction at } A + \left(\frac{M_A}{L} - \frac{M_B}{L}\right)$$

$$= 120 + 26.67 - 53.33 = 93.3 \text{ kN}$$

Similarly, the final shear force at B

$$= \text{the simple support reaction at } B + \left(\frac{M_B}{L} - \frac{M_A}{L}\right)$$

$$= 240 + 53.33 - 26.67 = 266.7 \text{ kN}$$

A simple rule to remember is that the shear force is greater than the simple support reaction at the fixed end which has the greater numerical fixing moment.

BUILT−IN BEAMS

LOADING / MOMENT / SHEAR / DEFLECTION

Case 1: Uniformly distributed load W over span L

$$M_A = M_B = -\frac{WL}{12}$$
$$M_C = \frac{WL}{24}$$
$$R_A = R_B = W/2$$

Deflection: $0.21L \mid 0.58L \mid 0.21L$

$$d_{max.} = \frac{WL^3}{384EI}$$

Case 2: Two equal loads W/2 at distance a from each support

$$M_A = M_B = -\frac{Wa}{12L}(3L-2a)$$

Moment at centre $= \frac{Wa}{4}$

$$R_A = R_B = W/2$$

$$d_{max.} = \frac{Wa^2}{48EI}(L-a)$$

Case 3: Distributed load W over portion b, with a and c on either side (d = a+b, e = b+c)

$$M_A = \frac{-W}{12L^2 b}\left[e^3(4L-3e) - c^3(4L-3c)\right]$$

$$M_B = \frac{-W}{12L^2 b}\left[d^3(4L-3d) - a^3(4L-3a)\right]$$

When r is the simple support reaction
$$R_A = r_A + \frac{M_A - M_B}{L} \qquad R_B = r_B + \frac{M_B - M_A}{L}$$

When $a = c$,
$$d_{max.} = \frac{W}{384EI}(L^3 + 2L^2 a + 4La^2 - 8a^3)$$

Case 4: Load W over portion b from support B, with a on support A side; $a/L = m$

$$M_A = -\frac{WL}{12}\cdot m\,(3m^2 - 8m + 6)$$

$$M_B = -\frac{WL}{12}\cdot m^2(4 - 3m)$$

$$+M_{max.} = \frac{WL}{12}m^2\left(-\frac{3}{2}m^5 + 6m^4 - 6m^3 + 6m^2 + 15m - 8\right)$$

When $x = \frac{a}{2}(m^3 - 2m^2 + 2)$

$$R_A = \frac{W(m^3 - 2m^2 + 2)}{2} \qquad R_B = \frac{W\cdot m^3(2-m)}{2m}$$

When $a = L/2$ and $x_1 = 0.445L$
$$d_{max.} = \frac{WL^3}{333EI}$$
$$d_C = \frac{WL^3}{384EI}$$

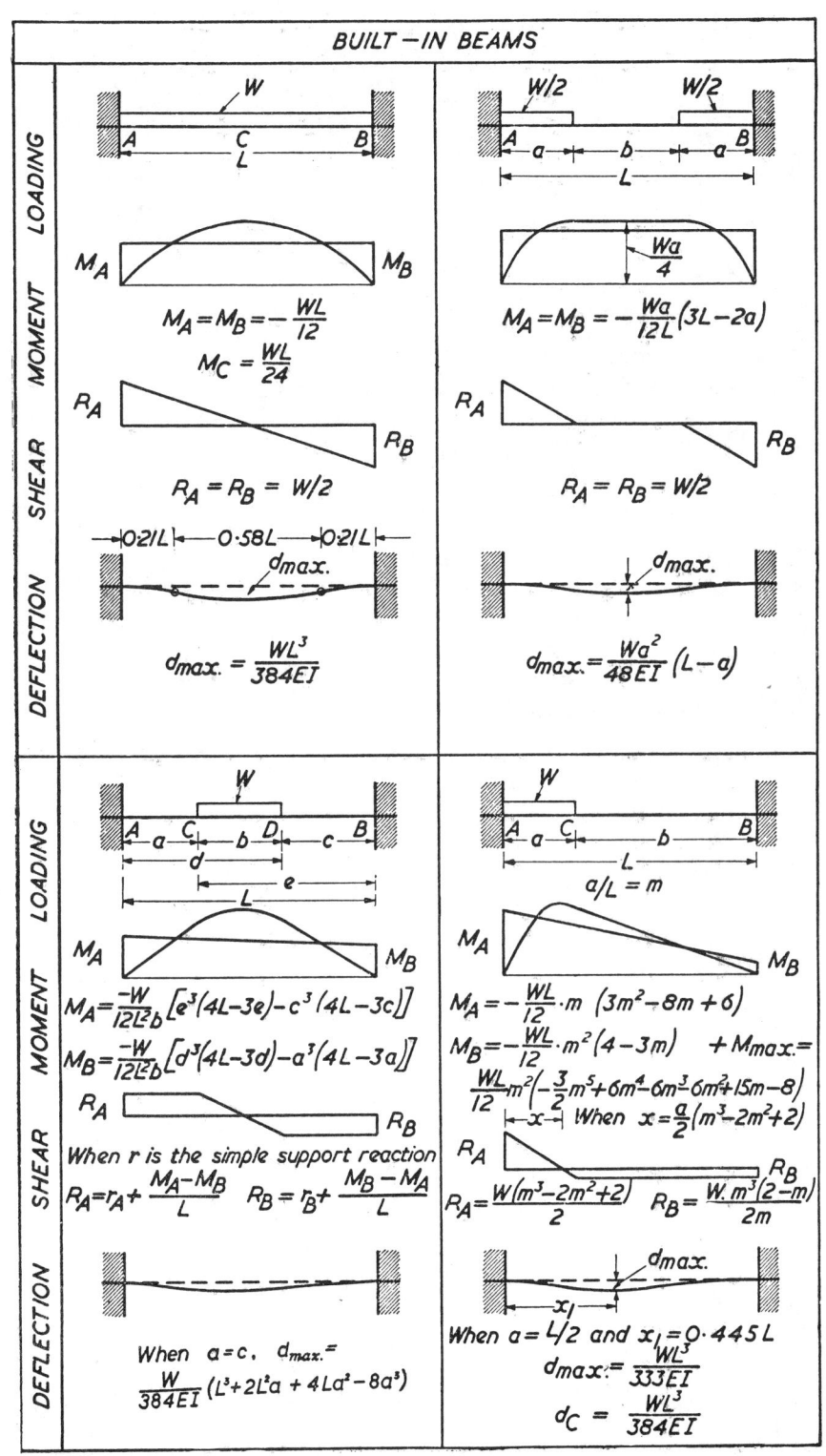

BUILT-IN BEAMS

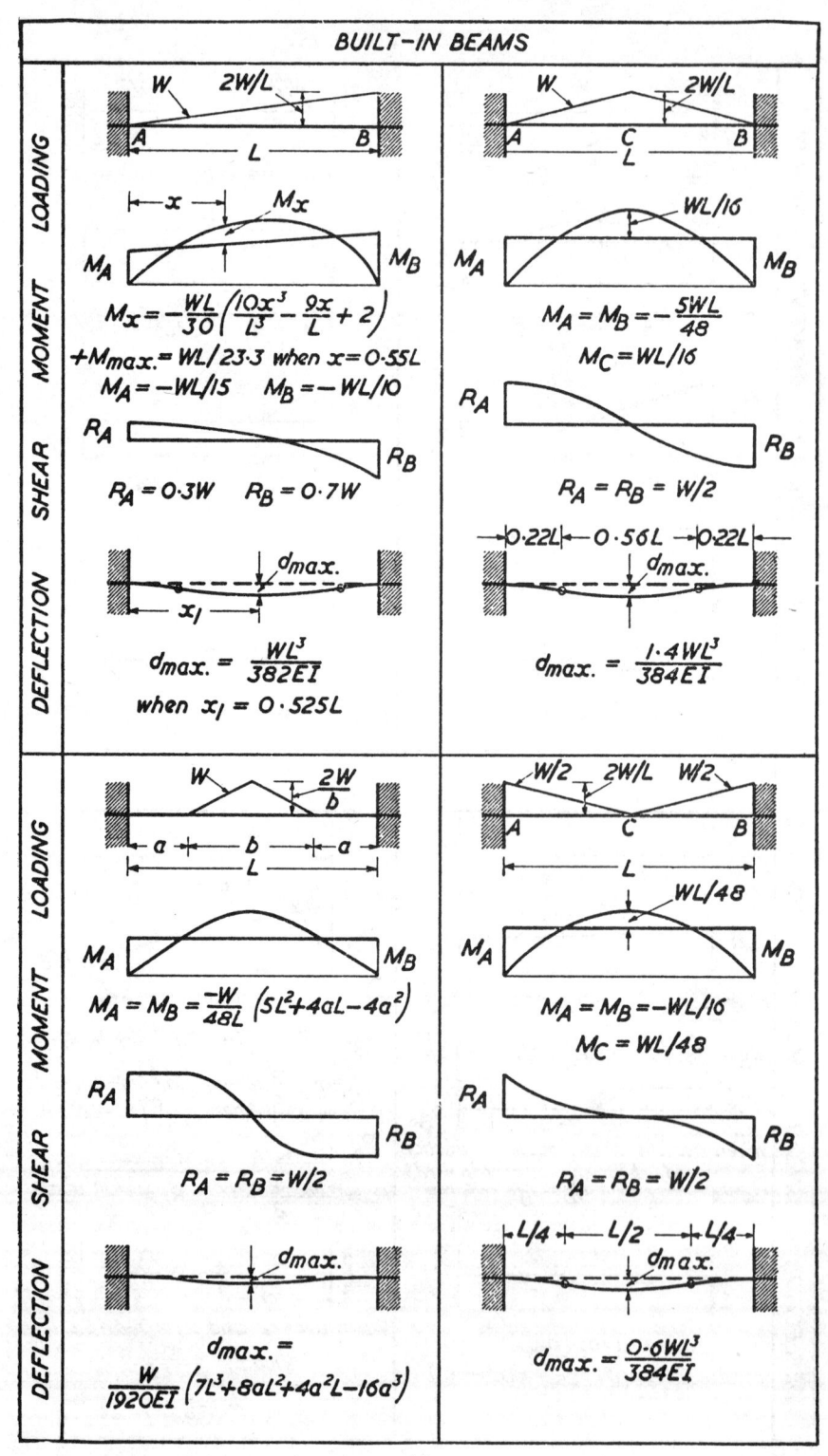

Case 1: Triangular load, max at B

LOADING: W, $2W/L$, span A to B, length L

MOMENT:
$$M_x = -\frac{WL}{30}\left(\frac{10x^3}{L^3} - \frac{9x}{L} + 2\right)$$
$+M_{max.} = WL/23.3$ when $x = 0.55L$
$M_A = -WL/15 \quad M_B = -WL/10$

SHEAR: $R_A = 0.3W \quad R_B = 0.7W$

DEFLECTION:
$$d_{max.} = \frac{WL^3}{382EI}$$
when $x_1 = 0.525L$

Case 2: Triangular load, symmetric peak at C

LOADING: W, $2W/L$ at centre C, span L

MOMENT: $WL/16$
$$M_A = M_B = -\frac{5WL}{48}$$
$M_C = WL/16$

SHEAR: $R_A = R_B = W/2$

DEFLECTION: $0.22L$ — $0.56L$ — $0.22L$
$$d_{max.} = \frac{1.4WL^3}{384EI}$$

Case 3: Triangular load over central length b

LOADING: W, peak $2W/b$, with $a + b + a = L$

MOMENT:
$$M_A = M_B = \frac{-W}{48L}\left(5L^2 + 4aL - 4a^2\right)$$

SHEAR: $R_A = R_B = W/2$

DEFLECTION:
$$d_{max.} = \frac{W}{1920EI}\left(7L^3 + 8aL^2 + 4a^2L - 16a^3\right)$$

Case 4: V-shaped load

LOADING: $W/2$, $2W/L$, $W/2$; span L, centre C

MOMENT: $WL/48$
$$M_A = M_B = -WL/16$$
$M_C = WL/48$

SHEAR: $R_A = R_B = W/2$

DEFLECTION: $L/4$ — $L/2$ — $L/4$
$$d_{max.} = \frac{0.6WL^3}{384EI}$$

BUILT-IN BEAMS

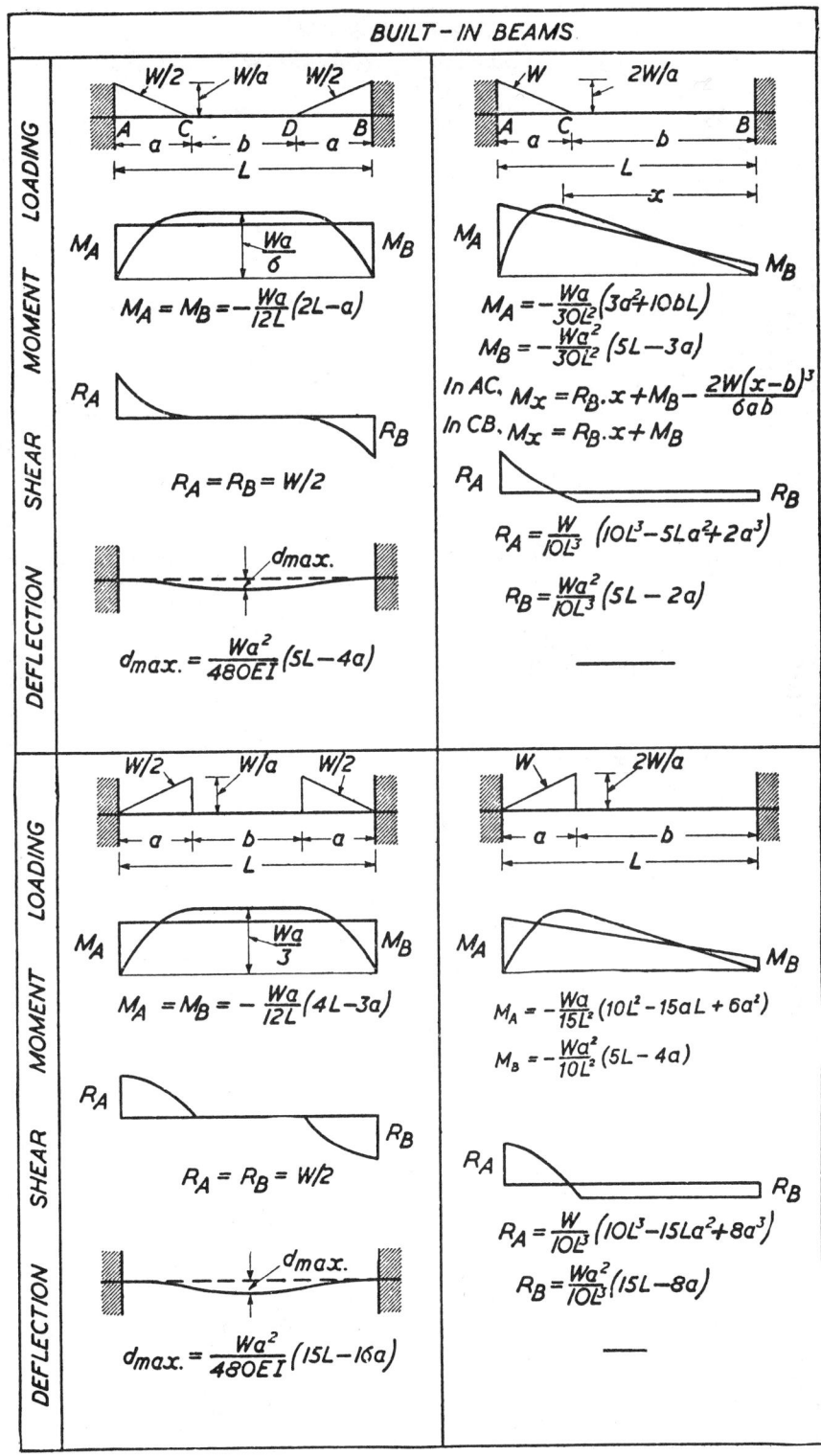

Top-left panel (Loading: symmetric triangular loads W/2 at ends, W/a peak at center, spans a-b-a):

$$M_A = M_B = -\frac{Wa}{12L}(2L-a)$$

Peak moment at center: $\frac{Wa}{6}$

$$R_A = R_B = W/2$$

$$d_{max.} = \frac{Wa^2}{480EI}(5L-4a)$$

Top-right panel (Loading: triangular load, W at A, peak 2W/a at C, span a then b to B):

$$M_A = -\frac{Wa}{30L^2}(3a^2+10bL)$$

$$M_B = -\frac{Wa^2}{30L^2}(5L-3a)$$

In AC, $M_x = R_B \cdot x + M_B - \frac{2W(x-b)^3}{6ab}$

In CB, $M_x = R_B \cdot x + M_B$

$$R_A = \frac{W}{10L^3}(10L^3 - 5La^2 + 2a^3)$$

$$R_B = \frac{Wa^2}{10L^3}(5L-2a)$$

Bottom-left panel (Loading: two triangular loads separated by gap b, each of base a, peak W/a):

$$M_A = M_B = -\frac{Wa}{12L}(4L-3a)$$

Peak moment: $\frac{Wa}{3}$

$$R_A = R_B = W/2$$

$$d_{max.} = \frac{Wa^2}{480EI}(15L-16a)$$

Bottom-right panel (Loading: triangular load at left, base a, peak 2W/a, then span b):

$$M_A = -\frac{Wa}{15L^2}(10L^2 - 15aL + 6a^2)$$

$$M_B = -\frac{Wa^2}{10L^2}(5L-4a)$$

$$R_A = \frac{W}{10L^3}(10L^3 - 15La^2 + 8a^3)$$

$$R_B = \frac{Wa^2}{10L^3}(15L-8a)$$

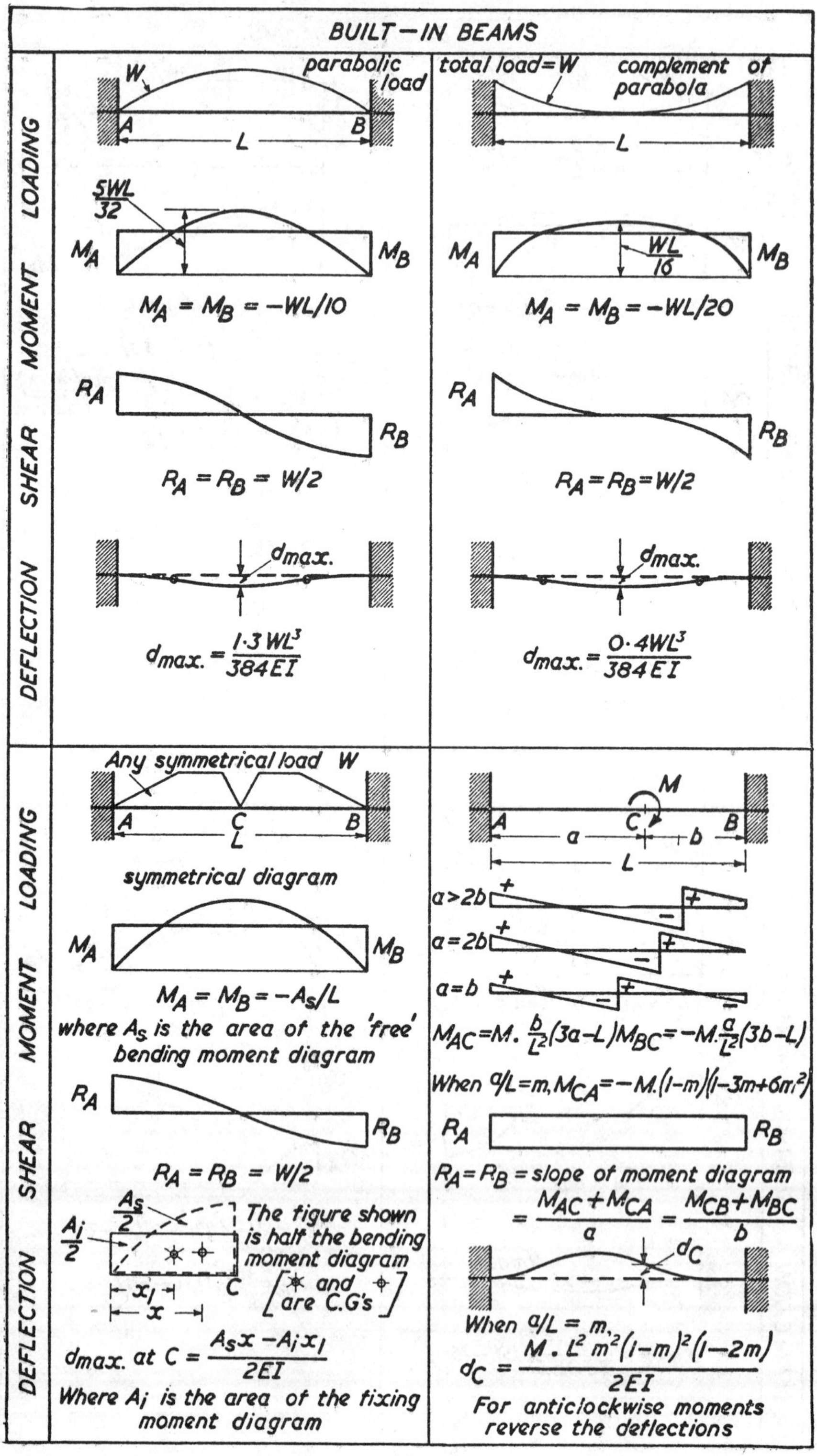

BUILT—IN BEAMS
LOADING
MOMENT
SHEAR
DEFLECTION
parabolic load
W
A
B
L
5WL/32
M_A
M_B
M_A = M_B = −WL/10
R_A
R_B
R_A = R_B = W/2
d_max.
d_max. = 1·3 WL³/384EI
total load=W
complement of parabola
WL/16
M_A = M_B = −WL/20
R_A = R_B = W/2
d_max. = 0·4WL³/384EI
Any symmetrical load W
C
symmetrical diagram
M_A = M_B = −A_s/L
where A_s is the area of the 'free' bending moment diagram
R_A = R_B = W/2
A_s/2
A_i/2
x_1
x
The figure shown is half the bending moment diagram
[✳ and ⌖ are C.G's]
d_max. at C = (A_s x − A_i x_1)/2EI
Where A_i is the area of the fixing moment diagram
M
a
b
a>2b
a=2b
a=b
+
−
M_AC = M. (b/L²)(3a−L) M_BC = −M.(a/L²)(3b−L)
When a/L = m, M_CA = −M.(1−m)(1−3m+6m²)
R_A = R_B = slope of moment diagram
= (M_AC + M_CA)/a = (M_CB + M_BC)/b
d_C
When a/L = m,
d_C = M . L² m²(1−m)²(1−2m)/2EI
For anticlockwise moments reverse the deflections

BUILT - IN BEAMS

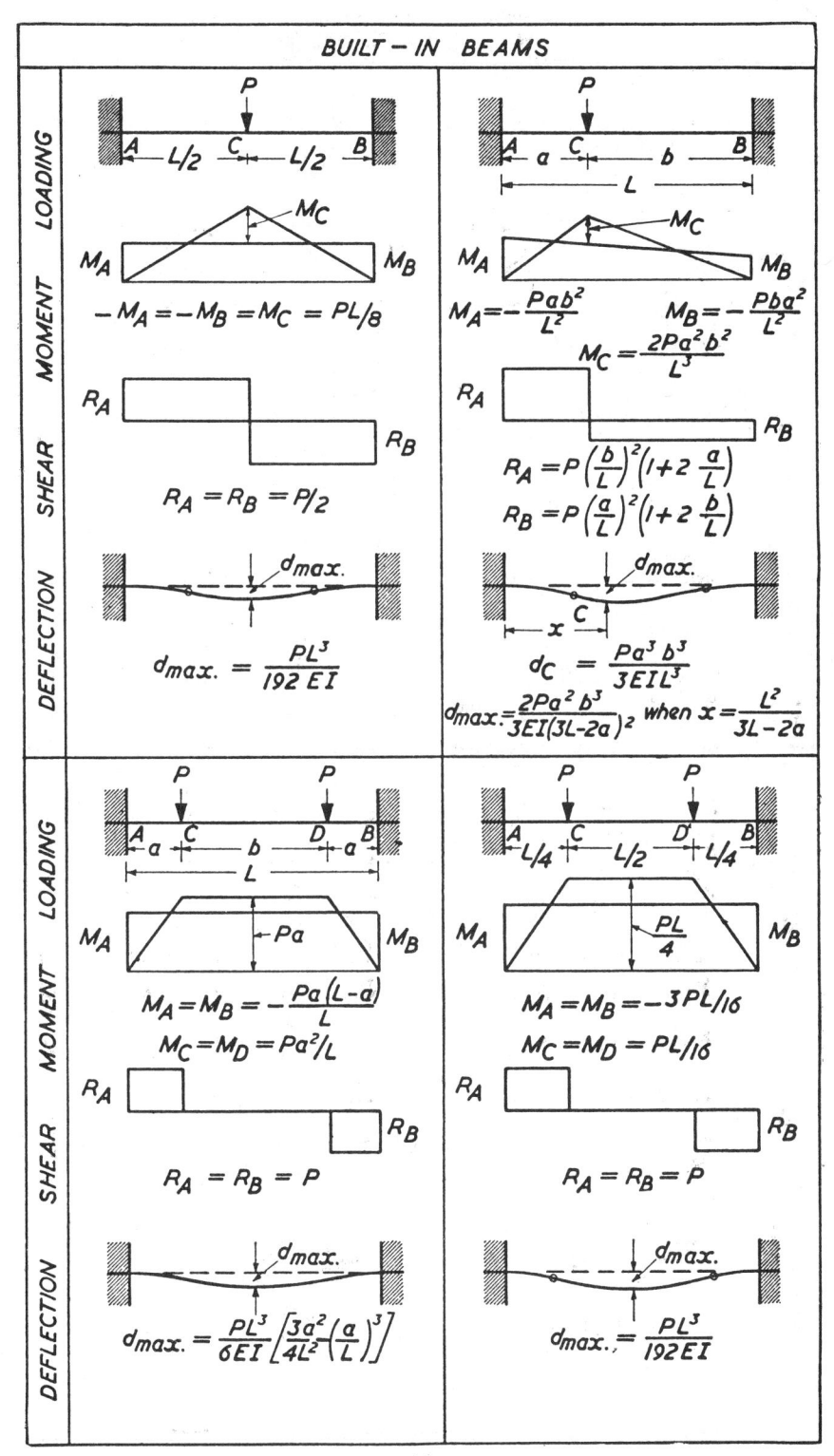

Case 1: Central point load

LOADING / MOMENT:

$-M_A = -M_B = M_C = PL/8$

SHEAR:

$R_A = R_B = P/2$

DEFLECTION:

$$d_{max.} = \frac{PL^3}{192\,EI}$$

Case 2: Point load at distance a

MOMENT:

$M_A = -\dfrac{Pab^2}{L^2}$, $M_B = -\dfrac{Pba^2}{L^2}$

$M_C = \dfrac{2Pa^2b^2}{L^3}$

SHEAR:

$R_A = P\left(\dfrac{b}{L}\right)^2 \left(1 + 2\,\dfrac{a}{L}\right)$

$R_B = P\left(\dfrac{a}{L}\right)^2 \left(1 + 2\,\dfrac{b}{L}\right)$

DEFLECTION:

$$d_C = \frac{Pa^3 b^3}{3EIL^3}$$

$$d_{max.} = \frac{2Pa^2 b^3}{3EI(3L-2a)^2} \quad \text{when } x = \frac{L^2}{3L-2a}$$

Case 3: Two symmetric point loads at distance a from supports

MOMENT:

$M_A = M_B = -\dfrac{Pa(L-a)}{L}$

$M_C = M_D = Pa^2/L$

SHEAR:

$R_A = R_B = P$

DEFLECTION:

$$d_{max.} = \frac{PL^3}{6EI}\left[\frac{3a^2}{4L^2}\left(\frac{a}{L}\right)^3\right]$$

Case 4: Two point loads at L/4 from supports

MOMENT:

$M_A = M_B = -3PL/16$

$M_C = M_D = PL/16$

SHEAR:

$R_A = R_B = P$

DEFLECTION:

$$d_{max.} = \frac{PL^3}{192\,EI}$$

BUILT — IN BEAMS

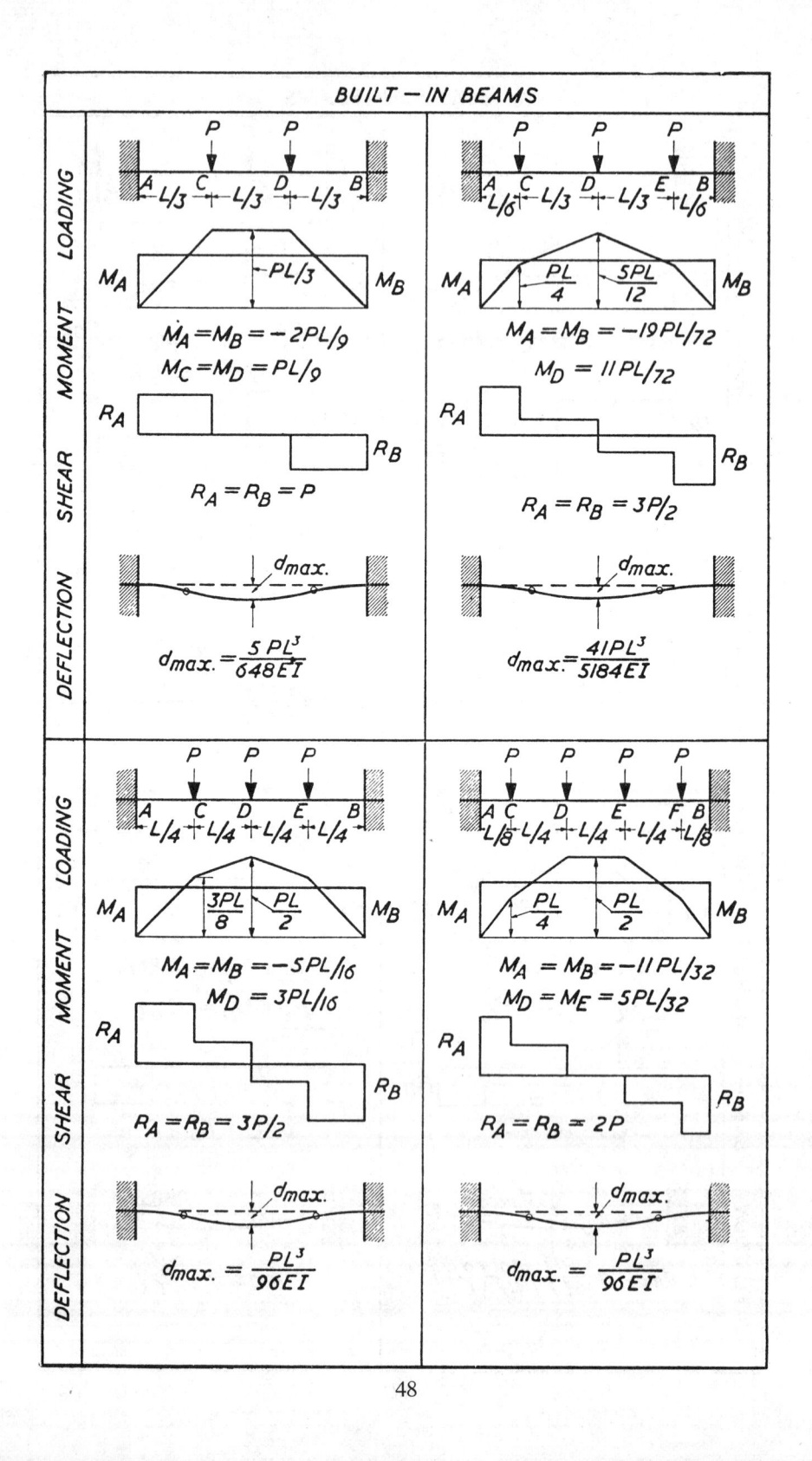

Loading: Two loads P at C and D, spaced $L/3$, $L/3$, $L/3$.

Moment:
$$M_A = M_B = -2PL/9$$
$$M_C = M_D = PL/9$$

Shear:
$$R_A = R_B = P$$

Deflection:
$$d_{max.} = \frac{5PL^3}{648EI}$$

Loading: Three loads P at C, D, E, spaced $L/6$, $L/3$, $L/3$, $L/6$.

Moment: $PL/4$ and $5PL/12$
$$M_A = M_B = -19PL/72$$
$$M_D = 11PL/72$$

Shear:
$$R_A = R_B = 3P/2$$

Deflection:
$$d_{max.} = \frac{41PL^3}{5184EI}$$

Loading: Three loads P at C, D, E, spaced $L/4$ each.

Moment: $3PL/8$ and $PL/2$
$$M_A = M_B = -5PL/16$$
$$M_D = 3PL/16$$

Shear:
$$R_A = R_B = 3P/2$$

Deflection:
$$d_{max.} = \frac{PL^3}{96EI}$$

Loading: Four loads P at C, D, E, F, spaced $L/8$, $L/4$, $L/4$, $L/4$, $L/8$.

Moment: $PL/4$ and $PL/2$
$$M_A = M_B = -11PL/32$$
$$M_D = M_E = 5PL/32$$

Shear:
$$R_A = R_B = 2P$$

Deflection:
$$d_{max.} = \frac{PL^3}{96EI}$$

BUILT – IN BEAMS

LOADING / MOMENT (left case): Four loads P at C, D, E, F between fixed supports A and B, spacing $L/5$.

$$M_A = M_B = -2PL/5$$
$$M_D = M_E = PL/5$$

Peak positive moment $= \dfrac{3PL}{5}$

SHEAR (left): $R_A = R_B = 2P$

DEFLECTION (left):
$$d_{max} = \dfrac{13\,PL^3}{1000\,EI}$$

LOADING / MOMENT (right case): $(n-1)$ forces P between fixed supports, spacing $= L/n$.

$$M_A = M_B = -\dfrac{PL(n^2-1)}{12n}$$

SHEAR (right): $R_A = R_B = (n-1)P/2$

DEFLECTION (right):

When n is odd,
$$d_{max} = \dfrac{PL^3}{192\,EI}\left[n - \dfrac{1}{n}\right]\left[1 - \dfrac{1}{2}\left(1 - \dfrac{1}{n^2}\right)\right]$$

When n is even,
$$d_{max} = \dfrac{PL^3}{192\,EI}\left\{\left[3 - \dfrac{1}{2}\left(1 + \dfrac{4}{n^2}\right)\right]n - 2\left(n - \dfrac{1}{n}\right)\right\}$$

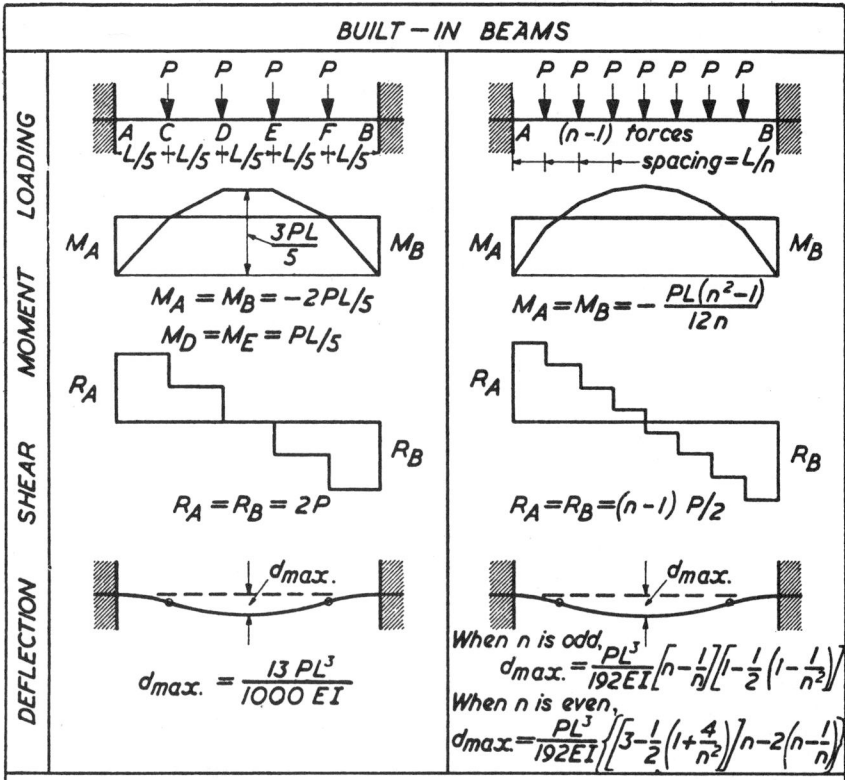

LOAD PER SPAN = W

CONTINUOUS BEAM with n equal spans of L/n, load W/n per span.

When $n > 10$, consider the load uniformly distributed.

The load on the outside stringers is carried directly by the supports
The continuous beam is assumed to be horizontal at each support
The reaction at the supports for each span $= W/2$, but the maximum
shear force in any span of the continuous beam $= \dfrac{W(n-1)}{2n} = A.W$
The value of the fixing moment at each support $= -B.WL$
The value of the maximum positive moment for each span $= C.WL$
The value of the maximum deflection for each span $\doteq 0.0026 \dfrac{WL^3}{EI}$

Value of n	A	B	C
2	0.2500	0.0625	0.0625
3	0.3333	0.0741	0.0370
4	0.3750	0.0781	0.0469
5	0.4000	0.0800	0.0400
6	0.4167	0.0811	0.0439
7	0.4286	0.0816	0.0408
8	0.4375	0.0820	0.0430
9	0.4444	0.0823	0.0413
10	0.4500	0.0825	0.0425

4. CONTINUOUS BEAMS

THE solution of this type of beam consists, in the first instance, of the evaluation of the fixing or negative moments at the supports.

The most general method is the use of Clapeyron's Theorem of Three Moments.

The theorem applies only to any two adjacent spans in a continuous beam and in its simplest form deals with a beam which has all the supports at the same level, and has a constant section throughout its length.

The proof of the theorem results in the following expression:

$$M_A \times L_1 + 2M_B(L_1 + L_2) + M_C \times L_2 = 6\left(\frac{A_1 \times x_1}{L_1} + \frac{A_2 \times x_2}{L_2}\right),$$

where M_A, M_B and M_C are the numerical values of the moments at the supports A, B and C respectively and the remaining terms are as illustrated in Fig. 1.

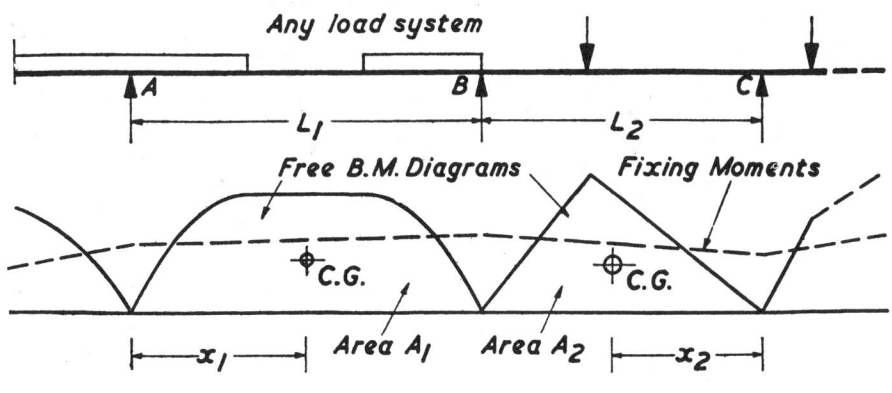

Fig. 1

In a continuous beam the conditions at the end supports are usually known, and these conditions provide starting points for the solution.

The types of end conditions are three in number, viz.:

1. Simply supported.
2. Partially fixed, e.g. a cantilever.
3. Completely fixed, i.e. the end of the beam is horizontal as in the case of a fixed beam.

One example of each of the above types will be worked out in full.

The S.F. at the end of any span is calculated after the support moments have been evaluated, in the same manner as for a fixed beam, each span being treated separately.

It is essential to note the difference between S.F. and reaction at any support, e.g. with reference to Fig. 1 the S.F. at support B due to span AB is a certain

CONTINUOUS BEAMS

amount, while that at support B due to span BC is another amount, but the reaction at the support is the sum of these two amounts.

If the section of the beam is not constant over its whole length, but remains constant for each span, the expression for the moments is rewritten as follows:

$$M_A \times \frac{L_1}{I_1} + 2M_B\left(\frac{L_1}{I_1} + \frac{L_2}{I_2}\right) + M_C \times \frac{L_2}{I_2} = 6\left(\frac{A_1 \times x_1}{L_1 \times I_1} + \frac{A_2 \times x_2}{L_2 \times I_2}\right)$$

in which I_1 = moment of inertia for span L_1 and I_2 = moment of inertia for span L_2.

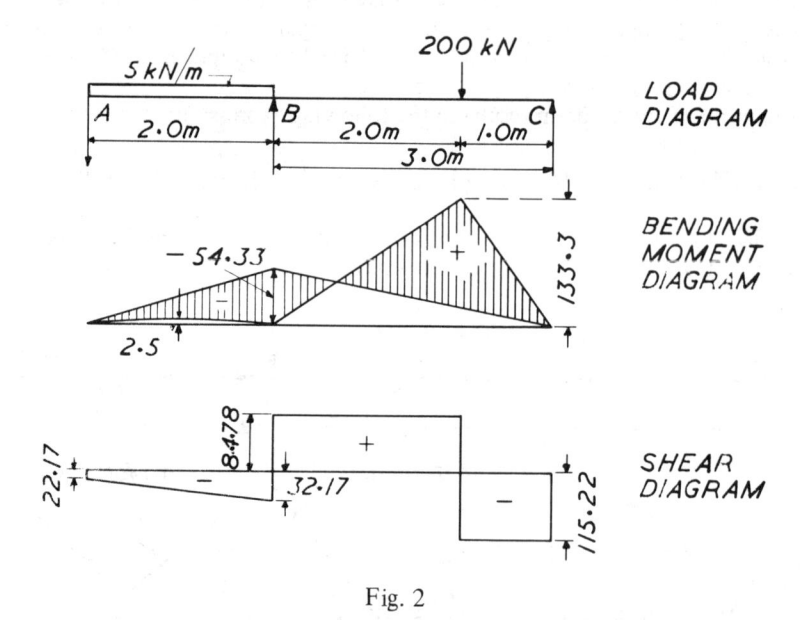

Fig. 2

Example 1. A two-span continuous beam ABC, of constant cross-section, is simply supported at A and C and loaded as shown in Fig. 2.

Applying Clapeyron's theorem,

$$L_1 = 2.0 \text{ m}$$

$$L_2 = 3.0 \text{ m}$$

$$A_1 = \frac{10 \times 2}{8} \times \frac{2 \times 2}{3} = \frac{10}{3} \text{ kNm}^2$$

$$A_2 = \frac{200 \times 1 \times 2}{3} \times \frac{3}{2} = 200 \text{ kNm}^2$$

$$x_1 = 1.0 \text{ m}$$

$$x_2 = \frac{4.0}{3} \text{ m}$$

Therefore $\quad M_A \times 2 + 2M_B(2 + 3) + M_C \times 3 = 6\left\{\dfrac{10 \times 1}{3 \times 2} + \dfrac{200 \times 4}{3 \times 3}\right\}.$

CONTINUOUS BEAMS

Since A and C are simple supports

$$M_A = M_C = 0.$$

Therefore $\quad M_B = \dfrac{6}{10}\left\{\dfrac{10}{6} + \dfrac{800}{9}\right\} = \dfrac{6}{10}\{90.56\} = 54.33 \text{ kNm}$

$$\text{S.F.}_A = 5 + \dfrac{0 - 54.33}{2} = 5 - 27.17 = -22.17 \text{ kN}$$

S.F.$_B$ for span AB = 5 + 27.17 = 32.17 kN

$$\text{S.F.}_C = \dfrac{200 \times 2}{3} + \dfrac{0 - 54.33}{3} = 133.33 - 18.11 = 115.22 \text{ kN}$$

S.F.$_B$ for span $BC = \dfrac{200}{3} + 18.11 = 66.67 + 18.11 = 84.78$ kN

Note that the negative reaction at A means that the end A will tend to lift off its support and will have to be held down.

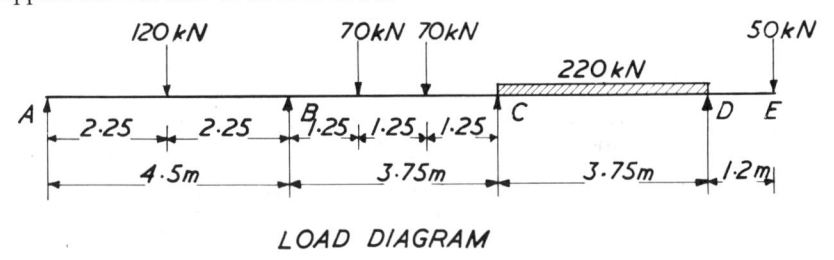

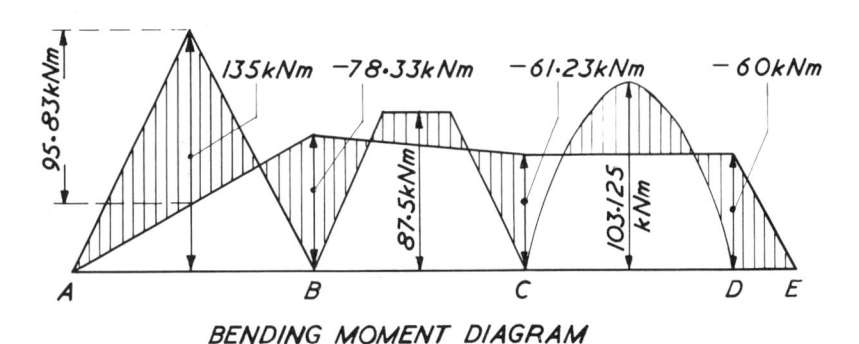

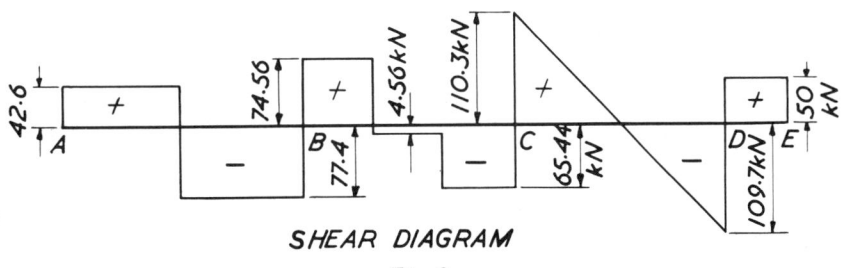

LOAD DIAGRAM

BENDING MOMENT DIAGRAM

SHEAR DIAGRAM

Fig. 3

CONTINUOUS BEAMS

Example 2. A three-span continuous beam $ABCDE$ is simply supported at A, cantilevered at the end D, and loaded as shown in Fig. 3. The section is not constant but varies as follows:

$$I_{AB} = 8\,500 \text{ cm}^4$$
$$I_{BC} = 6\,500 \text{ cm}^4$$
$$I_{CE} = 5\,500 \text{ cm}^4$$

Applying Clapeyron's theorem and noting that since the moments of inertia are given in cm units the B.M.s and spans must also be given in cm units

$$M_A = 0$$
$$M_D = 60 \text{ kNm} = 60 \times 10^2 \text{ kNcm}$$

$$A_1 = \frac{120 \times 4.5}{4} \times \frac{4.5}{2} = 303.75 \text{ kNm}^2 = 303.75 \times 10^4 \text{ kNcm}^2$$

$$A_2 = \frac{70 \times 3.75}{3} \times \frac{3.75 \times 2}{3} = 218.75 \text{ kNm}^2 = 218.75 \times 10^4 \text{ kNcm}^2$$

$$A_3 = \frac{220 \times 3.75}{8} \times \frac{3.75 \times 2}{3} = 257.81 \text{ kNm}^2 = 257.81 \times 10^4 \text{ kNcm}^2$$

First considering spans AB and BC

$$2M_B\left\{\frac{4.5}{8\,500} + \frac{3.75}{6\,500}\right\} \times 10^2 + M_C\left\{\frac{3.75}{6\,500}\right\} \times 10^2 =$$

$$= 6\left\{\frac{303.75 \times 2.25}{4.5 \times 8\,500} + \frac{218.75 \times 1.875}{3.75 \times 6\,500}\right\} \times 10^4$$

Therefore $\quad\quad 0.220M_B + 0.058M_C = 2\,076 \quad \ldots\ldots\ldots\ldots$ (i)

Next considering spans BC and CD

$$M_B\left\{\frac{3.75}{6\,500}\right\} \times 10^2 + 2M_C\left\{\frac{3.75}{6\,500} + \frac{3.75}{5\,500}\right\} \times 10^2 + 60\left\{\frac{3.75}{5\,500}\right\} \times 10^4$$

$$= 6\left\{\frac{218.75 \times 1.875}{3.75 \times 6\,500} + \frac{257.81 \times 1.875}{3.75 \times 5\,500}\right\} \times 10^4$$

Therefore $\quad\quad 0.058M_B + 0.252M_C = 2\,406 - 409 = 1\,997 \quad \ldots\ldots$ (2)

From equations (1) and (2)

$$M_B = 78.33 \text{ kNm}$$
$$M_C = 61.23 \text{ kNm}$$

$$\text{S.F.}_A = 60 + \frac{0 - 78.33}{4.5} = 60 - 17.4 = 42.6 \text{ kN}$$

S.F._B for span $AB \quad\quad = 60 + 17.4 = 77.4 \text{ kN}$

S.F._B for span $BC \quad\quad = 70 + \dfrac{78.33 - 61.23}{3.75} = 70 + 4.56 = 74.56 \text{ kN}$

S.F._C for span $BC \quad\quad = 70 - 4.56 = 65.44 \text{ kN}$

S.F._C for span $CD \quad\quad = 110 + \dfrac{61.23 - 60}{3.75} = 110 + 0.3 = 110.3 \text{ kN}$

S.F._D for span $CD \quad\quad = 110 - 0.3 = 109.7 \text{ kN}$

CONTINUOUS BEAMS

Example 3. A two-span continuous beam *ABC*, of constant cross-section, is built-in horizontally at *A* and *C*, supported at the same level at *B*, and loaded as shown in Fig. 4.

The effect of a fixed end is the same as that which would be produced if there had been, on the side of the support opposite to the span, another span of equal length to, and carrying the same load as, the span, except that this imaginary span is completely reversed from the actual span.

The effect can best be imagined if mirrors were to be placed at *A* and *C* when the span *AB* would be mirrored into the span AB_1 and the span *BC* into the span CC_1.

Interpreted, this means that the negative moment at each of the points B_1 and C_1 is equal to the negative moment at *B*.

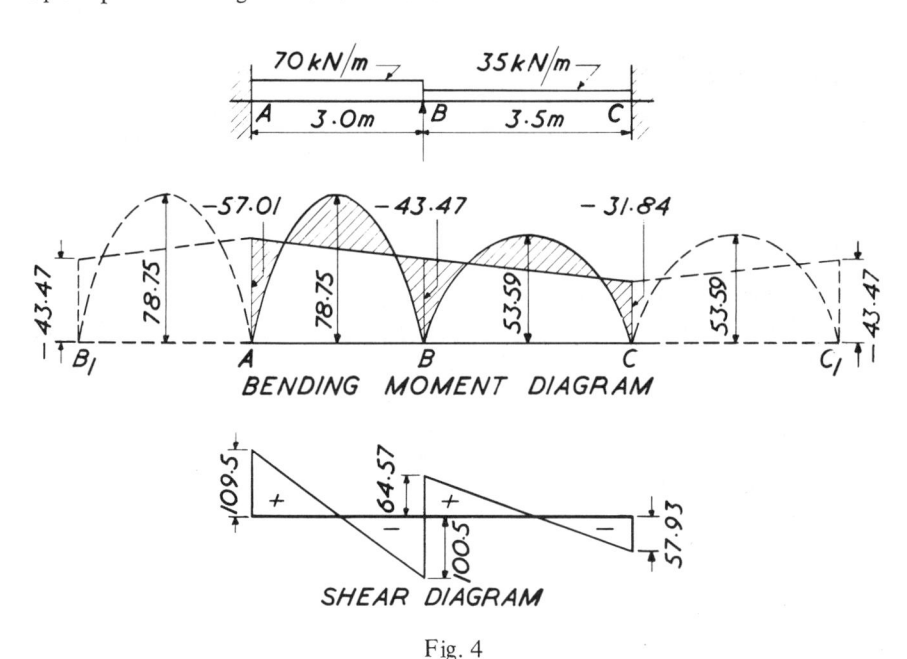

Fig. 4

Applying Clapeyron's theorem,

$$A_1 \text{ (area of free B.M. diagram on span } B_1 A) = \frac{210 \times 3.0}{8} \times \frac{3.0 \times 2}{3}$$

$$= \frac{630}{4} \text{ kNm}^2 = 157.5 \text{ kNm}^2$$

A_2 (,, ,, ,, $AB) = \frac{630}{4}$ kNm² = 157.5 kNm²

A_3 (,, ,, ,, $BC) = \frac{122.5 \times 3.5}{8} \times \frac{3.5 \times 2}{3}$

$$= 125 \text{ kNm}^2$$

A_4 (,, ,, ,, $CC_1) = 125$ kNm²

CONTINUOUS BEAMS

First considering spans $B_1 A$ and AB

$$M_{B_1} \times 3.0 + 2M_A(3.0 + 3.0) + M_B \times 3.0 = 6\left(\frac{630 \times 1.5}{4 \times 3.0} + \frac{630 \times 1.5}{4 \times 3.0}\right)$$

Note:— $M_{B_1} = M_B$

Therefore $\qquad M_B + 2M_A = 157.5$ kNm .... (i)

Secondly considering spans AB and BC.

$$M_A \times 3.0 + 2M_B(3.0 + 3.5) + M_C \times 3.5 = 6\left(\frac{630 \times 1.5}{4 \times 3.0} + \frac{125 \times 1.75}{3.5}\right)$$

$$M_A + 4.33M_B + 1.17M_C = 282.5 \ldots\ldots\ldots\ldots \text{(ii)}$$

Thirdly considering spans BC and CC_1.

$$M_B \times 3.5 + 2M_C(3.5 + 3.5) + M_{C_1} \times 3.5 = 6\left(\frac{125 \times 1.75}{3.5} + \frac{125 \times 1.75}{3.5}\right)$$

Note:— $M_{C_1} = M_B$

Therefore $\qquad M_B + 2M_C = 107.1$ kNm .... (iii)

From equations (i), (ii) and (iii)

$$M_A = 57.01 \text{ kNm}$$
$$M_B = 43.47 \text{ kNm}$$
$$M_C = 31.84 \text{ kNm}$$

$$\text{S.F.}_A = 105 + \frac{57.01 - 43.47}{3} = 105 + 4.513 = \text{say, } 109.5 \text{ kN}$$

S.F.$_B$ for span $AB = 105 - 4.513 = 100.5$ kN

$$\text{S.F.}_C = 61.25 + \frac{31.84 - 43.47}{3.5} = 61.25 - 3.32 = 57.93 \text{ kN}$$

S.F.$_B$ for span $BC = 61.25 + 3.32 = 64.57$ kN

It should be noted that further examples of continuous beams are dealt with in Chapter 13, "Methods of Structural Analysis."

Formulae for continuous beams are given in the pages immediately following this chapter.

EQUAL SPAN CONTINUOUS BEAMS
UNIFORMLY DISTRIBUTED LOADS

Moment = coefficient x W x L
Reaction = coefficient x W

where W is the U.D.L. on one span only and L is one span

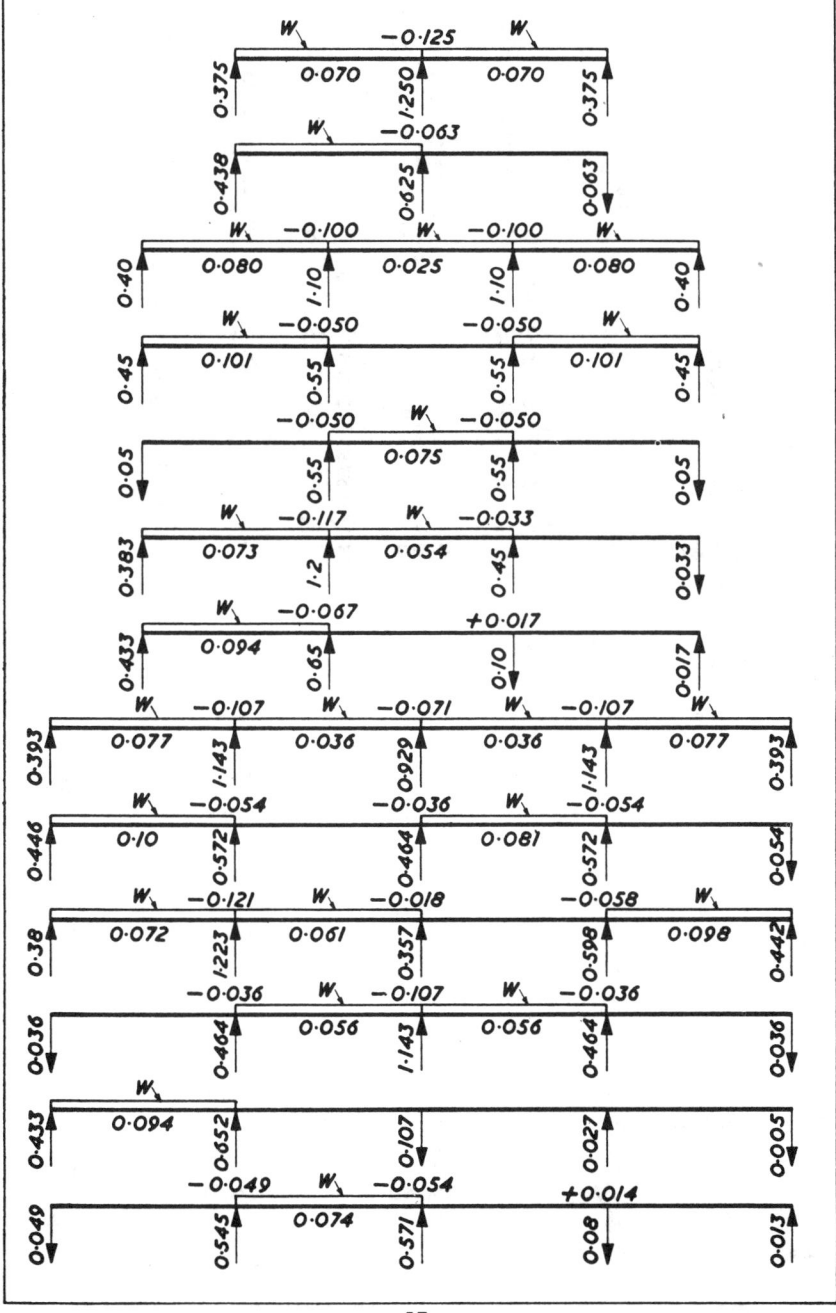

EQUAL SPAN CONTINUOUS BEAMS
CENTRAL POINT LOADS

Moment = coefficient x W x L
Reaction = coefficient x W
where W is the Load on one span only and L is one span

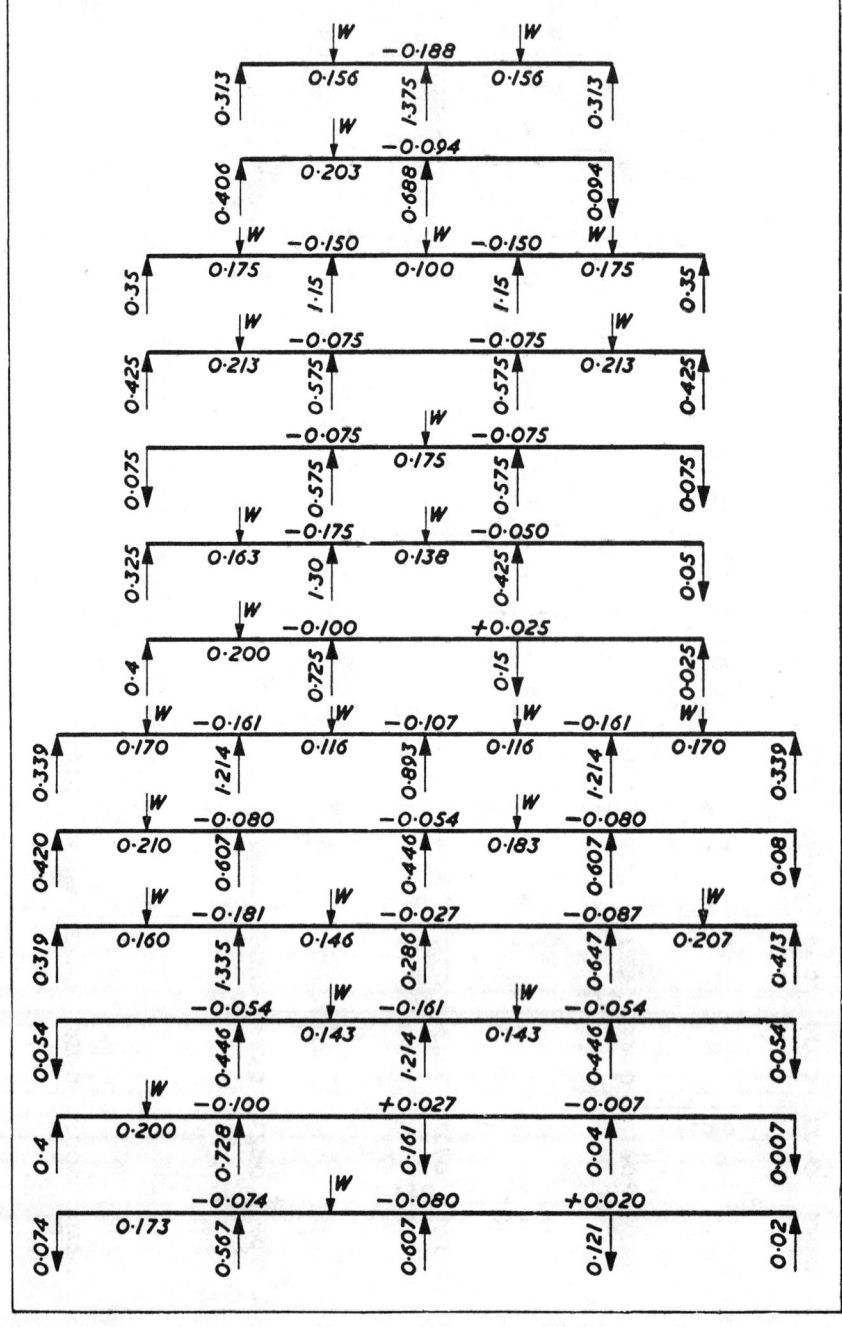

EQUAL SPAN CONTINUOUS BEAMS
POINT LOADS AT THIRD POINTS OF SPANS

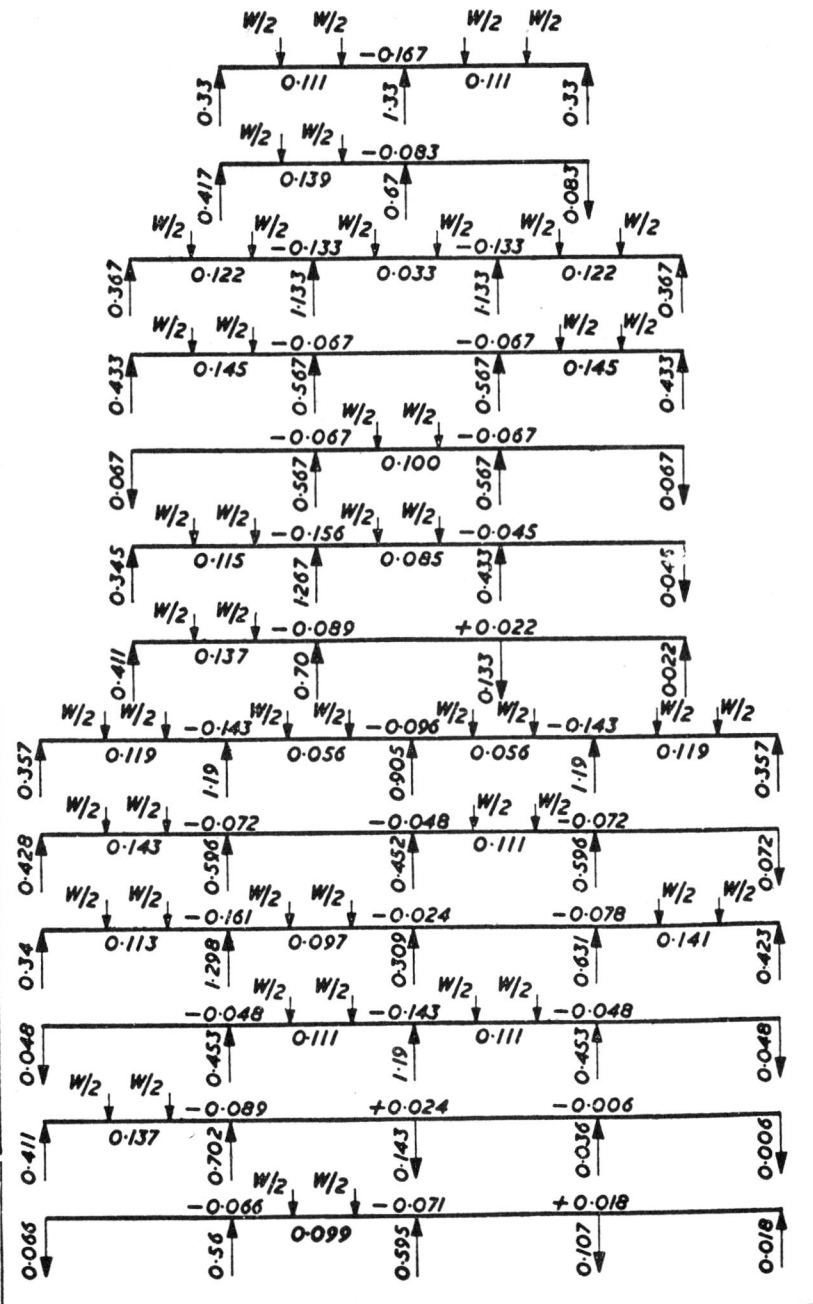

Moment = coefficient x W x L
Reaction = coefficient x W
where W is the <u>total</u> load on one span only & L is one span

5. CANTILEVER AND SUSPENDED-SPAN CONSTRUCTION

It is shown in the foregoing chapters and elsewhere in this manual that the general effect of the application of end-restraints to built-in beams and propped cantilevers, or of continuity in beams of more than one span, is to reduce the magnitude of the maximum bending moments. The notable exception to this statement is provided by the beam with a uniformly distributed load continuous over two equal spans, where the resulting internal support moment is $- WL/8$, numerically equal to the maximum free bending moment in a simply-supported beam. By contrast, when a beam with a central point load is built-in at its ends, the maximum bending moment is halved. For uniformly distributed loads, however, beams built-in or completely continuous over many spans do not provide the greatest economy as the support moments are $- WL/12$ while the moments at mid-span are $+ WL/24$. For other types of load the variations in bending moment can be even more erratic.

Only cantilever and suspended-span construction can achieve a more even distribution of moments. Although the principles of this form of construction have been known since they were patented and exploited by Gerber in Europe a century ago, this system could be said to have been sadly neglected in Great Britain since the erection of the Forth Railway Bridge.

The method offers distinct advantages where the loading pattern is constant, as for purlins and side-rails. Being statically determinate, the bending moments are unaffected by the sinking of supports, such as might occur in areas of mining subsidence.

Consider the simple example shown in Fig. 1. The beam *CB* cantilevers out to the point *h* where it is hinged to the suspended-span *Ah*.

Now the free moment under the point load *P*

$$M_P = \frac{Pab}{L_1}$$

and so the *BM* diagram for *Ah* may be drawn, as shown in Fig. 1b.

The reaction produced by the load P at the hinge

$$R_h = \frac{Pa}{L_1}$$

Therefore, the moment produced at the support *B*

$$M_B = -R_h \cdot c = -\frac{Pac}{L_1}$$

and the *BM* diagram for *hBC* may be drawn also as shown in Fig. 1b.

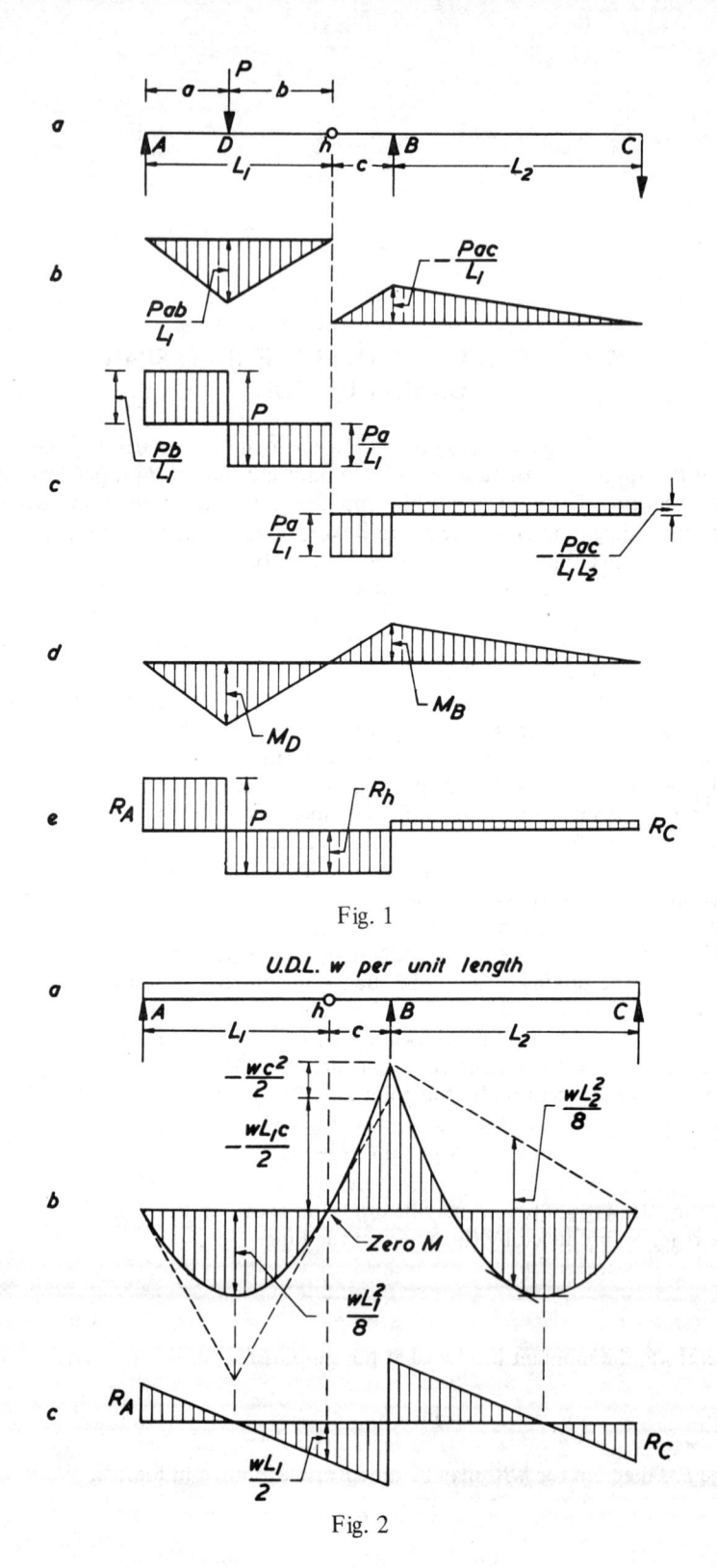

Fig. 1

Fig. 2

CANTILEVER AND SUSPENDED-SPAN CONSTRUCTION

To draw the *SF* diagrams it is necessary to calculate the reactions at *A* and *C*.

$$R_A = \frac{Pb}{L_1}$$

$$R_C = \frac{M_B}{L_2} = -\frac{Pac}{L_1 L_2}$$

The resulting diagrams for *Ah* and *hBC* are shown in Fig. 1c.

It should be particularly noted that the shearing forces on either side of the hinge *h* are of the same sign and magnitude from *D* to *B*. It may therefore be deduced that the *BM* diagram has a constant slope through the point of contraflexure formed by the hinge.

When the *BM* and *SF* diagrams are combined, as in Figs. 1d and e respectively, it will be seen how simple the calculations are for this form of construction.

Similar treatment may be applied to uniformly distributed loads. Consider the example shown in Fig. 2, where the load is *w* per unit length.

Then the max. free moment in the suspended span *Ah*

$$M = \frac{wL_1^2}{8}$$

and the *BM* diagram for *Ah* may be drawn, as shown in Fig. 2b.

The reaction produced by the suspended-span at *h*

$$R_h = \frac{wL_1}{2} = R_A$$

The moment produced at the support *B* is therefore

$$M_B = -R_h \cdot c - \frac{wc^2}{2}$$

The remainder of the *BM* diagram can now be drawn, as shown in Fig. 2b. The *SF* diagrams are shown in Fig. 2c.

Alternatively, the *BM* diagram may be drawn first by drawing the two free moment diagrams for the spans *AB* and *BC* and then merely drawing the restraint diagram through the point of contraflexure corresponding to the hinge *h*.

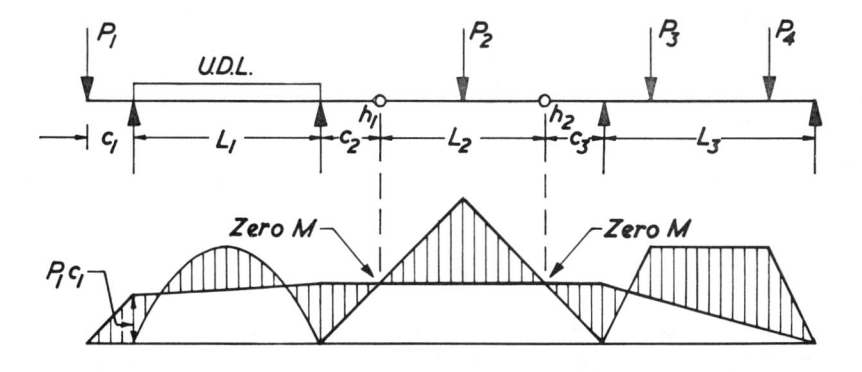

Fig. 3

64 CANTILEVER AND SUSPENDED-SPAN CONSTRUCTION

The same kind of procedure may be followed for any number of spans and any loading, typical examples being shown in Figs. 3 and 4.

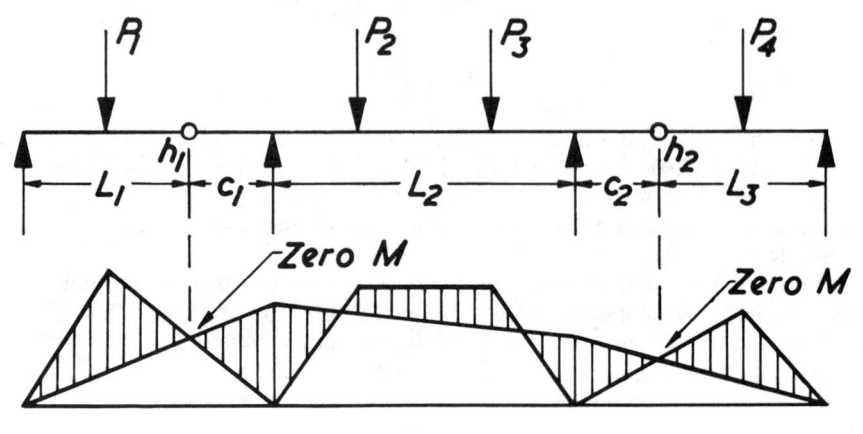

Fig. 4

For uniformly distributed loads, the greatest economy can be achieved when the maximum moments are reduced to $\pm WL/16$. This can be achieved in a structure, say with five spans, by reducing the lengths of the end spans to 0.8535 of the length of the three interior spans, as shown in Fig. 5. When the spans are all made equal,

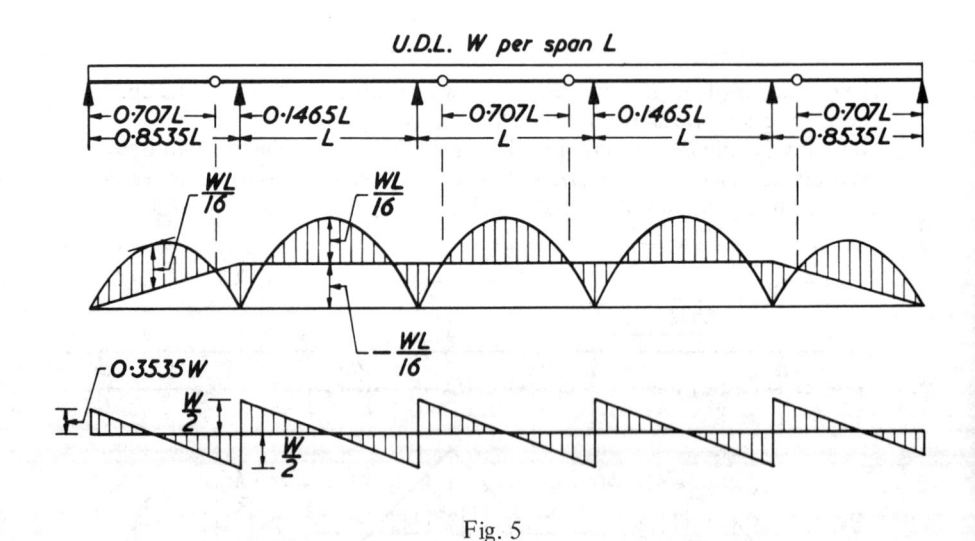

Fig. 5

as in Fig. 6, the free moments in the end suspended spans are greater than $WL/16$. The appropriate members can either be made stronger throughout or compounded in the zone of maximum moment.

CANTILEVER AND SUSPENDED-SPAN CONSTRUCTION

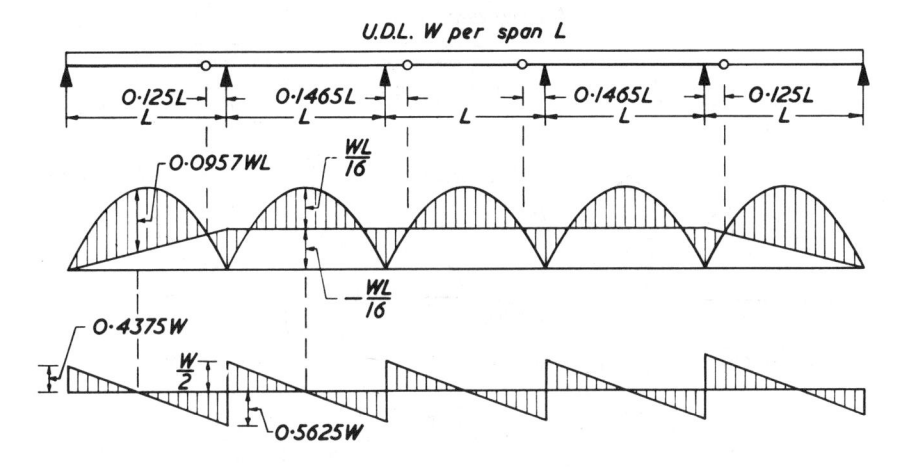

Fig. 6

Some typical arrangements for uniformly distributed loads on structures of multiple span are shown in Figs. 7 to 9.

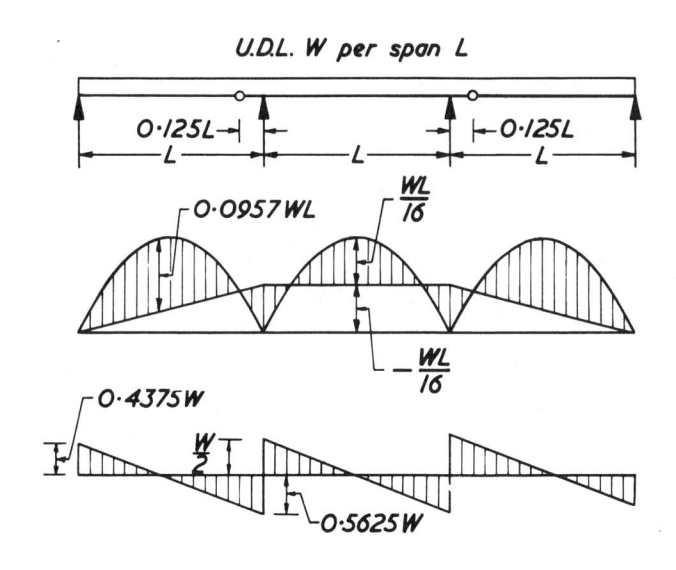

Fig. 7

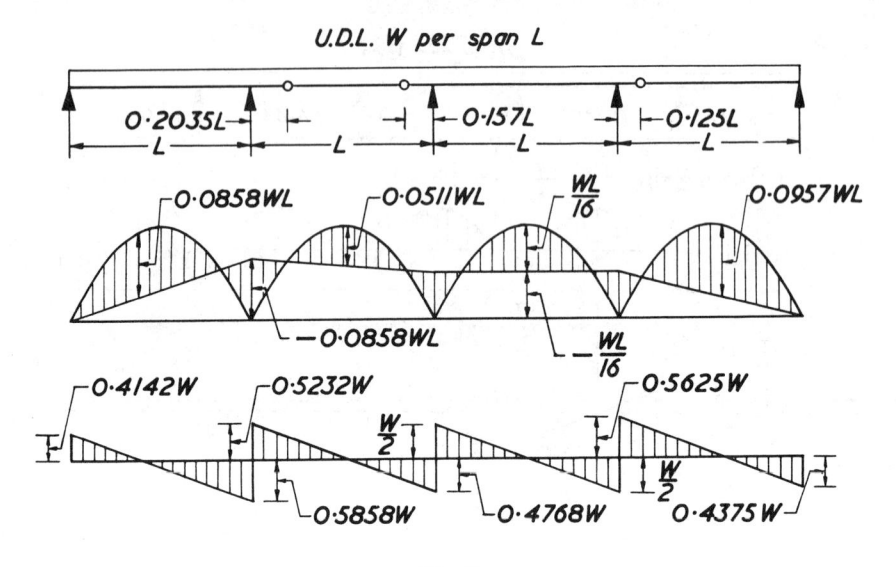

Fig. 8

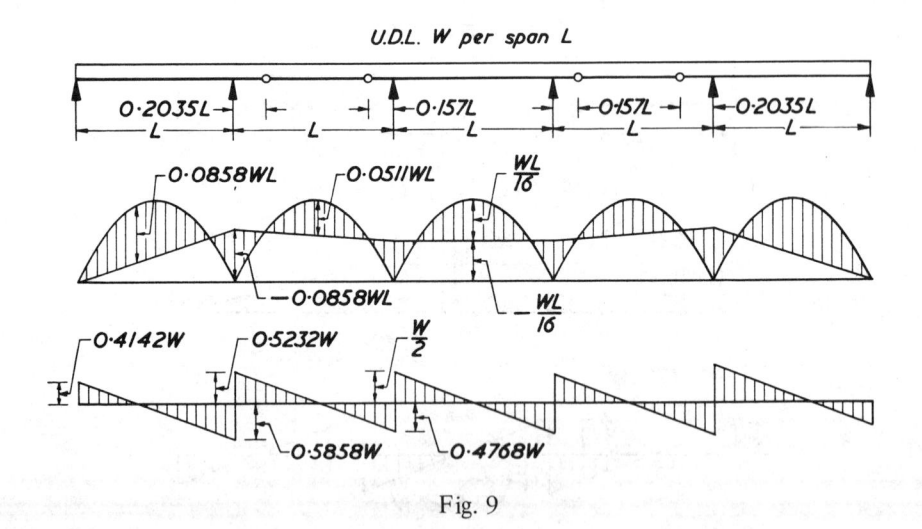

Fig. 9

6. PROPPED CANTILEVERS

BEAMS which are built-in at one end and simply supported at the other are known as propped cantilevers. Normally, the ends of the beams are on the same level, in which case bending moments and reactions may be derived in two ways, viz.: by employing the Theorem of Three Moments or by deflection formulae.

Solution by Theorem of Three Moments

Consider the propped cantilever AB in Fig. 1.

The bending moment at B may be found by using the Theorem of Three Moments, and assuming that AB is one span of a two-span continuous beam ABC which is symmetrical in every way about B.

Then the loads on AB and BC will produce free B.M.s whose areas are A_1 and A_2 respectively, the C.G.s of the areas being distances x_1 and x_2 from A and C respectively.

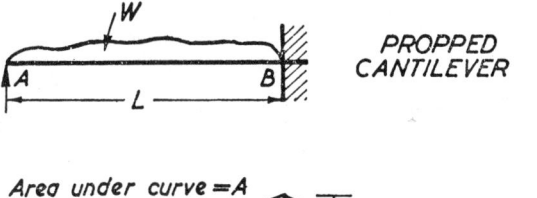

Fig. 1

Now $\quad M_A L_1 + 2M_B(L_1 + L_2) + M_C L_2 = 6\left(\dfrac{A_1 x_1}{L_1} + \dfrac{A_2 x_2}{L_2}\right).$

But $\quad L_1 = L_2 = \text{say}, L,$

$\quad M_A = M_C = 0,$

$\quad A_1 = A_2 = \text{say}, A$

and $\quad x_1 = x_2 = \text{say}, x.$

Hence, $\quad 2M_B(2L) = 6 \times 2\left(\dfrac{Ax}{L}\right)$

and $\quad M_B = \dfrac{3Ax}{L^2}.$

Therefore the moment at the fixed end of a propped cantilever = $3Ax/L^2$,

where A = the area of the free B.M. diagram, AB being considered as a simply-supported beam,

x = the distance from the prop to the C.G. of the free B.M. diagram

and L = the span.

The reactions at each support may be found by employing a modified form of the formula used for beams built-in at both ends, viz.:

$$\text{S.F.}_A = \text{the simple support reaction at } A - \frac{M_B}{L}$$

$$\text{S.F.}_B = \text{the simple support reaction at } B + \frac{M_B}{L}$$

where A is the propped end and B is built-in.

Example 1. The beam AB shown in Fig. 2 is propped at A and built-in at B. It is required to draw the B.M. and S.F. diagrams for the symmetrical triangular load of 100 kN.

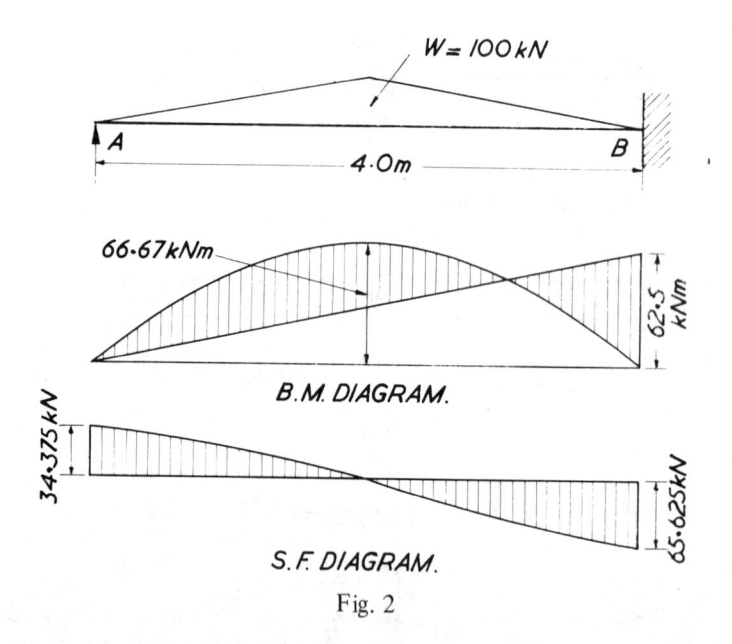

Fig. 2

Consider the free B.M. diagram. At any point x from A between A and the centre of the beam,

$$M_x = W\left(\frac{x}{2} - \frac{2x^3}{3L^2}\right)$$

Hence, the maximum B.M., at the centre,

$$= \frac{WL}{6} = \frac{100 \times 4}{6} = 66.67 \text{ kNm}$$

SOLUTION BY DEFLECTION FORMULAE

The area A of the free B.M. diagram

$$= 2\int_0^{L/2} W\left(\frac{x}{2} - \frac{2x^3}{3L^2}\right) dx$$

$$= 2W\left[\frac{x^2}{4} - \frac{2x^4}{12 \times 4^2}\right]_0^2 = 200(1 - 0.167)$$

$$= 166.7 \text{ kNm}^2$$

The C.G. of the free B.M. diagram is in the centre of the beam, i.e. 2 m. from A.

Hence, $$M_B = \frac{3Ax}{L^2} = \frac{3 \times 166.7 \times 2}{4 \times 4}$$

$$= 62.5 \text{ kNm}$$

Following normal convention this value is negative:

$$\text{S.F.}_A = \frac{W}{2} - \frac{M_B}{L} = 50 - \frac{62.5}{4} = 34.375 \text{ kN}$$

$$\text{S.F.}_B = 50 + \frac{M_B}{L} = 50 + 15.625 = 65.625 \text{ kN}$$

The B.M. and S.F. diagrams may now be drawn.

Solution by Deflection Formulae

When this method of solution is employed the propping force is found first and the B.M.s follow. The method is best explained by an example.

Example 2. Consider the propped cantilever shown in Fig. 3 which carries a uniformly distributed load of 40 kN.

Let the force in the prop be F.

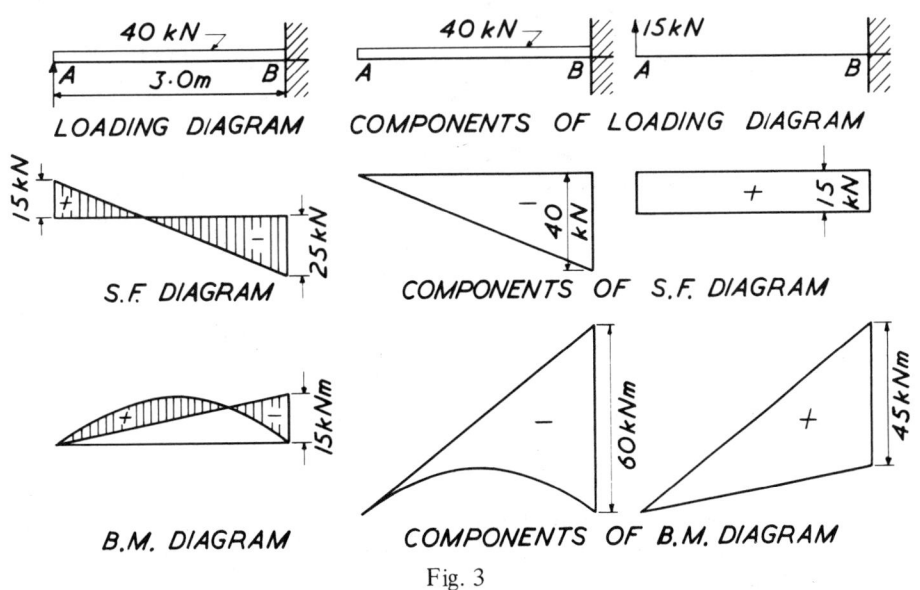

Fig. 3

Assume that this prop is removed. Then the deflection at A will equal $WL^3/8EI$, this being the standard deflection formula for a U.D.L. Now if the prop is assumed to be back in position then the upward deflection due to the prop will be $FL^3/3EI$.

Since A is on the same level as B,

$$\frac{FL^3}{3EI} = \frac{WL^3}{8EI},$$

whence
$$F = \text{S.F.}_A = \frac{3W}{8} = \frac{120}{8} = 15 \text{ kN}$$

and
$$\text{S.F.}_B = \frac{5W}{8} = \frac{200}{8} = 25 \text{ kN}$$

Without the prop the moment at B due to the load is $-WL/2$, whilst the moment at B due to the prop alone is $FL = 3WL/8$.

Hence the final moment at B

$$= M_B = WL\left(-\frac{1}{2} + \frac{3}{8}\right) = -\frac{WL}{8} = -\frac{40 \times 3}{8} \text{ kNm} = -15 \text{ kNm}$$

The S.F. and B.M. diagrams may now be drawn.

Encastré Beams and Propped Cantilevers

There is one further method in which the fixing moment may be determined in a propped cantilever, i.e. from a knowledge of the fixing moments in a similar encastré beam, as will be shown in the following example.

Example 3. Fig. 4 shows the B.M. diagram for an encastré beam AB, 3.0 m. long, carrying a point load of 250 kN at 1.25 m. from A.

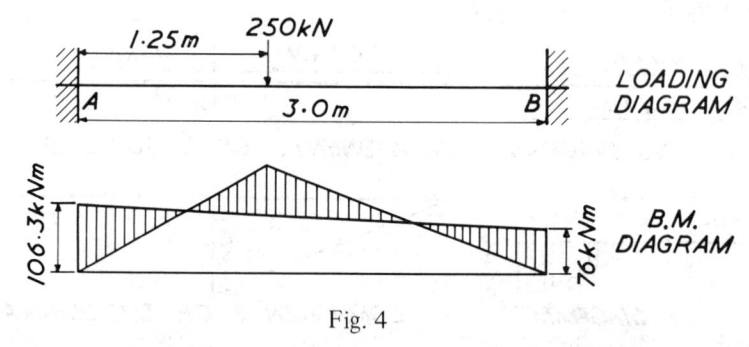

Fig. 4

Now if the fixing moment M_A is released, then it may be shown that the fixing moment M_B is increased by half the numerical value of M_A.

Therefore, for the propped cantilever shown in Fig. 5,

$$M_B = 76 + \frac{106.3}{2} = 129.2 \text{ kNm}$$

It will be appreciated that for symmetrical loading, the single fixing moment for a propped cantilever is one and a half times as great as the fixing moment at either end of a similar encastré beam.

Fig. 5

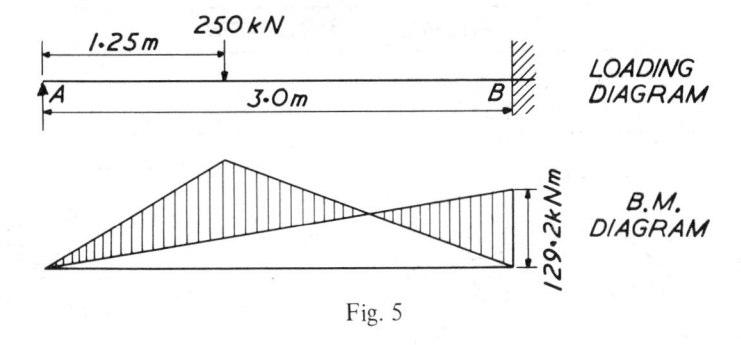

It should be noted that formulae for various propped cantilevers are given on pp. 72–79 inc.

Sinking of Supports

When the supports for a loaded propped cantilever do not maintain the same relative levels as in the unloaded condition, the B.M.s and S.F.s may be obtained by using the deflection method of Example 2. When the prop sinks the load which it takes is reduced, while the fixing moment at the other end is increased. Two special cases arise: the first when the prop sinks so much that no load is taken by the prop, and the second when the built-in end sinks so much that the fixing moment is reduced to zero, i.e. the cantilever resembles a simply supported beam.

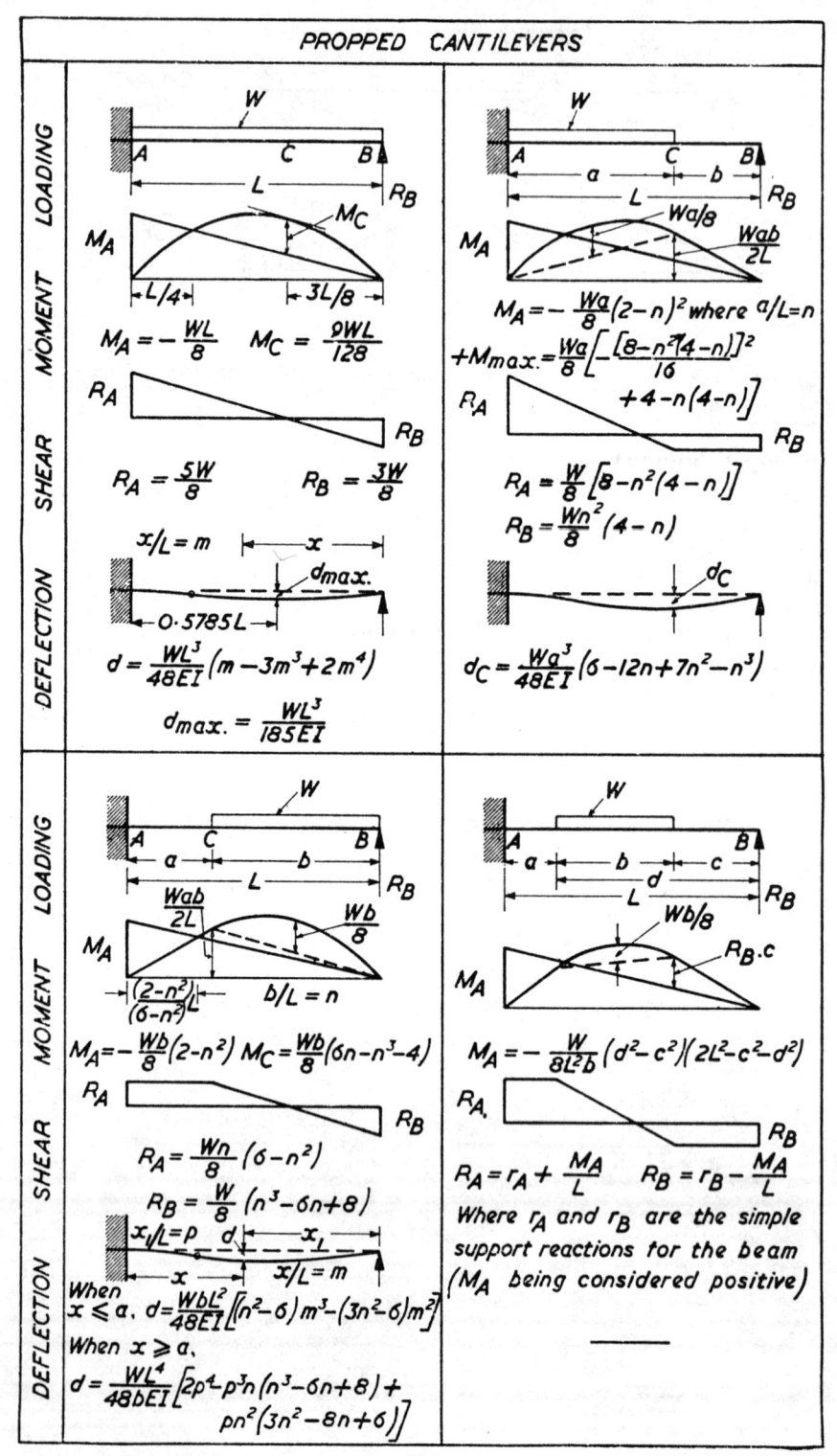

PROPPED CANTILEVERS

Case 1: Uniformly distributed load over full span

LOADING: Fixed at A, propped at B, UDL W over length L, reaction R_B.

MOMENT:
$$M_A = -\frac{WL}{8} \qquad M_C = \frac{9WL}{128}$$

Distances: $L/4$ and $3L/8$.

SHEAR:
$$R_A = \frac{5W}{8} \qquad R_B = \frac{3W}{8}$$

DEFLECTION: $x/L = m$, d_{max} at $0.5785L$.
$$d = \frac{WL^3}{48EI}(m - 3m^3 + 2m^4)$$
$$d_{max} = \frac{WL^3}{185EI}$$

Case 2: Point load W at distance a from fixed end

LOADING: Fixed at A, propped at B, load W at C, $a + b = L$.

MOMENT: $Wa/8$, $\frac{Wab}{2L}$
$$M_A = -\frac{Wa}{8}(2-n)^2 \text{ where } a/L = n$$
$$+M_{max} = \frac{Wa}{8}\left[-\frac{[8-n^2(4-n)]^2}{16} + 4 - n(4-n)\right]$$

SHEAR:
$$R_A = \frac{W}{8}[8 - n^2(4-n)]$$
$$R_B = \frac{Wn^2}{8}(4-n)$$

DEFLECTION:
$$d_C = \frac{Wa^3}{48EI}(6 - 12n + 7n^2 - n^3)$$

Case 3: Uniformly distributed load over part of span (from C to B)

LOADING: Fixed at A, propped at B, UDL W over length b, with $a + b = L$.

MOMENT: $\frac{Wab}{2L}$, $\frac{Wb}{8}$, $(2-n^2)/(6-n^2) \cdot L$, $b/L = n$
$$M_A = -\frac{Wb}{8}(2-n^2) \qquad M_C = \frac{Wb}{8}(6n - n^3 - 4)$$

SHEAR:
$$R_A = \frac{Wn}{8}(6 - n^2)$$
$$R_B = \frac{W}{8}(n^3 - 6n + 8)$$

DEFLECTION: $x_1/L = p$, $x/L = m$

When $x \leq a$:
$$d = \frac{WbL^2}{48EI}\left[(n^2 - 6)m^3 - (3n^2 - 6)m^2\right]$$

When $x \geq a$:
$$d = \frac{WL^4}{48bEI}\left[2p^4 - p^3n(n^3 - 6n + 8) + pn^2(3n^2 - 8n + 6)\right]$$

Case 4: Uniformly distributed load over interior portion

LOADING: Fixed at A, propped at B, UDL W over length b between distances a and c; d is distance from A to centre of load.

MOMENT: $Wb/8$, $R_B \cdot c$
$$M_A = -\frac{W}{8L^2 b}(d^2 - c^2)(2L^2 - c^2 - d^2)$$

SHEAR:
$$R_A = r_A + \frac{M_A}{L} \qquad R_B = r_B - \frac{M_A}{L}$$

Where r_A and r_B are the simple support reactions for the beam (M_A being considered positive)

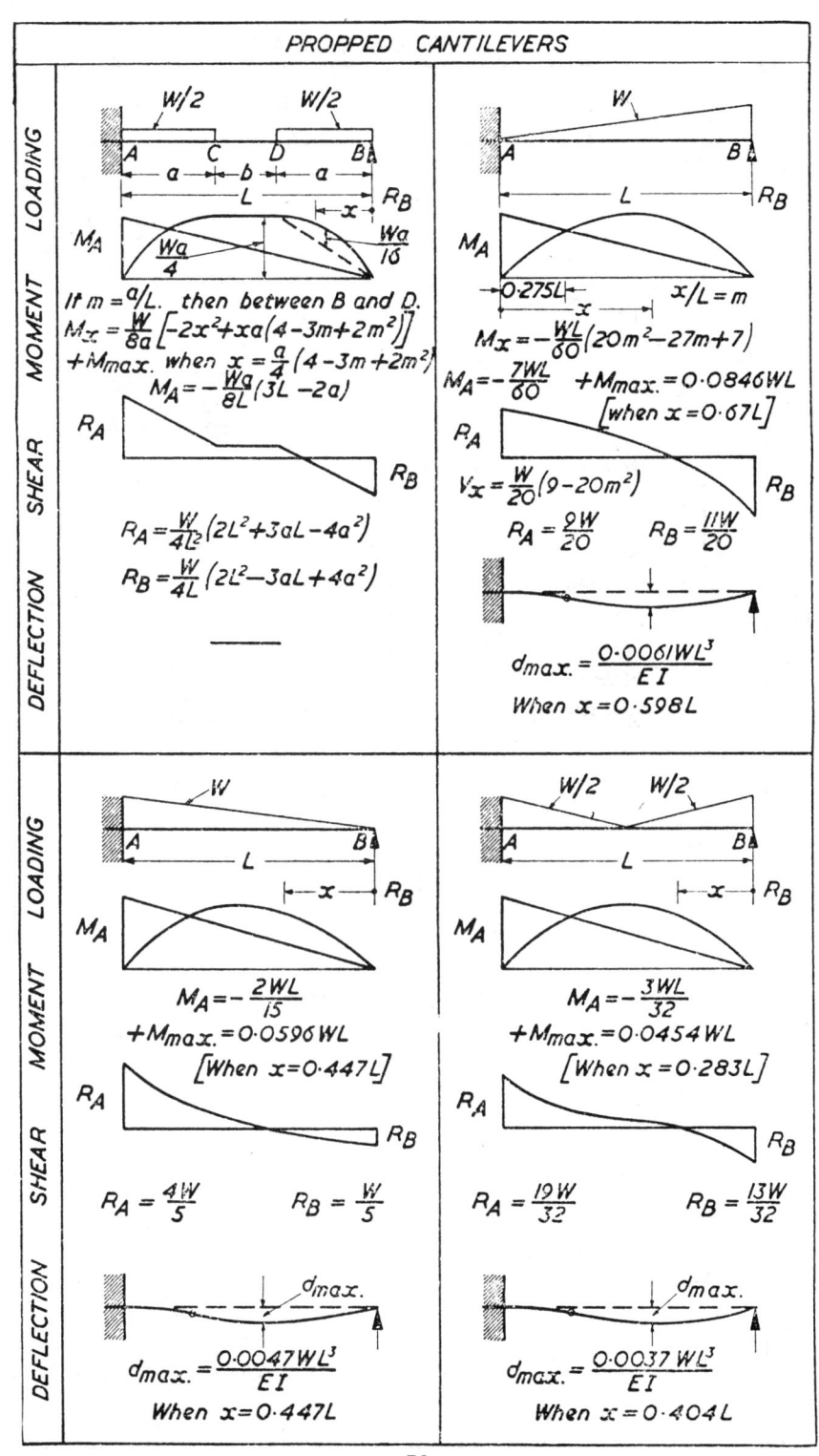

PROPPED CANTILEVERS

LOADING / MOMENT / SHEAR / DEFLECTION (Case 1: two partial UDLs W/2 each over length a, with gap b)

If $m = a/L$, then between B and D:
$$M_x = \frac{W}{8a}\left[-2x^2 + xa(4 - 3m + 2m^2)\right]$$
$+M_{max}$ when $x = \frac{a}{4}(4 - 3m + 2m^2)$
$$M_A = -\frac{Wa}{8L}(3L - 2a)$$

$$R_A = \frac{W}{4L^2}(2L^2 + 3aL - 4a^2)$$
$$R_B = \frac{W}{4L}(2L^2 - 3aL + 4a^2)$$

Case 2: Triangular load increasing from A to B, total W

$$M_x = -\frac{WL}{60}(20m^2 - 27m + 7), \quad x/L = m$$
$$M_A = -\frac{7WL}{60}$$
$+M_{max} = 0.0846 WL$ [when $x = 0.67L$]

$$V_x = \frac{W}{20}(9 - 20m^2)$$
$$R_A = \frac{9W}{20} \qquad R_B = \frac{11W}{20}$$

$$d_{max.} = \frac{0.0061 WL^3}{EI}$$
When $x = 0.598L$

Case 3: Triangular load increasing from B to A, total W

$$M_A = -\frac{2WL}{15}$$
$+M_{max.} = 0.0596 WL$ [When $x = 0.447L$]

$$R_A = \frac{4W}{5} \qquad R_B = \frac{W}{5}$$

$$d_{max.} = \frac{0.0047 WL^3}{EI}$$
When $x = 0.447L$

Case 4: Triangular loads peaking at mid-span, W/2 each side

$$M_A = -\frac{3WL}{32}$$
$+M_{max.} = 0.0454 WL$ [When $x = 0.283L$]

$$R_A = \frac{19W}{32} \qquad R_B = \frac{13W}{32}$$

$$d_{max.} = \frac{0.0037 WL^3}{EI}$$
When $x = 0.404L$

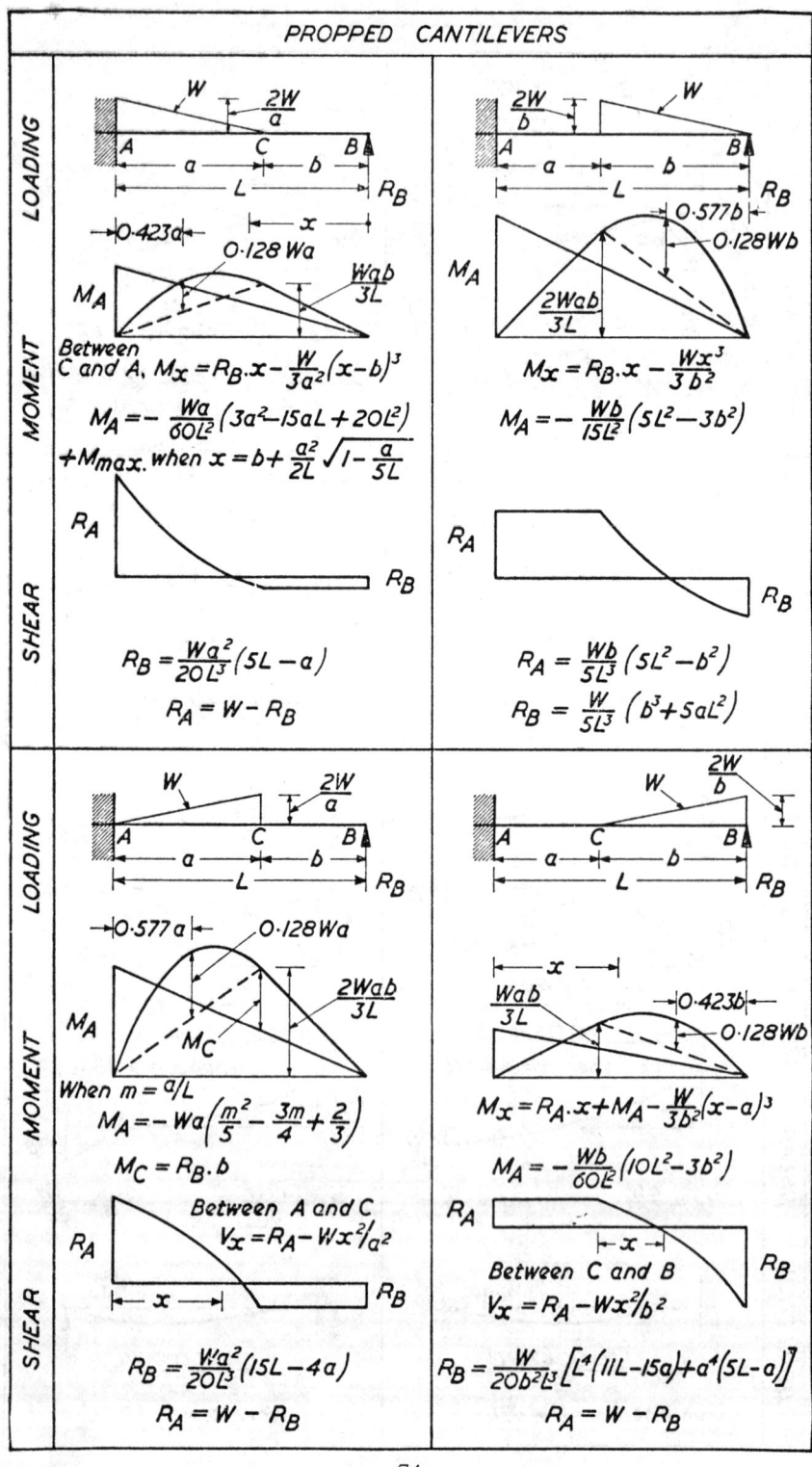

PROPPED CANTILEVERS

Case 1 (top left)

LOADING: Triangular load, max $2W/a$ at C, zero at A and B. Span A to C = a, C to B = b, total L. Reaction R_B at B.

MOMENT:
Distance $0.423a$ from C to point of max positive moment; $0.128Wa$ max; $Wab/3L$ shown.

Between C and A, $M_x = R_B \cdot x - \dfrac{W}{3a^2}(x-b)^3$

$M_A = -\dfrac{Wa}{60L^2}(3a^2 - 15aL + 20L^2)$

$+M_{max}$ when $x = b + \dfrac{a^2}{2L}\sqrt{1 - \dfrac{a}{5L}}$

SHEAR:
$R_B = \dfrac{Wa^2}{20L^3}(5L - a)$

$R_A = W - R_B$

Case 2 (top right)

LOADING: Triangular load, max $2W/b$ at C, zero at A and B. Span A to C = a, C to B = b.

MOMENT:
$0.577b$ location; $0.128Wb$; $2Wab/3L$.

$M_x = R_B \cdot x - \dfrac{Wx^3}{3b^2}$

$M_A = -\dfrac{Wb}{15L^2}(5L^2 - 3b^2)$

SHEAR:
$R_A = \dfrac{Wb}{5L^3}(5L^2 - b^2)$

$R_B = \dfrac{W}{5L^3}(b^3 + 5aL^2)$

Case 3 (bottom left)

LOADING: Triangular load, max $2W/a$ at C, zero at A, load on span A–C only.

MOMENT:
$0.577a$; $0.128Wa$; $2Wab/3L$; M_C.

When $m = a/L$,
$M_A = -Wa\left(\dfrac{m^2}{5} - \dfrac{3m}{4} + \dfrac{2}{3}\right)$

$M_C = R_B \cdot b$

SHEAR:
Between A and C, $V_x = R_A - Wx^2/a^2$

$R_B = \dfrac{Wa^2}{20L^3}(15L - 4a)$

$R_A = W - R_B$

Case 4 (bottom right)

LOADING: Triangular load, max $2W/b$ at B, zero at C, load on span C–B only.

MOMENT:
$Wab/3L$; $0.423b$; $0.128Wb$.

$M_x = R_A \cdot x + M_A - \dfrac{W}{3b^2}(x-a)^3$

$M_A = -\dfrac{Wb}{60L^2}(10L^2 - 3b^2)$

Between C and B, $V_x = R_A - Wx^2/b^2$

$R_B = \dfrac{W}{20b^2L^3}\left[L^4(11L - 15a) + a^4(5L - a)\right]$

$R_A = W - R_B$

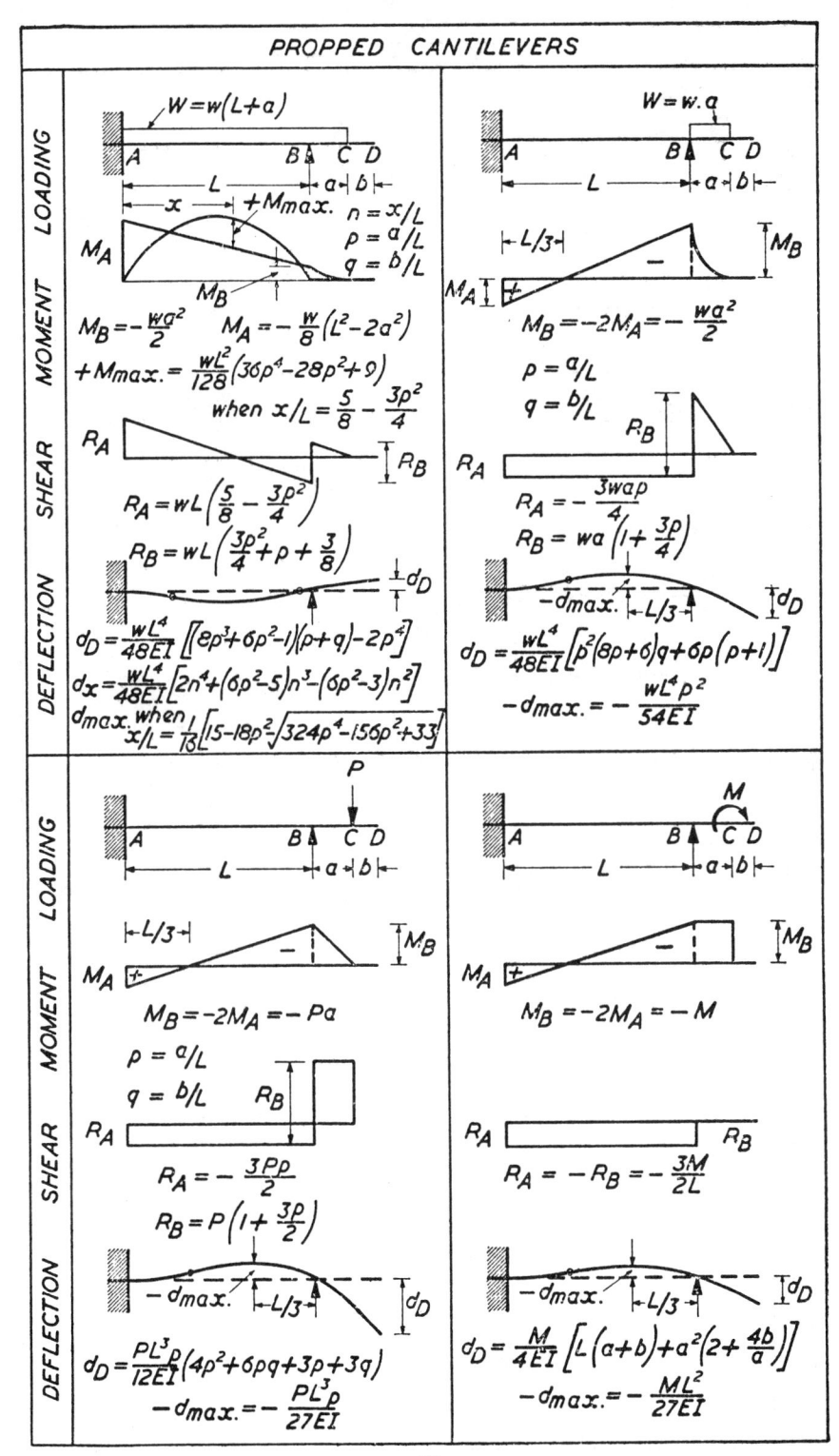

PROPPED CANTILEVERS

Case 1 (top-left): Distributed load $W = w(L+a)$

LOADING: Cantilever fixed at A, propped at B, with overhang to D. Distributed load w over length $L+a$. Distance L from A to B, a from B to C, b from C to D.

MOMENT:
$$M_B = -\frac{wa^2}{2} \qquad M_A = -\frac{w}{8}(L^2 - 2a^2)$$
$$+M_{max} = \frac{wL^2}{128}(36p^4 - 28p^2 + 9)$$
$$\text{when } x/L = \frac{5}{8} - \frac{3p^2}{4}$$
$$n = x/L, \quad p = a/L, \quad q = b/L$$

SHEAR:
$$R_A = wL\left(\frac{5}{8} - \frac{3p^2}{4}\right)$$
$$R_B = wL\left(\frac{3p^2}{4} + p + \frac{3}{8}\right)$$

DEFLECTION:
$$d_D = \frac{wL^4}{48EI}\left[(8p^3 + 6p^2 - 1)(p+q) - 2p^4\right]$$
$$d_x = \frac{wL^4}{48EI}\left[2n^4 + (6p^2 - 5)n^3 - (6p^2 - 3)n^2\right]$$
$$d_{max} \text{ when } x/L = \frac{1}{13L}\left[15 - 18p^2 - \sqrt{324p^4 - 156p^2 + 33}\right]$$

Case 2 (top-right): Distributed load $W = w \cdot a$ on overhang

MOMENT:
$$M_B = -2M_A = -\frac{wa^2}{2}$$
$$p = a/L, \quad q = b/L$$

SHEAR:
$$R_A = -\frac{3wap}{4}$$
$$R_B = wa\left(1 + \frac{3p}{4}\right)$$

DEFLECTION:
$$d_D = \frac{wL^4}{48EI}\left[p^2(8p+6)q + 6p(p+1)\right]$$
$$-d_{max} = -\frac{wL^4 p^2}{54EI}$$

Case 3 (bottom-left): Point load P at C

MOMENT:
$$M_B = -2M_A = -Pa$$
$$p = a/L, \quad q = b/L$$

SHEAR:
$$R_A = -\frac{3Pp}{2}$$
$$R_B = P\left(1 + \frac{3p}{2}\right)$$

DEFLECTION:
$$d_D = \frac{PL^3 p}{12EI}(4p^2 + 6pq + 3p + 3q)$$
$$-d_{max} = -\frac{PL^3 p}{27EI}$$

Case 4 (bottom-right): Applied moment M at C

MOMENT:
$$M_B = -2M_A = -M$$

SHEAR:
$$R_A = -R_B = -\frac{3M}{2L}$$

DEFLECTION:
$$d_D = \frac{M}{4EI}\left[L(a+b) + a^2\left(2 + \frac{4b}{a}\right)\right]$$
$$-d_{max} = -\frac{ML^2}{27EI}$$

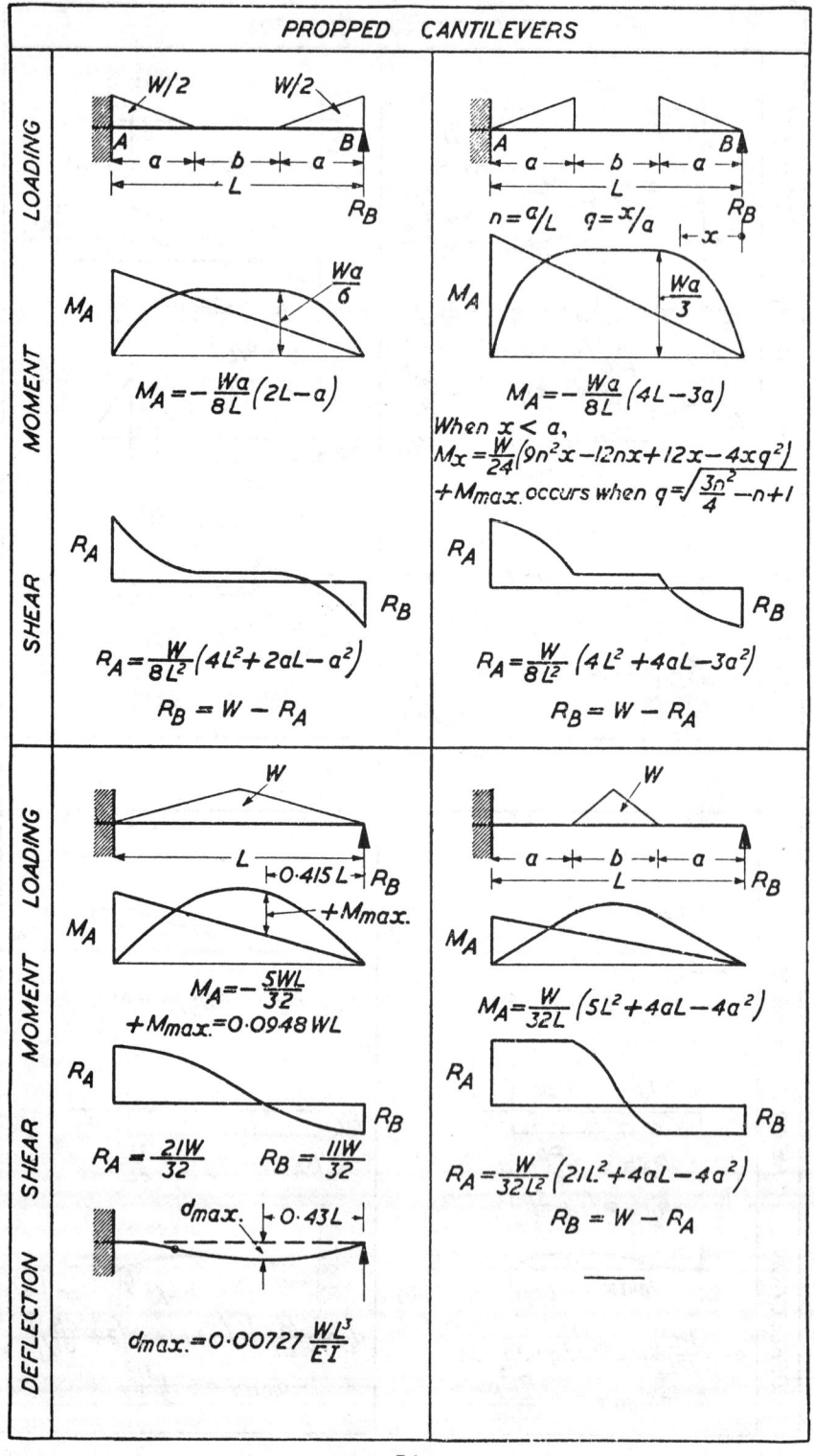

PROPPED CANTILEVERS

Top-left panel (Loading: triangular loads $W/2$ each over length a at both ends, span $L = a + b + a$):

$$M_A = -\frac{Wa}{8L}(2L - a)$$

Peak moment shown: $\frac{Wa}{6}$

$$R_A = \frac{W}{8L^2}(4L^2 + 2aL - a^2)$$

$$R_B = W - R_A$$

Top-right panel (Loading: triangular load rising to peak then falling, with $n = a/L$, $q = x/a$):

$$M_A = -\frac{Wa}{8L}(4L - 3a)$$

Peak moment shown: $\frac{Wa}{3}$

When $x < a$,
$$M_x = \frac{W}{24}(9n^2 x - 12nx + 12x - 4xq^2)$$

$+M_{max}$ occurs when $q = \sqrt{\frac{3n^2}{4} - n + 1}$

$$R_A = \frac{W}{8L^2}(4L^2 + 4aL - 3a^2)$$

$$R_B = W - R_A$$

Bottom-left panel (Loading: triangular load W over full span L, peak at $0.415L$ from R_B):

$$M_A = -\frac{5WL}{32}$$

$+M_{max} = 0.0948\,WL$

$$R_A = \frac{21W}{32} \qquad R_B = \frac{11W}{32}$$

d_{max} at $0.43L$

$$d_{max} = 0.00727\,\frac{WL^3}{EI}$$

Bottom-right panel (Loading: triangular load W centered, with dimensions $a + b + a = L$):

$$M_A = \frac{W}{32L}(5L^2 + 4aL - 4a^2)$$

$$R_A = \frac{W}{32L^2}(21L^2 + 4aL - 4a^2)$$

$$R_B = W - R_A$$

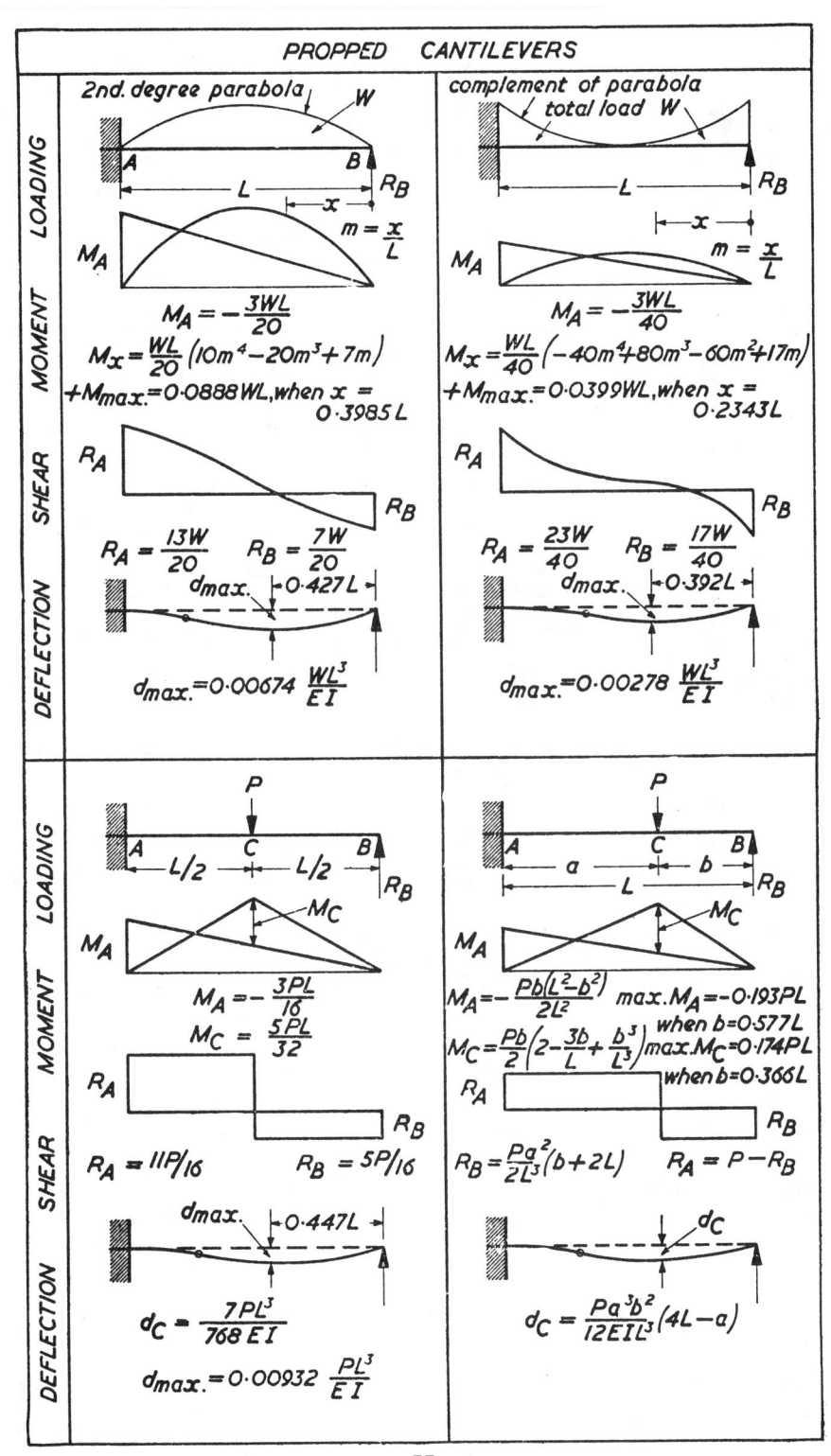

PROPPED CANTILEVERS

LOADING — 2nd. degree parabola, W; $m = x/L$

MOMENT:
$$M_A = -\frac{3WL}{20}$$
$$M_x = \frac{WL}{20}(10m^4 - 20m^3 + 7m)$$
$$+M_{max} = 0.0888\,WL, \text{ when } x = 0.3985L$$

SHEAR:
$$R_A = \frac{13W}{20} \quad R_B = \frac{7W}{20}$$

DEFLECTION: d_{max} at $0.427L$
$$d_{max} = 0.00674\,\frac{WL^3}{EI}$$

LOADING — complement of parabola, total load W; $m = x/L$

MOMENT:
$$M_A = -\frac{3WL}{40}$$
$$M_x = \frac{WL}{40}(-40m^4 + 80m^3 - 60m^2 + 17m)$$
$$+M_{max} = 0.0399\,WL, \text{ when } x = 0.2343L$$

SHEAR:
$$R_A = \frac{23W}{40} \quad R_B = \frac{17W}{40}$$

DEFLECTION: d_{max} at $0.392L$
$$d_{max} = 0.00278\,\frac{WL^3}{EI}$$

LOADING — P at C; A—$L/2$—C—$L/2$—B, R_B

MOMENT:
$$M_A = -\frac{3PL}{16}$$
$$M_C = \frac{5PL}{32}$$

SHEAR:
$$R_A = 11P/16 \quad R_B = 5P/16$$

DEFLECTION: d_{max} at $0.447L$
$$d_C = \frac{7PL^3}{768\,EI}$$
$$d_{max} = 0.00932\,\frac{PL^3}{EI}$$

LOADING — P at C; A—a—C—b—B, R_B; $L = a+b$

MOMENT:
$$M_A = -\frac{Pb(L^2 - b^2)}{2L^2} \quad \max M_A = -0.193\,PL \text{ when } b = 0.577L$$
$$M_C = \frac{Pb}{2}\left(2 - \frac{3b}{L} + \frac{b^3}{L^3}\right) \quad \max M_C = 0.174\,PL \text{ when } b = 0.366L$$

SHEAR:
$$R_B = \frac{Pa^2}{2L^3}(b + 2L) \quad R_A = P - R_B$$

DEFLECTION:
$$d_C = \frac{Pa^3 b^2}{12\,EI\,L^3}(4L - a)$$

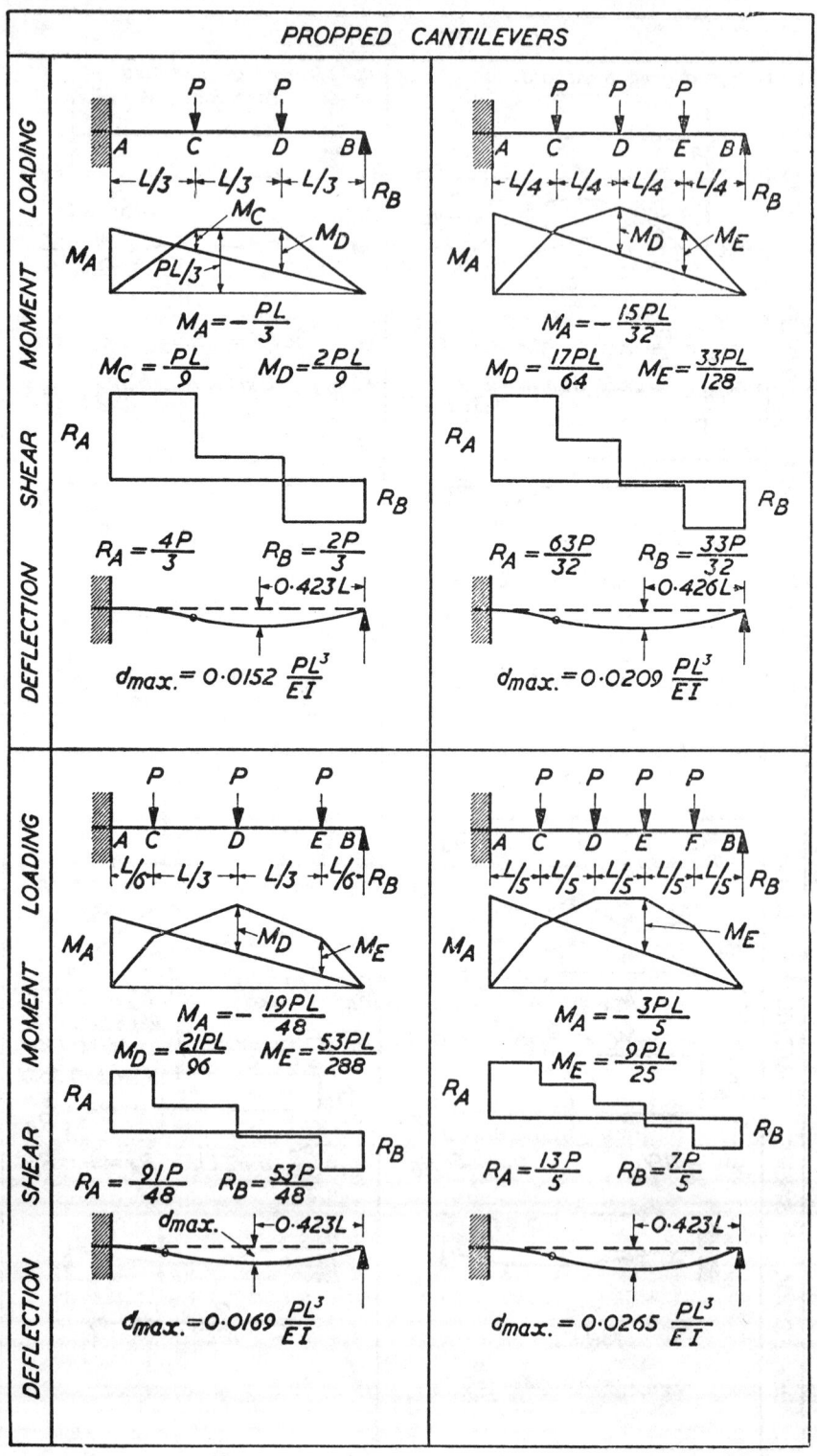

PROPPED CANTILEVERS

Case 1: Two equal loads at L/3 spacing

LOADING: Cantilever fixed at A, propped at B. Loads P at C and D. Spacing: L/3, L/3, L/3. Reaction R_B.

MOMENT:
$$M_A = -\frac{PL}{3}$$
$$M_C = \frac{PL}{9} \qquad M_D = \frac{2PL}{9}$$

SHEAR: R_A, R_B

$$R_A = \frac{4P}{3} \qquad R_B = \frac{2P}{3}$$

DEFLECTION: at $0.423L$ from B
$$d_{max.} = 0.0152 \frac{PL^3}{EI}$$

Case 2: Three equal loads at L/4 spacing

LOADING: Loads P at C, D, E. Spacing: L/4, L/4, L/4, L/4. Reaction R_B.

MOMENT:
$$M_A = -\frac{15PL}{32}$$
$$M_D = \frac{17PL}{64} \qquad M_E = \frac{33PL}{128}$$

SHEAR: R_A, R_B

$$R_A = \frac{63P}{32} \qquad R_B = \frac{33P}{32}$$

DEFLECTION: at $0.426L$ from B
$$d_{max.} = 0.0209 \frac{PL^3}{EI}$$

Case 3: Three equal loads at L/6, L/3, L/3, L/6

LOADING: Loads P at C, D, E. Spacing: L/6, L/3, L/3, L/6. Reaction R_B.

MOMENT:
$$M_A = -\frac{19PL}{48}$$
$$M_D = \frac{21PL}{96} \qquad M_E = \frac{53PL}{288}$$

SHEAR: R_A, R_B

$$R_A = \frac{91P}{48} \qquad R_B = \frac{53P}{48}$$

DEFLECTION: $d_{max.}$ at $0.423L$ from B
$$d_{max.} = 0.0169 \frac{PL^3}{EI}$$

Case 4: Four equal loads at L/5 spacing

LOADING: Loads P at C, D, E, F. Spacing: L/5, L/5, L/5, L/5, L/5. Reaction R_B.

MOMENT:
$$M_A = -\frac{3PL}{5}$$
$$M_E = \frac{9PL}{25}$$

SHEAR: R_A, R_B

$$R_A = \frac{13P}{5} \qquad R_B = \frac{7P}{5}$$

DEFLECTION: at $0.423L$ from B
$$d_{max.} = 0.0265 \frac{PL^3}{EI}$$

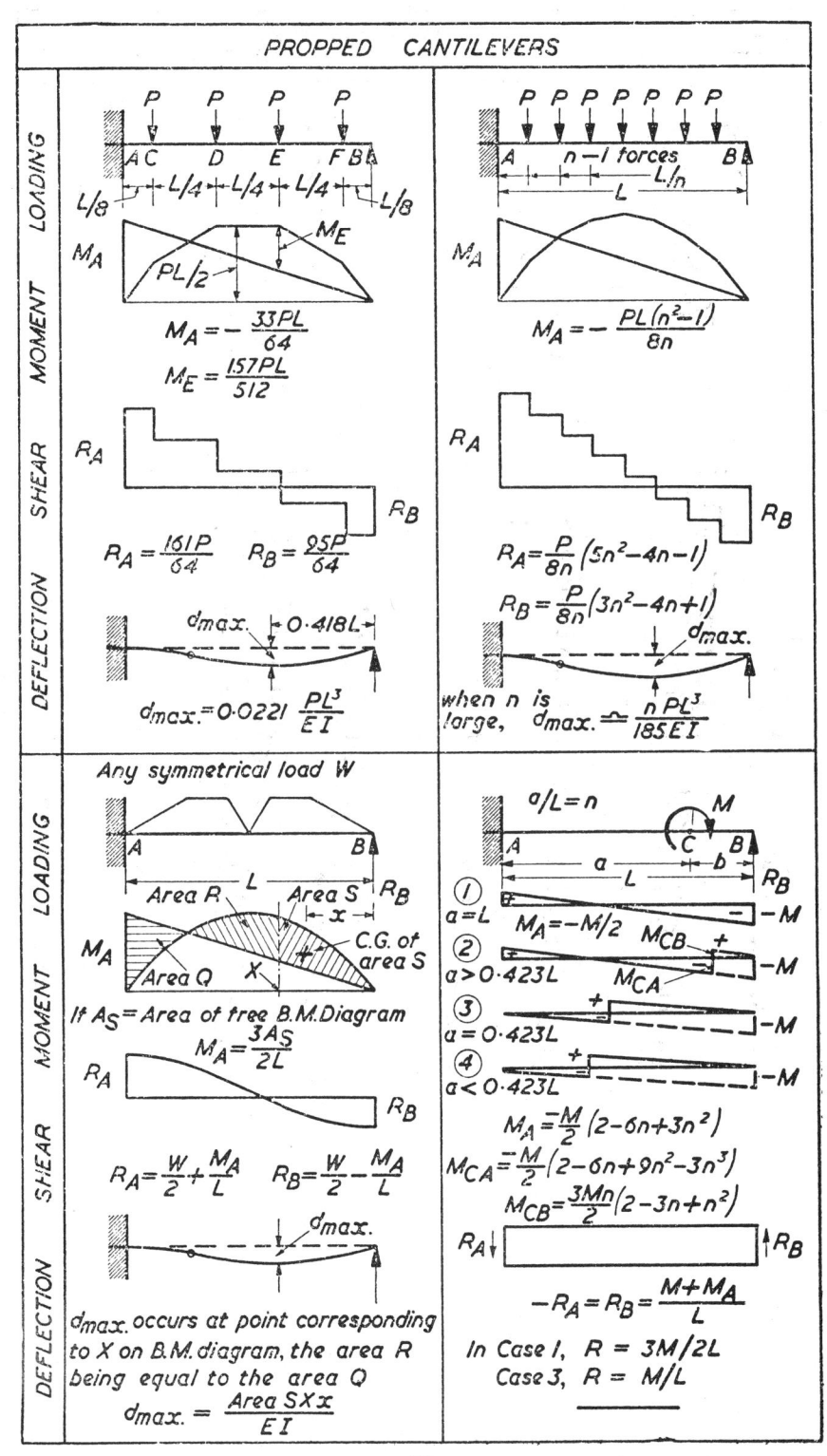

PROPPED CANTILEVERS

Case 1: Four equal point loads P at L/4 spacing, first at L/8 from A, last at L/8 from B

$$M_A = -\frac{33PL}{64}$$

$$M_E = \frac{157PL}{512}$$

$$R_A = \frac{161P}{64} \qquad R_B = \frac{95P}{64}$$

d_{max} at $0.418L$

$$d_{max} = 0.0221 \frac{PL^3}{EI}$$

Case 2: n−1 equal forces P spaced at L/n

$$M_A = -\frac{PL(n^2-1)}{8n}$$

$$R_A = \frac{P}{8n}(5n^2 - 4n - 1)$$

$$R_B = \frac{P}{8n}(3n^2 - 4n + 1)$$

when n is large, $d_{max} \simeq \dfrac{nPL^3}{185EI}$

Case 3: Any symmetrical load W

If A_S = Area of free B.M. Diagram

$$M_A = \frac{3A_S}{2L}$$

$$R_A = \frac{W}{2} + \frac{M_A}{L} \qquad R_B = \frac{W}{2} - \frac{M_A}{L}$$

d_{max} occurs at point corresponding to X on B.M. diagram, the area R being equal to the area Q

$$d_{max} = \frac{\text{Area } S \times x}{EI}$$

Case 4: Applied moment M at C, where $a/L = n$

① $a = L$, $M_A = -M/2$
② $a > 0.423L$
③ $a = 0.423L$
④ $a < 0.423L$

$$M_A = \frac{-M}{2}(2 - 6n + 3n^2)$$

$$M_{CA} = \frac{-M}{2}(2 - 6n + 9n^2 - 3n^3)$$

$$M_{CB} = \frac{3Mn}{2}(2 - 3n + n^2)$$

$$-R_A = R_B = \frac{M + M_A}{L}$$

In Case 1, $R = 3M/2L$
Case 3, $R = M/L$

7. THE DEFLECTION OF BEAMS

THE calculated deflection of a beam is limited by B.S. 449 to 1/360 of the span, calculated on loads other than the dead weight of the structural floors or roofs, together with steelwork and the weight of casing.

In the case of a simply supported beam, the uniformly distributed load W_D in tons, to produce this deflection is

$$W_D = CI$$

where C = tabulated load coefficient = $\dfrac{384E}{5 \times 360 \times L^2 \times 10^6}$

E = Young's Modulus taken as 2.1×10^5 N/mm^2

L = Length of span in metres

I = Moment of Inertia in cm^4

The load W_D will be less than the tabular load if the span exceeds 10 x the beam depth for Grade 55 steel, 12.17 x the beam depth for Grade 50 steel or 16.97 x the beam depth for Grade 43 steel.

In such cases it is necessary to confirm not only that the total load is within the capacity of the beam, but also that the loads to be considered for deflection purposes do not exceed W_D.

The table below gives limiting values of the span to depth ratio for uniformly loaded, simply supported beams for the different grades of steel for various ratios of Wd to Wt, where

Wd = load considered for deflection purposes, and
Wt = total load on beam.

Grade of Steel	Maximum Stress in N/mm^2	\multicolumn{6}{c}{Limiting Values of Span to Depth Ratio for Wd/Wt}						
		1.0	0.9	0.8	0.7	0.6	0.5
43	165	16.97	18.86	21.21	24.24	28.28	33.94
	155	18.06	20.07	22.58	25.80	30.10	36.12
	140	20.0	22.22	25.0	28.57	33.33	40.0
50	230	12.17	13.52	15.21	17.39	20.28	24.34
	215	13.02	14.47	16.28	18.60	21.70	26.04
55	280	10.0	11.11	12.50	14.29	16.67	20.0
	265	10.57	11.74	13.21	15.10	17.62	21.14

If the appropriate span to depth ratio is exceeded, then the relevant deflection will exceed 1/360th of the span unless the bending stress is reduced.

The most common methods of evaluating the deflection in any given case are three in number, although it is true to say that all these methods are variations of the same root principle.

They are:
(a) A mathematical method commonly called the method of slope-deflection.
(b) The area-moment method, first expounded by Mohr, and commonly called Mohr's theorem.
(c) A graphical method, which is really a graphical interpretation of the area-moment method.

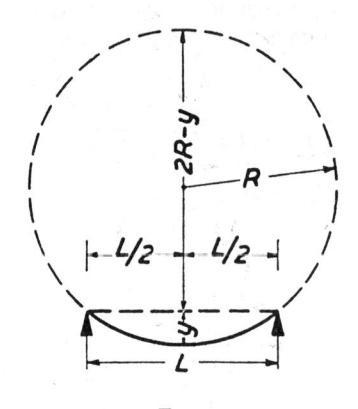

Fig. 1

Before starting to deal with these methods in detail, it is interesting to consider the deflection in the case of circular bending, which it will be remembered is the type of bending assumed in the Ordinary Beam Theory.

The portion AB of the beam shown in Fig. 1, on page 1, bends in a circular arc and by the geometry of the circle shown in Fig. 1 above.

$$y(2R - y) = \frac{L}{2} \times \frac{L}{2}$$

$$2Ry - y^2 = \frac{L^2}{4}$$

y^2 is extremely small compared with the other quantities, and may be neglected.

Therefore $$2Ry = \frac{L^2}{4}$$

and $$y = \frac{L^2}{8R}.$$

But $\dfrac{1}{R} = \dfrac{M}{EI}.$

Therefore $$y = \frac{ML^2}{8EI}$$

so that the maximum deflection in circular bending = $ML^2/8EI$.

MATHEMATICAL METHOD 83

If the section of the beam is constant, M must be constant for circular bending to occur. If, as is more usual, M varies throughout the span, then I must vary so as to keep the ratio M/I constant. This is not a practical proposition, although it is approached in the case of compound beams and plate girders where the plates are curtailed. A later example will illustrate this point.

The three methods will now be dealt with more fully.

Mathematical Method

It is necessary, first of all, to arrive at a convention of signs.

In Fig. 2, x is positive to the right and y is positive upwards. The slope dy/dx is positive, so that slopes upwards to the right are positive. The curvature shown is such that dy/dx increases as x increases, therefore, d^2y/dx^2 will be positive in this case. The type of bending illustrated in Fig. 1 will have been caused by what are considered positive B.M.s. Therefore positive B.M.s are associated with $+d^2y/dx^2$.

Now the rate of change of the slope, i.e. d^2y/dx^2, is the curvature and in the Beam Theory curvature, i.e. $1/R$, is equal to M/EI. Therefore,

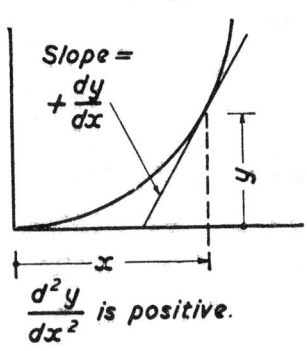

Fig. 2

$$\frac{d^2y}{dx^2} = \frac{M}{EI} \quad \text{or} \quad \frac{EI d^2y}{dx^2} = M.$$

This method is most easily applied to cases in which the B.M. at any point can be represented by a simple expression.

Example 1. Consider the case of the simply supported beam with a U.D.L. over the whole span, as in Fig. 3.

$$\text{B.M. at } x = \frac{wLx}{2} - \frac{wx^2}{2}$$

$$\frac{EI d^2y}{dx^2} = \frac{wLx}{2} - \frac{wx^2}{2}.$$

Integrating,

$$\frac{EI dy}{dx} = \frac{wLx^2}{4} - \frac{wx^3}{6} + C_1.$$

$dy/dx = 0$ at midspan, when $x = L/2$. Therefore,

$$0 = \frac{wL}{4} \times \left(\frac{L}{2}\right)^2 - \frac{w}{6}\left(\frac{L}{2}\right)^3 + C_1,$$

when

$$C_1 = -\frac{wL^3}{24}.$$

Therefore

$$\frac{EI dy}{dx} = \frac{wLx^2}{4} - \frac{wx^3}{6} - \frac{wL^3}{24}.$$

THE DEFLECTION OF BEAMS

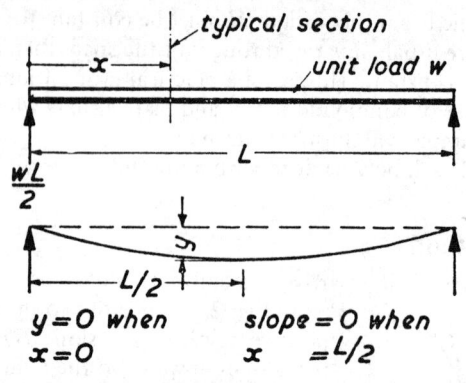

Fig. 3

From this expression the slope, dy/dx, at any point, can be found.

The slopes at the two ends of the beam are found by substituting $x = 0$ and $x = L$ in the expression.

In the first case, if $x = 0$, then

$$\frac{EIdy}{dx} = -\frac{wL^3}{24},$$

whence

$$\frac{dy}{dx} = -\frac{wL^3}{24EI}$$

which by the convention previously referred to, means that the slope is negative or upwards to the left.

In the second case, if $x = L$, then

$$\frac{EIdy}{dx} = \frac{wL^3}{4} - \frac{wL^3}{6} - \frac{wL^3}{24} = +\frac{wL^3}{24},$$

whence

$$\frac{dy}{dx} = +\frac{wL^3}{24EI}$$

i.e. the same value as in the first case, because the beam is symmetrically loaded, but of the opposite sign, indicating a slope upwards to the right.

Integrating the main expression again,

$$EIy = \frac{wLx^3}{12} - \frac{wx^4}{24} - \frac{wL^3 x}{24} + C_2$$

When $x = 0$, $y = 0$. Therefore $C_2 = 0$ and

$$EIy = \frac{wLx^3}{12} - \frac{wx^4}{24} - \frac{wL^3 x}{24},$$

for maximum deflection $x = L/2$. Whence,

$$y = -\frac{5wL^4}{384EI} = -\frac{5WL^3}{384EI}$$

the well-known result, although the minus sign is usually omitted. The diagrams shown on pages 29 to 38, 43 to 49 and 72 to 79 show downward deflections as positive in accordance with normal practice.

The Area-Moment Method

The area-moment method of analysis as used at the present time is usually attributed to Mohr who published his method of elastic loads in 1868. However, it was Professor C. E. Greene, of the University of Michigan who, in 1872, introduced the principles as they are now known. Subsequently, another German, Professor H. F. B. Müller-Breslau, extended the method to highly indeterminate structures.

A method of deriving the appropriate formulae will now be demonstrated.

Consider the cantilever ABC in Fig. 4 which is built in at A and carries a point load at C. Under the action of the load, the cantilever will no longer be horizontal except at A, the slope and consequent deflection varying from A to C. At B, for example, a short distance ds from A, the slope will be $d\theta$ and the deflection dy.

Now it is well known that

$$\frac{M}{I} = \frac{f}{y} = \frac{E}{R}$$

Ignoring the middle term,

$$\frac{M}{EI} = \frac{1}{R}$$

where M = the bending moment
E = the modulus of elasticity of the material
I = the moment of inertia
R = the radius of curvature of the member.

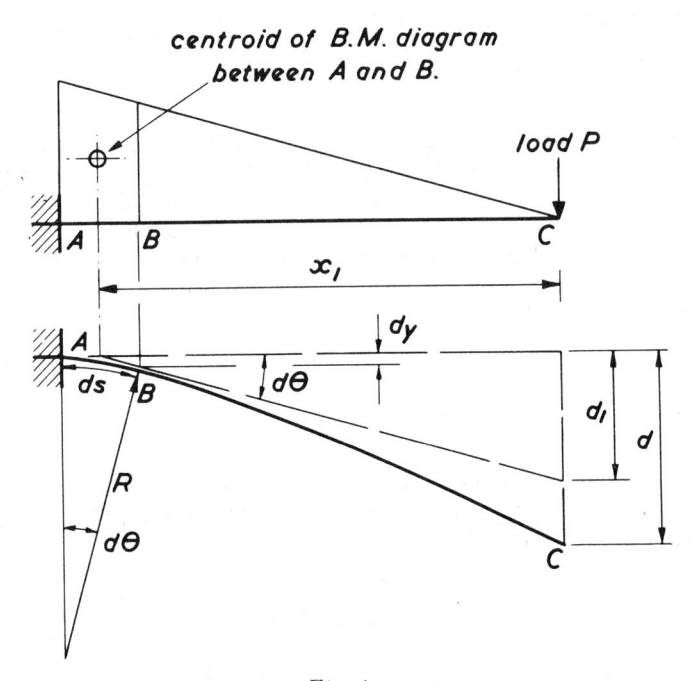

Fig. 4

Referring to Fig. 4, it will be seen that

$$ds = R \cdot d\theta \text{ (where } d\theta \text{ is measured in radians)}$$

or
$$d\theta = \frac{ds}{R} = \frac{1}{R} \times ds$$

$$= \frac{M \cdot ds}{EI}$$

Consequently, the total change in slope from A to C will be

$$\theta = \int_C^A \frac{M \cdot ds}{EI} \quad \ldots \ldots \ldots \ldots \ldots (a)$$

Returning to the short length ds between A and B, it will be seen that the deflection d_1 at C due to the bending of that short length alone may be found from the equation

$$d_1 = d\theta \cdot x_1 = \frac{M \cdot x_1 ds}{EI}$$

Consequently, the total vertical deflection d over the whole length of the cantilever may be found from the equation

$$d = \int_C^A \frac{M \cdot x \cdot ds}{EI} \quad \ldots \ldots \ldots \ldots \ldots (b)$$

In similar circumstances, where the deflection of a member is in a horizontal direction, the appropriate equation is

$$d = \int \frac{M \cdot y \cdot ds}{EI} \quad \ldots \ldots \ldots \ldots \ldots (c)$$

In equations (a) and (b) $M \cdot ds$ is the area of the B.M. diagram, whilst in equation (b) x is the lever arm between the centroid of the B.M. diagram and the point of deflection under consideration. It was from these data that Mohr and Greene developed the theorems of slope and deflection which may be expressed as follows:

Theorem I. The change in slope between any two points say, A and C in Fig. 4, in an *originally* straight member is equal to the area between corresponding points in the B.M. diagram, divided by EI.

i.e.
$$\theta = \frac{\Sigma A}{EI} \quad \ldots \ldots \ldots \ldots \ldots (d)$$

where ΣA = the area of the B.M. diagram.

Theorem II. The deflection of a point say, C in Fig. 4, in an originally straight member under flexure, in the direction perpendicular to the original axis of the member, measured from the tangent at a second point on the member say, A in Fig. 4 is equal to the statical moment of the B.M. diagram, divided by EI, taken about the first point C.

i.e.
$$d = \frac{\Sigma A x}{EI} \quad \ldots \ldots \ldots \ldots \ldots (e)$$

where x = the lever arm between the centroid of the B.M. diagram and C.

THE AREA-MOMENT METHOD　　　　　　87

Note that in cases where the B.M. diagram can be divided into convenient geometrical shapes, the formulae are more easily expressed in the form shown in equations (*d*) and (*e*) than (*a*) and (*b*) respectively.

B.M. diagram.

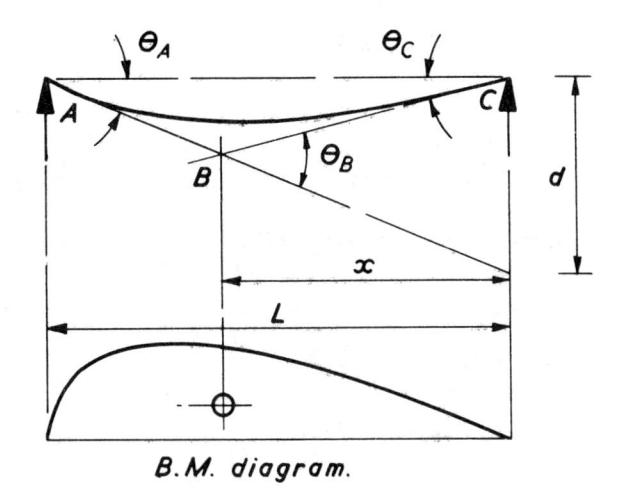

Fig. 5

The application of the theorems to a beam is demonstrated in Fig. 5, θ_B being the angular change between A and C measured in radians, which is equal to the area of the B.M. diagram, divided by EI; and d being the deflection, in this case upwards, measured perpendicular to the original axis of the member, of the point C from a tangent at A, which is equal to the statical moment of the B.M. diagram about C, divided by EI.

Now the angle θ_B is equal to the sum of the angles θ_A and θ_C at the ends of the beam. It will be observed that $\theta_A : \theta_C = x : (L - x)$ from which it is obvious that the angular change at either end of a beam can be obtained by calculating the support reaction at that end when the beam is loaded with its B.M. diagram, divided by EI.

Some standard slopes and moments for beams are shown in Fig. 6, the beams in each case being of constant section throughout.

Example 2. A 254 mm x 146 mm x 31 kg U.B. is loaded with a single point load as shown in Fig. 7. $E = 2.1 \times 10^5$ N/mm². $I = 4\,427$ cm⁴. It is required to find the deflection under the load, the maximum deflection and the deflection at the centre of the span.

$$\text{B.M. at } C = \frac{Wab}{L} = \frac{120 \times 1 \times 2}{3} = 80 \text{ kNm}$$

Considering the B.M. diagram as a load; taking moments about A and rearranging,

$$Rb = \frac{1}{3}\left(\frac{80 \times 2 \times 5}{2 \times 3} + \frac{80 \times 1 \times 2}{2 \times 3}\right) = 53.33 \text{ kNm}^2$$

88   THE DEFLECTION OF BEAMS

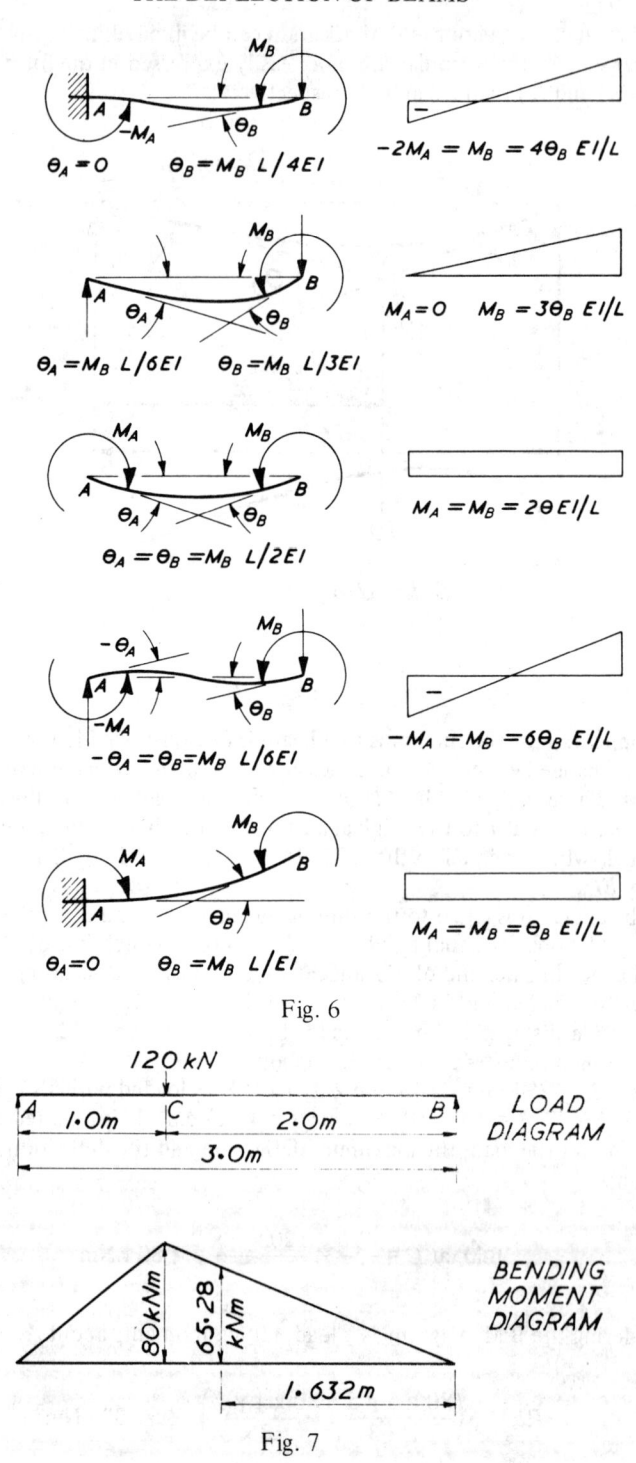

$\theta_A = 0$ $\theta_B = M_B L/4EI$ $-2M_A = M_B = 4\theta_B EI/L$

$\theta_A = M_B L/6EI$ $\theta_B = M_B L/3EI$ $M_A = 0$ $M_B = 3\theta_B EI/L$

$\theta_A = \theta_B = M_B L/2EI$ $M_A = M_B = 2\theta EI/L$

$-\theta_A = \theta_B = M_B L/6EI$ $-M_A = M_B = 6\theta_B EI/L$

$\theta_A = 0$ $\theta_B = M_B L/EI$ $M_A = M_B = \theta_B EI/L$

Fig. 6

LOAD DIAGRAM: 120 kN at C; A—1·0m—C—2·0m—B; total 3·0m

BENDING MOMENT DIAGRAM: 80 kNm, 65·28 kNm, 1·632 m

Fig. 7

THE AREA-MOMENT METHOD

Deflection at centre of span:

B.M. at centre of span = $\frac{1.5}{2} \times 80 = 60$ kNm

Secondary B.M. = $(53.3 \times 1.5) - \left(\frac{60 \times 1.5}{2} \times \frac{1.5}{3}\right)$

$= 79.95 - 22.5$

$= 57.45$ kNm³

Therefore deflection = $-\dfrac{57.45 \times 10^{12} \text{ Nmm}^3}{2.1 \times 10^5 \text{ N/mm}^2 \times 4\,427 \times 10^4 \text{ mm}^4} = 6.18$ mm

Deflection under the load:

Secondary B.M. = $(53.3 \times 2) - \left(\frac{80 \times 2}{2} \times \frac{2}{3}\right)$

$= 106.6 - 53.3 = 53.33$ kNm³

Therefore deflection = $-\dfrac{53.33 \times 10^{12} \text{ Nmm}^3}{2.1 \times 10^5 \text{ N/mm}^2 \times 4\,427 \times 10^4 \text{ mm}^4} = 5.75$ mm

Maximum deflection:
It is first necessary to find the position of maximum deflection. Since

$$\text{deflection} = -\frac{\text{Secondary } M}{EI}$$

it follows that the maximum deflection occurs where the secondary M is a maximum. As in all other cases of loading, the maximum B.M. occurs at the point of zero shear.

The point of zero shear will always occur between the load and the centre of the span.

Let x = the distance of the point X of zero secondary shear from B. Then:

B.M. at $X = \dfrac{x}{2} \times 80 = 40x$

Therefore $40x \times \dfrac{x}{2} = 20x^2 = 53.33$ whence $x = 1.632$ m

Secondary B.M. = $(53.33 \times 1.632) - \left(53.33 \times \dfrac{1.632}{3}\right)$

$= \dfrac{53.33 \times 1.632 \times 2}{3} = 57.99$ kNm³

Maximum deflection = $-\dfrac{57.99 \times 10^{12}}{2.1 \times 4\,427 \times 10^9} = 6.24$ mm

It will be seen that the maximum variation between all three values is approximately 8 per cent of the maximum deflection.

It can be proved that the maximum deflection, under any system of loading whatever, never occurs further than $0.0774L$ from the centre of the span, provided that the beam rests on simple supports.

If the slope at any point is required, use is made of the first proposition previously referred to. It is necessary to have a reference point from which all

THE DEFLECTION OF BEAMS

slopes can be measured, and this obviously is the point of zero slope, i.e. where the beam is horizontal. This is the point of maximum deflection.

The slopes at the ends of the beam will now be determined.

Slope at end B:

$$\text{Slope at } X = 0$$

$$\text{B.M. at } X = \frac{80 \times 1.632}{2} = 65.28 \text{ kNm}$$

Area of B.M. diagram between X and B

$$= \frac{65.28 \times 1.632}{2} = 53.33 \text{ kNm}^2$$

This is confirmed by the fact that this area must be equal to the reaction at B when the B.M. diagram is considered as a load.

Therefore:

$$\text{Slope at } B = +\frac{53.33}{EI} = +\frac{53.33 \times 10^9}{2.1 \times 10^5 \times 4\,427 \times 10^4} = +0.006 \text{ radian}$$

The slope is positive because the area is measured to the right of X and is therefore positive.

Slope at end A:

The total area of the B.M. diagram

$$= \frac{80 \times 3}{2} = 120 \text{ kNm}^2$$

Consequently the area of the B.M. diagram between X and A

$$= 120 - 53.33 = 66.67 \text{ kNm}^2$$

Therefore

$$\text{Slope at } A = -\frac{66.7}{EI} = -\frac{66.7 \times 10^9}{2.1 \times 10^5 \times 4\,427 \times 10^4}$$

$$= -0.0075 \text{ radian}$$

The slope is negative because the area is measured to the left of X.

Slope at C:

The area of the B.M. diagram between X and C

$$= \frac{80 + 65.28}{2} \times (2 - 1.632)$$

$$= 26.7 \text{ kNm}^2$$

C is to the left of X, hence

$$\text{Slope at } C = -\frac{26.7}{EI} = -\frac{26.7 \times 10^9}{2.1 \times 4\,427 \times 10^9}$$

$$= 0.0029 \text{ radian.}$$

A number of further examples will now be worked out to illustrate the application of the principles of the theory of deflection.

SIMPLE BEAMS 91

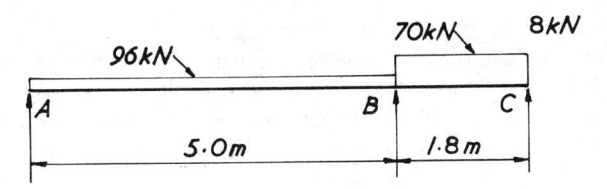

LOAD DIAGRAM

BENDING MOMENT DIAGRAM

DEFLECTED FORM

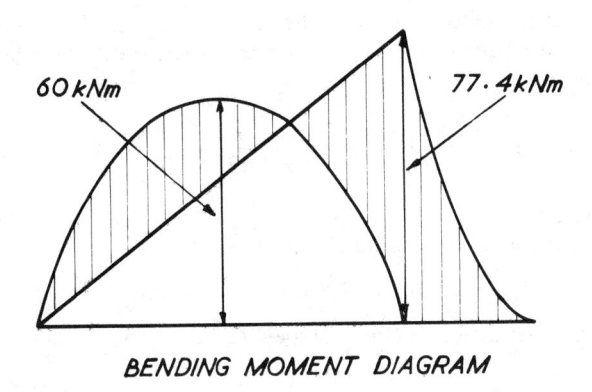

Fig. 8

Simple Beams

Example 3. To find the deflection at *C* of the cantilevered beam shown in Fig. 8.
$E = 2.1 \times 10^5$ N/mm²   *I* is constant and $= 4\,427$ cm⁴

The first step is to find the slope at *B* of the span *AB*, assuming that there is no load on the cantilevered portion.

$$\text{Slope} = \frac{\text{area of half B.M. diagram}}{EI}$$

$$= \tfrac{1}{2}\left(\frac{60 \times 5 \times \tfrac{2}{3} \times 10^9}{2.1 \times 4\,427 \times 10^9}\right)$$

$$= 0.0108 \text{ radian}$$

The slope is positive as will be seen from Fig. 2.

If there were no load on the cantilevered portion BC, then the slope at B would be continued along BC at a constant value. Since, in the case of relatively small angles of slope, the tangent of the angle is, for all practical purposes, equal to the angle, then the slope would cause an upward deflection at C equal to

$$0.0108 \times 1.8 \times 10^3 = 19.5 \text{ mm}$$

The load on the cantilever, however, causes the beam to rotate at B so that the deflection at C, with respect to B, is composed of two parts, viz. that due to the rotation and that due to the load.

The rotation at B due to the load on the cantilever is $ML/3EI$, where M is the bending moment at B due to the load on the cantilever, and L is the length of the span AB. Therefore:

$$\text{Slope at } B = \frac{77.4 \times 5}{3EI} = \frac{77.4 \times 5 \times 10^9}{3 \times 2.1 \times 4\,427 \times 10^9} = 0.0132 \text{ radian}$$

The slope is negative.

The downward deflection at C due to this slope

$$= 0.0132 \times 1.8 \times 10^3 = 23.8 \text{ mm}$$

The downward deflection at C with respect to B, due to the load on the cantilever.

$$= \left[\frac{70 \times 1.8^3}{8} + \frac{8 \times 1.8^3}{3} \right] \times \frac{10^{12}}{2.1 \times 4\,427 \times 10^9}$$

$$= 7.16 \text{ mm}$$

The net result is a downward deflection at C

$$= 19.5 - 23.8 - 7.16 = -11.46 \text{ mm}$$

Fixed Beams

The next example deals with a beam fixed at each end, but before working the example, the general application of the area-moment method to fixed beams will be considered.

The deflection at any point of a fixed beam is given by the difference between the downward deflection due to the load had the beam been simply supported, and the upward deflection due to the fixing moments.

Since, in any fixed beam, the centroids of the areas of the free moment and fixing moment diagrams lie on the same vertical line, it follows that when these areas are considered as loads, the reactions to the areas must be equal and opposite. Therefore when the areas are considered as loads, there are no resultant reactions and the maximum deflection will occur at a point where the area of the free moment diagram equals the area of the fixing moment diagram, working from the same end of the beam.

Example 4. Consider the simple case of a fixed beam of span L, carrying a total U.D.L. of W as shown in Fig. 9.

FIXED BEAMS

[Loading diagram: fixed beam with central point load W, span L]

LOADING DIAGRAM

[Bending moment diagram showing free moment WL/8 with fixing moment WL/12]

BENDING MOMENT DIAGRAM

Fig. 9

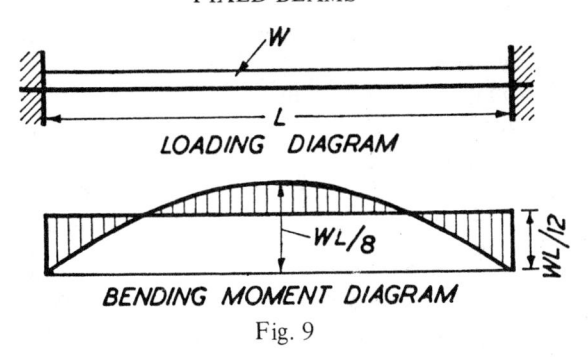

The maximum free B.M. = $WL/8$, while the fixing moments at each end = $WL/12$. The fixing moment diagram thus has a constant height of $WL/12$. The maximum deflection obviously occurs at the centre of the span.

Working from one end, the area of the free-moment diagram to this point

$$= \frac{1}{2} \times \frac{2}{3} \times \frac{WL}{8} \times L = \frac{WL^2}{24}$$

while the area of the fixing moment diagram

$$= \frac{WL}{12} \times \frac{L}{2} = \frac{WL^2}{24}$$

to the same point.

The maximum deflection in an encastré beam is given by the difference between the moments of the areas to one side of, and taken about, the point of maximum deflection, in this case the centre.

It must be remembered, however, that since, in all cases, the moment is equal to Reaction Moment minus Load Moment, where the Reaction and the Load refer to the areas of the appropriate B.M. diagrams, and as the reactions cancel out as previously shown, then the maximum deflection is given, numerically and with the correct sign, by deducting from the moment of the free moment diagram, the moment of the fixing moment diagram both taken about the point of maximum deflection.

Thus, in the case under consideration, the maximum deflection

$$= \frac{1}{EI} \left\{ \left(\frac{WL^2}{24} \times \frac{3L}{16} \right) - \left(\frac{WL^2}{24} \times \frac{L}{4} \right) \right\} = -\frac{WL^3}{384EI}.$$

In this particular case, and indeed for any case of symmetrical loading, the result could also have been arrived at in the following manner.

The central deflection due to the free moment diagram only

$$= -\frac{5}{384} \times \frac{WL^3}{EI},$$

while the deflection due to the fixing moment diagram (circular bending since M is constant) = $ML^2/8EI$, where $M = WL/12$, giving a deflection of $WL^3/96EI$, and the net deflection

$$= -\left(\frac{5}{384} \times \frac{WL^3}{EI} \right) + \left(\frac{1}{96} \times \frac{WL^3}{EI} \right) = -\frac{WL^3}{384EI}.$$

THE DEFLECTION OF BEAMS

Example 5. To find the position and value of the maximum deflection of the beam loaded as shown in Fig. 10.
$E = 2.1 \times 10^5$ N/mm². $I = 1\,522$ cm⁴.

Fig. 10

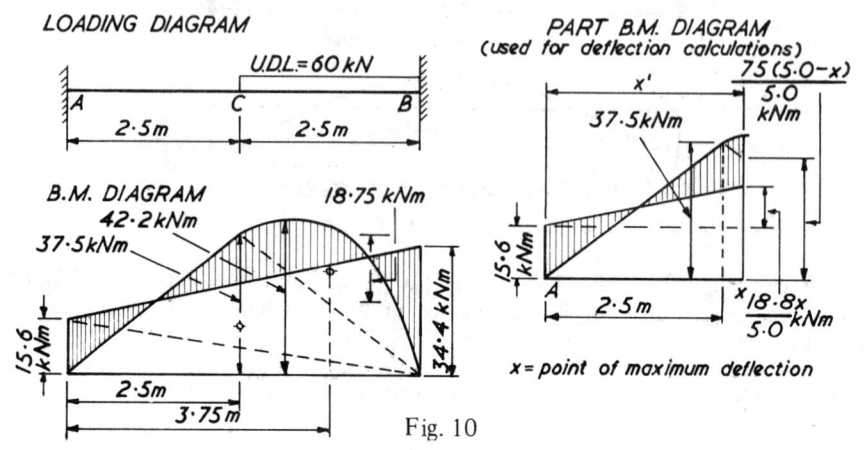

The free moment diagram can be divided into an isosceles triangle of height 37.5 kN at *C*, on which is superimposed over the length *CB* a parabola, of maximum height 18.75 kN.

The area of the free moment diagram

$$= \frac{37.5 \times 5}{2} + \frac{2 \times 18.75 \times 2.5}{3}$$

$$= 93.75 + 31.25 = 125 \text{ kNm}^2$$

Therefore
$$\frac{M_A + M_B}{2} \times 5 = 125 \text{ kNm}^2$$

whence $M_A + M_B = 50$ kNm . . . . (1)

Taking area moments about *A*, the moment of the fixing moment diagram, which can be divided into two triangles as shown, must equal the moment of the free moment diagram.

Hence $M_A \times \frac{L}{2} \times \frac{L}{3} + M_B \times \frac{L}{2} \times \frac{2L}{3} = (93.75 \times 2.5) + (31.25 \times 3.75)$

whence $M_A + 2M_B = 84.4$ kNm . . . . (2)

From the two equations $M_A = 15.6$ kNm and $M_B = 34.4$ kNm.

The maximum deflection will occur at *X*, at a distance *x* from *A* such that the areas of the free moment and fixing moment diagrams above the length *x* are equal. (See Fig. 10.)

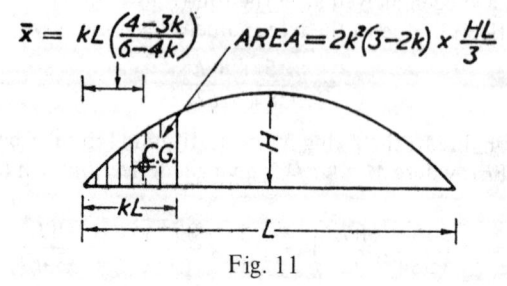

$$\bar{x} = kL\left(\frac{4-3k}{6-4k}\right) \quad AREA = 2k^2(3-2k) \times \frac{HL}{3}$$

Fig. 11

FIXED BEAMS

At this point it will be necessary to digress for a while in order to establish certain properties of a part of a parabola. See Fig. 11.

If a parabola of height H is constructed on a base L, then the area over a length kL measured from one end

$$= 2k^2(3-2k) \times \frac{HL}{3}$$

while the C.G. of this portion of area from the same end ($\bar{x}$ in the figure)

$$= kL\left[\frac{4-3k}{6-4k}\right]$$

Reverting to the parabola in the example

$H = 18.75$ kNm

$L = 2.5$ m (the half span on which the parabola is constructed)

and $kL = \left(x - \frac{5}{2}\right)$ or $k = \left(\frac{x}{2.5} - 1\right)$

The area of the fixing moment diagram on the length x

$$= 15.6x + \frac{18.8x^2}{10}$$

while the area of the free moment diagram on the same length

$$= \frac{37.5 \times 2.5}{2} + \left[\frac{37.5 + \frac{75(5.0-x)}{5.0}}{2}(x-2.5)\right] +$$

$$+ 2\left(\frac{x}{2.5} - 1\right)^2 \times \left[3 - 2\left(\frac{x}{2.5} - 1\right)\right]\left(\frac{18.75 \times 2.5}{3}\right)$$

$= 62.5 - 75x + 37.5x^2 - 4x^3$

The two areas must be equal, so that

$$15.6x + 1.88x^2 = -4x^3 + 37.5x^2 - 75x + 62.5$$

whence $x = 2.79$ m

The values of the individual small portions of areas are:

Free moment diagram (a) triangle = 46.875 kNm²
(b) trapezium = 10.24 ,,
(c) parabola = 1.16 ,,
       58.28 ,,

Fixing moment diagram (a) rectangle = 43.6 kNm²
(b) triangle = 14.68 ,,
       58.28 ,,

Taking moments about X, the maximum deflection

$$= \frac{10^3}{2.1 \times 1\,522}[46.87 \times 1.12 + 10.24 \times 0.14 + 1.16 \times 0.1 - 43.6 \times 1.395 - 14.68 \times 0.93]$$

$= 6.36$ mm.

THE DEFLECTION OF BEAMS

Example 6. A post is fixed at the base and hinged at the top. It carries an applied B.M. at an intermediate point. See Fig.12. Find the maximum deflection in the portion AB, the deflection at B, and the maximum deflection in the portion BC.

$E = 2.1 \times 10^5$ N/mm². $\quad I = 8\,500$ cm⁴.

The first step is to find the reaction at A.

Suppose the reaction at A to be removed, then the B.M. diagram would appear as shown in Fig 12 (b).

The deflection at A due to this B.M. diagram is given by

$$\frac{\text{Moment of B.M. diagram about } A}{EI} = \frac{60 \times 3.5 \times 4.25}{EI} = \frac{892.5}{EI}$$

If the reaction at A is now replaced, the deflection at A due to R_A

$$= \frac{1}{3} \times \frac{R_A L^3}{EI} = \frac{1}{3} \times \frac{R_A \times 216}{EI} = \frac{72 R_A}{EI}$$

Therefore for there to be no resultant deflection at A:

$$\frac{72 R_A}{EI} = \frac{892.5}{EI}$$

whence $\qquad R_A = 12.4$ kN

The final B.M. diagram is shown in Fig. 12 (c).

The B.M. diagram is now considered as a load and taking moments about C, the secondary reaction at A

$$= \frac{\left(14.4 \times \frac{1.16}{2} \times \frac{1.16}{3}\right) - \left(29 \times \frac{2.34}{2} \times 2.72\right) + \left(31 \times \frac{2.5}{2} \times 4.33\right)}{6}$$

$= 13.2$ kNm²

The maximum deflection between A and B is given at a distance x from A such that the area of the B.M. diagram on the length x is equal to the reaction at A.

Hence $\qquad \dfrac{31x}{2.5} \times \dfrac{x}{2} = 13.2, \quad$ whence $x = 1.46$ m

and the maximum deflection

$$= -\frac{\left[(13.2 \times 1.46) - \left(13.2 \times \frac{1.46}{3}\right)\right] \times 10^{12}}{(2.1 \times 10^5) \times (8\,500 \times 10^4)} = -0.72 \text{ mm}$$

i.e. a deflection to the right.

It is desirable to differentiate, by signs, between deflections to the right or to the left of the original vertical line of the post. If the post is viewed from the right, the secondary B.M. is of the type that corresponds to negative deflection. Accordingly, deflection to the right is considered negative and, conversely, that to the left is considered positive.

CONTINUOUS BEAMS 97

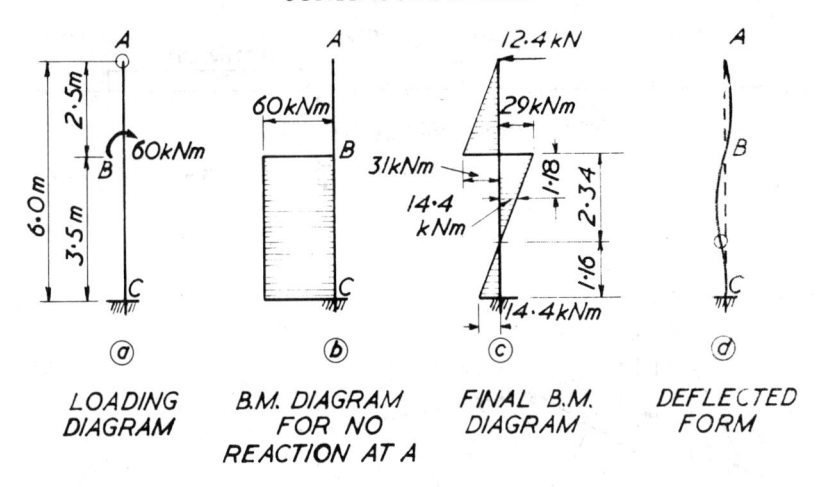

```
    A              A          12.4 kN          A
                60 kNm        29 kNm
     60 kNm   B                              B
   B          31 kNm
                14.4
                kNm
                          14.4 kNm
     C           C              C              C
```

(a) LOADING DIAGRAM   (b) B.M. DIAGRAM FOR NO REACTION AT A   (c) FINAL B.M. DIAGRAM   (d) DEFLECTED FORM

Fig. 12

The deflection at B

$$= -\frac{\left[(13.2 \times 2.5) - \left(31 \times \frac{2.5}{2} \times \frac{2.5}{3}\right)\right] \times 10^{12}}{(2.1 \times 10^5) \times (8\,500 \times 10^4)} = -0.04 \text{ mm}$$

i.e. again a deflection to the right.

The maximum deflection between B and C occurs at a distance of x_1 below B such that the area of the positive B.M. diagram on AB is equal to the sum of the reaction and the area of the negative B.M. diagram on the length x_1.

Hence $\qquad \dfrac{31 \times 2.5}{2} = 13.2 + \dfrac{x_1}{2}\left[29 + 29\left(\dfrac{2.34 - x_1}{2.34}\right)\right]$

whence $\qquad x_1^2 - 4.68x_1 + 4.13 = 0, \quad$ and $x_1 = 1.18$ m

and the maximum deflection

$$= -\frac{\left[(13.2 \times 3.68) - \left(31 \times \frac{2.5}{2} \times 2.01\right) + \left(14.4 \times \frac{1.18^2}{2}\right) + (29 - 14.4)\left(\frac{1.18^2}{3}\right)\right] \times 10^{12}}{(2.1 \times 10^5) \times (8\,500 \times 10^4)}$$

$= +0.68$ mm.

This deflection is to the left.

Continuous Beams

Example 7. To find the positions and values of the maximum deflections in the spans of a continuous beam.

The slopes and deflections in any span of a continuous beam can readily be determined by the use of the area-moment method, each span being treated separately after the negative moments at the points of support have been found.

98   THE DEFLECTION OF BEAMS

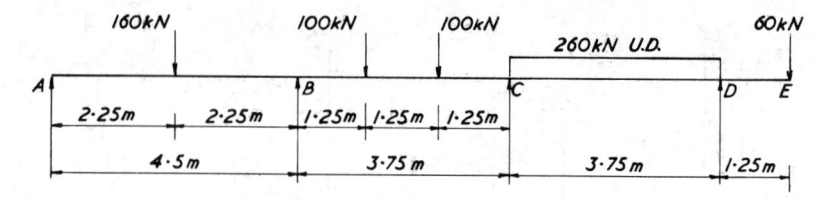

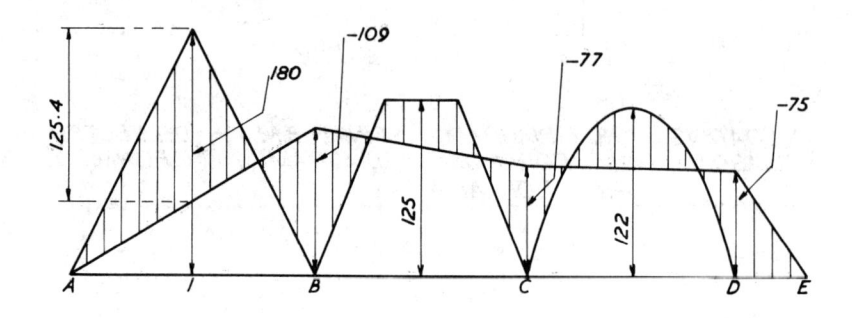

BENDING MOMENT DIAGRAM.

Fig. 13

Span AB:
[$I_{XX} = 8\,500$ cm^4]
   There are two B.M. diagrams in the span, one of which is positive and the other negative. Considering the B.M. diagrams as loads, the secondary reaction at A

$$= \left(\frac{180 \times 4.5}{2} \times \frac{1}{2}\right) - \left(\frac{109 \times 4.5}{2} \times \frac{1}{3}\right)$$

$$= 202.5 - 81.8 = 120.7 \text{ kNm}^2$$

The maximum deflection occurs at the point of maximum secondary B.M. which occurs at X, at a distance x from A such that the net area of the B.M. diagram on the length x equals the secondary reaction.

Therefore $\quad 120.7 = \left[\left(\frac{180}{2.25} \times \frac{x^2}{2}\right) - \left(\frac{109}{4.5} \times \frac{x^2}{2}\right)\right]$

whence $\quad x = 2.08$ m

The maximum deflection is found by the application of the principles previously set out and

$$= -\frac{\left[(120.7 \times 2.08) - \left(120.7 \times \frac{2.08}{3}\right)\right] \times 10^{12}}{(2.1 \times 10^5) \times (8\,500 \times 10^4)}$$

$$= -9.4 \text{ mm}$$

Span BC:

[I_{XX} = 6 195 cm^4]

Considering the B.M. diagrams as loads, the secondary reaction at B

$$= \left(\frac{125 \times 2.5}{2}\right) - \left(\frac{77 \times 3.75}{2}\right) - \left(\frac{32 \times 3.75}{2} \times \frac{2}{3}\right)$$

$$= -28.2 \text{ kNm}^2$$

while the secondary reaction at C

$$= \left(\frac{125 \times 2.5}{2}\right) - \left(\frac{77 \times 3.75}{2}\right) - \left(\frac{32 \times 3.75}{2} \times \frac{1}{3}\right)$$

$$= -8.2 \text{ kNm}^2$$

The fact that both secondary reactions are negative quantities means that the beam is hogging at each end, so that there will either be one point of maximum upward deflection or three points of maximum deflection: two upwards towards the ends of the span, and one downwards towards the centre. The latter circumstance will apply in the case under consideration.

The point of maximum deflection nearest to B will occur at a point X at a distance x from B such that

$$28.2 + \left(\frac{125}{1.25} \times \frac{x^2}{2}\right) = x\left(\frac{109 + 77 + \left(\frac{3.75 - x}{3.75}\right) \times 32}{2}\right)$$

$$28.2 + 50x^2 = x(109 - 4.3x)$$

Therefore $\qquad 54.3x^2 - 109x + 28.2 = 0$

whence $\qquad x = 0.3$ m

The corresponding upward deflection

$$= -\frac{\left[(-28.2 \times 0.3) - \left(\frac{125}{1.25} \times \frac{0.3^2}{6}\right) + \left(106.4 \times \frac{0.3^2}{2}\right) + \left(2.6 \times \frac{0.3^2}{2} \times \frac{2}{3}\right)\right] \times 10^{12}}{(2.1 \times 10^5) \times (6\ 195 \times 10^4)}$$

$$= +0.39 \text{ mm}$$

The point of maximum deflection nearest to C will occur at a point X_1, at a distance x_1 from C such that

$$8.2 + \left(\frac{125}{1.25} \times \frac{x_1^2}{2}\right) = x_1\left(\frac{77 + 77 + \frac{32x_1}{3.75}}{2}\right)$$

$$8.2 + 50x_1^2 = 77x_1 + 4.3x_1^2$$

Therefore $\qquad 54.3x_1^2 - 77x_1 + 8.2 = 0$

whence $\qquad x_1 = 0.11$ m.

The corresponding upward deflection

$$= -\frac{\left[(-8.2 \times 0.11) - \left(\frac{125}{1.25} \times \frac{0.11^2}{6}\right) + \left(77 \times \frac{0.11^2}{2}\right) + \left(0.9 \times \frac{0.11^2}{2} \times \frac{1}{3}\right)\right] \times 10^{12}}{(2.1 \times 10^5) \times (6\,195 \times 10^4)}$$

$= 0.05$ mm.

The point of maximum deflection nearest the centre is more easily found by working from the end C and will occur at a point X_2 at a distance x_2 from C such that

$$8.2 + \left(\frac{125 \times 1.25}{2}\right) + 125(x_2 - 1.25) = x_2 \left(\frac{77 + 77 + \frac{32x_2}{3.75}}{2}\right)$$

$$8.2 + 78.1 + 125x_2 - 156.2 = 77x_2 + 4.3x_2^2$$

Therefore $\qquad 4.3x_2^2 - 48x_2 + 70 = 0$

whence $\qquad x_2 = 1.75$ m

The corresponding maximum downward deflection

$$= -\frac{\left[(-8.2 \times 1.75) - \left(\frac{125 \times 1.25}{2} \times 0.875\right) - \left(\frac{125 \times 0.5^2}{2}\right) + \left(\frac{77 \times 1.75^2}{2}\right) + \left(\frac{14.9 \times 1.75^2}{2} \times \frac{1}{3}\right)\right] \times 10^{12}}{(2.1 \times 10^5) \times (6\,195 \times 10^4)}$$

$= -2.06$ mm

Span CD:

[$I_{xx} = 5\,749$ cm^4]

Since the negative moments at the ends of the span are so nearly equal, it may be assumed, for all practical purposes, that an average negative moment of 76 kNm obtains throughout the length of the span and that the maximum deflection occurs at the centre.

If that assumption is made, the maximum deflection

$$= -\left(\frac{5}{384} \times \frac{WL^3}{EI}\right) + \left(\frac{ML^2}{8EI}\right)$$

$$= -\frac{\left[-\frac{5 \times 260 \times 3.75^3}{384} + \frac{76 \times 3.75^2}{8}\right] \times 10^{12}}{(2.1 \times 10^5) \times (5\,749 \times 10^4)}$$

$= -3.75$ mm

Propped Cantilevers

It is required to find the position and amount of the maximum deflection in the propped cantilever shown in Fig. 14.

PROPPED CANTILEVERS

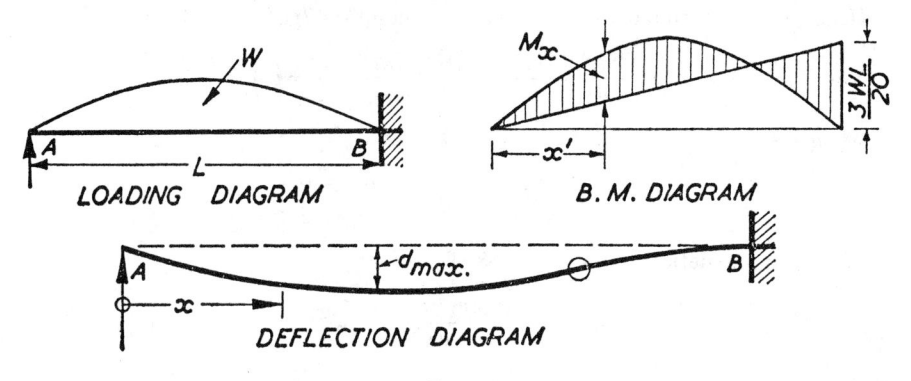

Fig. 14

The load W, which is distributed in the form of a second-degree parabola, produces the B.M. diagram illustrated.

Now $M_B = -3WL/20$ while the free B.M. at any point x ft. from A, considering AB to be a simply supported beam

$$= M_x = \frac{WL}{2}\left(\frac{x^4}{L^4} - \frac{2x^3}{L^3} + \frac{x}{L}\right)$$

Hence, for the propped cantilever,

$$M_x = \frac{WL}{2}\left(\frac{x^4}{L^4} - \frac{2x^3}{L^3} + \frac{x}{L}\right) - \left(\frac{3WL}{20} \times \frac{x}{L}\right)$$

$$= \frac{WL}{20}\left(\frac{10x^4}{L^4} - \frac{20x^3}{L^3} + \frac{7x}{L}\right)$$

Let $x/L = m$.

Now
$$EI\frac{d^2y}{dx^2} = M_x.$$

Hence
$$EI\frac{dy}{dx} = \frac{WL^2}{20}\left(\frac{10m^5}{5} - \frac{20m^4}{4} + \frac{7m^2}{2} + C\right)$$

When $x = L$,

the slope $\frac{dy}{dx} = 0$ and $m = 1$

Therefore $\quad 0 = (2 - 5 + 3\tfrac{1}{2} + C)$

whence $\quad C = -\tfrac{1}{2}$.

The deflection is a maximum when the slope is zero. Hence,

$$0 = 2m^5 - 5m^4 + \frac{7m^2}{2} - \frac{1}{2}$$

This equation is satisfied when $m = 0.4268$ or 1.

Hence, the maximum deflection occurs at a point $0.4268L$ from the prop A.

Now, $$EIy = \frac{WL^3}{20}\left(\frac{2m^6}{6} - \frac{5m^5}{5} + \frac{7m^3}{6} - \frac{m}{2} + D\right)$$

When $x = 0$ $y = 0$ and therefore $D = 0$.

$$y = d = \frac{WL^3}{20EI}\left(\frac{m^6}{3} - m^5 + \frac{7m^3}{6} - \frac{m}{2}\right)$$

For maximum deflection, $m = 0.4268$. Therefore,

$$d_{max} = -\frac{0.006742WL^3}{EI}$$

8. THE DEFLECTION OF COMPOUND GIRDERS WITH CURTAILED FLANGE PLATES

THE formula for the maximum deflection of a simply supported beam of prismatic section is

$$d = k \cdot \frac{WL^3}{EI}$$

where k is a constant depending on the form which the load W takes. When the centre of a beam is reinforced with curtailed flange plates the deflection is naturally not so great, but the reduction of deflection due to the extra plates requires special calculations. It can be shown that the appropriate formula for the deflection of a beam provided with one curtailed plate per flange as in Fig. 1, is

$$d = \frac{WL^3}{EI_0}\left[k - \frac{(n-1)}{n} \cdot f(m)\right]$$

where $n = I_1/I_0$ and $m = a/L$, I_0 being the original moment of inertia of the beam and I_1 being the total moment of inertia when the curtailed flange plates are added, while the function of m, $f(m)$, varies with the type of loading, k being the constant as above.

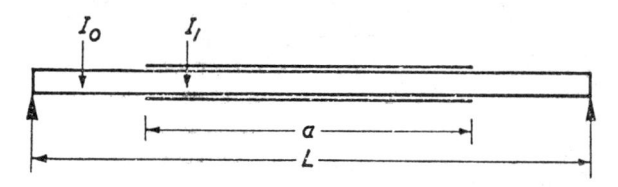

Fig. 1

Suppose that the beam of Fig. 1 is provided with a central point load W. Then the B.M. diagram, I diagram and M/I diagram appropriate to the beam are shown in Fig. 2, E being assumed to be constant and therefore ignored.

Considering the left half of the M/I diagram and taking moments about the left-hand support,

$$EId = \left(\frac{WL}{4} \times \frac{L}{4} \times \frac{L}{3}\right) - \frac{WL}{4}\left(\frac{n-1}{n}\right)\left(\frac{mL}{4}\right)\left(\frac{1}{2} - \frac{m}{6}\right)L$$

$$- \frac{WL}{4}(1-m)\left(\frac{n-1}{n}\right)\left(\frac{mL}{4}\right)\left(\frac{1}{2} - \frac{m}{3}\right)L$$

$$= \frac{WL^3}{48} - \frac{n-1}{n}\left(\frac{WL^3}{16}\right)m\left(\frac{1}{2} - \frac{m}{6}\right) - \frac{n-1}{1}\left(\frac{WL^3}{16}\right)(1-m)m\left(\frac{1}{2} - \frac{m}{3}\right)$$

$$= WL^3\left[\frac{1}{48} - \frac{n-1}{n} \cdot \frac{1}{16}\left(\frac{m}{2} - \frac{m^2}{6} + \frac{m}{2} - \frac{5m^2}{6} + \frac{m^3}{3}\right)\right]$$

103

THE DEFLECTION OF COMPOUND GIRDERS

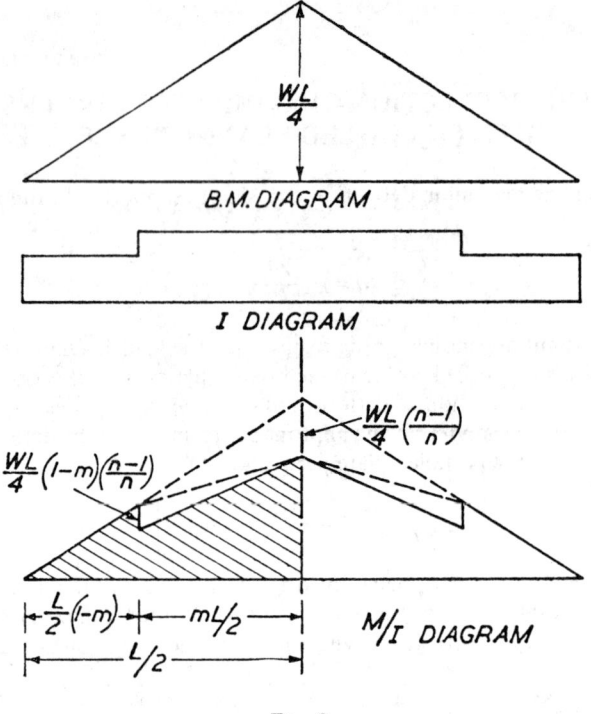

Fig. 2

Hence,

$$d = \frac{WL^3}{EI}\left[\frac{1}{48} - \frac{n-1}{n}\cdot\frac{1}{16}\left(m - m^2 + \frac{m^3}{3}\right)\right].$$

In a similar manner formulae for other loadings may be calculated. Five examples are shown in Fig. 3, each having one curtailed plate on each flange.

When an additional curtailed plate is added on each flange the amount of deflection is still further reduced and the appropriate formula can be expressed as:

$$d = \frac{WL^3}{EI_0}\left[k - \phi\cdot\frac{n-1}{n}\cdot f(m)\right].$$

It is desirable to make a small table as shown in Fig. 4 to calculate the value of d. In this table the values of n and m are functions of the beam. The values of D, E and F are calculated from the appropriate formulae for $f(m)$ given in Fig. 3. The values G, H and J are the deflection reduction coefficients for *each* individual plate.

It will be appreciated when considering the second plate that the first plate contributes to the value of I_2 and that the value G for the first plate has already been calculated. Allowance for this is made in calculating the value of $\phi(n-1)/n$. Thus the value for the second plate is $B - A$. Similarly, for the third plate the value is $C - B$, and so on.

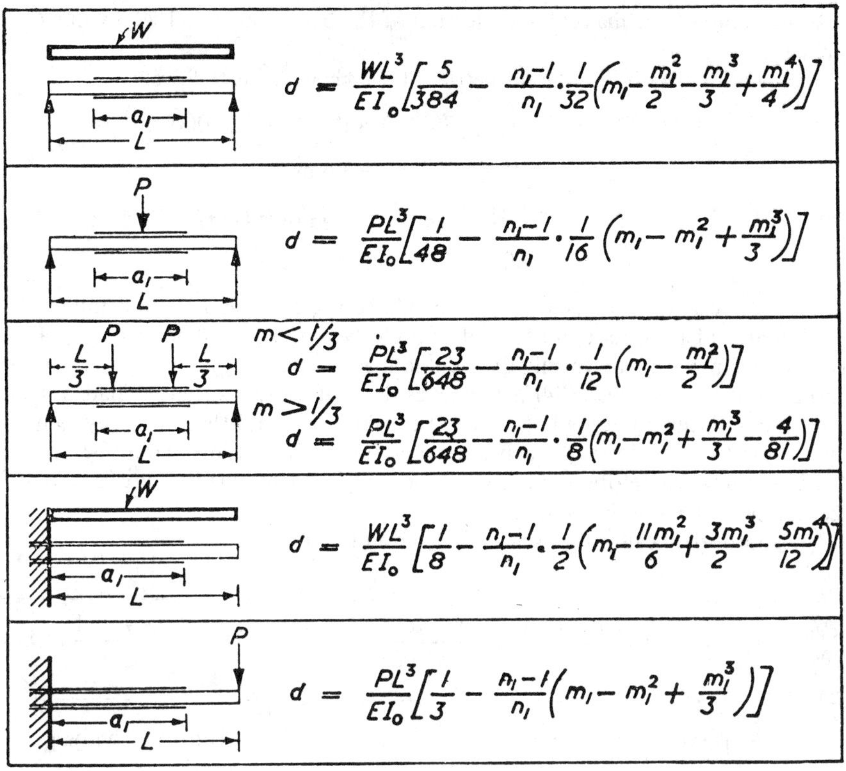

THE DEFLECTION OF COMPOUND GIRDERS

$$d = \frac{WL^3}{EI_0}\left[\frac{5}{384} - \frac{n_1-1}{n_1}\cdot\frac{1}{32}\left(m_1 - \frac{m_1^2}{2} - \frac{m_1^3}{3} + \frac{m_1^4}{4}\right)\right]$$

$$d = \frac{PL^3}{EI_0}\left[\frac{1}{48} - \frac{n_1-1}{n_1}\cdot\frac{1}{16}\left(m_1 - m_1^2 + \frac{m_1^3}{3}\right)\right]$$

$m < 1/3$
$$d = \frac{PL^3}{EI_0}\left[\frac{23}{648} - \frac{n_1-1}{n_1}\cdot\frac{1}{12}\left(m_1 - \frac{m_1^2}{2}\right)\right]$$

$m > 1/3$
$$d = \frac{PL^3}{EI_0}\left[\frac{23}{648} - \frac{n_1-1}{n_1}\cdot\frac{1}{8}\left(m_1 - m_1^2 + \frac{m_1^3}{3} - \frac{4}{81}\right)\right]$$

$$d = \frac{WL^3}{EI_0}\left[\frac{1}{8} - \frac{n_1-1}{n_1}\cdot\frac{1}{2}\left(m_1 - \frac{11m_1^2}{6} + \frac{3m_1^3}{2} - \frac{5m_1^4}{12}\right)\right]$$

$$d = \frac{PL^3}{EI_0}\left[\frac{1}{3} - \frac{n_1-1}{n_1}\left(m_1 - m_1^2 + \frac{m_1^3}{3}\right)\right]$$

Fig. 3

Plate No.	n	$\frac{n-1}{n}$	$\phi \cdot \frac{n-1}{n}$	m	$f(m)$	$\phi \cdot \frac{n-1}{n} \cdot f(m)$
1	I_1/I_0	A	A	a_1/L	D	$G = A \times D$
2	I_2/I_0	B	$B - A$	a_2/L	E	$H = E(B - A)$
3	I_3/I_0	C	$C - B$	a_3/L	F	$J = F(C - B)$

$$d = \frac{WL^3}{EI_0}\left[k - (G + H + J)\right].$$

Fig. 4

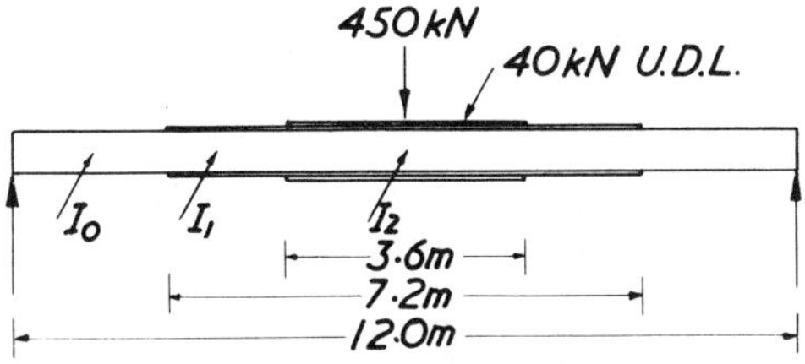

Fig. 5

THE DEFLECTION OF COMPOUND GIRDERS

As an example, the maximum deflection of the beam shown in Fig. 5 will be calculated.

Allowing for rivet holes, the properties of the beam are as follows:

$$I_0 = 164\,000 \text{ cm}^4 \quad I_1 = 239\,000 \text{ cm}^4 \quad I_2 = 320\,000 \text{ cm}^4$$

$$a_1 = 7.2 \text{ m} \quad a_2 = 3.6 \text{ m}$$

Hence
$$n_1 = I_1/I_0 = 1.455 \quad \text{and} \quad n_2 = I_2/I_0 = 1.947$$
$$m_1 = a_1/L = 0.6 \quad \text{and} \quad m_2 = a_2/L = 0.3.$$

The beam carries two loads, viz.: its own, assumed to be 40 kN uniformly distributed, and a central point load of 450 kN. Now the central deflection of a prismatic beam with a uniformly distributed load is $5WL^3/384EI$, while that for a central point load is $8WL^3/384EI$. Hence, as the uniformly distributed load is small compared with the point load, it could be considered with little inaccuracy to be concentrated at the centre of the beam, when the equivalent load would be $40 \times \tfrac{5}{8} = 25$ kN. Therefore, the beam will be analysed assuming that it carries only a central point load of 475 kN.

The appropriate calculations are tabulated in Fig. 6.

Plate No.	n	$\dfrac{n-1}{n}$	$\phi \cdot \dfrac{n-1}{n}$	m	$f(m)$	$\phi \cdot \dfrac{n-1}{n} \cdot f(m)$
1	1.455	0.313	0.313	0.6	0.0195	0.0061
2	1.947	0.486	0.173	0.3	0.0137	0.0024

$$\Sigma\phi \cdot \dfrac{n-1}{n} \cdot f(m) = \quad 0.0085$$

Fig. 6

The values of $f(m)$ for each plate is obtained from the appropriate formula in Fig. 3, viz.: $(1/16)[m_1 - m_1^2 + m_1^3/3]$.

$$d = \frac{WL^3}{EI_0}\left[\frac{1}{48} - 0.0085\right] = \frac{475 \times 10^3 \text{ N} \times 12^3 \times 10^9 \text{ mm}^3}{2.1 \times 10^5 \text{ N/mm}^2 \times 164\,000 \times 10^4 \text{ mm}^4}(0.02083 - 0.0085)$$

$$= 29.4 \text{ mm}.$$

9. BEAMS IN TORSION

Definitions

IN 1855, *Saint-Vénant*[1] showed that when the ends of a straight cylindrical bar are free to rotate under the action of two equal and opposite couples, the twist per unit length is constant throughout the bar. Consequently, this action is known as uniform torsion. If, however, some other section, such as a broad flanged beam, is restrained at the ends and a couple is applied somewhere along its length, the angle of twist varies. This action is known as non-uniform torsion. The two types are shown in Fig. 1.

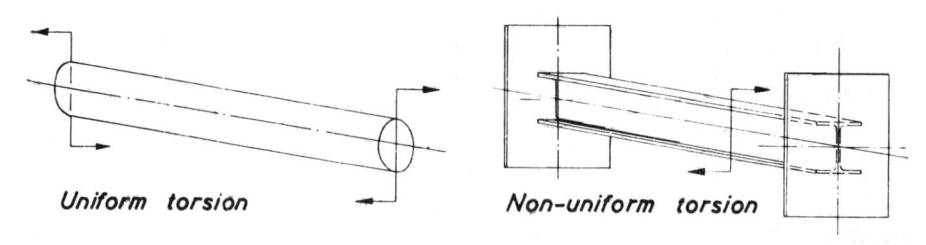

Fig. 1.—Uniform and non-uniform torsion

Saint-Vénant noted that all plane sections of cylinders remained plane after being twisted, while all radii remained straight. Now the fibres of such bars are deformed into helices, but for all practical purposes it may be assumed that they retain their original length. Consequently, the only stresses induced are shear stresses. As shown in Fig. 2, the tangential shear stresses, which vary uniformly in magnitude from zero at the longitudinal axis to a maximum at the circumference, are accompanied by equal shear stresses parallel with the longitudinal axis.

Saint-Vénant also discovered that even quite simple non-circular solid sections warp when twisted, some areas becoming concave and others convex. For example

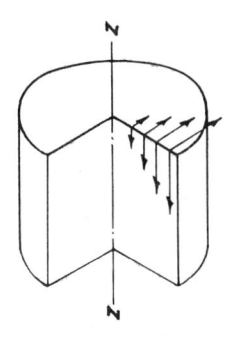

Fig. 2.—Types of shear stress

an equilateral triangle behaves as shown in Fig. 3. Under these conditions, radial lines from the centre of rotation do not remain straight and the distribution of shear stresses is not linear.

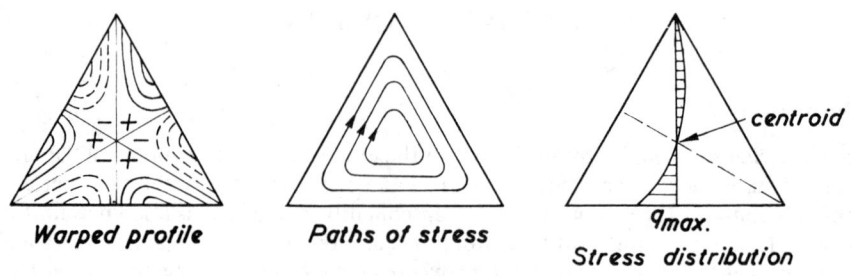

Fig. 3.—Warped triangular section

The sections which remain plane after twisting may be defined as follows:

1. Round bars or cylindrical tubes.
2. Open sections comprising *two* thin rectangles, the centre lines of which intersect at a point, e.g. angle or tee sections.
3. Thin-walled hollow sections, the resultants of the sides of which intersect in one point, as shown in Fig. 4.

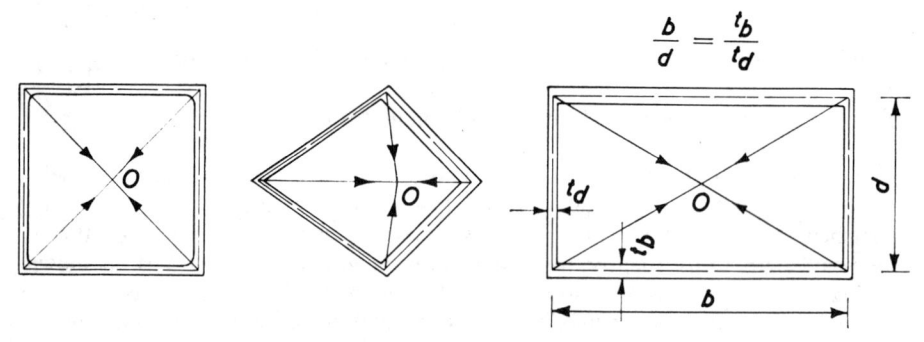

Fig. 4.—Hollow sections which remain plane

All other solid or hollow sections, including *rectangular* hollow sections of constant wall thickness, rolled steel beams and channels, warp when twisted.

Uniform Torsion

Straight round bars

Figure 5 shows a straight round bar subjected to pure torsion, which twists the outside fibres along AB through an angle ϕ to a position AB'.
Therefore,

$$\phi = \frac{q}{G} \qquad \qquad (1)$$

where q = the shear stress at the circumference
and G = the modulus of rigidity.

UNIFORM TORSION

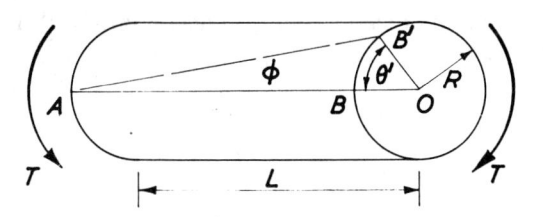

Fig. 5.—Twisted bar

The radius OB twists through an angle θ' to the position OB'. Hence,

$$\theta' = \frac{L\phi}{R} \text{ radian} \qquad \ldots (2)$$

where L = the length of the bar
and R = the radius.
From (1) and (2),

$$q = \phi G = \frac{R\theta'}{L} \times G \qquad \ldots (3)$$

The relationship between the applied torque T and the induced shear stresses q may be found from a consideration of Fig. 6.

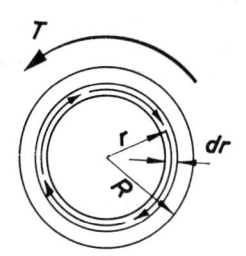

Fig. 6.—Shear-stresses induced by torque

The total shearing force on the elementary ring of radius r and thickness dr is

$$q' \times 2\pi r dr$$

The moment of this force about the polar axis is

$$dT = q' \times 2\pi r dr \times r$$

But
$$q' = qr/R$$

Hence, the moment of torsional resistance over the whole section is

$$T = 2\pi \frac{q}{R} \int_0^R r^3 dr = \frac{\pi}{2} q R^3$$

$$= \frac{\pi}{16} q D^3 \qquad \ldots (4)$$

Now the polar moment of inertia I_p of a cylinder is $(\pi/32) D^4$. Therefore equation (4) may be expressed as

$$T = q\frac{I_p}{R}$$

$$\text{or} \quad q = \frac{TR}{I_p} \quad \ldots \ldots (5)$$

Readers should note the close analogy which exists between this standard expression for torsion and that for bending, viz:

$$f_{bc} = f_{bt} = \frac{My}{I} = \frac{M}{Z}$$

Considering Fig. 5 and substituting from equation (3) in (5),

$$\theta' = \frac{TL}{GI_p}$$

Therefore, the angle of torsion per unit length of bar is

$$\theta = \frac{T}{GI_p} \quad \ldots \ldots (6)$$

This is the value given in Table 1.

TABLE 1

Type of section	Position of maximum shear stress	Expression for maximum shear stress	Expression for angular rotation per unit length in radians
Solid circular bar of diameter D	On the external surface	$\dfrac{16T}{\pi D^3}$	$\dfrac{32T}{G\pi D^4}$
Hollow shaft o.d. = D i.d. = d	On the external surface	$\dfrac{16DT}{\pi(D^4 - d^4)}$	$\dfrac{32T}{G\pi(D^4 - d^4)}$
Tube r = mean radius t = wall thickness	On the external surface	$\dfrac{T}{2\pi r^2 t}$	$\dfrac{T}{2G\pi r^3 t}$
Equilateral triangle s = side	In the middle of the sides	$\dfrac{20T}{s^3}$	$\dfrac{46.2T}{Gs^4}$
Square s = side	In the middle of the sides	$\dfrac{4.5T}{s^3}$	$\dfrac{7.2T}{Gs^4}$
Solid regular hexagon D = diameter of inscribed circle	In the middle of the sides	$\dfrac{5.32T}{D^3}$	$\dfrac{8.69T}{GD^4}$
Solid regular octagon D = diameter of inscribed circle	In the middle of the sides	$\dfrac{5.41T}{D^3}$	$\dfrac{9.26T}{GD^4}$

Hollow shafts

In considering hollow shafts, i.e. tubes with very thick walls, the same arguments can be used as for solid round bars. However, in this case, the polar moment of

inertia I_p is $(\pi/32)(D^4 - d^4)$, where D and d are the external and internal diameters of the shaft.

Consequently,

$$\theta = \frac{T}{G} \times \frac{32}{\pi(D^4 - d^4)}$$

$$\text{and } q = T \times \frac{16D}{\pi(D^4 - d^4)}$$

Hollow sections

Consider Fig. 7, showing a hollow section of any shape, the walls of which may be of variable or constant thickness but which are so thin compared with the other dimensions that the variation in stress between the inner and outer fibres may be ignored.

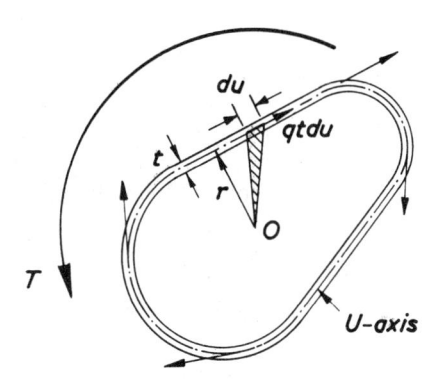

Fig. 7.—Shear flow in hollow section

The shear flow around the section, represented by the product qt is constant. Hence, where the wall is thin, the stress is high.

On an element tdu, the tangential force is $qtdu$. The moment of this force about the centre of rotation O is $qtrdu$, where r is the perpendicular dropped from the force to O. The sum of the moments of all the elements around the U-axis is equal to the torque T, that is

$$T = \oint qtrdu = qt \oint rdu$$

The integral represents twice the area A enclosed by the U-axis.

Therefore, $$T = 2Aqt$$

and $$q = \frac{T}{2At} \qquad \ldots \ldots (7)$$

This is known as Bredt's first formula.[2]

To calculate the angle of twist, use is made of Stokes's Law:

$$\oint qdu = 2G\theta A$$

Substituting from equation (7), the angle of twist per unit length,

$$\theta = \frac{T}{G}\left[\frac{\oint \frac{du}{t}}{4A^2}\right] \quad \ldots \ldots (8)$$

The reciprocal of the expression in brackets is analogous to the polar moment of inertia in equation (6). For non-circular sections, the symbol K is normally used instead of I_p.

Hence,
$$K = \frac{4A^2}{\oint \frac{du}{t}} \quad \ldots \ldots (9)$$

This is Bredt's second formula.[2]
It must be noted that Bredt's two formulas can be used only for hollow sections.

Tubes

For tubes, equation (7) becomes

$$q = \frac{T}{2\pi r^2 t} \quad \ldots \ldots (10)$$

where r = the mean radius, i.e. to the centre of the wall.

Also
$$K = I_p = \frac{4\pi^2 r^4}{2\pi r/t}$$

$$= 2\pi r^3 t$$

Hence,
$$\theta = \frac{T}{2G\pi r^3 t} \text{ per unit length} \quad \ldots \ldots (11)$$

Rectangular hollow sections

For rectangular hollow sections, some examples of which are shown in Fig. 8.

$$q_d = \frac{T}{2t_d bd} \quad \ldots \ldots (12a)$$

$$\text{and } q_b = \frac{T}{2t_b bd} \quad \ldots \ldots (12b)$$

$$K = \frac{4b^2 d^2}{\frac{2b}{t_b} + \frac{2d}{t_d}} = \frac{2b^2 d^2 t_b t_d}{bt_d + dt_b}$$

$$\text{and } \theta = \frac{T(bt_d + dt_b)}{2Gb^2 d^2 t_b t_d} \text{ per unit length} \quad \ldots \ldots (13)$$

The effect of corners. The internal angles of square or rectangular hollow sections should be rounded to give a free flow of stress. Angular external corners do not influence the stresses.

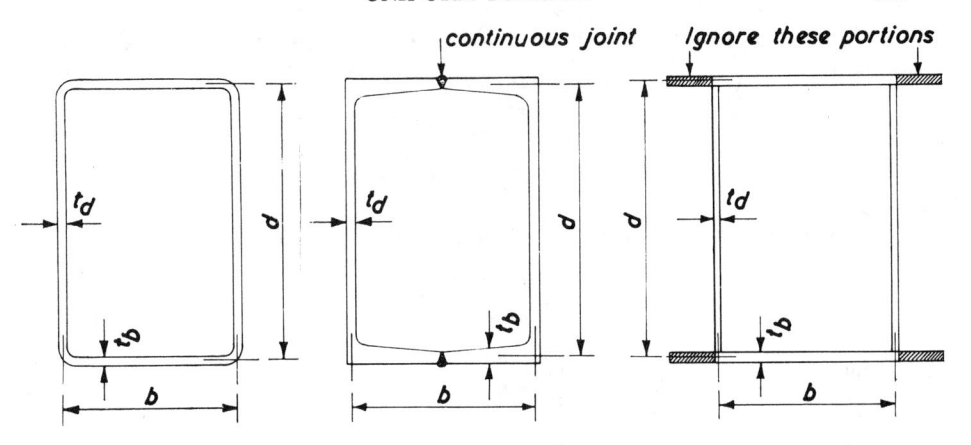

Fig. 8.—Types of rectangular hollow sections

Timoshenko[3] gives the following formula for the determination of the factor of concentration of stress, in terms of q, for internal angles:

$$C = \frac{t}{r_i} \left[\frac{1 - \frac{p}{4A}(r_e + r_i)}{\log_n \frac{r_e}{r_i}} \right] + \frac{pr_i}{2A} \quad \ldots \quad (14)$$

where p = the perimeter of the U-axis,

and r_e, r_i = the external and internal radii respectively of the corner.

Taper tubes

As taper tubes of constant wall thickness do not warp, they are comparatively easy to design.

Consider the short taper tube in Fig. 9. Let the radius at the origin be r_0 and at the free end r_1.

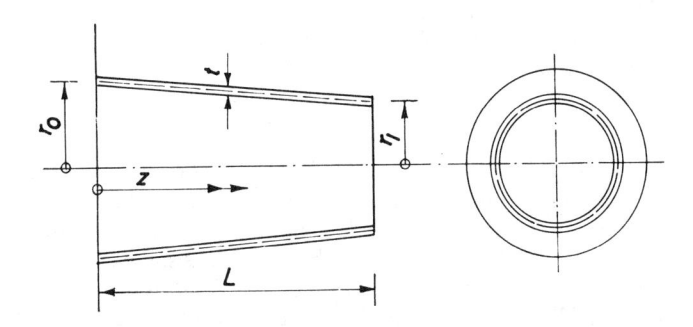

Fig. 9.—Taper tube

Then, using equations (10) and (11), at any point distance z from the origin,

$$q = \frac{T}{2\pi r^2 t}$$

and $\theta = \dfrac{T}{2G\pi r^3 t}$ per unit length

With constant torque, the total angle of twist at the free end will be

$$\theta' = \frac{T}{2G\pi t} \int \frac{dz}{r^3}$$

But, $\qquad dz = \dfrac{L \cdot dr}{r_0 - r_1}$

Hence $\qquad \theta' = \dfrac{TL}{2G\pi t(r_0 - r_1)} \int_{r_0}^{r_1} \dfrac{dr}{r^3}$

If the average radius is r_m, then

$$\theta' = \frac{TL r_m}{2G\pi r_0^2 r_1^2 t} \qquad \ldots \ldots (15)$$

Sections which warp

Saint-Vénant demonstrated that the differential equation of torsion could be applied to some of the more simple geometrical shapes, e.g. an equilateral triangle. Subsequently, *Weber* produced solutions for regular hexagons and octogons. Appropriate formulas are quoted in Table 1. Unfortunately, exact solutions do not exist for the sections normally rolled or extruded for use in structures. In Great Britain, however, a number of investigators, in particular Professor W. Fisher Cassie and his associates,[4,5] have used the membrane theory or relaxation methods to produce accurate torsional constants K for rolled steel sections. In the United States similar investigations have been carried out,[6] while the K values for universal beams, channels and Z-sections have been published by the *Bethlehem Steel Company*.[7]

The British Constructional Steelwork Association have profited especially from the American work in producing two publications[8] which deal exhaustively with design problems associated with the combined bending and torsion of beams and girders. While the torsional properties of British Standard sections are given in chapter 40, pp. 1020–1023, the reader is referred to the B.C.S.A. publications for detailed information on design.

In Germany, the K values for all standard rolled sections have been calculated by *Bornscheuer* and *Anheuser*.[9,10]

The Rectangle. When a square bar is subjected to torsion, plane sections warp into four concave and four convex zones, the shear stresses induced being of maximum value at the centre of the sides and zero at the corners, as shown in Fig. 10.

The effect of torsion on a rectangular bar or plate is similar, but the stresses induced in the middle of the long sides are greater than those in the middle of the short sides. In addition, if the ratio of depth d to breadth b exceeds 1.45, there are only two concave and two convex zones produced during warping.

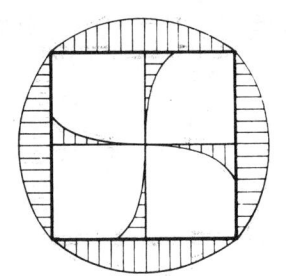

Fig. 10.—Shear stresses in twisted square bar

Saint-Vénant derived the following formulas for the maximum shear stresses:

In the long side, $$q_d = \frac{T}{\alpha b^2 d}$$ .... (16a)

In the short side, $$q_b = \frac{T}{\alpha b d^2}$$ .... (16b)

The corresponding formula for the angle of twist per unit length is

$$\theta = \frac{T}{\beta b^3 d G}$$ .... (17)

in which $K = \beta b^3 d$.

The values of α and β derived by Saint-Vénant are given in Table 2. When d/b exceeds 10, the values are always assumed to be $\frac{1}{3}$.

TABLE 2

d/b	1	1.5	1.75	2	2.5	3	4
α	0.208	0.231	0.239	0.246	0.258	0.267	0.282
β	0.141	0.196	0.214	0.229	0.249	0.263	0.281
d/b	5	6	7	8	9	10	∞
α	0.291	0.299	0.304	0.307	0.310	0.312	0.333
β	0.291	0.299	0.303	0.307	0.310	0.313	0.333

Comparison of sections

If a member will be subjected to heavy torque, a closed section such as a tube or rectangular hollow section should be used. It can readily be shown that a closed section is much more effective than an open member. Consider, for example, the tube and slit tube shown in Fig. 11.

For the tube, using equations (10) and (11),

$$q = \frac{T}{2\pi r^2 t}$$

and $$\theta = \frac{T}{2G\pi r^3 t}$$

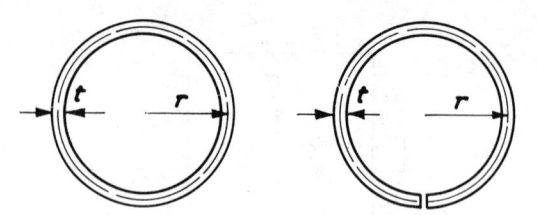

Fig. 11.—Comparison of closed and open sections

The slit tube is analysed as a rectangle, the 'depth' being the projected length of the perimeter.
Therefore,

$$q = \frac{T}{\frac{1}{3} \times b^2 d} = \frac{T}{\frac{1}{3} \times t^2 \times 2\pi r}$$

$$\theta = \frac{T}{\frac{1}{3} \times b^3 dG} = \frac{T}{\frac{1}{3} \times t^3 \times 2\pi rG}$$

Then the ratio of shear stress in the tube to that in the open section is:

$$\frac{T}{2\pi r^2 t} \times \frac{\frac{1}{3} \times t^2 \times 2\pi r}{T} = \frac{1}{3} \times \frac{t}{r}$$

while the ratio of twist is

$$\frac{T}{2G\pi r^3 t} \times \frac{\frac{1}{3} \times t^3 \times 2\pi rG}{T} = \frac{1}{3} \times \frac{t^2}{r^2}$$

It will be observed that if the thickness of the walls is one-tenth of the mean radius, then the shear stress in the slit tube is thirty times that in the entire tube, while the angle of twist is three hundred times as great.

Non-Uniform Torsion

All the foregoing formulas have been based on the assumptions that the members were subjected only to equal and opposite torques at the ends and that they could warp without hindrance, where appropriate. If conditions are such that these assumptions do not apply the members are subjected to non-uniform torsion.

The two publications already mentioned[8] should be consulted for information on this rather more complex subject.

BIBLIOGRAPHY

1. SAINT-VENANT, B. de. 'Mémoire sur la torsion des prismes.' *Mémoires des savants étrangers,* p. 253, (1855).
2. BREDT, R. 'Studien zur Dehnungselasticität'. *Z. VDI,* pp. 785 and 813, (1896).
3. TIMOSHENKO, S. *Strength of Materials,* Vol. 2. Macmillan.
4. CASSIE, Prof. W. F., and DOBIE, Dr. W. B. 'The torsional stiffness of structural sections.' *The Structural Engineer,* (March 1948).
5. DOBIE, Dr. W. B., and GENT, Dr. A. R. 'Accuracy of determination of the elastic torsional properties of non-circular sections using relaxation methods and the membrane analogy.' *The Structural Engineer,* (September 1952).
6. EL DARWISH, I. A., and JOHNSTON, B. G. 'Torsion of structural shapes.' *Journal of the Structural Division, Proceedings A.S.C.E.,* (February 1965).
7. HEINS, C. P., and SEABURG, P. A. *Torsion analysis of rolled steel sections.* Bethlehem Steel Corporation, 1963.
8. TERRINGTON, Dr. J. S. *Combined bending and torsion of beams and girders.* Publication No. 31 (First part), 1968, and Publication No. 31 (Second part), 1970. British Constructional Steelwork Association.
9. BORNSCHEUER, Prof. F. W., and ANHEUSER, L. 'Tafeln der Torsionskenngrössen für die Walzprofile der DIN 1025–1027.' *Der Stahlbau* (March 1961).
10. BORNSCHEUER, Prof. F. W. 'Schweissanschlüsse torsionsbeanspruchter Träger mit *I, U* und *Z* Querschnitten.' *Schweissen und Schneiden,* (March 1961).

10. FORCES IN PLANE FRAMES

A PERFECT frame is composed of straight bars between node points, assumed to be frictionless pin joints, arranged in such a manner as to give direct forces only in the several bars. This result will be achieved if the frame consists of an assemblage of triangles formed by the bars, so that the total number of bars is $2N - 3$, where N is the number of nodes, when all the external forces in the members of such a frame can be determined by simple statics in the following alternative ways:

(a) graphically by the use of the reciprocal diagram;
(b) by the method of sections or moments;
(c) by trigonometry or joint resolution;
(d) from B.M. and S.F. diagrams;
(e) by resolution;
(f) from the equilibrium polygon.

It is self-evident that the external forces acting on any frame, that is to say the loads and reactions, must form a system of forces in equilibrium. Further, since direct forces only are to be induced in the frame members all external forces must act at node points in the frame, and all the forces, internal and external, at each node also constitute separately systems of forces in equilibrium.

(a) Graphical Solution

Two figures consisting of lines and points lying in a plane are said to be reciprocal when: (1) to any node of one figure at which a given number of lines meet there is a corresponding polygon in the other figure bounded by the same number of sides, (2) to every line in one figure there corresponds a parallel line in the other figure, and (3) to every line in one figure joining the nodes there corresponds a line in the other figure separating the polygons corresponding to those nodes. Then, if the first figure represents a framework with loads acting on it, the second or reciprocal figure will give the forces acting on the frame and the forces in the members of the frame. The reciprocal figure is, in fact, an assembly of force polygons for the various nodes combined into one diagram, and is commonly called a force diagram. Each line in the figure, taken in opposite directions, represents two forces: one in each of two separate force polygons. Consider the truss shown in Fig. 1. The reactions at the supports, by the principle of symmetry, are 250 kN each, whence the load line is set down 1–6 and point O determined. Points in the force diagram are then found in the order *a, b, c, d, e* and *f*. The nature of the forces in the members is determined from the force polygons for the joints. For example, the force polygon for joint X is $O1, 1a, aO,$ and the arrows indicate the direction of the forces for this joint, being determined from the known force $O1$ acting upwards. Similarly for joint Y the force polygon is $a1, 12, 2b, ab,$ the known directions being $a1$, which is $1a$ reversed, and 12.

120   FORCES IN PLANE FRAMES

LOADED FRAME
6 panels of 7.5m = 45.0m
All loads 100 kN each

Members	LOAD in kN	Nature
1a, 5j.	250	T
2b, 6k.		
3d, 4g.	357	T
Oa, Ok.	354	C
Oc, Oh.	360	C
Oe, Of.	378	C
ab, jk.	100	T
bc, hj.	151	T
cd, gh.	7	C
de, fg.	29	T
ef	60	T

Reciprocal or force polygon

Force Polygon for joint X

Force Polygon for joint Y

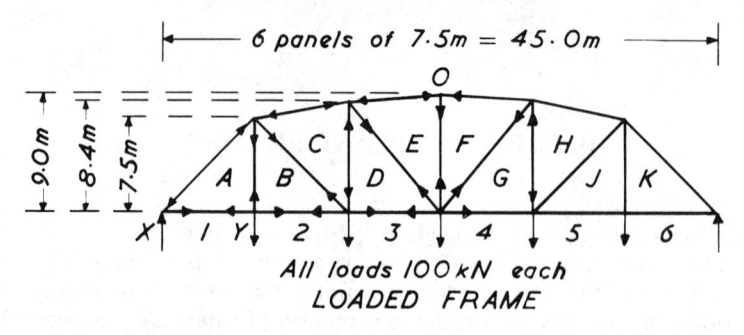

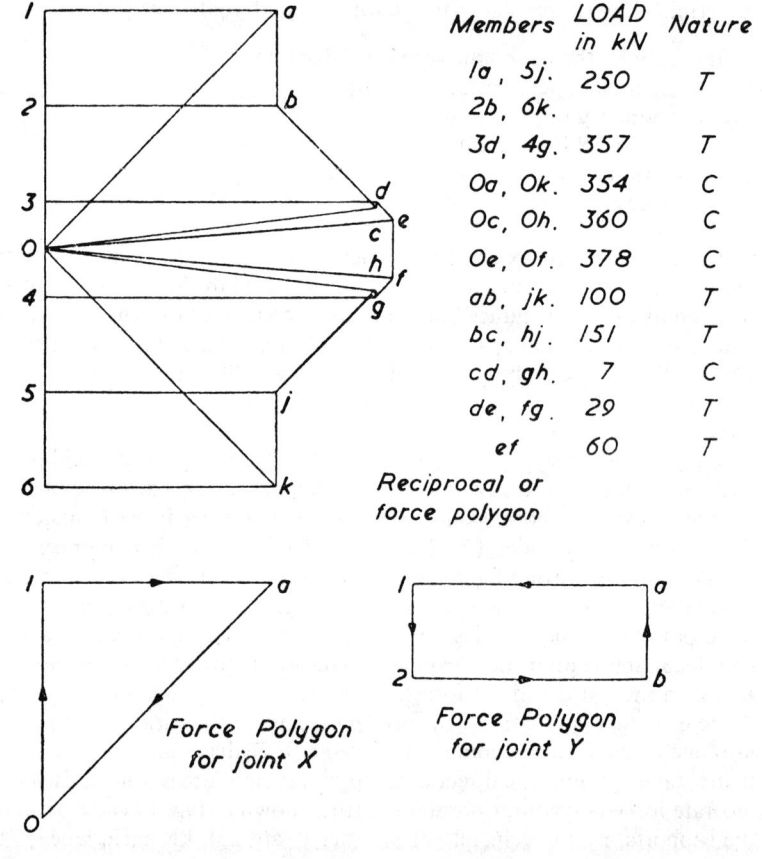

Fig. 1

It should be noted that no more than two unknowns, corresponding to the two conditions for equilibrium of a coplanar concurrent system of forces, can exist at any one node if the solution is to be by simple statics. Each line in the diagram, e.g. 1a, is used twice, once for each polygon and in opposite directions.

(b) *Solution by Method of Sections or Moments*

The basis of this method is the general case of equilibrium of non-concurrent forces. In Fig. 2 if *ABCDE* represents a portion of the frame

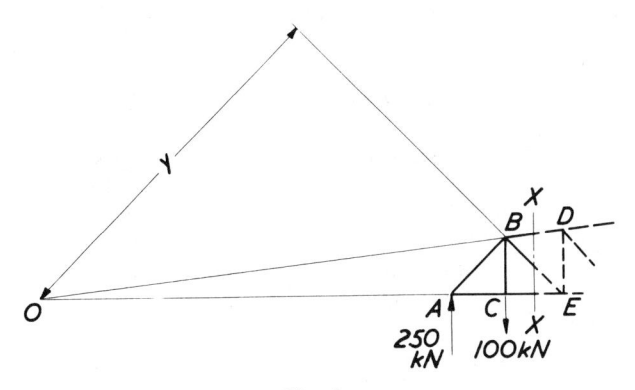

Fig. 2

shown in Fig. 1 cut by an imaginary section *XX*, then the forces in the bars cut by this section must hold in equilibrium all forces to one side of the section. All such sections must be drawn to cut three bars in conformity with

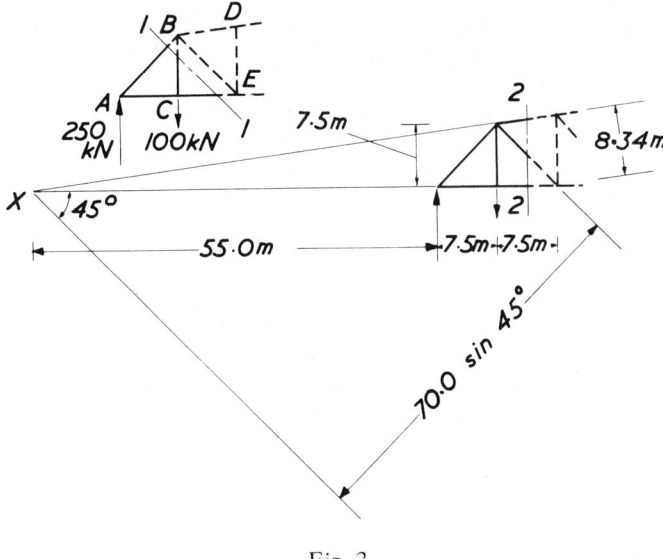

Fig. 3

the three conditions for equilibrium, $\Sigma V = 0$, $\Sigma H = 0$ and $\Sigma M = 0$, that is, the summation of the vertical components in the bars *BD*, *BE* and *CE* must be equal and opposite to the vertical force at the section; the horizontal components must summate to zero and the moments of the forces in the bars

taken about any convenient point must balance the moment of the external forces to one side of the section taken about the same point. Thus by taking moments about a point where two of the bars meet, e.g. point O, the moment of the force in the remaining bar BE also taken about O, $F_{BE} \times y$ must equal the moment about that point of the external forces at A and C.

Consider Fig. 3.

Section	Member	Moments about	Moments
1-1	AB	C (Intersection of BC and CE)	$+ 250$ kN $\times 7.5$ $- F_{AB} \times 0.707 \times 7.5\text{m} = 0$. $F_{AB} = 250 \times 1.414 = 354$ kN acting towards A, i.e. F_{AB} is compression.
	BC	A (Intersection of AB and CE)	$- 100$ kN $\times 7.5\text{m} + F_{BC} \times 7.5\text{m} = 0$. $F_{BC} = 100$ kN acting away from C, i.e. F_{BC} is tension.
	CE	B (Intersection of AB and BC)	$+ 250$ kN $\times 7.5\text{m} - F_{CE} \times 7.5\text{m} = 0$. $F_{CE} = 250$ kN acting away from C, i.e. F_{CE} is tension.
2-2	BD	E (Intersection of BE and CE)	$+ 250$ kN $\times 15\text{m} - 100$ kN $\times 7.5\text{m}$ $+ F_{BD} \times 8.34 = 0$. $F_{BD} = \dfrac{3\,000}{8.34} = 360$ kN compression.
	BE	X (Intersection of BD and CE)	$- 250$ kN $\times 55.0 + 100$ kN $\times 62.5$ $- F_{BE} \times 0.707 \times 70.0\text{m} = 0$ $F_{BE} = 151$ kN tension.

The remaining members may be calculated in a similar manner, but when dealing with the centre vertical it will be found impossible to draw a section other than to cut four bars, and in this case the loads in the other three bars must be known before the load in the centre vertical can be evaluated.

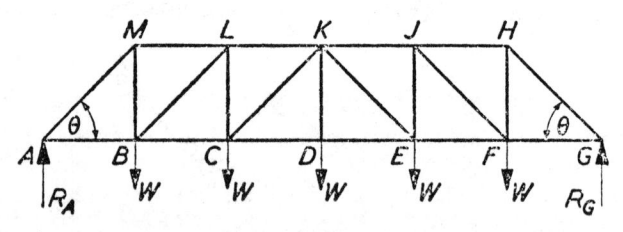

Fig. 4

(c) *Solution by Trigonometry or Joint Resolution*

This method is particularly useful for the solution of girders with parallel booms where the web members are inclined at constant angles. Consider the girder shown in Fig. 4. At joint A the known force is R_A, whence, for equilibrium, the force in AB, $F_{AB} = F_{FG} = R_A \cot \theta$, whilst the force in AM is $F_{AM} = F_{HG} = R_A \csc \theta$. Similarly, at joint M the known force is $F_{AM} = R_A \csc \theta$, whence for equilibrium $F_{ML} = F_{JH} = R_A \cot \theta$, whilst $F_{BM} = F_{HF} = R_A$.

At joint B there are two known forces F_{AB} and F_{BM}, whence the two unknown forces may be determined. The remaining joints are treated in a similar fashion and it will be noted that the method corresponds to the graphical construction in that the equilibrium is calculated separately for each joint.

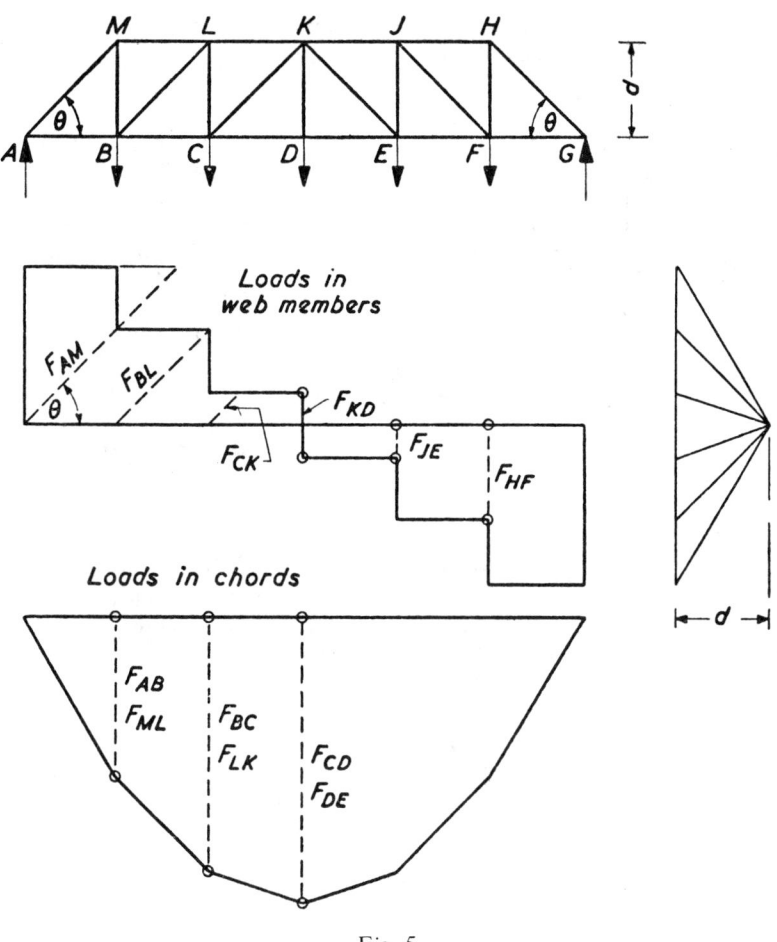

Fig. 5

(d) Solution by B.M. and S.F. Diagrams

This is an alternative method useful for the solution of parallel boom girders. Consider the girder shown in Fig. 5, then, by the method of sections, F_{CD} = B.M. at K divided by the depth between the booms whilst F_{LK} = B.M. at C divided by the depth. Hence the B.M. diagram to some scale will represent the boom forces or, if the B.M. diagram is drawn with a polar distance equal to the depth it will give the chord forces to the load scale.

The shear in any panel is resisted wholly by the web members since the chords are horizontal, thus the loads in the verticals are read directly from the S.F. diagram whilst the loads in the diagonals are found by resolving the shearing force in the direction of the diagonals.

(e) Solution by Resolution

This method consists of resolving the resultant force acting on any section into components in the directions of the three bars cut by the sections and is particularly useful in finding forces in web members in girders with non-parallel flanges subjected to varying loads.

Consider Section XX in Fig. 6. The resultant load on this section is $R_A - W_1$ acting where shown: the resultant of F_{DE}, F_{BD} and F_{BC} must be in

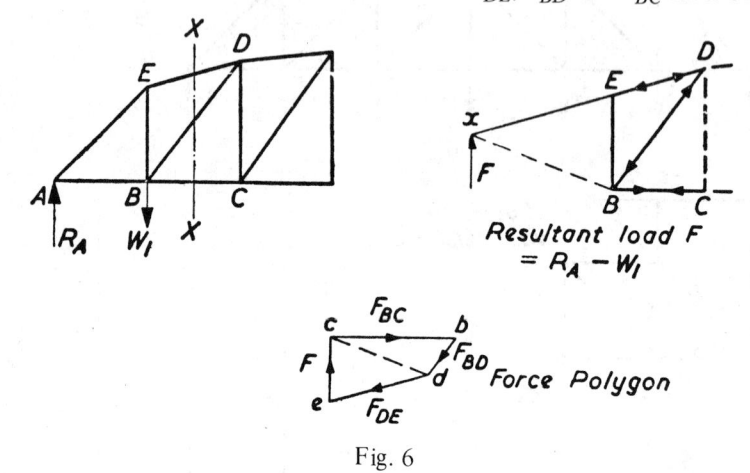

Fig. 6

equilibrium with this load. Produce DE to cut the line of action of F at point x and join this to B, the point of intersection of the other two bars cut by the section. Draw ec to represent F, ed parallel to DE, cd parallel to Bx to give point d. Then draw bd parallel to BD and bc parallel to BC, which gives the force polygon for the four forces concerned.

(f) Solution from the Equilibrium Polygon

This method is the application of the method of sections, but applied to the forces delineated by the equilibrium polygon in lieu of the original loads and

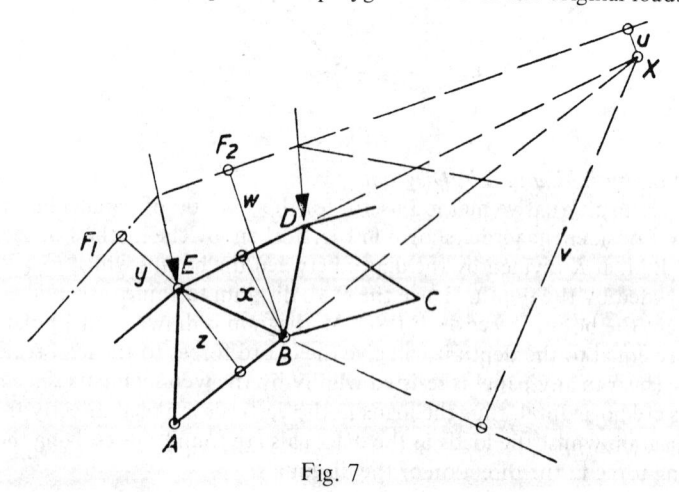

Fig. 7

reactions and in consequence is of particular application to three-hinged arch trusses. An example is illustrated in Fig. 7, where $ABCDE$ is part of a frame and F_1 and F_2 are the forces in the segments of the equilibrium polygon.

Then

$$F_{AB} = F_1 \times y/z \text{ (moments about } E\text{)}.$$

$$F_{DE} = F_2 \times w/x \text{ (moments about } B\text{)}.$$

$$F_{BE} = F_2 \times u/v \text{ (moments about } X\text{)}.$$

Forces in Knee-braced Frames

Figure 8 gives sections through a knee-braced shed (a) with columns hinged at their bases, (b) with columns fixed at their bases, with exaggerated diagrams of the deformed shape under side-wind loading with the resulting B.M. diagram on the columns, both knee-braces being assumed to act.

It is necessary first to find the reactions, and, considering case (a), there are four unknowns to be found, a vertical and a horizontal force at each hinge, so that the

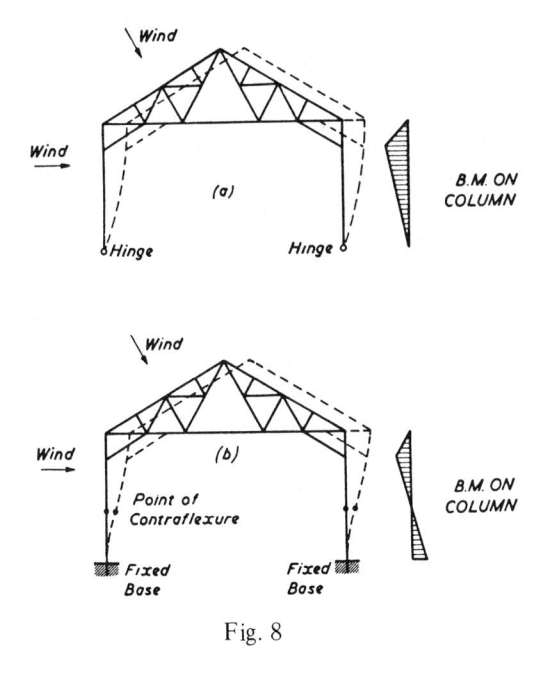

Fig. 8

external forces are an indeterminate system. The problem can only be rendered statically determinate by a reduction of the unknowns to three or less, and the usual practice in this matter is to assume a relationship between the horizontal forces at the hinges. These are usually proportioned according to the stiffness of the columns, and hence, if of equal section, length and moment of inertia, as is usually the case, the columns will share the horizontal load equally between them and hence the horizontal reactions are known, and the vertical reactions can be found by simple statics. Turning now to case (b), a further assumption is necessary regarding the

FORCES IN PLANE FRAMES

position of the points of contraflexure, normally taken as mid-way between the base of the column and the knee-brace. These points having been determined, the portion of the frame above them reduces to case (a).

In both cases the columns are subjected to direct and shear forces and bending moments, and therefore cannot be evaluated by simple statics. The method of calculating these values and the forces in the members of the frame is best explained by the examination of a typical problem, as shown in Fig. 9 (a).

The wind loading stipulated in British Standard Code of Practice, C.P. 3: Chapter V: Part 2: 1970 cannot be calculated without full knowledge of the particular building since this depends on relative length, breadth and height. For the purpose of this example, loads based on Chapter V (1952), but with metric equivalents, have been used.

Basic horizontal external wind pressure, $p = 720$ N/m² taken as positive blowing from left to right. This is considered as equally shared by both external walls, thus giving horizontal external wind pressures of $+0.5p$ on the windward face and $+0.5p$ on the leeward face. Suction pressures normal to the roof surfaces are assumed to be $-0.1p$ on the windward slope and $-0.45p$ on the leeward slope.

An internal pressure of $\pm 0.2p$ has been allowed for in addition.

The following calculations result, forming the basis of Fig. 9 (b).

$$W_1 = +0.5p - 0.2p = +0.3p$$
$$= 0.3 \times 720 \times 3.75 \times 4.5 \text{ N} = 3.65 \text{ kN}$$

$$W_2 = -0.1p - 0.2p = -0.3p$$
$$= 0.3 \times 720 \times 3.75 \times 8.66 \text{ N} = 7.02 \text{ kN}$$
in direction shown in Fig. 9

$$W_3 = -0.45p - 0.2p = -0.65p$$
$$= 0.65 \times 720 \times 3.75 \times 8.66 \text{ N} = 15.20 \text{ kN}$$
in direction shown in Fig. 9

$$W_4 = +0.5p + 0.2p = +0.7p$$
$$= 0.7 \times 720 \times 3.75 \times 4.5 \text{ N} = 8.50 \text{ kN}$$

Total horizontal load, left to right,

$$= 3.65 + 0.5(15.20 - 7.02) + 8.50$$
$$= 16.24 \text{ kN}$$

Whence $\qquad H_L = H_R = 8.12 \text{ kN}$ . . . .(a)

Total vertical load, upwards,

$$= 0.866(7.02 + 15.20)$$
$$= 19.24 \text{ kN}$$

Moment about left hinge,

$$V_R = \frac{(3.65 + 8.50) \times 2.25 - 7.02 \times 6.58 - 15.2 \times 6.41}{15}$$

$$= 7.75 \text{ kN (downwards)} \text{ . . . .(b)}$$

FORCES IN KNEE-BRACED FRAMES

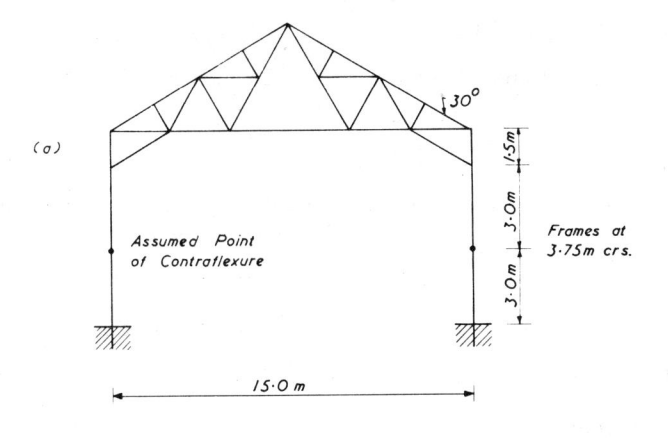

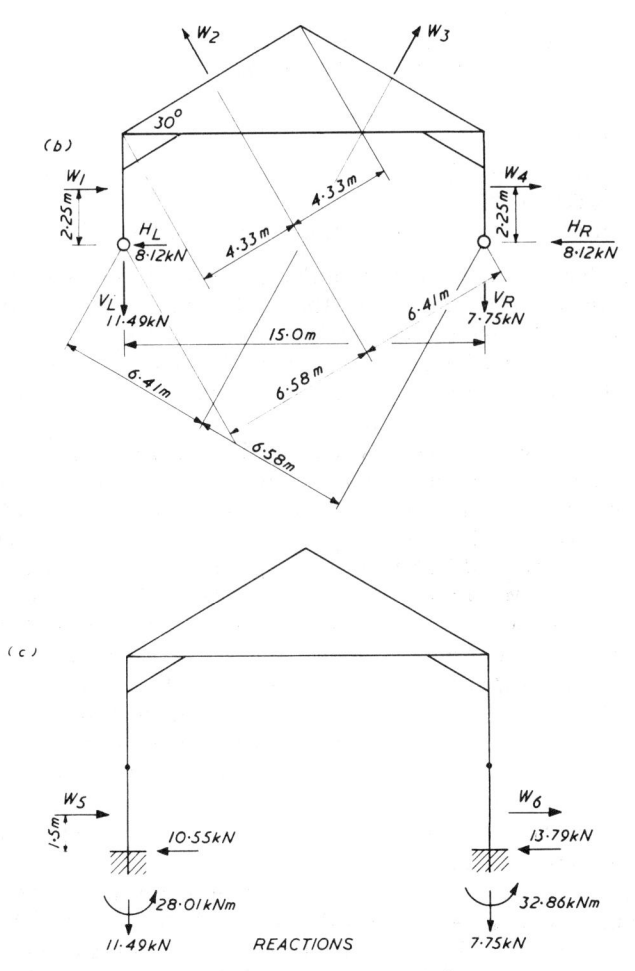

Fig. 9 (a) (b) and (c)

Moment about right hinge,

$$V_L = \frac{(3.65 + 8.50) \times 2.25 + 7.02 \times 6.41 + 15.2 \times 6.58}{15}$$

$= 11.49$ kN (downwards) . . . .(c)

These values (a), (b) and (c) are reactions at the points of contraflexure. The additional loads on the columns below these points are:

$$W_5 = 0.3 \times 720 \times 3.75 \times 3.0 \text{ N} = 2.43 \text{ kN}$$

$$W_6 = 0.7 \times 720 \times 3.75 \times 3.0 \text{ N} = 5.67 \text{ kN}$$

Taking these into consideration the reactions at the fixed bases will be:

Windward Column

Vertical reaction $\quad = -11.49$ kN
Horizontal reaction $= -8.12 - 2.43 = -10.55$ kN
Fixing moment $\quad = -(8.12 \times 3 + 2.43 \times 1.5) = -28.01$ kNm

Leeward Column

Vertical reaction $\quad = -7.75$ kN
Horizontal reaction $= -8.12 - 5.67 = -13.79$ kN
Fixing moment $\quad = -(8.12 \times 3 + 5.67 \times 1.5) = -32.86$ kNm

In order to draw the force diagram for the roof truss, it is necessary to make allowance for the B.M.s in the columns. This can be done in two ways: (a) by removal of the columns and substitution in their place of systems of forces applied at nodes, whose effect will be the same as that of the columns, and (b) by the addition of temporary framework arranged so that moments in the columns cause direct stresses only in the auxiliary frames.

The calculations for method (a) are shown in Fig. 9 (d) and (e). Considering the windward column, the side wind load on this, which is $W_1 = 3.65$ kN is divided proportionally to the column lengths, as to 0.61 kN at the eaves, 1.82 kN at the knee-brace level, and 1.22 kN at the hinge. These forces are then combined with the reactions at the point of contraflexure, resulting in final forces at the hinge of 11.49 kN vertically downwards and 6.9 kN horizontally. This last force must be transferred to the frame, and its effect about the knee-brace level is a horizontal shear of 6.9 kN and a moment of 20.7 kNm; the two forces shown of 13.8 kN at eaves and 20.7 kN at knee-brace level, will reproduce both the shear and the moment. To these forces are then added algebraically the appropriate wind loads, and the final truss reactions are shown on the extreme right of the diagram in Fig. 9 (d). Similarly, the forces on the leeward column are evaluated and the final truss reactions shown in Fig. 9 (e).

These six forces hold the truss in equilibrium against the applied wind loads, and the force diagram is reproduced in Fig. 10.

The direct forces in the columns above the knee-brace are given directly by the force diagram; the vertical forces below this are 11.49 kN and 7.75 kN tension (windward and leeward respectively), the shears are 8.12 kN each. In addition, there is a moment of 20.7 kNm at knee-brace level and a moment of opposite sign of

FORCES IN KNEE-BRACED FRAMES

WINDWARD STANCHION (d)

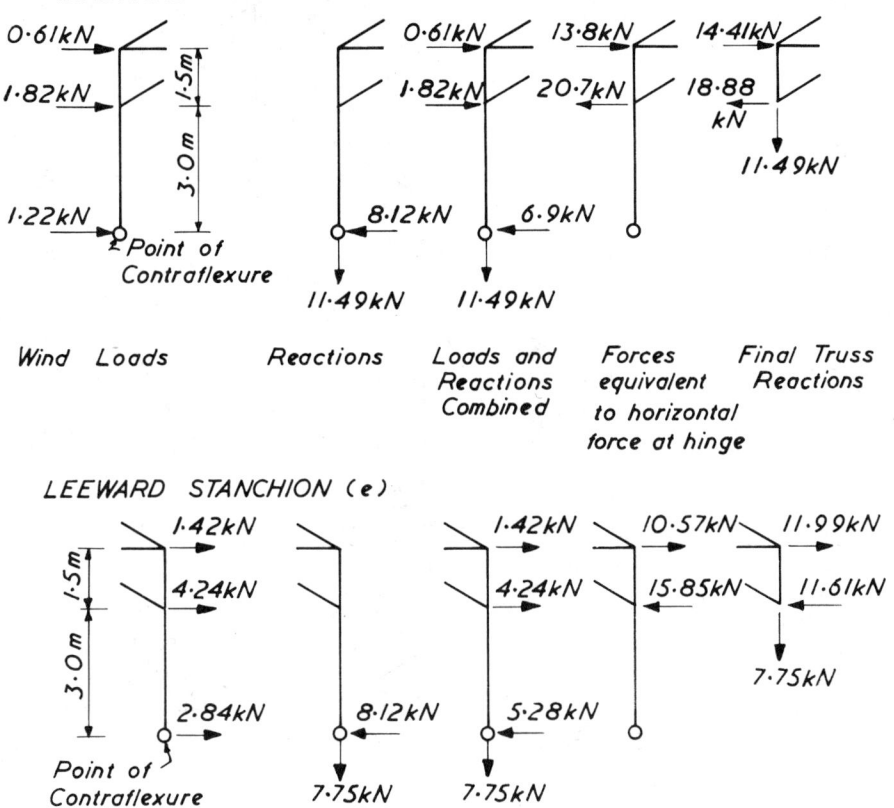

Wind Loads — Reactions — Loads and Reactions Combined — Forces equivalent to horizontal force at hinge — Final Truss Reactions

LEEWARD STANCHION (e)

Fig. 9 (d) and (e)

28.01 kNm at the base for the windward column, with corresponding figures 15.85 kNm and 32.86 kNm for the leeward column.

The solution for method (b) is shown in Fig. 11. Any convenient framework arrangement can be used as indicated by dotted lines: in this case the truss rafter has been prolonged to meet the horizontal through the foot of the knee-brace, to enable the wind loads previously calculated to be used in conjunction with the previously determined reactions at the points of contraflexure. The force diagram is then drawn in the usual manner.

The forces given in the diagram are correct for all members other than the columns, for these have been affected by the addition of the auxiliary framework. To determine the force in the columns, the auxiliary frames are removed and each column considered separately. Taking the windward column, then it is seen that below the knee-brace the direct force must equal the reaction, whilst in addition there is a moment of maximum value 20.7 kNm at knee-brace level. For the upper portion of the column the direct force must be such that it will hold in equilibrium the vertical reaction and the vertical component of the force in the knee-brace.

It will be found that the force diagram for method (a) can be constructed more easily than that for method (b), since there are four fewer joints to be considered

130

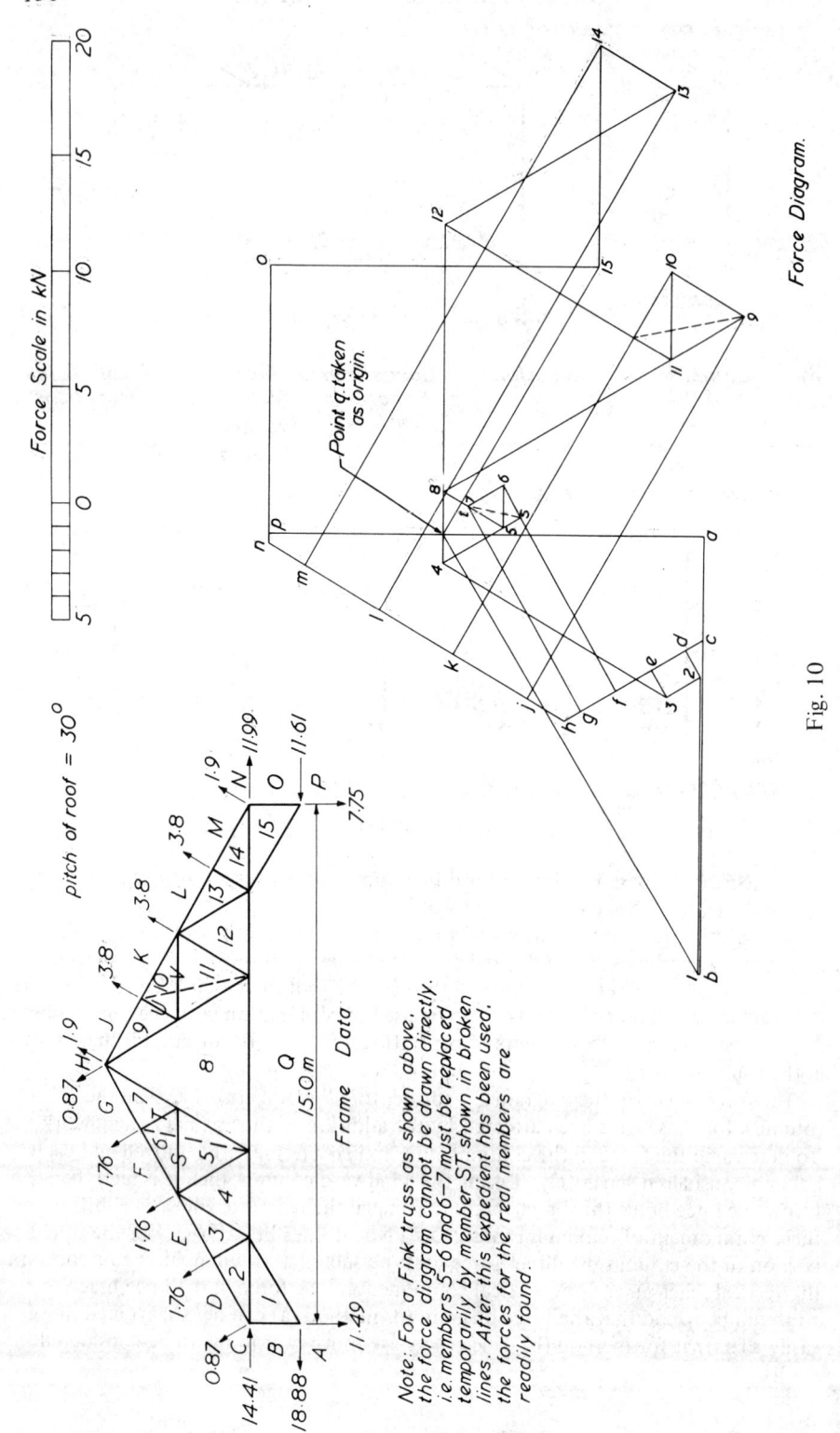

Force Scale in kN

pitch of roof = 30°

Frame Data

Note: For a Fink truss, as shown above, the force diagram cannot be drawn directly, i.e. members 5-6 and 6-7 must be replaced temporarily by member ST, shown in broken lines. After this expedient has been used, the forces for the real members are readily found.

Point q taken as origin.

Force Diagram.

Fig. 10

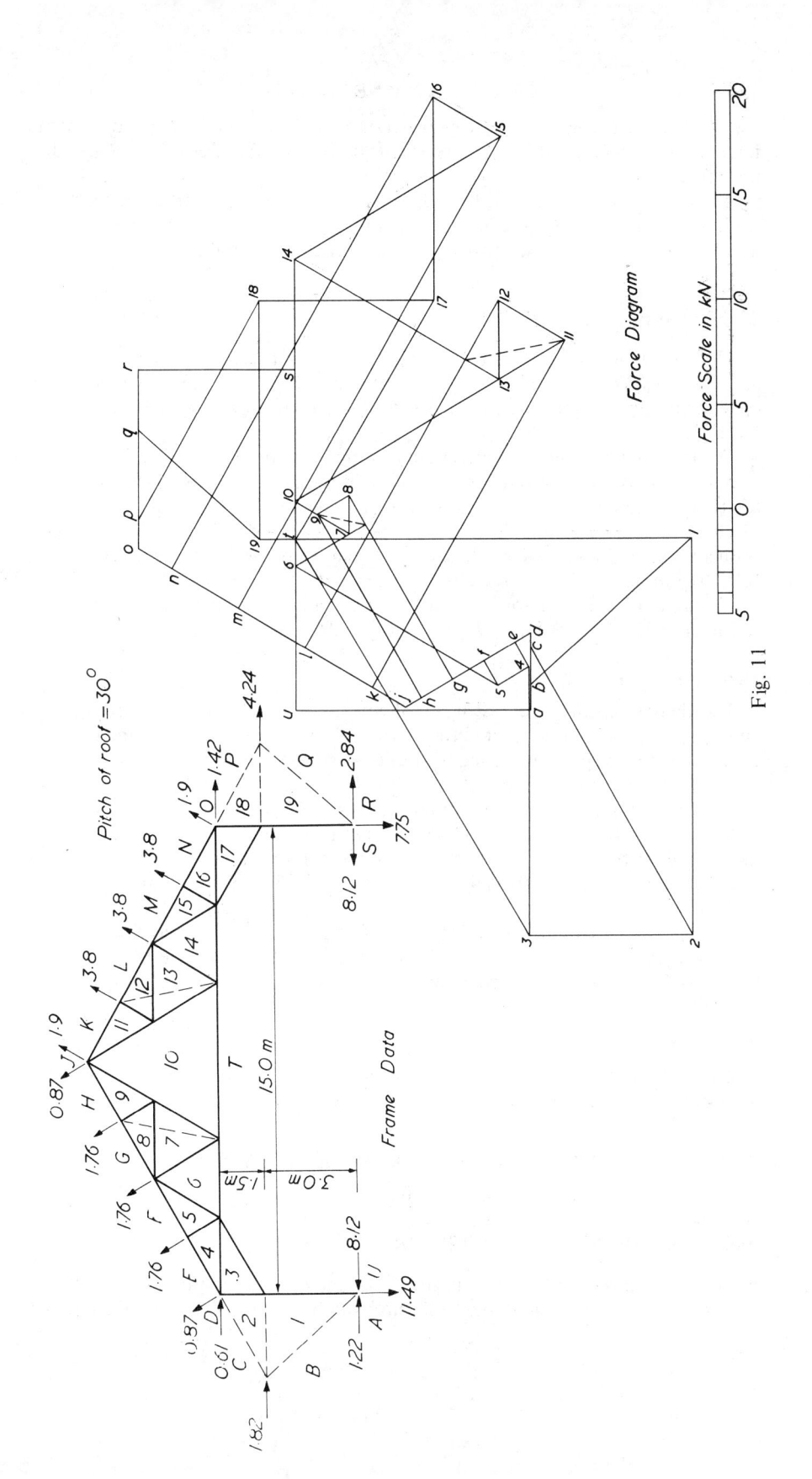

Fig. 11

132 FORCES IN PLANE FRAMES

and also since the diagram is more compact; moreover, the first method gives column forces directly, whereas the second does not. Both methods, of course, will give the same final answers.

The method of treating knee-braced frames described here is approximate, but it has been found satisfactory in practice. A more rigorous and accurate analysis would involve a consideration of the theory associated with portal frames.

The Three-hinged Arch

The three-hinged arch or portal is the only arch construction in which the horizontal reactions are statically determinate since there are three known points on the equilibrium polygon, or as it is more usually known in the case of an arch, the line of pressure.

The reactions and line of pressure can be found in three ways: (i) by drawing the equilibrium polygon to pass through the three hinges, (ii) by calculation of vertical reactions and horizontal thrust and constructing the ray diagram accordingly, or (iii) by superimposing upon the arch the free B.M. diagram caused by the loads acting as on a beam spanning between the two outer hinges, the scale being such that the B.M. diagram passes through the crown hinge.

The arch, XYZ, chosen for the examples is 18 m in span and carries 50 kN loads at 3 m intervals. The left hinge, X, is 3 m higher than the right hinge, Z.

Method (i) (see Fig. 12)

The load line *abcdef* is set down, a trial pole taken and the construction explained on page 133 is then followed, resulting in the pole of the ray diagram being fixed at point j in the force polygon, giving the reactions shown in Fig. 12, namely V_L = 100 kN, V_R = 150 kN and H = 150 kN. The hatched diagram, i.e. the difference between the equilibrium polygon (or line of pressure) and the line representing the arch axis, is the B.M. diagram for the arch, and the B.M. under load CD is given by the product of the polar distance jc of 150 kN and the vertical intercept between the line of pressure and arch axis at that point, namely 0.75 m giving a B.M. of 112.5 kNm.

It should be noted that the polar distance jc represents the horizontal thrust H, and consequently the B.M. at any point in the arch is given by the product of the horizontal thrust and the vertical intercept between the line of pressure and the arch axis at the point.

Method (ii)

The B.M. on the arch is zero at the hinge points X, Y and Z. Consequently, the following equations may be evolved by taking moments to the left of Y and Z respectively:

$$\Sigma M_Y = V_L \times 6\text{ m} - 50\text{ kN} \times 3\text{ m} - H \times 3\text{ m} = 0$$

$$\Sigma M_Z = V_L \times 18\text{ m} - 250\text{ kN} \times 9\text{ m} + H \times 3\text{ m} = 0$$

whence V_L = 100 kN, H = 150 kN and V_R = 150 kN.

With these figures known the ray diagram *abcdefgjka* can be constructed and the line of pressure drawn. The B.M.s on the arch can be found as for method (i) or may

THE THREE-HINGED ARCH   133

be obtained directly by calculation, e.g. by taking moments of all forces to the right of load CD we have

$$M_{CD} = 150 \text{ kN} \times 9 \text{ m} - 150 \text{ kN} \times 5.25 \text{ m} - (50 \text{ kN} \times 3 \text{ m} + 50 \text{ kN} \times 6 \text{ m})$$
$$= 112.5 \text{ kNm}$$

Method (iii)

The free B.M. diagram, or B.M. diagram for the five loads on a span of 18 m is as shown, and the B.M. at the crown hinge is 600 kNm. For the B.M. diagram to pass

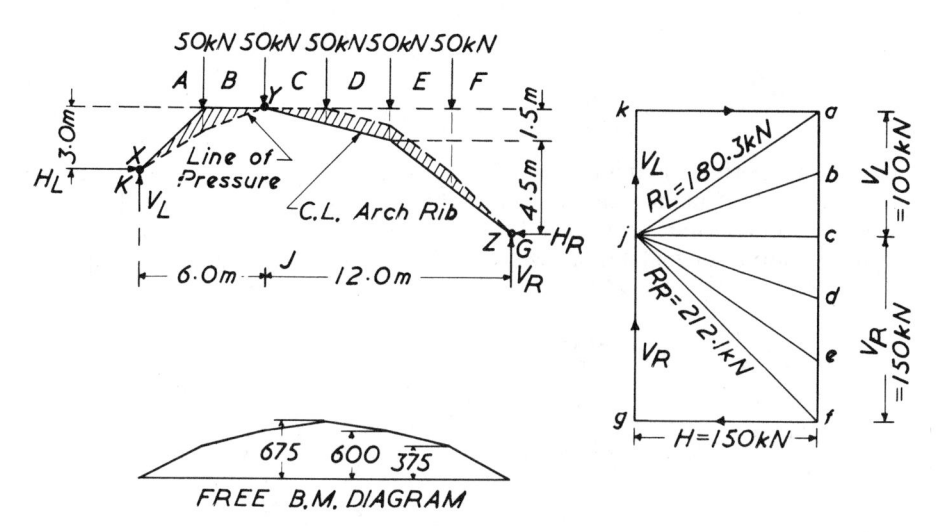

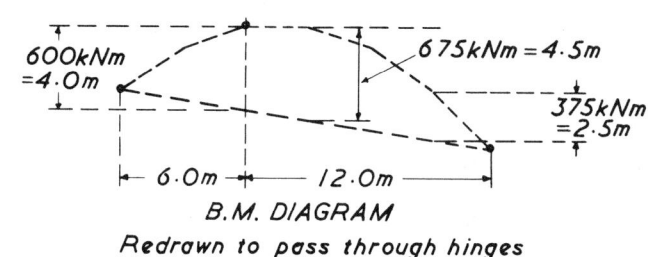

Fig. 12

through the crown hinge the moment of 600 kNm must be represented by the vertical distance of 4.0 m shown in the figure, that is, the space scale for moments is 150 kNm per metre on the scale layout of the arch. The B.M. diagram is then redrawn to this scale on the sloping base line between the two end hinges and superimposed on the line representing the arch axis, giving the same result as before.

The direct compression and shear in the rib at any point can be obtained by resolution from the line of pressure as, for example, between BC and CD the force is 150 kN which gives 145.5 kN rib compression and 36.4 kN shear as shown in Fig. 13. With regard to the B.M. it should be noted that tension on the underside of

the arch occurs when the line of pressure is above the arch axis and, conversely, tension occurs on the top of the arch when the line is below the axis.

It will be noted that the reactions at the supports are compounded of the horizontal thrust, equal at both ends, and the vertical reactions. The horizontal

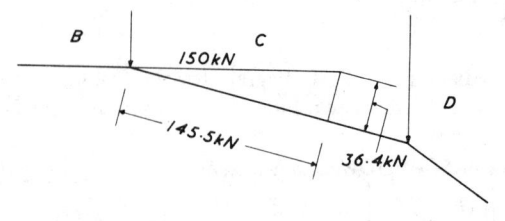

Fig. 13

thrusts at supports are in this case equal and opposite since there are no loads other than vertical loads on the arch; should there be horizontal loads or inclined loads having horizontal components then the horizontal reactions will not be equal and opposite and the horizontal thrust will not be constant across the arch. This can be

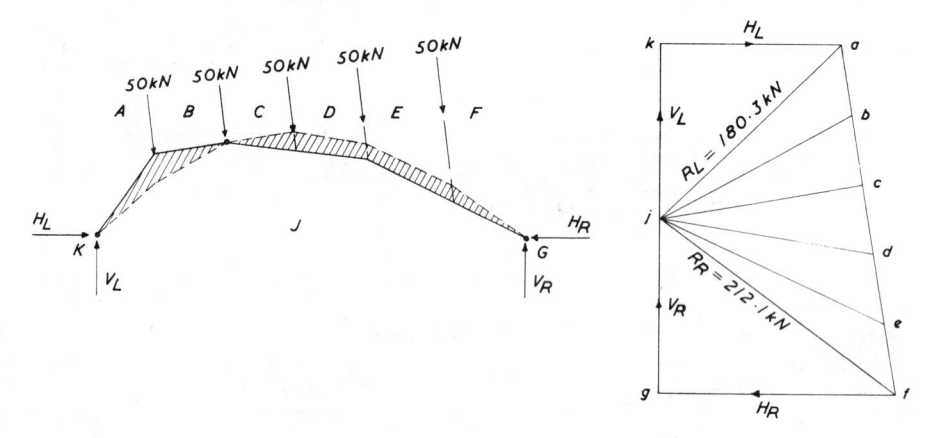

Fig. 14

shown if the arch previously considered is rotated so that the end hinges are in a horizontal line, all dimensions remaining unaltered as shown in Fig. 14. The magnitudes of the reactions will be unaltered, since the layout of the force system has not changed, but their vertical and horizontal components alter as indicated.

To deal with lattice or braced three-hinged arches, it is only necessary to calculate the reactions by any of the methods indicated, and, this having been done, the forces in the members of the frame can be found either graphically or analytically.

11. DEFLECTION OF FRAMED STRUCTURES

Mathematical Method

WHEN a structure is loaded, deflection will occur, and each of the loads on the structure will move through a certain distance thereby doing a certain amount of *external* work. Each of the members will become strained and in so doing will absorb a certain amount of *internal* work. According to the Principle of Work, the external work done on a structure must be equal to the internal work absorbed in straining the structure, and by applying this principle the deflection of the structure can be determined.

For the purpose of dealing with statically determinate plane frames in which all the loads are applied at node points and in which the supports are assumed not to move in the direction of the applied load, the mathematical treatment is relatively simple, since the internal work is done only in elongating and shortening the members of the frame.

Figure 1 shows a simple bracket consisting of two members AB and BC carrying a load W. The supports A and C are fixed in position.

Under the load W, AB will be in tension while BC will be in compression, and the point B will deflect by an amount d, the frame taking up the position shown by the dotted lines.

Let L_1 = length of member AB,

A_1 = the cross-sectional area of member AB,

F_1 = force in member AB (tension considered positive),

x_1 = extension of member AB (positive),

L_2 = length of member BC,

A_2 = cross-sectional area of member BC,

F_2 = force in member BC (compression considered negative),

and x_2 = shortening of member BC (negative).

Then, if E is constant for both members,

$$x_1 = \frac{F_1 L_1}{A_1 E}$$

and

$$x_2 = -\frac{F_2 L_2}{A_2 E}.$$

Now if the load is applied so as to increase gradually from zero to the value W the work done will be the average force acting × the distance moved through and therefore the total work done in straining the members

$$= \tfrac{1}{2} F_1 x_1 + \tfrac{1}{2} F_2 x_2$$
$$= \frac{1}{2} \cdot \frac{F_1^2 L_1}{A_1 E} + \frac{1}{2} \cdot \frac{F_2^2 L_2}{A_2 E}.$$

136 DEFLECTION OF FRAMED STRUCTURES

The plus sign is used because both F_2 and x_2 are negative.
Therefore, for any number of members, the total internal work done

$$= \frac{1}{2} \sum \frac{F^2 L}{AE}.$$

At the same time the external work done will be $\frac{1}{2} W \cdot d$.
Therefore, in the example

$$d = \frac{F_1^2 L_1}{W A_1 E} + \frac{F_2^2 L_2}{W A_2 E}$$

Now F_1 and F_2 will be proportional to W, and if U_1 and U_2 represent the

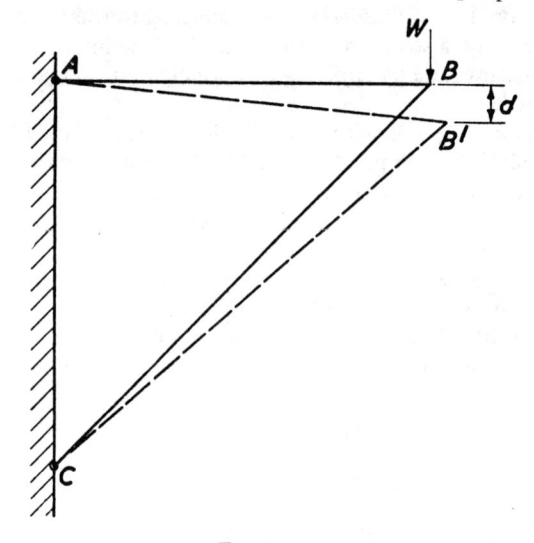

Fig. 1

forces caused in AB and BC respectively when a unit load is placed at B, then $U_1 = F_1/W$ and $U_2 = F_2/W$, consequently

$$d = W \frac{U_1^2 L_1}{A_1 E} + \frac{U_2^2 L_2}{A_2 E}$$

or in the general case

$$d = W \sum \frac{U^2 L}{AE}.$$

Figure 2 represents a more general case.
It is required to find the deflection of the point P in the direction C.
Consider the member 1 (AB) and imagine that, for the moment, all the other members are rigid, i.e. suffer no deformation.
Then, if a unit load is applied at P in the direction C and causes a deflection d_1 at P, the external work done

$$= \tfrac{1}{2} \times 1 \times d_1.$$

The internal work done $= \tfrac{1}{2} U_1 x_1$ where

U_1 = force in member 1 due to unit load at P

and $\qquad x_1$ = strain in member 1 due to unit load at P.

MATHEMATICAL METHOD

Therefore,
$$\tfrac{1}{2} \times 1 \times d_1 = \tfrac{1}{2} U_1 x_1$$
or
$$d_1 = U_1 x_1.$$

If, instead of placing unit load at P, the member 1 had been strained by an amount x_1, the deflection at P would have been d_1.

Accordingly the following rule is derived:

> The deflection of any point in a framed structure in a given direction due to an extension x in any one member of the structure is equal to the load in that member caused by unit load at the given point, acting in the given direction, multiplied by the extension x.

This rule can now be extended to cover all the members in the frame and consequently the total deflection = ΣUx.

In a loaded structure x will be the extension in any member due to the force F

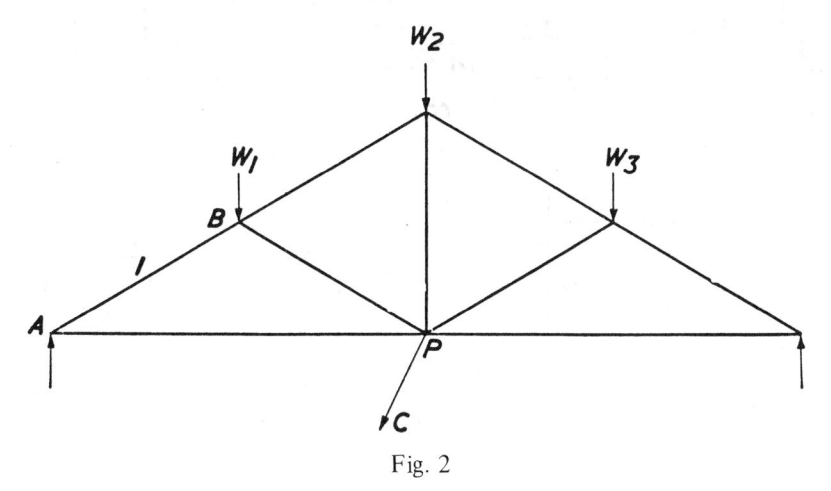

Fig. 2

in the member caused by the loading and $x = FL/AE$, where A is the cross-sectional area of the member.

Therefore the total deflection

$$d = \sum \frac{UFL}{AE}.$$

A complete determination of the movements of all node points in a girder requires two deflection calculations for each joint, vertical and horizontal. The graphical method referred to later gives a single displacement diagram from which the resultant movement of all the joints can be determined at the same time. The algebraic solution is exact while the graphical solution, although limited by the size of the drawing and the inherent inaccuracies in all graphical work, is a great saver of time.

Deflections due to temperature changes may be calculated by treating the change in length due to variation in temperature in a similar manner to strain caused by stress.

138   DEFLECTION OF FRAMED STRUCTURES

Deflections due to the forced distortion of members may be calculated by replacing FL/AE by the estimated or known change in length, increase in length being positive and decrease negative.

Example 1. Calculate the vertical deflection of the central point P of the lower boom of the Warren girder loaded symmetrically as shown in Fig. 3 (a). $E = 2.1 \times 10^5$ N/mm².

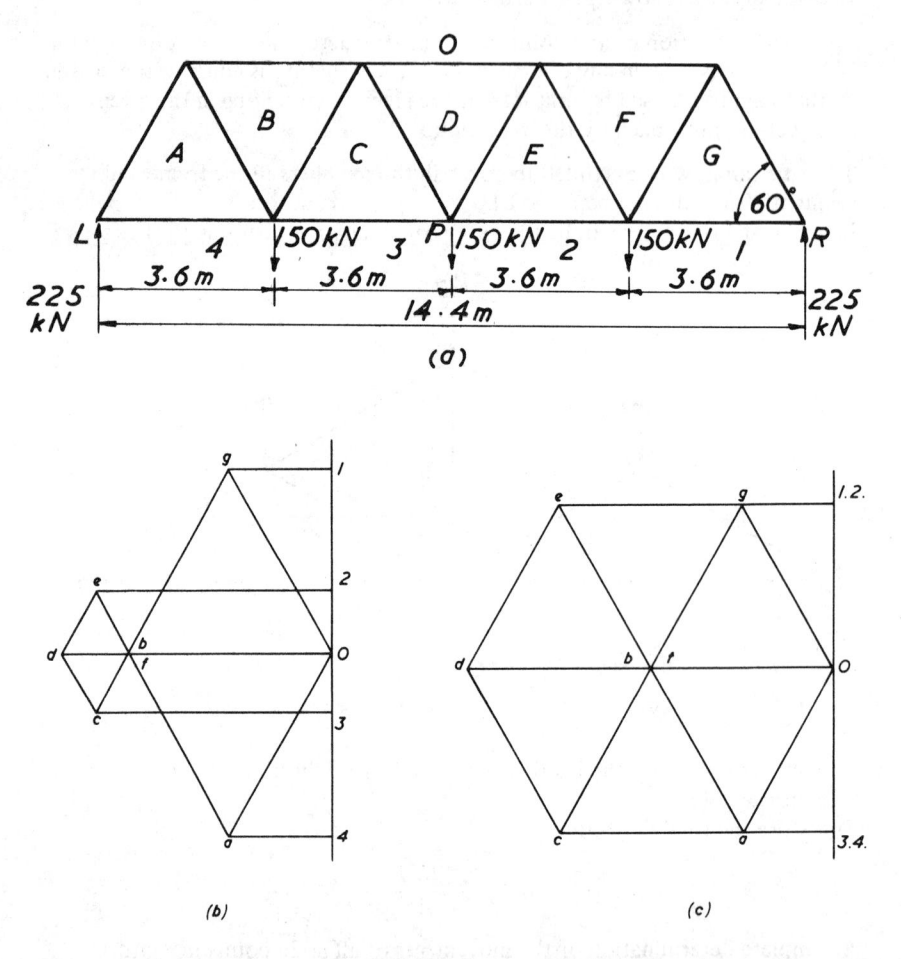

Fig. 3

Figure 3 (b) shows the force diagram for all external loads, while Fig. 3 (c) shows the force diagram due to unit vertical load at P.

In Fig. 4, for the purpose of this Example the members have not been designed in detail, but their areas have been assumed on the basis of using stresses of 202 and 135 N/mm² in the tension and compression members respectively, and, since the girder is symmetrically loaded, members having similar lengths and forces have been taken together, the length of each being multiplied by 2.

MATHEMATICAL METHOD

member	force F (kN)	area A (mm^2)	length L (m)	unit force U	$\frac{U \times F \times L}{10^4 A}$ N/mm
4A.IG.	130.0	621.3	2 x 3.6	.2887	43.5
3C.2E.	303.3	1449.7	2 x 3.6	.8662	130.5
OA.OG.	-260.0	1863.9	2 x 3.6	-.5775	58.0
OB.OF.	-260.0	1863.9	2 x 3.6	-.5775	58.0
OD.	-346.7	2485.2	3.6	-1.155	58.0
AB.FG.	260.0	1242.6	2 x 3.6	.5775	87.0
BC.EF.	-86.7	621.3	2 x 3.6	-.5775	58.0
CD.DE.	86.7	414.2	2 x 3.6	.5775	87.0
				TOTAL	= 580.0 x 10^4

– indicates compression

Fig. 4

Therefore the vertical deflection at *P*

$$\sum \frac{UFL}{AE} = \frac{580 \times 10^4 \text{ N/mm}}{2.1 \times 10^5 \text{ N/mm}^2} = 27.62 \text{ mm}$$

Example 2. Calculate the horizontal deflection of the central point *P* of the lower boom of the Warren Girder used in Example 1 and loaded as shown in Example 1. $E = 2.1 \times 10^5$ N/mm^2.

Bearing in mind that the girder is symmetrical and is symmetrically loaded and that the point at which the deflection is required is the central point of the bottom flange, it will be obvious that, in relation to the centre-line of the girder, this point will have no deflection in a horizontal direction.

There will, however, be some resultant horizontal movement between this point and the ends of the girder *L* and *R*, and it is really this horizontal movement that is required. Accordingly, if it is required to find the resultant horizontal movement between *L* and *P*, the point *L* can be assumed to be fixed in position. In this case all the horizontal reaction will be supplied at *L* and it will be obvious that with a horizontal unit load at *P* and the reaction to this load wholly supplied at *L*, and

therefore equal to unity, the only members of the truss subject to forces are the two members A4 and C3. If the unit load U is applied at P in the direction towards L, i.e. towards the left, then the forces in A4 and C3 under this unit load will be compressive.

The table corresponding to this unit loading is given below.

Member	Force F kN	Area A mm^2	Length L m	Unit Load U	$\dfrac{UFL}{A}$ 10^4 N/mm
A4	130.0	621.3	3.6	– 1.0	– 75.33
C3	303.3	1 449.7	3.6	– 1.0	– 75.33
				Total =	– 150.66 x 10^4

Therefore, the horizontal deflection at $P = -\dfrac{150.66 \times 10^4 \text{ N/mm}}{2.1 \times 10^5 \text{ N/mm}^2} = -7.17$ mm.

This means that the joint tends to move, in relation to L in the opposite direction to the applied load, i.e. the point P tends to move away from L, or conversely the point L tends to move away from the point P. The value of the calculated movement is confirmed by the measurement of the horizontal distance between a and e in Fig. 9 which equals 7.17 mm.

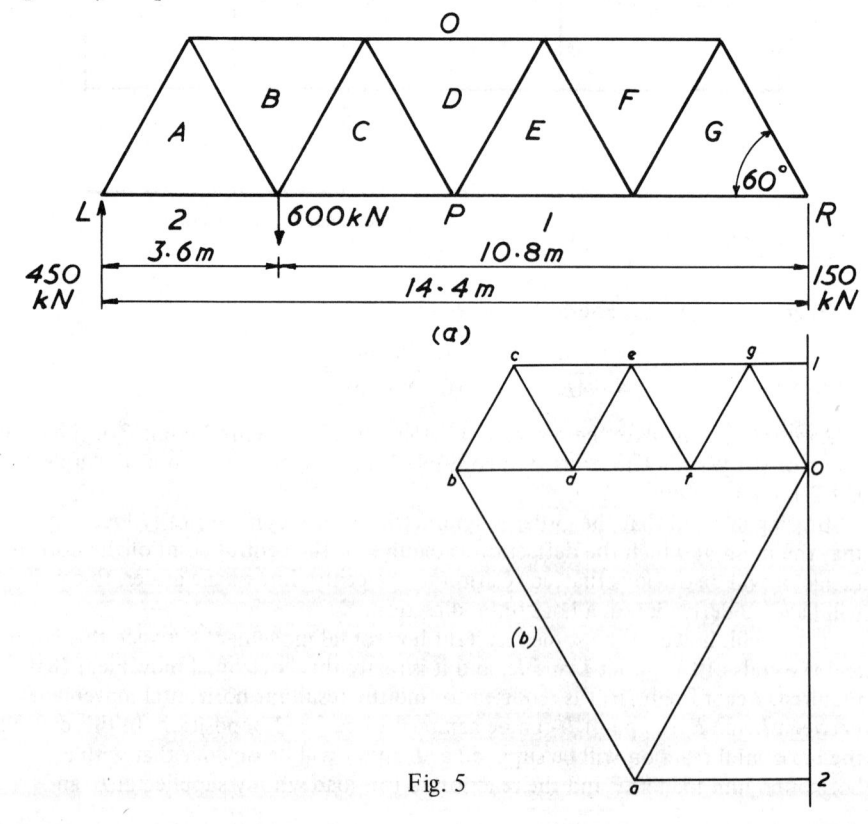

Fig. 5

MATHEMATICAL METHOD

member	force F (kN)	area A (mm^2)	length L (m)	unit force U	$\frac{U \times F \times L}{A}$ 10^4 N/mm
2A	259.8	621.3	3.6	.2887	43.46
1C	433.0	1449.7	3.6	.8662	93.14
1E	259.8	1449.7	3.6	.8662	55.88
1G	86.6	621.3	3.6	.2887	14.49
OA	-519.7	1863.9	3.6	-.5775	58.15
OB	-519.7	1863.9	3.6	-.5775	58.15
OD	-346.5	2485.2	3.6	-1.155	57.97
OF	-173.2	1863.9	3.6	-.5775	19.37
OG	-173.2	1863.9	3.6	-.5775	19.37
AB	519.7	1242.6	3.6	.5775	86.95
BC	173.2	621.3	3.6	-.5775	-57.96
CD	-173.2	414.2	3.6	.5775	-86.93
DE	173.2	414.2	3.6	.5775	86.93
EF	-173.2	621.3	3.6	-.5775	57.96
FG	173.2	1242.6	3.6	.5775	28.98

– indicates compression

TOTAL 535.91×10^4

Fig. 6

142   DEFLECTION OF FRAMED STRUCTURES

Example 3. Calculate the vertical deflection of the central point *P* of the lower boom of the Warren girder used in Example 1 but loaded asymmetrically as shown in Fig. 5 (a). $E = 2.1 \times 10^5$ N/mm^2.

It should be noted that, as the areas of the members are the same as in Example 1, the stresses in the members under this alternate load will be varied.

Figure 5 (b) shows the force diagram for the external load. The force diagram for vertical unit load at *P* is exactly as in Example 1 and consequently need not be redrawn. Figure 6 gives the summation UFL/A.

Therefore the vertical deflection at $P = \dfrac{535.91 \times 10^4 \text{ N/mm}}{2.1 \times 10^5 \text{ N/mm}^2} = 25.52$ mm

Graphical Method

The deflection of a framed structure may also be found by a graphical method which consists of plotting, to an enlarged scale, the movements of the joints of the structure. These diagrams are commonly referred to as Williot diagrams.

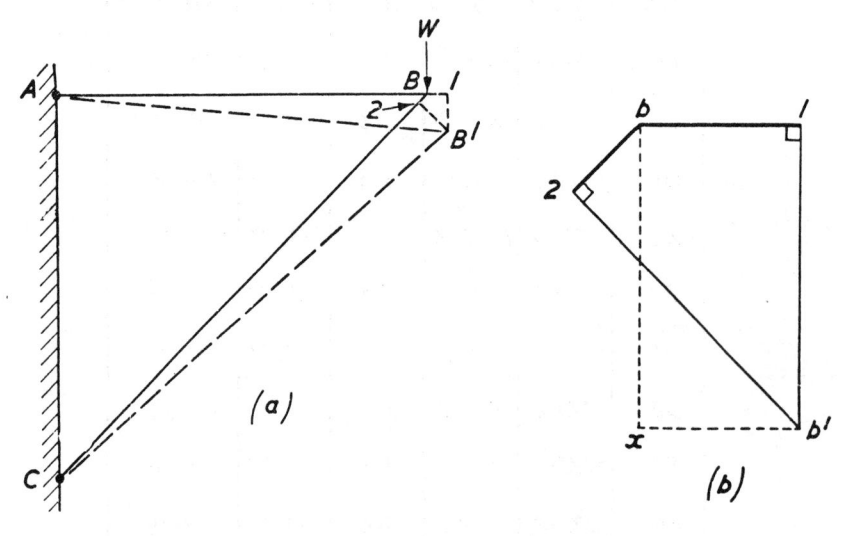

Fig. 7

As an illustration of the method, consider the simple bracket of Fig. 1, which is redrawn in Fig. 7 (a).

Under the load *W* the member *AB* stretches by an amount *B*1, while the member *BC* contracts by an amount *B*2.

Assuming that the member *AB* can rotate about *A*, and that member *BC* can rotate about *C*, then with centres *A* and *C* and radii *A*1 and *C*2 respectively, arcs can be drawn to intersect at B^1, which is the position taken up by the point *B* when the load *W* is applied.

In reality, the distances *B*1 and *B*2 are extremely small so that the arcs 1B^1 and 2B^1 can be replaced by straight lines at right angles to *A*1 and *C*2 respectively. The following construction, which is illustrated in Fig. 7 (b), will describe the method to be used.

GRAPHICAL METHOD 143

Set out, to a much enlarged scale, the lengths $b1$ and $b2$ parallel to their respective members AB and CB to represent the strains in those members, in value and direction. From 1 and 2 draw perpendiculars to $b1$ and $b2$ to intersect in b^1. b^1 will be the displaced position of the point b and if bx is vertical and b^1x is horizontal, then they represent, respectively, the vertical and horizontal deflections of the point B.

The method will now be extended to deal with a Warren girder and the three previous examples will be re-worked by the graphical method.

Example 4. The data is the same as in Example 1.

For the graphical construction it is required to know the strains in each member, and therefore another figure is required to arrive at these strains. See Fig. 8.

Member	length L (m)	area A (mm^2)	force F (kN)	strain $=\dfrac{FL}{AE}$ (mm)
2.14	3.6	621.3	130.0	3.59
6.10	3.6	1449.7	303.0	3.59
1.15	3.6	1863.9	−260.0	−2.39
4.12	3.6	1863.9	−260.0	−2.39
8	3.6	2485.2	−346.7	−2.39
3.13	3.6	1242.6	260.0	3.59
5.11	3.6	621.3	−86.7	−2.39
7.9	3.6	414.2	86.7	3.59

− indicates compression

Fig. 8

It is more convenient, when dealing with the graphical method, to letter the points of the girder and to number the members. For this reason the notation used in Fig. 9 (a) is different from that used in Fig. 3 (a).

The displacement diagram is shown in Fig. 9 (b).

Before commencing to draw the displacement diagram, it is necessary to relate all the displacements to one point on a member of the frame. This reference point must be a node point and it is assumed to be fixed in position. In addition this point

144   DEFLECTION OF FRAMED STRUCTURES

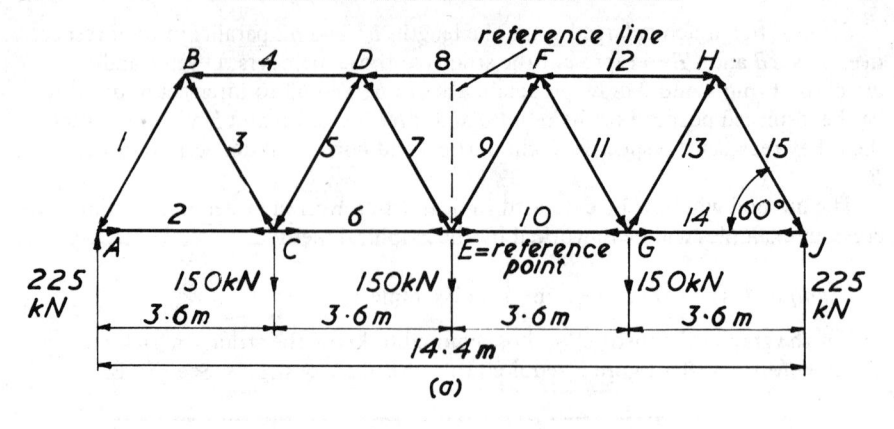

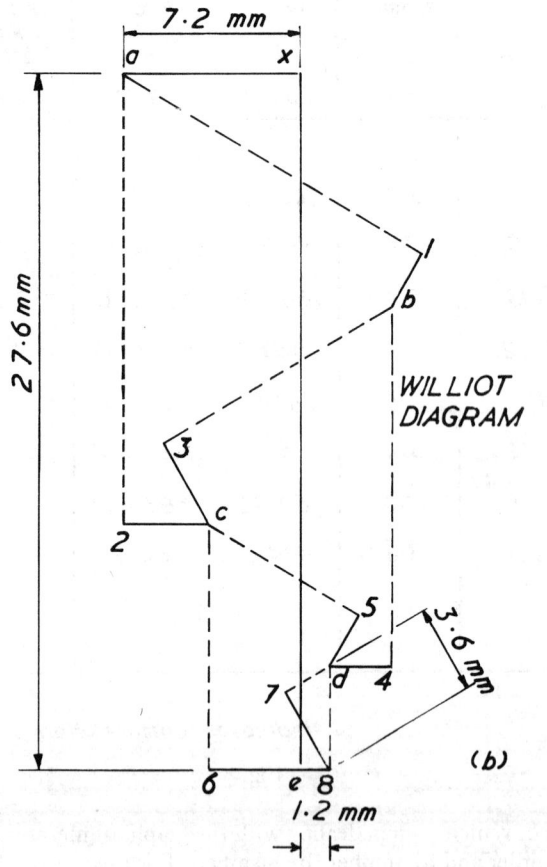

Fig. 9

must lie on a line which is assumed not to rotate. This line may or may not be a member of the frame.

For the purpose of this Example, which is for a symmetrical frame, symmetrically loaded, the point E is a particularly convenient reference point and the line which is assumed not to rotate is the line joining E to the mid-point of member 8. In this Example the principle of symmetry applies and it will only be necessary to deal with one half of the girder.

Accordingly the point e in Fig. 9 (b) is the starting point and the first operation is to find the displacement of the point D with respect to the point E. The analogy of the simple bracket will be apparent, because this displacement is concerned only with the strains in the members 7 and 8, and, more precisely, with only half the strain in member 8 since this member is bisected by the reference line.

Set out $e8$ parallel to member 8 and equal to half the strain in member 8, and set out $e7$ parallel to member 7 and equal to the strain in member 7. The values of the strains are taken from Fig. 8. Draw $8d$ and $7d$ at right angles to $e8$ and $e7$ respectively to meet at d. Then d is the displaced position of the point D with respect to the point E, and $d8$ will represent the vertical deflection of D with respect to E.

The only difficulty concerns the directions in which the lines $e8$ and $e7$ are to be drawn and the following simple rule will govern this matter.

Arrows are placed on the frame diagram, as shown in Fig. 9 (a), to represent the forces in the members. If the deflection of D with respect to E is required, the two members concerned are 7 and 8, and the directions of the forces in these members at the point D are considered. The lines $e8$ and $e7$ on Fig. 9 (b) are then drawn in the opposite directions to the respective arrows.

In other words, since member 8 is a strut, it will tend to shorten and consequently the point D will tend to move towards the right in relation to the point E, while since member 7 is a tie, the point D will tend to move away from the point E.

Having found the displaced position of D, it is now possible to find the displacement of C with respect to D. Draw $d5$ and $e6$ on Fig. 9 (b) parallel to members 5 and 6 respectively. The directions of $d5$ and $e6$ are governed by the arrows at C. Draw $5c$ and $6c$ at right angles to $d5$ and $e6$ respectively to meet in c, which is the displaced position of the point C.

By similar means the points b and a, which are the displaced positions of the points B and A respectively, are found.

The final vertical deflection of the point A with respect to the point E is the vertical distance between a and e and scales 27.6 mm as obtained by calculation in Example 1. Similarly, the horizontal distance between a and e = 7.2 mm as calculated in Example 2.

The scale used for the construction of the Williot diagram is too large to be applied directly to the truss drawing, but if a smaller scale is used the displacement may be indicated, although in an exaggerated manner, by the dotted lines in Fig. 10.

This outline represents the shape of the deformed truss on the basis that point E is fixed, i.e. the same basis on which the Williot diagram was drawn. It is obvious, however, that since points A and J remain fixed in elevation (although not laterally) a correction must be made by moving the broken outline downwards. The corresponding correction can be made on the Williot diagram by taking as a reference point the point X, which lies on the horizontal line through a and the vertical line through e. All displacements, both horizontal and vertical, are measured from X and thus all joints can be located from the one diagram.

146   DEFLECTION OF FRAMED STRUCTURES

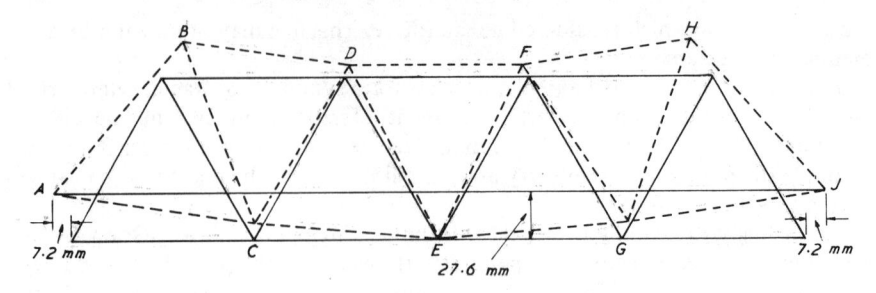

Fig. 10

Example 5. The data is the same as in Example 3.

In this case the loading is asymmetrical and accordingly the assumption that $E8$ remains vertical is not true. Therefore, the Williot diagram does not represent the true conditions and a suitable correction diagram must be devised.

Figure 12 (a) shows the frame diagram with the arrows representing the forces.

Figure 12 (b) shows the displacement diagram which is drawn by applying the principles described in Example 4, the strains being obtained from Fig. 11. The point e is the starting point, and the points a and j represent the displaced positions of the points A and J respectively. It will be seen that the vertical distances from a to e and from j to e are not equal, as, of course, they should be if the supports A and J are at the same level.

Figure 13 shows the deformed shape of the truss on the basis that the point e is fixed and the line $e8$ is vertical. In order that this deformed shape may correctly represent the deflection of the various joints it is necessary to rotate the dotted line figure about E in a clockwise direction until the points A and J are on a horizontal line.

In Fig. 13 let θ be the angle relative to the fixed point E through which the dotted outline must be rotated, then since the truss is a rigid frame all joints of the truss are rotated through the same angle and the amount of displacement of any joint is in direct proportion to the distance of that joint from the reference point. For example, due to rotation about E, points B and F move horizontally to the right, point B moving vertically upwards and point F vertically downwards.

The correction for each joint can be made graphically as follows:

Draw a vertical through e in Fig. 12 (b) to cut the horizontals through a and j in a_1 and j_1 respectively. On $a_1 j_1$ draw a replica of the truss to scale, as shown in chain lines. Denote each of the joints in the truss by letters corresponding to the original truss. This outline is a Mohr correction diagram and the final figure is a Williot-Mohr diagram.

The displacement of any joint may then be determined by direct measurement between points which correspond in the Mohr diagram and the Williot diagram, this being most conveniently done by measuring both horizontal and vertical components. For example the vertical deflection of the point E is given by the vertical distance between e_1 and e on the Williot-Mohr diagram and measures 25.6 mm which compares with 25.52 mm obtained by calculation in Example 3. Similarly the horizontal movement of B is the horizontal distance from b_1 to b.

GRAPHICAL METHOD

member	length L (m)	area A (mm²)	force F (kN)	strain $\frac{FL}{AE}$ (mm)
2	3.6	621.3	259.8	7.17
6	3.6	1449.7	433.0	5.12
10	3.6	1449.7	259.8	3.07
14	3.6	621.3	86.6	2.39
1	3.6	1863.9	-519.7	-4.79
4	3.6	1863.9	-519.7	-4.79
8	3.6	2485.2	-346.5	-2.39
12	3.6	1863.9	-173.2	-1.60
15	3.6	1863.9	-173.2	-1.60
3	3.6	1242.6	519.7	7.17
5	3.6	621.3	173.2	4.78
7	3.6	414.2	-173.2	-7.17
9	3.6	414.2	173.2	7.17
11	3.6	621.3	-173.2	-4.78
13	3.6	1242.6	173.2	2.39

– indicates compression

Fig. 11

148 DEFLECTION OF FRAMED STRUCTURES

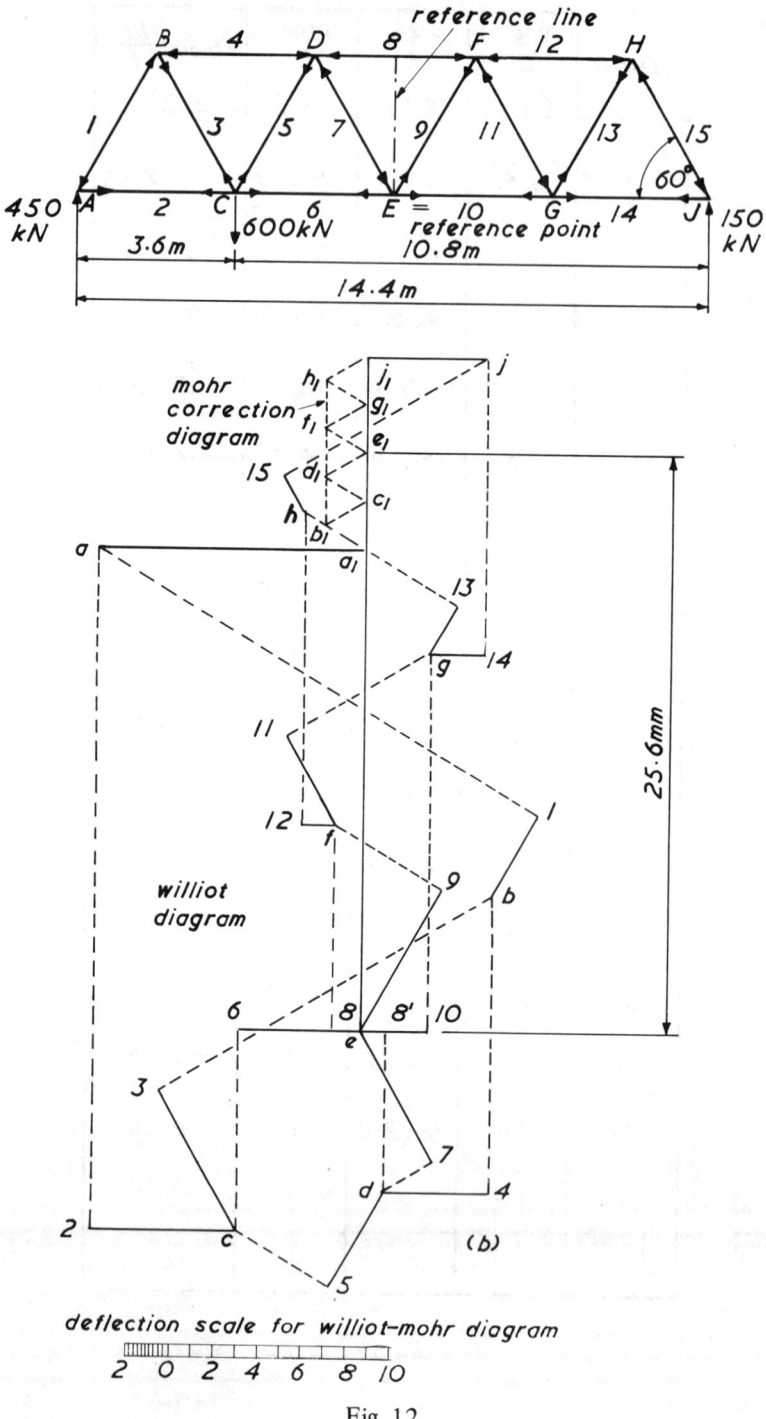

Fig. 12

GRAPHICAL METHOD

The principle of the Mohr correction diagram is further exemplified by the two additional Williot-Mohr diagrams for the same truss shown in Figs. 15 and 16.

In Fig. 15 the point A is assumed to be fixed in position and the member AB prevented from rotating (although allowed to shorten).

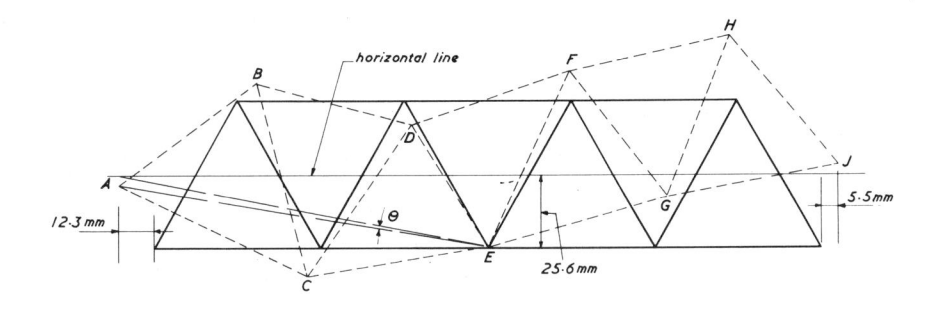

Fig. 13

If the exaggerated distorted frame were drawn, it would be apparent that the frame would have to be rotated in a clockwise direction about A to bring the point J down to the same level. The appropriate Mohr correction diagram is shown in Fig. 15. The deflected frame is shown in Fig. 14.

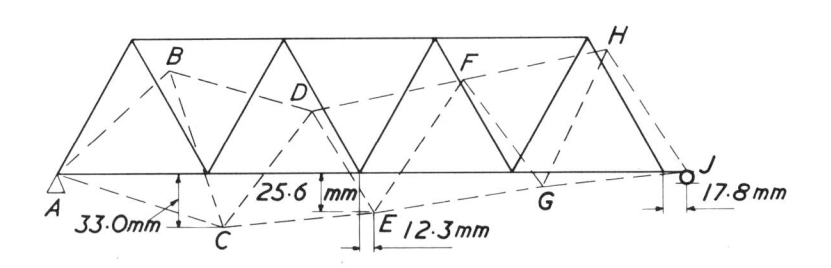

Fig. 14

In Fig. 16 the point E is once again assumed to be fixed, but this time the member EF is prevented from rotating. When the Williot diagram is completed the point a is higher than the point j. Hence, if the distorted frame were drawn it would have to be rotated in an anticlockwise direction about E to bring A and J to the same level. The Mohr diagram now appears on the right-hand side of the vertical through e.

A further frame is shown in Figs. 17 and 18 and vertical deflection of the centre node point of the lower boom is determined, both by calculation and by the drawing of a Williot-Mohr diagram. The values of FL/AE and FUL/AE are shown in Figs. 19 and 20 and the Williot-Mohr diagram is shown in Fig. 21.

150 DEFLECTION OF FRAMED STRUCTURES

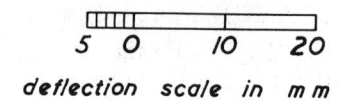

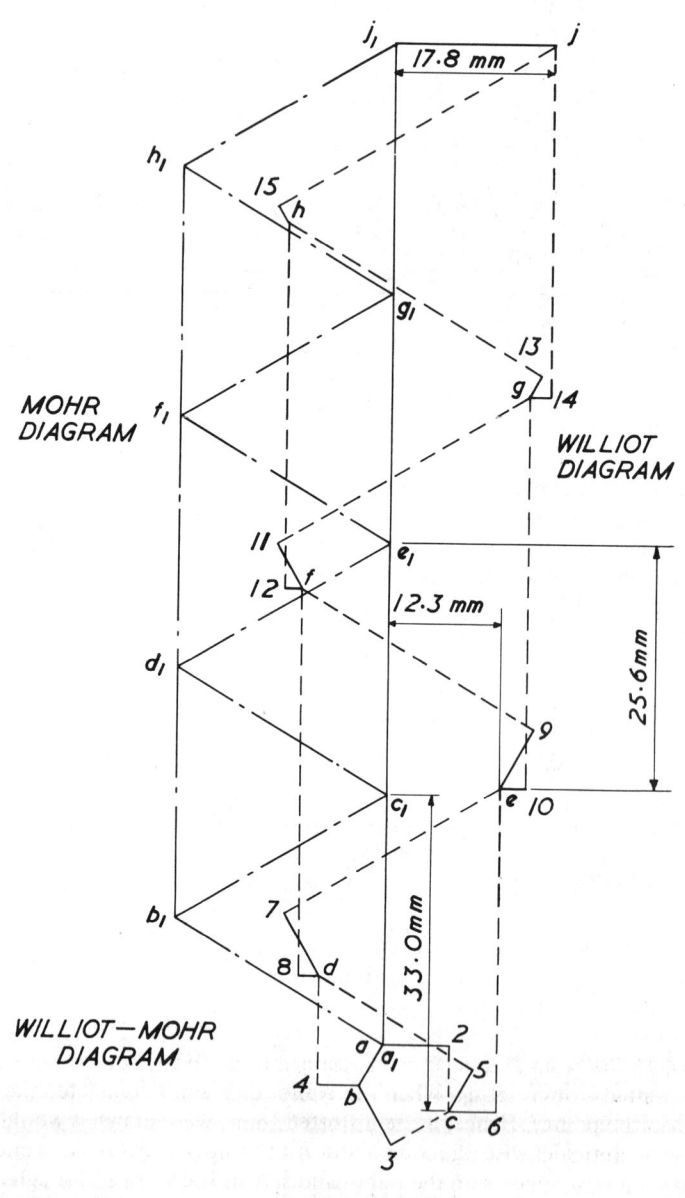

Fig. 15

GRAPHICAL METHOD

WILLIOT– MOHR DIAGRAM

reference point e
reference member ef

33.0mm

25.6mm

deflection scale in mm

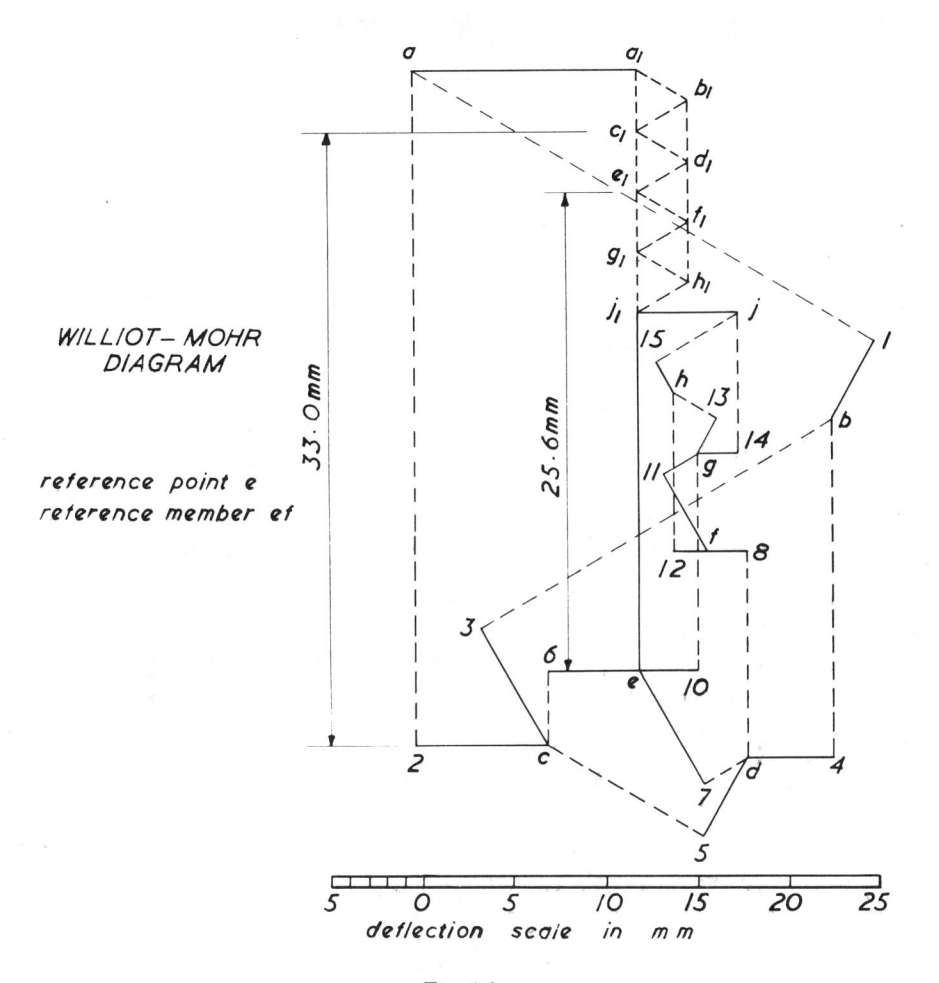

Fig. 16

frame as lettered for calculation method.

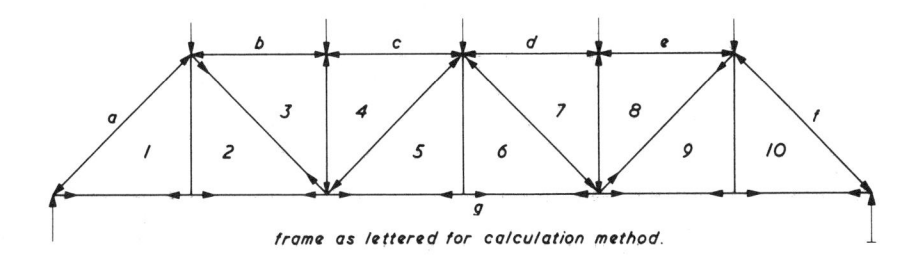

Fig. 17

152 DEFLECTION OF FRAMED STRUCTURES

frame as lettered for williot—mohr diagram.

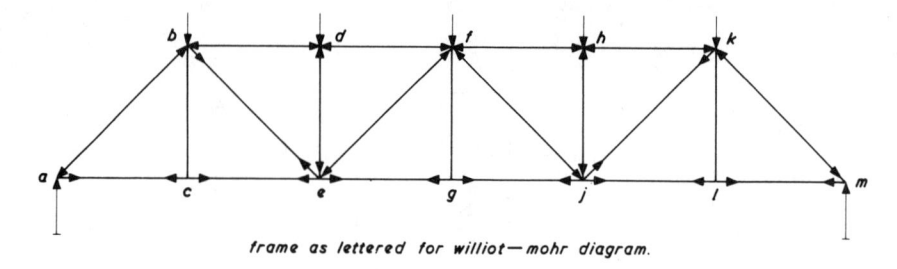

Fig. 18

member	F (kN)	L (m)	A (mm^2)	$\frac{E}{10^5 N/mm^2}$	$\frac{FL}{AE}$ (mm)
AB	−400·0	4·24	3225·8	2·1	−2·50
BD	−466·0	3·0	3741·9	··	−1·78
DF	−466·0	3·0	3741·9	··	−1·78
FH	−533·3	3·0	4322·6	··	−1·76
HK	−533·3	3·0	4322·6	··	−1·76
KM	−442·6	3·0	4129·0	··	−2·16
AC	283·3	3·0	3096·8	··	1·31
CE	283·3	3·0	3096·8	··	1·31
EG	550·0	3·0	3096·8	··	2·54
GJ	550·0	3·0	3096·8	··	2·54
JL	316·7	3·0	3096·8	··	1·46
LM	316·7	3·0	3096·8	··	1·46
BC	0	3·0	1806·4	··	0
BE	259·2	4·24	1806·4	··	2·90
DE	−100·0	3·0	1806·4	··	−0·79
EF	−117·9	4·24	1806·4	··	−1·32
FG	0	3·0	1806·4	··	0
FJ	23·56	4·24	1806·4	··	0·26
HJ	−200·0	3·0	1806·4	··	−1·58
JK	301·0	4·24	1806·4	··	3·37
KL	0	3·0	1806·4	··	0

Fig. 19

member	F (kN)	U	L (m)	A (mm²)	E 10⁵ N/mm²	FUL/A
A1	-400.0	-.707	4.24	3225.8	2.1	1.77
B3	-466.6	-1.0	3.0	3741.9	··	1.781
C4	-466.6	-1.0	3.0	3741.9	··	1.781
D7	-533.3	-1.0	3.0	4322.6	··	1.762
E8	-533.3	-1.0	3.0	4322.6	··	1.762
F10	-442.6	-.707	4.24	4129.0	··	1.529
G1	283.3	.5	3.0	3096.8	··	0.653
G2	283.3	.5	3.0	3096.8	··	0.653
G5	550.0	1.5	3.0	3096.8	··	3.806
G6	550.0	1.5	3.0	3096.8	··	3.806
G9	316.7	.5	3.0	3096.8	··	0.730
G10	316.7	.5	3.0	3096.8	··	0.730
1.2.	0	0	3.0	1806.4	··	0
2.3.	259.2	.707	4.24	1806.4	··	2.048
3.4.	-100.0	0	3.0	1806.4	··	0
4.5.	-117.9	-.707	4.24	1806.4	··	0.932
5.6.	0	1.0	3.0	1806.4	··	0
6.7.	-23.56	-.707	4.24	1806.4	··	0.186
7.8.	-200.0	0	3.0	1806.4	··	0
8.9.	301.6	.707	4.24	1806.4	··	2.383
9.10.	0	0	3.0	1806.4	··	0
				total deflection		26.31

Fig. 20

154 DEFLECTION OF FRAMED STRUCTURES

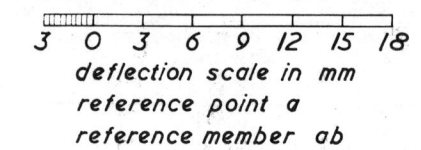

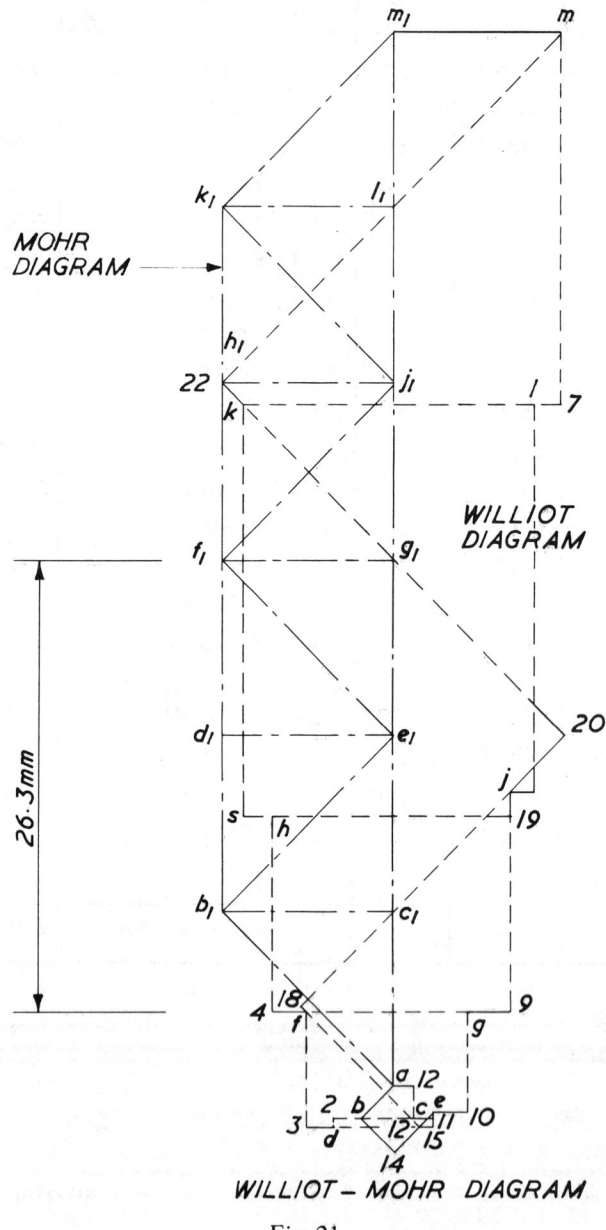

Fig. 21

12. INFLUENCE LINES

AN influence line for any given point P on a structure is such a line that its ordinate at any point Q gives the bending moment, shear force or similar quantity at P when a load is placed at Q.

It is important to note that an influence line is drawn for one point only and that while such a line will give the desired information at that point for any position of a load, it will not necessarily give the maximum value of the B.M., S.F., etc., on the structure as a whole. In any structure or for any system of loading, it may be necessary to construct several influence lines in order to obtain the absolute maxima on the structure. By comparison with an influence line, a B.M. diagram gives the B.M. at all points for one position of a load.

Influence lines can be drawn in three ways, viz.:

1. For unit loads and unit spans.
2. For unit loads and factual spans.
3. For factual loads and factual spans.

Each of these methods will be illustrated with reference to a simply supported beam.

Simply Supported Beams

Figure 1 (b) and (c) show the B.M. and S.F. influence lines for the point P on a simply supported beam of span AB, drawn for unit load and unit span. Note that when an influence line is drawn for unit span, both a and b are less than unity.

A. Point Loads

1. Bending Moments:

(a) Influence line drawn for unit load and unit span.

If a unit load is placed at Q, Fig. 1, then the B.M. at P is given by the product of the load and the ordinate of the influence line at the point of application of the load, viz.: CQ_1,

i.e. B.M. at $P = 1 \times CQ_1$.

If a series of unit loads are placed at Q, R and S respectively, then the B.M. at P is given by the sum of the products of the loads and their appropriate ordinates, e.g.:

B.M. at $P = 1 \times CQ_1 + 1 \times DR_1 + 1 \times ES_1$.

This particular influence line, i.e. that for unit load and unit span, can readily be used for the case of any loads on any span, but, in this case, the appropriate ordinates must be multiplied by the numerical values of both load and span. Thus if a series of loads W_1, W_2 and W_3 are placed on a span of

156 INFLUENCE LINES

Fig. 1

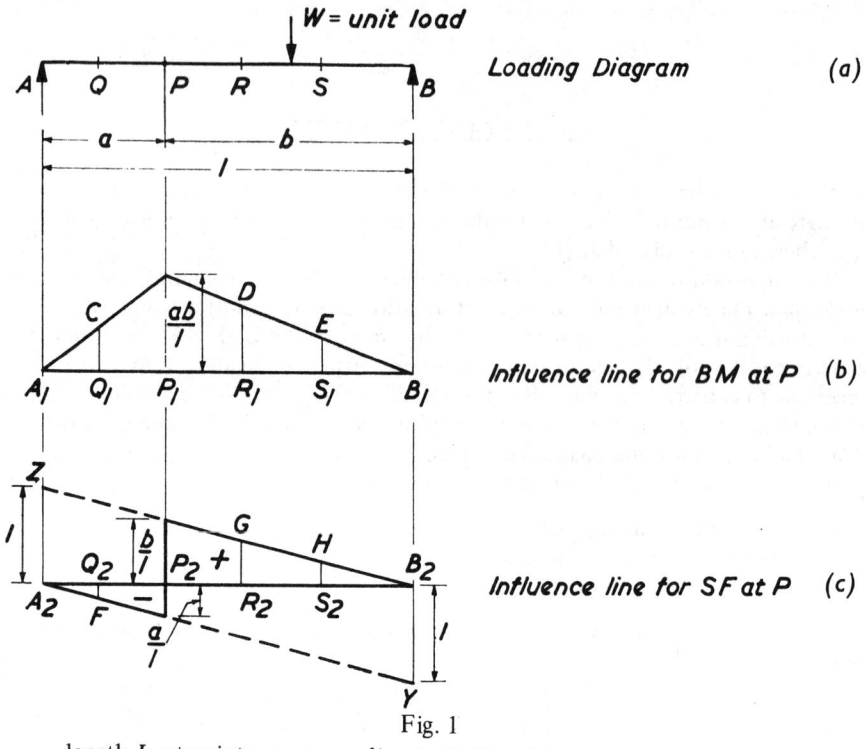

(a) Loading Diagram — W = unit load, points A, Q, P, R, S, B with distances a and b, total length l.

(b) Influence line for BM at P — triangle A₁–D–B₁ with peak $\frac{ab}{l}$ at D, points C, Q₁, P₁, R₁, S₁.

(c) Influence line for SF at P — with ordinates $\frac{b}{l}$ and $\frac{a}{l}$, points Z, A₂, Q₂, P₂, F, R₂, S₂, G, H, B₂, Y.

length L, at points corresponding to Q, R and S, then

B.M. at $P = W_1 \times CQ_1 \times L + W_2 \times DR_1 \times L + W_3 \times ES_1 \times L$
$= L(W_1 \times CQ_1 + W_2 \times DR_1 + W_3 \times ES_1)$.

(b) Influence line drawn for unit load and factual span.

In this case the ordinates will automatically be L times the values given in case 1 (a) as will be seen from the following example.

Suppose AP to be $\frac{1}{3} AB$.

Then in case 1 (a) the ordinate at P would be

$$\frac{\frac{1}{3} \times \frac{2}{3}}{1} = \frac{2}{9},$$

while in case 1 (b) if the span were 18 m the ordinate at P would be

$$\frac{6 \times 12}{18} = 4,$$

which is 18 times that in case 1 (a).

In this case, accordingly, if a series of unit loads are placed at Q, R and S respectively then

B.M. at $P = 1 \times CQ_1 + 1 \times DR_1 + 1 \times ES_1$,

and the values can be applied to the case of any loads by multiplying by the numerical values of the loads.

Thus, if loads W_1, W_2 and W_3 are placed at Q, R and S respectively,

B.M. at $P = W_1 \times CQ_1 + W_2 \times DR_1 + W_3 \times ES_1$.

SIMPLY SUPPORTED BEAMS 157

(c) Influence line drawn for factual load and factual span.

In this case the ordinates will automatically be WL times the values given in case 1 (a) and indeed will give directly the value of the B.M. It will be appreciated, however, that case 1 (c) can only be applied to a single load or a series of equal loads.

Summing up the three cases above, it will be seen that for any given loading and span, the B.M. is obtained as follows:

In case 1 (a) by multiplying the ordinates by the appropriate loads and by the span; in case 1 (b) by multiplying the ordinates by the appropriate loads only; and in case 1 (c) by direct measurement of the ordinates.

The three cases will be illustrated by the following example.

Example 1. Find, by the influence-line method, the B.M. at the left-hand quarter point of a span of 10 m when a single point load of 100 kN is placed at the other quarter point of the span.

The influence lines for cases 1 (a), (b) and (c) are shown in Fig. 2 (a), (b) and (c) respectively.

In Fig. 2 (a) the ordinate at P (the left-hand quarter point)

$$= \frac{0.25 \times 0.75}{1.0} = 0.1875,$$

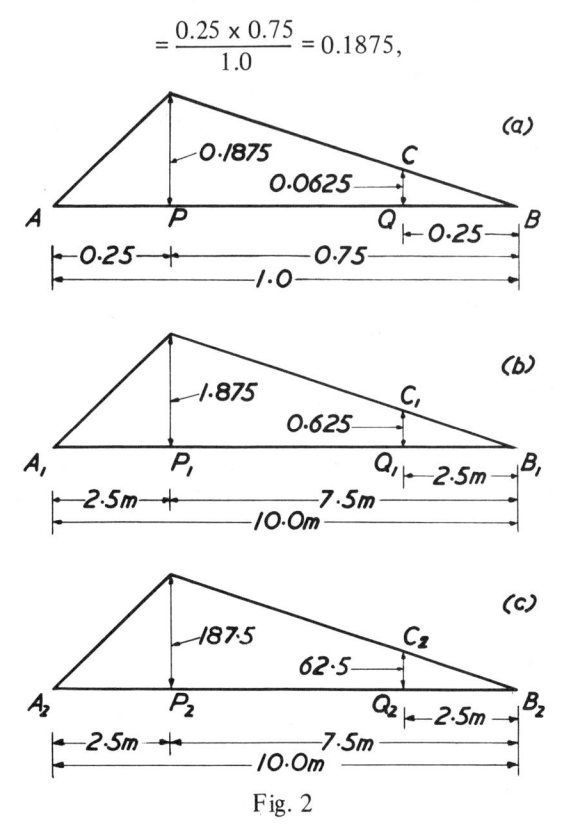

Fig. 2

and the corresponding ordinate at Q (the right-hand quarter point)

$$= \frac{0.1875}{3} = 0.0625 \quad \text{(which is merely a number)}.$$

Therefore, if a load of 100 kN is placed at Q and the span is 10 m

B.M. at P = 0.0625 × 100 kN × 10 m = 62.5 kNm

In Fig. 2 (b) the ordinate at P_1

$$= \frac{2.5 \text{ m} \times 7.5 \text{ m}}{10 \text{ m}} = 1.875 \text{ m}$$

and that at Q_1 = 0.625 m.

Therefore, B.M. at P_1 for the same conditions

= 0.625 × m × 100 kN = 62.5 kNm

In Fig. 2 (c) the ordinate at P_2 = 187.5 and that at Q_2 = 62.5 kNm. Therefore B.M. at P_2 for the same conditions = 62.5 kNm.

2. *Shear Forces:*

(a) Influence line drawn for unit load and unit span.

If a unit load is placed at Q, Fig. 1, then the S.F. at P is given by the product of the load and the appropriate ordinate FQ_2, i.e.:

S.F. at P = 1 × FQ_2.

If a series of unit loads are placed at Q, R and S respectively then

S.F. at P = 1 + FQ_2 + 1 × GR_2 + 1 × HS_2.

It will be noted that these quantities must be added *algebraically*, since FQ_2 is negative while the other two are positive.

If a series of loads W_1, W_2 and W_3 are placed at Q, R and S respectively then

S.F. at P = W_1 × FQ_2 + W_2 × GR_2 + W_3 × HS_2.

(b) Influence line drawn for unit load and factual span.

It will be seen from Fig. 1 (c) that, in the case of unit load, if the sloping lines are produced, intercepts of unity result at A and B.

Accordingly, there is no difference in application between cases 2 (a) and 2 (b). In other words, in the case of S.F. the span does not enter into consideration.

(c) Influence line drawn for factual load and factual span.

In this case, as in the case of B.M., the ordinates give directly the values required.

Once again, however, this case can only be applied to either a single load or a series of equal loads.

Summing up the three cases, it will be seen that for any given loading and span the S.F. is obtained as follows:

In cases 2 (a) and 2 (b), by multiplying the ordinates by the appropriate loads, and in case 2 (c) by direct measurement of the ordinates.

The three cases will be illustrated by the following example.

Example 2. Find, by the influence line method, the S.F. at the left-hand quarter point of a span of 10 m when a single point load of 100 kN is placed at the other quarter point of the span.

The influence lines for the three cases are shown in Figs. 3 (a), (b) and (c) respectively.

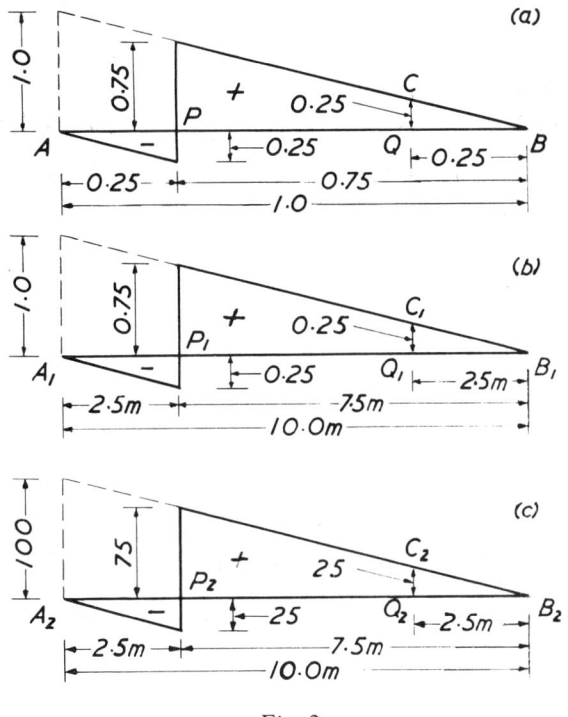

Fig. 3

In Fig. 3 (a) the appropriate ordinate at $P = 0.75$ and the corresponding ordinate at $Q = 0.25$ (a mere number).

In Fig. 3 (b) it will be seen that, since the intercept of the sloping line at A_1 is 1.0, i.e. exactly as in Fig. 3 (a), the ordinate at Q is also 0.25.

Therefore, irrespective of span, if a 100 kN load is placed at Q,

S.F. at $P = 0.25 \times 100$ kN $= 25$ kN

In Fig. 3 (c) the intercept at A_2 is 100 kN, i.e. the value of the load. Consequently the ordinate at $P_2 = 75$ and that at $Q_2 = 25$ kN.

Therefore, when 100 kN is placed at Q, the S.F. at $P = 25$ kN.

The difference in application between B.M. and S.F. will now be clear, i.e. that in the case of B.M., the span plays an important part, while in the case of S.F., it plays no part at all.

B. *Uniformly Distributed Loads Shorter than the Span*

If a U.D. load is considered to be made up of a large number of small point loads at very small distances apart, it will be appreciated that the ordinates of the point-

160   INFLUENCE LINES

load examples become a series of ordinates very close together, and in the limit the ordinates add up to an area.

1. *Bending Moments:*

   (a) Influence line drawn for unit intensity of load and unit span.

   If a U.D. load of unit intensity is placed over the length RS, Fig. 1 (a), then the B.M. at P will be given by the area R_1DES_1.

   This particular influence line can readily be used for the case of any load on any span, by multiplying the appropriate area by the actual loading intensity and by the square of the span. Thus, if a U.D. load of w per unit length is placed over the portion RS of a span of L units then the B.M. at P = area $R_1DES_1 \times wL^2$.

   It will be noticed that there is a difference between the multipliers in the cases of point and distributed loads. In the former, since the loads are the actual values, the unit ordinates are multiplied by the values of the load and the span, but in the latter, since w is only a load per unit length, the actual load is given in terms of wL and this quantity has still to be multiplied by L, thus giving the multiplier wL^2.

   (b) Influence line drawn for unit intensity of load and factual span.

   In this case the areas will automatically be L^2 times the values given in case 1 (a).

   If a U.D. load of unit intensity is placed over the portion RS then the B.M. at P will be given by the area R_1DES_1, while if the intensity of the loading is w per unit length, then the B.M. at P would be area $R_1DES_1 \times w$.

   (c) Influence line drawn for factual load and factual span.

   In this case the B.M. is given directly by the appropriate area.

The three cases will be illustrated by the following example.

Example 3. A U.D. load of 200 kN/m stands over a length of 2.5 m measured from the right-hand end of a span of 10.0 m. Find, by the influence-line method, the B.M. at the left-hand quarter point of the span.

The influence lines for the three cases are shown in Fig. 4 (a), (b) and (c) respectively.

In Fig. 4 (a) the ordinate at P, as previously shown, is 0.1875, while that at Q is 0.0625.

Therefore, the area under the B.M. influence line over the loaded length QB

$$= \frac{0.0625 \times 0.25}{2} = 0.0078125 \quad \text{(a mere number)}.$$

As the span is 10 m and the intensity of loading is 200 kN/m run over the length QB, then

B.M. at $P = 0.0078125 \times 200 \text{ kN/m} \times (10 \text{ m})^2 = 156.25$ kNm

In Fig. 4 (b) the ordinate at P_1 is 1.875 m, while that at $Q_1 = 0.625$ m.

Therefore, the area under the B.M. influence line over the loaded length QB

$$= \frac{0.625 \times 2.5}{2} = 0.78125 \text{ m}^2$$

SIMPLY SUPPORTED BEAMS   161

and with the intensity of loading of 200 kN/m run, the B.M. at P, the loaded length being QB

= 0.78125 m² × 200 kN/m = 156.25 kNm

Fig. 4

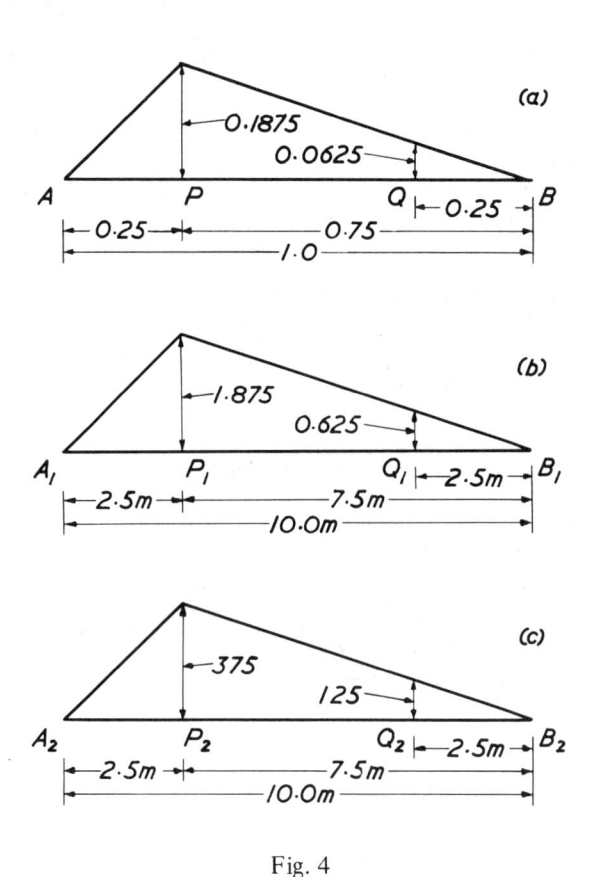

In Fig. 4 (c) the ordinate at P_2 is 375 kN and is found by multiplying the ordinate in case (b) by the intensity of loading. Consequently the ordinate at Q_2 is 125 kN, and the area over the loaded length QB

$$= \frac{125 \times 2.5}{2} = 156.25 \text{ kNm}$$

Therefore the B.M. at P due to a load of 200 kN/m over the length QB = 156.25 kNm.

2. *Shear Forces:*

(a) Influence line drawn for unit intensity of load and unit span.

If a load of unit intensity is placed over the length RS, Fig. 1 (a), then the S.F. at P will be given by the area R_2GHS_2.

162 INFLUENCE LINES

This particular influence line can readily be used for the case of any intensity of loading on any span, by multiplying the appropriate area by the actual loading intensity and by the span.

Thus if a U.D. load of w per unit length is placed over the portion RS of a span of L units, then the S.F. at P = area $R_2GHS_2 \times wL$.

Once again it will be seen that there is a difference between the applications depending on whether the loading is concentrated or U.D.

(b) Influence line drawn for unit intensity of load and factual span.

In this case, if the loading intensity is unity, the S.F. is again given by the area under the S.F. influence line standing over the loaded length.

In applying this case to any intensity of loading, it is only necessary to multiply the appropriate area by the intensity of loading w.

(c) Influence line drawn for factual load and factual span.

Once again, in this case, the S.F. is given directly by the appropriate area.

The three cases will be illustrated by the following example.

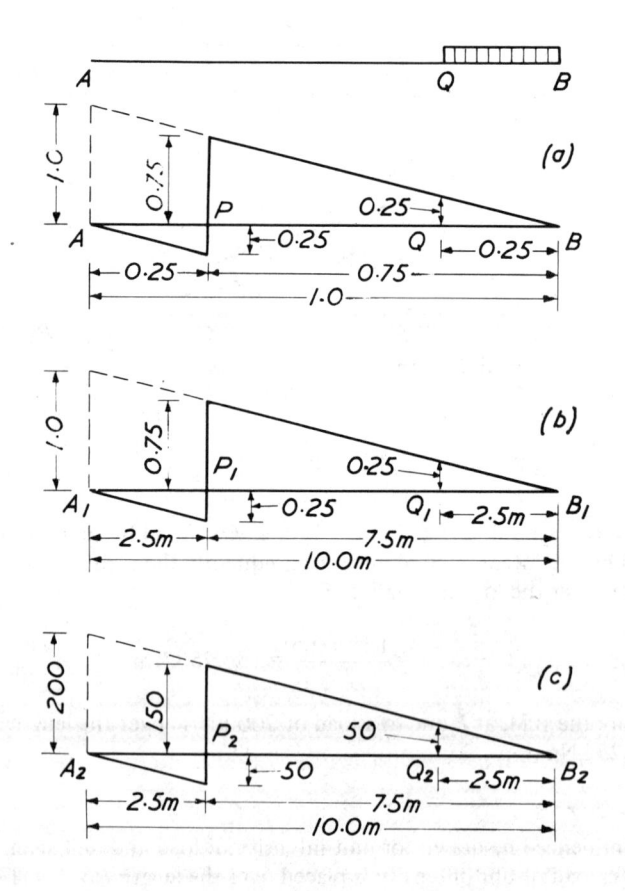

Fig. 5

SIMPLY SUPPORTED BEAMS

Example 4. For the same span and conditions of loading given in Example 3, find, by the influence-line method, the S.F. at the left-hand quarter point of the span.

The influence lines for the three cases are shown in Fig. 5 (a), (b) and (c) respectively.

In Fig. 5 (a) the ordinate at P is 0.75, while that at Q is 0.25. Therefore the area under the S.F. influence line over the loaded length QB

$$= \frac{0.25 \times 0.25}{2} = 0.03125 \quad \text{(a mere number)}.$$

As the span is 10 m, and the intensity of loading is 200 kN/m run over the length QB, then the S.F. at P

$$= 0.03125 \times 200 \text{ kN/m} \times 10 \text{ m} = 62.5 \text{ kN}.$$

In Fig. 5 (b) the ordinates are exactly as those in Fig. 5 (a). Therefore the area under the S.F. influence line over the loaded length QB

$$= \frac{0.25 \times 2.5}{2} = 0.3125 \text{ m}$$

With the intensity of loading of 200 kN/m run over the length QB, then the S.F. at P

$$= 0.3125 \text{ m} \times 200 \text{ kN/m} = 62.5 \text{ kN}.$$

In Fig. 5 (c) the ordinate at P is 150 kN/m, while that at Q is 50 kN/m. Therefore the area under the S.F. influence line over the loaded length QB

$$= \frac{50 \times 2.5}{2} = 62.5 \text{ kN}$$

Therefore the S.F. at P due to this loading = 62.5 kN.

In most practical cases, it is required to find the maximum B.M. or S.F. at any point due to the loading system.

In the case under consideration, i.e. that of a loaded length shorter than the span, the maximum B.M. at any point such as P in Fig. 1 (a) will be given when the point P divides the loaded length in the same proportions as it divides the span.

For S.F., two maxima are possible. The maximum positive (+ ve) S.F. at P is given when, if the load comes on from the right, the leading edge of the load had just reached P or if the load comes on from the left, the trailing edge of the load has just reached P, while the maximum negative (− ve) S.F. at P is given when, if the load comes on from the right, the trailing edge of the loading has just reached P, or if the load comes on from the left, the leading edge of the load has just reached P. (See Fig. 1 (c).)

C. Uniformly Distributed Loads Longer than the Span

This type of loading could be illustrated by a long train passing over a short span.

The methods of dealing with both B.M. and S.F. are exactly as given under the last heading, i.e. that for U.D. shorter than the span.

It only remains to illustrate the positions for maximum values.

In the case of B.M. it is obvious, by reference to Fig. 1 (b), that the maximum B.M. at P is given when the whole span is covered by the load.

Again in the case of S.F. there are two maxima and these are given when the load is in the positions previously indicated.

The use of influence lines is most helpful in the cases of continuous beams, arches and framed girders, when the values for B.M. and S.F., etc., for varying positions of loads can be much more rapidly evaluated than by any other means.

Influence Lines for Framed Structures

Figures 6–10 show typical influence lines for direct forces in members in Warren and Pratt or N-type frames.

The main points to be noted are detailed below:

1. The load has been assumed to be transmitted through deck stringers and cross beams so that load is applied to the main girders only at the panel points.

2. Considering Fig. 6 if the force in member QR is required, it could be obtained, by the method of sections, by taking moments about the point D. Consequently the influence line for the force in QR is, to some scale, the influence line for B.M. for the point D, with this difference: that, between Q and R the influence line is a straight line as shown. This truncating of the diagram is peculiar to the Warren-type girder and applies only to the loaded chord. Also, see the influence line for CD in Fig. 7. Similarly the influence line for CD is, to some scale, the influence line for B.M. for the point Q.

3. Considering Fig. 8 the influence line for the member QR is, to some scale, the influence line for B.M. for the point D. Here it should be noted that the influence line for member DE is the influence line for B.M. for the point R and is, evidently, the same as that for QR. This is substantiated by the fact that, for this type of frame, the forces in QR and DE are equal.

4. The influence lines for web members are influence lines for shear and are similar to that shown in Fig. 1 (c) with the difference that the line is drawn diagonally across the panel under consideration. It should be noted, that, in the case of the end diagonal in Fig. 6, the slope of the line $P_1 X_1$ depends on the position of X, i.e. the length of the approach span PX. X_1 is vertically below X. It should also be noted that, in this case, and indeed for all cases dealing with web members in the end panel, the shear-force influence line does not cross the base line, i.e. the shear force in the panel at the left-hand end of the girder is always positive. Note also the special case of the end vertical in Fig. 8.

5. If the influence line for any web member, e.g. the influence line for members QD and RD in Fig. 8, crosses the base line, then it is possible to get both positive and negative shear in the panel QR. This means that the member QD will, at some stage, be in compression. If the length of the member is such that it would be uneconomical to make it capable of resisting compression, then cross bracing is introduced. This cross bracing, usually comprised of members capable only of taking tension, introduces one or two unusual features in the influence lines for the members affected. For example, if, in Fig. 8, the panel $RSED$ is cross-braced and the members RE and SD are capable only of taking tension, then the influence line for SD will be as shown, while, at the same time, the influence line for RD will have the unusual form shown, since it will be possible for this member RD to take the whole load at the time when the load is at R.

INFLUENCE LINES FOR FRAMED STRUCTURES 165

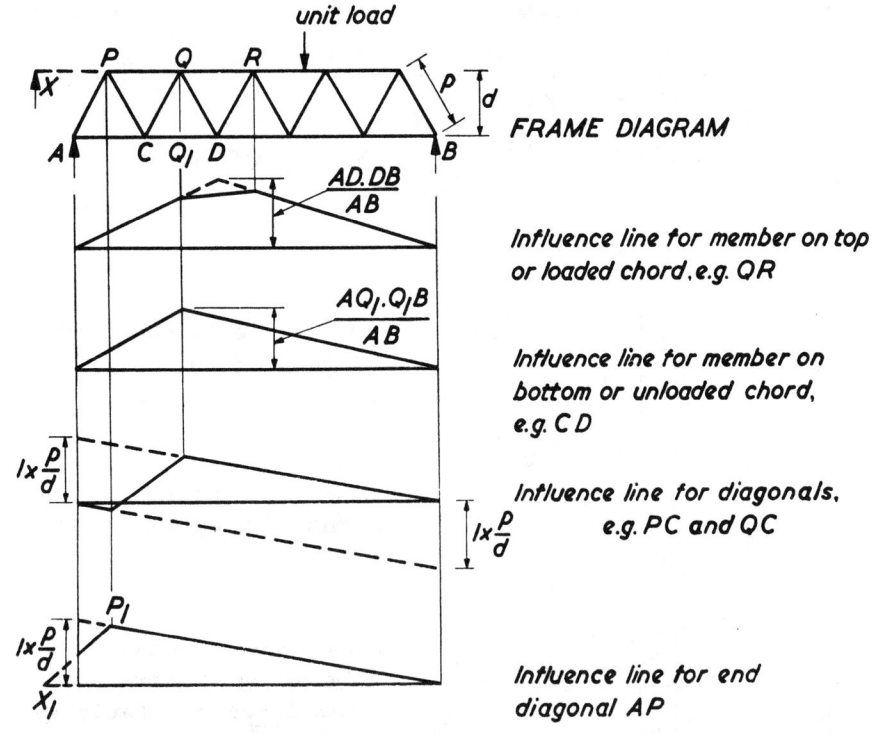

Fig. 6.—Warren girder, load on top chord

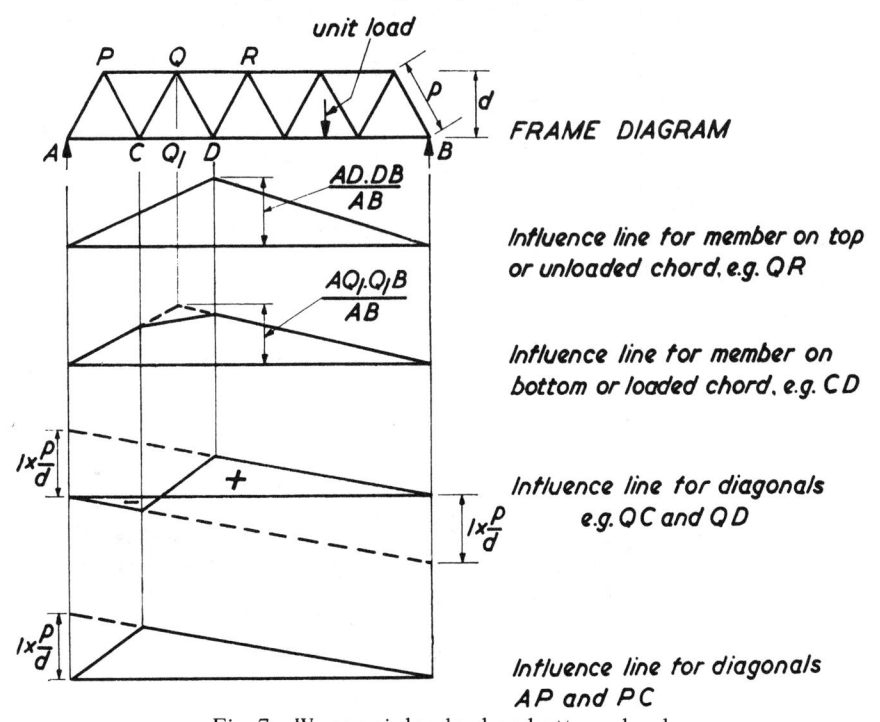

Fig. 7.—Warren girder, load on bottom chord

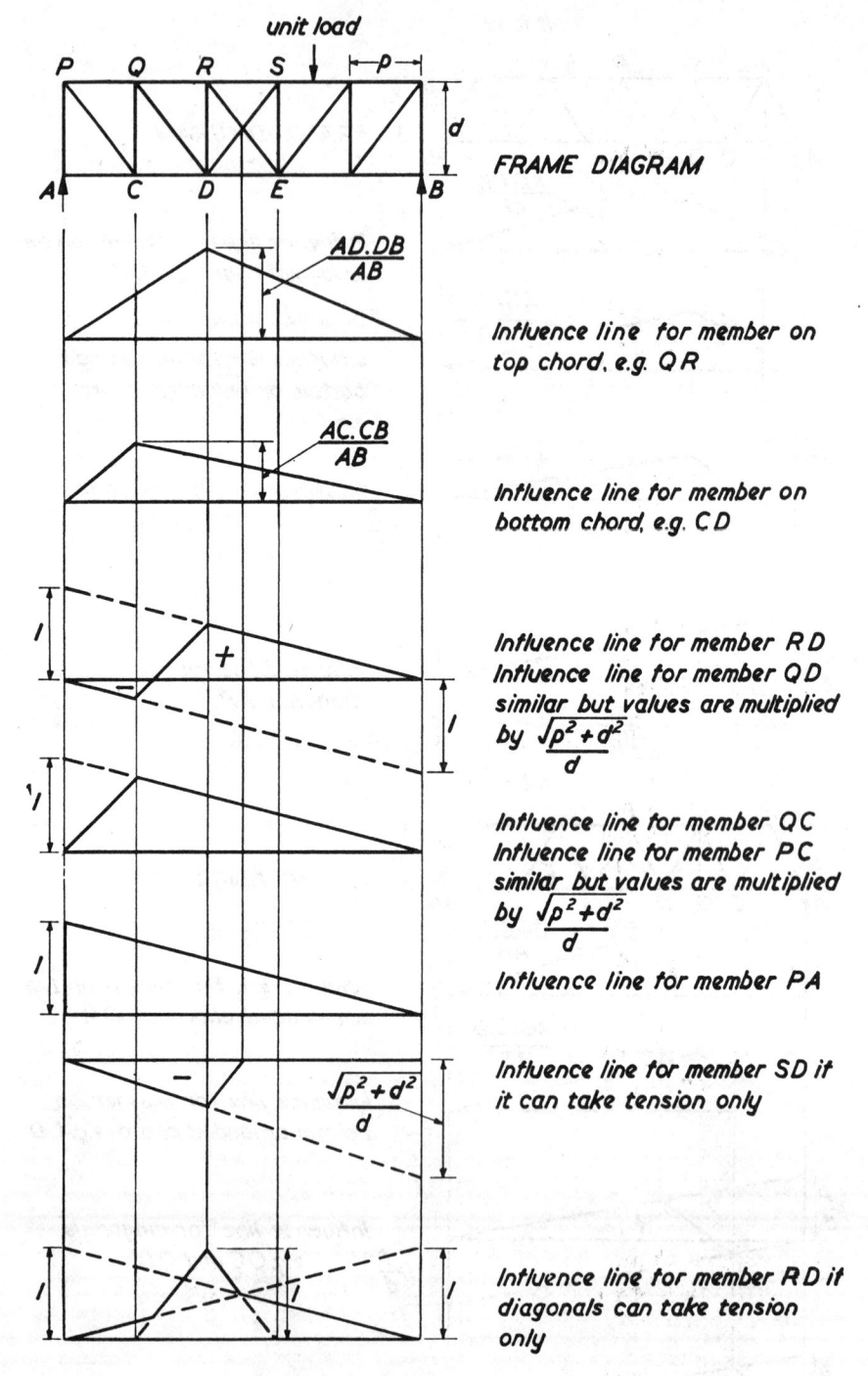

Fig. 8.—Pratt or N-truss, load on top chord

Influence line for member on top chord, e.g. QR

Influence line for member on bottom chord, e.g. CD

Influence line for member QC
Influence line for member QD
similar but values are multiplied by $\dfrac{\sqrt{p^2+d^2}}{d}$

Influence line for member PA
Influence line for member PC
similar but values are multiplied by $\dfrac{\sqrt{p^2+d^2}}{d}$

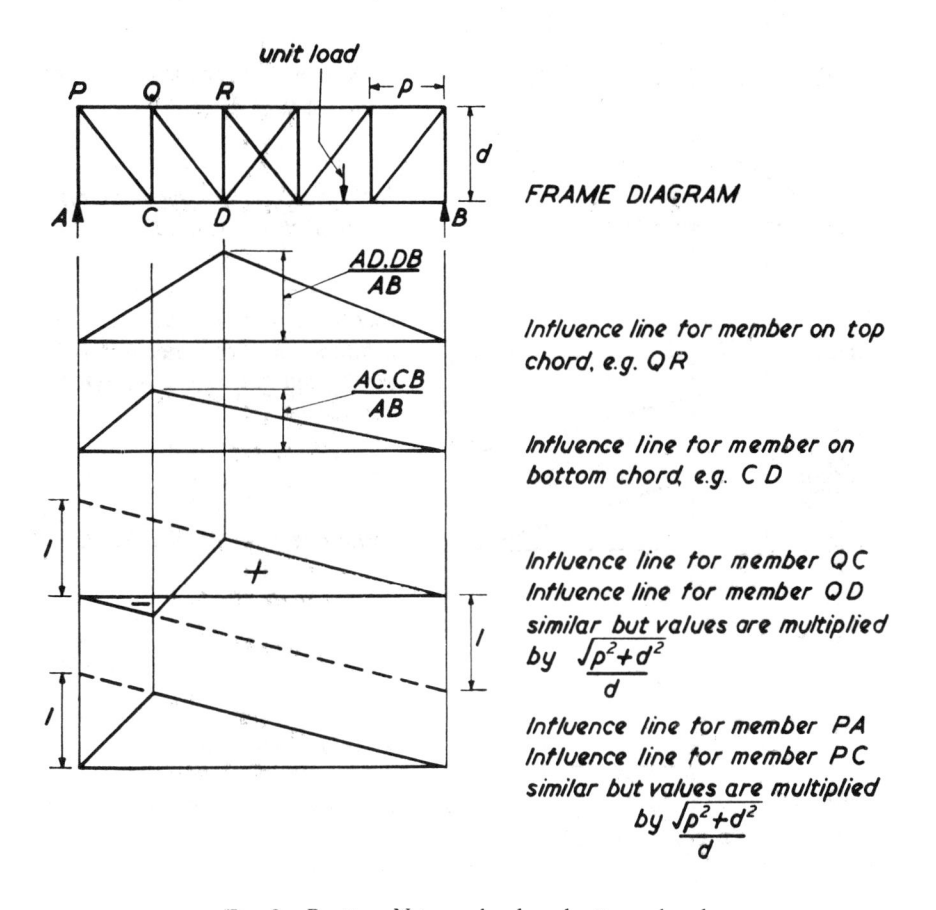

Fig. 9.—Pratt or N-truss, load on bottom chord

168   INFLUENCE LINES

FRAME DIAGRAM
LOAD ON TOP CHORD

Influence line for member on top chord, e.g. QR — $\dfrac{AC \cdot CB}{AB}$

Influence line for member on bottom chord, e.g. CD — $\dfrac{AD \cdot DB}{AB}$

Influence line for member QC
Influence line for member RC
similar but values are multiplied by $\dfrac{\sqrt{p^2+d^2}}{d}$

Influence line for member QA — $1 \times \dfrac{\sqrt{p^2+d^2}}{d}$

Influence line for member PA

FRAME DIAGRAM
LOAD ON BOTTOM CHORD
Influence lines for members on top and bottom chords are exactly as above

Influence line for member RD
Influence line for member RC
similar but values are multiplied by $\dfrac{\sqrt{p^2+d^2}}{d}$

Influence line for member QC
Influence line for member QA
similar but values are multiplied by $\dfrac{\sqrt{p^2+d^2}}{d}$

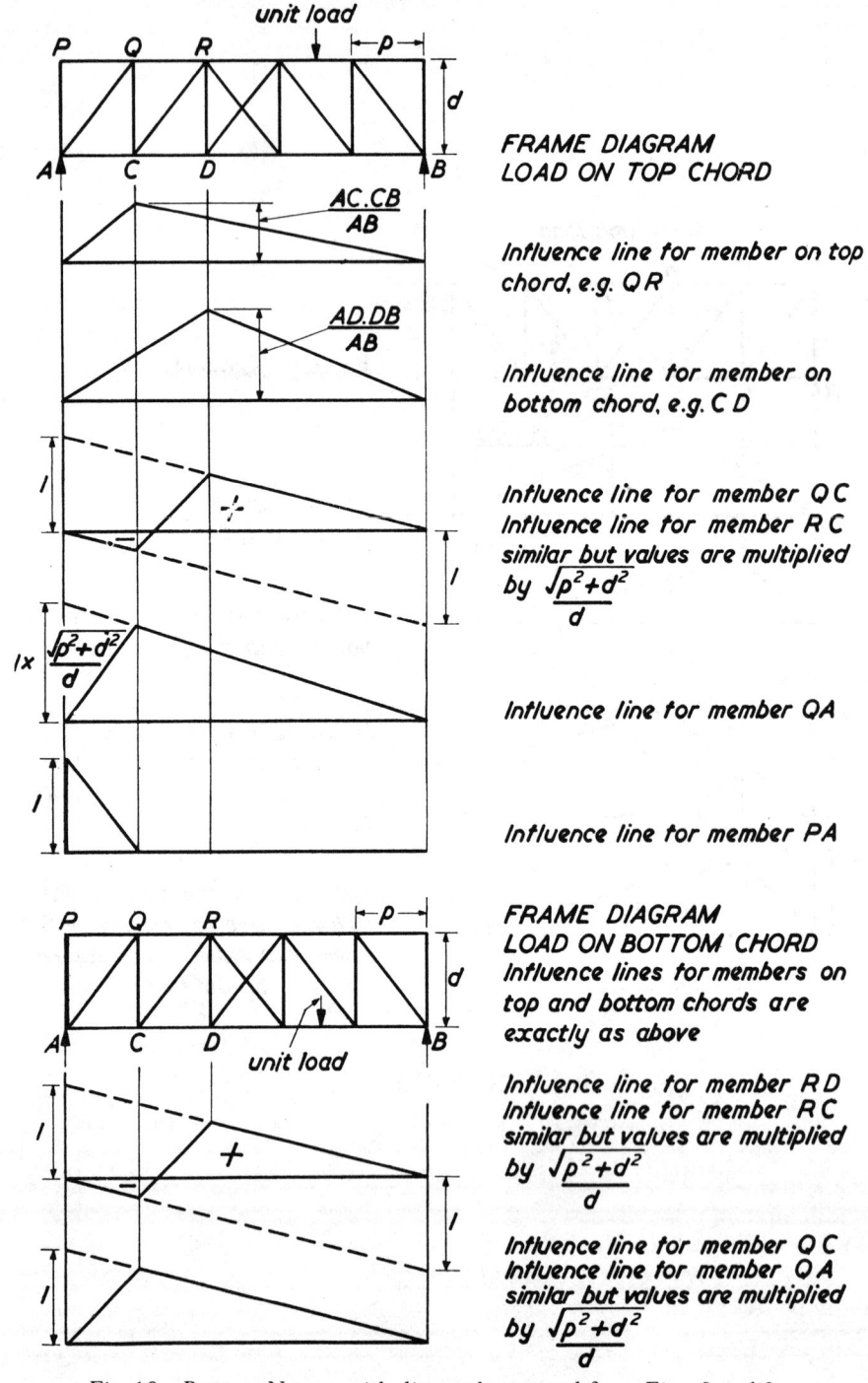

Fig. 10.—Pratt or N-truss with diagonals reversed from Figs. 8 and 9. Load on either top or bottom chord.

INFLUENCE LINES FOR FRAMED STRUCTURES

It will be clear from the foregoing diagrams that for a single point load the maximum B.M. and maximum S.F., both + ve and − ve, are given when the load is placed at clearly defined points.

For a U.D. load longer than the span, it will be clear that for maximum B.M. the whole span must be covered by the load, while for maximum S.F., the load should come up to the point X in Fig. 11.

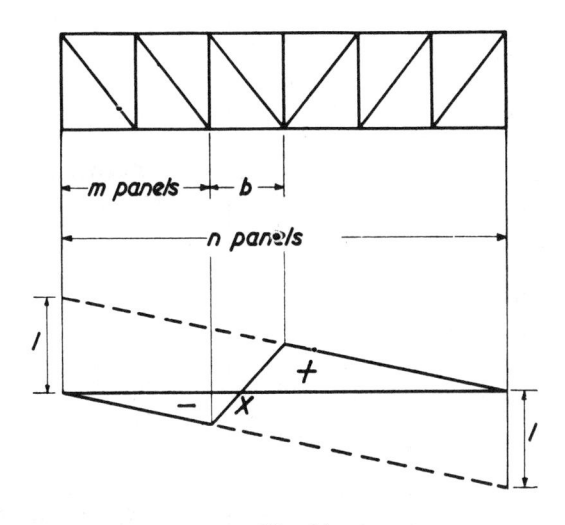

Fig. 11

The point X is found by the following means:

If there are n panels, each of length b, in a span, then the maximum S.F. in the $(m + 1)$ panel occurs when $\frac{m}{n-1}$ of that panel is covered by the load, in other words, the point X divides the panel as it divides the span.

Figures 12 to 16 show influence lines for some of the more uncommon types of frame.

Figure 12 shows a hog-backed girder.

Consider a unit load moving along the lower chord.

If it is required to draw the influence lines for the members PQ, QC and CD, i.e. those members cut by the line EE, then it should be obvious that the influence line for PQ is similar to that for member QR in Fig. 7, except that the scale is increased in proportion to the ratio of the length of QQ_1, i.e. the vertical height of Q, to the length SC, i.e. the normal from C to PQ, while that for CD is similar to that for member CD in Fig. 7.

Regarding member QC, the force in this member is a measure of the moment of all the external forces to the left of the section taken about the point X, this point being where QP produced and BA produced meet, i.e. Force in $QC = M_x/y$.

The influence line for QC is constructed as follows.

Set up from A_1 a distance x equal to the distance XA. Join B_1 to the point thus found and produce to meet the vertical through X in X_1. The influence line between

170   INFLUENCE LINES

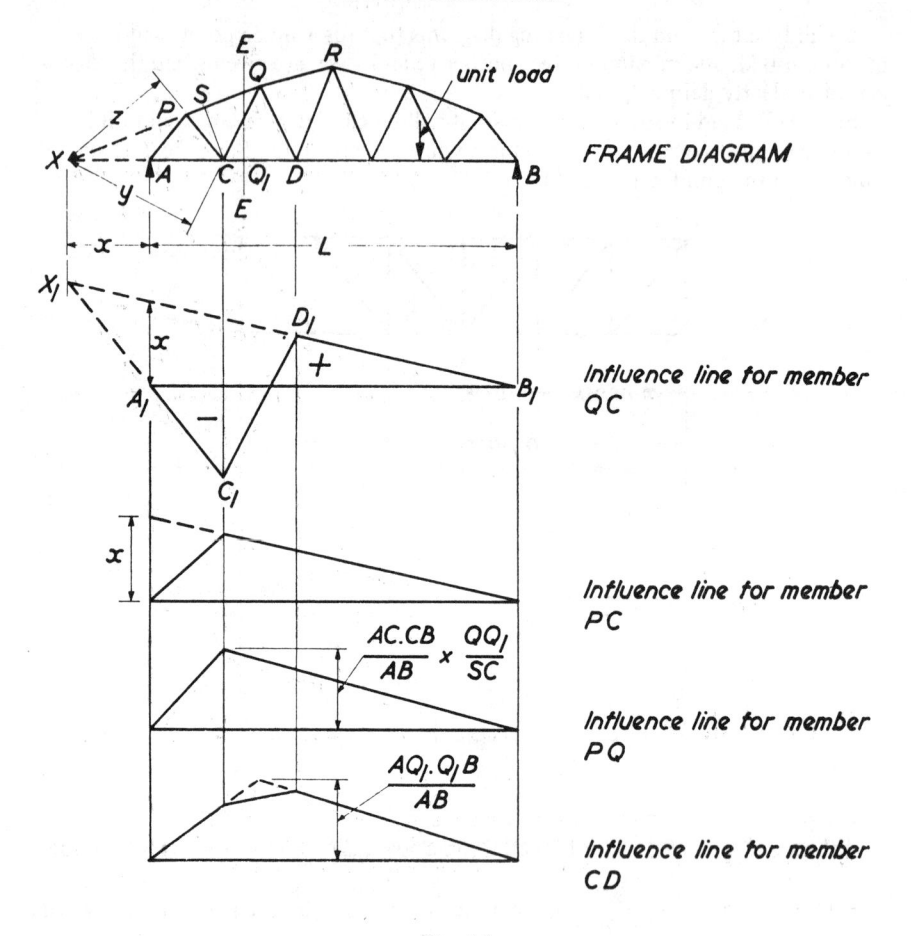

Fig. 12

B and D lies on this line, i.e. B_1D_1. Now join X_1A_1 and produce to C_1 which is vertically below C. Join C_1D_1. Then $A_1C_1D_1B_1$ is the complete influence line.

The force in member PC is also a measure of the moment about X, and consequently the influence line for PC is constructed in the same manner as that for QC, but, in this instance, the influence line does not cross the base line.

Figure 13 shows an N-truss with intermediate bracing. This type of bracing is used in long span trusses in order to avoid excessive panel lengths. In the example shown the load is on the top chord, in which case the intermediate bracing is needed only to reduce the unsupported lengths of the top chord in order to minimise or obviate local bending effects.

If the load is on the bottom chord then the bracing should be of the type shown in Fig. 14.

The influence lines for members on the bottom chord are exactly the same as shown in Fig. 8, but those for members in the top chord are affected by the intermediate bracing. For example, the influence line for members PQ and QR is

INFLUENCE LINES FOR FRAMED STRUCTURES

FRAME DIAGRAM

Influence line for members PQ and QR: $\dfrac{AC \cdot CB}{AB}$

Influence line for members RT and TU: $\dfrac{AD \cdot DB}{AB}$

Influence line for member CD: $\dfrac{AC \cdot CB}{AB}$

Influence line for member PS: $\dfrac{\sqrt{p^2+d^2}}{d}$

Influence line for member QS
Influence line for member RS
similar but values are multiplied by $\dfrac{\sqrt{p^2+d^2}}{2d}$

Influence line for member TV
Influence line for member UV
similar but values are multiplied by $\dfrac{\sqrt{p^2+d^2}}{2d}$

Influence line for member RC
Influence line for member SC
similar but values are multiplied by $\dfrac{\sqrt{p^2+d^2}}{d}$

Influence line for member RV

Influence line for member UD
Influence line for member VD
similar but values are multiplied by $\dfrac{\sqrt{p^2+d^2}}{d}$

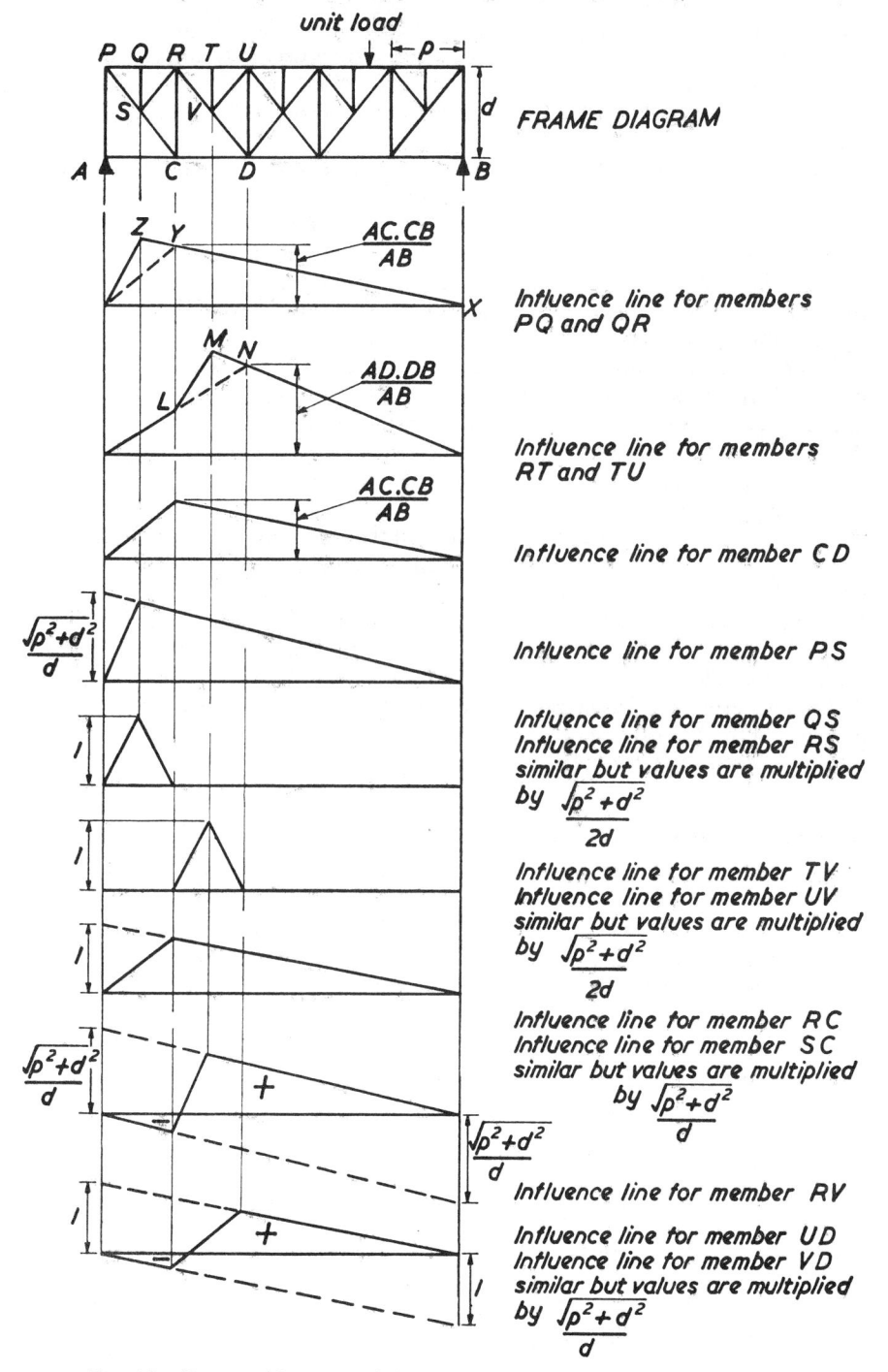

Fig. 13.—Pratt or N-truss with intermediate bracing. Load on top chord

similar to that for the corresponding member *PQ* in Fig. 8, i.e. that for the point *C*, except that in the panel *PR* a small triangle is superimposed on to the basic influence line. This triangle is formed by continuing the line *XY* until it reaches a point *Z* which is vertically under the point *Q* and then completing the influence line as shown.

Similarly the influence line for members *RT* and *TU* is the basic influence line for the point *D* with a similar triangle *LMN* superimposed upon it.

The influence lines for the main web members are of the usual form, except that the main diagonals, e.g. *RD*, have different influence lines for each subdivided part, while the influence lines for the intermediate members, e.g. *TV* and *UV*, have a special form, since no forces are developed in these members except when the load is in the appropriate panel, i.e. *RU*.

Figures 15 and 16 show the influence lines for lattice girders with K-type bracing. In general, the influence lines follow the usual pattern except that as far as the web members are concerned, any pair of diagonals, e.g. *EQ* and *EC*, share the shear in the panel equally and the actual forces in these members depend on the ratio of their lengths to the half-depth of the truss.

For the case of loading on the top chord, member *QF* and all other members which form the upper portions of the verticals will have a special form of influence line as shown in Fig. 15.

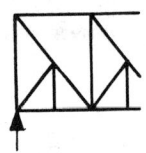

Fig. 14

Conversely, when the load is on the bottom chord, member *FC* and all other members which form the lower portions of the verticals will have the special form shown in Fig. 16.

As previously noted in the case of the N-girder, the influence lines for the web members in the end panel do not cross the base line, but in the cases of all other web members, reversals do take place.

Influence Lines for Built-in Beams

Figure 17 (a) shows the influence lines for the negative B.M.s M_A and M_B at the supports *A* and *B*.

The use of the influence lines in this case is illustrated by the example shown in Fig. 17 (b) and (c).

From the influence lines the fixing moment at *A* due to W_1

$$= 0.128 \, W_1 L = 0.128 \times 50 \times 10 = 64.0 \text{ kNm}$$

that due to W_2

$$= 0.144 \, W_2 L = 0.144 \times 60 \times 10 = 86.4 \text{ kNm}$$

and that due to W_3

$$= 0.063 \, W_3 L = 0.063 \times 40 \times 10 = 25.2 \text{ kNm}$$

Therefore the total fixing moment at *A*

$$= 64.0 \times 86.4 + 25.2 = 175.6 \text{ kNm}$$

Fixing moment at *B* due to W_1

$$= 0.032 \, W_1 L = 0.032 \times 50 \times 10 = 16.0 \text{ kNm}$$

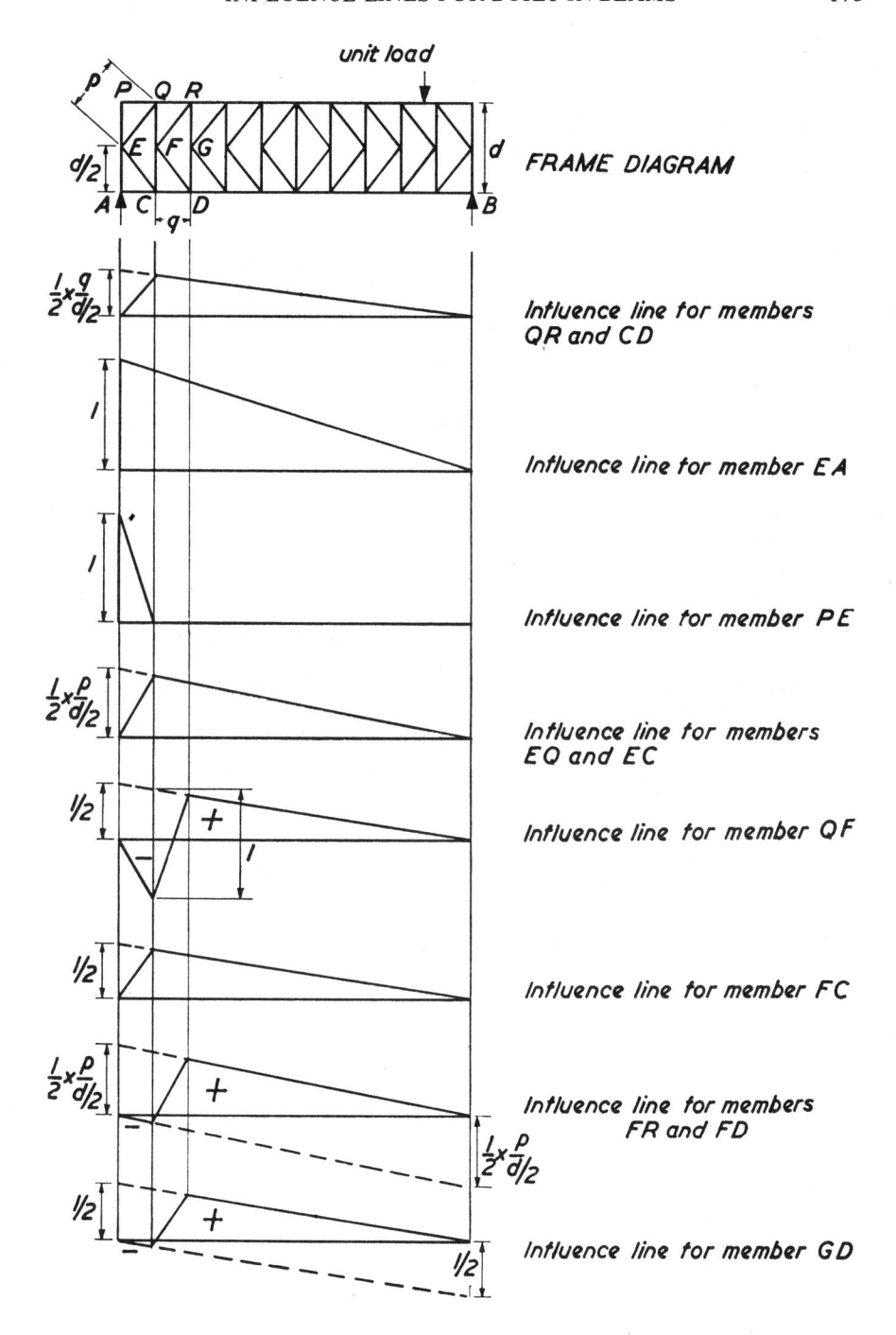

Fig. 15.—Lattice girder with K bracing. Load on top chord.

174 INFLUENCE LINES

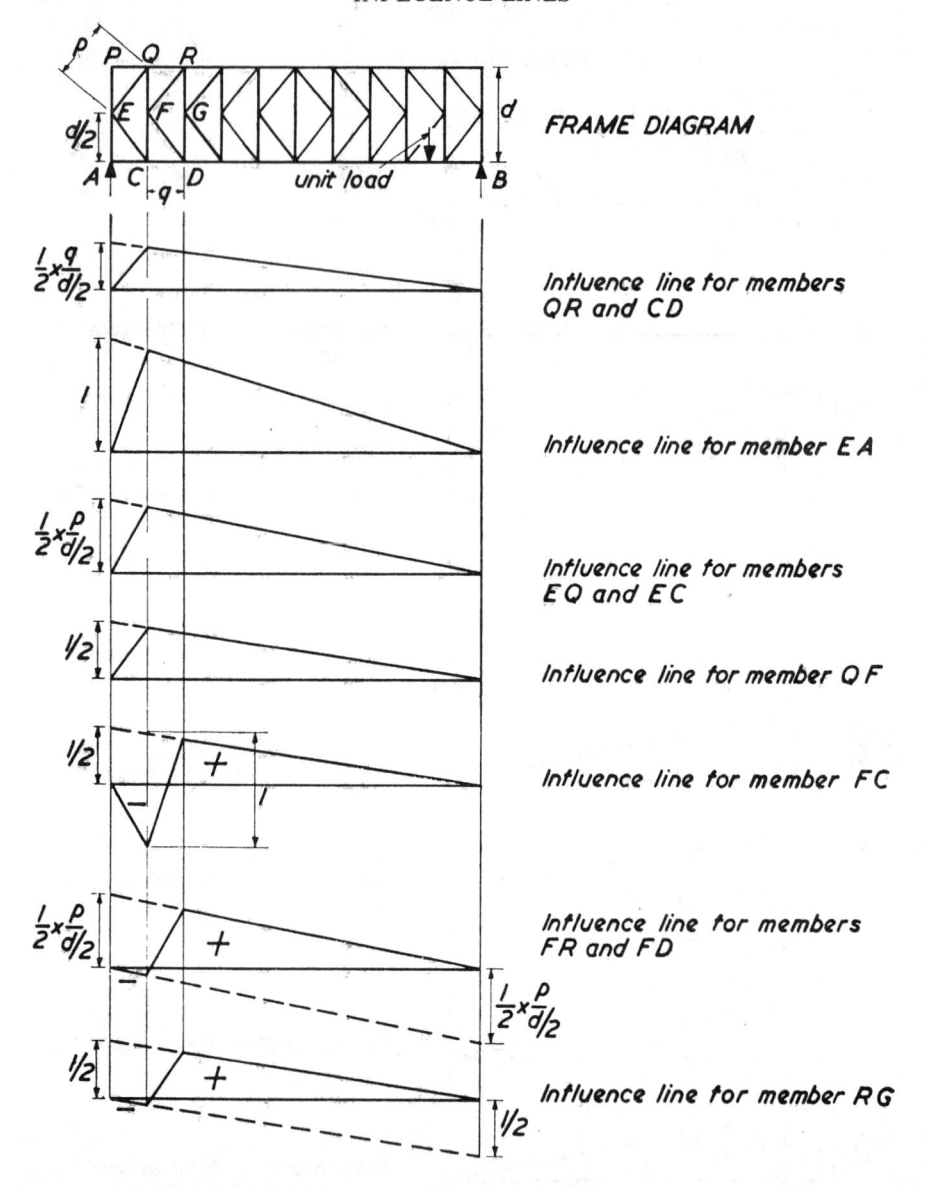

Fig. 16.—Lattice girder with K bracing. Load on bottom chord.

that due to W_2
$$= 0.096\ W_2 L = 0.096 \times 60 \times 10 = 57.6 \text{ kNm}$$
and that due to W_3
$$= 0.147\ W_3 L = 0.147 \times 40 \times 10 = 58.8 \text{ kNm}$$
Therefore the total fixing moment at B
$$= 16.0 + 57.6 + 58.8 = 132.4 \text{ kNm}$$

INFLUENCE LINES FOR BUILT-IN BEAMS 175

FIXED ENDED BEAM

(a) Influence lines for negative moments M_A and M_B at supports A and B

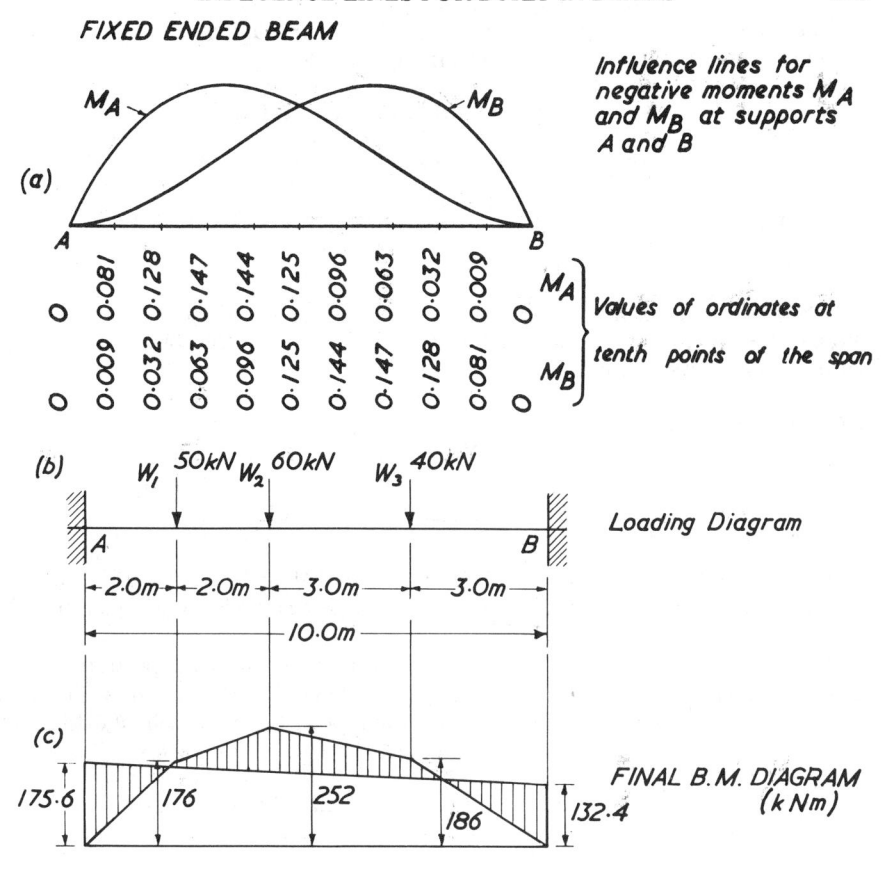

Values of ordinates at tenth points of the span:

M_A: 0, 0.009, 0.081, 0.128, 0.147, 0.144, 0.125, 0.096, 0.063, 0.032, 0.009, 0

M_B: 0, 0.009, 0.032, 0.063, 0.096, 0.125, 0.144, 0.147, 0.128, 0.081, 0

(b) Loading Diagram: W_1 50kN, W_2 60kN, W_3 40kN
Spans: 2.0m, 2.0m, 3.0m, 3.0m (total 10.0m)

(c) FINAL B.M. DIAGRAM (kNm): 175.6, 176, 252, 186, 132.4

Fig. 17

Influence lines for reactions at supports A and B

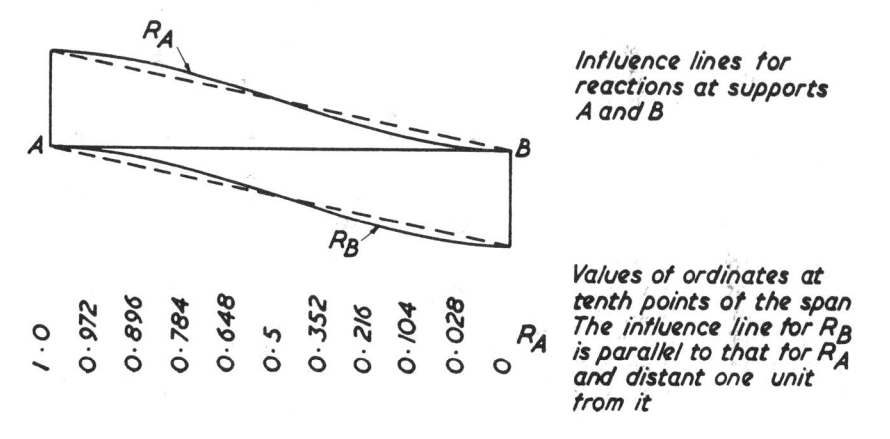

Values of ordinates at tenth points of the span. The influence line for R_B is parallel to that for R_A and distant one unit from it.

R_A: 1.0, 0.972, 0.896, 0.784, 0.648, 0.5, 0.352, 0.216, 0.104, 0.028, 0

Fig. 18

176   INFLUENCE LINES

The results will be checked by the principle of reciprocal moments which was mentioned in the chapter on Fixed Beams. See page 40.

$$M_A = \left(\frac{50 \times 2 \times 8}{10} \times \frac{8}{10}\right) + \left(\frac{60 \times 4 \times 6}{10} \times \frac{6}{10}\right) + \left(\frac{40 \times 7 \times 3}{10} \times \frac{3}{10}\right)$$

$$= 64.0 + 86.4 + 25.2 = 175.6 \text{ kNm}$$

$$M_B = \left(\frac{50 \times 2 \times 8}{10} \times \frac{2}{10}\right) + \left(\frac{60 \times 4 \times 6}{10} \times \frac{4}{10}\right) + \left(\frac{40 \times 7 \times 3}{10} \times \frac{7}{10}\right)$$

$$= 16.0 + 57.6 + 58.8 = 132.4 \text{ kNm}$$

The final B.M. diagram is as shown in Fig. 17 (c). Figure 18 shows the influence lines for the reactions at A and B. The dotted lines represent the corresponding influence lines for a simply supported beam.

Continuous Beams

All influence lines must be constructed from first principles and the following example will illustrate the method of construction.

Bending Moments

It is required to construct the influence line for B.M. at D, the mid-point of the span AB which is one span of a two-span continuous beam ABC, both spans being equal. See Fig. 19.

In any particular case, it is convenient to consider separately the portions of the beam to either side of the point for which the influence line is being drawn, because in these separate portions, the formulae for the B.M. at the point will take different forms.

Portion A to D. (See Fig. 19 (a).) If a load W is placed at the point X, which divides the span AB into two parts a and $(L - a)$, then the

$$\text{free B.M. at } X = \frac{W \times a \times (L - a)}{L}.$$

If the load is unity,

$$\text{the B.M. at } X = \frac{a(L - a)}{L}.$$

The area of the free B.M. diagram on the span AB

$$= \frac{a(L - a)}{L} \times \frac{L}{2} = \frac{a(L - a)}{2},$$

and the C.G. of this area from A, measured horizontally,

$$= AX + \tfrac{2}{3}XD = a + \frac{2}{3}\left(\frac{L}{2} - a\right) = \frac{L + a}{3}.$$

Applying the Theorem of Three Moments to spans AB and BC, and bearing in mind that $M_A = M_C = 0$, that there is no load on the span BC and that the two spans are equal, then

$$2M_B \times 2L = 6 \cdot \frac{a(L - a)}{2} \times \frac{L + a}{3} \times \frac{1}{L}$$

whence

$$M_B = \frac{a(L^2 - a^2)}{4L^2}.$$

CONTINUOUS BEAMS

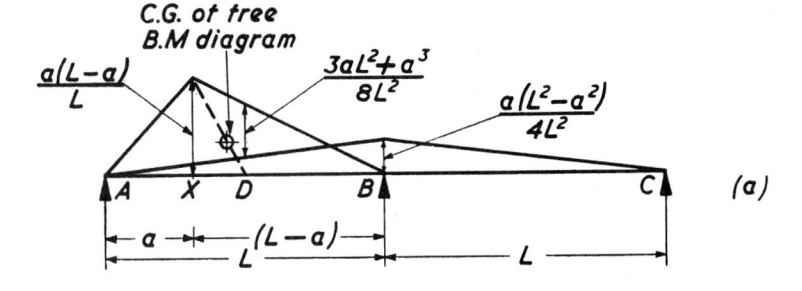

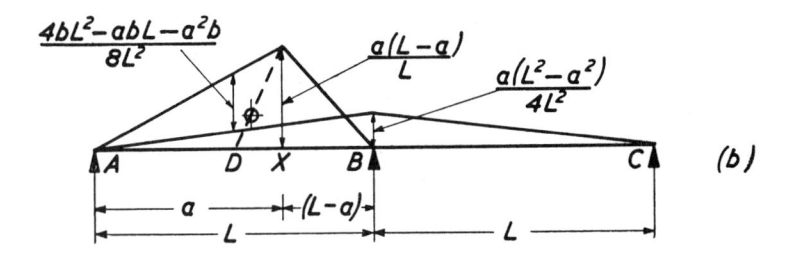

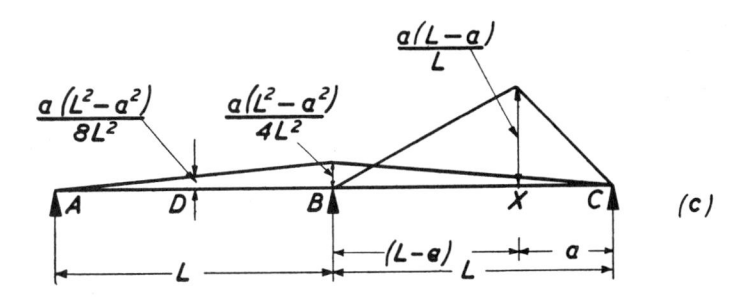

(figure with three bending moment diagrams labeled (a), (b), (c))

ALL THE B.M DIAGRAMS ARE FOR UNIT LOADS

Fig. 19

The height of the free B.M. diagram at D

$$= \frac{a(L-a)}{L} \times \frac{L/2}{L-a} = \frac{a}{2}$$

while the height of the negative B.M. diagram at D

$$= \frac{1}{2} \times \frac{a(L^2 - a^2)}{4L^2} = \frac{a(L^2 - a^2)}{8L^2},$$

therefore the net B.M. at $D = \dfrac{a}{2} - \dfrac{a(L^2 - a^2)}{8L^2} = \dfrac{3aL^2 + a^3}{8L^2}$,

and if the span is unity, this expression can be written:

$$\text{net B.M. at } D = \frac{3a + a^3}{8}.$$

From this expression the B.M. influence line for D from A to D can be built up by successive substitution of values for a. This is best done by means of a table as shown below:

a	$3a$	a^3	$3a + a^3$	$\dfrac{3a + a^3}{8}$
0.1	0.3	0.001	0.301	0.038
0.2	0.6	0.008	0.608	0.076
0.3	0.9	0.027	0.927	0.116
0.4	1.2	0.064	1.264	0.158
0.5	1.5	0.125	1.625	0.203

Portion D to B. (See Fig. 19 (b).) In this portion the horizontal distance from A to the C.G. of the free B.M. diagram is

$$\frac{L}{2} + \frac{1}{3}a - \frac{L}{2} = \frac{L+a}{3},$$

i.e. exactly as in the portion AD, and consequently

$$M_B = \frac{a(L^2 - a^2)}{4L^2} \quad \text{also as before.}$$

The height of the free B.M. diagram at D

$$= \frac{a(L - a)}{L} \times \frac{L/2}{a} = \frac{L - a}{2},$$

while the height of the negative B.M. diagram at D

$$= \frac{a(L^2 - a^2)}{8L^2} \quad \text{as before}$$

Therefore the net B.M. at D

$$= \frac{L - a}{2} - \frac{a(L^2 - a^2)}{8L^2} = \frac{4L^3 - 5aL^2 + a^3}{8L^2}$$

and again putting $L = 1$,

$$\text{net B.M. at } D = \frac{4 - 5a + a^3}{8}.$$

The corresponding table is given below:

a	$5a$	a^3	$4 - 5a + a^3$	$\dfrac{4 - 5a + a^3}{8}$
0.5	2.5	0.125	1.625	0.203
0.6	3.0	0.216	1.216	0.152
0.7	3.5	0.343	0.843	0.105
0.8	4.0	0.512	0.512	0.064
0.9	4.5	0.729	0.229	0.028
1.0	5.0	1.0	0	0

Portion B to C. (See Fig. 19 (c).) In this portion, if the distance a is measured from C, then the negative B.M. at B will be $[a(L^2 - a^2)]/4L^2$ for all values of a, and consequently the height of the B.M. diagram at D will be $[a(L^2 - a^2)]/8L^2$.

Putting $L = 1$, the B.M. at D will be $(a - a^3)/8$. Note that since there is no positive B.M. in the span AB when the load is on the span BC, all the values of B.M. at D will be negative.

The corresponding table is given below:

a	a^3	$a - a^3$	$\dfrac{a - a^3}{8}$
0.1	0.001	0.099	0.012
0.2	0.008	0.192	0.024
0.3	0.027	0.273	0.034
0.4	0.064	0.336	0.042
0.5	0.125	0.375	0.047
0.6	0.216	0.384	0.048
0.7	0.343	0.357	0.045
0.8	0.512	0.288	0.036
0.9	0.729	0.171	0.021
1.0	1.000	0	0

The complete influence line is shown in Fig. 21. (I.L. for point 5.)

Figures 20–22, 24–26 and 28–30 show B.M. influence lines for two-, three- and four-span continuous beams.

The diagrams are drawn to scale, but since it is inadvisable to scale the ordinates from small-scale diagrams the values of the ordinates at the tenth points of each span have been given so that the reader can construct his own influence lines from these ordinates. Influence lines have been drawn for each tenth point of the span and for the internal supports.

The diagrams have been drawn only for the cases of all spans being equal, but they can also be used for varying spans provided that the Moments of Inertia of the beams also vary so that the ratio I/L in each span is constant.

There are certain features of these influence lines to which particular attention must be drawn.

1. It will be seen, from a consideration of Figs. 24 to 26 that, in the case of a three-span beam, at the points for which the influence lines have been drawn, the moments due to loads in the end spans remote from the points under consideration are extremely small. Consequently if a fourth span had been added next to the remote span, the moments due to loads in this fourth span would be even smaller and could be neglected with little loss of accuracy. Moreover, the moments caused by loads in this fourth span would be of opposite sign to those obtaining in a three-span beam, and, accordingly, the discrepancy errs on the side of safety.

It follows that any further addition to the number of spans can make negligible differences to the values given, hence the influence lines for points in the end span of a three-span beam can be taken as applying to the end span of any series of spans exceeding two in number.

2. It will be seen, from Fig. 28 that the B.M. at the 1/5th point of the second span of a four-span beam is zero for all positions of the load in the third and fourth spans. This means that if the first and second spans are not

180   INFLUENCE LINES

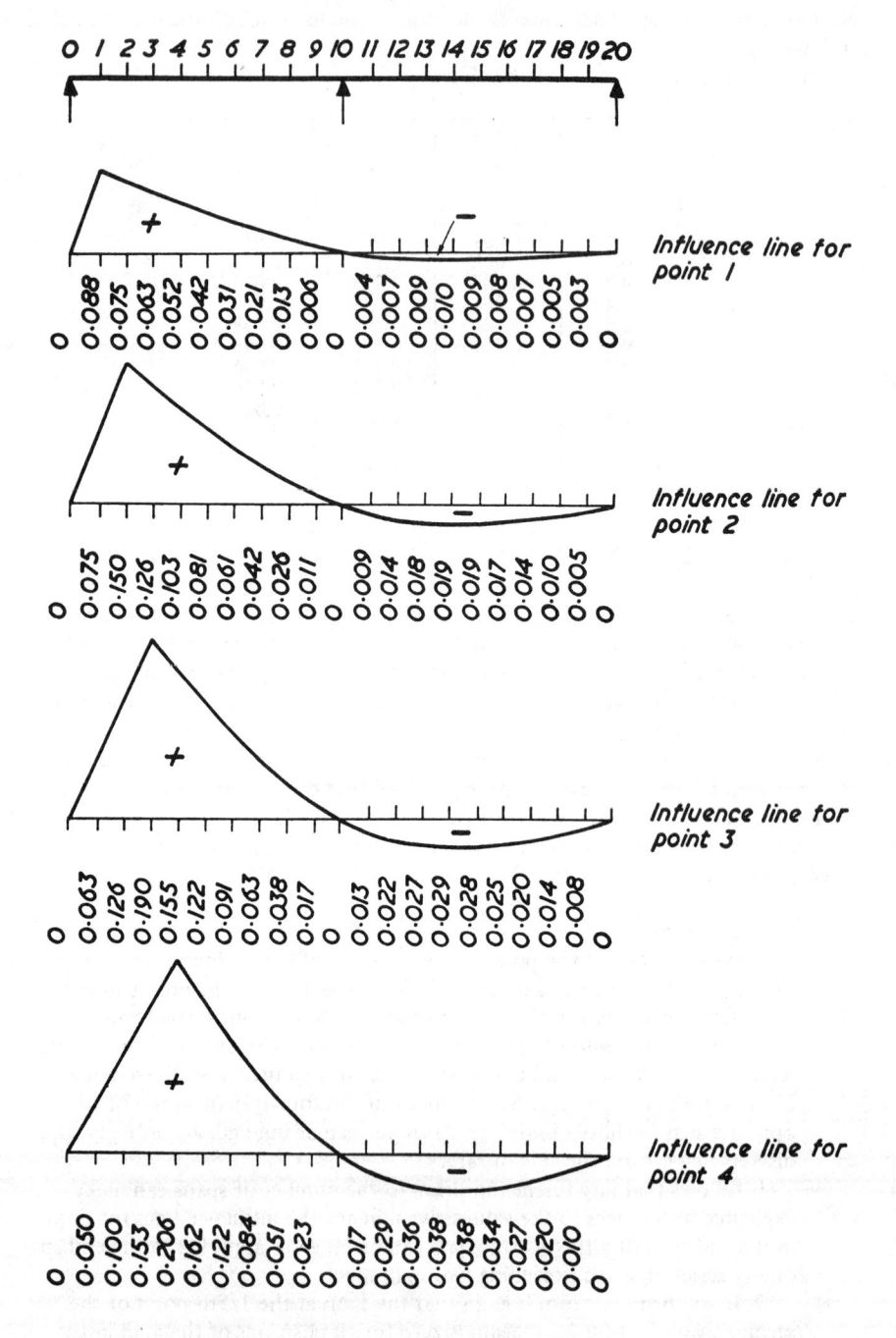

Fig. 20.—Influence lines for bending moments. Two span beam.

CONTINUOUS BEAMS 181

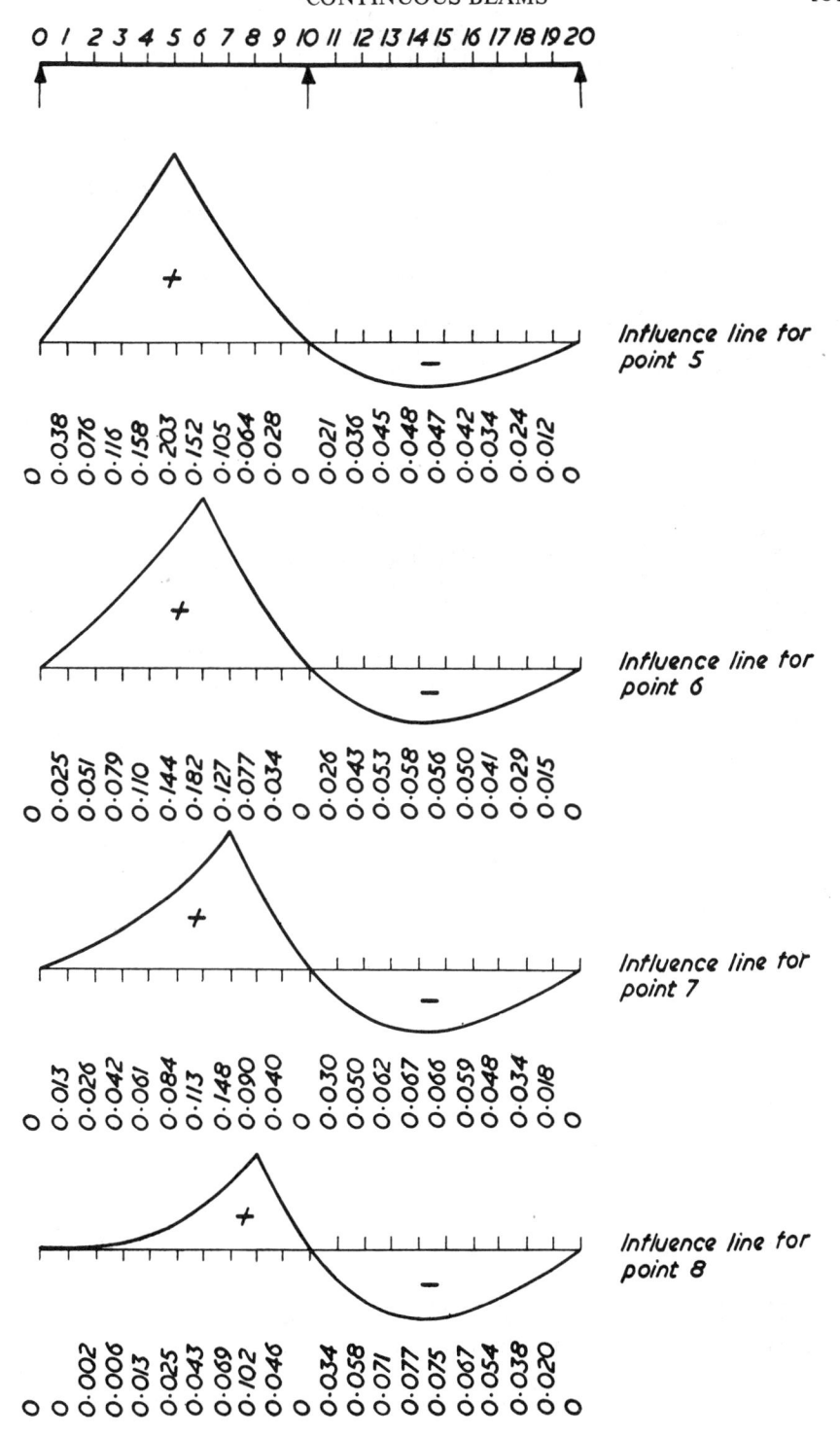

Fig. 21.—Influence lines for bending moments. Two span beam.

INFLUENCE LINES

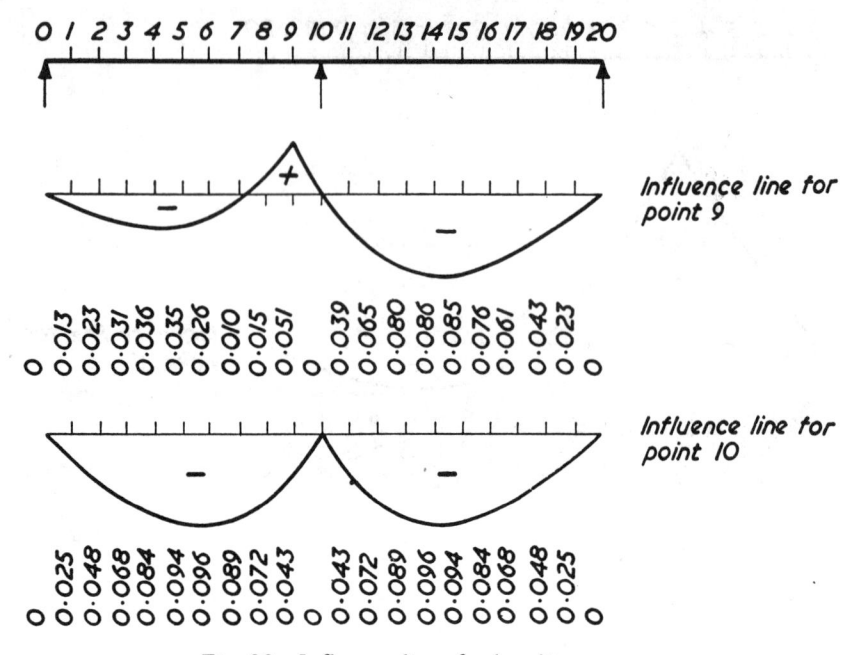

Fig. 22.—Influence lines for bending moments.

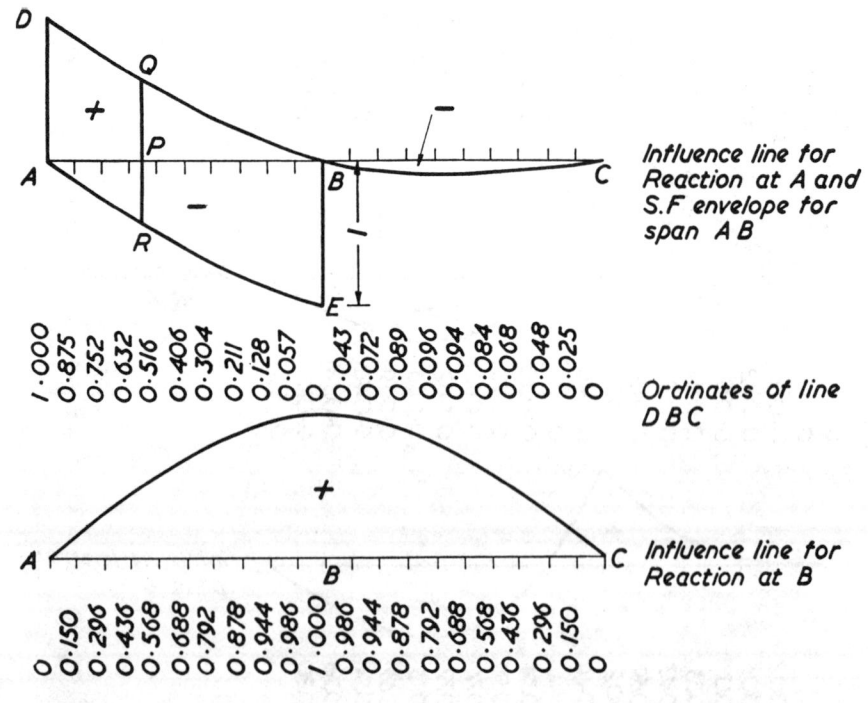

Fig. 23.—Influence lines for reactions and shear forces. Two span beam.

CONTINUOUS BEAMS

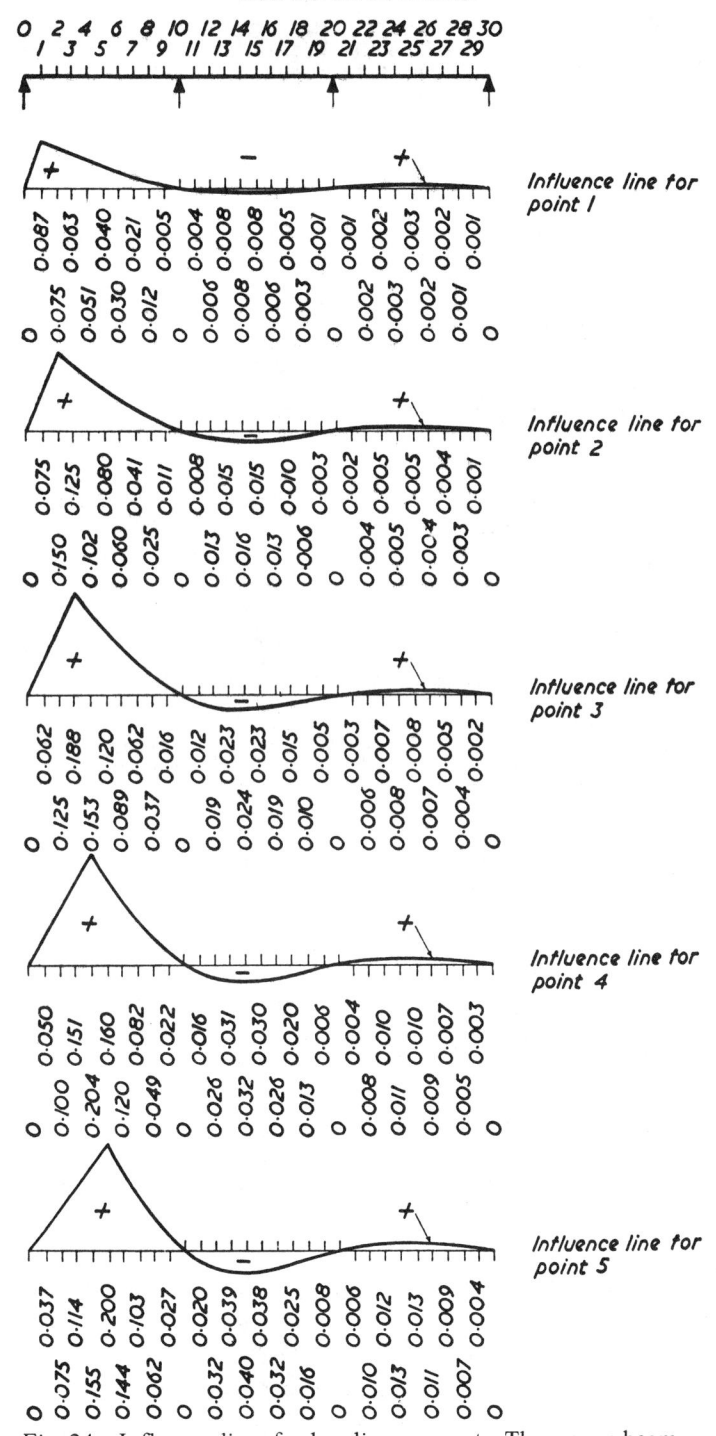

Fig. 24.—Influence lines for bending moments. Three span beam.

184   INFLUENCE LINES

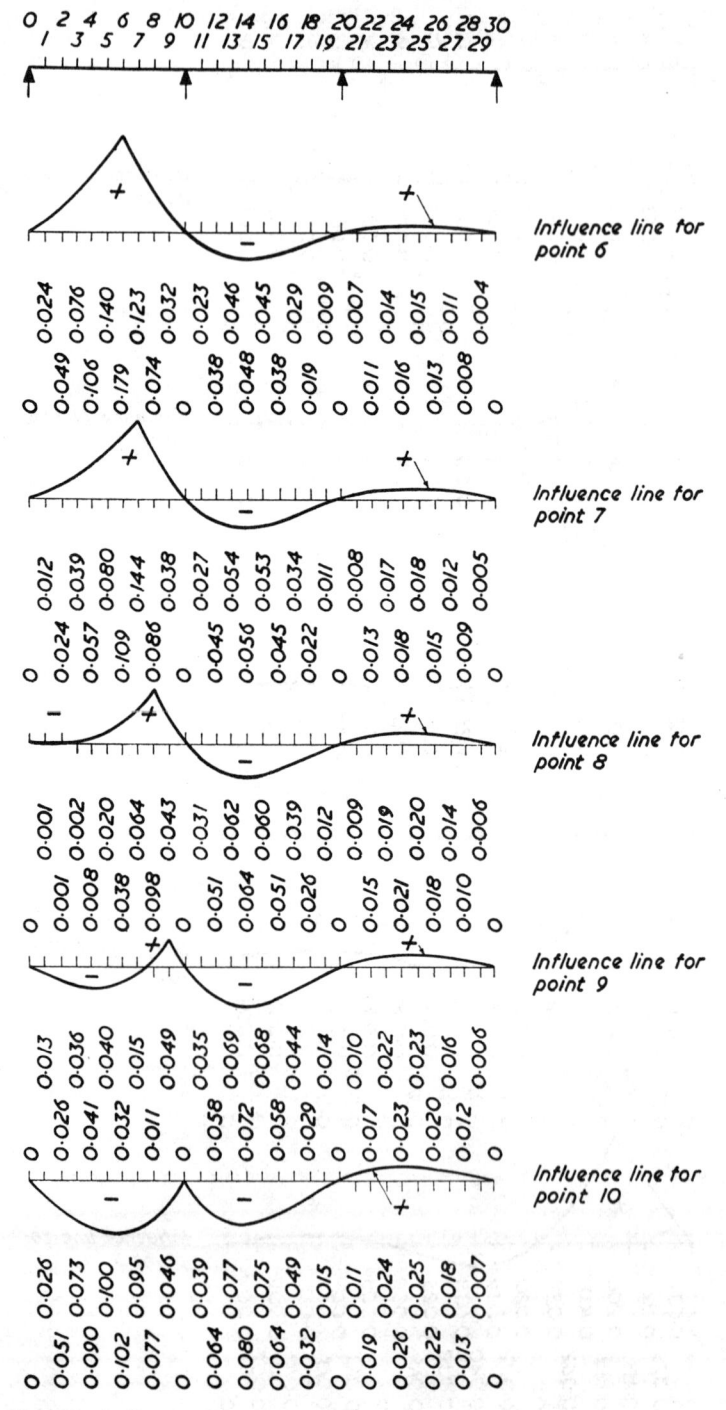

Fig. 25.—Influence lines for bending moments. Three span beam.

CONTINUOUS BEAMS

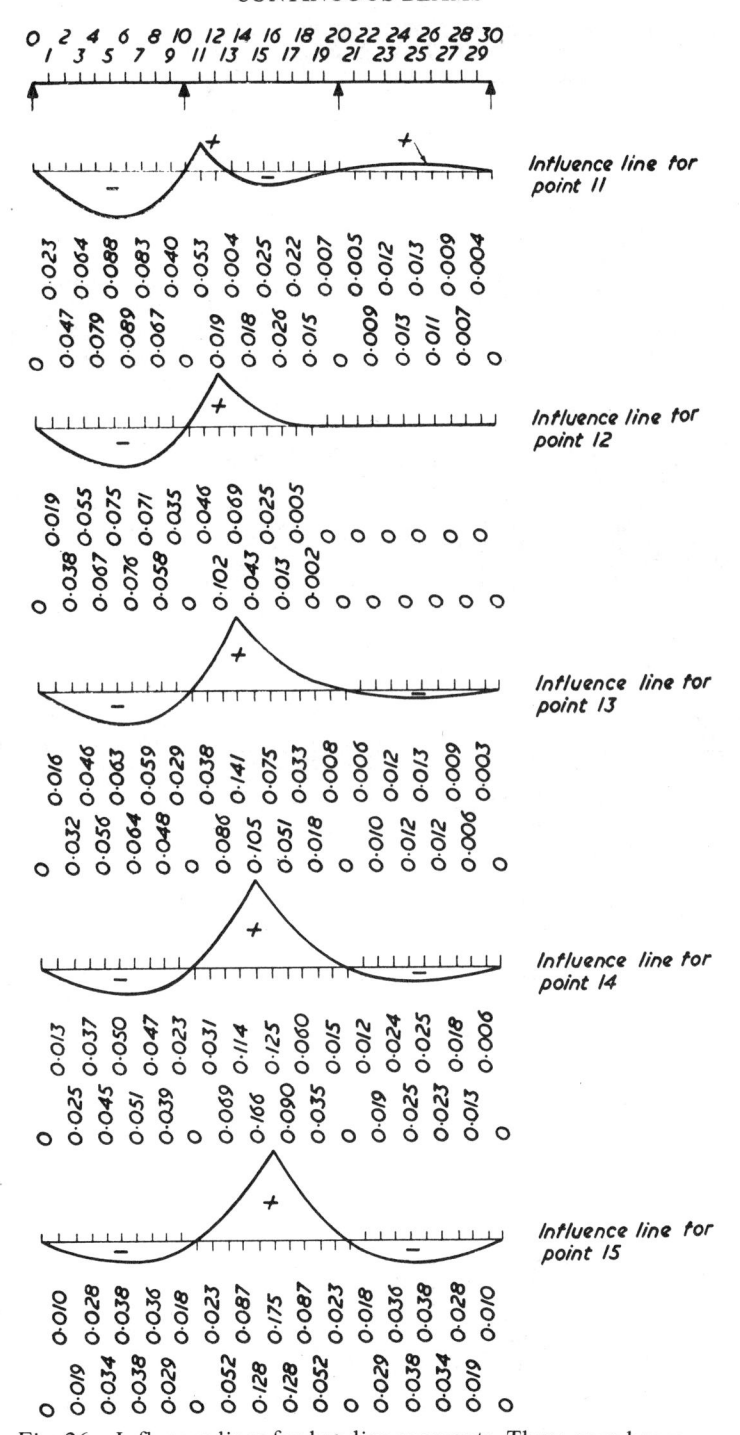

Fig. 26.—Influence lines for bending moments. Three span beam.

INFLUENCE LINES

Influence line for reaction at A and S.F. envelope for span AB

Ordinates of line EBCD:
1.000, 0.874, 0.749, 0.627, 0.510, 0.400, 0.298, 0.205, 0.123, 0.054, 0, 0.039, 0.064, 0.077, 0.080, 0.075, 0.064, 0.049, 0.032, 0.015, 0, 0.011, 0.019, 0.024, 0.026, 0.025, 0.022, 0.018, 0.013, 0.007, 0

Influence line for reaction at B

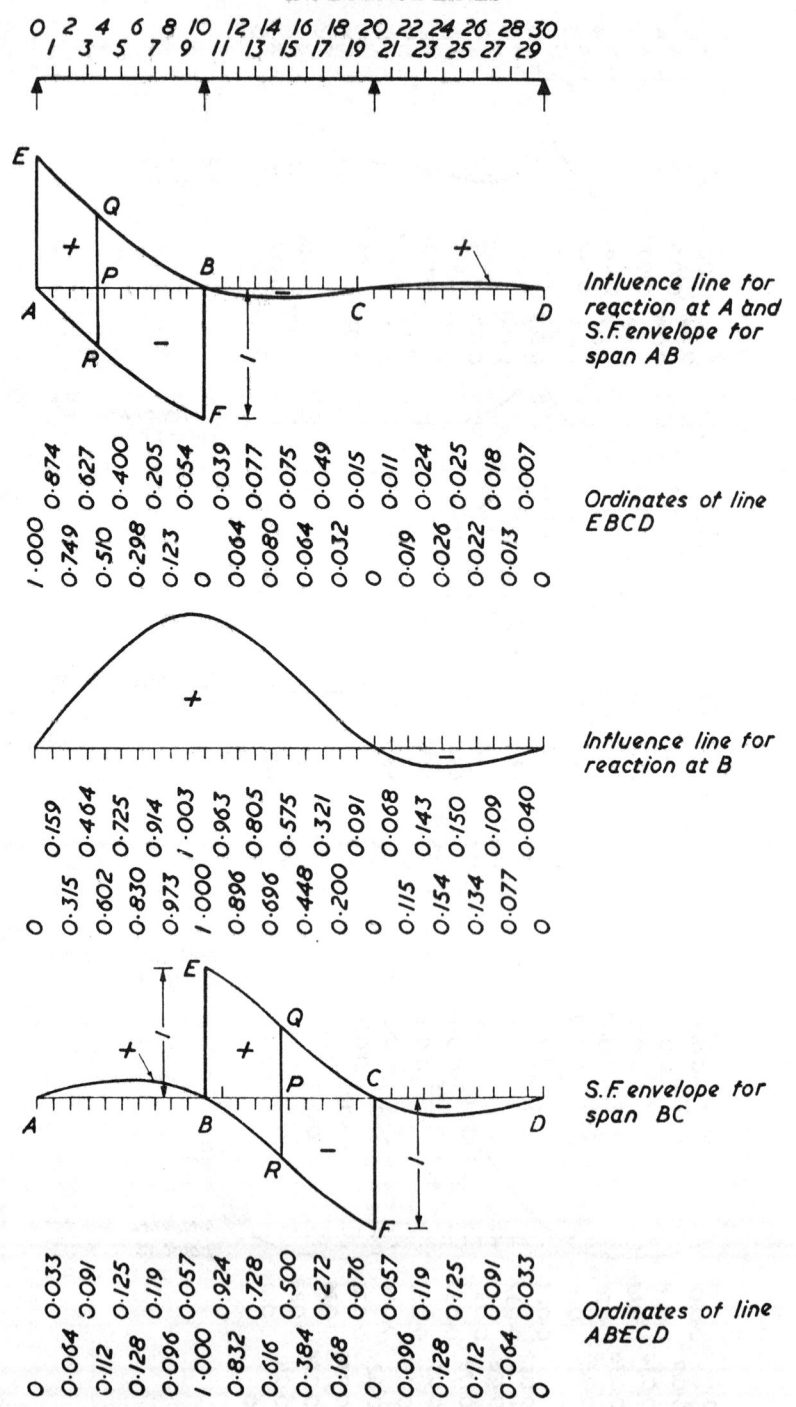

0, 0.159, 0.315, 0.464, 0.602, 0.725, 0.830, 0.914, 0.973, 1.003, 1.000, 0.963, 0.896, 0.805, 0.696, 0.575, 0.448, 0.321, 0.200, 0.091, 0, 0.068, 0.115, 0.143, 0.154, 0.150, 0.134, 0.109, 0.077, 0.040, 0

S.F. envelope for span BC

Ordinates of line ABECD:
0, 0.033, 0.064, 0.091, 0.112, 0.125, 0.128, 0.119, 0.096, 0.057, 0, 0.924, 0.832, 0.728, 0.616, 0.500, 0.384, 0.272, 0.168, 0.076, 0, 0.057, 0.096, 0.119, 0.128, 0.125, 0.112, 0.091, 0.064, 0.033, 0

Fig. 27.—Influence lines for reactions and shear forces. Three span beam.

CONTINUOUS BEAMS

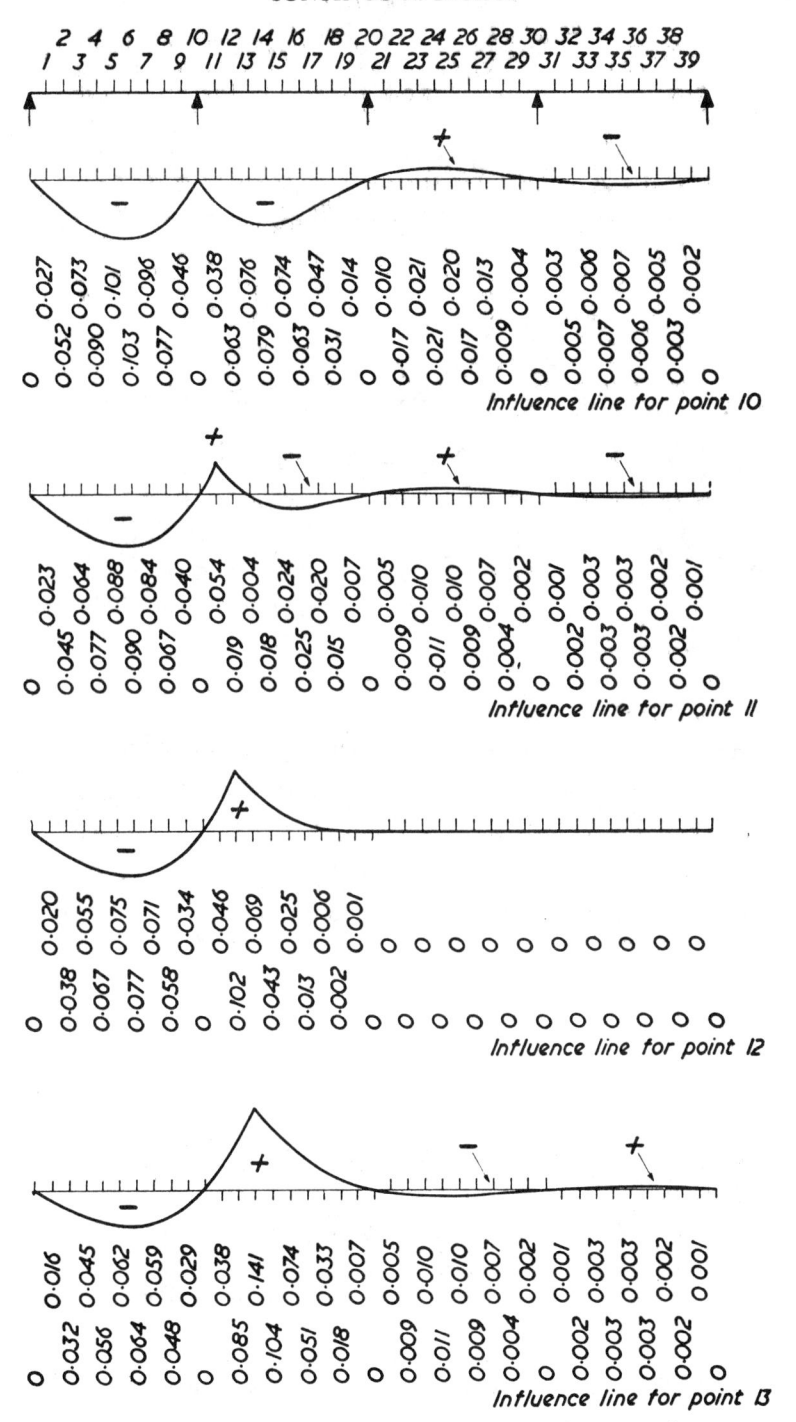

Fig. 28.—Influence lines for bending moments. Four span beam.

188   INFLUENCE LINES

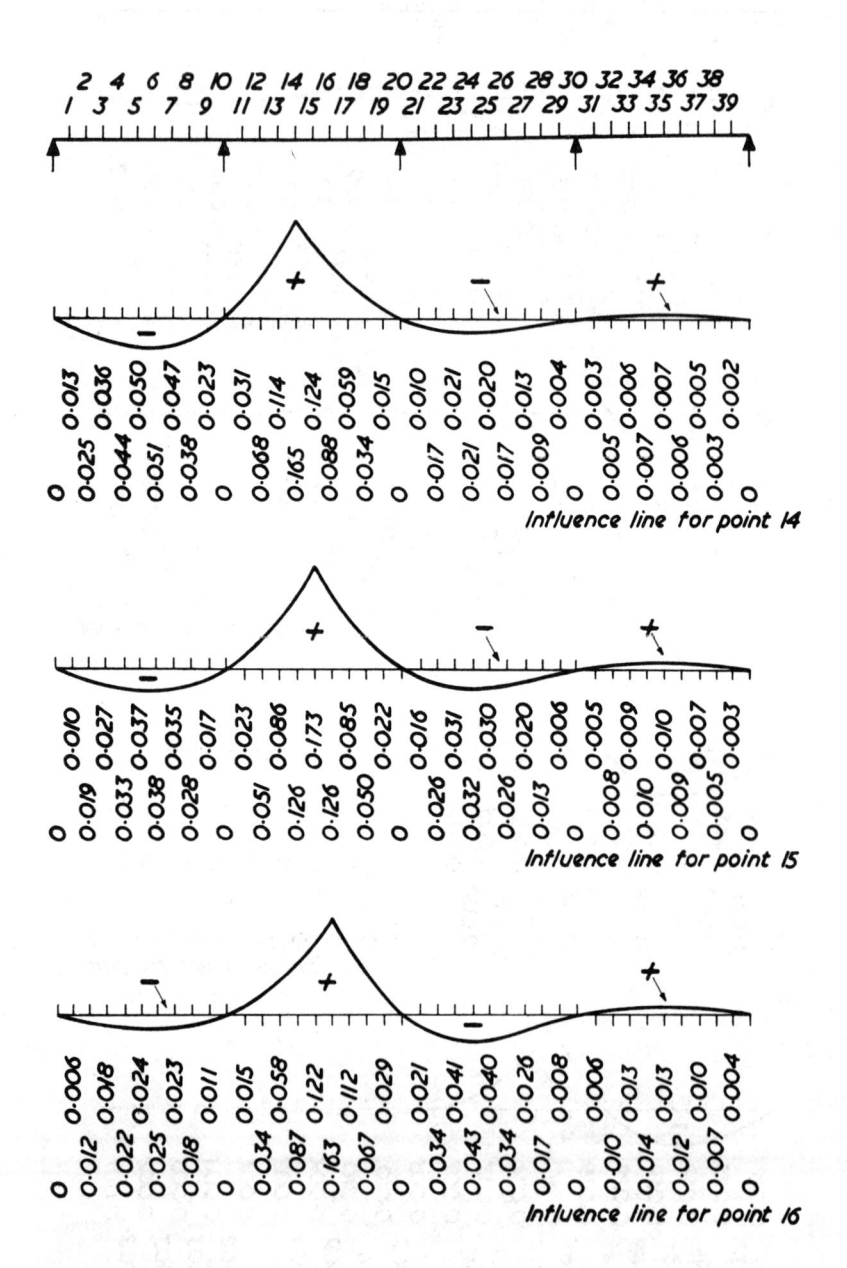

Fig. 29.—Influence lines for bending moments. Four span beam.

CONTINUOUS BEAMS

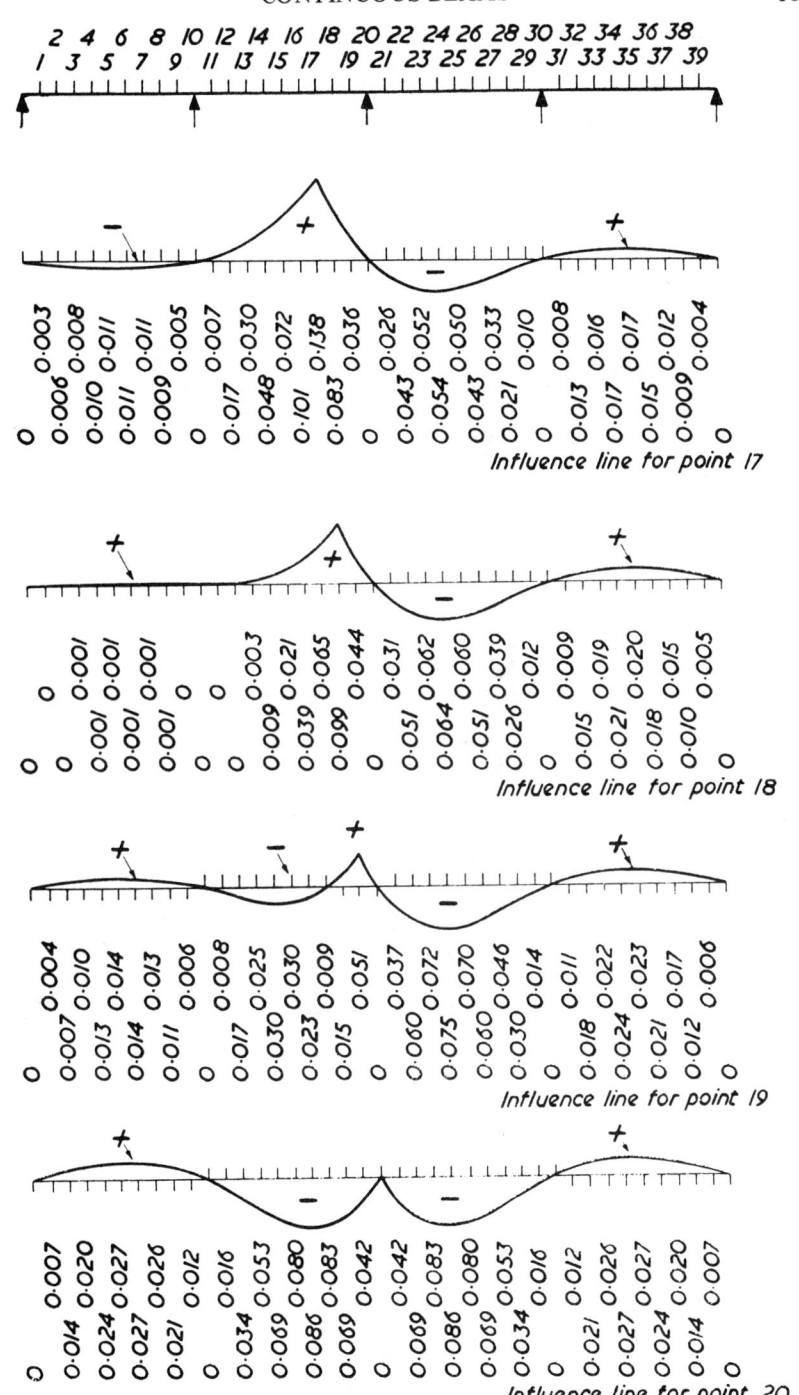

Fig. 30.—Influence lines for bending moments. Four span beam.

190 INFLUENCE LINES

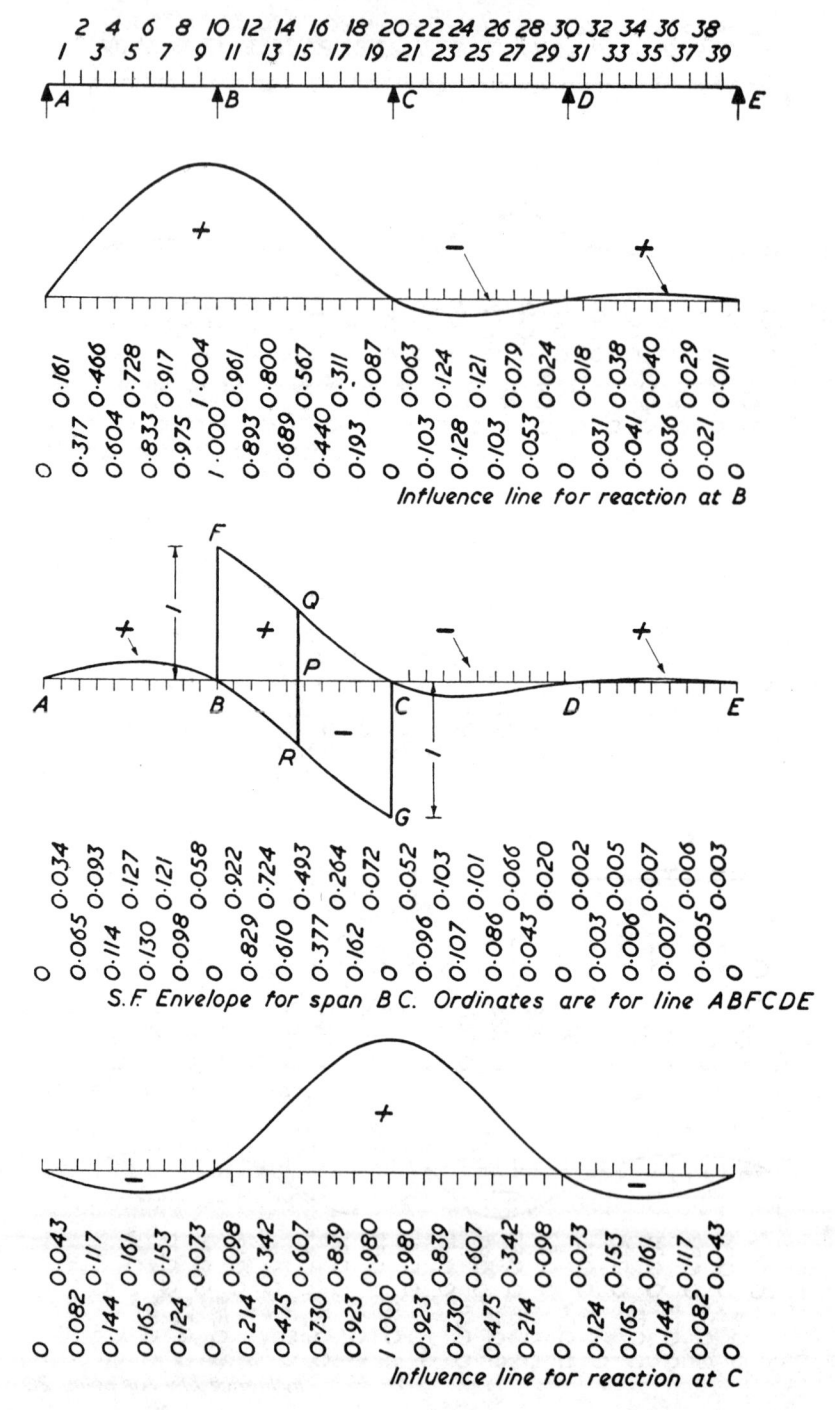

Fig. 31.—Influence lines for reactions and shear forces. Four span beam.

CONTINUOUS BEAMS

loaded, then for any case of loading on the third and fourth spans there will always be a point of contraflexure at the 1/5th point of the second span.

3. Again a comparison between Figs. 24 to 26 and Figs. 28 to 30 will show that there is little difference between the moments at the corresponding points in internal spans in the cases of three- and four-span beams. It follows that there will be even less discrepancy between corresponding points in four- and five-span beams.

Accordingly the influence lines shown in Figs. 28 to 30 may be taken as the influence lines for all internal spans of continuous beams of any series of spans exceeding three in number. The top diagram of Fig. 28 may be taken as the influence line for the end support but one, while the bottom diagram of Fig. 30 may be taken as the influence line for all other internal supports of a series of spans exceeding three in number.

Shear Force

The influence lines for S.F. are best constructed by means of envelopes. If reference is made to Fig. 1 (c), it will readily be seen that all the influence lines for S.F. in a simply supported span are contained between the two enveloping sloping lines $A_2 Y$ and $B_2 Z$, and that the influence line for any particular point, such as P_2 in the figure referred to, is constructed by drawing a vertical through P_2 to meet the two enveloping lines. Continuous beams may be treated in a similar manner.

Two-span Beam

It is required to construct the influence lines for the reactions at the supports of a two-span continuous beam. The following example will illustrate the method of construction.

Reaction at A. To construct the influence line for the reaction at A, one of the end supports of a two-span beam of equal spans. See Fig. 19.

If a unit load is placed at the point X distant a from A, then M_B, the negative moment at B, as shown in Fig. 19,

$$= \frac{a(L^2 - a^2)}{4L^2}.$$

The free reaction at A would be $1 \times \dfrac{L-a}{L}$, and the actual reaction in the given case would be

$$\frac{L-a}{L} - \frac{M_B}{L} = \frac{L-a}{L} - \frac{a(L^2 - a^2)}{4L^3}.$$

When $L = 1$, the reaction at $A = 1 - a - \dfrac{a - a^3}{4}$

INFLUENCE LINES

The values are plotted in the following table:

a	$1-a$	$\dfrac{a-a^3}{4}$	$1-a-\dfrac{a-a^3}{4}$
0	1.000	0	1.000
0.1	0.900	0.025	0.875
0.2	0.800	0.048	0.752
0.3	0.700	0.068	0.632
0.4	0.600	0.084	0.516
0.5	0.500	0.094	0.406
0.6	0.400	0.096	0.304
0.7	0.300	0.089	0.211
0.8	0.200	0.072	0.128
0.9	0.100	0.043	0.057
1.0	0	0	0

If the load is in the span BC then it is convenient to measure the distance a from C. In this case:

$$R_A = 0 - \frac{M_B}{L} = -\frac{a(L^2 - a^2)}{4L^3}$$

or $\qquad -\dfrac{a-a^3}{4}$ when $L = 1$.

The corresponding table is shown below:

a	$-\dfrac{a-a^3}{4}$
0	0
0.1	−0.025
0.2	−0.048
0.3	−0.068
0.4	−0.084
0.5	−0.094
0.6	−0.096
0.7	−0.089
0.8	−0.072
0.9	−0.043
1.0	0

It will be appreciated that these figures are applicable to R_C when the span AB is loaded.

The complete influence line for the reaction at A is drawn by the line DBC in Fig. 23.

If the line AE is drawn parallel to DB and distant one unit from it, then the lines DB and AE are the boundary lines of the S.F. envelope for all points in the span AB.

If the influence line for any point, e.g. P, is required, a vertical line QPR is drawn through P to meet DB and AE in Q and R respectively. The influence line for S.F. at P is $ARPQBC$.

CONTINUOUS BEAMS

Reaction at B. The influence line for R_B is built up as follows:

$$R_B = 1 - R_A - R_C$$

$$= 1 - \left(1 - a - \frac{a - a^3}{4}\right) - \left(-\frac{a - a^3}{4}\right)$$

$$= 1 - \left(1 - a - \frac{a - a^3}{4}\right) + \left(\frac{a - a^3}{4}\right)$$

The corresponding table is shown below:

a	$1 - a - \dfrac{a - a^3}{4}$	$\dfrac{a - a^3}{4}$	$1 - \left(1 - a - \dfrac{a - a^3}{4}\right) + \dfrac{a - a^3}{4}$
0	1.000	0	0
0.1	0.875	0.025	0.150
0.2	0.752	0.048	0.296
0.3	0.632	0.068	0.436
0.4	0.516	0.084	0.568
0.5	0.406	0.094	0.688
0.6	0.304	0.096	0.792
0.7	0.211	0.089	0.878
0.8	0.128	0.072	0.944
0.9	0.057	0.043	0.986
1.0	0	0	1.000

The complete influence line for the reaction at B is shown in Fig. 23.

It should be noted that the reaction at B is greater than the S.F. at B in the span AB, because the reaction is composed of the sum of the shears at B in both spans.

Three-span Beam

Consider the top diagram in Fig. 27.

The line *EBCD* shows the influence line for the reaction at A. Once again, if *AF* is drawn parallel to *EB* and distant one unit from it, then the lines *EB* and *AF* are the boundary lines of the S.F. envelope for all points in the span AB. For example, *ARPQBCD* is the influence line for shear at the point P.

As in the case of the influence lines for B.M. for points in the end span AB, the addition of any number of spans beyond D can make negligible difference to the S.F. in the span AB, and so the envelope shown can be taken as the envelope for the end span of any series of spans exceeding two in number.

The influence line for the reaction at B and the S.F. envelope for the centre span BC are also shown in Fig. 27. In the bottom diagram, *BF* is parallel to *EC* and distant one unit from it, and the lines *EC* and *BF* are the boundary lines of the S.F. envelope for all points in the span BC. Therefore, *ABRPQCD* is the influence line for shear at the point P.

Four or More Spans

The influence line for the reaction at B, the end but one support of a four-span beam, is shown in Fig. 31 and may be taken as that for the end support but one of any number of spans greater than three.

The S.F. envelope for an internal span of a four-span beam is also shown in Fig. 31, *FC* and *BG* being the boundary lines. Hence, *ABRPQCDE* is the influence line for shear at the point *P*. The envelope shown may be taken to be that for any internal span of a continuous beam of more than three spans.

The bottom diagram of the same figure shows the influence line for the reaction at *C*, the central support of a four-span beam. This may be taken as the influence line for any internal support of a continuous beam of more than three spans, other than the end support but one.

13. METHODS OF STRUCTURAL ANALYSIS

Introduction

IN the opening chapters of this book, methods of calculation are given for built-in and continuous beams, the methods being based on the theorems of Mohr and Clapeyron respectively, both of which were developed in the nineteenth century.

The methods of analysis of statically indeterminate structures to be described here are as follows:

1. The Area-Moment Method.
2. Moment Distribution.
3. Slope-Deflection.

As higher mathematics are not required when using these systems they should be of general appeal.

The Area-Moment Method has recently regained its former popularity, although it often appears in the guise known as Semi-Graphical Integration, Moment Distribution was evolved by Professor Hardy Cross, an American, details being first published in 1929. Slope Deflection was developed and made widely known by Professors Maney and Wilson of Minnesota University in 1915, but like Moment Distribution the method owes much to the work of Mohr.

Where convenient or practical, the same numerical examples have been employed to illustrate the various methods of analysis, although the examples have been chosen to cover as many aspects of structural engineering as possible. By so doing, it is hoped that readers will be able to form some idea of the relative merits of the methods for a particular type of problem. Generalising, Moment Distribution and Slope Deflection are of universal application, but the work involved is lengthy where sidesway or asymmetrical frames are being treated.

The Area-Moment Method may be used most effectively for dealing with members of variable moment of inertia.

This section of the book is followed by formulae and charts for the design of certain standard types of rigid frames. Readers may find it useful to consider the notes in the preamble to these tables and charts in conjunction with the following methods of analysis.

(a) The Area-Moment Method

The Reciprocal Theorem

The basis of the Area-Moment Method is stated in Chapter 7, but, in order to extend the scope of the examples given in this section, a knowledge of the Reciprocal Theorem is essential. Although this theorem, which was first enunciated by Professor Clerk-Maxwell in 1864, is fairly well known, it is perhaps advisable to lay down the conditions under which it may be used.

Consider the identical cantilevers shown in Figs. 1 and 2. When the load P_C is applied at C in Fig. 1, a deflection d_B is produced at B. If the same load, P_B is applied at B in Fig. 2, a deflection d_C is produced at C. It can be shown that the values of the two deflections are equal.

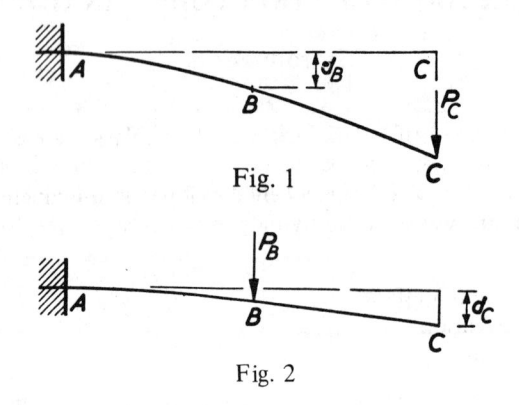

Fig. 1

Fig. 2

In the general case, where the loads are unequal, the work done by the force P_C on the corresponding deflection d_C is equal to that done by the force P_B on the corresponding deflection d_B.

i.e.
$$P_B d_B = P_C d_C \quad \quad \quad \quad \quad (1)$$

The theorem is not confined to linear deflections. The same reasoning can be applied to moments and their corresponding angular deflections.

Consider the identical cantilevers in Figs. 3 and 4. When the moment M_C is

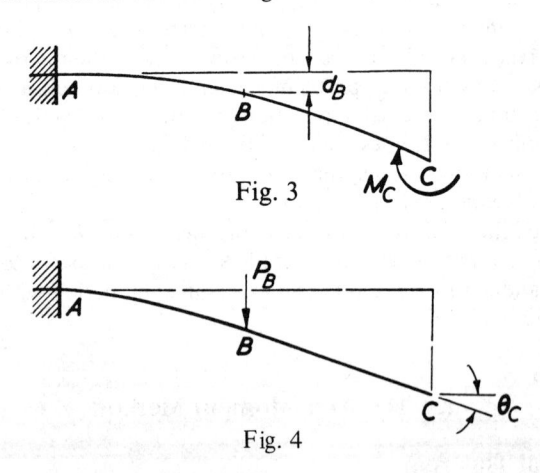

Fig. 3

Fig. 4

applied at C in Fig. 3, a deflection d_B is produced at B. If a load of the same numerical value is applied at B in Fig. 4, a slope θ_C is produced at C. It can be shown that the numerical values of the two deflections d_B and θ_C are equal.

In the general case:
$$M_C \theta_C = P_B d_B \quad \quad \quad \quad \quad (2)$$

THE AREA-MOMENT METHOD

In a similar way, the slope of the cantilever at B produced by a moment applied at C is equal to the slope at C produced by an equal moment applied at B.

In the general case:

$$M_B \theta_B = M_C \theta_C \quad \ldots \ldots \ldots \ldots \ldots \ldots \quad (3)$$

The theorem applies only to the three cases stated, whether the member concerned is a cantilever, beam or part of a frame. Hence, it is *incorrect* to assume that the slope of a member at B caused by a load at C is equal to the slope at C when the load is placed at B.

The reciprocal theorem may be used to reduce the amount of calculation involved in structures with more than one degree of indeterminacy. The calculations involve the solution of Maxwell-Mohr work equations which are of the general form shown below:

$$X_1 d_{1-1} + X_2 d_{2-1} + X_3 d_{3-1} \ldots + d_{P-1} = 0$$
$$X_1 d_{1-2} + X_2 d_{2-2} + X_3 d_{3-2} \ldots + d_{P-2} = 0$$
$$X_1 d_{1-3} + X_2 d_{2-3} + X_3 d_{3-3} \ldots + d_{P-3} = 0$$
$$\ldots \ldots \ldots \ldots \ldots \ldots \ldots \ldots = 0$$
$$\ldots \ldots \ldots \ldots \ldots \ldots \ldots \ldots = 0$$

now, from the reciprocal theorem,

$$d_{2-1} = d_{1-2}$$
$$d_{3-1} = d_{1-3}$$
and $$d_{3-2} = d_{2-3}$$

Hence, the terms to the left of and below the diagonal $X_1 d_{1-1}$, $X_2 d_{2-2}$, $X_3 d_{3-3}$, etc., need not be calculated. They merely repeat the corresponding figures on the other side of the diagonal.

Application of the Area-Moment Method

The application of area-moments is carried out in two stages. In the first stage the redundants are removed while in the second unit redundants are applied. At each stage the linear and, if necessary, angular deflections of the structure are calculated. By equating the deflections calculated at each stage the actual values of the redundants may be found.

Example 1. The procedure may be demonstrated by considering the loaded propped cantilever shown in Fig. 5. The redundant is either the restraint moment M_A or the prop B. Proceed assuming that the prop is the redundant. When this redundant is removed, the end B deflects, as shown in Fig. 6, the appropriate deflection formula being

$$d_B = \frac{\Sigma A x}{EI}$$

$$= \frac{(-1.8 \times 1.8 \times \tfrac{1}{2}) 3.0}{2}$$

$$d_B = -2.43 \text{ units.}$$

198 METHODS OF STRUCTURAL ANALYSIS

Fig. 5

Fig. 6

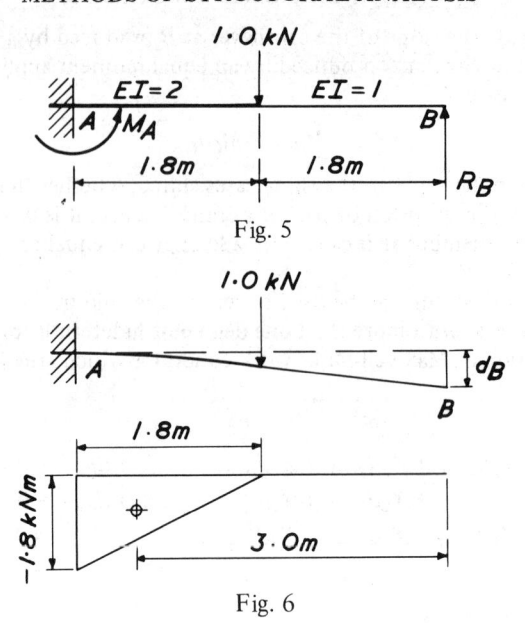

Now apply a unit upward propping force at B, as shown in Fig. 7. Then the upward deflection,

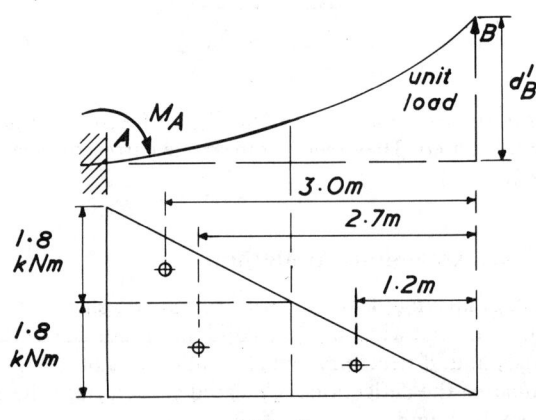

Fig. 7

$$d_B' = \frac{\Sigma Ax}{EI}$$

$$= \frac{(1.8 \times 1.8)2.7}{2} + \frac{(1.8 \times 1.8 \times \frac{1}{2})3.0}{2} + \frac{(1.8 \times 1.8 \times \frac{1}{2})1.2}{1}$$

$$= 4.37 + 2.43 + 1.94$$

$$= 8.74 \text{ units.}$$

Now the actual deflection at $B = 0$.

Therefore $\qquad d_B + R_B d_B' = 0$

THE AREA-MOMENT METHOD

or
$$R_B = \frac{2.43}{8.74}$$
$$= 0.2777 \text{ kN}$$

The positive result indicates that the prop acts in the direction assumed, i.e. upwards.

Therefore
$$M_A = (-1.0 \times 1.8) + (0.2777 \times 3.6)$$
$$= -0.8 \text{ kNm}$$

It is interesting to note that had the cantilever been prismatic, instead of varying in section, the force of the prop R_B would have been 0.3125 kN and M_A would have been −0.675 kNm. With no prop at all M_A would be −1.8 kNm.

As an alternative solution, assume that the redundant is the restraint moment M_A. If M_A is removed, as in Fig. 8, then the *slope* at A is equal to the reaction at A

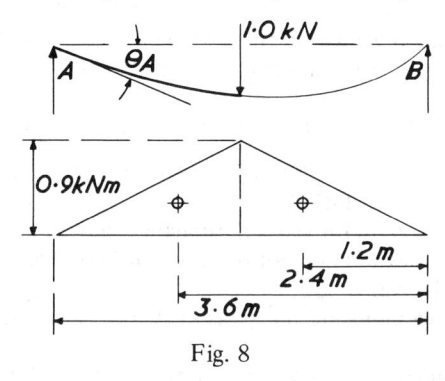

Fig. 8

assuming that the beam AB is loaded with its static B.M. diagram, divided by EI. Taking moments about B,

$$\theta_A \times 3.6 = \frac{(0.9 \times 1.8 \times \tfrac{1}{2})2.4}{2} + \frac{(0.9 \times 1.8 \times \tfrac{1}{2})1.2}{1}$$

$$\theta_A = \frac{1.944}{3.6} = 0.54 \text{ units}$$

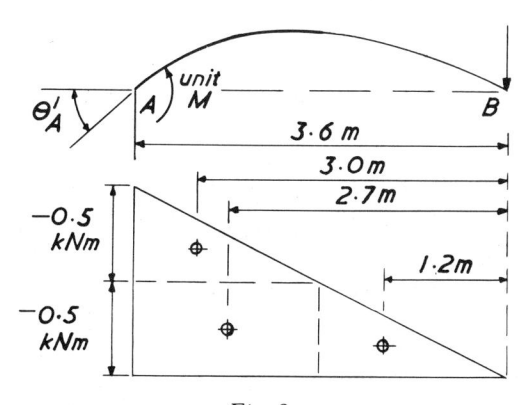

Fig. 9

If a unit moment is applied at A, as shown in Fig. 9, then the upward slope at A can be found once again by taking moments about B,

$$\theta_A' \times 3.6 = \frac{(-0.5 \times 1.8)2.7}{2} + \frac{(-0.5 \times 1.8 \times \tfrac{1}{2})3.0}{2} + \frac{(-0.5 \times 1.8 \times \tfrac{1}{2})1.2}{1}$$

$$\theta_A' = -\left(\frac{1.215 + 0.675 + 0.54}{3.6}\right)$$

$$= -0.675 \text{ units.}$$

But there is actually zero slope at A. Therefore

$$\theta_A + (M_A \times \theta_A') = 0$$

From which

$$M_A = \frac{0.54}{0.675}$$

$$= 0.8 \text{ kNm.}$$

This value is numerically equal to that obtained by the first method. The positive result indicates that the moment is applied as assumed, i.e. it is anti-clockwise.

The steps in the solution of a statically indeterminate structure are as follows:

1. Reduce the structure to a statically determinate condition by removing redundant forces or moments.
2. Draw the B.M. diagram for this determinate condition.
3. Divide the B.M. diagram into convenient geometrical areas and locate the centroid of each area.
4. Calculate the area-moments and areas to find the linear and angular displacements.
5. Apply unit redundants to the structure and calculate the relevant B.M. diagrams.
6. Repeat steps 3 and 4 for these diagrams.
7. Equate the results in steps 4 and 6 to find the values of the actual redundants.
8. Using these values, calculate the final B.M. diagram.

The method of reducing a structure to a statically determinate condition is often a matter of choice on the part of the designer. Various methods are adopted in the examples which follow.

Portal Frames

Example 2. Figure 10 shows the outline of a portal frame for a small warehouse

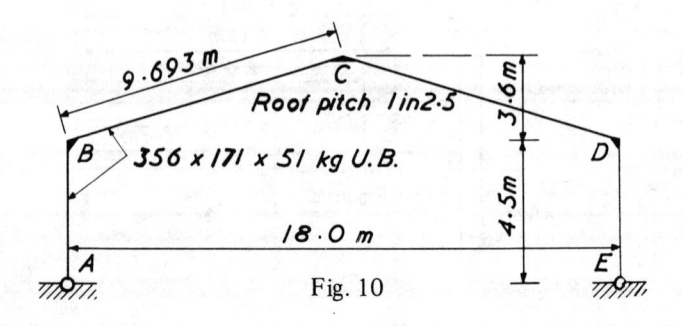

Fig. 10

building, the frames being set at 4.5 m centres. It will be assumed that the frame is composed throughout of 356 × 171 × 51 kg U.B.s, with small triangular haunches at the eaves and ridge joints.

Calculate the B.M.s induced at *B, C* and *D* by the dead load.

Dead Load (on slope of rafter):

Average weight of sheeting and thermal insulation
   or glazing . . . . 250 N/m²
   Purlins, say . . . . 100 N/m²

                                                                                 350 N/m²

Total dead load to be supported per frame
   = 4.5 × 350 = ... 1 575 N/m
Weight of rafters, say . . . . 550 N/m

                                                                                 2 125 N/m

The total dead load over one frame, allowing for the slope of the rafters and the slight overhang of the sheeting is:

$$20 \times 2\,125 = 42\,500 \text{ N} = 42.5 \text{ kN}$$

Although most of the load is actually applied at the purlin points, the error is negligible if it is assumed that it is uniformly distributed over the rafters.

Static B.M. Diagram:

This frame is statically indeterminate to the first degree and may be reduced to a determinate condition by fixing the *position* of one foot, say *A*, and removing the horizontal reaction at the other foot, say *E*, when that foot can move outwards without restraint.

The static B.M. diagram for the dead load, shown in Fig. 11, will be parabolic

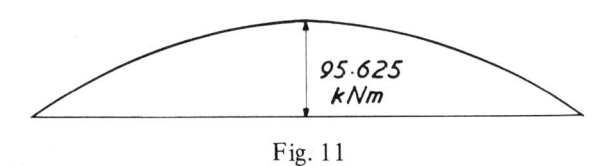

95·625 kNm

Fig. 11

with a base *L* of length *BD* and a height equal to $WL/8 = \dfrac{42.5 \times 18}{8} = 95.625$ kNm.

This diagram is projected on to the rafters as shown in Fig. 12.

Area-Moments for Statically Determinate Condition:

The rafters being 9.693 m long, measured along the centre-line, the area of the diagram on each rafter = $9.693 \times 95.625 \times \tfrac{2}{3}$ = 617.93 units. The centroid of the area is normal to a point situated $\tfrac{5}{8}$ of the distance up the rafter, or at a height 2.25 m above the eaves and 6.75 m above floor level.

202   METHODS OF STRUCTURAL ANALYSIS

Fig. 12

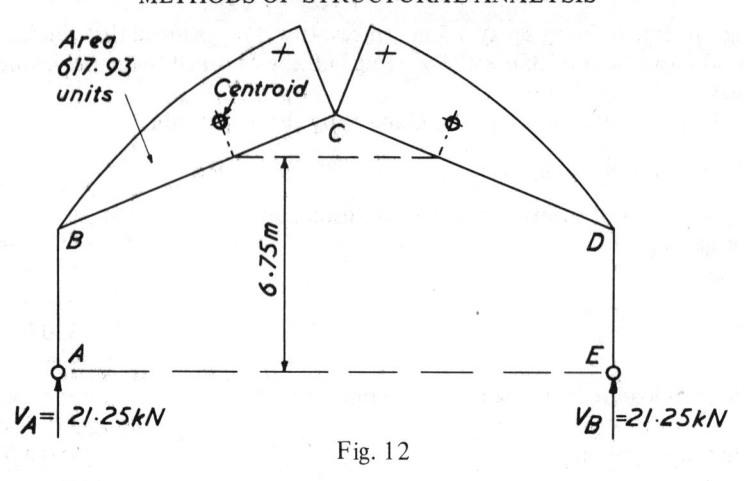

Taking area-moments about AE, the outward deflection at E

$$d_E = \frac{\Sigma Ay}{EI} = \frac{2 \times 617.93 \times 6.75}{EI}$$

$EId_E = 8\,342.06$ units.

Area-Moments for Unit Horizontal Thrust:

When unit horizontal thrusts are applied at the feet of the frame, the B.M. diagram shown in Fig. 13 is obtained.

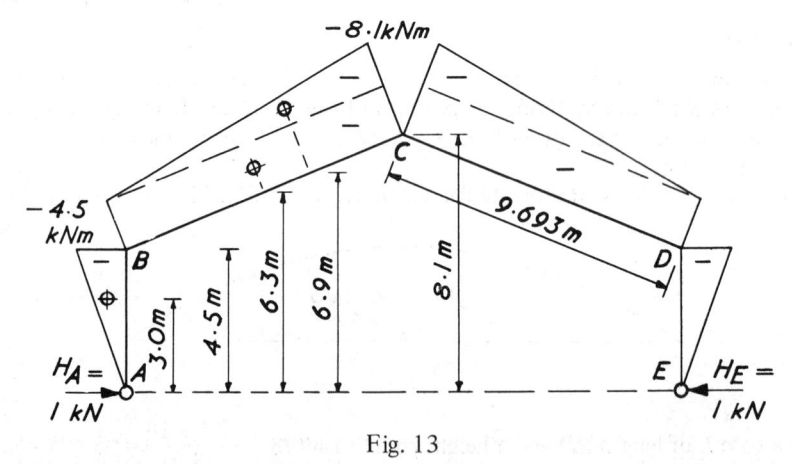

Fig. 13

$M_B = M_D = -1 \times 4.5 = -4.5$ kNm
$M_C = -1 \times 8.1 = -8.1$ kNm

Taking area-moments about the base AE, the deflection at E

$$d_E = \frac{2(-4.5 \times 4.5 \times \tfrac{1}{2})3.0}{EI} + \frac{2(-4.5 \times 9.693)6.3}{EI} + \frac{2(-3.6 \times 9.693 \times \tfrac{1}{2})6.9}{EI}$$

$EId_E = -851.18$ units.

THE AREA-MOMENT METHOD

Final Horizontal Thrusts:
Now the actual deflection at $E = 0$
Therefore $\quad 8\,342.06 - 851.18\,H_E = 0$
or $\quad\quad\quad\quad\quad\quad H_E = 9.8$ kN
The final B.M. diagram is shown in Fig. 14.

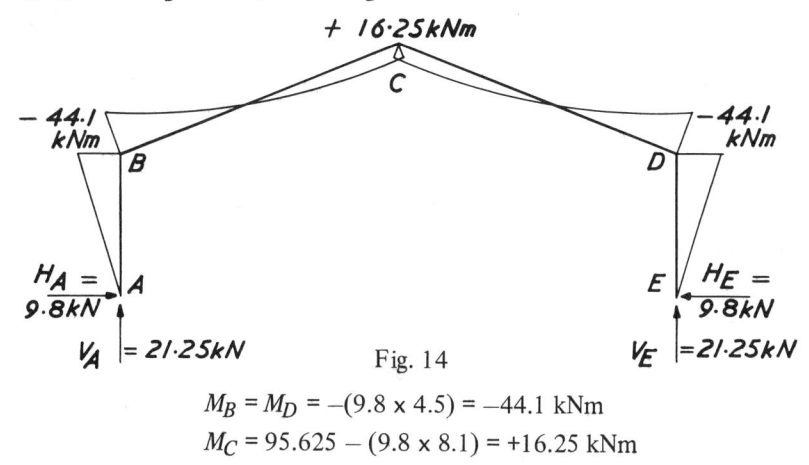

Fig. 14

$M_B = M_D = -(9.8 \times 4.5) = -44.1$ kNm
$M_C = 95.625 - (9.8 \times 8.1) = +16.25$ kNm

Snow Load

Having calculated the moments induced at B, C and D by the dead loads, which were assumed to be uniformly distributed, it is easy to calculate the effect of snow on the structure. Clause 6.3 of CP3–Chapter V Part 1 (1967) states, *inter alia,* that on a roof having a slope greater than 10 and less than 30 degrees, allowance shall be made for an imposed load of 0.75 kN/m² measured on plan. This unit load is equivalent to 60.75 kN uniformly distributed over the whole frame. Therefore, the moments induced by the snow load may be found by multiplying those due to dead load by $60.75/42.5 = 1.43$, with the following results:

$$M_B = M_D = -63.1 \text{ kNm}$$
$$M_C = +23.2 \text{ kNm}$$
$$H_A = H_E = +14.0 \text{ kN}$$
while $\quad\quad\quad V_A = V_E = +30.39$ kN

Deflection of Eaves Joints

It was not necessary to calculate the actual deflections at E for the statically determinate condition or for the unit horizontal thrust and the term EI cancelled out in the final calculations. Nevertheless, the actual deflection at any point can be calculated by inserting values for E, the modulus of elasticity, and I, the moment of inertia.

Suppose it is necessary to know the spread of the eaves joints B and D under dead and snow loads. The appropriate combined B.M. diagram is shown in Fig. 15. The horizontal deflection of D with respect to B may be found by taking area-moments of the diagram along BCD about the axis BD.

METHODS OF STRUCTURAL ANALYSIS

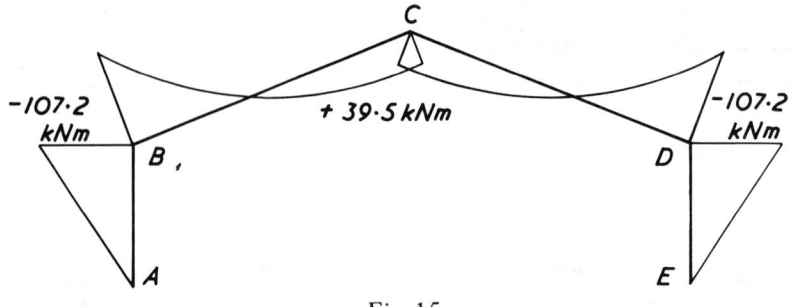

Fig. 15

Now $I = 14\,118$ cm^4 for a 356 x 171 x 51 kg U.B.

Assume that $E = 2.1 \times 10^5$ N/mm^2.

The B.M. diagram can be split into its components as shown in Fig. 16.

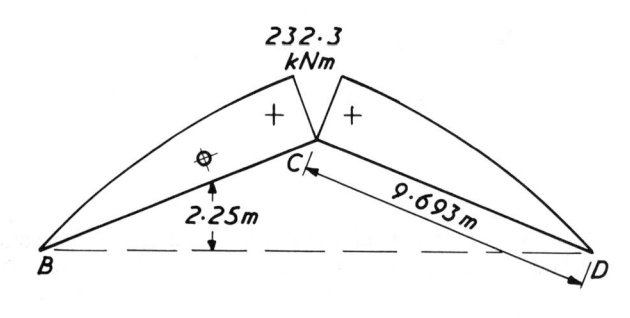

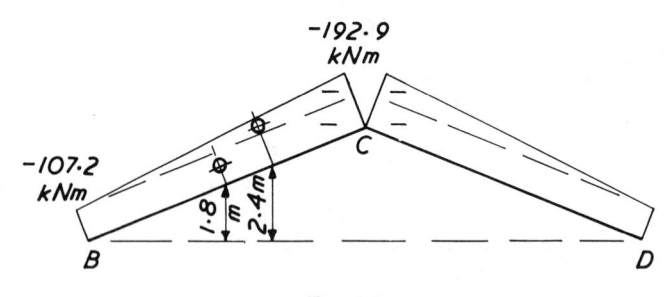

Fig. 16

Then the deflection of D, relative to B,

$$d_D = \frac{\Sigma Ay}{EI}$$

$EI d_D = 2(232.3 \times 9.693 \times \tfrac{2}{3})2.25 - 2(107.2 \times 9.693)1.8 - 2(85.7 \times 9.693 \times \tfrac{1}{2})2.4$

$$d_D = \frac{(6\,755.1 - 3\,740.7 - 1\,993.7) \times 10^{12}}{2.1 \times 10^5 \times 14\,118 \times 10^4} = 34.4 \text{ mm}$$

Hence, each eaves joint moves 17.2 mm outwards.

Wind Loading

In the next two examples the effect of wind loads will be studied, using the same frame as in Example 2.

Example 3

If the basic wind speed is 40 m/s, the dynamic pressure of the wind, q, in accordance with CP3—Chapter V Part 2 (1970), will be 385 N/m² if $S1 = 1.0$, $S2 = 0.62$ and $S3 = 1.0$. It can be assumed that $0.7q$ is the pressure on the windward face and $0.25q$ the suction on the leeward face. Internal pressure or suction will be ignored.

Consider only the windward load on AB, as shown in Fig. 17.

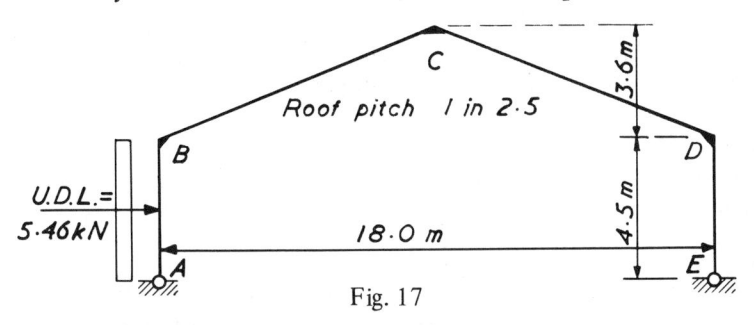

Fig. 17

AB is 4.5 m high and the frames are spaced at 4.5 m centres. Hence the total load on AB, which can be assumed, with negligible error, to be uniformly distributed over the full height, is

$$0.7 \times 385 \times 4.5 \times 4.5 = 5\,460 \text{ N} = 5.46 \text{ kN}.$$

Static B.M. Diagram:

As in the previous example, it will be assumed that the foot A is fixed in position but that E is free to move outwards under the effect of the load.

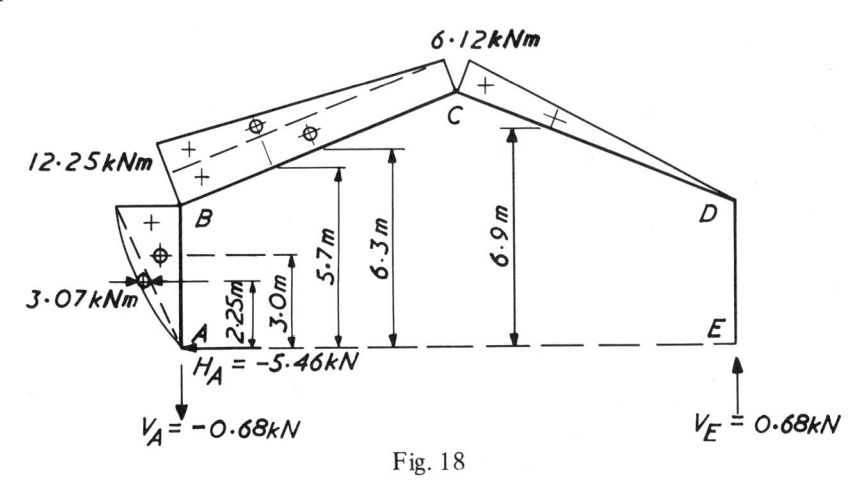

Fig. 18

Then the static B.M. diagram is shown in Fig. 18. Taking moments about A,

$$V_E \times L = \frac{W \times h}{2}$$

$$V_E = \frac{5.46 \times 4.5}{2 \times 18} = 0.68 \text{ kN}$$

Similarly
$$V_A = -0.68 \text{ kN}$$

Then
$$M_B = 0.68 \times 18 = 12.2 \text{ kNm}$$
$$M_C = 0.68 \times 9 = 6.12 \text{ kNm}$$
$$M_D = 0$$

The B.M. diagram for AB will be a triangle of height 12.25 kNm at B and zero at A, on which is superimposed a parabola of height $Wh/8 = 5.46 \times 4.5/8 = 3.07$ kNm.

Area-Moments for Statically Determinate Condition:

Taking moments of the static B.M. diagram about the base AE and working from D round to A, the deflection at E

$$d_E = \frac{\Sigma Ay}{EI}$$

$$= \frac{(6.12 \times 9.693 \times \tfrac{1}{2})6.9}{EI} + \frac{(6.12 \times 9.693)6.3}{EI} + \frac{(6.13 \times 9.693 \times \tfrac{1}{2})5.7}{EI}$$

$$+ \frac{(12.25 \times 4.5 \times \tfrac{1}{2})3.0}{EI} + \frac{(3.07 \times 4.5 \times \tfrac{2}{3})2.25}{EI}$$

$EI d_E = 851.13$ units.

Area-Moments for Unit Horizontal Thrust:

These will be identical with those shown in Fig. 13.
Therefore $\quad EId_E = -851.18$ units.

Final Horizontal Thrust:

Now $\quad$ the actual deflection at $E = 0$

Therefore $\quad 851.13 - 851.18 H_E = 0$

or $\quad H_E = +1.00$ kN

Consequently, $\quad H_A = -5.46 + 1.00 = -4.46$ kN

Hence the B.M.s induced in the frame by the wind load on AB are found as follows:

$$M_B = -(+1.00 \times 4.5) + (0.68 \times 18) = +7.74 \text{ kNm}$$
$$M_C = -(+1.00 \times 8.1) + (0.68 \times 9) = -1.98 \text{ kNm}$$
$$M_D = -(+1.00 \times 4.5) \qquad\qquad = -4.5 \text{ kNm}$$

The final B.M. diagram is shown in Fig. 19.

THE AREA-MOMENT METHOD

Fig. 19

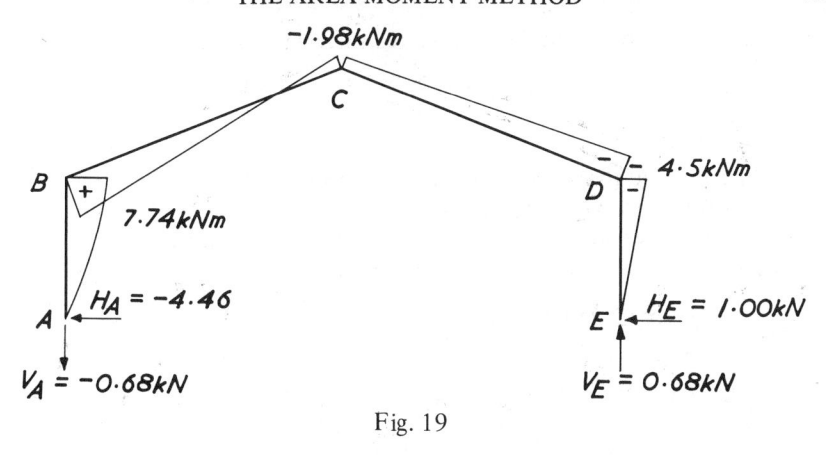

The B.M.s induced by the suction load of $0.25q$ on DE may be calculated by multiplying the corresponding values just calculated by -0.357.

Therefore the values will be as follows:

$$M_B = -0.357(+7.74) = -2.76 \text{ kNm}$$
$$M_C = -0.357(-1.98) = +0.71 \text{ kNm}$$
$$M_D = -0.357(-4.5) = +1.61 \text{ kNm}$$

Example 4. Calculate the B.M.s induced at B, C and D by the wind load on the rafter BC, shown in Fig. 20.

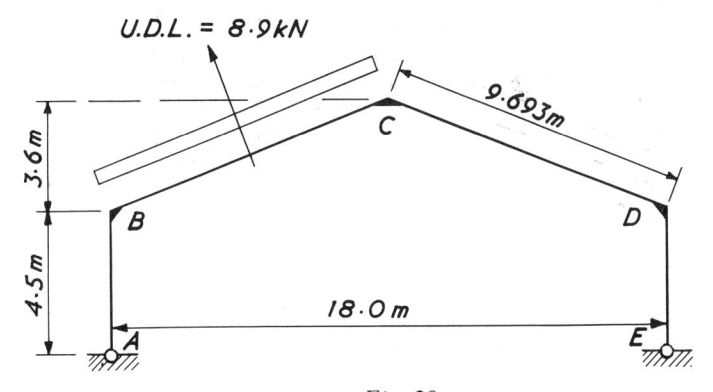

Fig. 20

Table 8 of CP3—Chapter V Part 2 (1970) gives the wind pressures to be taken on roofs normal to the surface, due to wind blowing at right angles to the eaves, in terms of the unit pressure q.

A roof slope of 1 in 2.5 is equivalent to an angle of $21° 48'$. The pressure for this slope is interpolated as $-0.33q$ (i.e. a suction) for the windward rafter BC and equals $-0.4q$ for the leeward rafter CD. If allowance is also made for an internal pressure of $0.2q$, the total suction on the rafter BC is $-0.33q - 0.2q = -0.53q$.

208 METHODS OF STRUCTURAL ANALYSIS

Hence, the suction load on *BC*, considered as being uniformly distributed, is

$$-0.53 \times 385 \times 4.5 \times 9.693 = -8.9 \text{ kN}.$$

Static B.M. Diagram:

It will be assumed that *A* is fixed in position and that *E* is free to move horizontally *inwards* under the action of the wind.

As the load acts obliquely, there will be a horizontal reaction at *A*.

Hence $H_A = 8.9 \times \sin 21° 48' = 3.31 \text{ kN}.$

Fig. 21

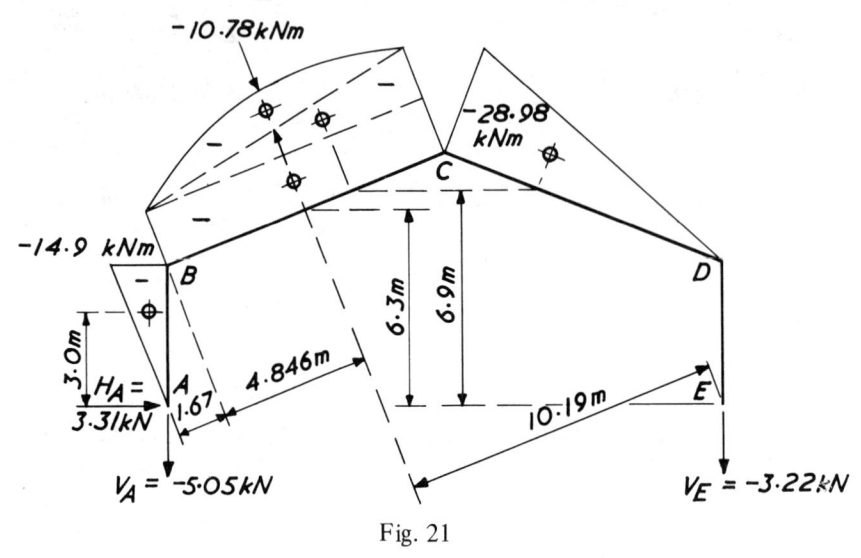

Figure 21 shows not only the static B.M. diagram but also the appropriate lever arms for the load.

Taking moments about *A* and *E* respectively

$$V_E = \frac{-8.9 \times 6.516}{18} = -3.22 \text{ kN}$$

$$V_A = \frac{-8.9 \times 10.19}{18} = -5.05 \text{ kN}$$

Then
$M_D = V_E \times 0 = 0$
$M_C = V_E \times 9.0 = -28.98 \text{ kNm}$
$M_B = -H_A \times 4.5 = -14.9 \text{ kNm}.$

Area-Moments for Statically Determinate Condition:

Taking area-moments about *AE* and working from *D* round to *A*

$$d_E = \frac{\Sigma Ay}{EI}$$

$EI d_E = (-28.98 \times 9.693 \times \tfrac{1}{2})6.9 + (-14.9 \times 9.693)6.3 + (-14.08 \times 9.693 \times \tfrac{1}{2})6.9$
$\qquad\qquad + (-10.78 \times 9.693 \times \tfrac{2}{3})6.3 + (-14.9 \times 4.5 \times \tfrac{1}{2})3.0$

$\qquad = -2\,889.3 \text{ units.}$

THE AREA-MOMENT METHOD

Area-Moments for Unit Horizontal Thrust:

These will be the same as in the two previous examples.

Hence $EId_E = -851.18$ units.

Final Horizontal Thrusts:

The actual deflection at $E = 0$

Hence $-2\,889.3 - 851.18H_E = 0$

or $H_E = -3.39$ kN

The thrust $H_A = 3.31 + H_E = -0.08$ kN.

Final B.M.s:

The final B.M. diagram is shown in Fig. 22.

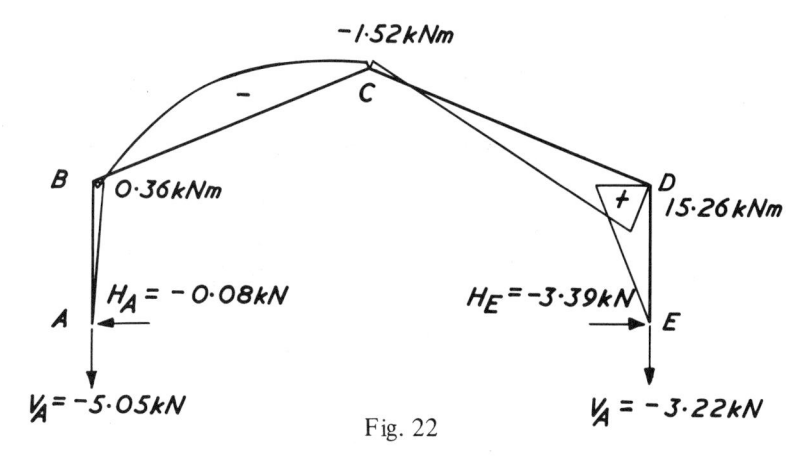

Fig. 22

$M_B = 0.08 \times 4.5 = +0.36$ kNm

$M_C = (3.39 \times 8.1) - (3.22 \times 9) = -1.52$ kNm

$M_D = 3.39 \times 4.5 \qquad = +15.26$ kNm

B.M.s for Suction on Leeward Rafter, CD:

The suction on the rafter CD is $-0.4q - 0.2q = -0.6q$ which is $0.6/0.53 = 1.13$ times the suction on the rafter BC.

Hence, the corresponding final thrusts and B.M.s will be as follows:

$H_A = 1.13(-3.39) = -3.83$ kN
$H_E = 1.13(-0.08) = -0.09$ kN
$V_A = 1.13(-3.22) = -3.64$ kN
$V_E = 1.13(-5.05) = -5.71$ kN
$M_B = 1.13(+15.26) = +17.24$ kNm
$M_C = 1.13(-1.52) = -1.72$ kNm
$M_D = 1.13(+0.36) = +0.41$ kNm

Summary of Bending Moments:

The B.M.s in kNm for Examples 2 to 4 may be summarised as follows:

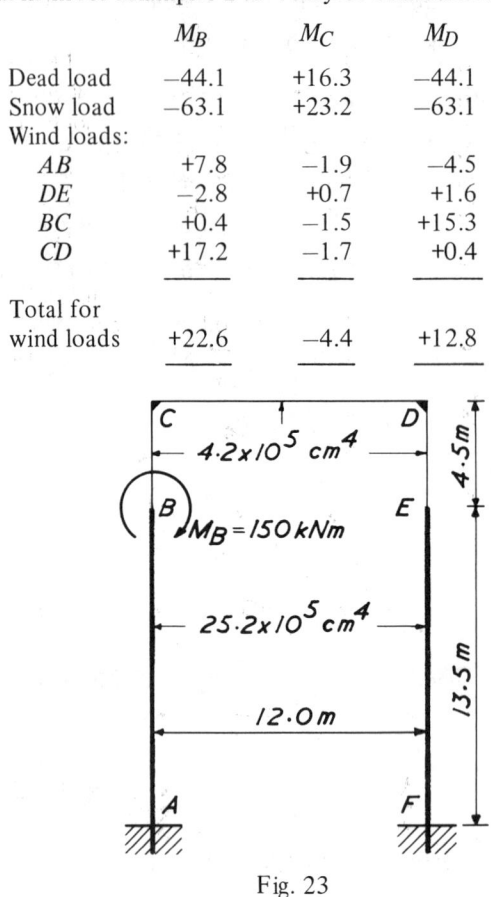

	M_B	M_C	M_D
Dead load	−44.1	+16.3	−44.1
Snow load	−63.1	+23.2	−63.1
Wind loads:			
AB	+7.8	−1.9	−4.5
DE	−2.8	+0.7	+1.6
BC	+0.4	−1.5	+15.3
CD	+17.2	−1.7	+0.4
Total for wind loads	+22.6	−4.4	+12.8

Fig. 23

Industrial Building Frame

Example 5. The industrial building frame shown in Fig. 23, which has fixed feet, is indeterminate to the third degree. The frame can be reduced to a statically determinate condition in several ways, but for this example the horizontal and vertical reactions and the fixing moment at F will be removed so that the frame can be treated as a cantilever.

In the free state, Fig. 24, the end F undergoes linear deflections in a horizontal direction, 1, and in a vertical direction, 2, and angular rotation, 3. As these movements are due to the moment M_B at B, they are defined as d_{M-1}, d_{M-2} and θ_{M-3} as shown in Fig. 24.

Now the redundants H_F, V_F and M_F have the effect of bringing F back to its original position. It is more convenient, however, to apply a unit horizontal force, 1, a unit vertical force, 2, and a unit moment, 3, at F, each of which will produce linear and angular movements at F. But the actual linear and angular movements

at F are zero. Then the combined effect of the moment M_B and the reactions may be expressed by the following Maxwell-Mohr work equations:

$$H_F d_{1 \to 1} + V_F d_{2 \to 1} + M_F d_{3 \to 1} + d_{M \to 1} = 0$$
$$H_F d_{1 \to 2} + V_F d_{2 \to 2} + M_F d_{3 \to 2} + d_{M \to 2} = 0$$
$$H_F \theta_{1 \to 3} + V_F \theta_{2 \to 3} + M_F \theta_{3 \to 3} + d_{M \to 3} = 0.$$

In each term the first subscript for d or θ relates to the force or moment that produced the deflection or rotation respectively, while the second indicates the direction of the deflection or rotation.

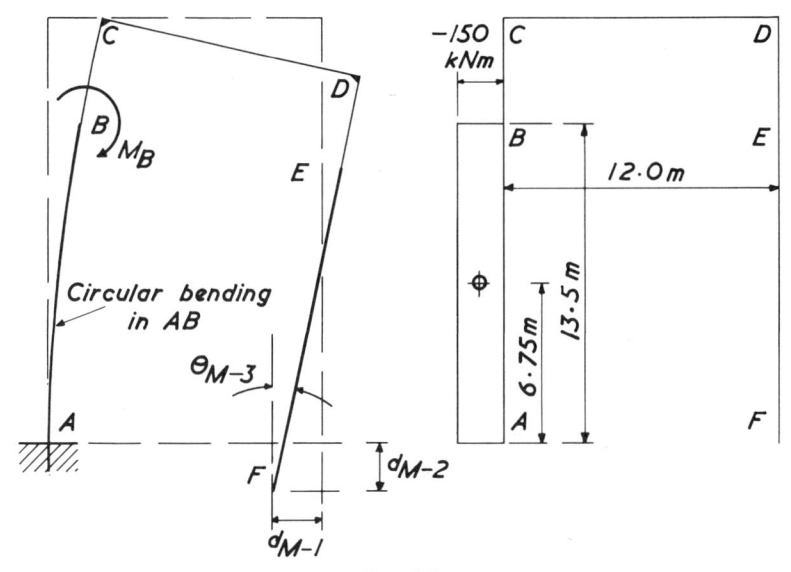

Fig. 24

It is advisable to adopt a definite system for the application of the unit redundants. As shown in Fig. 24, the moment applied at B tends to rotate the frame in a clockwise direction. Consequently, it can be assumed that H_F will act outwards, V_F upwards and M_F in an anti-clockwise direction. Therefore, the unit redundants are shown applied in these directions.

Cases will constantly arise where it is not obvious in which direction the redundants act. Assumptions must be made and, where these are wrong, the final answer will have a negative value.

In the following calculations the modulus of elasticity E, being constant, is ignored, while the values of the moment of inertia of the members are comparative only, viz.: 6 for the columns below crane level and 1 for the frame above crane level.

This arrangement results in simplification of the calculations and while the 'deflections' can be correctly compared they are not expressed in any real linear or angular unit.

Unit Horizontal Load 1 at F:

Consider Fig. 25, which shows the unit horizontal load applied at F.

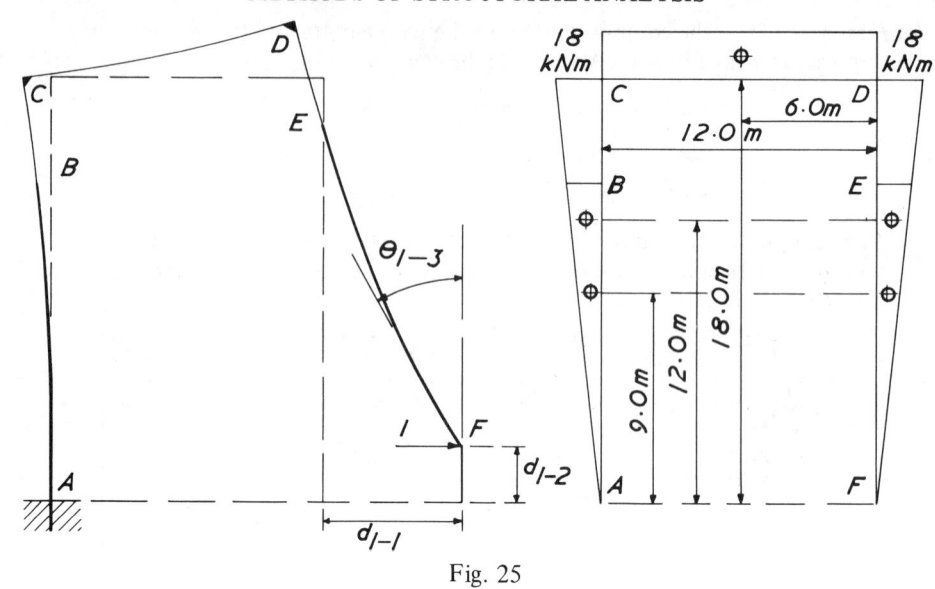

Fig. 25

Taking area-moments about the axis AF and considering the columns first

$$d_{1-1} = \frac{\Sigma Ay}{I}$$

$$= \frac{2(13.5 \times 13.5 \times \frac{1}{2})9.0}{6} + \frac{2[(18.0 \times 18.0 \times \frac{1}{2})12.0 - (13.5 \times 13.5 \times \frac{1}{2})9.0]}{1} +$$

$$+ \frac{(18.0 \times 12.0)18.0}{1}$$

$$= \frac{2(91.125)9.0}{6} + \frac{2[(162)12.0 - (91.125)9.0]}{1} + \frac{(216)(18.0)}{1}$$

$$= 6\,409.125 \text{ units}$$

Taking area-moments about the axis DF and working from A round to F

$$d_{1-2} = \frac{\Sigma Ax}{I}$$

$$= \frac{(91.125)12.0}{6} + \frac{(162 - 91.125)12.0}{1} + \frac{(216)6}{1} + 0 + 0$$

$$= 2\,328.75 \text{ units.}$$

Considering the rotation at F

$$\theta_{1-3} = \frac{\Sigma A}{I}$$

$$= \frac{2(91.125)}{6} + \frac{2(162 - 91.125)}{1} + \frac{216}{1}$$

$$= 388.125 \text{ units.}$$

Unit Vertical Load 2 at F:

Consider Fig. 26, which shows the unit vertical load applied at F.

THE AREA-MOMENT METHOD 213

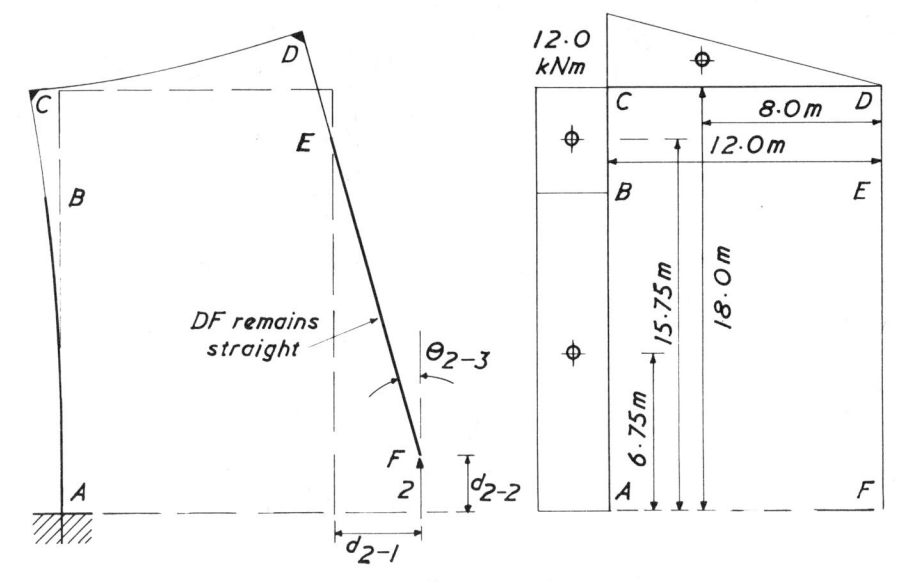

Fig. 26

Taking area-moments about AF

$$d_{2-1} = \frac{\Sigma Ay}{I}$$

$$= \frac{(12.0 \times 13.5)6.75}{6} + \frac{(12.0 \times 4.5)15.75}{1} + \frac{(12.0 \times 12.0 \times \frac{1}{2})18.0}{1}$$

$$= \frac{(162)6.75}{6} + \frac{(54) \times 15.75}{1} + \frac{(72)18.0}{1}$$

$$= 2\ 328.75 \text{ units.}$$

Note: $d_{2-1} = d_{1-2}$.

Taking area-moments about DF

$$d_{2-2} = \frac{\Sigma Ax}{I}$$

$$= \frac{(162)12.0}{6} + \frac{(54)12.0}{1} + \frac{(72)8.0}{1}$$

$$= 1\ 548 \text{ units.}$$

Rotation at F:

$$\theta_{2-3} = \frac{(162)}{6} + \frac{(54)}{1} + \frac{(72)}{1}$$

$$= 153 \text{ units.}$$

Unit Moment 3 at F:

Considering Fig. 27, which shows the unit moment 3 applied at F, and taking area-moments about AF

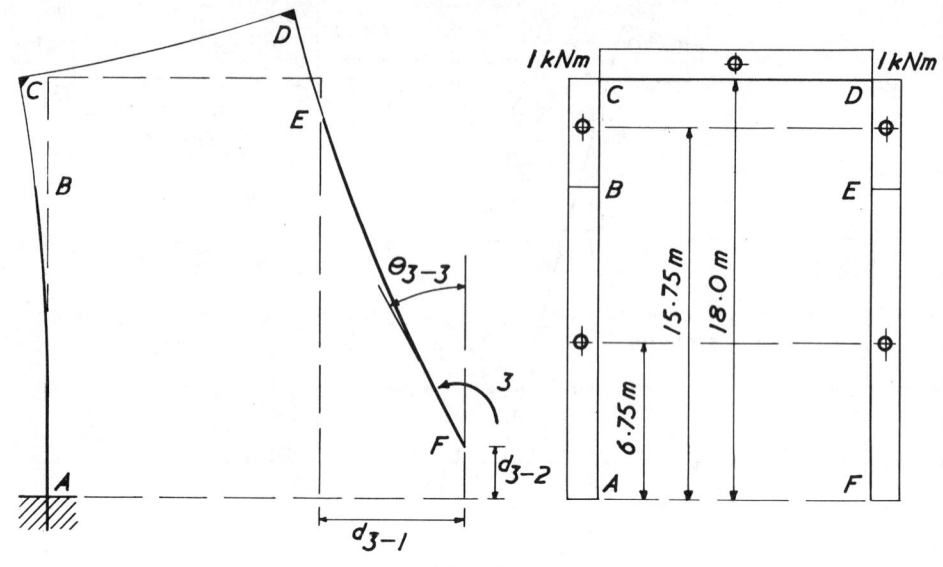

Fig. 27

$$d_{3\to1} = \frac{\Sigma Ay}{I}$$
$$= \frac{2(1.0 \times 13.5)6.75}{6} + \frac{2(1.0 \times 4.5)15.75}{1} + \frac{(1.0 \times 12.0)18.0}{1}$$
$$= \frac{2(13.5)6.75}{6} + \frac{2(4.5)15.75}{1} + \frac{(12.0)18.0}{1}$$
$$= 388.125 \text{ units.}$$

Note: $d_{3\to1} = \theta_{1\to3}$.

Taking area-moments about DF,
$$d_{3\to2} = \frac{(13.5)12}{6} + \frac{(4.5) \times 12}{1} + \frac{(12.0)6}{1}$$
$$= 153 \text{ units.}$$

Note: $d_{3\to2} = \theta_{2\to3}$.

$$\theta_{3\to3} = \frac{\Sigma A}{I}$$
$$= \frac{2(13.5)}{6} + \frac{2(4.5)}{1} + \frac{(12.0)}{1}$$
$$= 25.5 \text{ units.}$$

Static B.M. Diagram:

Referring back to Fig. 24 and taking area-moments about AF
$$d_{M\to1} = \frac{\Sigma Ay}{I}$$
$$= \frac{(-150 \times 13.5)6.75}{6}$$
$$= \frac{(-2\,025)6.75}{6}$$
$$= -2\,278.125 \text{ units.}$$

Taking area-moments about DF

$$d_{M-2} = \frac{\Sigma Ax}{I}$$

$$d_{M-2} = \frac{(-2\,025)12.0}{6}$$

$$= -4\,050 \text{ units.}$$

Rotation at F:

$$\theta_{M-3} = \frac{\Sigma A}{I}$$

$$= \frac{(-2\,025)}{6}$$

$$= -337.5 \text{ units.}$$

Sufficient calculations have been made to complete the work equations which can be rearranged, taking account of signs, as follows:

$6\,409.125\ H_F + 2\,328.75\ V_F + 388.125\ M_F = 2\,278.125$

$2\,328.75\ H_F + 1\,548.0\ V_F + 153.0\ M_F = 4\,050.0$

$388.125\ H_F + 153.0\ V_F + 25.5\ M_F = 337.5$

(Note the symmetry of the equations.)

From which
$H_F = -\ 5.70 \text{ kN}$
$V_F = +\ 3.21 \text{ kN}$
$M_F = +80.69 \text{ kNm.}$

Therefore, H_F, being negative, actually acts inwards, not outwards as assumed.

The final B.M. diagram is shown in Fig. 28. The calculations are completed in full to demonstrate that the values for d_{2-1}, d_{3-1} and d_{3-2} are numerically equal

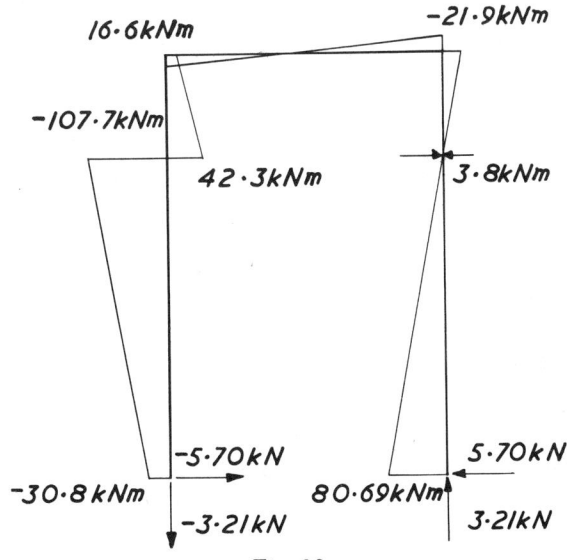

Fig. 28

to d_{1-2}, θ_{1-3} and θ_{2-3} respectively. It will be obvious, however, that the work may be greatly simplified by making use of the symmetrical form of the equations. Furthermore, if the frame is now loaded in another manner, the expressions for H_F, V_F and M_F remain in the exact form given above. Only the figures to the right of the equations, which relate to the actual loads, are changed.

Symmetrical Multi-Storey Frames

Side Loading

When horizontal loads are applied at the joints of symmetrical building frames as shown in Figs. 29 (a) and 30 (a) points of contraflexure form at the centre of the cross-beams and the moments induced in the right-hand half of the frame are numerically equal to those in the left-hand half. Consequently, such a frame may be

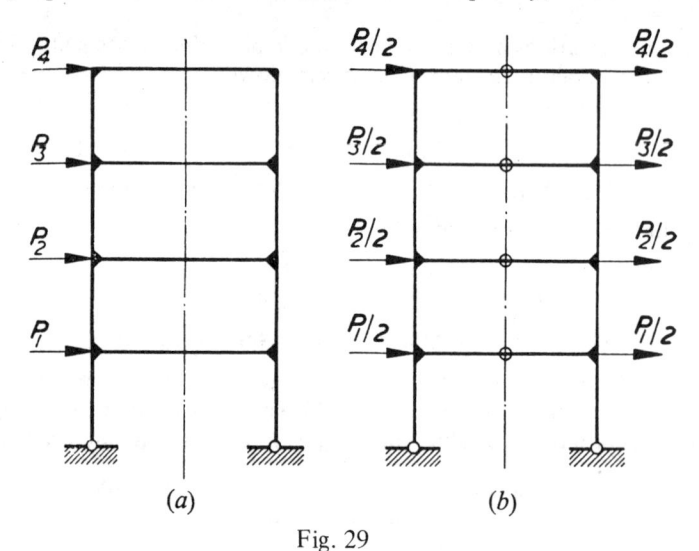

Fig. 29

Fig. 30

analysed by considering one-half of the frame subjected to one-half of the loads, as shown in Figs. 29 (b) and 30 (b).

Readers may note that a similar approach was adopted by Mr. N. Naylor, B.Sc., A.M.I.Struct.E., in his article on the Moment Distribution method of analysis mentioned on page 256. However, Naylor's method may be applied only to frames with parallel columns, as shown in Fig. 29. Area-moments may be applied also to frames with sloping columns, as shown in Fig. 30.

Example 6. Calculate the bending moments induced in the frame shown in Fig. 31. The frame may be analysed by considering the left-hand half of the frame

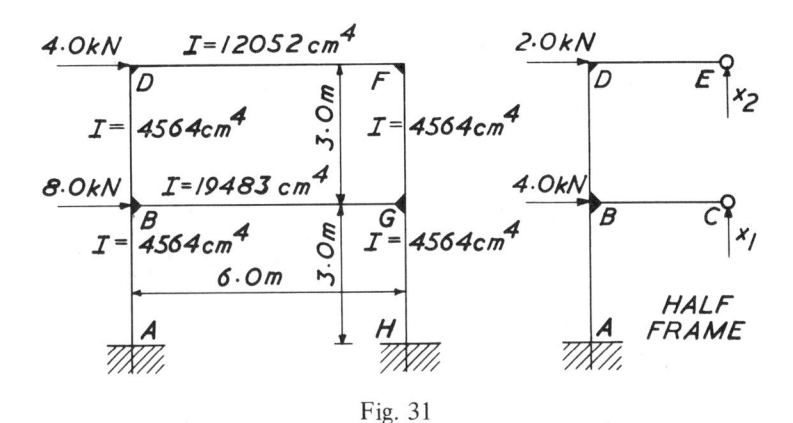

Fig. 31

under half the loads, as shown. When this is done the only unknowns are the vertical propping forces x_1 and x_2. Consequently, the appropriate Maxwell-Mohr work equations are as follows:

$$x_1 d_{1 \to 1} + x_2 d_{2 \to 1} + d_{P \to 1} = 0$$
$$x_1 d_{1 \to 2} + x_2 d_{2 \to 2} + d_{P \to 2} = 0$$

As before, relative moments of inertia can be used and whilst the actual inertias of the sections are shown, in the half frame these have been reduced to proportions taking the column sections as 1.0.

Area-Moments for Props at C and E:

Considering Fig. 32, ignoring the modulus of elasticity E, which is constant, and taking area-moments about the vertical axis through CE

$$d_{1 \to 1} = \frac{\Sigma A x}{I}$$

$$= \frac{(3.0 \times 3.0)3.0}{1.0} + \frac{(3.0 \times 3.0 \times \frac{1}{2})2.0}{4.27}$$

$$= 27.0 + 2.11$$

$$= 29.11 \text{ units.}$$

Similarly, considering Figs. 32 or 33,

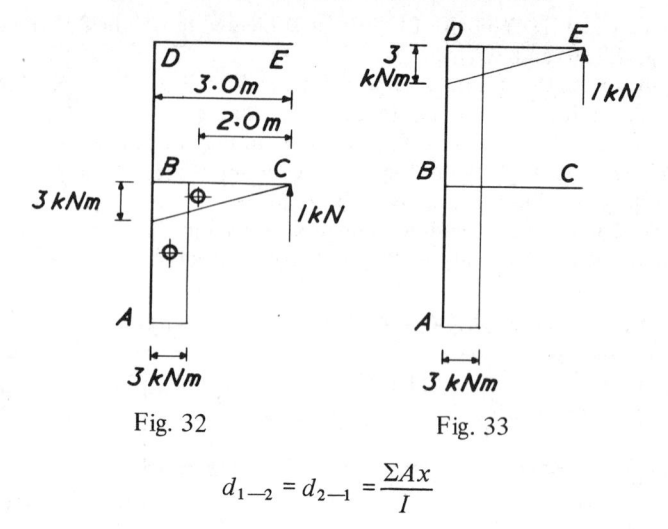

METHODS OF STRUCTURAL ANALYSIS

Fig. 32

Fig. 33

$$d_{1-2} = d_{2-1} = \frac{\Sigma Ax}{I}$$

(where the area A applies to the member AB only)

$$= \frac{(3.0 \times 3.0)3.0}{1.0}$$

$$= 27.0 \text{ units.}$$

Considering Fig. 33

$$d_{2-2} = \frac{(3.0 \times 6.0)3.0}{1.0} + \frac{(3.0 \times 3.0 \times \frac{1}{2})2.0}{2.64}$$

$$= 54.0 + 3.41$$

$$= 57.41 \text{ units.}$$

Static B.M.

The static B.M. diagram is shown in Fig. 34.

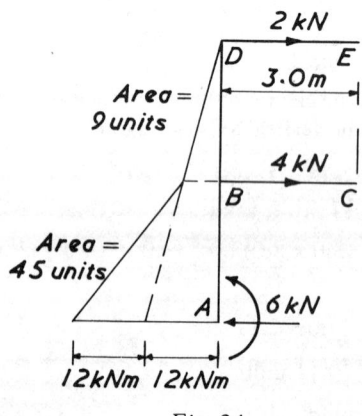

Fig. 34

THE AREA-MOMENT METHOD

Taking area-moments as before

$$d_{P-1} = \frac{\Sigma Ax}{I}$$
$$= \frac{-(45.0 \times 3.0)}{1}$$
$$= -135 \text{ units.}$$

$$d_{P-2} = \frac{\Sigma Ax}{I}$$
$$= \frac{-(45+9)3.0}{1}$$
$$= -162 \text{ units.}$$

Work Equations:

The work equations may now be rearranged and written down as follows:

$$29.11 x_1 + 27.0 \; x_2 = 135.0$$
$$27.0 \; x_1 + 57.41 x_2 = 162.0$$

From which
$$x_1 = 3.58 \text{ kN}$$
$$x_2 = 1.14 \text{ kN.}$$

Final B.M.s

The final B.M.s, shown in Fig. 35, are calculated as follows:

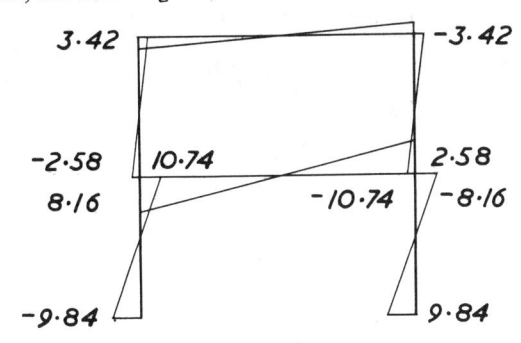

Fig. 35

$M_{DE} = M_{DB} = 1.14 \times 3.0 \quad = 3.42 \text{ kNm}$

$M_{BD} = -(2.0 \times 3.0) + 3.42 = -2.58 \text{ kNm}$

$M_{BC} = 3.58 \times 3.0 \quad\quad\quad = 10.74 \text{ kNm}$

$M_{BA} = 10.74 - 2.58 \quad\quad = 8.16 \text{ kNm}$

$M_{AB} = -(2.0 \times 6.0) - (4.0 \times 3.0) + 3.42 + 10.74$
$\quad\quad = -9.84 \text{ kNm.}$

Example 7. The frame shown in Fig. 36 has hinged feet which necessitate a different method of treatment from that in the previous Example.

When this frame is cut down the middle it requires a support, such as a vertical prop applied at *C*, to hold it in position.

220 METHODS OF STRUCTURAL ANALYSIS

Fig. 36

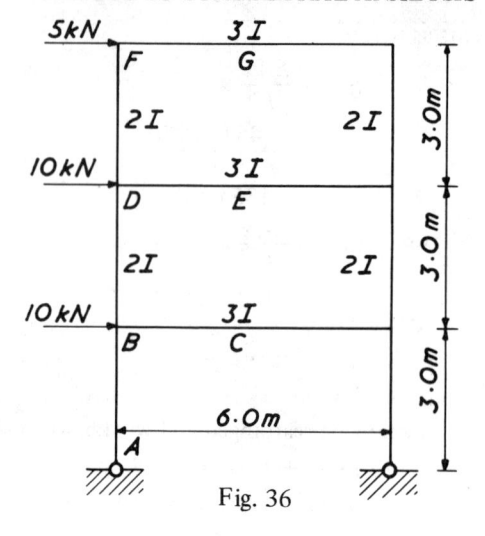

Taking moments about A, the force in the prop

$$V_C = \frac{(5.0 \times 3.0) + (5.0 \times 6.0) + (2.5 \times 9.0)}{3.0}$$

$$= 22.5 \text{ kN}$$

Then the appropriate static moment diagram is shown in Fig. 37.

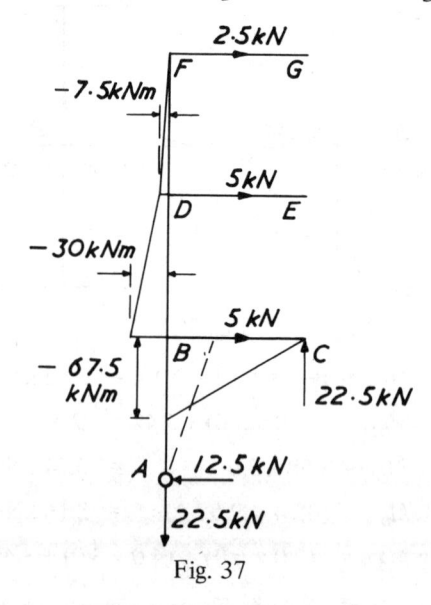

Fig. 37

As before, the redundants are x_1 and x_2.
Then

$$x_1 d_{1 \to 1} + x_2 d_{2 \to 1} + d_{P \to 1} = 0$$

$$x_1 d_{1 \to 2} + x_2 d_{2 \to 2} + d_{P \to 2} = 0$$

THE AREA-MOMENT METHOD 221

Considering Fig. 38, and taking area-moments about the vertical axis through CG

$$d_{1-1} = \frac{\Sigma Ax}{EI}$$

$$= \frac{(3.0 \times 3.0)3.0}{2.0} + \frac{2(3.0 \times 3.0 \times \frac{1}{2})2.0}{3.0}$$

$$= 13.5 + 6.0$$

$$= 19.5 \text{ units.}$$

Similarly, considering Fig. 38 or 39,

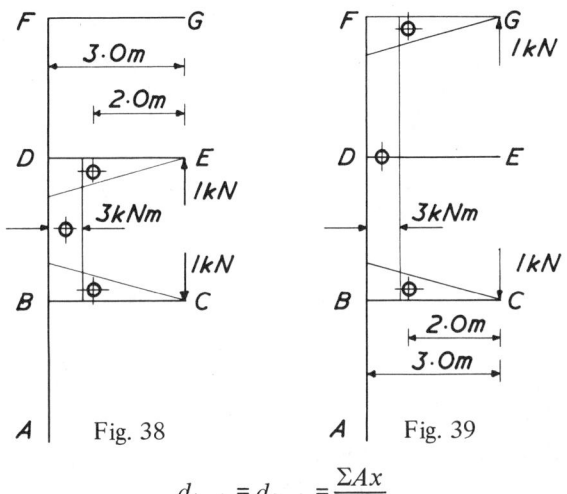

Fig. 38   Fig. 39

$$d_{1-2} = d_{2-1} = \frac{\Sigma Ax}{EI}$$

(where the area A applies to the members BD and BC only)

$$= \frac{(3.0 \times 3.0)3.0}{2.0} + \frac{(3.0 \times 3.0 \times \frac{1}{2})2.0}{3.0}$$

$$d_{1-2} = d_{2-1} = 13.5 + 3.0 = 16.5 \text{ units.}$$

Considering Fig. 39,

$$d_{2-2} = \frac{(3.0 \times 6.0)3.0}{2.0} + \frac{2(3.0 \times 3.0 \times \frac{1}{2})2.0}{3.0}$$

$$= 27.0 + 6.0 = 33.0 \text{ units.}$$

The static B.M. diagram is shown in Fig. 37:

$$d_{P-1} = -\left[\frac{(7.5 \times 3.0)3.0}{2.0} + \frac{(22.5 \times 3.0 \times \frac{1}{2})3.0}{2.0} + \frac{(67.5 \times 3.0 \times \frac{1}{2})2.0}{3.0}\right]$$

$$= -(33.75 + 50.625 + 67.5) = -151.875 \text{ units.}$$

$$d_{P-2} = -\left[\frac{151.875}{1.0} + \frac{(7.5 \times 3.0 \times \frac{1}{2})3.0}{2}\right]$$

$$= -(151.875 + 16.875) = -168.75 \text{ units.}$$

Then the work equations become:
$$19.5x_1 + 16.5x_2 = 151.875$$
$$16.5x_1 + 33\ x_2 = 168.75$$
From which
$$x_1 = 6.000 \text{ kN}$$
$$x_2 = 2.114 \text{ kN}$$

Now the final B.M. diagram may be drawn, as shown in Fig. 40.

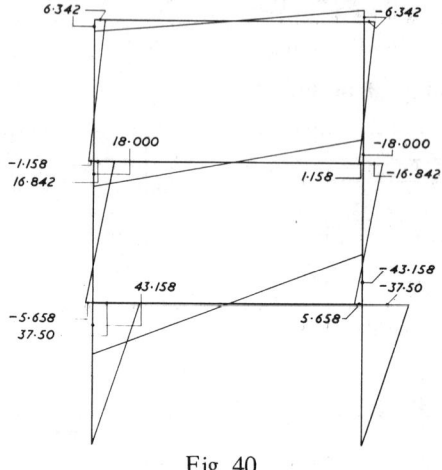

Fig. 40

Vertical Loading

Area-moments compare favourably with other methods of solution of symmetrical frames subjected to side loading at the joints. However, unless the members are non-prismatic, vertical loading is more quickly treated by some other method.

A frame vertically loaded may be cut down the middle as shown in Fig. 41. For symmetrical loads there will be no vertical forces V and the number of redundants per cross-beam will be reduced from three to two.

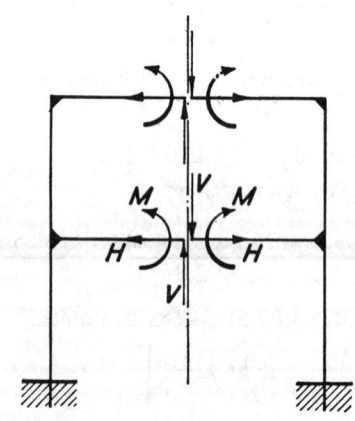

Fig. 41

Closed Frames

Similar treatment to that shown in Fig. 41 may be applied to a closed frame under any type of load. Fig. 42 shows a typical frame together with the appropriate B.M. diagrams.

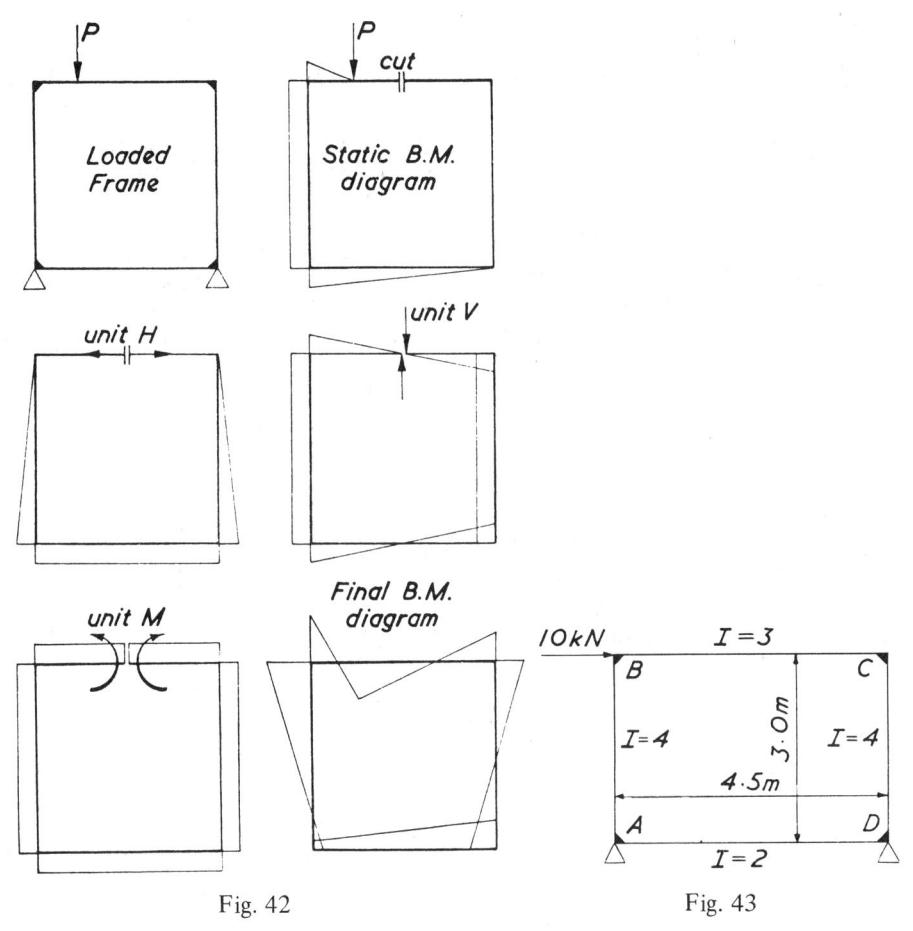

Fig. 42

Fig. 43

Example 8. The frame *ABCD* shown in Fig. 43 is subjected to a horizontal load of 1 kN applied at *B*. Such a load is analogous to those in Examples 6 and 7. There will be points of contraflexure at the mid-points of the cross-beams *BC* and *AD* while the forces *H* at the mid-point of *BC* will be 0.5 kN. Hence the only unknown is *V*.

Unit Vertical Loads:

Consider Fig. 44.

$$d_{2-2} = \frac{\Sigma Ax}{I}$$

$$= \frac{(2.25 \times 2.25 \times \tfrac{1}{2})1.5 - (-2.25 \times 2.25 \times \tfrac{1}{2})1.5}{2}$$

224   METHODS OF STRUCTURAL ANALYSIS

$$+ \frac{(2.25 \times 2.25 \times \frac{1}{2})1.5 - (-2.25 \times 2.25 \times \frac{1}{2})1.5}{3}$$

$$+ \frac{(2.25 \times 3.0)2.25 - (-2.25 \times 3.0)2.25}{4}$$

$$= 13.922 \text{ units.}$$

Static B.M.s

The appropriate static B.M. diagram is shown in Fig. 45.

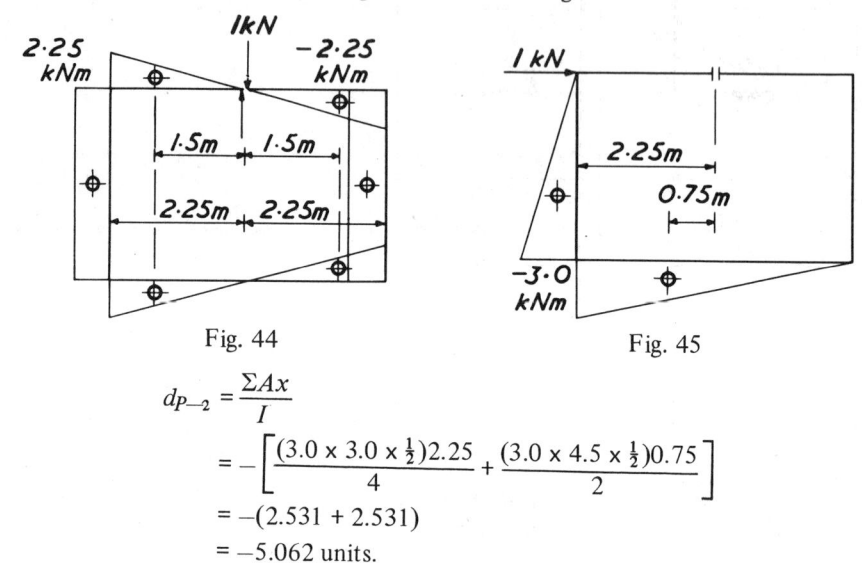

Fig. 44   Fig. 45

$$d_{P-2} = \frac{\Sigma Ax}{I}$$

$$= -\left[\frac{(3.0 \times 3.0 \times \frac{1}{2})2.25}{4} + \frac{(3.0 \times 4.5 \times \frac{1}{2})0.75}{2}\right]$$

$$= -(2.531 + 2.531)$$

$$= -5.062 \text{ units.}$$

Now there is no vertical deflection at the mid-point of *BC*. Hence

$$Vd_{2-2} + d_{P-2} = 0$$

or $13.922 V = 5.062$

and $V = 0.364$ kN.

Final B.M.s

$H = 0.5$ kN and $V = 0.364$ kN.

Therefore, the final B.M.s, shown in Fig. 46, are derived as follows:

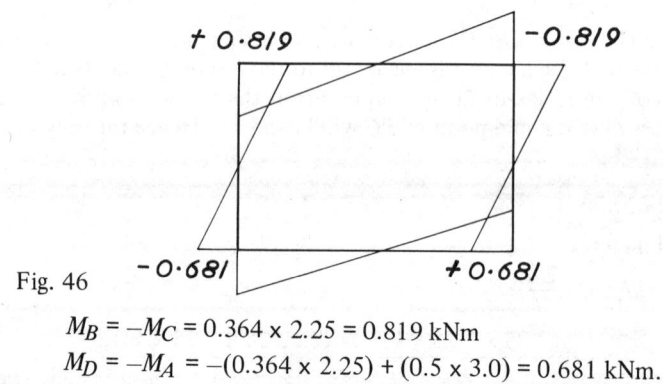

Fig. 46

$$M_B = -M_C = 0.364 \times 2.25 = 0.819 \text{ kNm}$$
$$M_D = -M_A = -(0.364 \times 2.25) + (0.5 \times 3.0) = 0.681 \text{ kNm.}$$

(b) Moment Distribution

Moment Distribution is a mechanical process of dealing with indeterminate structures by means of successive approximations in which the moments themselves are treated directly, the calculations involved being purely arithmetical.

The method is unique in that all joints are initially considered to be fixed against rotation. The fixed end moments are determined for each member as though it were an encastré beam and then the joints are allowed to rotate, either separately or all at once, the moments induced by the rotations being distributed among the members until the algebraic sum of the moments at each internal joint is zero.

The sign convention most commonly adopted for Moment Distribution is that all moments acting on individual members from supports or other members of a frame

MOMENT DISTRIBUTION CONVENTION

NORMAL CONVENTION

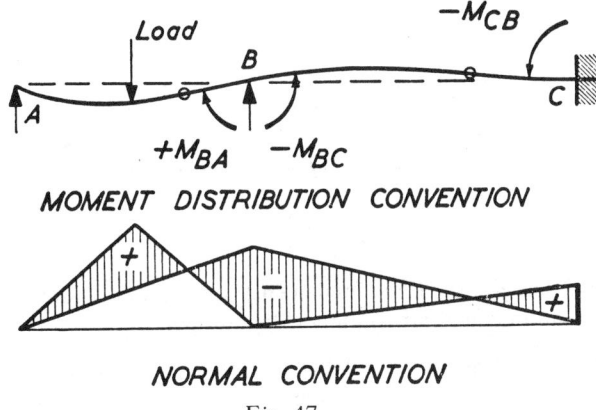

Fig. 47

are positive if clockwise in application and negative if anti-clockwise. Before a B.M. diagram is drawn, this convention must be translated into the normal convention whereby in continuous beams, for example, sagging moments are positive and hogging moments are negative. The two conventions are compared in Fig. 47.

It will be found that the operations of Moment Distribution are more readily understood and checked if the reader considers initially how the structure deflects under load. Consequently, deflection diagrams are incorporated in many of the examples.

Although the structural principles on which Moment Distribution is based are well known, it is advisable to consider them in a definite sequence.

Figure 48 shows a beam AB of constant cross-section, i.e. a prismatic beam, fixed in position and direction at A and fixed in position, but not in direction at B. When the moment M_{BA} is applied at B a moment M_{AB} is induced at A. It was mentioned on page 88 that

$$M_{AB} = \tfrac{1}{2} M_{BA} \quad \ldots \ldots \ldots \ldots \ldots \ldots \ldots \ldots \text{(i)}$$

and $M_{BA} = 4E \tan\theta \dfrac{I}{L}$

$$= 4E\theta \dfrac{I}{L} \quad \text{(for small values of } \theta \text{)} \ldots \ldots \ldots \text{(ii)}$$

Similarly, if A is fixed in position but not in direction, as in Fig. 49, then

$$M_{BA} = 3E\theta \frac{I}{L} \quad \text{(for small values of } \theta\text{)} \quad \ldots \ldots \ldots \text{(iii)}$$

Fig. 48

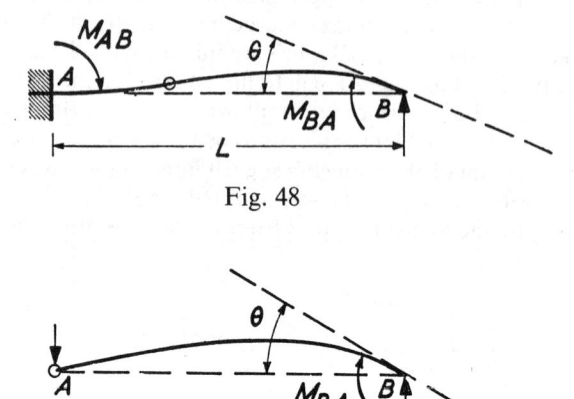

Fig. 49

These equations give the three fundamental principles of Moment Distribution applicable to continuous beams on unyielding supports.

Principle I (Equation (i)). When a moment is applied at one end of a prismatic beam, that end remaining fixed in position but not in direction, the other end being fixed both in position and direction, a moment of half the amount and the same sign is induced at the second end.

Principle II (Equation (ii)). When one end of a beam remains fixed in position and direction, the moment required to produce a rotation of a given angle at the other end of the beam, which remains fixed in position, is proportional to the value I/L for the beam, provided that E is constant. The value I/L, known by the symbol K, is the stiffness factor for the particular beam in question.

Principle III (Equation (iii)). When one end of a beam is rotated through a given angle, remaining fixed in position, and the other end remains fixed in position but not in direction, the moment required at the first end is $\frac{3}{4}$ of that required if the second end were fixed both in position and direction, i.e. the equivalent stiffness factor for the beam is $\frac{3}{4}I/L = \frac{3}{4}K$.

The three foregoing principles alone are applied when the supports do not yield. However, when the joints change their positions B.M.s have to be modified accordingly. Consider Fig. 50 in which the end A of an encastré beam AB, of span L, has settled an amount d, the ends A and B remaining parallel in direction.

It is shown on page 41 that

$$M_{AB} = M_{BA} = \frac{6EId}{L^2} \quad \ldots \ldots \ldots \ldots \ldots \text{(iv)}$$

Similarly, in Fig. 51, where the end A is hinged, i.e. not fixed in direction,

$$M_{BA} = \frac{3EId}{L^2} = \frac{6EId}{2L^2} \quad \ldots \ldots \ldots \ldots \ldots \text{(v)}$$

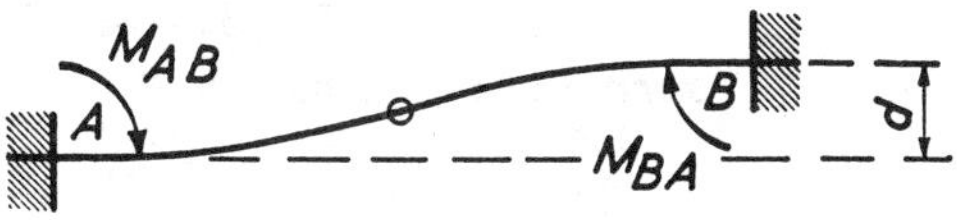

Fig. 50

Fig. 51

From these equations, the following further principles may be derived.

Principle IV (Equation (iv)). When one end of a beam is deflected through a given distance, that end remaining parallel to its original position and the other remaining fixed in position and direction, equal moments of the same sign are induced at each end, proportional to the I/L^2 value of the beam.

Principle V (Equation (v)). When a hinged end of a beam is deflected through a given distance, the other end remaining fixed in position and direction, a moment is induced at the second end, proportional to the $I/2L^2$ value of the beam.

Having stated the principles, the moment distribution processes may be explained by considering some simple examples.

Continuous Beams

Example 1. Figure 52 shows a continuous beam ABC, of constant cross-section, which is fixed in position and direction at A and C and simply supported at B and which carries uniformly distributed loads of 65.0 kN/m on AB and 32.0 kN/m on BC.

Under these loads the beam will rotate in an anti-clockwise direction at B and, as it is a fundamental assumption in the theory of continuous beams that the slope does not change over a support, the beam will rotate the same amount θ on either side of B. However, assume that the beam does not rotate at B, but through some locking device remains horizontal after the loads are applied. Then AB and BC are in effect two separate encastré beams and the moments at the end of each span are fixed-end moments (F.E.M.s) depending only on the functions of the span and the loading.

The functions of a span are (1) L, its length, and (2) I, its moment of inertia.

When I is constant throughout all spans, the process is straightforward. Hence, in this example I may be ignored.

METHODS OF STRUCTURAL ANALYSIS

LOAD DIAGRAM: Beam fixed at A and C, with 65.0 kN per m over span AB (3.0 m) and 32.0 kN per m over span BC (3.6 m); $I = 5000 \text{ cm}^4$.

DEFLECTION: zero slope at A, slope at B, zero slope at C, with points of contraflexure shown.

DISTRIBUTION TABLE

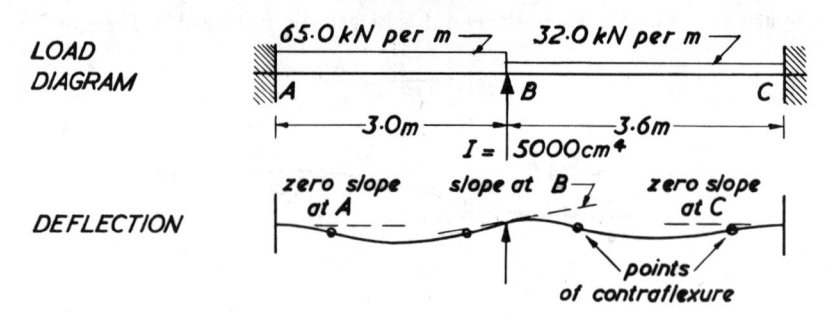

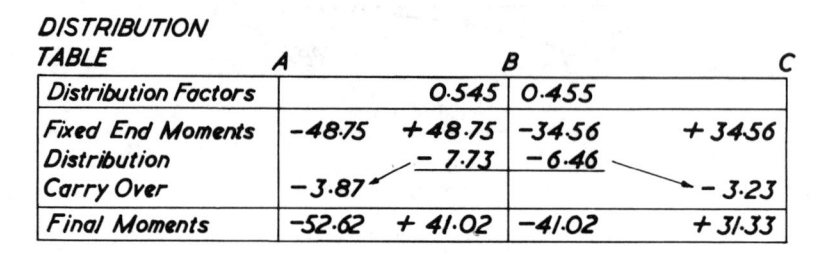

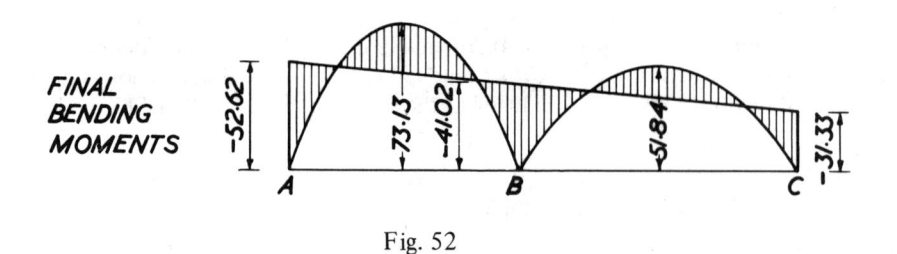

	A	B		C
Distribution Factors		0.545	0.455	
Fixed End Moments	−48.75	+48.75	−34.56	+34.56
Distribution		−7.73	−6.46	
Carry Over	−3.87			−3.23
Final Moments	−52.62	+41.02	−41.02	+31.33

FINAL BENDING MOMENTS: −52.62 at A; −73.13, −41.02 near B left; −51.84 on span BC; −31.33 at C.

Fig. 52

Consider the F.E.M.s for the span *AB*. From the tables on pp. 271 to 275.

F.E.M.$_{AB}$, (being anti-clockwise)

$$= -\frac{WL}{12} = \frac{-65.0 \times 3.0 \times 3.0}{12}$$

$$= -48.75 \text{ kNm.}$$

F.E.M.$_{BA}$ (being clockwise)

$$= +48.75 \text{ kNm.}$$

Similarly, for the span *BC*

$$\text{F.E.M.}_{BC} = -\frac{WL}{12} = \frac{-32 \times 3.6 \times 3.6}{12}$$

$$= -34.56 \text{ kNm}$$

F.E.M.$_{CB}$ = +34.56 kNm

Having made these calculations the beam can be released at B and allowed to rotate in an anti-clockwise direction. Now the algebraic sum of the moments on either side of this support must be zero. However, when the beam was horizontal at B, the value of M_{AB} was +48.75 kNm and that of M_{BC} was −34.56 kNm. Therefore, to produce equilibrium at B, the total moment induced by the rotation of the beam there must be −14.19 kNm, since the moments are out of balance by +48.75 − 34.56 = +14.19 kNm.

But the span ends meeting at B rotate through the same angle θ. Consequently, the moments induced by rotation on either side of B are proportional to the stiffness of AB and BC (Principle II). In other words, the moment of −14.19 kNm is distributed between AB and BC in proportion to their stiffness, i.e. in the proportion $K_{AB}/(K_{AB} + K_{BC})$ to the left and $K_{BC}/(K_{AB} + K_{BC})$ to the right.

These proportions are known as the distribution factors (D.F.) for the spans. Although it may sometimes be more accurate to employ fractions, these factors are usually expressed in decimals, but, in any case, the factors for a support or joint must always add up to unity.

Now $I = 5\,000$ cm^4, $AB = 3.0$ m, $BC = 3.6$ m.

Hence $$K_{AB} = \frac{5\,000 \times 10^4}{3.0 \times 10^3} = 16.67 \times 10^3$$

and $$K_{BC} = \frac{5\,000 \times 10^4}{3.6 \times 10^3} = 13.89 \times 10^3$$

Therefore $$\text{D.F.}_{AB} = \frac{16.67 \times 10^3}{16.67 \times 10^3 + 13.89 \times 10^3} = 0.545$$

$$\text{D.F.}_{BC} = \frac{13.89 \times 10^3}{16.67 \times 10^3 + 13.89 \times 10^3} = 0.455$$

The operation of moment distribution is shown in the distribution table in Fig. 52, 0.545 × −14.19 = −7.73 kNm being added to the end BA and 0.455 × −14.19 = −6.46 kNm being added to the end BC.

From a consideration of Principle I moments are induced at the outer ends of the beam at A and C, equal to half the moments distributed between the spans at B and of the same signs.

Hence, 0.5 × −7.73 = −3.865 kNm must be transferred to the end A and 0.5 × −6.46 = −3.23 kNm to the end C. This process, which is known as the 'carry-over' process, is shown in Fig. 52.

The final moments in the beam are found by adding each column algebraically. When constructing a B.M. diagram it is convenient to remember that the moment to the right of a support in a distribution table bears the same sign as the support moment in the B.M. diagram (in the normal sign convention). Therefore, the final moments at A, B and C are respectively −52.62, −41.02 and −31.33 kN.

The maximum static or 'free' B.M.s for AB and BC are obtained from the formula $+WL/8$ and equal 73.13 and 51.84 kNm respectively.

Example 2. The continuous beam $ABCDE$, which is shown in Fig. 53 and which has been analysed on p. 54, is simply supported at A and overhangs the other outside support D. Consequently, the beam is free to rotate at A and D, although it is restrained to a certain extent at D by the load at E, and when deriving F.E.M.s the beam is assumed to be fixed in a horizontal position at B and C only. Therefore,

230   METHODS OF STRUCTURAL ANALYSIS

Principle III applies to spans AB and CD, and the stiffness factors for these spans equal $\tfrac{3}{4}K$. The stiffness factor for $BC = K$.

Now the moment of inertia I differs for each span, although it is constant throughout a span, as shown in Fig. 53.

Hence
$$\tfrac{3}{4}K_{AB} = \tfrac{3}{4} \times \frac{8\,500 \times 10^4}{3.0 \times 10^3} = 14.17 \times 10^3$$

$$K_{BC} = \frac{6\,500 \times 10^4}{3.75 \times 10^3} = 17.33 \times 10^3$$

$$\tfrac{3}{4}K_{CD} = \tfrac{3}{4} \times \frac{5\,500 \times 10^4}{3.75 \times 10^3} = 11.00 \times 10^3$$

$$\text{D.F.}_{BA} = \tfrac{3}{4}K_{AB}/(\tfrac{3}{4}K_{AB} + K_{BC}) = \frac{14.17 \times 10^3}{17.33 \times 10^3 + 14.17 \times 10^3} = 0.45$$

D.F.$_{BC}$ = 1 − 0.45 = 0.55

$$\text{D.F.}_{CB} = K_{BC}/(K_{BC} + \tfrac{3}{4}K_{CD}) = \frac{17.33 \times 10^3}{17.33 \times 10^3 + 11.00 \times 10^3} = 0.612$$

D.F.$_{CD}$ = 1 − 0.612 = 0.388.

Now AB and CD are treated as fixed at one end only.
Consider the tables of F.E.M.s (Pages 271 to 275)

$$\text{F.E.M.}_{BA} = +\frac{3PL}{16} = \frac{3 \times 120 \times 4.5}{16} = 101.25 \text{ kNm}$$

$$\text{F.E.M.}_{BC} = -\frac{2PL}{9} = -\frac{2 \times 70 \times 3.75}{9} = -58.33 \text{ kNm}$$

F.E.M.$_{CB}$ = +58.33 kNm

$$\text{F.E.M.}_{CD} = -\frac{WL}{8} = -\frac{220 \times 3.75}{8} = -103.13 \text{ kNm}$$

Now under any circumstances in the remainder of the beam the moment at D can only be that due to the load on the cantilever.

The cantilever moment

$$M_{DE} = -M_{DC} = -50 \times 1.2 = -60 \text{ kNm.}$$

Sufficient data have been accumulated to analyse the beam by moment distribution as shown in Fig. 53.

The inexperienced may prefer to deal with the cantilever first. The moments M_{DE} and M_{DC} are inserted in the appropriate columns, and half M_{DC} is carried over to the other end of the span CD. Subsequently the support D is ignored until the final moments are summated.

The preliminary operations at support C demand some explanation. When the beam is released the unbalanced moment

$$= +30 + (58.33 - 103.13) = -14.8 \text{ kNm.}$$

To balance this moment +14.8 kNm must be distributed between the ends CB and CD.

MOMENT DISTRIBUTION

LOAD DIAGRAM

Loads: 120 kN at B (4.5 m from A), 70 kN and 70 kN between B and C (3.75 m), 220 kN distributed between C and D (3.75 m), 50 kN at E (1.2 m beyond D).

$I = 8500\ cm^4$ (A–B), $I = 6500\ cm^4$ (B–C), $I = 5500\ cm^4$ (C–D)

DEFLECTION DIAGRAM

DISTRIBUTION TABLE

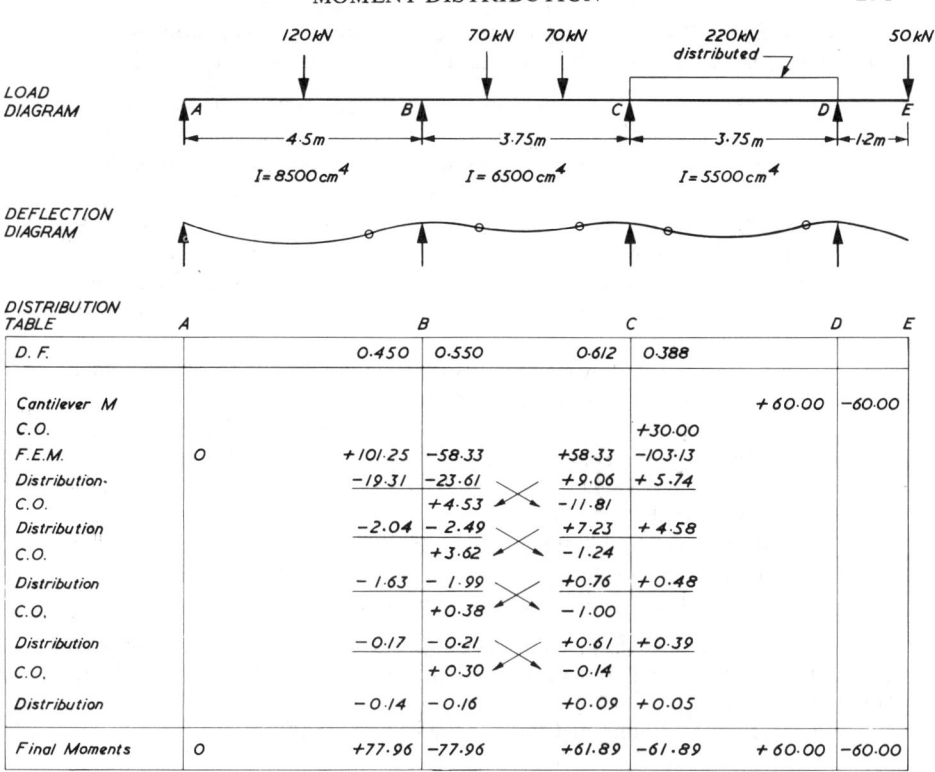

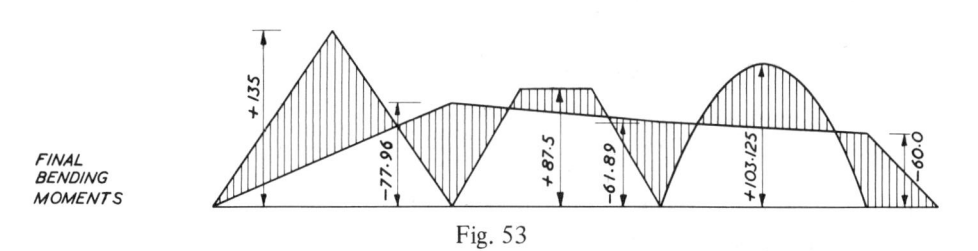

	A	B		C		D	E
D. F.		0.450	0.550	0.612	0.388		
Cantilever M						+60.00	−60.00
C.O.				+30.00			
F.E.M.	0	+101.25	−58.33	+58.33	−103.13		
Distribution		−19.31	−23.61	+9.06	+5.74		
C.O.			+4.53	−11.81			
Distribution		−2.04	−2.49	+7.23	+4.58		
C.O.			+3.62	−1.24			
Distribution		−1.63	−1.99	+0.76	+0.48		
C.O.			+0.38	−1.00			
Distribution		−0.17	−0.21	+0.61	+0.39		
C.O.			+0.30	−0.14			
Distribution		−0.14	−0.16	+0.09	+0.05		
Final Moments	0	+77.96	−77.96	+61.89	−61.89	+60.00	−60.00

FINAL BENDING MOMENTS

Values: +135, −77.96, +87.5, −61.89, +103.125, −60.0

Fig. 53

It will be observed that the work is not completed in one cycle of distribution and carry-over as in the previous example. After one operation of distribution the moments are balanced at the supports B and C. Unfortunately, the carry-over moments from B to C and from C to B throw the support moments out of balance, and further cycles of distribution and carry-over are required until the moments distributed are so small that the process may be halted.

As the beam is free to rotate at A, there can be no moment there, neither can there be at D, other than the cantilever moment. Hence, no moments are carried over from B to A or from C to D.

It is correct to carry-over moments to an outside fixed-ended support after distribution, as in Example 1, but for all internal supports a distribution table is finished with a line of distribution. Otherwise, the carry-over moments would leave these integral supports out of balance.

METHODS OF STRUCTURAL ANALYSIS

It is convenient at this stage to summarise some of the points relating to continuous beams which have been brought out by the foregoing examples.

1. *Cantilever Ends.* The only moment which can occur at the supports is that due to loads on the cantilever. After carry-over to the adjacent interior support, treat the end span as being simply supported at its outer end.
2. *Simply Supported End Spans.* As the beam is free to rotate about an end support, no moment can occur there and, consequently, there is no carry over to an end support.
3. *Fixed Ends.* Moments must be carried over to the fixed end from the adjacent interior support, but not in the opposite direction as a fixed end cannot rotate.
4. *Distribution Table.* After each distribution draw a line to signify that at that stage the moments at the support are in equilibrium.
5. *Completion.* Always finish at a fixed-ended support with a carry-over. Always finish at an interior support with a line of distribution.

Any degree of accuracy may be obtained, but the longer the cycles of moment distribution and carry-over are continued the nearer will the results be to those obtained by one of the classical methods.

In practice, however, it is sufficiently accurate to stop the process when the moment distributed is about 2 per cent of the original F.E.M.

Example 3. Figure 54 shows a continuous beam *ABC* which is freely supported at *A* and fixed at *C*. The B.M.s induced by the settlement of *B* can be calculated by Moment Distribution.

Let $E = 2.1 \times 10^5$ N/mm², and the I of the beam = 5 000 cm⁴.

Then
$$\tfrac{3}{4}K_{AB} = \frac{3 \times 5\,000 \times 10^4}{4 \times 3.0 \times 10^3} = 1.250 \times 10^4$$

while
$$K_{BC} = \frac{5\,000 \times 10^4}{4.5 \times 10^3} = 1.111 \times 10^4$$

whence
$$D.F._{DA} = \frac{1.25 \times 10^4}{1.25 \times 10^4 + 1.111 \times 10^4} = 0.53$$

and
$$D.F._{BC} = 1 - 0.53 = 0.47.$$

Now *A* is simply supported. Hence, remembering the derivation of Principle V,

$$F.E.M._{BA} = -\frac{3EId}{L^2}$$
$$= -\frac{3 \times 2.1 \times 10^5 \times 5\,000 \times 10^4 \times 25}{3\,000^2}$$
$$= -87.50 \text{ kNm.}$$

The beam is fixed at *C*. Therefore, remembering the derivation of Principle IV,

$$F.E.M._{BC} = F.E.M._{CB} = +\frac{6EId}{L^2}$$
$$= \frac{6 \times 2.1 \times 10^5 \times 5\,000 \times 10^4 \times 25}{4\,500^2}$$
$$= 77.78 \text{ kNm.}$$

MOMENT DISTRIBUTION   233

The distribution table and final moments are shown in Fig. 54.

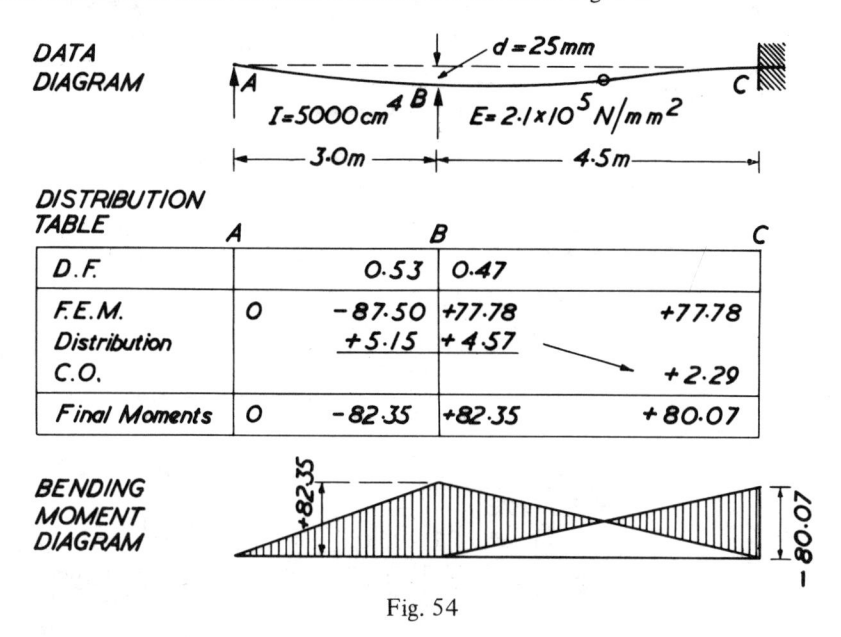

DATA
DIAGRAM

$I = 5000 \text{ cm}^4$, $d = 25 \text{ mm}$, $E = 2.1 \times 10^5 \text{ N/mm}^2$, 3.0 m, 4.5 m

DISTRIBUTION TABLE

	A	B		C
D.F.		0.53	0.47	
F.E.M.	0	−87.50	+77.78	+77.78
Distribution		+5.15	+4.57	
C.O.				+2.29
Final Moments	0	−82.35	+82.35	+80.07

BENDING MOMENT DIAGRAM

Fig. 54

Example 4. Figure 55 shows a beam fixed in position and direction at A and C and in which a clockwise moment of 10 kNm is applied at B.

The beam may be analysed by employing two stages. In the first stage the beam is propped at B so that it cannot change its position under the action of the moment. Consequently, the beam may be treated as a continuous beam. In the second stage a load equal in size and opposite in direction to the prop is applied at B. When the moments resulting from the two stages are added algebraically, the final moments are obtained.

Moment Distribution is employed only in the first stage. The beam being of constant EI, the stiffness of BC is three times that of AB. The load on the prop is found by considering the shears associated with the Stage I moments. Thus,

$$R_B = \frac{1.25 + 2.5}{4.5} - \frac{7.5 + 3.75}{1.5}$$

$$= -6.667 \text{ kN (acting downwards)}.$$

The moments resulting from the application of an upward load of 6.667 kN at B can be calculated by formula.

The 'free' bending moment under the load

$$= -\frac{Wab}{L} = -\frac{6.667 \times 4.5 \times 1.5}{6.0} = -7.5 \text{ kNm}$$

$$\text{F.E.M.}_{AB} = \frac{Wab^2}{L^2} = 1.875 \text{ kNm}$$

$$\text{F.E.M.}_{CB} = \frac{Wa^2b}{L^2} = 5.625 \text{ kNm}$$

234 METHODS OF STRUCTURAL ANALYSIS

DATA DIAGRAM

Fig. 55

STAGE I LOAD DIAGRAM

STAGE II LOAD DIAGRAM

DISTRIBUTION TABLE

MOMENT DIAGRAM

MOMENT DIAGRAM

FINAL BENDING MOMENTS

Fig. 56

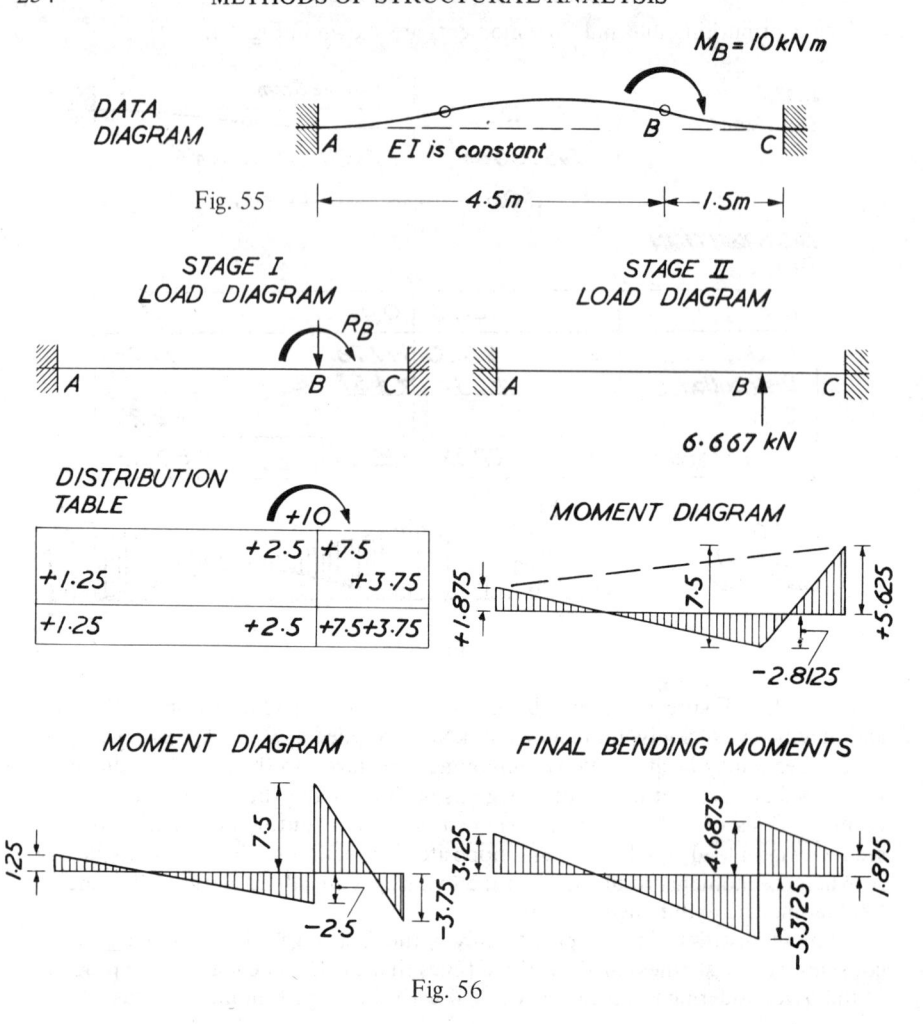

The Stage I, Stage II and final B.M.s are shown in Fig. 56.

Portal Frames

Example 5. *ABCD* in Fig. 57 is a symmetrical portal frame with fixed feet and with columns of equal stiffness. When *BC* is loaded symmetrically as shown the frame does not side-sway (i.e. *B* and *C* remain symmetrical about the vertical axis, although both sink a small amount and each moves slightly towards the vertical axis). Consequently, the moments may be found in the frame by flattening it out into a continuous beam *ABCD*.

There is justification for this action because it is a fundamental assumption that the connections between the beams and columns are rigid, just as it is assumed that there is no sudden change of slope in a deformed continuous beam at its supports.

MOMENT DISTRIBUTION

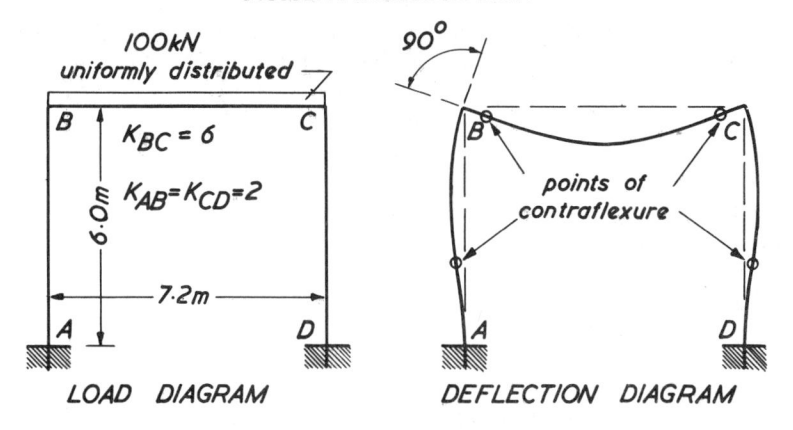

LOAD DIAGRAM

100kN uniformly distributed
$K_{BC} = 6$
$K_{AB} = K_{CD} = 2$
6.0 m
7.2 m

DEFLECTION DIAGRAM

90°
points of contraflexure

DISTRIBUTION TABLE

D.F.	A		0.25	B	0.75	0.75	C	0.25		D
F.E.M.					−60.00	+60.00				
Dist.			+15.00		+45.00	−45.00		−15.00		
C.O.	+7.50				−22.50	+22.50				−7.50
Dist.			+ 5.62		+16.88					
C.O.	+2.81				− 8.44	Figures can be				
Dist.			+ 2.11		+ 6.33	repeated here with				
C.O.	+1.06				− 3.16	opposite signs				
Dist.			+ 0.79		+2.36					
C.O.	+0.39				− 1.18					
Dist.			+ 0.30		+0.89					
C.O.	+0.15									
Final Moments	+11.91		+ 23.82		−23.82	+23.82		−23.82		−11.91

Fig. 57

Now $\text{F.E.M.}_{BC} = -\text{F.E.M.}_{CB} = -\dfrac{WL}{12} = -\dfrac{100 \times 7.2}{12}$

$= -60 \text{ kNm.}$

$K_{AB} : K_{BC} : K_{CD} = 1 : 3 : 1.$

Then $\text{D.F.}_{BA} = \text{D.F.}_{CD} = \dfrac{1}{1+3} = 0.25$

and $\text{D.F.}_{BC} = \text{D.F.}_{CB} = \dfrac{3}{1+3} = 0.75$

The distribution table is shown in Fig. 57.

It should be noted that it is not essential to carry over from B to A or from C to D after each distribution. It would suffice to carry over half the final moments at B and C to A and D respectively.

236 METHODS OF STRUCTURAL ANALYSIS

It is appropriate at this stage to include a practical 'short cut' to reduce the amount of work in the distribution table. Just as it is convenient to consider that the equivalent stiffness of a simply supported end span is $\frac{3}{4}K$, so can one modify the stiffness factors of other members under certain conditions.

The portal frame of Fig. 57 and its load are symmetrical about the centre of the beam BC.

If, in this case, the stiffness factor for BC is taken as $K/2$, then there is no need to carry-over between B and C.

As $K_{AB} : K_{BC}/2 : K_{CD} = 1 : 1.5 : 1$.

$$\text{D.F.}_{BA} = \text{D.F.}_{CD} = \frac{1}{1+1.5} = 0.4$$

and

$$\text{D.F.}_{BC} = \text{D.F.}_{CB} = 1 - 0.4 = 0.6$$

Employing this method a very short distribution table results as shown in Fig. 58.

DISTRIBUTION TABLE

	A		B		C		D
D.F.		0.4	0.6		0.6	0.4	
F.E.M.			−60.00	+60.00			
Dist.		+24.00	+36.00	−36.00	−24.00		
C.O.	+12.00						−12.00
Final Moments	+12.00	+24.00	−24.00	+24.00	−24.00	−12.00	

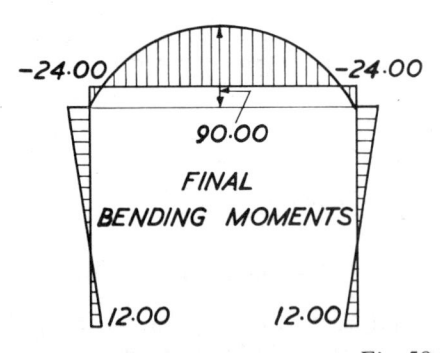

Fig. 58

The final moments are exact.

The above procedure may be adopted for all members which are subject to *equal* end rotations in opposite directions. The principle will be used again later.

Example 6. When a portal frame is asymmetrical in shape or is asymmetrically loaded, it tends to sway to one side, and analysis by Moment Distribution has to be carried out in two stages. In the first stage the moments are derived assuming

MOMENT DISTRIBUTION 237

that the frame is propped against sway, while in the second moments induced by the sway are calculated.

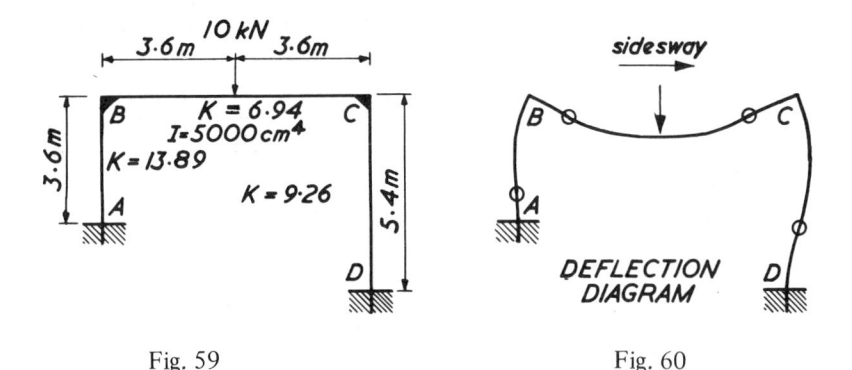

Fig. 59

Fig. 60

Consider the frame shown above, which has a constant I of 5 000 cm⁴.

$$K_{AB} = 5\ 000 \times 10^4/3.6 \times 10^3 = 1.389 \times 10^4$$
$$K_{BC} = 5\ 000 \times 10^4/7.2 \times 10^3 = 0.694 \times 10^4$$
$$K_{CD} = 5\ 000 \times 10^4/5.4 \times 10^3 = 0.926 \times 10^4$$

Hence

$$\text{D.F.}_{BA} = \frac{1.389 \times 10^4}{1.389 \times 10^4 + 0.694 \times 10^4} = 0.667$$

$$\text{D.F.}_{BC} = 0.333$$

$$\text{D.F.}_{CB} = \frac{0.694 \times 10^4}{0.694 \times 10^4 + 0.926 \times 10^4} = 0.428$$

$$\text{D.F.}_{CD} = 0.572.$$

Now

$$\text{F.E.M.}_{BC} = -\text{F.E.M.}_{CB} = -\frac{WL}{8} = -\frac{10 \times 7.2}{8}$$

$$= -9.00 \text{ kNm.}$$

When the frame is prevented from swaying, the moments are those obtained in Fig. 61.

Now a frame sways because of unbalanced horizontal thrust.
The thrust at A

$$= \frac{+3.75 + 7.50}{3.6} = 3.125$$

while the thrust at D

$$= \frac{-6.18 + 3.09}{5.4} = -1.717$$

Hence, the propping force equals 1.41 kN and acts in a horizontal direction from C towards B.

238 METHODS OF STRUCTURAL ANALYSIS

The second stage of the calculations is to find what moments result when a force of 1.41 kN acts in a horizontal direction from B towards C.

Unfortunately, there is no direct method of achieving this object. Nevertheless, within the elastic range of the material, the moments produced in a frame are

DISTRIBUTION TABLE FOR STAGE I MOMENTS.

	A		B		C		D
D.F.		$2/3$		$1/3$	$3/7$	$4/7$	
F.E.M.				−9.00	+9.00		
Dist.		+6.00		+3.00	−3.86	−5.14	
C.O.	+3.00			−1.93	+1.50		−2.57
Dist.		+1.29		+0.64	−0.64	−0.86	
C.O.	+0.64			−0.32	+0.32		−0.43
Dist.		+0.21		+0.11	−0.14	−0.18	
C.O.	+0.11						−0.0
Stage I Moments	+3.75		+7.50	−7.50	+6.18	−6.18	−3.09

Fig. 61

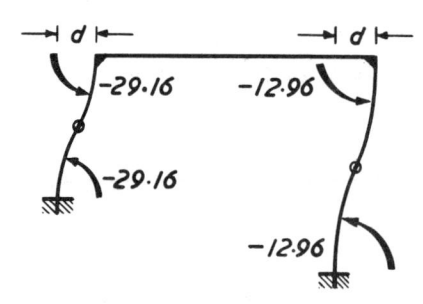

Fig. 62

proportional to the applied forces. Hence, if it can be calculated that a certain B.M. produces a known lateral force, then the bending moment resulting from another lateral force in the same place may be calculated by proportion.

Let the frame sway an amount d along the line BC, the joints B and C being prevented from rotation, as shown in Fig. 62.

By Principle IV the moments induced in AB and CD are proportional to their I/L^2 values.

Hence,

F.E.M.$_{AB}$: F.E.M.$_{BA}$: F.E.M.$_{CD}$: F.E.M.$_{DC}$ = $1/3.6^2$: $1/3.6^2$: $1/5.4^2$: $1/5.4^2$

$$= -29.16 : -29.16 : -12.96 : -12.96.$$

Using these arbitrary moments, release the joints B and C and calculate the resulting moments in the frame, as shown in Fig. 63.

MOMENT DISTRIBUTION

DISTRIBUTION TABLE FOR SIDESWAY

	A	B		C	D
D.F.		2/3	1/3	3/7	4/7
F.E.M.	−29·16	−29·16		−12·96	−12·96
Dist.		+19·44	+9·72	+5·55	+7·41
C.O.	+9·72		+2·78	+4·86	+3·70
Dist.		−1·85	−0·93	−2·08	−2·78
C.O.	−0·93		−1·04	−0·46	−1·39
Dist.		+0·69	+0·35	+0·20	+0·26
C.O.	+0·35				+0·13
Final Moments	−20·02	−10·88	+10·88	+8·07	−8·07	−10·50

Fig. 63

The resulting shears

$$= \frac{-20.02 - 10.88}{3.6} + \frac{-8.07 - 10.52}{5.4}$$

$$= -12.03 \text{ kN.}$$

This force is 12.03/1.41 = 8.532 times as great as the propping force in Stage I. Hence, the Stage II moments are 1/8.532 of those calculated above.

	A	B	C	D
Stage I Moments	+3·75	−7·50	−6·18	−3·09
Stage II Moments	−2·35	+1·28	−0·95	−1·23
Final Moments	+1·40	−6·22	−7·13	−4·32

Fig. 64

The Stage I and Stage II moments are shown in Figs. 64, 65–67. When added algebraically they provide the final B.M.s.

Example 7. In general, rigid frames with hinged feet are a little more easy to analyse than those with fixed feet, and the analysis is particularly easy with frames symmetrical in shape and in loading. However, the frame shown in Fig. 68 is somewhat complicated as it is asymmetrical and requires analysis in two stages as in the last example.

The approach is new because the legs are sloping and account must be taken of the vertical reactions V_A and V_D which affect the moments at B and C.

Taking moments to the left of B,

$$2.7V_A + M_{BA} - 6.6H_A = 0 \quad \ldots \ldots \ldots \ldots \ldots \text{(i)}$$

Taking moments to the left of C,

$$5.7V_A + M_{CB} - 6.6H_A - (100 \times 1.5) = 0 \quad \ldots \ldots \ldots \text{(ii)}$$

240 METHODS OF STRUCTURAL ANALYSIS

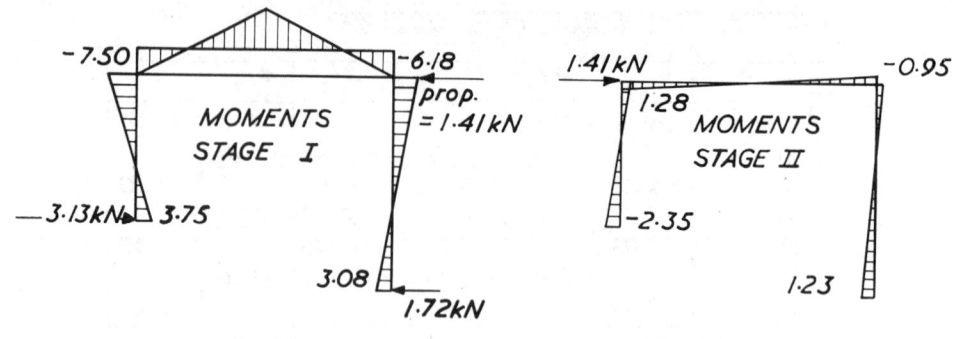

Fig. 65

Fig. 66

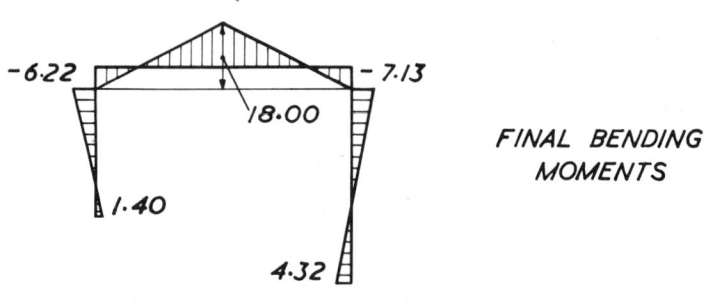

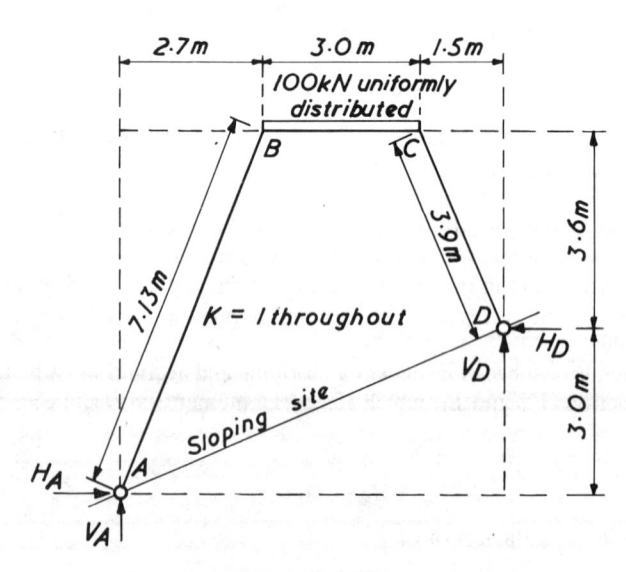

Fig. 67 — FINAL BENDING MOMENTS

Fig. 68

MOMENT DISTRIBUTION 241

Taking moments to the left of D,

$$7.2V_A - 3.0H_A - (100 \times 3.0) = 0 \quad\ldots\ldots\ldots\ldots \text{(iii)}$$

Subtracting (i) from (ii),

$$3.0V_A + M_{CB} - M_{BA} - 150 = 0 \quad\ldots\ldots\ldots\ldots \text{(iv)}$$

Multiplying (iii) by 6.6/3,

$$15.84V_A - 6.6H_A - 660 = 0 \quad\ldots\ldots\ldots\ldots \text{(v)}$$

Subtracting (i) from (v),

$$13.14V_A - M_{BA} - 660 = 0 \quad\ldots\ldots\ldots\ldots \text{(vi)}$$

Dividing (vi) by 13.14/3.00,

$$3.0V_A - 0.228M_{BA} - 150.69 = 0 \quad\ldots\ldots\ldots\ldots \text{(vii)}$$

Subtracting (vii) from (iv),

$$M_{CB} - 0.772M_{BA} + 0.69 = 0$$

or, as $-M_{BA} = +M_{BC}$, $\quad 0.772M_{BC} + M_{CB} = -0.69.$

This is the fundamental equation of equilibrium of the frame.
The Stage I moments are found as shown in Fig. 69.

	A	B		C		D
D.F.		0.6	0.4	0.4	0.6	
F.E.M.	0	−25.0		+25.0		0
Distribution		+15.0	+10.0	−10.0	−15.0	
Stage I Moments	0	+15.0	−15.0	+15.0	−15.0	0

Fig. 69

Note that the stiffness coefficients are symmetrical about the central member and that the load is symmetrical. Providing that the frame is propped, the K value of the central member can be halved and carry-over dispensed with between B and C for this Stage, as in Example 5.

From these moments

$$0.772M_{BC} + M_{CB} = (0.772 \times -15.00) + 15.00 = 3.42.$$

As this equation does not give the same result as the fundamental equation, the frame side-sways with a resulting modification of moments so that

$$0.772M_{BC} + M_{CB} = -4.11.$$

If it is assumed that the member BC side-sways a horizontal amount d to the left, as shown in Fig. 70, then

$$d = 6.6\phi_1 = 3.6\phi_3$$

i.e. $\quad\quad \phi_3 = 1.833\phi_1.$

METHODS OF STRUCTURAL ANALYSIS

The vertical movement of B with regard to C

$$= 3.0\phi_2 = 2.7d/6.6 + 1.5d/3.6$$
$$= (2.7/6.6 + 1.5/3.6)6.6\phi_1$$
$$= 5.450\phi_1$$

i.e. $\phi_2 = 1.816\phi_1$.

From Principles IV and V,

$$\text{F.E.M.}_{BA} = \frac{3EKd}{L} = +3EK_1\phi_1$$

$$\text{F.E.M.}_{BC} = \text{F.E.M.}_{CB} = -6EK_2\phi_2$$

$$\text{F.E.M.}_{CD} = +3EK_3\phi_3$$

The K values all equal 1.

If $\phi_1 = 1$, then $\phi_2 = 1.816$ and $\phi_3 = 1.833$.

Hence

$\text{F.E.M.}_{BA} : \text{F.E.M.}_{BC} : \text{F.E.M.}_{CB} : \text{F.E.M.}_{CD}$

$$= 1 : -2(1 \times 1.816) : -2(1 \times 1.816) : (1 \times 1.833)$$
$$= 1 : -3.632 : -3.632 : 1.833.$$

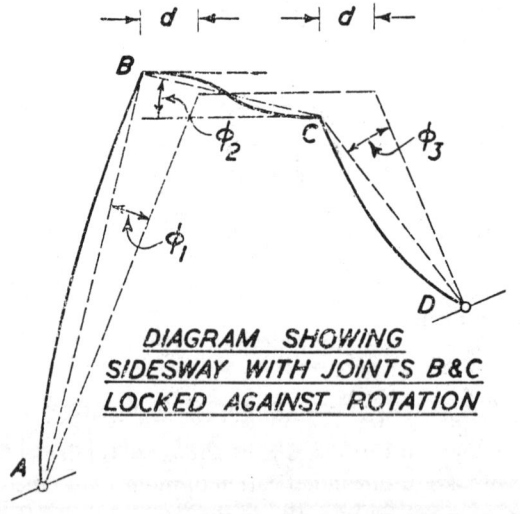

DIAGRAM SHOWING SIDESWAY WITH JOINTS B & C LOCKED AGAINST ROTATION

Fig. 70

Employing those moments as arbitrary F.E.M.s the Stage II distribution is carried out as shown in Fig. 71.

From these moments,

$$0.772M_{BC} + M_{CB} = -3.848.$$

MOMENT DISTRIBUTION

But the Stage II moments should be such that

$$0.772 M_{BC} + M_{CB} = -4.11.$$

Hence, the appropriate correction coefficient for the Stage II moments

$$= \frac{-4.11}{-3.848} = 1.07.$$

STAGE II DISTRIBUTION TABLE

	A	B			C	D
D.F.		0·43	0·57	0·57	0·43	
F.E.M.	0	+1·000	−3·632	−3·632	+1·833	0
Dist.		+1·128	+1·504	+1·028	+0·771	
C.O.			+0·514	+0·752		
Dist.		−0·220	−0·294	−0·430	−0·322	
C O			−0·215	−0·147		
Dist.		+0·092	+0·123	+0·084	+0·063	
C.O.			+0·042	+0·062		
Dist.		−0·018	−0·024	−0·035	−0·027	
Final Moments	0	+1·982	−1·982	−2·318	+2·318	0

Fig. 71

Correcting the Stage II moments and adding them to the Stage I moments, the final moments are obtained as follows:

$$M_{BC} = 1.07 \times (-1.982) - 15.00$$
$$= -17.12 \text{ kNm}$$
$$M_{CB} = 1.07 \times (-2.318) + 15.00$$
$$= +12.52 \text{ kNm}.$$

The final B.M. diagram is shown in Fig. 72.

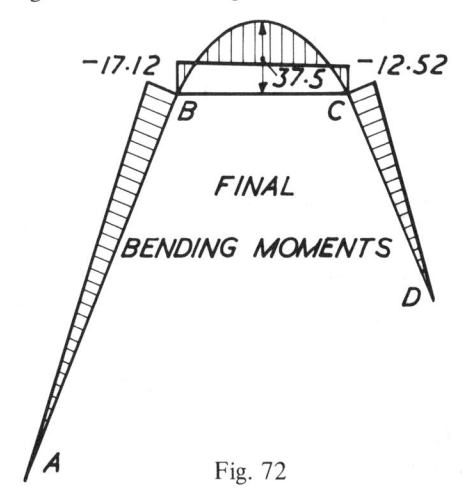

Fig. 72

Frames with Pitched Roofs

Example 8. When one joint of a three-member frame, such as that in Example 6, is displaced the displacement of other joints can be found from the geometry of the frame. When, however, four member frames, such as those in Fig. 73, are under

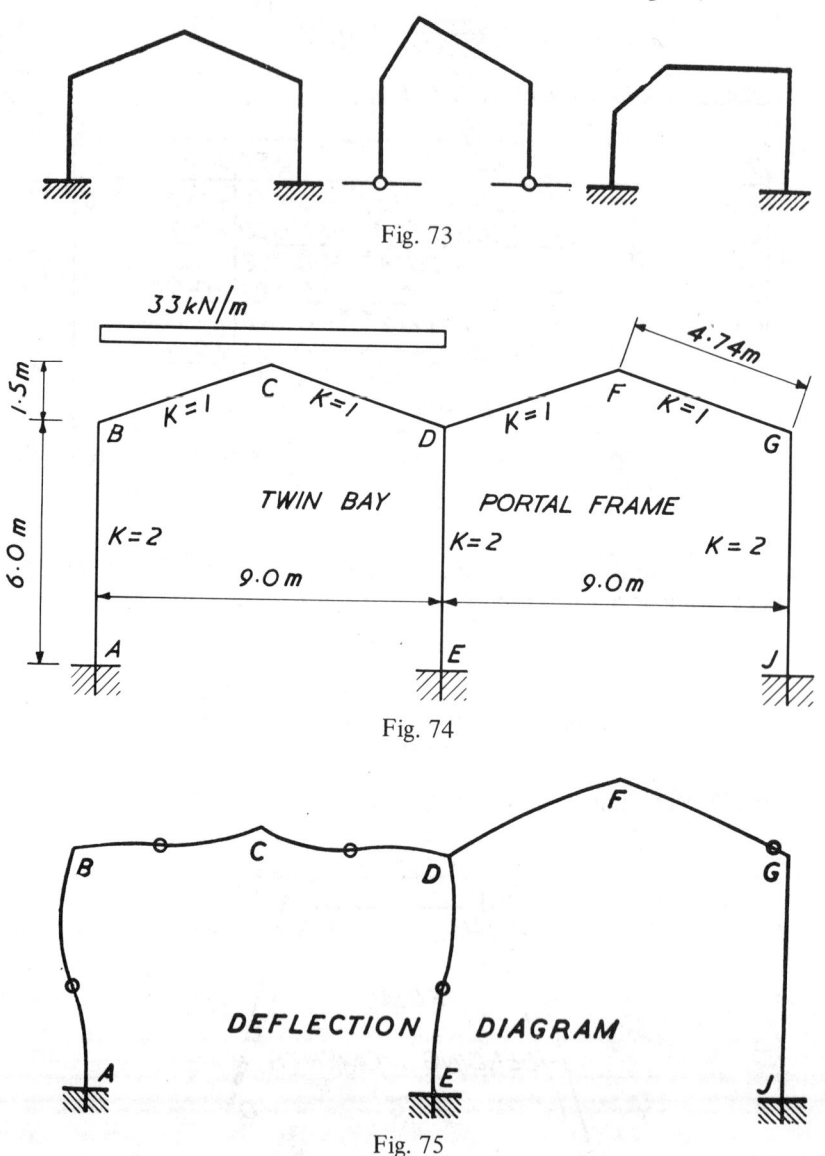

Fig. 73

Fig. 74

Fig. 75

consideration it is necessary to know the movements of two joints before those of the remaining joints can be calculated.

Similarly when the twin portal in Fig. 74–5 is loaded the effect of the movement of the three joints *B*, *D* and *G* must be known before the frame can be analysed.

MOMENT DISTRIBUTION

Analysis is carried out in stages as follows:

I. By assuming that the joints B, D and G are fixed in position but free to rotate.
II. By allowing each joint separately to sway under arbitrary loading while the other two remain fixed in position.
III. By proportioning the thrusts resulting from the previous stages so that no induced horizontal forces remain at B, D and G.
IV. By adjusting the moments at the various stages in proportion to the final thrusts.

Care is needed with signs. When *comparing* thrusts, those acting from right to left are considered to be negative. Similarly, downward vertical reactions are negative. As before, clockwise moments are positive.

Stage I:
$$K_{AB} = K_{DE} = K_{GJ} = 2$$
$$K_{BC} = K_{CD} = K_{DF} = K_{FG} = 1.$$

Hence
$$\text{D.F.}_{BA} = \text{D.F.}_{GJ} = 0.67$$
$$\text{D.F.}_{BC} = \text{D.F.}_{GF} = 0.33$$
$$\text{D.F.}_{CB} = \text{D.F.}_{CD} = \text{D.F.}_{FD} = \text{D.F.}_{FG} = 0.5$$
$$\text{D.F.}_{DC} = \text{D.F.}_{DF} = 0.25$$
$$\text{D.F.}_{DE} = 0.50$$

$$\text{F.E.M.}_{BC} = -\text{F.E.M.}_{CB} = \text{F.E.M.}_{CD} = -\text{F.E.M.}_{DC} = -\frac{WL}{12}$$

$$= -\frac{33 \times 4.5^2}{12} = -55.69 \text{ kNm.}$$

The distribution table for Stage I is shown in Fig. 76, and the appropriate B.M. diagram in Fig. 77.

$$H_A = \frac{18.75 + 37.52}{6} = \frac{56.29}{6} = 9.38 \text{ kN}$$

$$H_E = \frac{-14.21 - 28.43}{6} = \frac{-42.64}{6} = -7.11 \text{ kN}$$

$$H_J = \frac{-0.58 - 1.16}{6} = \frac{-1.74}{6} = -0.29 \text{ kN.}$$

Taking moments to the left of D

$$9V_A + 18.75 + 40.90 - 6H_A - \frac{(297 \times 9)}{2} = 0$$

$$V_A = \frac{-18.75 - 40.90 + 56.29 + 1\,336.5}{9}$$

$$= 148.13 \text{ kN}$$

DISTRIBUTION TABLE FOR STAGE I MOMENTS

	AB	BA	BC	CB	CD	DC	DE	ED	DF	FD	FG	GF	GJ	JG
D.F.	0	0.67	0.33	0.5	0.5	0.25	0.5	0	0.25	0.5	0.5	0.33	0.67	0
F.E.M.			−55.69	+55.69	−55.69	+55.69								
Dist.		+37.13	+18.56			−13.92	−27.85	−13.92	−13.92					
C.O.	+18.56			+9.28	−6.96					−6.96				
Dist.				−1.16	−1.16				+1.74	+3.48	+3.48	+1.74		
C.O.		−0.58			−0.58									
Dist.		+0.39	+0.19		−0.29	−0.29	−0.58	−0.29	−0.29			−0.58	−1.16	
C.O.	+0.19			+0.10	−0.15					−0.15	−0.29			
Dist.				+0.02	+0.03					+0.22	+0.22			
	+18.75	+37.52	−37.52	−63.93	+63.93	+40.90	−28.43	−14.21	−12.47	−3.41	+3.41	+1.16	−1.16	−0.58

Fig. 76

MOMENT DISTRIBUTION

Taking moments to the left of C

$$4.5V_A + 18.75 + 63.93 - 7.5H_A - \frac{(148.5 \times 4.5)}{2} - 1.5H_B = 0$$

$$H_B = \frac{666.57 + 18.75 + 63.93 - 70.36 - 334.13}{1.5}$$

$$= 229.87 \text{ kN}.$$

Taking moments to the right of D

$$6H_J - 9V_J - 0.58 - 12.47 = 0$$

$$V_J = \frac{1.74 - 0.58 - 12.47}{9}$$

$$= -1.26 \text{ kN (i.e. } V_J \text{ acts downwards)}.$$

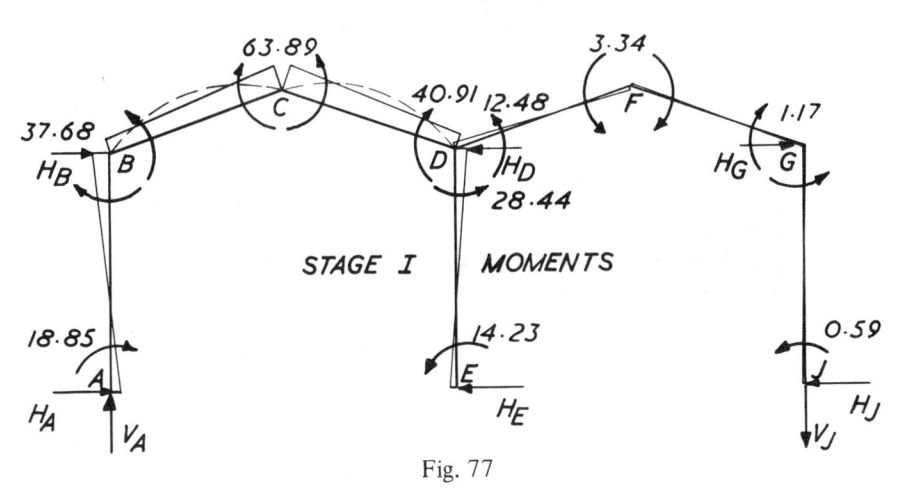

Fig. 77

Taking moments to the right of F

$$7.5H_J + 4.5V_J - 0.58 + 3.41 - 1.5H_G = 0$$

$$H_G = \frac{2.175 + 5.655 - 0.58 + 3.41}{1.5}$$

$$= 7.11 \text{ kN}.$$

Now $\Sigma H = 0$

Therefore, $-H_D = H_A + H_B + H_E + H_G + H_J$

$-H_D = 9.38 + 229.87 - 7.11 + 7.11 - 0.29$

$H_D = -238.96$ kN (i.e. H acts from right to left).

Summarising the Stage I induced thrusts at eaves and valley level:

$$H_B = 229.87 \text{ kN}$$
$$H_D = -238.96 \text{ kN}$$
$$H_G = 7.11 \text{ kN}.$$

248   METHODS OF STRUCTURAL ANALYSIS

Stage II:

During this stage the props are removed from joints B, D and G and the moments due to sway are calculated for each joint.

Consider joint B in Fig. 78. If D and G remain fixed in position and B deflects an amount 2d to the left, then the ridge C will move a horizontal distance d. But C will also drop vertically. Provided d is small then C will drop an amount equal to

$$\frac{\text{span}}{2 \times \text{rise}} \times d = \frac{90}{30} \times d = 3d.$$

Now the angle of pitch of the roof θ is $18°\ 26'$. Hence, provided the ends of the rafters do not change their slopes, the deflection of the ends with respect to one another = $3d \cos \theta + d \sin \theta$.

Therefore, the deflection = $3.163d$.

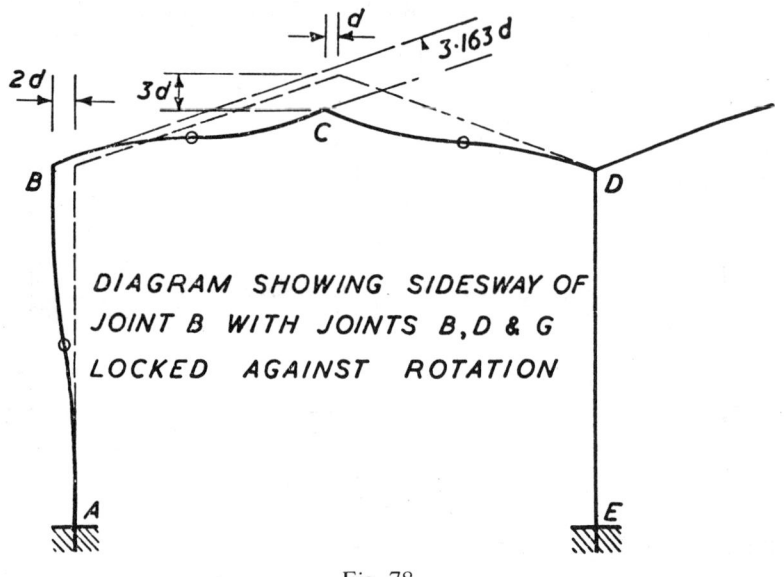

DIAGRAM SHOWING SIDESWAY OF JOINT B WITH JOINTS B, D & G LOCKED AGAINST ROTATION

Fig. 78

Providing that the joints B and C do not rotate, the deflection diagram is as shown in Fig. 78.

The F.E.M.s are proportional to K/L and to the amount of deflection.

Hence $\quad$ F.E.M.$_{AB}$ = F.E.M.$_{BA}$: $\dfrac{K2d}{L} = \dfrac{4d}{6}$

and $\quad$ F.E.M.$_{BC}$ = F.E.M.$_{CB}$ = $-$F.E.M.$_{CD}$ = $-$F.E.M.$_{DC}$: $-\dfrac{K3.163d}{L}$

$$= -\frac{3.163d}{4.743}.$$

By coincidence $\quad \dfrac{4d}{6} = \dfrac{3.163d}{4.743}$

MOMENT DISTRIBUTION

DISTRIBUTION TABLE WHEN JOINT B SIDESWAYS

	AB	BA	BC	CB	CD	DC	DE	ED	DF	FD	FG	GF	GJ	JG
D.F.	0	0.67	0.33	0.5	0.5	0.25	0.5	0	0.25	0.5	0.5	0.33	0.67	0
F.E.M.		+20.00	+20.00	−20.00	−20.00	+20.00	+20.00							
Dist.					−2.50	−5.00	−10.00	−5.00	−5.00	−2.50				
C.O.			+1.25	+1.25						+1.25	+1.25			
Dist.		+0.62			+0.62	+0.62			+0.62			+0.62		
C.O.		−0.21			−0.31	−0.31		−0.31	−0.31			−0.21		
Dist.			−0.10	−0.10	−0.15		−0.62			−0.15	−0.10			
C.O.	−0.21		+0.13	+0.12	+0.12					+0.12	+0.13	−0.21		
Dist.														
	+19.79	+19.59	−19.59	−18.72	+18.72	+15.31	−10.62	−5.31	−4.69	−1.28	+1.28	+0.41	−0.41	−0.21

Fig. 79

METHODS OF STRUCTURAL ANALYSIS

If arbitrary moments of, say, 20 kNm are applied to the ends of the member AB, then the same amount, with appropriate signs, can be applied to the members BC and CD.

If all the joints previously fixed against rotation are allowed to rotate while D and G remain fixed in position, then the distribution table is as shown in Fig. 79. The appropriate B.M. diagram is shown in Fig. 80.

$$H_A = \frac{19.59 + 19.79}{6} = \frac{39.38}{6} = 6.56 \text{ kN}$$

$$H_E = \frac{-10.62 - 5.31}{6} = \frac{-15.93}{6} = -2.66 \text{ kN}$$

$$H_J = \frac{-0.41 - 0.21}{6} = \frac{-0.62}{6} = -0.10 \text{ kN.}$$

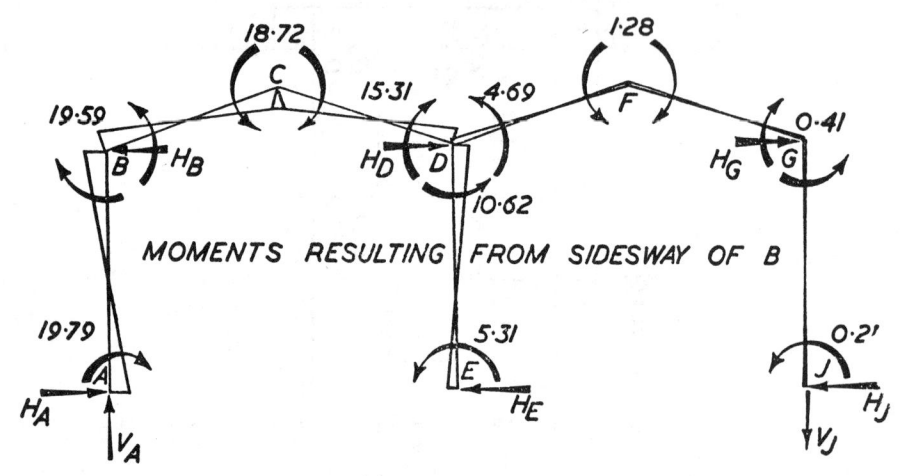

Fig. 80

Taking moments to the left of D

$$9V_A + 19.79 + 15.31 - 6H_A = 0$$

$$V_A = \frac{39.38 - 19.79 - 15.31}{9}$$

$$= 0.476 \text{ kN.}$$

Taking moments to the left of C

$$1.5H_B + 4.5V_A - 7.5H_A + 19.79 - 18.72 = 0$$

$$H_B = \frac{49.22 + 18.72 - 2.14 - 19.79}{1.5}$$

$$= \frac{45.99}{1.5} = 30.67 \text{ kN.}$$

H_B acts from right to left.

Taking moments to the right of D

$$9V_J + 6H_J - 4.69 - 0.21 = 0$$

$$V_J = \frac{4.69 + 0.21 - 0.62}{9}$$

$$= 0.476 \text{ kN.}$$

V_J acts downwards.

Taking moments to the right of F

$$-1.5H_G + 4.5V_J + 7.5H_J - 0.21 + 1.28 = 0$$

$$H_G = \frac{2.14 + 0.78 - 0.21 + 1.28}{1.5}$$

$$= 2.66 \text{ kN.}$$

Now $\Sigma H = 0$.

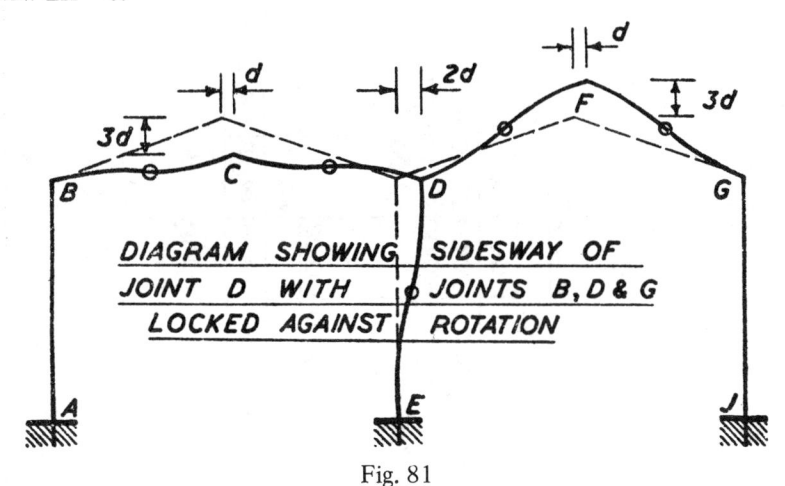

Fig. 81

Then $-H_D = 6.56 - 2.66 - 0.10 - 30.67 + 2.66$

$H_D = 24.21$ kN.

Summarising the induced thrusts at the eaves and valley due to the side-sway of D:

$$H_B = -30.67 \text{ kN}$$
$$H_D = 24.21 \text{ kN}$$
$$H_G = 2.66 \text{ kN.}$$

Now consider joint D. When B and G remain fixed in position and D moves $2d$ to the right, C drops $3d$ vertically and F rises $3d$ vertically. Both C and F move to the right a horizontal amount d. Hence the deflection diagram is as shown in Fig. 81.

The types of deflection being similar to those when B side-swayed, the arbitrary F.E.M.s chosen are 20 kNm. The distribution table is shown in Fig. 82, and the appropriate B.M. diagram is shown in Fig. 83.

DISTRIBUTION TABLE WHEN JOINT D SIDESWAYS

	AB	BA	BC	CB	CD	DC	DE	ED	DF	FD	FG	GF	GJ	JG
D.F.	0	0.67	0.33	0.5	0.5	0.25	0.5	0	0.25	0.5	0.5	0.33	0.67	0
F.E.M.		−20.00		−20.00	+20.00	+20.00	−20.00	−20.00	+20.00	+20.00	−20.00	−20.00		
Dist.		+13.33	+6.67	+3.33	−2.50	−5.00	−10.00		−5.00	−2.50	+3.33	+6.67	+13.33	
C.O.	+6.67							−5.00						+6.67
Dist.				−0.42	−0.41	−0.21	+0.22		−0.21	−0.41	−0.42			
C.O.		−0.21						+0.11					−0.21	
Dist.		+0.14	+0.07	+0.10	+0.10					+0.10	+0.10	+0.07	+0.14	
C.O.	+0.07													+0.07
	+6.74	+13.47	−13.47	−17.09	+17.09	+14.89	−29.78	−24.89	+14.89	+17.09	−17.09	−13.47	+13.47	+6.74

Fig. 82

MOMENT DISTRIBUTION

$$H_A = \frac{13.47 + 6.74}{6} = \frac{20.21}{6} = 3.37 \text{ kN}$$

$$H_E = \frac{-29.78 - 24.89}{6} = \frac{-54.67}{6} = -9.11 \text{ kN}$$

$$H_J = \frac{13.47 + 6.74}{6} = 3.37 \text{ kN}.$$

Taking moments to the left of D,

$$9V_A + 6.74 + 14.89 - 6H_A = 0$$

$$V_A = \frac{20.21 - 6.74 - 14.89}{9}$$

$$= -0.158 \text{ kN}.$$

V_A acts downwards.

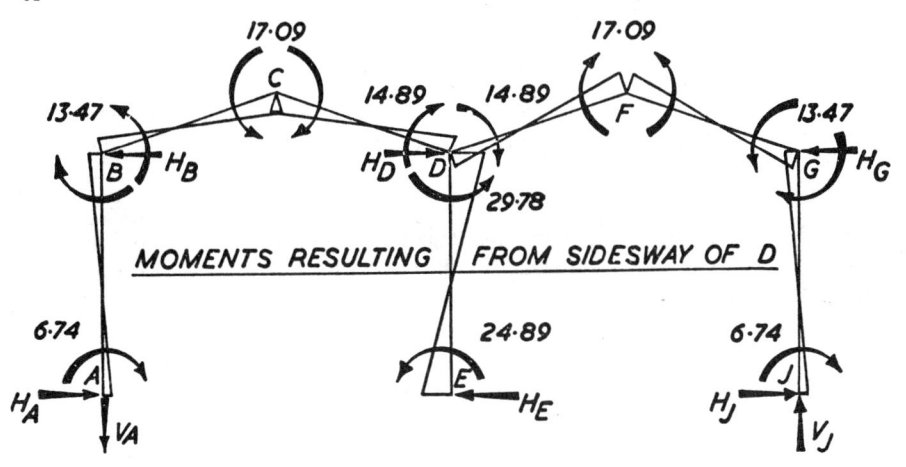

Fig. 83

MOMENTS RESULTING FROM SIDESWAY OF D

Similarly

$$V_J = 0.157 \text{ kN}.$$

V_J acts upwards.

Taking moments to the left of C

$$1.5H_B + 6.74 - 17.09 - 7.5H_A - 4.5V_A = 0$$

$$H_B = \frac{17.09 + 25.26 + 0.72 - 6.74}{1.5}$$

$$= -24.22 \text{ kN}.$$

Similarly

$$H_G = -24.22 \text{ kN}.$$

Now $\Sigma H = 0$.

METHODS OF STRUCTURAL ANALYSIS

Hence $\quad -H_D = 3.37 + 3.37 - 9.11 - 24.22 - 24.22$

and $\quad H_D = 50.81$ kN.

Summarising the induced thrusts at the eaves and valley due to the side-sway of D:

$$H_B = -24.22 \text{ kN}$$
$$H_D = 50.81 \text{ kN}$$
$$H_G = -24.22 \text{ kN}.$$

Now consider the side-sway of joint G. When joint G moves inwards an amount $2d$, the joints B and D being fixed in position, the effect is similar to that when B moved outwards an amount $2d$ with joints D and G fixed in position. The final figures for bending moment on the right side of the frame are of the same amount but of reverse sign to those on the left side of the frame when B moved outwards. Consequently, the corresponding thrusts are of the same amount but are opposite in direction.

Therefore, the thrusts at the eaves and valley due to the side-sway of G are:

$$H_B = 2.66 \text{ kN}$$
$$H_D = 24.21 \text{ kN}$$
$$H_G = -30.67 \text{ kN}.$$

Stage III:

For the equilibrium of the frame the Stage I induced thrusts at the eaves and valley must be balanced by the Stage II induced thrusts. Let x be the correction coefficient for the Stage II thrusts due to the side-sway of joint B, and y and z be the coefficients for the side-sway of joints D and G respectively.

Then at joint B: $\quad 229.87 = 30.67x + 24.22y - 2.66z$

at joint D: $\quad 238.96 = 24.21x + 50.81y + 24.21z$

at joint G: $\quad 7.11 = -2.66x + 24.22y + 30.67z$.

When these simultaneous equations are solved

$$x = 4.272$$
$$y = 3.817$$
$$z = -2.412.$$

As the value of z is negative, the joint G moves outwards when the frame is loaded, not inwards as originally supposed from a consideration of the Stage I thrusts.

Stage IV:

When the Stage II moments are multiplied by the appropriate coefficient they may be added algebraically to the Stage I moments as shown in Fig. 84.

The final B.M. diagram is shown in Fig. 85.

MOMENT DISTRIBUTION

COMPUTATION OF FINAL MOMENTS

	AB	BC	CD	DC	DE
Stage I Moments	+ 18.76	− 37.52	− 63.93	+ 40.90	− 28.43
Sidesway of B × 4.272	+ 84.53	− 83.68	+ 79.96	+ 65.40	− 45.36
Sidesway of D × 3.817	+ 25.73	− 51.42	+ 65.23	+ 56.84	−113.67
Sidesway of G × −2.412	+ 0.51	− 0.99	+ 3.09	+ 11.31	+ 25.62
Final Moments	+129.52	−173.60	+ 84.35	+174.44	−161.85

	ED	DF	FG	GJ	JG
Stage I Moments	− 14.21	− 12.47	+ 3.41	− 1.16	− 0.58
Sidesway of B × 4.272	− 22.68	− 20.03	+ 5.47	− 1.75	− 0.90
Sidesway of D × 3.817	− 95.01	+ 56.84	− 65.23	+ 51.42	+ 25.73
Sidesway of G × −2.412	+ 12.81	− 36.93	+ 45.15	− 47.25	− 47.73
Final Moments	−119.09	− 12.60	− 11.20	+ 1.25	− 23.48

Fig. 84

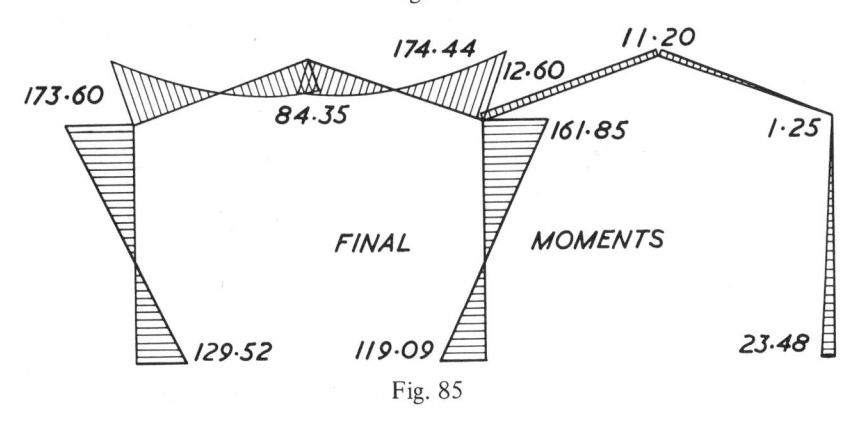

Fig. 85

It is interesting to compare these final moments with those which would result if the frame were loaded with 33 kN/m over the whole roof and also with those for a single frame loaded over the whole roof, as shown in Fig. 86. It will be observed that there is little variation of moment at joints analogous to A, B, C and D.

Multi-storey Frames

Example 9. Consider the symmetrical building frame in Fig. 87. When the uniformly distributed load is applied to the first floor the frame deflects in such a manner that the ends of the beams are subjected to equal rotation in opposite directions. Hence, we may reduce the stiffness of the beams to $K/2$ and dispense with carry-over between B and E and between C and D, as was done with the symmetrical rectangular frames.

256   METHODS OF STRUCTURAL ANALYSIS

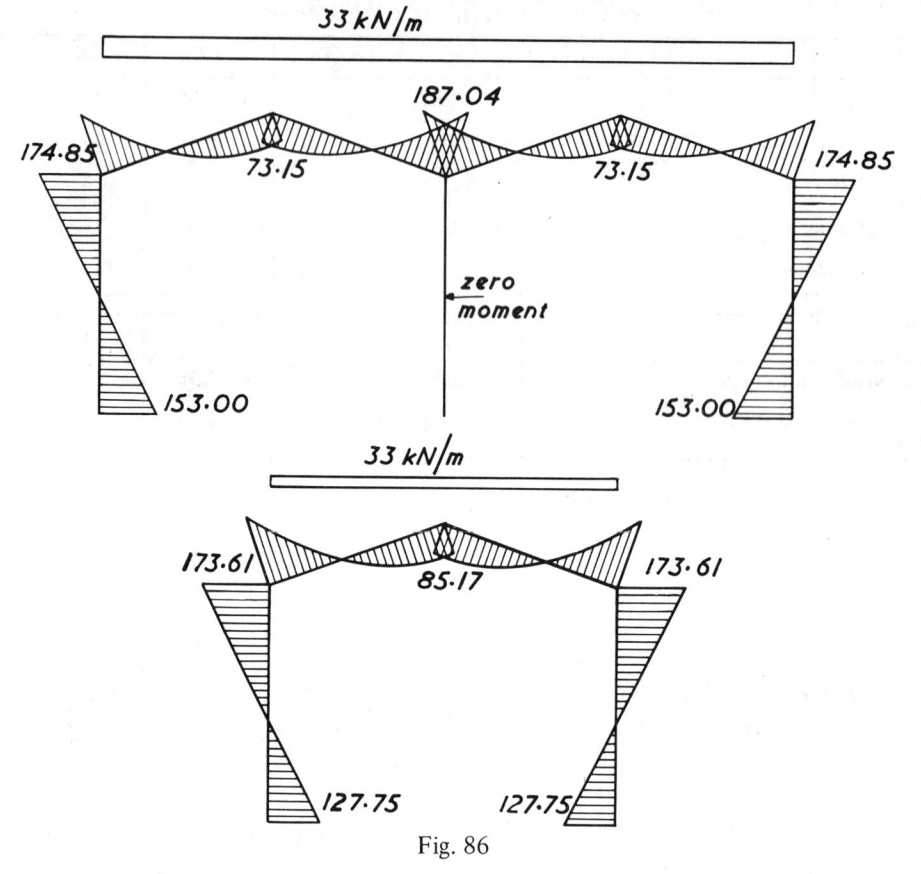

Fig. 86

As *EI* is constant

$$K_{AB} : K_{BC} : K_{BE} : K_{CD} = 2 : 2 : 1 : 1.$$

Now $\quad$ F.E.M.$_{BE}$ = F.E.M.$_{EB}$ = $-\dfrac{WL}{12}$ = $-\dfrac{33 \times 6^2}{12}$ = -99.00 kNm.

There is no necessity to consider more than one-half of the frame and the distribution may be carried out as shown in Fig. 87.

The final B.M. diagram is shown in Fig. 88.

Side Loading on Symmetrical Frames

In an article entitled 'Side Sway in Symmetrical Building Frames', published in the *Structural Engineer* of April 1950, Mr. N. Naylor, B.Sc., A.M.I.Struct.E., presented a very useful modification of Moment Distribution which will be derived and illustrated by Examples 10 to 15.

Figure 89 shows a continuous column *PQR* which is fixed in position and direction at *P*, fixed in direction but not in position at *R* and rigidly attached to the end *Q* of the beam *QS*, the other end of the beam being hinged and free to move in a horizontal direction.

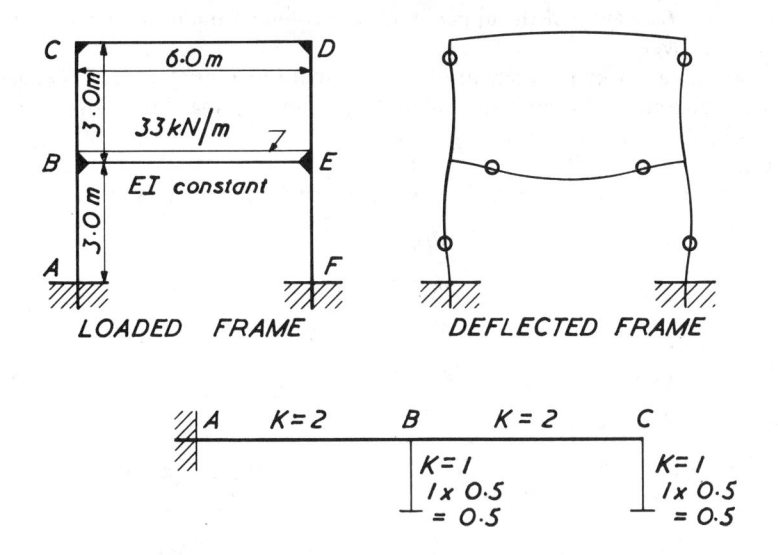

MOMENT DISTRIBUTION

[Loaded frame: portal frame CDEF with C-D top beam 6.0m, BE middle beam, 33 kN/m load on BE, EI constant, heights 3.0m each, supports at A and F]

[Deflected frame shown]

	A	K=2	B	K=2	C
			K=1, 1×0.5 = 0.5		K=1, 1×0.5 = 0.5

Distribution Factors	A		B			C	
		4/9	1/9	4/9		4/5	1/5
F.E.M.			−99.00				
Distribution		+44.00	+11.00	+44.00			
C.O.	+22.00				+22.00		
Distribution				−17.60			−4.4
C.O.				−8.80			
Distribution		+3.91	+0.98	+3.91			
C.O.	+1.96				+1.96		
Distribution					−1.57		−0.39
Final Moments	+23.96	+47.91	−87.02	+39.11		+4.79	−4.79

Fig. 87

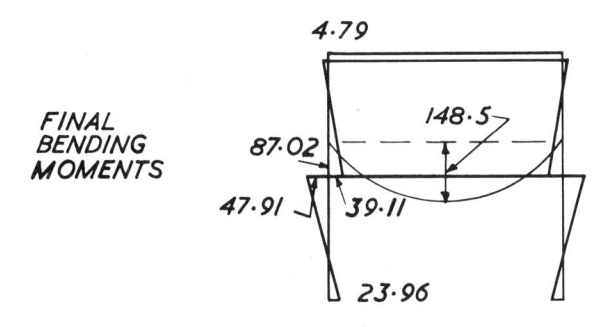

FINAL BENDING MOMENTS

4.79
87.02
148.5
47.91 39.11
23.96

Fig. 88

258   METHODS OF STRUCTURAL ANALYSIS

Let the I/L values of the upper and lower columns and beams be K_U, K_L and K_B respectively.

Now let a clockwise moment M be applied at the joint Q. Then R, Q and S will move horizontally to the right, and each member meeting at Q will slope an amount θ at that joint. Each column will behave like a cantilever and the general effect will be as shown in Fig. 90.

It can be shown that M is distributed among the members, so that

$$M_{QP} = E \cdot K_L \cdot \theta$$
$$M_{QR} = E \cdot K_U \cdot \theta$$
$$M_{QS} = 3E \cdot K_B \cdot \theta,$$

that is, in the proportion $K_L : K_U : 3K_B$.

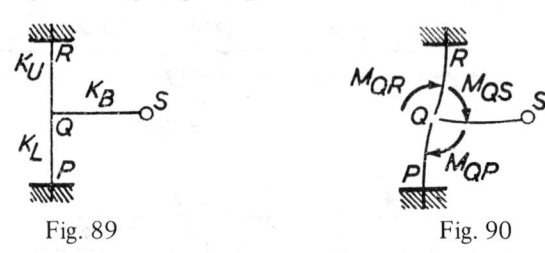

Fig. 89

Fig. 90

As each column acts as a cantilever, the carry-over factor from Q to P or from Q to R is -1. Since the beam is hinged at S, there is no carry-over to S. Moment distribution can be applied to the numerical example shown in Fig. 91.

The joint Q is initially assumed to be locked so that the column ends remain vertical and the beam end remains horizontal. Under these conditions the side loads

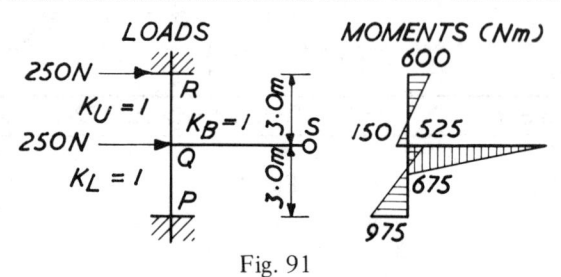

Fig. 91

induce moments at each end of the column lengths, their magnitude being $Ph/2$, where P is the appropriate load and h the column height.

Hence, $\text{F.E.M.}_{QR} = \text{F.E.M.}_{RQ} = Ph/2 = \dfrac{250 \times 3}{2} = 375$ Nm

while $\text{F.E.M.}_{QP} = \text{F.E.M.}_{PQ} = (250 + 250) \times \dfrac{3}{2} = 750$ Nm.

The D.F. for QR and $QP = 1/(1 + 3 + 1) = 0.2$, while that for $QS = 3/(1 + 3 + 1) = 0.6$.

The joint Q can now be unlocked when distribution takes place as shown in Fig. 92.

MOMENT DISTRIBUTION

Figure showing beam P-Q-R with $K_L = 1$, $K_U = 1$, support at S with $K_B = 1$, and $3 \times 1 = 3$.

DISTRIBUTION TABLE

	P		Q		R
D. F.		0.2	0.6	0.2	
F. E.M.	−750	−750		−375	−375
Distribution		+225	+675	+225	
C.O. (factor = −1)	−225				−225
Final Moments	−975	−525	+675	−150	−600

Fig. 92

Note the value and sign of the moments carried over. The final B.M. diagram is as shown in Fig. 91.

The principles employed in this example may be applied to symmetrical frames of the type shown in Fig. 93.

Under the side loading shown points of contraflexure exist in the centres of the beams, and the moments induced in the right-hand half of the frame are numerically equal to those in the left-hand half.

Consequently, the moments may be calculated in a symmetrical frame subjected to side loads by considering one-half only of the frame. It should be noted that the beam length will be halved, and therefore the beam to be considered will be twice as stiff. Hence, the original K value must be multiplied by 6 when considering D.F.s for half the frame.

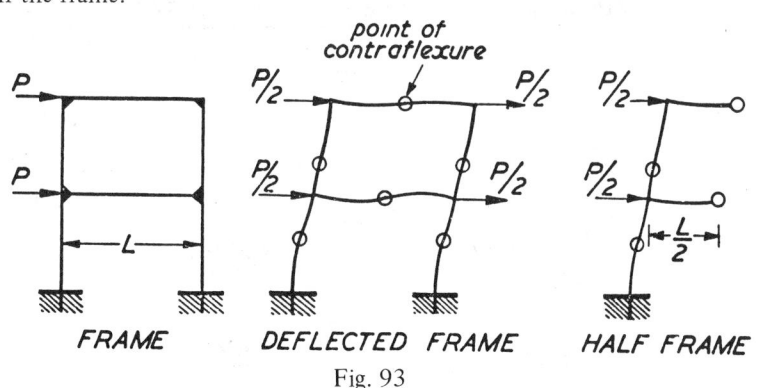

FRAME   DEFLECTED FRAME   HALF FRAME

Fig. 93

Single-bay Portal Frame

Example 10. The most simple application is to the ordinary symmetrical portal frame with fixed feet, as shown in Fig. 94.

The stiffness factor for $AB = K = 1$, while that for $BC = 6 \times K = 6$.

$$\text{D.F.}_{BA} = 1/7; \quad \text{D.F.}_{BC} = 6/7$$

$$\text{F.E.M.}_{AB} = \text{F.E.M.}_{BA} = \frac{P}{2} \times \frac{h}{2} = \frac{500}{2} \times \frac{3}{2} = 375 \text{ Nm.}$$

260   METHODS OF STRUCTURAL ANALYSIS

The distribution table is shown in Fig. 95, and the final B.M. diagram in Fig. 94

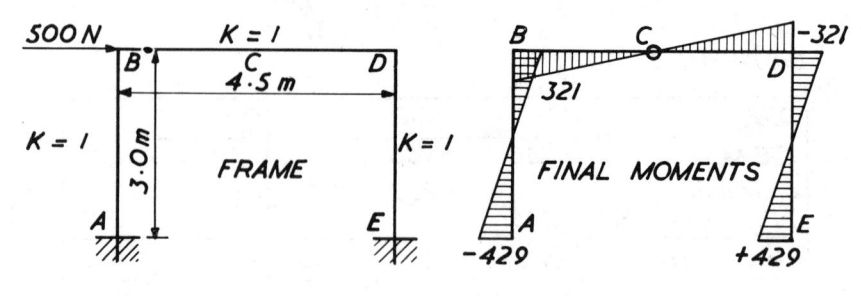

Fig. 94

DISTRIBUTION TABLE

D.F.	A		B
		1/7	6/7
F.E.M.	−375	−375	
Distribution		+54	+321
C.O. (factor = −1)	−54		
Final Moments	−429	−321	+321

Fig. 95

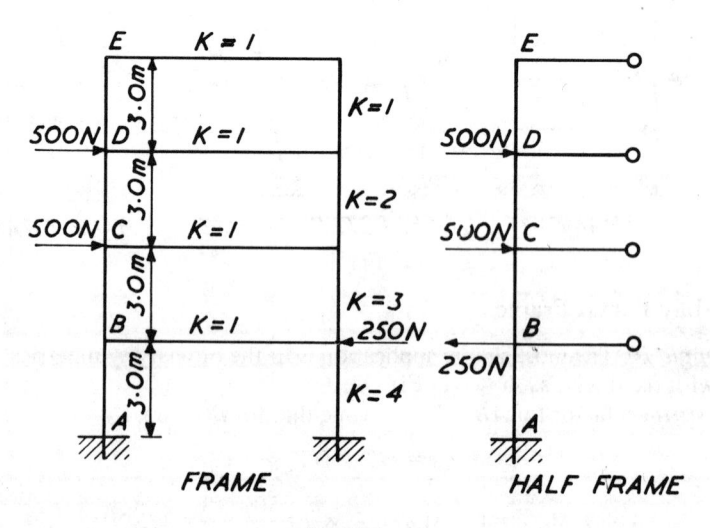

Fig. 96

MOMENT DISTRIBUTION

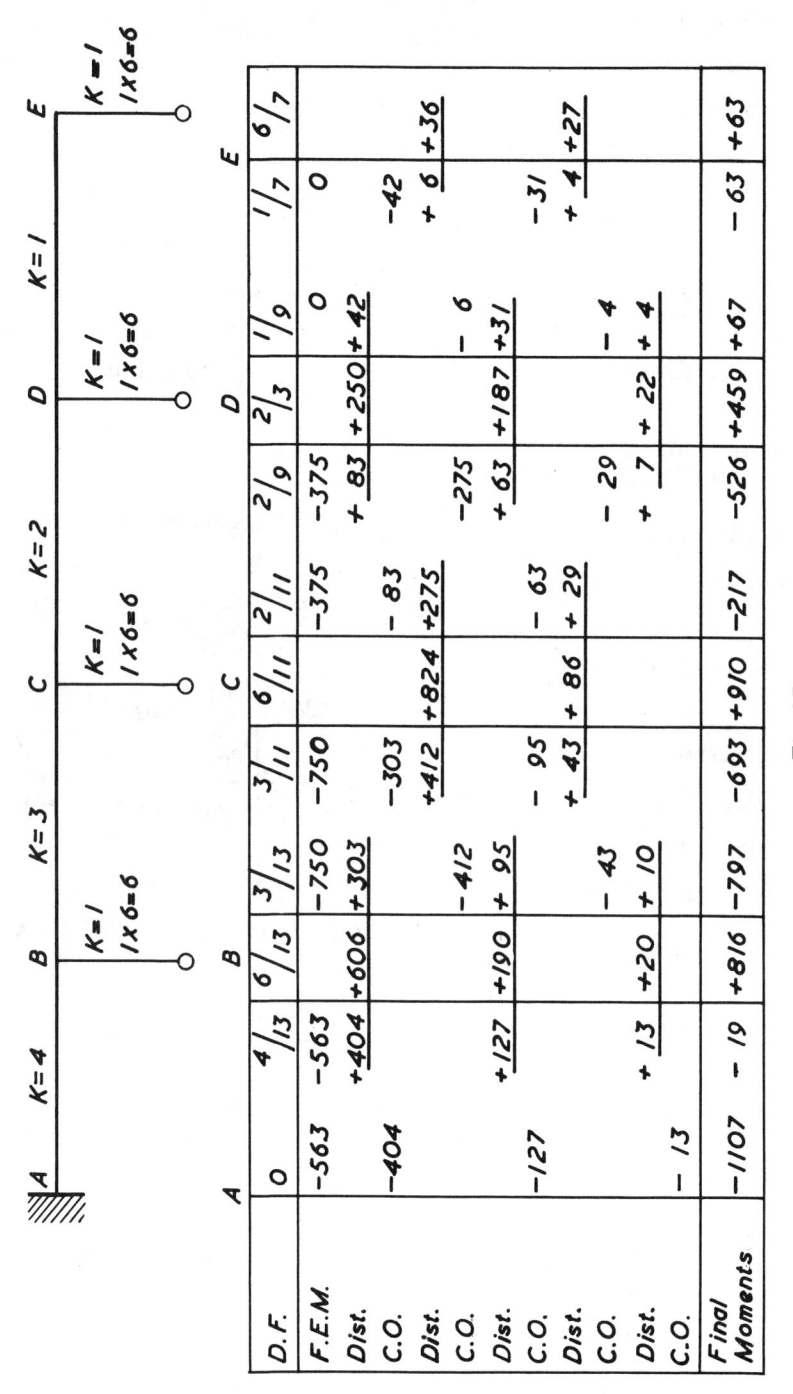

Structure (top): A (K=4, fixed) — B (K=3, K=1, 1×6=6) — C (K=2, K=1, 1×6=6) — D (K=1, K=1, 1×6=6) — E

D.F.	A 0	4/13	B 6/13	3/13	3/11	C 6/11	2/11	2/9	D 2/3	1/9	1/7	E 6/7
F.E.M.	−563	−563		−750	−750		−375	−375		0	0	
Dist.		+404	+606	+303				+83	+250	+42		
C.O.	−404				−303	+412	−83				−42	+36
Dist.				−412	+412	+824	+275	−275			+6	
C.O.	−127									−6		
Dist.		+127	+190	+95				+63	+187	+31		
C.O.				−95	+43	+86	−63				−31	
Dist.							+29	−29			+4	+27
C.O.	−13			−43						−4		
Dist.		+13	+20	+10				+7	+22	+4		
C.O.	−13											
Final Moments	−1107	−19	+816	−797	−693	+910	−217	−526	+459	+67	−63	+63

Fig. 97

261

Multi-storey Frames

Frames with Fixed Feet

  Example 11. See Fig. 96.

  There are no F.E.M.s for *DE* but

$$F.E.M._{DC} = F.E.M._{CD} = 250 \times \frac{3}{2} = 375 \text{ Nm}$$

$$F.E.M._{CB} = F.E.M._{BC} = (250 + 250) \times \frac{3}{2} = 750 \text{ Nm}$$

$$F.E.M._{BA} = F.E.M._{AB} = (250 + 250 - 125) \times \frac{3}{2} = 562.5 \text{ Nm}.$$

The distribution table is shown in Fig. 97, and the final B.M. diagram in Fig. 98.

Frames with Hinged Feet

  Example 12. The frame shown in Fig. 99 is a symmetrical frame having hinges at the feet *A* and *F*. Hence, *AB* and *EF* behave like cantilevers loaded at *A* and *F* with a force $H/2 = 5$ kN.

  Consequently

$$M_{BA} = M_{EF} = 5 \times 3 = 15 \text{ kNm}.$$

  In the upper storey

$$F.E.M._{CB} = F.E.M._{BC} = P/2 \times h/2$$
$$= 5 \times \frac{3}{2} = 7.5 \text{ kNm}.$$

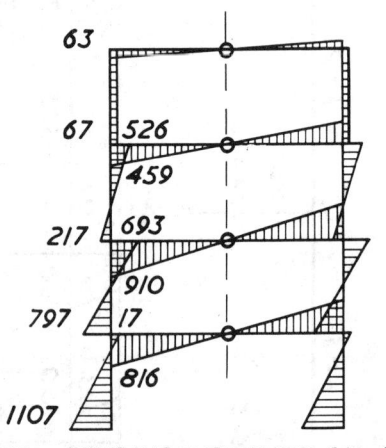

FINAL BENDING MOMENTS (Nm)
Fig. 98

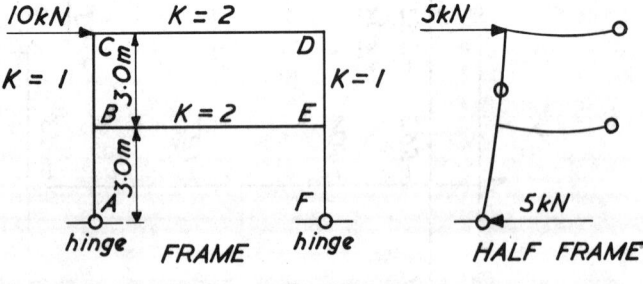

Fig. 99

The cantilever moment *M* is applied to joint *B* and distributed between the upper column and beam, there being an appropriate carry-over to *C* which is absorbed in the first distribution at that joint. Then the F.E.M.s for the upper

MOMENT DISTRIBUTION 263

column are introduced at *B* and *C* and distribution takes place between the beams and upper column in the normal way, as shown in Fig. 100, the lower column being completely ignored. Of course, the cantilever moment and F.E.M.s may be introduced on the same line in the distribution table, but initially in calculations of this kind it may assist to separate them as shown above. The final moments are shown in Fig. 101.

Frames with Non-prismatic Columns

Example 13. In industrial buildings containing crane gantries it is usual to have columns of varying section. A typical example is shown in Fig. 102.

DISTRIBUTION TABLE

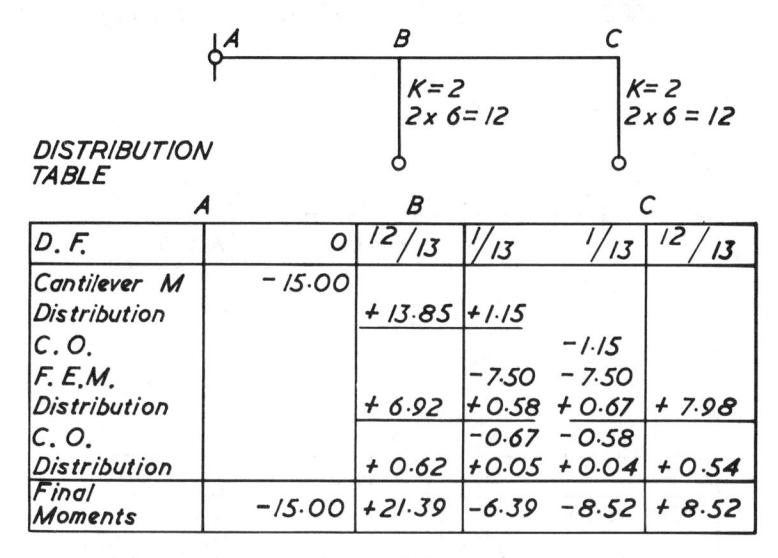

D.F.	A	B			C
	0	12/13	1/13	1/13	12/13
Cantilever M	−15.00				
Distribution		+13.85	+1.15		
C.O.				−1.15	
F.E.M.			−7.50	−7.50	
Distribution		+6.92	+0.58	+0.67	+7.98
C.O.			−0.67	−0.58	
Distribution		+0.62	+0.05	+0.04	+0.54
Final Moments	−15.00	+21.39	−6.39	−8.52	+8.52

Fig. 100

Under the side loading shown the columns undergo similar deflections and the crane beam positions *B* and *E* remain the same distance apart.

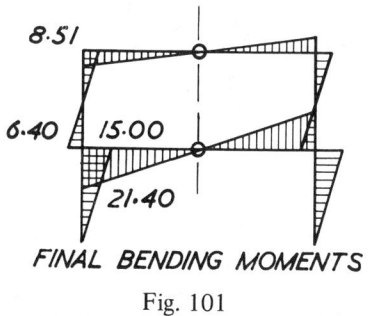

Therefore it can be assumed that *B* and *E* are joined by a beam of zero stiffness, and the frame analysed like a two-storey frame.

$$F.E.M._{CB} = F.E.M._{BC}$$
$$= -\frac{900 \times 3}{2 \times 2} = -675 \text{ Nm}$$

$$F.E.M._{BA} = F.E.M._{AB}$$
$$= -\frac{900 \times 9}{2 \times 2} = -2\,025 \text{ Nm}.$$

FINAL BENDING MOMENTS

Fig. 101

The distribution table is shown in Fig. 103, while the final B.M. diagram is shown in Fig. 102.

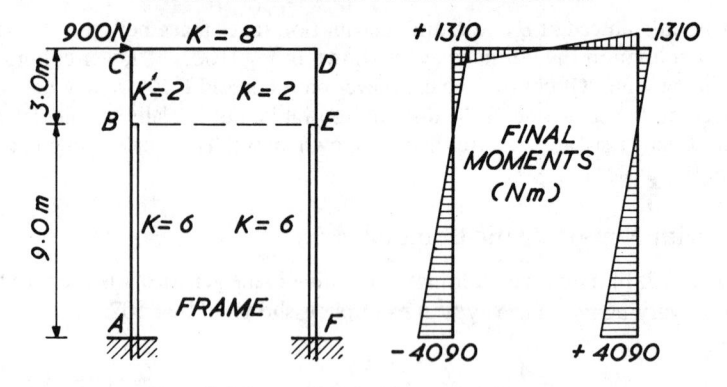

Fig. 102

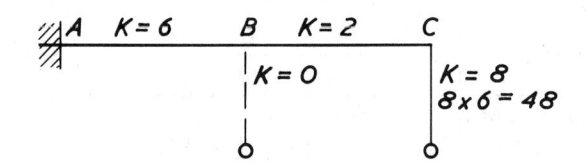

Fig. 103

Example 14. The frame in Example 13 could be analysed in a similar manner if loads of the same magnitude and direction were applied at *B* and *E*. However, the problem shown in Fig. 104 is not so simple because the two columns do not distort in the same manner. Nevertheless, the problem can be solved by carrying out the analysis in two stages as shown in Figs. 104–7.

In Stage I the work follows the same lines as in the last four examples, but in Stage II the calculations are similar to those in Examples 1 to 10.

Stage I:

$$F.E.M._{CB} = F.E.M._{BC} = 0$$

$$F.E.M._{BA} = F.E.M._{AB} = 1\,125 \times \frac{12}{2}$$

$$= 6\,750 \text{ Nm.}$$

The distribution table is shown in Fig. 105.

Fig. 104

STAGE I DISTRIBUTION
NAYLOR SYSTEM

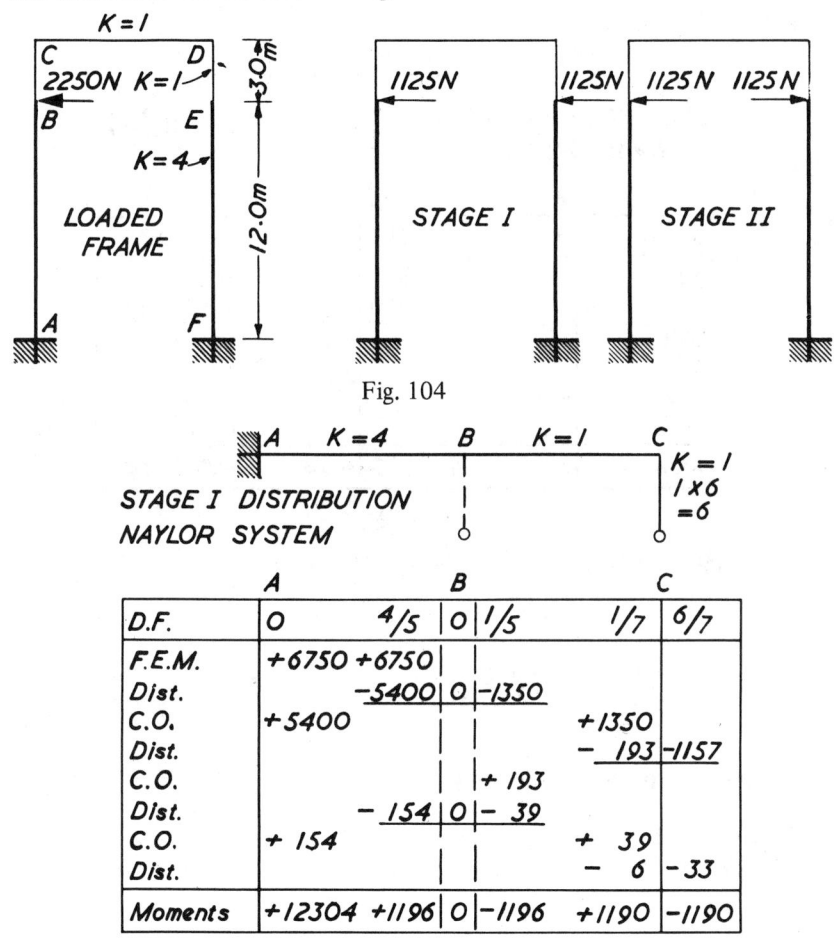

D.F.	A			B			C	
	0	4/5	0	1/5			1/7	6/7
F.E.M.	+6750	+6750						
Dist.		−5400	0	−1350				
C.O.	+5400						+1350	
Dist.							−193	−1157
C.O.				+193				
Dist.		−154	0	−39				
C.O.	+154						+39	
Dist.							−6	−33
Moments	+12304	+1196	0	−1196			+1190	−1190

Fig. 105

Stage II:

Because of symmetry it is possible to make these calculations by considering only one-half of the frame. The K value of CD is halved and carry-over between C and D is omitted.

The F.E.M.s for AB and BC caused by a deflection at B will vary with the $I/L^2 = K/L$ value of the column. If arbitrary moments are applied at each end of the members, the propping force can be calculated. As a result the moments induced by the given load of 1 125 N can be calculated.

Now K/L for $AB = 4/12 = 0.333$
and K/L for $BC = 1/3 = 0.333$.

Assume that the arbitrary moments for $AB = 100$ Nm.

Then those for $BC = -100 \times \dfrac{0.333}{0.333} = -100$ Nm.

The distribution table is shown in Fig. 106.

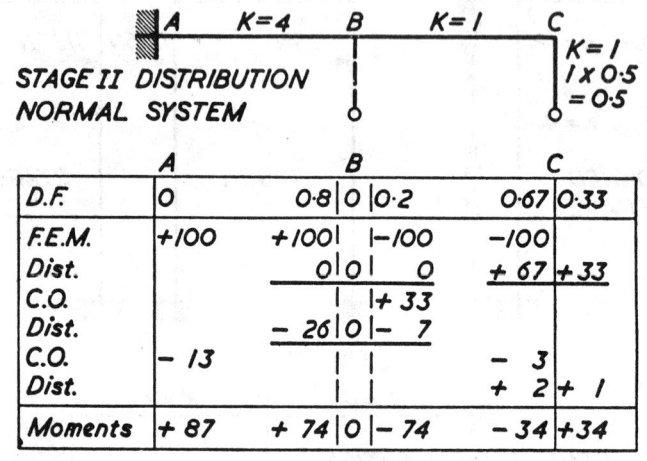

```
        A    K=4    B    K=1    C
                                  K=1
STAGE II DISTRIBUTION              1×0·5
NORMAL SYSTEM                      =0·5
```

D.F.	A	B	C
	0	0·8 | 0·2	0·67|0·33
F.E.M.	+100	+100 |-100	-100
Dist.		0 | 0	+67 |+33
C.O.		| +33	
Dist.		-26 | -7	
C.O.	-13		-3
Dist.			+2 | +1
Moments	+87	+74 | -74	-34 |+34

Fig. 106

The propping force required to avoid any deflection at B

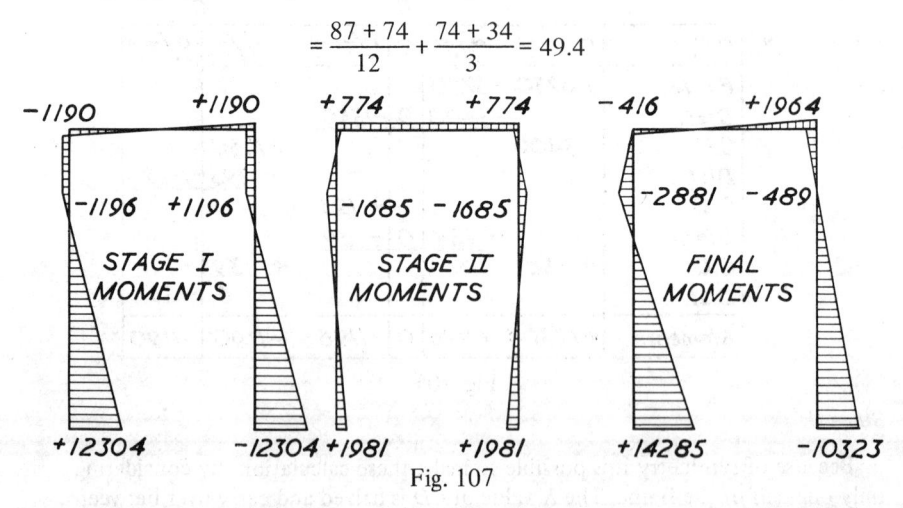

$$= \frac{87 + 74}{12} + \frac{74 + 34}{3} = 49.4$$

```
-1190    +1190    +774     +774    -416     +1964

   -1196 +1196      -1685 -1685       -2881  -489

   STAGE I          STAGE II          FINAL
   MOMENTS          MOMENTS           MOMENTS

+12304    -12304 +1981    +1981    +14285    -10323
```
Fig. 107

But the load on the frame = 1 125 N. Therefore, the moments resulting from the foregoing distribution table must be multiplied by 1 125/49.4 giving:
$M_{AB} = +1\,981$, $M_{BC} = -1\,685$ and $M_{CD} = +774$ Nm.

The B.M. diagrams for Stages I and II are shown in Fig. 107 together with the final moments resulting from a combination of the stages.

MOMENT DISTRIBUTION

Vierendeel Girders

Example 15. When the top and bottom chords of any panel of Vierendeel girders are of the same section, the girders may be analysed in the same way as multi-storey frames. The F.E.M.s are derived from the panel shears.

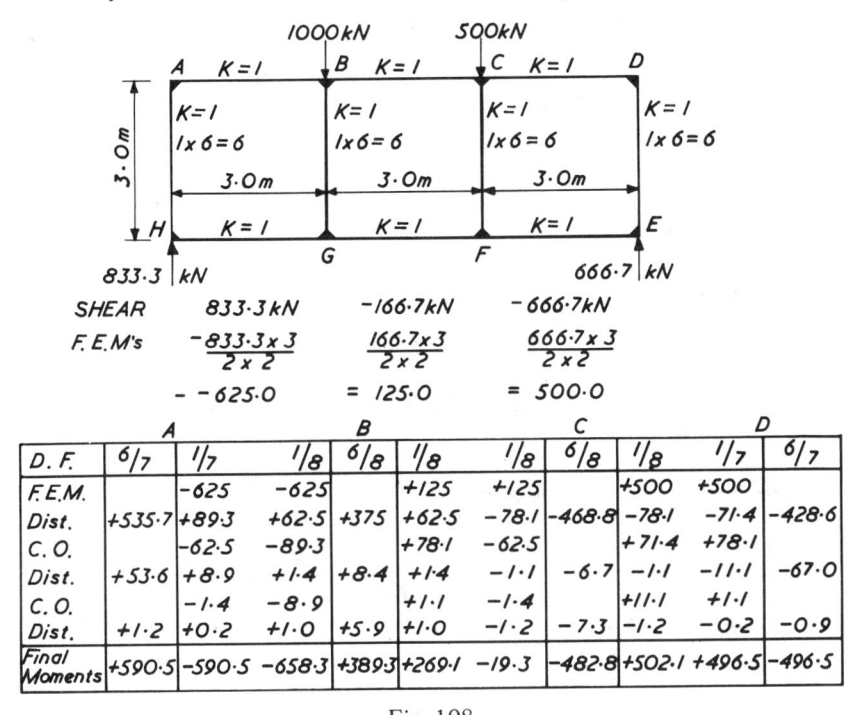

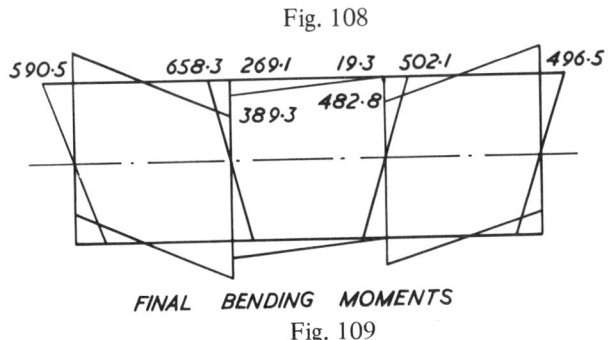

	A			B			C			D
D.F.	6/7	1/7	1/8	6/8	1/8	1/8	6/8	1/8	1/7	6/7
F.E.M.		−625	−625		+125	+125		+500	+500	
Dist.	+535·7	+89·3	+62·5	+375	+62·5	−78·1	−468·8	−78·1	−71·4	−428·6
C.O.		−62·5	−89·3		+78·1	−62·5		+71·4	+78·1	
Dist.	+53·6	+8·9	+1·4	+8·4	+1·4	−1·1	−6·7	−1·1	−11·1	−67·0
C.O.		−1·4	−8·9		+1·1	−1·4		+11·1	+1·1	
Dist.	+1·2	+0·2	+1·0	+5·9	+1·0	−1·2	−7·3	−1·2	−0·2	−0·9
Final Moments	+590·5	−590·5	−658·3	+389·3	+269·1	−19·3	−482·8	+502·1	+496·5	−496·5

Fig. 108

FINAL BENDING MOMENTS
Fig. 109

The girder in Fig. 108 is analysed as shown, and the final B.M. diagram is shown in Fig. 109.

Interpanel Loading

In most multi-storey buildings wind loads are transmitted to the columns at floor level, but occasionally in industrial buildings, and often with Vierendeel girders, loads are applied between the panel points.

268 METHODS OF STRUCTURAL ANALYSIS

Although Naylor's method involves two stages of analysis, the work is simplified compared with the normal method. An example will demonstrate the method of treatment.

FRAME DATA
C———D U.D.L = 3.0 kN/m
B K constant E
A————F 7.5m

STAGE II + STAGE I (3.0m + 3.0m)

FINAL B.M. DIAGRAM

STAGE II − STAGE I

STAGE I
1.5 kN/m on left column; 9 kN at A; 1.5 kN at F; 9 kN

F.E.M's
$$\frac{4.5 kN \times 3}{2 \times 2} = -3.375 \text{ kNm}$$

$$\frac{WL}{12} = \frac{4.5 \times 3}{12} = \pm 1.125 \text{ kNm}$$

$$M_{BA} = -(9 \times 3) + (1.5 \times 3 \times 1.5)$$
$$= -20.25 \text{ kNm}$$

STAGE I DISTRIBUTION NAYLOR SYSTEM

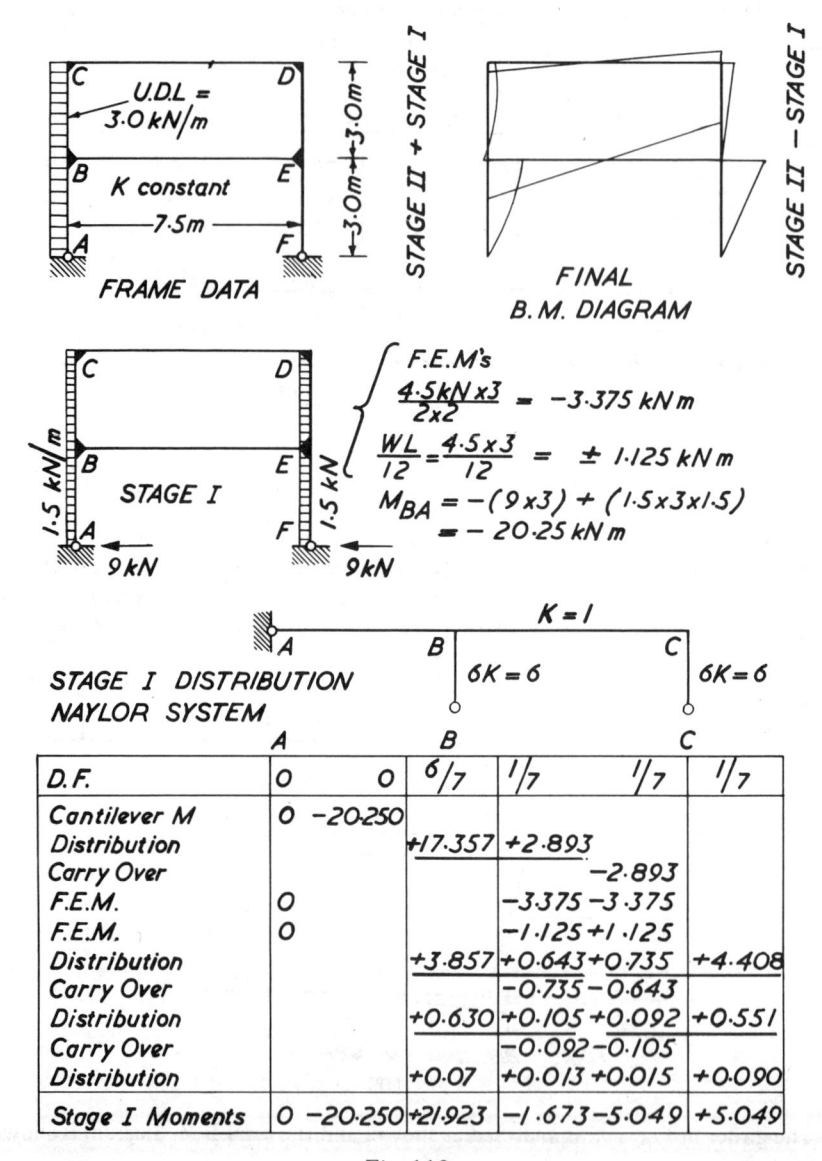

$K = 1$ between B and C; $6K = 6$ at B (down); $6K = 6$ at C (down)

D.F.	A		B $6/7$	$1/7$	$1/7$	C $1/7$
Cantilever M	0	−20.250				
Distribution			+17.357	+2.893		
Carry Over					−2.893	
F.E.M.	0			−3.375	−3.375	
F.E.M.	0			−1.125	+1.125	
Distribution			+3.857	+0.643	+0.735	+4.408
Carry Over				−0.735	−0.643	
Distribution			+0.630	+0.105	+0.092	+0.551
Carry Over				−0.092	−0.105	
Distribution			+0.07	+0.013	+0.015	+0.090
Stage I Moments	0	−20.250	+21.923	−1.673	−5.049	+5.049

Fig. 110a

Example 16. The U.D.L. applied to the left column of the frame in Fig. 110 is divided in a manner similar to that adopted for Example 14, and the distribution is carried out in two stages as shown. Note that it is only necessary to consider the left half of the frame in each case.

MOMENT DISTRIBUTION

$$\text{F.E.M's}$$
$$\frac{WL}{12} = \frac{4.5 \times 3}{12} = \pm 1.125 \text{ kNm}$$

$$\frac{WL}{8} = \frac{4.5 \times 3}{8} = +1.688 \text{ kNm}$$

STAGE II DISTRIBUTION
NORMAL SYSTEM

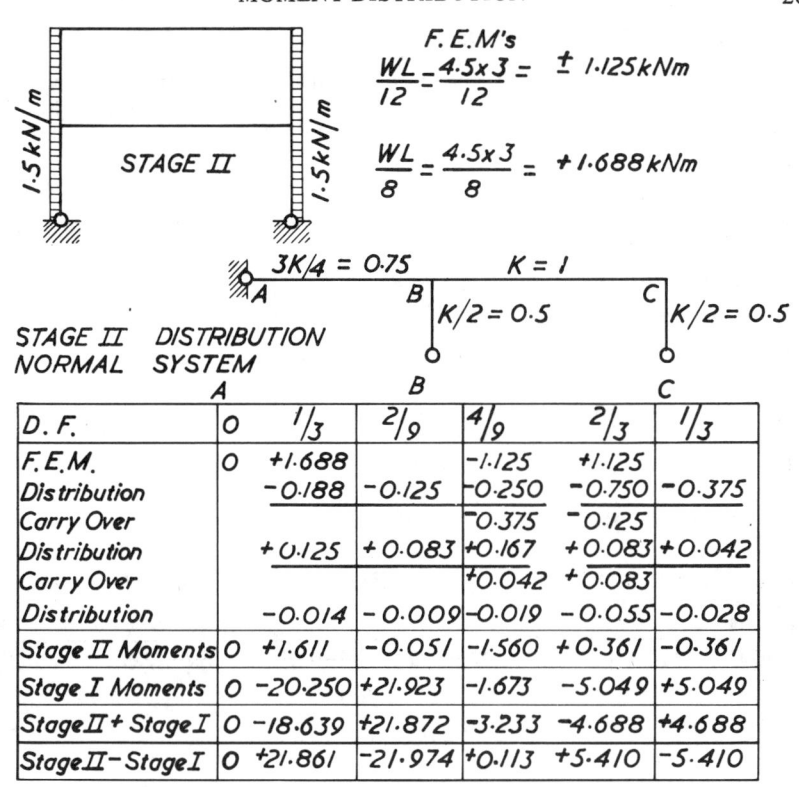

D.F.	A		B		C	
D.F.	0	1/3	2/9	4/9	2/3	1/3
F.E.M.	0	+1.688		−1.125	+1.125	
Distribution		−0.188	−0.125	−0.250	−0.750	−0.375
Carry Over				−0.375	−0.125	
Distribution		+0.125	+0.083	+0.167	+0.083	+0.042
Carry Over				+0.042	+0.083	
Distribution		−0.014	−0.009	−0.019	−0.055	−0.028
Stage II Moments	0	+1.611	−0.051	−1.560	+0.361	−0.361
Stage I Moments	0	−20.250	+21.923	−1.673	−5.049	+5.049
Stage II + Stage I	0	−18.639	+21.872	−3.233	−4.688	+4.688
Stage II − Stage I	0	+21.861	−21.974	+0.113	+5.410	−5.410

Fig. 110b

The final B.M. diagram is shown in detail in Fig. 110c.

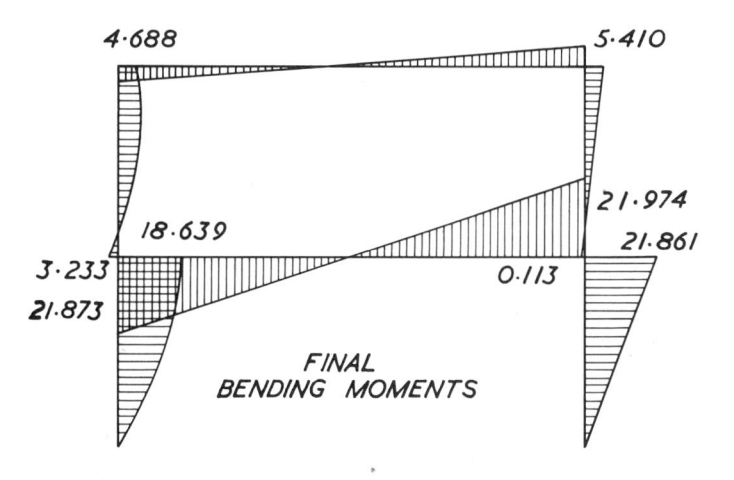

FINAL BENDING MOMENTS

4.688   5.410
18.639   21.974
3.233   21.861
21.873   0.113

Fig. 110c

Unsymmetrical Vertical Loads

Unsymmetrical vertical loads can be divided in a similar manner, as shown in Fig. 111, the Naylor method being employed for the first stage and the normal method for the second.

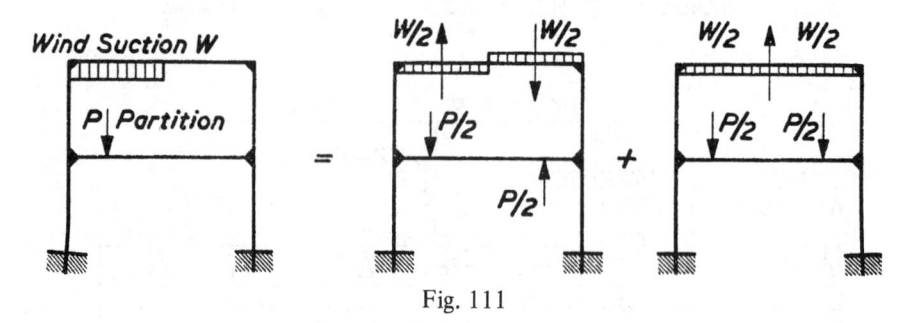

Fig. 111

As a further aid, the F.E.M.s may be divided as shown in Fig. 112.

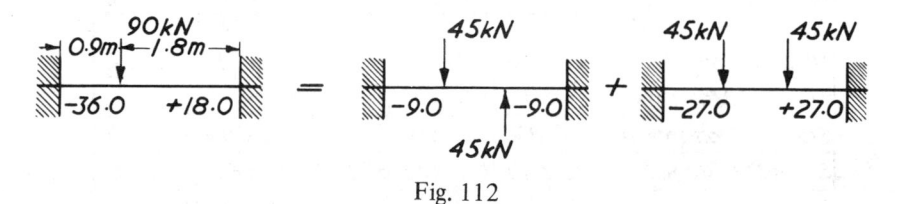

Fig. 112

FIXED-END MOMENTS

BUILT-IN BEAMS OF CONSTANT CROSS-SECTION

Symmetrical loadings	Values of $+M_A$ and $-M_B$
Total U.D.L.$= W$	$-\dfrac{WL}{12}$
$W/2$, $W/2$ at ends, a, $L-2a$, a	$-\dfrac{Wa}{12L}(3L-2a)$
W over central b, a, b, a	$-\dfrac{W}{24L}(3L^2-b^2)$
Triangular W over L	$-\dfrac{5WL}{48}$
Triangular W, a, $L-2a$, a	$-\dfrac{W}{48L}(5L^2+4aL-4a^2)$
$W/2$, $W/2$ triangular full	$-\dfrac{WL}{16}$
$W/2$, $W/2$ triangular, a, $L-2a$, a	$-\dfrac{Wa}{12L}(2L-a)$
$W/2$, $W/2$ triangular at ends, a, a	$-\dfrac{Wa}{12L}(4L-3a)$
W 2nd degree parabola	$-\dfrac{WL}{10}$
$W/2$, $W/2$ complement of parabola	$-\dfrac{WL}{20}$
P central	$-\dfrac{PL}{8}$
P, P, a, $L-2a$, a	$-\dfrac{Pa}{L}(L-a)$
$(n-1)$ forces, L/n	$-\dfrac{PL}{12n}(n^2-1)$
n forces, L/n, $L/2n$	$-\dfrac{PL}{24n}(2n^2+1)$
Any symmetrical loading	$-A_S/L$ Where A_S is the area of the 'free' bending moment diagram

METHODS OF STRUCTURAL ANALYSIS

FIXED-END MOMENTS

BUILT-IN BEAMS OF CONSTANT CROSS-SECTION

Asymmetrical loadings	Values of M_A and M_B
W over left half (L/2), A to B	$M_A = -\dfrac{11}{96}WL$ $M_B = +\dfrac{5}{96}WL$
W over length a, $a/L = m$	$M_A = -\dfrac{WL}{12}\cdot m(3m^2 - 8m + 6)$ $M_B = +\dfrac{WL}{12}\cdot m^2(4 - 3m)$
W over central portion b; a, b, c segments; $d = a+b$, $e = b+c$	$M_A = -\dfrac{W}{12L^2 b}\left[e^3(4L-3e) - c^3(4L-3c)\right]$ $M_B = +\dfrac{W}{12L^2 b}\left[d^3(4L-3d) - a^3(4L-3a)\right]$
Triangular load W, max at B	$M_A = -\dfrac{WL}{10}$ $M_B = +\dfrac{WL}{15}$
Triangular load W over a, max at A, then b unloaded	$M_A = -\dfrac{Wa}{30L^2}(3a^2 + 10bL)$ $M_B = +\dfrac{Wa^2}{30L^2}(5L - 3a)$
Triangular load W over a, max at end of a	$M_A = -\dfrac{Wa}{15L^2}(10L^2 - 15aL + 6a^2)$ $M_B = +\dfrac{Wa^2}{10L^2}(5L - 4a)$
Applied moment $+M$ at distance a from A, b from B	$M_A = +M\dfrac{b}{L^2}(3a - L)$ $M_B = +M\dfrac{a}{L^2}(3b - L)$
Point load P at distance a from A, b from B	$M_A = -\dfrac{Pab^2}{L^2}$ $M_B = +\dfrac{Pa^2 b}{L^2}$
Two crane loads P, spaced a, at distance x from A	When $x = \dfrac{4L - 3a - \sqrt{4L^2 - 9a^2}}{6}$ $M_{A\,max.} = -\dfrac{P}{54L^2}\left[8L^3 + (4L^2 - 9a^2)^{3/2}\right]$ corresponding $M_B = +\dfrac{P}{54L^2}\left[4L^3 + (2L^2 + 9a^2)\sqrt{4L^2 - 9a^2}\right]$
General loading W, free B.M. diagram area A_S, centroid at $\bar{x}$ from A	$M_A = -\left[\dfrac{4A_S}{L} - \dfrac{6A_S\bar{x}}{L^2}\right]$ $M_B = +\left[\dfrac{6A_S\bar{x}}{L^2} - \dfrac{2A_S}{L}\right]$

Where A_S is the area of the 'free' B.M. Diagram and $\bar{x}$ is the distance from A to its centroid

MOMENT DISTRIBUTION
FIXED-END MOMENTS

PROPPED CANTILEVERS OF CONSTANT CROSS-SECTION	
Symmetrical loadings	Values of fixing moment M_A
Total U.D.L. = W	$-\dfrac{WL}{8}$
W/2, W/2; a, $L-2a$, a	$-\dfrac{Wa}{8L}(3L-2a)$
W; a, b, a	$-\dfrac{W}{16L}(3L^2-b^2)$
W (triangular, length L)	$-\dfrac{5WL}{32}$
W (triangular); a, $L-2a$, a	$-\dfrac{W}{32L}(5L^2+4aL-4a^2)$
W/2, W/2 (triangles meeting at centre)	$-\dfrac{3WL}{32}$
W/2, W/2; a, $L-2a$, a	$-\dfrac{Wa}{8L}(2L-a)$
W/2, W/2; a, a	$-\dfrac{Wa}{8L}(4L-3a)$
W 2nd degree parabola	$-\dfrac{3WL}{20}$
W/2, W/2 complement of parabola	$-\dfrac{3WL}{40}$
P (at B)	$-\dfrac{3PL}{16}$
P, P; a, $L-2a$, a	$-\dfrac{3Pa}{2L}(L-a)$
(n-1) forces, spacing L/n	$-\dfrac{PL}{8n}(n^2-1)$
n forces, spacing L/n, $L/2n$ at ends	$-\dfrac{PL}{16n}(2n^2+1)$
A Any symmetrical loading B	$-3A_s/2L$ Where A_s is the area of the 'free' bending moment diagram

For cantilevers of opposite hand, the fixing moments M_B are of opposite sign.

METHODS OF STRUCTURAL ANALYSIS
FIXED-END MOMENTS

PROPPED CANTILEVERS OF CONSTANT CROSS-SECTION	
Asymmetrical loadings	Values of fixing moment M_A
W at centre of left half, spans $L/2$, $L/2$	$-\dfrac{9}{64}WL$
W at centre of right half, spans $L/2$, $L/2$	$-\dfrac{7}{64}WL$
W over length a from A, $a/L = m$, remainder $L-a$	$-\dfrac{Wa}{8}(2-m)^2$
W over length b from B, $b/L = m$, with a, b	$-\dfrac{Wb}{8}(2-m^2)$
W over length d, with segments a, b, c	$-\dfrac{W}{8L^2 b}(d^2-c^2)(2L^2-c^2-d^2)$
Triangular load W, max at A, over L	$-\dfrac{2}{15}WL$
Triangular load W, max at B, over L	$-\dfrac{7}{60}WL$
Triangular W over a from A, then b	$-\dfrac{Wa}{60L^2}(3a^2-15aL+20L^2)$
Triangular W over b from B	$-\dfrac{Wb}{15L^2}(5L^2-3b^2)$
Triangular W over a from A, $a/L = m$	$-Wa\left(\dfrac{m^2}{5} - \dfrac{3m}{4} + \dfrac{2}{3}\right)$
Triangular W over b from B	$-\dfrac{Wb}{60L^2}(10L^2-3b^2)$
Moment $+M$ applied, $a/L = n$	$+\dfrac{M}{2}(2-6n+3n^2)$
Point load P at distance a from A, b from B	$-\dfrac{Pb}{2L^2}(L^2-b^2)$
General load W, with $\bar{x}$ from B	$-\dfrac{3A_S \bar{x}}{L^2}$

Where A_S is the area of the 'free' B.M. diagram, considering AB as a beam, and $\bar{x}$ is the distance from B to its centroid

For cantilevers of opposite hand, the fixing moments M_B are of opposite sign.

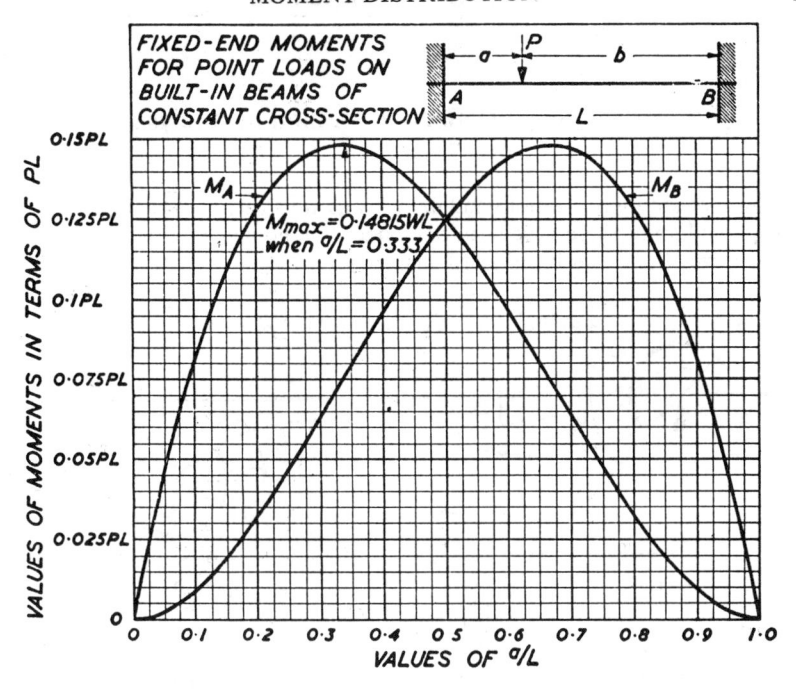

MOMENT DISTRIBUTION

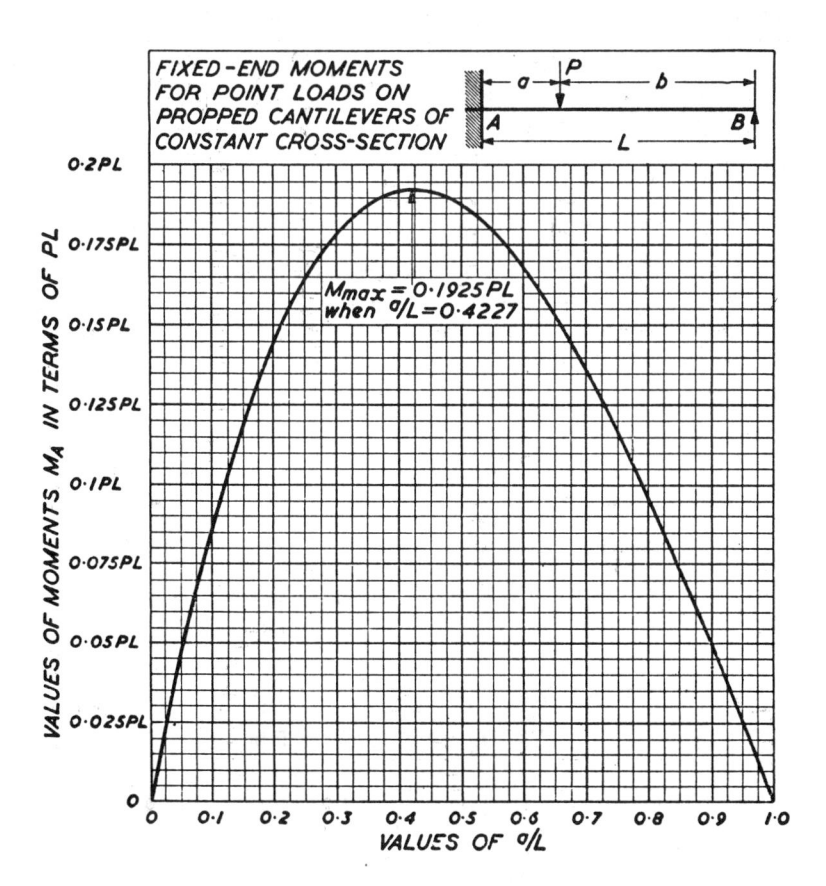

FIXED-END MOMENTS FOR POINT LOADS ON BUILT-IN BEAMS OF CONSTANT CROSS-SECTION

$M_{max} = 0.14815 WL$ when $a/L = 0.333$

FIXED-END MOMENTS FOR POINT LOADS ON PROPPED CANTILEVERS OF CONSTANT CROSS-SECTION

$M_{max} = 0.1925 PL$ when $a/L = 0.4227$

(c) The Slope Deflection Method of Analysis

In this method joint rotations and deflections are treated as the unknown quantities and, once these have been evaluated, the moments follow automatically by substituting the values in standard equations.

SLOPE DEFLECTION SYMBOLS

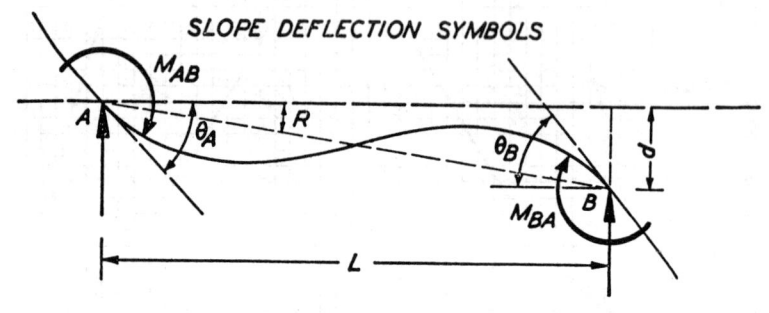

Fig. 113

Suppose that the member AB in Fig. 113 is one unloaded span of a continuous beam and that the member is of constant moment of inertia. Then, for the conditions shown

$$M_{AB} = 2EK(2\theta_A + \theta_B - 3R)$$
$$M_{BA} = 2EK(2\theta_B + \theta_A - 3R),$$

where E and K have the normal significance,

θ_A and θ_B are the angles the joints make with the horizontal

and R is the angle of rotation of B with respect to A when B sinks an amount d (i.e. $R = d/L$).

With regard to sign convention,

θ is positive when the tangent to the beam rotates in a clockwise direction,

R is positive when the beam rotates in a clockwise direction

and M is positive when the moment acts in a clockwise direction on the beam.

Therefore, the various values of M, θ and R in Fig. 113 are all positive.

Suppose that the span AB carries a load acting downwards in the normal fashion. Then

$$M_{AB} = 2EK(2\theta_A + \theta_B - 3R) - \text{F.E.M.}_{AB}$$
$$M_{BA} = 2EK(2\theta_B + \theta_A - 3R) + \text{F.E.M.}_{BA}.$$

F.E.M.$_{AB}$ and F.E.M.$_{BA}$ are the fixed-end moments which would exist if AB were a fixed-end beam. The values and signs used are precisely the same as in the Moment Distribution Method and the tables given on pages 271 to 275 are equally of use for the Slope Deflection Method.

When the end A of a beam AB is hinged, the formula for the moment at the other end is modified as follows:

$$M_{BA} = EK(3\theta_B - 3R) \quad \text{(unloaded condition)}$$

or $\qquad M_{BA} = EK(3\theta_B - 3R) + \text{F.E.M.}_{BA} \quad \text{(loaded condition)}$

An analogy for this modification exists in Moment Distribution where the stiffness factor for a beam hinged at one end is reduced to $\frac{3}{4}K$. It should be noted that the value of the F.E.M. is that applicable to beams hinged at one end and fixed in direction and position at the other.

The standard formulae will be applied to some of the examples which appeared in the section on Moment Distribution. The reader should examine especially the signs which are given to the rotation R.

As in most other methods of analysis the value of the modulus of elasticity E can be ignored in nearly every example.

When calculating the values of F.E.M.s for loads acting downwards in the normal fashion the appropriate signs can be ignored because the fundamental formulae automatically provide the correct signs.

The final B.M. diagram is prepared by considering all hogging moments as negative and all sagging moments as positive.

When the method of Slope Deflection is used to find the moments in a continuous beam, the slope of the beam over each internal support is calculated. The values of the slopes may be useful in calculating the deflections in interior spans, but great care is needed with signs. In the slope-deflection calculations a positive value for the slope means that the beam has rotated in a clockwise direction. In the purely mathematical sense a positive slope is one 'going upwards to the right', i.e. dy/dx is positive, see Fig. 2, Page 83. Furthermore, it is essential that the units employed throughout the Slope Deflection calculations should be the same as those for E and I. Otherwise the values of the slopes will not be related to the units employed in the deflection calculations.

Continuous Beams

Example 1. Consider the continuous beam in Fig. 114. Under the action of the loads, the joint B rotates in an anti-clockwise direction. When this rotation θ_B is calculated the whole beam can be analysed.

Let the suffixes 1 and 2 be applied to AB and BC respectively.

A and C are fixed in direction as well as position. Hence,

$$\theta_A = 0 = \theta_C.$$

278   METHODS OF STRUCTURAL ANALYSIS

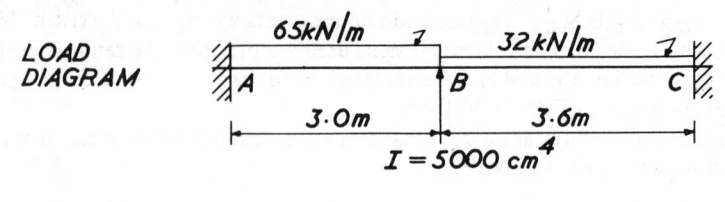

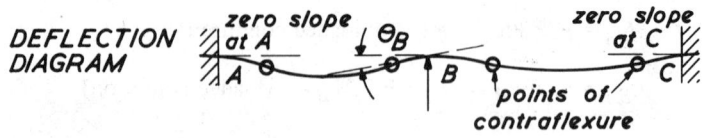

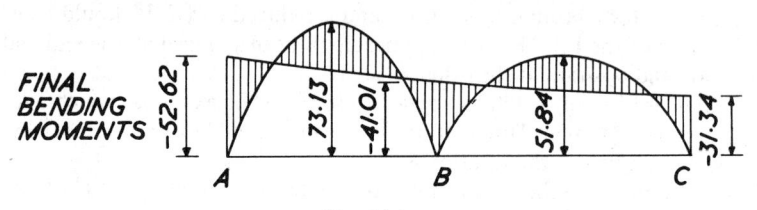

LOAD DIAGRAM: 65 kN/m over span AB (3.0 m), 32 kN/m over span BC (3.6 m), $I = 5000$ cm^4

DEFLECTION DIAGRAM: zero slope at A, θ_B at B, zero slope at C, points of contraflexure

FINAL BENDING MOMENTS: -52.62 at A, 73.13, -41.01 at B, 51.84, -31.34 at C

Fig. 114

A, B and C are on the same level. Therefore,

$$R_1 = 0 = R_2.$$

$$\text{F.E.M.}_{AB} = \text{F.E.M.}_{BA} = \frac{WL}{12} = 48.75 \text{ kNm}$$

$$\text{F.E.M.}_{BC} = \text{F.E.M.}_{CB} = \frac{WL}{12} = 34.56 \text{ kNm}.$$

Now $$M_{BA} + M_{BC} = 0.$$

Hence, $2EK_1(2\theta_B + \theta_A - 3R_1) + 48.75 + 2EK_2(2\theta_B + \theta_C - 3R_2) - 34.56 = 0$.

But $$\theta_A = 0 = \theta_C,$$
$$R_1 = 0 = R_2,$$

$$K_1 = \frac{I}{L} = \frac{5\,000 \times 10^{-8}}{3.0} \frac{\text{m}^4}{\text{m}}$$

and $$K_2 = \frac{I}{L} = \frac{5\,000 \times 10^{-8}}{3.6} \frac{\text{m}^4}{\text{m}}$$

Therefore

$$\left(2E \times \frac{5\,000 \times 10^{-8}}{3.0} \times 2\theta_B\right) + \left(2E \times \frac{5\,000 \times 10^{-8}}{3.6} \times 2\theta_B\right) = 34.56 - 48.75$$

$$E\theta_B = -11.61 \times 10^4.$$

THE SLOPE DEFLECTION METHOD OF ANALYSIS

Using the basic formulae

$$M_{AB} = 2EK_1(\theta_B) - 48.75$$
$$= \left(-2 \times \frac{5\,000 \times 10^{-8}}{3.0} \times 11.61 \times 10^4\right) - 48.75 = 52.62 \text{ kNm}$$

$$M_{BA} = -M_{BC}$$
$$= 2EK_1(2\theta_B) + 48.75$$
$$= \left(-2 \times \frac{5\,000 \times 10^{-8}}{3.0} \times 2 \times 11.61 \times 10^4\right) + 48.75 = 41.01 \text{ kNm}$$

$$M_{CB} = 2EK_2(\theta_B) + 34.56$$
$$= \left(-2 \times \frac{5\,000 \times 10^{-8}}{3.6} \times 11.61 \times 10^4\right) + 34.56 = 31.34 \text{ kNm}.$$

Example 2. The continuous beam *ABCDE* in Fig. 115 involves the treatment of simply supported end spans and a cantilever.

Let the suffixes 1, 2 and 3 apply to *AB*, *BC* and *CD* respectively.

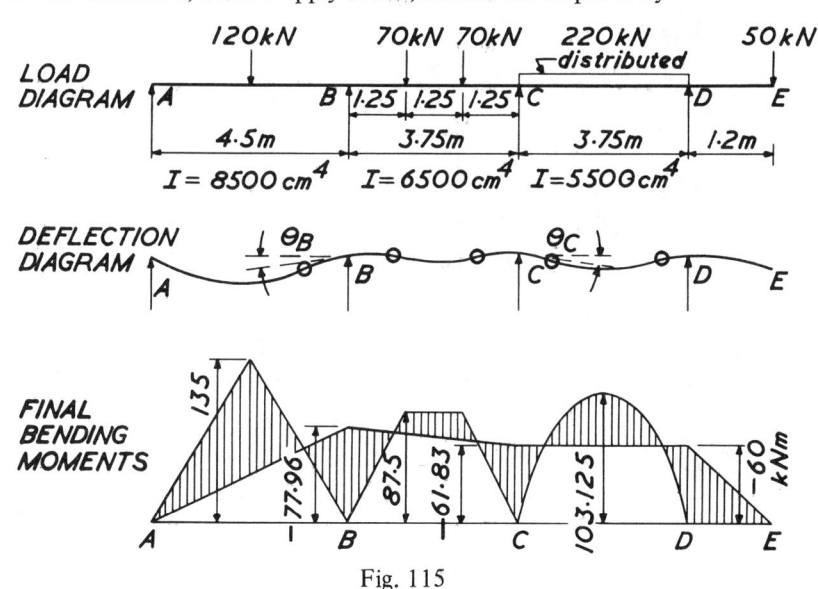

Fig. 115

Now the supports *A*, *B*, *C* and *D* are on the same level. Hence

$$R_1 = 0 = R_2 = R_3.$$

Also
$$M_{BA} + M_{BC} = 0$$
$$M_{CB} + M_{CD} = 0$$
$$M_{DC} = 60 \text{ kNm.}$$

The effect at *C* of the cantilever *DE* is that F.E.M.$_{CD}$ is reduced in value by half the amount of M_{DC}, i.e. by 30 kNm (cf. Principle I in Moment Distribution).

280   METHODS OF STRUCTURAL ANALYSIS

$$EK_1(3\theta_B - 3R_1) + F.E.M._{BA} + 2EK_2(2\theta_B + \theta_C - 3R_2) - F.E.M._{BC} = 0$$
$$2EK_2(2\theta_C + \theta_B - 3R_2) + F.E.M._{CB} + EK_3(3\theta_C - 3R_3) - F.E.M._{CD} + M_{DC}/2 = 0$$

Hence $E \times 1.89 \times 10^{-5}(3\theta_B) + 101.25 + 2E \times 1.733 \times 10^{-5}(2\theta_B + \theta_C) - 58.33 = 0$

$$12.6 \times 10^{-5} E\theta_B + 3.466 \times 10^{-5} E\theta_C + 42.92 = 0 \quad \ldots \ldots \text{(i)}$$

$2E \times 1.733 \times 10^{-5}(2\theta_C + \theta_B) + 58.33 + E \times 1.467 \times 10^{-5}(3\theta_C) - 103.1 + 30 = 0$

$$3.466 \times E\theta_B \times 10^{-5} + 11.333 E\theta_C \times 10^{-5} - 14.77 = 0.$$

or $12.6 E\theta_B \times 10^{-5} + 41.199 E\theta_C \times 10^{-5} - 53.69 = 0 \quad \ldots \ldots \text{(ii)}$

Subtracting (i) from (ii)

$$37.733 E\theta_C \times 10^{-5} = 96.61$$
$$E\theta_C = 2.56 \times 10^5.$$

Substituting in equation (i)

$$E\theta_B = -4.11 \times 10^5.$$

Employing these values of $E\theta_B$ and $E\theta_C$

$$M_{BA} = -M_{BC}$$
$$= EK_1(3\theta_B) + 101.25$$
$$= -1.89 \times 10^{-5} \times 3 \times 4.11 \times 10^5 + 101.25$$
$$= 77.96 \text{ kNm}.$$

$$M_{CB} = -M_{CD}$$
$$= 2EK_2(2\theta_C + \theta_B) + 58.33$$
$$= 2 \times 1.733 \times 10^{-5}(2.56 \times 2 - 4.11) \times 10^5 + 58.33$$
$$= 61.83 \text{ kNm}.$$

Example 3. Figure 116 shows a continuous beam *ABC*, freely supported at *A* and fixed at *C*. The B.M. induced by the settlement of 25 mm at *B* will be calculated.

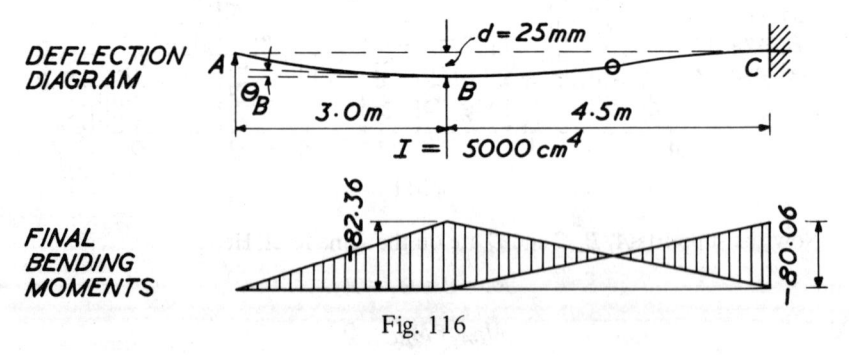

Fig. 116

Assume that $E = 2.1 \times 10^5 \text{ N/mm}^2$.
Let the suffixes 1 and 2 be applied to spans *AB* and *BC* respectively.
A being simply supported, $M_{AB} = 0$.
C being fixed in direction, $\theta_C = 0$.
B sinks 25 mm = *d*.

$$R_1 = \frac{d}{L_1} = \frac{25}{3\,000}$$

$$R_2 = -\frac{d}{L_2} = -\frac{25}{4\,500}$$

Now $M_{BA} + M_{BC} = 0.$

Hence $EK_1(3\theta_B - 3R_1) + 2EK_2(2\theta_B + \theta_C - 3R_2) = 0$

$$2.1 \times 10^5 \times \frac{5\,000 \times 10^4}{3.0 \times 10^3} \left(3\theta_B - \frac{3 \times 25}{3.0 \times 10^3}\right)$$

$$+ 2 \times 2.1 \times 10^5 \times \frac{5\,000 \times 10^4}{4.5 \times 10^3}\left(2\theta_B + \frac{3 \times 25}{4.5 \times 10^3}\right) = 0$$

$$1.889\theta_B = 0.926 \times 10^{-3}$$

$$\theta_B = 0.00049.$$

Using this value of θ_B

$$M_{BA} = EK_1(3\theta_B - 3R_1)$$

$$= 2.1 \times 10^5 \times \frac{5\,000 \times 10^4}{3.0 \times 10^3}\left(0.00147 - \frac{3 \times 25}{3 \times 10^3}\right)$$

$$= -82.36 \text{ kNm}.$$

Similarly

$$M_{CB} = 2EK_2(\theta_B - 3R_2)$$

$$= 2 \times 2.1 \times 10^5 \times \frac{5\,000 \times 10^4}{4.5 \times 10^3}\left(0.00049 + \frac{3 \times 25}{4.5 \times 10^8}\right)$$

$$= 80.06 \text{ kNm}.$$

Example 4. Figure 117 shows a beam ABC which is fixed in position and direction at A and C and in which a clockwise moment of 10 kNm is applied at B. The beam is analysed by finding the rotation and displacement of B.

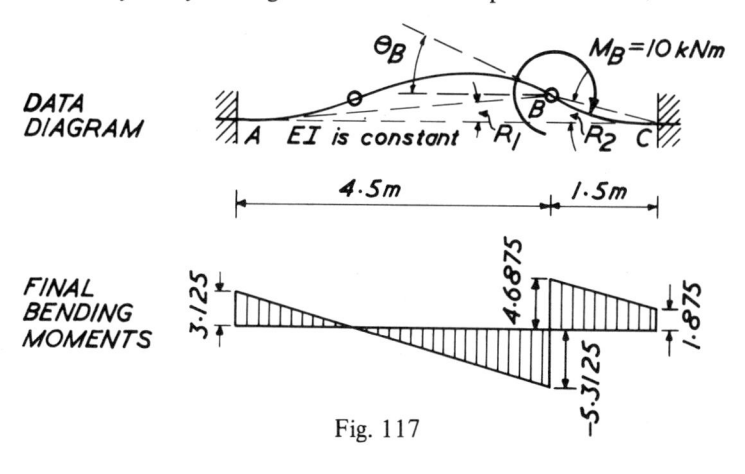

Fig. 117

Now $\theta_A = \theta_C = 0$.

Let the suffixes 1 and 2 be applied to AB and BC respectively.

By virtue of the fact that θ_B is positive it is reasonable to assume that B moves upwards and, consequently, that R_1 is negative. If these assumptions are correct the numerical value calculated for R_1 will be positive. Alternatively it could be assumed that R_1 was positive and the sign of the numerical value would provide the correct sign for R_1.

EI being constant $\qquad K_2 = 3K_1$,

also $\qquad R_2 = -3R_1$.

Now $\qquad M_{BA} + M_{BC} = 10 \text{ kNm}$ . . . . (i)

The moment at B induces vertical shear forces at A and C and for equilibrium these forces must be equal and opposite.

Let the shear forces be S.

Then $\qquad M_{AB} + M_{BA} + 4.5S = 0$

$\qquad M_{BC} + M_{CB} + 1.5S = 0$,

or $\qquad (M_{AB} + M_{BA}) - 3(M_{BC} + M_{CB}) = 0$ . . . . (ii)

Equation (i) can be rewritten:

$$2EK_1(2\theta_B + 3R_1) + 2E(3K_1)(2\theta_B - 3 \times 3R_1) = 10$$

$$16EK_1\theta_B - 48EK_1R_1 = 10 . . . . \text{(iii)}$$

Similarly, equation (ii) becomes:

$$2EK_1(\theta_B + 3R_1) + 2EK_1(2\theta_B + 3R_1) - 3\{2E(3K_1)(2\theta_B - 3 \times 3R_1) +$$
$$2E(3K_1)(\theta_B - 3 \times 3R_1)\} = 0$$

$$-48EK_1\theta_B + 336EK_1R_1 = 0 . . . . \text{(iv)}$$

Multiplying equation (iii) by 3

$$48EK_1\theta_B - 144EK_1R_1 = 30 . . . . \text{(v)}$$

Adding equations (iv) and (v)

$$192EK_1R_1 = 30$$

$$EK_1R_1 = 10/64.$$

Substituting in equation (iii)

$$16EK_1\theta_B - \frac{48 \times 10}{64} = 10$$

$$EK_1\theta_B = 70/64.$$

Using these values of $EK_1\theta_B$ and EK_1R_1 in the fundamental equations,

$$M_{AB} = 2EK_1(\theta_B + 3R_1)$$

$$= \frac{140 + 60}{64}$$

$$= 3.125 \text{ kNm}.$$

THE SLOPE DEFLECTION METHOD OF ANALYSIS 283

$$M_{BA} = 2EK_1(2\theta_B + 3R_1)$$
$$= \frac{280 + 60}{64}$$
$$= 5.3125 \text{ kNm}$$
$$M_{BC} = 2E(3K_1)(2\theta_B - 3 \times 3R_1)$$
$$= \frac{840 - 540}{64}$$
$$= 4.6875 \text{ kNm}.$$

(Note $M_{BA} + M_{BC} = 10$ kNm.)
$$M_{CB} = 2E(3K_1)(\theta_B - 3 \times 3R_1)$$
$$= \frac{420 - 540}{64}$$
$$= -1.875 \text{ kNm}.$$

The final B.M.s are shown in Fig. 117.

It will be appreciated that this problem is similar to that for a column provided with a bracket load.

Symmetrical Portal Frames

Example 5. Rigid frames which are symmetrical in shape and symmetrically loaded are easy to analyse by the Slope Deflection Method. The portal frame in Fig. 118 will be analysed as an example.

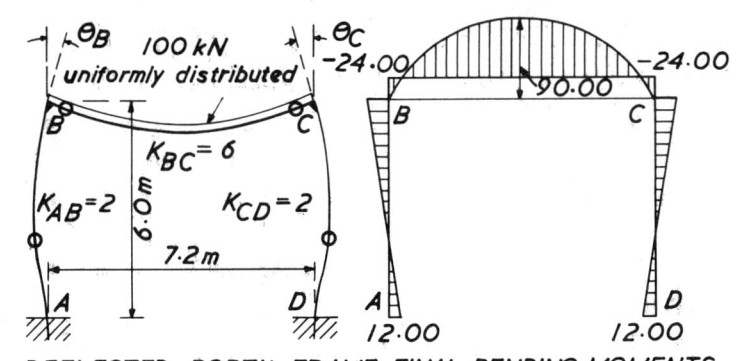

DEFLECTED PORTAL FRAME : FINAL BENDING MOMENTS

Fig. 118

$$\text{F.E.M.}_{BC} = \text{F.E.M.}_{CB} = \frac{100 \times 7.2}{12}$$
$$= 60 \text{ kNm}.$$

Let the suffixes 1, 2 and 3 be applied to the members AB, BC and CD respectively.

Then $K_1 : K_2 : K_3 = 1 : 3 : 1.$

Being symmetrical the frame does not sway under load. In addition, it is assumed that the lengths of the members do not change.

284   METHODS OF STRUCTURAL ANALYSIS

Hence
$$R_1 = 0 = R_2 = R_3$$
$$\theta_B = -\theta_C.$$

As A and D are fixed in direction as well as position
$$\theta_A = 0 = \theta_D.$$

Now
$$M_{BA} + M_{BC} = 0,$$

i.e.
$$2EK_1(2\theta_B) + 2EK_2(2\theta_B + \theta_C) - \text{F.E.M.}_{BC} = 0$$
$$2EK_1(2\theta_B) + 2E(3K_1)(\theta_B) - 60 = 0$$
$$10EK_1\theta_B = 60$$
$$EK_1\theta_B = 6.$$

Using the fundamental formula
$$M_{AB} = 2EK_1(\theta_B)$$
$$= 12 \text{ kNm}$$
$$M_{BA} = 2EK_1(2\theta_B)$$
$$= 24 \text{ kNm}.$$

Similarly
$$M_{CD} = -24 \text{ kNm}$$
$$M_{DC} = -12 \text{ kNm}.$$

Asymmetrical Portal Frames

Example 6. As the frame shown in Fig. 119 is not symmetrical it is less easy to analyse than the previous example.

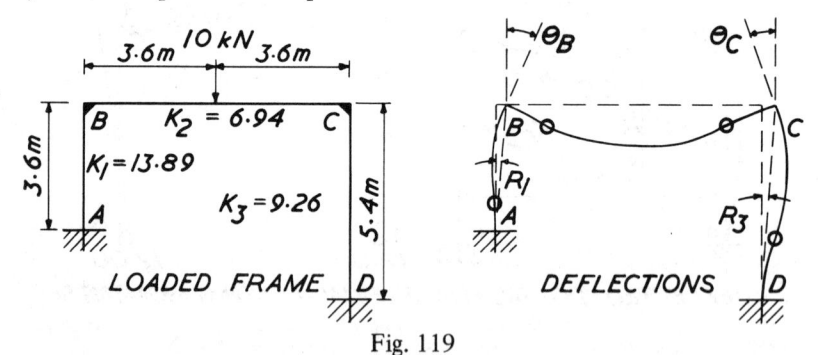

Fig. 119

Let the suffixes 1, 2 and 3 be applied to the members AB, BC and CD respectively.

From the data there is no slope at A and D, i.e.
$$\theta_A = 0 = \theta_D.$$

B and C are at the same height and under load it is assumed that they retain their positions relative to one another, i.e.
$$R_2 = 0.$$
$$\text{F.E.M.}_{BC} = \text{F.E.M.}_{CB} = 9 \text{ kNm}.$$

THE SLOPE DEFLECTION METHOD OF ANALYSIS 285

The conditions of equilibrium require that

$$\Sigma M_B = 0, \quad \Sigma M_C = 0 \quad \text{and} \quad \Sigma H = 0.$$

Hence
$$M_{BA} + M_{BC} = 0$$
$$M_{CB} + M_{CD} = 0$$
$$\frac{M_{AB} + M_{BA}}{L_1} + \frac{M_{CD} + M_{DC}}{L_3} = 0.$$

The side-sway of B equals that of C. Hence
$$R_1 L_1 = R_3 L_3.$$

Using the basic formulae
$$2EK_1(2\theta_B - 3R_1) + 2EK_2(2\theta_B + \theta_C) - 9 = 0$$
$$2EK_2(2\theta_C + \theta_B) + 2EK_3(2\theta_C - 3R_3) + 9 = 0$$
$$\frac{2EK_1(\theta_B - 3R_1) + 2EK_1(2\theta_B - 3R_1)}{L_1} + \frac{2EK_3(2\theta_C - 3R_3) + 2EK_3(\theta_C - 3R_3)}{L_3} = 0.$$

Now $\quad K_1 = 13.889, \quad K_2 = 6.944, \quad K_3 = 9.259$
and $\quad R_1 = 1.5 R_3$

Hence
$$2 \times 13.889 E(2\theta_B - 4.5R_3) + 2 \times 6.944 E(2\theta_B + \theta_C) - 9 = 0$$
$$2 \times 6.944 E(2\theta_C + \theta_B) + 2 \times 9.259 E(2\theta_C - 3R_3) + 9 = 0$$
$$\frac{2 \times 13.889 E}{3.6}(3\theta_B - 9R_3) + \frac{2 \times 9.259 E}{5.4}(3\theta_C - 6R_3) = 0$$

That is:
$$83.333 E\theta_B + 13.889 E\theta_C - 125.001 ER_3 - 9 = 0 \quad \ldots \ldots \text{(i)}$$
$$13.889 E\theta_B + 64.815 E\theta_C - 55.556 ER_3 + 9 = 0 \quad \ldots \ldots \text{(ii)}$$
$$13.889 E\theta_B + 6.173 E\theta_C - 54.012 ER_3 = 0 \quad \ldots \ldots \text{(iii)}$$

Multiplying (ii) by 2.25
$$31.250 E\theta_B + 145.834 E\theta_C - 125.001 ER_3 + 20.25 = 0 \quad \ldots \text{(iv)}$$

Subtracting (iv) from (i)
$$52.083 E\theta_B - 131.945 E\theta_C - 29.25 = 0 \quad \ldots \ldots \ldots \text{(v)}$$

Multiplying (iii) by 2.314
$$32.143 E\theta_B + 14.286 E\theta_C - 125.001 ER_3 = 0 \quad \ldots \ldots \text{(vi)}$$

Subtracting (vi) from (i)
$$51.191 E\theta_B - 0.397 E\theta_C - 9 = 0. \quad \ldots \ldots \ldots \text{(vii)}$$

Multiplying (vii) by 1.017
$$52.083 E\theta_B - 0.404 E\theta_C - 9.157 = 0 \quad \ldots \ldots \ldots \text{(viii)}$$

Subtracting (viii) from (v)
$$-131.541 E\theta_C - 20.093 = 0$$

Whence
$$E\theta_C = -0.15275$$
$$E\theta_B = 0.17463$$
$$ER_3 = 0.02744$$
and
$$ER_1 = 0.04116.$$

Using these values in the basic formulae:

$$M_{AB} = 2EK_1(\theta_B - 3R_1)$$
$$= 2 \times 13.889(0.17463 - 0.12348)$$
$$= 1.42 \text{ kNm}$$

$$M_{BA} = 2EK_1(2\theta_B - 3R_1)$$
$$= 2 \times 13.889(0.34926 - 0.12348)$$
$$= 6.27 \text{ kNm}$$

$$M_{CD} = 2EK_3(2\theta_C - 3R_3)$$
$$= 2 \times 9.259(-0.30550 - 0.08232)$$
$$= -7.18 \text{ kNm}$$

$$M_{DC} = 2EK_3(\theta_C - 3R_3)$$
$$= 2 \times 9.252(-0.15275 - 0.08232)$$
$$= -4.35 \text{ kNm}.$$

The final B.M. diagram is shown in Fig. 120.

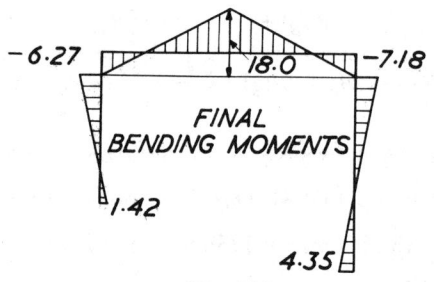

−6·27   18·0   −7·18

FINAL BENDING MOMENTS

1·42

4·35

Fig. 120

Example 7. The frame shown in Fig. 121 is analysed in a similar manner to that of the last example, but the analysis is complicated by the side-sway of the joints B and C. As the legs slope and are of different length, the relative values of R for the three members must be obtained by geometry or other means. This was done for the Stage II moments for the same frame in the section on Moment Distribution and will not be repeated here.

Let the suffixes 1, 2 and 3 apply to members AB, BC and CD respectively.
Now EK is constant throughout the frame.
Also
$$R_2 = -1.816R_1$$
and
$$R_3 = 1.833R_1.$$
The frame may be analysed knowing that:
$$M_{BA} + M_{BC} = 0$$
$$M_{CB} + M_{CD} = 0$$
$$0.772M_{BA} + M_{CD} = 0.69 \text{ kNm (see p. 241, line 13)}.$$

THE SLOPE DEFLECTION METHOD OF ANALYSIS

Fig. 121

Hence

$$EK(3\theta_B + 3R_1) + 2EK(2\theta_B + \theta_C - 3 \times 1.816 R_1) - \text{F.E.M.}_{BC} = 0$$

$$7EK\theta_B + 2EK\theta_C - 7.896 EKR_1 - 25 = 0 \quad \ldots \ldots \ldots (i)$$

$$2EK(2\theta_C + \theta_B - 3 \times 1.816 R_1) + \text{F.E.M.}_{CB} + EK(3\theta_C + 3 \times 1.833 R_1) = 0$$

$$2EK\theta_B + 7EK\theta_C - 5.396 EKR_1 + 25 = 0$$

or $\quad 7EK\theta_B + 24.5 EK\theta_C - 18.886 EKR_1 + 87.5 = 0. \ldots \ldots (ii)$

Subtracting equation (i) from (ii)

$$22.5 EK\theta_C - 10.99 EKR_1 + 112.5 = 0 \ldots \ldots \ldots (iii)$$

Also $\quad 0.772 EK(3\theta_B + 3R_1) + EK(3\theta_C + 5.5 R_1) = 0.69$

Hence

$$2.316 EK\theta_B + 3EK\theta_C + 7.816 EKR_1 = 0.69$$

or $\quad 7.0 EK\theta_B + 9.067 EK\theta_C + 23.623 EKR_1 = 2.085 \ldots \ldots (iv)$

Subtracting equation (i) from (iv)

$$7.067 EK\theta_C + 31.519 EKR_1 + 22.915 = 0$$

or $\quad 22.5 EK\theta_C + 100.350 EKR_1 + 72.957 = 0 \ldots \ldots \ldots (v)$

Subtracting equation (iii) from (v)

$$111.340 EKR_1 = 39.543$$

$$EKR_1 = 0.3552.$$

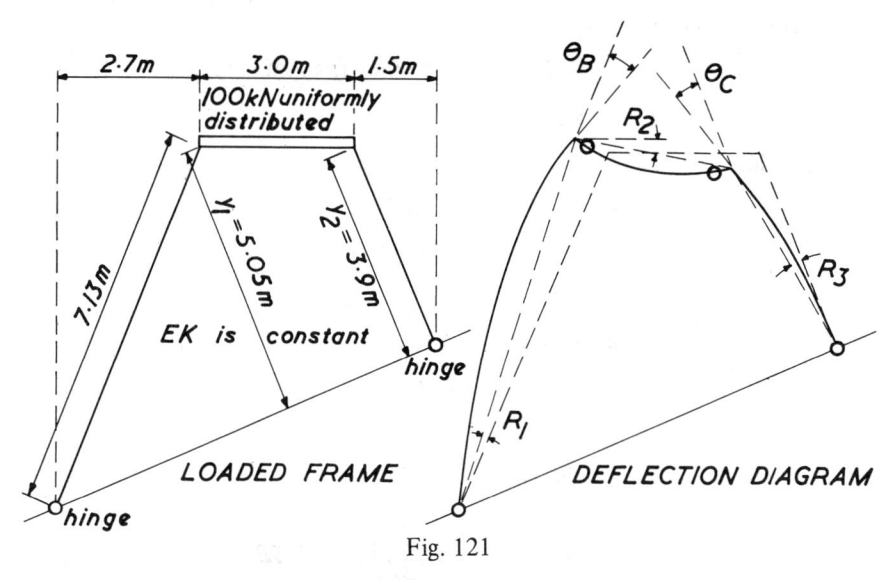

Substituting in equation (v)

$$22.5 EK\theta_C = -72.957 - (100.35 \times 0.3552)$$

$$EK\theta_C = -4.826.$$

Substituting in equation (i)

$$7EK\theta_B = (2 \times 4.826) + (7.896 \times 0.3552) + 25$$
$$EK\theta_B = 5.351.$$

Employing the fundamental formulae

$$M_{BA} = EK(3\theta_B + 3R_1)$$
$$= 3(5.351 + 0.3552)$$
$$= 17.12 \text{ kNm.}$$
$$M_{CD} = EK(3\theta_C + 5.5R_1)$$
$$= 3(-4.826) + 5.5(0.3552)$$
$$= -12.52 \text{ kNm.}$$

The final B.M. diagram is shown in Fig. 122.

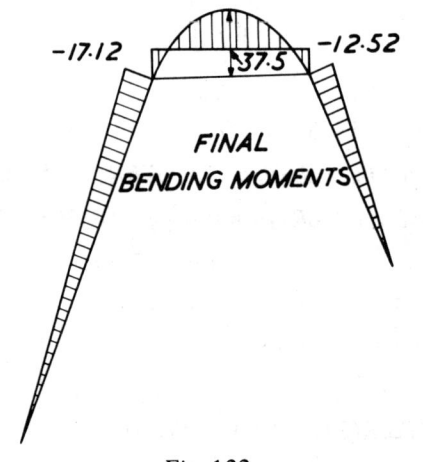

Fig. 122

14. SLOPES AND DEFLECTIONS IN RIGID FRAMES

WHEN a rigid frame has been analysed by employing the Slope Deflection Method, it is possible to find the displacement and angle of rotation of every joint in the structure directly from the analysis computations provided, of course, that the actual, rather than comparative, units have been used throughout.

It is normal to assume that there is no longitudinal strain in any member and that deflections of any point can be measured at right angles to the original position of the member concerned. This being so, rotations and deflections can be computed by using the Area-Moment Method described on pages 85 *et seq.* (Figs. 4, 5 and 6 in particular).

Example 1. It is not always necessary to carry out any detailed calculations because, in cases where there is symmetry, formulae can frequently be used. Suppose that it is desired to find the rotation of the joints B and D in Fig. 1 and

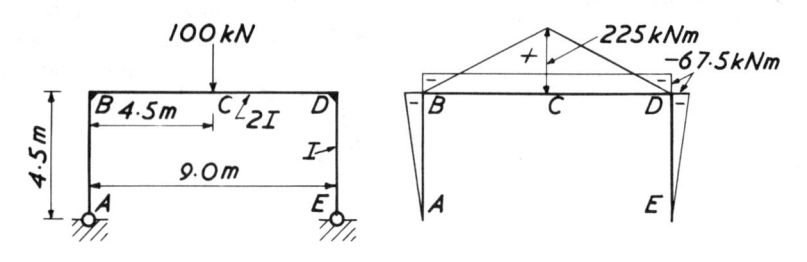

Fig. 1

to find the maximum deflections in the legs and cross-beam. This information can be obtained from the final B.M. diagram and a knowledge of certain formulae dispersed through various parts of this book.

Neglecting any shortening of the members due to strain or curvature, then the points B and D can be assumed not to change their position under the symmetrical loading shown. Hence, AB or DE can be considered as simply supported beams, at the ends of which, at B and D respectively, are applied moments M equal to 67.5 kNm. Considering Fig. 1 and the associated calculations,

$$\theta = \frac{ML}{3EI}$$

Therefore

$$\theta_B = -\theta_B = \frac{67.5 \times 4.5}{3EI} \text{ radian}$$

where clockwise moments and rotations are considered positive.

On page 89, in the section dealing with the deflection of beams, it was shown that the deflection

$$d = -\frac{\text{Secondary B.M.}}{EI}$$

Consider the B.M. for AB shown in Fig. 1 and the properties of a similarly shaped triangular load, as shown on page 32.

Then the maximum secondary B.M. occurs at a point $(0.5774 \times 4.5) = 2.598$ m from A as shown in Fig. 2, and the deflection

$$d = 0.128 WL/EI \quad \text{(where } W = \text{the area of the B.M. diagram for } AB\text{)}$$

$$= \frac{0.128 \times 67.5 \times 4.5 \times 4.5}{2EI}$$

$$= \frac{87.48}{EI} \text{ (m}^3 \text{ units)}$$

i.e. the deflection is outwards from the frame.

Considering the cross-beam BD, the rotations of B and D are already known and it remains to find the maximum deflection, which occurs under the load.

The B.M. diagram consists of a rectangle of height -67.5 kNm and an isosceles triangle of height $+225$ kNm and base BD. Consulting the appropriate properties for U.D.L.s and triangular loads on pages 31 and 32,

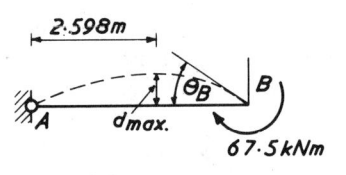

Fig. 2

$$d = -\frac{(-W_1 L/8) + (+W_2 L/6)}{E(2I)}$$

$$= -\frac{(-67.5 \times 9) \times 9/8 + (225 \times 9/2) \times 9/6}{E(2I)}$$

$$= -\frac{417.66}{EI} \text{ (m}^3 \text{ units)},$$

i.e. the deflection is inwards.

It will be noted that the sign convention usual in mathematics has been adopted in this particular example, but any system may be adopted provided it is used throughout a specific example.

Example 2. The frame in Fig. 3 is subjected to a horizontal point load applied at B, which results in the B.M. and thrust diagram shown. Calculate the rotations and deflections of B and D.

It will be observed that a point of contraflexure exists at C, the mid-point of BD, as M_B and M_D are equal and opposite. Consequently, C is in line with B and D and all three points can be assumed to have retained their relative height, each moving horizontally by the same amount.

Then the portion BC may be considered to be a beam simply supported at B and C with a moment of 225 kNm applied at B.

$$\theta_B = \theta_D = \frac{M_B L}{3EI} = \frac{225 \times 4.5}{3EI} = \frac{337.5}{EI} \text{ radian}$$

SLOPES AND DEFLECTIONS IN RIGID FRAMES

Since the slope of the cantilever at A is unknown it is convenient to consider the displacement of A relative to B. Hence, consider B fixed in position and A free to move to the left. Then the relative displacement is in two parts, one due to the rotation of B and the other due to the thrust at A when BA acts as a cantilever, as shown in Fig. 4.

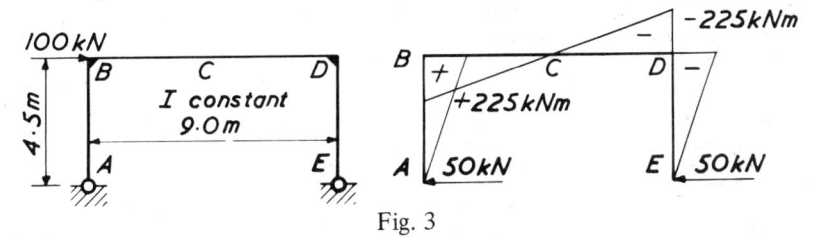

Fig. 3

Therefore the horizontal deflection of A due to the rotation of B,

$$d_\theta = AB \times \theta_B = \frac{4.5 \times 337.5}{EI}$$

$$= \frac{1\,518.75}{EI} \text{ (m}^3 \text{ units)}$$

The horizontal deflection due to the thrust at A,

$$d_H = \frac{PL^3}{3EI} = \frac{50 \times 4.5^3}{3EI} = \frac{1\,518.75}{EI}$$

Therefore the total horizontal deflection of A relative to B

$$= \frac{1\,518.75 + 1\,518.75}{EI} = \frac{3\,037.5}{EI} \text{ (m}^3 \text{ units)}.$$

To determine the deflection in millimetres it is necessary to resolve into mm units throughout

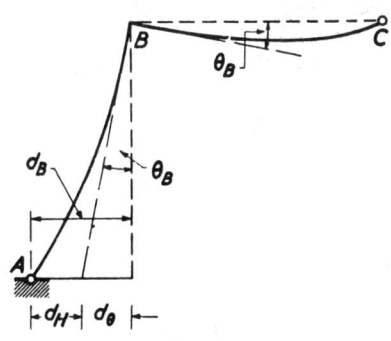

Fig. 4

Suppose the framework is composed of 533 mm × 210 mm × 82 kg U.B. sections, for which $I_{xx} = 47\,363$ cm^4.

Then the deflection of B or D,

$$d = \frac{3\,037.5 \text{ kNm}^3}{EI} = \frac{3\,037.5 \times 10^3 \times 10^9}{2.1 \times 10^5 \times 47\,363 \times 10^4} \text{ mm}$$

$$= 30.5 \text{ mm}.$$

SLOPES AND DEFLECTIONS IN RIGID FRAMES

Example 3. The frame shown in Fig. 5 is of uniform section throughout, and it is required to find the deformed shape of the frame under the action of the crane loads only. The crane bracket loads produce moments Pc of 30 kNm and 12 kNm at Q and R respectively. Then the B.M. diagram may be constructed by consulting pages 373 and 374.

In terms of Pc, the following moments may be read off the charts:

Joint	B.M. due to equal loads at Q and R	B.M. due to a single load at Q
A	+0.319	−0.032
B	+0.218	+0.417
C	−0.148	−0.074
D	+0.218	−0.199
E	+0.319	+0.352

After multiplication by the appropriate coefficients of Pc (12 and 18 respectively for the equal and single loads) and adding algebraically the values for the two types of load, the moments shown in the Figure are obtained.

The values at Q are, respectively, above the bracket

$$10.122 - \left(\frac{10.122 - 3.252}{5}\right) + \frac{30}{5} = +14.748$$

and below the bracket

$$10.122 - \left(\frac{10.122 - 3.252}{5}\right) - \frac{30 \times 4}{5} = -15.252.$$

Values at R are obtained in a similar manner.

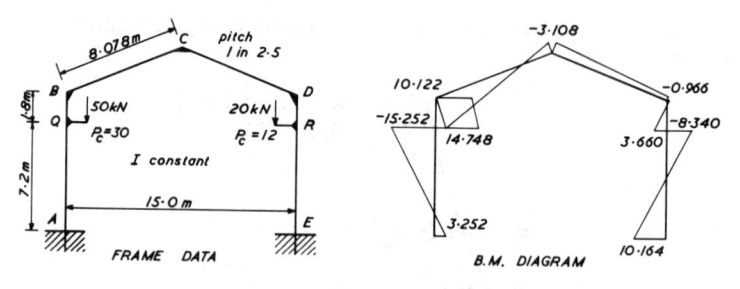

Fig. 5

From the B.M. diagram it is possible to calculate both the slope and the displacement of any point in the frame. The ends A and E are fixed in position and direction, i.e. $\theta_A = \theta_E = 0$, and hence form useful points from which to commence operations.

Now

$$\theta = \frac{\Sigma A}{EI}$$

where A = area of B.M. diagram, and

$$d = \frac{\Sigma Ay}{EI}$$

where Ay = area moment of B.M. diagram.

SLOPES AND DEFLECTIONS IN RIGID FRAMES

The work can be done by commencing at A and working through to E, but it is more convenient to work from A to C and then from E to C; the results should be identical and this provides a ready check upon the accuracy of the calculations.

Commencing from A, see Fig. 6, and assuming deflections into the frame to be positive,

$$d_Q = \frac{(+15.252 \times 7.2^2 \times \tfrac{1}{2} \times \tfrac{1}{3}) - (3.252 \times 7.2^2 \times \tfrac{1}{2} \times \tfrac{2}{3})}{EI}$$

$$= \frac{75.58}{EI} \text{ (m}^3 \text{ units)}.$$

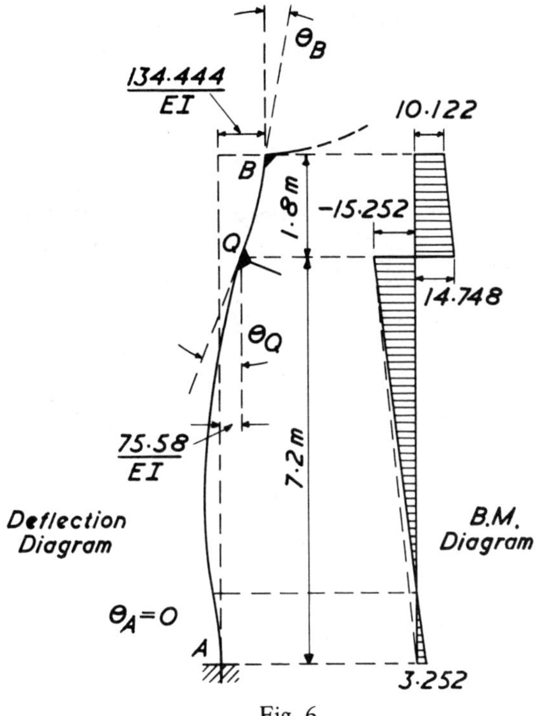

Fig. 6

The slope at Q,

$$\theta_Q = \frac{(15.252 \times 7.2/2) - (3.252 \times 7.2/2)}{EI}$$

$$= \frac{43.2}{EI} \text{ radian (clockwise rotation positive)}.$$

The deflection of B,

$$d_B = +\frac{75.58}{EI} + \frac{43.2 \times 1.8}{EI} - \frac{(10.122 \times 1.8 \times 0.9) + (14.748 - 10.122) \times 1.8^2 \times \tfrac{1}{2} \times \tfrac{1}{3}}{EI}$$

$$= \frac{134.444}{EI} \text{ (m}^3 \text{ units)}.$$

SLOPES AND DEFLECTIONS IN RIGID FRAMES

The slope at B,

$$\theta_B = \text{slope at } Q + \text{rotation between } Q \text{ and } B.$$

$$= \frac{43.2}{EI} - \frac{(14.748 + 10.122) \times 1.8}{2 \times EI}$$

$$= \frac{20.817}{EI} \text{ radian.}$$

Now calculate the deflection of C with respect to B, measuring normal to the slope of the rafter BC, as shown in Fig. 7.

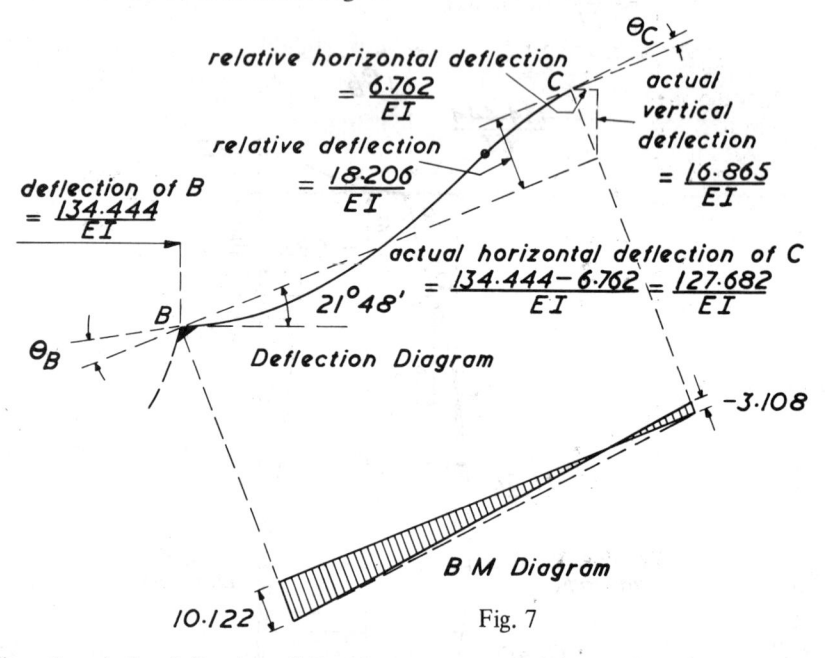

Fig. 7

Then the relative deflection of C with respect to B = the deflection due to the rotation of B + the deflection in BC,

$$= \frac{20.817 \times 8.078}{EI} - \frac{(10.122 \times 8.078^2 \times \frac{1}{2} \times \frac{2}{3})}{EI}$$

$$+ \frac{3.108 \times 8.078^2 \times \frac{1}{2} \times \frac{1}{3}}{EI}$$

$$= -\frac{18.206}{EI} \text{ (m}^3 \text{ units).}$$

Now the rafter has an angle of slope of $21° \, 48'$.

$$\sin 21° \, 48' = 0.3714.$$

$$\cos 21° \, 48' = 0.9285.$$

Then the relative horizontal deflection of C

$$= -\frac{18.206 \times 0.3714}{EI} = -\frac{6.762}{EI} \text{ (m}^3 \text{ units)}$$

SLOPES AND DEFLECTIONS IN RIGID FRAMES

while the actual vertical deflection of C

$$= -\frac{18.206 \times 0.9285}{EI} = -\frac{16.904}{EI} \text{ (m}^3 \text{ units)}$$

Hence, the horizontal displacement of C = displacement at B + displacement of C relative to B

$$= \frac{134.444 - 6.762}{EI} = \frac{127.682}{EI} \text{ (m}^3 \text{ units)}$$

and the vertical deflection

$$= -\frac{16.904}{EI} \text{ (m}^3 \text{ units)}.$$

Therefore, the ridge C has moved upwards and to the right.

The slope at the ridge = the rotation at B + the rotation between B and C

$$\theta_C = \frac{20.817}{EI} - \frac{(10.122 - 3.108) \times 8.078 \times \frac{1}{2}}{EI}$$

$$= \frac{-7.513}{EI} \text{ radian.}$$

Therefore, the ridge C has rotated in an anti-clockwise direction.

Now consider the right half of the frame.

Commencing from E, the deflection at R

$$d_R = \frac{(-10.164 \times 7.2^2 \times \frac{1}{2} \times \frac{2}{3}) + (8.340 \times 7.2^2 \times \frac{1}{2} \times \frac{1}{3})}{EI}$$

$$= \frac{-103.576}{EI} \text{ (m}^3 \text{ units)}.$$

The slope at R

$$\theta_R = \frac{(10.164 - 8.340) \times 7.2 \times \frac{1}{2}}{EI}$$

$$= \frac{6.566}{EI}.$$

The deflection of D

$$d_D = -\frac{103.576}{EI} - \frac{6.566 \times 1.8}{EI} - \frac{3.660 \times 1.8^2 \times \frac{1}{2} \times \frac{2}{3}}{EI} + \frac{0.966 \times 1.8^2 \times \frac{1}{2} \times \frac{1}{3}}{EI}$$

$$= -\frac{118.826}{EI} \text{ (m}^3 \text{ units)}.$$

The slope at D

$$\theta_D = \frac{6.566 + (3.660 - 0.966) \times 1.8 \times \frac{1}{2}}{EI}$$

$$= \frac{8.990}{EI} \text{ radian.}$$

Considering the deflection of C with respect to D and normal to DC,

relative $d_C = -\frac{8.990 \times 8.078}{EI} + \frac{0.966 \times 8.078^2 \times \frac{1}{2}}{EI}$

$$+ \frac{(3.108 - 0.966) \times 8.078^2 \times \frac{1}{2} \times \frac{1}{3}}{EI}$$

$$= -\frac{17.808}{EI} \text{ (m}^3 \text{ units)}$$

which compares reasonably well with the figure of $-18.206/EI$ for the relative deflection of C with regard to B.

Then the relative horizontal deflection of C

$$= -\frac{17.808 \times 0.3714}{EI} = -\frac{6.614}{EI}$$

while the actual vertical deflection of C

$$= -\frac{17.808 \times 0.9285}{EI} = -\frac{16.535}{EI}$$

Therefore the actual horizontal deflection of C, calculated along the right half of the frame

$$= -\frac{118.826 - 6.614}{EI} = -\frac{125.440}{EI}$$

compared with $127.682/EI$ when calculated along the left half of the frame.

The calculated vertical deflection is $-16.535/EI$ compared with $-16.904/EI$ when calculated along the left half of the frame.

The slope at C

$$\theta_C = \frac{8.990}{EI} - \frac{(3.108 + 0.966) \times 8.078 \times \frac{1}{2}}{EI}$$

$$= -\frac{7.465}{EI} \text{ radian,}$$

which compares favourably with the value of $-7.513/EI$ obtained from the previous calculations.

The slight differences in values for deflections and rotations can be due to slight inaccuracies both in the evaluation of the original B.M.s and from the calculations given here. It is likely that the greatest inaccuracy here could be traced to the values of the B.M.s.

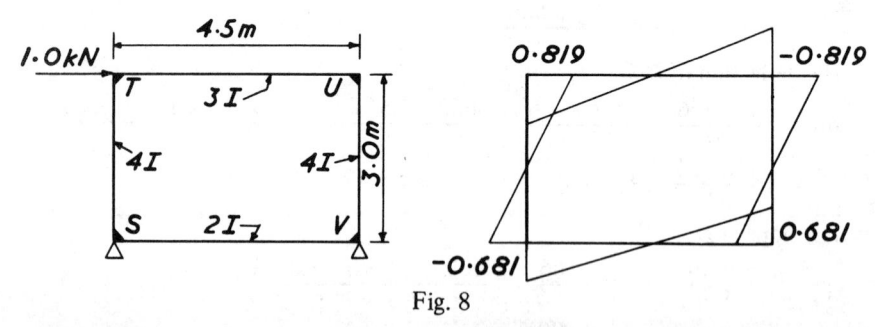

Fig. 8

Example 4. Figure 8 shows the frame to be analysed. The slopes and deflections of the joints will be calculated.

Both TU and SV are members at the ends of which have been applied moments of the same value and the same kind of rotation, i.e. all the rotations are clockwise. Then each member is like the member BD in Fig. 3 and the slopes at each end could be calculated as in Example 2. The easier way is to use the formula, applicable in such a case,

$$\theta = \frac{ML}{6EI}$$

SLOPES AND DEFLECTIONS IN RIGID FRAMES

Therefore

$$\theta_T = \theta_U = \frac{0.819 \times 4.5}{6E(3I)}$$

$$= \frac{0.205}{EI} \text{ radian,}$$

while

$$\theta_S = \theta_V = \frac{0.681 \times 4.5}{6E(2I)}$$

$$= \frac{0.255}{EI} \text{ radian.}$$

Now S and V are fixed in position, so the deflection of both T and U can be found by working upwards either from S or V.

Commencing from S, the deflection of T = deflection due to θ_S plus the deflection due to the moments in ST. Therefore,

$$d_T = \frac{0.255 \times 3.0}{EI} + \frac{0.681 \times 3^2 \times \tfrac{1}{2} \times \tfrac{2}{3}}{E(4I)} - \frac{0.819 \times 3^2 \times \tfrac{1}{2} \times \tfrac{1}{3}}{E(4I)}$$

$$= +\frac{0.969}{EI} \text{ (m}^3 \text{ units).}$$

The deflection of U is the same amount.

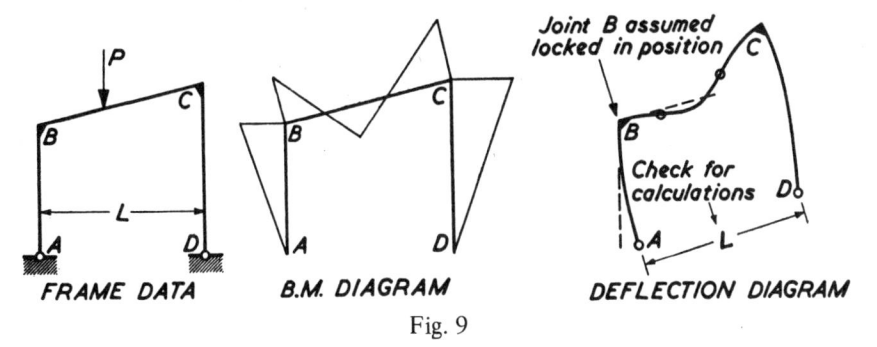

FRAME DATA   B.M. DIAGRAM   DEFLECTION DIAGRAM

Fig. 9

Frames where no Symmetry Exists

Cases will arise where none of the expedients already stated can be employed. In this event, the method of procedure is to assume that one of the joints is locked in position and direction and to allow the frame to rotate and deflect about it. The method is outlined in Fig. 9.

The foregoing Area-Moment principles apply, but the work is a little more tedious. The deflection and rotation of A, with respect to B, due to the horizontal thrust at A are calculated and the relative position of A is plotted. Then the deflection and rotation of C relative to B are calculated and the position and slope of C are plotted. From C the relative deflection of D is due to the rotation of C and to the deflection of CD resulting from the horizontal thrust at D. All these operations have been demonstrated in the preceding Examples. When the calculations are complete, the distance between the points A and D should be L.

15. FORMULAE FOR RIGID FRAMES

THE formulae given in this section are based on Professor Kleinlogel's Rahmenformeln and Mehrstielige Rahmen, published by Wilhelm Ernst & Sohn of Berlin, to whom grateful acknowledgment is made. The formulae are applicable to frames which are symmetrical about a central vertical axis, with the single exception of the triangular frame, and in which each member is of constant moment of inertia.

Formulae are given for the following types of frame:

Single-storey Frame

Frame I. Hingeless rectangular portal frame.
II. Two-hinged rectangular portal frame.
III. Hingeless gable frame with vertical legs.
IV. Two-hinged gable frame with vertical legs.
V. Hingeless frame with skew corners.
VI. Two-hinged frame with skew corners.
VII. Two-hinged triangular frame.

Multi-bay Frame

VIII. Twin Gable Frame with hinged feet.

The loadings are so arranged that dead, snow and wind loads may be reproduced on all the frames. For example, wind suction acting normal to the sloping rafters of a building may be divided into horizontal and vertical components, for which appropriate formulae are given, although all the signs must be reversed because the loadings shown in the tables act inwards, not outwards as in the case of suction. Crane loads, including surge, are also shown in a number of the single-storey frames.

It should be noted that, with few exceptions, the loads between node or panel points are uniformly distributed over the *whole* member. It is appreciated that it is normal practice to impose loads on frames through purlins, side rails or beams. By using the coefficients in Fig. 1, however, allowance can be made for many other symmetrically placed loads on the cross-beams of frames I and II shown above, where the difference in effect is sufficient to warrant the corrections being made. The indeterminate B.M.s in the whole frame are calculated as though the loads were uniformly distributed over the beam being considered, and then all are adjusted by multiplying by the appropriate coefficient in Fig. 1. It may be of interest to state why these adjustments can be made. In any statically indeterminate structure the indeterminate moments vary directly with the value of the following quantity:

$$\frac{\text{Area of the free B.M. diagram}}{EI}$$

Where the loaded member is of constant cross-section, EI may be ignored.

FORMULAE FOR RIGID FRAMES

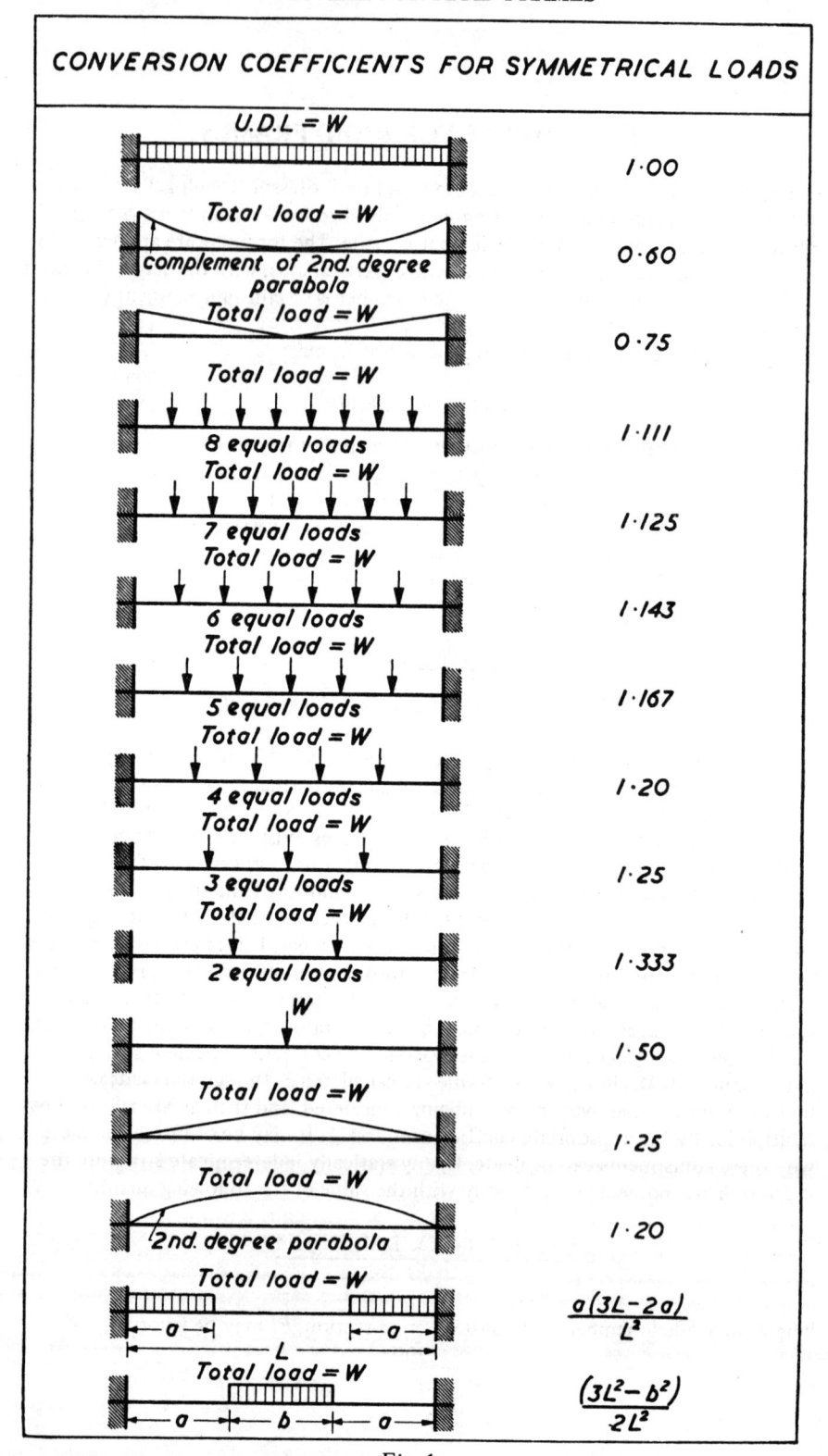

CONVERSION COEFFICIENTS FOR SYMMETRICAL LOADS

Loading	Coefficient
U.D.L = W	1·00
Total load = W, complement of 2nd. degree parabola	0·60
Total load = W (triangular, inverted)	0·75
Total load = W, 8 equal loads	1·111
Total load = W, 7 equal loads	1·125
Total load = W, 6 equal loads	1·143
Total load = W, 5 equal loads	1·167
Total load = W, 4 equal loads	1·20
Total load = W, 3 equal loads	1·25
Total load = W, 2 equal loads	1·333
W (central point load)	1·50
Total load = W (triangular)	1·25
Total load = W, 2nd. degree parabola	1·20
Total load = W, two end U.D.L. of length a, span L	$\dfrac{a(3L-2a)}{L^2}$
Total load = W, central U.D.L. of length b, ends a	$\dfrac{(3L^2 - b^2)}{2L^2}$

Fig. 1

Consider, as an example, the case of an encastré beam of constant cross-section and of length L carrying a U.D.L. of W. Then the area of the free B.M. diagram

$$= \frac{WL}{8} \times \frac{2L}{3} = \frac{WL^2}{12}.$$

If, however, W were a central point load, the area of the free B.M. diagram would be

$$\frac{WL}{4} \times \frac{L}{2} = \frac{WL^2}{8}.$$

The F.E.M.s due to the two types of loadings are $WL/12$ and $WL/8$ respectively, thus demonstrating that the indeterminate moments vary with the area of the free B.M. diagram and proving that the indeterminate moments are in the proportion of 1 : 1.5.

No rules can be laid down for the effect on the reactions of a change in the mode of application of the load, although sometimes they will vary with the indeterminate moments. Consider a simple rectangular portal with hinged feet. If a U.D.L. placed over the whole of the beam is replaced by a central point load of the same magnitude, then the knee moments will increase by 50 per cent with a corresponding increase in the horizontal thrusts H, while the vertical reactions V will remain the same.

Although the foregoing remarks relating to the indeterminate moments resulting from symmetrical loads apply to all rectangular portals, the rule applies for asymmetrical loads imposed upon the cross-beam of a rectangular portal frame with hinged feet. If a vertical U.D.L. on the cross-beam is replaced by any vertical load of the same magnitude, then the indeterminate moments vary with the areas of the respective free B.M. diagrams.

No doubt readers who use the tables frequently will learn short cuts, but it is not inappropriate to mention some. For example, if a U.D.L. of W over the whole of a single-bay symmetrical frame is replaced by a U.D.L. *of the same magnitude of W* over either the left-hand or right-hand half of the frame, the horizontal thrust at the feet is unaltered. If the frame has a pitched roof then the ridge moments will also be unaltered.

It will be noted that the formulae for the load P on a single crane bracket are related to those for loads P on both crane brackets. Consider Fig. 2.

Then M_A ($= M_E$) and M_P ($= M_D$) in Fig. 2 (b) are equal to $(M_A + M_E)$ and $(M_B + M_D)$ respectively in Fig. 2 (a), while M_C in Fig. 2 (b) is double the value of M_C in Fig. 2 (a).

When frames have hinged feet, the moments resulting from surge loads P can be written down without calculation, although the frames are nominally statically indeterminate. The moments at both the loads and at the knees are equal to Pa, where a is the height of the point of application of the loads above the feet of the frame.

The charts on pages 345 to 397 have been prepared to assist in the design of rectangular frames or frames with roof pitches of 1 in 5 or 1 in 2.5. Results for intermediate pitches may be interpolated with reasonable accuracy.

The charts on pages 398 to 411 are for two-bay portal frames with a roof pitch of 1 in 2.5 only.

Fig. 2

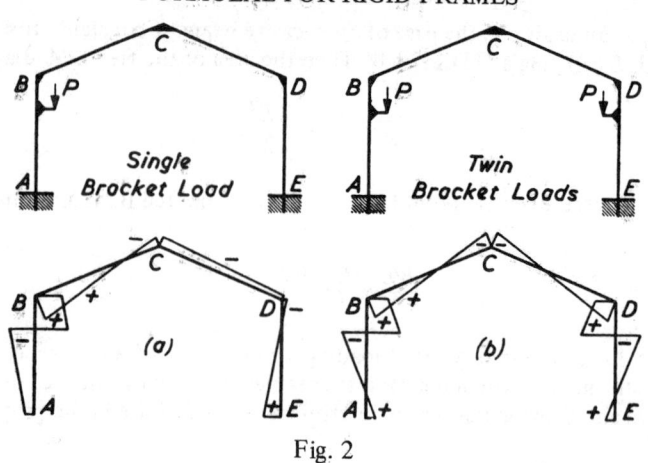

Arrangement of Formulae

Each set of formulae is treated as a separate chapter. The data required for each frame, together with the constants to be used in the various formulae, are given on the first page of each chapter. This general information is followed by the detailed formulae for the various loading conditions, each of which is illustrated by two diagrams placed side by side, the left-hand diagram giving a loading condition and the right-hand one giving the appropriate B.M. and reaction diagram. It should be noted, however, that some B.M.s change their signs as the frames change their proportions. This will be appreciated by examining the charts.

For simple frames, i.e. for single-bay, single-storey frames, the formulae for reactions immediately follow the formulae for B.M.s for each load. For multi-storey or multi-bay frames the formulae for B.M.s are given first in a group and are followed by formulae for reactions, shears and thrusts, also in a group.

Considering the simple frames only, the kind of formula depends on the degree of indeterminacy and the shape of the frame. Auxiliary Coefficients X are introduced whenever the direct expressions become complicated or for other reasons of expediency.

No hard and fast rules can be laid down for the nomenclature and it must be noted that each set of symbols and constants applies only to the particular frame under consideration, although, of course, an attempt has been made to produce similarity in the types of symbols.

The formulae for multi-storey or multi-bay frames may seem less complicated than for simple frames, but they are based on numerous constants and composite coefficients which must be accurately computed.

Sign Conventions

All computations must be carried out algebraically, hence every quantity must be given its correct sign. The results will then be automatically correct in sign and magnitude.

The direction of the load or applied moment shown in the left-hand diagram for each load condition is considered to be positive. If the direction of the load or

FORMULAE FOR RIGID FRAMES

moment is reversed, the signs of all the results obtained from the formulae as printed must be reversed.

For simple frames, the moments causing tension on the inside faces of the frame are considered to be positive. Upward vertical reactions and inward horizontal reactions are also positive.

For multi-storey or multi-bay frames the same general rules apply to moments and vertical reactions. However, in the case of a two-span portal frame, for example, a problem arises with the central column. It is assumed in the formulae that the central column belongs to the left-hand bay, so that if the column bends inwards towards this bay, the moment is positive. Similarly, in a multi-storey frame, a cross-beam is associated with the storey below it, tension on the lower face providing a positive B.M.

Horizontal reactions at the feet of multi-storey or multi-bay frames have been given signs in the diagrams but the appropriate rules are given on the first page dealing with reactions in each chapter. In general it may be said that these thrusts bear the same sign as the moment which they create in the joint at the top of the column upon which they act. It should be noted that this system is opposite to that which operates for simple frames.

Checking Calculations for Indeterminate Frames

Calculations for indeterminate frames may be checked by using some other method of analysis, but it is also possible to check any frame or portion of a frame, such as that above the line AB in Fig. 3 by ensuring that

1. The three fundamental statical equations, i.e. $\Sigma H = 0$, $\Sigma V = 0$ and $\Sigma M = 0$, have been satisfied and, in addition, either that
2. The sum of the areas of the M/EI diagram above any line, such as AB, is zero if A and B are fully fixed; or
3. The sum of the moments, with respect to the base AB, of the areas of the M/EI diagram above the line AB is zero if A and B are partially restrained (as shown in the Figure) or are hinged.

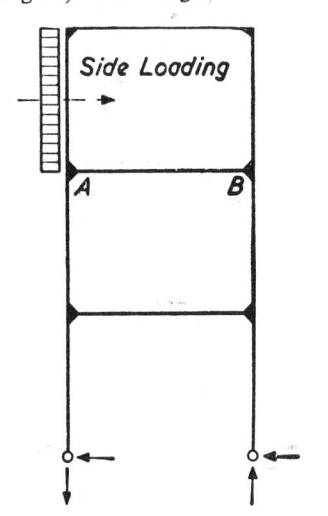

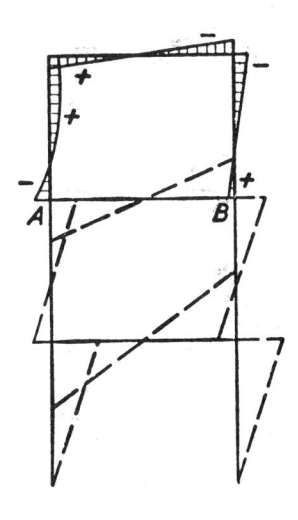

Fig. 3

FORMULAE FOR RIGID FRAMES

It is of interest to note that the underlying principles in Rules 2 and 3 above are those used in the application of the Column Analogy method of analysis.

As an example of Rule 2, consider the frame in Fig. 4, where EI is constant.

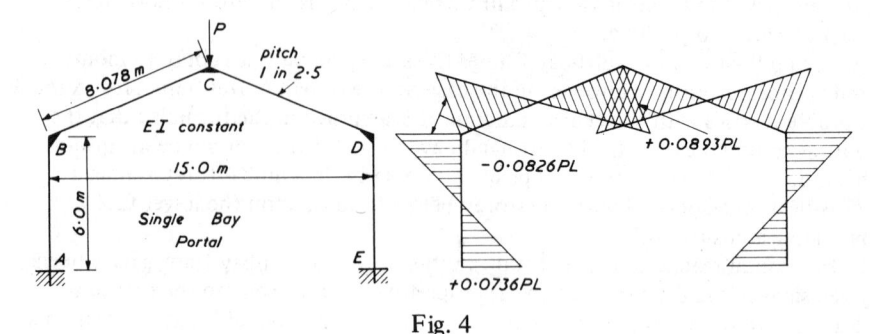

Fig. 4

Then the sum of the areas of the M/EI diagram, considering the legs first, is

$$\frac{2}{EI}\left[\frac{(+0.0736 - 0.0826) \times 6.0}{2}\right] + \frac{2}{EI}\left[\frac{(-0.0826 + 0.0893) \times 8.078}{2}\right]$$

$$= \frac{-0.054 + 0.054}{EI} = 0.$$

Thus demonstrating that the moments calculated are correct.

Now consider the frame in Fig. 5, an example for Rule 3.

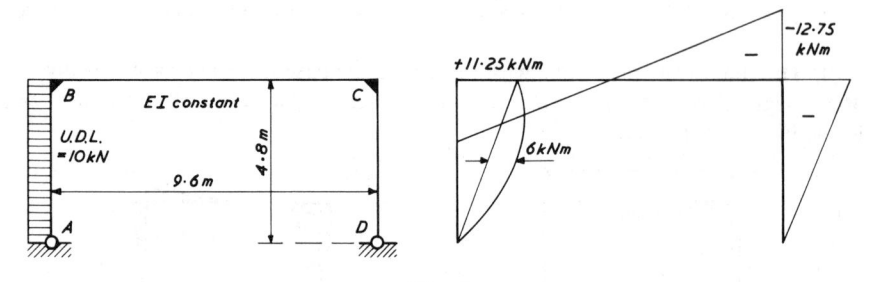

Fig. 5

Then the sum of the moments of the areas of the M/EI diagram, working from A round to D, is

$$\frac{1}{EI}\left\{\left[\frac{11.25 \times 4.8}{2} \times \frac{2 \times 4.8}{3}\right] + \left[\frac{6 \times 4.8 \times 2}{3} \times \frac{4.8}{2}\right]\right.$$

$$\left. + \left[\left(\frac{11.25 - 12.75}{2}\right) 9.6 \times 4.8\right] + \left[\frac{-12.75 \times 4.8}{2} \times \frac{2 \times 4.8}{3}\right]\right\} = 0$$

$$= \frac{1}{EI}\left[\frac{4.8^2 \times 2}{2 \times 3}(11.25 + 6 - 4.5 - 12.75)\right] = 0$$

Demonstrating again that the calculations are correct.

Frame I

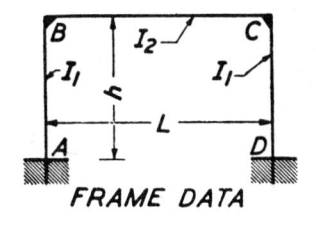

FRAME DATA

Coefficients:
$$k = \frac{I_2}{I_1} \cdot \frac{h}{L}$$
$$N_1 = k + 2 \qquad N_2 = 6k + 1$$

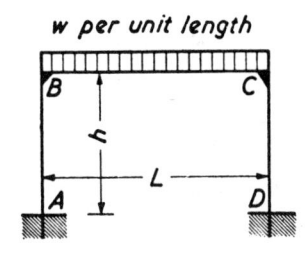

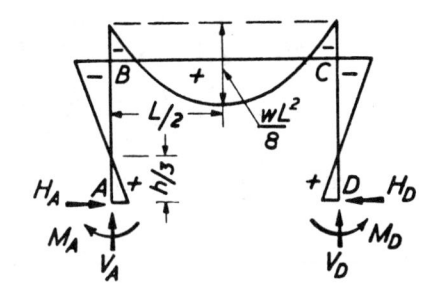

---

w per unit length

$$M_A = M_D = \frac{wL^2}{12N_1} \qquad M_B = M_C = -\frac{wL^2}{6N_1} = -2M_A$$

$$M_{max} = \frac{wL^2}{8} + M_B \qquad V_A = V_D = \frac{wL}{2} \qquad H_A = H_D = \frac{3M_A}{h}$$

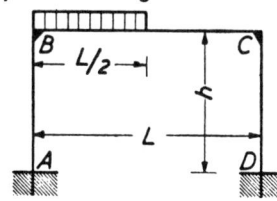

---

w per unit length

$$M_A = \frac{wL^2}{8}\left[\frac{1}{3N_1} - \frac{1}{8N_2}\right] \qquad M_B = -\frac{wL^2}{8}\left[\frac{2}{3N_1} + \frac{1}{8N_2}\right]$$

$$M_D = \frac{wL^2}{8}\left[\frac{1}{3N_1} + \frac{1}{8N_2}\right] \qquad M_C = -\frac{wL^2}{8}\left[\frac{2}{3N_1} - \frac{1}{8N_2}\right]$$

$$V_D = \frac{wL}{8}\left[1 - \frac{1}{4N_2}\right] \qquad V_A = \frac{wL}{2} - V_D \qquad H_A = H_D = \frac{wL^2}{8hN_1}$$

Extract: 'Kleinlogel, Rahmenformeln' 11. Auflage Berlin—Verlag von Wilhelm Ernst & Sohn.

FORMULAE FOR RIGID FRAMES

$$M_A = \frac{wh^2}{4}\left[-\frac{k+3}{6N_1} - \frac{4k+1}{N_2}\right] \qquad M_B = \frac{wh^2}{4}\left[-\frac{k}{6N_1} + \frac{2k}{N_2}\right]$$

$$M_D = \frac{wh^2}{4}\left[-\frac{k+3}{6N_1} + \frac{4k+1}{N_2}\right] \qquad M_C = \frac{wh^2}{4}\left[-\frac{k}{6N_1} - \frac{2k}{N_2}\right]$$

$$H_D = \frac{wh(2k+3)}{8N_1} \qquad H_A = -(wh - H_D) \qquad V_A = -V_D = -\frac{wh^2 k}{LN_2}$$

Constants: $a_1 = \dfrac{a}{h} \qquad b_1 = \dfrac{b}{h}$

$$X_1 = \frac{Pc}{2N_1}[1 + 2b_1 k - 3b_1^2(k+1)] \qquad X_2 = \frac{Pcka_1(3a_1 - 2)}{2N_1}$$

$$X_3 = \frac{3Pcka_1}{N_2}$$

$$M_A = +X_1 - \left(\frac{Pc}{2} - X_3\right) \qquad M_B = +X_2 + X_3$$

$$M_D = +X_1 + \left(\frac{Pc}{2} - X_3\right) \qquad M_C = +X_2 - X_3$$

$$H_A = H_D = \frac{Pc}{2h} + \frac{X_1 - X_2}{h} \qquad V_D = \frac{2X_3}{L} \qquad V_A = P - V_D$$

$$M_1 = M_A - H_A a \qquad M_2 = M_B + H_D b$$

Extract: 'Kleinlogel, Rahmenformeln' 11. Auflage Berlin—Verlag von Wilhelm Ernst & Sohn.

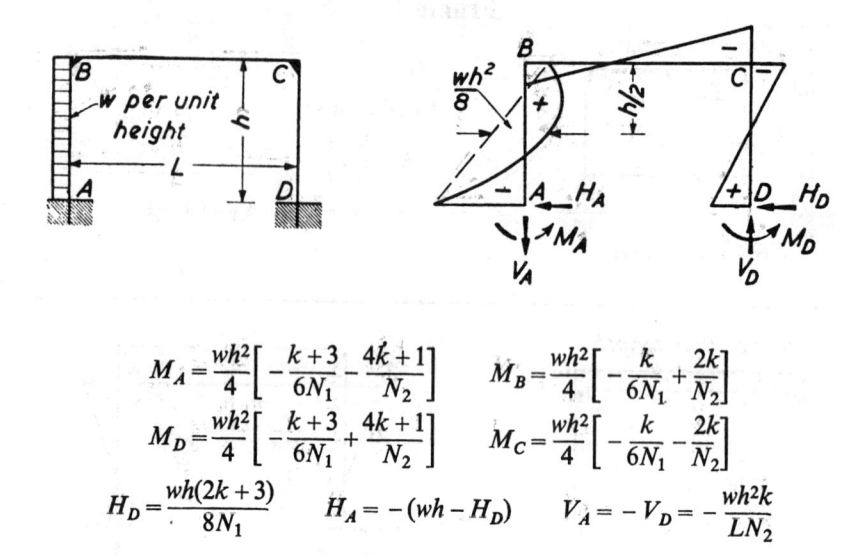

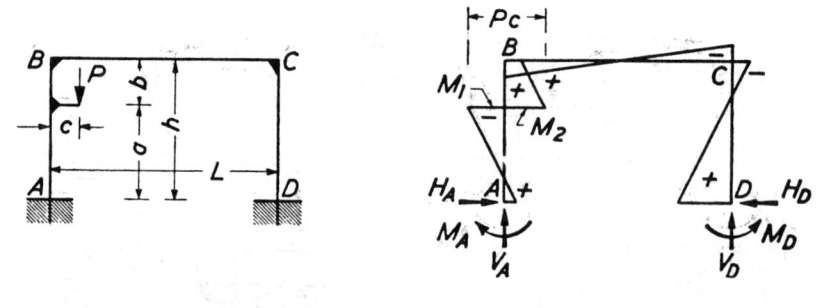

FRAME I

Constants: $a_1 = \dfrac{a}{h}$ $b_1 = \dfrac{b}{h}$

$$X_1 = \frac{Pc}{2N_1}[1 + 2b_1k - 3b_1^2(k+1)] \qquad X_2 = \frac{Pcka_1(3a_1 - 2)}{2N_1}$$

$$M_A = M_D = \frac{Pc}{N_1}[1 + 2b_1k - 3b_1^2(k+1)] = 2X_1$$

$$M_B = M_C = \frac{Pcka_1(3a_1 - 2)}{N_1} = 2X_2$$

$$V_A = V_D = P \qquad H_A = H_D = \frac{Pc + M_A - M_B}{h}$$

$$M_1 = M_A - H_A a \qquad M_2 = M_B + H_D b$$

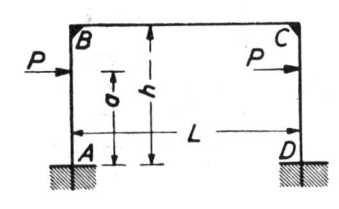

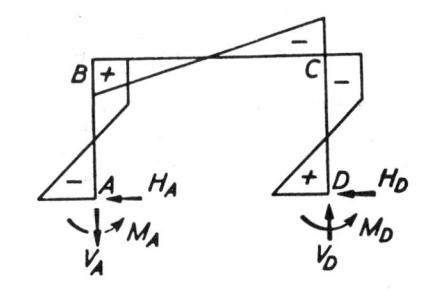

Constants: $a_1 = \dfrac{a}{h}$ $X_1 = \dfrac{3Paa_1 k}{N_2}$

$$M_A = -Pa + X_1 \qquad M_B = X_1$$

$$M_D = +Pa - X_1 \qquad M_C = -X_1$$

$$V_A = -V_D = -\frac{2X_1}{L} \qquad H_A = -H_D = -P$$

Extract: 'Kleinlogel, Rahmenformeln' 11. Auflage Berlin—Verlag von Wilhelm Ernst & Sohn.

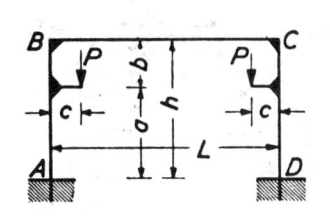

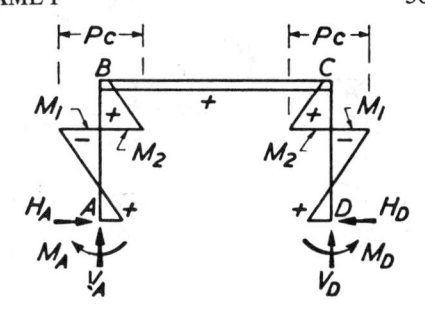

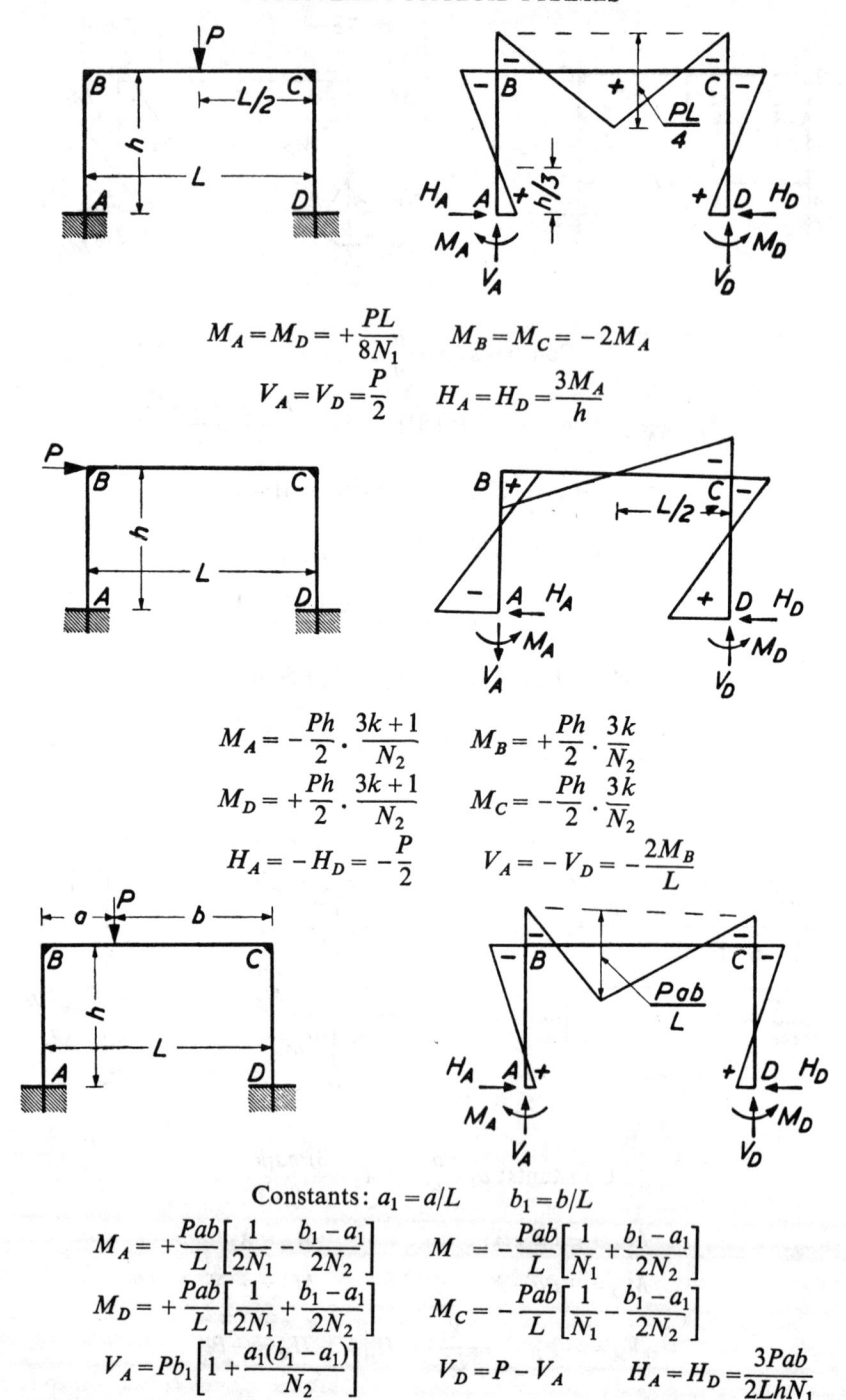

FORMULAE FOR RIGID FRAMES

$$M_A = M_D = +\frac{PL}{8N_1} \qquad M_B = M_C = -2M_A$$

$$V_A = V_D = \frac{P}{2} \qquad H_A = H_D = \frac{3M_A}{h}$$

$$M_A = -\frac{Ph}{2} \cdot \frac{3k+1}{N_2} \qquad M_B = +\frac{Ph}{2} \cdot \frac{3k}{N_2}$$

$$M_D = +\frac{Ph}{2} \cdot \frac{3k+1}{N_2} \qquad M_C = -\frac{Ph}{2} \cdot \frac{3k}{N_2}$$

$$H_A = -H_D = -\frac{P}{2} \qquad V_A = -V_D = -\frac{2M_B}{L}$$

Constants: $a_1 = a/L \qquad b_1 = b/L$

$$M_A = +\frac{Pab}{L}\left[\frac{1}{2N_1} - \frac{b_1 - a_1}{2N_2}\right] \qquad M_B = -\frac{Pab}{L}\left[\frac{1}{N_1} + \frac{b_1 - a_1}{2N_2}\right]$$

$$M_D = +\frac{Pab}{L}\left[\frac{1}{2N_1} + \frac{b_1 - a_1}{2N_2}\right] \qquad M_C = -\frac{Pab}{L}\left[\frac{1}{N_1} - \frac{b_1 - a_1}{2N_2}\right]$$

$$V_A = Pb_1\left[1 + \frac{a_1(b_1 - a_1)}{N_2}\right] \qquad V_D = P - V_A \qquad H_A = H_D = \frac{3Pab}{2LhN_1}$$

Extract: 'Kleinlogel, Rahmenformeln' 11. Auflage Berlin—Verlag von Wilhelm Ernst & Sohn.

Frame II

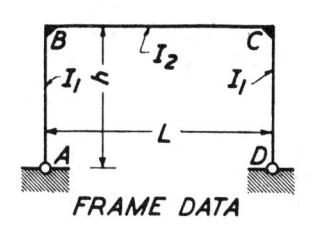

FRAME DATA

Coefficients:

$$k = \frac{I_2}{I_1} \cdot \frac{h}{L}$$

$$N = 2k + 3$$

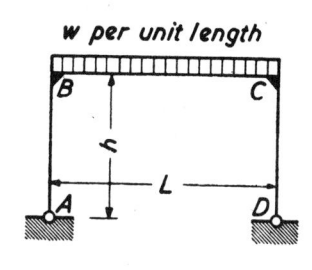

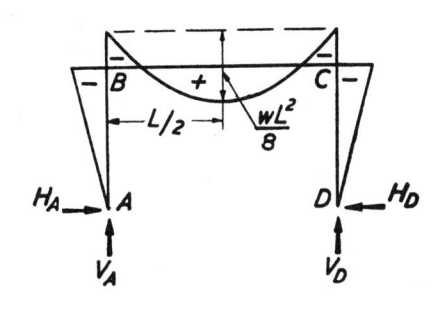

w per unit length

$$M_B = M_C = -\frac{wL^2}{4N} \qquad M_{\max} = \frac{wL^2}{8} + M_B$$

$$V_A = V_D = \frac{wL}{2} \qquad H_A = H_D = -\frac{M_B}{h}$$

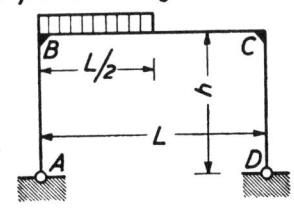

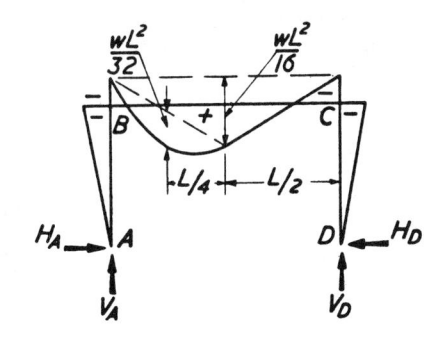

w per unit length

$$M_B = M_C = -\frac{wL^2}{8N}$$

$$V_A = \frac{3wL}{8} \qquad V_D = \frac{wL}{8} \qquad H_A = H_D = -\frac{M_B}{h}$$

Extract: 'Kleinlogel, Rahmenformeln' 11. Auflage Berlin—Verlag von Wilhelm Ernst & Sohn.

FORMULAE FOR RIGID FRAMES

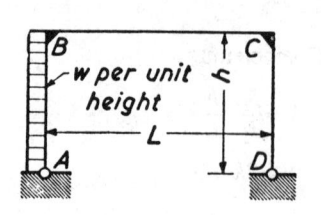

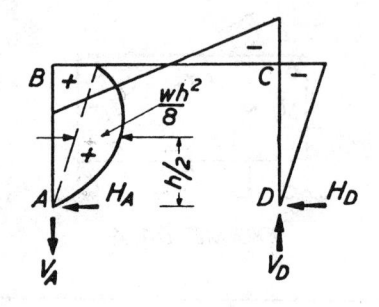

$$M_B = \frac{wh^2}{4}\left[-\frac{k}{2N}+1\right] \qquad H_D = -\frac{M_C}{h}$$

$$M_C = \frac{wh^2}{4}\left[-\frac{k}{2N}-1\right] \qquad H_A = -(wh - H_D)$$

$$V_A = -V_D = -\frac{wh^2}{2L}$$

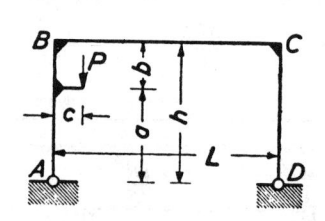

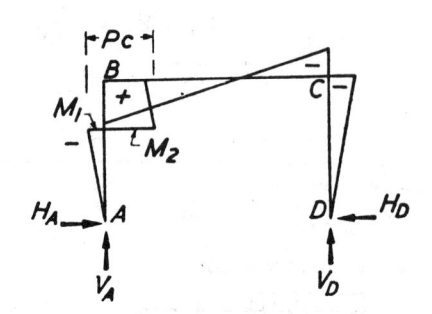

Constant: $a_1 = \frac{a}{h}$

$$M_B = \frac{Pc}{2}\left[\frac{(3a_1^2-1)k}{N}+1\right]$$

$$M_C = \frac{Pc}{2}\left[\frac{(3a_1^2-1)k}{N}-1\right] \qquad H_A = H_D = -\frac{M_C}{h}$$

$$V_D = \frac{Pc}{L} \qquad V_A = P - V_D$$

$$M_1 = -H_A a \qquad M_2 = Pc - H_A a$$

Extract: 'Kleinlogel, Rahmenformeln' 11. Auflage Berlin—Verlag von Wilhelm Ernst & Sohn.

FRAME II

Constant: $a_1 = \dfrac{a}{h}$

$$M_B = M_C = \dfrac{Pc(3a_1^2 - 1)k}{N}$$

$$H_A = H_D = \dfrac{Pc - M_B}{h} \qquad V_A = V_D = P$$

$$M_1 = -H_A a \qquad M_2 = Pc - H_A a$$

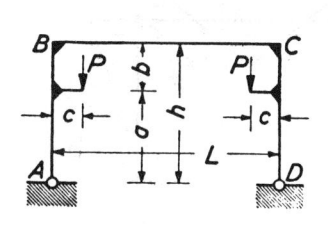

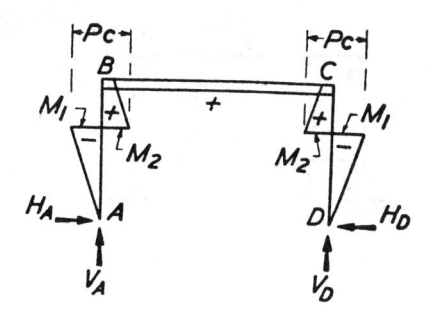

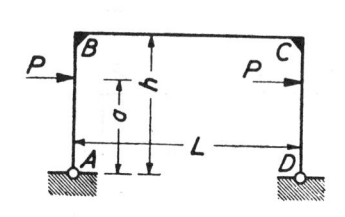

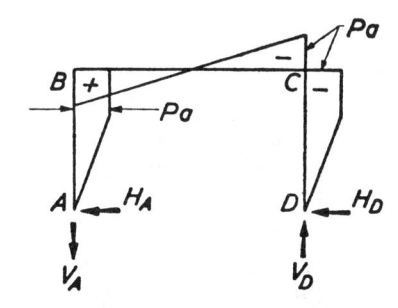

$$M_B = -M_C = Pa \qquad H_A = H_D = P$$

$$V_A = -V_D = -\dfrac{2Pa}{L}$$

Moment at loads $= \pm Pa$

Extract: 'Kleinlogel, Rahmenformeln' 11. *Auflage Berlin—Verlag von Wilhelm Ernst & Sohn.*

FORMULAE FOR RIGID FRAMES

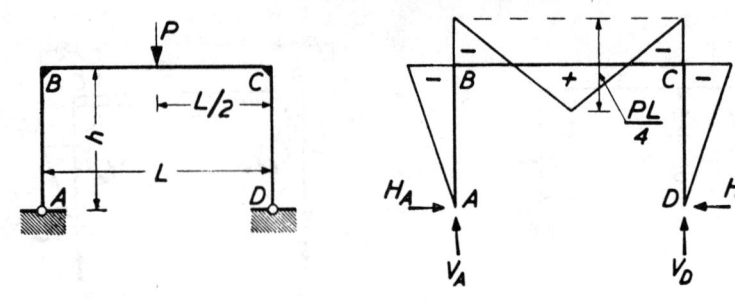

$$M_B = M_C = -\frac{3PL}{8N} \quad V_A = V_D = \frac{P}{2} \quad H_A = H_D = -\frac{1M_B}{h}$$

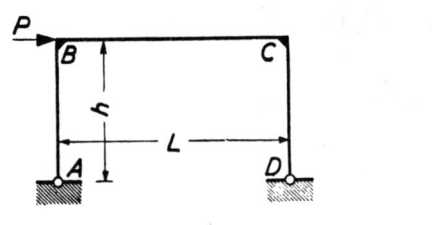

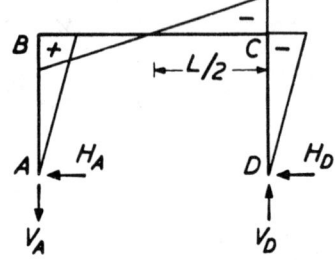

$$M_B = -M_C = +\frac{Ph}{2}$$

$$V_A = -V_D = -\frac{Ph}{L} \quad H_A = -H_D = -\frac{P}{2}$$

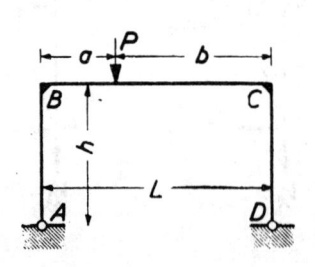

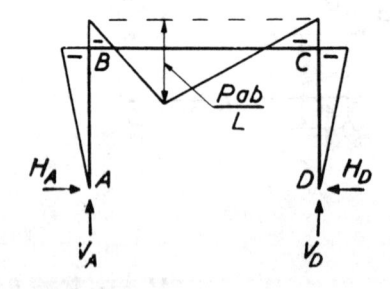

$$M_B = M_C = -\frac{Pab}{L} \cdot \frac{3}{2N}$$

$$V_A = \frac{Pb}{L} \quad V_D = \frac{Pa}{L} \quad H_A = H_D = -\frac{M_B}{h}$$

Extract: 'Kleinlogel, Rahmenformeln' 11. Auflage Berlin—Verlag von Wilhelm Ernst & Sohn.

Frame III

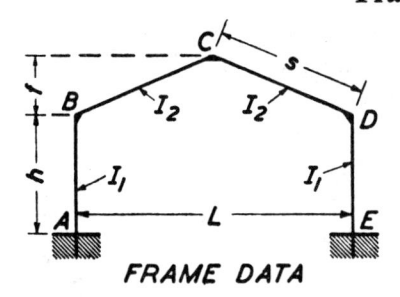

Coefficients:

$$k = \frac{I_2}{I_1} \cdot \frac{h}{s} \qquad \phi = \frac{f}{h}$$

$$m = 1 + \phi$$

$$B = 3k + 2 \qquad C = 1 + 2m$$

FRAME DATA

$$K_1 = 2(k + 1 + m + m^2) \qquad K_2 = 2(k + \phi^2)$$

$$R = \phi C - k \qquad N_1 = K_1 K_2 - R^2 \qquad N_2 = 3k + B$$

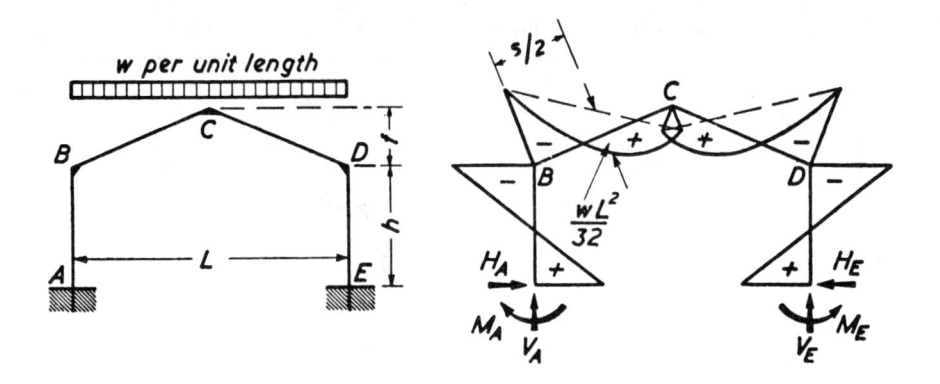

$$M_A = M_E = \frac{wL^2}{16} \cdot \frac{k(8 + 15\phi) + \phi(6 - \phi)}{N_1}$$

$$M_B = M_D = -\frac{wL^2}{16} \cdot \frac{k(16 + 15\phi) + \phi^2}{N_1}$$

$$M_C = \frac{wL^2}{8} - \phi M_A + m M_B$$

$$V_A = V_E = \frac{wL}{2} \qquad H_A = H_E = \frac{M_A - M_B}{h}$$

Extract: 'Kleinlogel, Rahmenformeln' 11. Auflage Berlin—Verlag von Wilhelm Ernst & Sohn.

FORMULAE FOR RIGID FRAMES

w per unit length

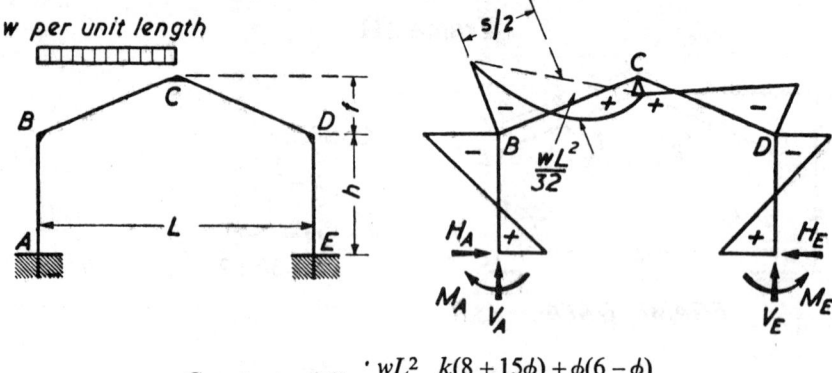

Constants: $^*X_1 = \dfrac{wL^2}{32} \cdot \dfrac{k(8+15\phi)+\phi(6-\phi)}{N_1}$

$^*X_2 = \dfrac{wL^2}{32} \cdot \dfrac{k(16+15\phi)+\phi^2}{N_1}$ $\quad X_3 = \dfrac{wL^2}{32N_2}$

$M_A = +X_1 - X_3 \quad M_B = -X_2 - X_3 \quad M_E = +X_1 + X_3 \quad M_D = -X_2 + X_3$

$^*M_C = \dfrac{wL^2}{16} - \phi X_1 - mX_2$

$V_E = \dfrac{wL}{8} - \dfrac{2X_3}{L} \quad V_A = \dfrac{wL}{2} - V_E \quad H_A = H_E = \dfrac{X_1 + X_2}{h}$

* Note that X_1, $-X_2$ and M_C are respectively half the values of $M_A (=M_E)$, $M_B (=M_D)$ and M_C from the previous set of formulæ where the whole span was loaded.

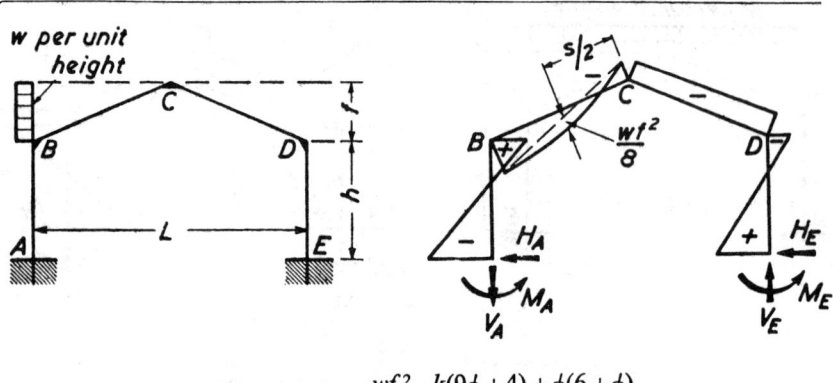

---

w per unit height

Constants: $X_1 = \dfrac{wf^2}{8} \cdot \dfrac{k(9\phi+4)+\phi(6+\phi)}{N_1}$

$X_2 = \dfrac{wf^2}{8} \cdot \dfrac{k(8+9\phi)-\phi^2}{N_1} \quad X_3 = \dfrac{wfh}{8} \cdot \dfrac{4B+\phi}{N_2}$

$M_A = -X_1 - X_3 \quad M_B = +X_2 + \left(\dfrac{wfh}{2} - X_3\right)$

$M_E = -X_1 + X_3 \quad M_D = +X_2 - \left(\dfrac{wfh}{2} - X_3\right)$

$M_C = -\dfrac{wf^2}{4} + \phi X_1 + mX_2$

$V_A = -V_E = -\dfrac{wfh(2+\phi)}{2L} + \dfrac{2X_3}{L} \quad H_E = \dfrac{wf}{2} - \dfrac{X_1+X_2}{h} \quad H_A = -(wf - H_E)$

FRAME III

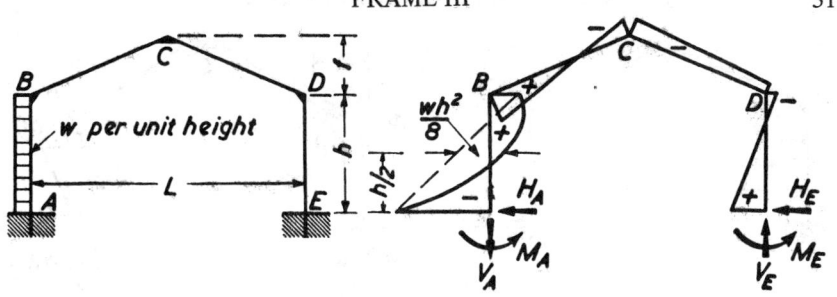

Constants: $X_1 = \dfrac{wh^2}{8} \cdot \dfrac{k(k+6) + k\phi(15+16\phi) + 6\phi^2}{N_1}$

$$X_2 = \dfrac{wh^2 k(9\phi + 8\phi^2 - k)}{8N_1} \qquad X_3 = \dfrac{wh^2(2k+1)}{2N_2}$$

$$M_A = -X_1 - X_3 \qquad M_B = +X_2 + \left(\dfrac{wh^2}{4} - X_3\right)$$

$$M_E = -X_1 + X_3 \qquad M_D = +X_2 - \left(\dfrac{wh^2}{4} - X_3\right)$$

$$M_C = -\dfrac{whf}{4} + \phi X_1 + m X_2$$

$$V_A = -V_E = -\dfrac{wh^2}{2L} + \dfrac{2X_3}{L} \qquad H_E = \dfrac{wh}{4} - \dfrac{X_1 + X_2}{h} \qquad H_A = -(wh - H_E)$$

Constants: $a_1 = \dfrac{a}{h} \qquad b_1 = \dfrac{b}{h}$

$$Y_1 = Pc[2\phi^2 - (1 - 3b_1^2)k] \qquad Y_2 = Pc[\phi - (3a_1^2 - 1)k]$$

$$X_1 = \dfrac{Y_1 K_1 - Y_2 R}{2N_1} \qquad X_2 = \dfrac{Y_2 K_2 - Y_1 R}{2N_1} \qquad X_3 = \dfrac{Pc}{2} \cdot \dfrac{B - 3(a_1 - b_1)k}{N_2}$$

$$M_A = -X_1 - X_3 \qquad M_B = +X_2 + \left(\dfrac{Pc}{2} - X_3\right)$$

$$M_E = -X_1 + X_3 \qquad M_D = +X_2 - \left(\dfrac{Pc}{2} - X_3\right) \qquad M_C = -\dfrac{\phi Pc}{2} + \phi X_1 + m X_2$$

$$M_1 = M_A - H_A a \qquad M_2 = M_B + H_E b$$

$$V_E = \dfrac{Pc - 2X_3}{L} \qquad V_A = P - V_E \qquad H_A = H_E = \dfrac{Pc}{2h} - \dfrac{X_1 + X_2}{h}$$

Extract: 'Kleinlogel, Rahmenformeln' 11. Auflage Berlin—Verlag von Wilhelm Ernst & Sohn.

FORMULAE FOR RIGID FRAMES

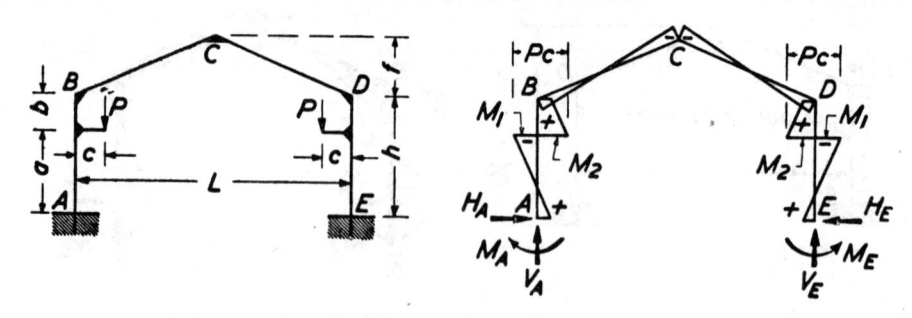

Constants: $a_1 = \dfrac{a}{h}$ $b_1 = \dfrac{b}{h}$

$$Y_1 = Pc[2\phi^2 - (1 - 3b_1^2)k]$$
$$Y_2 = Pc[\phi C + (3a_1^2 - 1)k]$$

$$M_A = M_E = \frac{Y_2 R - Y_1 K_1}{N_1} \qquad M_B = M_D = \frac{Y_2 K_2 - Y_1 R}{N_1}$$

$$M_C = -\phi(Pc + M_A) + mM_B$$

$$V_A = V_D = P \qquad H_A = H_E = \frac{Pc + M_A - M_B}{h}$$

$$M_1 = M_A - H_A a \qquad M_2 = M_B + H_E b$$

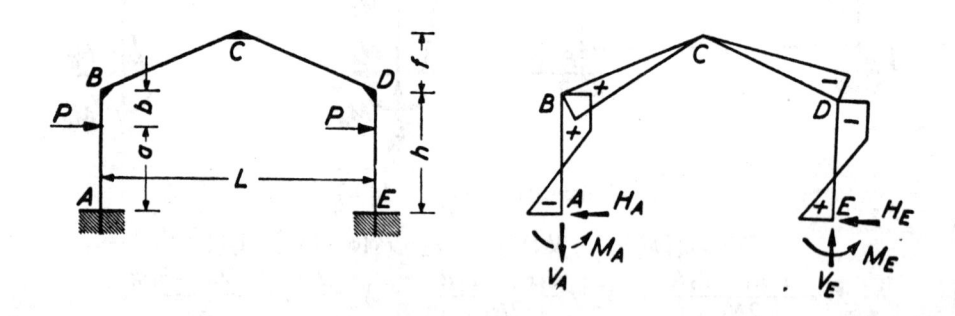

Constant: $X_1 = \dfrac{Pa(B + 3b_1 k)}{N_2}$

$$M_A = -M_E = -X_1 \qquad M_B = -M_D = Pa - X_1 \qquad M_C = 0$$

$$V_A = -V_E = -2\left[\frac{Pa - X_1}{L}\right] \qquad H_A = -H_E = -P$$

Extract: 'Kleinlogel, Rahmenformeln' 11. Auflage Berlin—Verlag von Wilhelm Ernst & Sohn.

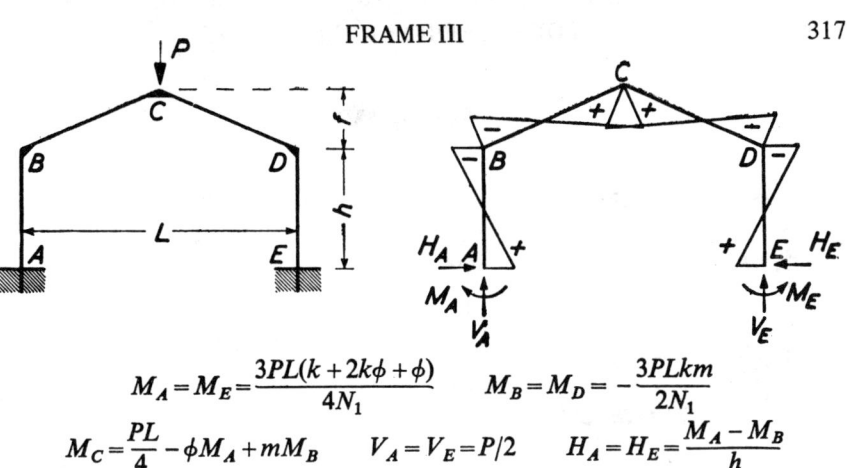

FRAME III

$$M_A = M_E = \frac{3PL(k+2k\phi+\phi)}{4N_1} \qquad M_B = M_D = -\frac{3PLkm}{2N_1}$$

$$M_C = \frac{PL}{4} - \phi M_A + mM_B \qquad V_A = V_E = P/2 \qquad H_A = H_E = \frac{M_A - M_B}{h}$$

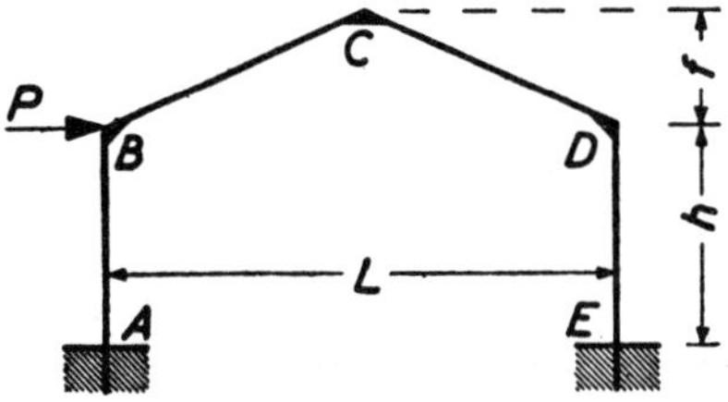

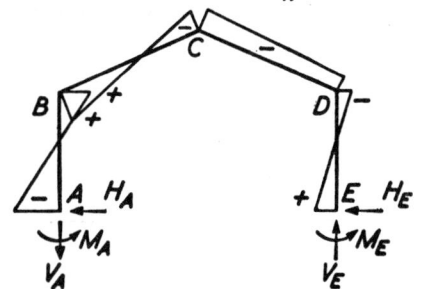

Constants: $X_1 = \dfrac{3Pf(k+2\phi k+\phi)}{2N_1}$ $\qquad X_2 = \dfrac{3Pfmk}{N_1} \qquad X_3 = \dfrac{PhB}{2N_2}$

$$M_A = -X_1 - X_3 \qquad M_B = +X_2 + \left(\frac{Ph}{2} - X_3\right)$$
$$M_C = -\frac{Pf}{2} + \phi X_1 + mX_2$$
$$M_E = -X_1 + X_3 \qquad M_D = +X_2 - \left(\frac{Ph}{2} - X_3\right)$$

$$V_A = -V_E = -\frac{Ph - 2X_3}{L} \qquad H_E = \frac{P}{2} - \frac{X_1 + X_2}{h} \qquad H_A = -(P - H_E)$$

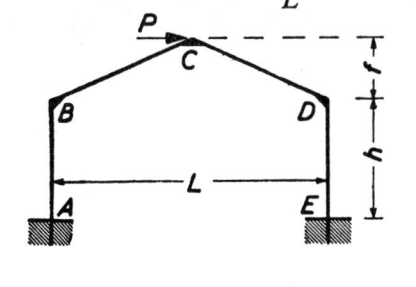

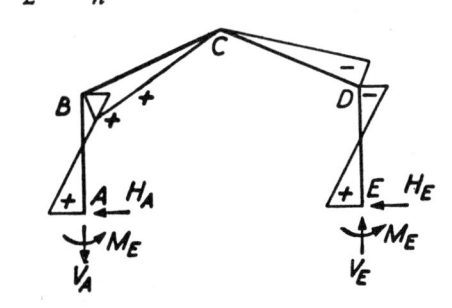

$$M_A = -M = -\frac{PhB}{2N_2} \qquad M_B = -M_D = +\frac{3Phk}{2N_2} \qquad M_C = 0$$

$$V_A = -V_E = -\frac{P(h+f) + 2M_A}{L} \qquad H_A = -H_E = -\frac{P}{2}$$

Extract: '*Kleinlogel, Rahmenformeln*' 11. Auflage Berlin—Verlag von Wilhelm Ernst & Sohn.

Frame IV

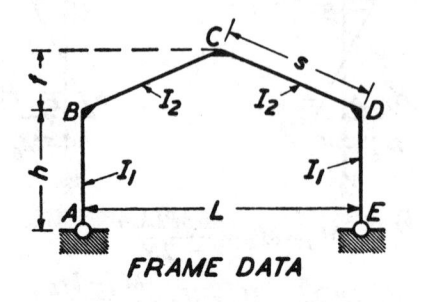

FRAME DATA

Coefficients:

$$k = \frac{I_2}{I_1} \cdot \frac{h}{s}$$

$$\phi = \frac{f}{h}$$

$$m = 1 + \phi$$

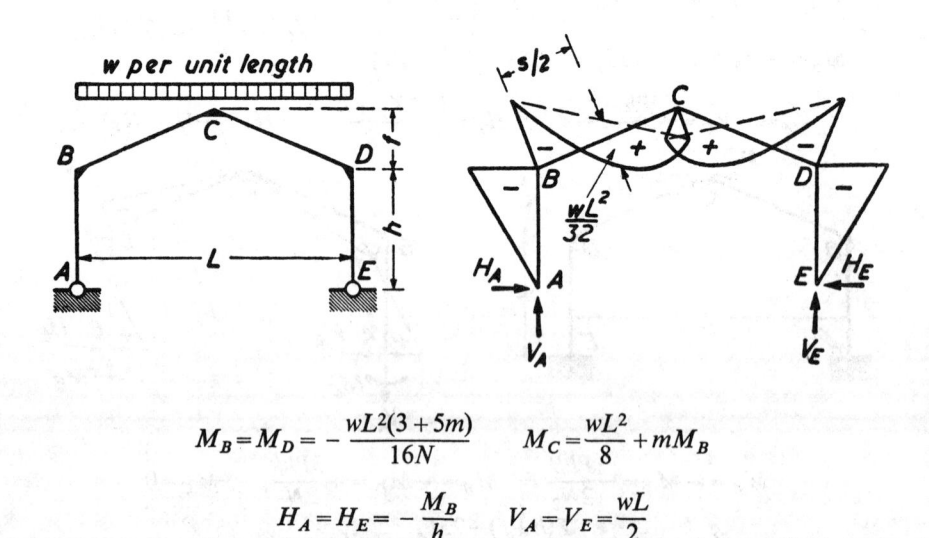

$$B = 2(k+1) + m \qquad C = 1 + 2m \qquad N = B + mC$$

$$M_B = M_D = -\frac{wL^2(3+5m)}{16N} \qquad M_C = \frac{wL^2}{8} + mM_B$$

$$H_A = H_E = -\frac{M_B}{h} \qquad V_A = V_E = \frac{wL}{2}$$

Extract: 'Kleinlogel, Rahmenformeln' 11. Auflage Berlin—Verlag von Wilhelm Ernst & Sohn.

FRAME IV

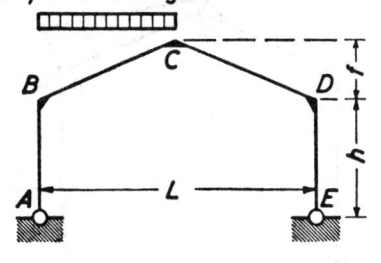

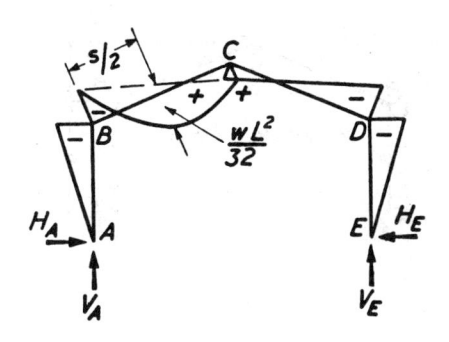

$$M_B = M_D = -\frac{wL^2(3+5m)}{32N} \qquad M_C = \frac{wL^2}{16} + mM_B$$

$$H_A = H_E = -\frac{M_B}{h} \qquad V_A = \frac{3wL}{8} \qquad V_E = \frac{wL}{8}$$

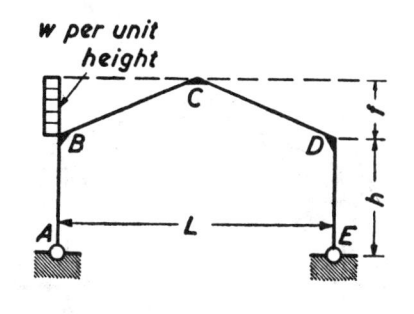

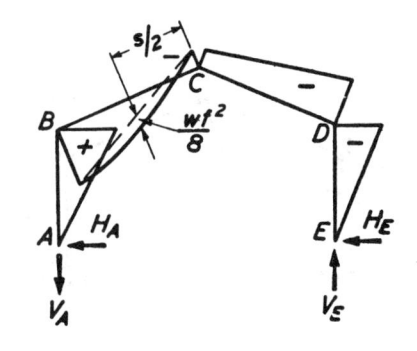

Constant: $X = \dfrac{wf^2(C+m)}{8N}$

$$M_B = +X + \frac{wfh}{2} \qquad M_C = -\frac{wf^2}{4} + mX$$

$$M_D = +X - \frac{wfh}{2} \qquad V_A = -V_E = -\frac{wfh(1+m)}{2L}$$

$$H_A = -\frac{X}{h} - \frac{wf}{2} \qquad H_E = -\frac{X}{h} + \frac{wf}{2}$$

Extract: 'Kleinlogel, Rahmenformeln' 11. Auflage Berlin—Verlag von Wilhelm Ernst & Sohn.

FORMULAE FOR RIGID FRAMES

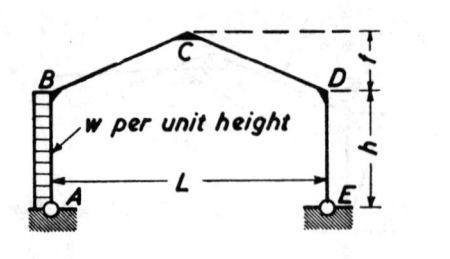

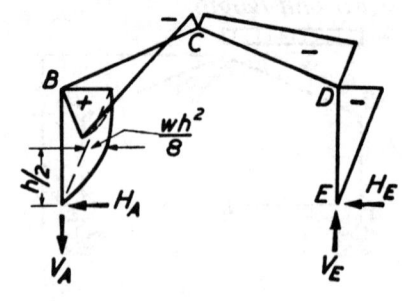

$$M_D = -\frac{wh^2}{8} \cdot \frac{2(B+C)+k}{N} \qquad M_B = \frac{wh^2}{2} + M_D$$

$$M_C = \frac{wh^2}{4} + mM_D$$

$$V_A = -V_E = -\frac{wh^2}{2L} \qquad H_E = -\frac{M_D}{h} \qquad H_A = -(wh - H_E)$$

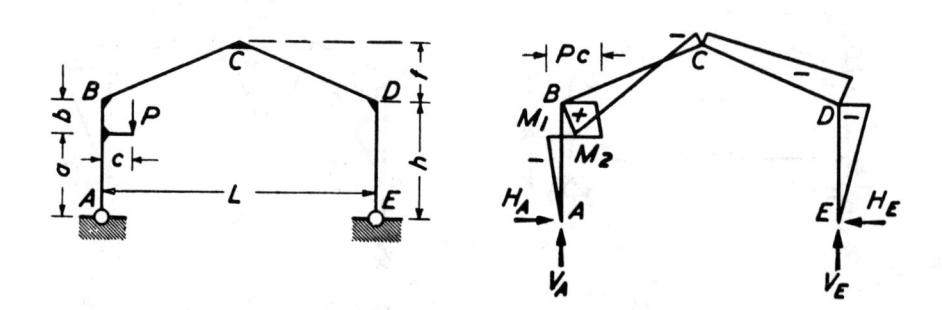

Constants: $a_1 = \dfrac{a}{h}$ $\qquad X = \dfrac{Pc}{2} \cdot \dfrac{B+C-k(3a_1^2-1)}{N}$

$$M_B = Pc - X \qquad M_D = -X \qquad M_C = \frac{Pc}{2} - mX$$

$$M_1 = -a_1 X \qquad M_2 = Pc - a_1 X$$

$$V_E = \frac{Pc}{L} \qquad V_A = P - V_E \qquad H_A = H_E = \frac{X}{h}$$

Extract: 'Kleinlogel, Rahmenformeln' 11. Auflage Berlin—Verlag von Wilhelm Ernst & Sohn.

FRAME IV

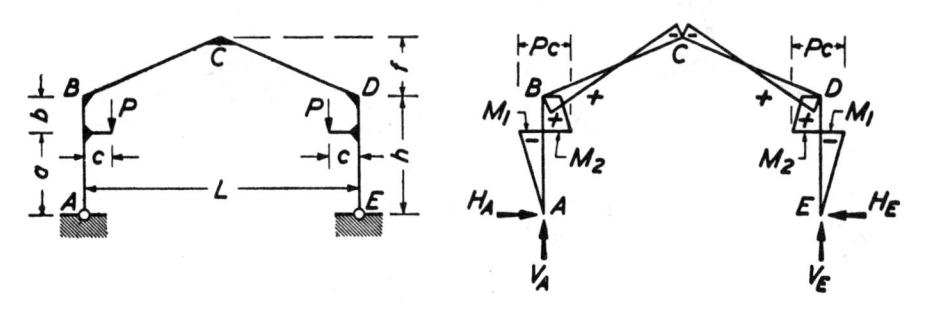

Constant: $a_1 = \dfrac{a}{h}$

$$M_B = M_D = Pc \cdot \frac{\phi C + k(3a_1^2 - 1)}{N} \qquad M_C = -\phi Pc + mM_B$$

$$H_A = H_E = \frac{Pc - M_B}{h} \qquad V_A = V_E = P$$

$$M_1 = -a_1(Pc - M_B) \qquad M_2 = (1 - a_1)Pc + a_1 M_B$$

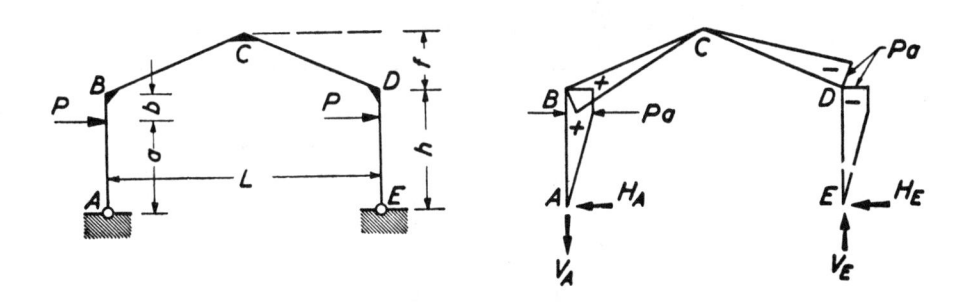

$$M_B = -M_D = Pa \qquad M_C = 0$$

$$H_A = -H_E = -P \qquad V_A = -V_E = -\frac{2Pa}{L}$$

Moment at loads $= \pm Pa$

Extract: 'Kleinlogel, Rahmenformeln' 11. Auflage Berlin—Verlag von Wilhelm Ernst & Sohn.

FORMULAE FOR RIGID FRAMES

$$M_B = M_D = -\frac{PL}{4} \cdot \frac{C}{N} \qquad M_C = +\frac{PL}{4} \cdot \frac{B}{N}$$

$$V_A = V_E = \frac{P}{2} \qquad H_A = H_E = -\frac{M_B}{h}$$

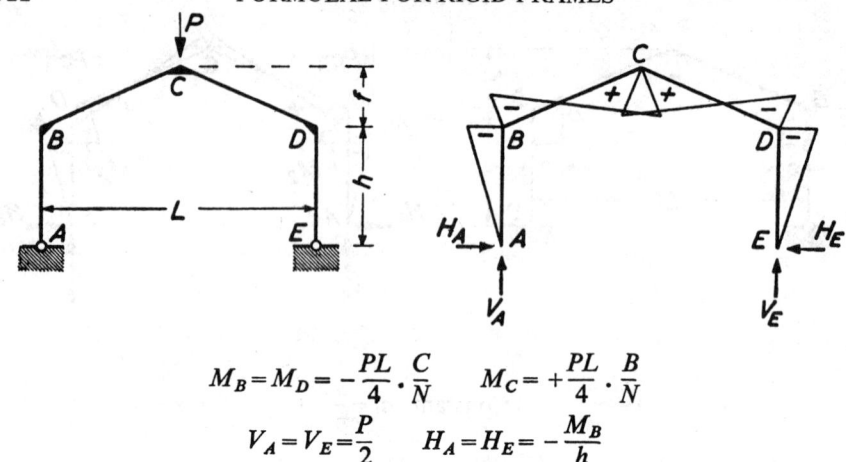

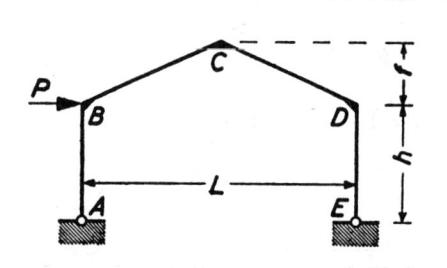

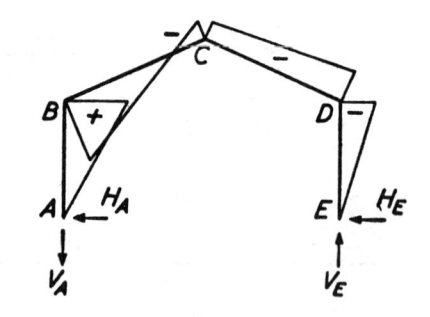

$$M_D = -\frac{Ph(B+C)}{2N} \qquad M_B = Ph + M_D \qquad M_C = \frac{Ph}{2} + mM_D$$

$$V_A = -V_E = -\frac{Ph}{L} \qquad H_E = -\frac{M_D}{h} \qquad H_A = -(P - H_E)$$

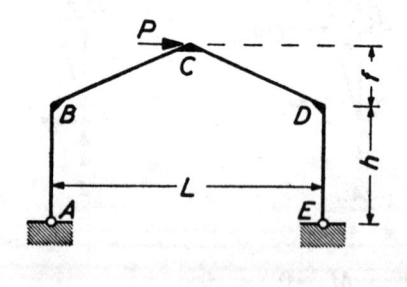

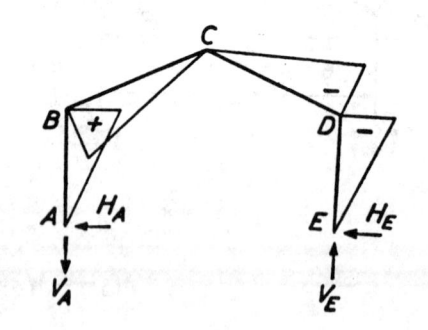

$$M_B = -M_D = +\frac{Ph}{2} \qquad M_C = 0 \qquad V_A = -V_E = -\frac{Phm}{L} \qquad H_A = -H_E = -\frac{P}{2}$$

Extract: 'Kleinlogel, Rahmenformeln' 11. Auflage Berlin—Verlag von Wilhelm Ernst & Sohn.

Frame V

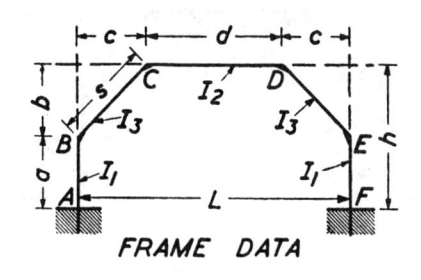

FRAME DATA

Coefficients:

$$k_1 = \frac{I_3}{I_1} \cdot \frac{a}{s} \qquad k_2 = \frac{I_3}{I_2} \cdot \frac{d}{s}$$

$$c_1 = \frac{c}{L} \qquad d_1 = \frac{d}{L}$$

$$\phi = \frac{b}{a} \qquad m = \frac{h}{a} = 1 + \phi$$

$$(2c_1 + d_1 = 1)$$

$C_1 = \phi(2 + 3k_2)$

$C_2 = 1 + m(2 + 3k_2)$

$R = \phi C_2 - k_1$

$B = 3k_1 + 2 + d_1 \qquad C_3 = 1 + d_1(2 + k_2)$

$K_1 = 2(k_1 + 1) + m(1 + C_2)$

$K_2 = 2k_1 + \phi C_1$

$N_1 = K_1 K_2 - R^2$

$N_2 = 3k_1 + B + d_1 C_3$

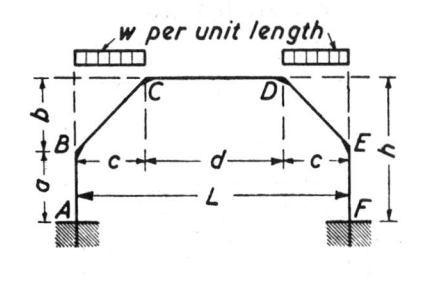

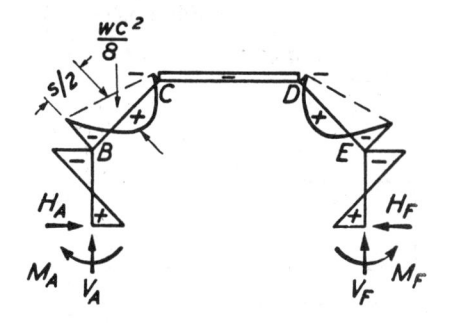

Constants: $Y_1 = \dfrac{wc^2}{4}(2C_1 + \phi) \qquad Y_2 = \dfrac{wc^2}{4}(2C_2 + 1 + m)$

$$M_A = M_F = \frac{Y_1 K_1 - Y_2 R}{N_1} \qquad M_B = M_E = -\frac{Y_2 K_2 - Y_1 R}{N_1}$$

$$M_C = M_D = \frac{wc^2}{2} - \phi M_A + m M_B$$

$$V_A = V_F = wc \qquad H_A = H_F = \frac{M_A - M_B}{a}$$

Extract: '*Kleinlogel, Rahmenformeln*' 11. *Auflage Berlin—Verlag von Wilhelm Ernst & Sohn.*

FORMULAE FOR RIGID FRAMES

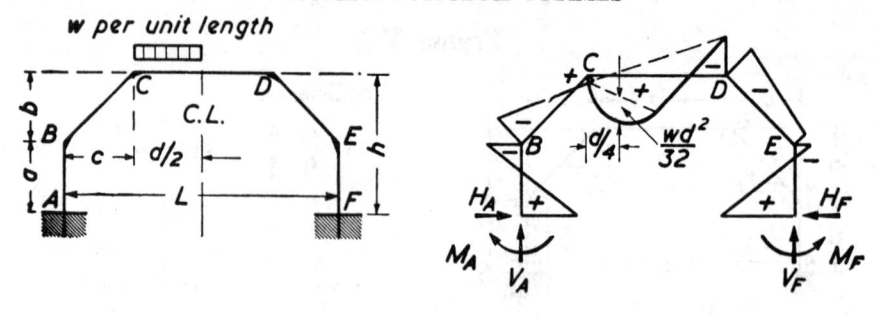

Constants: $Y_3 = \dfrac{wd^2}{32}(8c_1C_3 + d_1k_2)$

$Y_1 = \dfrac{wd}{4}(2cC_1 + d\phi k_2)$ $\qquad Y_2 = \dfrac{wd}{4}(2cC_2 + dmk_2)$

$X_1 = \dfrac{Y_1K_1 - Y_2R}{2N_1}$ $\qquad X_2 = \dfrac{Y_2K_2 - Y_1R}{2N_1}$ $\qquad X_3 = \dfrac{Y_3}{2N_2}$

$M_A = +X_1 - X_3$ $\qquad M_B = -X_2 - X_3$ $\qquad H_A = H_F = \dfrac{X_1 + X_2}{a}$

$M_F = +X_1 + X_3$ $\qquad M_E = -X_2 + X_3$

$M_C = \dfrac{wdc}{4} - \phi X_1 - mX_2 + \left(\dfrac{c_1 wd^2}{8} - d_1 X_3\right)$

$M_D = \dfrac{wdc}{4} - \phi X_1 - mX_2 - \left(\dfrac{c_1 wd^2}{8} - d_1 X_3\right)$

$V_A = \dfrac{wd(4c + 3d) + 16X_3}{8L}$ $\qquad V_F = \dfrac{wd(4c + d) - 16X_3}{8L}$

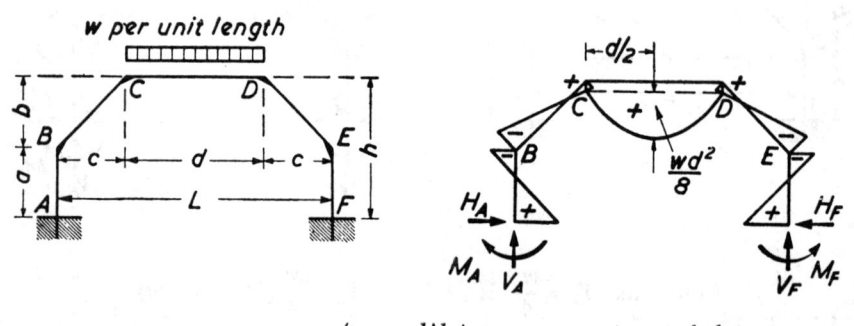

Constants: $Y_1 = wd\left(cC_1 + \dfrac{d\phi k_2}{2}\right)$ $\qquad Y_2 = wd\left(cC_2 + \dfrac{dmk_2}{2}\right)$

$X_1 = \dfrac{Y_1K_1 - Y_2R}{2N_1}$ $\qquad X_2 = \dfrac{Y_2K_2 - Y_1R}{2N_1}$

$M_A = M_F = X_1$ $\qquad M_B = M_E = -X_2$ $\qquad M_C = M_D = \dfrac{wdc}{2} - \phi X_1 - mX_2$

$V_A = V_F = \dfrac{wd}{2}$ $\qquad H_A = H_F = \dfrac{X_1 + X_2}{a}$

Extract: 'Kleinlogel, Rahmenformeln' 11. Auflage Berlin—Verlag von Wilhelm Ernst & Sohn.

FRAME V

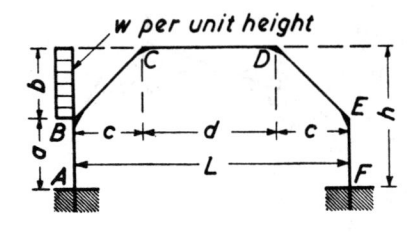

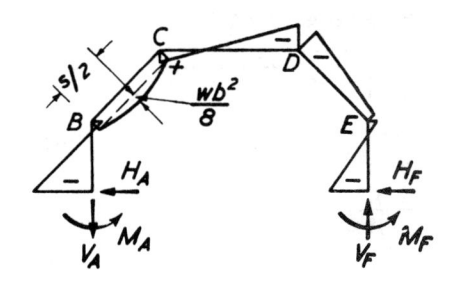

Constants: $Y_1 = \dfrac{wb^2}{4}(2C_1 - \phi)$

$Y_2 = \dfrac{wb^2}{4}(2C_2 - 1 - m)$ $\qquad Y_3 = wab(B + d_1 C_3) + \dfrac{wb^2}{4}(2d_1 C_3 + 1 + d_1)$

$$X_1 = \dfrac{Y_1 K_1 - Y_2 R}{2N_1} \qquad X_2 = \dfrac{Y_2 K_2 - Y_1 R}{2N_1} \qquad X_3 = \dfrac{Y_3}{2N_2}$$

$$M_A = -X_1 - X_3 \qquad M_B = +X_2 + \dfrac{wab}{2} - X_3$$

$$M_F = -X_1 + X_3 \qquad M_E = +X_2 - \dfrac{wab}{2} + X_3$$

$$M_C = -\dfrac{wb^2}{4} + \phi X_1 + m X_2 + \dfrac{d_1}{2}\left(wab + \dfrac{wb^2}{2} - 2X_3\right)$$

$$M_D = -\dfrac{wb^2}{4} + \phi X_1 + m X_2 - \dfrac{d_1}{2}\left(wab + \dfrac{wb^2}{2} - 2X_3\right)$$

$$V_A = -V_F = \dfrac{2wab + wb^2 - 4X_3}{2L}$$

$$H_F = \dfrac{wb}{2} - \dfrac{X_1 + X_2}{a} \qquad H_A = -(wb - H_F)$$

Extract: '*Kleinlogel, Rahmenformeln*' 11. Auflage Berlin—Verlag von Wilhelm Ernst & Sohn.

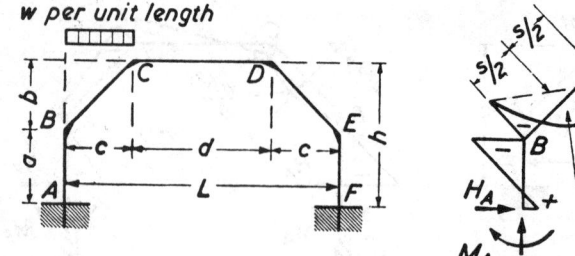

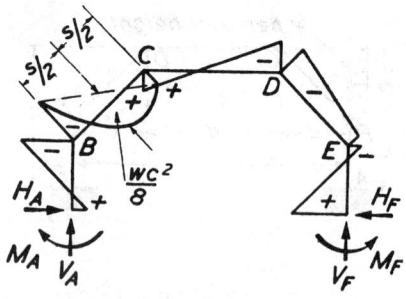

Constants: $Y_1 = \dfrac{wc^2}{4}(2C_1 + \phi)$ $\quad Y_2 = \dfrac{wc^2}{4}(2C_2 + 1 + m)$ $\quad Y_3 = \dfrac{wc^2}{4}(2d_1 C_3 + 1 + d_1)$

$$X_1 = \frac{Y_1 K_1 - Y_2 R}{2N_1} \qquad X_2 = \frac{Y_2 K_2 - Y_1 R}{2N_1} \qquad X_3 = \frac{Y_3}{2N_2}$$

$$M_A = +X_1 - X_3 \qquad M_B = -X_2 - X_3 \qquad M_F = +X_1 + X_3 \qquad M_E = -X_2 + X_3$$

$$M_C = \frac{wc^2}{4} - \phi X_1 - m X_2 + \frac{d_1}{2}\left(\frac{wc^2}{2} - 2X_3\right)$$

$$M_D = \frac{wc^2}{4} - \phi X_1 - m X_2 - \frac{d_1}{2}\left(\frac{wc^2}{2} - 2X_3\right)$$

$$V_F = \frac{wc^2 - 4X_3}{2L} \qquad V_A = wc - V_F \qquad H_A = H_F = \frac{X_1 + X_2}{a}$$

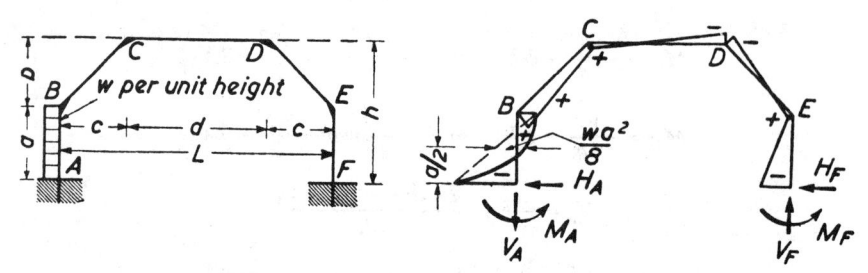

Constants: $Y_1 = \dfrac{wa^2}{4}(2\phi C_1 + k_1)$ $\quad Y_2 = \dfrac{wa^2}{4}(2\phi C_2 - k_1)$ $\quad Y_3 = \dfrac{wa^2}{2}(B + d_1 C_3 + k_1)$

The formulæ for X_1, X_2 and X_3 are the same as above.

$$M_A = -X_1 - X_3 \qquad M_B = +X_2 + \left(\frac{wa^2}{4} - X_3\right)$$

$$M_F = -X_1 + X_3 \qquad M_E = +X_2 - \left(\frac{wa^2}{4} - X_3\right)$$

$$M_C = -\frac{wa^2 \phi}{4} + \phi X_1 + m X_2 + d_1\left(\frac{wa^2}{4} - X_3\right)$$

$$M_D = -\frac{wa^2 \phi}{4} + \phi X_1 + m X_2 - d_1\left(\frac{wa^2}{4} - X_3\right)$$

$$V_A = -V_F = -\frac{2}{L}\left(\frac{wa^2}{4} - X_3\right) \qquad H_F = \frac{wa}{4} - \frac{X_1 + X_2}{a} \qquad H_A = -(wa - H_F)$$

Extract: 'Kleinlogel, Rahmenformeln' 11. Auflage Berlin—Verlag von Wilhelm Ernst & Sohn.

Frame VI

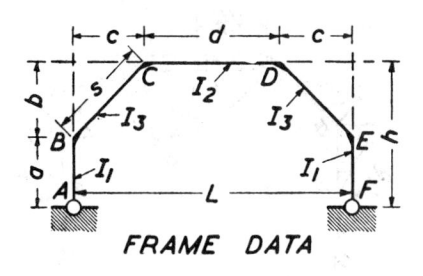

Coefficients:

$$k_1 = \frac{I_3}{I_1} \cdot \frac{a}{s} \qquad k_2 = \frac{I_3}{I_2} \cdot \frac{d}{s}$$

$$a_1 = \frac{a}{h} \qquad c_1 = \frac{c}{L}$$

$$B = 2a_1(k_1+1)+1 \qquad C = a_1 + 2 + 3k_2$$

$$N = a_1 B + C$$

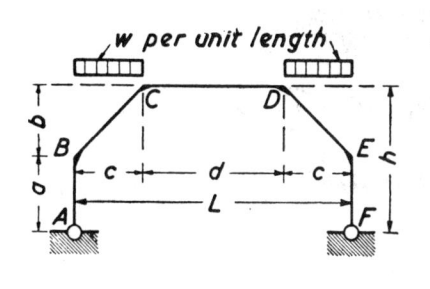

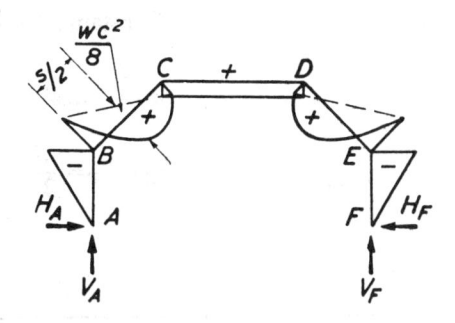

FRAME DATA

Constant: $X = \dfrac{wc^2}{4} \cdot \dfrac{3a_1 + 5 + 6k_2}{N}$

$$M_B = M_E = -a_1 X \qquad M_C = M_D = \frac{wc^2}{2} - X \qquad V_A = V_F = wc \qquad H_A = H_F = \frac{X}{h}$$

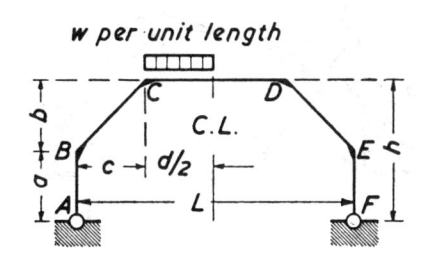

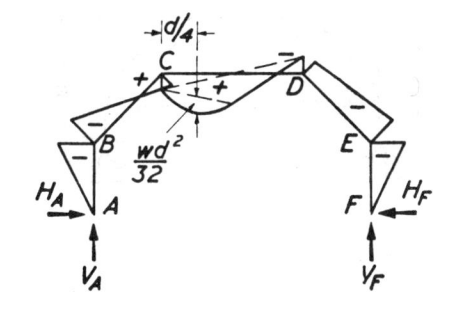

Constant: $X = \dfrac{wd(2cC + dk_2)}{8N}$

$$M_C = \frac{c_1 wd}{8}(3d+4c) - X \qquad M_D = \frac{c_1 wd}{8}(4c+d) - X \qquad M_B = M_E = -a_1 X$$

$$V_A = \frac{wd(3d+4c)}{8L} \qquad V_F = \frac{wd(4c+d)}{8L} \qquad H_A = H_F = \frac{X}{h}$$

Extract: 'Kleinlogel, Rahmenformeln' 11. Auflage Berlin—Verlag von Wilhelm Ernst & Sohn.

FORMULAE FOR RIGID FRAMES

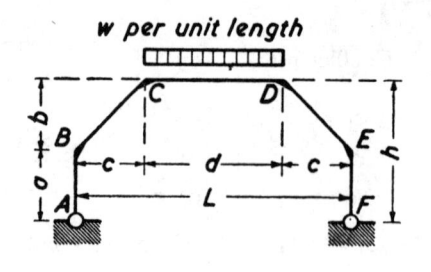

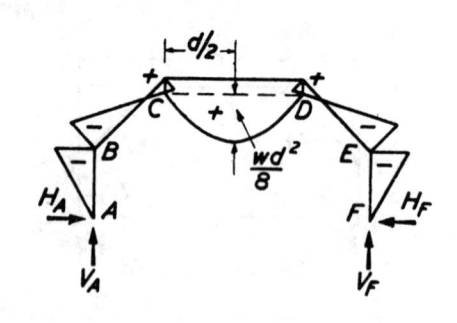

Constant: $X = \dfrac{wd}{4} \cdot \dfrac{2cC + dk_2}{N}$

$M_B = M_E = -a_1 X \qquad M_C = M_D = \dfrac{wdc}{2} - X$

$V_A = V_F = \dfrac{wd}{2} \qquad H_A = H_F = \dfrac{X}{h}$

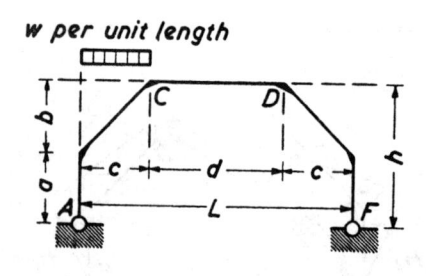

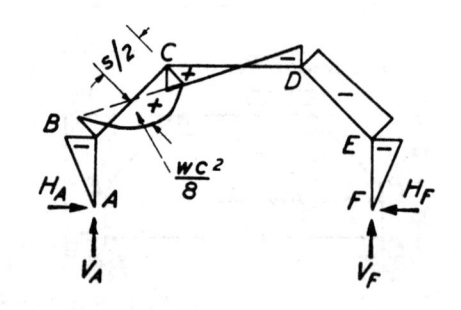

Constant: $X = \dfrac{wc^2(2C + a_1 + 1)}{8N}$

$M_B = M_E = -a_1 X \qquad M_C = (1 - c_1)\dfrac{wc^2}{2} - X$

$V_F = \dfrac{wc^2}{2L} \qquad V_A = wc - V_F \qquad H_A = H_F = \dfrac{X}{h}$

Extract: '*Kleinlogel, Rahmenformeln*' 11. *Auflage Berlin—Verlag von Wilhelm Ernst & Sohn.*

FRAME VI

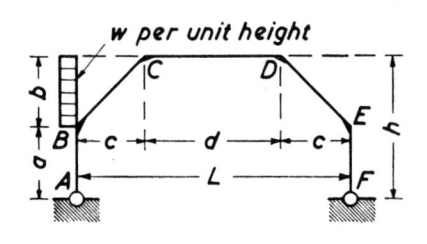

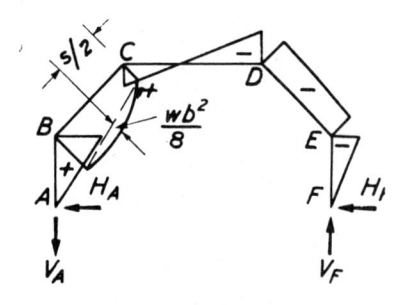

Constant: $X = \dfrac{wb}{8} \cdot \dfrac{4a(B+C) + b(2C + a_1 + 1)}{N}$

$M_B = wba - a_1 X \qquad M_E = -a_1 X$

$M_C = V_F(L - c) - X \qquad M_D = -X + V_F \cdot c$

$V_A = -V_F = -\dfrac{wb(a+h)}{2L} \qquad H_F = \dfrac{X}{h} \qquad H_A = -(wb - H_F)$

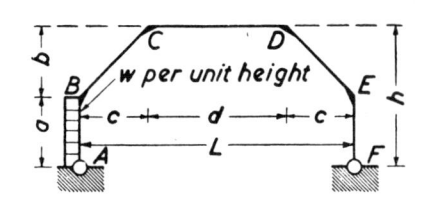

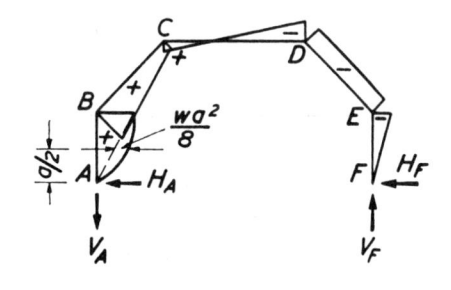

Constant: $X = \dfrac{wa^2}{8} \cdot \dfrac{2(B+C) + a_1 k_1}{N}$

$M_B = \dfrac{wa^2}{2} - a_1 X \qquad M_E = -a_1 X$

$M_C = V_F(L - c) - X \qquad M_D = -X + V_F \cdot c$

$V_A = -V_F = -\dfrac{wa^2}{2L} \qquad H_F = \dfrac{X}{h} \qquad H_A = -(wa - H_F)$

Extract: '*Kleinlogel, Rahmenformeln*' 11. *Auflage Berlin—Verlag von Wilhelm Ernst & Sohn.*

Frame VII

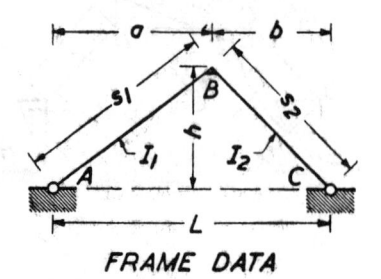

FRAME DATA

Coefficients:

$$k = \frac{I_1}{I_2} \cdot \frac{s_2}{s_1} \qquad N = 1 + k$$

$$a_1 = \frac{a}{L} \qquad b_1 = \frac{b}{L}$$

$$(a_1 + b_1 = 1)$$

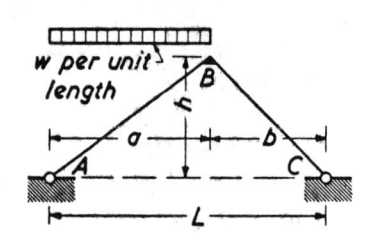

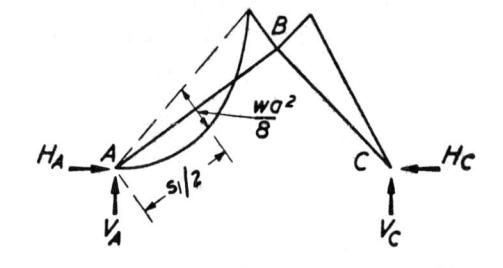

$$M_B = -\frac{wa^2}{8N} \qquad V_C = \frac{wa^2}{2L} \qquad V_A = wa - V_C$$

$$H_A = H_C = \frac{wa^2 b_1}{2h} - \frac{M_B}{h}$$

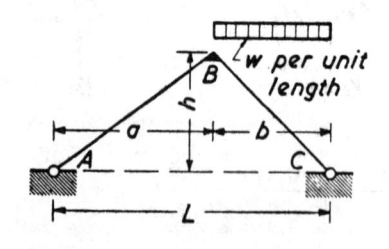

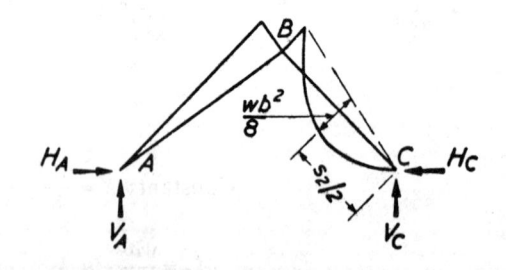

$$M_B = -\frac{wb^2 k}{8N} \qquad V_A = \frac{wb^2}{2L} \qquad V_C = wb - V_A$$

$$H_A = H_C = \frac{wb^2 a_1}{2h} - \frac{M_B}{h}$$

Extract: 'Kleinlogel, Rahmenformeln' 11. Auflage Berlin—Verlag von Wilhelm Ernst & Sohn.

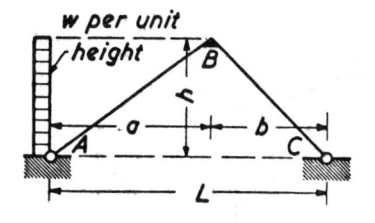

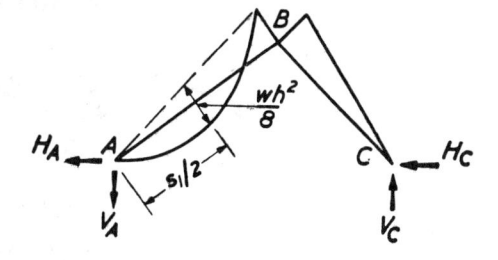

FRAME VII

$$M_B = -\frac{wh^2}{8N} \qquad V_A = -V_C = -\frac{wh^2}{2L}$$

$$H_C = \frac{whb_1}{2} - \frac{M_B}{h} \qquad H_A = -(wh - H_C)$$

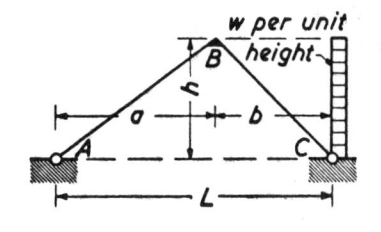

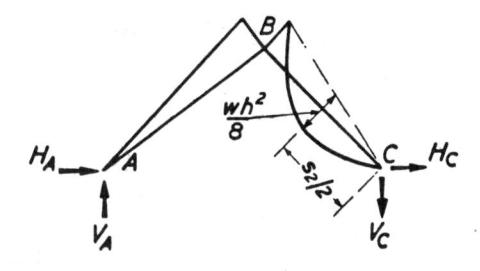

$$M_B = -\frac{wh^2 k}{8N} \qquad V_A = -V_C = \frac{wh^2}{2L}$$

$$H_A = \frac{wha_1}{2} - \frac{M_B}{h} \qquad H_C = -(wh - H_A)$$

Extract: '*Kleinlogel, Rahmenformeln*' 11. Auflage Berlin—Verlag von Wilhelm Ernst & Sohn.

Frame VIII

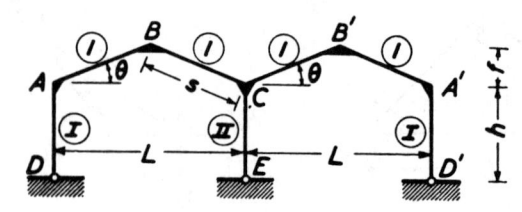

FRAME DATA

Constants:

$$\phi = \frac{f}{h} \quad x_1 = \frac{I_1 h}{I_I s} \quad x_{II} = \frac{I_1 h}{I_{II} s} \quad \cos\theta = \frac{L}{2s} \quad \sin\theta = \frac{f}{s}$$

$$N_1 = 8x_I + 12(1+\phi) + 7\phi^2 \qquad N_2 = 2x_I + 12(1+\phi) + 4\phi^2 + 4x_{II}$$

Influence Coefficients:

$$n_{11} = \frac{2}{N_1} \qquad n_{12} = n_{21} = \frac{2+3\phi}{2N_1} \qquad n_{22} = \frac{x_I + 2 + 3\phi + 2\phi^2}{N_1}$$

$$L_1 = \frac{2+5\phi}{4N_1} \qquad c_1 = \frac{6+7\phi}{4N_1} \qquad y_{11} = 2\phi c_1$$

$$L_2 = \frac{2x_I + 2 - \phi - 2\phi^2}{4N_1} \qquad c_2 = \frac{6x_I + 6 + \phi}{4N_1} \qquad y_{12} = 2\phi c_2$$

$$n = \frac{1}{2N_2} \qquad L' = \frac{1+\phi}{2N_2} \qquad c = \frac{3+2\phi}{2N_2} \qquad r = \frac{1}{N_2}$$

$$y_{13} = \frac{6 + 9\phi + 4\phi^2 + 4x_{II}}{2N_2} \qquad y_{23} = \frac{6 + 3\phi + 4x_{II}}{2N_2}$$

$$y_{14} = \frac{2x_I + 6 + 3\phi}{2N_2} \qquad y_{24} = \frac{2x_I + 6 + 9\phi + 4\phi^2}{2N_2}$$

$$(y_{13} + y_{14} = 0.5) \qquad (y_{23} + y_{24} = 0.5)$$

Composite Influence Coefficients:

$$s_1 = +n_{11} + L_1 \qquad s'_1 = -L_1 + n_{21} \qquad r_1 = -n_{11} + n_{21}$$
$$s_2 = +n_{12} - L_2 \qquad s'_2 = +L_2 + n_{22} \qquad r_2 = -n_{12} + n_{22}$$
$$s' = L' + n \qquad r = 2n$$

$$m_1 = \frac{+s_1 - s'_1}{2} = L_1 - \frac{r_1}{2} \qquad m_2 = \frac{-s_2 + s'_2}{2} = L_2 + \frac{r_2}{2}$$

$$y_3 = y_{23} + c\phi \qquad y_{13} - y_{23} - 2c\phi = 0$$
$$y_4 = y_{24} - c\phi \qquad y_{14} - y_{24} + 2c\phi = 0$$

Note.—The four rafters of equal length AB, BC, CB' and $B'A'$ with the member number 1 are allocated another series of member numbers according to the figure on page 334 as a distinguishing mark for the static values referring to these rafters.

Extract: 'Kleinlogel, *Mehrstielige Rahmen*', Band I + II. Berlin—Verlag von Wilhelm Ernst & Sohn.

FRAME VIII

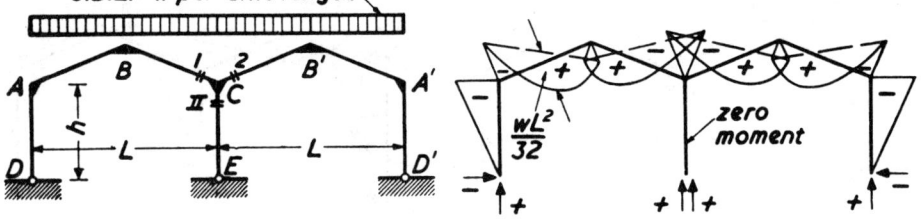

$$M_A = M'_A = -\frac{wL^2}{4}(m_1 + 2c_1)$$

$$M_{C1} = M_{C2} = -\frac{wL^2}{4}(m_2 + 2c_2)$$

$$M_{CII} = 0$$

$$M_B = M'_B = +\frac{wL^2}{8} + \frac{M_A(1+2\phi) + M_{C1}}{2}$$

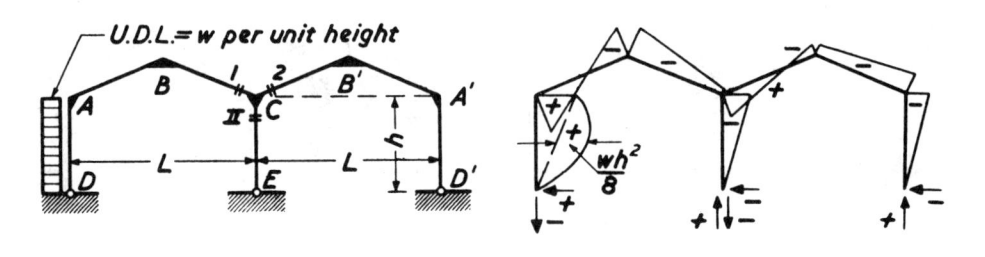

$$M_A = \frac{wh^2}{4} \cdot x_I(-n_{11} - n) + \frac{wh^2}{2}(+y_{11} + y_{13})$$

$$M'_A = \frac{wh^2}{4} \cdot x_I(-n_{11} + n) + \frac{wh^2}{2}(+y_{11} - y_{13})$$

$$M_{C1} = \frac{wh^2}{4} \cdot x_I(+n_{12} - n) + \frac{wh^2}{2}(+y_{12} - y_{14})$$

$$M_{C2} = \frac{wh^2}{4} \cdot x_I(+n_{12} + n) + \frac{wh^2}{2}(+y_{12} + y_{14})$$

$$M_{CII} = -2\left[\frac{wh^2}{4} \cdot x_I n + \frac{wh^2}{2} \cdot y_{14}\right]$$

$$M_B = -\frac{wh^2\phi}{2} + \frac{M_A(1+2\phi) + M_{C1}}{2}$$

$$M'_B = \frac{M_{C2} + M'_A(1+2\phi)}{2}$$

Extract: 'Kleinlogel, Mehrstielige Rahmen', Band I+II. Berlin—Verlag von Wilhelm Ernst & Sohn.

FORMULAE FOR RIGID FRAMES

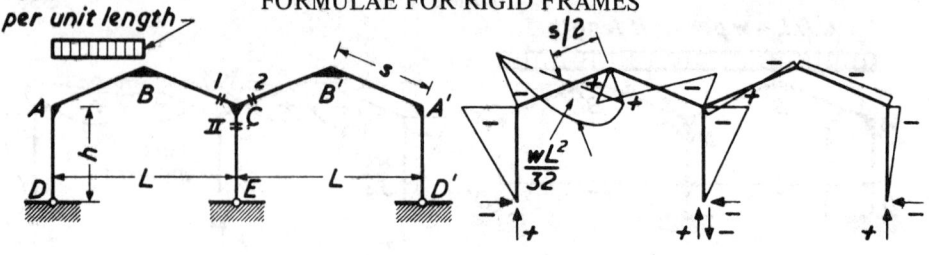

$$M_A = \frac{wL^2}{16}(-s_1 - s' - 2c_1 - 2c) \qquad M'_A = \frac{wL^2}{16}(-s_1 + s' - 2c_1 + 2c)$$

$$M_{C1} = \frac{wL^2}{16}(+s_2 - s' - 2c_2 - 2c) \qquad M_{C2} = \frac{wL^2}{16}(+s_2 + s' - 2c_2 + 2c)$$

$$M_{CII} = -\frac{wL^2}{8}(s' + 2c)$$

$$M_B = \frac{wL^2}{16} + \frac{M_A(1+2\phi) + M_{C1}}{2} \qquad M'_B = \frac{M_{C2} + M'_A(1+2\phi)}{2}$$

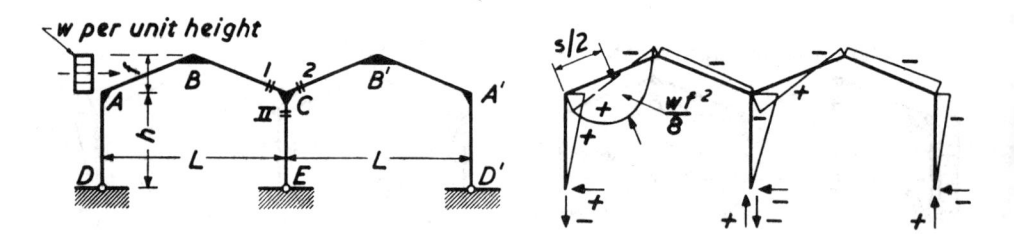

Constant: $W_1 = wfh$

$$M_A = \frac{wf^2}{4}(-s_1 - s') + \frac{3wf^2}{2}(+c_1 + c) + W_1 y_{23}$$

$$M'_A = \frac{wf^2}{4}(-s_1 + s') + \frac{3wf^2}{2}(+c_1 - c) - W_1 y_{23}$$

$$M_{C1} = \frac{wf^2}{4}(+s_2 - s') + \frac{3wf^2}{2}(+c_2 + c) - W_1 y_{24}$$

$$M_{C2} = \frac{wf^2}{4}(+s_2 + s') + \frac{3wf^2}{2}(+c_2 - c) + W_1 y_{24}$$

$$M_{CII} = -\frac{wf^2 s'}{2} + 3wf^2 c - 2W_1 y_{24}$$

$$M_B = -\frac{3wf^2}{4} + \frac{M_A(1+2\phi) + M_{C1}}{2}$$

$$M'_B = \frac{M_{C2} + M'_A(1+2\phi)}{2}$$

Extract: 'Kleinlogel, Mehrstielige Rahmen', Band I+II. Berlin—Verlag von Wilhelm Ernst & Sohn.

FRAME VIII

$$M_A = \frac{wL^2}{16}(+s'_1 - s' - 2c_1 - 2c) \qquad M'_A = \frac{wL^2}{16}(+s'_1 + s' - 2c_1 + 2c)$$

$$M_{C1} = \frac{wL^2}{16}(-s'_2 - s' - 2c_2 - 2c) \qquad M_{C2} = \frac{wL^2}{16}(-s'_2 + s' - 2c_2 + 2c)$$

$$M_{CII} = -\frac{wL^2}{8}(s' + 2c)$$

$$M_B = \frac{wL^2}{16} + \frac{M_A(1+2\phi) + M_{C1}}{2} \qquad M'_B = \frac{M_{C2} + M'_A(1+2\phi)}{2}$$

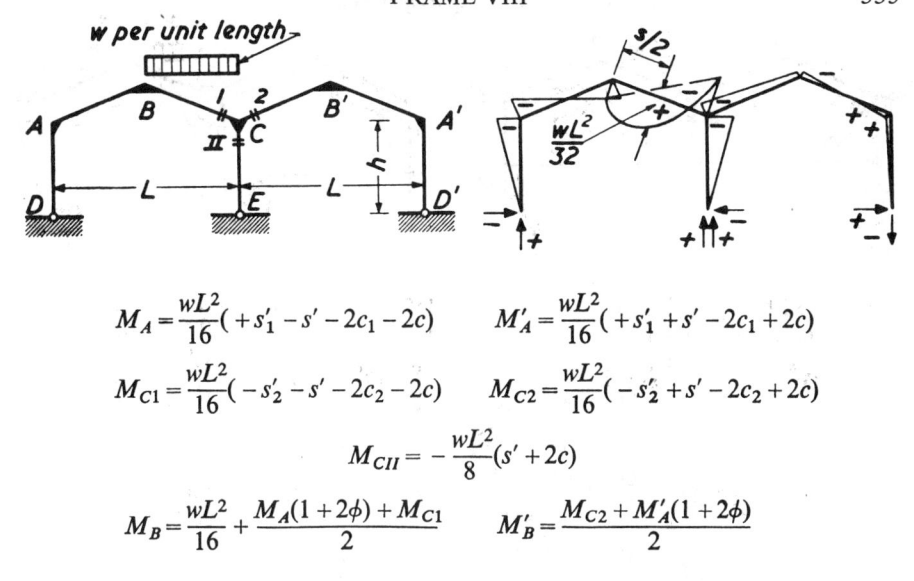

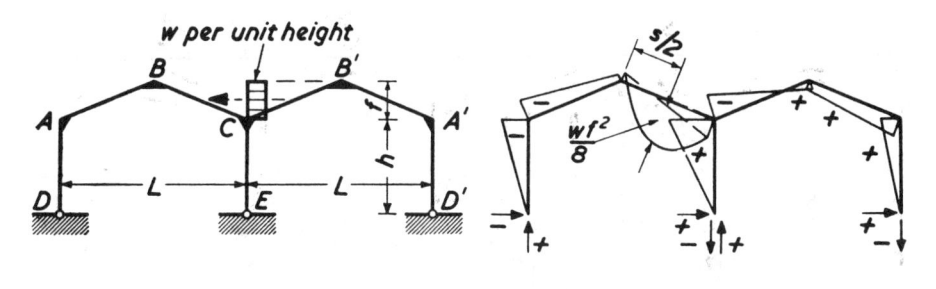

Constant: $W_1 = wfh$

$$M_A = \frac{wf^2}{4}(+s'_1 - s' - 2c_1 - 2c) - W_1 y_{23}$$

$$M'_A = \frac{wf^2}{4}(+s'_1 + s' - 2c_1 + 2c) + W_1 y_{23}$$

$$M_{C1} = \frac{wf^2}{4}(-s'_2 - s' - 2c_2 - 2c) + W_1 y_{24}$$

$$M_{C2} = \frac{wf^2}{4}(-s'_2 + s' - 2c_2 + 2c) - W_1 y_{24}$$

$$M_{CII} = -\frac{wf^2}{2}(s' + 2c) + 2W_1 y_{24}$$

$$M_B = \frac{wf^2}{4} + \frac{M_A(1+2\phi) + M_{C1}}{2} \qquad M'_B = \frac{M_{C2} + M'_A(1+2\phi)}{2}$$

Extract: 'Kleinlogel, Mehrstielige Rahmen', Band I + II. Berlin—Verlag von Wilhelm Ernst & Sohn.

FORMULAE FOR RIGID FRAMES

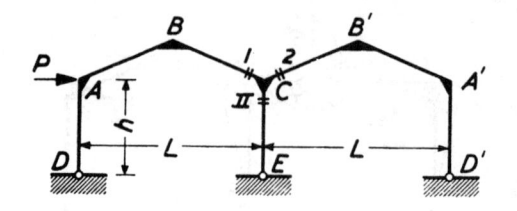

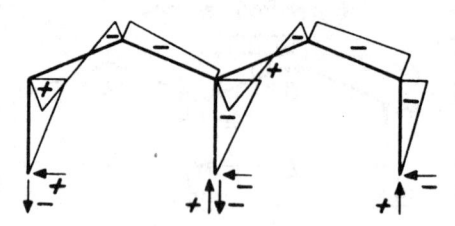

$$M_A = M(+y_{11} + y_{13})$$ where $M = Ph$
$$M'_A = M(+y_{11} - y_{13})$$
$$M_{C1} = M(+y_{12} - y_{14})$$
$$M_{C2} = M(+y_{12} + y_{14}) \qquad M_{CII} = -2My_{14}$$
$$M_B = \frac{M_A(1+2\phi) + M_{C1}}{2} - Pf$$
$$M'_B = \frac{M_{C2} + M'_A(1+2\phi)}{2}$$

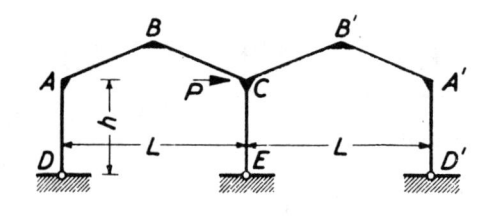

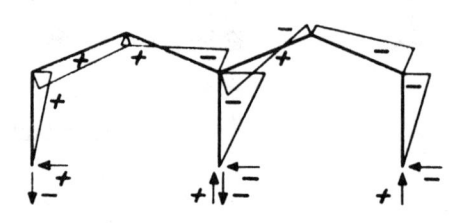

$$M_A = +My_{23}$$ where $M = Ph$
$$M'_A = -My_{23}$$
$$M_{C1} = -My_{24} \qquad M_{CII} = -2My_{24}$$
$$M_{C2} = +My_{24}$$
$$M_B = \frac{M_A(1+2\phi) + M_{C1}}{2}$$
$$M'_B = \frac{M_{C2} + M'_A(1+2\phi)}{2}$$

Extract: 'Kleinlogel, *Mehrstielige Rahmen*', Band I + II. Berlin—Verlag von Wilhelm Ernst & Sohn.

FRAME VIII

$$M_A = M'_A = -Mc_1$$
$$M_{C1} = M_{C2} = -Mc_2 \quad \text{where} \quad M = PL$$
$$M_{CII} = 0$$
$$M_B = M'_B = \frac{PL}{4} + \frac{M_A(1+2\phi) + M_{C1}}{2}$$

$$M_A = \frac{M}{2}(-c_1 - c)$$

$$M'_A = \frac{M}{2}(-c_1 + c) \qquad \text{where} \quad M = PL$$

$$M_{C1} = \frac{M}{2}(-c_2 - c)$$

$$M_{C2} = \frac{M}{2}(-c_2 + c) \qquad M_{CII} = -Mc$$

$$M_B = \frac{PL}{4} + \frac{M_A(1+2\phi) + M_{C1}}{2}$$

$$M'_B = \frac{M_{C2} + M'_A(1+2\phi)}{2}$$

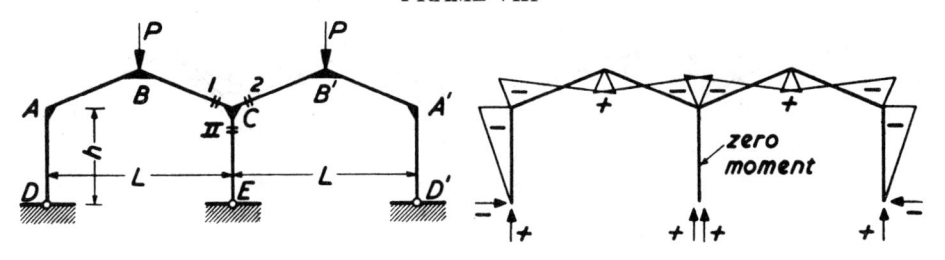

Extract: 'Kleinlogel, Mehrstielige Rahmen', Band I + II. Berlin—Verlag von Wilhelm Ernst & Sohn.

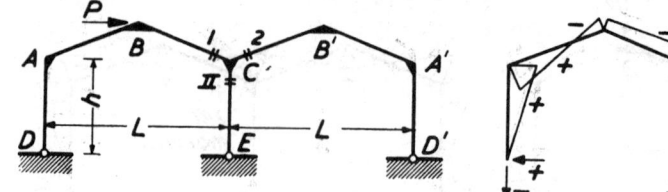

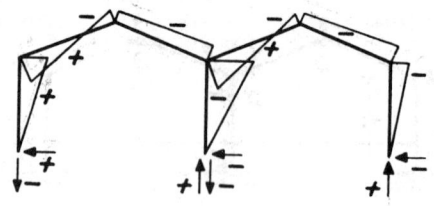

$$M_A = M(+c_1\phi + y_3)$$
$$M'_A = M(+c_1\phi - y_3)$$
$$M_{C1} = M(+c_2\phi - y_4)$$
$$M_{C2} = M(+c_2\phi + y_4)$$
$$M_B = \frac{M_A(1+2\phi) + M_{C1} - Pf}{2}$$
$$M'_B = \frac{M_{C2} + M'_A(1+2\phi)}{2}$$

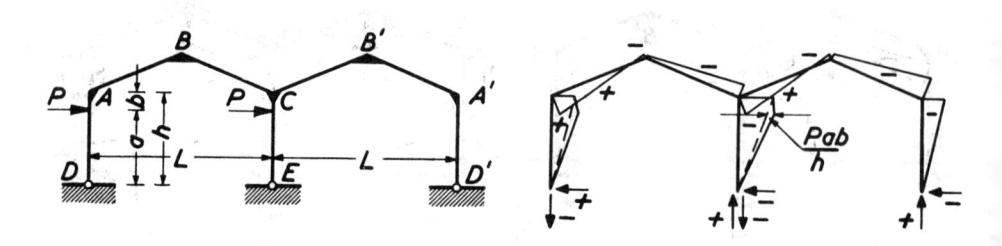

where $M = Ph$

$$M_{CII} = -2My_4$$

Constants: $a_1 = a/h \quad b_1 = b/h$

$$R_I = Pba_1(1+a_1)x_I \qquad L_{II} = Pab_1(1+b_1)x_{II}$$
$$M_A = R_I(-n_{11} - n) + Pa(+y_{11} + y_{13}) + L_{II}r + Pay_{23}$$
$$M'_A = R_I(-n_{11} + n) + Pa(+y_{11} - y_{13}) - L_{II}r - Pay_{23}$$
$$M_{C1} = R_I(+n_{12} - n) + Pa(+y_{12} - y_{14}) + L_{II}r - Pay_{24}$$
$$M_{C2} = R_I(+n_{12} + n) + Pa(+y_{12} + y_{14}) - L_{II}r + Pay_{24}$$
$$M_{CII} = -2(R_I n + Pay_{14} - L_{II}r + Pay_{24})$$
$$M_B = \frac{M_A(1+2\phi) + M_{C1}}{2} - Pa\phi$$
$$M'_B = \frac{M_{C2} + M'_A(1+2\phi)}{2}$$

Extract: 'Kleinlogel, Mehrstielige Rahmen', Band I+II. Berlin—Verlag von Wilhelm Ernst & Sohn.

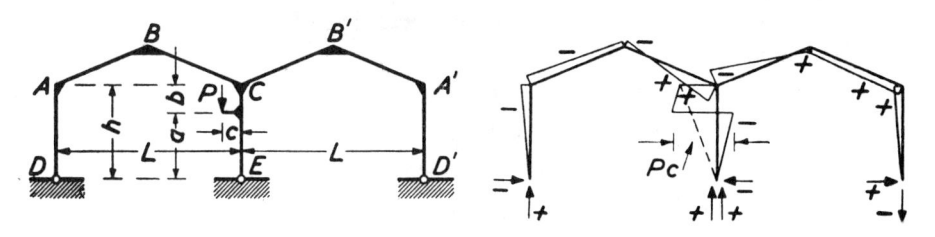

FRAME VIII

Constants: $a_1 = a/h \quad b_1 = b/h$

$$L_{II} = Pc(1 - 3a_1^2)x_{II}$$
$$M_A = -L_{II}r - Pcy_{23}$$
$$M'_A = +L_{II}r + Pcy_{23}$$
$$M_{C1} = -L_{II}r + Pcy_{24} \qquad M_{CII} = 2(-L_{II}r + Pcy_{24})$$
$$M_{C2} = +L_{II}r - Pcy_{24}$$
$$M_B = \frac{M_A(1 + 2\phi) + M_{C1}}{2}$$
$$M'_B = \frac{M_{C2} + M'_A(1 + 2\phi)}{2}$$

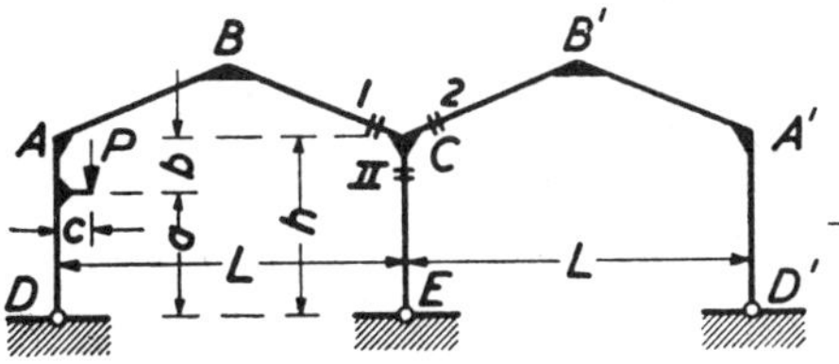

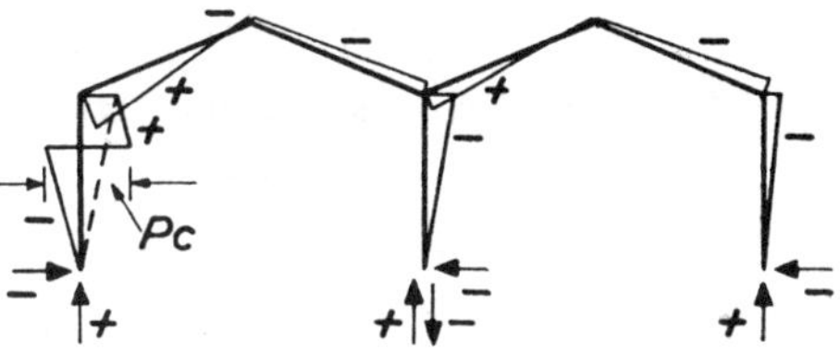

Constants: $a_1 = a/h \quad b_1 = b/h$

$$R_I = Pc(1 - 3a_1^2)x_I$$
$$M_A = R_I(-n_{11} - n) + Pc(+y_{11} + y_{13})$$
$$M'_A = R_I(-n_{11} + n) + Pc(+y_{11} - y_{13})$$
$$M_{C1} = R_I(+n_{12} - n) + Pc(+y_{12} - y_{14})$$
$$M_{C2} = R_I(+n_{12} + n) + Pc(+y_{12} + y_{14})$$
$$M_{CII} = -2(R_I n + Pcy_{14})$$
$$M_B = \frac{M_A(1 + 2\phi) + M_{C1}}{2} - Pc\phi$$
$$M'_B = \frac{M_{C2} + M'_A(1 + 2\phi)}{2}$$

Extract: 'Kleinlogel, *Mehrstielige Rahmen*', Band I + II. Berlin—Verlag von Wilhelm Ernst & Sohn.

FORMULAE FOR RIGID FRAMES

Formulæ for Support Reactions, Shear and Axial Forces.
Nomenclature:

V = vertical reaction or axial force in columns.
N = axial force in rafters or columns.
T = shear force at ends of rafters or columns.
H = horizontal reactions in columns.

(a) For all loads:

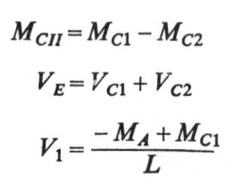

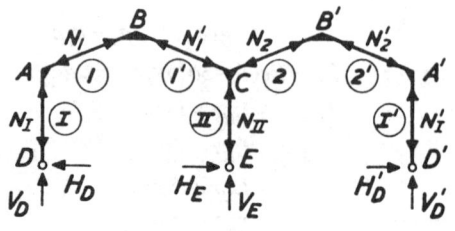

$M_{CII} = M_{C1} - M_{C2}$

$V_E = V_{C1} + V_{C2}$

$V_1 = \dfrac{-M_A + M_{C1}}{L}$

$V_2 = \dfrac{-M_{C2} + M'_A}{L}$

All the values of H and V shown are positive

$$H_I = \dfrac{M_A}{h} \qquad H_{II} = \dfrac{M_{CII}}{h} \qquad H'_I = \dfrac{M'_A}{h}$$

The axial thrusts in the columns are:
$$N_I = V_D \qquad N_{II} = V_E \qquad N'_I = V'_D$$

The horizontal thrusts in unloaded columns are:
$$H_D = H_I \qquad H_E = H_{II} \qquad H'_D = H'_I$$

The shear forces in unloaded columns are:
$$T_I = +H_I \qquad T_{II} = -H_{II} \qquad T'_I = -H'_I$$

(b) For vertical U.D.L. over the whole frame:

U.D.L. = w per unit length

$V_D = +V_1 + \dfrac{wL}{2}$

$V_{C2} = +V_2 + \dfrac{wL}{2}$

$V_{C1} = -V_1 + \dfrac{wL}{2}$

$V'_D = -V_2 + \dfrac{wL}{2}$

Reactions shown are those for total U.D.L.

For vertical U.D.L. over the extreme left rafter:

$$V_D = +V_1 + \dfrac{3wL}{8} \qquad V_{C2} = +V_2$$

$$V_{C1} = -V_1 + \dfrac{wL}{8} \qquad V'_D = -V_2$$

Extract: 'Kleinlogel, Mehrstielige Rahmen', Band I+II. Berlin—Verlag von Wilhelm Ernst & Sohn.

FRAME VIII

For vertical U.D.L. over the second rafter from the left:

$$V_D = +V_1 + \frac{wL}{8} \qquad V_{C2} = +V_2$$

$$V_{C1} = -V_1 + \frac{3wL}{8} \qquad V'_D = -V_2$$

The shear forces in the rafters for *all* vertical U.D.L.s are:

$$T_{L1} = +V_D \cos\theta + H_I \sin\theta \qquad T_{R1} = T_{L1} - W_1 \cos\theta$$
$$T'_{R1} = -V_{C1} \cos\theta - H_I \sin\theta \qquad T'_{L1} = T'_{R1} + W'_1 \cos\theta$$
$$T_{L2} = +V_{C2} \cos\theta + H'_I \sin\theta \qquad T_{R2} = T_{L2} - W_2 \cos\theta$$
$$T'_{R2} = -V'_D \cos\theta - H'_I \sin\theta \qquad T'_{L2} = T'_{R2} + W'_2 \cos\theta$$

where the suffix $L1$ refers to the left-hand end of the first rafter from the left, $R1$ refers to the right-hand end of the first rafter, and so on, and where W_1, W'_1, W_2 and W'_2 are the U.D.L.s on the first, second, third and fourth rafters from the left.

The axial thrusts for *all* vertical U.D.L.s are:

$$N_{1A} = +V_D \sin\theta - H_I \cos\theta \qquad N_{1B} = N_{1A} - W_1 \sin\theta$$
$$N'_{1C} = +V_{C1} \sin\theta - H_I \cos\theta \qquad N'_{1B} = N'_{1C} - W'_1 \sin\theta$$
$$N_{2C} = +V_{C1} \sin\theta - H'_I \cos\theta \qquad N_{2B} = N_{2C} - W_2 \sin\theta$$
$$N'_{2A} = +V'_D \sin\theta - H'_I \cos\theta \qquad N'_{2B} = N'_{2A} - W'_2 \sin\theta$$

where W_1, W'_1, W_2 and W'_2 are the U.D.L.s on the first, second, third and fourth rafters respectively from the left.

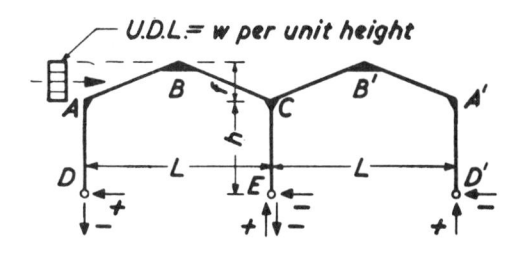

(c) For horizontal U.D.L. applied to the extreme left rafter:

$$V_D = +V_1 - \frac{wf^2}{2L} = -V_{C1}$$

$$V_{C2} = +V_2 = -V'_D$$

$$H_{C1} = H_I - wf \qquad H_{C2} = H'_I$$

The shear forces in the rafters are:

$$T_{L1} = +V_D \cos\theta + H_D \sin\theta \qquad T_{L2} = -V'_D \cos\theta + H_{C2} \sin\theta$$
$$T_{R1} = T_{L1} - wf \sin\theta \qquad T_{R2} = T_{L2}$$
$$T'_1 = +V_D \cos\theta - H_{C1} \sin\theta \qquad T'_2 = -V'_D \cos\theta - H'_D \sin\theta$$

The axial thrusts are:

$$N_{1A} = +V_D \sin\theta - H_D \cos\theta \qquad N_{2C} = -V'_D \sin\theta - H_{C2} \cos\theta$$
$$N_{1B} = N_{1A} + wf \cos\theta \qquad N_{2B} = N_{2C}$$
$$N'_1 = -V_D \sin\theta - H_{C1} \cos\theta \qquad N'_2 = +V'_D \sin\theta - H'_D \cos\theta$$

Extract: 'Kleinlogel, Mehrstielige Rahmen', Band I+II. Berlin—Verlag von Wilhelm Ernst & Sohn.

(d) For horizontal U.D.L. acting to the left on the second rafter from the left:

$$V_D = +V_1 + \frac{wf^2}{2L} = -V_{C1}$$

$$V_{C2} = +V_2 = -V'_D$$

$$H_{C1} = H_I + wf$$

$$H_{C2} = H'_I$$

U.D.L.= w per unit height

The shear forces in the rafters are:

$$T'_{R1} = +V_D \cos\theta - H_{C1}\sin\theta \qquad T'_{R2} = -V'_D\cos\theta - H'_D\sin\theta$$

$$T'_{L1} = T'_{R1} + wf\sin\theta \qquad T'_{L2} = T'_{R2}$$

$$T_1 = +V_D\cos\theta + H_D\sin\theta \qquad T_2 = -V'_D\cos\theta + H_{C2}\sin\theta$$

The axial thrusts are:

$$N'_{1C} = -V_D\sin\theta - H_{C1}\cos\theta \qquad N'_{2A} = +V'_D\sin\theta - H'_D\cos\theta$$

$$N'_{1B} = N'_{1C} + wf\cos\theta \qquad N'_{2B} = N_{2A}$$

$$N_1 = +V_D\sin\theta - H_D\cos\theta \qquad N_2 = -V'_D\sin\theta - H_{C2}\cos\theta$$

(e) For horizontal U.D.L. applied to extreme left column:

U.D.L.= w per unit height

$$V_D = +V_1 \qquad V_{C2} = +V_2$$

$$V_{C1} = -V_1 \qquad V'_D = -V_2$$

$$H_D = H_I + \frac{wh}{2} \qquad H_E = H_{II}$$

$$H'_D = H'_I$$

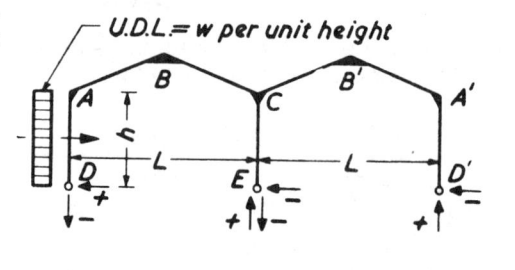

The shear forces in the columns are:

$$T_D = +H_D \qquad T_E = -H_E \qquad T'_D = -H'_D$$

$$T_A = +H_D - wh \qquad T_C = -H_E \qquad T'_A = -H'_D$$

The shear forces in the rafters are:

$$T_1 = +V_1\cos\theta + T_A\sin\theta \qquad T_2 = +V_2\cos\theta - T'_A\sin\theta$$

$$T'_1 = +V_1\cos\theta - T_A\sin\theta \qquad T'_2 = +V_2\cos\theta + T'_A\sin\theta$$

The axial thrusts in the rafters are:

$$N_1 = +V_1\sin\theta - T_A\cos\theta \qquad N_2 = +V_2\sin\theta + T'_A\cos\theta$$

$$N'_1 = -V_1\sin\theta - T_A\cos\theta \qquad N'_2 = -V_2\sin\theta + T'_A\cos\theta$$

Extract: 'Kleinlogel, Mehrstielige Rahmen', Band I+II. Berlin—Verlag von Wilhelm Ernst & Sohn.

FRAME VIII

(f) Gantry crane loads:

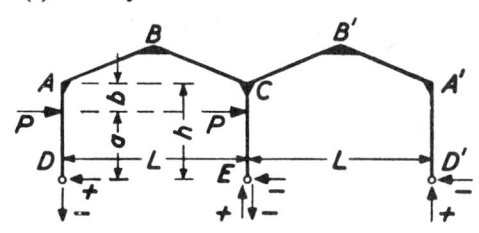

(1) Surge loads:

$$V_D = +V_1 \qquad V_{C2} = +V_2$$
$$V_{C1} = -V_1 \qquad V'_D = -V_2$$
$$H_D = H_I + \frac{Pb}{h} \qquad H_E = H_{II} - \frac{Pb}{h}$$
$$H'_D = H'_I$$

The shear forces in the column are:

$$T_D = +H_D \qquad T_E = -H_E \qquad T'_D = -H'_D$$
$$T_A = +H_D - P \qquad T_C = -H_E - P \qquad T'_A = -H'_D$$

(2) Bracket load on central column CE:

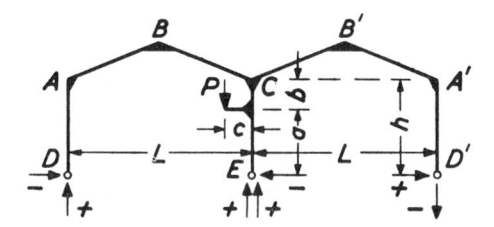

$$V_D = +V_1 \qquad V_{C2} = +V_2$$
$$V_{C1} = P - V_1 \qquad V'_D = -V_2$$
$$H_D = H_I \qquad H_E = H_{II} - \frac{Pc}{h}$$
$$H'_D = H'_I$$

(3) Bracket load on column AD:

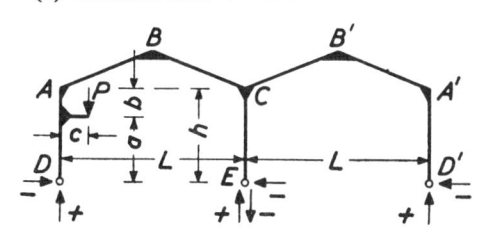

$$V_D = P + V_1 \qquad V_{C2} = +V_2$$
$$V_{C1} = -V_1 \qquad V'_D = -V_2$$
$$H_D = H_I - \frac{Pc}{h} \qquad H_E = H_{II}$$
$$H'_D = H'_I$$

(g) For vertical point loads on the ridges:

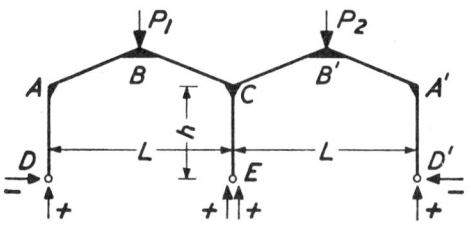

$$V_D = +V_1 + \frac{P_1}{2}$$
$$V_{C1} = -V_1 + \frac{P_1}{2}$$
$$V_{C2} = +V_2 + \frac{P_2}{2}$$
$$V'_D = -V_2 + \frac{P_2}{2}$$

Reactions shown are applicable when $P_1 = P_2$

Extract: 'Kleinlogel, Mehrstielige Rahmen', Band I+II. Berlin—Verlag von Wilhelm Ernst & Sohn.

FORMULAE FOR RIGID FRAMES

$$T_1 = +V_D \cos\theta + H_I \sin\theta \qquad N_1 = +V_D \sin\theta - H_I \cos\theta$$
$$T_1' = -V_{C1} \cos\theta - H_I \sin\theta \qquad N_1' = +V_{C1} \sin\theta - H_I \cos\theta$$
$$T_2 = +V_{C2} \cos\theta + H_I' \sin\theta \qquad N_2 = +V_{C2} \sin\theta - H_I' \cos\theta$$
$$T_2' = -V_D' \cos\theta - H_I' \sin\theta \qquad N_2' = +V_D' \sin\theta - H_I' \cos\theta$$

(h) For horizontal point loads applied at joints:

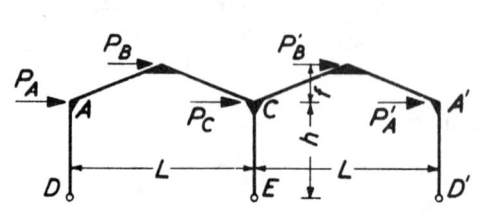

Reactions vary with the number and value of the loads

$$V_D = +V_1 - \frac{P_B f}{L}$$
$$V_{C2} = +V_2 - \frac{P_B' f}{L}$$
$$V_{C1} = -V_D$$
$$V_D' = -V_{C2}$$
$$H_{C1} = H_I - P_A - P_B$$
$$H_{C2} = H_I' + P_B' + P_A'$$

$$T_1 = +V_D \cos\theta + (H_I - P_A) \sin\theta \qquad N_1 = +V_D \sin\theta - (H_I - P_A) \cos\theta$$
$$T_1' = +V_D \cos\theta - H_{C1} \sin\theta \qquad N_1' = -V_D \sin\theta - H_{C1} \cos\theta$$
$$T_2 = -V_D' \cos\theta + H_{C2} \sin\theta \qquad N_2 = -V_D' \sin\theta - H_{C2} \cos\theta$$
$$T_2' = -V_D' \cos\theta - (H_I' + P_A') \sin\theta \qquad N_2' = +V_D' \sin\theta - (H_I' + P_A') \cos\theta$$

Extract: 'Kleinlogel, Mehrstielige Rahmen', Band I+II. Berlin—Verlag von Wilhelm Ernst & Sohn.

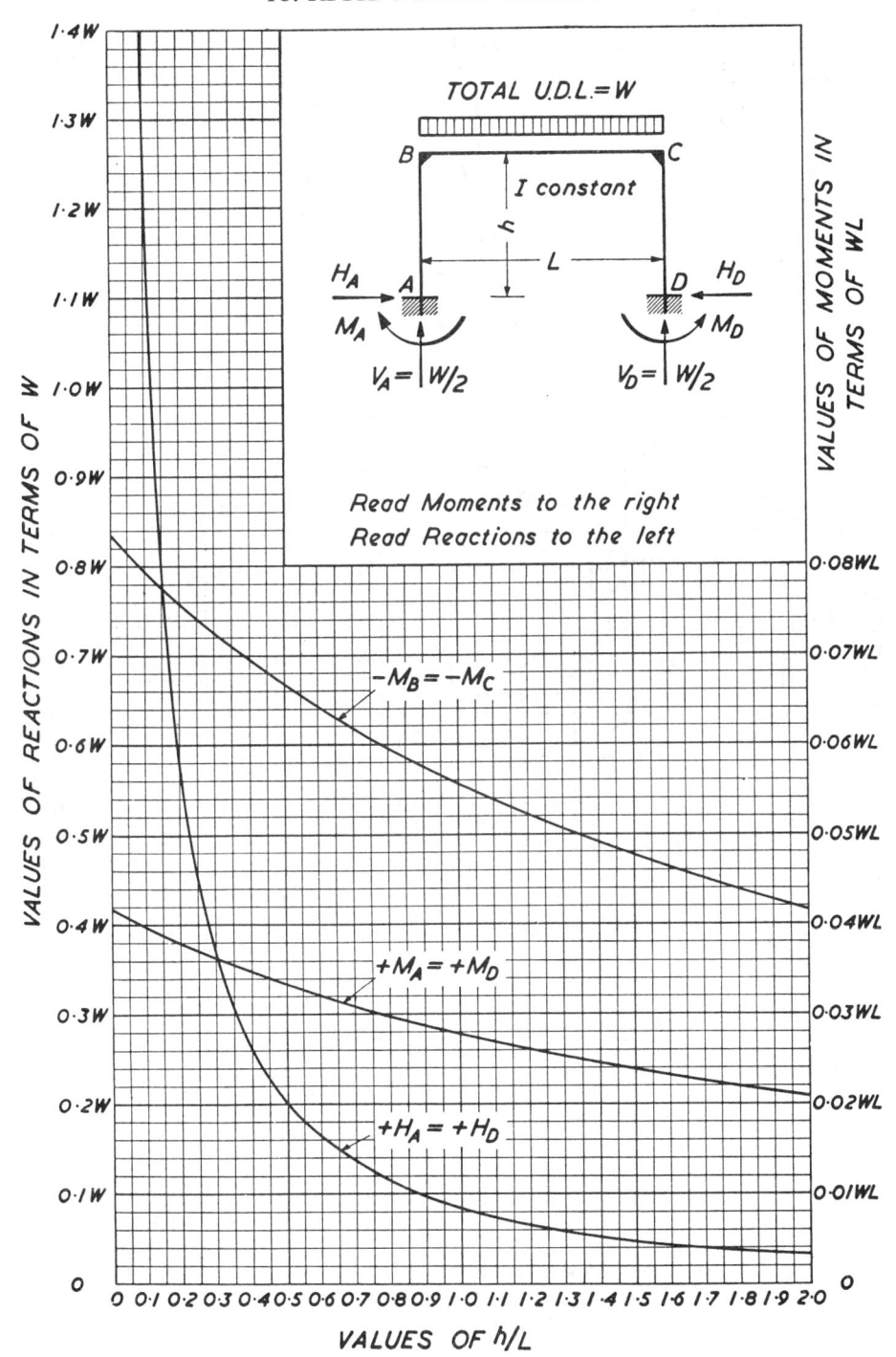

16. RIGID FRAME CHARTS

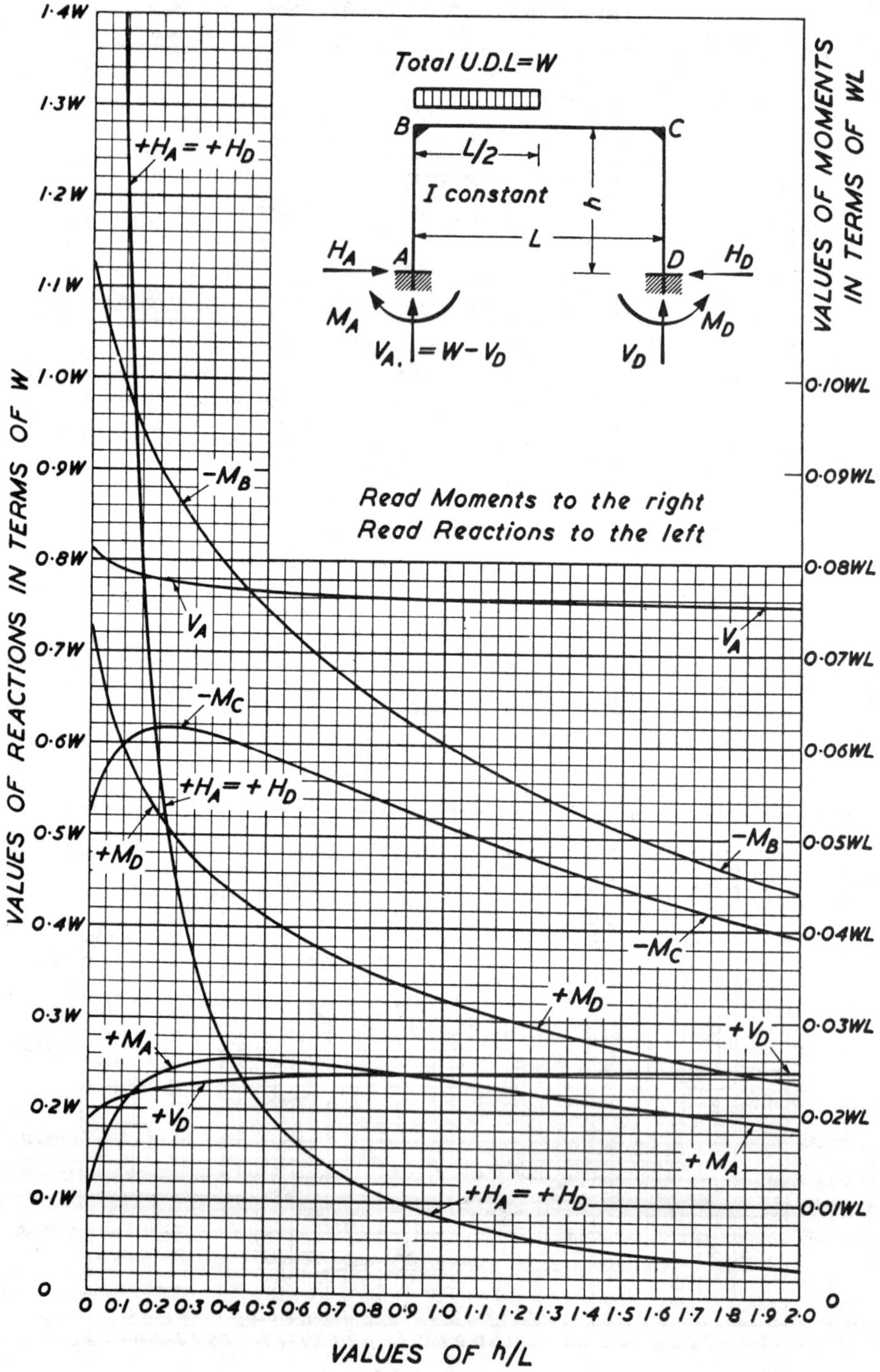
Total U.D.L=W
B
C
L/2
I constant
h
L
H_A
A
D
H_D
M_A
M_D
$V_A = W - V_D$
V_D
Read Moments to the right
Read Reactions to the left
$+H_A = +H_D$
$-M_B$
V_A
$-M_C$
$+M_D$
$+M_A$
$+V_D$
VALUES OF REACTIONS IN TERMS OF W
VALUES OF MOMENTS IN TERMS OF WL
VALUES OF h/L
0
0·1W
0·2W
0·3W
0·4W
0·5W
0·6W
0·7W
0·8W
0·9W
1·0W
1·1W
1·2W
1·3W
1·4W
0·01WL
0·02WL
0·03WL
0·04WL
0·05WL
0·06WL
0·07WL
0·08WL
0·09WL
0·10WL
0 0·1 0·2 0·3 0·4 0·5 0·6 0·7 0·8 0·9 1·0 1·1 1·2 1·3 1·4 1·5 1·6 1·7 1·8 1·9 2·0

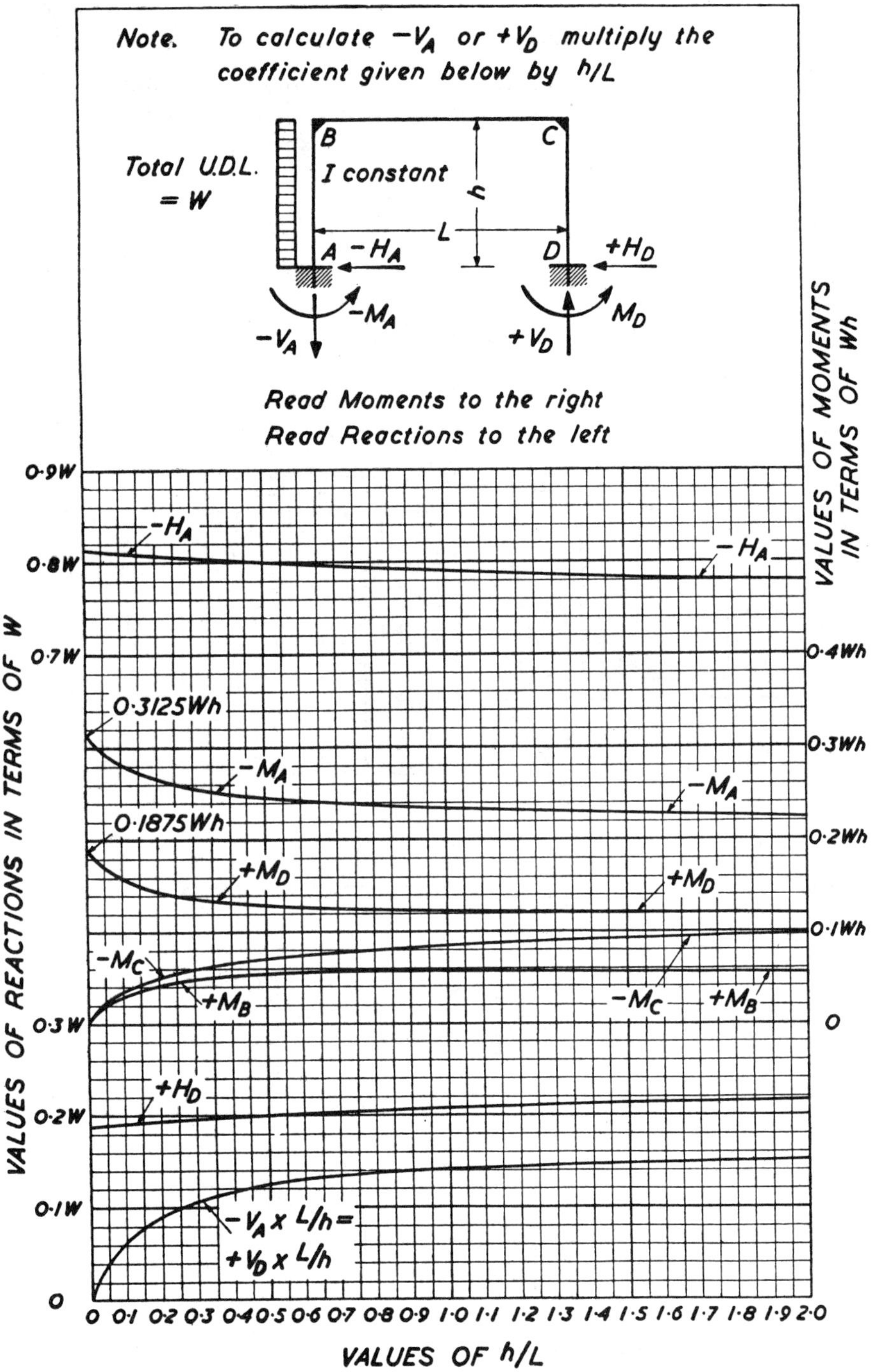
Note. To calculate $-V_A$ or $+V_D$ multiply the coefficient given below by h/L
Total U.D.L. = W
I constant
B
C
h
L
A
$-H_A$
D
$+H_D$
$-M_A$
M_D
$-V_A$
$+V_D$
Read Moments to the right
Read Reactions to the left
$-H_A$
0·3125Wh
$-M_A$
0·1875Wh
$+M_D$
$-M_C$
$+M_B$
$+H_D$
$-V_A \times L/h = +V_D \times L/h$
0·9W
0·8W
0·7W
0·3W
0·2W
0·1W
0
0·4Wh
0·3Wh
0·2Wh
0·1Wh
0
VALUES OF REACTIONS IN TERMS OF W
VALUES OF MOMENTS IN TERMS OF Wh
0 0·1 0·2 0·3 0·4 0·5 0·6 0·7 0·8 0·9 1·0 1·1 1·2 1·3 1·4 1·5 1·6 1·7 1·8 1·9 2·0
VALUES OF h/L

348

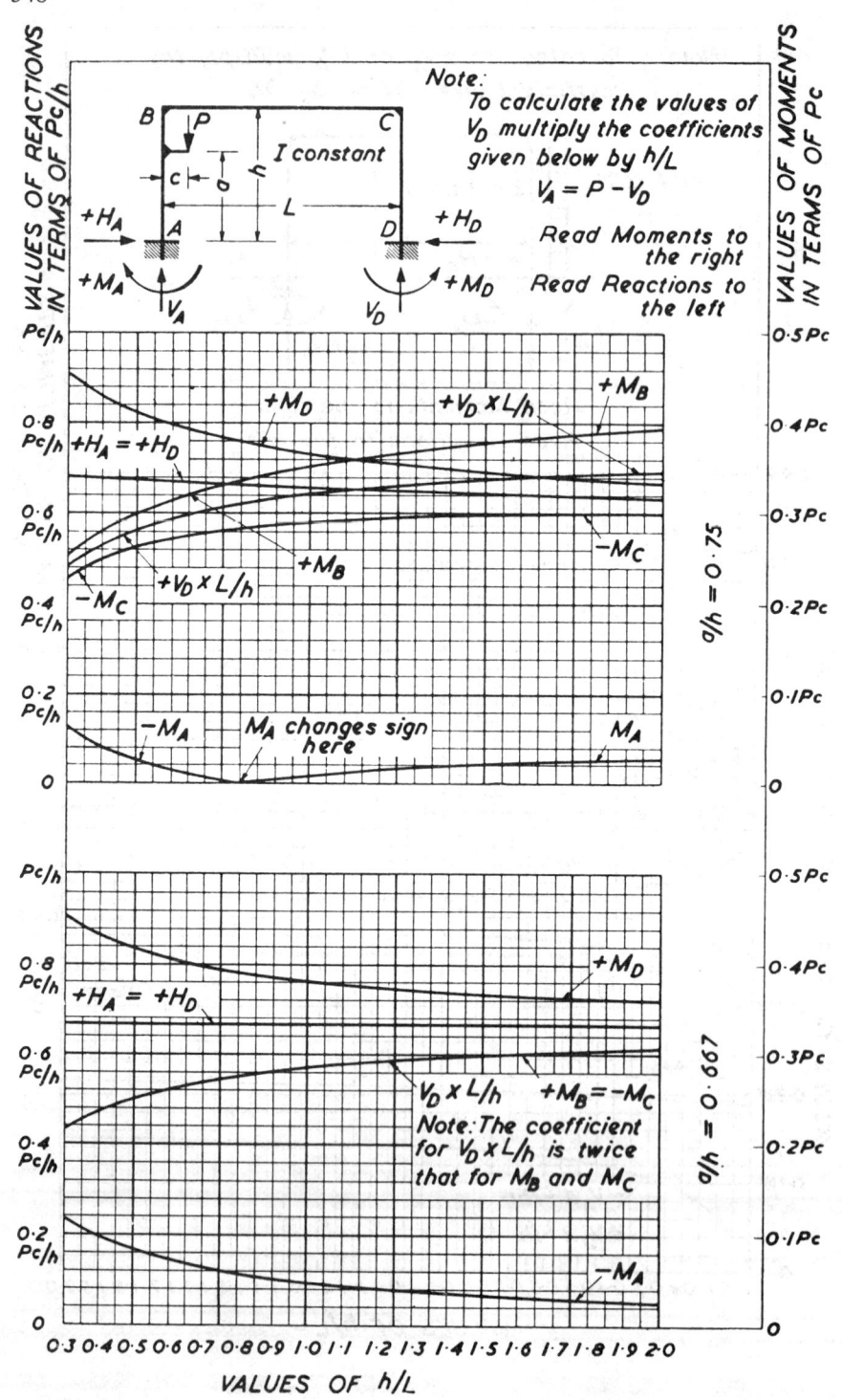

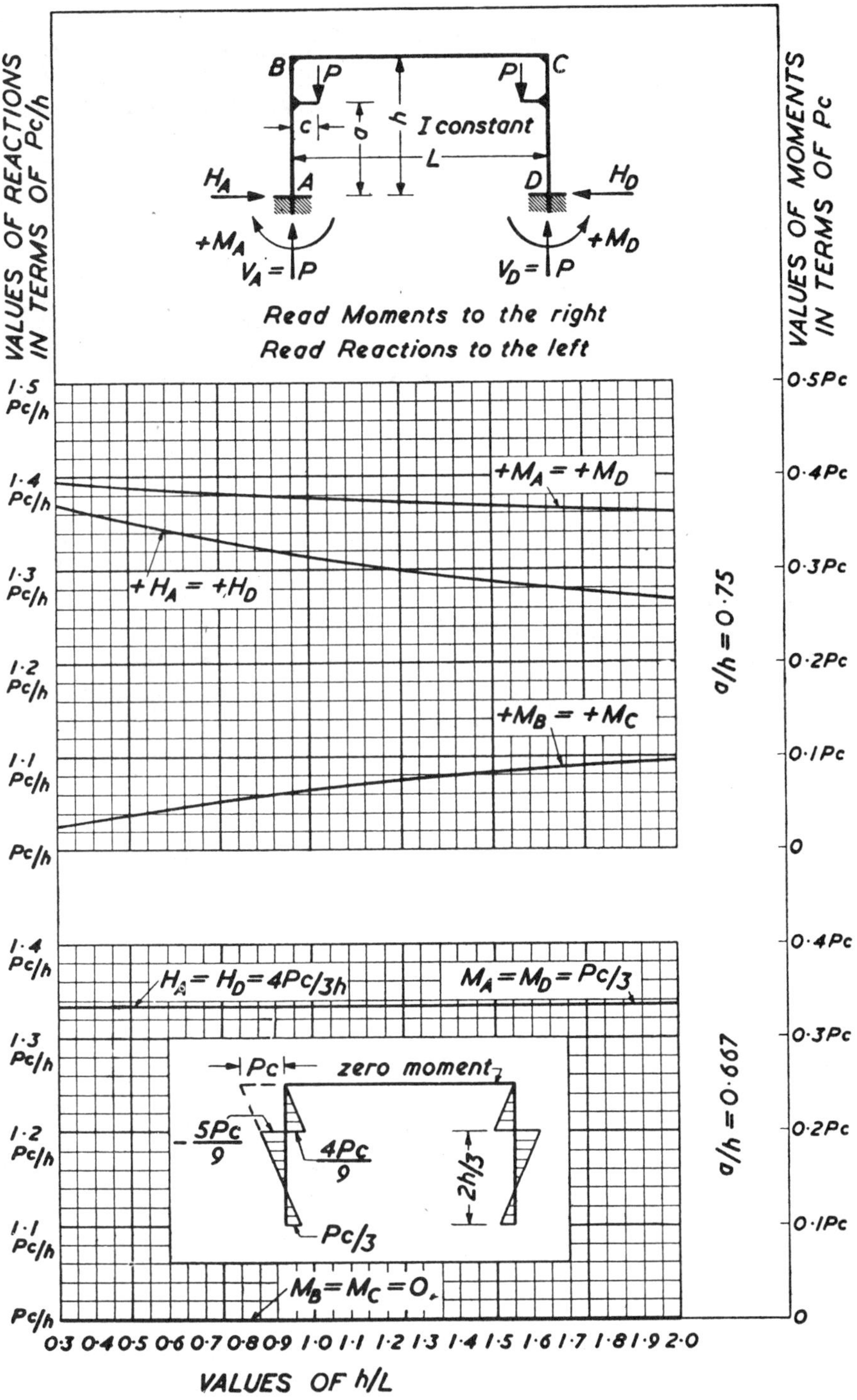
VALUES OF REACTIONS IN TERMS OF Pc/h
VALUES OF MOMENTS IN TERMS OF Pc
B
C
P
P
c
a
h
I constant
L
HA
A
D
HD
+MA
+MD
VA = P
VD = P
Read Moments to the right
Read Reactions to the left
1·5 Pc/h
1·4 Pc/h
1·3 Pc/h
1·2 Pc/h
1·1 Pc/h
Pc/h
0·5Pc
0·4Pc
0·3Pc
0·2Pc
0·1Pc
0
+MA = +MD
+HA = +HD
+MB = +MC
a/h = 0·75
HA = HD = 4Pc/3h
MA = MD = Pc/3
a/h = 0·667
Pc
zero moment
-5Pc/9
4Pc/9
2h/3
Pc/3
MB = MC = 0
0·3 0·4 0·5 0·6 0·7 0·8 0·9 1·0 1·1 1·2 1·3 1·4 1·5 1·6 1·7 1·8 1·9 2·0
VALUES OF h/L

350

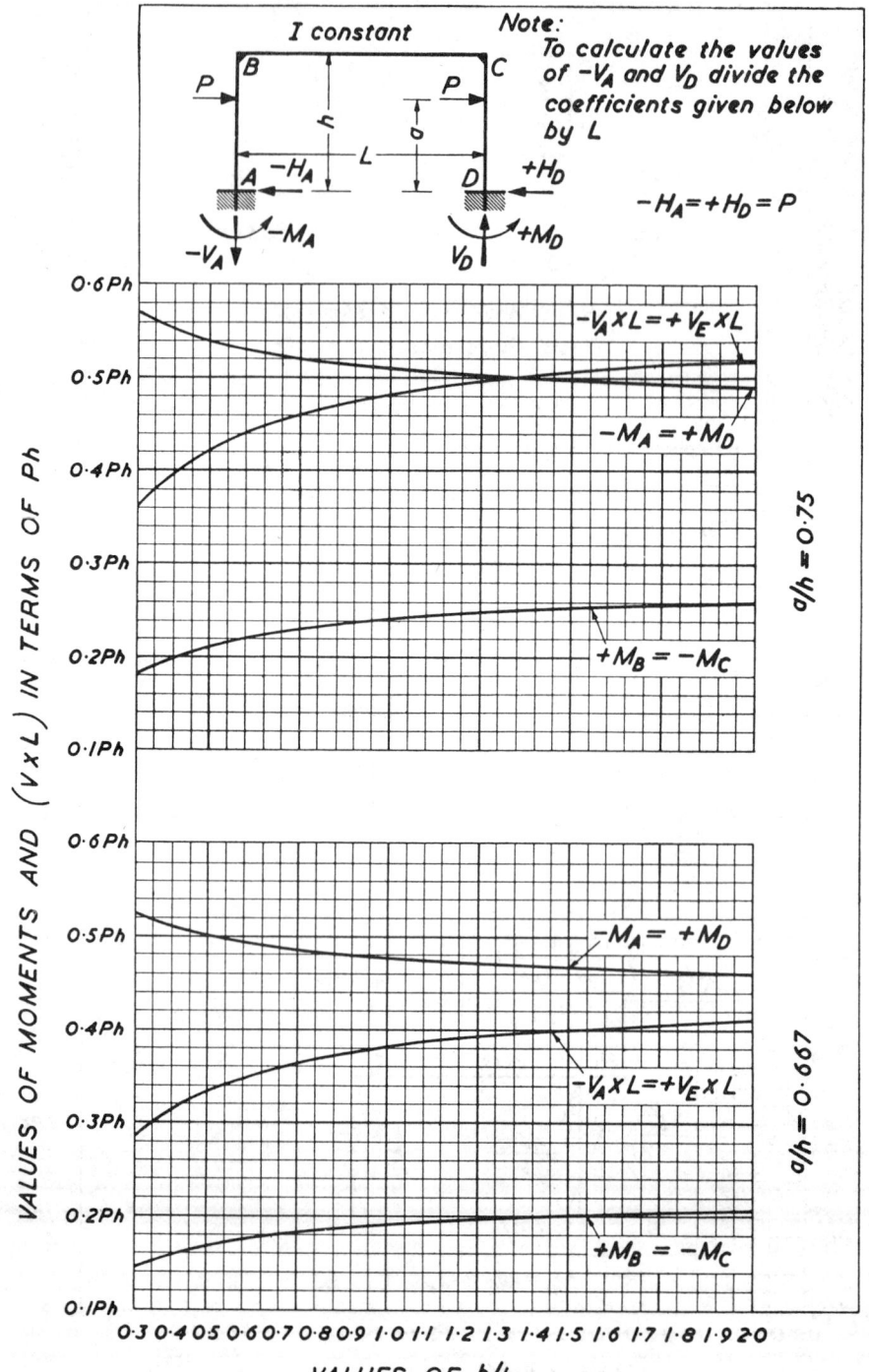

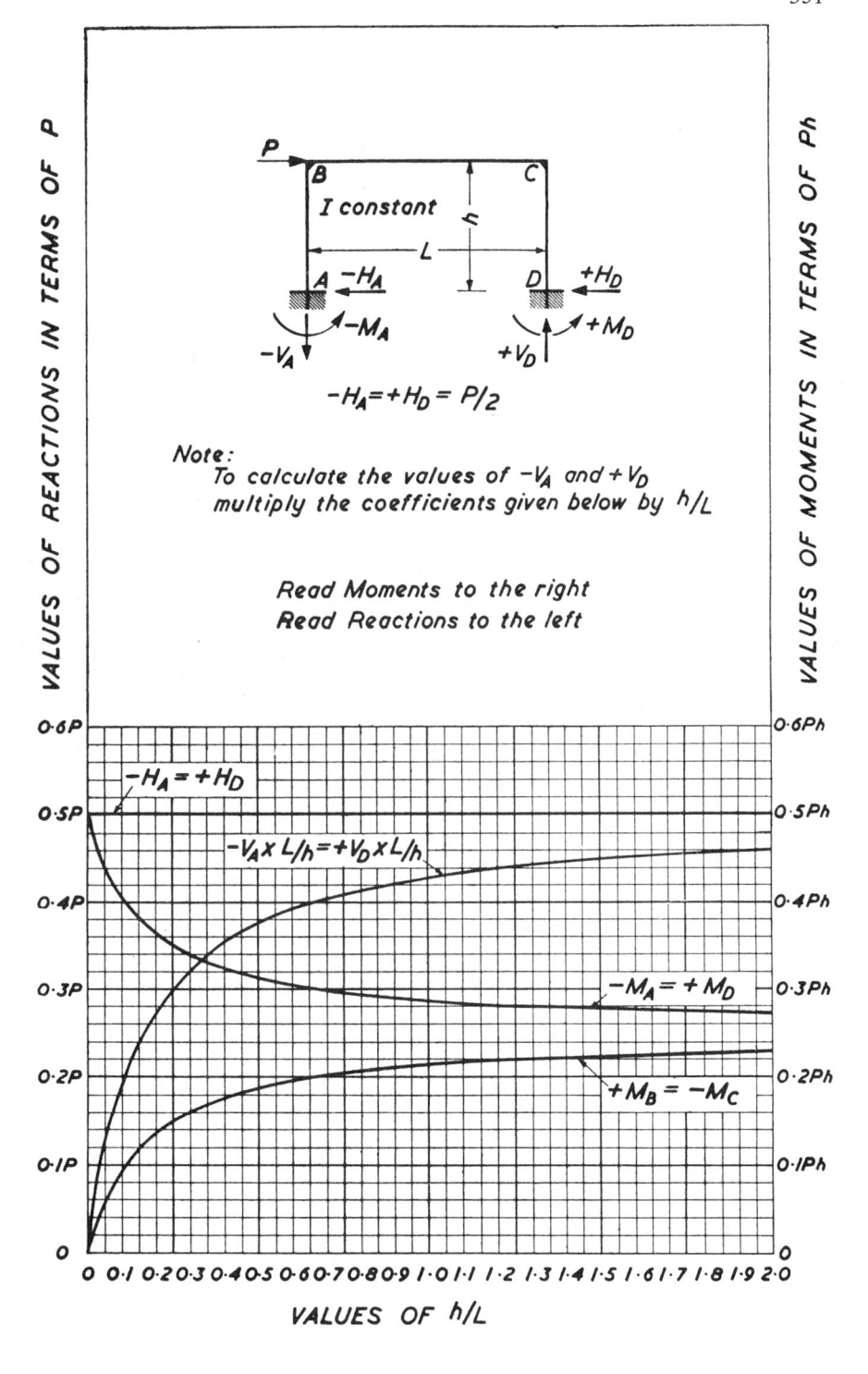

351

VALUES OF REACTIONS IN TERMS OF P

VALUES OF MOMENTS IN TERMS OF Ph

$-H_A = +H_D = P/2$

Note:
To calculate the values of $-V_A$ and $+V_D$ multiply the coefficients given below by h/L

Read Moments to the right
Read Reactions to the left

$-H_A = +H_D$

$-V_A \times L/h = +V_D \times L/h$

$-M_A = +M_D$

$+M_B = -M_C$

VALUES OF h/L

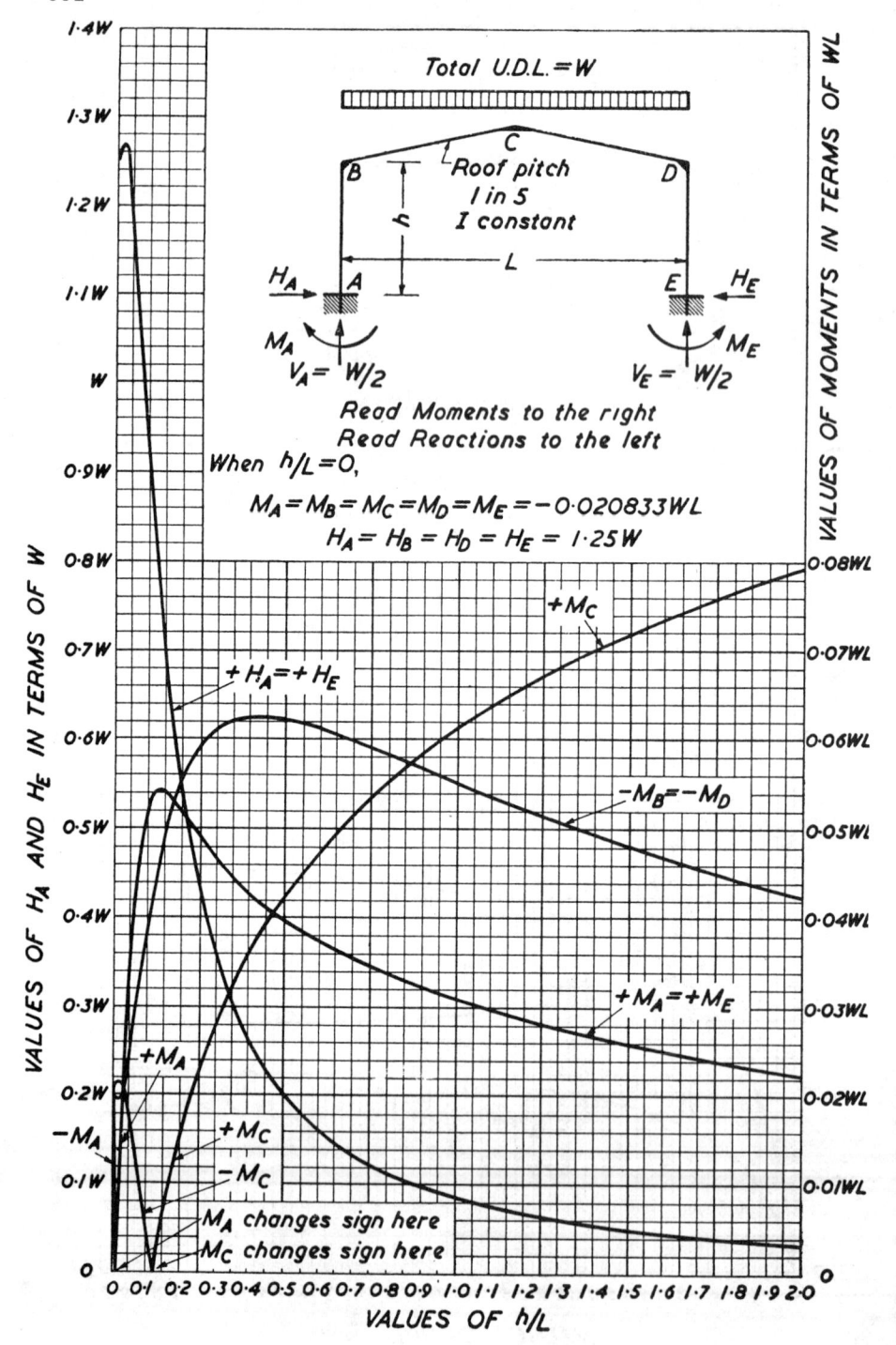

Total U.D.L. = W

Roof pitch 1 in 5
I constant

Read Moments to the right
Read Reactions to the left

When $h/L = 0$,

$M_A = M_B = M_C = M_D = M_E = -0.020833WL$
$H_A = H_B = H_D = H_E = 1.25W$

VALUES OF H_A AND H_E IN TERMS OF W

VALUES OF MOMENTS IN TERMS OF WL

VALUES OF h/L

$+M_C$
$+H_A = +H_E$
$-M_B = -M_D$
$+M_A = +M_E$
$+M_A$
$-M_A$
$+M_C$
$-M_C$

M_A changes sign here
M_C changes sign here

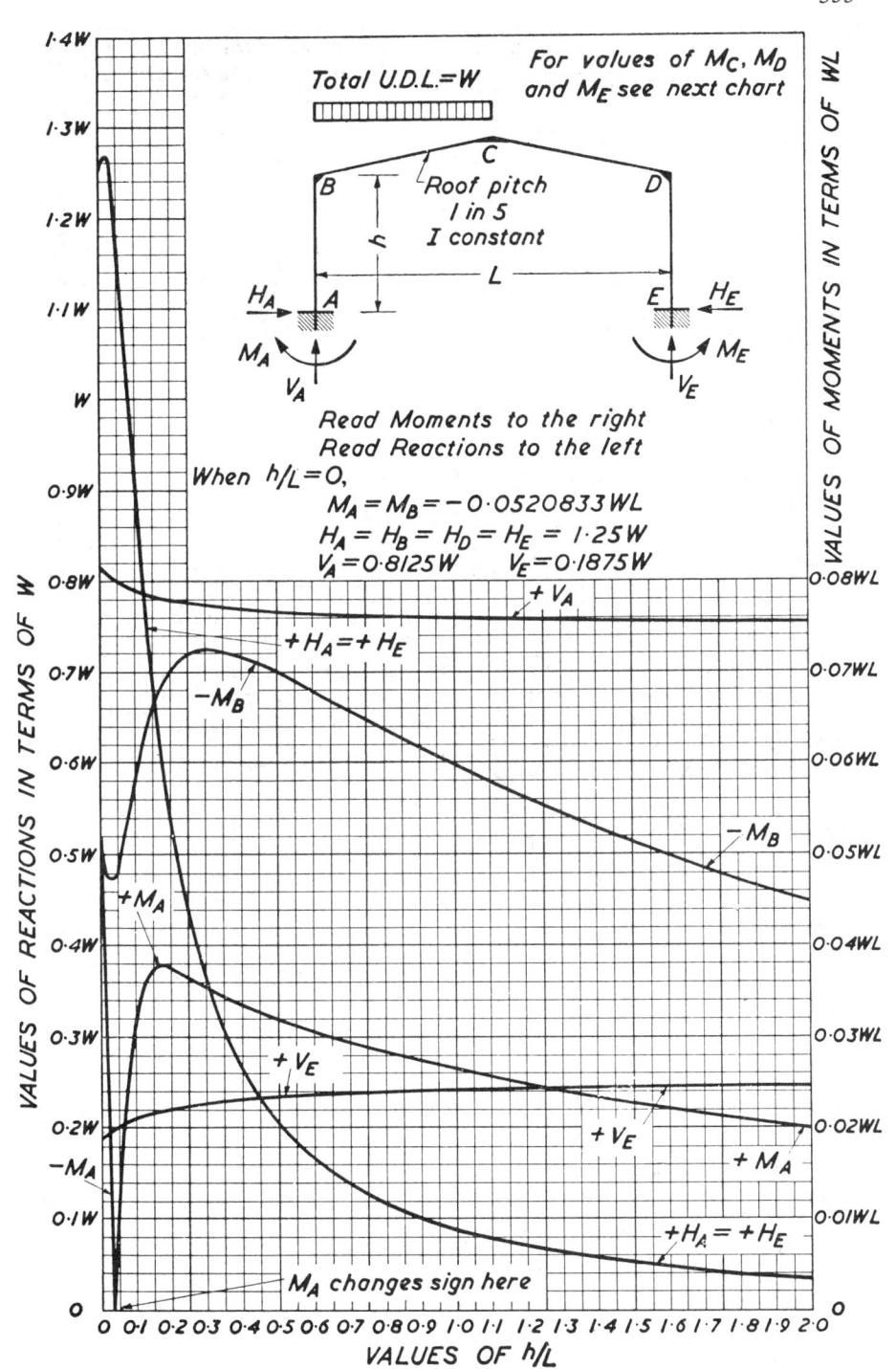

353

Total U.D.L. = W

For values of M_C, M_D and M_E see next chart

Roof pitch 1 in 5
I constant

Read Moments to the right
Read Reactions to the left

When $h/L = 0$,
$M_A = M_B = -0.0520833\,WL$
$H_A = H_B = H_D = H_E = 1.25\,W$
$V_A = 0.8125\,W \qquad V_E = 0.1875\,W$

$+V_A$
$+H_A = +H_E$
$-M_B$
$-M_B$
$+M_A$
$+V_E$
$+V_E$
$+M_A$
$-M_A$
$+H_A = +H_E$

M_A changes sign here

VALUES OF h/L

VALUES OF REACTIONS IN TERMS OF W

VALUES OF MOMENTS IN TERMS OF WL

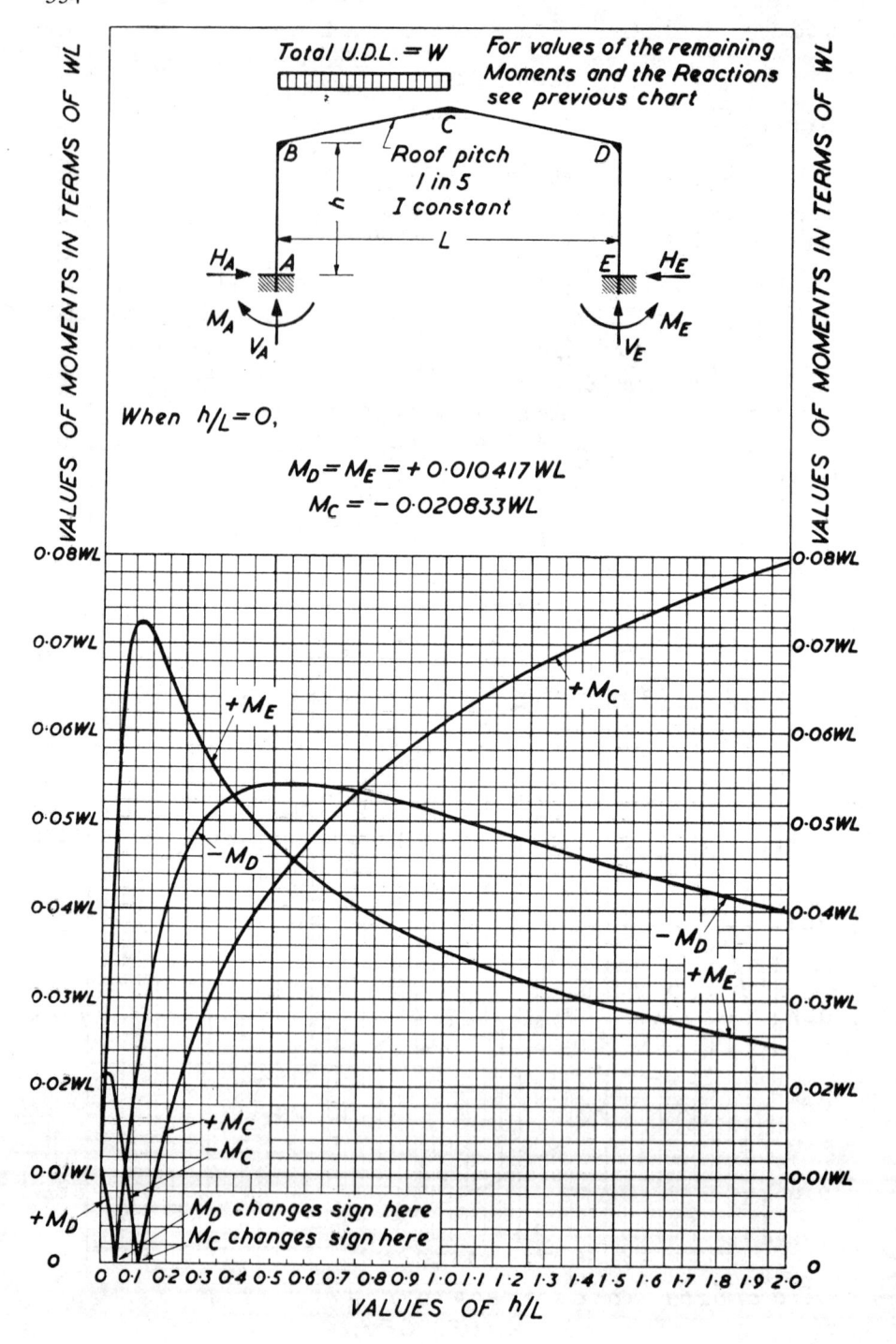

VALUES OF MOMENTS IN TERMS OF WL

Total U.D.L. = W

For values of the remaining Moments and the Reactions see previous chart

Roof pitch 1 in 5
I constant

When $h/L = 0$,

$$M_D = M_E = +0.010417\,WL$$
$$M_C = -0.020833\,WL$$

M_D changes sign here
M_C changes sign here

VALUES OF h/L

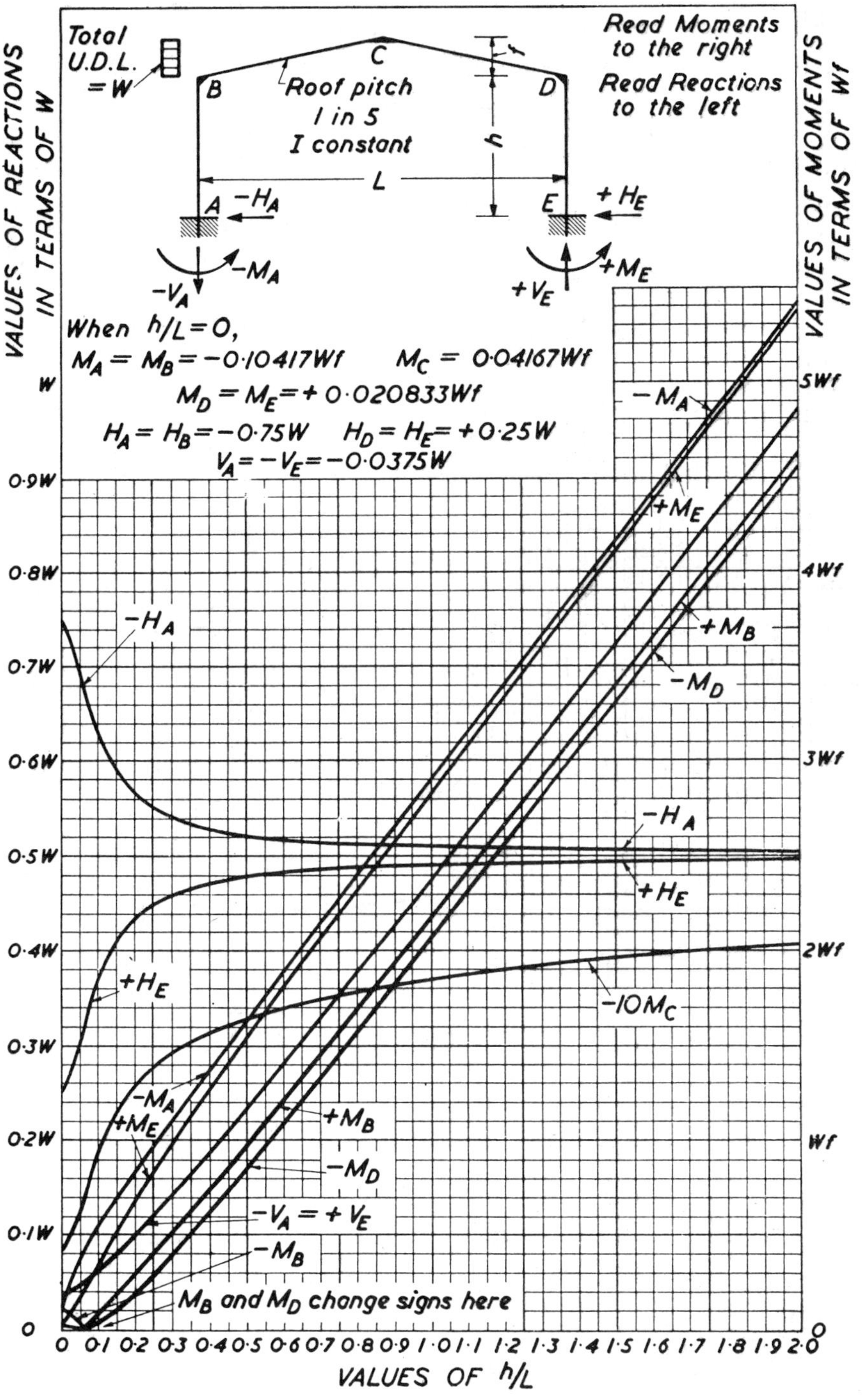

Total U.D.L. = W
Roof pitch 1 in 5
I constant
B
C
D
A
E
L
h
f
−H_A
+H_E
−M_A
+M_E
−V_A
+V_E
Read Moments to the right
Read Reactions to the left
VALUES OF REACTIONS IN TERMS OF W
VALUES OF MOMENTS IN TERMS OF Wf
When h/L = 0,
M_A = M_B = −0·10417Wf   M_C = 0·04167Wf
M_D = M_E = + 0·020833Wf
H_A = H_B = −0·75W   H_D = H_E = +0·25W
V_A = −V_E = −0·0375W
W
0·9W
0·8W
0·7W
0·6W
0·5W
0·4W
0·3W
0·2W
0·1W
0
5Wf
4Wf
3Wf
2Wf
Wf
0
−M_A
+M_E
+M_B
−M_D
−H_A
+H_E
−10M_C
−V_A = + V_E
−M_B
M_B and M_D change signs here
0 0·1 0·2 0·3 0·4 0·5 0·6 0·7 0·8 0·9 1·0 1·1 1·2 1·3 1·4 1·5 1·6 1·7 1·8 1·9 2·0
VALUES OF h/L

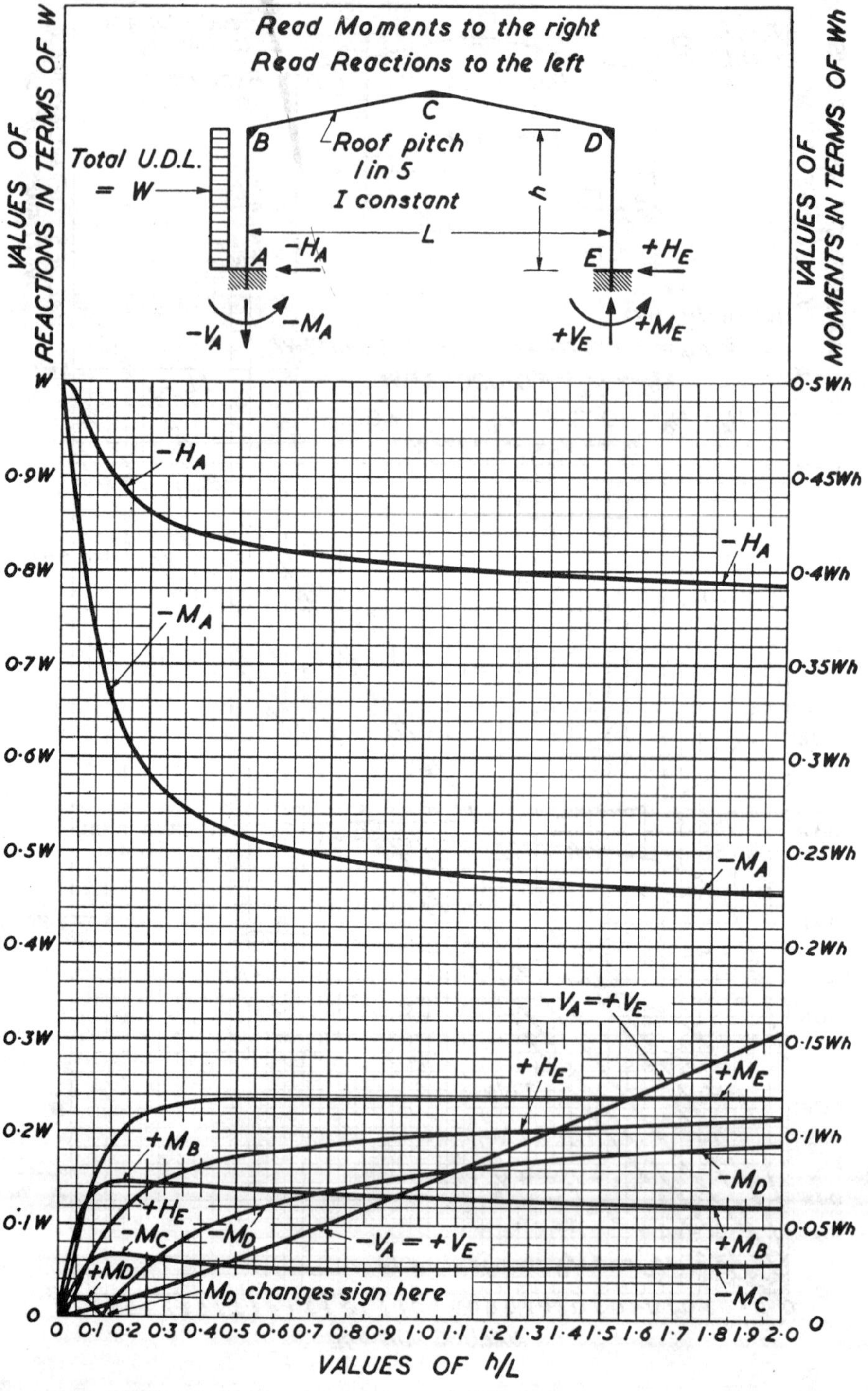
Read Moments to the right
Read Reactions to the left
Total U.D.L. = W
Roof pitch 1 in 5
I constant
B
C
D
A
E
L
h
$-H_A$
$+H_E$
$-V_A$
$-M_A$
$+V_E$
$+M_E$
VALUES OF REACTIONS IN TERMS OF W
VALUES OF MOMENTS IN TERMS OF Wh
W
0·9W
0·8W
0·7W
0·6W
0·5W
0·4W
0·3W
0·2W
0·1W
0
0·5Wh
0·45Wh
0·4Wh
0·35Wh
0·3Wh
0·25Wh
0·2Wh
0·15Wh
0·1Wh
0·05Wh
0
$-H_A$
$-M_A$
$-V_A = +V_E$
$+H_E$
$+M_E$
$+M_B$
$-M_D$
$-M_C$
$+M_D$
M_D changes sign here
0 0·1 0·2 0·3 0·4 0·5 0·6 0·7 0·8 0·9 1·0 1·1 1·2 1·3 1·4 1·5 1·6 1·7 1·8 1·9 2·0
VALUES OF h/L

357

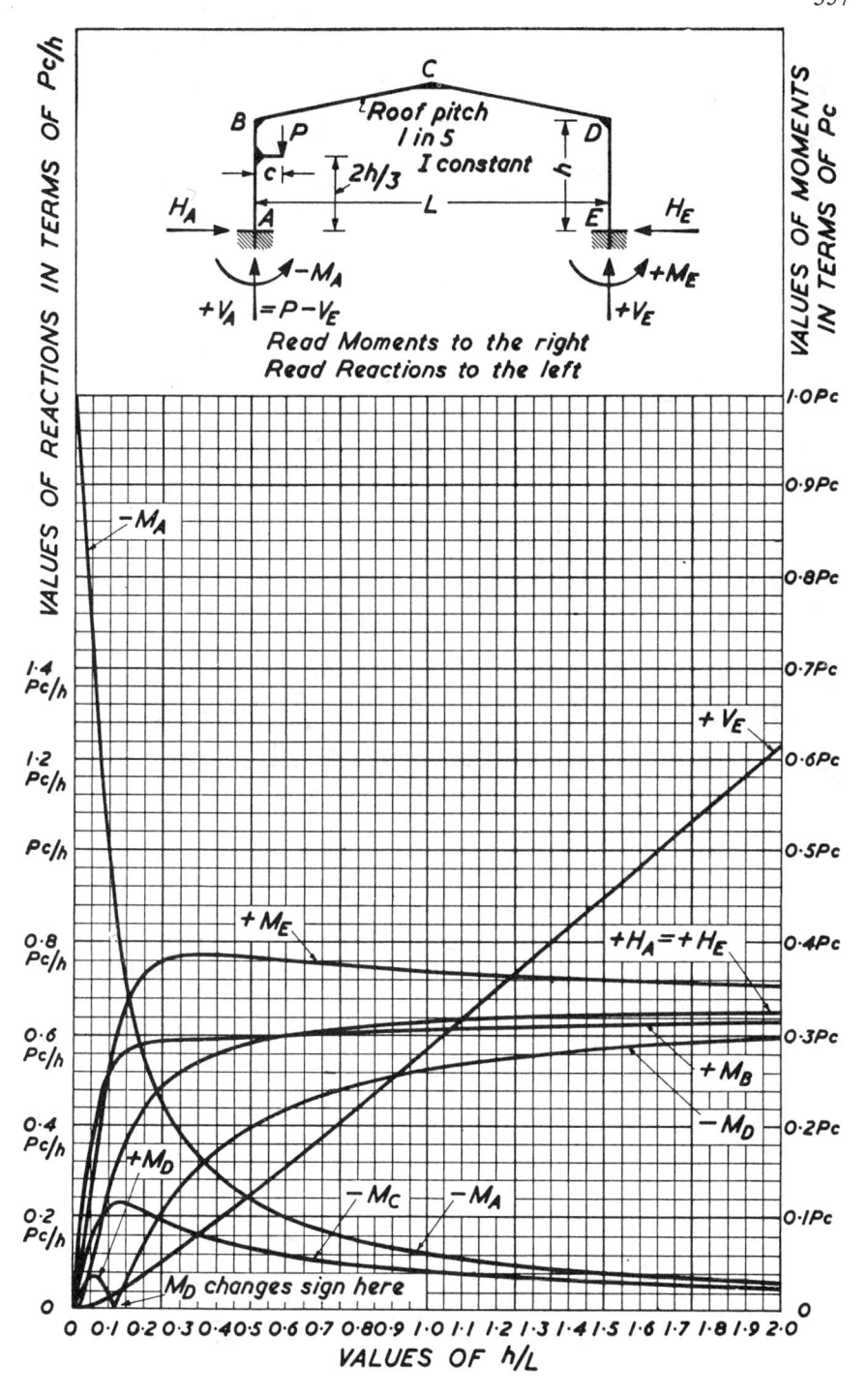

358

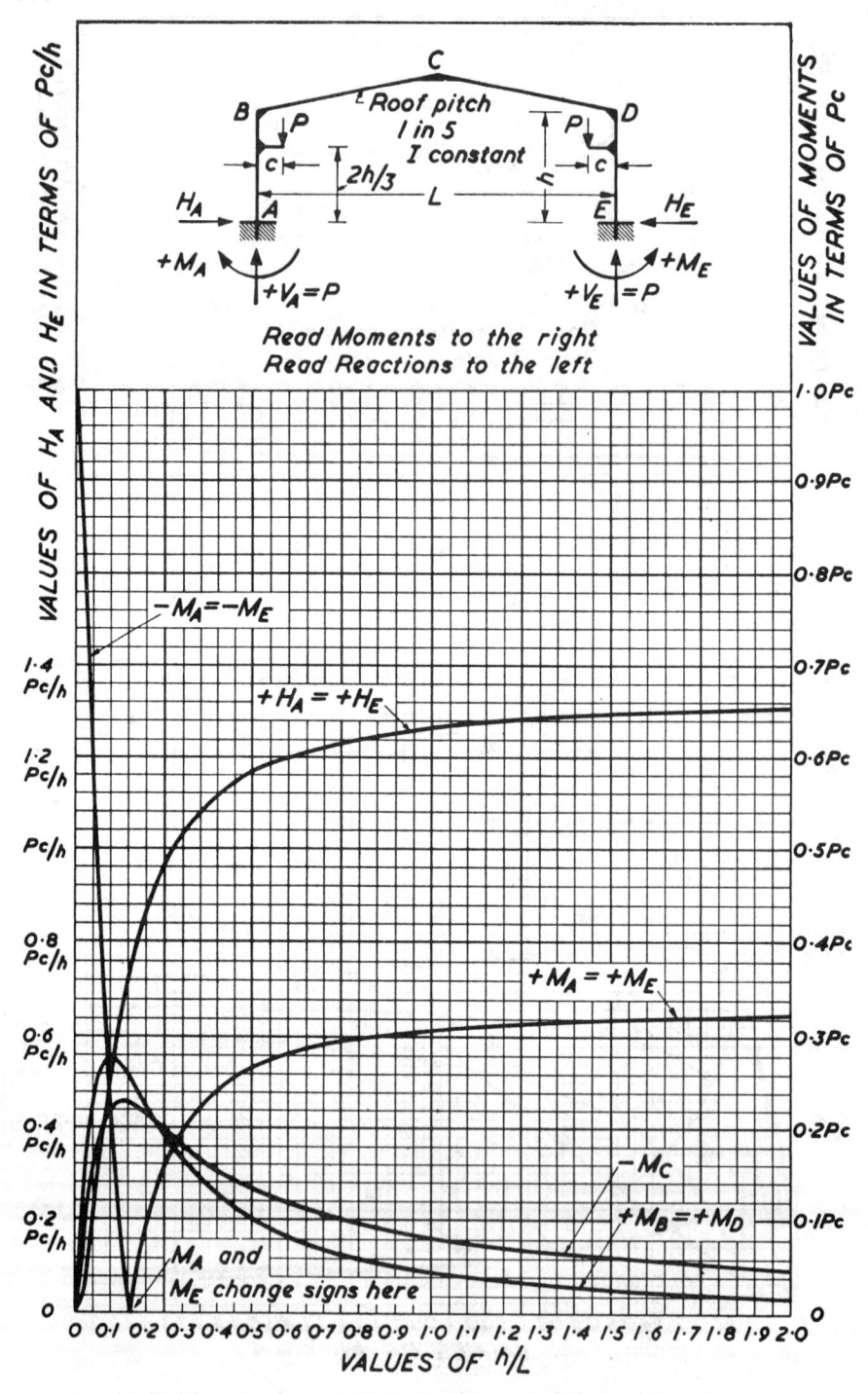

359

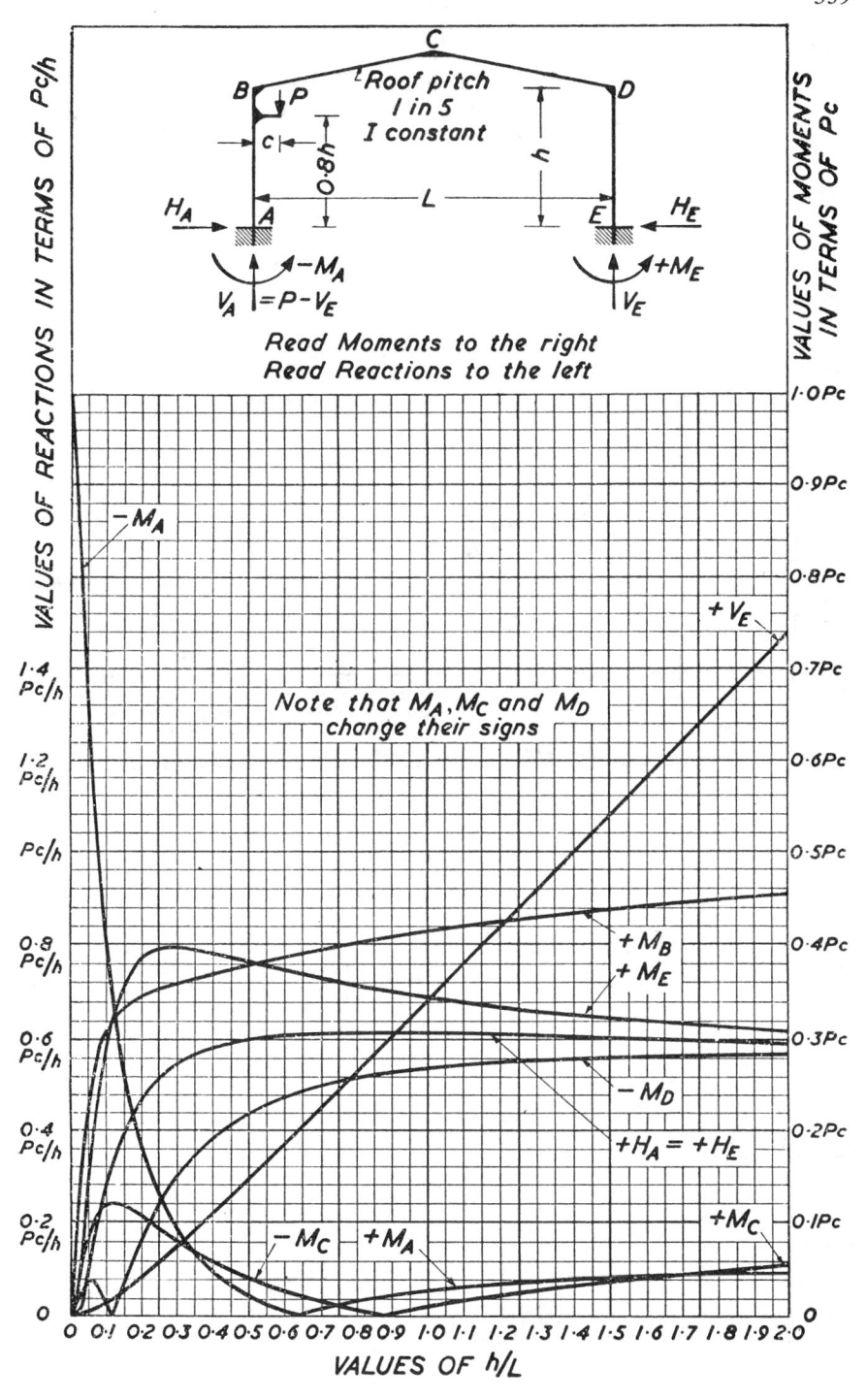

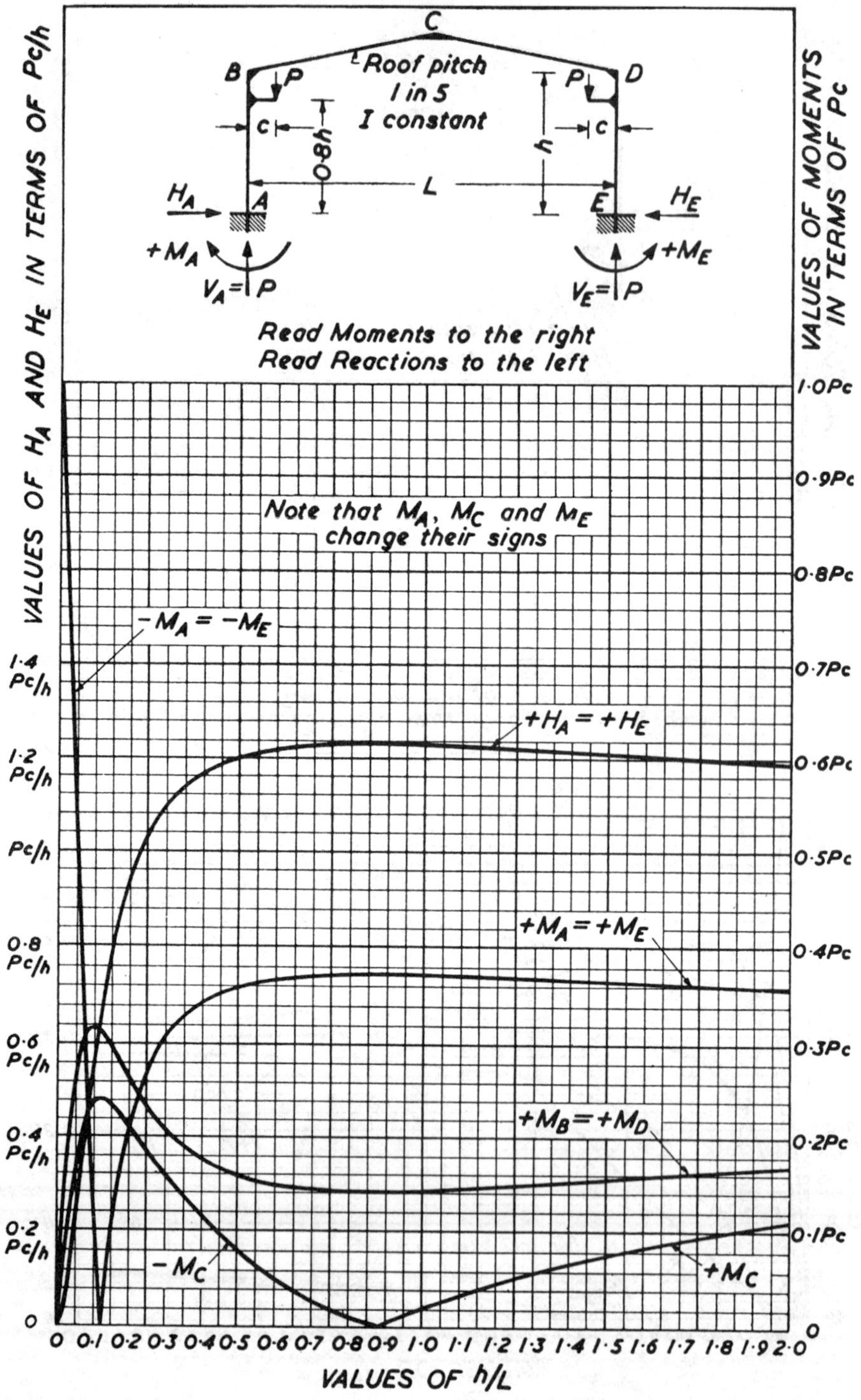
C
B
P
Roof pitch
1 in 5
I constant
P
D
c
0·8h
h
c
L
H_A
A
E
H_E
$+M_A$
$V_A = P$
$V_E = P$
$+M_E$
Read Moments to the right
Read Reactions to the left
Note that M_A, M_C and M_E change their signs
$-M_A = -M_E$
$+H_A = +H_E$
$+M_A = +M_E$
$+M_B = +M_D$
$-M_C$
$+M_C$
VALUES OF H_A AND H_E IN TERMS OF Pc/h
VALUES OF MOMENTS IN TERMS OF Pc
1·4 Pc/h
1·2 Pc/h
Pc/h
0·8 Pc/h
0·6 Pc/h
0·4 Pc/h
0·2 Pc/h
0
1·0Pc
0·9Pc
0·8Pc
0·7Pc
0·6Pc
0·5Pc
0·4Pc
0·3Pc
0·2Pc
0·1Pc
0
0 0·1 0·2 0·3 0·4 0·5 0·6 0·7 0·8 0·9 1·0 1·1 1·2 1·3 1·4 1·5 1·6 1·7 1·8 1·9 2·0
VALUES OF h/L

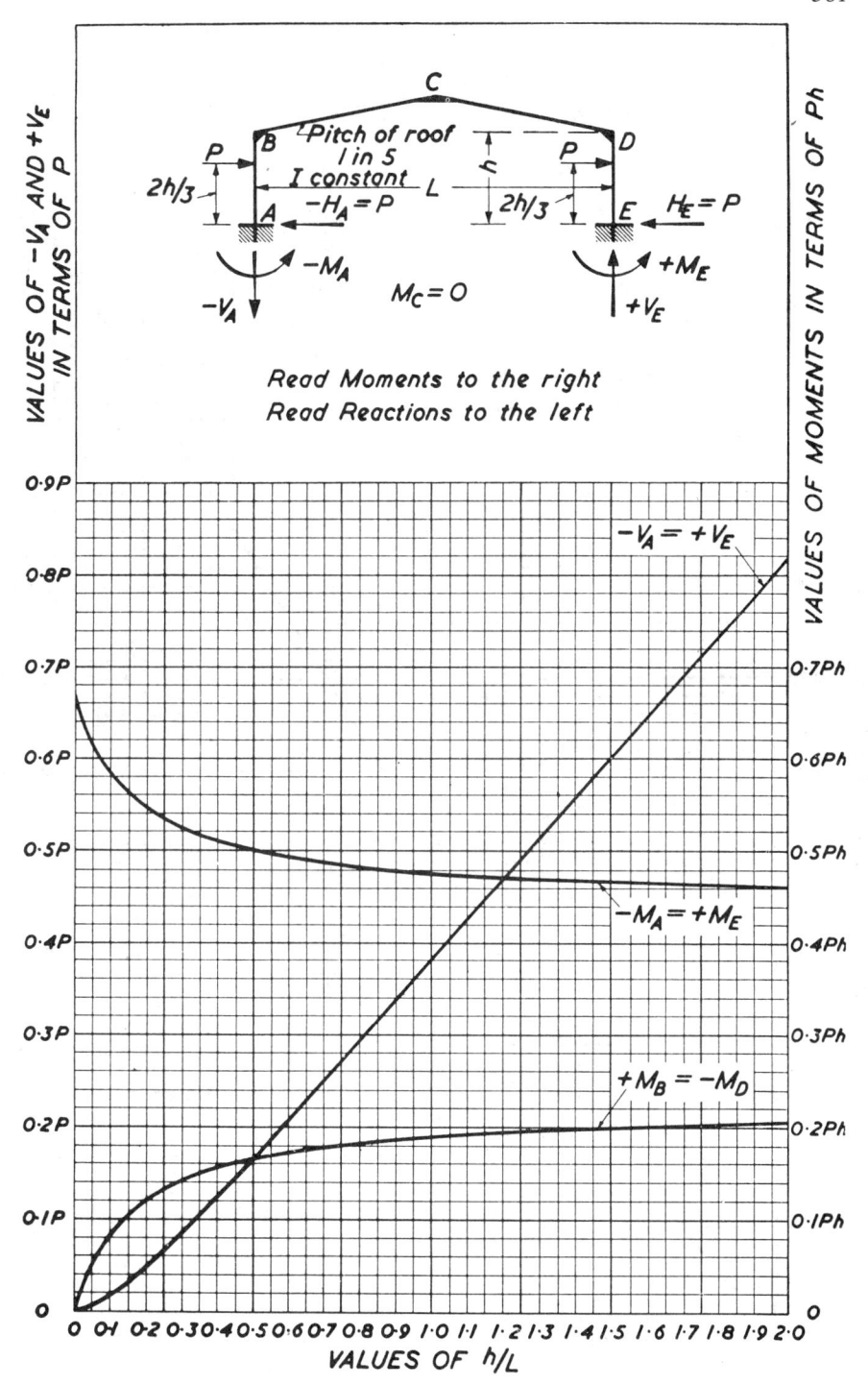

361

VALUES OF $-V_A$ AND $+V_E$ IN TERMS OF P

VALUES OF MOMENTS IN TERMS OF Ph

Pitch of roof 1 in 5
I constant
$-H_A = P$
$H_E = P$
$-M_A$
$+M_E$
$-V_A$
$+V_E$
$M_C = 0$

Read Moments to the right
Read Reactions to the left

$-V_A = +V_E$

$-M_A = +M_E$

$+M_B = -M_D$

VALUES OF h/L

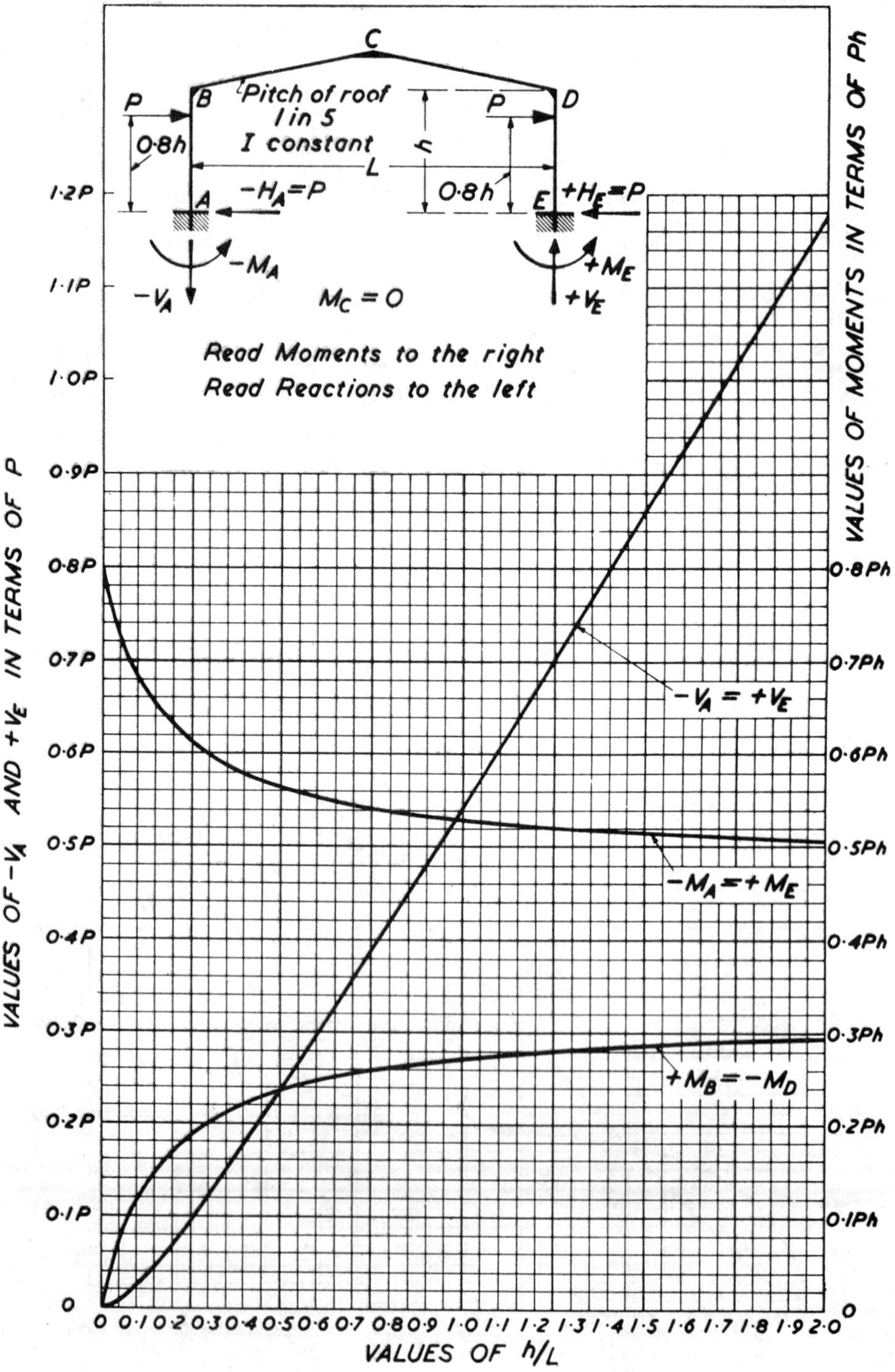
C
B
D
P
P
Pitch of roof
1 in 5
I constant
0·8h
0·8h
h
L
A
E
$-H_A = P$
$+H_E = P$
$-M_A$
$+M_E$
$-V_A$
$+V_E$
$M_C = 0$
Read Moments to the right
Read Reactions to the left
$-V_A = +V_E$
$-M_A = +M_E$
$+M_B = -M_D$
VALUES OF $-V_A$ AND $+V_E$ IN TERMS OF P
VALUES OF MOMENTS IN TERMS OF Ph
VALUES OF h/L
0 0·1 0·2 0·3 0·4 0·5 0·6 0·7 0·8 0·9 1·0 1·1 1·2 1·3 1·4 1·5 1·6 1·7 1·8 1·9 2·0
0 0·1P 0·2P 0·3P 0·4P 0·5P 0·6P 0·7P 0·8P 0·9P 1·0P 1·1P 1·2P
0 0·1Ph 0·2Ph 0·3Ph 0·4Ph 0·5Ph 0·6Ph 0·7Ph 0·8Ph

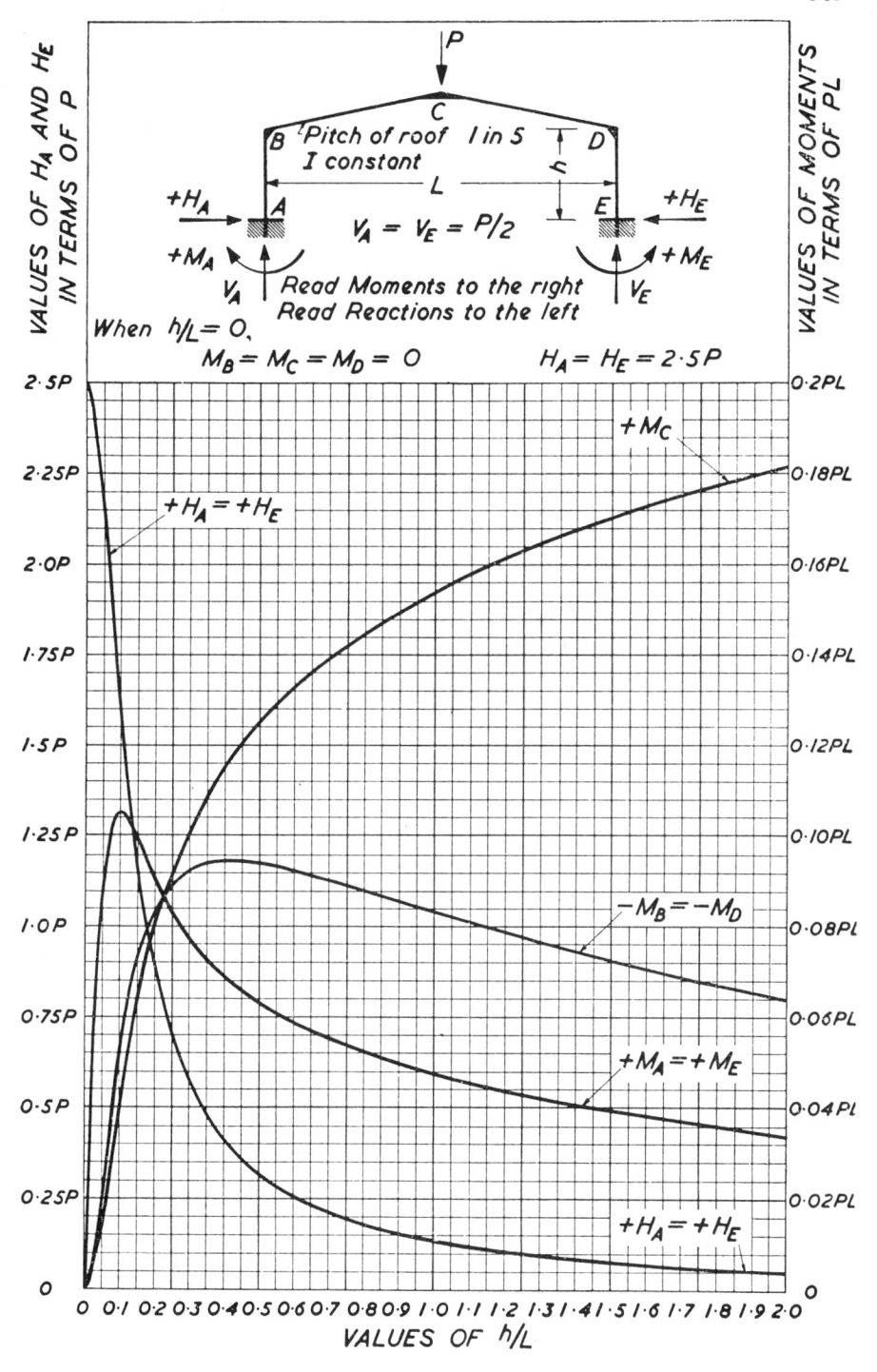

363

VALUES OF H_A AND H_E IN TERMS OF P

VALUES OF MOMENTS IN TERMS OF PL

Pitch of roof 1 in 5
I constant

$V_A = V_E = P/2$

Read Moments to the right
Read Reactions to the left

When $h/L = 0$,
$M_B = M_C = M_D = 0$ $H_A = H_E = 2.5P$

$+M_C$

$+H_A = +H_E$

$-M_B = -M_D$

$+M_A = +M_E$

$+H_A = +H_E$

VALUES OF h/L

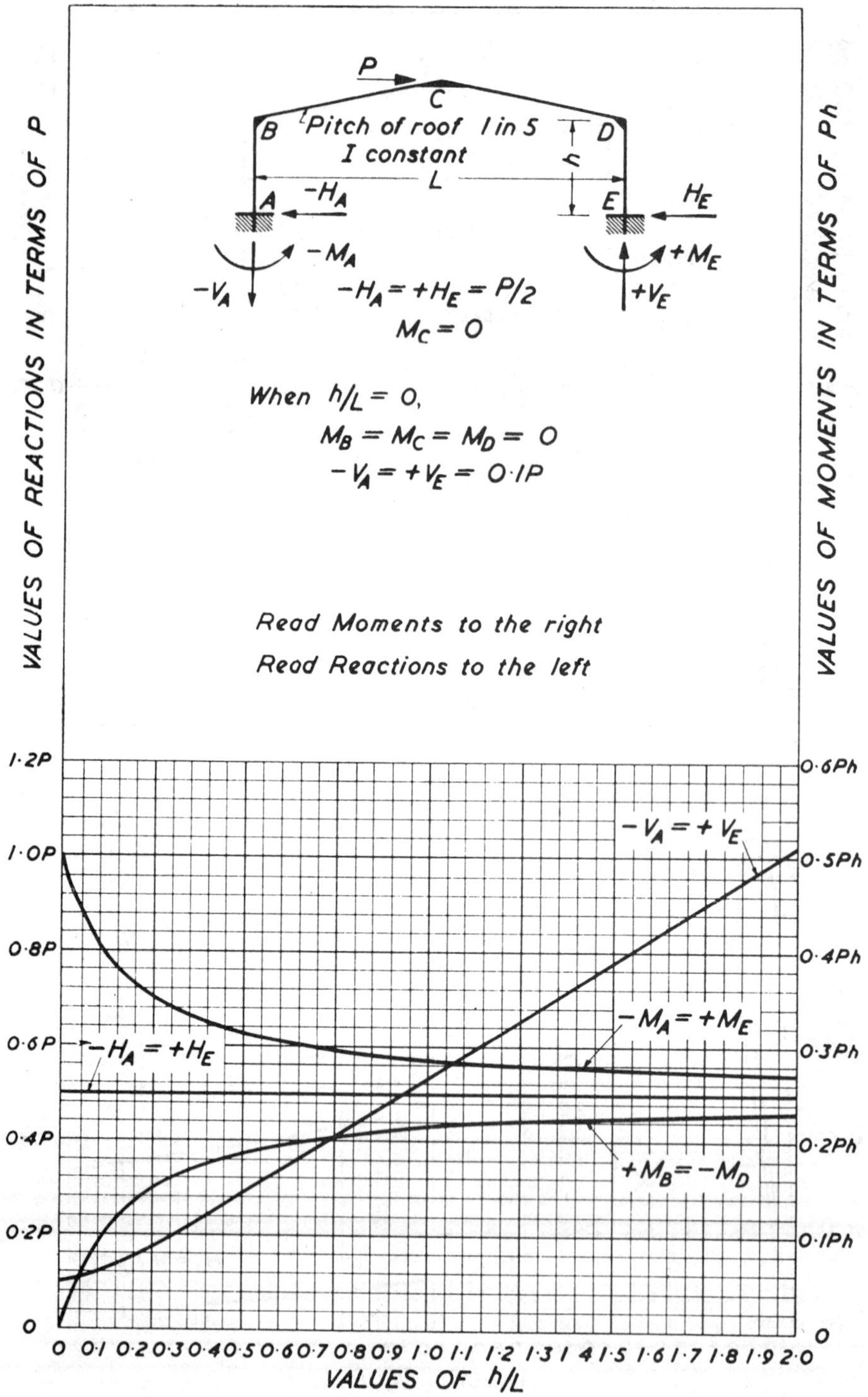

P
C
B
D
Pitch of roof 1 in 5
I constant
h
L
A
E
$-H_A$
H_E
$-M_A$
$+M_E$
$-V_A$
$+V_E$
$-H_A = +H_E = P/2$
$M_C = 0$
When $h/L = 0$,
$M_B = M_C = M_D = 0$
$-V_A = +V_E = 0{\cdot}1P$
Read Moments to the right
Read Reactions to the left
VALUES OF REACTIONS IN TERMS OF P
VALUES OF MOMENTS IN TERMS OF Ph
1·2P
1·0P
0·8P
0·6P
0·4P
0·2P
0
0·6Ph
0·5Ph
0·4Ph
0·3Ph
0·2Ph
0·1Ph
0
$-V_A = +V_E$
$-M_A = +M_E$
$-H_A = +H_E$
$+M_B = -M_D$
0 0·1 0·2 0·3 0·4 0·5 0·6 0·7 0·8 0·9 1·0 1·1 1·2 1·3 1·4 1·5 1·6 1·7 1·8 1·9 2·0
VALUES OF h/L

365

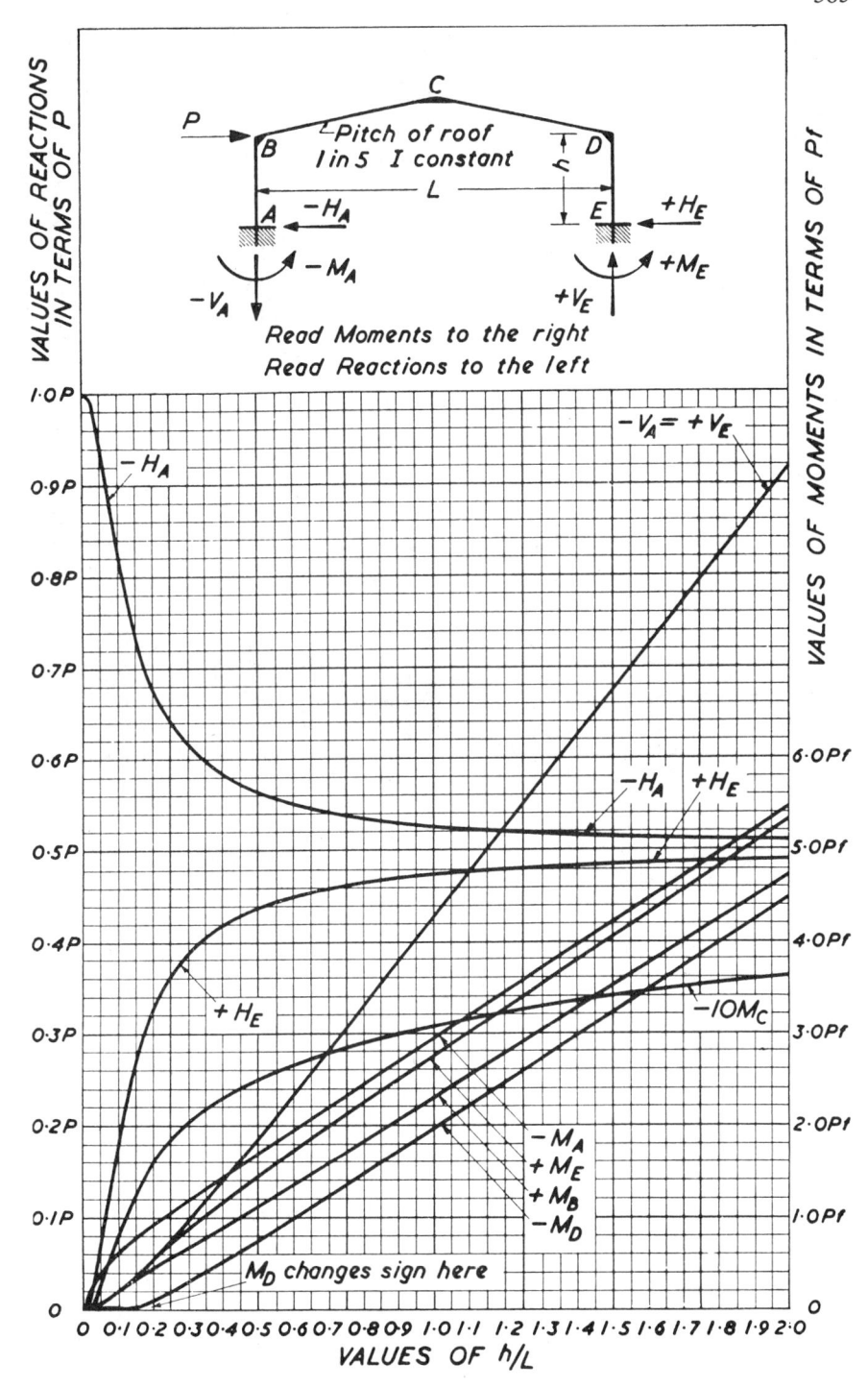

Total U.D.L. = W
Roof pitch 1 in 2·5
I constant

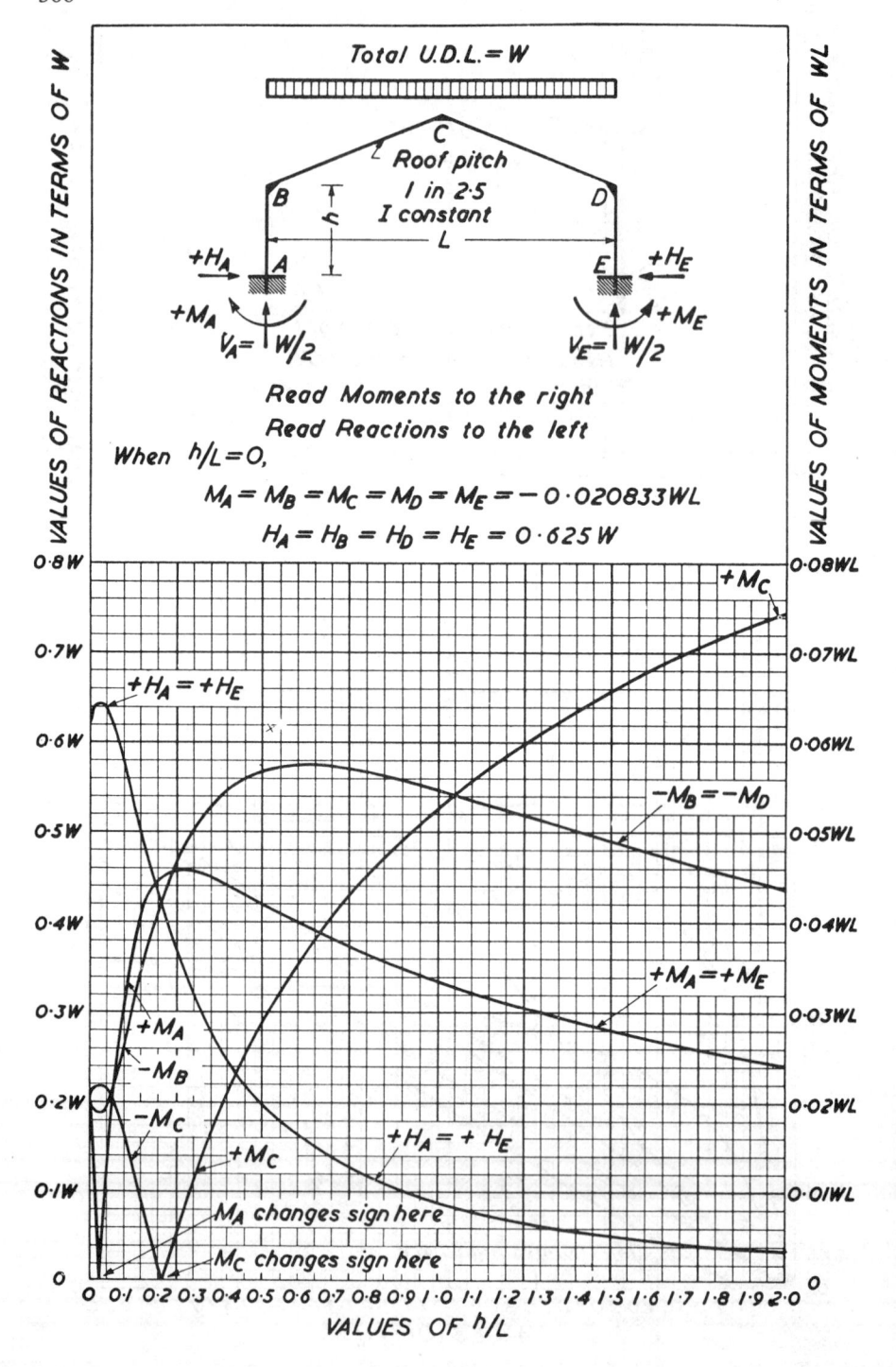

Read Moments to the right
Read Reactions to the left

When $h/L = 0$,

$M_A = M_B = M_C = M_D = M_E = -0.020833 WL$

$H_A = H_B = H_D = H_E = 0.625 W$

367

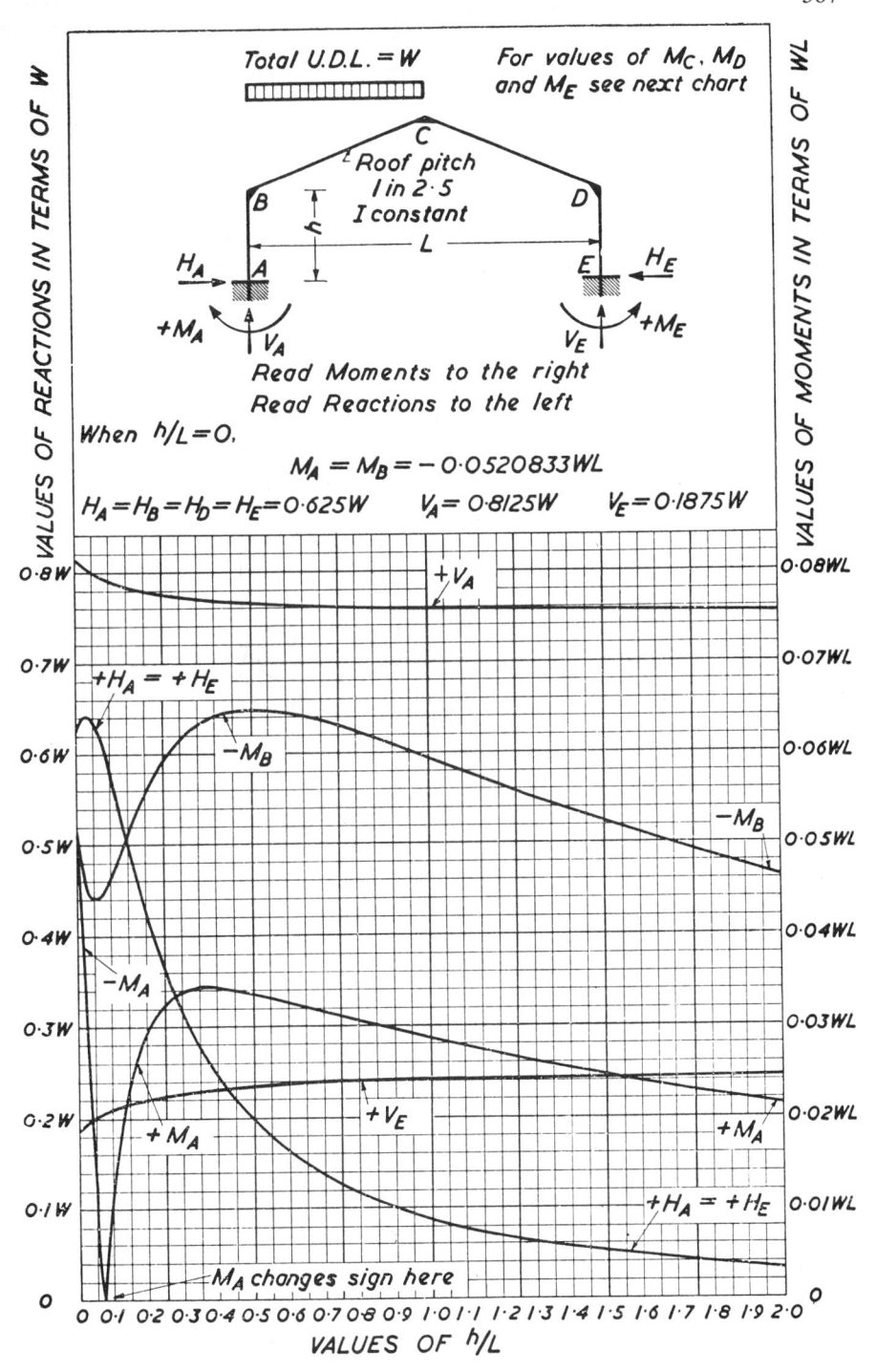

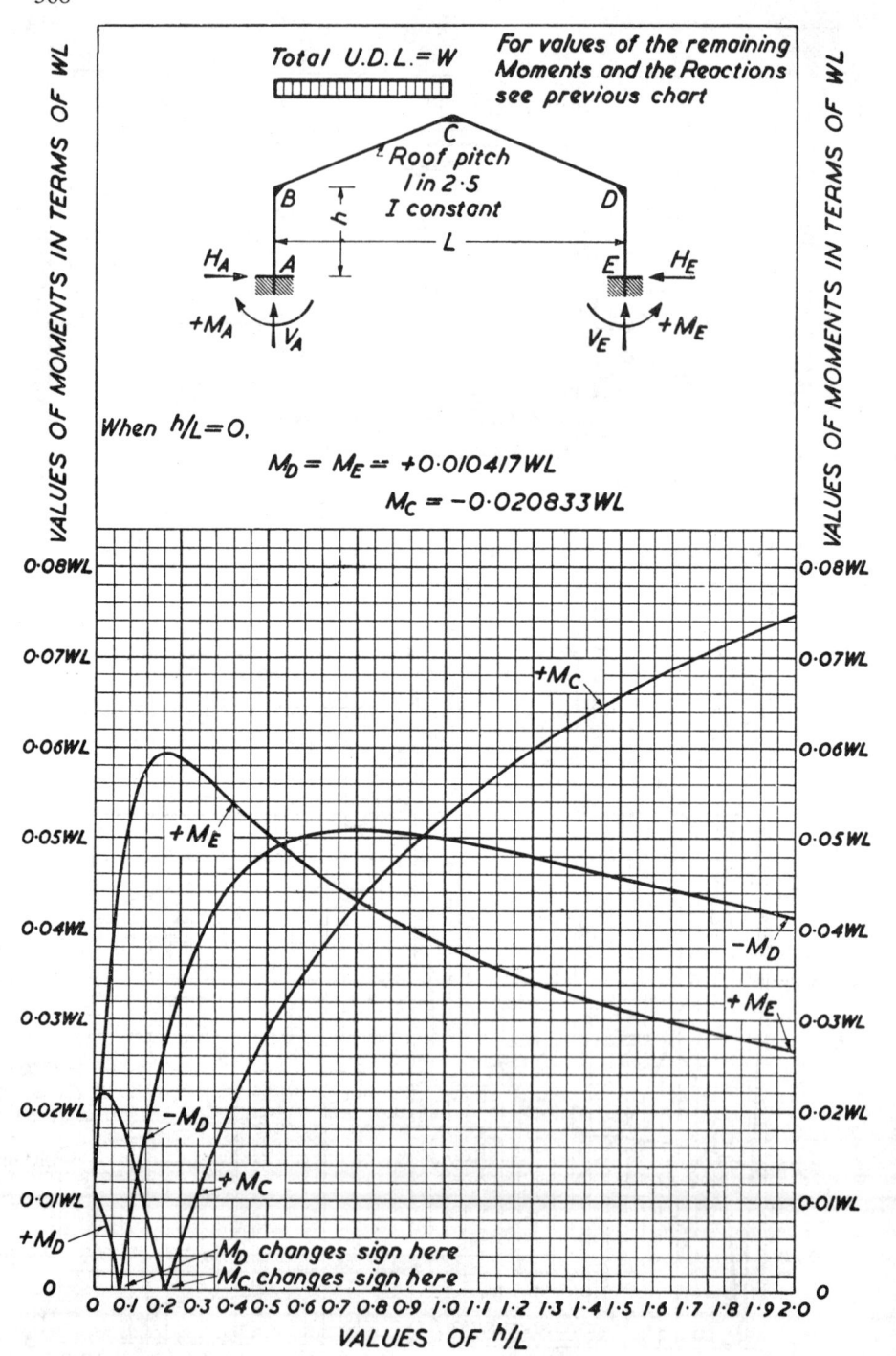

VALUES OF MOMENTS IN TERMS OF WL

Total U.D.L. = W

For values of the remaining Moments and the Reactions see previous chart

Roof pitch 1 in 2·5
I constant

When $h/L = 0$.

$M_D = M_E = +0.010417 WL$

$M_C = -0.020833 WL$

VALUES OF h/L

369

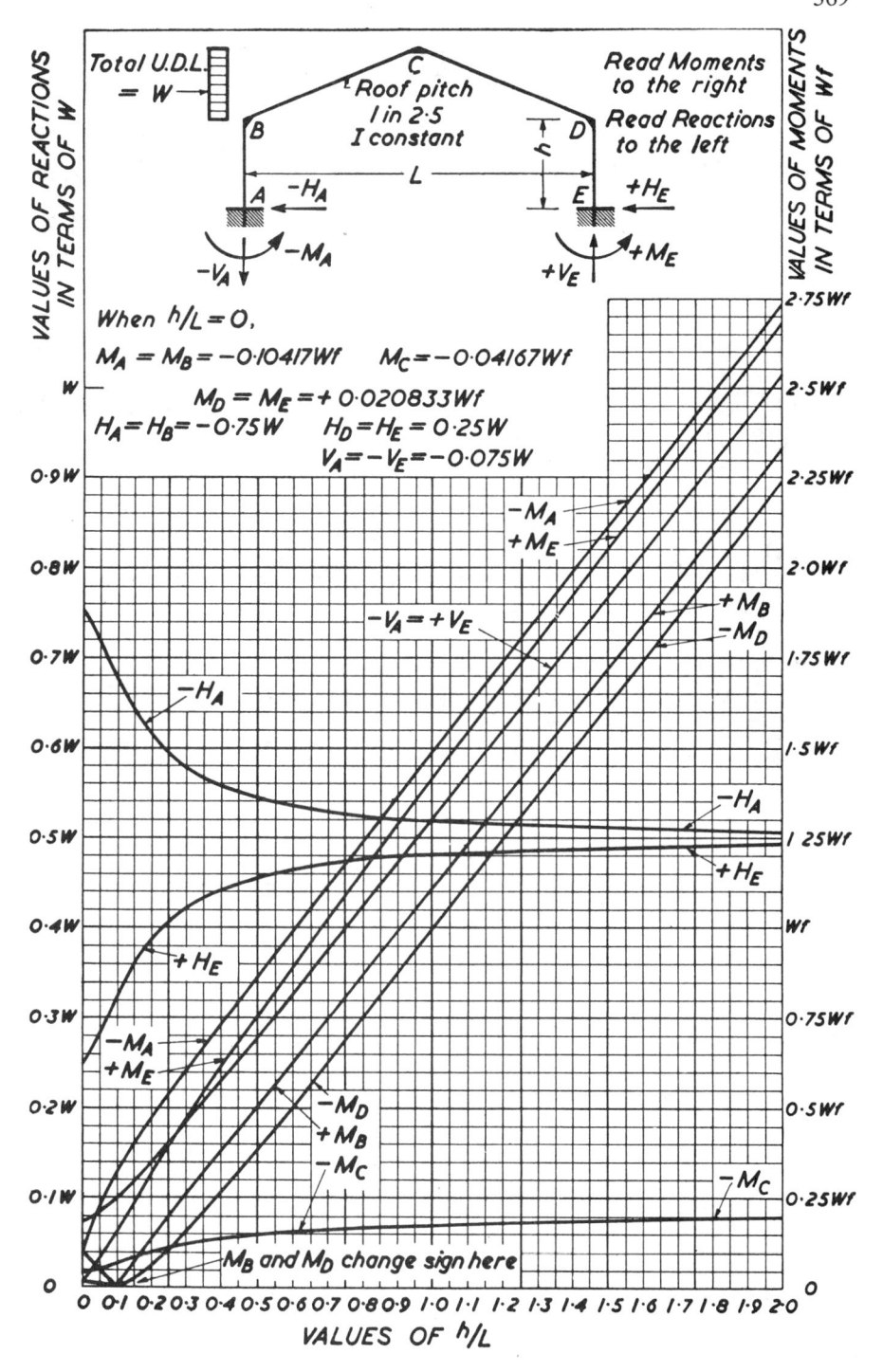

When $h/L = 0$,

$M_A = M_B = -0.10417Wf$ $M_C = -0.04167Wf$

$M_D = M_E = +0.020833Wf$

$H_A = H_B = -0.75W$ $H_D = H_E = 0.25W$

$V_A = -V_E = -0.075W$

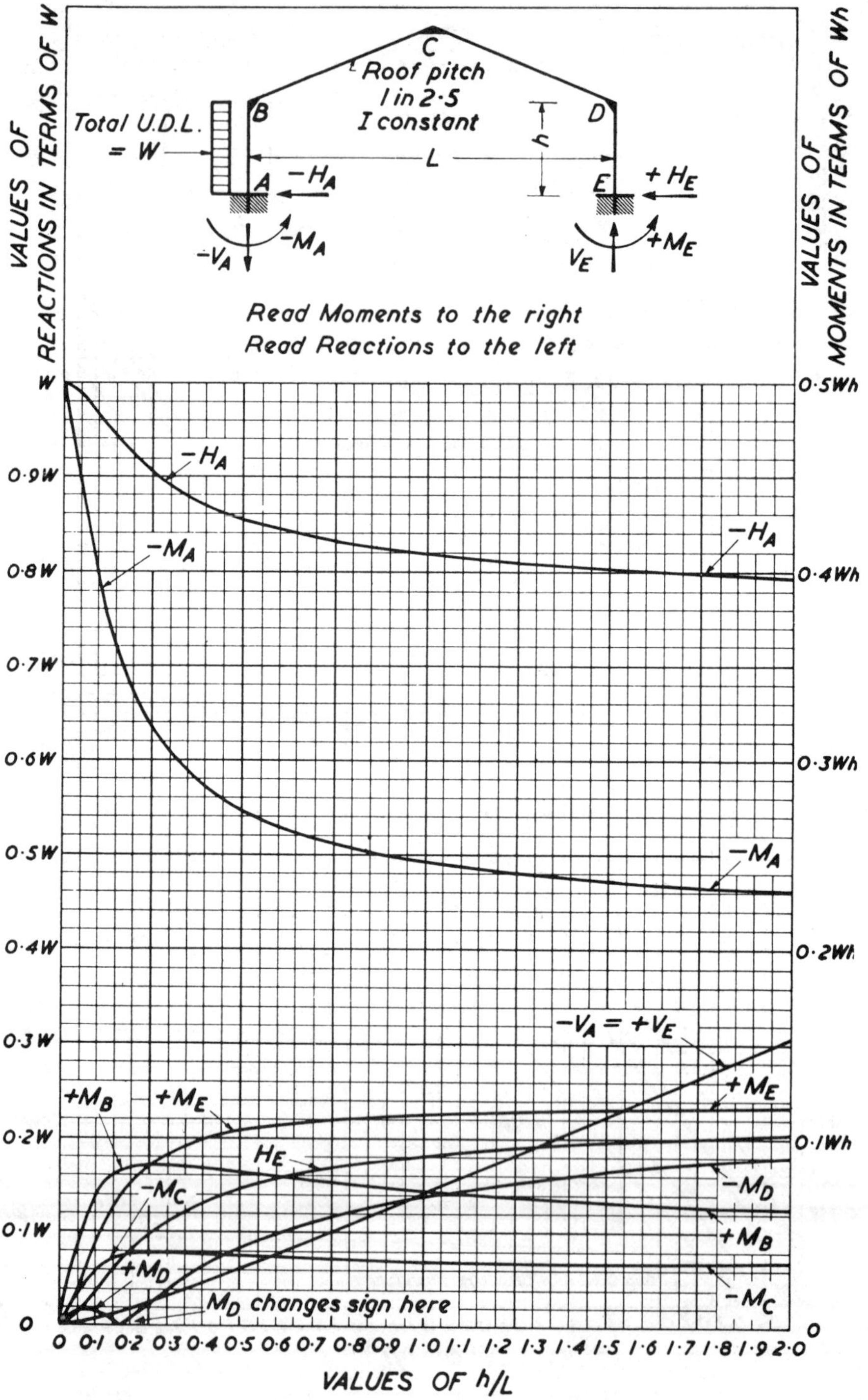
VALUES OF REACTIONS IN TERMS OF W
VALUES OF MOMENTS IN TERMS OF Wh
C
Roof pitch
1 in 2·5
I constant
B
D
Total U.D.L.
= W
h
L
A
−HA
E
+HE
−VA
−MA
VE
+ME
Read Moments to the right
Read Reactions to the left
W
0·9W
0·8W
0·7W
0·6W
0·5W
0·4W
0·3W
0·2W
0·1W
0
0·5Wh
0·4Wh
0·3Wh
0·2Wh
0·1Wh
0
−HA
−MA
−VA = +VE
+ME
+MB
HE
−MC
−MD
+MD
MD changes sign here
0 0·1 0·2 0·3 0·4 0·5 0·6 0·7 0·8 0·9 1·0 1·1 1·2 1·3 1·4 1·5 1·6 1·7 1·8 1·9 2·0
VALUES OF h/L

371

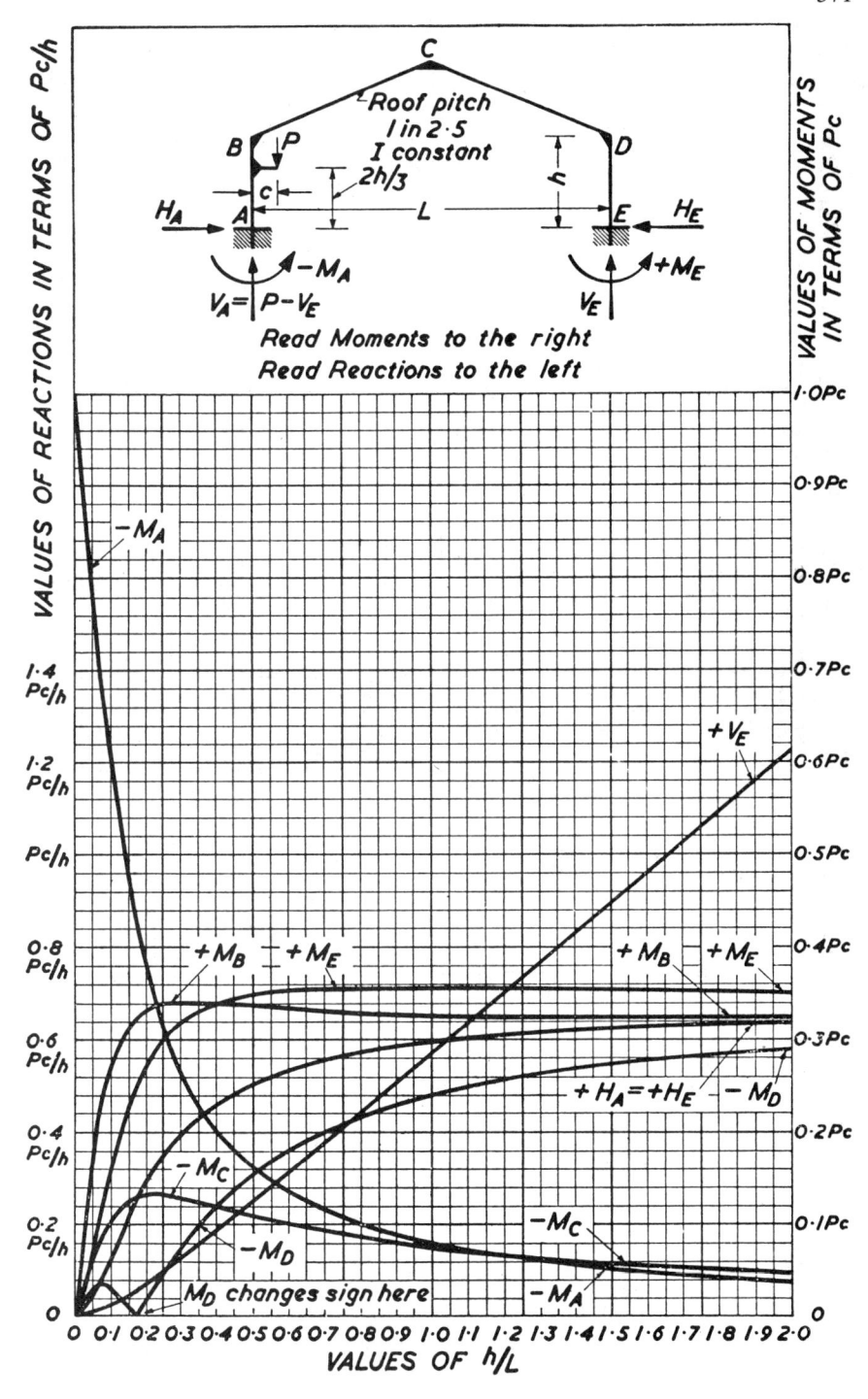

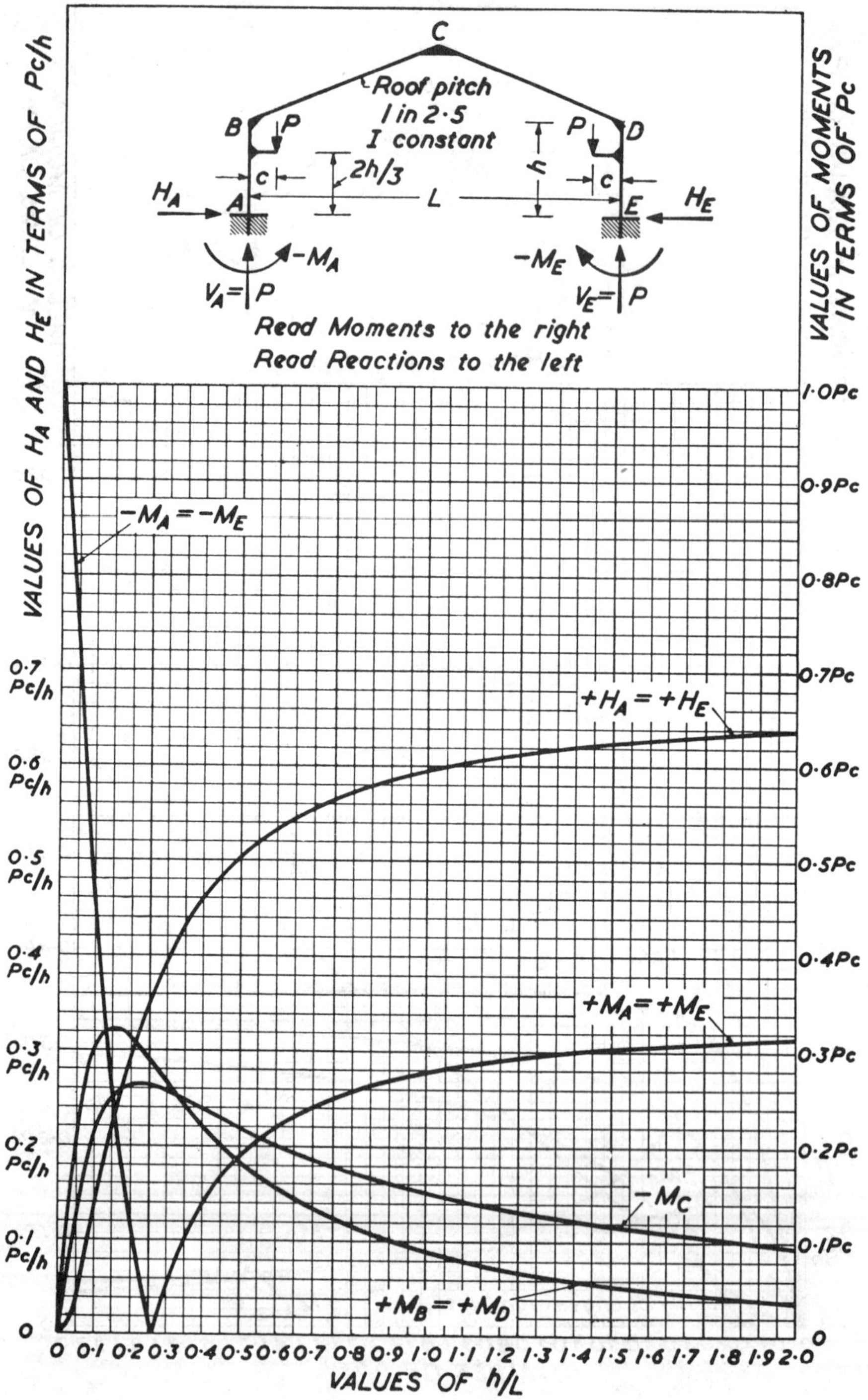
C
Roof pitch
I in 2·5
I constant
B
P
P
D
c
2h/3
h
c
H_A
A
L
E
H_E
$-M_A$
$-M_E$
$V_A = P$
$V_E = P$
Read Moments to the right
Read Reactions to the left
VALUES OF H_A AND H_E IN TERMS OF Pc/h
VALUES OF MOMENTS IN TERMS OF Pc
$-M_A = -M_E$
$+H_A = +H_E$
$+M_A = +M_E$
$-M_C$
$+M_B = +M_D$
0.7 Pc/h
0.6 Pc/h
0.5 Pc/h
0.4 Pc/h
0.3 Pc/h
0.2 Pc/h
0.1 Pc/h
0
1.0Pc
0.9Pc
0.8Pc
0.7Pc
0.6Pc
0.5Pc
0.4Pc
0.3Pc
0.2Pc
0.1Pc
0
0 0.1 0.2 0.3 0.4 0.5 0.6 0.7 0.8 0.9 1.0 1.1 1.2 1.3 1.4 1.5 1.6 1.7 1.8 1.9 2.0
VALUES OF h/L

373

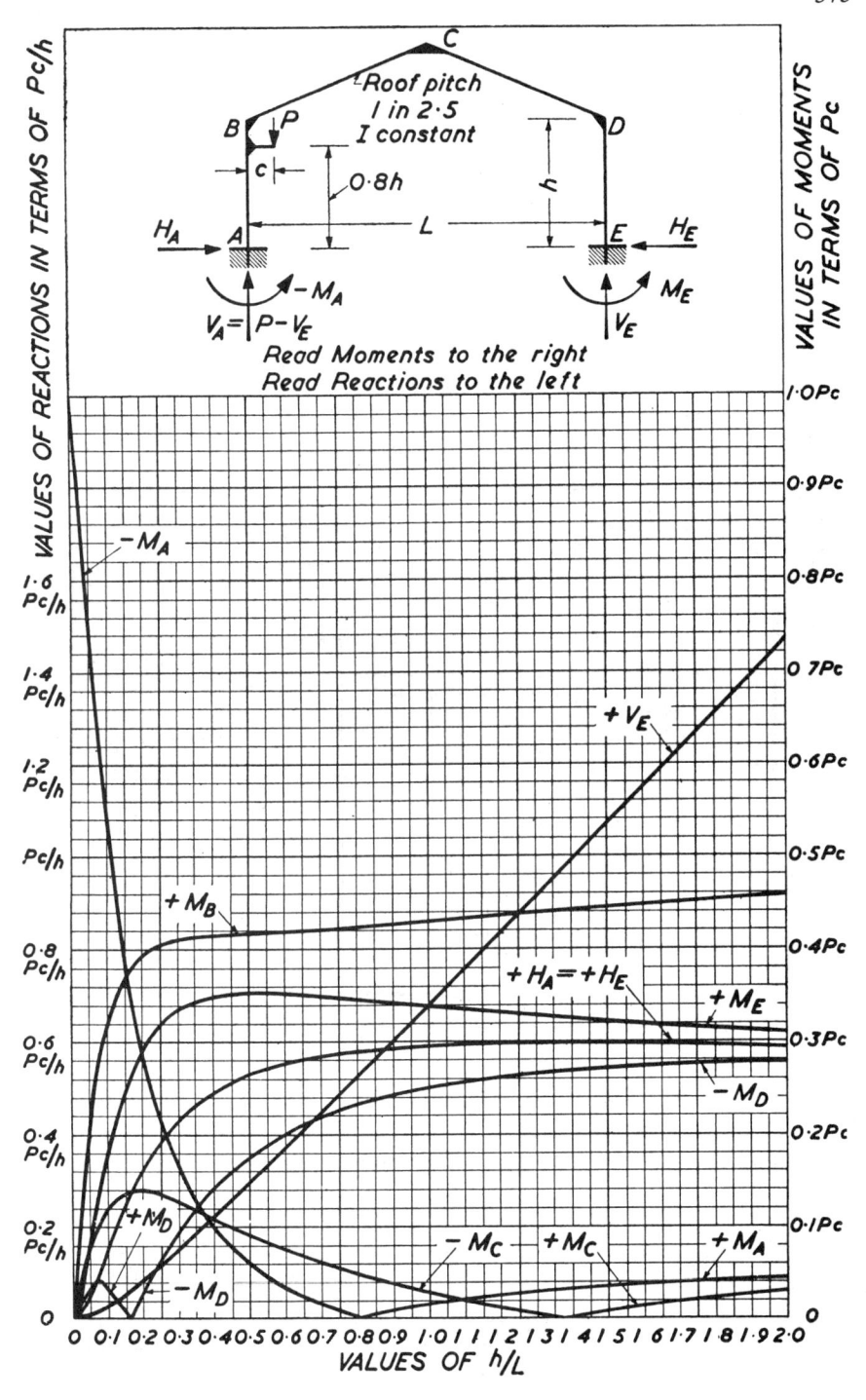

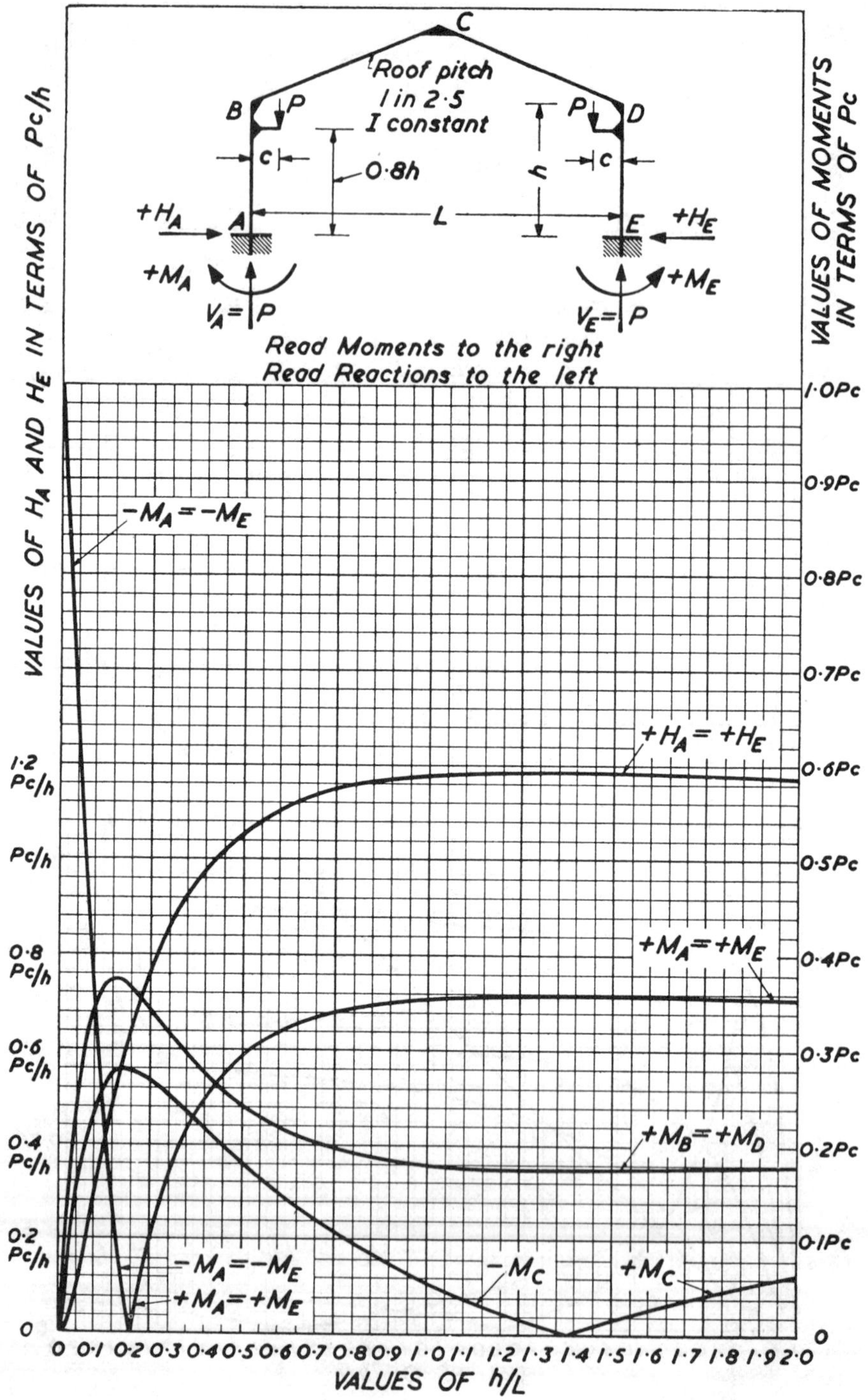

375

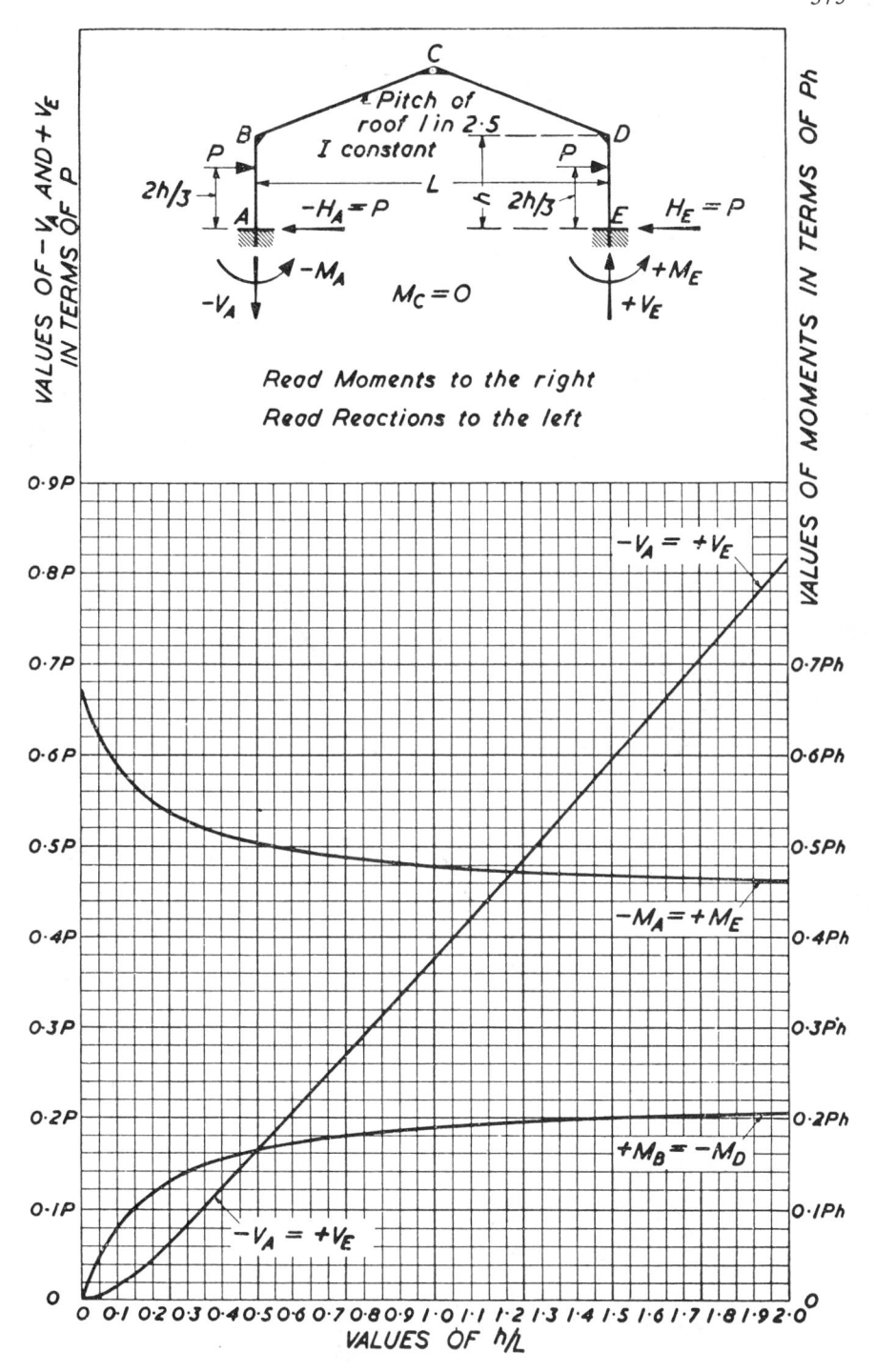

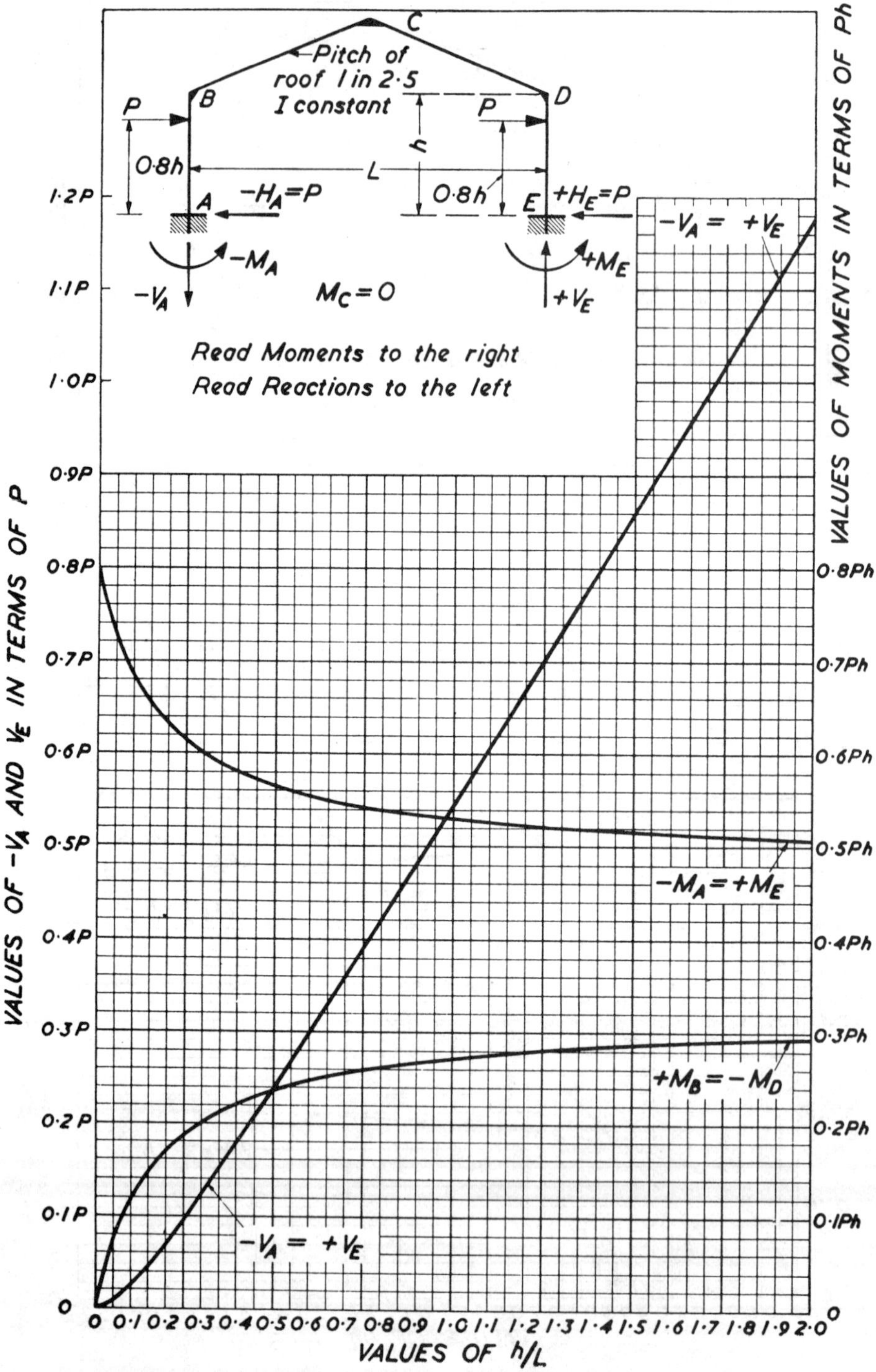

C
Pitch of roof 1 in 2·5
I constant
B
D
P
P
h
0·8h
L
0·8h
A
E
$-H_A=P$
$+H_E=P$
$-M_A$
$+M_E$
$-V_A$
$+V_E$
$M_C=0$
Read Moments to the right
Read Reactions to the left
$-V_A = +V_E$
$-M_A=+M_E$
$+M_B=-M_D$
$-V_A = +V_E$
VALUES OF $-V_A$ AND V_E IN TERMS OF P
VALUES OF MOMENTS IN TERMS OF Ph
0
0·1P
0·2P
0·3P
0·4P
0·5P
0·6P
0·7P
0·8P
0·9P
1·0P
1·1P
1·2P
0
0·1Ph
0·2Ph
0·3Ph
0·4Ph
0·5Ph
0·6Ph
0·7Ph
0·8Ph
0 0·1 0·2 0·3 0·4 0·5 0·6 0·7 0·8 0·9 1·0 1·1 1·2 1·3 1·4 1·5 1·6 1·7 1·8 1·9 2·0
VALUES OF h/L

377

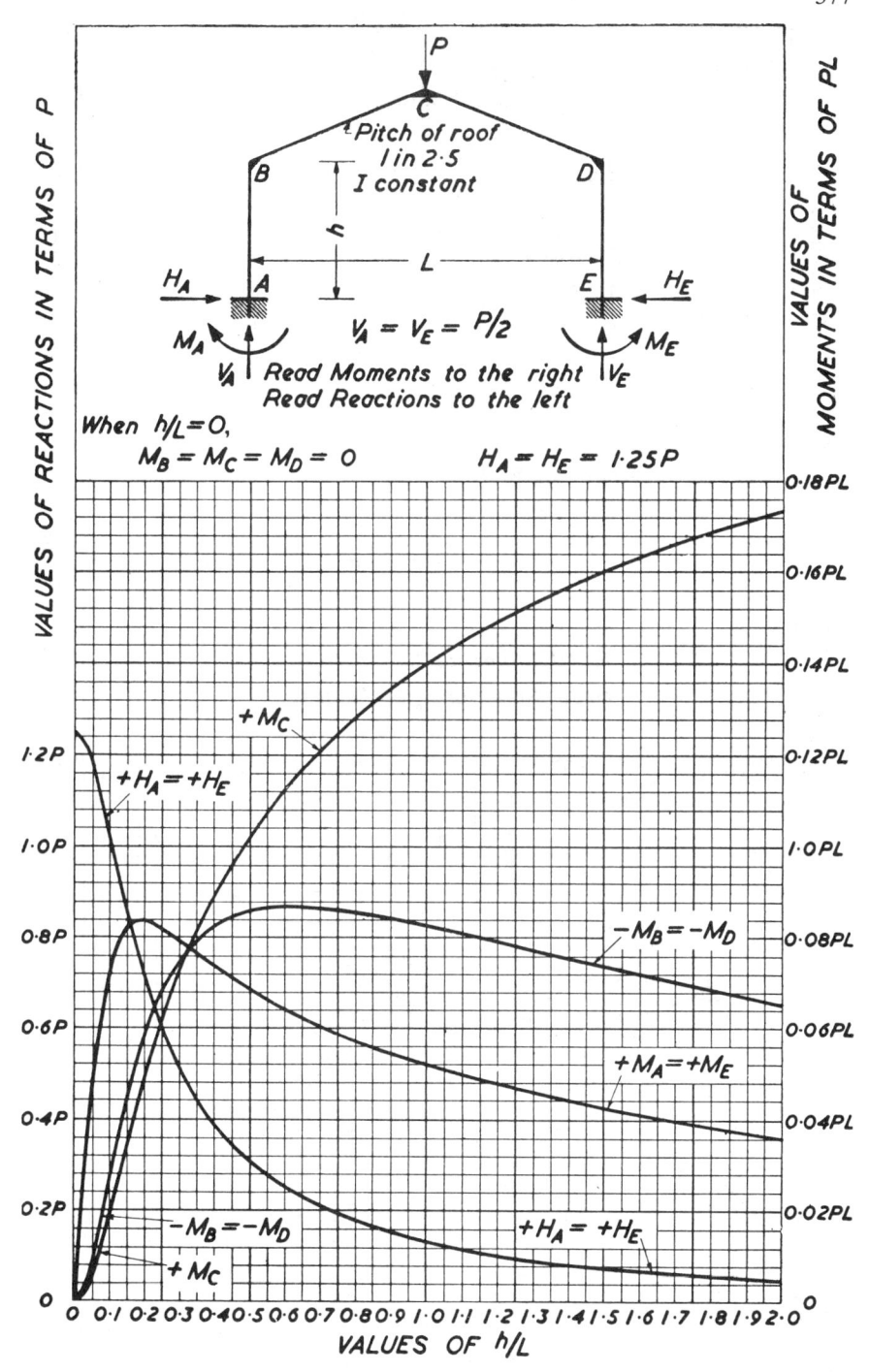

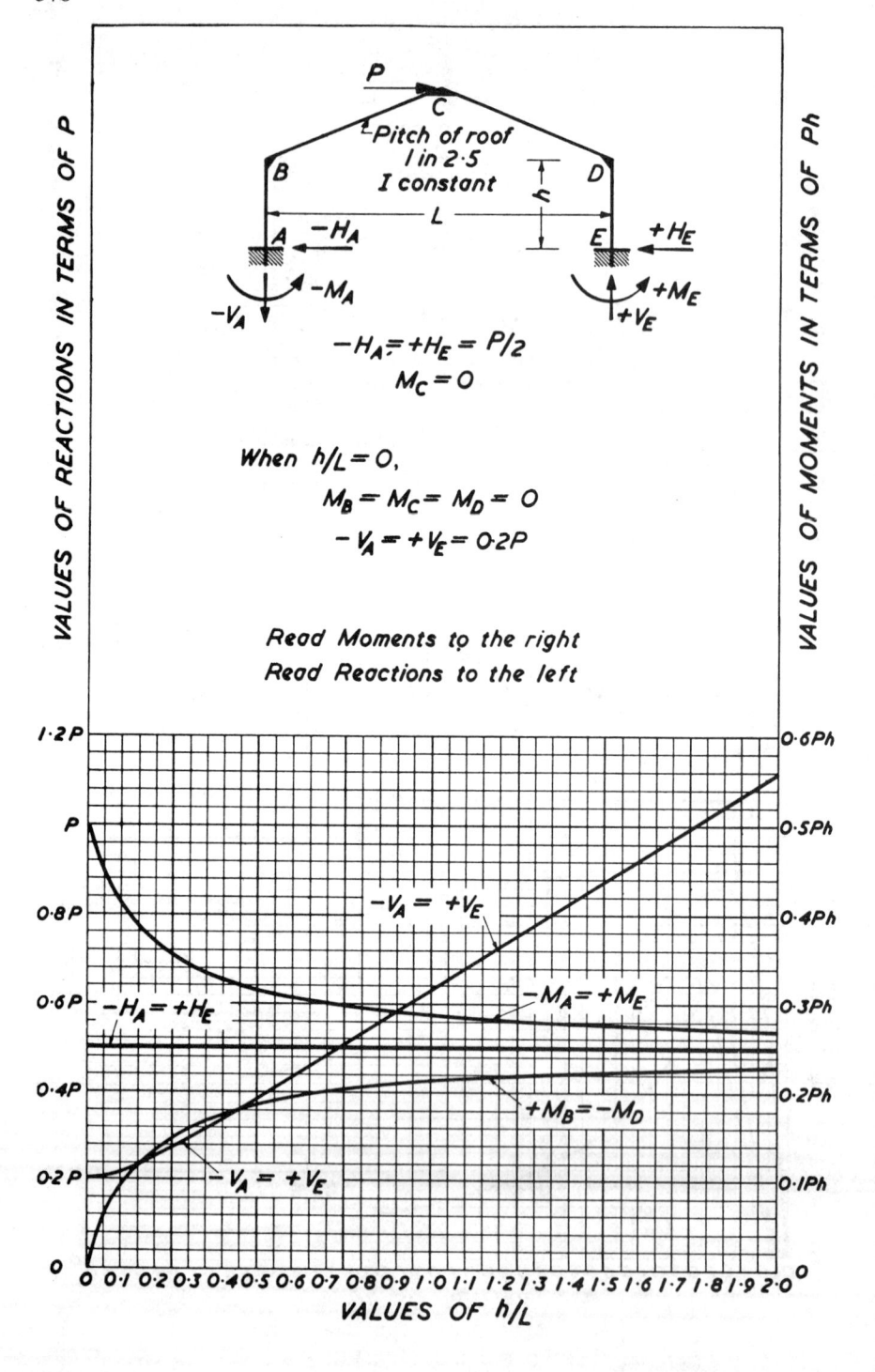

VALUES OF REACTIONS IN TERMS OF P

VALUES OF MOMENTS IN TERMS OF Ph

Pitch of roof 1 in 2·5
I constant

$-H_A = +H_E = P/2$
$M_C = 0$

When $h/L = 0$,
$M_B = M_C = M_D = 0$
$-V_A = +V_E = 0.2P$

Read Moments to the right
Read Reactions to the left

$-V_A = +V_E$

$-H_A = +H_E$

$-M_A = +M_E$

$+M_B = -M_D$

$-V_A = +V_E$

VALUES OF h/L

379

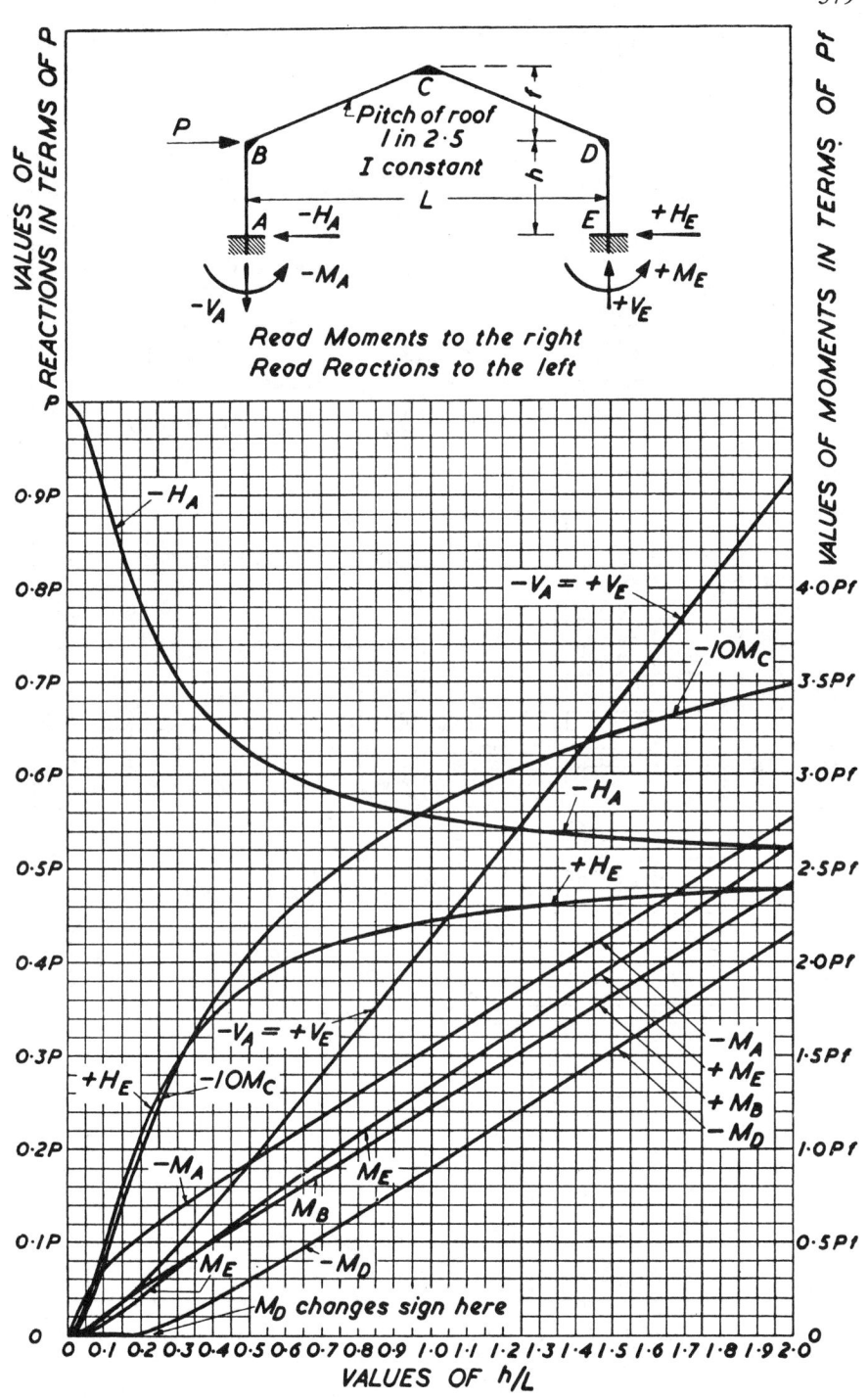

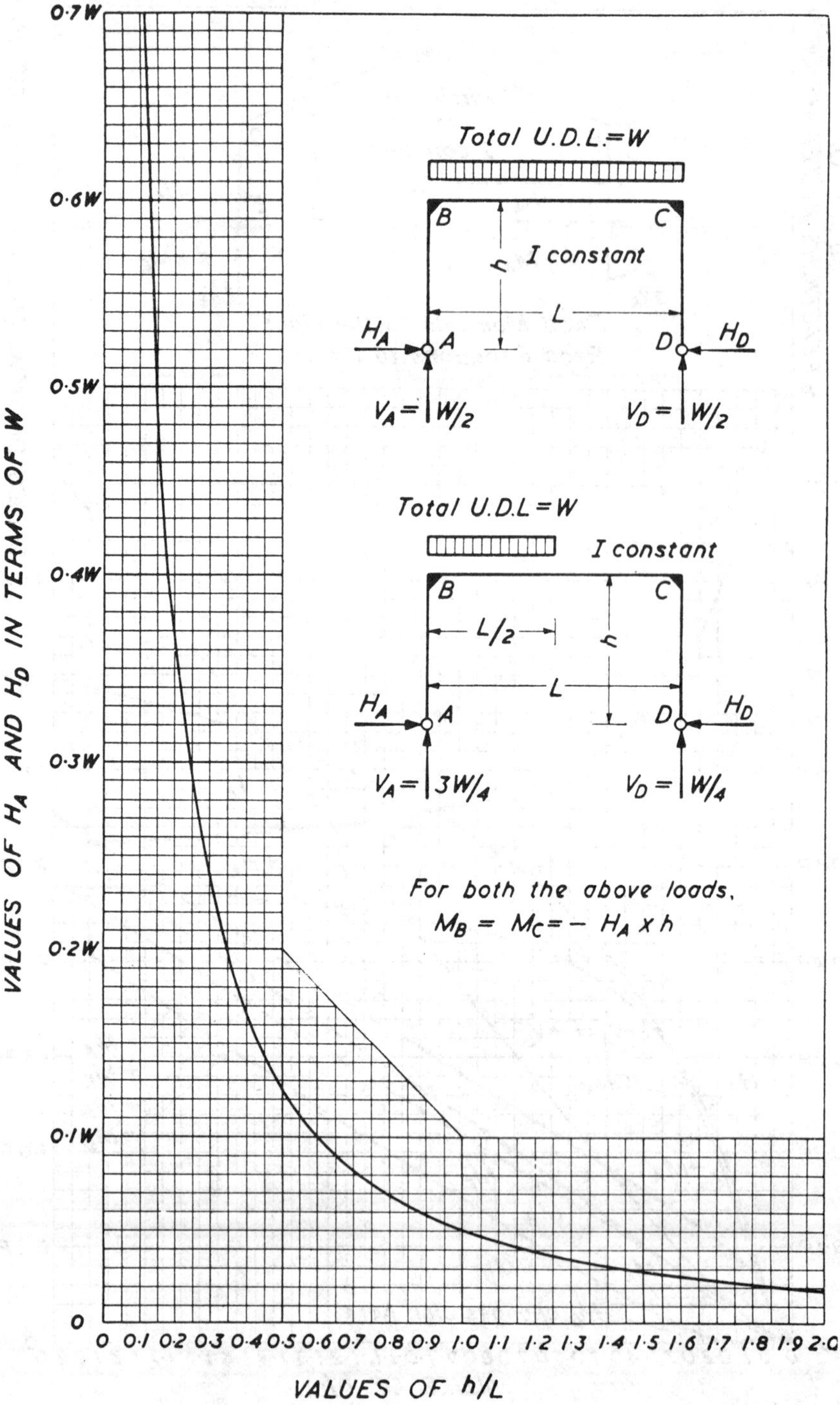
Total U.D.L.=W
B
C
h
I constant
L
H_A
A
D
H_D
V_A = W/2
V_D = W/2
Total U.D.L = W
I constant
B
C
L/2
h
L
H_A
A
D
H_D
V_A = 3W/4
V_D = W/4
For both the above loads,
$M_B = M_C = - H_A \times h$
0.7W
0.6W
0.5W
0.4W
0.3W
0.2W
0.1W
0
0 0.1 0.2 0.3 0.4 0.5 0.6 0.7 0.8 0.9 1.0 1.1 1.2 1.3 1.4 1.5 1.6 1.7 1.8 1.9 2.0
VALUES OF H_A AND H_D IN TERMS OF W
VALUES OF h/L

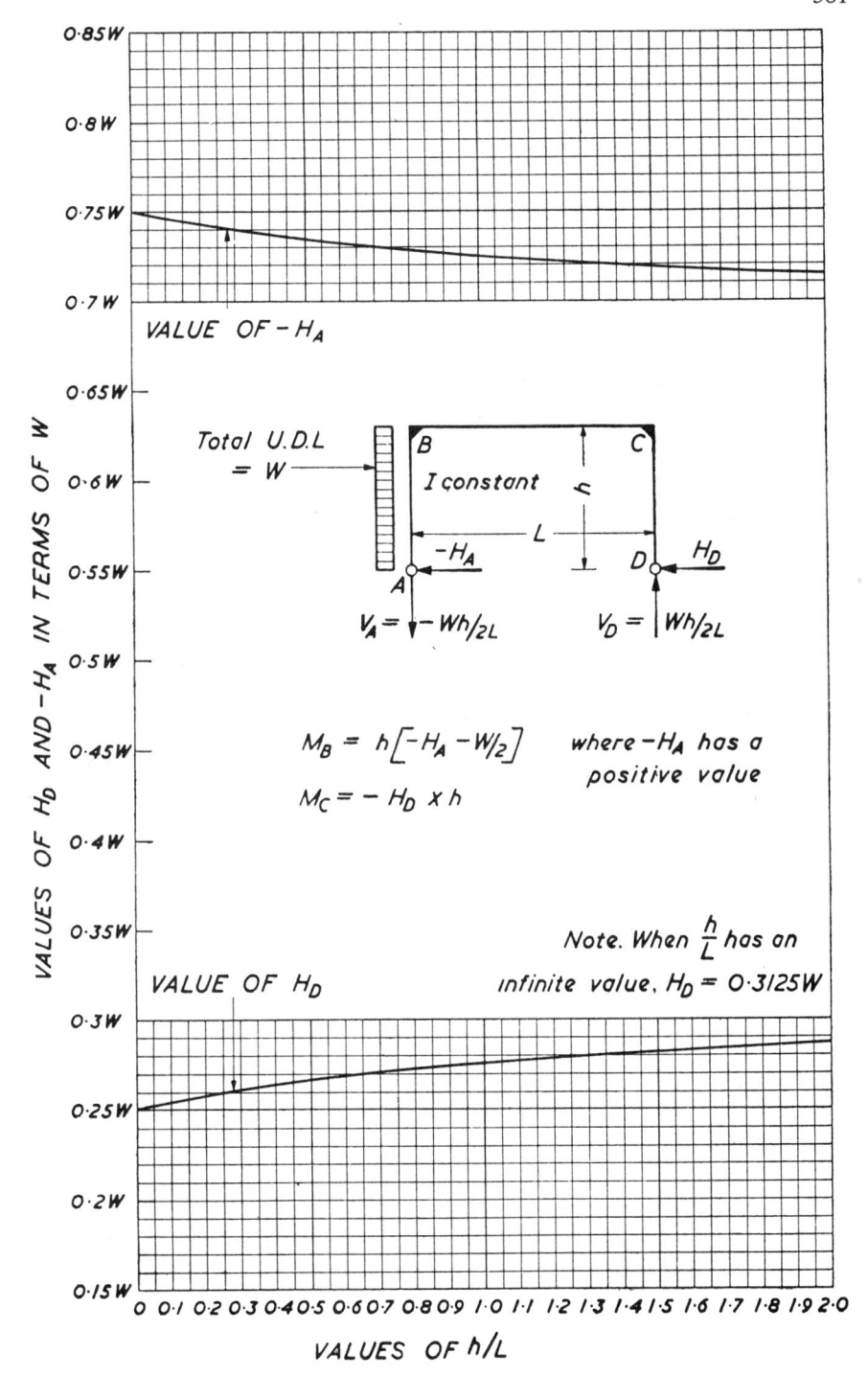

Chart showing values of H_D and $-H_A$ in terms of W versus values of h/L.

VALUE OF $-H_A$

VALUE OF H_D

Total U.D.L = W, I constant

$V_A = -Wh/2L$, $V_D = Wh/2L$

$M_B = h\left[-H_A - W/2\right]$ where $-H_A$ has a positive value

$M_C = -H_D \times h$

Note. When $\dfrac{h}{L}$ has an infinite value, $H_D = 0.3125W$

382

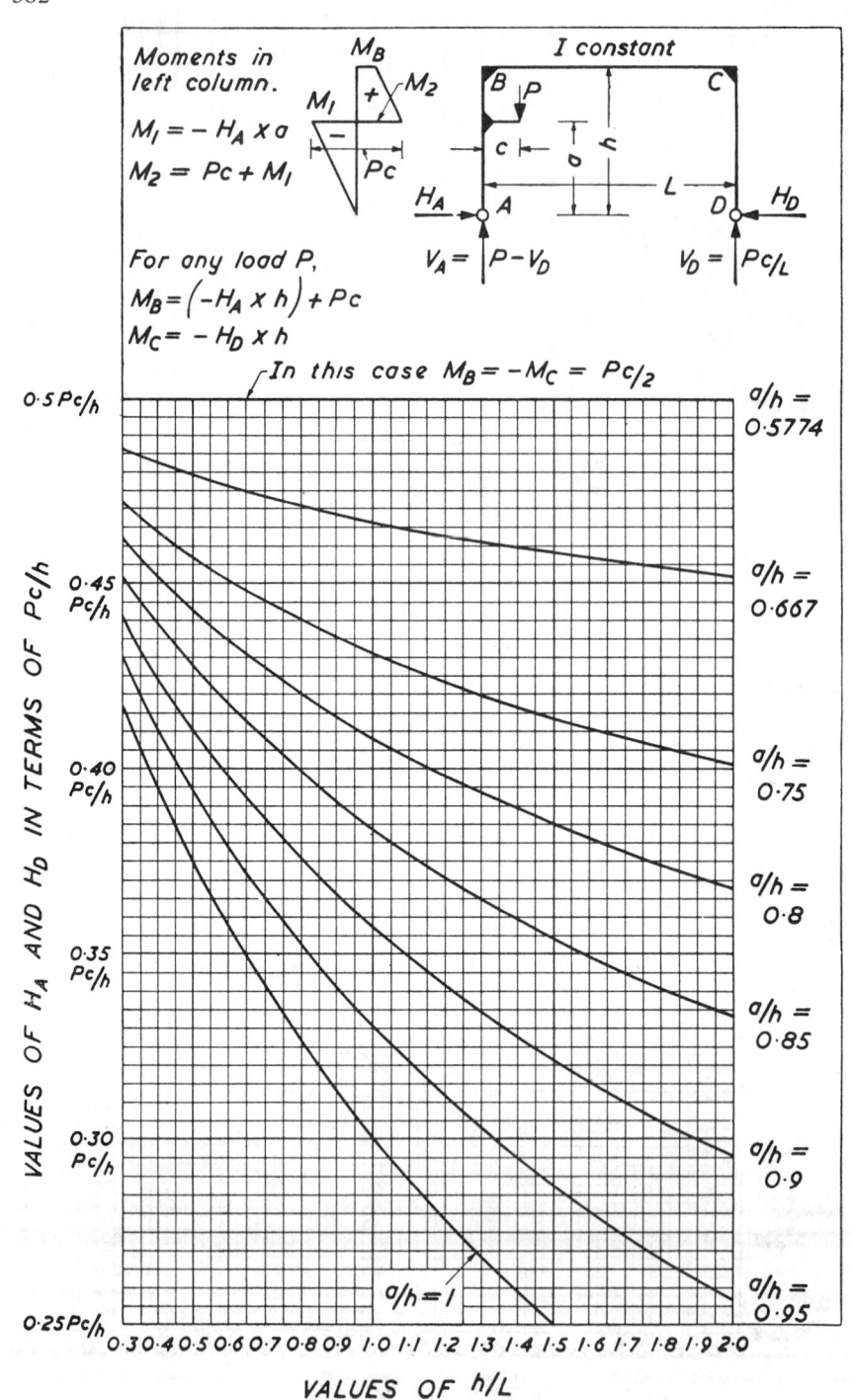

Moments in left column.

$M_1 = -H_A \times a$

$M_2 = Pc + M_1$

For any load P,

$M_B = (-H_A \times h) + Pc$

$M_C = -H_D \times h$

In this case $M_B = -M_C = Pc/2$

I constant

$V_A = P - V_D$ $\quad V_D = Pc/L$

VALUES OF H_A AND H_D IN TERMS OF Pc/h

- $0.5\,Pc/h$ — $a/h = 0.5774$
- $0.45\,Pc/h$ — $a/h = 0.667$
- $0.40\,Pc/h$ — $a/h = 0.75$
- — $a/h = 0.8$
- $0.35\,Pc/h$ — $a/h = 0.85$
- $0.30\,Pc/h$ — $a/h = 0.9$
- $a/h = 1$
- $0.25\,Pc/h$ — $a/h = 0.95$

VALUES OF h/L : 0.3 0.4 0.5 0.6 0.7 0.8 0.9 1.0 1.1 1.2 1.3 1.4 1.5 1.6 1.7 1.8 1.9 2.0

Moments in left column:

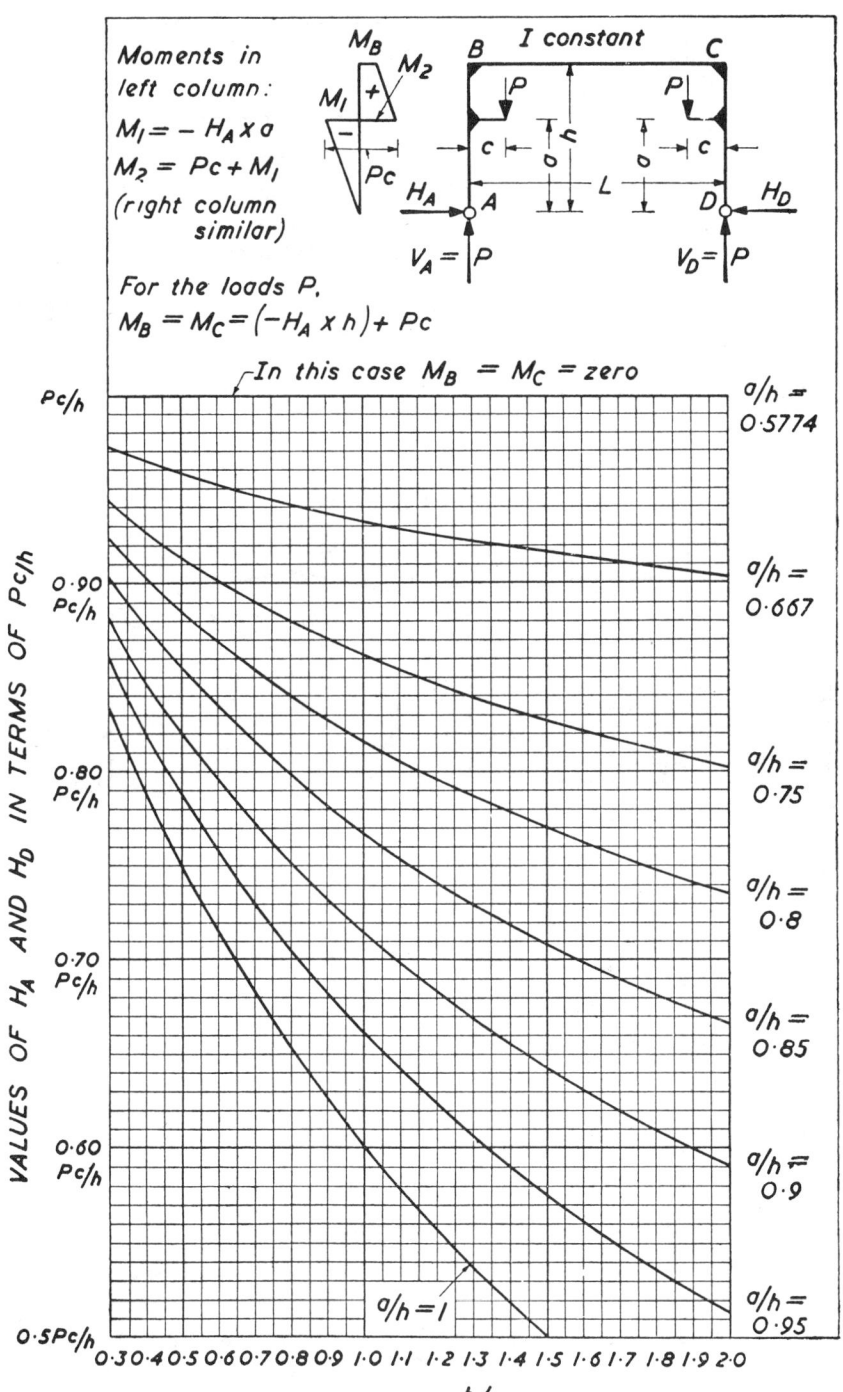

$M_1 = -H_A \times a$

$M_2 = Pc + M_1$

(right column similar)

For the loads P,

$M_B = M_C = (-H_A \times h) + Pc$

In this case $M_B = M_C$ = zero

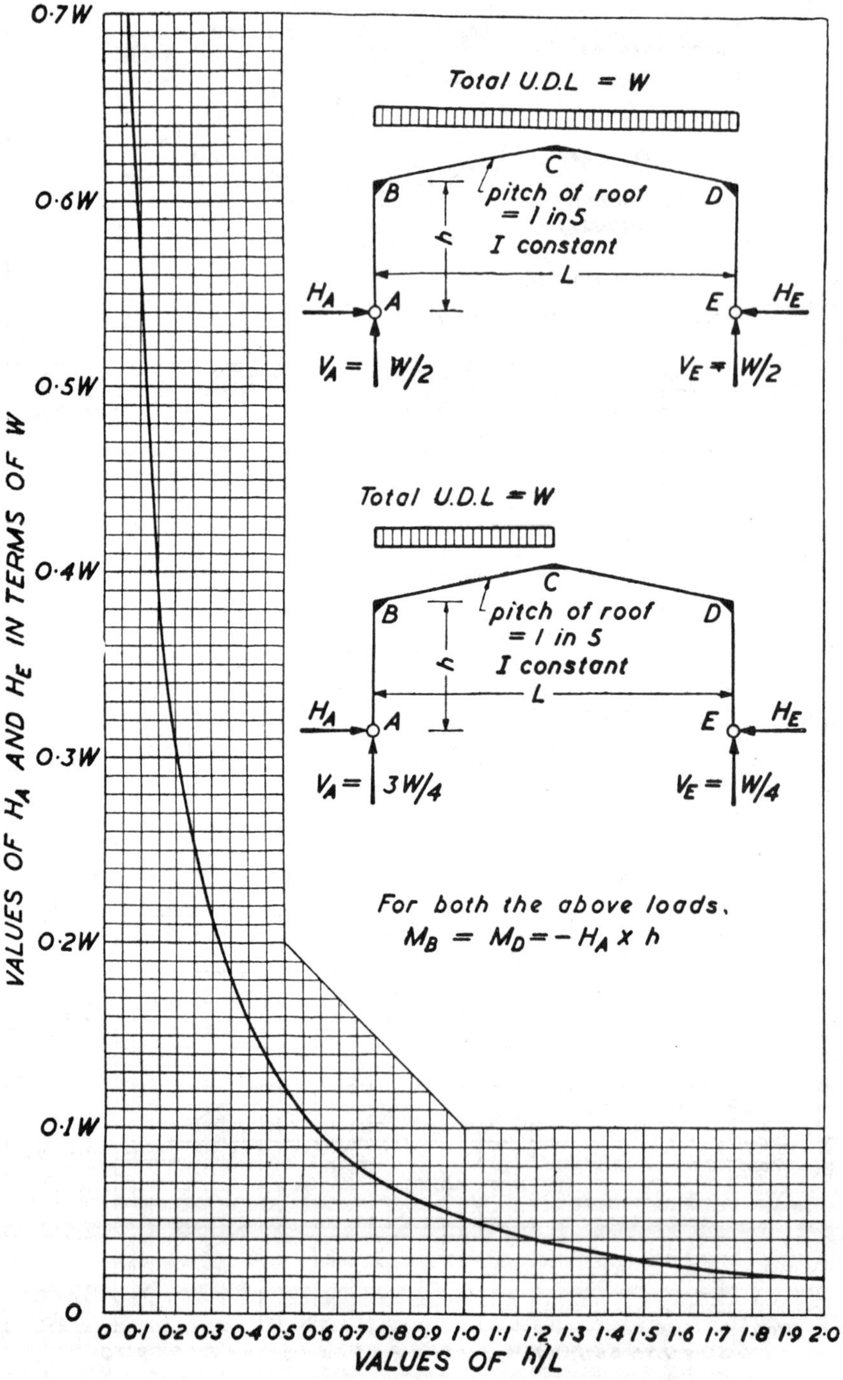
0·7W
0·6W
0·5W
0·4W
0·3W
0·2W
0·1W
0
VALUES OF H_A AND H_E IN TERMS OF W
Total U.D.L = W
C
B
D
pitch of roof = 1 in 5
I constant
h
L
H_A
A
E
H_E
V_A = W/2
V_E = W/2
Total U.D.L = W
C
B
D
pitch of roof = 1 in 5
I constant
h
L
H_A
A
E
H_E
V_A = 3W/4
V_E = W/4
For both the above loads,
$M_B = M_D = -H_A \times h$
0 0·1 0·2 0·3 0·4 0·5 0·6 0·7 0·8 0·9 1·0 1·1 1·2 1·3 1·4 1·5 1·6 1·7 1·8 1·9 2·0
VALUES OF h/L

Total U.D.L. = W

pitch of roof = 1 in 5
I constant

$V_A = \downarrow -W\left(\dfrac{h+0.5f}{L}\right)$

$V_E = \uparrow W\left(\dfrac{h+0.5f}{L}\right)$

$M_B = -H_A \times h$ (resulting value positive)

$M_D = -H_E \times h$ (resulting value negative.)

CURVE FOR H_E

CURVE FOR $-H_A$

VALUES OF $-H_A$ AND H_E IN TERMS OF W

VALUES OF h/L

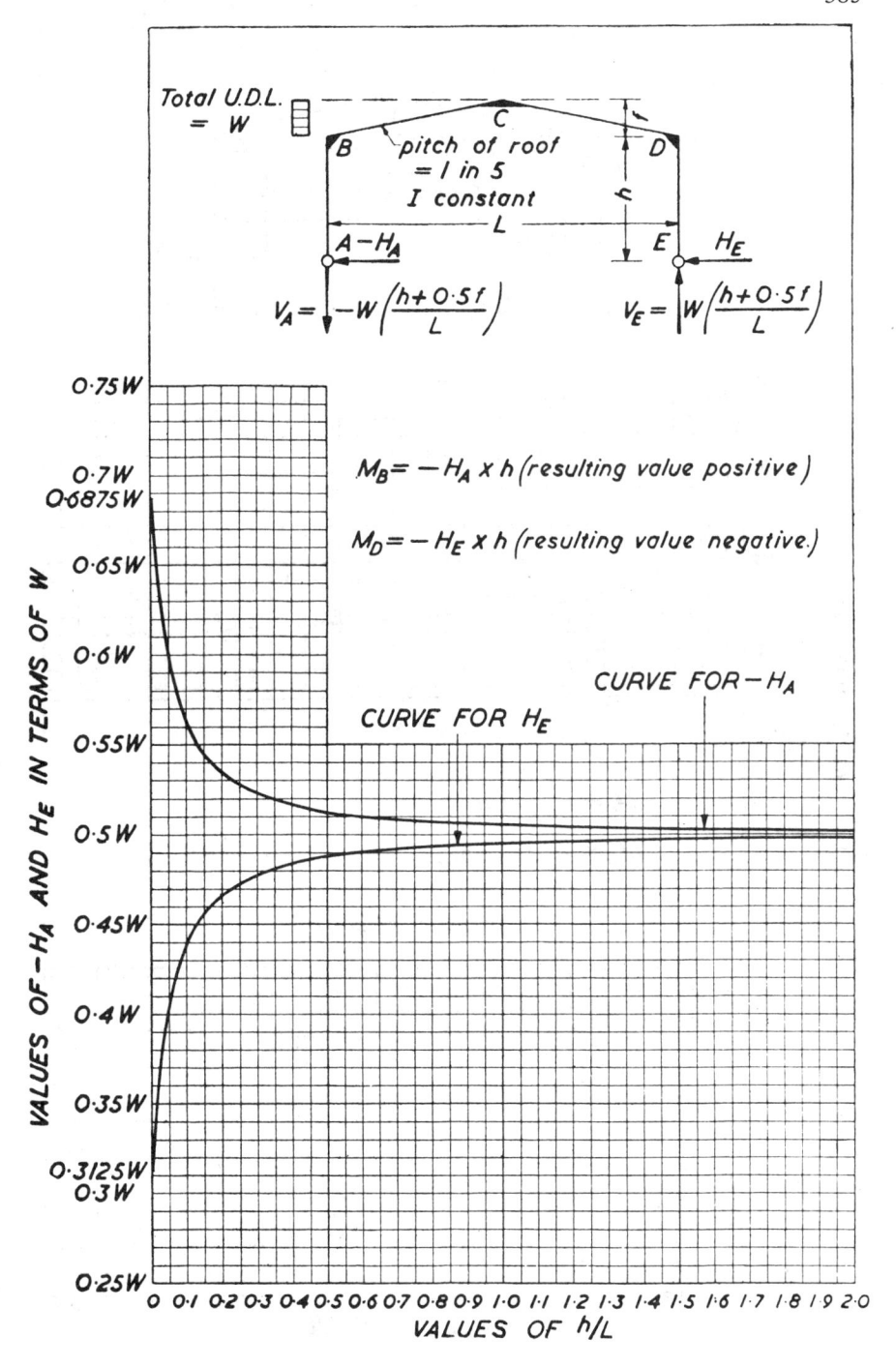

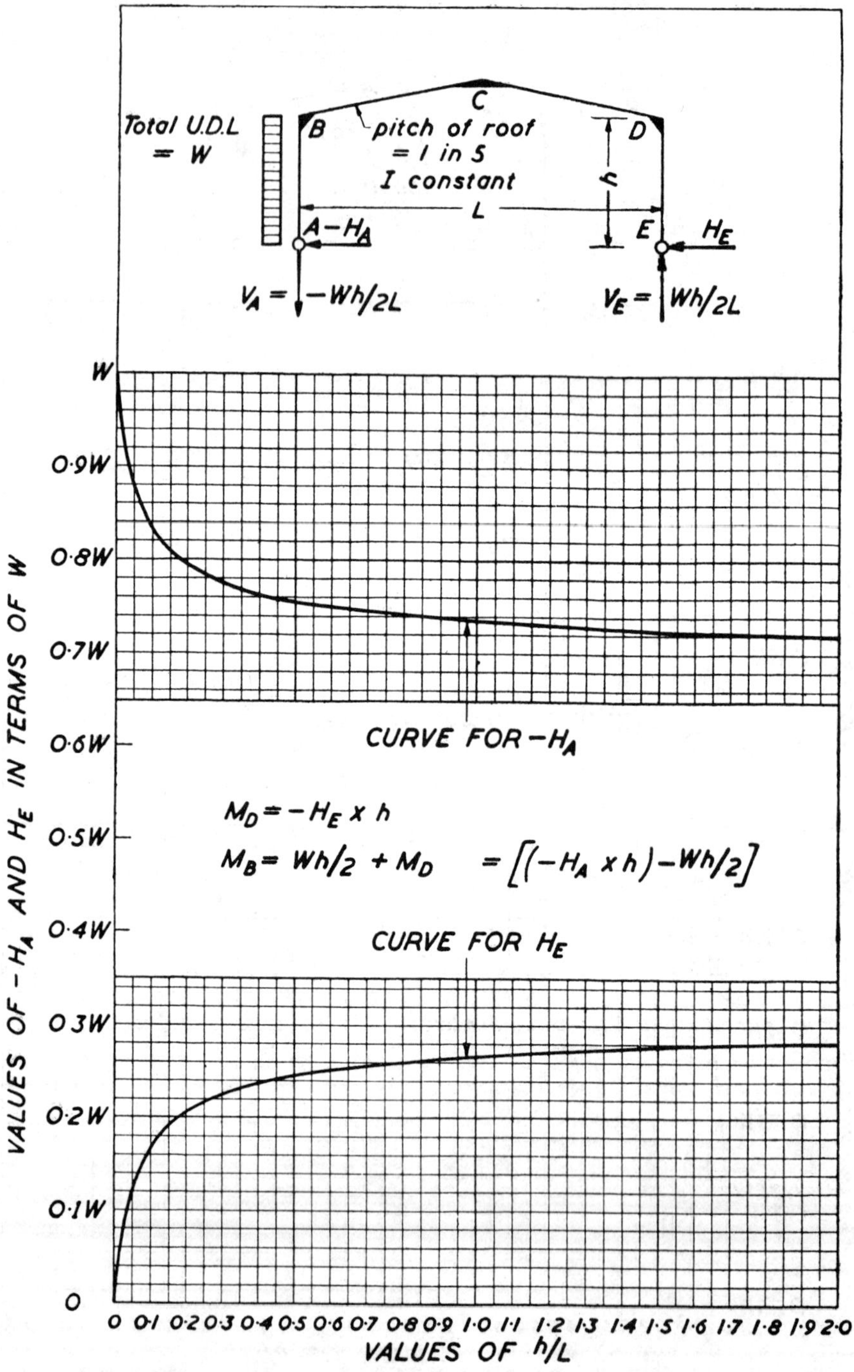

Total U.D.L = W
B
C
D
pitch of roof = 1 in 5
I constant
h
L
A − H_A
E
H_E
V_A = −Wh/2L
V_E = Wh/2L
W
0·9W
0·8W
0·7W
0·6W
0·5W
0·4W
0·3W
0·2W
0·1W
0
VALUES OF −H_A AND H_E IN TERMS OF W
CURVE FOR −H_A
M_D = −H_E x h
M_B = Wh/2 + M_D = [(−H_A x h) − Wh/2]
CURVE FOR H_E
0 0·1 0·2 0·3 0·4 0·5 0·6 0·7 0·8 0·9 1·0 1·1 1·2 1·3 1·4 1·5 1·6 1·7 1·8 1·9 2·0
VALUES OF h/L

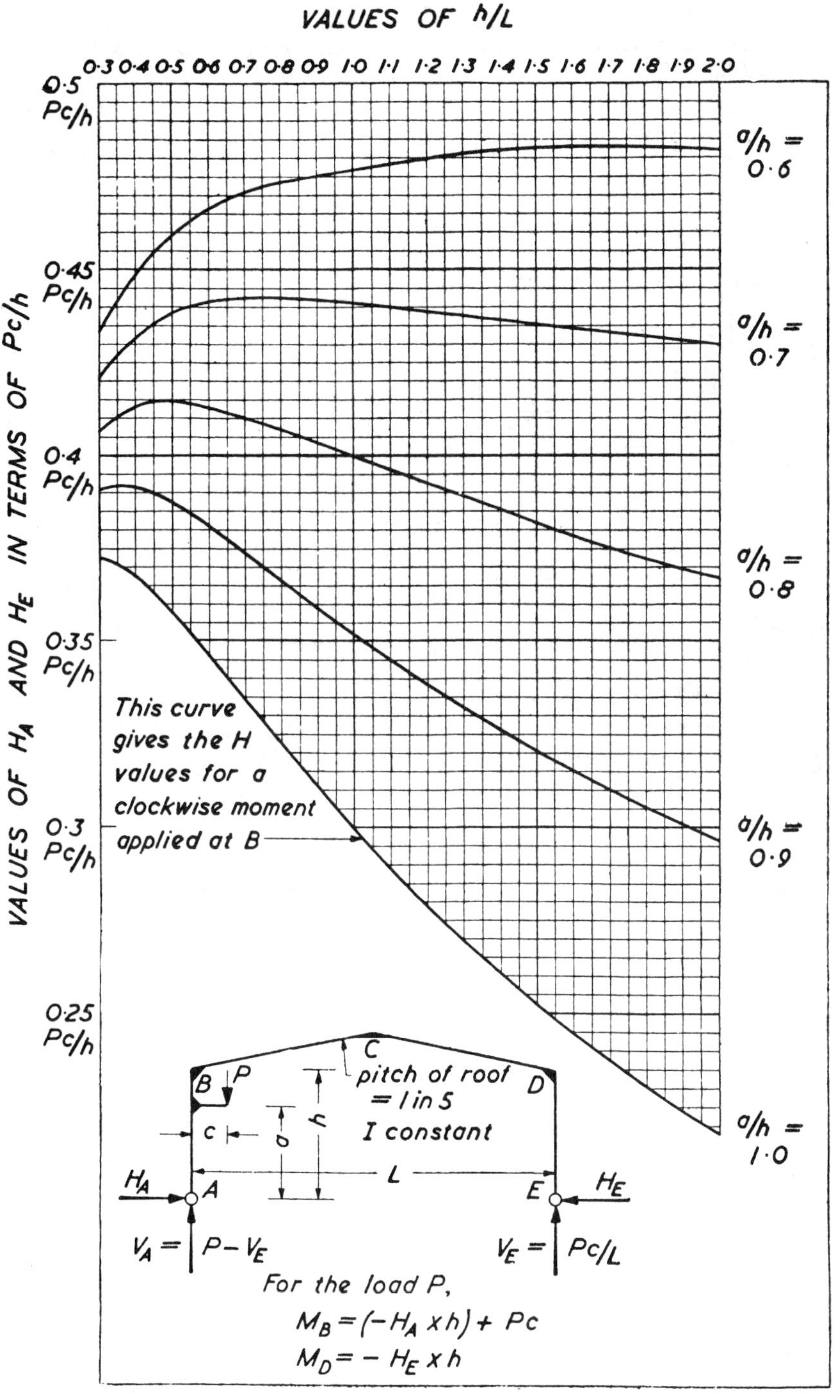
VALUES OF h/L
0·3 0·4 0·5 0·6 0·7 0·8 0·9 1·0 1·1 1·2 1·3 1·4 1·5 1·6 1·7 1·8 1·9 2·0
VALUES OF H_A AND H_E IN TERMS OF Pc/h
0·5 Pc/h
0·45 Pc/h
0·4 Pc/h
0·35 Pc/h
0·3 Pc/h
0·25 Pc/h
a/h = 0·6
a/h = 0·7
a/h = 0·8
a/h = 0·9
a/h = 1·0
This curve gives the H values for a clockwise moment applied at B
B
P
C
pitch of roof = 1 in 5
D
c
a
h
I constant
L
H_A
A
E
H_E
V_A = P − V_E
V_E = Pc/L
For the load P,
$M_B = (-H_A \times h) + Pc$
$M_D = -H_E \times h$

VALUES OF h/L

VALUES OF H_A AND H_E IN TERMS OF Pc/h

$a/h = 0.6$
$a/h = 0.7$
$a/h = 0.8$
$a/h = 0.9$
$a/h = 1.0$

This curve gives the H values for equal & opposite moments applied at B & D

pitch of roof = 1 in 5
I constant

$V_A = P$ $V_E = P$

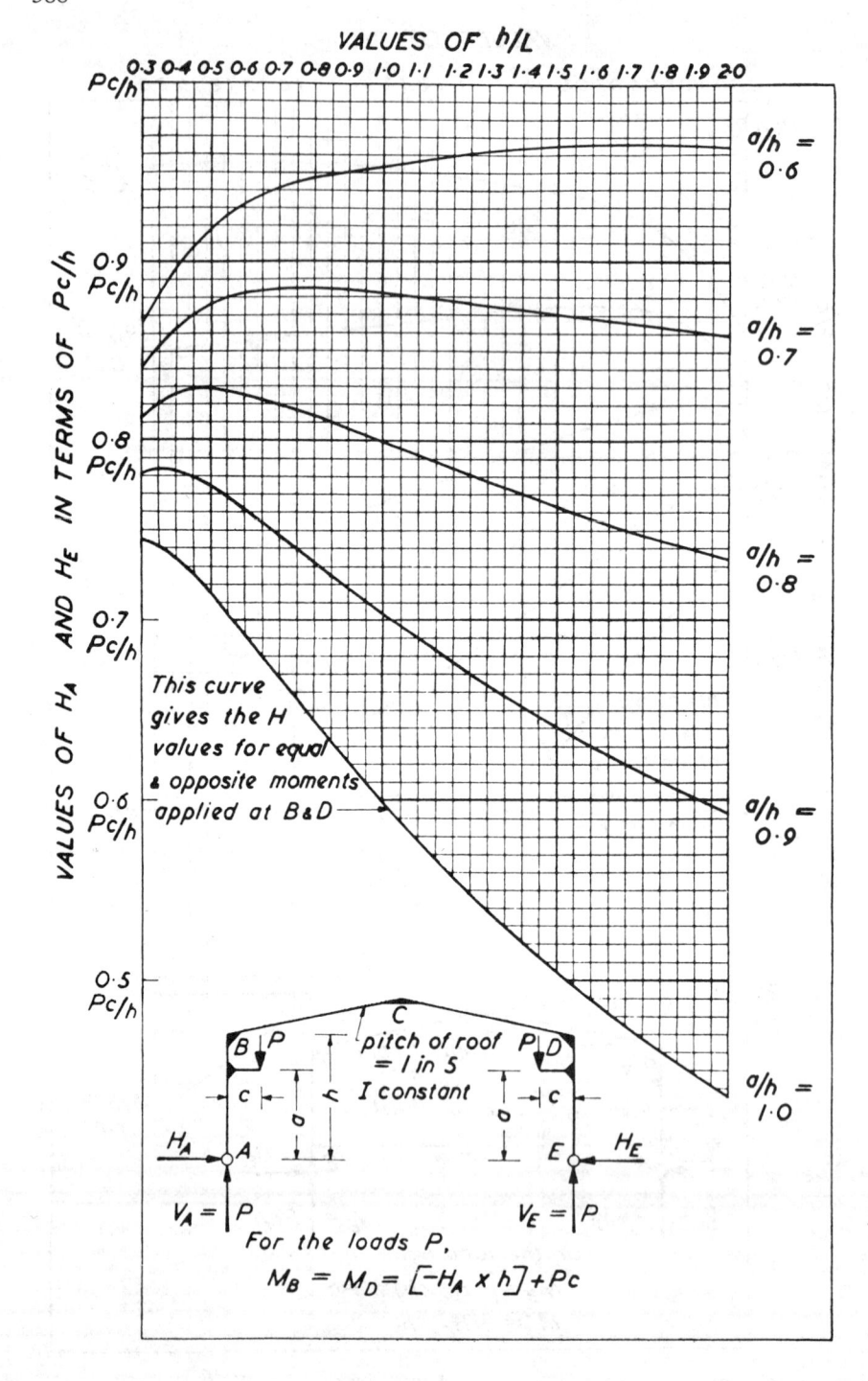

For the loads P,
$M_B = M_D = [-H_A \times h] + Pc$

VALUES OF H_A AND H_E IN TERMS OF P

Value when $h/L = 0$

pitch of roof 1 in 5
I constant

For the load P,

$$M_B = M_D = -H_A \times h$$

VALUES OF h/L

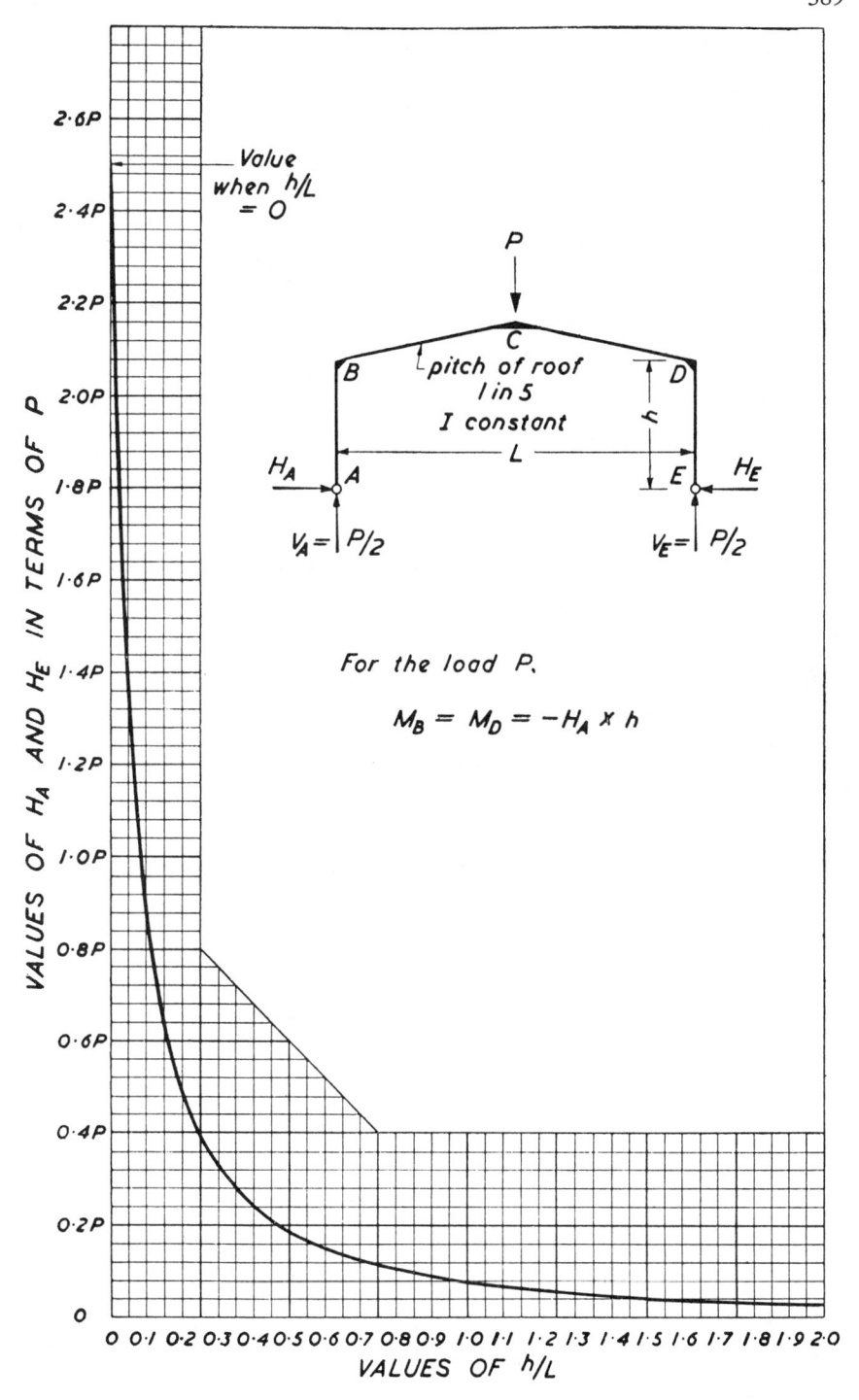

For the load P,

$M_B = -H_A \times h$ (resulting value positive)

$M_D = -H_E \times h$ (resulting value negative)

CURVE FOR $-H_A$

CURVE FOR H_E

VALUES OF $-H_A$ AND H_E IN TERMS OF P

VALUES OF h/L

pitch of roof 1 in 5
I constant

$V_A = -Ph/L$

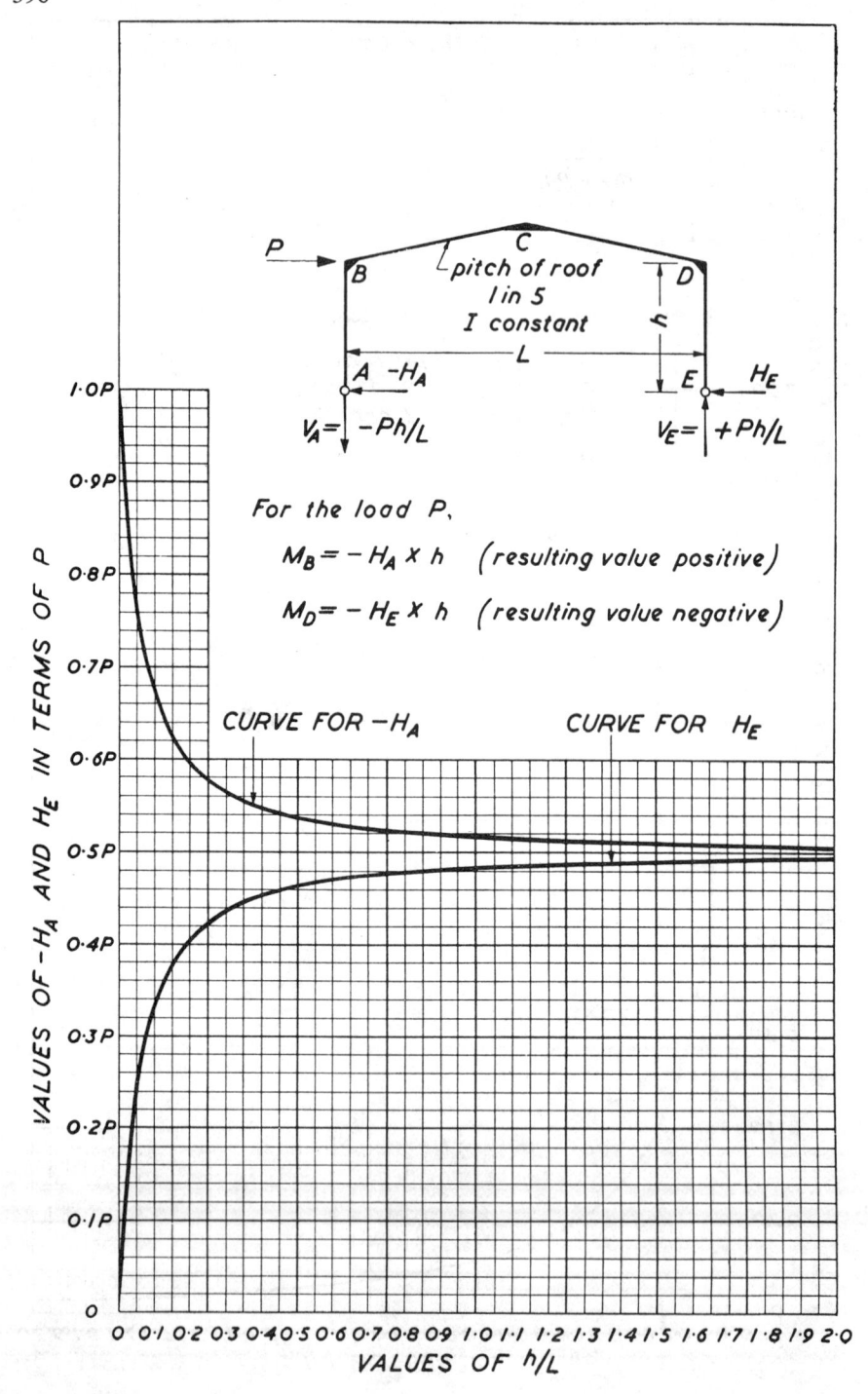

$V_E = +Ph/L$

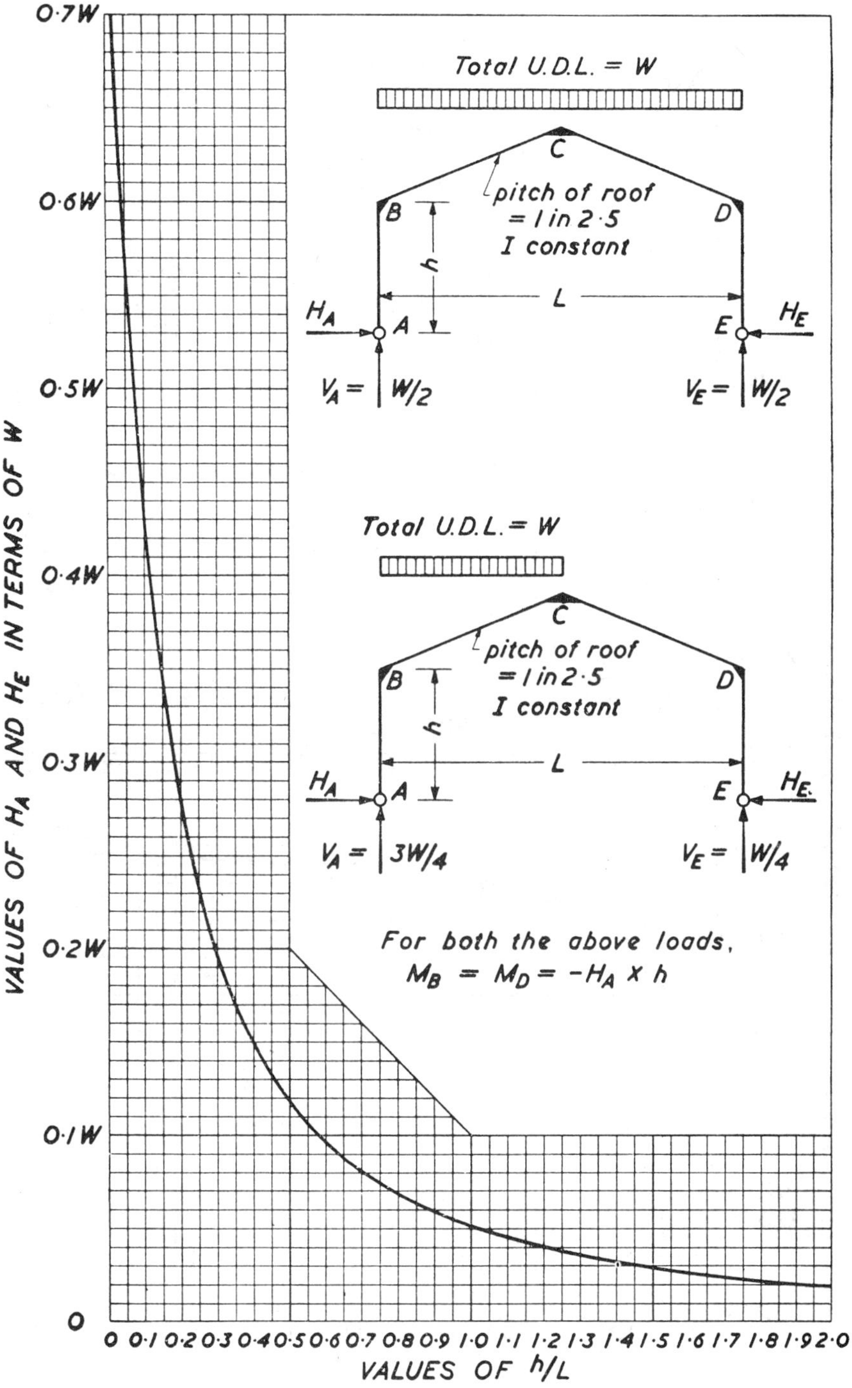
Total U.D.L. = W
pitch of roof = 1 in 2·5
I constant
B
C
D
h
L
H_A
A
E
H_E
V_A = W/2
V_E = W/2
Total U.D.L. = W
pitch of roof = 1 in 2·5
I constant
B
C
D
h
L
H_A
A
E
H_E
V_A = 3W/4
V_E = W/4
For both the above loads,
$M_B = M_D = -H_A \times h$
0·7W
0·6W
0·5W
0·4W
0·3W
0·2W
0·1W
0
VALUES OF H_A AND H_E IN TERMS OF W
0 0·1 0·2 0·3 0·4 0·5 0·6 0·7 0·8 0·9 1·0 1·1 1·2 1·3 1·4 1·5 1·6 1·7 1·8 1·9 2·0
VALUES OF h/L

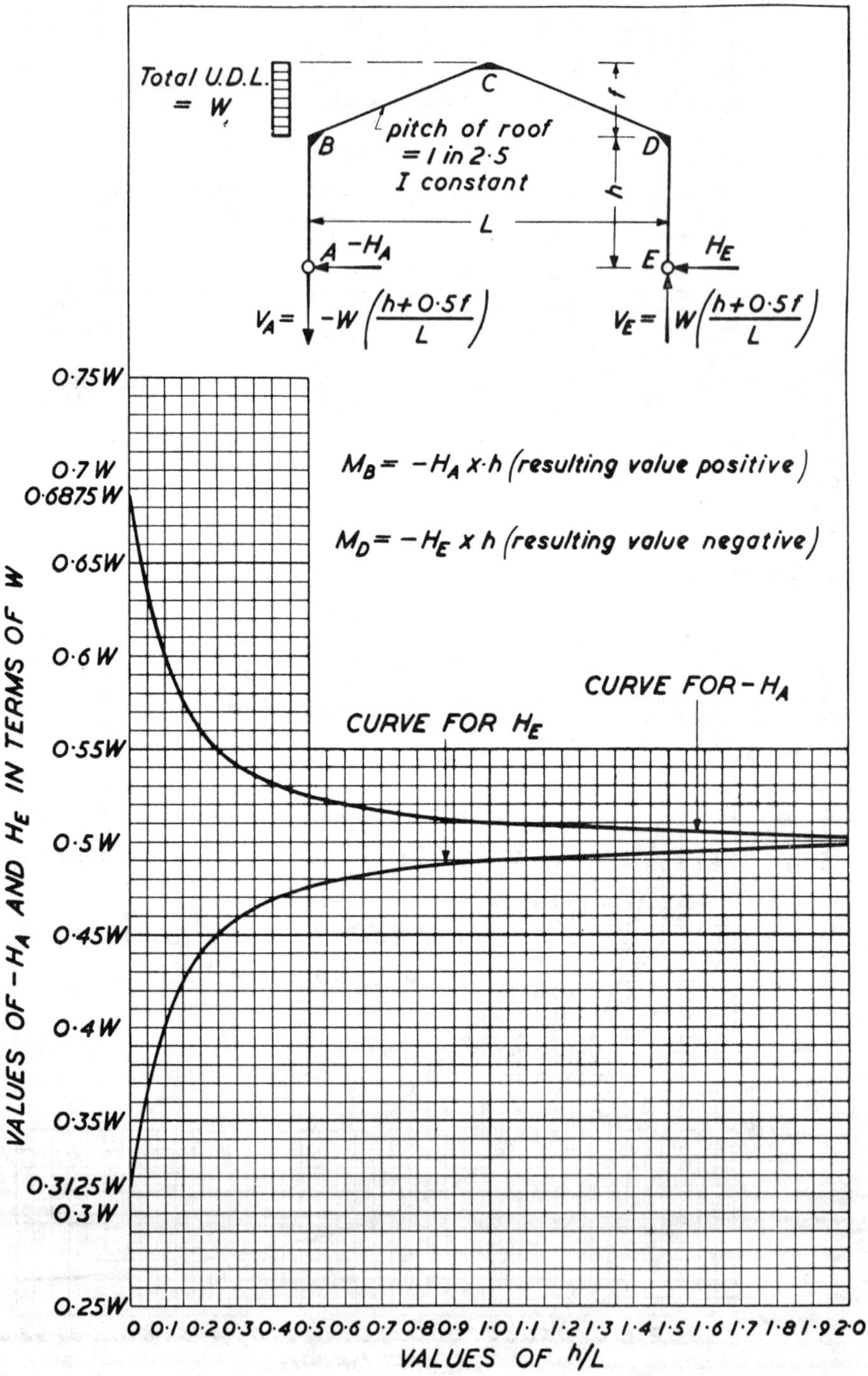
Total U.D.L. = W
C
f
pitch of roof = 1 in 2·5
I constant
B
D
h
L
A
$-H_A$
E
H_E
$V_A = -W\left(\frac{h+0.5f}{L}\right)$
$V_E = W\left(\frac{h+0.5f}{L}\right)$
$M_B = -H_A \times h$ (resulting value positive)
$M_D = -H_E \times h$ (resulting value negative)
CURVE FOR $-H_A$
CURVE FOR H_E
0·75W
0·7W
0·6875W
0·65W
0·6W
0·55W
0·5W
0·45W
0·4W
0·35W
0·3125W
0·3W
0·25W
VALUES OF $-H_A$ AND H_E IN TERMS OF W
0 0·1 0·2 0·3 0·4 0·5 0·6 0·7 0·8 0·9 1·0 1·1 1·2 1·3 1·4 1·5 1·6 1·7 1·8 1·9 2·0
VALUES OF h/L

Total U.D.L. = W

pitch of roof = 1 in 2·5
I constant

$V_A = -Wh/2L$
$V_E = Wh/2L$

CURVE FOR $-H_A$

$M_D = -H_E \times h$
$M_B = Wh/2 + M_D = \left[(-H_A \times h) - Wh/2\right]$

CURVE FOR H_E

VALUES OF $-H_A$ AND H_E IN TERMS OF W

VALUES OF h/L

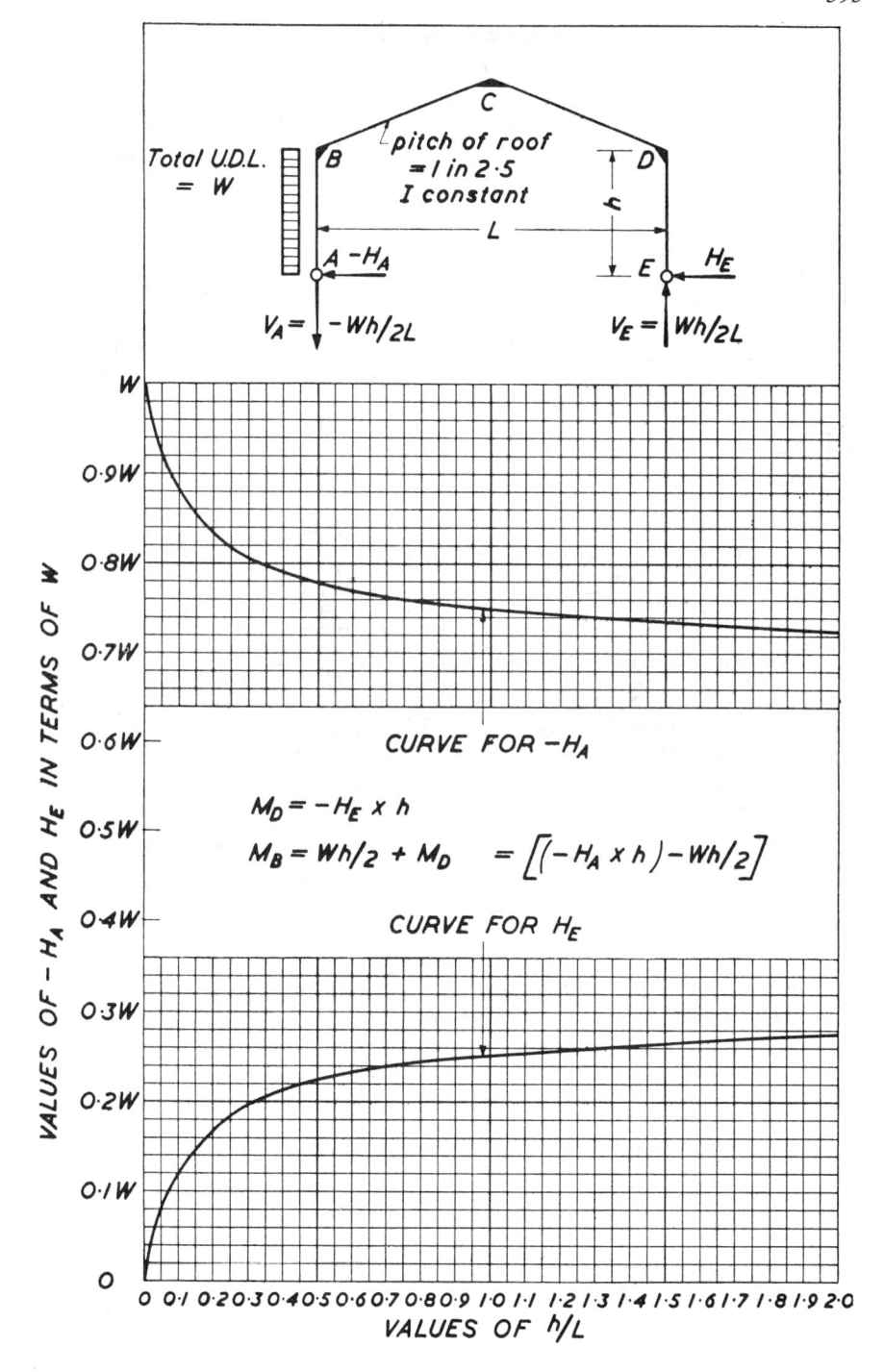

394

VALUES OF h/L

0.3 0.4 0.5 0.6 0.7 0.8 0.9 1.0 1.1 1.2 1.3 1.4 1.5 1.6 1.7 1.8 1.9 2.0

VALUES OF H_A AND H_E IN TERMS OF Pc/h

$a/h = 0.6$
$a/h = 0.7$
$a/h = 0.8$
$a/h = 0.9$
$a/h = 1.0$

This curve gives the H values for a clockwise moment applied at B

pitch of roof = 1 in 2.5
I constant

$V_A = P - V_E$
$V_E = Pc/L$

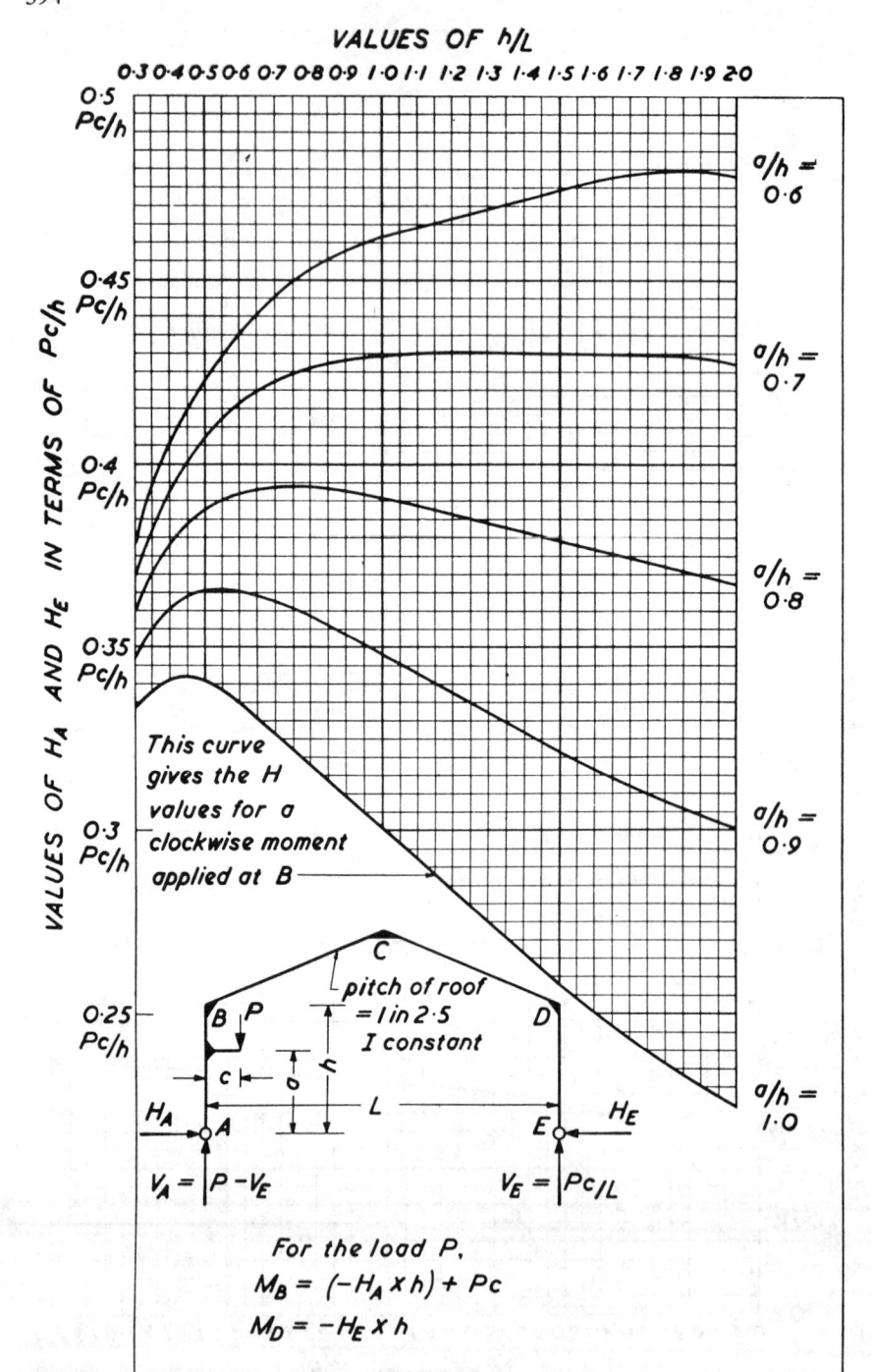

For the load P,
$M_B = (-H_A \times h) + Pc$
$M_D = -H_E \times h$

395

VALUES OF h/L

0·3 0·4 0·5 0·6 0·7 0·8 0·9 1·0 1·1 1·2 1·3 1·4 1·5 1·6 1·7 1·8 1·9 2·0

VALUES OF H_A AND H_E IN TERMS OF Pc/h

$a/h = 0.6$
$a/h = 0.7$
$a/h = 0.8$
$a/h = 0.9$
$a/h = 1.0$

This curve gives the H values for equal & opposite moments applied at B & D

pitch of roof = 1 in 2·5
I constant

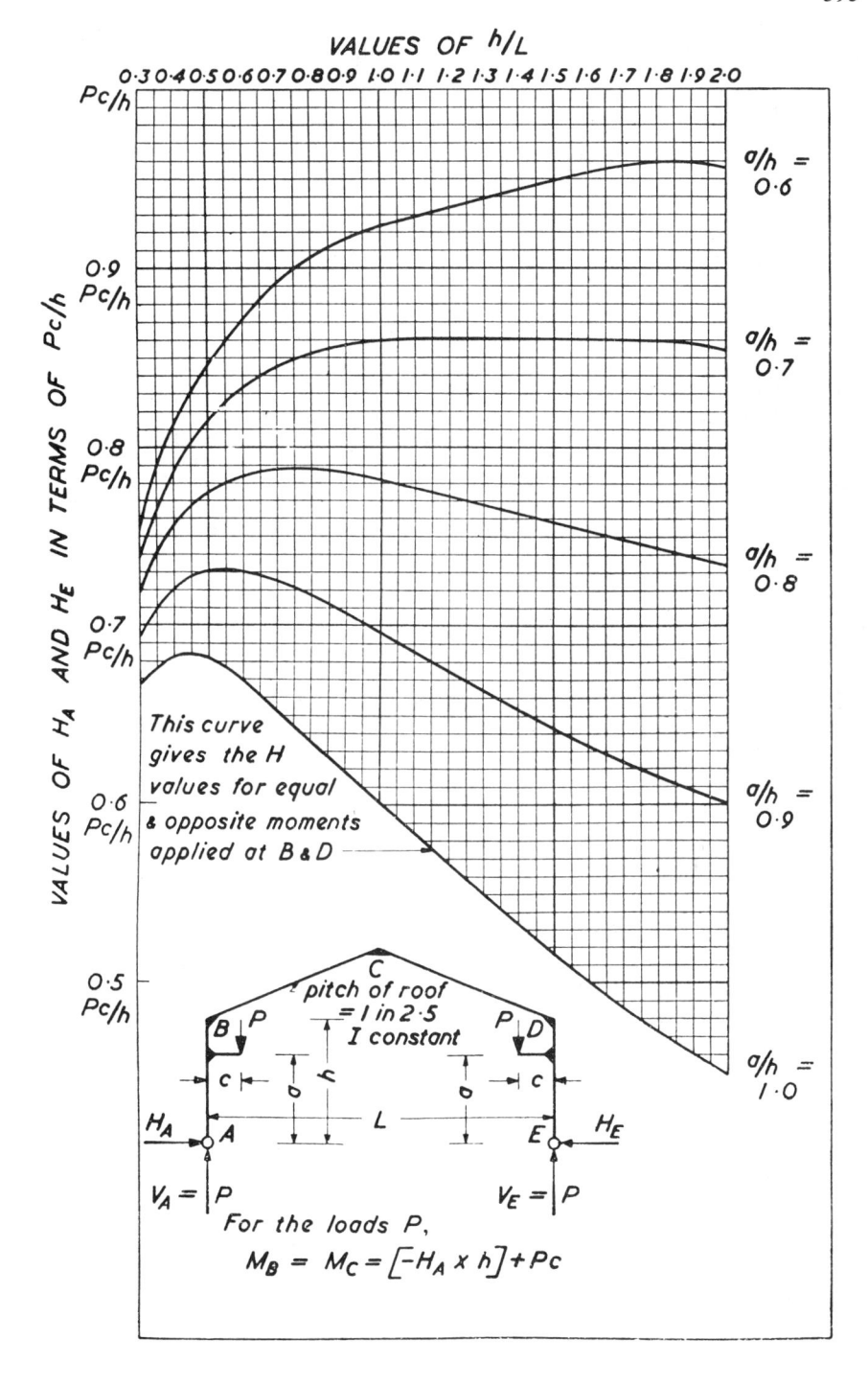

For the loads P,
$M_B = M_C = [-H_A \times h] + Pc$

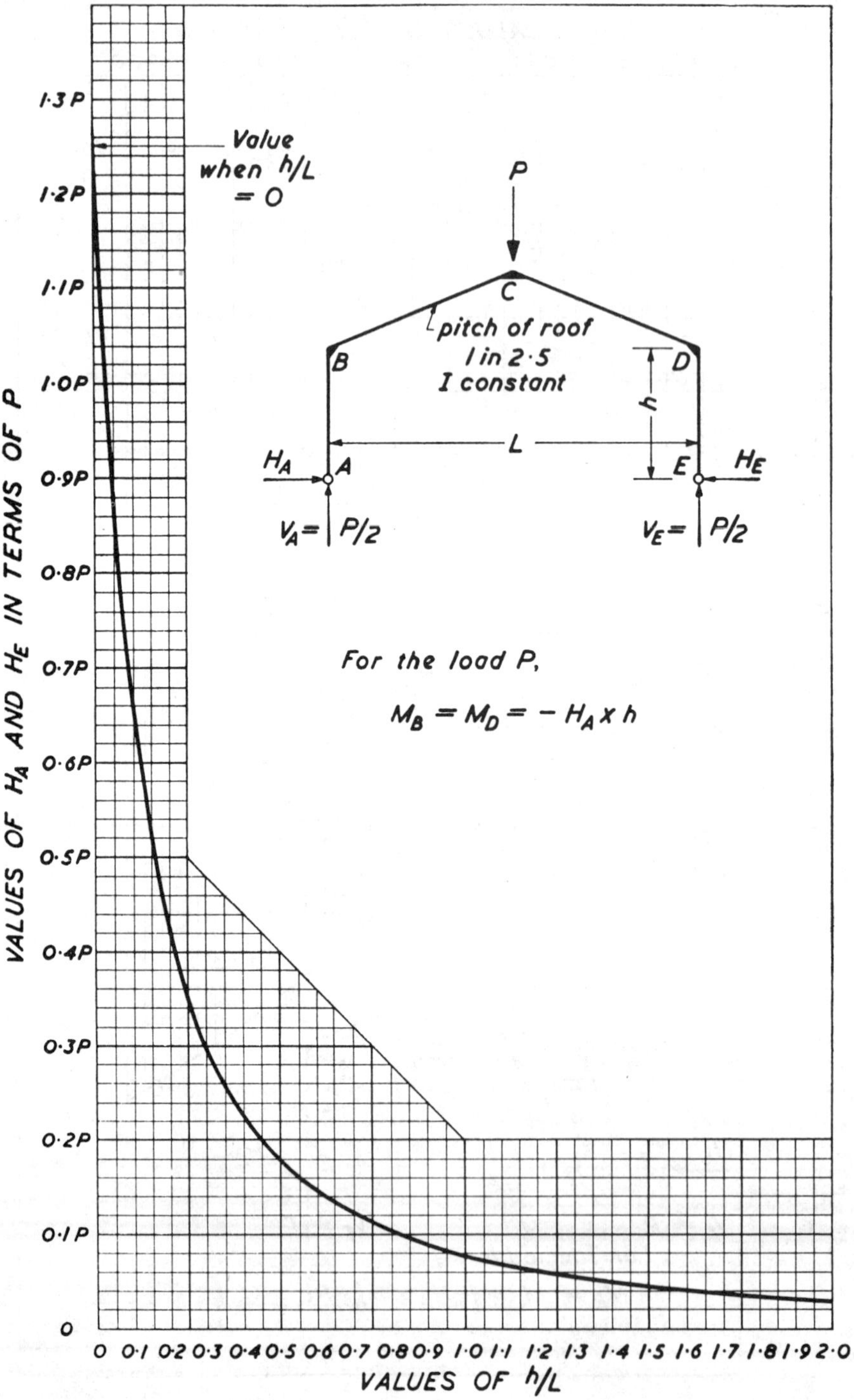
Value when h/L = 0
P
C
pitch of roof
1 in 2·5
I constant
B
D
h
L
A
E
H_A
H_E
V_A = P/2
V_E = P/2
For the load P,
M_B = M_D = − H_A × h
1·3P
1·2P
1·1P
1·0P
0·9P
0·8P
0·7P
0·6P
0·5P
0·4P
0·3P
0·2P
0·1P
0
VALUES OF H_A AND H_E IN TERMS OF P
0 0·1 0·2 0·3 0·4 0·5 0·6 0·7 0·8 0·9 1·0 1·1 1·2 1·3 1·4 1·5 1·6 1·7 1·8 1·9 2·0
VALUES OF h/L

pitch of roof
1 in 2.5
I constant

For the load P,

$M_B = -H_A \times h$ (resulting value positive)

$M_D = -H_E \times h$ (resulting value negative)

CURVE FOR $-H_A$

CURVE FOR H_E

VALUES OF $-H_A$ AND H_E IN TERMS OF P

VALUES OF h/L

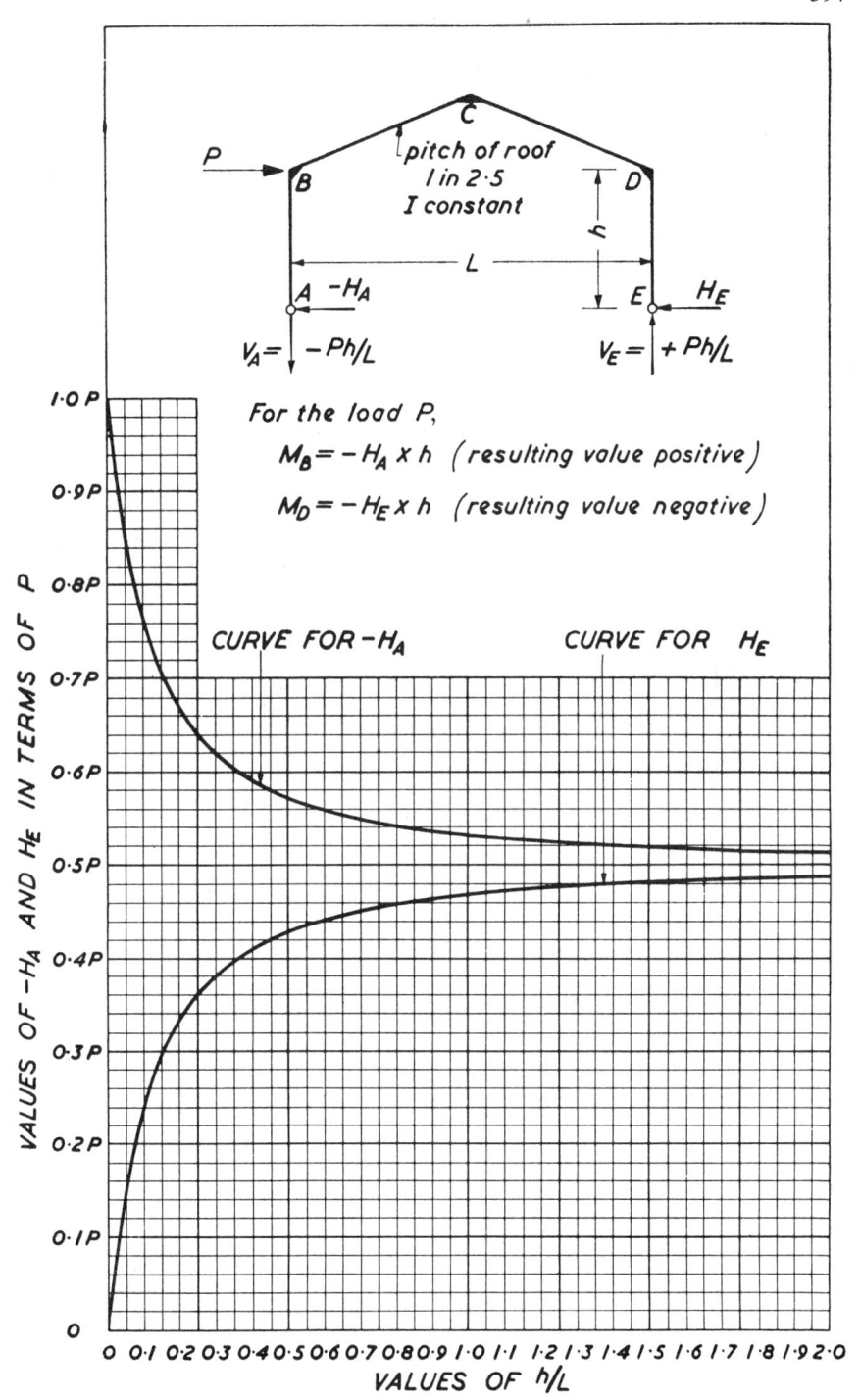

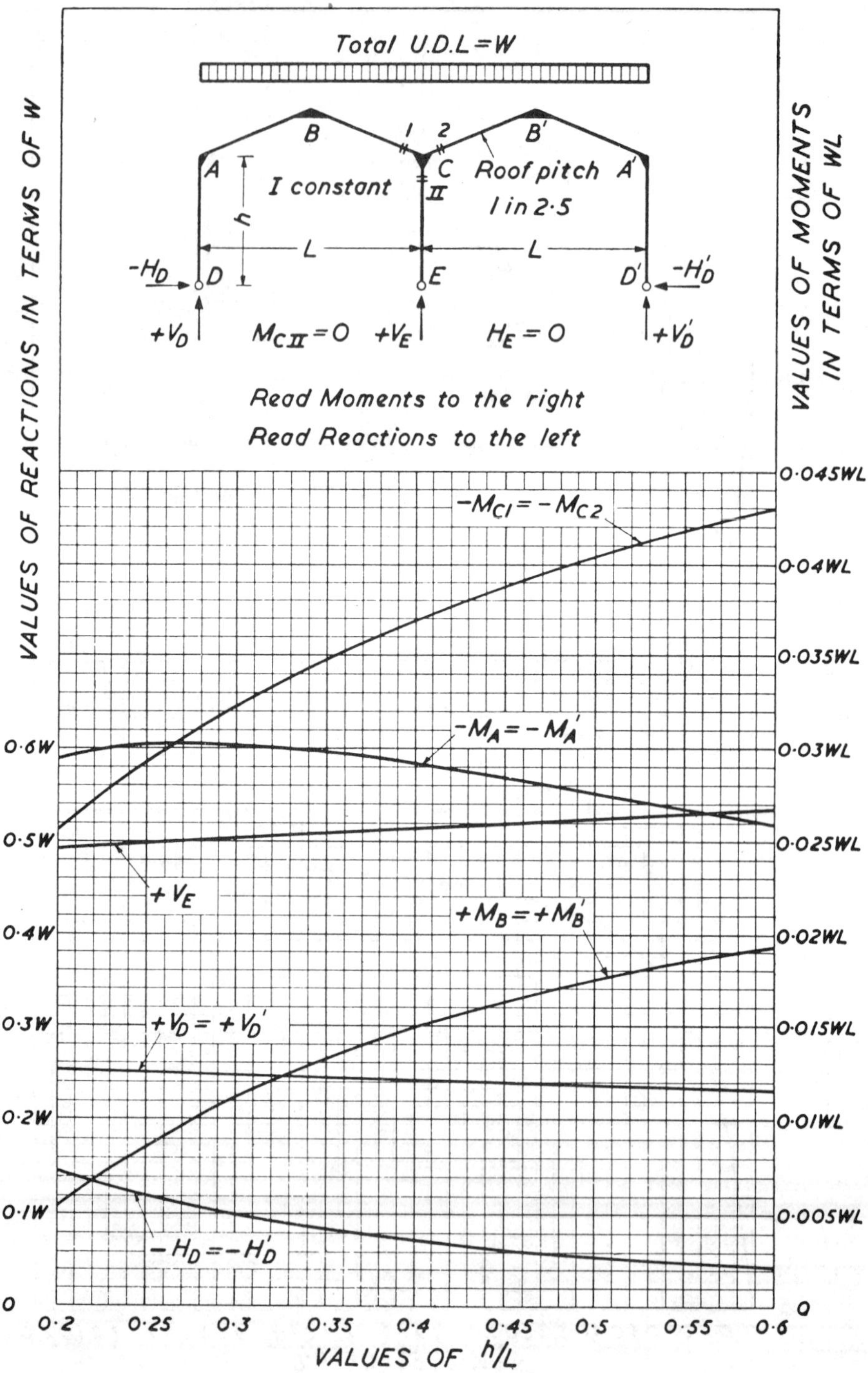

Total U.D.L=W
B
B'
1
2
A
C
Roof pitch
A'
I constant
II
1 in 2·5
h
L
L
$-H_D$
D
E
D'
$-H'_D$
$+V_D$
$M_{CII}=0$
$+V_E$
$H_E=0$
$+V'_D$
Read Moments to the right
Read Reactions to the left
VALUES OF REACTIONS IN TERMS OF W
VALUES OF MOMENTS IN TERMS OF WL
$-M_{C1}=-M_{C2}$
$-M_A=-M'_A$
$+V_E$
$+M_B=+M'_B$
$+V_D=+V'_D$
$-H_D=-H'_D$
0·6W
0·5W
0·4W
0·3W
0·2W
0·1W
0
0·045WL
0·04WL
0·035WL
0·03WL
0·025WL
0·02WL
0·015WL
0·01WL
0·005WL
0
0·2
0·25
0·3
0·35
0·4
0·45
0·5
0·55
0·6
VALUES OF h/L

399

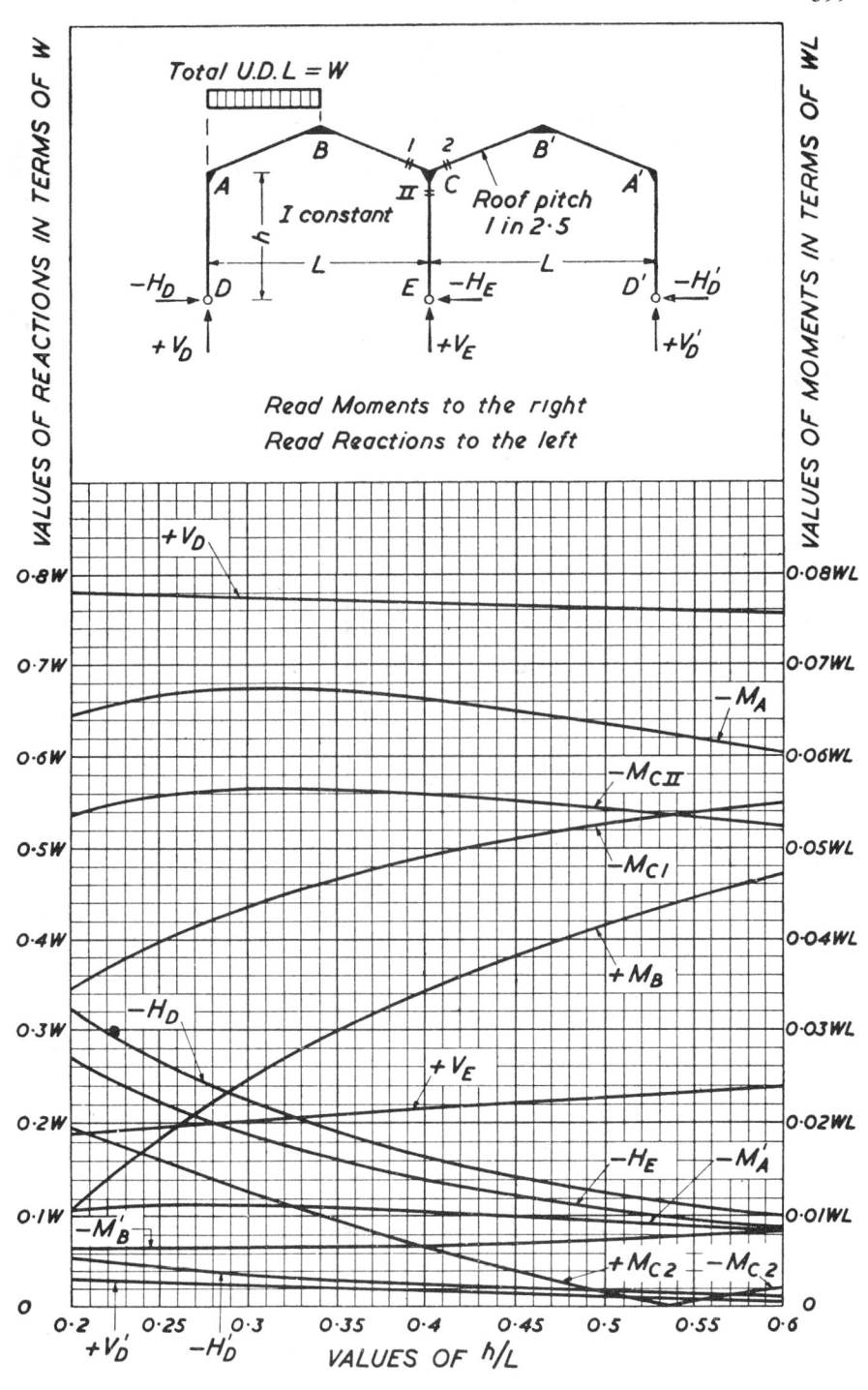

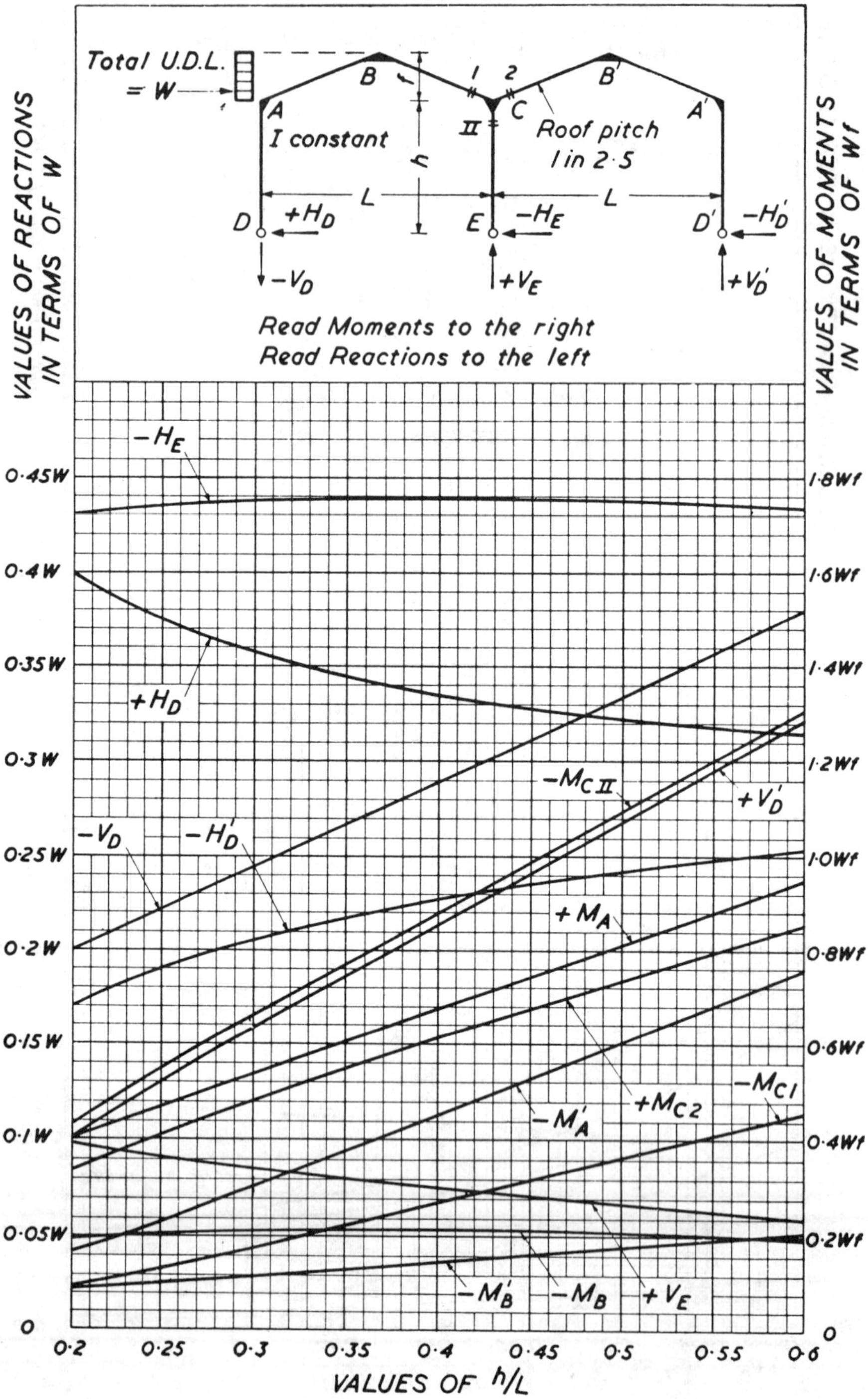
Total U.D.L. = W
I constant
Roof pitch 1 in 2·5
Read Moments to the right
Read Reactions to the left
VALUES OF REACTIONS IN TERMS OF W
VALUES OF MOMENTS IN TERMS OF Wf
VALUES OF h/L
0·45W
0·4W
0·35W
0·3W
0·25W
0·2W
0·15W
0·1W
0·05W
0
1·8Wf
1·6Wf
1·4Wf
1·2Wf
1·0Wf
0·8Wf
0·6Wf
0·4Wf
0·2Wf
0
0·2
0·25
0·3
0·35
0·4
0·45
0·5
0·55
0·6
−H_E
+H_D
−M_C II
+V'_D
−V_D
−H'_D
+M_A
+M_C2
−M_C1
−M'_A
−M'_B
−M_B
+V_E

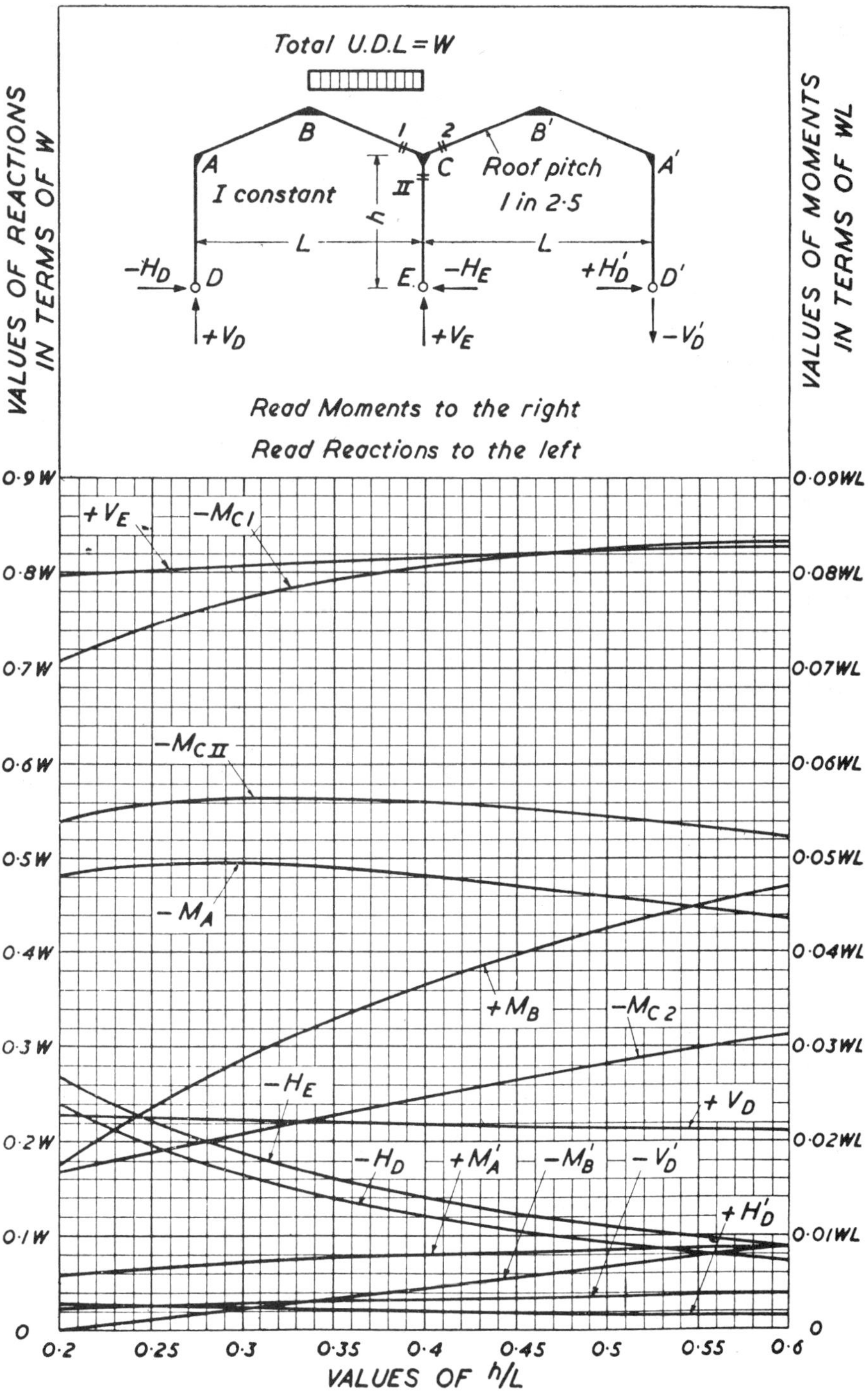
Total U.D.L = W
A
B
1
2
B'
A'
C
I constant
II
Roof pitch
1 in 2·5
h
L
L
−H_D
D
E
−H_E
+H'_D
D'
+V_D
+V_E
−V'_D
Read Moments to the right
Read Reactions to the left
VALUES OF REACTIONS IN TERMS OF W
VALUES OF MOMENTS IN TERMS OF WL
+V_E
−M_C1
−M_CII
−M_A
+M_B
−M_C2
−H_E
+V_D
−H_D
+M'_A
−M'_B
−V'_D
+H'_D
0·9W
0·8W
0·7W
0·6W
0·5W
0·4W
0·3W
0·2W
0·1W
0
0·09WL
0·08WL
0·07WL
0·06WL
0·05WL
0·04WL
0·03WL
0·02WL
0·01WL
0
0·2
0·25
0·3
0·35
0·4
0·45
0·5
0·55
0·6
VALUES OF h/L

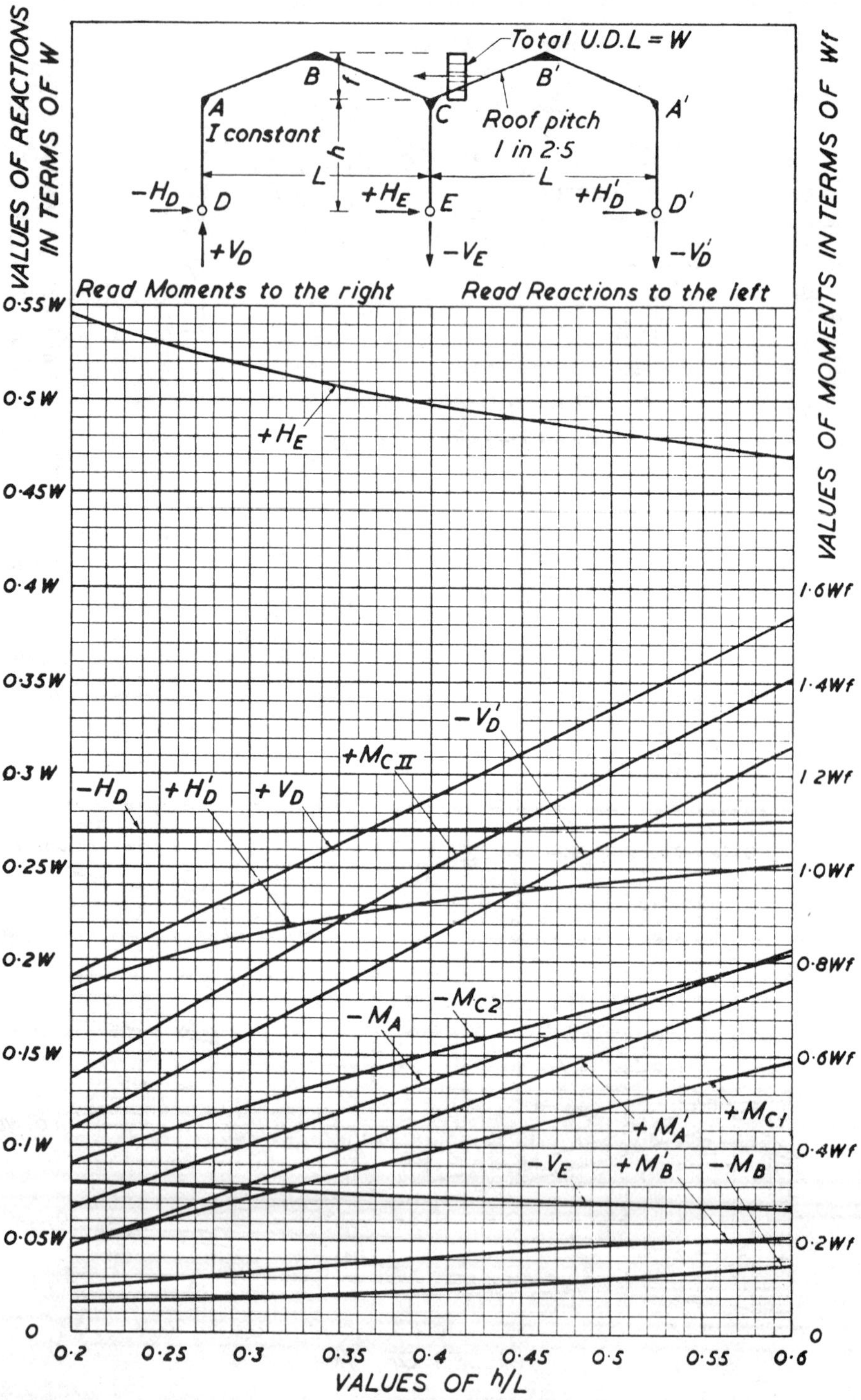
Total U.D.L = W
B
B'
A
C
A'
I constant
Roof pitch
1 in 2·5
h
f
L
L
−H_D
D
+H_E
E
+H'_D
D'
+V_D
−V_E
−V'_D
VALUES OF REACTIONS IN TERMS OF W
VALUES OF MOMENTS IN TERMS OF Wf
Read Moments to the right
Read Reactions to the left
0·55W
0·5W
0·45W
0·4W
0·35W
0·3W
0·25W
0·2W
0·15W
0·1W
0·05W
0
1·6Wf
1·4Wf
1·2Wf
1·0Wf
0·8Wf
0·6Wf
0·4Wf
0·2Wf
0
+H_E
−V'_D
+M_C II
−H_D
+H'_D
+V_D
−M_A
−M_C2
+M'_A
+M_C1
−V_E
+M'_B
−M_B
0·2
0·25
0·3
0·35
0·4
0·45
0·5
0·55
0·6
VALUES OF h/L

403

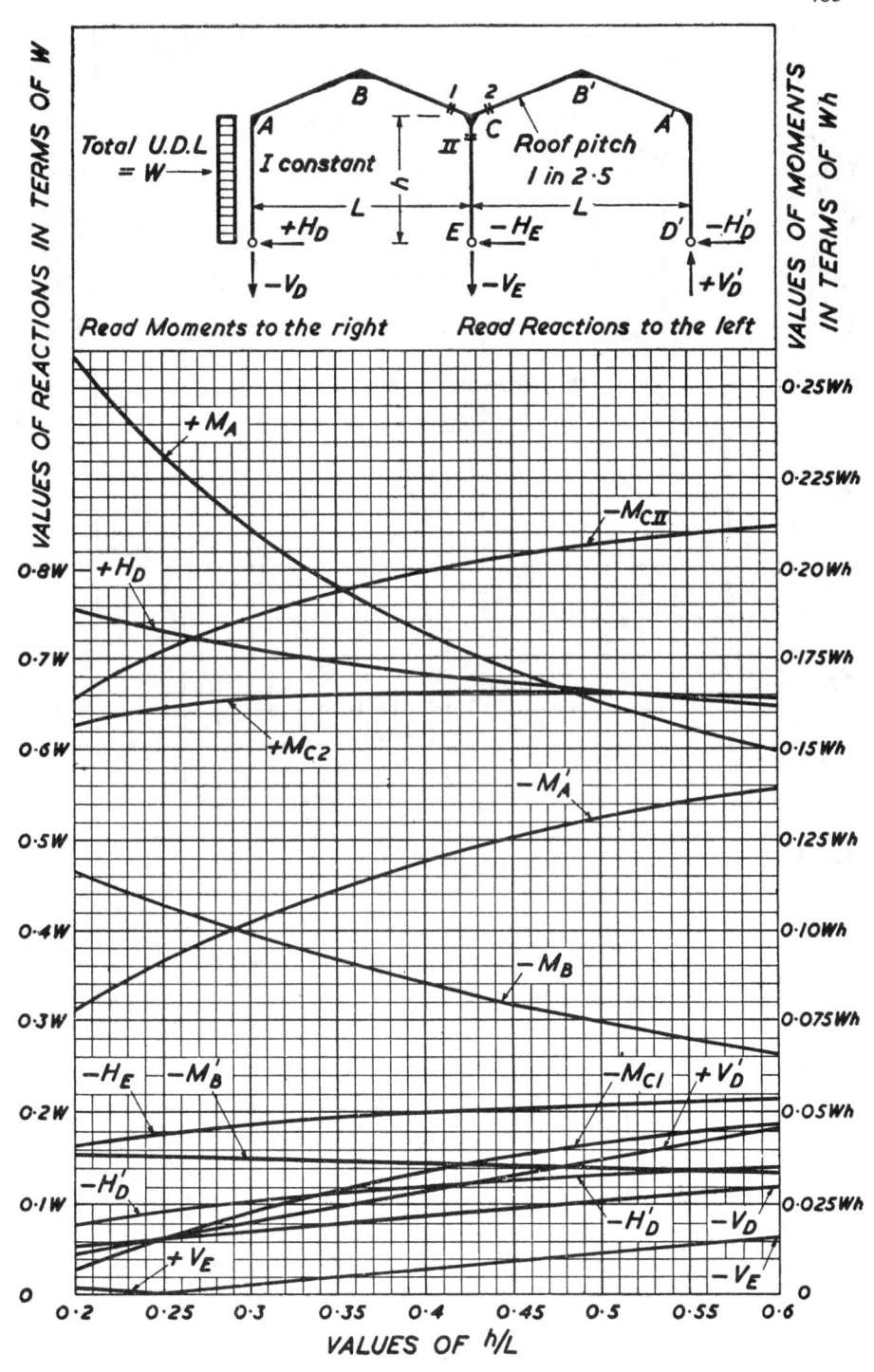

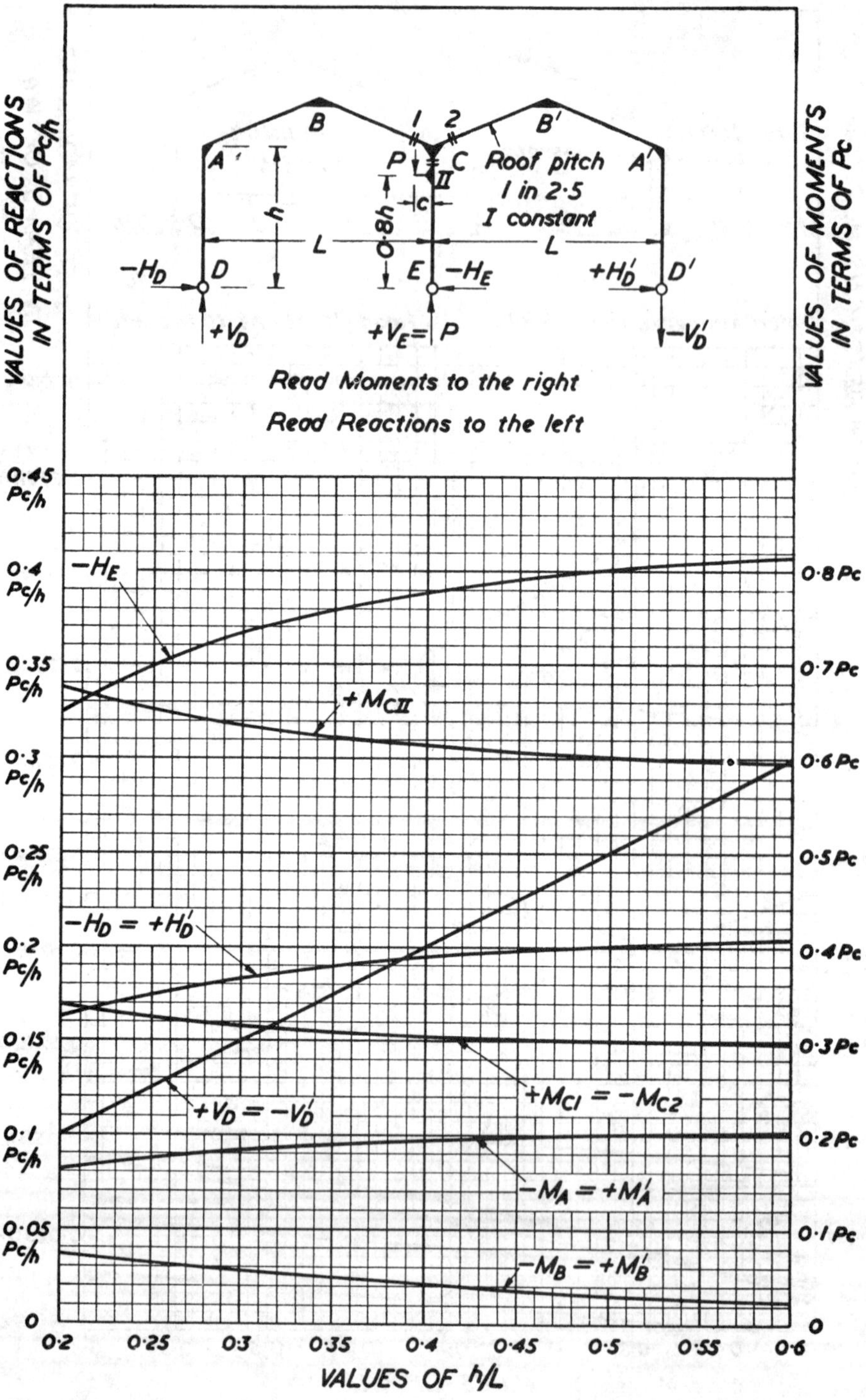
VALUES OF REACTIONS IN TERMS OF Pc/h
VALUES OF MOMENTS IN TERMS OF Pc
B
B'
1
2
A
A'
P
C
II
Roof pitch
1 in 2·5
I constant
h
0·8h
c
L
L
−H_D
D
E
−H_E
+H'_D
D'
+V_D
+V_E = P
−V'_D
Read Moments to the right
Read Reactions to the left
0·45 Pc/h
0·4 Pc/h
0·35 Pc/h
0·3 Pc/h
0·25 Pc/h
0·2 Pc/h
0·15 Pc/h
0·1 Pc/h
0·05 Pc/h
0
0·8 Pc
0·7 Pc
0·6 Pc
0·5 Pc
0·4 Pc
0·3 Pc
0·2 Pc
0·1 Pc
0
−H_E
+M_CII
−H_D = +H'_D
+M_C1 = −M_C2
+V_D = −V'_D
−M_A = +M'_A
−M_B = +M'_B
0·2
0·25
0·3
0·35
0·4
0·45
0·5
0·55
0·6
VALUES OF h/L

VALUES OF REACTIONS IN TERMS OF Pc/h

VALUES OF MOMENTS IN TERMS OF Pc

Roof pitch 1 in 2.5
I constant

Read Moments to the right
Read Reactions to the left

Note: $V_D = P - (V_E + V_D')$

Curves labelled: $+M_A$, $-H_D$, $-H_E$ or $-M_{C\mathrm{II}}$, $+M_{C2}$, $+V_D'$, $-M_B$, $-M_A'$ or $-H_D'$, $-M_{C1}$, $-M_B'$, $+V_E$

VALUES OF h/L

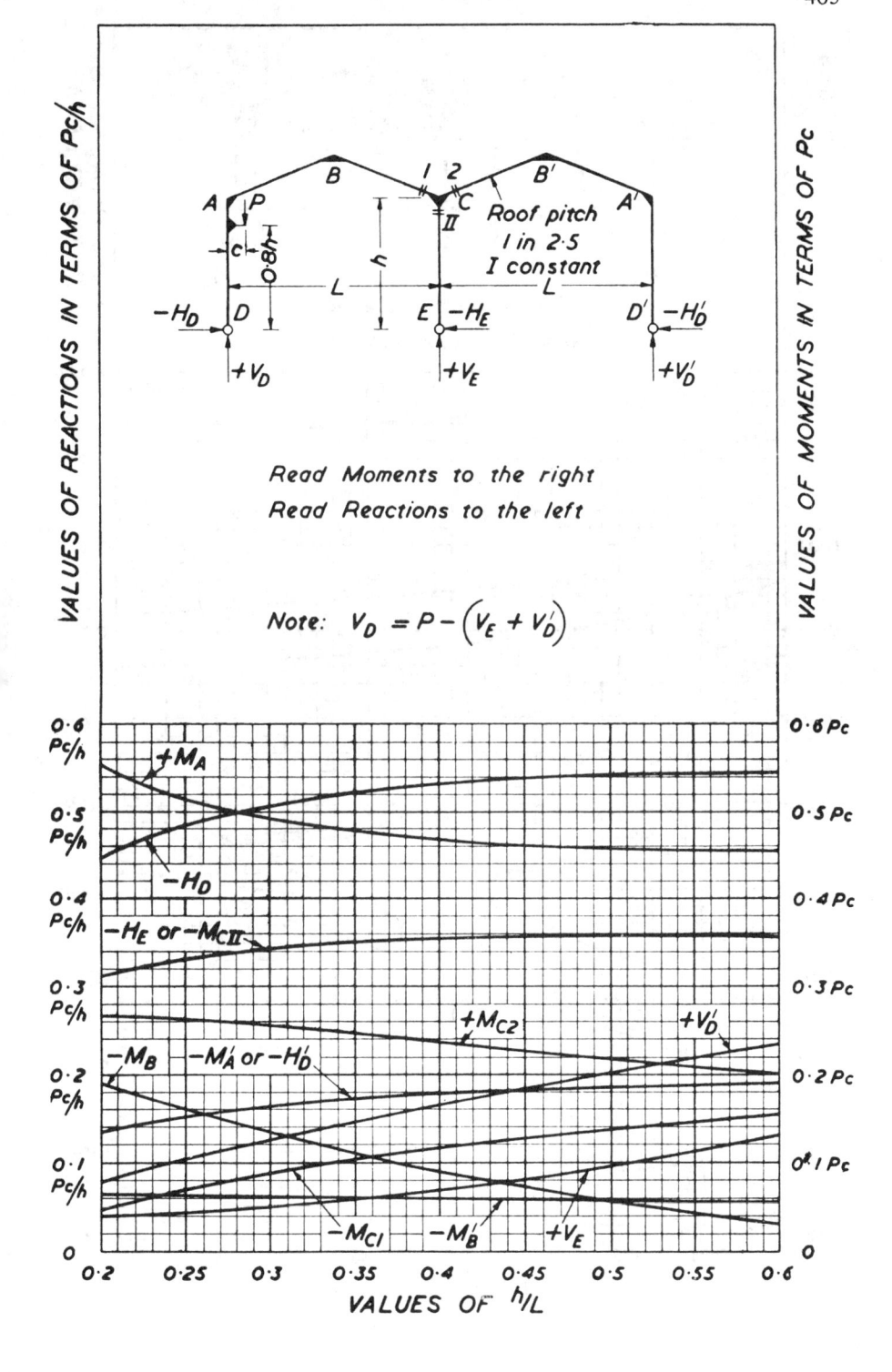

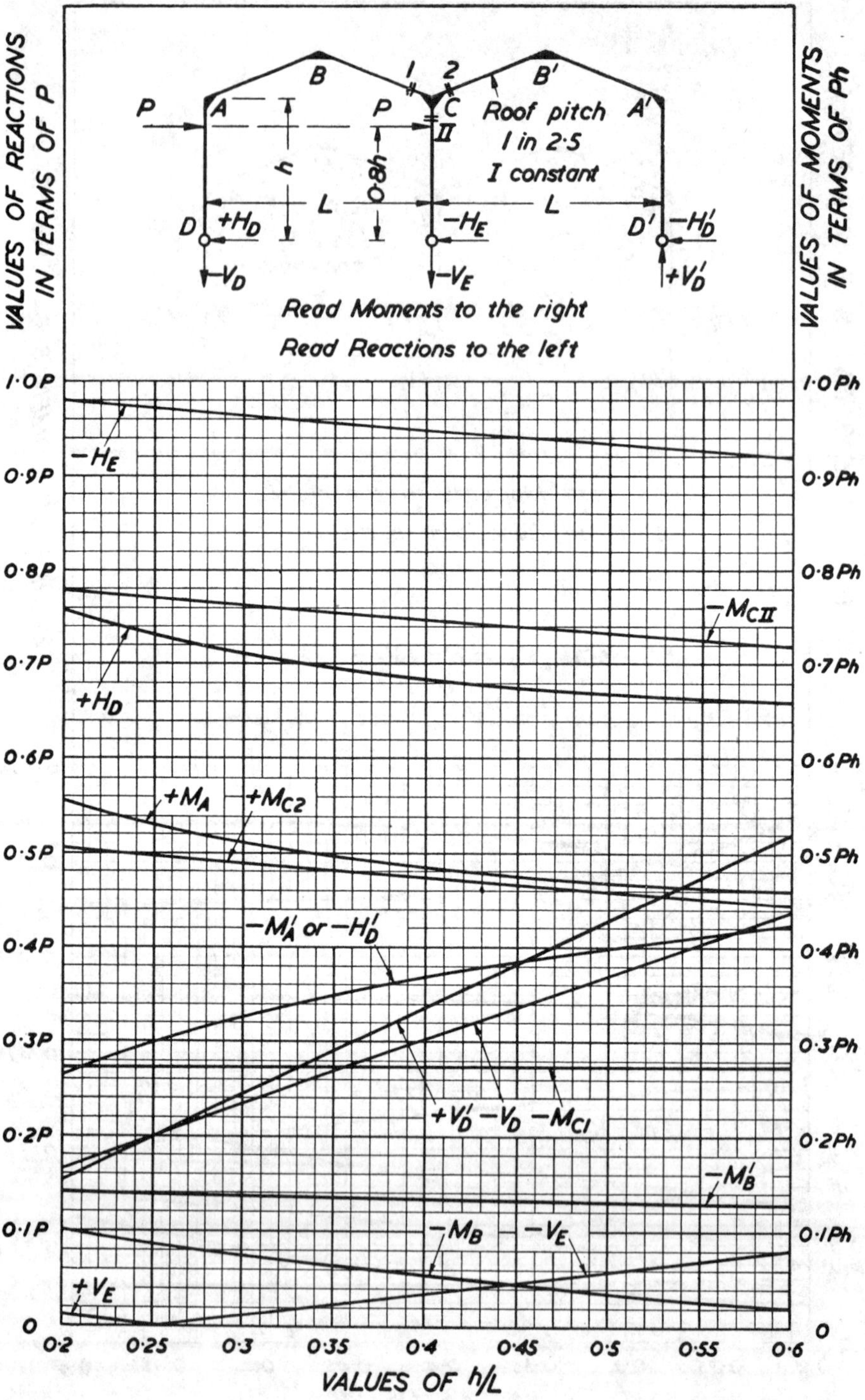
VALUES OF REACTIONS IN TERMS OF P
VALUES OF MOMENTS IN TERMS OF Ph
P
A
B
1
2
C
II
B'
A'
P
h
0·8h
Roof pitch
1 in 2·5
I constant
L
L
D
+H_D
-H_E
D'
-H'_D
-V_D
-V_E
+V'_D
Read Moments to the right
Read Reactions to the left
1·0P
0·9P
0·8P
0·7P
0·6P
0·5P
0·4P
0·3P
0·2P
0·1P
0
1·0Ph
0·9Ph
0·8Ph
0·7Ph
0·6Ph
0·5Ph
0·4Ph
0·3Ph
0·2Ph
0·1Ph
0
-H_E
-M_CII
+H_D
+M_A
+M_C2
-M'_A or -H'_D
+V'_D -V_D -M_CI
-M'_B
-M_B
-V_E
+V_E
0·2
0·25
0·3
0·35
0·4
0·45
0·5
0·55
0·6
VALUES OF h/L

407

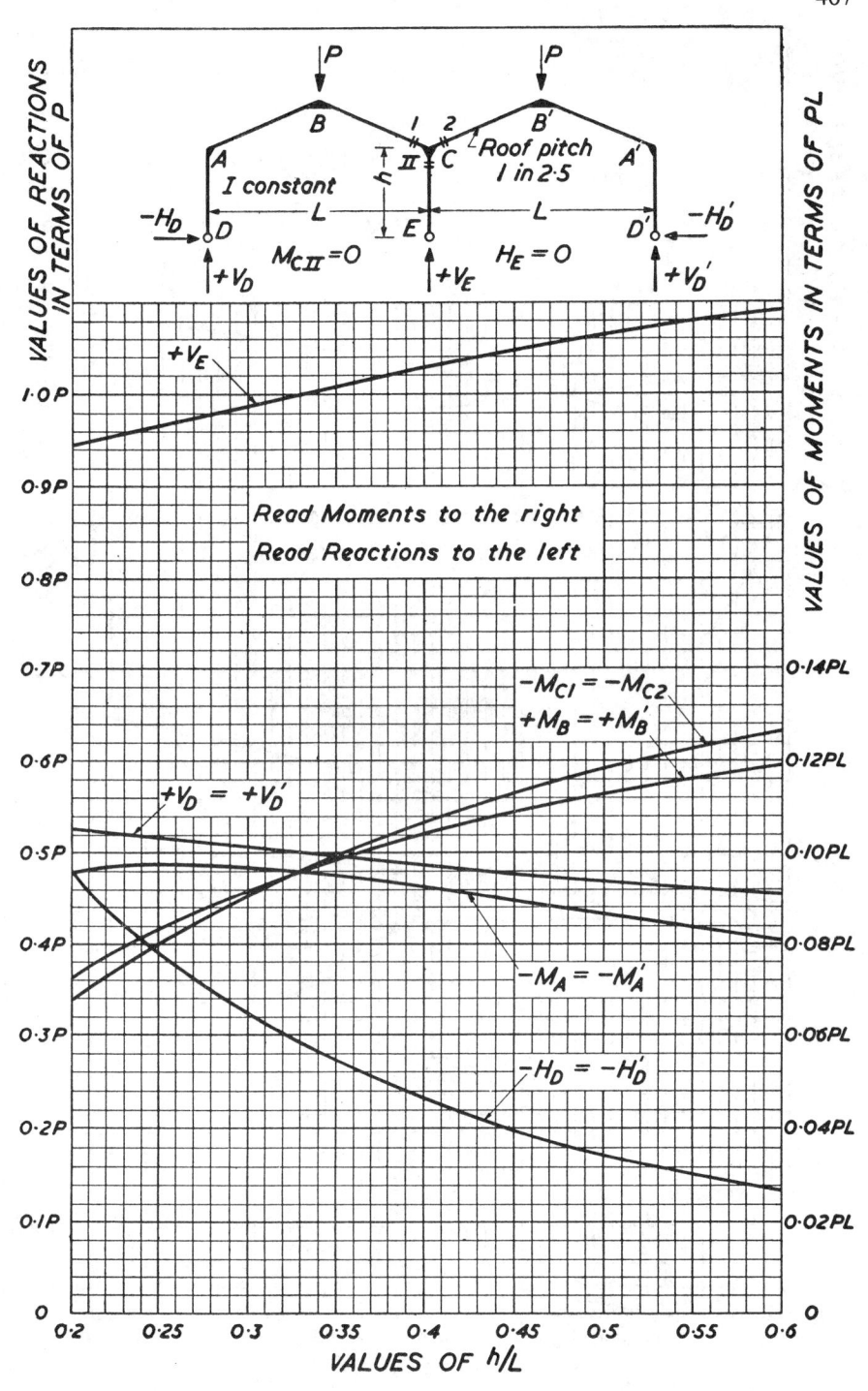

408

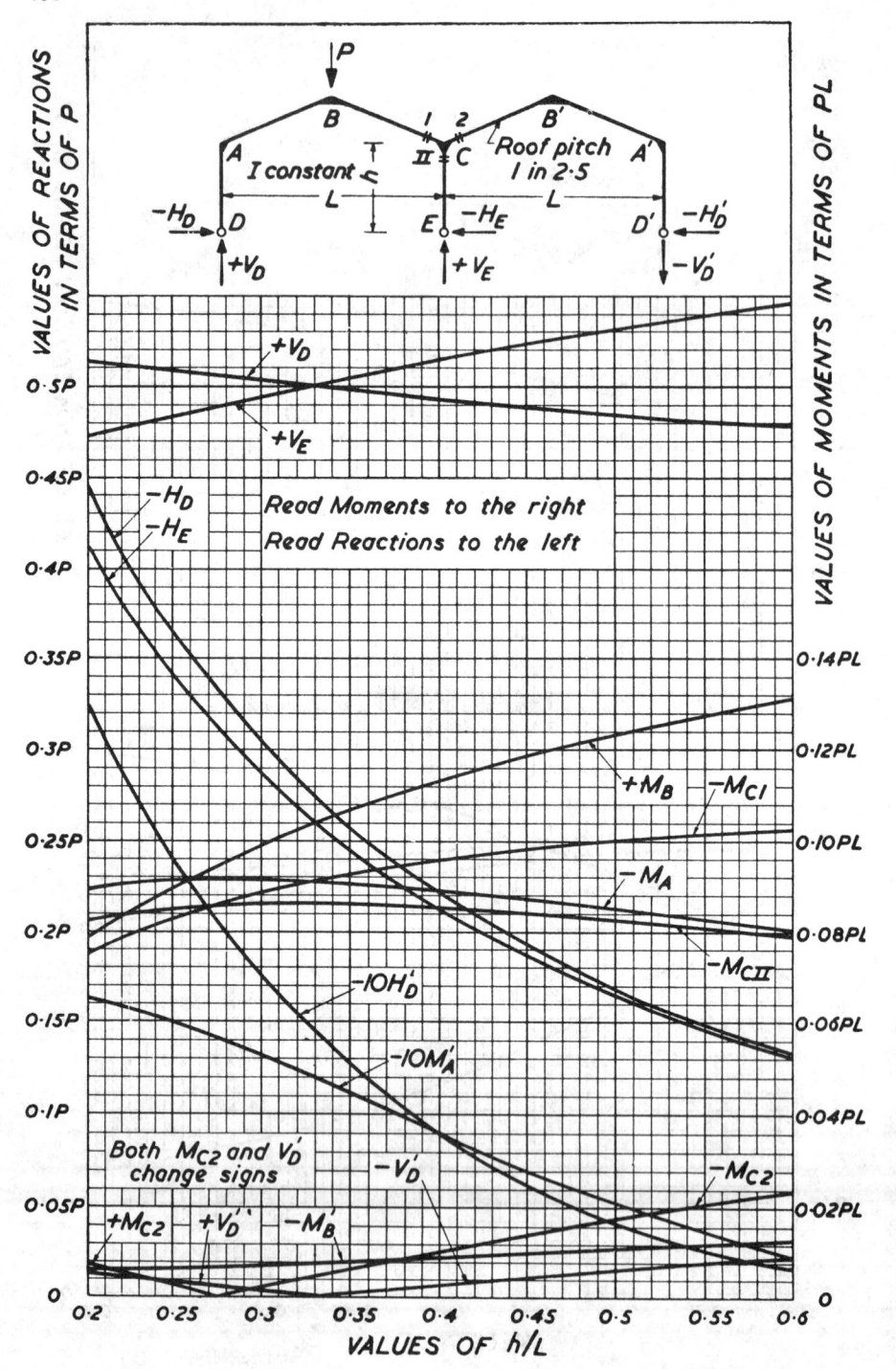

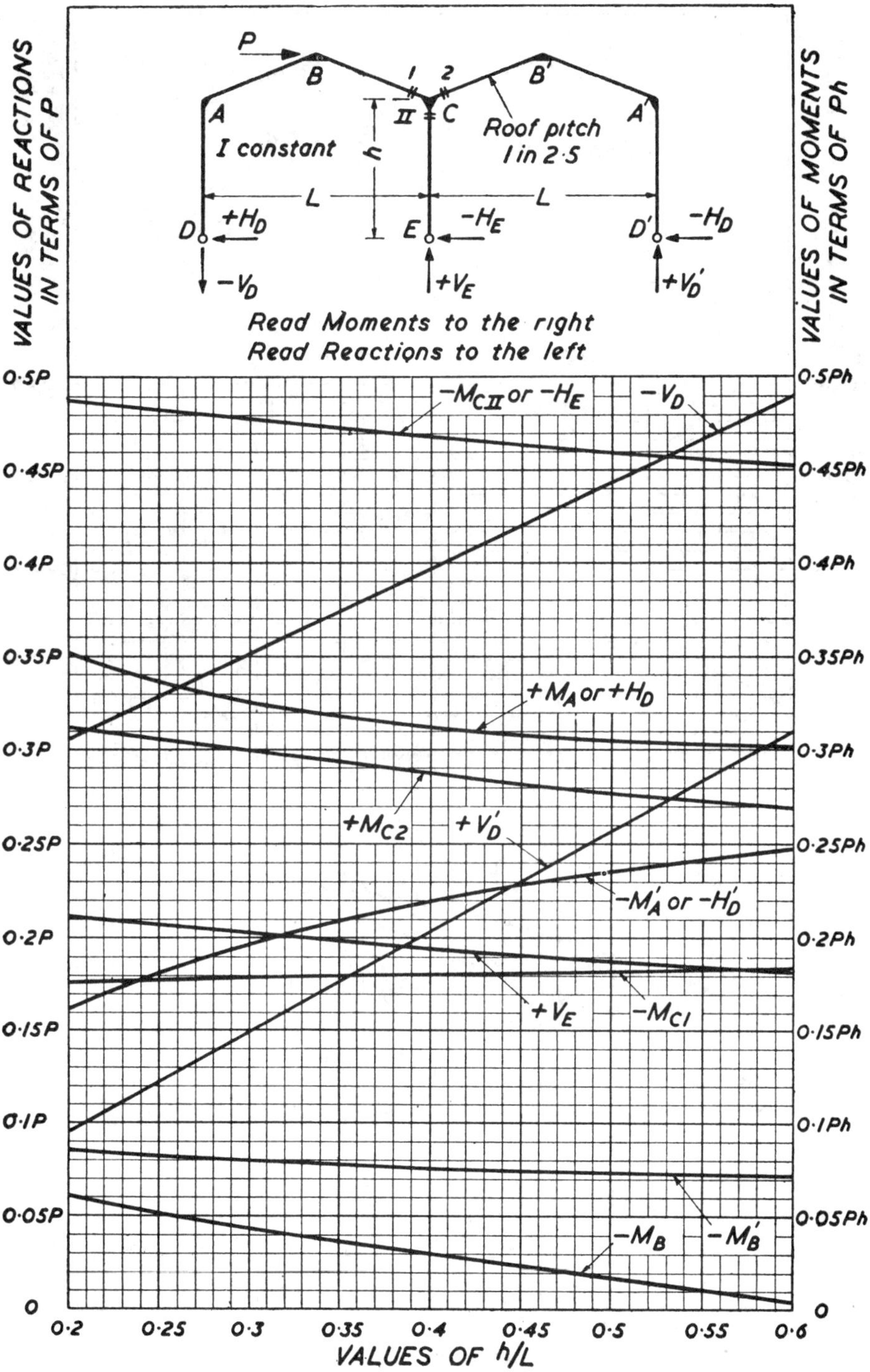
VALUES OF REACTIONS IN TERMS OF P
VALUES OF MOMENTS IN TERMS OF Ph
P
B
B'
1
2
A
A'
II
C
I constant
Roof pitch
1 in 2·5
h
L
L
D
+H_D
E
-H_E
D'
-H_D
-V_D
+V_E
+V_D'
Read Moments to the right
Read Reactions to the left
0·5P
0·45P
0·4P
0·35P
0·3P
0·25P
0·2P
0·15P
0·1P
0·05P
0
0·5Ph
0·45Ph
0·4Ph
0·35Ph
0·3Ph
0·25Ph
0·2Ph
0·15Ph
0·1Ph
0·05Ph
0
-M_CII or -H_E
-V_D
+M_A or +H_D
+M_C2
+V_D'
-M_A' or -H_D'
+V_E
-M_C1
-M_B
-M_B'
0·2
0·25
0·3
0·35
0·4
0·45
0·5
0·55
0·6
VALUES OF h/L

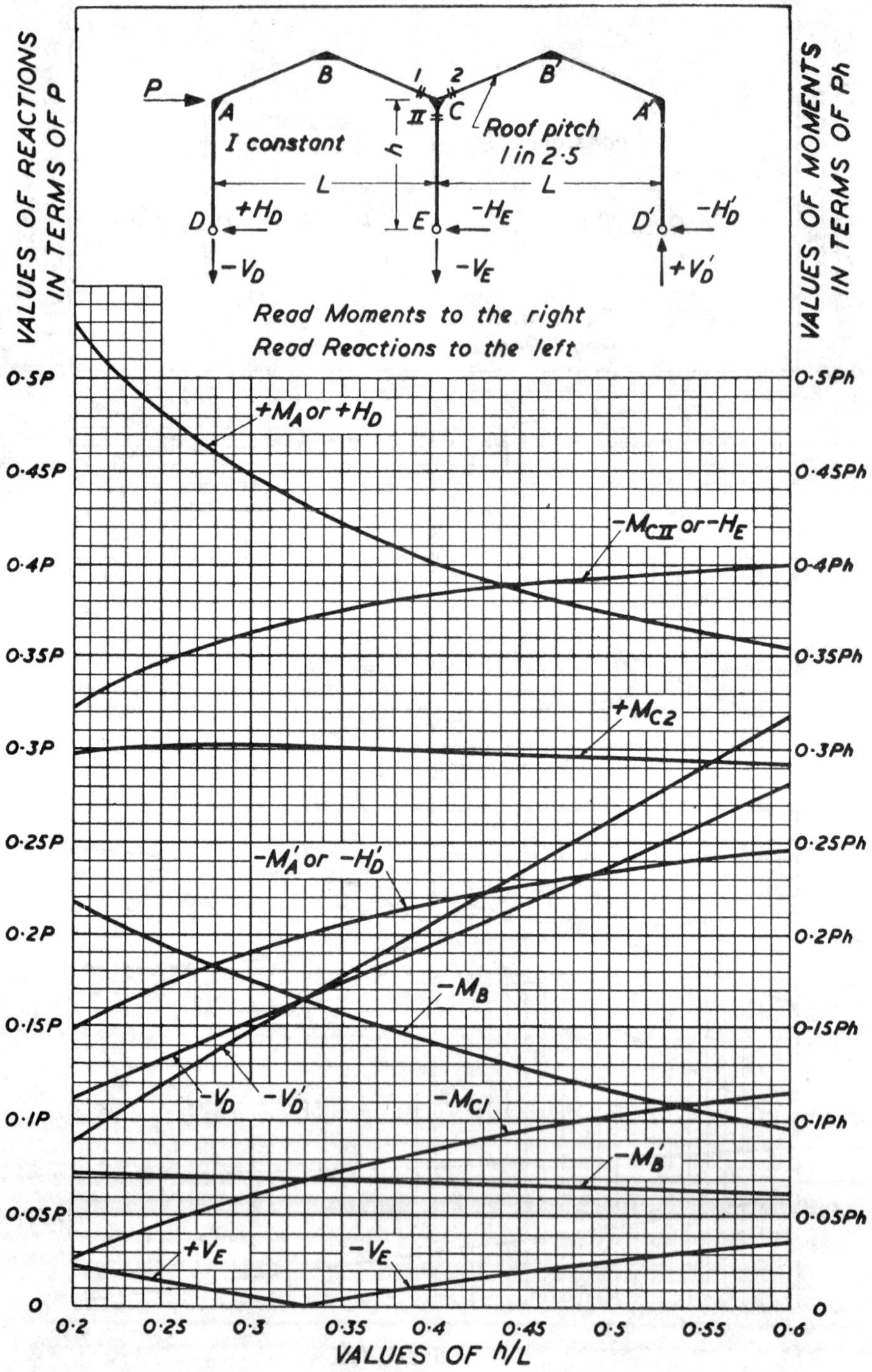
P
A
B
B'
A'
C
1
2
II
I constant
h
Roof pitch
1 in 2·5
L
L
D
+H_D
E
-H_E
D'
-H'_D
-V_D
-V_E
+V'_D
Read Moments to the right
Read Reactions to the left
VALUES OF REACTIONS IN TERMS OF P
VALUES OF MOMENTS IN TERMS OF Ph
0·5P
0·45P
0·4P
0·35P
0·3P
0·25P
0·2P
0·15P
0·1P
0·05P
0
0·5Ph
0·45Ph
0·4Ph
0·35Ph
0·3Ph
0·25Ph
0·2Ph
0·15Ph
0·1Ph
0·05Ph
0
+M_A or +H_D
-M_CII or -H_E
+M_C2
-M'_A or -H'_D
-M_B
-V_D
-V'_D
-M_C1
-M'_B
+V_E
-V_E
0·2
0·25
0·3
0·35
0·4
0·45
0·5
0·55
0·6
VALUES OF h/L

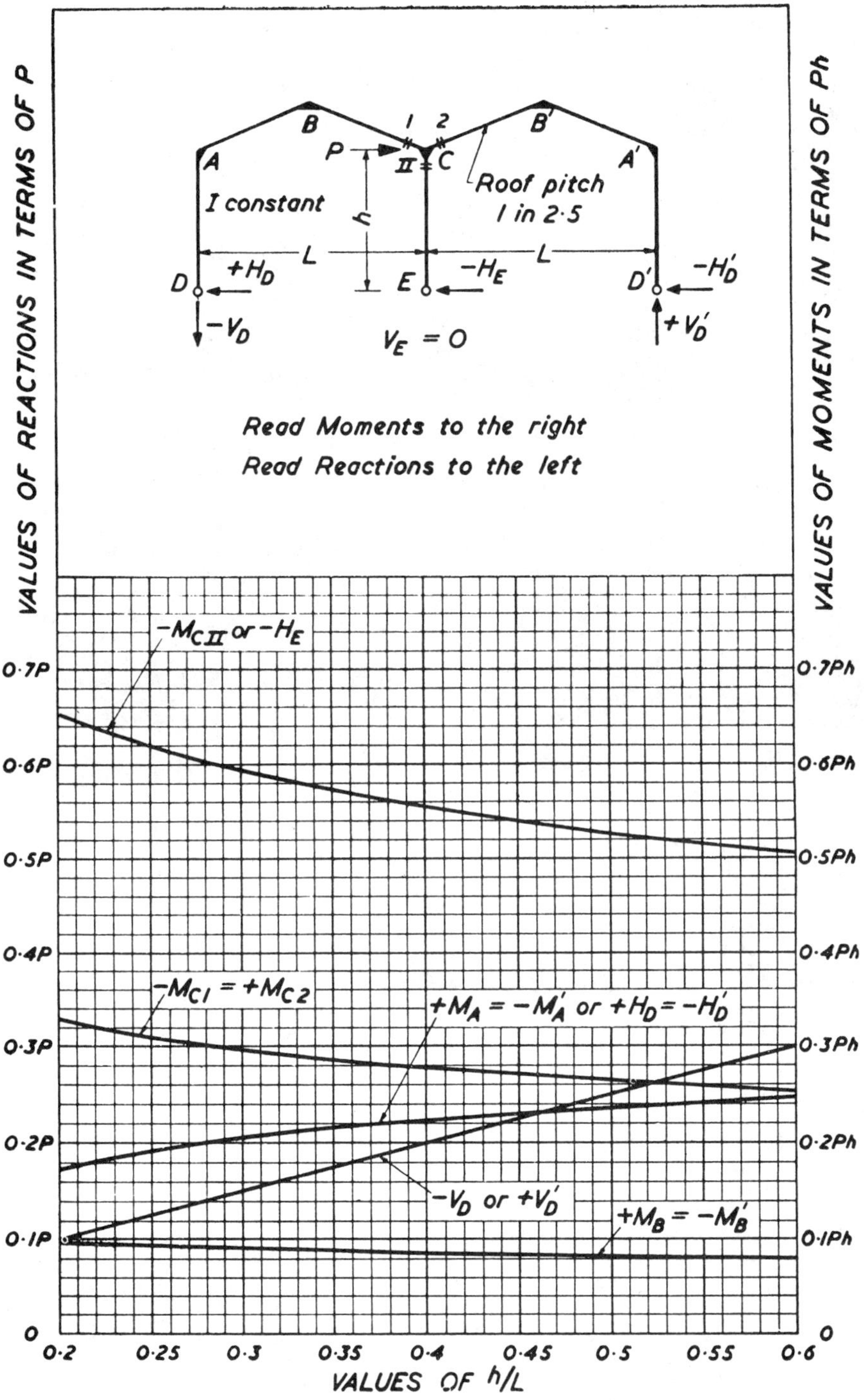
VALUES OF REACTIONS IN TERMS OF P
VALUES OF MOMENTS IN TERMS OF Ph
A
B
1
2
B'
A'
P
II
C
Roof pitch
1 in 2·5
I constant
h
L
L
+H_D
-H_E
-H'_D
D
E
D'
-V_D
V_E = O
+V'_D
Read Moments to the right
Read Reactions to the left
-M_CII or -H_E
-M_C1 = +M_C2
+M_A = -M'_A or +H_D = -H'_D
-V_D or +V'_D
+M_B = -M'_B
0·7P
0·6P
0·5P
0·4P
0·3P
0·2P
0·1P
0
0·7Ph
0·6Ph
0·5Ph
0·4Ph
0·3Ph
0·2Ph
0·1Ph
0
0·2
0·25
0·3
0·35
0·4
0·45
0·5
0·55
0·6
VALUES OF h/L

17. VIERENDEEL GIRDERS

IN 1896, a Belgian engineer, Professor Arthur Vierendeel, suggested a method of constructing an 'open' web girder with rigid joints, comprised of a top and a bottom chord with vertical members only between the booms. This type of girder is known as a Vierendeel girder.

Unlike a truss with diagonal members which are generally designed for direct stresses only, the Vierendeel girder members are subjected to bending, axial and shear stresses. Typical forms of Vierendeel girder are shown in Fig. 1.

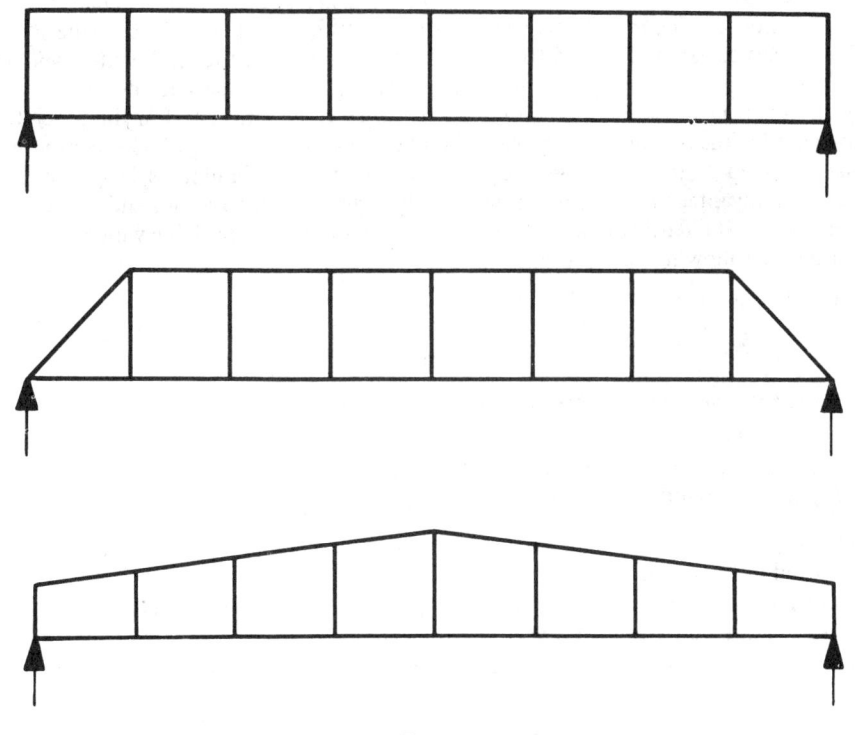

Fig. 1

Many uses can be made of this type of rigid frame, especially for bridges, as in Belgium, though in Great Britain Vierendeel girders are more commonly seen in church, school and industrial structures where clerestory lighting is required and the absence of diagonal frame members is desirable.

Methods of Analysis

The Vierendeel girder is a statically indeterminate structure but a simple static analysis can be adopted. The structure will become statically determinate if three pins are introduced for each panel and the characteristic action of the structure is maintained by placing the pins at the mid-lengths of the chord members and mid heights of the verticals.

The statically determinate analysis is not generally suitable for Vierendeels with:

(a) inclined members,
(b) chords of radically different stiffness,
(c) posts of variable depth or with loads applied away from node points.

The limitations of a statically determinate analysis necessitate consideration of other methods. If the top and bottom chords of any panel of a Vierendeel girder are of the same section, Naylor's application of moment distribution can be used. A worked example is given in Chapter 13.

Modified moment distribution methods have been developed for the analysis of Vierendeel girders. The substitute frame method (1) can be used for the analysis of parallel chorded Vierendeel bents with chords of different stiffness in the panels. The stiffnesses can be adjusted to allow for curved gussets at the ends of the members.

A further modification of moment distribution procedure has been developed for the analysis of a Vierendeel girder with inclined top chords (2). For the general case with inclined members and loading applied away from node points, a computer analysis based on the generalised slope deflection equations minimises design time. A further advantage is that the computer output gives joints rotations and translations. Standard computer application programmes are available which require the following data for input:

(a) The geometry of the frame.
(b) The physical properties of the members of the frame.
(c) The alternative loadings on the frame.

After calculation, the computer will provide the following results:

(a) The deformations at each joint in the frame.
(b) The axial and shear forces in each member.
(c) The moments at each end of each member.

The results of a computer (see reference (3)) and statically determinate analysis are compared in the design example given later.

The plastic theory may be applied to the design of Vierendeel girders. Typical modes of failure are shown in Fig. 2 and it should be noted that failure of the structure, as a whole, generally results from local failure of a small number of its members. The failure mode indicated in Fig. 2 (a) is due to plastic hinges formed at the end of chord members. In Fig. 2 (b), plastic hinges are formed at the ends of two of the verticals and at one section in each chord. Figure 2 (c) indicates failure by formation of hinges in one or more of the verticals and at two sections in each chord.

Detailed information of the plastic analysis of Vierendeel girders is given in reference (4).

The approach would be to design the chord members against failure by mode shown in Fig. 2 (a) and then to design the verticals against failure by modes shown

Fig. 2

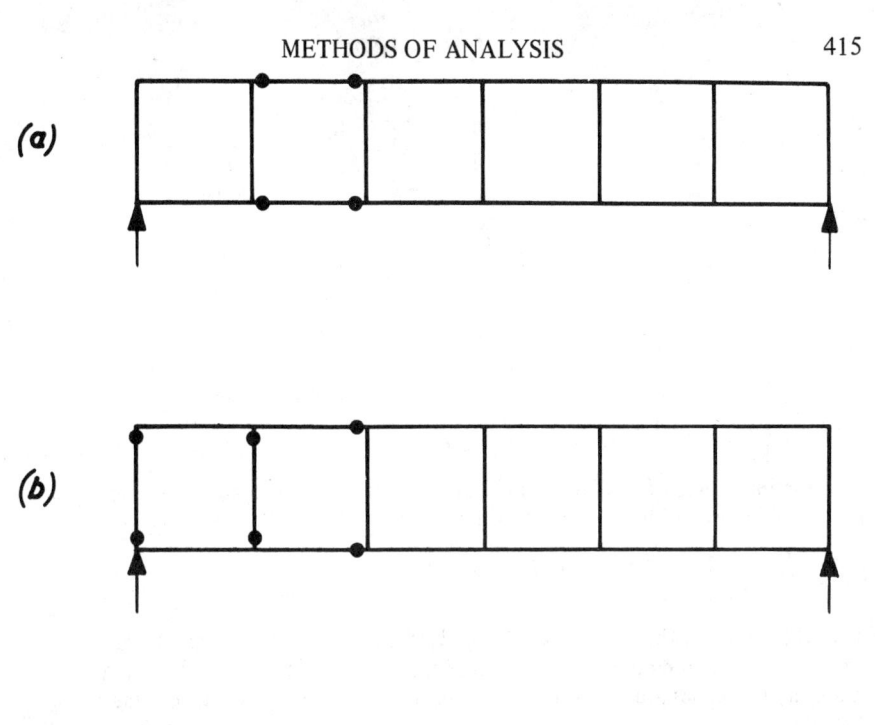

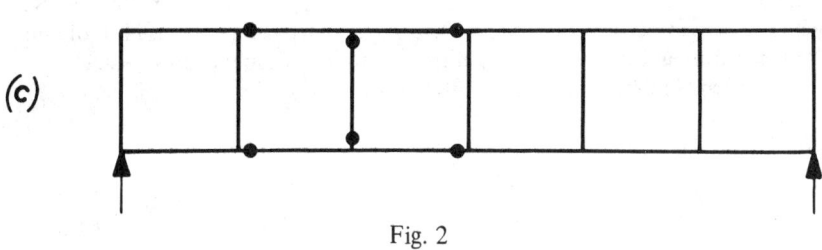

in Figs. 2 (b) and 2 (c). Standard programmes are available for the plastic analysis of plane frameworks which no doubt could be modified to give the collapse conditions for Vierendeels.

Joints

Although considerable increase in strength can be obtained by reinforcing the joints with curved flanges, Fig. 3, this type of joint is rather expensive and is now rarely used. Similar strength can be obtained by the use of simple diagonal stiffeners and critical sections are easier to locate. Detailed information on joint design is given in the earlier part of this chapter. Calculations for a typical joint without stiffening are included in a subsequent design example.

Example

The Vierendeel girder shown in Fig. 4 forms part of the roof structure for an industrial building and is subjected to node point loading only. (For simplicity, top

VIERENDEEL GIRDERS

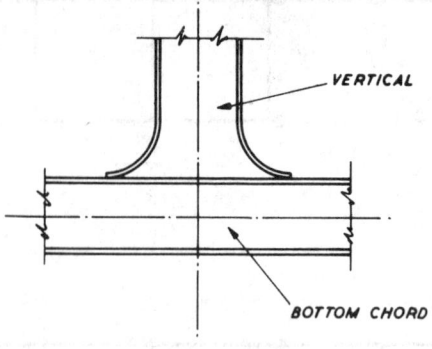

Fig. 3

and bottom chords of similar section have been used.) The bending moments, axial and shear forces will be compared using:

(1) The statically determinate analysis.
(2) A computer analysis based on slope deflection theory.

It should be noted that for the statically determinate analysis, the sectional properties of the members are not required, whereas for the computer analysis, sectional properties are necessary. The section sizes obtained from the statically determinate analysis will be used for the computer analysis. However, for frames which are not symmetrical, a statically determinate analysis can be used to obtain preliminary sizes and the computer can then be used for rapid analysis using variations of the preliminary sizes to obtain optimum results.

(1) *Statically Determinate Analysis*

As the frame is symmetrical, only half will be considered and imaginary pins are placed at the mid-span of the chords and mid-height of the verticals as indicated in Fig. 4 (a).

Considering the vertical equilibrium of the structure,

$$\text{total downward load} = 10(30.2 + 33.8)$$
$$= 640 \text{ kN}$$

Hence $RL = RR = 320 \text{ kN}$

Vertical shear at $X - X = 320 - (15.1 + 16.9)$
$$= 288 \text{ kN.}$$

This is divided equally between the top and bottom chords to give a shear at the pins of 144 kN.
Similarly,

Vertical shear at $Y - Y = 320 - (15.1 + 16.9 + 30.2 + 33.8)$
$$= 224 \text{ kN}$$

$$\text{Shear at pins} = \frac{224}{2}$$
$$= 112 \text{ kN.}$$

METHODS OF ANALYSIS

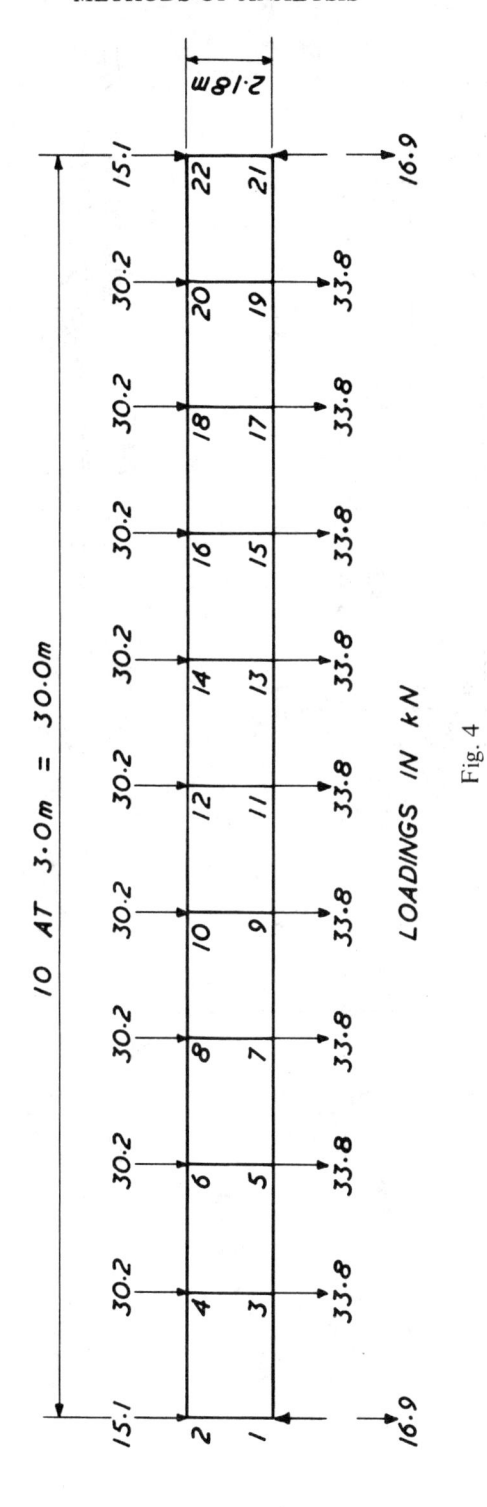

Fig. 4

LOADINGS IN kN

418 VIERENDEEL GIRDERS

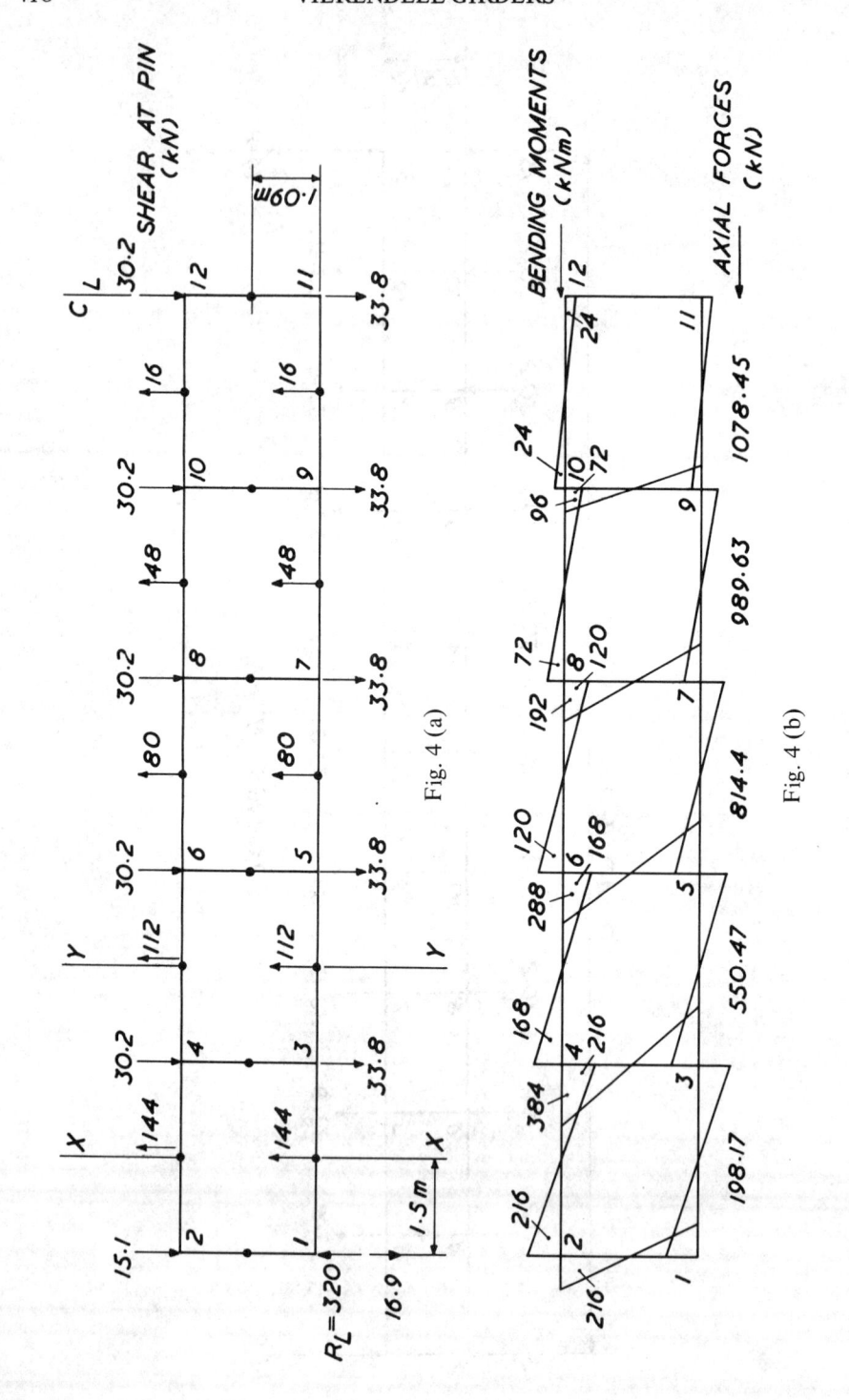

Fig. 4 (a)

Fig. 4 (b)

This procedure is repeated for the remaining chord pins and the shears are indicated in Fig. 4 (a). The chord bending moments are obtained by multiplying the pin shears by half the length of the panel.

$$M_{2-4} = 144 \times \frac{3.0}{2} = 216 \text{ kNm}$$

$$M_{4-6} = 112 \times \frac{3.0}{2} = 168 \text{ kNm}.$$

Considering the joint equilibrium

$$M_{2-1} = 216 \text{ kNm}$$

$$M_{4-3} = (216 + 168) = 384 \text{ kNm}.$$

The shear in the verticals is obtained by dividing the moment in the verticals at the joints by half their height.

For member 1–2 shear = $\frac{216 \times 2}{2.18} = 198.2$ kN

For member 3–4 shear = $\frac{384 \times 2}{2.18} = 352.3$ kN.

The bending moment diagram for the frame is indicated in Fig. 4 (b).
The axial force in the chords is obtained by summing the horizontal shears.

$$\text{Member } 1-3 = 198.2 \text{ kN}$$

$$\text{Member } 3-5 = (198.2 + 352.3) = 550.5 \text{ kN}.$$

The axial forces are indicated in Fig. 4 (b).

(2) *Computer Analysis*

From the statically determinate analysis, the verticals and chords can be designed. These are as follows:

Top and bottom chords 2/381 x 102 x 55 kg channels placed back to back 255 mm apart.

Verticals:

Member	Section
1, 2	457 x 190 x 89 kg U.B.
3, 4	533 x 210 x 122 kg U.B.
5, 6	457 x 190 x 98 kg U.B.
7, 8	457 x 190 x 74 kg U.B.
9, 10	457 x 190 x 74 kg U.B.
11, 12	457 x 190 x 74 kg U.B.

420 VIERENDEEL GIRDERS

The following information (normally written on a standard form) is required as input data for the computer.

1. Frame data, i.e. joint co-ordinates and section identifier.
2. Section data, i.e. area and moment of inertia.
3. Young's modulus of elasticity.
4. Constraint data. Note:— At joint 1 (see Fig. 5) there are two constraints, one in the X-direction and one in the Y-direction. At joint 21 there is a constraint in the Y-direction only.
5. Loading data. In this example the loading consists of point loads in the Y-direction only.

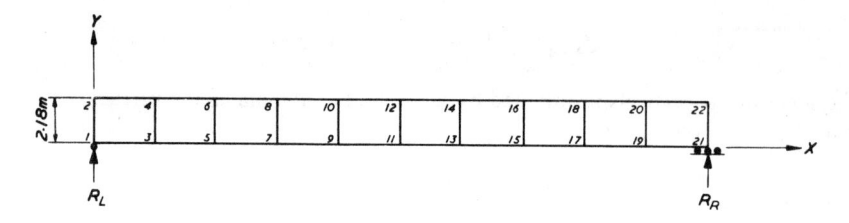

Fig. 5

Zero loads and moments must be entered where zero deformations have been specified.

The printed output from the computer is of the form shown in Tables 1 and 2. Table 1 gives the joint translation in the X and Y directions and rotations. Table 2 gives the axial forces, shear forces and bending moments at the ends of each member.

The bending moments rounded off to the nearest whole number are indicated in Fig. 6. Comparing these values with those in Fig. 4 (b), it can be seen that the largest variation is in chord members. For a symmetrical Vierendeel of this type, the statically determinate analysis appears to be adequate since these chord members are designed for combined bending and direct loads. This would not be true for other types of Vierendeel and the use of a computer will facilitate optimisation of member sizes. The computer analysis also gives joint translations and rotations, the maximum vertical translation being 51.85 mm or approximately 1/575 span (see Table 1).

Member Sizing and Joint Design

Once the moment and axial forces are known, the members can be designed in the normal manner. A typical calculation for a joint will now be given using the moments derived from computer analysis.

COMPUTER RESULTS

TABLE 1

Displacements and Rotations at Joints

Joint No.	X mm	Y mm	Z-Rotation Radians
1	0.000000	0.000000	−0.004191
2	7.160910	−0.144021	−0.004153
3	0.195999	−15.888290	−0.003978
4	6.964911	−15.888163	−0.003961
5	0.746764	−30.096067	−0.003631
6	6.414146	−30.094192	−0.003634
7	1.552041	−41.722214	−0.002783
8	5.608869	−41.720336	−0.002782
9	2.523080	−49.273330	−0.001450
10	4.637830	−49.271344	−0.001450
11	3.580455	−51.849164	−0.000000
12	3.580455	−51.847201	0.000000
13	4.637830	−49.273330	0.001450
14	2.523080	−49.271344	0.001450
15	5.608869	−41.722214	0.002783
16	1.552041	−41.720336	0.002782
17	6.414146	−30.096067	0.003631
18	0.746764	−30.094192	0.003634
19	6.964911	−15.888290	0.003978
20	0.195999	−15.888163	0.003961
21	7.160910	0.000000	0.004191
22	0.000000	−0.144021	0.004153

TABLE 2

Loads and Moments in Members Designated by Joint Numbers at each end

Joint No.	Axial Load kN	Shear kN	Moment kNm
1	157.88	−192.60	−211.44
2	−157.88	192.60	−208.42
1	−192.60	145.22	211.44
3	192.60	−145.22	224.21
2	192.60	142.78	208.42
4	−192.60	−142.78	219.92
3	−0.19	−348.61	−381.25
4	0.19	348.61	−378.73
3	−541.21	111.61	157.04
5	541.21	−111.61	177.79
4	541.21	112.39	158.81
6	−541.21	−112.39	178.36
5	−2.26	−250.10	−272.48
6	2.26	250.10	−272.74
5	−791.31	80.07	94.69
7	791.31	−80.07	145.52
6	791.31	79.93	94.37
8	−791.31	79.93	145.42
7	−1.72	−162.89	−177.57
8	1.72	162.89	−177.53

TABLE 2 (*continued*)

Joint No.	Axial Load kN	Shear kN	Moment kNm
7	−954.20	47.99	32.05
9	954.20	−47.99	111.91
8	954.20	48.01	32.11
10	−954.20	−48.01	111.93
9	−1.82	−84.84	−92.47
10	1.82	84.84	−92.48
9	−1039.04	16.00	−19.44
11	1039.04	−16.00	67.45
10	1039.04	16.00	−19.45
12	−1039.04	−16.00	67.44
11	−1.79	−0.00	−0.00
12	1.79	0.00	−0.00
11	−1039.04	−16.00	−67.45
13	1039.04	16.00	19.44
12	1039.04	−16.00	−67.44
14	−1039.04	16.00	19.45
13	−1.82	84.84	92.47
14	1.82	−84.84	92.48
13	−954.20	−47.99	−111.91
15	954.20	47.99	−32.05
14	954.20	−48.01	−111.93
16	−954.20	48.01	−32.11
15	−1.72	162.89	177.57
16	1.72	−162.89	177.53
15	−791.31	−80.07	−145.52
17	791.31	80.07	−94.69
16	791.31	−79.93	−145.42
18	−791.31	79.93	−94.37
17	−2.26	250.10	272.48
18	2.26	−250.10	272.74
17	−541.21	−111.61	−177.79
19	541.21	111.61	157.04
18	541.21	−112.39	−178.36
20	−541.21	112.39	−158.81
19	−0.19	348.61	381.25
20	0.19	−348.61	378.73
19	−192.60	−145.22	−224.21
21	192.60	145.22	−211.44
20	192.60	−142.78	−219.92
22	−192.60	142.78	−208.42
21	157.88	192.60	211.44
22	−157.88	−192.60	208.42

METHODS OF ANALYSIS 423

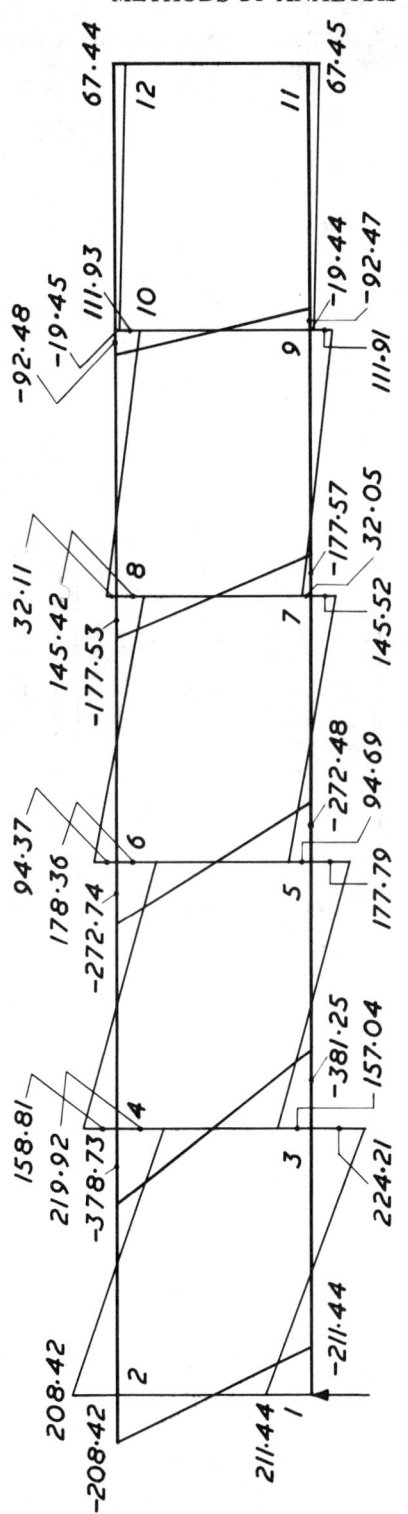

Fig. 6

424   VIERENDEEL GIRDERS

Consider Joint 4, see Fig. 7; the vertical member is connected to the chords via 2 no. 250 x 22 mm plates x 680 mm long.

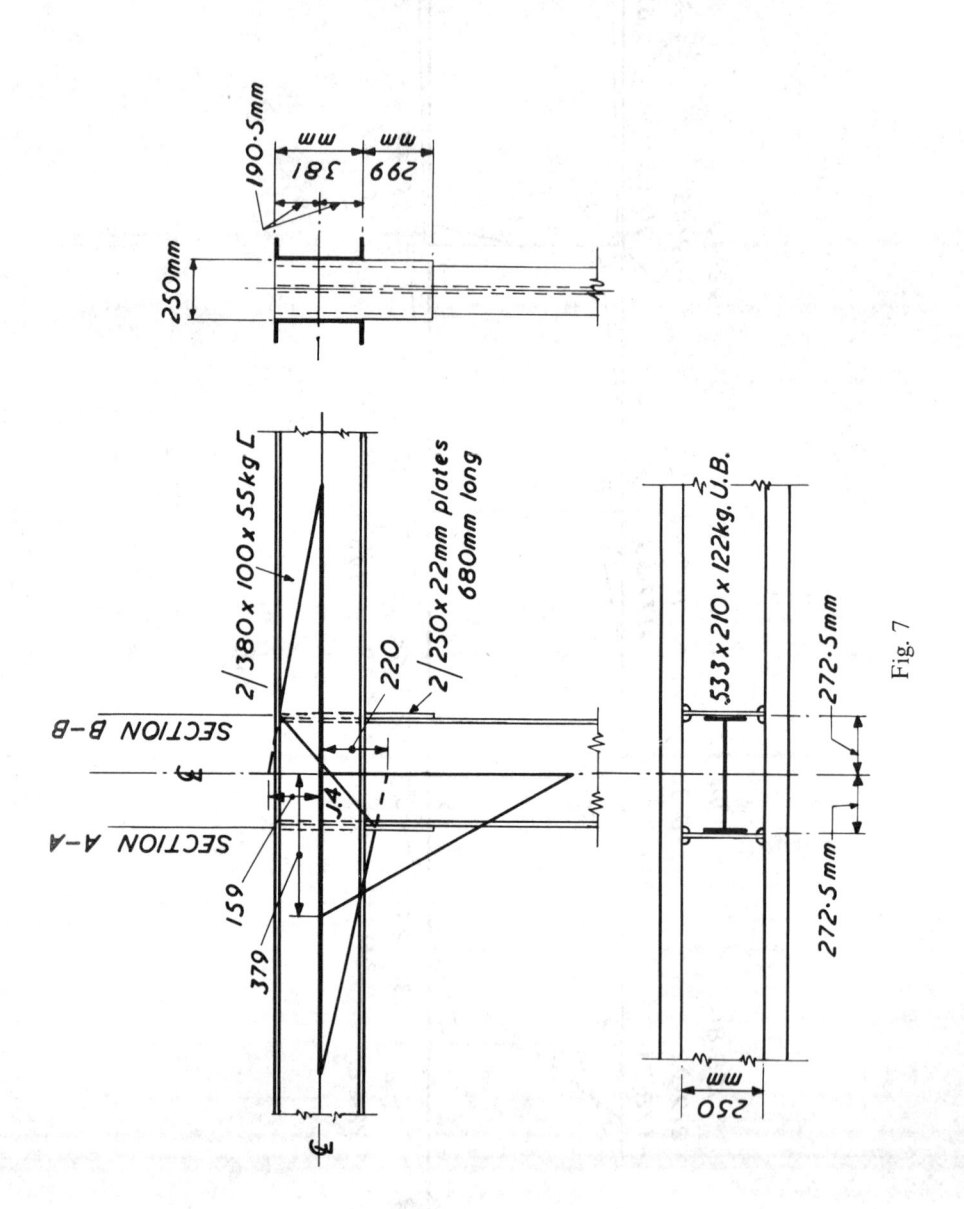

Fig. 7

CALCULATIONS FOR DETAILS

Weld between Connection Plate and the Chord

Force in each weld:

due to end moment in vertical	$= \dfrac{379 \times 10^3}{567}$	$= 668$ kN
due to end moment in top chord	$= \dfrac{(208 + 220) - (159 + 178)}{2 \times 3.0}$	$= 15$ kN
		$\overline{683\ \text{kN}}$
	(1) in each weld $= \dfrac{683}{4}$	$= 171$ kN
due to shear in vertical member	$= \dfrac{379 + 381}{2 \times 2.18}$	$= 174$ kN
	(2) in each weld $= \dfrac{174}{4}$	$= 43.5$ kN
resulting from moment due to shear force	$= \dfrac{87.2 \times 19}{22}$	$= 75$ kN
	(3) in each weld	$= 75$ kN

See Fig. 9 for forces acting on weld.

Resultant

$$= \sqrt{171^2 + 43.5^2 + 75^2} \qquad = \underline{192\ \text{kN}}$$

Force per mm assuming effective length 330 mm

$$= \dfrac{192 \times 10^3}{330} \qquad = \underline{58\ \text{N}}$$

Use 8 mm F.W. at 115 N/mm²

Connection Plate

Assume effective length of 380 mm:

(1) Longitudinal shear force:

$$f_{S1} = \dfrac{684 \times 10^3}{2 \times 380 \times 22} \qquad = \underline{40.9\ \text{N/mm}^2}$$

(2) Shear stress:

$$f_S = \dfrac{87.2 \times 10^3}{380 \times 22} \qquad = \underline{10.4\ \text{N/mm}^2}$$

VIERENDEEL GIRDERS

Bending moment due to shear force:

$$-\frac{87.2 \times 19}{380} \qquad = 4.4 \text{ kNmm}$$

(3) Bending moment stress:

$$fb = \frac{4.4 \times 10^3 \times 6}{1 \times (22)^2} \qquad = 54.0 \text{ N/mm}^2$$

See Fig. 10 for stress acting on plate.

Equivalent B.M. stress, see B.S. 449, Cl.14 (c),

$$fb_c = \sqrt{54^2 + 3(10.4)^2} \qquad = 56.9 \text{ N/mm}^2$$

Principal Stress*

$$= \frac{56.9}{2} + \sqrt{\left(\frac{56.9}{2}\right)^2 + 40.9^2} \qquad = 78.3 \text{ N/mm}^2$$

The shear in the webs of the boom across the joint can be found by calculating the rate of change of the bending moment. The points of contraflexure are determined from Fig. 8. By similar triangles:

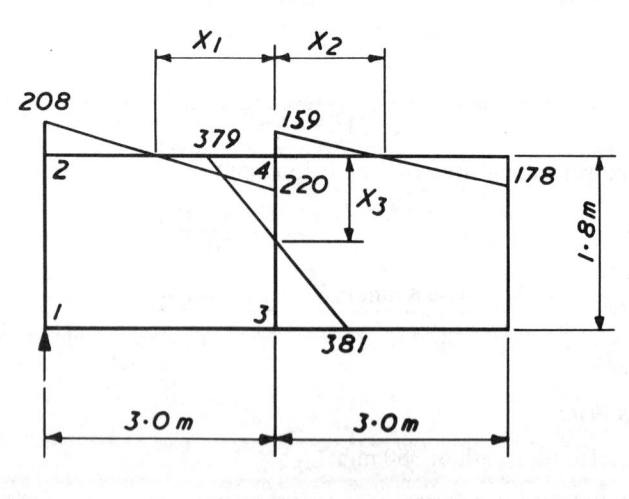

Fig. 8

---

* Rigorous inspection of the part of the joint under consideration will reveal that the applied loads result in local effects not taken into account in the design. In fact, these will be found to have only an insignificant effect on the final result, and the calculated stresses in the plate are considered sufficiently accurate.

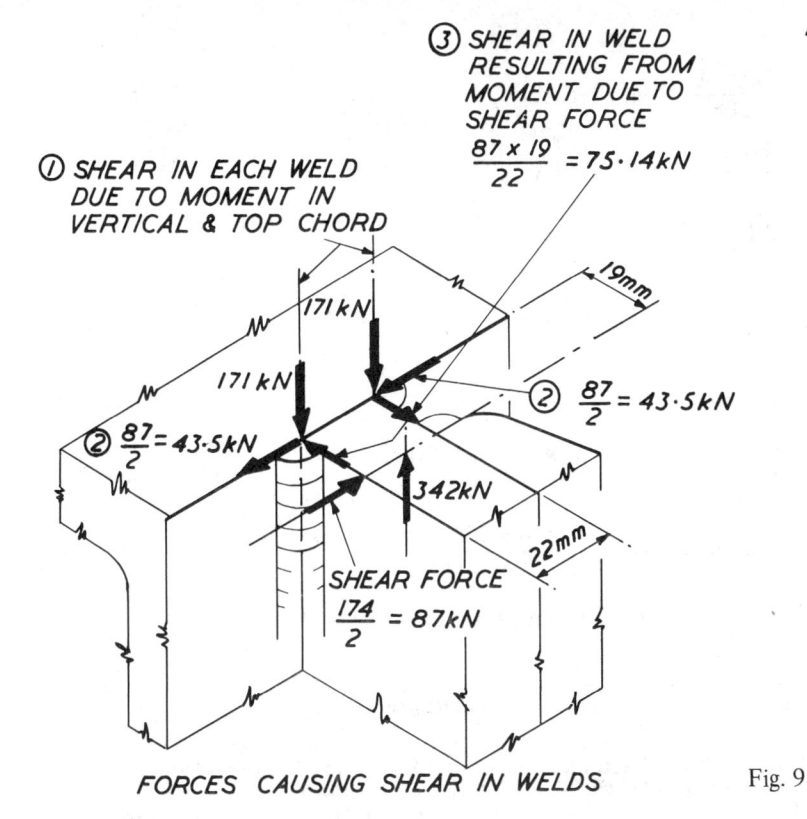

427

Fig. 9 — FORCES CAUSING SHEAR IN WELDS

① SHEAR IN EACH WELD DUE TO MOMENT IN VERTICAL & TOP CHORD — 171 kN, 171 kN

③ SHEAR IN WELD RESULTING FROM MOMENT DUE TO SHEAR FORCE
$$\frac{87 \times 19}{22} = 75.14 \text{ kN}$$

② $\frac{87}{2} = 43.5$ kN

② $\frac{87}{2} = 43.5$ kN

342 kN

SHEAR FORCE $\frac{174}{2} = 87$ kN

19 mm, 22 mm

Fig. 10 — STRESSES ACTING ON PLATE

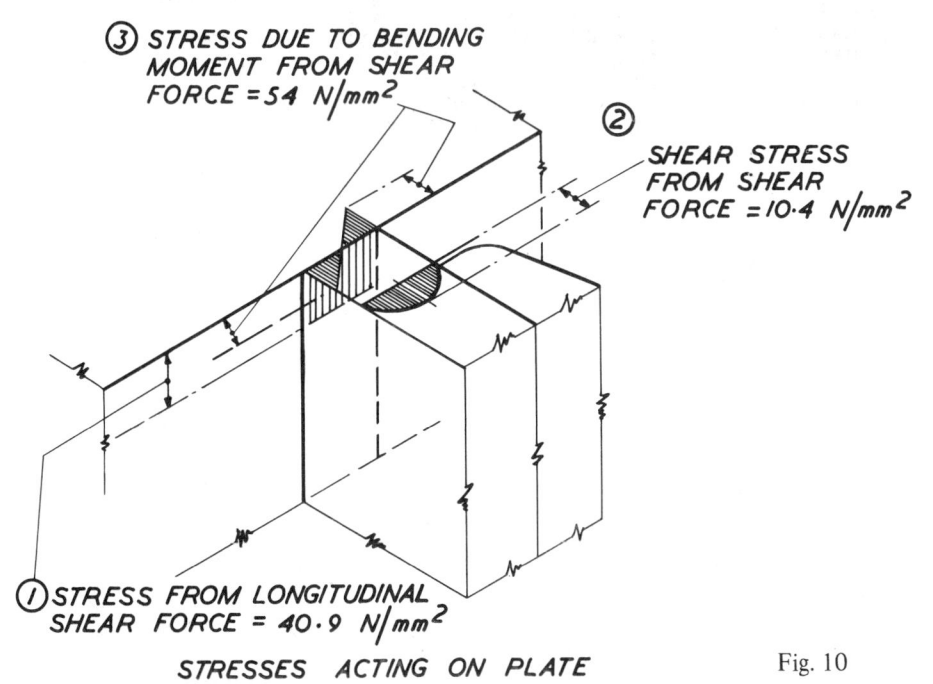

③ STRESS DUE TO BENDING MOMENT FROM SHEAR FORCE = 54 N/mm²

② SHEAR STRESS FROM SHEAR FORCE = 10.4 N/mm²

① STRESS FROM LONGITUDINAL SHEAR FORCE = 40.9 N/mm²

$$\frac{X_1}{220} = \frac{3.0 - X_1}{208} \text{ from which } X_1 \qquad = 1.54 \text{ m}$$

$$\frac{X_2}{159} = \frac{3.0 - X_2}{178} \text{ from which } X_2 \qquad = 1.42 \text{ m}$$

$$\frac{X_3}{379} = \frac{2.18 - X_3}{381} \text{ from which } X_3 \qquad = 1.09 \text{ m}.$$

B.M. at $A - A$ (Fig. 7) $= 220 \times \dfrac{1\,268}{1\,540}$ $\qquad = 181$ kNm

B.M. at $B - B$ (Fig. 7) $= 159 \times \dfrac{1\,148}{1\,540}$ $\qquad = 128$ kNm

Shear $\qquad = \dfrac{(181 + 128) \times 10^3}{545}$ $\qquad = 568$ kN

Average shear stress $= \dfrac{568 \times 10^3}{2 \times 10.4 \times 381}$ $\qquad = 72$ N/mm^2

Max. permissible shear stress = 115 N/mm^2.

REFERENCES

1. LIGHTFOOT, E. 'Substitute Frames in the Analysis of Rigid Jointed Structures', *Civil Engineering and Public Works Review,* Vol. 52, December, 1957.
2. LIGHTFOOT, E. 'A Moment Distribution Method for Vierendeel Bents and Girders with Inclined Chords', *I.C.E. Paper* No. 6284.
3. BARIC COMPUTING SERVICES LTD. – Structural Analysis Programme.
4. HENDRY, A. 'Plastic Analysis and Design of Mild Steel Vierendeel Girders', *Journal of Institution of Structural Engineers,* July, 1955.
5. HENDRY, A. 'An Investigation of the Strength of Welded Portal Frame Connections', *Journal of Institution of Structural Engineers,* October, 1950.

18. FRAMING FOR SINGLE-STOREY SHEDS

THE framing for single-span single-storey sheds can be arranged in a number of ways dependent upon the design assumptions made and those in general use are shown in Figs. 1 to 6. (a) Typical B.M. diagrams for the stanchions are shown in the Figures and, in addition, the dotted lines indicate, in an exaggerated manner, the deflected form of the structure under side-wind loading.

The cross-section shown in Fig. 1 (a) is perhaps more widely used than any other and consists basically of roof trusses and stanchions. The trusses may be carried directly by stanchions as shown in Fig. 1 (b) or there may be in addition intermediate trusses supported by eaves beams, Fig. 1 (c). In each case the stanchions must be fixed at their bases and will act as vertical cantilevers under

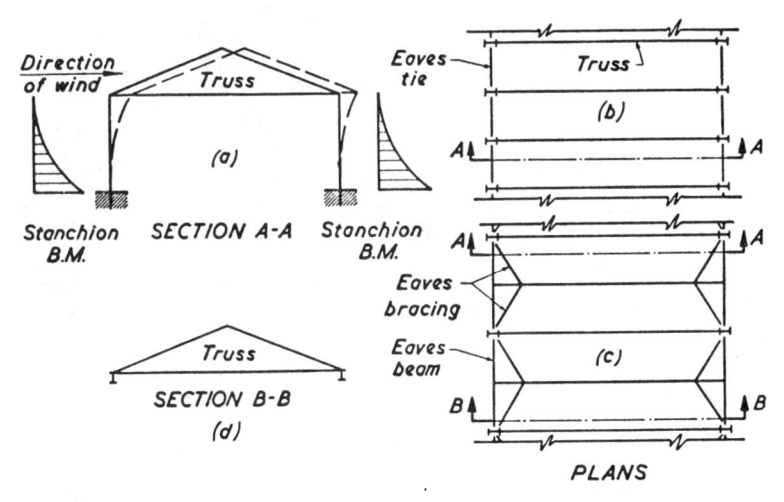

Fig. 1

the action of side-wind loading. The wind B.M. in the stanchions must be taken up by the foundations. For the design of suitable stanchion bases and foundations see Chapters 28 and 30. Where intermediate trusses are carried by eaves beams, lateral bracing, as shown in Fig. 1 will be required to transmit to the stanchions the horizontal reactions of the trusses unless the beams are designed for the combined vertical and horizontal loads. Bracing is normally the more economical.

The approximate weights of single span roof trusses of bolted construction, for spans between 10 m and 30 m can be calculated based on the area carried times 100 to 150 N/m². Welded trusses are somewhat lighter and weigh between 80 and 120 N/m² of area carried.

The imposition of moments on the foundations can be avoided by the use of knee braces between the trusses and stanchions as shown in cross-section and plan in Figs. 2 (a) and (b). Where eaves beams are used, as in Figs. 2 (c) and (d), the

intermediate trusses will be of simple form and, again, lateral bracing at eaves level will be required to transmit the horizontal loads to the stanchion caps. For a shed otherwise identical with that shown in Fig. 1 the maximum B.M. in the stanchions will be reduced and, in addition its position will be transferred from the stanchion base to the junction of knee brace and stanchion.

If head room or other conditions so dictate, the braces may be on one side only, the other end of the truss being simply supported, in which case the knee-braced stanchion takes up all the moment due to side wind on the frame as a

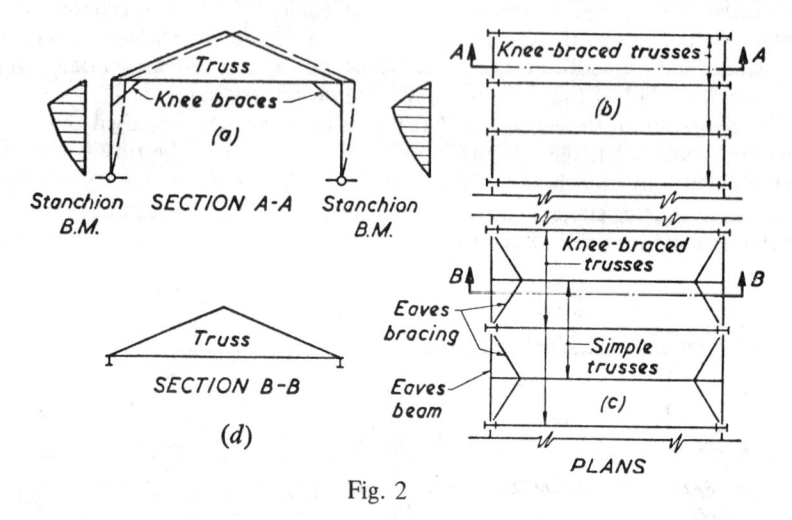

Fig. 2

whole whilst the other stanchion acts as a vertical post carrying direct load but subjected to a wind B.M. as a beam simply supported at base and cap levels. (See Fig. 3.)

A common variant is that shown in Fig. 4, which still further reduces the stanchion bending moments, other conditions remaining unaltered, but at the

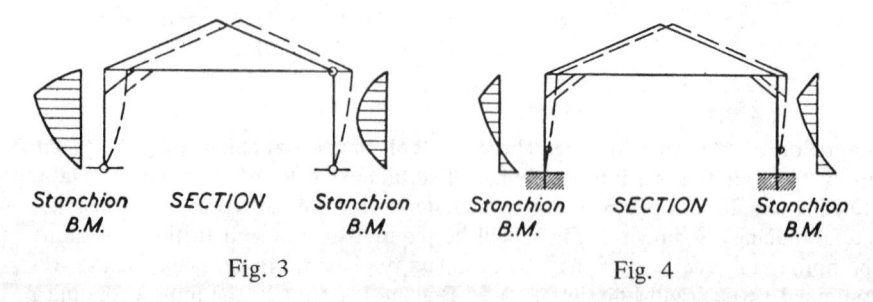

Fig. 3   Fig. 4

expense of imposing moments on the foundations approximately half those shown in Fig. 1.

Another form of construction, shown in Fig. 5, employs horizontal eaves girders of lattice construction to transmit all horizontal loads from trusses and intermediate posts to the ends of the building and thence via bracing to the foundations. This type of construction is particularly economical when the building is short enough to enable the wind loads to be taken to the end frames

FRAMING FOR SINGLE-STOREY SHEDS 431

by eaves girders of reasonable proportions, as the additional bracing in these girders will be relatively light. The trusses may be supported directly by stanchions as shown in plan at Fig. 5 (b) or by eaves beams and stanchions as depicted in Fig. 5 (c). Alternatively, if the building is too long to give reasonably light eaves girders, the spans of these girders may be reduced by the insertion, at appropriate intermediate points, of frames designed to give the necessary horizontal resistance.

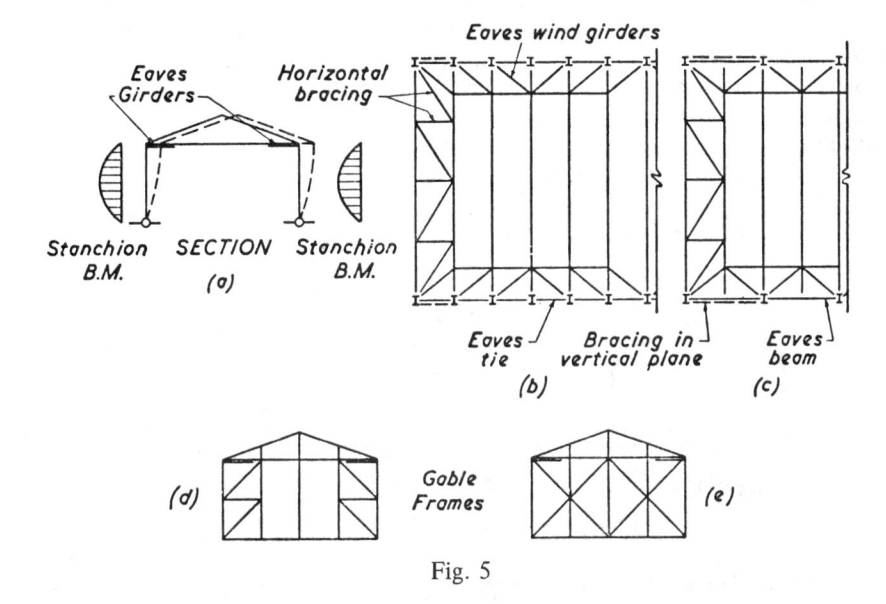

Fig. 5

The horizontal eaves girders may be replaced by girders in the planes of the truss rafters and in which the purlins can be made to serve as the chord members of the lattice girders.

Alternative arrangements of the gable frames are shown in Figs. 5 (d) and (e).

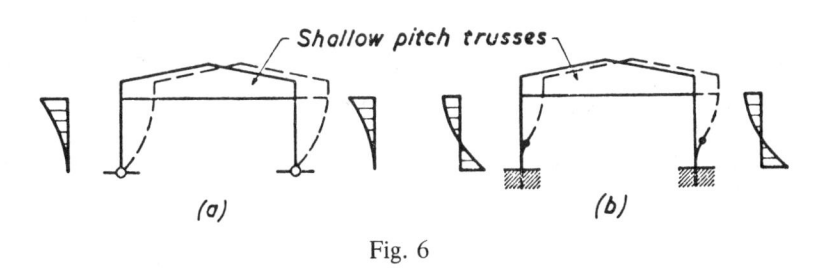

Fig. 6

In Fig. 5 (d) stanchions are braced together in pairs with diagonals in tension only, and hence each pair is assumed to act in resisting the wind effect on one side only, whilst in Fig. 5 (e) stiff bracing is used, capable of acting both in tension and compression so that all the stanchions in the gable act in conjunction.

If it is required to use a roof of shallow pitch and in consequence there is a reasonable depth of truss at the shoe, the effect of a knee brace will be simulated as shown in Figs. 6 (a) and (b), showing stanchions hinged and fixed at their

bases respectively. The horizontal eaves lattice girders referred to previously may be used with this type of roof if required.

Arched or portal construction may also be used, the principal types being shown in Fig. 7; they are respectively three-hinged, two-hinged, and hingeless or fixed arches. The first has the advantage of being statically determinate although the moments and crown deflection when loaded are greater than for the other two types. The moments in the fixed-ended arch or portal are the least of the three. All three types have the disadvantage of imposing horizontal thrusts of some magnitude upon the foundations, whilst the hingeless type also causes moments on the foundations. The calculations for the two-hinged and hingeless portals are necessarily based upon the normal arch theory, which presupposes the existence of the appropriate horizontal resistance at the feet of the portals to provide the arch thrust and it should be noted that comparatively small horizontal displacements at the footings will cause considerable redistribution of the moments. These types of framing should not therefore be used in cases where the

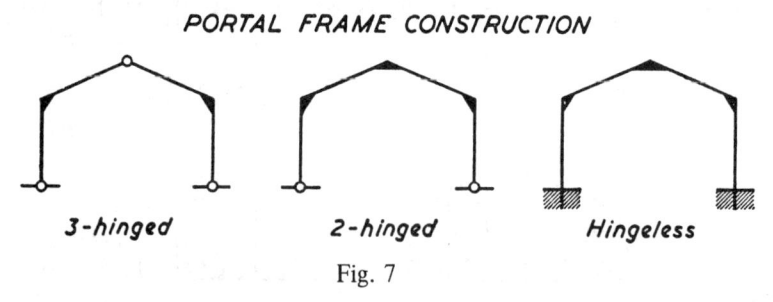

PORTAL FRAME CONSTRUCTION

3-hinged   2-hinged   Hingeless

Fig. 7

existence of adequate horizontal resistance is suspect unless the effect of arch spread is included in the design calculations or unless positive means are taken to determine the extent of the lateral movement of the portal feet, for example, by the insertion of ties at base level.

The calculations for a three-pinned frame also call for the provision of an adequate horizontal thrust, but the effect of footing spread in this case is of little importance: all forces and moments in the frame are increased in inverse ratio to that of the reduced crown height compared with the original crown height, but no redistribution effects occur.

It is generally found that the weight of a portal frame designed on the elastic theory is greater than that of the lightest comparable construction utilising trusses and columns, but this effect may be offset by savings and advantages in other directions.

The portal frames may be of solid I-section (joist or welded plate) or of lattice type. Formulæ dealing with two-hinged and hingeless frames are given on pages 305 to 329. These may be used to calculate moments, shears and thrusts when a solid section is used, but if lattice work is employed the vertical and horizontal reactions can be calculated from the formulæ given and a stress diagram can be drawn in the normal manner utilising the reactions so found. In the case of the three-pinned arch the vertical reactions at the feet may be found by taking moments about them in the normal manner whilst the horizontal reactions can be found by utilising the fact that the moment at the crown hinge is zero. For examples of the calculations for a three-pinned arch see pages 132–134.

Multi-span Ridged Roof Sheds

Multi-span sheds may be constructed of a series of simple truss or portal spans provided that a relatively large number of internal columns can be permitted. The principal advantage of multi-bay construction in truss work is, however, the ease with which large floor areas may be covered with a minimum number of supports. Fig. 8 (a) indicates a conventional layout in which apex lattice girders support

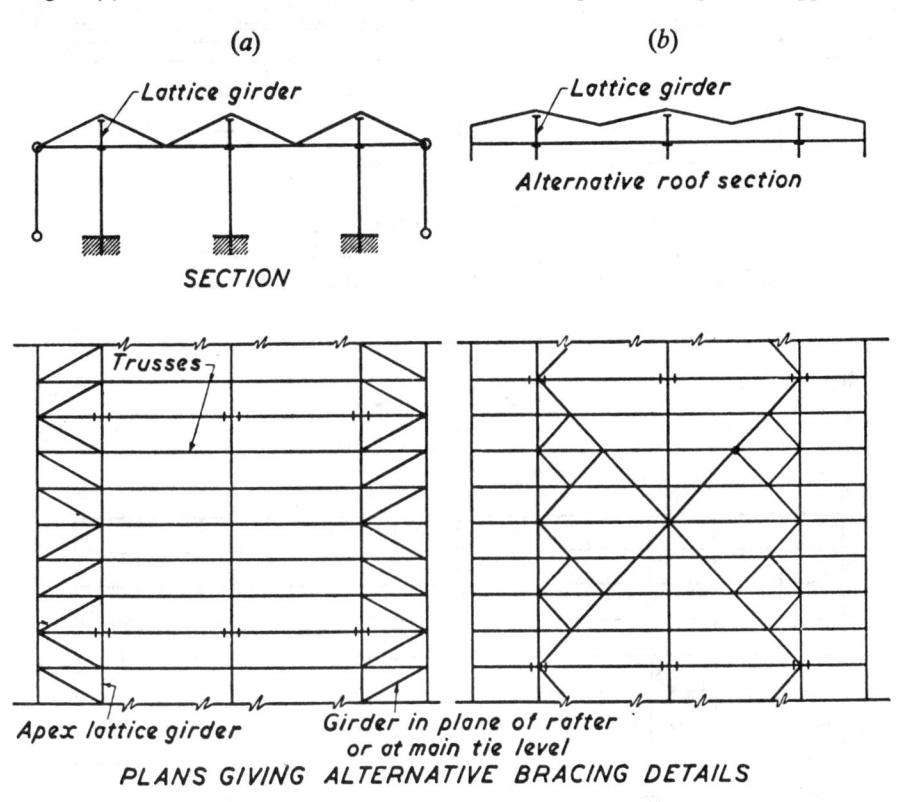

Fig. 8

umbrella-type roof trusses. The lattice girders are normally designed as simple spans between columns but may also be designed as continuous girders. The construction is economical for lattice girders with spans up to 36–45 m at spacings not exceeding 18–21 m. A stiffer structural frame results from the use of trusses having appreciable depth at the valley as indicated in the alternative section, Fig. 8 (b). The same type of construction may be adopted with a north-light truss outline, and it may be convenient to replace the vertical apex lattice girder by an inclined girder in the plane of the north slopes of the slung trusses.

Should it be necessary still further to reduce the number of columns, the apex lattice girders may be carried by main lattice girders simply supported at their ends, or continuous over intermediate supports if desired, as shown in Fig. 9. Such girders will project through the roof covering, and weathering details are apt to be somewhat complicated.

SECTION A-A (a)

PLAN (b)

SECTION B-B (c)

Fig. 9

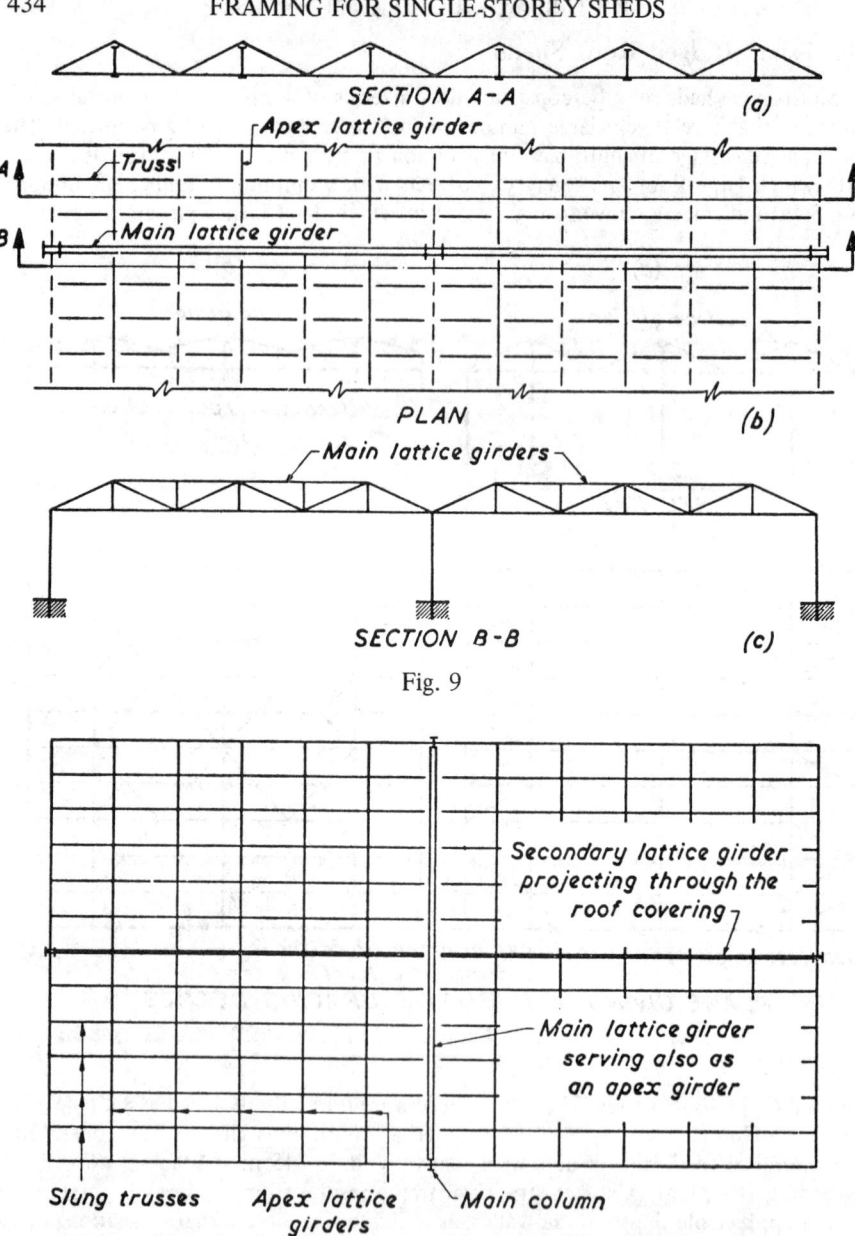

Fig. 10

The sub-division of framing may be carried still further, as shown in Fig. 10 which gives a layout which dispenses entirely with internal columns.

A form of construction which has also found favour is shown in Fig. 11.

It consists of lattice girders carrying valley beams which, in turn, support north-light trusses running parallel to the lattice girders. The top chords of these

MULTI-SPAN RIDGED ROOF SHEDS

(a) SECTION A-A
MAIN LATTICE GIRDER

(b) SECTION B-B
ROOF TRUSSES

(c)

Fig. 11

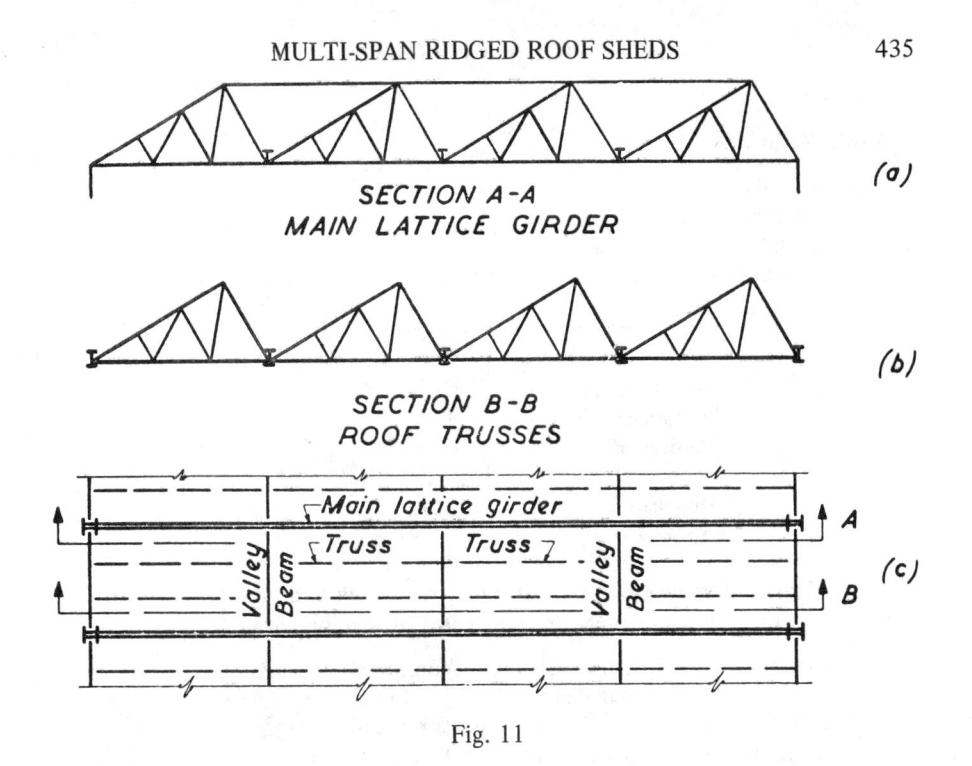

girders are thus above the roof covering and elaborate flashing arrangements are required.

Flat-roofed Single-storey Sheds

These roofs may be of simple joist construction for moderate spans, but if required with monitor lights advantage may be taken of the upstand of the monitor to incorporate lattice girders inside the vertical glazing to reduce the number of supporting columns as shown in Fig. 20 (b), Chapter 19. Alternatively the upper and lower roof levels together with the monitor upstand can be designed as a cranked beam as shown in Fig. 20 (c), Chapter 19.

Lattice Girders — Stress Data

In the following pages of this section stress coefficients are tabulated for the members of various standard types of lattice girders by which the forces in them may be readily obtained under the loading conditions indicated.

The coefficients are calculated for equal loads at all panel points other than end supports and they are to be multiplied by the panel point load qualifier as indicated.

Throughout the tables compression is indicated as negative ($-$) and tension as positive ($+$).

Parallel Flange Lattice Girders

In the following tables,

W = the load per panel,
d = constant depth of girder
and p = constant width of panel.

The stress coefficients are to be multiplied by qualifiers as follows:

Type of Member	Multiply by Qualifier
Top chords TC ⎫ Bottom chords BC ⎭	$\dfrac{W.p}{d}$
Verticals V	W
Diagonals D	$\dfrac{W}{d}\sqrt{(d^2 + p^2)}$

Figs. 12 and 13 show two types of girder with even and odd numbers of panels respectively. In both types there are four cases of loading, viz.:

1. Vertical struts, diagonal ties, load on compression flange.
2. Vertical struts, diagonal ties, load on tension flange.
3. Vertical ties, diagonal struts, load on compression flange.
4. Vertical ties, diagonal struts, load on tension flange.

The tables of stress coefficients are given on pages 438–439.

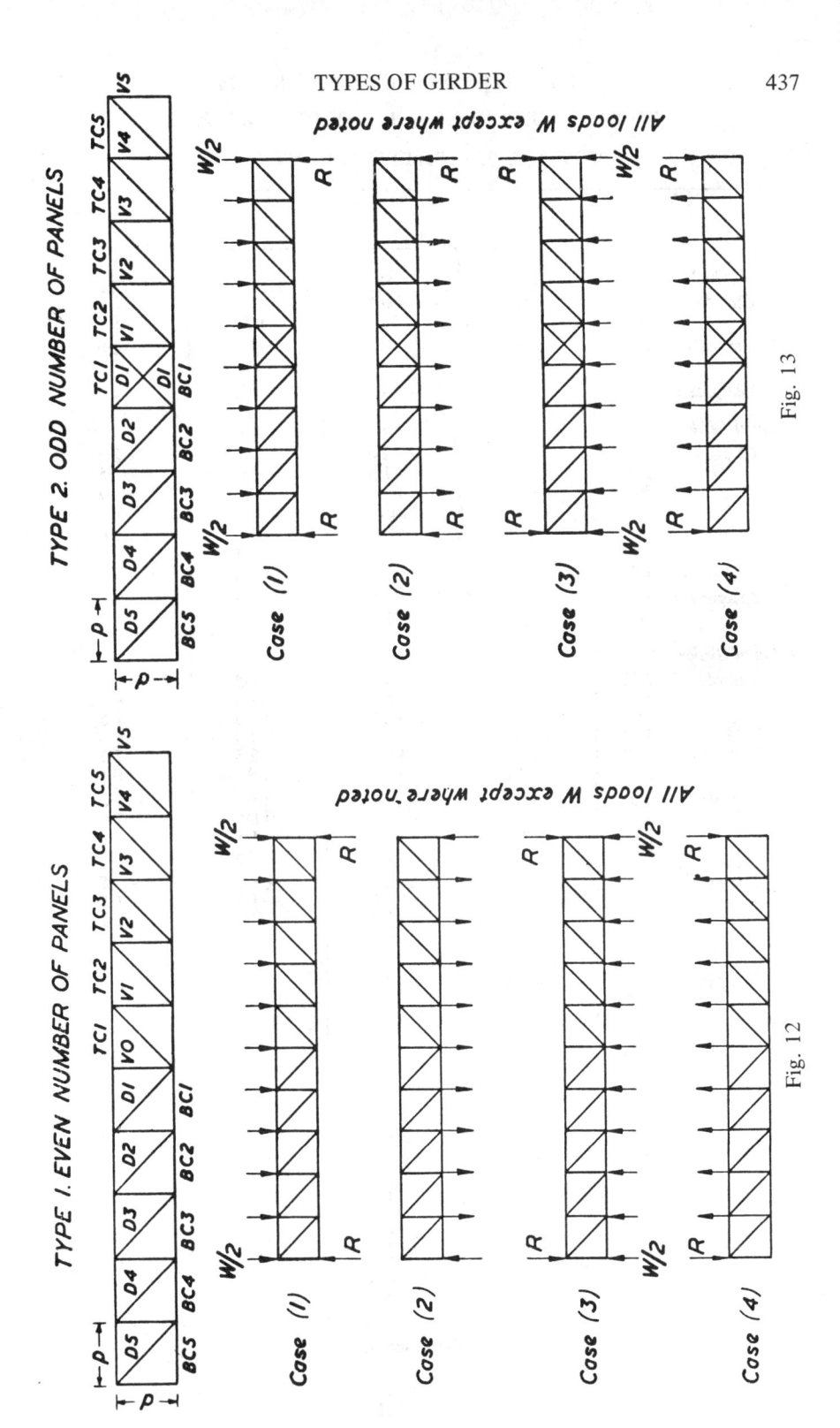

TYPES OF GIRDER 437

Fig. 13

Fig. 12

STRESS COEFFICIENTS FOR LATTICE GIRDERS
TYPE 1 EVEN NUMBER OF PANELS

		Case (1)	Case (2)	Case (3)	Case (4)
2-Panel Girder					
Chords	TC1	−0·5	−0·5	+0·5	+0·5
	BC1	nil	nil	nil	nil
Verticals	V0	−1·0	nil	nil	+1·0
	V1	−1·0	−0·5	−0·5	nil
Diagonals	D1	−0·5	+0·5	−0·5	−0·5
4-Panel Girder					
Chords	TC1	−2·0	−2·0	+2·0	+2·0
	TC2	−1·5	−1·5	+1·5	+1·5
	BC1	+1·5	+1·5	−1·5	−1·5
	BC2	nil	nil	nil	nil
Verticals	V0	−1·0	nil	nil	+1·0
	V1	−1·5	−0·5	+0·5	+1·5
	V2	−2·0	−1·5	−0·5	nil
Diagonals	D1	+0·5	+0·5	−0·5	−0·5
	D2	+1·5	+1·5	−1·5	−1·5
6-Panel Girder					
Chords	TC1	−4·5	−4·5	+4·5	+4·5
	TC2	−4·0	−4·0	+4·0	+4·0
	TC3	−2·5	−2·5	+2·5	+2·5
	BC1	+4·0	+4·0	−4·0	−4·0
	BC2	+2·5	+2·5	−2·5	−2·5
	BC3	nil	nil	nil	nil
Verticals	V0	−1·0	nil	nil	+1·0
	V1	−1·5	−0·5	+0·5	+1·5
	V2	−2·5	−1·5	+1·5	+2·5
	V3	−3·0	−2·5	−0·5	nil
Diagonals	D1	+0·5	+0·5	−0·5	−0·5
	D2	+1·5	+1·5	−1·5	−1·5
	D3	+2·5	+2·5	−2·5	−2·5
8-Panel Girder					
Chords	TC1	−8·0	−8·0	+8·0	+8·0
	TC2	−7·5	−7·5	+7·5	+7·5
	TC3	−6·0	−6·0	+6·0	+6·0
	TC4	−3·5	−3·5	+3·5	+3·5
	BC1	+7·5	+7·5	−7·5	−7·5
	BC2	+6·0	+6·0	−6·0	−6·0
	BC3	+3·5	+3·5	−3·5	−3·5
	BC4	nil	nil	nil	nil
Verticals	V0	−1·0	nil	nil	+1·0
	V1	−1·5	−0·5	+0·5	+1·5
	V2	−2·5	−1·5	+1·5	+2·5
	V3	−3·5	−2·5	+2·5	+3·5
	V4	−4·0	−3·5	−0·5	nil
Diagonals	D1	+0·5	+0·5	−0·5	−0·5
	D2	+1·5	+1·5	−1·5	−1·5
	D3	+2·5	+2·5	−2·5	−2·5
	D4	+3·5	+3·5	−3·5	−3·5

For qualifiers, see p. 436

LATHE GIRDERS – STRESS DATA

STRESS COEFFICIENTS FOR LATTICE GIRDERS—continued

TYPE 1—continued

		Case			
		(1)	(2)	(3)	(4)
10-*Panel Girder*					
Chords	TC1	−12·5	−12·5	+12·5	+12·5
	TC2	−12·0	−12·0	+12·0	+12·0
	TC3	−10·5	−10·5	+10·5	+10·5
	TC4	−8·0	−8·0	+8·0	+8·0
	TC5	−4·5	−4·5	+4·5	+4·5
	BC1	+12·0	+12·0	−12·0	−12·0
	BC2	+10·5	+10·5	−10·5	−10·5
	BC3	+8·0	+8·0	−8·0	−8·0
	BC4	+4·5	+4·5	−4·5	−4·5
	BC5	nil	nil	nil	nil
Verticals	V0	−1·0	nil	nil	+1·0
	V1	−1·5	−0·5	+0·5	+1·5
	V2	−2·5	−1·5	+1·5	+2·5
	V3	−3·5	−2·5	+2·5	+3·5
	V4	−4·5	−3·5	+3·5	+4·5
	V5	−5·0	−4·5	−0·5	nil
Diagonals	D1	+0·5	+0·5	−0·5	−0·5
	D2	+1·5	+1·5	−1·5	−1·5
	D3	+2·5	+2·5	−2·5	−2·5
	D4	+3·5	+3·5	−3·5	−3·5
	D5	+4·5	+4·5	−4·5	−4·5

TYPE 2 ODD NUMBER OF PANELS

		Case			
		(1)	(2)	(3)	(4)
3-*Panel Girder*					
Chords	TC1	−1·0	−1·0	+1·0	+1·0
	TC2	−1·0	−1·0	+1·0	+1·0
	BC1	+1·0	+1·0	−1·0	−1·0
	BC2	nil	nil	nil	nil
Verticals	V1	−1·0	nil	nil	−1·0
	V2	−1·5	−1·0	−0·5	nil
Diagonals	D1	nil	nil	nil	nil
	D2	+1·0	+1·0	−1·0	−1·0
5-*Panel Girder*					
Chords	TC1	−3·0	−3·0	+3·0	+3·0
	TC2	−3·0	−3·0	+3·0	+3·0
	TC3	−2·0	−2·0	+2·0	+2·0
	BC1	+3·0	+3·0	−3·0	−3·0
	BC2	+2·0	+2·0	−2·0	−2·0
	BC3	nil	nil	nil	nil
Verticals	V1	−1·0	nil	nil	+1·0
	V2	−2·0	−1·0	+1·0	+2·0
	V3	−2·5	−2·0	−0·5	nil
Diagonals	D1	nil	nil	nil	nil
	D2	+1·0	+1·0	−1·0	−1·0
	D3	+2·0	+2·0	−2·0	−2·0

For qualifiers, see p. 436

STRESS COEFFICIENTS FOR LATTICE GIRDERS—*continued*

TYPE 2—*continued*

		Case (1)	Case (2)	Case (3)	Case (4)
7-Panel Girder					
Chords	$TC1$	−6·0	−6·0	+6·0	+6·0
	$TC2$	−6·0	−6·0	+6·0	+6·0
	$TC3$	−5·0	−5·0	+5·0	+5·0
	$TC4$	−3·0	−3·0	+3·0	+3·0
	$BC1$	+6·0	+6·0	−6·0	−6·0
	$BC2$	+5·0	+5·0	−5·0	−5·0
	$BC3$	+3·0	+3·0	−3·0	−3·0
	$BC4$	nil	nil	nil	nil
Verticals	$V1$	−1·0	nil	nil	+1·0
	$V2$	−2·0	−1·0	+1·0	+2·0
	$V3$	−3·0	−2·0	+2·0	+3·0
	$V4$	−3·5	−3·0	−0·5	nil
Diagonals	$D1$	nil	nil	nil	nil
	$D2$	+1·0	+1·0	−1·0	−1·0
	$D3$	+2·0	+2·0	−2·0	−2·0
	$D4$	+3·0	+3·0	−3·0	−3·0
9-Panel Girder					
Chords	$TC1$	−10·0	−10·0	+10·0	+10·0
	$TC2$	−10·0	−10·0	+10·0	+10·0
	$TC3$	−9·0	−9·0	+9·0	+9·0
	$TC4$	−7·0	−7·0	+7·0	+7·0
	$TC5$	−4·0	−4·0	+4·0	+4·0
	$BC1$	+10·0	+10·0	−10·0	−10·0
	$BC2$	+9·0	+9·0	−9·0	−9·0
	$BC3$	+7·0	+7·0	−7·0	−7·0
	$BC4$	+4·0	+4·0	−4·0	−4·0
	$BC5$	nil	nil	nil	nil
Verticals	$V1$	−1·0	nil	nil	+1·0
	$V2$	−2·0	−1·0	+1·0	+2·0
	$V3$	−3·0	−2·0	+2·0	+3·0
	$V4$	−4·0	−3·0	+3·0	+4·0
	$V5$	−4·5	−4·0	−0·5	nil
Diagonals	$D1$	nil	nil	nil	nil
	$D2$	+1·0	+1·0	−1·0	−1·0
	$D3$	+2·0	+2·0	−2·0	−2·0
	$D4$	+3·0	+3·0	−3·0	−3·0
	$D5$	+4·0	+4·0	−4·0	−4·0

For qualifiers, see p. 436

19. SPACE FRAMES

SKELETAL space frames were first used by primitive peoples who found by instinct and experience that the most efficient way of using the materials available was to build a cone or dome shaped structure of tree branches and cover this with a flexible weather-proof material. The buffalo hide covered American Indian tepee and the African thatched round house are typical of this simple structural method.

The arguments in favour of the efficiency of this form are as valid for large structures as for the small dwellings which initiated their use, but the rate of their development has been slow for the following reasons:

The lack of materials which were strong enough to form a self-supporting skeletal structure over large spans. Monolithic space structures were first developed in masonry which has fair compressive but negligible tensile strength. Skeletal structures which were built in these media required heavy applied dead loads at their crown or at points liable to develop tension to keep the material in compression. Flying buttresses are common examples in plane structures but the applications of such materials to space frames are limited. Timber was used more extensively but as the jointing technique then used could only develop the full compression load of the material it tended to be used in the same manner as stone. A few types of timber skeletal space structures were developed before the advent of steel, these were mostly hipped and polygonal roof structures sometimes tied at eaves level with wrought iron ties.

At the beginning of the nineteenth century steel, wrought iron and cast iron began to be produced in quantity and the first difficulty was overcome.

Methods of predicting the sizes of members required was the next difficulty to be faced and through the nineteenth and this century until a few years ago, many of the greatest brains in science tried to develop analytical methods and structural systems which could be resolved simply. In spite of their efforts the calculation and design of such structures remained so complex that they stayed the preserve of the most accurate and brilliant brains. Today we stand on the threshold of the solution of this difficulty; the electronic digital computer provides such a powerful mode of calculation that programmes have been devised for it by which most forms of construction can be accurately analysed.

Methods of connecting members has remained another problem which has not yet been solved for every type of construction. For very large structures site welding can be the most economical and satisfactory, for smaller structures many ingenious mechanical joints have been devised which allow for the simple connection of members meeting in three planes. The bibliography provides a short section dealing with patented and other jointing methods. So much work is being carried out, so many new materials (including adhesives) are being developed that we are again on the brink of a big advance towards comprehensive solutions of this problem in many different ways and should be ready to take advantage of these new methods as they are developed.

442   SPACE FRAMES

The calculation of the geometry and the detailing of members in space once represented a painstaking, highly skilled and tedious task. Most of the tedium has been removed by the advent of the computer. Standardised methods of fabrication of details to set jointing techniques can eventually lead to mass production of prefabricated components to a degree of accuracy and finish not yet general in the structural engineering industry.

The economic and other advantages accruing from the use of space frames have been clouded by their use for purposes for which they are not necessarily suited or by the use of the wrong type of frame for a purpose for which a space frame could be used. Generally there are good reasons, other than economical ones, for the selection of a particular type of space frame, but sometimes the selection of a frame economically unsuited for its purpose leads people to consider them uneconomical in general. The descriptive part of the chapter is intended to provide some guidance on suitable uses for various types of frame and the examples are intended to illustrate methods of analysis for simple frames. The more complex frames can best be selected and designed with specialist advice.

Types of Space Structures in General Use

1. Grids

These are generally used to support floors or flat roofs but may also replace an array of simple members in folded plate or arch construction. They consist of a series of interconnected beams or lattice girders spanning in two or more directions. Their great advantage is that, as they act in a manner analogous to a plate supported on all sides (therefore taking bending loads in more than one direction) for a given weight of structural material it is possible to reduce the construction depth to span ratio appreciably compared to that required by systems of simple main and secondary beams. To obtain full benefit of this advantage it is important to use the best grid form for the particular case of span in each direction, load and form of support. Generally speaking the best situation is when the distance between supports or lines of supports is approximately equal in each direction, but special forms of grid can be used where this ideal is not achievable.

Two basic forms exist: (a) single layer grids whose members are composed of single beams or latticed girders which have their major axes horizontal (i.e. their chords or top and bottom flanges are disposed vertically one above the other) and (b) double layer grids, which are invariably of lattice construction, whose top and bottom chords are not disposed in the same vertical planes and in many cases do not even follow the same geometrical pattern (see Fig. 2). Some of the geometrical grid patterns used are as follows:

(i) *Square or Rectangular Grids*

(a) *Single layer:* for simple structures uniformly loaded little advantage can be gained from this form over primary and secondary beam construction. Where heavy incidental or concentrated alternative loadings need to be carried (Bridge

TYPES OF SPACE STRUCTURES 443

decks carrying *HA* and *HB* loading for example) rectangular grids using the minimum number of cross or distributional beams can provide an economical structure (Fig. 1).

(b) *Double layer:* can show three fold advantage for structures carrying U.D. loads, they give much better load distribution than the single layer type, enabling the use of smaller sections, are also amenable to mass production methods of

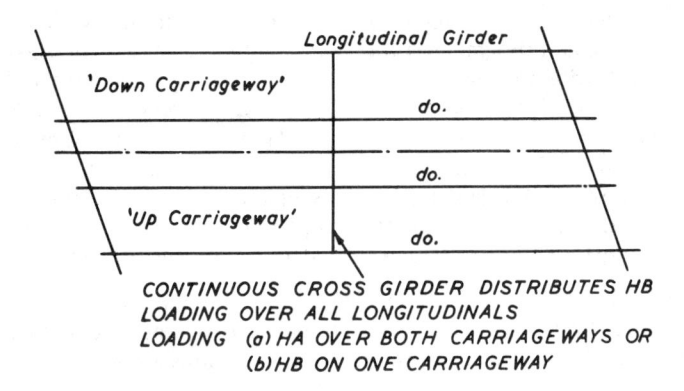

CONTINUOUS CROSS GIRDER DISTRIBUTES HB LOADING OVER ALL LONGITUDINALS
LOADING (a) HA OVER BOTH CARRIAGEWAYS OR (b) HB ON ONE CARRIAGEWAY

LINE PLAN OF BRIDGE GIRDERS

Fig. 1

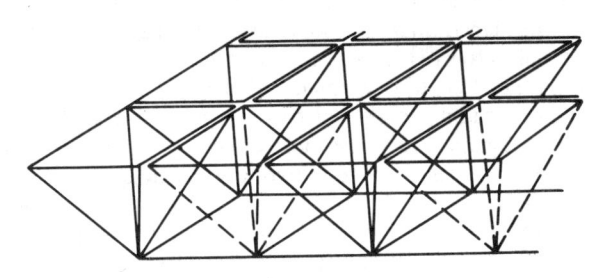

AXOMETRIC VIEW OF GRID
(ALTERNATE PYRAMIDS IN BROKEN LINE FOR CLARITY)

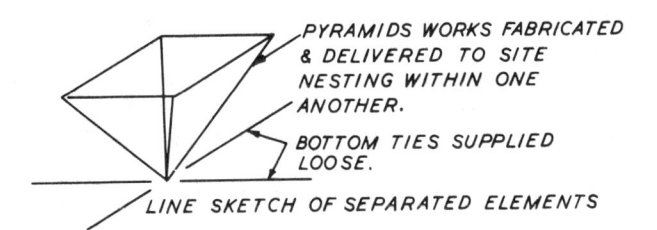

PYRAMIDS WORKS FABRICATED & DELIVERED TO SITE NESTING WITHIN ONE ANOTHER.
BOTTOM TIES SUPPLIED LOOSE.
LINE SKETCH OF SEPARATED ELEMENTS

DOUBLE LAYER GRID

Fig. 2

fabrication, the units thus produced stack very compactly for storage and transport (Fig. 2). Several patented forms of roof construction exist in this form, both in this country and on the continent. Some of these are capable of spanning up to 45 m in each direction.

(ii) *Diagonal Grids*

(a) *Single layer:* are fairly simple to design and fabricate (many computer bureaux have standard programmes for their structural analysis) and show excellent load distribution for square or nearly square bays. Example 1 shows a grid of this type and it is noticeable that the shearing forces and bending moments have maximum values which are similar for each beam. They are the most efficient form of single layer grid for spans of up to 22.5 m.

The example shows this grid type in its simplest form, even so it demonstrates clearly that its deflection to construction depth and span to weight characteristics are markedly superior to primary and secondary beam construction.

(b) *Double layer type:* Because the single layer type gives such good distribution, double layer diagonal grids are rarely used, where additional span or distributional characteristics are required other grid forms are used.

(iii) *Three Way Grids*

For wide spans, irregular shapes and areas in which internal supports cannot be placed equidistantly in each direction this form is often used. They consist of beam or lattice beam members running in three different directions, forming a triangular pattern in plan from which some members may be omitted to form a secondary hexagonal pattern (Fig. 3 (a), (b) and (c)).

The stiffness or the strength of elements of the grid or of single members may be altered to suit the loading conditions or spans prevailing at different points of the whole structure. This makes it the most flexible type of grid structure but also the most difficult to design. Simple two-way, single-layer grids can be designed in any good general design office and fabricated by any shop which has the initiative to take on work which varies only slightly from plane frame and beam fabrication. Double layer two-way grids are not so simple to design but can be the simplest form to fabricate.

Three-way grids of any form, though, can show the greatest advantage over simple, single spanning beams for the cases noted above, but must remain the preserve of the specialist consultant to design and the most accurate and best equipped shops to fabricate.

Figures 3 (a), (b) and (c) illustrate some of the devices by which material and fabrication time may be saved by using this form of construction. From these sketches some of the complexities in the analysis of such structures can also be envisaged.

To sum up the merits of the various grid forms, it may be said that generally:

(a) Rectangular two-way single-layer grids can only show to advantage under special loading conditions.
(b) Rectangular two-way double-layer grids and single-layer diagonal grids are best used over areas of equal span or equal bays in each direction.

TYPES OF SPACE STRUCTURES

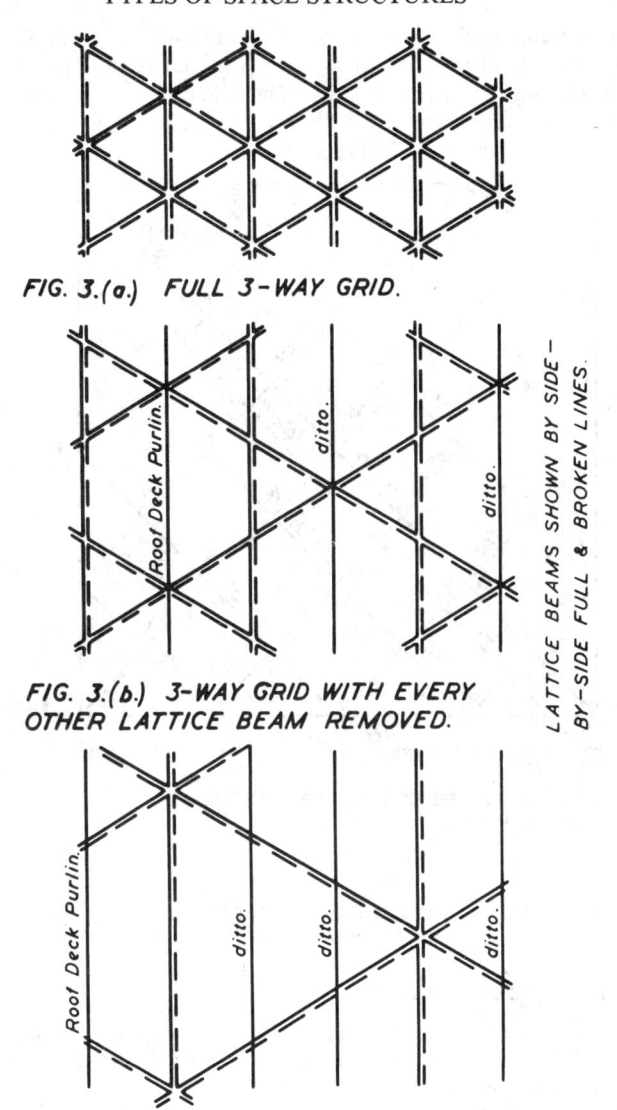

FIG. 3.(a.) FULL 3-WAY GRID.

FIG. 3.(b.) 3-WAY GRID WITH EVERY OTHER LATTICE BEAM REMOVED.

FIG. 3.(c.) 3-WAY GRID WITH 2 OUT OF 3 LATTICE BEAMS REMOVED.

Fig. 3

(c) Three-way single or double-layer grids provide the most economical method of spanning over very large or irregularly shaped areas, but are so much more difficult to design that they are worth while only for major projects designed by specialist consultants.

Example 1. 18 m Square diagonal grid, edges simply supported, dimensions and sizes as shown in Fig. 4.

446   SPACE FRAMES

The Beams are of negligible torsional compared with bending rigidity. The calculations are derived in terms of unit loads placed at the beam intersections, the distance a being also taken as unity. The calculations may therefore be used for any similar arrangements having the inertia's of the beams in each direction in a ratio 1.5:1. Let I = inertia of smaller section.

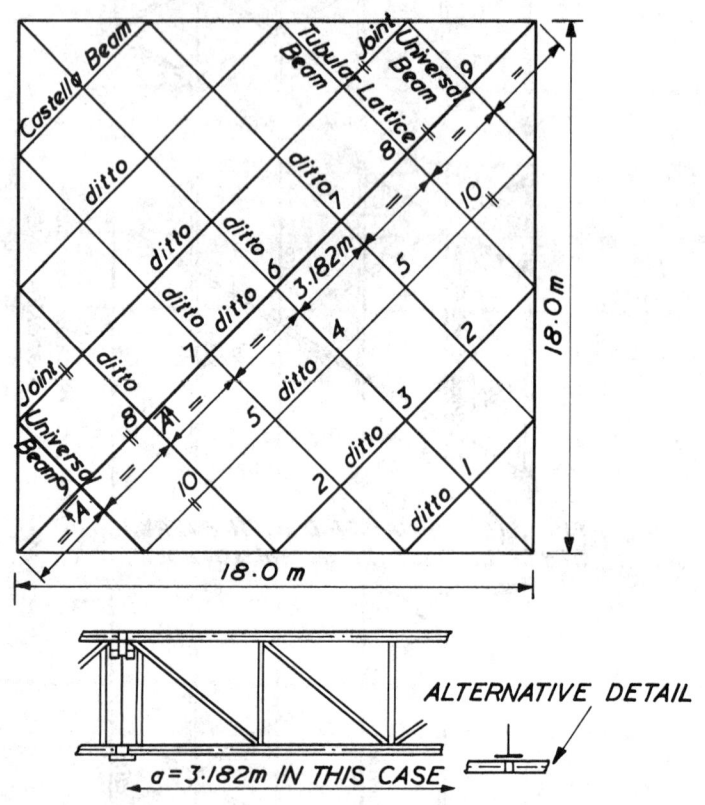

HALF SECTION A-A

THE TUBULAR BOTTOM CHORDS MAY BE PLACED ABOVE OR BELOW THE BOTTOM FLANGE OF THE CASTELLA BEAMS AS SHOWN
LATTICE BEAMS JOINTED AT EACH INTERSECTION
CASTELLA BEAMS JOINTED AS NOTED

DETAILS OF GRID IN EXAMPLE No.1

Fig. 4

The vertical deflections of each numbered point are worked out in terms of the applied unit loads and the reaction or 'link force' between the lattice and castellated beam at that point. These link forces are the unknown X_1, X_2, X_3 ... derived in the equations below. As the deflection for the castellated beam and the lattice beam must be identical at each separate point they may be equated. For example the deflection at point 1 is:

EXAMPLE OF SPACE FRAME

Castellated Beam *Lattice Beam*

$$I\delta_1 = \frac{0.167a^3 X_1}{1.5E} = \frac{a^3}{E}[2.04167(1 - X_1) + 3.375(1 - X_3) + 3.9583(1 - X_4)$$
$$+ 3.9167(1 - X_6) + 1.2917(1 - X_1) + 2.4583(1 - X_3)$$
$$+ 3.375(1 - X_4)]$$

By grouping the terms together the following expression is obtained:

$$I\delta_1 = \frac{0.111a^3 X_1}{E} = \frac{a^3}{E}[20.417 - 3.333X_1 - 5.833X_2 - 7.333X_4 - 3.917X_6]$$

Dividing through by a^3/E and rearranging the two right-hand equations:

$$\frac{\delta_1 EI}{a^3} = 20.416 = 3.444X_1 + 5.833X_2 + 7.333X_4 + 3.917X_6$$

Similar expressions are derived for each point, from which the following set of simultaneous equations are obtained.

	X_1	X_2	X_3	X_4	X_5	X_6	X_7	X_8	X_9	X_{10}
20·416 =	3·44 +	0 +	5·833 +	7·33 +	0 +	3·916 +	0 +	0 +	0 +	0
8·33 =	0 +	3·22 +	0·611 +	0 +	3·83 +	0 +	2·16 +	0 +	0 +	0
37·5 =	5·83 +	1·22 +	11·55 +	13·66 +	0 +	7·33 +	0 +	0 +	0 +	0
48·75 =	7·33 +	0 +	13·66 +	20·96 +	5·11 +	9·75 +	0 +	0 +	0 +	2·88
14·33 =	0 +	3·83 +	0 +	2·55 +	11·11 +	0 +	3·833 +	0 +	0 +	2·55
52·66 =	7·83 +	0 +	14·66 +	19·5 +	0 +	17·77 +	13·0 +	9·77 +	5·22 +	0
16·45 =	0 +	4·33 +	0 +	0 +	7·66 +	6·5 +	16·45 +	9·11 +	4·88 +	0
3·16 =	0 +	0 +	0 +	0 +	0 +	4·88 +	9·11 +	8·44 +	3·88 +	1·835
0·16 =	0 +	0 +	0 +	0 +	0 +	2·611 +	4·88 +	3·88 +	2·388 +	0
2·25 =	0 +	0 +	0 +	1·44 +	2·55 +	0 +	0 +	0·916 +	0 +	2·88

The solution of these equations by computer gives

$X_1 = 4.4994$ $X_5 = 0.2498$ $X_8 = 1.237$
$X_2 = 2.0463$ $X_6 = 0.3802$ $X_9 = -3.4586$
$X_3 = -0.6244$ $X_7 = 0.5370$ $X_{10} = -0.3168$
$X_4 = 0.9644$

From these the bending moment diagrams and shearing force diagrams for each beam may be derived (as they give the same values on each side of the centre line, only half of each diagram is drawn (see Figs. 5 and 6)).

448 SPACE FRAMES

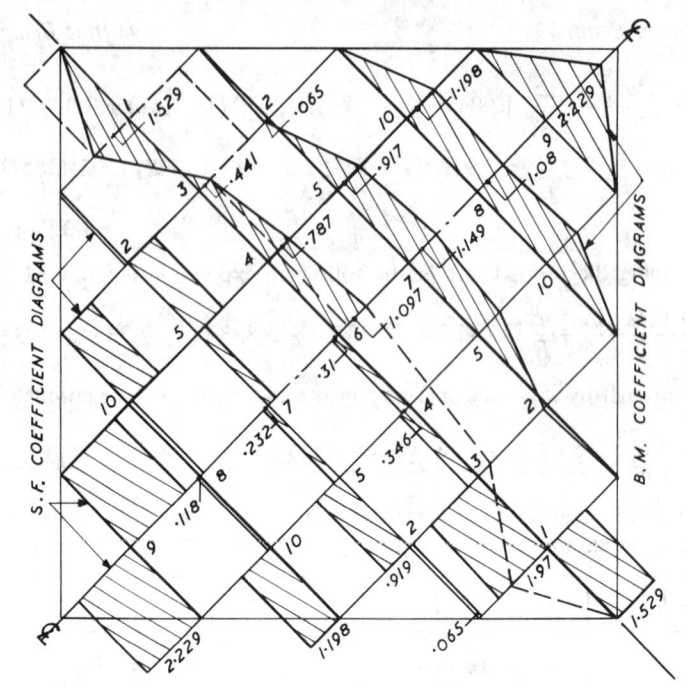

LATTICE BEAMS, B.M. & S.F. COEFFICIENTS
FOR B.M.'s x by wa^3, FOR S.F.'s x by wa^2

Fig. 5

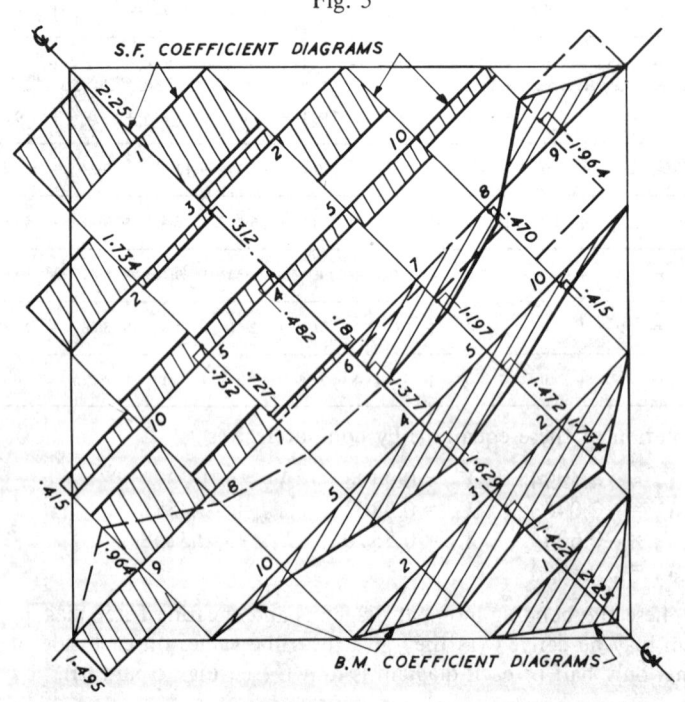

CASTELLA BEAMS B.M. & S.F. COEFFICIENTS
FOR B.M.'s x by wa^3, FOR S.F.'s x by wa^2

Fig. 6

EXAMPLE OF SPACE FRAME

These diagrams are for unit load at unit spacing. For the particular example, allowing a total load of 2.0 kN/m² over the plan area of the roof, the shear forces must be multiplied by

$$3.182^2 \times 2.0 = 20.25 \text{ kN}$$

The bending moments by

$$3.182^3 \times 2.0 = 64.44 \text{ kNm}$$

Selecting the worst cases

Lattice Beams

Point 9. B.M. = 64.44 × 2.229 = 143.6 kNm
Use 406 × 152 × 59 kg U.B.
I = 20 620 cm⁴ Z = 1 011 cm³
f_{bc} = 142 N/mm²
(Locate through castellation of other beam)

Point 1. B.M. = 64.44 × 1.529 = 98.5 kNm
Chord load = $\dfrac{98.5}{0.75}$ = 131.4 kN
Bottom use 76 × 76 × 3.25 mm R.H.S. (− 133.1 kN)
Top use 76 × 51 × 3.25 mm R.H.S. (157.4 kN)

Point 10. B.M. = 64.44 × 1.198 = 77.2 kNm
Chord load = $\dfrac{77.2}{0.75}$ = 102.9 kN
Effective lengths = 2.1 m (XX) and 0.525 m (YY)
Top use 76 × 51 × 3.25 mm R.H.S. (− 101.8 kN)
Bottom use 64 × 38 × 3.25 mm R.H.S. (122.8 kN)
All R.H.S. chords to B.S. 4360 Grade 50

Castellated Beams

Point 1. B.M. = 64.44 × 2.25 = 145 kNm
Use 610 × 140 × 39 kg castellated U.B.
I = 29 007 cm⁴ Z = 966.8 cm³
f_{bc} = 150 N/mm²
(This B.M. is at point of lateral support
∴ p_{bc} = 165 N/mm²)

Point 9. B.M. = 64.44 × 1.96 = 126 kNm

Section as above

Point 2. B.M. = 64.44 × 1.73 = 112 kNm

Section as above

Consider chords at 0.75 m centres; check second moment of area at end of calculation

Deflection at point 6 = $\dfrac{3.718 \times 3.182^3 \times 20.25}{2.1 \times 10^5 \times 19\,300}$ = 60 mm = $\dfrac{L}{300}$

$= \dfrac{L}{800}$ for 0.75 kN/m² imposed load.

The maximum shear force to be taken by a diagonal bracing

$$= 1.97 \times 20.25 = 39.9 \text{ kN}$$

Diagonal length = 1.093 m

$$\text{Max. load} = \frac{1.093}{0.75} \times 39.9 = 58.1 \text{ kN} \quad \text{Use } 43 \times 3.25 (59.4 \text{ kN})$$

For shears less than 24.5 kN, use 27 × 3.25 (35.6 kN)

The maximum compression load to be taken by a vertical

$$= -39.9 \text{ kN}$$
$$\text{Effective length} = 0.525 \text{ m} \quad \text{Use } 34 \times 3.25 (-40.2 \text{ kN})$$

For shears less than 27 kN   Use 27 × 3.25 (−27.0 kN)

−ve loads compressive, +ve loads tensile, bracing all in C.H.S.

Loads in brackets after section size are safe loads to B.S. 449 Part 2 1969 from Stewarts and Lloyds Ltd., 'Safe Load Tables and Section Properties'.

Second Moment of Area of Lattice beams, considering chords only

$$= 33.52^2 \times 7.55 + 42.7^2 \times 5.91 + 31.09 + 13.36$$
$$= 19\ 300 \text{ cm}^4, \text{ Ex. } 76 \times 51 \times 3.25 \text{ and } 64 \times 38 \times 3.25 \text{ R.H.S.}$$

All beams as this except in areas near points 1 and 9.

The weight of steel per m² is approximately 17 kg of which 75 per cent is in Castellated sections and 25 per cent tubular sections.

2. Space Structures Resoluble into a Series of Plane Frames

Simple examples of this form are bridges, conveyor bridges, crane girders and crane jibs, electricity supply transmission towers and radio aerial masts and towers of rectangular or square cross-section subjected to loads which induce torsion

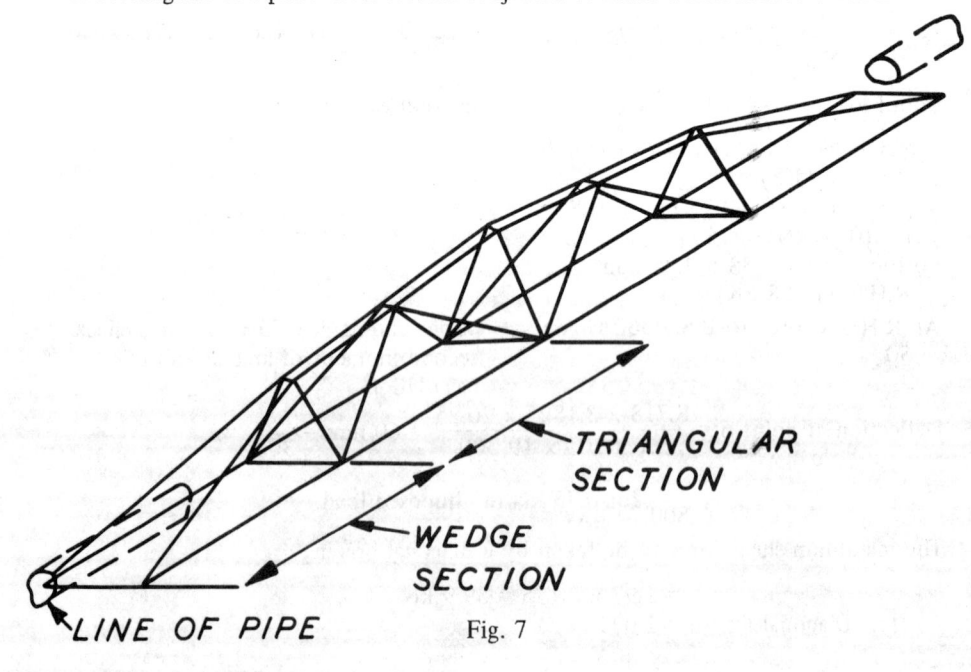

Fig. 7

into the complete structure. The cross-section need not be uniform throughout, in fact pipe and conveyor bridges are frequently designed as changing from rectangular to triangular cross-section through their length, the rectangular and the triangular sections each being of varying section (see Fig. 7). It is important, for reasonably accurate analysis by this method and for ease of fabrication, that the structure can be resolved into a series of plane frames. Even if the structure is so resolute, the analysis by this method is often an approximation.

Taking the case of a transmission tower under the broken conductor condition (i.e. one of the conductor wires is assumed broken on one side of the tower, the continuation of this conductor is assumed unbroken and exerting a heavy horizontal tension on the crossarm in the direction of the line), the heavy torsion load thus induced in the tower trunk is generally assumed to be divided between the parallel faces in line with the crossarms and those normal to the crossarms, and shears are applied at the level of the crossarm in each of these four planes to correspond with the applied torque – i.e. the shear to each plane is equal to $PL/2B$ (see Fig. 8 for derivation of P, L and B).

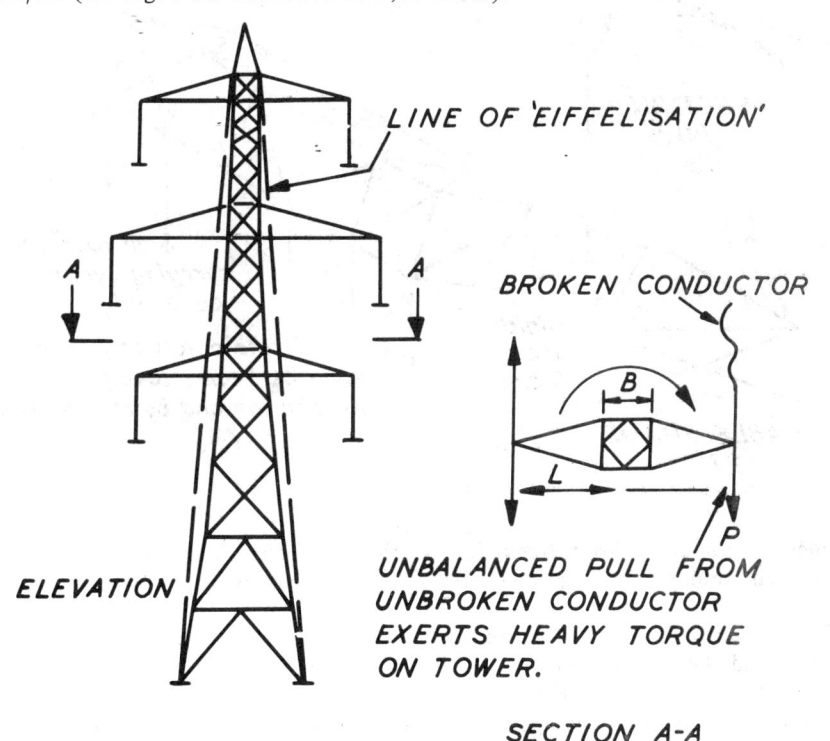

Fig. 8

Now, if the legs of the tower are 'eiffelised' (that is, if they have an inward camber as has the Eiffel Tower), the sides of the tower do not form a plane frame, thus the solution of this frame must be no more than a good approximation. In addition to the loads due to torsion the unbroken conductor exerts an horizontal load on the tower body in the direction of the line, this is usually

452   SPACE FRAMES

taken as being shared between the faces of the tower normal to the crossarm axis. All the other forms of loading on transmission towers can be reasonably considered as plane frame loadings and are not considered in this chapter.

Space frame roof structures are often designed by this method of resolution into plane frames.

To demonstrate the principle in its simplest form consider the single-span north-light roof, shown in Fig. 9, in each slope of which there is a lattice girder. The 'verticals' or struts in these girders act as rafters spanning from eaves to apex levels, as shown in Fig. 9 (a) and (c), and receive the dead and wind loads from the purlins, which have been omitted from the Figure for the sake of clarity.

The eaves reaction of each rafter is postulated to be vertical and this is shown

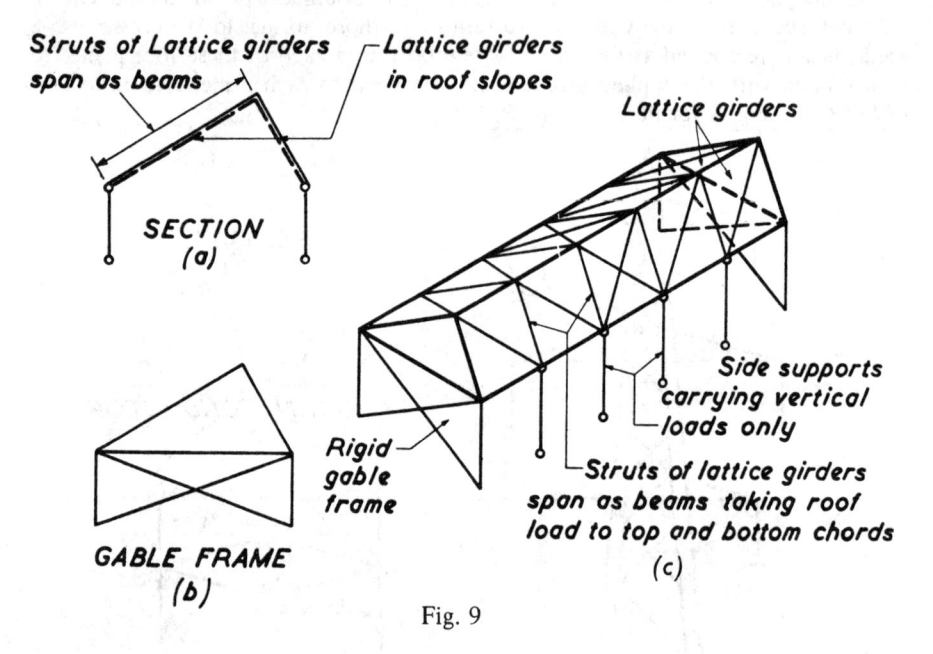

Fig. 9

provided by side posts. From this it follows that the direction of the reaction at the apex end under dead load, wind pressure or wind suction can be found as shown in Fig. 10 (a), (b) and (c).

The load on each rafter is then resolved into components in the lines of these reactions, and the original load on each is replaced by the equivalent loads at eaves and apex, as shown in Fig. 11.

At each panel point at the apex these equivalent loads from the rafters are then resolved into components in the planes of the slopes, to give the panel point loads for the design of the lattice girders.

Consider, for example, the north-light roof shown in outline in Fig. 12, and assume that the rafters are at 3.75 m centres, thereby fixing the panel lengths for both girders. With the horizontal wind pressure taken at 0.5 kN/m^2, and assuming that the sides of the building and roof covering are such that the flow of air through them is practically negligible, the loads for design purposes will be as shown.

SPACE FRAME ROOF STRUCTURE 453

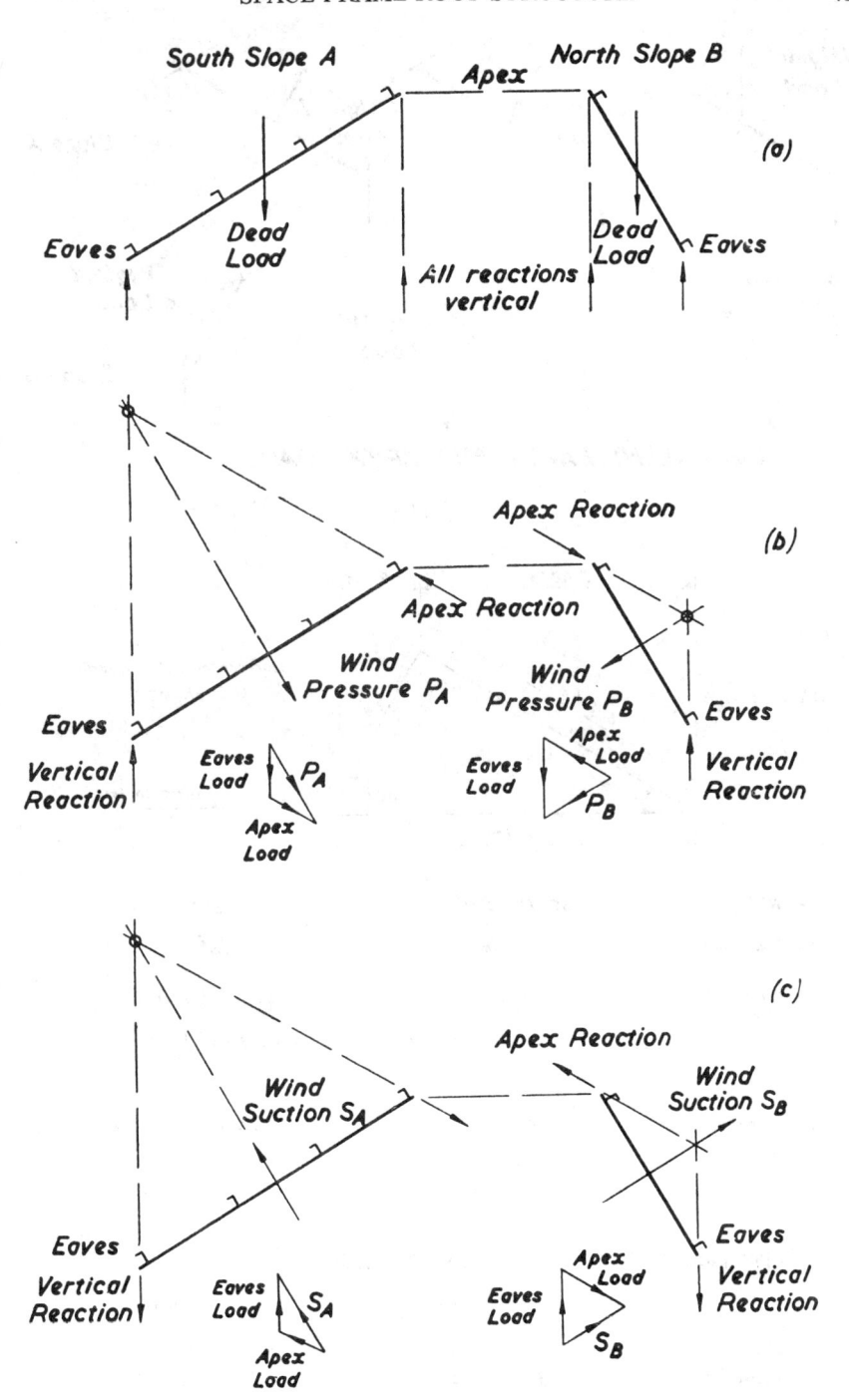

Fig. 10

EQUIVALENT EAVES AND APEX LOADS

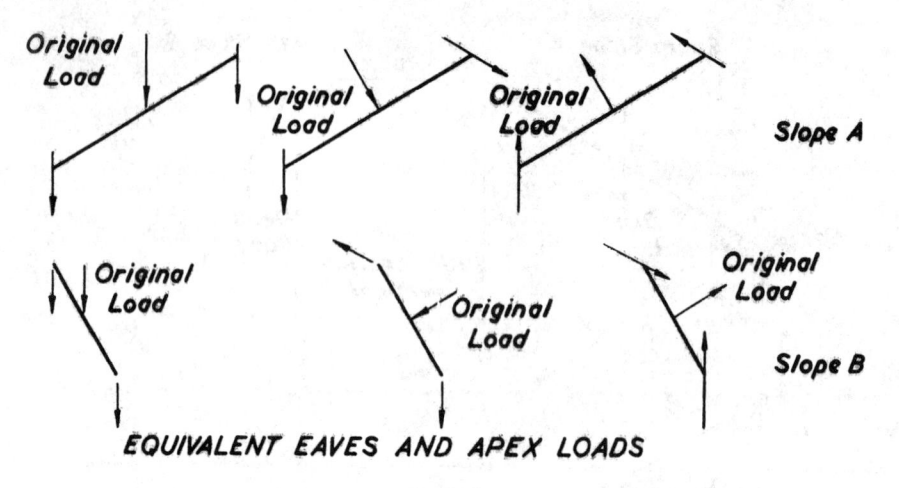

Fig. 11

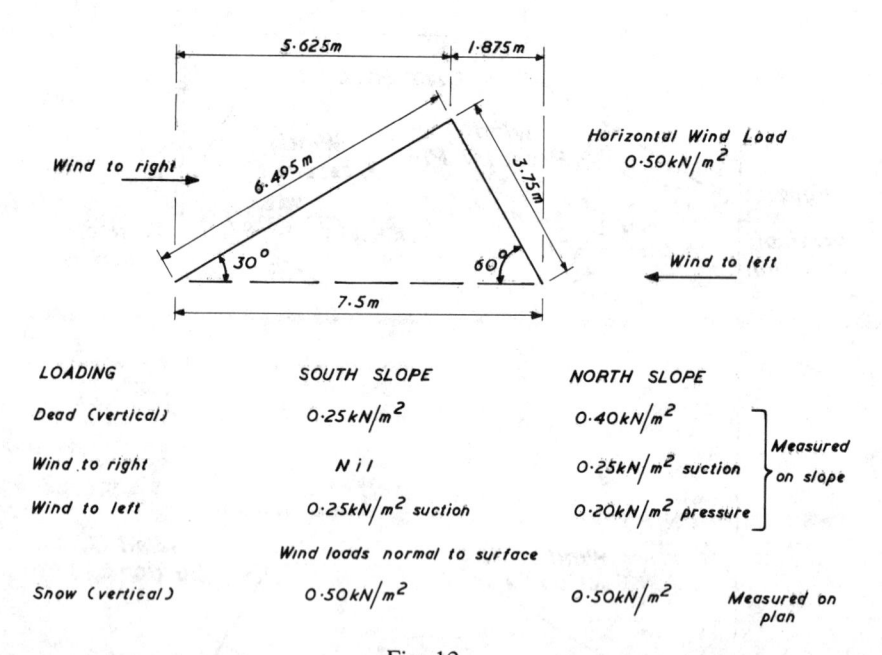

LOADING	SOUTH SLOPE	NORTH SLOPE	
Dead (vertical)	$0.25 kN/m^2$	$0.40 kN/m^2$	⎫
Wind to right	Nil	$0.25 kN/m^2$ suction	⎬ Measured on slope
Wind to left	$0.25 kN/m^2$ suction	$0.20 kN/m^2$ pressure	⎭
Wind loads normal to surface			
Snow (vertical)	$0.50 kN/m^2$	$0.50 kN/m^2$	Measured on plan

Fig. 12

The loads per panel will therefore be as shown in Fig. 13.

	South slope	North slope
Dead load	6.089 kN	5.625 kN
Snow load	10.547 kN	3.516 kN
Wind to right	Nil	3.516 kN suction
Wind to left	6.089 kN suction	2.813 kN

SPACE FRAME ROOF STRUCTURE

Fig. 13

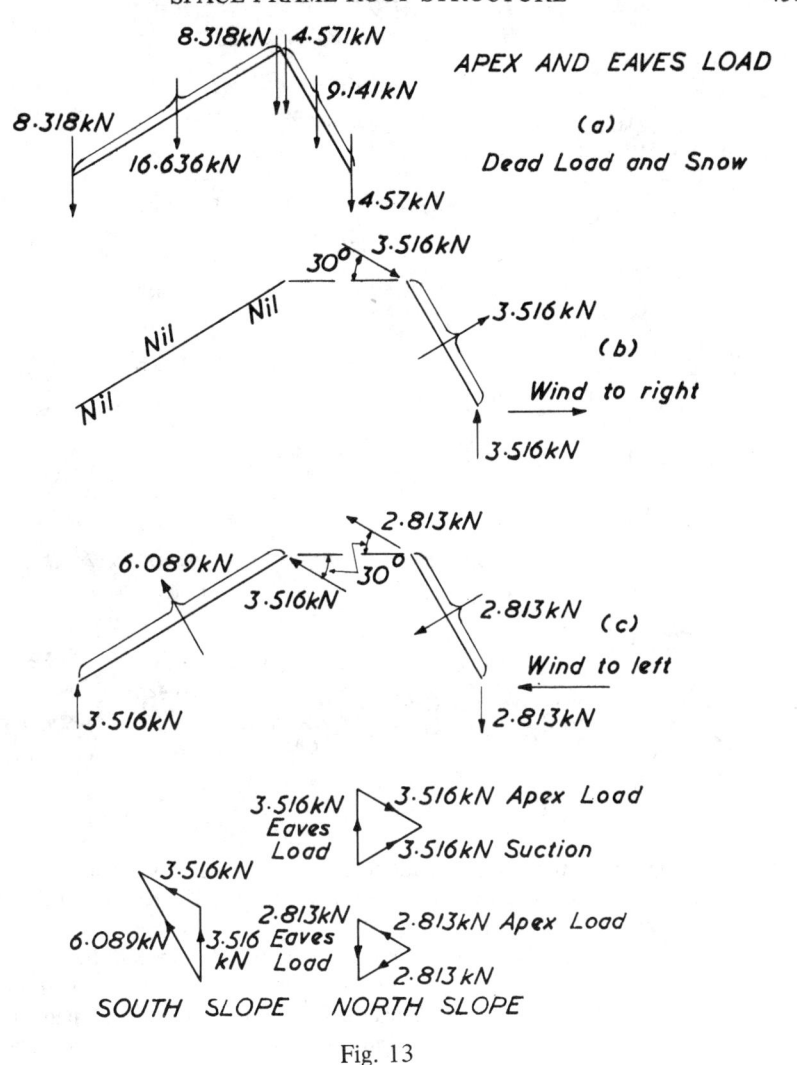

The apex loads are also shown in this Figure and are tabulated below:

	South slope	North slope
Dead and snow	8.318 kN	4.571 kN
Wind to right	Nil	3.516 kN
Wind to left	3.516 kN	2.813 kN

These loads are then resolved into components in the respective slopes as shown in Fig. 14, from which it is obvious that the maximum panel point load for the south-slope girder is given with the wind blowing to the left and for the north-slope girder with the wind blowing to the right, and the following figures result:

Maximum Panel Point Loads

South slope

Dead and snow	12.89 × sin 30° = 6.445 kN
Wind to left	6.33 × sin 30° = 3.165 kN
	9.610 kN

North slope

Dead and snow	12.89 × cos 30° = 11.161 kN
Wind to right	3.52 × cos 30° = 3.043 kN
	14.204 kN

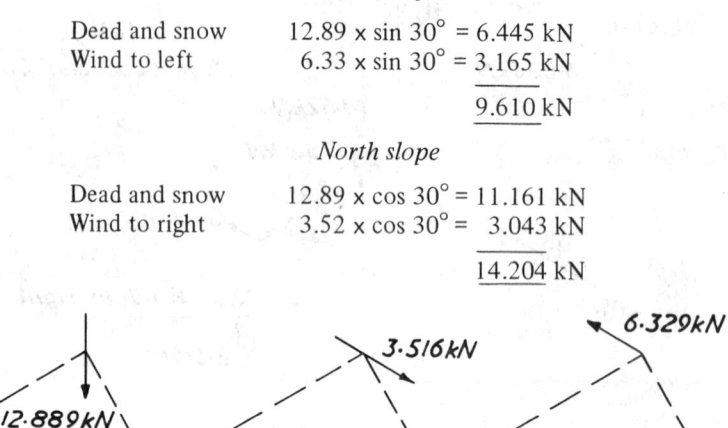

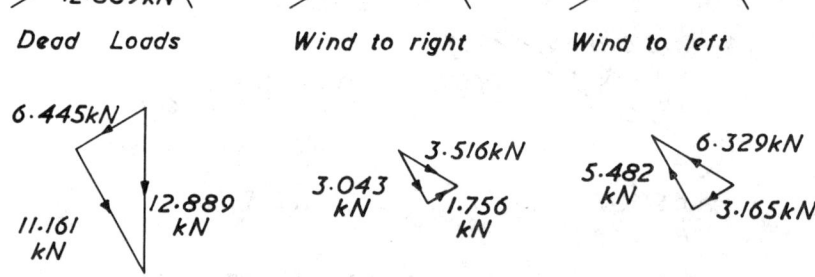

Fig. 14

These are maximum panel point loads to be applied at the top chords of the lattice girders, provided that independent girders are used. The case of girders with a common top chord is considered later.

It now remains to find the forces to be applied at panel points in the bottom chords of the girders due to wind effect on the sides of the buildings. Assuming that the shed is 3.0 m high to eaves with posts at 3.75 m centres to suit the panel length, and that there is suction and pressure as indicated in Fig. 15, then the side posts may be assumed to act as vertical beams spanning from ground to eaves level. Their

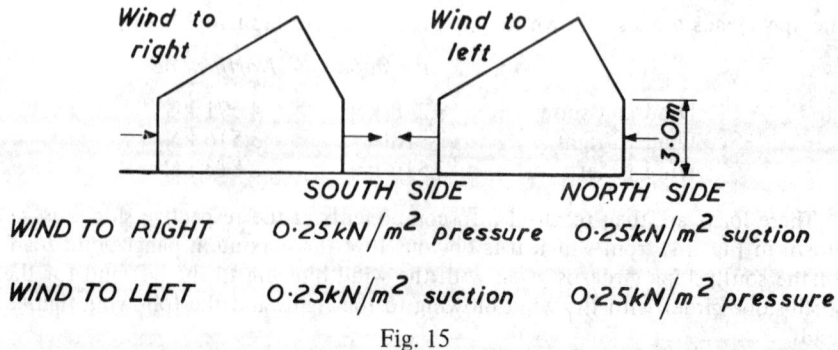

WIND TO RIGHT $0.25 kN/m^2$ pressure $0.25 kN/m^2$ suction

WIND TO LEFT $0.25 kN/m^2$ suction $0.25 kN/m^2$ pressure

Fig. 15

SPACE FRAME ROOF STRUCTURE

horizontal reactions on each of the lattice girders will therefore be half the total load on each side, i.e. 1.406 kN. These horizontal reactions are then resolved into the planes of the lattice girders, resulting in panel point loads in the planes of the girders of 1.621 kN in the south slope and 2.812 kN in the north slope. Considering wind blowing to the left, then for the south slope the load due to suction on the south side acts in the same direction as the wind effect at the apex, whilst wind blowing to the right increases the loads in the north slope girder.

The final loadings for the two lattice girders are thus as shown in Fig. 16 (a) and (b).

The vertical forces in the side posts are combined from the vertical reactions shown in Fig. 13, and the additional vertical component caused by resolving the wind loads on the sides of the building into the planes of the lattice girders, and are:

	South side posts	North side posts
Dead and snow	+ 8.318 kN	+ 4.570 kN
Wind to left,		
from rafter	− 3.516 kN	+ 2.813 kN
from side wind	− 0.812 kN	+ 2.432 kN
Wind to right,		
from rafter	Nil	− 3.516 kN
from side wind	+ 0.812 kN	− 2.432 kN
Maximum value	+ 9.130 kN	+ 9.815 kN

If the lattice girders in the two slopes have a common top chord, some reduction can be made in the panel point loads at the apex, since the values given in Fig. 16 (a) and (b) are the maximum which can exist in either slope and are not possible at the same time; it is therefore necessary to consider the load which can exist in both slopes simultaneously. See Fig. 16 (c) and (d).

Apex loads	South slope	North slope
Dead and snow	6.445 kN	11.161 kN
Wind to left	3.165 kN	− 5.482 kN
	9.610 kN	5.679 kN
Dead and snow	6.445 kN	11.161 kN
Wind to right	− 1.760 kN	3.043 kN
	4.699 kN	14.204 kN
Eaves loads		
Wind to left	1.621 kN	−2.813 kN
Wind to right	−1.621 kN	2.813 kN

It will be seen that the web members and the bottom chord members of the lattice girders in both slopes should be designed for the loads previously given, but that the common top chord can be designed for the reduced total loads given in Fig. 16 (d).

From this example it will be apparent that the basic principle of this method of design is that advantage is taken of the line common to two inclined planes, since

458   SPACE FRAMES

(a) *PART SOUTH SLOPE GIRDER*

Panel Point Loads at Apex 9.610kN each

Panel Point Loads at Eaves 1.621kN each

(dimensions: 3.75m, 3.75m, 6.495m)

(b) *PART NORTH SLOPE GIRDER*

Panel Point Loads at Apex 14.204kN each

Panel Point Loads at Eaves 2.813kN each

(dimensions: 3.75m, 3.75m)

LOADS FOR INDEPENDENT GIRDERS

(c)
- 9.610 kN each / 1.621 kN each — SOUTH SLOPE
- 5.679 kN each / 2.813 kN each — NORTH SLOPE

COINCIDENT LOADINGS – WIND TO LEFT

(d)
- 4.699 kN each / 1.621 kN each — SOUTH SLOPE
- 14.204 kN each / 2.813 kN each — NORTH SLOPE

COINCIDENT LOADINGS – WIND TO RIGHT

Fig. 16.

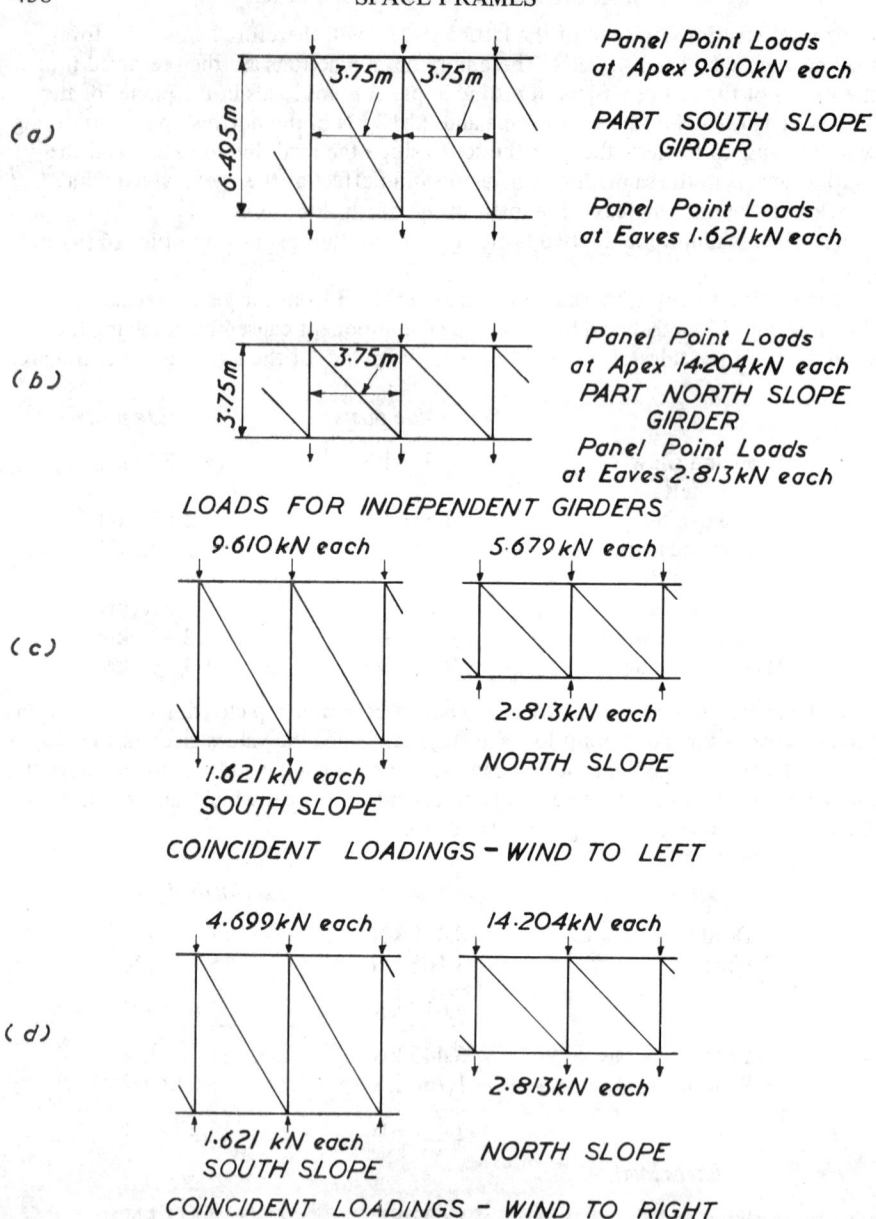

any loads acting at this line can be resolved into components acting in these planes, and any such line is regarded as the boom of two lattice girders. (See *Example 2* for a fully worked example).

The end reactions of the lattice girders are provided by triangulated or other stiff frames at the ends of the building or intermediate points. These frames transfer the loads to the foundations.

SPACE FRAME ROOF STRUCTURE 459

The system can easily be extended to the case of multi-bay roofs as shown in Fig. 17, which depicts, under the action of dead loads only, a four-bay roof without internal supports. It will be noticed that lattice girders extend the full depth of each slope, and in consequence, for any given overall width, an increase in the number

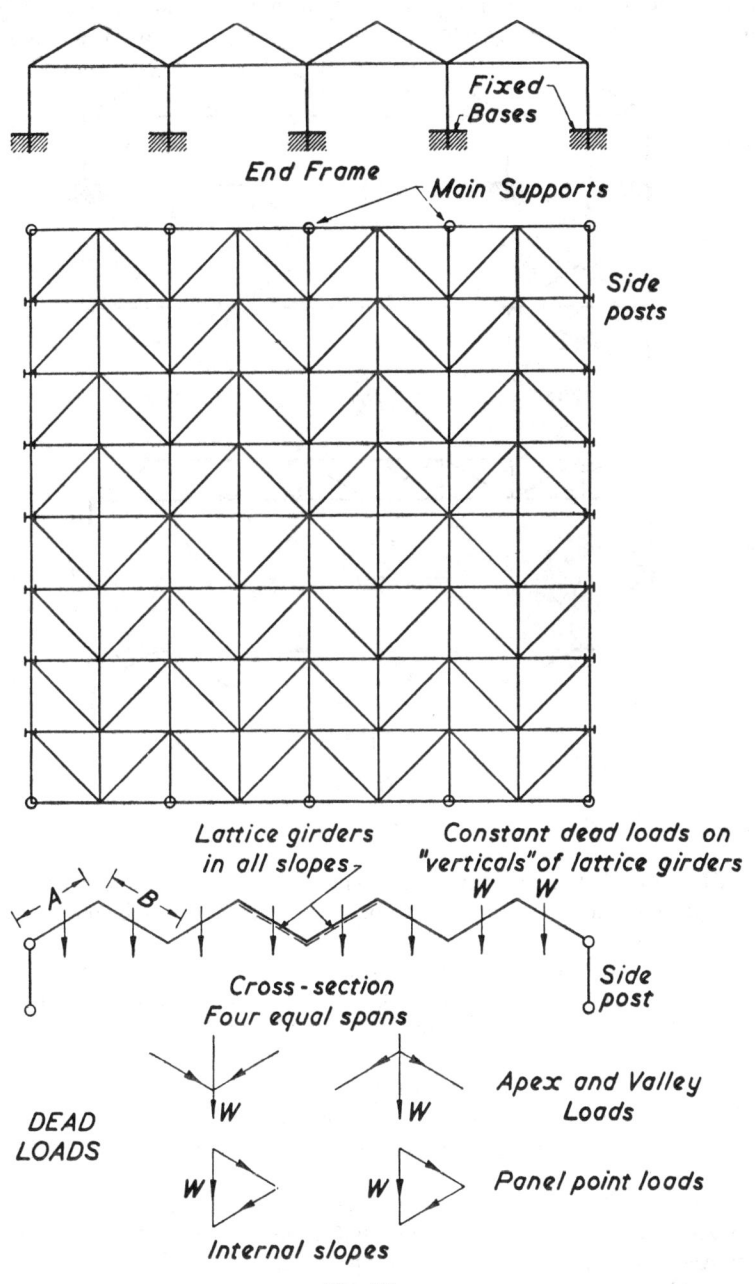

Fig. 17

of lattice girders will result in a reduction in the length of the 'verticals' of the girders, with a consequent reduction of the B.M.s in these members. It is thus economical in framing to utilise a fairly large number of slopes. Under the action of dead loads only, all lattice girders, other than those in the two outer slopes, receive equal panel point loads at both top and bottom chords provided that the slopes are of the same length and the dead load per unit of area is constant. The two outer

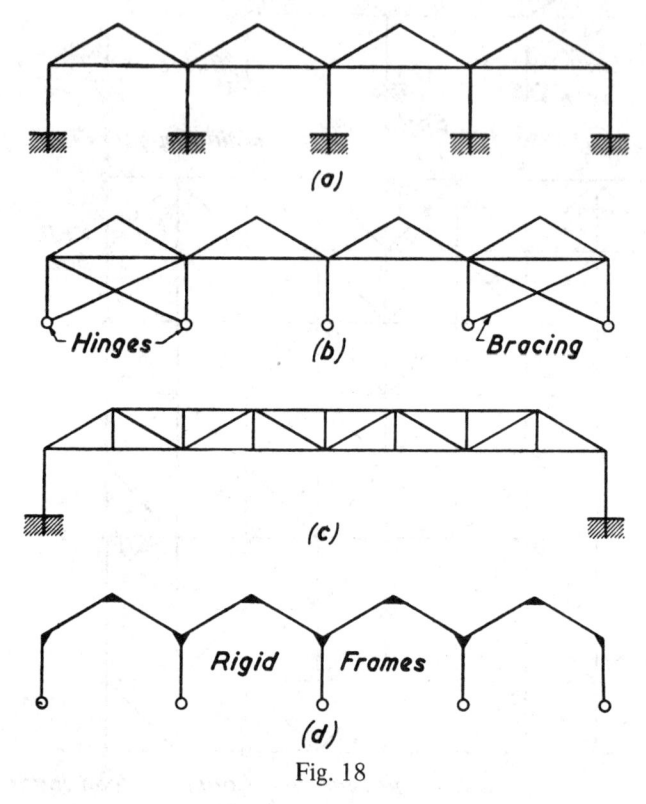

Fig. 18

girders (A) are loaded at their top chords only, the reaction at the eaves end of each 'vertical' being supplied by a post. This top-chord load is the same amount as for all other top-chord loads in the structure.

The action of the roof framing under wind loads is analysed as explained in connection with the example of the north-light roof, but it must be remembered when finding the panel point loads that the reactions at the eaves are vertical, which thus affects the loads in the two slopes A and B.

The example in Fig. 18 shows several types of end frame. In diagram (a) a tie at eaves level ensures stability of the sloping lattice girders, and the horizontal components of their reactions are taken to foundation level by columns fixed at their bases.

In diagram (b) the three inner columns serve to take vertical components only, and horizontal forces are transferred through the eaves level tie to the braced frames. Alternatively, as shown in diagram (c), supports may be provided at the eaves only, the inclined lattice girders being supported by another lattice girder in a vertical

SPACE FRAME ROOF STRUCTURE 461

plane. Rigid frames can also be utilised as sketched in diagram (d).

The examples so far discussed have utilised lattice girders of equal depth to the length of the roof slope, but this is not essential. The rafters used in the lattice girders must in all cases span from valley to apex as beams, in addition to forming

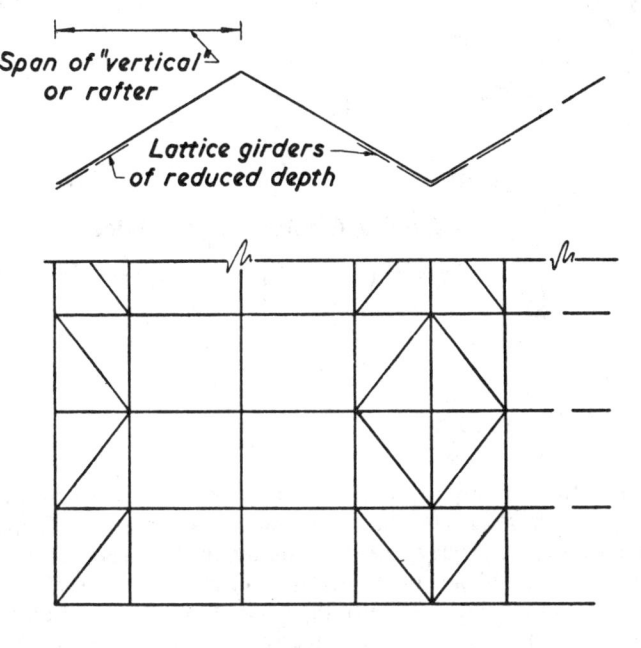

Fig. 19

the struts of the lattice girders of reduced depth as shown in Fig. 19. There must always be at least one lattice girder per slope.

The construction can also be extended to mansard roofs, and Fig. 20 demonstrates this application. The loads from the roof are carried to the node points of the lattice girders by the rafters which also act as the struts in the girders. The loads at the nodes are resolved into the planes of the girders which meet at the nodes as previously explained. Horizontal lattice girders in Fig. 20 (b) and (c) will take no load other than that due to unsymmetrical loading such as wind, as the dead loads are self-cancelling.

Example 2 illustrates the design of a folded plate type roof in skeletal construction resolved into plane frames, the general arrangement being shown in Fig. 21. The calculations are annotated to be self-explanatory, but the following points are also relevant to this structural form:

(a) The roof slopes in the example are equal about the ridges, but the method can be applied to the completion of the design of north light construction as illustrated in pages 452 to 458, with the exception that two roof plane frames, two thrust girders and two rafters need to be designed.
(b) The method can also be applied to monitor roofs but in this case the cheeks of each monitor must be braced to give stability to the upper roof.

SPACE FRAMES

Fig. 20

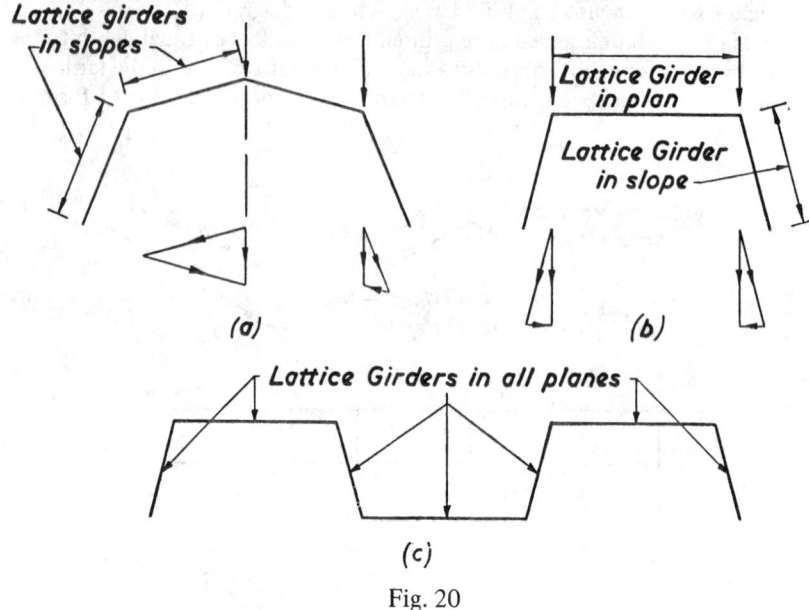

(c) It is extremely easy to analyse the separate plane frames for this kind of structure, it is almost as easy to forget the fact that these frames are compounded into a single space structure which must have overall stability. For example the omission of the end thrust girders or the ties in Example (b) would lead to instability, even though each of the plane frames have been properly designed. This is particularly important in the case of north light roofs of this form, where bracing in the south slope sometimes tends to be forgotten, leading to instability and bowing of the chords of the north light girder. In monitor roofs both upper and lower roofs require bracing, or the rafters made continuous between braced planes.

(d) Where more than three plane frames are linked together to form a space structure similar to that illustrated in example (b), the load taken by each frame cannot be assessed, even approximately, from consideration of statics. Furthermore under some conditions of loading such structures can become unstable if considered as pin jointed, even though they may be apparently stable under uniformly distributing loading. Consequently structures in this category are dealt within the next section as braced vaults.

It must be emphasised that the methods of analysis outlined in this section are in general only approximate. They can be used effectively for preliminary designs and for comparatively small structures but for major structures or repetitive designs more accurate methods of final analysis should be used. With the growing use and development of programmes for the analysis of space structures by means of the digital computer, it is safe to say that in the next five years, final analysis will always be made by this means. The case for the analysis of torsional loads in transmission towers and other structures of box section by means of the computer is particularly strong as this is the only means of obtaining accurate results economically.

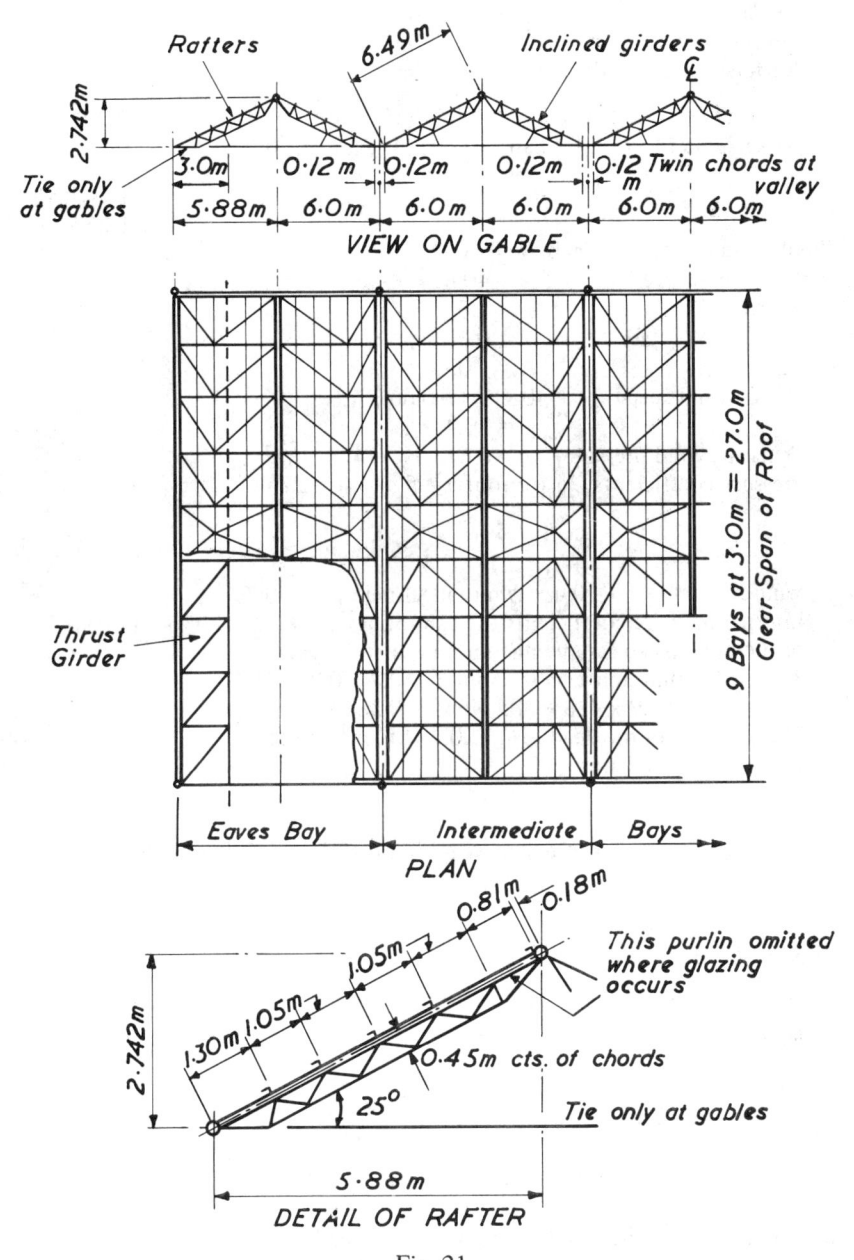

Fig. 21

Example 2.

Loading

Dead Load:

Decking, etc.	$= 275$ N/m^2
Purlins	$= 65$ N/m^2
Rafters and inclined girders	$= 150$ N/m^2

Total dead load on plan $= 490$ N/m^2 $= 0.49$ kN/m^2
Super Load (To C.P.3. Chap. V. Part 1): On plan $= 0.75$ kN/m^2

$\qquad\qquad\qquad\qquad\qquad\qquad\qquad\qquad\qquad\qquad$ 1.24 kN/m^2
$\qquad\qquad\qquad\qquad\qquad\qquad\qquad\qquad\qquad$ say 1.25 kN/m^2

Wind Load (To C.P.3. Chap. V. Part 2):
Basic wind speed V say $\quad= 40$ m/s
Factor $S1 \quad\quad\quad\quad\quad\quad= 1.0$
Factor $S2$ (class 3.C.) $\quad= 0.66$
Factor $S3 \quad\quad\quad\quad\quad\quad= 1.0$
Design wind speed $V_s = 40 \times 1.0 \times 0.66 \times 1.0$
$\qquad\qquad\qquad\qquad\quad = 26.4$ m/s
Dynamic wind pressure $\quad q = 0.613\, V_s^2 = 428$ N/m^2
Pressure coefficients C_{pe} for roof $h = 5$ m $\quad w = 27$ m

$$\frac{h}{w} = \frac{5}{27} = \frac{1}{5.4} \leqslant \frac{1}{2}, \text{ slope} = 25°$$

Wind normal $C_{pe} = -0.4q$ Wind tangential $C_{pe} = -0.7q$
Internal pressure coefficients C_{pi} cannot be determined without more details of the building so assume maximum value $= 0.2q$
Then maximum uplift on roof will be $(0.7 + 0.2)q = 0.9q$
Wind uplift $= 0.9 \times 428 = 385$ N/m^2 $= 0.385$ kN/m^2
This is less than the dead load of 0.49 kN/m^2 therefore the effect on the roof structure can be ignored.

Purlins

These are designed in accordance with Clause 45 B.S. 449:Part 2 1969.

Rafter

Type 1. Latticed tubular girder

Load on rafter $= 6.0 \times 3.0 \times 1.25 = 22.5$ kN

Reactions $\quad = \dfrac{22.5}{2} = 11.25$ kN

Force in chords due to load on rafter $= \dfrac{22.5 \times 6}{8 \times 0.45} = \pm 37.5$ kN

Maximum shear resisted by internal bracings $= 11.25 \times \dfrac{5}{6} = 9.375$ kN

Maximum force in internal bracing (by vector diag. Fig. 21(a)) $= 12.0$ kN Sections:

(a) For Upper chord size see Main Girder calculations.

SPACE FRAME ROOF STRUCTURE

(b) For Lower chord use 34 x 3.25 mm C.H.S. (circular hollow section)
 Actual load + 37.5 kN.
 Allowable load 315 x 155 = 48 800 N = 48.8 kN
(c) For Internal Bracings use 27 x 3.25 mm C.H.S.

Actual load = 12 kN. Effective $\dfrac{l}{r} = \dfrac{100 \times 0.7}{8.4} = 58$

Allowable load = 243 x 127 = 30.8 kN.

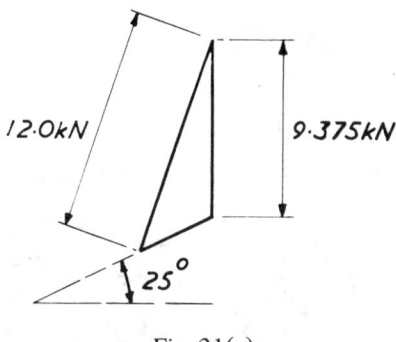

Fig. 21(a)

Type 2. Castellated Joist.

Bending Moment = $\dfrac{22.5 \times 6}{8}$ = 16.88 kNm

Axial Load (from Main Girder calcs.) = −128.6 kN (3, 5).
                                        −149.3 kN (A2).

(A2 only combined with half B.M.)

Try 152 x 89 mm x 17 kg joist castellated to 228 x 89 mm

$\dfrac{l}{r} = \dfrac{10.67 \times 10^2}{20.1} = 53.1 \qquad p_c = 131 \text{ N/mm}^2$

$\dfrac{D}{T} = \dfrac{229}{8.28} = 27.6 \qquad p_{be} = 165 \text{ N/mm}^2$

$\sum \dfrac{f}{p} = \dfrac{128.6 \times 10^3}{18.1 \times 10^2 \times 131} + \dfrac{16.88 \times 10^6}{181.9 \times 10^3 \times 165} = 0.543 + 0.563 = 1.106$

No good.
 (d) For rafter use 267 x 102 x 22 kg Castella (ex 178 x 102 joist)
   or 254 x 102 x 22 kg U.B.

Main Inclined Girders

Consider forces acting on girders due to dead load plus super load.

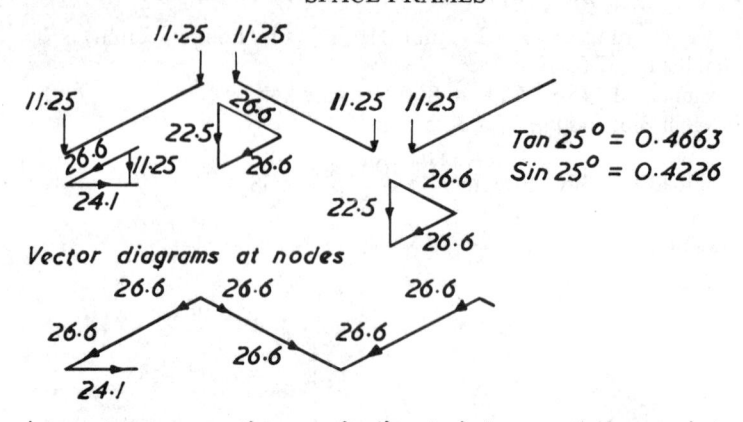

466   SPACE FRAMES

Vector diagrams at nodes

$Tan\ 25° = 0.4663$
$Sin\ 25° = 0.4226$

Internal forces acting on inclined girders and thrust girder

Fig. 21 (b) shows internal forces acting on inclined girders and thrust girders

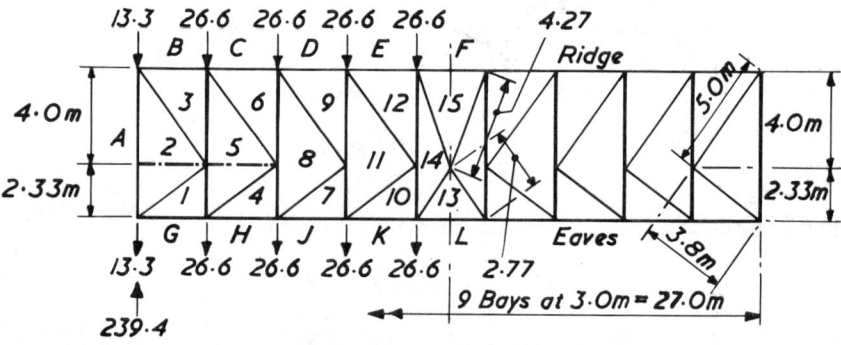

General arrangement and forces on inclined girders
Fig. 21(c)

Length of diagonals:

Member 2–3. $4.0^2 = 16.0$
              $3.0^2 = 9.0$
              $\overline{25.0} = 5.0^2$

Member 1–2. $2.33^2 = 5.43$
              $3.0^2 = 9.0$
              $\overline{14.43} = 3.8^2$

Member 14–15. $4.0^2 = 16.0$
              $1.5^2 = 2.25$
              $\overline{18.25} = 4.27^2$

Member 13–14. $2.33^2 = 5.43$
              $1.5^2 = 2.25$
              $\overline{7.68} = 2.77^2$

SPACE FRAME ROOF STRUCTURE

The diagonals of the girders have been arranged to node on a purlin line so that the members in tension are longer than those in compression.

Loads in chords and sections of chords:

Members F15 & L13: Bending Moment = 212.8 x 13.5 = 2 873
− 53.2 x 4 x 6.0 = 1 277
_____
1 596 kNm

$$\text{Load in member} = \frac{1\,596}{6.33} = \pm 252.1 \text{ kN}$$

Sections: Top chord F15

Axial Load = 2 x −252.1 = −504.2 kN

(Note: Compressive loads in members are considered negative and tensile loads considered positive.)

Effective length = 3.0 x 0.85 = 2.55 m

(e) Try 219 mm x 5.39 mm C.H.S.

$$\frac{l}{r} = \frac{2.55 \times 10^2}{7.57} = 33.69 \qquad p_c = 141 \text{ N/mm}^2$$

Allowable load = 36.12 x 10² x 141 = 509 300 N = 509.3 kN

Bottom chord L13.

Axial load = +252.1 kN

(f) Try 114 mm x 5.39 mm C.H.S.

Allowable load = 18.45 x 10² x 155 = 286 kN

Members C6 & H4: Bending Moment = 212.8 x 6.0 = 1 276.8
− 52.3 x 3.0 = 159.6
_____
1 117.2 kNm

$$\text{Load in member} = \frac{1\,117.2}{6.33} = \pm 176.5 \text{ kN}$$

Depending upon the relationship between fabrication and material costs it may prove economical to 'curtail' the chord sections. To enable this detail to be checked by the fabricator the sections for C6 and H4 will be calculated.

Sections: Top chord C6.

Axial load = 2 x −176.5 = −353 kN
Effective length = 2.55 m

From steelmakers' tables safe load for

(g) 168 mm x 5.39 mm C.H.S. = 372 kN

Bottom chord H4.
Axial load = + 176.5 kN

468   SPACE FRAMES

(h) Use 114 mm x 5.39 mm C.H.S. The details will become awkward if a tube of smaller diameter is used and it is dangerous to use the same diameter tube with a different wall thickness in the one component, i.e. bottom chord.

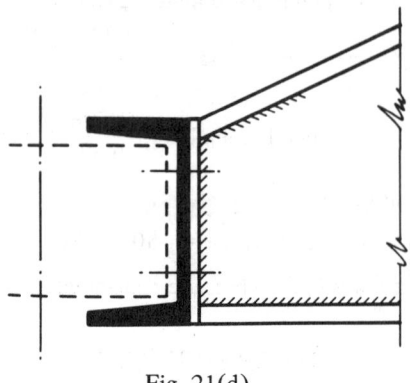

Fig. 21(d)

Consider alternative section for bottom chord using rolled steel channel (Fig. 21 (d)). This would produce a neat detail at the valley and if castellated rafters are used the structure could be site bolted as against site welding using tubes. The latter would obviously be neater but more expensive.

Consider Bottom chord L13.

Axial load + 252.1 kN

(j) Try 152 mm x 76 mm x 18 kg R.S.C.

Gross Area = 22.77 cm^2
Less holes 2 x 2.1 x 0.635 = 20.67 cm^2

Net area 20.10 cm^2
Allowable load = 20.1 x 10^2 x 155 = 311 600 N = 311.6 kN

Loads in bracings and sections of bracings.

The safe loads shown in the following table are taken direct from the steelmakers' tables.

The factor for effective length is taken as 0.7 for site welded construction and 0.85 for site bolting. For the purpose of the table site welding has been considered.

Shear coefficients:

$$\text{Members} \quad 2-3, 5-6, 8-9, 11-12 = \frac{5.0}{6.33} = 0.79$$

$$\text{''} \quad 1-2, 4-5, 7-8, 10-11 = \frac{3.8}{6.33} = 0.60$$

$$\text{''} \quad 3-5, 6-8, 9-11, 12-14 = \frac{4.0}{6.33} = 0.63$$

$$\text{''} \quad 1-5, 4-8, 7-11, 10-14 = \frac{2.33}{6.33} = 0.37.$$

SPACE FRAME ROOF STRUCTURE

Member	Shear in Panel kN	Load in Member kN — Due to Shear	Due to Lattice Rafter	Total	Effective Length (0.7 L) m	Section mm	r cm	l/r	Allowable Load kN
A–2		(212.8 × .63) + 13.3 = −147.4	−18.8	−166.2	1.49*	114 × 3.66 C.H.S.	3.91	38.1	−177.9
1–2	212.8	212.8 × .60 = −127.7		−127.7	2.66	"		68.1	−150.0
2–3	212.8	212.8 × .79 = +168.1		+168.1	3.50	"		89.6	+197.0
1–5		(159.6 × .37) + 26.6 = + 85.7		+ 85.7	1.63	"		41.7	+197.0
3–5		(159.6 × .63) + 26.6 = −127.1	−37.5	−163.9	1.49*	"		38.1	−177.9
4–5	159.6	159.6 × .60 = − 95.8	−37.5	− 95.8	2.66	89 × 4.06 C.H.S.	3.02	88.1	− 99.4
5–6	159.6	159.6 × .79 = +126.0		+126.0	3.50	"		116.0	+167.4
4–8		(106.4 × .37) + 26.6 = + 66.0		+ 66.0	1.63	114 × 3.66 C.H.S.	3.91	41.7	+197.0
6–8		(106.4 × .63) + 26.6 = − 93.6	−37.5	−131.4	1.49*	"		38.1	−177.0
7–8	106.4	106.4 × .60 = − 63.8	−37.5	−101.4	2.66	89 × 4.06 C.H.S.	3.02	88.1	−101.4
8–9	106.4	106.4 × .79 = + 84.1		+ 84.1	3.50	60 × 3.25 C.H.S.	2.01	175.0	+ 90.4
7–11		(53.2 × .37) + 26.6 = + 46.3	−37.5	+ 46.3	1.63	114 × 3.68 C.H.S.	3.91	41.7	+197.0
9–11		(53.2 × .63) + 26.6 = − 60.1	−37.5	− 97.6	1.49*	"		38.1	−177.9
10–11	53.2	53.2 × .60 = − 31.9		− 31.9	2.66	76 × 3.25 C.H.S.	2.50	102.6	− 56.4
11–12	53.2	53.2 × .79 = + 42.0		+ 42.0	3.50	60 × 3.25 C.H.S.	2.01	175.0	+ 90.4
10–14		+ 26.6		+ 26.6	1.63	114 × 3.66 C.H.S.	3.91	41.7	+197.0
12–14		− 26.6	−37.5	− 64.1	1.49*	114 × 3.66 C.H.S.	3.91	38.1	−177.9
13–14		—		—	1.94	60 × 3.25 C.H.S.	2.01	96.8	− 48.3
14–15		—		—	2.99	"		149.0	+ 90.4

*Two bays used to calculate effective length to cater for glazing bar 2.13 m long.

470

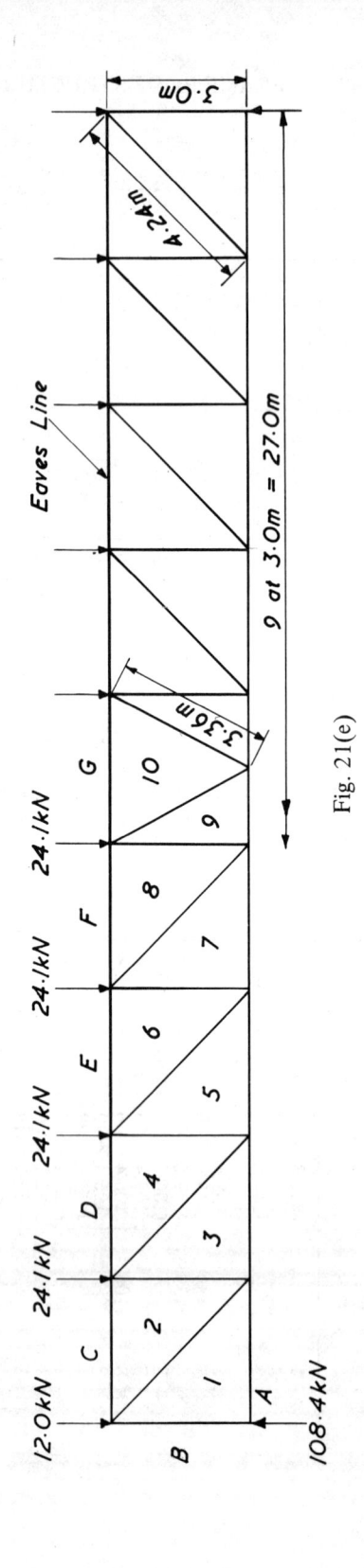

Fig. 21(e)

SPACE FRAME ROOF STRUCTURE

Thrust Girder at Eaves (shown on Fig. 21(e))

The chord load at the eaves line reduces the effect of the inclined girder forces so this can be ignored.

The chord load in member A9 = $\dfrac{(96.4 \times 12.0) - (24.1 \times 18.0)}{3.0}$

$$= \dfrac{1\,157 - 434}{3.0} = +241 \text{ kN}$$

(k) Use 114 x 5.39 mm C.H.S. (Allowable load = +286 kN)

Bracing Member	Shear kN	Load kN	Eff. Length m (0.7 L)	Section mm	Allowable Load
B–1	108.4	–108.4	2.10	89 x 4.06 C.H.S.	–124.6
1–2	96.4	+136.2	2.97	– ,, –	+168.0
2–3	,,	– 96.4	2.10	– ,, –	–124.6
3–4	72.3	+102.2	2.97	76 x 3.25 C.H.S.	+115.0
4–5	,,	– 72.3	2.10	– ,, –	– 75.7
5–6	48.2	+ 68.1	2.97	60 x 3.25 C.H.S.	+ 90.4
6–7	,,	– 48.2	2.10	76 x 3.25 C.H.S.	– 75.7
7–8	24.1	+ 34.1	2.97	60 x 3.25 C.H.S.	+ 90.4
8–9	,,	– 24.1	2.10	– ,, –	– 43.7
9–10	–	–	2.35	– ,, –	+ 90.4

Tie at Gables

Resultant load in tie = 193 – 96.4 = +96.6 kN (see Fig. 21(f)).
Use 60 x 4.06 mm C.H.S. tie.
Gable peak verge rails and sheeting rails would be added to carry sheeting if required.

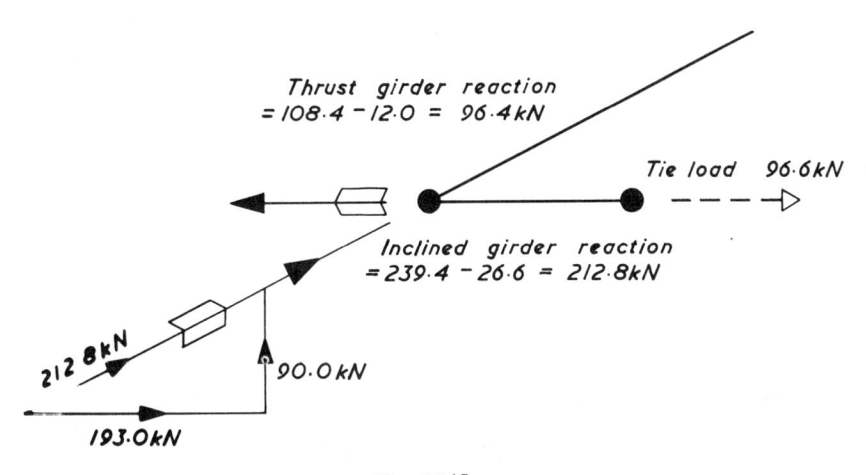

Fig. 21(f)

3. Braced Vault Construction

As mentioned in the previous section, roof structures built from four or more skeletal plane frames to form an arched vault are considered in this section. Two way arched roof structures covering circular, elliptical, regularly polygonal (including square) areas in plan are dealt with as domes in the next section. These arbitrary definitions have been made to avoid overlap between matters discussed in the separate sections.

The most common types of braced vault are:

(i) *Rectangular vaults of prismatic cross section with vertical end diaphragms.* The cross sectional shape is generally that of a segment of a circle, but may be semi-elliptical or parabolic. The circular section allows for simpler detailing and fabrication but the last two are probably more efficient structurally.

(ii) *Rectangular hipped vaults* (to which class Example 3 belongs) in which the cross section is as for type (i) but the ends are shaped away in curved or flat hips. This form has two advantages over type (i):

   (a) The hips have the effect of stiffening the central area of the vault; this is an important consideration for vaults which have very light cladding and may be subjected to heavy differential loading from wind, etc.
   (b) Junctions between the ends and sides of adjacent roofs are simplified and rainwater collection can be unified into a single system (gutters can be carried all round each bay) thus providing an insurance against temporary blockage of individual down pipes.

(iii) *North light vaults.* Generally these are supported along their edges by north-light girders and are thus not two-way spanning vaults in the strict sense. A number of pure vaults have been constructed in which the framing to the north lights merely serves to separate the north and south edges of adjacent vaults and to support the glazing. However, provided a north-light girder has slender bracing members which do not obstruct the light unduly, it appears more logical to provide girders rather than spacing members only in this position.

North-light vaults provide an unobstructed soffit to the south roof slopes, by selection of the right shape and surface to this soffit the uniformity of light distribution within the building can be enhanced without detracting from the quality of the natural light.

(iv) *Double curved vaults.* Though this type are generally more difficult to fabricate and to detail, they are probably the most suited to skeletal construction, the double curvature providing extra stiffness enabling lighter members and connections to be used. Anticlastic surface shapes are suited for use as auditorium roofs and may be designed to provide all straight members apart from the transverse ribs. The division between arched vault and suspended structures occurs within this range of surface shapes.

Generally, if the curvature in both directions is upward to the centre, the structure is obviously an arched vault; if the curvature is downwards in both directions it is a suspended structure. If the structural shape is curved in one direction upwards and one downwards (i.e. anticlastic) then the structure may be partly suspended, partly arched, or fully suspended or fully arched depending on its surface shape and boundary conditions.

BRACED VAULT CONSTRUCTION

Under grossly unequal loading conditions braced vaults can give rise to large local deformations even to the point of instability, though the same structure under uniformly distributed loads may show very small deflections and be quite stable.

Where unequal loading can only be due to sub-hurricane wind effects (for example the loads tabulated in C.P.3 Chapter V Part 2, 1970), provided the jointing system employed allows a high degree of joint rigidity and the periphery is held in position by a stiff edge beam or columns, these and the distributional effect of the roof covering are generally deemed sufficient to restrain gross deformation in vaults of normal proportions.

Where the unequal loading is due to hurricane or greater wind effects or particularly severe ice and snow loadings vaults are often stiffened by intermediate deep ribs in addition to a stiff ring beam.

For a series of vaults arranged in bays, with their frames connected structurally in the valleys, the edge beams need not extend across each valley line but should carry all round the periphery of the building.

The computer programmes so far available for the analysis of vault structures are based on the assumption of small linear deflections under load. It is therefore important that the deflections and axial loads and moments should be printed out in the analysis to see if they are acceptable in service and compatible with the small deflection theory assumed.

Example 3 illustrates the application of the computer to a simple hipped vault. Fig. 22 shows the general dimensions of the structure. The data arrangement fed into the computer, the output of loads, moments and deflections under unit load per square metre of plan area are annotated to refer to each member. From this the worst loads for each type of member have been abstracted, factored by the dead plus imposed load of 1.25 kN/m^2 and checked for combined stress under B.S. 449: Part 2, 1970. This vault would be quite stiff under all forms of loading and though having too few slopes to be of the ideal structural shape, has a steelwork content of 11.5 kg/m^2 of area covered. This could undoubtedly be improved on, but is still a reasonably economical proposition for any type of pitched roof construction.

Such a vault provides greatly increased usable volume within a building for a given eaves height when compared with truss or portal construction. Its appearance can be immeasurably superior, and this analysis indicates that its cost can be comparable with any other form of construction.

Example 3. Braced Hipped Vault. The axial loads and moments in the members of the vault shown in Fig. 22 have been calculated for an applied load of 1.0 kN/m^2 of plan area. From these unit loads and moments the member forces for the actual imposed loads are derived by direct proportion, by multiplying by the plan load — 1.25 kN/m^2 in this case.

Section Properties x elastic moduli in Nmm units

EA	EI_x	EI_y	GI_p	
3.23×10^8	4.89×10^{11}	4.89×10^{11}	3.64×10^{11}	114 x 4.47 mm C.H.S.
1.56×10^8	1.04×10^{11}	1.04×10^{11}	0.79×10^{11}	76 x 3.25 mm C.H.S.
2.27×10^8	1.97×10^{11}	1.97×10^{11}	1.50×10^{11}	89 x 4.06 mm C.H.S.

End of Section Properties.

474 SPACE FRAMES

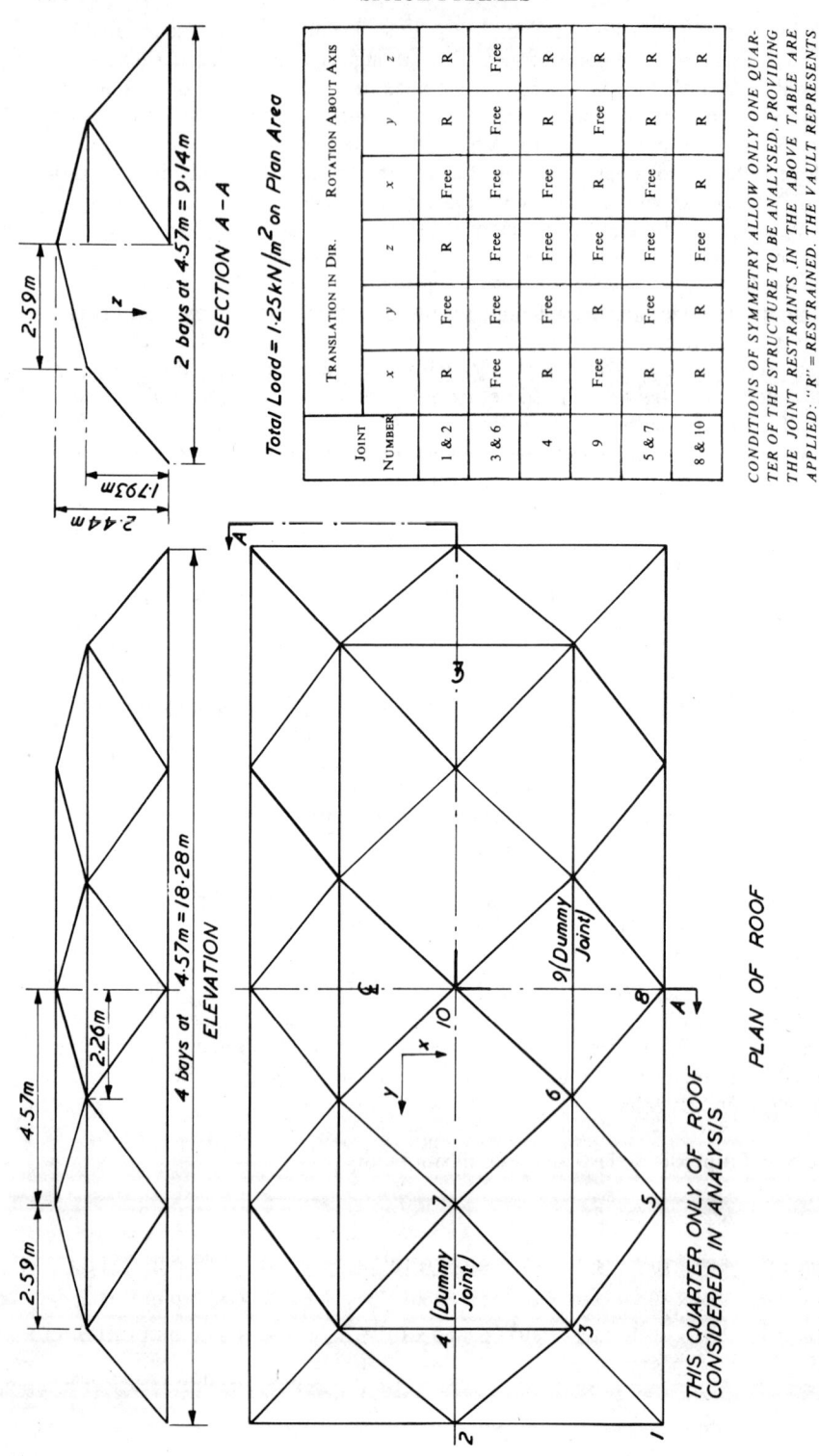

Fig. 22

BRACED VAULT CONSTRUCTION

Joint	x	y	z
1	4 570	9 140	2 440
2	0	9 140	2 440
3	2 590	7 160	647
4	0	7 160	647
5	4 570	4 570	2 440
6	2 590	2 260	647
7	0	4 570	0
8	4 570	0	2 440
9	2 590	0	647
10	0	0	0

Ordinates of joints from 1 to 10 consecutively arranged in x, y, z order.

End of ordinates.

Number of equations to be solved in each group of joints, joints 1 to 4 on first line, 5 to 7 on second line, 8 to 10 on third line (see Fig. 22).

```
2 2 6 3
3 6 3
1 3 1
```

End of equation data.

Joint	P_x	P_y	P_z	M_x	M_y	M_z
1	0	0	0	0	0	0
2	0	0	0	0	0	0
3	0	0	5.23	0	0	0
4	0	0	0	0	0	0
5	0	0	4.54	0	0	0
6	0	0	10.44	0	0	0
7	0	0	5.92	0	0	0
8	0	0	2.26	0	0	0
9	0	0	0	0	0	0
10	0	0	2.96	0	0	0
z						

External loads and moments applied from 1 to 10 consecutively, arranged in x, y, z order loads first, moments last. In this case only vertical loads are applied.

End of External load data,

All the physical data required for the analysis of the particular vault is contained in the print out from the data tape shown in the box above. Some small amount of additional data is required concerning the sub-division of the actual analysis programme, but this is not shown.

The deformations of each joint in terms of translations and rotations are obtained and printed out by the computer programme. Note that the joints subjected to restraint by virtue of the symmetry of loading and geometry of the structure have no deflection or rotation in the restrained direction and therefore have no print out in these directions.

From these deformations the member loads and moments are calculated and printed out on the next page.

The axial loads and moments obtained from the computer programme result print out reproduced must be multiplied by 1.25 to them to kN and kNm.

In this case only the axial load shown in the first column and the worst of the moments shown in the two right-hand columns are important. The other columns are taken up with shear-forces and torsional moments which are too small to be significant for this type of structure.

Checking the stresses for the worst loaded case of each member gives the following:

SPACE FRAMES

Members	Joint Numbers	Axial Load kN P_x		Shearing Forces kN P_y		P_z		Torsional Moment kNm M_x		Bending Moments kNm M_y		M_z	
End 1	1												
	3	5.9	1	3.8	−2	−5.7	−2	−4.5	−2	1.9	−2	9.9	−2
End 2	5	−3.4	1	0.0	−52	−8.7	−3	0.0	−52	6.4	−3	0.0	−52
	2	0.0	−52	2.0	−2	0.0	−52	9.7	−3	0.0	−52	4.5	−2
End 1	2												
End 2	3	2.3	−2	−6.8	−3	7.5	−3	−9.2	−3	−1.5	−3	−5.9	−3
	1	0.0	−52	−2.0	−2	0.0	−52	−9.7	−3	0.0	−52	4.5	−2
End 1	3												
	2	−2.3	−2	6.8	−3	−7.5	−3	9.2	−3	−2.6	−2	−1.9	−2
	4	−1.2	1	1.1	−7	−1.8	−6	0.0	−52	9.7	−2	−7.8	−2
End 2	7	3.5	1	4.0	−2	2.2	−1	1.3	−1	−5.0	−1	5.0	−2
	6	3.9	1	4.2	−2	−1.9	−1	−5.9	−2	4.0	−1	9.7	−2
	5	−4.2	1	−5.4	−3	1.6	−2	2.3	−3	−4.0	−2	−2.6	−2
	1	−5.9	1	−3.8	−2	5.7	−2	4.5	−2	1.7	−1	2.8	−2
End 1	4												
End 2	3	1.2	1	−1.1	−7	1.8	−6	0.0	−52	−9.7	−2	7.8	−2
End 1	5												
	1	3.4	1	0.0	−52	8.7	−3	0.0	−52	3.3	−2	0.0	−52
End 2	8	−8.4	1	0.0	−52	4.9	−3	0.0	−52	−1.1	−1	0.0	−52
	6	3.1	1	8.8	−2	−1.2	−1	−1.5	−2	1.8	−1	1.6	−1
	3	4.2	1	5.4	−3	−1.6	−2	−2.3	−3	−1.9	−2	5.0	−3
End 1	6												
	7	1.4	0	3.4	−2	2.0	−1	−8.5	−3	−3.1	−1	4.7	−2
	10	1.8	1	−4.1	−2	2.8	−1	4.5	−2	−5.0	−1	−7.1	−2
End 2	9	5.2	1	0.0	−52	0.0	−52	−4.7	−8	−3.1	−1	−7.6	−2
	8	4.7	0	−7.9	−2	1.8	−1	−3.8	−3	−3.5	−1	−1.5	−1
	5	−3.1	1	−8.8	−2	1.2	−1	1.5	−2	2.6	−1	1.5	−1
	3	−3.9	1	−4.2	−2	1.9	−1	5.9	−2	5.4	−1	1.1	−1
End 1	7												
	3	−3.5	1	−4.0	−2	−2.2	−1	−1.3	−1	−3.5	−1	9.9	−2
End 2	6	−1.4	0	−3.4	−2	−2.0	−1	8.5	−3	−4.0	−1	7.3	−2
	10	4.6	1	0.0	−52	2.3	−2	0.0	−52	−1.5	−1	0.0	−52

Notation of forces: A x 10^n where A = the first number and n = the integer following
(e.g. 5.9 1 = 5.9 x 10 = 59)
Cases which are criteria for member types are underlined.

114 x 4.47 C.H. Sections.

Member 1, 3 load = −59 x 1.25 = 73.8 kN, moment = 0.09 x 1.25
= 0.124 kNm

Length = 3 325 mm
effective length = 2 325 mm

$$\text{Unity eq.} = \frac{73.8}{192} + \frac{0.124 \times 10^3}{40.7 \times 165} = 0.40$$

Member 6, 9 load = −52 x 1.25 = 65.0 kN, moment = 0.31 x 1.25
= 0.388 kNm

Length = 4 572 mm
effective length = 3 200 mm

$$\text{Unity eq.} = \frac{65.0}{154} + \frac{0.388 \times 10^3}{40.7 \times 165} = 0.48$$

89 x 4.06 mm C.H. Sections

Member 5, 8 load = 84 x 1.25 = 105 kN, moment = 0.011 x 1.25
= 0.014 kNm

Length = 4 572 mm
effective length = 3 200 mm

$$\text{Unity eq.} = \frac{105}{168} + \frac{0.014 \times 10^3}{22 \times 165} = 0.63$$

76 x 3.25 mm C.H. Sections

Member 5, 6 load = −31 x 1.25 = 39 kN, moment = 0.18 x 1.25
= 0.225 kNm

Length = 3 325 mm
effective length = 2 325 mm

$$\text{Unity eq.} = \frac{39.0}{67.5} + \frac{0.225 \times 10^3}{13.2 \times 165} = 0.68$$

Member 3, 5 load = +42 x 1.25 = 52.5 kN, moment = 0.04 x 1.25
= 0.05 kNm

$$\text{Unity eq.} = \frac{52.5}{115} + \frac{0.05 \times 10^3}{13.2 \times 165} = 0.48$$

Maximum downward deflection at centre

= 14.8 x 1.25 at joint 8
= 18.5 mm
= Span divided by 986.

All the sections are well in hand and the construction is somewhat over-designed. Even so, the weight of structural steel is only 11.5 kg/m² of covered area in plan.

Domes

By far the most advanced and familiar form of skeletal space structure is the dome, though its range is perhaps the most limited. Probably the reason for this public cognizance is that such large domes have been built; one in Austin, Texas, has a diametral span in plan of over 180 m and spans of up to 450 m are being considered in this country. Only suspension structures can compete economically over such large spans, and these suffer from the major defects that in their most efficient form storm water must be collected from the centre of the span and their stability can be suspect under wind loading.

Domes, however, have a fundamentally stable shape which in any case tends to streamline windflow, thus precluding the possibility of sharp changes of pressure or suction round their surface. (Sharp changes of pressure, particularly if they can develop rhythmically, can allow even low velocity winds to produce far worse effects on a structure than uniform or slowly changing pressures at high velocities.) Storm water disposal can only be a problem of volume and then only in the case of very large domes.

Though the most efficient shape in plan is circular, domes can be efficient structures in elliptical, polygonal and even square plan shape. Their prime use is probably to cover sports arenas, large auditoriums and prestige exhibition halls, but they have been used with conspicuous success for comparatively small spans by continental architects and engineers. For auditoriums the most efficient structural shape may not match the best acoustic form; for such uses both criteria must be examined in the design.

The simplest form of dome comprises a series of radial arch ribs linked by purlins. Where the spacing of the ribs may be shortened as shown in Fig. 23 (a), the thrust of the shortened ribs must be transferred to the through ribs by means of a braced

ring. The natural cross-sectional shape of a ribbed dome is probably elliptical, in order to allow the dead loads to be taken axially through the ribs without undue eccentricity. For large span domes dead loads and snow loads are almost certainly the most important. Dead loads per rib increase in magnitude in direct proportion to the distance from the centre with a secondary increase in magnitude due to the slope of the surface. Snow loads also increase in magnitude in direct proportion to the distance from the centre but have a secondary decrease in magnitude due to the slope of the surface. The line of thrust through the ribs will depend on the balance between these two loads.

In spite of the fact that a segment of a sphere can rarely be the correct shape to conform with the line of thrust derived from the worst loading condition, most domes are built of this shape to simplify detailing.

If all the ribs of a dome are shortened (i.e. do not reach the centre) diagonal bracing must be introduced between ribs and purlins; such domes are generally statically determinate as pin-jointed structures and are known as Schwëdler domes.

All ribbed domes induce an horizontal thrust at their feet which must be taken up by a ring beam or by the foundations. They are comparatively simple to analyse and the tendency has been to take advantage of this simplicity; the simplicity of analysis is not matched by ease of detailing, fabrication or covering, for the following reasons:

(a) The ribs are always subjected to some bending action, therefore must be deep in comparison with their width. The fabrication of a deep curved section, even if braced, is not straightforward.
(b) Purlins to support the roof covering must also be curved, which again is an operation not welcomed by fabricators.
(c) Roof covering cannot easily be made in prefabricated units owing to the constantly changing profile of the ribs.

More sophisticated methods of analysis have led to the development of other skeletal forms of dome. The principal advantages which all of these new forms offer are:

(i) Covering techniques are greatly simplified.
(ii) Components may be fabricated in straight, short lengths generally connected by patent connectors or site welding, or by light continuous curved members which are often out of plane with each other and connected by loop shear connectors.

A development of the ribbed dome which simplifies covering is the square grid dome shown in Fig. 23 (b) in which members act as an interconnected arched rectangular grid. Domes of up to 30 m span have been built in this form.

Triangular or three-way grid domes are the most used today. They lend themselves to production in short, straight units and are adaptable to the polygonal and square forms shown in Figs. 23 (c) and (d).

The skeletal dome can be an economical and visually pleasing structure provided the function of the area it covers demands domical form either by reason of large clear diametrical span (e.g. a covered arena) or architectural effect (e.g. the vaulted roof of a church or assembly hall). In the first case it will be a most economical form, in the second case it can be a not uneconomical form. In both cases its appearance, both internally and externally, can justify its selection, where the economical difference between it and other forms are marginal.

DOMES

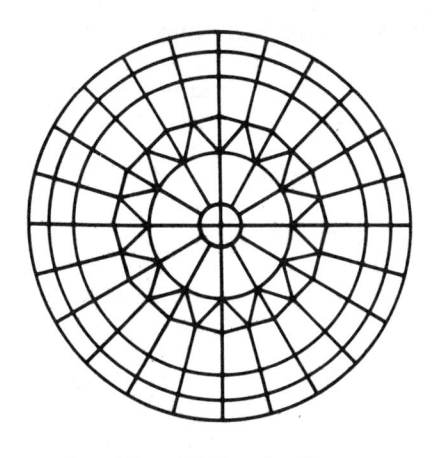

(a) PLAN OF RIBBED DOME.
SHOWING SHORTENED RIBS.

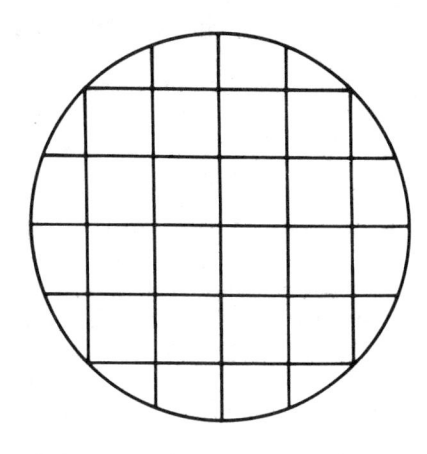

(b) PLAN OF GRID DOME.

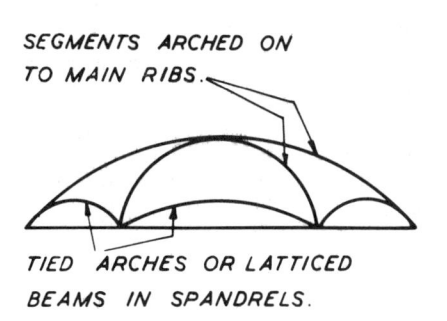

SEGMENTS ARCHED ON
TO MAIN RIBS.

TIED ARCHES OR LATTICED
BEAMS IN SPANDRELS.

SECTION A-A.

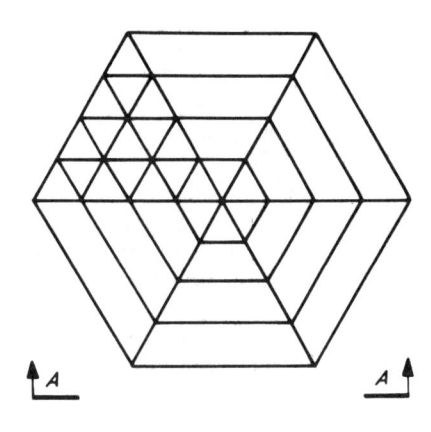

(c) PLAN OF HEXAGONAL DOME.

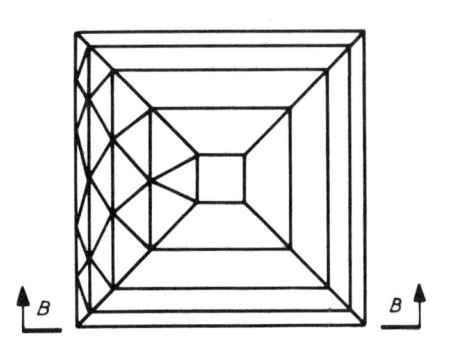

(d) PLAN OF SQUARE PLATE DOME.

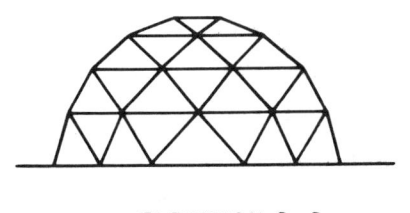

ELEVATION B-B.

Fig. 23

Suspension and Tension Supported Structure

Some forms of suspended structure appear to the practical engineer as mere evidence of the desire of a certain minority of the profession to use an unusual structure at any price. It is true that some of the ideas put forward for this type of structure are outside the realms of reason and have tended to discredit the logic on which they are based.

In fact they can be used logically over a wider range of structures than domes or vaults and their application has a long history. Suspension bridges leap to the mind as being the most efficient way of bridging wide gaps where no intermediate supports are possible. Such bridges have been used from very early times, notably in China where a very old chain suspension bridge is still in use and, perhaps better known, in Peru where stood the bridge of St. Luis Rey, built at least 600 years ago and bridging an 45 m span chasm.

Today suspension bridges are often space structures by the most meticulous definition and notable examples of pipe bridges in Italy and Austria have become too well known to structural engineers to dwell on in this chapter.

Figure 24 (a) and (b) shows two applications of the structural form to smaller structures.

See references under suspension structures of the bibliography.

Another form of tension supported structure in more common use is the guyed mast. Guyed masts have been built up to 450 m in height and are projected to much greater heights. Their design and analysis are complicated by the following factors:

(a) The wind load varies in intensity over the height of the mast, and must be considered as approaching from any point of the compass. As wind is the principal source of load this makes assessment of the worst loading case laborious and complex.

(b) The strain in each of the guys is sufficiently great to require special consideration in deriving the bending and shear effects in the mast itself. The mast must therefore be considered as a continuous beam on elastic supports.

Three sets of guys are usual, being connected to a braced mast of triangular cross-section. The guys are usually taken to anchorage points diametrically in line with the mast chords, two or three guys being attached to each anchorage. A very tall mast may have up to three sets of anchorage points at successively increasing distances out from the base. The mast foot is usually mounted on a steel ball set in an hemispherical socket packed in grease and contained in a weather-proof box.

So far no computer programme for their analysis is in use in this country, but the problem is one particularly amenable to solution by this means.

Materials of Construction

Naturally the same rules apply to the selection of members for space frames as apply to any other form of construction. For example, I sections or latticed beams are best for spanning members and tubular sections for compression members. In addition to this criterion of selection by structural shape, for space frames some special factors related to detailing and calculation must be borne in mind.

For members which must be detailed in double angles, details can become complicated for flanged and unsymmetrical sections, analysis by computer becomes

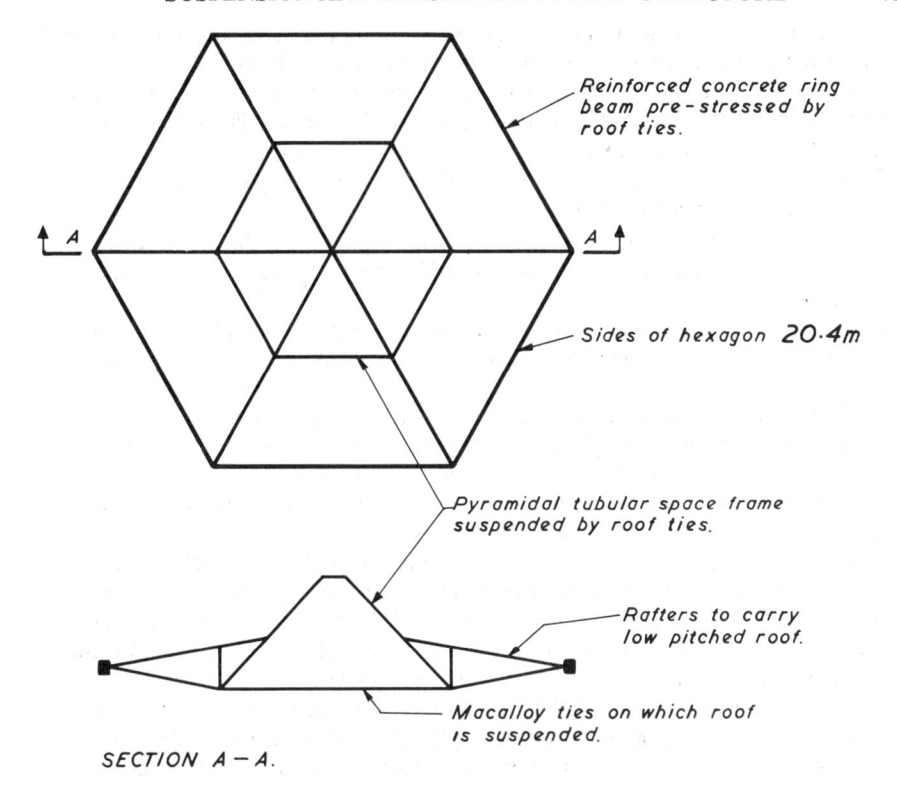

SUSPENSION AND TENSION SUPPORTED STRUCTURE 481

Reinforced concrete ring beam pre-stressed by roof ties.

Sides of hexagon 20·4m

Pyramidal tubular space frame suspended by roof ties.

Rafters to carry low pitched roof.

Macalloy ties on which roof is suspended.

SECTION A−A.

(a) SKETCH SHOWING ROOF CONSTRUCTION OF CHICHESTER FESTIVAL THEATRE

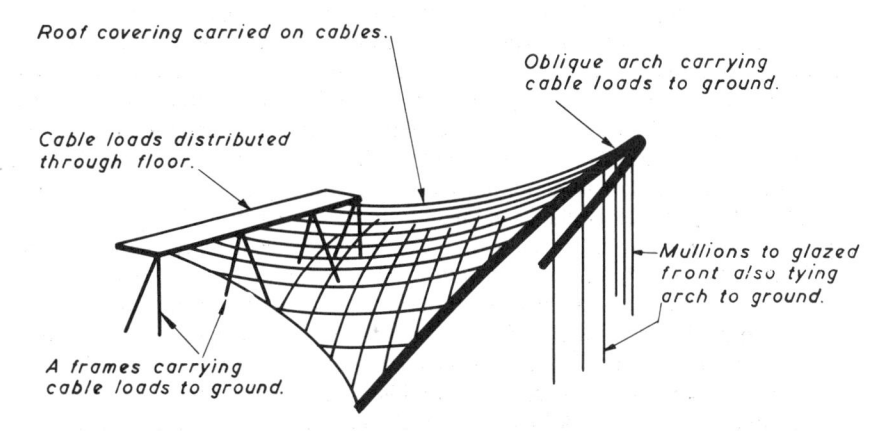

Roof covering carried on cables.

Oblique arch carrying cable loads to ground.

Cable loads distributed through floor.

Mullions to glazed front also tying arch to ground.

A frames carrying cable loads to ground.

(b) PERSPECTIVE OF CONSTRUCTION OF A THEATRE IN PRAGUE.

Fig. 24

slightly more complex as additional changes of co-ordinate system are needed to reduce members to the single Cartesian system required for such analyses.

The bending or curving of unsymmetrical sections is not a simple operation and usually needs to be carried out by hand by a skilled blacksmith. The bending of a solid web I section about its main axis is also not a good production proposition.

Tubular and tubular latticed members show to advantage for this type of operation for the following reasons:

(a) a circular section presents the same profile to any line drawn to its centre; therefore the detailing of double angles becomes a simple matter of solid geometry.
(b) a circular section can easily be curved to any shape required between rollers and curved lattice beams can be fabricated by welding bracings to pre-bent chords.
(c) hollow sections are made to a higher degree of accuracy than other hot finished sections, making accurate manipulation and assembly far easier.

Cold formed sections, so far not mentioned in the text or the examples, could be a useful addition to hot rolled steel and hollow sections, particularly when the sections are designed to make simple connections.

Considering the three examples in the foregoing and analysing the choice of section may help to illustrate the reasons governing the selection. In Example 1 castellated sections are used for the major part of the grid as they form a compact stiff section at comparatively low cost. It would not be advisable to have castellated sections running in the other direction as the joints between the intersecting members would be expensive and difficult to make. Lattice tubular sections can be jointed easily out of plane with the castellated flanges, using flange plates and high tensile bolts. As the ratio of stiffness to strength cannot be made the same as for the castellated beams the calculations must account for this variation of stiffness.

For the end lattice beam a universal beam has been passed through one of the castellations of the intersecting beam in order to provide a strength/stiffness ratio able to carry additional moment but fitting into the general stiffness pattern. A purlin would have to be provided over this beam to carry the decking. By passing this beam through a castellation a difficult site joint has been avoided.

A lattice beam could be provided in this position to suit the stiffness and strength requirements but would not be as economical as the simple beam suggested.

In Example 2 castellated or lightweight universal beams are the obvious choice for rafters. Channels allow simple connections for the valley main tension chords to the inclined beams. The purlins could be either rolled steel angle or cold formed sections, the latter named section needing to be approximately two fifths of the weight of the equivalent rolled steel section to be competitive. The ridge member forming the top chord of the inclined beams and the diagonal bracings are probably best in circular hollow sections. Thus the sections requiring to be strong in bending or having to provide an area of steel in tension are selected from the cheapest material formed to the most suitable section to resist bending or tension and to make simple connections. Members which are in pure compression and will be connected by welding are in the most suitable shape for these purposes.

Example 3. As axial load dominates in all members and practically every member is involved in compound angles, tubular sections are the obvious choice. An additional factor is the simplified computer programme for tubular sections which is not apparent from the data and print out provided.

Concluding Notes

The examples are intended to illustrate three of the possible methods by which the design and analysis of skeletal space frames may be approached.

The simplest approach is that the structure may be conceived in such a manner that it may be reduced to a series of statically determinate plane frames. Often, in order to accomplish this reduction, the structure must be made to deviate from its most natural and economical form. Even when the frame is so stretched out of its best shape the statical analysis of the resulting structure is often still only an approximation.

The second method is useful where the deflections of interacting components may easily be calculated in terms of the external loads and the redundant reactions between the members. Generally these deflections are best calculated with the help of special tables of standard cases. The deflections thus derived can be formed, by hand, into a set of simultaneous equations which are then solved by the digital computer.

Lastly, where the deflections cannot be readily expressed in terms of the external loads and redundancies, the structure can be analysed by a special computer programme which derives deflections, axial loads, shears, bending moments and torsional moments in members from the geometry of the structure and its members, the elastic constants of the material and the externally applied loads.

Obviously the design conception for the last two approaches is the most difficult and the initial form of the structure must be evolved by experience, approximate analysis and logic before the computer makes the final analysis.

The grid and vault examples (which represent the second and third approaches respectively) are rather elementary and crude illustrations of each of their forms: nevertheless their construction depth and the weight of constructional steel used per square metre for the span and total load sustained compares favourably with that of any system of plane frames beams and purlins which may be envisaged to fulfil the same purpose. An equally favourable strength to weight ratio could be established for space frame compared to plane frame structures over a very wide range of construction, provided the best form of space structure is used for the specific case.

BIBLIOGRAPHY

General Analysis

SOUTHWELL, R. V. 'Primary stress determination in space frames', *Engineering*, 1920, p. 165.
GRINTER, L. E. *Theory of Modern Steel Structures*, Vol. 2, *Statically Indeterminate Structures and Space Frames.* (New York: Macmillan Co., 1937, 285 pp.)
TIMOSHENKO, S. and YOUNG, D. H. *Theory of Structures.* New York and London: McGraw Hill, 1945.
LIVESLEY, R. V. *Matrix Methods of Structural Analysis 1964.* Pergamon Press.
MAKOWSKI, Z. S. *Steel Space Structures*, Michael Joseph Ltd. London 1965.
MATHESON, J. L. *Hyperstatic Structures,* Volumes 1 and 2. Butterworth, 1959.
MAKOWSKI, Z. S. *Räumliche Tragwerke auf Stahl.* Verlag Stahleisen, Düsseldorf 1963.
DAVIS, R. M. (Ed.) *Space Structures*, Blackwell Scientific Publications Ltd., Oxford 1966.

Grids

BAER, O. A. 'Steel frame folded plate roof', *Journal of the Structural Division, Proc. A.S.C.E.*, June 1961, p. 35.

HENDRY, A. W. and JAEGER' L. G. 'The load distribution in highway bridge decks', *Proceedings, A.S.C.E.*, July 1956.
The Analysis of Grid Frameworks and Related Structures. Chatto and Windus, 1958.
LIGHTFOOT, E., and SAWKO, F. 'Grid frameworks resolved by generalised slope-deflection', *Engineering*, January 2, 1959, pp. 18–20.
'The analysis of grid frameworks and floor systems by the electronic computer' *Structural Engineer*, March 1960, pp. 79–87.
MAKOWSKI, Z. S. 'Interconnected systems, two-and three-dimensional grids', *The Guild's Engineer*, 1955, pp. 11–28.
MAKOWSKI, Z. S. and RAMIREZ, R. 'Modern grid frameworks of a regular hexagonal layout'. *Technika i Nauka*, Journal of the Institutions of Polish Engineers abroad, No. 5, 1959, pp. 1–41.
MARTIN, J. and HERNANDEZ, J. 'Orthogonal gridworks loaded normally to their planes', *Proc. A.S.C.E.*, Journal of the Structural Division, January 1960, St. 1.
MATHESON, J. L. 'Moment distribution applied to rectangular rigid space frames', *Journal of I.C.E.*, No. 3, 1947–48.

Domes

CHATEAU DU S.* 'Structure spatiale spherique en trame tridirectionelle', *L'architecture d'aujourd'hui*, No. 81, 1959.
'Coupoles reticules', *Proceedings, IASS Colloquium,* Paris 1962.
GONDIKAS, J. M. and SALVADORI, M. G. 'Wind stresses in domes', *Journal of the Engineering Mechanics Division, A.S.C.E.*, October 1960, pp. 13–29.
MAKOWSKI, Z. S. and PIPPARD, A. J. S. 'Experimental analysis of space structures, with particular reference to braced domes', *Proc. I.C.E.*, Part III, Dec. 1952, pp. 420–441.
MAKOWSKI, Z. S. and PALMER, D. 'Domes – their history and development', *The Guild's Engineer*, 1956, pp. 50–61.
MAKOWSKI, Z. S. and GOGATE, M. N. 'Stress analysis of three-pinned arch-ribbed domes', *Proc. I.C.E.*, Part III, Vol. 5, pp. 824–844.
MITCHELL, L. H. 'A shell analogy for framed domes', *Research and Development Branch, Aeronautical Research Labs.* Note ARL/SM/208, December 1953, Melbourne, Australia.
MITCHELL, L. H. 'A shell Analogy for Framed Domes', *Engineering*, Vol. 183, June 14th 1957, pp. 754–5.
VENANZI, U. and VANNACCI, G. F. 'Tubular Lattice Girder Dome for the Sports Palace of Bologna, Italy', *Acier, (Steel)* Vol. 22, Nov. 1957, pp. 447–51.
ANON. 'The Pittsburgh Public Auditorium (U.S.A.)', *Acier (Steel)*, Vol. 25, Feb. 1960, pp. 71–73.
VÖLKEL, 'Assembly Hall in Steel Construction', *Acier (Steel)*, Vol. 24, Dec. 1959, pp. 518–19.
COHEN, E. and GOLDSMITH, R. 'Cantilever frame for retractable roof of Pittsburgh Public Auditorium (U.S.A.)', *Acier (Steel)*, Vol. 27, July/Aug. 1962, pp. 315–22.
LEDERER, F. 'Developments in tubular domes', *Tubular Structures*, Issue No. 3. Stewarts & Lloyds Ltd. 1964.
BENJAMIN, B. S. *The Analysis of Braced Domes*, Asia Publishing House. 1965.

Braced Vaults

MAKOWSKI, Z. S. and HOWLEY, M. 'The analysis of braced barrel vaults', *Proc. Institution of Polish Engineers Abroad in Great Britain*, March 1957, pp. 341–347.
MATSUSHITA, F., SATO, M. and HAYASHI, T. 'An experimental study of non-uniformly trussed steel shells consisting of triangular elements of varying dimensions', *Proceedings of the symposium of steel structures, Japan Society for the Promotion of Science*, Tokyo, October 1961, pp. 1–10.
PAGANO, M.* 'Theoretical and experimental research on triangulated steel vaults', *Proceedings, IASS Colloquium*, Paris 1962.

Suspension Structures

WEISS, C. 'The design and construction of the Chichester Festival Theatre', *Structural Engineer*, Vol. XL, No. 12, Dec. 1962. pp. 389–405.
SAMUELY, F. J. 'Structural Pre-stressing', *The Structural Engineer*, Feb. 1955.
ESQUILLAN, N. and SAILLARD, Y. *Hanging Roofs*. North Holland Publishing Co., Amsterdam. 1963. (Papers from I.A.S.S. Colloquium, Paris 1962).

*Contained in this volume together with a paper dealing with the theatre in Prague mentioned in the text and illustrated in Fig. 10 (b).

Connections

le tube d'acier dans la construction mètalique. Chambre syndicale des fabricants de tubes d'acier. 37 avenue George V, Paris.
Glued metal joints. I.A.B.S.E. Rio de Janeiro. 1964.
Proceedings of the symposium on new ideas in structural design. Japanese Society for the Promotion of Science.

Acier-Stahl-Steel	Brussels	
Tubular Steel Construction	Stewarts & Lloyds Limited, London	Periodicals in which new jointing methods are frequently reviewed.
Der Stahlbau	Düsseldorf	

Column. Issue No. 12, 1964. Yawata Iron & Steel Co. Ltd., 1, 1-chome, Marunouchi, Chiyada, Tokyo, Japan.

20. DESIGN OF ANGLE STRUTS

As in the case of stanchions, the carrying capacity of angle struts carrying axial loads depends to some extent on the slenderness ratio of the strut. According to B.S. 449: Part 2: 1969, angle struts may be divided into five groups. These five groups are given below:

1. Single-angle discontinuous struts with single-bolted or riveted connections at each end.
2. Single-angle discontinuous struts connected at each end by not less than two bolts or rivets in line or their equivalents in welding.
3. Double-angle discontinuous struts, back to back, connected to one side only of a gusset. The end connections are not considered to affect the permissible stresses on this type of strut even if they are only single-bolted or riveted.
4. Double-angle discontinuous struts, back to back, connected to both sides of a gusset by not less than two bolts or rivets in line or their equivalents in welding.
5. Continuous angle struts such as those forming the rafters of trusses, etc.

It should be noted that in dealing with groups 1 to 4 the eccentricity of the connection with respect to the centroid of the strut may be ignored and the struts designed as axially-loaded members.

Group 1

The effective length must be taken as the full length centre to centre of intersections and the allowable stress must not exceed 80 per cent of the values given in Tables 17a, 17b or 17c for steel grades 43, 50 and 55 respectively.

Example 1. Single-angle discontinuous strut with single-riveted connections, having a length of 2.4 m between intersections and carrying a load of 34 kN.

$$L = 2.4 \text{ m} = 2.4 \times 10^3 \text{ mm}$$

Try 76 × 76 × 9.4 mm angle of grade 43 steel.

$$\text{Area} = 13.47 \text{ cm}^2$$

$$r_v = 1.48 \text{ cm}$$

$$l/r_v = \frac{2.4 \times 10^3}{1.48 \times 10^2} = 162 \qquad p_c = 35 \text{ N/mm}^2$$

$$\text{Allowable stress} = 0.8 \times 35 = 28 \text{ N/mm}^2$$

$$\text{Actual stress} = \frac{34.0 \times 10^3}{13.47 \times 10^2} = 25.2 \text{ N/mm}^2$$

The section is adequate.

DESIGN OF ANGLE STRUTS

It should be noted that *in no case* shall the slenderness ratio of this type of strut exceed 180, whereas the slenderness ratio of other types of struts carrying loads *resulting from wind forces only,* may be as high as 250.

Group 2

The effective length may be taken as 0.85 times the length of the strut, centre-to-centre of intersections and the allowable stress must not exceed the values given in Tables 17a, etc., of B.S. 449.

Example 2. Single-angle discontinuous strut with double-riveted connections having the same length between intersections and carrying the same load as in Example 1.

Length centre-to-centre of intersections = 2.4 m = 2.4×10^3 mm.

Try 76 x 76 x 6.2 mm angle of grade 43 steel.

$$\text{Area} = 9.12 \text{ cm}^2$$

$$r_y = 1.49 \text{ cm}$$

$$\frac{l}{r_y} = \frac{0.85 \times 2.4 \times 10^3}{1.49 \times 10} = 137 \qquad p_c = 48 \text{ N/mm}^2$$

$$\text{Actual stress} = \frac{34 \times 10^3}{9.12 \times 10^2} = 37.3 \text{ N/mm}^2.$$

The section is adequate.

It will be noted that the section required in Example 1 weighs 10.57 kg/m run, while that required in Example 2 weighs 7.16 kg/m run, representing a saving of 3.41 kg/m run or 32 per cent; a saving gained by the addition of one more rivet at each end of the strut.

Group 3

The effective length may be taken as 0.85 times the length of the strut, centre-to-centre of intersections and the allowable stress must not exceed the values given in Tables 17a, etc., of B.S. 449.

Group 4

The effective length may be taken as between 0.7 and 0.85 times the distance between intersections, depending on the degree of restraint and the allowable stress must not exceed the value given in Tables 17a, etc., of B.S. 449.

Examples 3, 4 and 5 deal with a discontinuous strut having a length of 3.0 m between intersections and carrying a load of 100 kN. It will be designed first as a single-angle discontinuous strut with double-riveted connections; secondly as a double-angle discontinuous strut, back to back, connected to one side only of a gusset; and thirdly as a double-angle discontinuous strut, back to back, connected to both sides of a gusset with double riveting.

Example 3. Single-angle discontinuous strut with double-riveted connections.

Length centre-to-centre of intersections = 3.0 m = 3.0×10^3 mm.

ECCENTRICITY OF LOADING

Try 102 × 102 × 9.4 mm angle of grade 50 steel.

$$\text{Area} = 18.39 \text{ cm}^2$$

$$r_y = 1.99 \text{ cm}$$

$$\frac{l}{r_y} = \frac{0.85 \times 3.0 \times 10^3}{1.99 \times 10} = 128 \qquad p_c = 60 \text{ N/mm}^2$$

$$\text{Actual stress} = \frac{100 \times 10^3}{18.39 \times 10^2} = 54.4 \text{ N/mm}^2.$$

The section is adequate.

Example 4. Double-angle discontinuous strut, back to back, connected to one side of a gusset.

Length centre-to-centre of intersections = 3.0 m = 3.0 × 10³ mm.

Try two angles 76 × 51 × 6.2 mm, 8 mm back to back, of grade 50 steel.

$$\text{Area} = 15.18 \text{ cm}^2$$

$$r_y = 2.18 \text{ cm}$$

$$\frac{l}{r_y} = \frac{0.85 \times 3.0 \times 10^3}{2.18 \times 10} = 117 \qquad p_c = 70 \text{ N/mm}^2$$

$$\text{Actual stress} = \frac{100 \times 10^3}{15.18 \times 10^2} = 65.9 \text{ N/mm}^2.$$

The section is adequate.

Example 5. Double-angle discontinuous strut, back to back, connected to both sides of a gusset with not less than two rivets in line.

Try two angles 63 × 51 × 6.2 mm, 8 mm back to back, of grade 50 steel.
Length centre-to-centre of intersections = 3.0 m = 3.0 × 10³ mm.

$$\text{Area} = 13.64 \text{ cm}^2$$

$$r_x = 2.0 \text{ cm}$$

$$\frac{l}{r_x} = \frac{0.7 \times 3.0 \times 10^3}{2.0 \times 10} = 105 \qquad p_c = 85 \text{ N/mm}^2$$

$$\text{Actual stress} = \frac{100 \times 10^3}{13.64 \times 10^2} = 73.3 \text{ N/mm}^2.$$

The section is adequate.

The weights of the sections in Examples 3, 4 and 5 are 14.4, 11.9 and 10.7 kg/m respectively. To the last two of these weights, an allowance should be made for intermediate fastenings.

The next two examples, Nos. 6 and 7, deal with a continuous angle strut, i.e. the main rafter of a roof truss.

The loads and B.M.s are as shown in Fig. 1. It should be noted that the continuous member is assumed to have fixed ends at the eaves and apex.

It will be designed first as a single-angle strut and then as a double-angle strut.

490 DESIGN OF ANGLE STRUTS

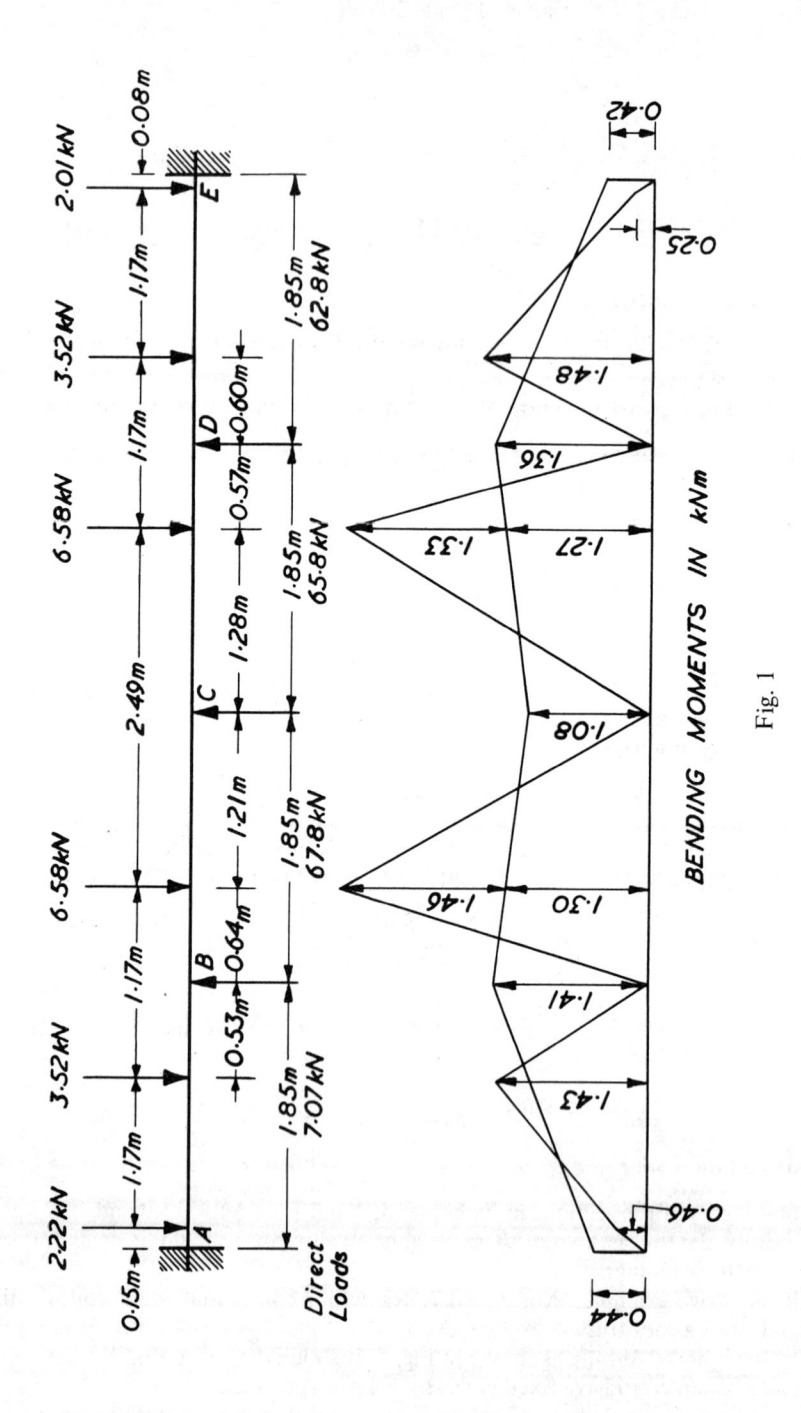

Fig. 1

ECCENTRICITY OF LOADING

Example 6. Single-angle continuous strut.

An inspection of the B.M. diagram in conjunction with the direct loads in the members and the positions of the purlins, will show that the design will be governed by the conditions at one of the following points:

(a) The support B.
(b) The support C.
(c) The purlin point between these two supports.

A single angle subjected to bending, if freely supported, will not bend about the horizontal axis *xx* but about some other axis, as shown in Example 2 in Section 1, Bending and Axial Stresses.

It will have been noted, from that section, that the determination of the axis of bending is not an easy matter.

If the rafter of a roof truss is a single angle, it will tend to bend about some other axis than the horizontal axis *xx,* but it will be appreciated that, due to the effect of the gussets at the panel points and also due to intermediate connections at the purlin points, the determination of the axis of bending, which is not necessarily the same at both the points mentioned, is an even more complicated problem than in the case of the simply supported angle beam.

It is necessary, moreover, to take into account the effect of the gusset plate in applying the direct load eccentrically to the member.

A practical design method should take these facts into consideration and should be, at the same time, convenient to use in the design office.

The following calculations are based on such a method, which has been found, from experience, to give reasonably satisfactory results.

In ascertaining the relevant properties of the section chosen it is necessary to evaluate certain bending stress coefficients, as follows:

1. Those due to eccentrically applied direct loads.
2. Those due to transverse bending.

Figure 2 shows the method of determining the first of these coefficients.

$$\text{Coefficient for point } B = \frac{OP \times \sin\theta \times BV}{I_v} - \frac{OP \times \cos\theta \times BU}{I_u}$$

$$\text{Coefficient for point } C = \frac{OP \times \cos\theta \times CU}{I_u} - \frac{OP \times \sin\theta \times CV}{I_v}$$

Maximum compressive bending stress = Load × max. coefficient.

Note. P is the point of application of the load assumed to be at the centre of gusset thickness and on the centre line of the connecting rivets or bolts.

In this example one angle, 102 × 89 × 7.8 mm of grade 43 steel will be investigated. The properties of the angle are:

$$\text{Area} = 14.41 \text{ cm}^2 \qquad r_x = 3.17 \text{ cm}$$
$$I_u = 199 \text{ cm}^4 \qquad r_y = 2.68 \text{ cm}$$
$$I_v = 49.1 \text{ cm}^4 \qquad r_v = 1.85 \text{ cm}$$

For the purpose of calculating the first coefficient, the dimension x will be taken as 55 m and the gusset thickness as 10 mm, whence from Fig. 3,

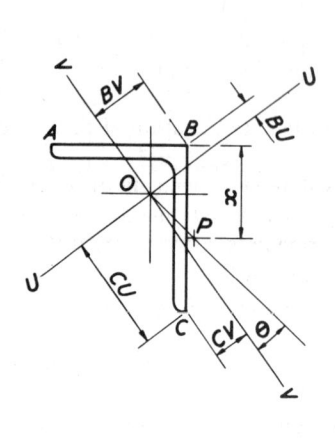

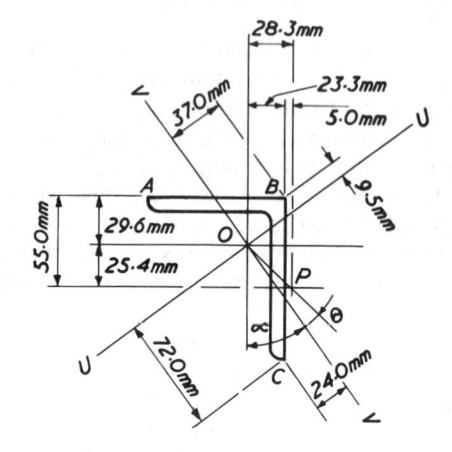

$OP = 38.0$ mm $\qquad CV = 24.0$ mm

$\theta = 11° \ 20'$ $\qquad BU = 9.5$ mm

$CU = 72.0$ mm $\qquad BV = 37.0$ mm

Fig. 2   Fig. 3

The coefficient for point B

$$= \frac{OP \times \sin\theta \times BV}{I_v} - \frac{OP \times \cos\theta \times BU}{I_u}$$

$$= \frac{38.0 \times 0.197 \times 37.0}{49.1 \times 10^4} - \frac{38.0 \times 0.980 \times 9.5}{199 \times 10^4}$$

$$= (5.64 - 1.78) \times 10^{-4}$$

$$= 3.86 \times 10^{-4}/\text{mm}^2.$$

The coefficient for point C

$$= \frac{OP \times \cos\theta \times CU}{I_u} - \frac{OP \times \sin\theta \times CV}{I_v}$$

$$= \frac{38.0 \times 0.980 \times 72.0}{199 \times 10^4} - \frac{38.0 \times 0.197 \times 24.0}{49.1 \times 10^4}$$

$$= (13.47 - 3.66) \times 10^{-4}$$

$$= 9.81 \times 10^{-4}/\text{mm}^2.$$

Therefore the maximum coefficient is given at the point C. It will generally be found that such is the case, although it must not necessarily be taken that it is always so.

Figure 4 shows the method of determining the second of these coefficients.

Coefficient for point B

$$= +\frac{BV \times \sin \alpha}{I_v} + \frac{BU \times \cos \alpha}{I_u}$$

Coefficient for point A

$$= -\frac{AV \times \sin \alpha}{I_v} + \frac{AU \times \cos \alpha}{I_u}$$

Coefficient for point C

$$= -\frac{CV \times \sin \alpha}{I_v} - \frac{CU \times \cos \alpha}{I_u}$$

(+ indicates compression; – indicates tension.)
The angle $\alpha = 36°\ 54'$ and the values of CU, CV, BU and BV are as given previously.
$AV = 32.0$ mm and $AU = 61.0$ mm.

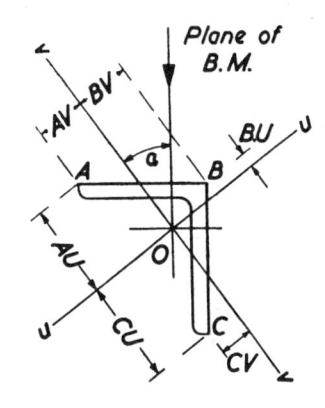

Fig. 4

The coefficients will be found to be as follows:

$$\text{Coefficient for point } B = +0.41 \times 10^{-4}/\text{mm}^3$$
$$\text{„ „ „ } C = -0.58 \times 10^{-4}/\text{mm}^3$$
$$\text{„ „ „ } A = -0.15 \times 10^{-4}/\text{mm}^3$$

(a) Support B

(i) Axis xx:

Effective length = 0.7 × panel length
= 0.7 × 1.85 = 1.295 m

$$\frac{l}{r_x} = \frac{1.295 \times 10^3}{3.17 \times 10} = 40.8$$

(ii) Axis yy:

Effective length = 1.0 × distance between purlins
= 1.0 × 1.17 = 1.17 m

$$\frac{l}{r_y} = \frac{1.17 \times 10^3}{2.68 \times 10} = 43.6$$

(iii) Axis vv:

Effective length = 1.0 × maximum distance between support and adjacent purlin

= 1.0 × 0.64 = 0.64 m

$$\frac{l}{r_v} = \frac{0.64 \times 10^3}{1.85 \times 10} = 34.5$$

Therefore greatest value of $l/r = 43.6$.

$$p_c = 137 \text{ N/mm}^2$$

$$f_c = \frac{70.7 \times 10^3 \text{ N}}{14.41 \times 10^2 \text{ mm}^2} = 49 \text{ N/mm}^2$$

Bending moment at support B = 1.41 kNm negative.

At this point, it will be the point C which has the maximum compressive bending stress.

$$f_{bc} = (70.7 - 67.8) \times \text{coef. from Fig. 3} + 1.41 \times \text{coef. from Fig. 4}$$
$$= 2.9 \text{ kN} \times (9.81 \times 10^{-4})/\text{mm}^2 + (1.41 \times 10^3)\text{kNmm} \times (0.58 \times 10^{-4})/\text{mm}^2$$
$$= 0.0028 \text{ kN/mm}^2 + 0.0818 \text{ kN/mm}^2$$
$$= 0.0846 \text{ kN/mm}^2 = 84.6 \text{ N/mm}^2$$

It may be assumed that the provisions of Clause 19.c of B.S. 449 apply only to sections remote from the gusset plate and that in respect of angles subjected to combined bending and compression the allowable bending stress p_{bc} can be taken to be 165 N/mm² at the gusset plate.

$$\frac{f_c}{p_c} + \frac{f_{bc}}{p_{bc}} = \frac{49}{137} + \frac{84.6}{165} = 0.36 + 0.53 = 0.89$$

Therefore the section is adequate at support B.

(b) Support C
 (i) Axis *xx*:

$$l/r_x \text{ as for support B} = 40.8$$

(ii) Axis *yy*:

$$\text{Effective length} = 1.0 \times 2.49 = 2.49 \text{ m}$$
$$\frac{l}{r_y} = \frac{2.49 \times 10^3}{2.68 \times 10} = 93$$

(iii) Axis *vv*:

$$\frac{l}{r_v} = \frac{1.28 \times 10^3}{1.85 \times 10} = 69$$

Therefore the greatest value of $l/r = 93$.

$$p_c = 87 \text{ N/mm}^2$$
$$f_c = \frac{67.8 \times 10^3}{14.41 \times 10^2} = 47 \text{ N/mm}^2.$$

Bending moment at support C = 1.08 kNm negative.
At this point, it will again be the point C which has the maximum compressive stress.

$$f_{bc} = (67.8 - 65.8) \times (9.81 \times 10^{-4}) + (1.08 \times 10^3) \times (0.58 \times 10^{-4})$$
$$= 0.0029 \text{ kN/mm}^2 + 0.0626 \text{ kN/mm}^2$$
$$= 0.0655 \text{ kN/mm}^2 = 65.5 \text{ N/mm}^2$$

$$\frac{f_c}{p_c} + \frac{f_{bc}}{p_{bc}} = \frac{47.0}{87.0} + \frac{65.5}{165.0} = 0.54 + 0.40 = 0.94.$$

Therefore the section is adequate at support C.

ECCENTRICITY OF LOADING

(c) At the purlin point between B and C.

(i) Axis xx:
$$l/r_x \text{ as for support B} = 40.8$$

(ii) Axis yy:
$$l/r_y \text{ as for support C} = 93.0$$

(iii) Axis vv:
$$l/r_v = \frac{1.21 \times 10^3}{1.85 \times 10} = 66.$$

Therefore the greatest value of $l/r = 93$.

$$p_c = 87 \text{ N/mm}^2$$

$$f_c = \frac{67.8 \times 10^3}{14.41 \times 10^2} = 47 \text{ N/mm}^2.$$

Bending moment at purlin point = 1.46 kNm positive.

At this point it will be the point B which has the maximum compressive stress, but it is reasonable to assume, from the location of the purlin point between the supports B and C, that the bending stress due to the eccentricity of the end connections is negligible.

$$f_{bc} = (1.46 \times 10^3) \times (0.41 \times 10^{-4}) = 0.06 \text{ kN/mm}^2 = 60 \text{ N/mm}^2$$

$$\frac{f_c}{p_c} + \frac{f_{bc}}{p_{bc}} = \frac{47.0}{87.0} + \frac{60.0}{165.0} = 0.54 + 0.37 = 0.91.$$

Therefore the section is adequate at the purlin point.
Accordingly, the section selected is adequate.

Example 7. Double-angle continuous strut.

In this case, there will be bending only about the axis xx, i.e. in a vertical plane. Moreover, it will be obvious, by consideration of Example 6, that only support C need be investigated.

Try two 63 x 51 x 6.2 mm angles of grade 50 steel, 10 mm back to back.

$$\text{Area} = 13.64 \text{ cm}^2$$
$$r_x = 1.98 \text{ cm}$$
$$r_y = 2.37 \text{ cm}$$
$$\text{Min. } Z_x = 12.2 \text{ cm}^3.$$

Support C

(i) Axis xx:
$$l/r_x = \frac{0.7 \times 1.85 \times 10^3}{1.98 \times 10} = 65$$

(ii) Axis yy:
$$l/r_y = \frac{2.49 \times 10^3}{2.37 \times 10} = 105$$

DESIGN OF ANGLE STRUTS

Therefore the greatest value of $l/r = 105$.

$$p_c = 85 \text{ N/mm}^2$$

$$f_c = \frac{67.8 \times 10^3}{13.64 \times 10^2} = 49.5 \text{ N/mm}^2.$$

Bending moment at C = 1.08 kNm negative.

$$f_{bc} = \frac{1.08 \times 10^6}{12.2 \times 10^3} = 88.5 \text{ N/mm}^2$$

$$\frac{f_c}{p_c} + \frac{f_{bc}}{p_{bc}} = \frac{49.5}{85.0} + \frac{88.5}{230.0} = 0.58 + 0.38 = 0.96.$$

Therefore the section is adequate.

The comparative weights of the sections in Examples 6 and 7 are 11.3 and 10.7 kg/m respectively.

There is probably not a great deal to choose between the sections.

The single angle has the advantage that there is less fabrication than with the double angle because the latter requires washer-riveting at intervals. It also has the advantage from the point of view of maintenance, since it is virtually impossible to paint the inner vertical surfaces of a double angle.

Against this, the double angle has two principal advantages: first, since the rivets at the joints will be in double shear or bearing, fewer will be needed than in the single angle; secondly, the purlin cleats, which should be connected to the rafters by a minimum of two rivets, will be of smaller section, since in the double angle the two rivets can be placed one in each angle, whereas in the single angle they must be placed in line down the rafter back.

21. ENGINEERING WORKSHOP DESIGN

THE following calculations for a steel-framed workshop building with an electric overhead travelling crane are such as would be done in the project stage for the preparation of an estimate of cost and would have to be finally checked in detail in accordance with BS 449 Part 2 1969.

Since this type of building is often subject to alteration to suit changes in work type and other owners requirements, the simple construction adopted lends itself more readily to alteration than the more sophisticated designs frequently used nowadays. Though three main grades of structural steelwork, viz.: Grades 43, 50 and 55 are now available, there is little merit in using anything better than Grade 43 for buildings of this size. However, where larger dimensions or heavier crane loads apply, Grade 50 might be used with advantage provided that deflection is watched.

Crane gantries, by virtue of the loading which they are required to sustain, can be subject to fatigue, particularly if the cranes are part of a production cycle, and in continuous operation at maximum capacity. It is therefore advisable to investigate this problem as recommended in BS 153, Steel Girder Bridges. This has not been done in the example which follows.

Data for building

Overall length	= 48.5 m
Length c-c gable columns	= 48.0 m say
Overall width	= 18.5 m
Width c-c roof columns	= 18.0 m say
Height ground to crane rail	= 9.0 m
Height crane rail to u/s roof	= 2.5 m
Height ground to u/s roof	= 11.5 m

Roof covering: protected metal sheeting with fibreboard lining and one stretch of patent glazing 2.0 m deep on each slope.
Side covering: similar sheets and lining above 2.0 m brick wall on permanent side and above ground on temporary side. Gable cladding: similar with provision for sliding door 3.5 m wide x 5.0 m high each gable.
One side prepared for future extension of a similar bay.
One end prepared for future extension.
Building to house a gantry to carry one electric overhead travelling crane of 30 metric tonne capacity.

Crane span	= 17.0 m
End clearance	= 0.3 m
Minimum hook approach	= 1.0 m
Length of end carriage	= 4.0 m
Wheel centres in end carriage	= 3.5 m
Weight of crane without crab	= 25.0 Mg or tonne
Weight of crab	= 5.0 Mg or tonne

Steelwork designed to BS 449: Part 2: 1969.
Wind loading to CP 3: Chap. V: Part 2: 1970.

Framing to building

Length c-c end stanchions = 48.0 m.
Roof trusses at 4.0 m centres gives 12 bays.
Main stanchions at 8.0 m centres gives 6 bays.
Trusses carried on main and intermediate stanchions on permanent side.
Trusses carried on main stanchions and valley beams on temporary side.
Gable frame on stanchions framed to suit door at permanent end.
Truss and false framing carried on main stanchions at temporary end with framing arranged to suit door below.
Purlins and rails to sheeting at 2.0 m centres.

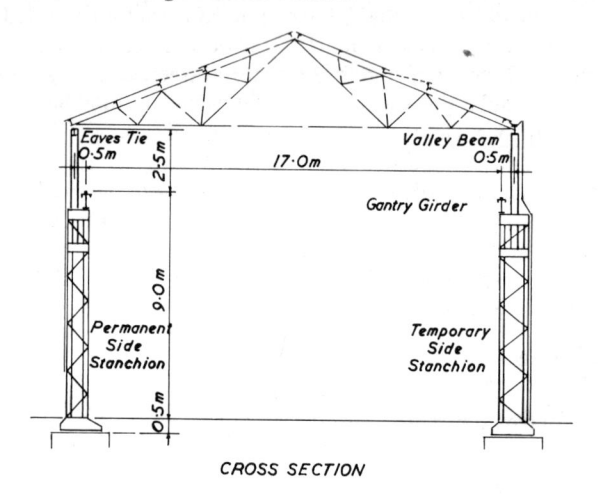

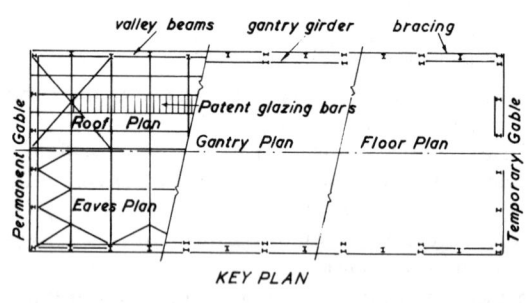

Fig. 1

Wind on Building

Calculations in conformity with requirements of CP 3 Chapter V Part 2 1970.

In order to arrive at the loading to be considered it is necessary to know the location of the site as well as other conditions. For the purpose of this example the following assumptions are made:—

'Building situated in town in S.E. England with no unusual topological conditions. Building is 18.5 m wide x 48.5 m long x 15.0 m high.'

MAIN STANCHIONS

Fig. 2

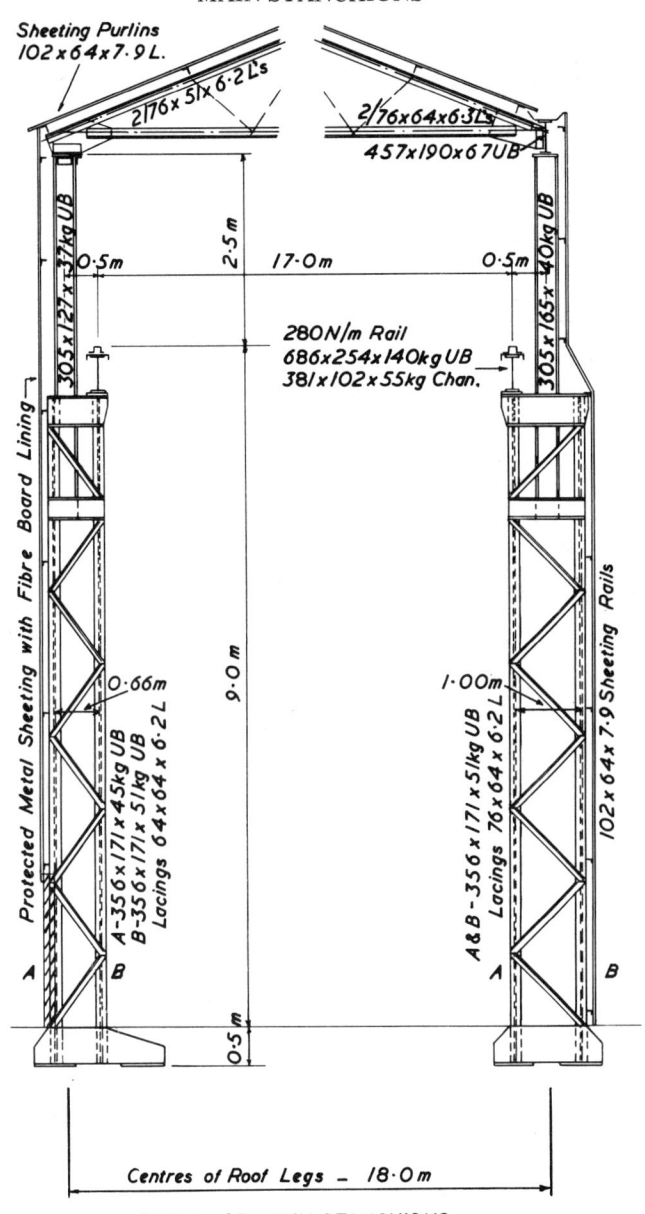

DETAIL OF MAIN STANCHIONS.

Basic wind speed V = 40 m/s
Topography factor S_1 = 1.0
Ground roughness category 3
Building size class B
Factor S_2 varies with height as shown in Fig. 3.

ENGINEERING WORKSHOP DESIGN

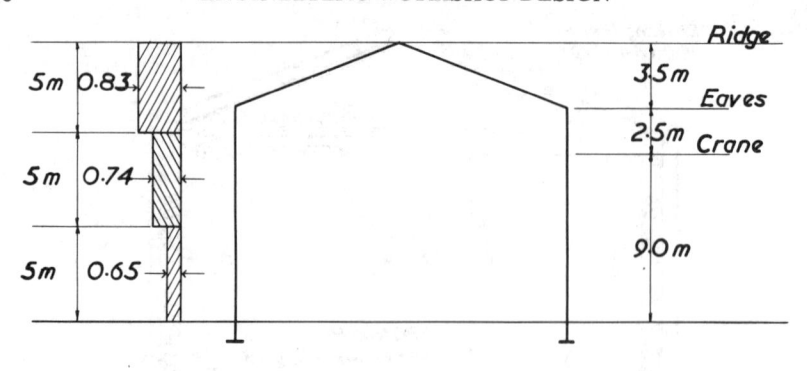

VALUES OF FACTOR S2

Fig. 3

Statistical factor S3 = 1.0

Note. The loading shown applies to the steelwork. A separate set of values must be used for the cladding and its connections to the steelwork.

From the assumed information

$$\text{Design wind speed } V_s = V \times S1 \times S2 \times S3$$
$$= 40 \times 1 \times S2 \times 1$$
$$= 40\ S2$$

Dynamic wind pressure $\quad q = KV_s^2$

$$= 0.613\ V_s^2\ \text{N/m}^2$$

Then for three stages of height

$$0\text{–}5\ \text{m} \quad q = 0.613(40 \times 0.65)^2 = 414\ \text{N/m}^2$$
$$5\ \text{m}\text{–}10\ \text{m} \quad q = 0.613(40 \times 0.74)^2 = 537\ \text{N/m}^2$$
$$10\ \text{m}\text{–}15\ \text{m} \quad q = 0.613(40 \times 0.83)^2 = 676\ \text{N/m}^2$$

Consider wind load on roof

$$\text{Height to eaves} = 11.5\ \text{m} = h$$
$$\text{Width of building} = 18.5\ \text{m} = w$$
$$\text{Then } h/w = 11.5/18.5 = 0.62$$
$$\text{Slope of roof} = 22°$$

Then pressure coefficients are

C_{pe} wind ⟷ Windward = −0.7 Leeward = −0.5

wind ↓ = −0.8 maximum both slopes

Internal pressure coefficient C_{pi}
Maximum effect on roof is when $C_{pi} = +0.2$

SHEETING PURLINS

Then maximum uplift on both slopes
$$= (-0.8 - (+0.2))q$$
$$= -1.0q$$
$$q = 676 \text{ N/m}^2$$
Then load each slope $= -676 \text{ N/m}^2$

Note. Local areas of cladding and its connections must be designed for increased loads.

For general stability consider load on sides and roof without internal pressure.

Height to eaves $= h = 11.5$ m
Width of building $= w = 18.5$ m
Length of building $= l = 48.5$ m

Ratio
$h/w = \dfrac{11.5}{18.5} = 0.6$
$l/w = \dfrac{48.5}{18.5} = 2.6$

Then pressure coefficients for sides are

Wind ⟷ Windward $= +0.7$ Leeward $= -0.3$ Local $= -1.1$
↕ Both sides $= -0.5$

The wind on the end does not affect stability across the building since the two values cancel each other.

Then total force transversely $= 1.0q$ N/m².

Design of members

Sheeting purlins 4.0 m span 2.0 m centres

Dead Load Sheets $= 4 \times 2 \times 110$ N/m² $= 880$ N
Lining $= 4 \times 2 \times 50$ N/m² $= 400$ N
Purlin   say   400 N
                   ─────────
                   1 680 N $= 1.68$ kN

Live Load to CP 3 Chap. V Part 1 1967.
Load $= 4 \times 1.8 \times 0.75$ kN/m² $= 5.4$ kN

Wind Load $= 4 \times 2 \times -676$ N/m² $= -5.4$ kN

Then load is either $+7.08$ kN or -3.72 kN
Design for $+7.08$ kN

From BS 449, Part 2, 1969 $L/45 = 89$ mm $L/60 = 67$ mm

and $Z = \dfrac{WL}{1.8 \times 10^3} = \dfrac{7.08 \text{ kN} \times 4\,000 \text{ mm}}{1.8 \times 10^3}$

$= 15.75$ cm³

Use   102 mm $\times$ 63 mm $\times$ 7.8 mm angle
$Z = 19.0$ cm³

N.B. Whilst the effect of the wind is to reverse the direction of the net load it does not alter the basic design of the purlin.

Glazing Purlins   4.0 m span   2.0 m centres.

Dead Load   Sheets   = 4 x 1 x 110 N/m² =   440 N
              Lining   = 4 x 1 x 50 N/m² =   200 N
              Glazing = 4 x 1 x 300 N/m² = 1 200 N
              Purlin   say   560 N
                                         ─────────
                                         2 400 N = 2.4 kN

Live Load   As before $= \dfrac{5.4 \text{ kN}}{7.8 \text{ kN}}$ total

It dimensional requirements are met,

$$Z \text{ reqd.} = \frac{7.8 \text{ kN} \times 4\,000 \text{ mm}}{1.8 \times 10^3} = 17.4 \text{ cm}^3$$

Use 102 mm x 63 mm x 7.8 mm angle with 51 mm x 51 mm x 6.3 mm shelf angle to glazing, bolted or battened.

Roof Truss   18.0 m span   4.0 m centres.

Dead Load per truss

   Sheets   = 16 x 4 x 110 N/m² = 7 040 N
   Lining   = 16 x 4 x 50 N/m² = 3 200 N
   Glazing = 4 x 4 x 300 N/m² = 4 800 N
   Purlins = 8 x 4 x 100 N/m   = 3 200 N
            + 4 x 4 x 140 N/m   = 2 240 N
   Truss   say   9 000 N
                                 ──────────
                                 29 480 N = 29.5 kN

Live Load per truss

   = 18 x 4 x 0.75 kN/m² $= \dfrac{54.0 \text{ kN}}{83.5 \text{ kN}}$

        Say 10 panels at 8.4 kN = 84.0 kN
        Reactions   = 42.0 kN

Wind Load per truss (including internal pressure)

        = 20 x 4 x − 676 N/m = − 54 000 N = − 54.0 kN

This represents 10 panels @ − 5.4 kN each acting normal to the slopes and can act in conjunction with dead load only. Hence, the net effect will give reversals of stress in all truss members.

ROOF TRUSS

Frame Diagram

Force Diagram D.L.+ LL

Force Diagram W.L.+ Int. press.

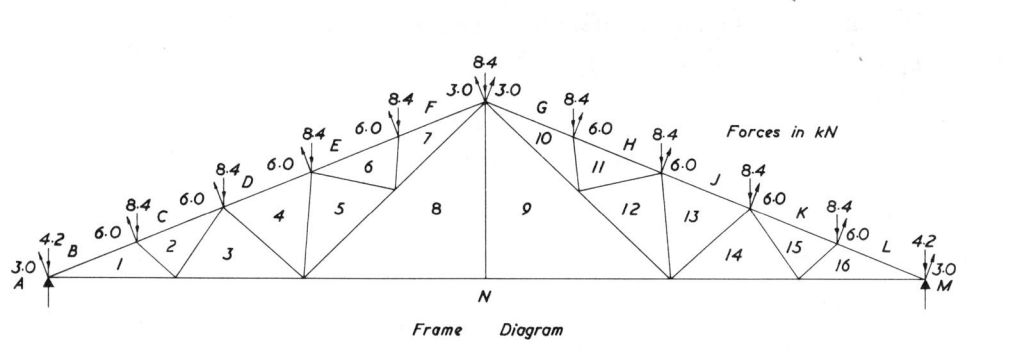

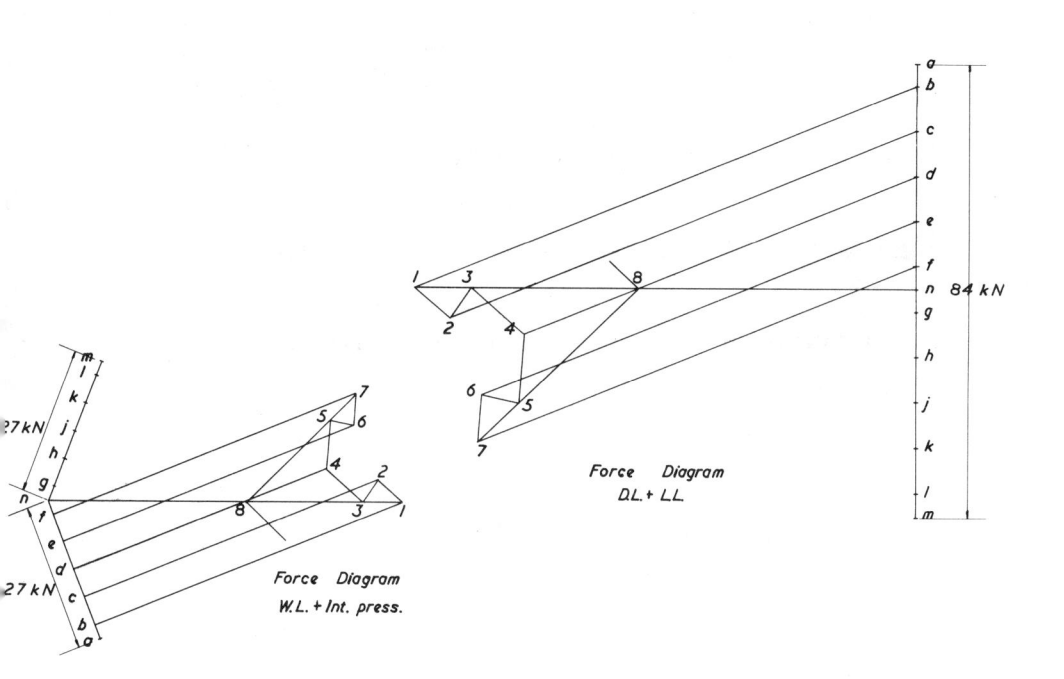

Member	Mark	D.L.+L.L.	D.L only	W.L.	D.L.+ W.L.
Rafter	b/ etc	+ 130 kN	+ 37 kN	- 63 kN	-26 kN
Main Tie	n/ etc	- 98 kN	- 35 kN	+ 68 kN	+33 kN
Main Tie Centre	n8 n9	- 55 kN	-20 kN	- 40 kN	-20 kN
Main Strut	3.4 etc	+15 kN	+5 kN	- 10 kN	- 5 kN
Crown Tie	5.8 etc	-43 kN	-15 kN	+ 28 kN	+ 13 kN

Fig. 4

Design for truss members (Grade 43 steel)

Rafter Panel length xx = 2.0 m

Purlin centres yy = 2.0 m

Load from force diagrams = + 103 kN

or − 33 kN

Try 2/angles 76 mm x 51 mm x 6.2 mm as minimum for practical purposes.

Length xx = 2.0 m Coef. 0.85 Effective l_{xx} = 1.7 m = 170 cm

Length yy = 2.0 m Coef. 1.0 Effective l_{yy} = 2.0 m = 200 cm

$$l/r_{yy} = \frac{200}{2.18} = 92 \qquad p_c = 85 \text{ N/mm}^2$$

$$\text{Actual stress } f_c = \frac{103 \text{ kN}}{15.2 \text{ cm}^2} = \frac{103 \times 10^3 \text{ N}}{15.2 \times 10^2 \text{ mm}^2} = 68 \text{ N/mm}^2$$

Ample margin as strut and therefore as tie for reversal,

∴ Section is OK.

Main Tie Load from force diagrams = − 98 kN

or + 40 kN

Since wind causes reversal section suitable as strut must be used.

Try 2/angles 76 mm x 51 mm x 6.2 mm similar to rafter.

Gross area = 2 x 7.59 = 15.18 cm² = 1 518 mm²

Area of connected legs = 2 x 73.1 x 6.22 = 910 mm²

Area of outstanding legs = 2 x 47.7 x 6.22 = 594 mm²

$a1$ = nett area of connected legs = 910 − 2 x 18 x 6.22

= 910 − 224   = 686 mm²

$a2$ = nett area of outstanding legs   = 594 mm²

Then effective area of pair of angles

$$= 686 + \frac{5 \times 686}{5 \times 686 + 594} \times 594$$

$$= 686 + \frac{3\,430}{4\,024} \times 594$$

= 686 + 504 = 1 190 mm²

$$\text{Stress in tension} = \frac{98 \times 10^3 \text{ N}}{1\,190 \text{ mm}^2} = 83 \text{ N/mm}^2$$

Permissible stress p_t = 155 N/mm²   ∴ OK for tension.

ROOF TRUSS

Consider reversal of stress placing tie in compression.

    Length xx = 2.5 m   Coef. 0.85   Effective l_{xx} = 2.12 m = 212 cm

    Length yy = 5.0 m   Coef. 1.0   Effective l_{yy} = 5.0 m = 500 cm

    (Assuming longitudinal tie at main node points.)

$$l/r_{xx} = \frac{212}{2.42} = 88$$

$$l/r_{yy} = \frac{500}{2.18} = 229 \quad p_c = 18 \text{ N/mm}^2 + 25\% \text{ for wind} = 22.5 \text{ N/mm}^2$$

Actual stress on gross area $f_c = \dfrac{33 \times 10^3 \text{ N}}{15.2 \times 10^2 \text{ mm}^2} = 21.8 \text{ N/mm}^2$

This is within, but increase to next size to improve stiffness of truss.

    Use 2/angles 76 mm x 63 mm x 6.2 mm

Main Tie centre   Load from force diagrams =   – 55 kN

                                                   or + 22 kN

    Length xx = 3.75 m   Coef. 0.85   Effective l_{xx} = 3.18 m = 318 cm

    Length yy = 7.5 m   Coef. 1.0   Effective l_{yy} = 7.5 m = 750 cm

Try section suggested for outer sections.

$$l/r_{yy} = 750/2.78 = 268 \qquad p_c = 13.5 \text{ N/mm}^2 + 25\% = 17.0 \text{ N/mm}^2$$

Act stress, $f_c = \dfrac{20 \times 10^3 \text{ N}}{16.7 \times 10^2 \text{ mm}^2} = 12.0 \text{ N/mm}^2$

It should be noted that the l/r ratio of 268 is just in excess of the value of 250 mentioned in BS 449, Clause 33, but since the actual length of the member will be reduced slightly by the end connections the size will be accepted.

    2/angles 76 mm x 63 mm x 6.2 mm will be used throughout.

Crown Tie   Load from force diagram =   – 43 kN

                                               or + 13 kN

         Length xx = 2.5 m   Length yy = 5.0 m

From calculations made for main tie it can be shown that 2/angles 76 x 51 x 6.2 will be satisfactory.

Main strut Load from force diagram = +15 kN or –5 kN
                        Use 1/angle 63 mm x 63 mm x 6.2 mm

Remainder of members – Use 1/angle 63 mm x 51 mm x 6.2 mm.

Design for valley beam on temporary side

Span between main stanchions = 8.0 m

Load:—
Centre point load from truss = 42 kN
Centre ,, ,, ,, future truss = 42 kN
Dist. load from own wt and gutter = 16 kN say
$$\overline{100 \text{ kN}}$$

$$\text{Maximum B.M.} = \frac{84 \times 8}{4} + \frac{16 \times 8}{8}$$

$$= 168 + 16 = 184 \text{ kNmm.}$$

Try 406 x 178 x 67 kg UB.

Beam restrained at centre by truss connection

$$l/r_{yy} = \frac{400}{3.86} = 104. \qquad D/T = 409/14.3 = 28.6$$

$p_{bc} = 152 \text{ N/mm}^2$

$$\text{Act. } f_{bc} = \frac{184 \text{ kNm}}{1\,186 \text{ cm}^3} = \frac{184 \times 10^6 \text{ Nmm}}{1\,186 \times 10^3 \text{ mm}^2} = 155 \text{ N/mm}^2 \text{ high.}$$

Use beam 457 x 190 x 67 kg UB in Grade 43 steel.

Beam should be checked against deflection from live load, i.e. centre point load of 54 kN.

$$\text{Deflection} = \frac{WL^3}{48EI}$$

$$= \frac{54 \text{ kN} \times (8 \text{ m})^3}{48 \times 2.1 \times 10^5 \text{ N/mm}^2 \times 29\,337 \text{ cm}^4}$$

$$= \frac{54 \times 10^3 \text{ N} \times 512 \times 10^9 \text{ mm}^3}{48 \times 2.1 \times 10^5 \text{ N/mm}^2 \times 29\,337 \times 10^4 \text{ mm}^4}$$

$$= 9.4 \text{ mm}$$

Span = 8 000 mm $\frac{1}{360}$ span = 22.2 mm

Then beam well in for deflection.

Design for crane gantry girders

Span of girder = 8.0 m.

Maximum Loads on gantry (N.B.:— For design purposes 1 kg f is taken as 10 N — actually 9.807 N).

CRANE GANTRY GIRDERS

Vertical load

$$\text{From crane} = \tfrac{1}{2}\{25.0 \text{ Mg}\} = 12\,500 \text{ kg} = 125 \text{ kN}$$

$$\text{From crab + lift} = \tfrac{16}{17}\{5 + 30\} \text{ Mg} = 33\,000 \text{ kg} = 330 \text{ kN}$$

$$\overline{455 \text{ kN}}$$

Add 25% for impact = 114 kN

$$\overline{569 \text{ kN}}$$

say 570 kN = 285 kN each on two wheels at 3.5 m cts
From weight of gantry   say 16 kN

Co-existent Vertical load

From crane	= 125 kN
From crab + lift	= 20 kN
	145 kN
From wt of gantry say	16 kN

Horizontal load

Crab + lift = 35 Mg = 350 kN

Cross-surge 10%. = 35 kN

on 4 wheels   = say 9 kN per wheel

Maximum Reactions on stanchions

Vertical

$$\text{From crane} = 285\left[1 + \frac{4.5}{8}\right] = 1.5625 \times 285 = 444 \text{ kN}$$

From girder   = 16 kN

460 kN

Horizontal

From cross-surge = 1.5625 × 9   = 14 kN

Minimum co-existent reactions

Vertical

From crane = 1.5625 × 122.5   = 192 kN

From girder   = 16 kN

208 kN

Horizontal

From cross-surge   = 14 kN

Maximum bending moment — vertical

From crane

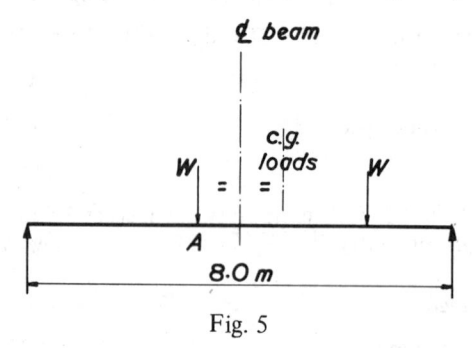

Fig. 5

For two equal loads W.

$$\text{Maximum bending moment at } A = \frac{2W}{L}\left(\frac{L}{2}-\frac{c}{4}\right)^2$$

$$= \frac{2W}{8}(4-0.875)^2$$

$$= 2.44W$$

$W = 285$ kN $\therefore$ B.M. $= 695$ kNm

From girder

If dead load $= 16.0$ kN distributed,

Bending moment at A

$$= 8 \times 3.125 - \frac{16}{8}\left(\frac{3.125}{2}\right)^2$$

$$= 15 \text{ kNm}$$

Total vertical bending moment $= 710$ kNm

Maximum bending moment — horizontal

From crane cross-surge $W = 9$ kN

Then bending moment $= 9 \times 2.44$

$$= 22 \text{ kNm}$$

Selection of size for crane gantry girders

Except in cases where horizontal surge girders are adopted it is usual to make the gantry girders out of compound sections where the top flange is much stronger than the bottom flange. This serves two purposes, i.e.

(a) To provide more effective resistance to instability of the compression flange.
(b) To give a large resistance to horizontal bending from the crane cross-surge.

The properties of suitable sections for gantry girders can be obtained from structural steelwork handbooks, but an example of calculating these properties is given for reference.

For this example, a section composed of a 610 x 229 x 113 kg UB with a 305 x 89 x 42 kg channel on the top flange, all in Grade 50 steel, will first be examined. Calculations for the properties are as follows:

Properties of compound sections for gantry girders

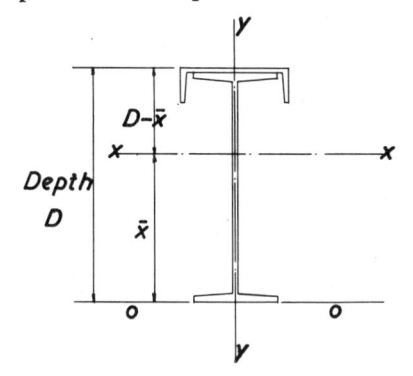

Section comprising:—

610 x 229 x 113 kg UB

and 305 x 89 x 42 kg channel forming a welded section.

Fig. 6

Vertical

Part	Depth	Area a	Lever arm y	ay	ay^2	Own Inertia	Inertia $0-0$
	cm	cm^2	cm	cm^3	cm^4	cm^4	cm^4
Channel	1.02	53.11	59.54	3 162	188 276	325	188 600
UB	60.70	144.30	30.35	4 380	132 918	87 262	220 280
Σ	61.72	197.41	—	7 542	—	—	408 880

$$\bar{x} = \frac{\Sigma ay}{\Sigma a} = \frac{7\,542}{197.41} = 38.2 \text{ cm} : D - \bar{x} = 23.52 \text{ cm}$$

$$I_{oo} = I_{xx} + ay^2 \quad \therefore I_{xx} = I_{oo} - ay^2$$

$$= 408\,880 - 7\,542 \times 38.2$$

$$= 408\,880 - 288\,104 = 120\,766 \text{ cm}^4$$

$$Z_{xx} \text{ top} = \frac{I_{xx}}{D - \bar{x}} = \frac{120\,776}{23.52} = 5\,135 \text{ cm}^3$$

$$Z_{xx} \text{ bottom} = \frac{I_{xx}}{\bar{x}} = \frac{120\,776}{38.2} = 3\,162 \text{ cm}^3$$

Horizontal

Full section

$$I_{yy} = I_{xx} \text{ for channel} + I_{yy} \text{ for UB}$$

$$= 7\,061 + 3\,184$$

$$= 10\,245 \text{ cm}^4$$

$$r_{yy} = \sqrt{\frac{I_{yy}}{a}} = \sqrt{\frac{10\,245}{197.41}} = 7.2 \text{ cm}$$

Top flange only I_{yy} part = I_{xx} for channel + $\dfrac{TB^3}{12}$ for UB flange

$$= 7\,061 + \dfrac{1.73 \times 22.8^3}{12}$$

$$= 7\,061 + 1\,709 = 8\,770 \text{ cm}^4$$

$$Z_{yy} \text{ top} = \dfrac{8\,770}{15.25} = 575 \text{ cm}^3$$

Horizontal area of top flange,

$$= tD \text{ for channel} + TB \text{ for UB}$$

$$= 1.02 \times 30.5 + 1.73 \times 22.8$$

$$= 31.11 + 39.44 = 70.55 \text{ cm}^2$$

Width of top flange = depth of channel = 30.5 cm.
Then mean thickness of top flange T_m

$$= \dfrac{70.55}{30.5}$$

$$= 2.31 \text{ cm}$$

To determine p_{bc} and p_{bt}

Effective length of compression flange = span

$$= 8.0 \text{ m}$$

$$l/r_{yy} = \dfrac{8.0 \times 10^2 \text{ cm}}{7.2 \text{ cm}}$$

$$= 111$$

$$D/T_m = \dfrac{61.72}{2.31}$$

$$= 26.7$$

From table 3b $\qquad p_{bc} = 188 \text{ N/mm}^2$

table 2 $\qquad p_{bt} = 230 \text{ N/mm}^2$

Maximum vertical bending moment = 710 kNm

Maximum horizontal bending moment = 22 kNm

Top flange

$$f_{bc} = \dfrac{710 \times 10^6 \text{ Nmm}}{5\,135 \times 10^3 \text{ mm}^3} + \dfrac{22 \times 10^6 \text{ Nmm}}{575 \times 10^3 \text{ mm}^3}$$

$$= 138 + 38 = 176 \text{ N/mm}^2$$

$$p_{bc} = 188 + 10\% = 207 \text{ N/mm}^2$$

DEFLECTION OF GANTRY GIRDER

Bottom flange

$$f_{bt} = \frac{710 \times 10^6 \text{ Nmm}}{3\,162 \times 10^3 \text{ mm}^3} = 225 \text{ N/mm}^2$$

Then section is O.K. for stress.

Check deflection of gantry girder selected

The maximum deflection will vary as the loads roll across the span, but exact calculations are lengthy and unnecessary. A useful assumption is that the maximum deflection occurs at the centre of the span when the two wheels are equidistant on either side of the centreline. The deflection can then be calculated using the standard formula, viz.:

$$\delta\max = \frac{PL^3}{6EI}\left[\frac{3a}{4L} - \left(\frac{a}{L}\right)^3\right]$$

where
P = value of one load

a = distance of one load from adjacent reaction

For this example,

$$P = 285 \text{ kN} \quad a = \left(\frac{L-c}{2}\right) = \frac{8-3.5}{2} = 2.25 \text{ m}$$

Then deflection due to crane,

$$= \frac{285 \text{ kN} \times (8 \text{ m})^3}{6EI}\left[\frac{3 \times 2.25}{4 \times 8} - \left(\frac{2.25}{8}\right)^3\right]$$

$$= \frac{285 \times 10^3 \text{ N} \times 512 \times 10^9 \text{ mm}^3}{6 \times 2.1 \times 10^5 \text{ N/mm}^2 \times 120\,766 \times 10^4 \text{ mm}^4}[0.208 - 0.022]$$

$$= \frac{285 \times 512 \times 0.186}{6 \times 2.1 \times 120.766} \text{ mm}$$

$$= 18 \text{ mm}$$

And deflection due to gantry girder,

$$= \frac{5 \times 16 \text{ kN} \times (8 \text{ m})^3}{384 \times 2.1 \times 10^5 \text{ N/mm}^2 \times 120\,766 \times 10^4 \text{ mm}^4}$$

$$= 0.5 \text{ mm}$$

Total deflection = 18.5 mm — say 20.0 mm

If span = 8.0 m = 8 000 mm

Deflection = 1/400 span

This could be too much. Often a deflection of 1/500 span is specified. A deeper girder in Grade 43 steel would be a more suitable section.

Try girder composed of 686 x 254 x 140 kg UB with a 381 x 102 x 55 kg channel on top flange. Grade 43 steel. Calculating properties as before.

$$I_{xx} = 190\ 590 \text{ cm}^4$$
$$Z_{xx} \text{ top} = 7\ 328 \text{ cm}^3$$
$$Z_{xx} \text{ bottom} = 4\ 388 \text{ cm}^4$$
$$r_{yy} \text{ full section} = 8.89 \text{ cm}$$
$$Z_{yy} \text{ top flange} = 918 \text{ cm}^3$$
$$D/T = 30.2$$

On span of 8.0 m,

$$l/r_{yy} = 90$$

Then
$$p_{bc} = 165 \text{ N/mm}^2$$
$$p_{bt} = 165 \text{ N/mm}^2$$

Actual stresses,

$$f_{bc} = \frac{710 \times 10^6 \text{ Nmm}}{7\ 328 \times 10^3 \text{ mm}^3} + \frac{22 \times 10^6 \text{ Nmm}}{918 \times 10^3 \text{ mm}^3}$$

$$= 96 + 24 = 120 \text{ N/mm}^2$$

$$f_{bt} = \frac{710 \times 10^6 \text{ Nmm}}{4\ 388 \times 10^3 \text{ mm}^3}$$

$$= 162 \text{ N/mm}^2$$

Then section is O.K. for stress and since Inertia is approximately 50 per cent more than before, deflection will be within required limit.

Therefore, use section

$$\left.\begin{array}{l} 686 \times 254 \times 140 \text{ kg UB} \\ 381 \times 102 \times\ \ 55 \text{ kg channel} \end{array}\right\} \text{ in Grade 43 steel.}$$

Design for side and gable rails

Maximum span say 4.0 m   Maximum spacing say 2.0 m.

Vertical Load

Sheets and lining = 4 x 2 x 160 N/m² = 1.28 kN

Own weight   say = 0.40 kN
                        ─────────
                         1.68 kN

Horizontal Load

Apart from the eaves rail which takes only half a panel, the value of c_{pe} is a maximum of ±0.7 which should be increased by 0.3 for internal pressure. Therefore the rails should be designed for a wind load equal to q.

SIDE AND GABLE RAILS

Taking the maximum value of q at 676 N/m² is conservative since this only applies to the upper portion of the building.
Then,

$$\text{Horizontal load} = 4 \times 2 \times 676 \text{ N/m}^2 = 5.4 \text{ kN say}$$

Bending Moments

Allowing for continuity take B.M. = $WL/10$.

$$\text{Vertical B.M.} = \frac{1.68 \times 4}{10} = 0.67 \text{ kNm}$$

$$\text{Horizontal B.M.} = \frac{5.4 \times 4}{10} = 2.16 \text{ kNm}$$

Selection of section

Try 102 mm x 63 mm x 7.8 mm angle similar to purlin with long leg horizontal

$I_{xx} = 129$ cm⁴ Z_{xx} max = 38.2 cm³ Z_{xx} min = 19.0 cm³
$I_{yy} = 39$ cm⁴ Z_{yy} max = 26.2 cm³ Z_{yy} min = 8.0 cm³

$$\text{Max. } f_{bc} = \frac{M_V}{Z_{yy} \text{ max}} + \frac{M_H}{Z_{xx} \text{ min}}$$

$$f_{bt} = \frac{M_V}{Z_{yy} \text{ min}} + \frac{M_H}{Z_{xx} \text{ min}}$$

$$= \frac{0.67 \times 10^6 \text{ Nmm}}{8 \times 10^3 \text{ mm}^3} + \frac{2.16 \times 10^6 \text{ Nmm}}{19 \times 10^3 \text{ mm}^3}$$

$$= 84 + 114 = 198 \text{ N/mm}^2$$

Permissible p_{bt} = 165 N/mm² + 25 per cent for wind = 207 N/mm²

Section is O.K. Use 102 mm x 63 mm x 7.8 mm angle with long leg horizontal.

Design for intermediate side posts – permanent side

Height from floor to u/s roof = 11.5 m
Height from base to u/s roof = 12.0 m
Rail spacing = 2.0 m

Assume battened rail at 6 m from base.

Vertical Load

Reaction from truss = 42.0 kN
Side sheeting, etc. 9.5 x 4 x 0.2 =   7.6 kN
Own weight   say   =   4.4 kN
                         ─────────
                          54.0 kN

Horizontal Load from wind

Take $1.0q$. This load varies but it is on the top side to take an average of 537 N/m².
Then,
$$\text{Horizontal load} = 11.5 \times 4 \times 537 \text{ N/m}^2 = 24.8 \text{ kN}$$
$$\text{Horizontal moment as a beam} = \frac{24.8 \times 12}{8} = 37.2 \text{ kNm}$$

Selection of size

Actual length $xx = 12.0$ m Coef. 1.5 = 18.0 m effective
Actual length $yy = 6.0$ m Coef. 0.75 = 4.5 m effective

Try 305 × 127 × 37 kg UB in Grade 43 steel.

$$l/r_{xx} = \frac{18 \times 10^2}{12.3} = 146$$

$$l/r_{yy} = \frac{4.5 \times 10^2}{2.58} = 175 \quad p_c = 30 \text{ N/mm}^2$$

increased for wind by 25 per cent = 37.5 N/mm²

For bending effective length of compression flange is rail spacing, i.e. 2.0 m.

Then,
$$l/r_{yy} = \frac{2 \times 10^2}{2.58} = 78 \quad \frac{D}{T} = 28.4$$

$$p_{bc} = p_{bt} = 165 \text{ N/mm}^2 + 25\% = 207 \text{ N/mm}^2$$

Then actual stress
$$f_c = \frac{54 \times 10^3 \text{ N}}{47.4 \times 10^2 \text{ mm}^2} = 11.5 \text{ N/mm}^2$$

and actual stress
$$f_{bt} \text{ or } f_{bc} = \frac{37.2 \times 10^6 \text{ Nmm}}{470 \times 10^3 \text{ mm}^3} = 79 \text{ N/mm}^2$$

Ratio
$$\left. \begin{array}{l} \dfrac{f_c}{p_c} = \dfrac{11.5}{37.5} = 0.31 \\[2ex] \dfrac{f_{bc}}{p_{bc}} = \dfrac{79}{207} = 0.37 \end{array} \right\} = 0.68$$

Section O.K.

Use 305 × 127 × 37 kg UB.

Similar section should be used for temporary posts and for permanent gable posts.

Design for main stanchions

One of the main factors to be considered is the horizontal load from the wind.

The previous calculations indicated the need to allow for a load equal to $1.0\,q$ N/m² on the vertical height ground to eaves, and nothing on the roof slopes since the net effect of the various coefficients cancels out for all practical purposes.

Wind drag can have an effect on a large area of roof, but since the ratios of width to height and width to length are both less than 4, it can be ignored in the present case.

MAIN STANCHIONS

The value $1.0\, q$ N/m² is based on $+0.7\, q$ on windward face and $-0.3\, q$ on leeward face. Since, however, the two faces are connected together by the roof structure, the proportions will be adjusted in accordance with the relative stiffness of the main steelwork in each face.

The following treatment of the wind load is considered sufficiently accurate for design of main stanchions.

Consider side wind on 8.0 m cts. of stanchions.
Referring to Fig. 3 and values of q as calculated

$$
\begin{aligned}
\text{Wind load Grd} - 5.0\text{ m} &= 5 \times 8 \times 414 &= 16.56\text{ kN} \\
\text{\textquotedbl}\quad\text{\textquotedbl}\quad 5\text{ m} - 10\text{ m} &= 5 \times 8 \times 537 &= 21.48\text{ kN} \\
\text{\textquotedbl}\quad\text{\textquotedbl}\quad 10\text{ m} - \text{eaves} &= 1.5 \times 8 \times 676 &= 8.11\text{ kN} \\
&\text{Total} &= 46.15\text{ kN}
\end{aligned}
$$

Taking moment about base, reaction at eaves level

$$
\begin{aligned}
&= \tfrac{1}{12}[16.56 \times 3 + 21.48 \times 8 + 8.11 \times 11.25] \\
&= \tfrac{1}{12}[49.68 + 171.84 + 91.24] \\
&= \tfrac{1}{12}[312.76] = 26.06\text{ kN.}
\end{aligned}
$$

If 26.06 kN represents the force at the truss level to be resisted by the steelwork

Overturning moment at base

$$= 26.06 \times 12 = 312.76\text{ kNm}$$

$$\text{Forces in stanchions} = \pm \frac{312.76}{18} = \pm 17.4\text{ kN}$$

The moment due to the wind, i.e. 312.76 kNm, will be distributed between the stanchions on either side which, if similar sections are used, is a function of the centres of the two legs forming the stanchions, viz.:—

Permanent side cts of legs   say 0.66 m: $\dfrac{0.66}{2} = 0.33: 0.33^2 = 0.11$

Temporary side   ,, ,, ,,   ,,   0.9 m: $\dfrac{0.9}{2} = 0.45: 0.45^2 = \underline{0.20}$

$$\hspace{10cm} 0.31$$

Then on permanent side take $\dfrac{0.11}{0.31}$ or say one third

and on temporary side take $\dfrac{0.20}{0.31}$ or say two thirds.

So for permanent side stanchion

$$\text{Shear} = \frac{26.06}{3} = 8.7\text{ kN}$$

$$\text{Moment} = \frac{312.76}{3} = 104.25\text{ kNm}$$

and for temporary side stanchion

$$\text{Shear} = 17.4\text{ kN}$$

$$\text{Moment} = 208.5\text{ kNm.}$$

ENGINEERING WORKSHOP DESIGN

Permanent side stanchions

Loading details

Roof leg above crane cap

Vertical   Maximum   Minimum

Reaction from truss   42.0 kN   15.0 kN
Load from side cladding say   3.0 kN   3.0 kN
Own weight   1.5 kN   1.5 kN
                                                  46.5 kN   19.5 kN

Horizontal

From wind at roof level   8.7 kN
From surge at rail level   14.0 kN

Roof leg below crane cap

Vertical   Maximum   Minimum

From above $\dfrac{.48}{.66}$ =   34.0 kN   14.0 kN

Load from cladding say   7.0 kN   7.0 kN
Own weight say   5.0 kN   5.0 kN
                                                  46.0 kN   26.0 kN

50 per cent of wind overturning = ± 8.7 kN.

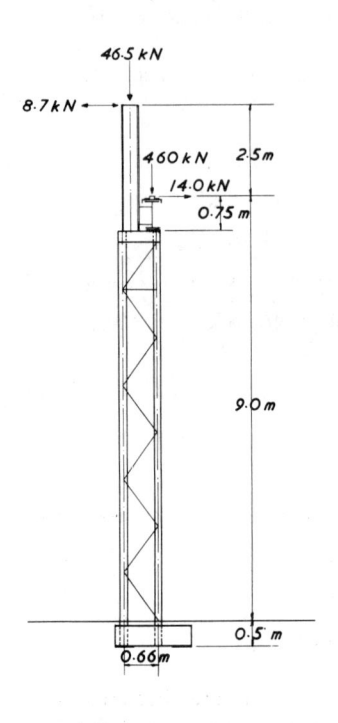

Fig. 7

Crane leg below crane cap

Vertical

From above $\dfrac{.18}{.66}$ =   12.5 kN   1.0 kN
From crane   444.0 kN   192.0 kN
From girder   16.0 kN   16.0 kN
Own weight say   5.0 kN   5.0 kN
                                   467.5 kN   22.0 kN no crane
                                                    214.0 kN crane remote

50 per cent of wind overturning = ± 8.7 kN.

Horizontal loads as above giving base moments
of Wind 104 kNm and Surge 133 kNm.

MAIN STANCHIONS 517

Roof leg above crane cap

Length = 3.25 m
Effective length xx = 1.5 x 3.25 = 4.875 m
yy = 1.0 x 3.25 = 3.25 m
Vertical load = 46.5 kN
Bending moment from wind = 8.7 x 3.25 = 28.3 kNm ⎫
from surge = 14.0 x 0.75 = 10.5 kNm ⎬ 38.8 kNm

Try 305 x 165 x 40 kg UB in Grade 43 steel.

$l/r_{xx} = \dfrac{4.875 \times 10^2}{12.9} = 38$ $p_c = 92$ N/mm² + 25% for wind = 125 N/mm²

$l/r_{yy} = \dfrac{3.25 \times 10^2}{3.66} = 89$ $p_{bc} = 165$ N/mm² + 25% for wind = 207 N/mm²

$f_c = \dfrac{46.5 \times 10^3 \text{ N}}{51.4 \times 10^2 \text{mm}^2} = 9.1$ N/mm² ratio = 0.073

$f_{bc} = \dfrac{38.8 \times 10^6 \text{ Nmm}}{560 \times 10^3 \text{ mm}^3} = 70.0$ N/mm² ratio = 0.34

0.413

Try 305 x 127 x 37 kg UB in Grade 43 steel.

$l/r_{xx} = \dfrac{4.875 \times 10^2}{12.3} = 40$ $p_c = 55$ N/mm² + 25% = 69 N/mm²

$l/r_{yy} = \dfrac{3.25 \times 10^2}{2.58} = 126$

$D/T = 28$ $p_{bc} = 130$ N/mm² + 25% = 163 N/mm²

$f_c = \dfrac{46.5 \times 10^3}{47.4 \times 10^2} = 9.8$ N/mm² ratio = 0.143

$f_{bc} = \dfrac{38.8 \times 10^6}{470 \times 10^3} = 83.0$ N/mm² ratio = 0.61

0.753

Section O.K.

Use 305 x 127 x 37 kg UB in Grade 43 steel.

Roof leg below crane cap

Length = 8.75 m.   Spacing of lacings 2.0 m.
Effective length xx = 0.85 x 8.75 = 7.5 m
yy = 1.0 x 2.0 = 2.0 m

Vertical load (maximum) = $46.0 + 8.7 + \dfrac{237}{.66} = 429$ kN.

518 ENGINEERING WORKSHOP DESIGN

Try 356 × 171 × 45 kg UB in Grade 43 steel.

$$l/r_{xx} = \frac{7.5 \times 10^2}{14.6} = 51$$
$$l/r_{yy} = \frac{2 \times 10^2}{3.58} = 56$$
$$p_c = 129 \text{ N/mm}^2 + 25\% = 162 \text{ N/mm}^2$$

$$f_c = \frac{429 \times 10^3}{56.9 \times 10^2} = 75 \text{ N/mm}^2 \qquad \text{Section O.K.}$$

Crane leg below crane cap

Length = 8.75 m.   Spacing of lacings 2.0 m.
Effective length xx = 0.85 × 8.75 = 7.5 m
yy = 1.0 × 2.0 = 2.0 m
Vertical load (maximum) = 467 + 8.7 + 374 = 850 kN.

Using same section as before,

$$f_c = \frac{850 \times 10^3}{56.9 \times 10^2} = 150 \text{ N/mm}^2 \qquad p_c \text{ inc.} = 162 \text{ N/mm}^2.$$

Section proves adequate but increase to next heavier section to give margin against eccentricity of crane loading.

Use 356 × 171 × 51 kg UB in Grade 43 steel.

Lacings

Total vertical load on combined stanchion,

From construction = 46 + 467.5 = 513.5 kN } 531.0 kN
From wind O.T.M.   = 17.5 kN }

Take 2½% as shear on lacings = 13.3 kN
Wind shear   = 8.7 kN
Surge   = 14.0 kN
                               ─────────
Load carried by lacings based on   36.0 kN horizontal.

|0.66 m|

Length = $\sqrt{1^2 + .66^2}$ = 1.2 m

Force = $\frac{1.2}{0.66}$ × 36.0 kN = 65 kN on pair

= 32.5 kN each.

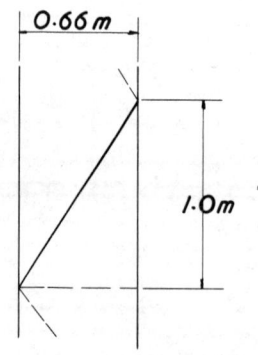

1·0m   Try angle 63 mm × 63 mm × 6.2 mm.

$$l/r_{min} = \frac{1.2 \times 10^2}{1.25} = 96 \qquad p_c = 84 + 25\% = 105 \text{ N/mm}^2$$

$$f_c = \frac{32.5 \times 10^3}{7.59 \times 10^2} = 43 \text{ N/mm}^2$$

Fig. 8   Use lacings 63 mm × 63 mm × 6.2 mm angle.

Temporary side stanchions

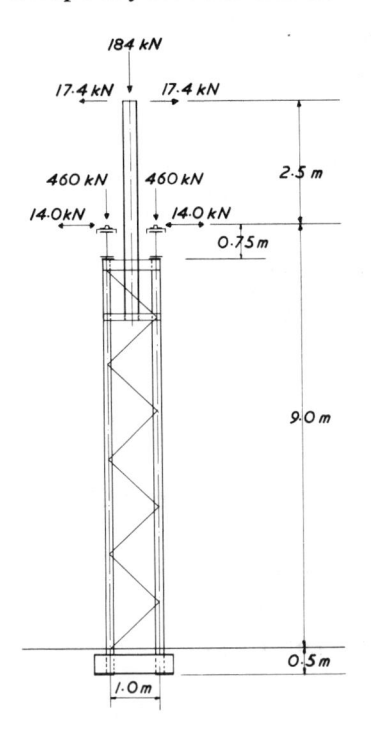

Loading details

Roof leg above crane cap

Vertical

	Future Maximum	Present Minimum
Reaction from trusses	84.0 kN	15.0 kN
Reaction from valley	100.0 kN	31.0 kN
Temporary cladding		3.0 kN
Own weight	2.0 kN	2.0 kN
	186.0 kN	51.0 kN

Horizontal

From wind at roof 17.4 kN

From surge *one* side 14.0 kN.

N.B. When future bay added, wind load could reduce but extra surge.

Fig. 9

Crane legs below crane cap

Vertical

		Maximum	Minimum
From above	=	93 kN	26 kN
From crane	=	444 kN	192 kN
From girder	=	16 kN	16 kN
Own weight	=	6 kN	6 kN
		559 kN	32 kN present outer leg. No crane.

50 per cent wind O.T.M. ± 8.7 kN.

Base moments

　Wind 17.4 × 12 = 209 kNm

　Surge 14.0 × 9.5 = 133 kNm

　　　　　　　　　342 kNm

　　　Forces in legs from moments = ± 342 kN

　　∴ Maximum load in lower shaft = 910 kN

Roof leg above crane cap

$$\text{Length} = 3.25 \text{ m}$$
$$\text{Effective length } xx = 1.5 \times 3.25 = 4.875 \text{ m}$$
$$yy = 1.0 \times 3.25 = 3.25 \text{ m}$$
$$\text{Vertical load} = 186 \text{ kN}$$

$$\text{Bending moment from wind} = 57 \text{ kNm}$$
$$\text{from surge} = \underline{10 \text{ kNm}}$$
$$67 \text{ kNm}$$

Try 305 x 165 x 40 kg UB in Grade 43 steel.

$$\left. \begin{array}{l} l/r_{xx} = \dfrac{4.875 \times 10^2}{12.9} = 38 \\[6pt] l/r_{yy} = \dfrac{3.25 \times 10^2}{3.67} = 89 \end{array} \right\} \quad \begin{array}{l} p_c = 92 \text{ N/mm}^2 \\[6pt] p_{bc} = 165 \text{ N/mm}^2 \end{array} \quad \begin{array}{l} p_c \text{ inc.} = 125 \text{ N/mm}^2 \\[6pt] p_{bc} \text{ inc.} = 207 \text{ N/mm}^2 \end{array}$$

$$f_c = \frac{186 \times 10^3}{51.4 \times 10^2} = 36.3 \text{ N/mm}^2 \qquad \text{ratio} = 0.29$$

$$f_{bc} = \frac{67 \times 10^6}{560 \times 10^3} = 120 \text{ N/mm}^2 \qquad \text{ratio} = \underline{0.58}$$

$$0.87$$

Section O.K.

Use 305 x 165 x 40 kg UB in Grade 43 steel.

Crane legs below crane cap

$$\text{Length 8.75 m.} \qquad \text{Spacing of lacings 2.0 m.}$$
$$\text{Effective length } xx = 7.5 \text{ m}$$
$$yy = 2.0 \text{ m}$$
$$\text{Vertical load (maximum)} = 910 \text{ kN.}$$

Try 356 x 171 x 51 kg UB.

$$\left. \begin{array}{l} l/r_{xx} = \dfrac{7.5 \times 10^2}{14.8} = 51 \\[6pt] l/r_{yy} = \dfrac{2 \times 10^2}{3.71} = 54 \end{array} \right\} \quad p_c = 130 \text{ N/mm}^2 + 25\% = 162.5 \text{ N/mm}^2$$

$$f_c = \frac{910 \times 10^3}{64.5 \times 10^2} = 142 \text{ N/mm}^2$$

Section O.K.

Use 356 x 171 x 51 kg UB in Grade 43 steel.

MAIN STANCHIONS

Lacings

Total vertical load on combined stanchion

From construction = 2 x 559 = 1 118 kN

From wind O.T.M. = 17.4 kN but ignore since above load includes future bay.

Take 2½ per cent as shear carried by lacings = 28 kN

Wind shear	= 17 kN
Surge (two cranes)	= 28 kN
Load carried by lacings based on	73 kN horizontal

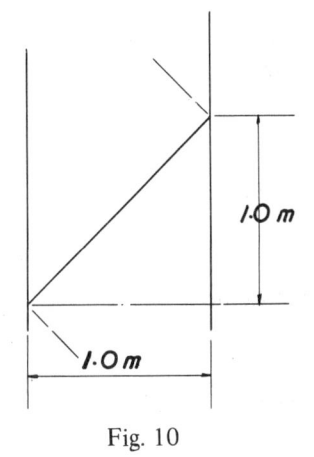

Length of bar = $\sqrt{2}$ = 1.414 m

Force = $\frac{1.414}{1}$ x 73 = 104 kN on pair

= 52 kN each.

Try angle 76 mm x 76 mm x 6.2 mm.

$l/r_{min} = \frac{1.414 \times 10^2}{1.49} = 95$

$p_c = 85 + 25\% = 106$ N/mm^2

$f_c = \frac{52 \times 10^3}{9.12 \times 10^2} = 57$ N/mm^2 good margin

Try angle 76 mm x 63 mm x 6.2 mm.

$l/r_{min} = \frac{1.414 \times 10^2}{1.33} = 106 \qquad p_c = 73 + 25\% = 91$ N/mm^2

$f_c = \frac{52 \times 10^3}{8.36 \times 10^2} = 63$ N/mm^2 \qquad O.K.

Use lacings 76 mm x 63 mm x 6.2 mm angle.

Fig. 10

Design for bracings

Gable wind girder at truss tie level

From previous calculations pressure coefficient for end of building is – 0.5 q. Allowing for internal pressure increase to – 0.8 q.

Then reaction at tie level for wind from below will be 80% of that from side wind.

On full effective width of 18.0 m.

Load on wind girder = 18/8 x 0.8 x 26.06 kN

= 47.0 kN.

ENGINEERING WORKSHOP DESIGN

Wind on apex
$$= 18 \times 3.5 \times \tfrac{1}{2} \times 0.8 \times 676 \text{ N} = 17 \text{ kN}$$

Reaction at wind girder level
$$= 12.9/12 \times 17 = 18.2 \text{ kN}$$

Then total load on wind girder $\qquad = 65.2$ kN

Assume distributed between stanchion positions as shown in Fig. 11.

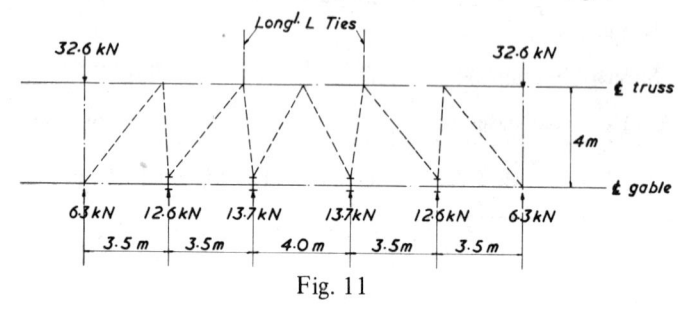

Fig. 11

Length of end diagonal approximately $= \sqrt{4^2 + 3.5^2} = 4.7$ m

Force in end diagonal $= \dfrac{4.7}{4.0} \times 26.3 = 31.0$ kN

Effective length of member, say $4.7 \times 0.85 = 4.0$ m.

Try 2/angles 63 mm × 51 mm × 6.2 mm.

$l/r_{min} = \dfrac{4.0 \times 10^2}{1.95} = 205 \qquad p_c = 22$ N/mm² increased for wind by 25% = 27.5 N/mm²

Actual stress $\qquad f_c = \dfrac{31 \times 10^3}{13.6 \times 10^2} = 23$ N/mm² $\qquad$ O.K.

Use 2/angles 63 mm × 51 mm × 6.2 mm for all members.

Eaves bracing

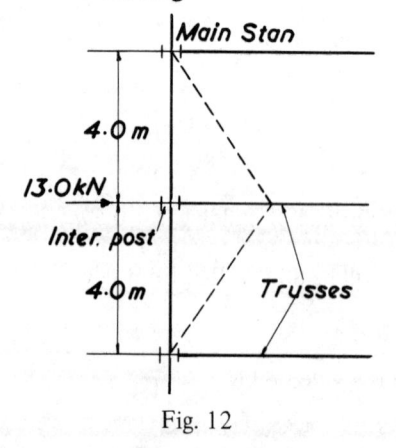

Fig. 12

Bracing required to transfer wind on intermediate post to main stanchions.

Wind on post from previous calculations
$$= 24.8 \text{ kN}$$

Reaction at top $= \dfrac{6.25}{12} \times 24.8$

$$= 13.0 \text{ kN}$$

Force in brace $= \dfrac{4.7}{2.5} \times \dfrac{13}{2} = 12.3$ kN

Use single angle 89 mm × 76 mm × 6.3 mm.

Vertical bracing

Bracing, or some other form of longitudinal stiffness is required to transfer the reaction from the gable wind girder, and the crane longitudinal surge, to base level.

Horizontal load from wind,

Permanent side 32.6 kN

Temporary side 65.2 kN on two legs = 32.6 kN each.

Longitudinal surge from crane = 5% x 455 kN = 91 kN

Take half at each end   = 45.5 kN

Design bracing above crane for horizontal load of 32.6 kN and below crane for horizontal load of 78.1 kN.

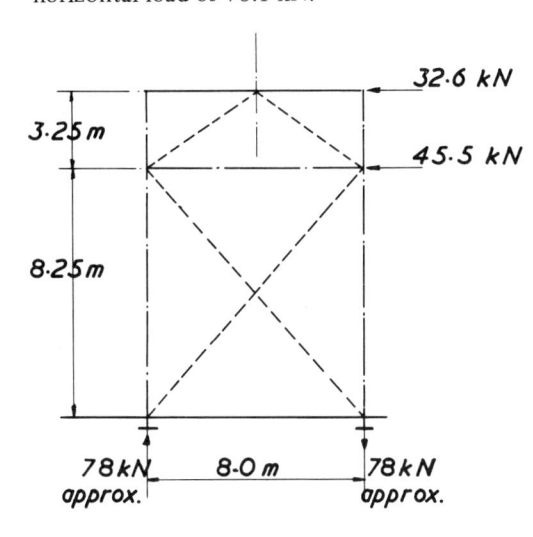

Above crane

Length of bar $= \sqrt{4^2 + 3.25^2}$

$= 5.15$ m

Force $= \dfrac{5.15}{3.25} \times 32.6 = 52.0$ kN

Effective length $0.85 \times 5.15 = 4.4$

Try 2/angles 89 mm x 63 mm x 6.2 mm.

$l/r_{min} = \dfrac{4.4 \times 10^2}{2.96} = 165$

$p_c = 34$ N/mm^2

$+ 25\% = 42.5$ N/mm^2

$f_c = \dfrac{52 \times 10^3 \text{ N}}{18.2 \times 10^2 \text{ mm}^2}$

$= 28.6$ N/mm^2

Section O.K. Use 2/angles 89 mm x 63 mm x 6.2 mm.

VERTICAL BRACING

Fig. 13

Below crane level

Assume all load taken on one bar in tension

Load = 78.1 kN

Try single angle 102 mm x 76 mm x 7.9 mm

Gross area = 13.48 cm^2

Net area of connected leg,

$$= (101.6 - 4) \times 7.9 - 18 \times 7.9$$
$$= 740 - 140 = 600 \text{ mm}^2 = 6 \text{ cm}^2$$

Area of outstanding leg,

$$= (76.2 - 4) \times 7.9$$
$$= 570 \text{ mm}^2 = 5.7 \text{ cm}^2$$

Then effective area,

$$= 6 + \left(\frac{3 \times 6}{3 \times 6 + 5.7}\right) 5.7$$
$$= 6 + 4.3$$
$$= 10.3 \text{ cm}^2$$

Permissible stress = 155 N/mm² + 25% = 193 N/mm²

Actual stress,

$$f_t = \frac{78.1 \times 10^3 \text{ N}}{10.3 \times 10^2 \text{ mm}^2} = 76 \text{ N/mm}^2$$

Ample margin but section used on account of length.

Use single angle 102 mm x 76 mm x 7.9 mm counterbraced.

Design for stanchion base
Permanent side
Loading conditions

Case 1. Maximum

Roof leg vertical	=	46.0 kN
Crane leg vertical	=	467.5 kN
Wind overturning	=	17.4 kN
Wind moment	=	104 kNm
Surge moment	=	133 kNm

Case 2. minimum no crane

Roof leg vertical	=	26.0 kN
Crane leg vertical	=	22.0 kN
Wind overturning	=	− 17.4 kN
Wind moment	=	104 kNm

Case 3. minimum with crane

Roof leg vertical	=	26.0 kN
Crane leg vertical	=	214.0 kN
Wind overturning	=	− 17.4 kN
Wind moment	=	104 kNm
Surge moment	=	133 kNm

STANCHION BASE

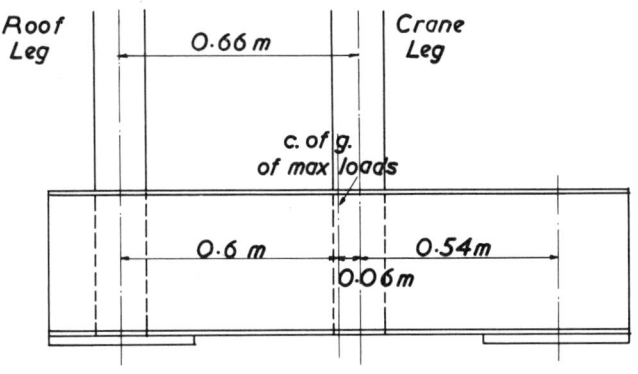

Fig. 14

Consider maximum loads.

$$\text{C. of G. of vertical loads} = 467.5 \times 0.66 \times \frac{1}{513.5} = 0.6 \text{ m}$$

If base made symmetrical about this centre line.
Effective centres of base plates = 1.2 m
Total vertical load = 532 kN
On each plate = 266 kN
Total overturning moment = 237 kNm
On each plate = 237/1.2 = 200 kN
Then total vertical load on baseplate = 266 + 200 = 466 kN
Assuming baseplate size 0.4 m × 0.6 m = 0.24 m²

$$\text{Maximum pressure} = \frac{466}{0.24} = 1\,960 \text{ kN/m}^2$$

This is satisfactory.

For maximum uplift consider *case 2*.

Uplift on baseplate *B*.

From vertical load, moment equals 26 × 0.6 − 22 × 0.06 = 15.6 − 1.3 = 14.3 kNm

$$\text{Uplift} = \frac{14.3}{1.2} = 12 \text{ kN}$$

From overturning = $\frac{17.4}{2}$ = 9 kN

From wind moment = $\frac{104}{1.2}$ = 87

$$\begin{array}{rl} \text{Total uplift} & = 108 \text{ kN} \\ \text{Less vertical} & = 24 \text{ kN} \\ \hline \text{Nett uplift} & = 84 \text{ kN} \end{array}$$

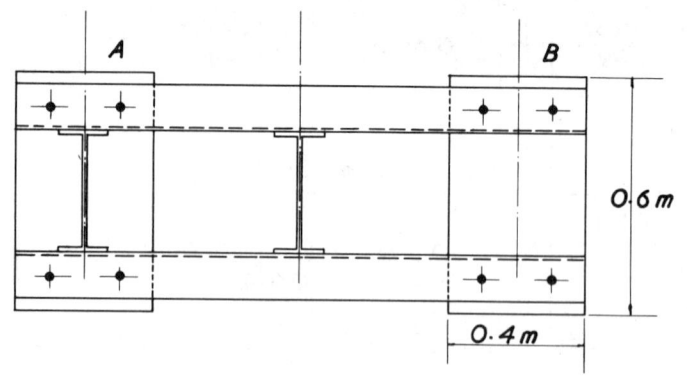

Fig. 15

Foundation bolts

Bolts strength designation 4.6.

BS 449 Part 2 1969 gives p_t = 130 N/mm²   Add 25% wind = 162.5 N/mm²

Approximate value of nett area = 70% gross area.

22 mm dia. bolt.   gross area = 380 mm²   nett area approximately = 266 mm²

Then using 2 bolts,

$$\text{Tensile stress} = \frac{84 \times 10^3 \text{ N}}{2 \times 266} = 160 \text{ N/mm}^2$$

Using 4 bolts,

$$\text{Tensile stress} = 80 \text{ N/mm}^2$$

∴ Use 4 H.D. bolts 22 mm dia. in each baseplate.

Temporary Side

Loading conditions

Case 1. Maximum future

Present crane leg vertical	= 559 kN
Future crane leg vertical	= 559 kN
Wind overturning	= 17.4 kN
Wind moment	= 209 kNm
Surge moment	= 133 kNm

Case 2. Minimum present no crane

Present crane leg vertical	= 48 kN
Future crane leg vertical	= 32 kN
Wind overturning	= − 17.4 kN
Wind moment	= 209 kNm

TEMPORARY SIDE BASE 527

Case 3. Minimum present with crane

$$\begin{aligned}
\text{Present crane leg vertical} &= 238 \text{ kN} \\
\text{Future crane leg vertical} &= 32 \text{ kN} \\
\text{Wind overturning} &= -17.4 \text{ kN} \\
\text{Wind moment} &= 209 \text{ kNm} \\
\text{Surge moment} &= 133 \text{ kNm}
\end{aligned}$$

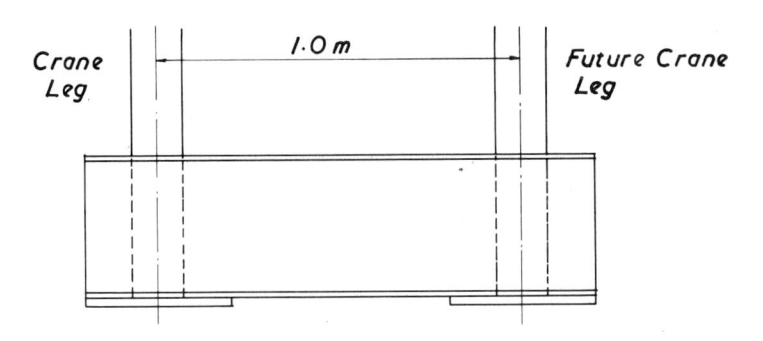

Fig. 16

Consider maximum loads.

Using baseplates under each leg.
Maximum vertical load on each

$$= 559 + 8.7 + 342/1 = 910 \text{ kN}$$

Using plate $0.6 \times 0.6 = 0.36 \text{ m}^2$

Pressure $= 910/0.36 = 2\,530 \text{ kN/m}^2$

which is satisfactory.

Consider maximum uplift.

Ecc. moment from vertical load $= (238 - 32)(0.5) = 103 \text{ kNm}$

Moment from wind and surge $\qquad = 342 \text{ kNm}$

Total moment $\qquad = 445 \text{ kNm}$

Then max. uplift $= \dfrac{445}{1} + \dfrac{17.4}{2} - \dfrac{270}{2}$

$= 445 - 135 = 319 \text{ kN}$

32 mm dia. bolts.   gross area $= 800 \text{ mm}^2$ $70\% = 560 \text{ mm}^2$

Tensile stress for 4 bolts $= \dfrac{319 \times 10^3 \text{ N}}{4 \times 560} = 142 \text{ N/mm}^2$ say O.K.

Use 4 H.D. bolts 32 mm dia. to each plate.

Design of sundry items

The permanent gable will be constructed as shown in Fig. 17.

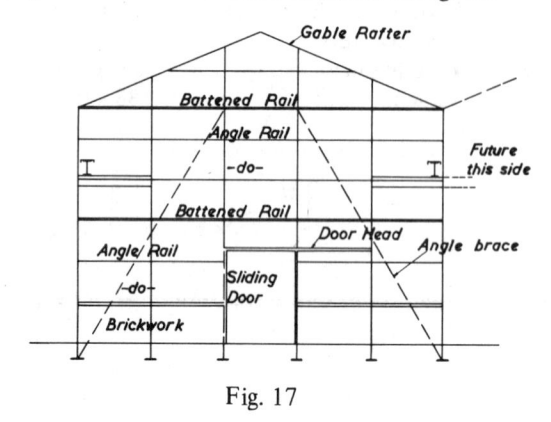

Fig. 17

Gable rafter

Maximum span 4.0 m.   Use 152 x 89 x 24 kg channel.

Beams carrying gantry girder

Span 3.5 m. Point load from gantry at 0.5 m from one end, say 450 kN to include own weight.

Then reaction on corner stanchion,

$$= \frac{3.0}{3.5} \times 450$$

$$= 364 \text{ kN.}$$

Max B.M. = 364 x 0.5 = 182 kNm

$$Z \text{ reqd.} = \frac{182 \times 10^6 \text{ Nmm}}{165 \text{ N/mm}^2} = 1.1 \times 10^6 \text{ mm}^3$$

$$= 1\,100 \text{ cm}^3$$

Use 457 x 190 x 67 kg UB

web to be stiffened under gantry girder.

Corner stanchions

Permanent side/permanent end

Load roof and side say	10 kN
Reaction from crane	364 kN
Own weight	10 kN
	384 kN

CORNER STANCHIONS

Temporary side/permanent end

Load from valley beam	50 kN
Load roof and side say	20 kN
Load from present and future cranes	728 kN
Own weight	10 kN
	808 kN

Both stanchions will be made same size so design for maximum load.

Length xx = 12.0 m   Coef. 0.85   Effective = 10.2 m

Length yy = 6.0 m   Coef. 1.0   Effective = 6.0 m

Try 254 x 254 x 73 kg UC.

$$l/r_{yy} = \frac{6.0 \times 10^2 \text{ cm}}{6.45 \text{ cm}} = 93 \qquad p_c = 87 \text{ N/mm}^2$$

Actual stress, $f_c = \dfrac{808 \times 10^3 \text{ N}}{93 \times 10^2 \text{ mm}^2} = 87 \text{ N/mm}^2$

Since full load from both present and future cranes has been taken, this would appear to be satisfactory. Again, the effective length could no doubt be reduced.

∴ Use 254 x 254 x 73 kg UC in Grade 43 steel.

Note: These two stanchions will require substantial anchorage against uplift from vertical bracing.

Rafter bracing

Used mainly for erection purposes. Use 76 mm x 63 mm x 6.2 mm angle clipped to purlins for support.

Longitudinal Ties to trusses

Use 76 mm x 76 mm x 6.2 mm angle.

Battened rails in gable and side

Use 2/angles 89 mm x 63 mm x 6.2 mm with 89 mm leg vertical.

Vertical bracing in permanent gable

Use angle 89 mm x 63 mm x 6.2 mm.

Penultimate truss at both ends
Gable truss at temporary end

Since these form booms of wind girders main tie should be increased.
Use 2/angles 76 mm x 76 mm x 7.8 mm to allow for extra loading conditions.

22. PLASTIC THEORY AND DESIGN

Foreword

THE attempt to present in a few pages the basis of the plastic method of design of steel frames has led to some inevitable omissions. Further, the compression in these pages is such that the designer unfamiliar with the method may be left with the impression that plastic design is very complicated. This is not so. Plastic theory may not be a universal panacea, but for those problems for which it is suited it will be found to be far easier than elastic analysis; further, plastic design methods are simple and rational, and the designer can be confident that his calculations are reflected in actual behaviour.

The following numbered references are suggested for further reading; they are quoted in the following pages in the appropriate places:

1. BAKER, J. F. *The Steel Skeleton,* Vol. 1. Cambridge U.P. 1954.
2. BAKER, J. F., HORNE, M. R., HEYMAN, J. *The Steel Skeleton,* Vol. 2. Cambridge U.P. 1956.
3. NEAL, B. G. *Plastic Methods of Structural Analysis.* Chapman and Hall 1956.
4. HEYMAN, J. *Beams and Framed Structures.* Pergamon 1964.
  The pitched roof portal frame is dealt with at length in
5. HEYMAN, J. *Plastic Design of Portal Frames.* Cambridge U.P. 1957.
6. BAKER, J. F. BCSA Publication No. 21: 'Plastic Design in Steel to B.S. 968'.
  Stability problems are discussed in 2 above and column design curves are given in
7. HORNE, M. R. BCSA Publication No. 23: 'Plastic Design of Columns'.
  Multi-storey design is covered by a report from the Joint Committee of the Institute of Welding and the Institution of Structural Engineers:
8. *Fully Rigid Multi-storey Welded Steel Frames.* Inst. of Structural Engineers.

Introduction (1, 2*)

The work of the Steel Structures Research Committee, in particular their tests on actual buildings, made it clear that a steel frame behaves very differently from the way assumed by the conventional designer. In certain cases 'elastic design' may hide so much real behaviour that it is dangerous; the real factor of safety of a structural element can be much less than the designer thinks. In other cases, the real factor of safety may be excessive, and an uneconomical structure will result. Indeed, elastic methods give a poor indication of the strength of even the simplest redundant structure, although reasonable estimates can be made of deflections.

After the publication of the Final Report of the Steel Structures Research Committee, in 1936, it was evident that some other method of design would have to be found if the steel-framed building was to advance. As a result, work started on plastic theory, and the plastic method of design was first permitted in this country in 1948, when a clause was inserted in the new edition of B.S. 449.

* The numbers refer to the references listed in the Foreword.

Simple plastic theory is concerned with the *strength* of steel framed structures. Thus, if a given structure is analysed, plastic theory will estimate the values of the loads which will cause collapse of that structure. In the *design* process, where a structure is required to carry given loads, attention is again concentrated on the collapse state. The required reserve of strength against collapse is obtained by designing the structure to fail, not of course under the working loads, but under those loads increased by a *load factor*.

Before embarking on a plastic design, therefore, the designer must be satisfied that the main design criterion for his structure is that of strength. No firm rules can be given as to whether a particular structure is likely to be suitable for plastic design. As a rough guide, however, if maximum permitted stresses govern a conventional elastic design, then plastic methods can probably be used; alternatively, if deflection limitations govern an elastic design, then they will probably preclude the use of plastic theory for that particular structure. Similarly, other considerations, such as fatigue, may make it essential to use an elastic design method.

The plastic design process therefore consists essentially in proportioning a structure so that it is on the point of collapse under factored working loads. Stress distributions under the actual working loads are not normally computed, and working load deflections will be calculated only if it is required to check that these do not influence the design.

Plastic theory has been developed to deal with a specific class of structure, the ductile rigid frame. A framed structure carries applied loads mainly by bending of the members, and plastic collapse analysis is undertaken in essence by the examination only of bending moment diagrams, although more sophisticated methods of calculation may mask this fact. The effects of axial loads and shear forces on the members of a frame are assumed to be small, although allowance can be made for these in a particular design. The potential instability of compression members does not fit easily into simple plastic theory, and special methods have been developed to deal with columns of building frames, which will be referred to later on. In no case, however, is primary plastic collapse of a frame allowed to occur by instability of a member; it is assumed that all members remain stable at collapse of a frame, and the design of columns consists in checking that they are indeed stable.

It will be seen, therefore, that trusses, whose members are subjected to large axial forces rather than bending, cannot be dealt with by simple plastic theory; special methods exist for the design of trusses which are outside the scope of this chapter. For the purpose of this exposition of plastic theory, it will be assumed that the structure satisfies the following requirements:

1. Loads are carried mainly by bending, and the effects of axial load and shear force on a member are small.
2. The designer is satisfied that strength is the main design criterion; checks on deflections may have to be made if these are suspected to be significantly large.
3. The design is fabricated in a ductile steel to B.S. 4360.

Subject to these limitations, plastic theory makes the design process easy and rational. Since the design is rational, in the sense that it deals with an accurately ascertainable criterion, that of collapse, it will also be economical; greatest economy will be achieved if a rigid frame is used.

Basis of the Theory (2, 3, 4)

If a short length of rolled steel joist is subjected to a gradually increasing bending moment, and the values of curvature measured, then a curve similar to that shown in Fig. 1 (a) will be obtained. Elastic behaviour for small values of bending moment is followed by inelastic behaviour, greatly increased curvatures being observed for

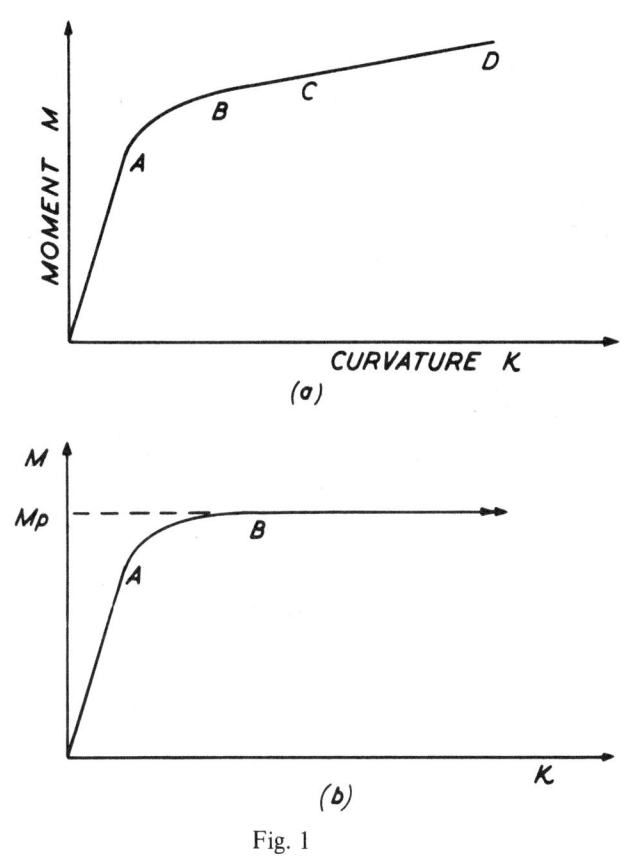

Fig. 1

relatively small increases in bending moment. An actual test on an as-received specimen will show some strain-hardening as shown by the rising curve CD; for the purposes of simple plastic theory, it is assumed, conservatively, that a maximum moment M_p is reached (Fig. 1 (b)) at which curvature can increase indefinitely. The maximum moment M_p is known as the *full plastic moment*; the cross-section of a member which is fully plastic can sustain large local increases of curvature, and is known as a *plastic hinge*. It will be assumed for the time being that the value M_p of the full plastic moment is unaffected by axial loads and shear forces; this point is discussed below, and corrections made to the value of M_p.

Consider a simply supported beam, Fig. 2 (a), subjected to some system of working loads which can be specified in terms of one of their number, W. The bending moment diagram can be drawn and the maximum bending moment M determined. Suppose now that all the loads acting on the beam are multiplied by a

PLASTIC THEORY AND DESIGN

common load factor λ; clearly the bending moment diagram is simply increased in size in the ratio λ, and the maximum bending moment will have value λM, Fig. 2 (b). The load factor λ may be increased until the bending moment λM just reaches the full plastic moment M_p of the cross-section of the beam; the *collapse* condition is therefore

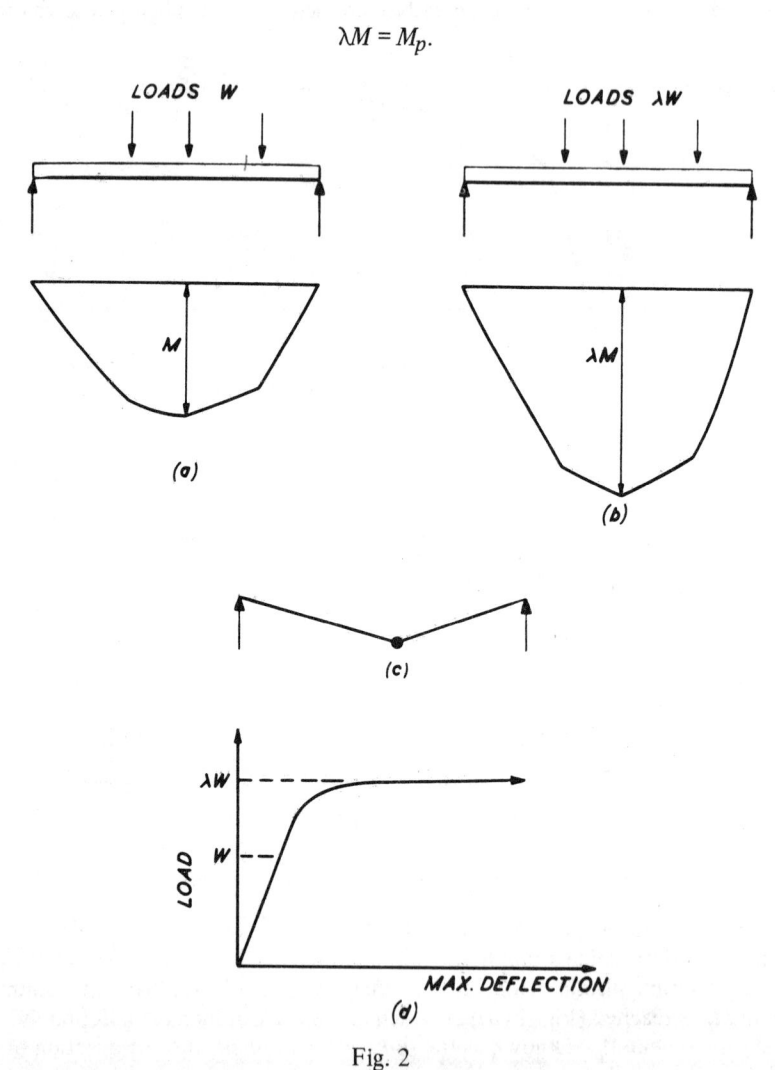

$$\lambda M = M_p.$$

Fig. 2

At this condition, the cross-section at the point of maximum moment will become a plastic hinge, and a *mechanism* movement will be possible as shown in Fig. 2 (c), the load-deflection curve for the beam being as shown in Fig. 2 (d).

This simple example has illustrated one of the basic features of plastic collapse, namely, the formation of a mechanism permitting gross deformation of the structure without any increase in the applied collapse load.

Load Factors

The primary function of a load factor is, of course, to ensure that a structure will be safe under service conditions. In fact, a load factor performs several duties, since a reserve of strength in a structure is required to cover uncertainty of the values of the loads, imperfections in workmanship, errors in design and fabrication, and so on. Under these circumstances, it is difficult to fix a value of load factor on a theoretical basis, although some probability studies have been made.

However, there is a hard core of practical experience with existing structures which can be used. Put in the simplest terms, steel structures designed in accordance with B.S. 449 have shown themselves to be safe, and a plastic design incorporating the minimum load factor implied by B.S. 449 should therefore also be safe.

This minimum load factor is found in a simply supported beam and has the value 1.75 for dead plus superimposed loading; the value may be derived from a study of the beam in Fig. 2. If the bending moment diagram in Fig. 2 (a) corresponds to working values W of the loads, with a maximum bending moment M, then an elastic design in steel to B.S. 4360 grade 43 would be made on the basis that the greatest bending stress should not exceed 165 N/mm².

$$M = 165 Z_e$$

where Z_e is the elastic modulus of cross-section of the beam.

Corresponding to elastic moduli, plastic moduli are given in the section tables from which the full plastic moments M_p can be calculated. (These plastic moduli are first moments of area, and will be referred to again below.) If the collapse state of the beam is as shown in Fig. 2 (b), then the collapse condition can be written

$$\lambda M = M_p = 245 Z_p$$

where 245 N/mm² is the guaranteed yield stress of steel to B.S. 4360 grade 43, and Z_p is the plastic modulus of the beam.

Comparing the elastic design with the collapse condition, it will be seen that the load factor λ is given by

$$\lambda = \frac{245}{165} \times \frac{Z_p}{Z_e} = 1.48 \frac{Z_p}{Z_e}$$

The ratio Z_p/Z_e for any cross-section is known as the *shape factor*; for joists, universal sections, and built-up I sections of similar proportions, the shape factor has a value very close to 1.15. Thus the collapse load factor λ for a simply supported beam of I section is 1.48 x 1.15 = 1.71 or say 1.75.

B.S. 449 permits an increase in working stresses of 25 per cent if such increase is solely due to the effect of wind. It is easy to show that this implies a reduction in collapse load factor to 1.4. Such a reduction in load factor is, of course, a measure of the statistical unlikelihood of full wind and superload acting at the same time. Thus the designer will use a load factor of 1.75 on dead plus superimposed loading, and carry out a second analysis at a load factor of 1.4 on dead plus superimposed plus wind loads. Naturally, the more critical of the two cases will give the actual design.

The load factor 1.75 (reduced to 1.4 for the wind case) will be used in all examples given here. It must be emphasised that, as far as can be judged by experience with structures of elastic design in steel to B.S. 4360 grade 43, the

value of 1.75 is safe, and there might be a case for reducing the value for certain classes of building. For example, for multi-storey office blocks, braced against wind, where rigorous load patterns are prescribed, (8) the load factor might well be reduced to 1.5.

Redundant Beams

As a very simple example of a redundant structure, which however will illustrate the ease and power of plastic methods, consider the propped cantilever carrying a central load W, shown in Fig. 3 (a). This beam has one redundancy, denoted by the bending moment M, and the complete bending moment diagram is sketched in Fig. 3 (a). This bending moment diagram consists essentially of two parts: a diagram

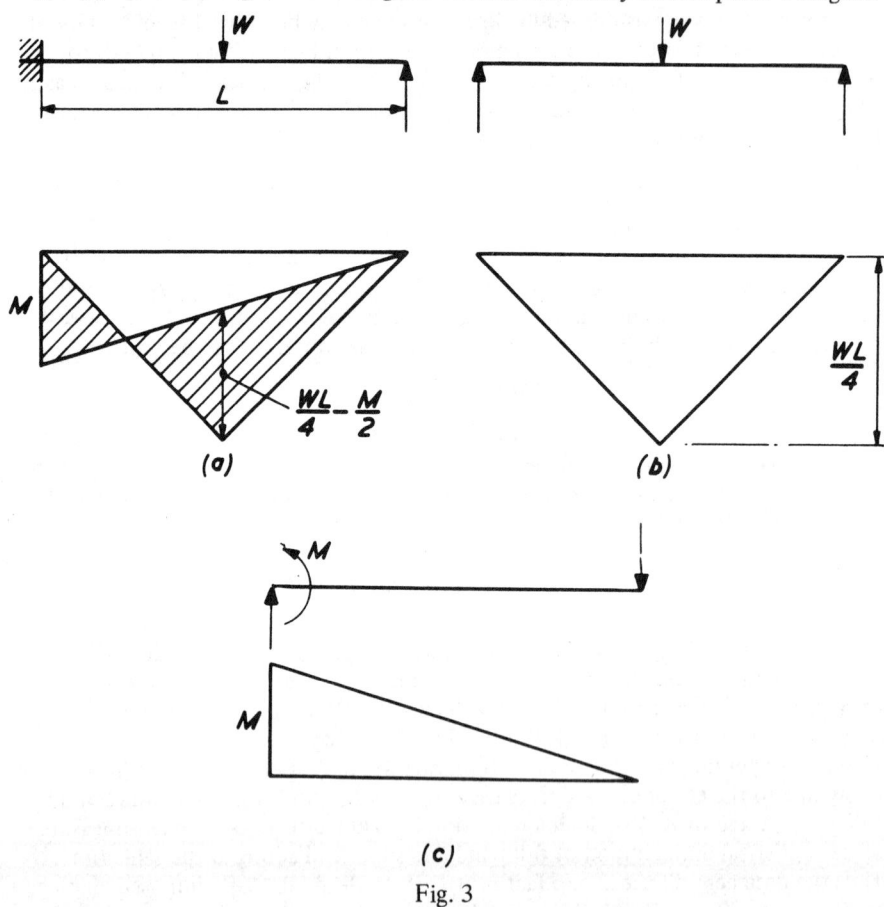

Fig. 3

due to the external loading acting on a statically determinate beam, as shown in Fig. 3 (b), and a diagram due to the redundant bending moment M, shown in Fig. 3 (c). Fig. 3 (b) shows the *free bending moments*, and Fig. 3 (c) the *reactant moments*. It will be seen that the diagram of Fig. 3 (a) results by superimposing Figs. 3 (b) and (c).

In general, any structural system may be broken down in this way into free and reactant components. The general form of the bending moment diagrams is independent of material properties and can be determined without stating whether the problem is elastic or plastic. The whole problem of structural analysis lies in the determination of the values of the redundancies; for the propped cantilever, the value M must be found.

In order to determine values of redundancies, the moment-curvature curve must be used, and it is instructive to follow the arguments leading to an *elastic* solution to the problem. In Fig. 3 (a), the elastic value M will be found by using the conditions that the beam has zero deflections at both ends and zero slope at one end. Only one value of M can be found which will lead to elastic curvatures of the beam satisfying these boundary conditions. The use of slope-deflection equations or other techniques of analysis may render the elastic problem relatively easy, at least if the number of redundancies is few. However, the final solution obtained may have little relation to reality, since the boundary conditions are implicit, and these are difficult to satisfy in practice. For example, the supports of the propped cantilever may sink slightly, and the slope at the 'fixed' end will almost certainly not be truly zero; these anomalies can have a great effect on the elastic distribution of bending moments.

By contrast, the plastic solution to the same problem is both easier to obtain and bears a close relation to reality. The 'compatibility' boundary conditions are not used in plastic theory; supports may sink, and an end of a beam may be imperfectly fixed. Nevertheless, the collapse load of an actual beam will be predicted with great accuracy by plastic methods, since the collapse load is independent of such practical imperfections as sinking of supports, flexibility of connections, and so on.

Returning to Fig. 3 (a), it was seen that this bending moment diagram is general, being valid both for the elastic and for the plastic problem. The key to the determination of the value of M for the plastic problem is the fact that, at collapse, the beam must be capable of deforming as a mechanism. Thus plastic hinges must be formed within the length of the beam.

Moving directly to the correct solution, consider the bending moment diagram of Fig. 4 (a), where the full plastic moment M_p is reached at the two cross-sections shown. Figure 4 (b) shows the beam in its collapse state, with two plastic hinges, and the end support acting as a third hinge; three hinges in a straight line permit the formation of an elementary mechanism. Thus the mechanism condition is satisfied by the bending moment distribution of Fig. 4 (a). Further, it is clear by inspection that the values of the bending moments reach, but do not exceed anywhere in the beam, the value M_p.

Calculation is very simple. Comparing Figs. 3 (a) and 4 (a), it will be seen that

$$M = M_p$$
$$\frac{WL}{4} - \tfrac{1}{2}M = M_p.$$

Hence $M_p = M = WL/6$. Thus, if W represents the *working* value of the load, and a load factor of 1.75 is required, a beam section must be provided whose full plastic moment is at least $(1.75)WL/6$.

It will have been noted that, in arriving at this design of the beam, calculations have been based directly on the bending moment diagram. No reference has been

made to slopes or deflections, and the supports, for example, may settle slightly without affecting the value of M_p.

These observations on the propped cantilever can be stated more formally as three master requirements of plastic design. First, equilibrium must be satisfied; that is, proper bending moment diagrams must be constructed which are in equilibrium with the applied loads. Secondly, a structure at collapse must be capable

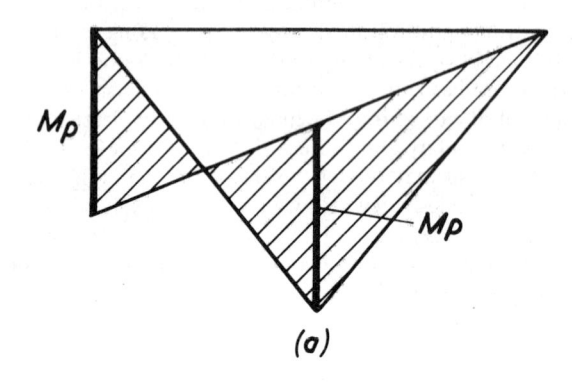

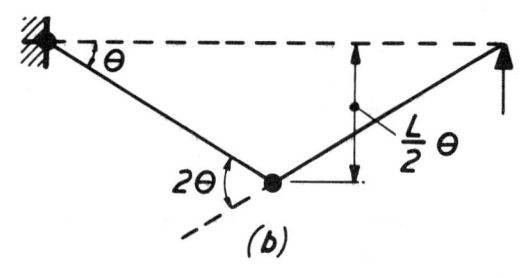

Fig. 4

of deforming as a mechanism, due to the formation of plastic hinges. Thirdly, the yield condition must be satisfied; the bending moments must not exceed the full plastic values. These three conditions form the whole basis of plastic design; a structure at collapse is required to satisfy the conditions of

1. Equilibrium.
2. Mechanism.
3. Yield.

It may be shown that these conditions are all that are required. If a design satisfies the conditions, then the designer can be certain that he has the correct solution to his problem (providing, of course, the fundamental assumptions of simple plastic theory are obeyed; that is, there are no instability or other extraneous effects, referred to above).

Returning to the example of the propped cantilever, the collapse equation can be found even more easily if the work equation is used to replace calculations based on the bending moment diagram. In the plastic collapse mechanism of Fig. 4 (b), it will be seen that if a rotation θ occurs at the plastic hinge at the fixed end, the

rotation at the sagging hinge under the load must be 2θ. The corresponding deflection of the load is $(L/2)\theta$. (The collapse mechanism has been drawn with the beam straight between hinge points. Although the beam will in fact have elastic curvatures, these do not affect the calculations.)

The work dissipated at a plastic hinge is equal to the product of the full plastic moment and the hinge rotation. Thus the total work dissipated in the mechanism of Fig. 4 (b) is $M_p(\theta) + M_p(2\theta) = 3M_p\theta$. The work done by the load acting on the plastic collapse mechanism is $W([L/2]\theta)$, and these two work quantities must be equal, i.e. $WL/2 = 3M_p$, which is the collapse equation previously derived.

It may be shown that the writing of the work equation is exactly equivalent to writing an equilibrium equation; thus, if the work equation is correctly formulated, the equilibrium condition will be satisfied automatically, and the mechanism condition is, of course, also satisfied. However, if the collapse equation ($WL = 6M_p$) is derived from work considerations, the bending moment diagram must be constructed to check that the yield condition is not violated. An example of such a check is given in the next example.

It will have been noted that the rotation θ in Fig. 4 (b) disappeared from the calculations when the work equation was written. The mechanism has, in fact, one degree of freedom, specified in terms of the arbitrary θ, and this parameter occurred as a common multiplier in all the work terms. It will have been noted also that the bending moment diagram of Fig. 4 (a) could be determined directly from the collapse condition, the original statical indeterminacy of the propped cantilever also disappearing. This feature of plastic design recurs in all structures at the point of collapse; that part of the structure concerned in the collapse mechanism becomes statically determinate, and this greatly simplifies the working compared with elastic analysis.

A rule can be derived concerning the number of hinges required to turn a structure into a mechanism of one degree of freedom. The simply supported beam of Fig. 2, which is, of course, statically determinate, required the formation of a single hinge to make it into a mechanism. For a redundant structure, the insertion of one hinge, whether real or plastic, will reduce the degree of redundancy by one. Thus if a structure has originally a number N redundancies, the insertion of N hinges, properly placed, will make that structure statically determinate. One further hinge will turn the statically determinate structure into a collapse mechanism of one degree of freedom.

It is to be expected, then, that a structure with N redundancies will require the formation of $(N + 1)$ plastic hinges at collapse. Thus the simply supported beam needs one hinge; the propped cantilever, having one redundancy, will need two hinges (see Fig. 4 (b)); and a beam having both ends fixed will need three hinges, corresponding to the two redundancies. The rule of $(N + 1)$ hinges is always obeyed for *regular* collapse mechanisms involving the whole of a structure. However, in complex frames, an incomplete or partial mechanism may be formed involving only a portion of the frame. In this case, that portion of the frame involved in the collapse will become statically determinate, and the collapse load can be calculated; the rest of the frame, however, will remain statically indeterminate, and the bending moment distribution cannot be found by plastic methods alone. Nevertheless, simple theorems are available, and are referred to below, whose use enables the designer to be *certain* of his collapse load calculations, even though the bending moments in part of a structure have not been determined.

PLASTIC THEORY AND DESIGN

In thinking of an actual structure loaded slowly up to the point of collapse, it will be clear that certain plastic hinges will form before others; if necessary, a complete elastic-plastic history could be worked out. For the purpose of calculating a plastic collapse load, however, such a history is quite unnecessary. The designer can concentrate directly on the collapse state and satisfy the three master conditions of mechanism, equilibrium, and yield. He then has absolute certainty that the structure cannot possibly collapse at loads lower than those for which the structure is designed.

Upper and Lower Bounds

Consider a propped cantilever under the action of a uniformly distributed load W, Fig. 5 (a). The general bending moment diagram is again composed of free and reactant components, Fig. 5 (b); note that the reactant line due to the unknown bending moment M is exactly the same as before, Fig. 3 (c), being purely a function of the structure and not of the loading. Fig. 5 (c) evidently gives the collapse condition of the propped cantilever; a hogging hinge forms at the fixed end, and a sagging hinge at some internal cross-section, leading to the collapse mechanism of Fig. 5 (d). By drawing, or by analysis, the value of M_p in Fig. 5 (c) is determined as

$$M_p = 0.686 \left(\frac{WL}{8}\right)$$

The difficulty in this simple problem lies in locating the sagging hinge, and the opportunity may be taken to introduce upper and lower bound theorems which may be applied to approximate solutions. Suppose, for example, that a first guess for the collapse mechanism is the configuration shown in Fig. 4 (b), where the sagging hinge is located at midspan. If the work equation is written for this mechanism, then, since the uniformly distributed load descends an average distance $\frac{L}{4}\theta$,

$$W\left(\frac{L}{4}\theta\right) = 3M_p\theta$$

or

$$M_p = \frac{2}{3}\left(\frac{WL}{8}\right)$$

Of the three master conditions, only those of equilibrium and mechanism have so far been satisfied; a plastic collapse mechanism has been guessed, and the writing of the work equation ensures that equilibrium has been satisfied. It may be shown that such a solution, not necessarily satisfying the yield condition, is unsafe. That is, the actual value of M_p must be larger, or at best equal to, the value derived from a guessed mechanism, i.e.

$$M_p \geqslant \frac{2}{3}\left(\frac{WL}{8}\right)$$

The bending moment diagram corresponding to the assumed mechanism may now be examined; it is sketched in Fig. 6. It will be seen that the derived value of

UPPER AND LOWER BOUNDS

(figures a, b, c, d)

Fig. 5

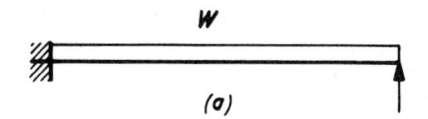

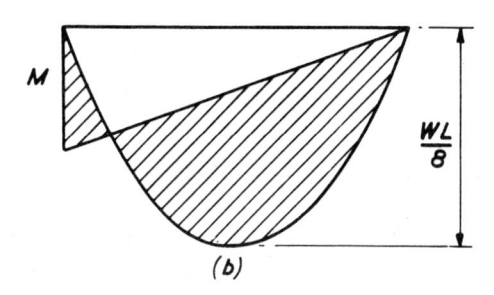

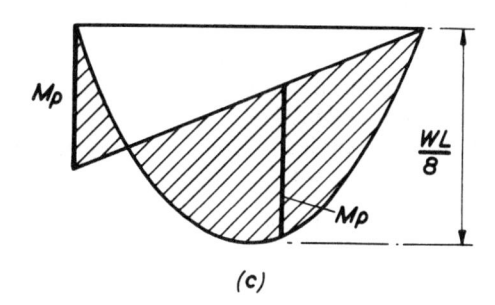

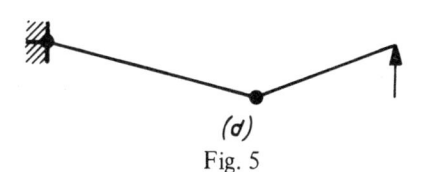

$M_p = \frac{2}{3}(WL/8)$ is exceeded in the ratio 25/24 (i.e. 4 per cent) at the most critical section, so that the yield condition is not satisfied.

Consider, however, a new design of the beam with

$$M_p = \frac{25}{24} \cdot \frac{2}{3} \cdot \frac{WL}{8} = 0.694 \left(\frac{WL}{8}\right),$$

in conjunction with the bending moment diagram of Fig. 6. This set of bending moments satisfies equilibrium; it also satisfies the yield condition since $M_p = 0.694(WL/8)$ is not exceeded. It may be shown that any solution satisfying equilibrium and yield, but not necessarily the mechanism condition, is safe; that is,

$$M_p \leq 0.694 \left(\frac{WL}{8}\right).$$

For this particular problem, therefore, the design value of M_p has been bracketed between fairly close limits (4 per cent apart), and this is certainly good enough for design purposes. In general, close limits cannot be found easily for more complicated structures, but the unsafe solution forms the basis of a valuable technique for

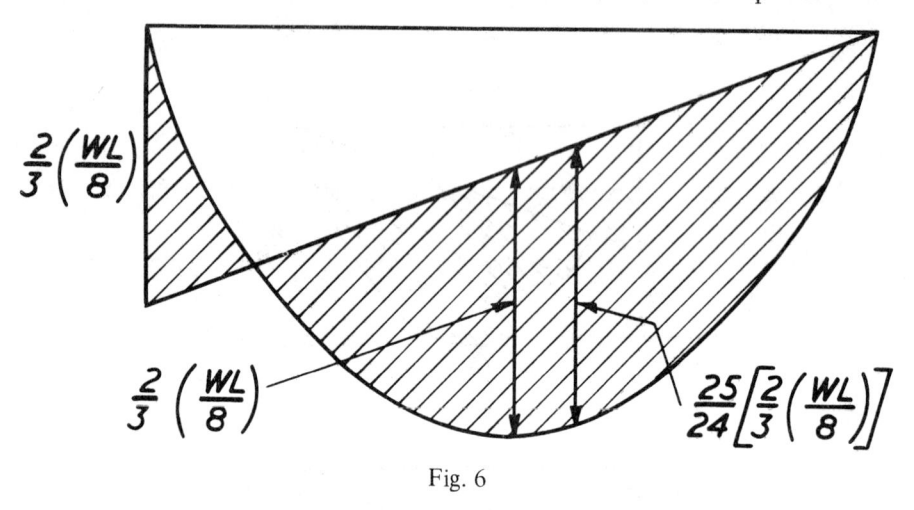

$\frac{2}{3}\left(\frac{WL}{8}\right)$

$\frac{2}{3}\left(\frac{WL}{8}\right)$ $\quad\quad\quad\quad\quad\quad\quad\quad\quad\quad\quad\quad\quad\quad\frac{25}{24}\left[\frac{2}{3}\left(\frac{WL}{8}\right)\right]$

Fig. 6

deriving a quick answer to a problem. This will be commented on when the method of combination of mechanisms is described.

For the moment it may be recorded that bounds can be determined by satisfying the three conditions in pairs, as shown:

$$\text{Exact} \begin{cases} \text{Mechanism} \\ \text{Equilibrium} \\ \text{Yield} \end{cases} \begin{matrix} \text{Unsafe} \\ \\ \text{Safe} \end{matrix}$$

Continuous Beams

The ideas so far presented will now be applied to a more complicated design problem, that of the continuous beam. A continuous purlin, for example, is a beam resting on a number of supports; the supports are provided by the rafters of the main frames in the building. The fact that the rafters are flexible and will deflect slightly is irrelevant from the point of view of plastic design.

Consider first a building of four main frames, spaced uniformly at a distance l apart; then the beam to be designed will be as shown in Fig. 7 (a). A free bending moment diagram may be drawn as in Fig. 7 (b), consisting simply of a set of identical parabolas of height $wl^2/8$; this diagram has been drawn by considering each span as if it were simply supported. Actually, a bending moment will exist at each internal support, and the reactant line consists therefore of a continuous line which is straight for each span of the beam.

If a uniform section purlin is used, then it is reasonable to suppose that the end spans, behaving like propped cantilevers, will be weaker than the internal span which will behave plastically like a fixed-ended beam. To test this assumption, a collapse mechanism will be tried for the end spans, and then it will be seen whether the central span is safe.

CONTINUOUS BEAMS

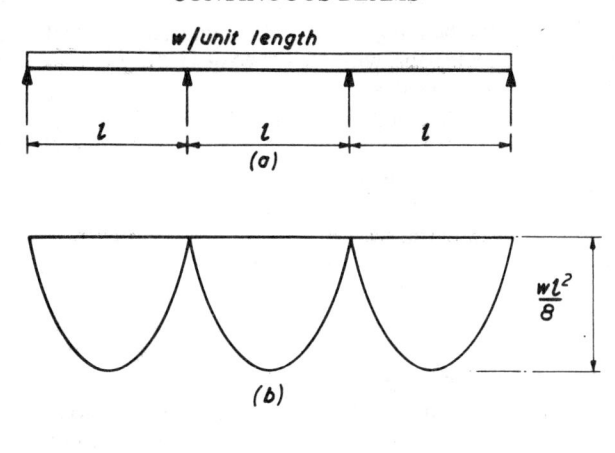

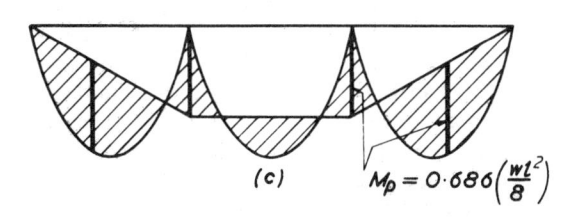

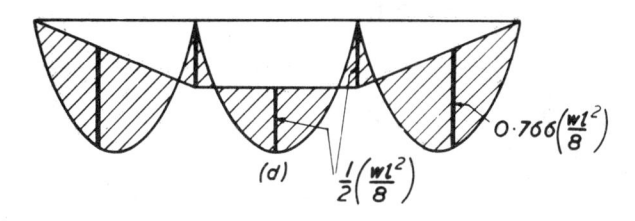

Fig. 7

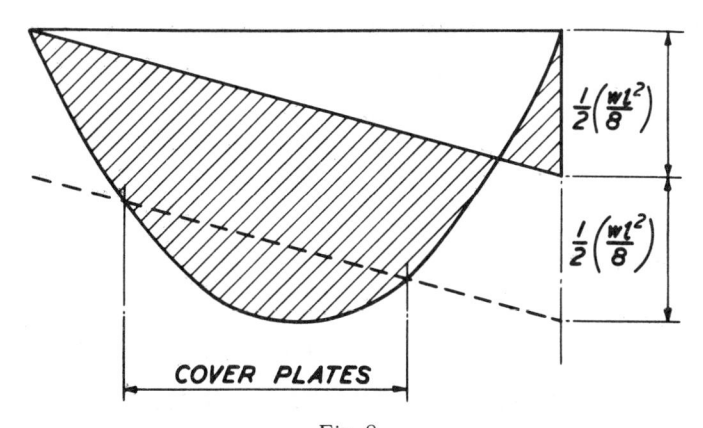

Fig. 8

Figure 7 (c) shows the bending moment diagram with the reactant line adjusted to give collapse in the end spans, with $M_p = 0.686(wl^2/8)$. It is seen immediately that the bending moment at the centre of the internal span is less than M_p; in fact, it has value $0.314(wl^2/8)$. Thus the bending moment diagram of Fig. 7 (c) satisfies the equilibrium condition, since it is a proper combination of free and reactant diagrams. It satisfies the yield condition, since the moment $M_p = 0.686(wl^2/8)$ is nowhere exceeded. And the mechanism condition is satisfied by the end spans. Thus the correct solution has been obtained.

As a practical design, the uniform section purlin may be wasteful of material, although it may be less costly than to vary the section along the length of the purlin. However, suppose the internal span were designed to have as small a section as possible, so that it collapses as a fixed-ended beam with $M_p = \frac{1}{2}(wl^2/8)$. The bending moment diagram for the whole purlin is shown in Fig. 8 (d). By drawing or calculation it is found that the end spans are called upon to carry maximum sagging moments of $0.766(wl^2/8)$. The end spans must therefore be strengthened, either by providing a uniform section having this value of full plastic moment, or by using the section for the internal span and strengthening it locally with cover plates. The extent of such cover plates may be determined quickly by drawing the bending moment diagram for an end span, Fig. 8. It will be seen from the next section on 'Full Plastic Moments' that the size of the cover plates can be calculated easily.

A purlin of more than three spans can be designed in the same way, but the problem of incomplete collapse arises. Consider, for example, a five-span purlin of uniform section, collapsing as before in the end spans, Fig. 9. The reactant line is fixed for the end spans by the collapse condition, but the collapse analysis will not determine the bending moments for the internal spans. However, in Fig. 9 is shown

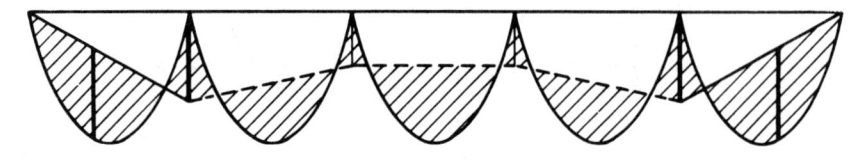

Fig. 9

dotted a reactant line for the internal spans which is drawn in the proper way; that is, the reactant line is continuous, and straight between supports. It has been sketched by eye to satisfy clearly the yield criterion; by inspection, the bending moments in the internal spans shown in Fig. 9 do not exceed the value $M_p = 0.686(wl^2/8)$.

The bending moment diagram in Fig. 9, therefore, satisfies the equilibrium condition; it is a proper combination of free and reactant moments. Further, a collapse mechanism is formed in each end span. Further still, the yield condition is satisfied by the bending moments in the diagram. Since the three master conditions are satisfied, the correct solution has been obtained.

Other variations in design of this continuous beam can be investigated readily; for example, the spacing of the gable frames could be reduced so that the propped cantilever ends of the continuous beam are of shorter span. It may be shown that if the end spans are approximately 85 per cent of the internal spans, a uniform section purlin can be used which will just be on the point of collapse both in the end and in the internal spans.

Full Plastic Moments (2)

The basic theory has been presented, and more complex structures, such as the portal frame, can now be discussed. Before moving on, however, detail design of a cross-section will be considered, and the effects of axial load and shear force determined.

The ideal elastic/plastic behaviour in bending of Fig. 1 (b) stems from the ideal elastic/plastic stress–strain curve of steel in tension or compression shown in Fig. 10. From O to A the material is elastic; when the yield stress σ_0 is reached, indefinite extension of the material can take place. (The actual stress–strain curve will show some strain hardening, and the ideal curve of Fig. 10 is 'safe'.) If a rectangular cross-section is subjected to pure bending, then the normal elastic distribution of stress at yield of the section will be as shown in Fig. 11 (b), the corresponding bending moment being $\frac{1}{6}bd^2\sigma_0$. The elastic modulus of a rectangular section is $\frac{1}{6}bd^2$.

If the bending moment is increased above this value yield will occur in the outer fibres, where, however, the stress will remain constant at σ_0. The stress distribution

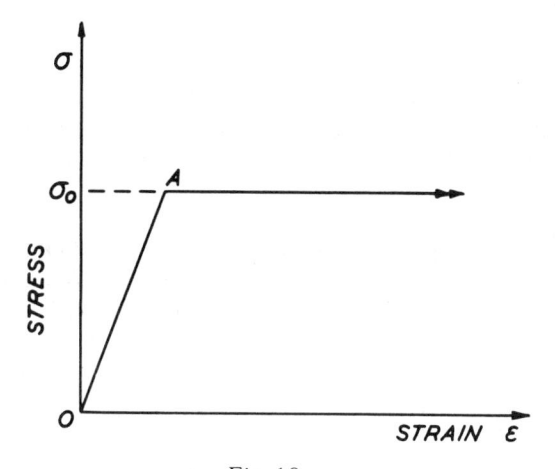

Fig. 10

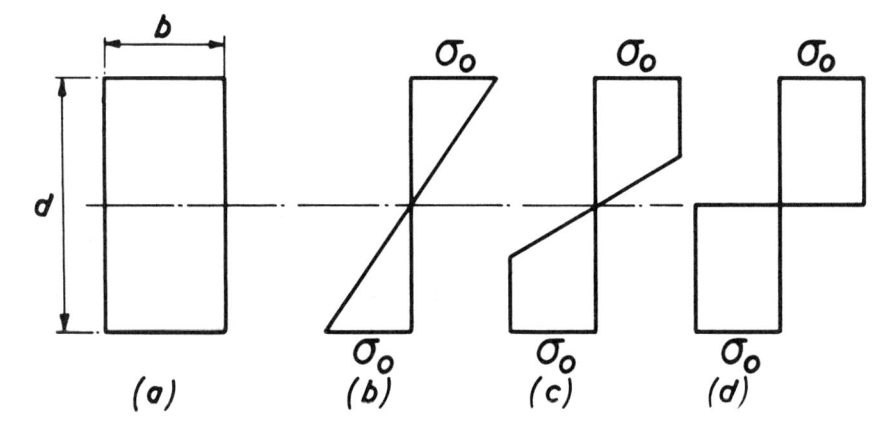

Fig. 11

will be modified to that shown in Fig. 11 (c), and, as the bending moment is further increased, the limit will be reached when the whole cross-section is plastic, as shown in Fig. 11 (d). Figure 12 reproduces this full plastic distribution, and it will be seen that $M_p = \frac{1}{4}bd^2\sigma_0$. The plastic modulus, $\frac{1}{4}bd^2$, is the first moment of area of the cross-section about the zero stress axis.

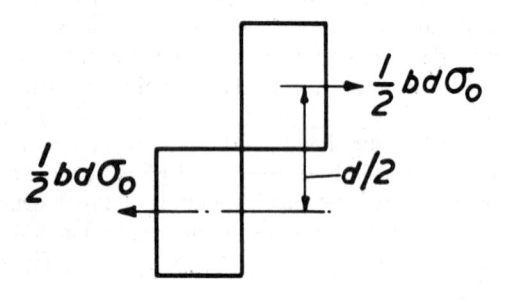

Fig. 12

Plastic moduli may be worked out in this way for cross-sections of any shape. For example, the plate girder of Fig. 13 will have a plastic modulus equal to

$$\tfrac{1}{4}bd^2 + BT(d + T)$$

where the first term corresponds to the web and the second to the flanges.

Care must be taken when dealing with bending about an axis which is not an axis of symmetry. For example, the T section of Fig. 14 (a) will have the fully

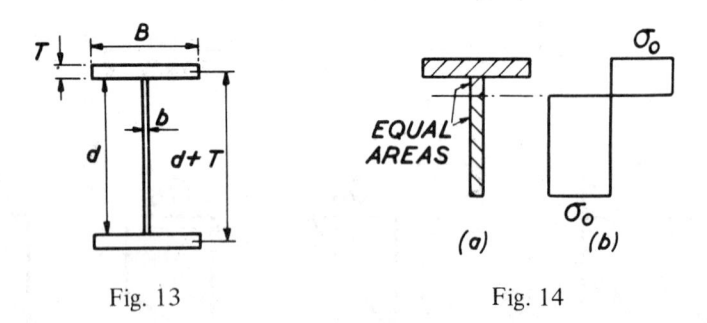

Fig. 13   Fig. 14

plastic stress distribution of Fig. 14 (b), where the zero stress axis does not coincide with the usual neutral axis for elastic bending. Instead, the zero stress axis is fixed from consideration of the fact that, in the absence of axial load, there must be no net axial force on the cross-section. Since the stresses in tension and compression are uniform and numerically equal, the zero stress axis must divide the cross-section into two equal areas. Having fixed the zero stress axis in this way, it is a simple matter to calculate the first moments of area about that axis in order to determine the section modulus.

Axial Load (2)

If a cross-section is acted upon by an axial load as well as a bending moment, then the zero stress axis will shift so that the net force across the section is equal to the axial load. Consider, for example, the rectangular cross-section of Fig. 15 (a) subjected to a bending moment M and an axial load P (say compressive). If M and P have such magnitudes that the cross-section is fully plastic, then the stress distribution will be as shown in Fig. 15 (b). It will be seen that this stress distribution can be divided into three 'blocks', the central block corresponding to the axial load P and the two outside blocks to the bending moment M. The following equations may be written:

$$P = \alpha b d \sigma_0 = \alpha P_p \text{ say}$$

$$M = (1 - \alpha^2) \frac{bd^2}{4} \sigma_0 = (1 - \alpha^2) M_p$$

Eliminating α,

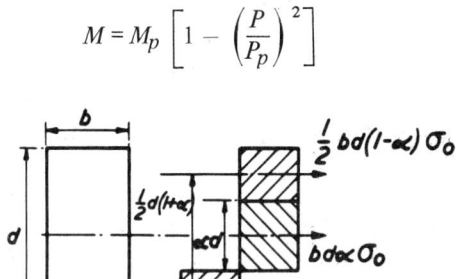

$$M = M_p \left[1 - \left(\frac{P}{P_p} \right)^2 \right]$$

Fig. 15

Thus the full plastic moment is reduced by the presence of the axial load.

A similar analysis may be carried out for Universal Beam and Column sections; indeed, the analysis is virtually unchanged for small axial loads, since the zero stress axis will remain within the rectangular cross-section web. The section tables give the reductions to be made in the plastic section modulus due to the presence of axial load; separate formulae are given for each section in the tables for the cases of low and high axial load. In the case of high axial load, the zero stress axis moves out of the web into one of the flanges, and a new analysis leading to a slightly different formula has to be made.

For low axial loads, the reduction in full plastic moment is small. Figure 16 shows the plastic modulus of a 203 × 203 × 71 kg U.C. plotted against mean axial stress (it is assumed that the yield stress of the material is 245 N/mm^2); an axial stress of 15 N/mm^2 produces a drop in full plastic moment of only about 1 per cent.

It may be mentioned here that web buckling may occur if the ratio d/b (Fig. 13) is high, and precautions must be taken when designing built-up sections. In addition, certain of the Universal Beam sections are liable to web instability, and some are not suitable for plastic design on account of *flange* instability (see below), these

sections are noted in Tables 1 and 2 at the end of this chapter. For a section in steel to B.S. 4360 grade 43, no web buckling problems arise if the web depth to thickness ratio d/b is less than 53. If the ratio exceeds 53, premature web buckling occurs if the mean axial stress computed on the whole area of cross-section, is greater than $\frac{15b^2}{A}\left(2\,125 - 25\frac{d}{b}\right)$ N/mm², where A is the total area of cross-section in mm². The web depth to thickness ratio should, in any case, not exceed 85.

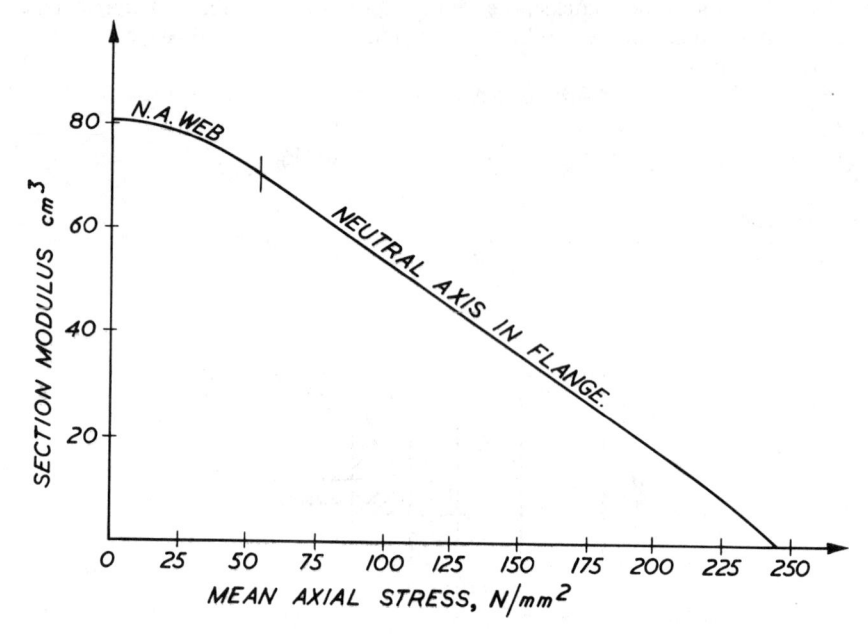

Fig. 16

For a section in steel to B.S. 4360 grade 50, if d/b exceeds 44 premature buckling will occur at a mean axial stress greater than $\frac{15b^2}{A}\left(2\,600 - 37\frac{d}{b}\right)$; d/b should not exceed 70.

Shear Force (2, 7)

Shear force acting on a cross-section will also reduce the effective full plastic moment at that section, and allowance may have to be made in some designs. For an I section, Fig. 13, it was seen that the full plastic moment could be expressed in the form

$$M_p = \tfrac{1}{4}bd^2\sigma_0 + BT(d+T)\sigma_0$$
$$= M_w + M_f$$

where M_w and M_f are the contributions from the web and flanges respectively. Theoretical and experimental studies of the problem have led to the following

design rules. The shear force at a cross-section is assumed to be resisted by a uniform shear stress τ acting on the web alone; i.e. if the shear force is F, then $F = bd\tau$. Then the full plastic moment of an I section may be written

$$M_p = M_w \sqrt{1 - 3\left(\frac{\tau}{\sigma_0}\right)^2} + M_f$$

The same limiting depth to thickness ratio must be observed in order that premature buckling does not occur; for steel to B.S. 4360 grade 43, d/b should not exceed 85, the corresponding ratio being 70 for steel to B.S. 4360 grade 50.

Flange Stability

Flange breadth to thickness ratios B/T, Fig. 15, must also be limited for stability reasons; the limits are 18 and 15 for steel to B.S. 4360 grade 43 and 50 respectively.

Design Example 1. The two-span continuous girder shown in Fig. 17 will be designed plastically. The loads shown are working values, and the design load factor is 1.75. Due to symmetry, one-half only of the girder need be considered (Fig. 18 (a)). Figure 18 (b) indicates the expected collapse bending moment diagram.

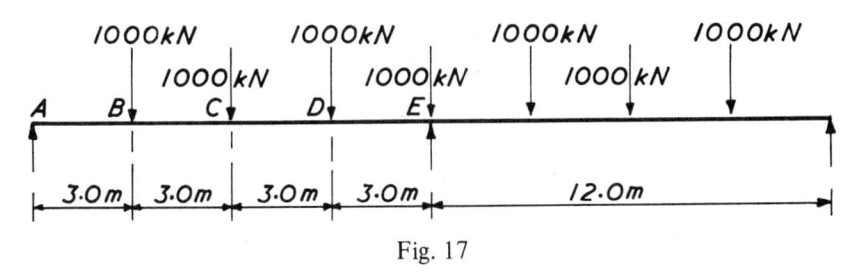

Fig. 17

There is great flexibility in the design of this girder, since by variation of the flange areas, and possibly of the web, the hogging moment M at E can take a large number of values. For the purposes of this example, it will be assumed that the girder has a uniform cross-section throughout its length.

Ignoring for the present the fact that the full plastic moment at E will be reduced by the effect of shear force, a uniform section design, from Fig. 18 (b), would be given by $M = \frac{2}{3}(10\,500) = 7\,000$ kNm.

Assume as a first trial that the actual value of M at E is 6 000 kNm, then the design bending moments for the girder will be as shown in Fig. 19 (a), and the shear force diagram at collapse will be as shown in Fig. 19 (b). The greatest shear force is 3 125 kN; it will be seen from the formula for reduction in full plastic moment that the shear force on the web cannot exceed $\sigma_0/\sqrt{3}$. Using steel to B.S. 4360 grade 43, with a yield stress of 245 N/mm², $\frac{16}{\sqrt{3}} = 141.5$, so that a minimum web area of

$$\frac{3\,125 \times 10^3}{141.5} = 22.1 \times 10^3 \text{ mm}^2 = 221 \text{ cm}^2$$

is required. Remembering that d/b should not exceed 85, a 1 500 mm x 20 mm web will be tried, having area 300 cm² and

$$M_w = \frac{1}{4} \times 20 \times (1\,500)^2 \times 245 = 2\,756 \times 10^6 \text{ Nmm} = 2\,756 \text{ kNm}.$$

PLASTIC THEORY AND DESIGN

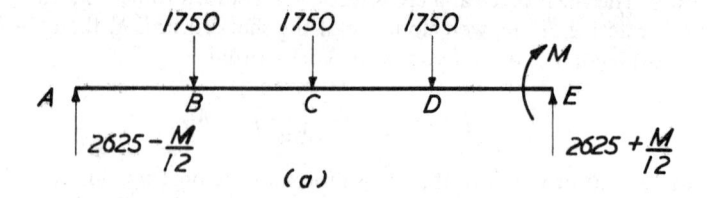

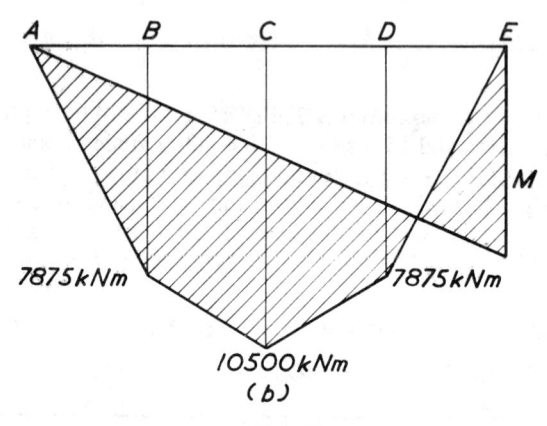

Fig. 18

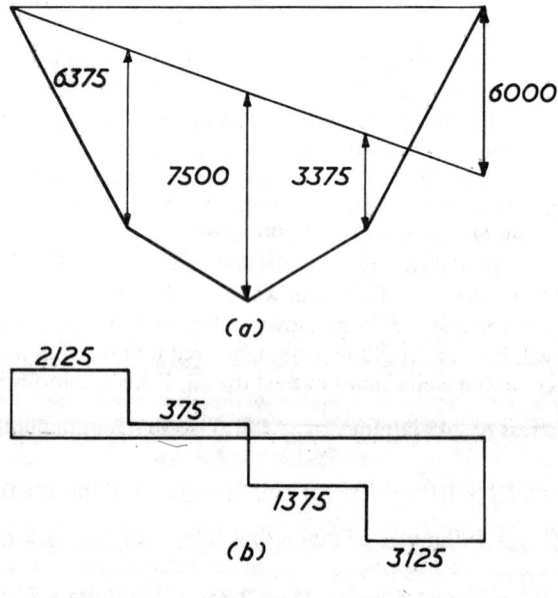

Fig. 19

FLANGE STABILITY

Now, at E, the mean shear stress on the web is $\dfrac{3\,125 \times 10^3}{300 \times 10^2} = 104.3 \text{ N/mm}^2$

hence, at E

$$M_w = 2\,756\sqrt{1 - 3\left(\dfrac{104.3}{245}\right)^2} = 1\,860 \text{ kNm}.$$

Thus the flanges at E are required to contribute $6\,000 - 1\,860 = 4\,140$ kNm to the full plastic moment. At C, where the shear force is small, the flanges are required to contribute approximately $7\,500 - 2\,756 = 4\,744$ kNm. Evidently the first trial of $6\,000$ kNm as the bending moment at E was slightly in error, and an almost exact estimate may now be made. Assuming the same web, suppose that M_w at E is $1\,800$ kNm, and that M_w at C is $2\,700$ kNm. Then, from Fig. 18 (b),

for E: $\qquad M = 1\,800 + M_f$

and for C: $\qquad \tfrac{1}{2}M + 2\,700 + M_f = 10\,500$

From these equations, $M_f = 4\,600$ kNm, $M = 6\,400$ kNm, and a revised design bending moment diagram may be constructed as shown in Fig. 20 (a), the corresponding shear force diagram being that of Fig. 20 (b).

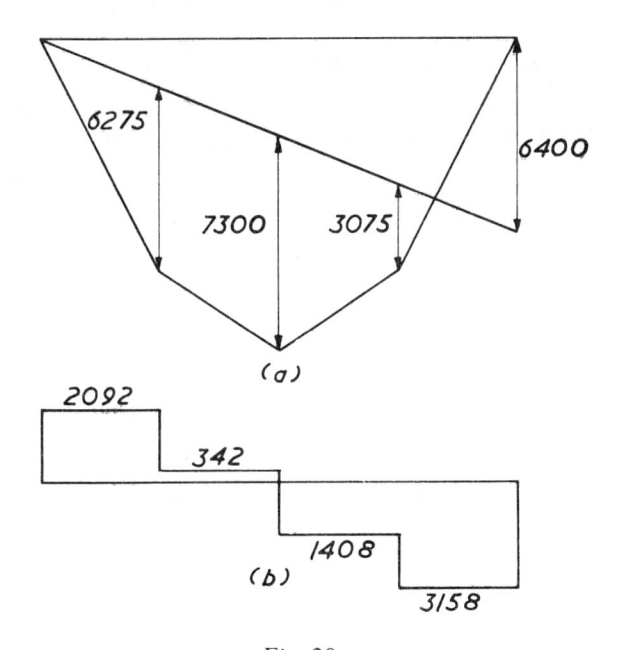

Fig. 20

For the cross-section at E

$$M_w = 2\,756\sqrt{1 - 3\left(\dfrac{105.3}{245}\right)^2} = 1\,830 \text{ kNm}$$

required $\qquad M_f = 6\,400 - 1\,830 = 4\,570$ kNm.

For the cross-section at C

$$M_w = 2\,756\sqrt{1 - 3\left(\frac{47.0}{245}\right)^2} = 2\,630 \text{ kNm}$$

required $\quad M_f = 7\,300 - 2\,630 = 4\,670$ kNm.

Assuming flanges each of area A and of thickness 30 mm, with $\sigma_0 = 240$ N/mm², then

$$240 \times A \times 1\,530 = 4\,670 \times 10^6$$

i.e. $A = \dfrac{4\,670 \times 10^6}{36.7 \times 10^4}$ mm² $= 130 \times 10^2$ mm² say 440 × 30 mm. Thus the uniform girder will have a 1 500 mm × 20 mm web and 440 mm × 30 mm flanges; the total area is 564 cm².

The calculations may be repeated for a girder made of steel to B.S. 4 360 grade 50, with a nominal $\sigma_0 = 345$ N/mm². For the web, the maximum shear stress is $\dfrac{345}{\sqrt{3}} = 200$ N/mm². Using Fig. 20 for the first trial, the minimum web area is $\dfrac{3\,158 \times 10^3}{200} = 15.8 \times 10^3$ mm² $= 158$ cm². Hence, remembering that d/b is now limited to 70, a 1 200 mm × 20 mm web will be tried, having

$$M_w = \frac{1}{4} \times 20 \times 1\,200^2 \times 345 = 2\,484 \times 10^6 \text{ Nmm} = 2\,484 \text{ kNm}.$$

At E, $M_w = 2\,484\sqrt{1 - 3\left(\dfrac{132}{345}\right)^2} = 1\,860$ kNm; $M_f = 6\,400 - 1\,860 = 4\,540$ kNm

At C, $M_w = 2\,484\sqrt{1 - 3\left(\dfrac{59}{345}\right)^2} = 2\,370$ kNm; $M_f = 7\,300 - 2\,370 = 4\,930$ kNm

Fig. 21

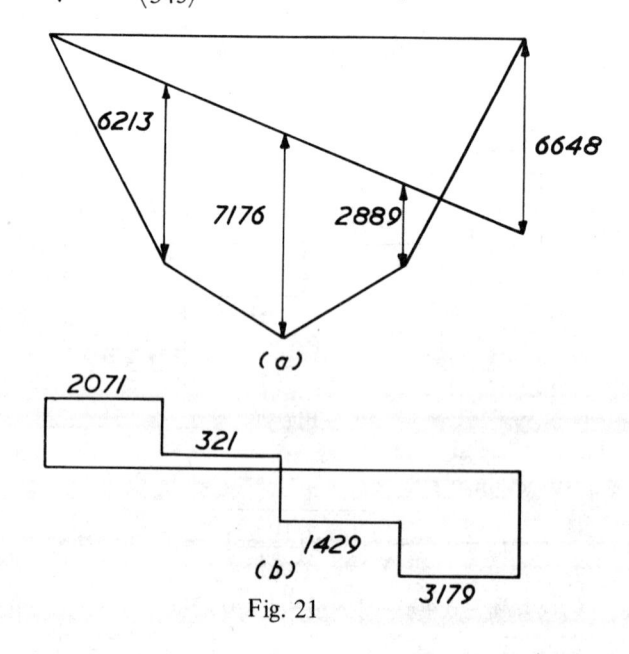

Adjusting this trial solution as before, the more accurate design conditions of Fig. 21 are arrived at.

At E, $M_w = 2\,484\sqrt{1 - 3\left(\dfrac{132.5}{345}\right)^2} = 1\,855$ kNm; $M_f = 6\,648 - 1\,855 = 4\,793$ kNm

At C, $M_w = 2\,484\sqrt{1 - 3\left(\dfrac{59.5}{345}\right)^2} = 2\,371$ kNm; $M_f = 7\,176 - 2\,371 = 4\,805$ kNm

Assuming 28 mm flanges, $345 \times A \times 1\,228 = 4\,805 \times 10^6$, i.e. $A = 114 \times 10^2\,\text{mm}^2$, say 420 mm x 28 mm. Thus the uniform girder in steel to B.S. 4360 grade 50, will have a 1 200 mm x 20 mm web and 420 mm x 28 mm flanges; the total cross-sectional area is 475 cm², a saving of about 20 per cent over the grade 43 girder.

Frame Analysis (2, 3, 4)

The beam is an essentially stable structural element, although precautions may have to be taken to guard against lateral instability. As such, the stable beam satisfies the basic assumptions of simple plastic theory, and the plastic theorems are directly applicable. A frame, however, consists of beams attached to columns, and the column is a potentially unstable structural element. As mentioned in the introduction, unstable column behaviour will not be permitted in a plastic design. It will be assumed that all columns remain stable, and checks will be made finally to justify this assumption for any particular design.

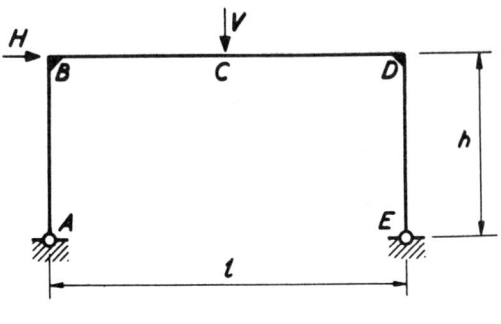

Fig. 22

The simple rectangular portal frame of Fig. 22 has pinned feet, and thus has one redundancy. The vertical load V and horizontal load H are highly idealised representations of dead plus superimposed load and of wind load; the basic properties of the frame may be demonstrated, however, under this ideal loading. The same techniques that were used for beams may be used for the plastic design of frames. In Fig. 23 the frame has been made statically determinate by freeing one column foot, and Fig. 23 (a) shows the condition which will produce free bending moments. The single redundancy S in Fig. 23 (b) will lead to the reactant diagram, and the free and reactant diagrams are sketched separately in Fig. 24.

554   PLASTIC THEORY AND DESIGN

Fig. 23

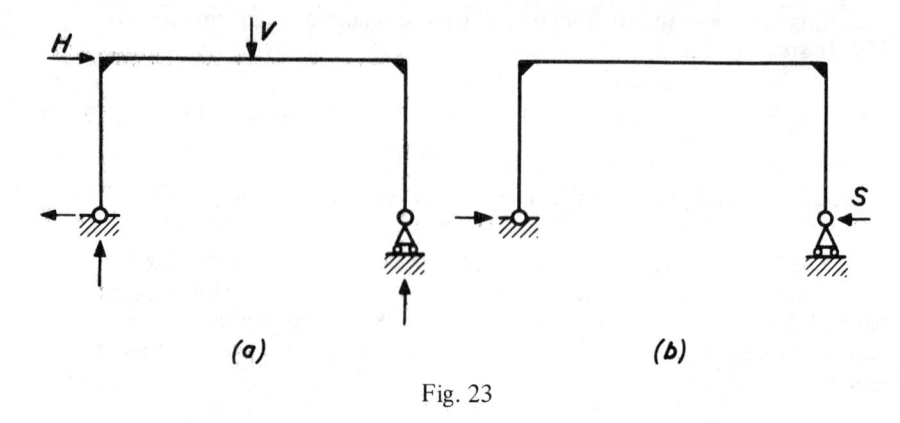

Since the frame has one redundancy, two plastic hinges will be required for a regular plastic collapse mechanism; the only two mechanisms possible are sketched in Fig. 25. Figures 24 (a) and (b) are superimposed in Fig. 26 to give the bending moment diagrams corresponding to the two mechanisms.

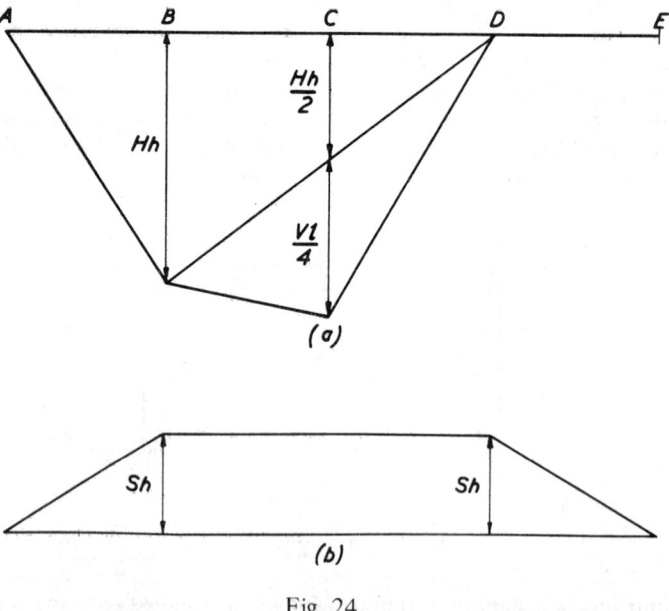

Fig. 24

It will be seen that the mechanism which occurs depends entirely on the relative magnitudes of the free bending moments at B and C in Fig. 24 (a). If the free moment at C, $Vl/4 + Hh/2$, exceeds that at B, Hh (i.e. $Vl/4 > Hh/2$), then Figs. 25 (a) and 26 (a) are appropriate, and M_p can be determined directly as $Vl/8 + Hh/4$. Similarly, if the wind load H is relatively high compared with V (more precisely, if $Hh/2 > Vl/4$), then the free bending moment at B will exceed that at C, and Fig. 26 (b) gives $M_p = Hh/2$.

FRAME ANALYSIS   555

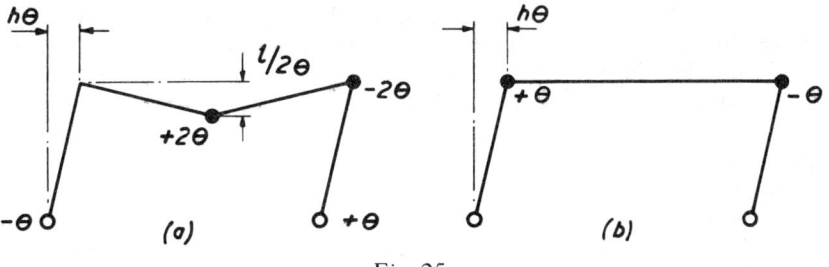

Fig. 25

These design values of M_p may be derived directly by writing the work equations corresponding to the mechanisms of collapse. For example, in Fig. 25 (a), the side load H moves through a distance $h\theta$, the vertical load V moving through $(l/2)\theta$, so that

$$Hh\theta + \frac{Vl}{2}\theta = 4M_p\theta,$$

i.e.
$$M_p = \frac{Vl}{8} + \frac{Hh}{4}, \text{ as before}$$

Similarly, for the pure sidesway mechanism, Fig. 25 (b),

$$Hh\theta = 2M_p\theta,$$

i.e.
$$M_p = \frac{Hh}{2}.$$

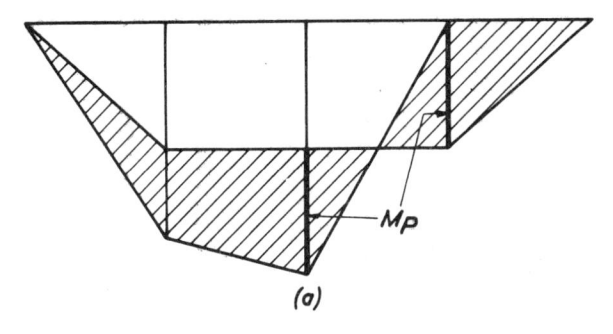

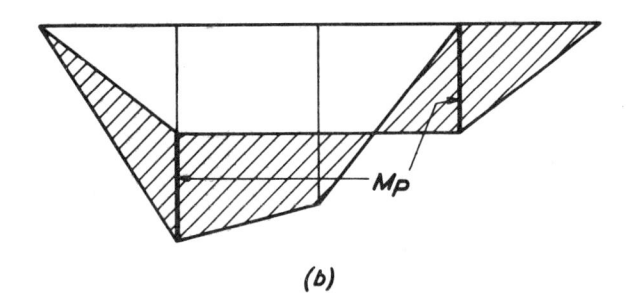

Fig. 26

In Figs. 25, plus and minus signs are associated with the hinge rotations. For example, the sagging hinge at the centre of the beam in Fig. 25 (a) has rotation $+2\theta$, while the hogging hinge at the top of the right-hand column has rotation -2θ. It is of great importance that the sign convention chosen for bending moments should be the same as that chosen for the corresponding hinge rotations. Thus at the central sagging hinge of Fig. 25 (a) where the rotation is $+2\theta$, the bending moment at collapse will be $+M_p$; at the hogging hinge, where the rotation is -2θ, the bending moment has value $-M_p$. Thus the work done in the hinges is $(+2\theta)(+M_p) + (-2\theta)(-M_p) = 4M_p\theta$; in all cases, plastic work dissipated at a hinge is positive.

For determining the work done in the hinges, therefore, it is necessary only to take the numerical value of the hinge rotation and multiply by the appropriate value of M_p. Signs of hinge rotations are crucial for the full use of the principle of virtual work, discussed below.

In the design of this simple frame, it has been assumed implicitly that the frame had uniform section M_p. For a large range of frame sizes and of loading, the uniform frame is likely to give the most economical design, but non-uniform designs can be investigated very quickly. Suppose, for example, that the two columns have full plastic moment M_C which is greater than M_B, the full plastic moment of the beam. The two plastic collapse mechanisms will be as shown in Fig. 27, where the plastic hinge at the top of a column forms in the beam rather than the column. The collapse bending moment diagrams will be as shown in Fig. 26, except that M_p is now replaced by M_B.

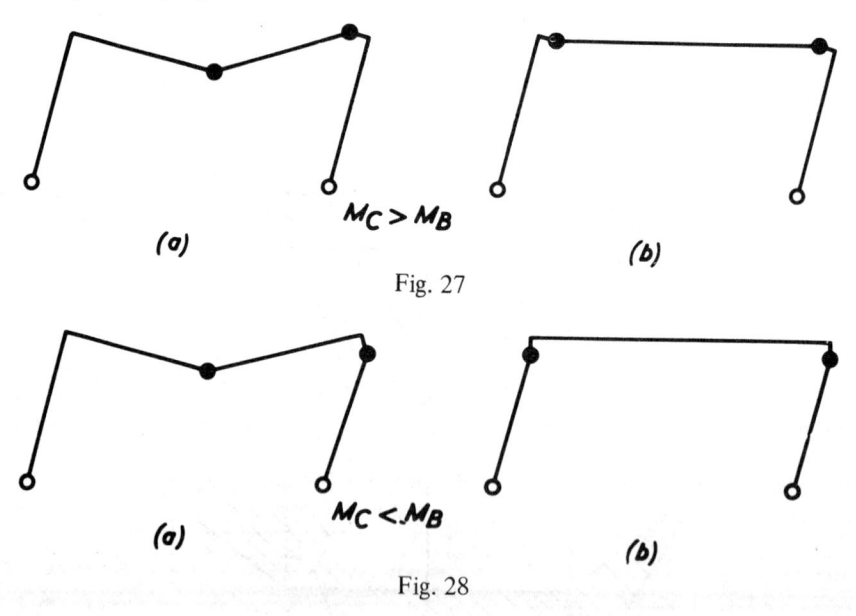

(a) $M_C > M_B$ (b)

Fig. 27

(a) $M_C < M_B$ (b)

Fig. 28

Should the columns be weaker than the beam, then the collapse mechanisms change slightly to those of Fig. 28, the bending moment diagrams being shown in Fig. 29. From Fig. 29 (a), comparing with Fig. 24 (a), it will be seen that

$$M_B + M_C = \frac{Hh}{2} + \frac{Vl}{4};$$

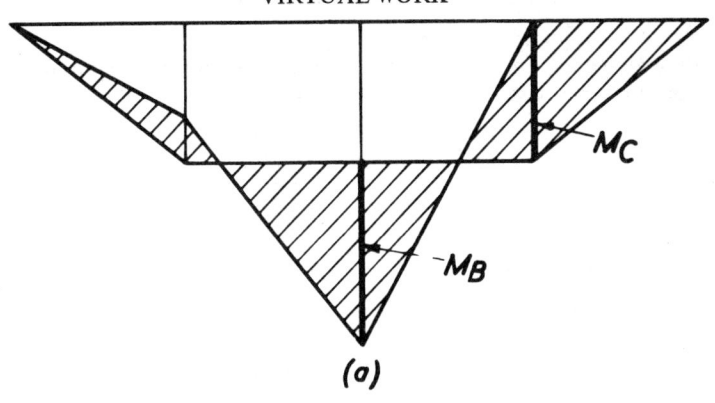

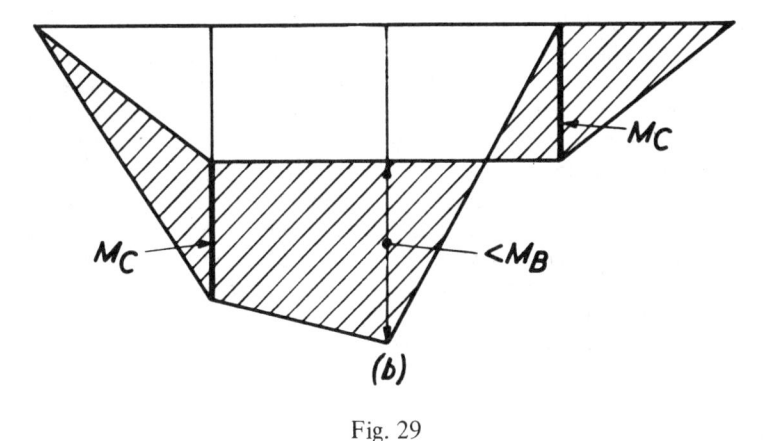

similarly, from Fig. 29 (b), $M_C = Hh/2$, and M_B must be greater than $Vl/4$ as well as greater than M_C.

Virtual Work (2, 3, 4)

In writing the work equation for the two simple collapse mechanisms discussed above, it was easy to evaluate the work done by the external loading. For practical loading cases, involving distributed and/or a large number of point loads, this is not always so since the work terms due to the loading are more tedious to calculate. However, a method, based upon virtual work, is of very great use and power when dealing with practical loading cases. It may be shown that, if M_i are bending moments in equilibrium with the collapse loads acting on a frame, $(M_p)_i$ are the full plastic moments at sections i, and ϕ_i are hinge rotations of a plastic collapse mechanism, then, by virtual work,

$$\Sigma M_i \phi_i = \Sigma (M_p)_i \phi_i$$

558   PLASTIC THEORY AND DESIGN

The moments M_i can be *any* equilibrium set, and it is convenient to take M_i as the free bending moments. Thus, from Fig. 24 (a) for the pinned-base portal frame, a possible set of moments M_i is

$$M_i \equiv [M_A, M_B, M_C, M_D, M_E] \equiv \left[0, Hh, \frac{Hh}{2} + \frac{Vl}{4}, 0, 0\right]$$

Now Fig. 25 (a), for example, gives the rotations of a plastic collapse mechanism:

$$\phi_i \equiv [\phi_A, \phi_B, \phi_C, \phi_D, \phi_E] \equiv [-\theta, 0, 2\theta, -2\theta, 0]$$

Using the virtual work equation,

$$(0)(-\theta) + (Hh)(0) + \left(\frac{Hh}{2} + \frac{Vl}{4}\right)(2\theta) + (0)(-2\theta) + (0)(0) = 4M_p\theta$$

which gives $4M_p = Hh + \dfrac{Vl}{2}$, as before.

The importance of this approach to the problem lies in the fact that the moments M_i, for which free bending moments may be used, need be determined just once at the start of the calculations. Alternative collapse mechanisms can then be tried quickly; since each trial corresponds to an unsafe solution, the largest value of M_p will give the correct solution. The problem now becomes one of finding the proper mechanism for a complex structure; the simple portal frame could be studied almost by inspection, but more complicated frames require a logical technique for deriving the correct mechanism of collapse. Such a technique may be found in the method of combination of mechanisms, which is mentioned later in this chapter.

Design Example 2. The portal frame of Fig. 30 will be designed to a load factor of 1.75 on the vertical loading, and 1.4 on the vertical loading together with the wind load of 40 kN. It seems obvious that since the wind load is relatively small, this combination will not be critical, but a check will be made. In this check, reductions in full plastic moment due to shear force and axial load will be ignored; these effects will be investigated later.

Fig. 30

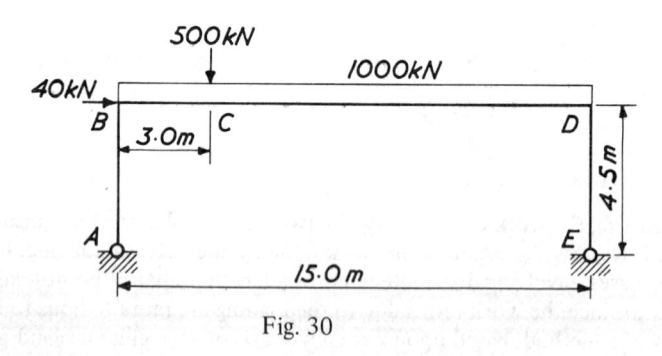

Figure 31 shows the free bending moment diagram for the beam under vertical loading only. The maximum free moment has value 2 700 kNm (unfactored), so that, since the beam collapses effectively as a fixed-ended beam, the required M_p (again unfactored) is 1 350 kNm. Thus a section must be provided having $M_p = 1.75 \times 1\,350 = 2\,363$ kNm, in order to achieve the design load factor of 1.75.

VIRTUAL WORK

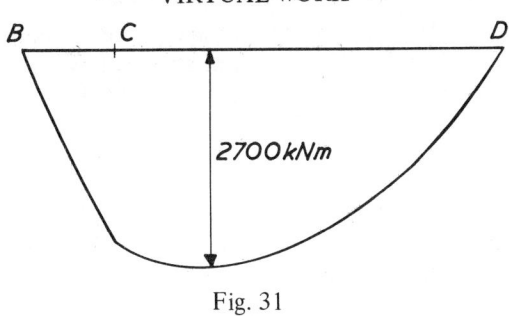

Fig. 31

The collapse bending moment diagram under vertical plus wind loading is shown in Fig. 32; again assuming a uniform frame, it will be seen that M_p = 1 404 kNm, so that the section must provide

$$M_p = 1.4 \times 1\ 404 = 1\ 966 \text{ kNm}.$$

Clearly the vertical loading only, at a load factor of 1.75, is critical. Using steel to B.S. 4360 grade 43, a plastic modulus of $2\ 363 \times 10^6/245 \times 10^3 = 9\ 645$ cm^3 is required; a 914 x 305 x 253 kg U.B. has Z_p = 10 930 cm^3. The effects of shear force and axial load on this section will now be investigated, and typical calculations will be made for corner B, Fig. 30.

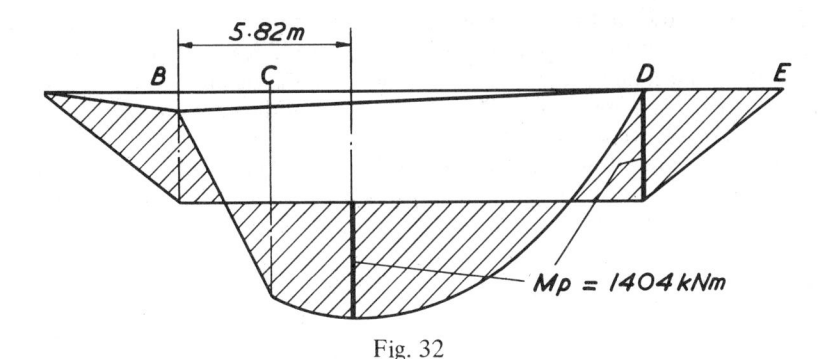

Fig. 32

Column AB carries at collapse an axial thrust of 900 x 1.75 = 1 575 kN together with a shear force of 1 350 x 1.75/4.5 = 525 kN. The axial stress is therefore $1\ 575 \times 10^3/323 \times 10^2 = 49$ N/mm^2; thus n = 49/245 = 0.2, and, from the section tables, the reduction in plastic modulus is 15 040 x 0.2^2 = 602 cm^3. This last calculation neglects the effect of the shear force acting together with the axial load. The combined effect is small, but may be calculated from a formula given by Horne (*British Welding Journal,* April 1958, p. 170): the combined reduction in section modulus is 703 cm^3, thus reducing the initial 10 930 cm^3 to an effective 10 227 cm^3. This value exceeds the figure of 9 645 cm^3 required for the design.

A greater reduction in section modulus at the corner B is however realised if the end of the beam rather than the top of the column is considered. This section is subjected to a shear force of 1 575 kN (together with an axial thrust of 525 kN). The effect of the shear force may be calculated by the method of Design Example 1 above, using nominal dimensions for the 914 x 305 x 253 kg U.B. Thus the web

of this section has area 918 × 17.3 = 159 cm² and plastic modulus ¼ × 17.3 × 918² = 3 644 × 10³ mm³ = 3 644 cm³. The mean shear stress is 1 575 × 10³/159 × 10² = 99 N/mm², and the reduction in section modulus is therefore

$$3\,644\left[1 - \sqrt{1 - 3\left(\frac{99}{245}\right)^2}\right] = 1\,046 \text{ cm}^3$$

If combined axial load and shear are considered, the reduction is 1 112 cm³. The initial section modulus of 10 930 cm³ is reduced to 9 818 cm³ at the end of the beam at B, but this value still exceeds 9 645 cm³, and the section chosen is thus satisfactory.

The columns must be checked for stability, and this will be done later. In fact column AB is just stable, so that a uniform section 914 × 305 × 253 kg U.B. represents one possible design for the frame of Fig. 30. There is, of course, no point in increasing the column section, since the beam has minimum section. The beam could however be increased and the columns reduced in section in an attempt to achieve a more economical design; there is little scope for such a move, however, for reasons of column stability. The uniform section frame is likely to be the most economical in this example.

The Fixed-base Rectangular Frame (2)

The frame shown in Fig. 33 (a) has three basic modes of collapse, two of them identical with those for the pinned-base frame (Figs. 33 (c) and (d)), except that the hinges at the column feet are plastic rather than simple pins. Since the fixed-base frame, however, is more capable of resisting side load, then a third mode, Fig. 33 (b), is possible for relatively small values of the side load H.

A uniform section frame carrying the loads of Fig. 33 (a) will collapse as shown in Fig. 33 (b) if $Vl/2 = 4M_p$; the corresponding collapse equations for Figs. 33 (c) and (d) are $Hh + (Vl/2) = 6M_p$ and $Hh = 4M_p$, respectively.

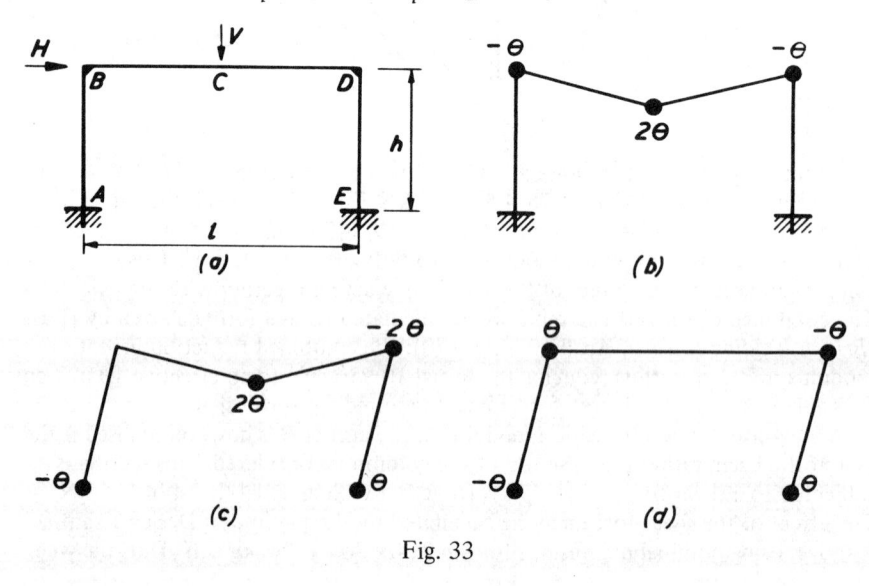

Fig. 33

THE FIXED-BASE RECTANGULAR FRAME

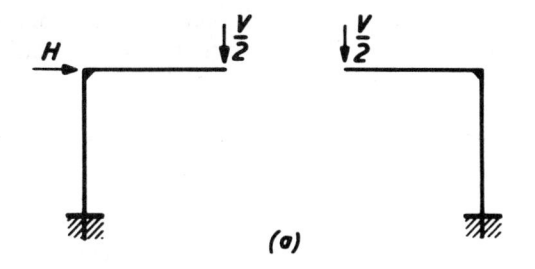

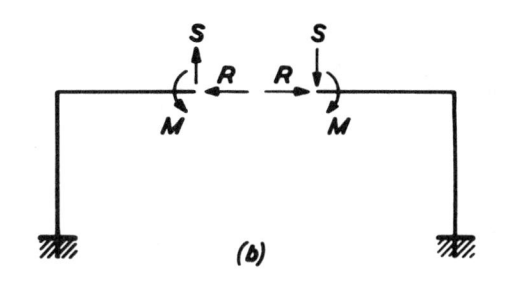

Fig. 34

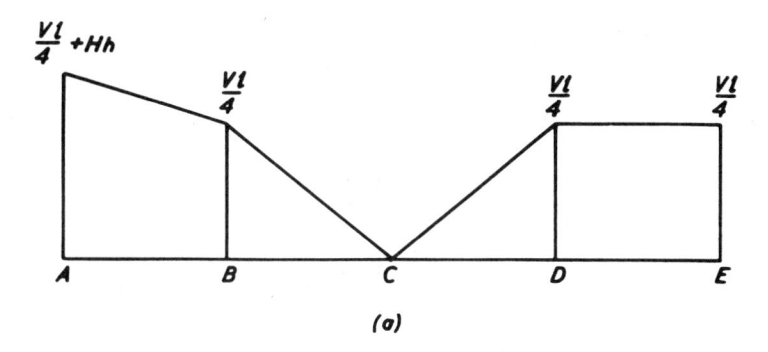

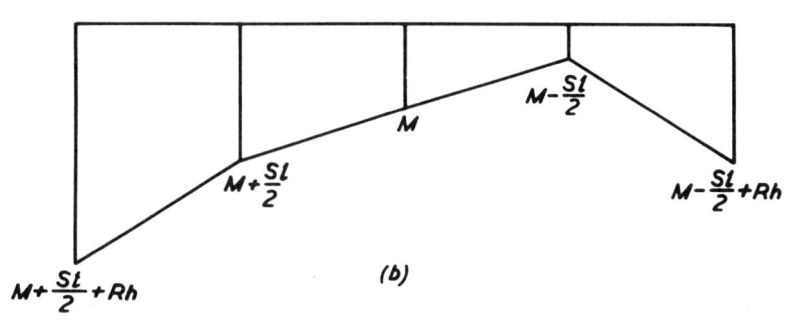

Fig. 35

Bending moment diagrams are drawn conveniently if the frame is split at the centre of the beam, Fig. 34. Each half of the frame becomes statically determinate, and the free bending moments may be determined from Fig. 34 (a), and the reactant moments (in terms of the redundancies M, R and S) from Fig. 34 (b). The two bending moment diagrams are shown separately in Fig. 35. Note that the reactant diagram, Fig. 35 (b), consists of three straight lines, the slopes of these lines for the columns being equal and opposite (equal numerically to the shear force R across the column).

Figure 36 illustrates the collapse bending moment diagram for (in arbitrary units) $l = 2$, $h = 1$, $V = 10$, $H = 8$. By calculation or by drawing, the value of M_p is found to be 3 units. Consider now the more practical loading of Fig. 37, in which the beam carries a uniformly distributed load of 20 units. The free bending moment diagram for such loadings can be drawn in the same way, by splitting the frame at the centre. The collapse bending moment diagram for Fig. 37 is shown in Fig. 38, and, again by drawing or calculation, the value of M_p is 3.016 units.

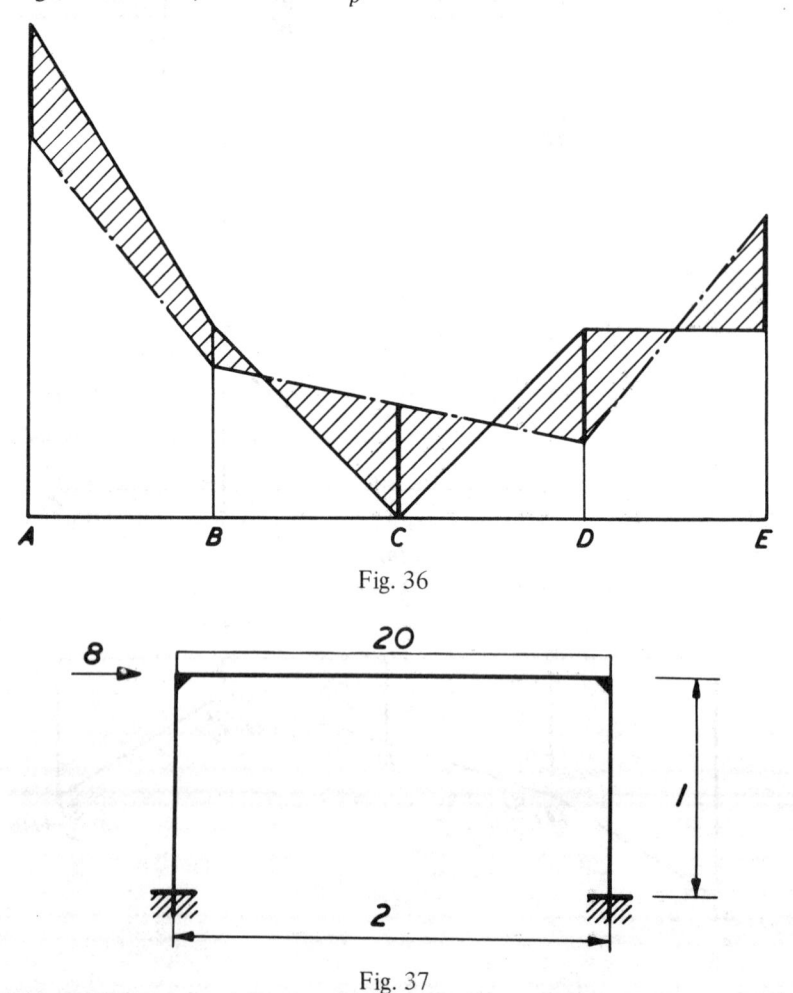

Fig. 36

Fig. 37

THE FIXED-BASE RECTANGULAR FRAME

Figure 38 may be compared with Fig. 36 to show the small difference introduced by replacing a distributed load by a point load of the appropriate value. The maximum free bending moment for the beam alone in Fig. 37 is $Wl/8 = (20)(2)/8 = 5$ units. The same value of free bending moment would be obtained by a point load of 10 units; $Vl/4 = (10)(2)/4 = 5$ units. The analysis of the point load case is, of course, much simpler, and led to $M_p = 3$ compared with $M_p = 3.016$ for the distributed loading.

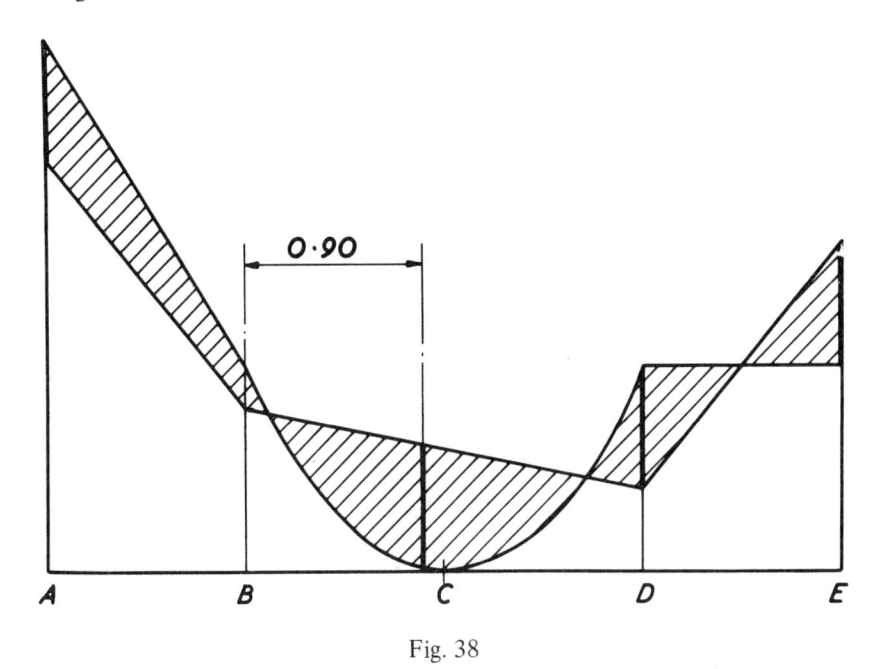

Fig. 38

This technique, of replacing actual loading on a beam by a single point load of magnitude sufficient to give the same maximum free bending moment, is sometimes of great value. The resulting design values of M_p are likely to be in error by only 1 or 2 per cent ($\frac{1}{2}$ per cent in the above example), and the designer is able more easily to appreciate the plastic behaviour of the frame if he is dealing with simple loading conditions.

There are, of course, many ways of making the simple portal frame statically determinate in order that the free bending moment diagram may be drawn. One convenient way is to insert three pins into the original portal, as shown in Fig. 39 (a), leading to the free bending moments of Fig. 39 (b). The reactant line, being a function of the frame, has exactly the same form as before, and Fig. 40 corresponds exactly to Fig. 36. The advantage of making the frame statically determinate in this way is well illustrated by the case of the multi-storey, multi-bay frame, Fig. 41. It will be seen that each beam is effectively isolated from its neighbours, so that the free bending moment diagram can be found for each beam separately. Similarly, each internal column length is hinged at both ends, and can carry no shear; all wind loads are carried on the windward column. The free bending moment diagram for the entire frame can thus be drawn quickly.

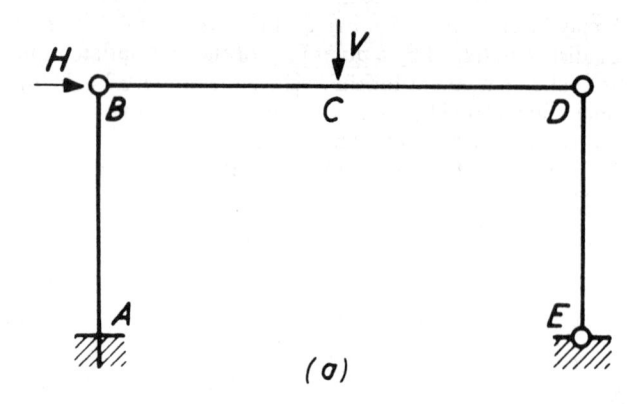

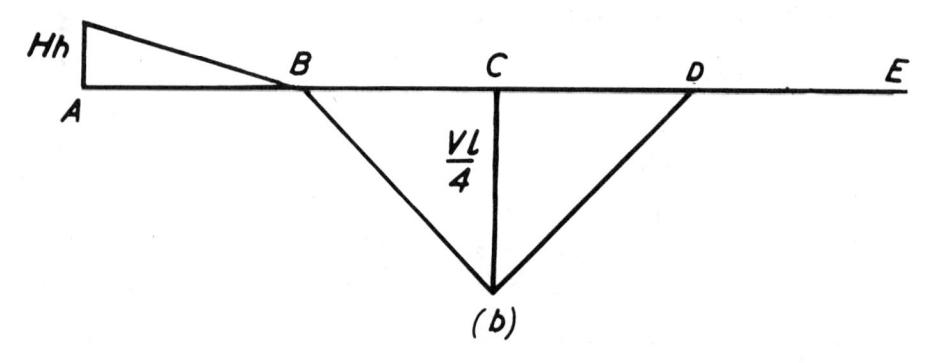

Fig. 39

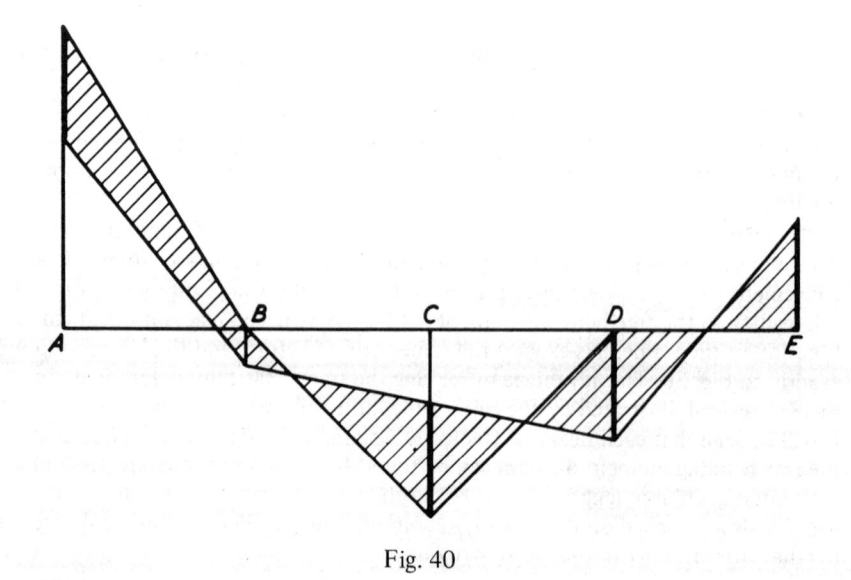

Fig. 40

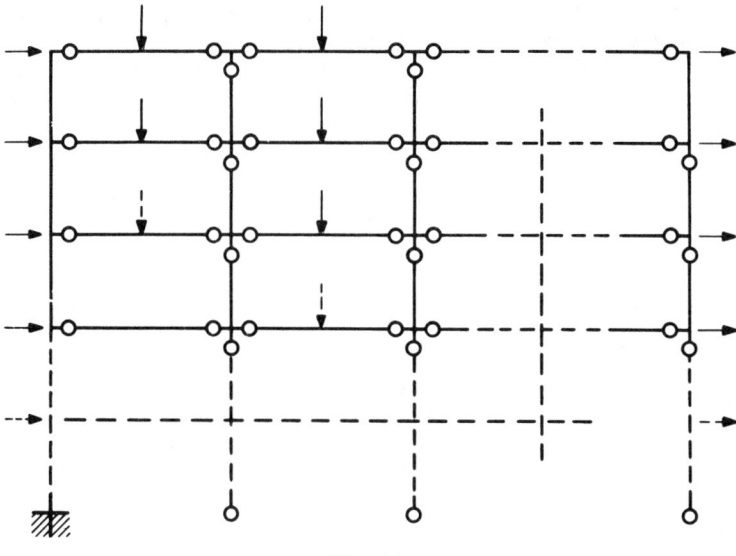

Fig. 41

Pitched Roof Frames (5, 6)

The pitched roof portal frame may, under certain conditions, develop deflections large enough to invalidate the assumptions of simple plastic theory. Thus a plastic design must be checked to ensure that deflections are not troublesome. To this end, rules have been developed which indicate when a particular frame is likely to be unsuitable for plastic design. Following the usual philosophy, it will be assumed in the first instance that deflections are small; the final design will then be checked by the rules.

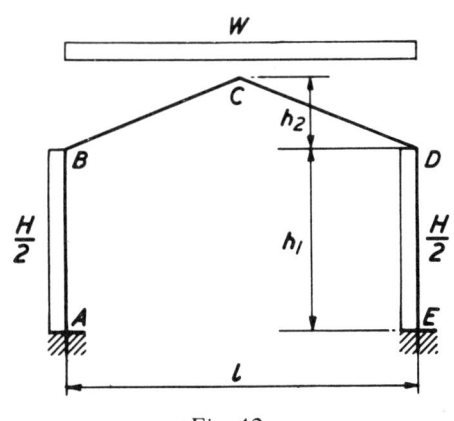

Fig. 42

Consider first the pitched portal frame of Fig. 42 carrying the uniformly distributed loads shown. (For the sake of clarity, drag and suction on the roof are neglected.) An approximate loading is shown in Fig. 43 (a), and it will be found that

for normal frame sizes this approximate loading will give very accurate design estimates. Indeed, the even simpler approximate loading of Fig. 43 (b) is also very accurate, and will serve to illustrate the main features of pitched roof design. The free bending moment diagram may be drawn, as for the rectangular portal, by splitting the frame at the apex C; Fig. 44 shows the two free bending moment diagrams corresponding to the two approximate loadings of Fig. 43. In both of Figs. 44, the free bending moments at the five cardinal points A to E have the same corresponding values.

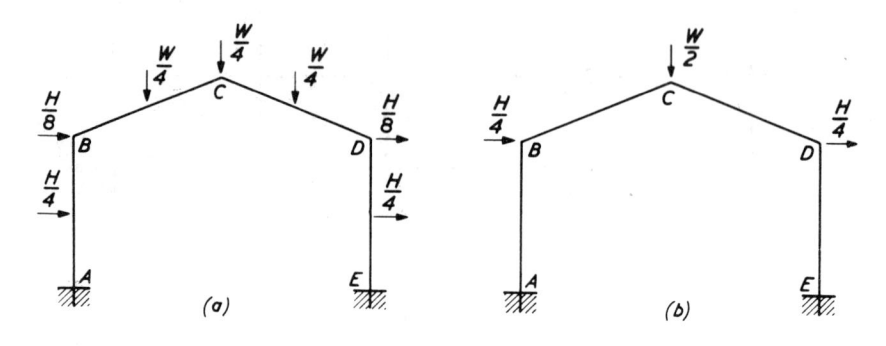

Fig. 43

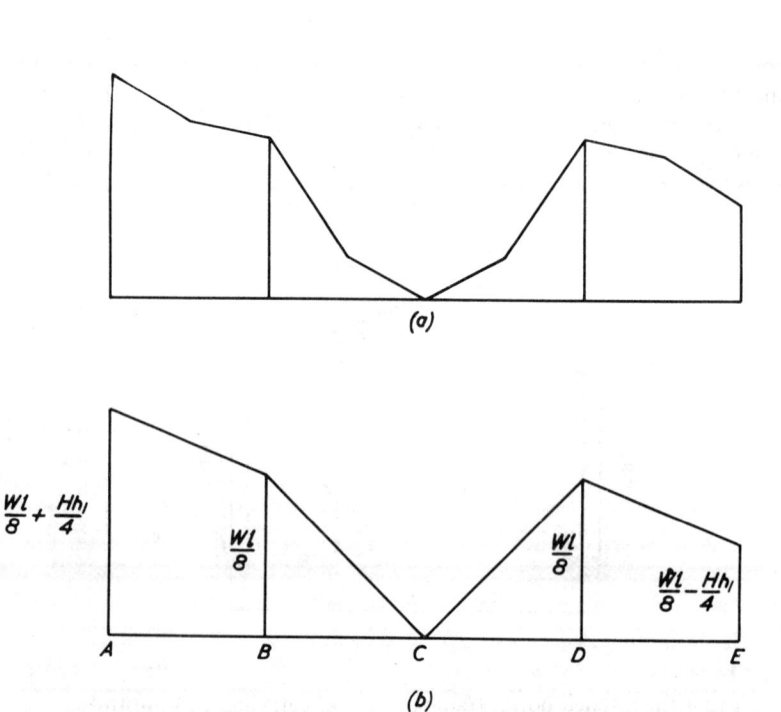

Fig. 44

PITCHED ROOF FRAMES

The reactant line may be found by considering the effects of the redundancies M, R and S shown in Fig. 45; it will be seen that the values of the reactant moments at the cardinal points are

$$A: \quad M + R(h_1 + h_2) + \frac{Sl}{2}$$

$$B: \quad M + Rh_2 + \frac{Sl}{2}$$

$$C: \quad M$$

$$D: \quad M + Rh_2 - \frac{Sl}{2}$$

$$E: \quad M + R(h_1 + h_2) - \frac{Sl}{2}$$

Thus the actual bending moment at any section can be expressed easily as the sum of the free and reactant moments; at B, for example, from Fig. 43 (b) and the above expressions for the reactant moments, the actual bending moment is given by

$$-\frac{Wl}{8} + \left(M + Rh_2 + \frac{Sl}{2}\right)$$

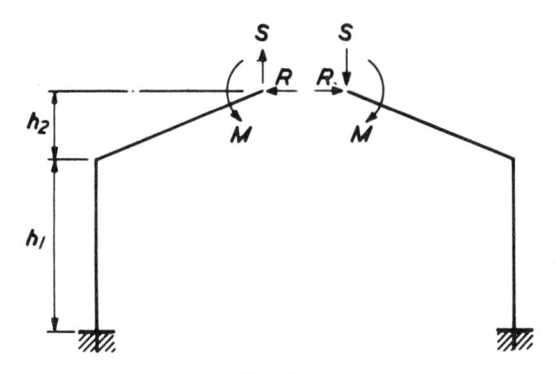

Fig. 45

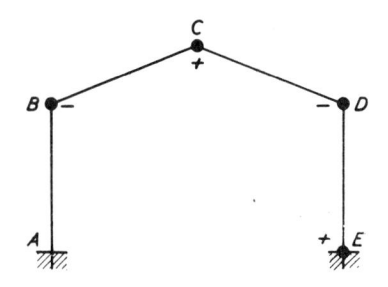

Fig. 46

PLASTIC THEORY AND DESIGN

Consider now the common collapse mode shown in Fig. 46, which involves hinges at B, C, D and E. If the frame has uniform section, then the occurrence of these hinges can be expressed by equating the actual moment at each hinge to $\pm M_p$:

$$B: \quad -\frac{Wl}{8} + \left(M + Rh_2 + \frac{Sl}{2}\right) = -M_p$$

$$C: \quad 0 + M = M_p$$

$$D: \quad -\frac{Wl}{8} + \left(M + Rh_2 - \frac{Sl}{2}\right) = -M_p$$

$$E: \quad -\left(\frac{Wl}{8} - \frac{Hh_1}{4}\right) + \left(M + R(h_1 + h_2) - \frac{Sl}{2}\right) = M_p$$

It is instructive to note that the frame has three redundancies, M, R and S, and therefore requires four hinges for the formation of a regular mechanism. The above four equations serve to determinate the values of the three redundancies, and also to determine the single collapse equation from which M_p may be found.

The equations solve to give

$$S = 0$$

$$R = \frac{1}{(1+k)}\frac{Wl}{8h_1} - \frac{H}{4}$$

$$M = M_p = \frac{1}{(1+k)}\left[\frac{Wl}{16} + \frac{Hh_2}{8}\right]$$

where $k = h_2/h_1$, and the final bending moment diagram is as shown in Fig. 47. (The general shape of the reactant line may be noted; it consists of four straight lines, with the slopes for columns AB and DE equal and opposite.)

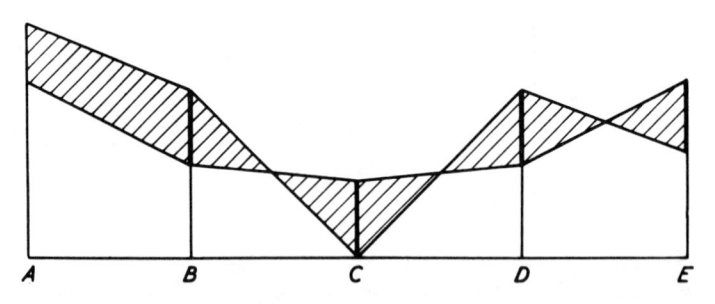

Fig. 47

Inspection of Fig. 47 reveals that the mechanism and equilibrium conditions are satisfied. The yield condition will also be satisfied if the moment at A is less than M_p. This can be checked by drawing, for numerical values of the loads. Alternatively, the actual moment at A has value:

$$-\left(\frac{Wl}{8} + \frac{Hh_1}{4}\right) + M + R(h_1 + h_2) + \frac{Sl}{2}$$

PITCHED ROOF FRAMES

and substituting the now known values of M, R and S, M_A is found to have the value $M_p - (Hh_1/2)$. Thus for relatively small values of H, M_A is certainly less than M_p, and Hh_1 must equal $4M_p$ before M_A becomes equal to $-M_p$.

It will be seen that for $H = 0$ (i.e. no wind load), the above analysis gives the correct mode of collapse with the moment at column foot A just equal to, but not exceeding, M_p. The same mode, Fig. 46, occurs for reasonably shaped frames subjected to wind loads, and is the most commonly occurring basic mode. For tall frames, and for small angles of pitch, the mode of Fig. 48 (a) may sometimes be encountered; Fig. 48 (b) represents a very unusual mode.

(a)

(b)

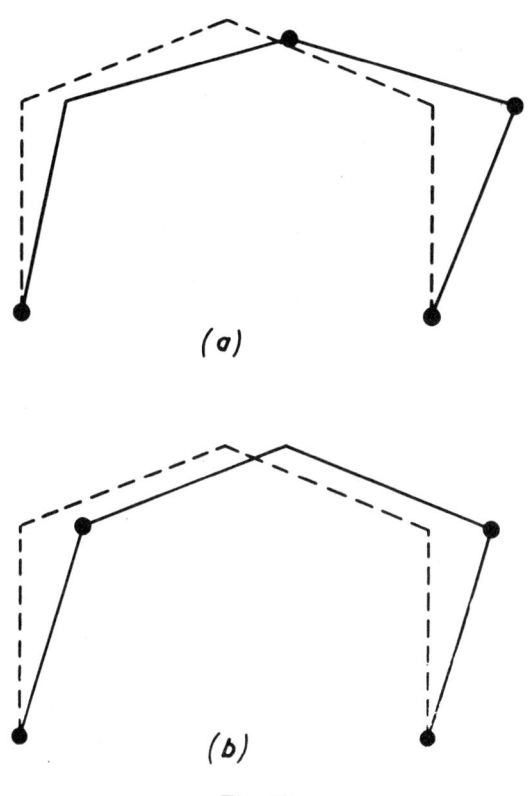

Fig. 48

The writing of four simultaneous equations to express collapse, as was done above, is one way of expressing simultaneously the equilibrium and the mechanism conditions. Designers who have become completely familiar with this method may wish, by using virtual work, to avoid solving such equations; the virtual work equation gives just one equation, and does away with the need to determine the values of the redundancies. It does not, however, necessarily guarantee that the yield condition is satisfied, but this condition can be checked most elegantly and easily by using virtual work again.

Figure 49 shows the basic collapse mode, with hinges at B, C, D and E, with the single degree of freedom specified in terms of a rotation θ about the instantaneous

PLASTIC THEORY AND DESIGN

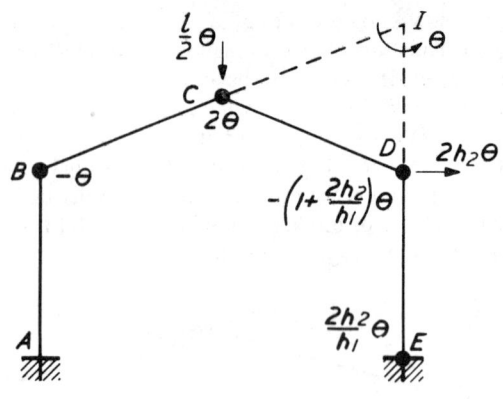

Fig. 49

centre I of rafter CD. Consideration of the mechanism motion leads to the hinge rotations shown, namely

$$\phi_i \equiv (-\theta, 2\theta, -(1+2k)\theta, 2k\theta)$$

for hinges B, C, D, E. The free bending moments at the four sections are, from Fig. 44 (b),

$$M_i \equiv \left(-\frac{Wl}{8}, 0, -\frac{Wl}{8}, -\frac{Wl}{8} + \frac{Hh_1}{4}\right),$$

and the collapse bending moments are, of course,

$$(M_p)_i \equiv (-M_p, M_p, -M_p, M_p).$$

Thus the virtual work equation gives

$$\left(-\frac{Wl}{8}\right)(-\theta) + (0)(2\theta) + \left(-\frac{Wl}{8}\right)(-\{1+2k\}\theta) + \left(-\frac{Wl}{8} + \frac{Hh_1}{4}\right)(2k\theta)$$

$$= M_p[\theta + 2\theta + (1+2k)\theta + 2k\theta]$$

i.e.
$$M_p = \frac{1}{(1+k)}\left[\frac{Wl}{16} + \frac{Hh_2}{8}\right]$$

which is, of course, the correct collapse equation.

Suppose now that it is required to check the yield condition at column foot A, i.e. the value M_A at collapse is required. For this purpose, a virtual mechanism is chosen involving a hinge rotation at A; for example, the mechanism of Fig. 50 has rotations at A, B, D, E of values $(-\theta, \theta, -\theta, \theta)$. These hinge rotations can be multiplied by any set of equilibrium moments, and the sum equated to the sum of the same rotations multiplied by any other set of equilibrium moments. Now the collapse bending moments are certainly an equilibrium set; the values at A, B, D and E are $(M_A, -M_p, -M_p, M_p)$. (For the signs of M_p, see Fig. 46.) The free bending moments form another equilibrium set:

$$\left(-\frac{Wl}{8} - \frac{Hh_1}{4}, -\frac{Wl}{8}, -\frac{Wl}{8}, -\frac{Wl}{8} + \frac{Hh_1}{4}\right)$$

Thus

$$(M_A)(-\theta) + (-M_p)(\theta) + (-M_p)(-\theta) + (M_p)(\theta)$$

$$= \left(-\frac{Wl}{8} - \frac{Hh_1}{4}\right)(-\theta) + \left(-\frac{Wl}{8}\right)(\theta) + \left(-\frac{Wl}{8}\right)(-\theta) + \left(-\frac{Wl}{8} + \frac{Hh_1}{4}\right)(\theta)$$

i.e. $$M_A = M_p - \frac{Hh_1}{2}, \text{ as before.}$$

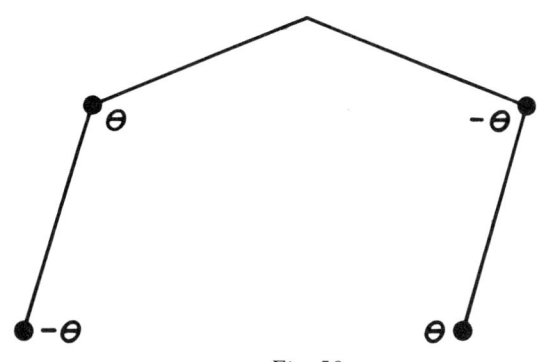

Fig. 50

Design Example 3. The pitched roof frame of Fig. 51 will be designed in steel to B.S. 4360 grade 43. The frame spacing is 6.0 m, the roof loading (dead plus superimposed) 1.5 kN/m², and the load factor 1.75. Purlins are spaced at 1.2 m on plan. (It will be assumed that the wind loading is such that an approximate analysis, based upon the loading of Fig. 43 (b), indicates clearly that vertical loading at a load factor of 1.75 is more critical than vertical loading plus wind at a factor of 1.4.)

In the absence of wind load, the approximate analysis gave

$$M_p = \frac{1}{(1+k)} \frac{Wl}{16},$$

where W is the total load on the frame; the mode of collapse was that of Fig. 46. (With the loading symmetrical, a fifth hinge will of course form at the column foot A.) For the particular dimensions of the frame, $k = h_2/h_1 = 2.48/4.2 = 0.59$,

$$W = 12 \times 6 \times 1.5 = 108 \text{ kN (unfactored)},$$

and $$\frac{Wl}{16} = \frac{108 \times 12}{16} = 81 \text{ kNm.}$$

Hence $M_p = 81/1.59 = 51$ kNm. A section must be provided, therefore, having M_p at least equal to $51 \times 1.75 = 89$ kNm, in order that the design load factor should be achieved.

However, before selecting a section, an accurate analysis will first be made using the actual loading. Figure 52 shows half the frame, and the free bending moments may be found:

A	B	B_1	B_2	B_3	B_4	C
−162.0	−162.0	−103.7	−58.3	−25.9	−6.5	0 kNm

PLASTIC THEORY AND DESIGN

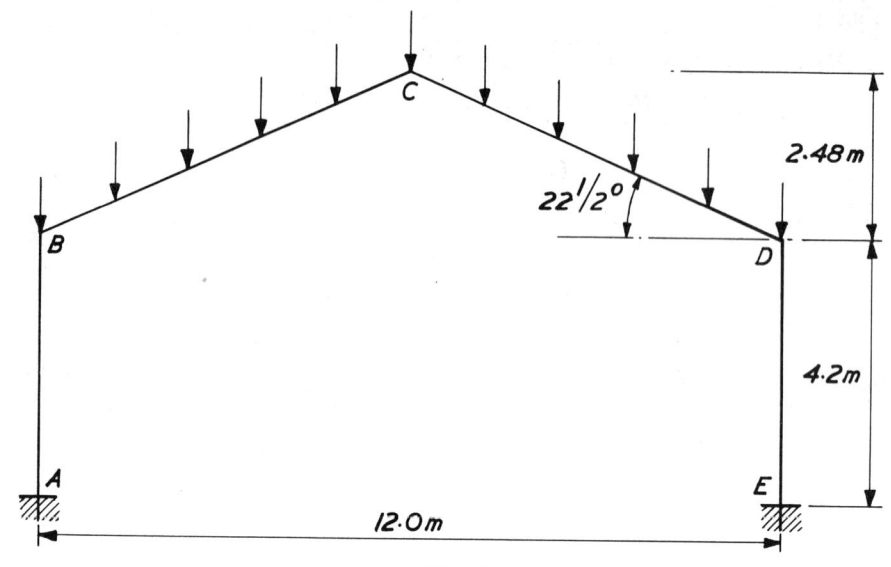

Fig. 51

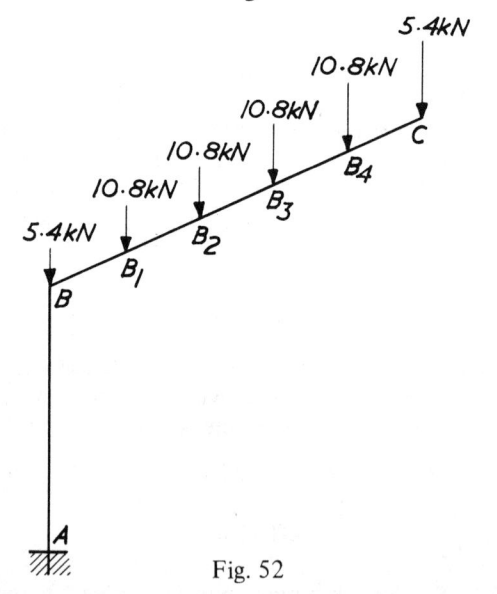

Fig. 52

Using the virtual work equation (or the simultaneous equations) the value of M_p is confirmed as 51.0 kNm, and Fig. 53 shows the net bending moments, with hinges at A, B, C, D and E. From the figure, or by calculation, the actual bending moments for the assumed collapse mode are:

	A	B	B_1	B_2	B_3	B_4	C
	51.0	−51.0	−4.6	28.7	49.1	56.5	51.0 kNm

PITCHED ROOF FRAMES

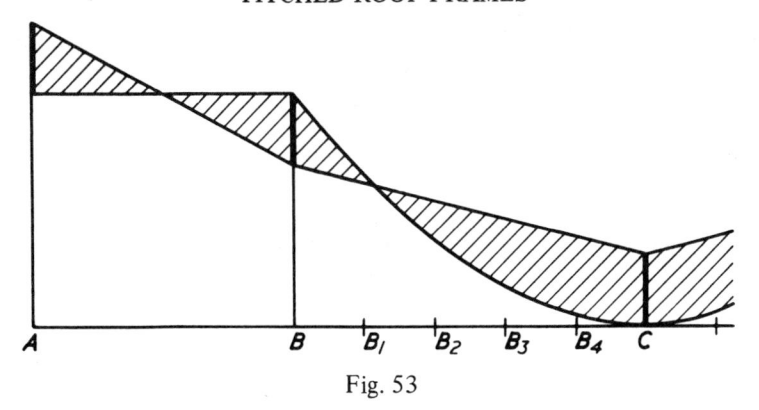

Fig. 53

It will be seen that the yield condition is violated at section B_4, where the bending moment exceeds $M_p = 51.0$ kNm. This means that the mechanism of Fig. 46 is incorrect, and it seems reasonable to try the mechanism of Fig. 54. The following calculations will be done first by writing the simultaneous equations, and then by the equation of virtual work.

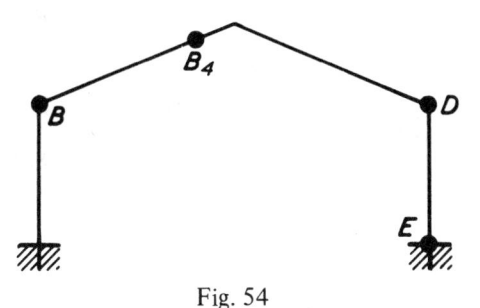

Fig. 54

With the given dimensions of the frame, the reactant moments (in terms of the redundancies M, R, S, Fig. 45) have values:

$$A: \quad M + 6.68R + 6S$$
$$B: \quad M + 2.48R + 6S$$
$$C: \quad M$$
$$D: \quad M + 2.48R - 6S$$
$$E: \quad M + 6.68R - 6S.$$

Between B and C, and between C and D, the reactant line is straight. Thus the following equations may be written for the formation of hinges at B, B_4, D and E:

$$B: \quad -162.0 + M + 2.48R + 6S = -M_p$$
$$B_4: \quad -\ 6.5 + M + 0.5R + 1.2S = M_p$$
$$D: \quad -162.0 + M + 2.48R - 6S = -M_p$$
$$E: \quad -162.0 + M + 6.68R - 6S = M_p$$

574 PLASTIC THEORY AND DESIGN

These equations solve to give

$$S = 0$$
$$R = 25.05 \text{ kN}$$
$$M = 47.1 \text{ kNm}$$
$$M_p = 52.9 \text{ kNm}.$$

The following table may now be drawn up, giving the total bending moments throughout the frame:

	A	B	B_1	B_2	B_3	B_4	C
Free	−162.0	−162.0	−103.7	−58.3	−25.9	−6.5	0 ...
Reactant	214.9	109.1	96.7	84.2	71.8	59.4	47.1 ...
Total	52.9	−52.9	−7.0	25.9	45.9	52.9	47.1 ...

It will be seen that the correct solution has now been obtained, since the value M_p = 53.0 kNm is not exceeded anywhere in the frame.

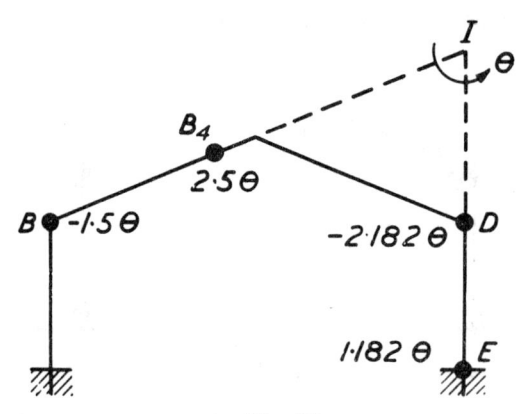

Fig. 55

To obtain these same results by virtual work, consider the mechanism of Fig. 55, in which the hinge rotations are again referred to a rotation θ about the instantaneous centre of rafter CD. The hinge rotations shown in the figure may be checked geometrically; in fact, they may be determined quickly by combining mechanisms, as shown later. The following table may be drawn up:

	A	B	B_1	B_2	B_3	B_4	C	D	E
Free	−162.0	−162.0	−103.7	−58.3	−25.9	−6.5	0	−162.0	−162.0
Collapse		$-M_p$				$+M_p$		$-M_p$	$+M_p$
Mechanism Fig. 55		-1.5θ				$+2.5\theta$		-2.182θ	-1.182θ
Mechanism Fig. 56		-0.5ϕ				$+2.5\phi$	-2ϕ		
Mechanism Fig. 57		$-\psi$				$+2.5\psi$		-1.5ψ	

The first three lines of this table give the free bending moments, the collapse bending moments (i.e. M_p at the assumed hinge points), and the hinge rotations of

the assumed collapse mechanism, Fig. 56. By virtual work:

$$(-1.5)(-162.0) + (2.5)(-6.5) + (-2.182)(-162.0) + (1.182)(-162.0)$$
$$= (1.5 + 2.5 + 2.182 + 1.182)(M_p)$$

from which $M_p = 52.9$ kNm.

As yet, this value of M_p is an unsafe estimate of the correct value, since it has not been shown that the yield condition is satisfied. The values of the bending moments must be found throughout the frame and this can now be done by drawing, or again by using virtual work. Consider, for example, the virtual mechanism of Fig. 56, involving hinge discontinuities at B, B_4, and C; this mechanism is shown in the fourth line of the above table. The hinge rotations may be written down quickly by noting that from B to B_4, Fig. 52, is a distance of four purlin points (4.8 m), while from B_4 to C is one purlin point (1.2 m). Thus the rotation at C in Fig. 56 must be four times that at B, while the rotation at B_4 is the numerical sum of the rotations at B and C.

Using this mechanism with the first two lines of the table:

$$(-0.5)(-162) + (2.5)(-6.5) + (-2.0)(0)$$
$$= (-0.5)(-M_p) + (2.5)(M_p) + (-2.0)(M_C)$$

where M_C is the bending moment at collapse at the apex C. Substituting in the value $M_p = 52.9$ kNm, M_C is found to be 47.0 kNm. Similarly, using the mechanism of Fig. 57,

$$(-1.0)(-162) + (2.5)(-25.9) + (-1.5)(0)$$
$$= (-1.0)(-M_p) + (2.5)(M_{B_3}) + (-1.5)(M_C)$$

On substituting $M_p = 52.9$ kNm, $M_C = 47.0$ kNm, M_B is found to be 45.9 kNm. Continuing in this way, a complete static check may be made of the whole frame.

Combination of Mechanisms (2, 3, 4)

The hinge rotations of Fig. 55 are determined quite easily, but the work becomes tedious if each new trial mechanism has to be examined directly. Fortunately, the technique of combining mechanisms, while being logical, also saves a considerable amount of labour.

The basic collapse mechanism for the pitched roof frame is shown in Fig. 49, and, on substituting the value $k = h_2/h_1 = 0.59$, the hinge rotations for design Example 3 become, at B, C, D and E respectively, $-\theta$, 2θ, -2.182θ, 1.182θ. It was found that this mechanism was incorrect, in that the hinge did not form at the apex C but at the first purlin point B_4. To change Fig. 49 to Fig. 55, hinge C must be suppressed, and hinge B_4 introduced. Now the mechanism of Fig. 56 involves a hinge at B_4; consider, therefore, the following table of hinge rotations:

	A	B	B_1	B_2	B_3	B_4	C	D	E
Fig. 49		-1					2	-2.182	1.182
Fig. 56		-0.5				2.5	-2		
Fig. 55		-1.5				2.5		-2.182	1.182

It will be seen that the hinge rotations of Fig. 55 are simply the sums of the rotations of Figs. 49 and 56. This usage of elementary mechanisms to build up by superposition more complex mechanisms is one of the basic tools of plastic design.

The rule for determining the number of independent mechanisms required for a given structure is simple. If N is the number of 'critical sections', at which hinges might form, and if R is the number of redundancies of the frame, then the number

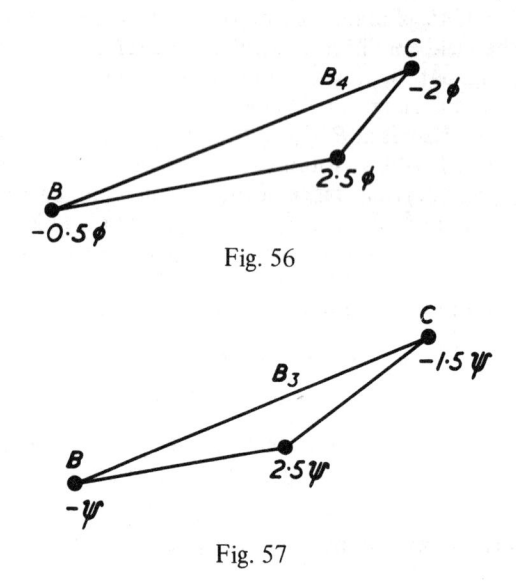

Fig. 56

Fig. 57

of independent mechanisms is $(N - R)$. Thus, for the pitched roof frame under the loading of Fig. 43 (b), hinges can form only at the five sections A, B, C, D, E. Since the frame has 3 redundancies, all possible mechanisms of collapse can be built up from 2 independent mechanisms, and these may well be taken as the mechanisms of Figs. 49 and 50. For the real loading, it is clear that each purlin point is a potential critical section; further, each purlin point requires one extra independent mechanism, and these may well be taken as those of Figs. 56, 57, etc.

The method of combination of mechanisms will not be amplified further here; the method is discussed at length in the literature. It may be noted that the technique sketched above differs slightly from that presented in the standard texts.

Deflections (6)

In common with elastic structural analysis, plastic theory assumes that deflections have little effect on equilibrium equations. Thus additional bending moments in a column, for example, due to axial loads combined with frame sway, are usually ignored. Plastic designs, no less than elastic designs, must be checked to see that the design assumptions are obeyed.

The pitched roof portal frame may develop uncomfortably large deflections, and semi-empirical rules have been developed to estimate the reduction in load factor that might be expected. Should this reduction exceed 10 per cent, then plastic design is not recommended. B.C.S.A. Publication 21 1963 gives the following

percentage reduction (Δ) in the collapse load factor for single-span pitched roof frames:

$$\Delta = \beta \frac{M_p l}{EI}$$

In this expression, M_p is the full plastic moment of the section and I the second moment of area; E is the elastic modulus (2.1×10^5 N/mm^2) and l the frame span. The parameter β is a function of the frame geometry only, and is given for fixed-base frames in Fig. 58. (Similar curves are given in the B.C.S.A. Publication for pinned-base frames.)

A trial section is selected to give the required nominal load factor for a particular design. Using the section properties and the appropriate value of β, the percentage reduction Δ in the load factor is calculated.

The formula expressing the percentage reduction in load factor, while semi-empirical, is conservative in that strain-hardening of the steel is neglected. It is in fact improbable that the full calculated reduction in load factor would be observed in a test to destruction, since strain-hardening will strengthen the frame.

Design Example 3 (continued). Accurate analysis of the pitched roof frame gave $M_p = 52.9$ kNm. For a nominal load factor of 1.75, the required plastic section modulus in steel to B.S. 4360 grade 43 is thus

$$Z_p = \frac{52.9 \times 1.75 \times 10^6}{245} = 377 \times 10^3 \text{ mm}^3$$

A 254 x 146 x 31 kg U.B. section is therefore selected, having $Z_p = 395$ cm^3, and the uncorrected value of load factor is thus

$$\lambda = \frac{245 \times 395 \times 10^3}{52.9 \times 10^6} = 1.83$$

With the given dimensions (Fig. 51), $k = h_2/h_1 = 0.59$, $p = h_2/l = 0.207$, and Fig. 58 gives $\beta = 42$. Thus

$$\Delta = \frac{\beta M_p l}{EI} = \frac{42 \times 245 \times 395 \times 10^3 \times 12 \times 10^3}{2.1 \times 10^5 \times 4\,427 \times 104} = 5.25$$

It is to be expected, therefore, that the collapse load factor of 1.83 might be reduced by deflections by 5.25 per cent to 1.73.

Column Design (7)

The stability of a column carrying a given axial load depends not only on the slenderness ratio about the minor axis, but also on the slenderness ratio about the major axis, and, most importantly, on the torsional properties of the section of the column. In addition, stability depends on the ratio of end moments acting on the column; other things being equal, a column bent in single curvature is more likely to become unstable than one bent in double curvature.

Horne has published curves (B.C.S.A. Publication 23) dealing with the plastic design of columns, in which the bending moment acting at one end of the column has the full plastic value M_p, while the bending moment at the other end can have

578   PLASTIC THEORY AND DESIGN

any value βM_p. Double curvature is indicated by $\beta = -1$, a pinned end by $\beta = 0$, and single curvature by $\beta = 1$. For a column having certain specified torsional properties, a chart, similar to that of Fig. 59, relates mean axial stress to permitted slenderness ratio about the minor axis for various values of β, the ratio of end moments. (Slenderness ratio about the major axis is of relatively slight importance, and is ignored for these *plastic* charts.)

The torsional property in question is the T value, which is tabulated at the end of this chapter.

Design Example 2 (continued). Column AB of Fig. 30 will be checked for stability at collapse. A 914 × 305 × 253 kg U.B. was selected for the portal frame, and the column is subjected to the full plastic moment at the top. Since the portal is pin-based, the ratio of end moments β is zero. At a load factor of 1.75, the column is subjected to a mean axial stress

$$p = \frac{(900)(1.75)(10^3)}{323 \times 10^2} = 49 \text{ N/mm}^2$$

Now $l/r_y = 4.5 \times 10^2/6.23 = 72.2$, and, from the tables, $T = 181 \text{ N/mm}^2$. From Fig. 59 for $T = 185$, the permitted axial stress p for $l/r_y = 72.2$ and for $\beta = 0$ is 50 N/mm², so that the column is just stable.

Design Example 3 (continued). For the uniform pitched roof portal frame of Fig. 51, a 254 × 146 × 31 kg U.B. was selected; the stability of column DE will be checked (see Fig. 54). For this column, $T = 238$, $l/r_y = 4.2 \times 10^2/3.18 = 132$, $\beta = -1$, and the mean axial stress at collapse is $54 \times 10^3 \times 1.75/3\,990 = 23.7 \text{ N/mm}^2$. Charts similar to Fig. 59 for $T = 215$ and $T = 245$ give the permitted axial stress as 21 and 24 N/mm² respectively; by interpolation, the permitted axial stress for $T = 238$ is 23.5 N/mm², so that the column can be considered just stable.

Further Stability Problems (2, 7)

The charts similar to Fig. 59 show a 'limiting slenderness ratio'. For slenderness ratios smaller than the limiting value, any ratio of end moments will effectively permit the development of full carrying capacity. A column which is found to be unsatisfactory can be brought into this 'safe' region by the provision of lateral supports, thus reducing the slenderness ratio, which is based on the unsupported length. As such, the limiting slenderness ratio curve on the charts thus gives design rules for lateral stability of a section.

The B.C.S.A. report gives also charts for the *elastic* design of columns. For example if the 914 × 305 × 253 kg U.B. had been retained for the *beam* of the portal frame of Design Example 2, but the column sections increased, then those columns would have remained elastic at collapse of the frame. The extra charts in the B.C.S.A. report enable these columns to be checked for stability, but these charts do of course require converting to metric units.

Repeated Loading (2, 3, 4)

Repeated loading on a steel frame does not necessarily preclude the use of plastic methods of design. If, however, the number of reversals of load during the estimated life of the structure can be numbered in the millions, then certainly an elastic design should be made, with the main design criterion that of fatigue.

CONNECTIONS

Work is proceeding at Cambridge and elsewhere on the loading of a structure into the plastic range by a few thousand (rather than million) reversals. It is too early yet to draw definite conclusions, but there seems at present to be no reason why plastic methods should not be used, without modification.

There remains the problem of a very few (tens or hundreds) applications of random combinations of live load, which may lead to incremental collapse, or to shakedown. Incremental collapse occurs in the following way. A certain combination of loads acting on a given frame may cause plastic hinges to form, but not in sufficient number to transform the frame into a mechanism. However, small plastic deformations could occur at these hinges. Another, different combination of loads may lead to the formation of other hinges, and so on. If all the hinges, which are formed at different stages, would transform the frame into a mechanism if they occurred simultaneously, then the small plastic deformations at the hinges can build up into large deformations over repeated cycles of loading. If this occurred, then, although an actual mechanism was not formed at any one stage in the loading history, the frame would nevertheless distort incrementally into the collapse state.

Shakedown occurs when, despite some initial plastic deformation, the frame eventually resists all further variations and combinations of load by purely elastic action.

For usual ratios of live to dead load, the probability is small of critical combinations of load leading to incremental collapse. However, for unusually high live loads, the designer ought to consult the literature for techniques of shakedown analysis.

Connections

It will have been seen that the problem of joining together two members of a frame is mainly that of assessing the *strength* of the connection. From the point of view of simple plastic theory, the deformation characteristics of the connection are not of prime importance. For example, in a continuous beam involving a change of section, the connection must be such that the full plastic moment of the weaker section can be developed without failure of the connection. In some design problems, of course, a flexible connection, even if of adequate strength, may lead to objectionable deflections, and the designer must then ensure that the structure as a whole is serviceable.

In other problems, the designer may deliberately use over-strong or very stiff connections, in order to satisfy some particular design requirement. A good example is haunching at the eaves and apex of pitched roof portal frames. Consider the collapse bending moment diagram of Fig. 53. (It will be remembered that the hinge forms not at the apex C but at the purlin point B_4.) It is clear that if the designer introduces an apex haunch, perhaps extending from B_4 to the corresponding purlin point on the other rafter, then the collapse bending moment diagram and the collapse load will be unaffected. However, the working load bending moment diagram will have the same general features as Fig. 53, and, in particular, there will be a region of high and nearly constant bending moment in the neighbourhood of the apex C. Thus, although the *strength* of the frame is not affected by the apex haunch, deflections are likely to be reduced significantly.

By contrast, if a haunch were introduced at the eaves, B, where the bending moment diagram is 'peaky', little improvement would be expected in deflection behaviour; the collapse load, however, will be markedly increased by an eaves haunch.

For whatever reason a connection is made, the design requirement for the connection is the same. The connection must be strong enough to permit the sections meeting at a joint to develop their required strengths. That is, if a hinge should form at collapse adjacent to a connection, then the connection must be capable of taking, without failure, whatever bending moment is transmitted through it to the other members.

This does not mean that every connection should necessarily be full strength. For example, returning to Fig. 53, a site connection could be made near purlin point B_1, the connection being subjected to only very small bending moments.

In general, however, full-strength connections will be required. Greatest economy of material, and cleanness of the connection, will almost certainly be achieved if welding, whether site or shop, is used. The design criterion for the welds themselves is clear; failure of an individual weld cannot be tolerated, and the welds must have sufficient throat and run so that the full strength of the structural elements can be developed.

Full strength connections can also be made by bolting, and high strength bolts; will lead to the greatest economy.

For general methods of calculation of connections, the reader is referred to Chapter 28.

Special Note

The examples shown are in steel to B.S. 4360 grades 43 and 50 only. Grade 55 steel has limited use for plastic design because of local flange instability.

TABLE 1
UNIVERSAL BEAMS
T Values and Suitability for Plastic Action

Serial Size mm	Wt. per m kg	T Value N/mm²	Steel to B.S. 4360 Grade 43	Steel to B.S. 4360 Grade 50
914 x 419	387	287	Yes	Yes
,, ,,	343	227	,,	,,
914 x 305	289	233	Yes	Yes
,, ,,	253	181	,,	,,
,, ,,	225	143	,,	,,
,, ,,	201	118	,,	,,
838 x 292	226	189	Yes	Yes
,, ,,	194	140	,,	,,
,, ,,	176	117	,,	No
762 x 267	196	211	Yes	Yes
,, ,,	173	165	,,	,,
,, ,,	147	123	,,	,,
686 x 254	170	223	Yes	Yes
,, ,,	152	181	,,	,,
,, ,,	140	155	,,	,,
,, ,,	125	126	,,	No
610 x 305	238	451	Yes	Yes
,, ,,	179	264	,,	,,
,, ,,	149	189	,,	No
610 x 229	140	239	Yes	Yes
,, ,,	125	193	,,	,,
,, ,,	113	159	,,	,,
,, ,,	101	130	,,	No
533 x 330	211	470	Yes	Yes
,, ,,	189	383	,,	,,
,, ,,	167	303	,,	,,
533 x 210	122	289	Yes	Yes
,, ,,	109	231	,,	,,
,, ,,	101	203	,,	,,
,, ,,	92	170	,,	,,
,, ,,	82	137	,,	No
457 x 190	98	321	Yes	Yes
,, ,,	89	267	,,	,,
,, ,,	82	227	,,	,,
,, ,,	74	189	,,	,,
,, ,,	67	156	,,	,,
457 x 152	82	309	Yes	Yes
,, ,,	74	257	,,	,,
,, ,,	67	210	,,	,,
406 x 178	74	277	Yes	Yes
,, ,,	67	228	,,	,,
,, ,,	60	184	,,	,,
,, ,,	54	150	,,	No
406 x 152	74	347	Yes	Yes
,, ,,	67	284	,,	,,
,, ,,	59	227	,,	,,
406 x 140	46	151	Yes	Yes
,, ,,	39	110	,,	No
381 x 152	67	332	Yes	Yes
,, ,,	60	265	,,	,,
,, ,,	52	208	,,	,,
356 x 171	67	335	Yes	Yes
,, ,,	57	243	,,	,,
,, ,,	51	197	,,	,,
,, ,,	45	156	,,	No
356 x 127	39	180	Yes	Yes
,, ,,	33	134	,,	,,
305 x 165	54	342	Yes	Yes
,, ,,	46	259	,,	,,
,, ,,	40	201	,,	No
305 x 127	48	408	Yes	Yes
,, ,,	42	323	,,	,,
,, ,,	37	262	,,	,,
305 x 102	33	234	Yes	Yes
,, ,,	28	180	,,	,,
,, ,,	25	135	,,	,,
254 x 146	43	424	Yes	Yes
,, ,,	37	326	,,	,,
,, ,,	31	238	,,	No
254 x 102	28	294	Yes	Yes
,, ,,	25	234	,,	,,
,, ,,	22	196	,,	,,
203 x 133	30	408	Yes	Yes
,, ,,	25	311	,,	No

TABLE 2
UNIVERSAL COLUMNS
T Values and Suitability for Plastic Action

Serial Size mm	Wt. per m kg	T Value N/mm²	Steel to B.S. 4360 Grade 43	Steel to B.S. 4360 Grade 50
396 x 381	634	6 380	Yes	Yes
,, ,,	550	5 040	,,	,,
,, ,,	467	3 800	,,	,,
,, ,,	393	2 800	,,	,,
,, ,,	339	2 150	,,	,,
,, ,,	287	1 581	,,	,,
Column Core	476	4 100	,,	,,
356 x 368	202	891	,,	,,
,, ,,	177	693	,,	No
,, ,,	153	526	,,	No
,, ,,	129	381	No	No
305 x 305	283	3 040	Yes	Yes
,, ,,	240	2 270	,,	,,
,, ,,	198	1 600	,,	,,
,, ,,	158	1 050	,,	,,
,, ,,	137	804	,,	,,
,, ,,	117	602	,,	No
,, ,,	97	417	No	No
254 x 254	167	2 390	Yes	Yes
,, ,,	133	1 575	,,	,,
,, ,,	107	1 063	,,	,,
,, ,,	89	754	,,	,,
,, ,,	73	515	,,	No
203 x 203	86	1 618	Yes	Yes
,, ,,	71	1 144	,,	,,
,, ,,	59	807	,,	,,
,, ,,	52	633	,,	No
,, ,,	46	497	No	No
152 x 152	37	956	Yes	Yes
,, ,,	30	628	,,	No
,, ,,	23	420	No	No

Information based on tables prepared by Professor M. R. Horne.

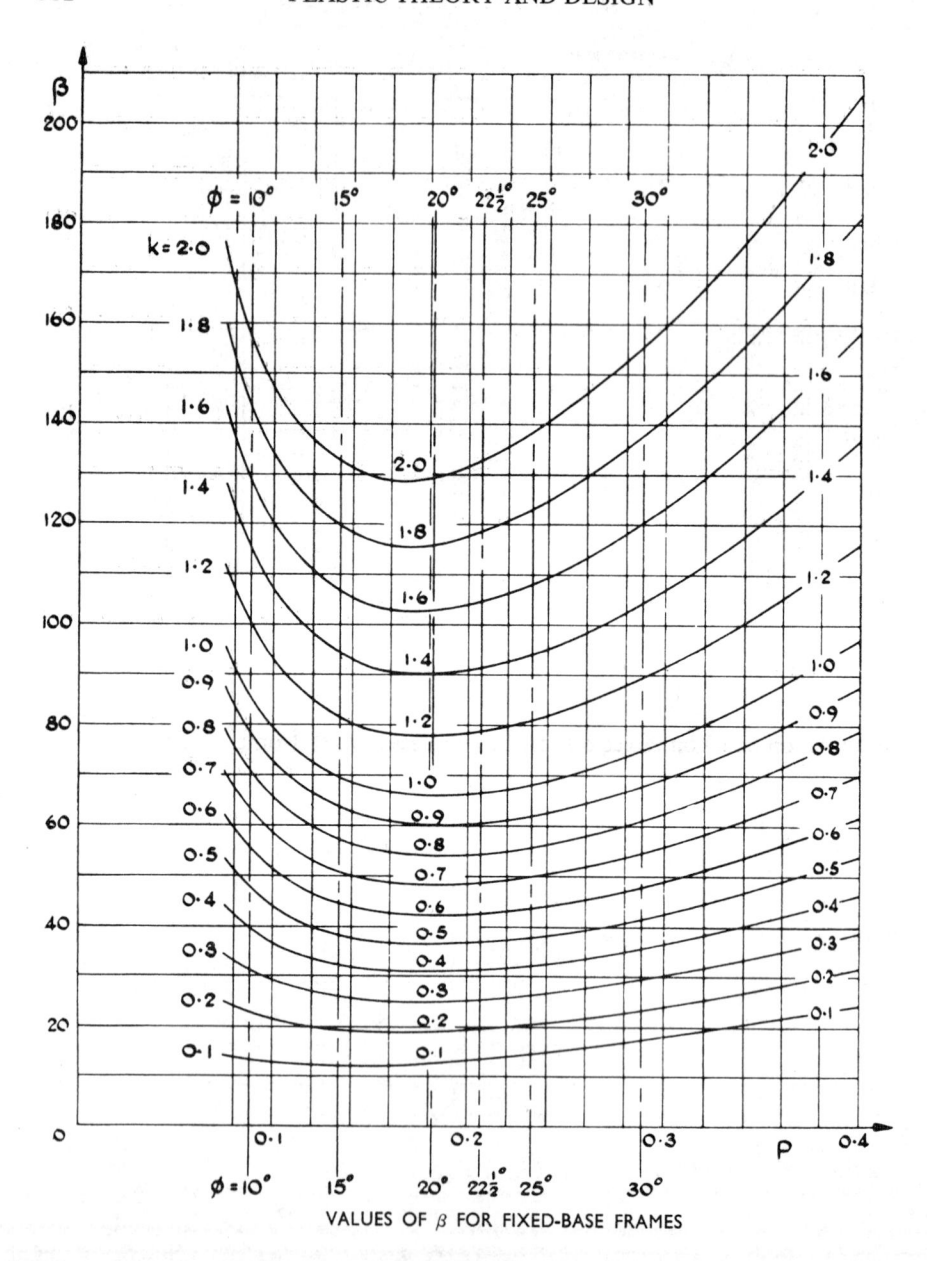

Fig. 58

VALUES OF β FOR FIXED-BASE FRAMES

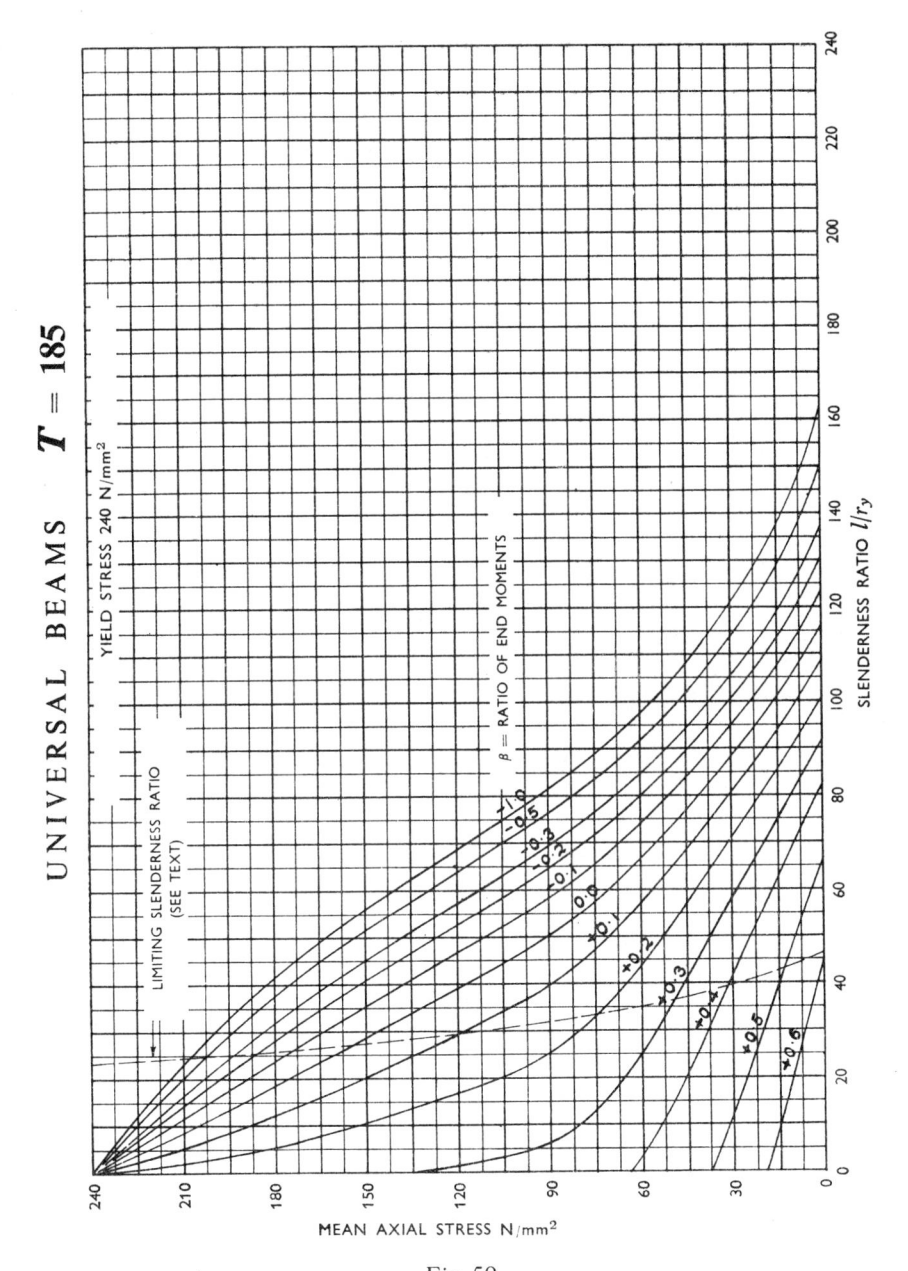

Fig. 59

23. COMPOSITE CONSTRUCTION

AS the metric version of C.P. 117 is under review and is not likely to be available for some time this chapter has been metricated from the original, and should, therefore, be regarded only as a guide to the principles involved. The various references in the text are to codes and publications which are in imperial terms and any necessary conversions must be made by the reader.

Figure 1 shows a steel beam supporting a concrete slab, there being no mechanical connection between them and bond stress being taken as zero. When the slab is loaded, both slab and beam deflect individually; there is relative movement at the common interface and the whole of the load is taken by the steel joist acting alone, the resulting stress diagram being as shown.

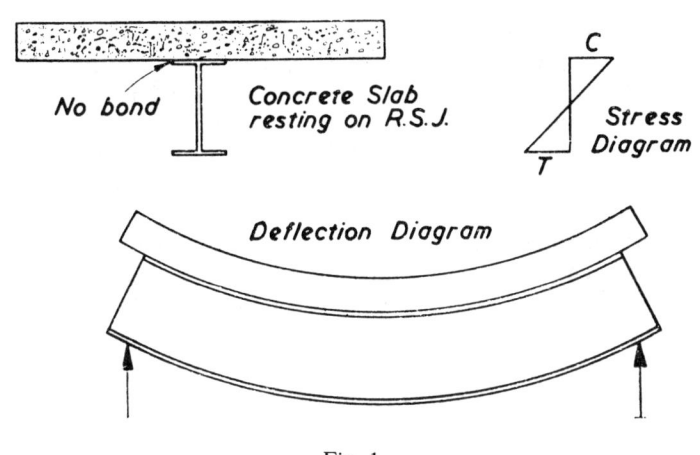

Fig. 1

If, however, there is some form of mechanical bond between the two elements to transfer the horizontal shear from slab to beam across the common interface, so that relative slip between slab and beam is obviated, then together the 'composite section' will behave as a T-beam, in which all or most of the compression will be taken by the concrete and all the tension will be taken by the steel.

Although composite action designs have been used in Europe for many years, they have only recently been used for building work in Great Britain, although the method has been widely adopted for bridgework in this country and is practically universal for plate girder or beam bridges with reinforced concrete decks. This position has changed radically since 1960, following the publication by the British Constructional Steelwork Association of three brochures presenting the sectional properties of an extensive range of composite sections suitable for use in steel framed buildings. These brochures, based on elastic design, set out a provisional basis for design, since, at the date of issue, official regulations did not exist in Great Britain, although both European and American regulations were available.

A considerable further step forward was taken with the publication, in January 1965, of C.P. 117, Part 1, 'Composite Construction in Structural Steel and Concrete: Simply-supported beams in Buildings.' This sets out the detailed requirements for the design of composite sections and includes both elastic and load-factor design methods. Designs made to the requirements of this Code will satisfy all local byelaw requirements and the contents of the B.C.S.A. brochures have been revised in accordance with Code requirements, and embodied in one brochure, No. 25, published in 1965.

There are a number of ways in which a composite beam can be constructed. The steel beam, simply supported at its ends, can carry the shuttering and concrete during the construction period. In this case the construction loads are carried solely by the steel beam and the super load and weight of finishes applied after the concrete has set and attained the requisite strength, will be carried by the composite section.

Alternatively, the beam supporting the concrete and shuttering can be continuously propped during construction so that it does not deflect, thus remaining unstressed. In this case, after the concrete has attained the requisite strength, the props are removed, after which both dead and superimposed loads are carried by the composite section. For all practical purposes the beam may be assumed to be continuously propped if it has three supports equally spaced between the ends.

From the point of view of strength of member, the second method is the more advantageous but against this must be set ease of construction in the first case, which leaves the floor below entirely clear of props. The resulting stress diagrams are shown in the top two diagrams in Fig. 2, and they are the two most commonly used methods.

Other arrangements of props are possible and a single support at the centre of the span has advantages. If this is used, then there will be a negative moment over the central prop due to the construction loads and the stresses induced in the steel beam at this point will remain when the concrete sets; they are opposite in nature to those developed subsequently. The forces due to the removal of the central prop and subsequent application of the superimposed load will be taken by the composite section of the full span. The final stresses will be summation of those due to the superimposed load and those due to the positive and negative moments induced during construction and in certain cases economy will be shown over the two previous methods but at the cost of extra work in design.

It is also possible to secure further economy in steel by jacking a predetermined upward deflection into the beam, or by prestressing the beam by other methods, before casting the slab. The effect of this on final stresses is also shown in Fig. 2.

The most commonly used shear connector is undoubtedly the headed stud, welded electrically to the steel joist by means of a 'stud gun' and the relative merits of these and others forms are dealt with later on pages 592-5 and 608-9, when detail design is investigated. The duty of shear connectors is primarily to resist horizontal movement between the concrete slab and steel beam and so transfer the horizontal shear. It is also necessary to restrain the slab, which is under compression, from lifting off the beam and it is for this reason that studs are headed and that loops are added to bar connectors. If a solid concrete casing is used, then the shear connection is most conveniently made by rod stirrups bent to a rectangular outline, such as are used in conventional reinforced concrete T-beams: the material costs for these will be less than for studs, since the stirrups can be of mild steel, whereas it is universal to make studs of a higher quality steel.

COMPOSITE CONSTRUCTION

Typical Stress Diagrams

Stresses when the dead load is taken by the steel only

Stresses when the dead load is taken by the composite beam

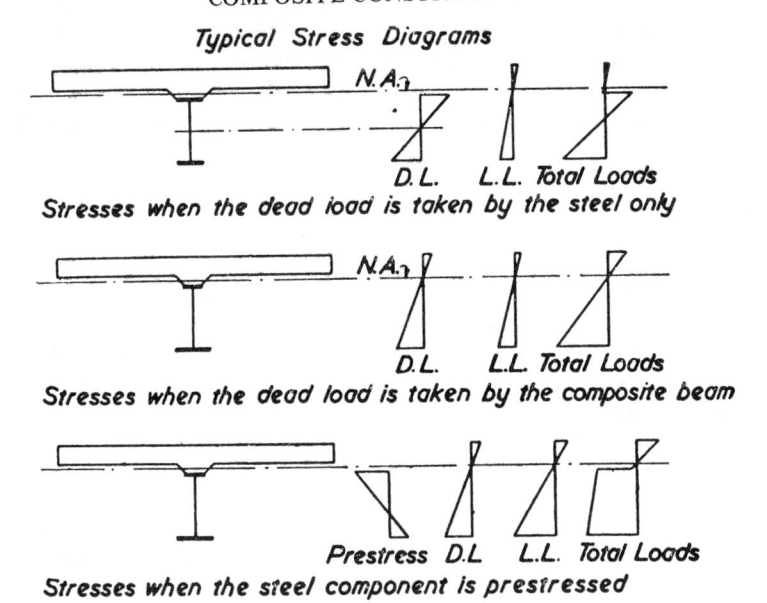

Stresses when the steel component is prestressed

Fig. 2

Before proceeding to investigate the design procedure it is interesting to study briefly the economy which can be obtained by the adoption of composite action when compared with what may be called conventional design. All multi-storey buildings are required to have a standard fire resistance period, which varies with the use, size and location of the building, so that the adoption of some kind of concrete floor is practically universal. The thickness of this floor, spans being the same, will be unaltered, whether the steel frame be designed conventionally or on the basis of composite action. There will, however, be a considerable reduction in both weight and cost of steelwork if used compositely.

The steel members will require some form of fire protection, and this can be given by the use of solid casing in concrete, or hollow lightweight casings, such as vermiculite or similar board. It is obvious that, for any given layout of the frame, the least weight of steelwork will result from the use of hollow lightweight casings to the beams and solid concrete casings to the stanchions, allowance being made in the latter for the load carrying capacity of the concrete in accordance with Clause 30b of B.S. 449. However, it will often be found that the least overall cost of steel and casing will be given by the use of lightweight hollow casings throughout, notwithstanding the extra weight of steel in the stanchions.

Composite beam sections can be designed elastically, using the method of transformed sections, final stresses in concrete and steel being restricted to those given in the appropriate documents, C.P. 114 and B.S. 449. Alternatively, they can be designed by the ultimate load, or load factor, method using a factor of 1.75. Steel sections obtained by this mode of design will be smaller than those obtained elastically.

Composite stanchion sections can be designed on the basis of Clause 30b, of B.S. 449, or they may be designed as reinforced concrete columns on the lines of

the requirements of C.P. 114, pending publication of a further part of C.P. 117 dealing with columns. The use of B.S. 449 is recommended.

Savings in weights of steel beams designed in mild steel with elastic composite action can exceed 25 per cent, whilst if high yield stress steel is used in conjunction with load factor design the saving can be up to 40 per cent, compared with the weight of the steel beam to carry the same load. Savings in stanchion weight when designed compositely vary; the greatest saving is on the smaller sizes, where the relative load carried by the concrete is the larger and for information on this point the reader is referred to Chapter 30.

The economy resulting from the use of composite action in a conventional office block, 10 storeys high was investigated by a Joint Working Party of the Ministry of Public Building and Works and the British Constructional Steelwork Association, and the main results are given in a paper by Mr. L. R. Creasy, M.O.P.B.W., published in *The Structural Engineer*, December 1964. This paper shows, in the section dealing with relative costs, that when allowance has been made for the increased column stresses permitted by the 1964 Addendum to B.S. 449, and later incorporated in the 1965 edition of the Standard, a reduction of approximately 20 per cent in the weight of the steel framework and about the same percentage reduction in the overall cost of the carcase of the building including steelwork, casing, floors and foundations is obtainable. Greater savings will be made on buildings more heavily loaded.

A further and real advantage to the steelwork designer is that the adoption of composite action for beams reduces the deflection criterion, as the stiffness of the composite section is many times that of the steel beam required to carry the same load. Thus one of the handicaps to the use, in beams, of more highly stressed steel is lessened and Grade 50 steel can be used with advantage.

Design Procedure

(i) Slabs

The design of the reinforced concrete floor slab is independent of composite action and the thickness will be determined by the span and loading conditions, or alternatively, by the minimum thickness required for fire resistance.

To ensure compliance with local By-laws, slabs should be designed in accordance with the requirements of C.P. 114, 'The Structural Use of Reinforced Concrete in Buildings'. The effective span of the slab should be taken as the lesser of:

(*a*) the distance between centres of bearings; or
(*b*) the clear distance between edges of supports plus the effective depth of the slab.

Thus in the case of composite sections utilising solid concrete encasures, to which reference is made later in this chapter, the effective span is usually that given under heading (*b*) above.

Slabs may be designed by the elastic method, or by the load-factor method. Design factors for slabs reinforced in tension only, utilising the elastic method are given in Table D on page 628 for the three most commonly used mixes, and for design by the load-factor method, the reader is referred to Clause 306 of C.P. 114.

DESIGN PROCEDURE

It is important to note that, as the stresses caused by composite action need not be added to the normal bending stresses in the slabs, this implies slabs spanning in one direction only, i.e. at right angles to the beam span. Slab bending moments, where continuous, can be calculated for the following arrangements of superimposed load:

(a) alternate spans loaded and all other spans unloaded; or
(b) any two adjacent spans loaded and all other spans unloaded.

In such cases, the negative moments so calculated may be increased or decreased by not more than 15 per cent, provided that these modified negative moments are used for the calculation of the corresponding span moments.

Generally, however, slabs uniformly loaded and spanning over three or more approximately equal spans, may be assumed to have bending moment values as set out in Table A below. Two spans may be considered as approximately equal when they do not differ by more than 15 per cent of the longer span.

TABLE A

	Near Middle of End Span	At Support Next to End Support	Middle of Interior Spans	At Other Interior Supports
Dead load moment	$+\dfrac{W_d l}{12}$	$-\dfrac{W_d l}{10}$	$+\dfrac{W_d l}{24}$	$-\dfrac{W_d l}{12}$
Superimposed load moment	$+\dfrac{W_s l}{10}$	$-\dfrac{W_s l}{9}$	$+\dfrac{W_s l}{12}$	$-\dfrac{W_s l}{9}$

W_d = total dead load
W_s = total superimposed load $\bigg\}$ per span l

The use of precast shuttering as an integral part of the composite section is permitted by C.P. 114, subject to some minor constructional requirements which are set out in Clause 503 of that document. Hollow tile floors can also be used provided that over the beams there is solid concrete of sufficient width to develop the required moment.

(ii) Composite Sections

(a) Elastic Method

Stresses must be computed on the basis of a fully composite section with a modular ratio, $m = 15$, the concrete being assumed to take no tension. If the steel beam is unpropped during construction and thus carries the surrounding concrete casing, if any, and the floor, it must be designed in accordance with B.S. 449 and the stresses at this stage added to those resulting from full composite action when the superload is taken into account. The final stresses must not exceed the permissible stresses for steel and concrete given in B.S. 449 and C.P. 114 respectively. The

590   COMPOSITE CONSTRUCTION

reference to B.S. 449 at the construction stage indicates that the allowable stresses will be those given in Tables 3a, etc., of the Standard, making allowance for the unsupported length as set out in Clause 26a. The stresses in the final condition will be those given in Table 2 of B.S. 449 since continuous lateral support is given by the floor.

The effective width, b, of the concrete compression flange must not exceed the least of the following (see Fig. 3 and 4):

    For T beams   : one third of the span of the beam
(slabs on both sides) : the centres of the steel beams
    Fig. 3   : twelve times the slab thickness plus the width of the support.

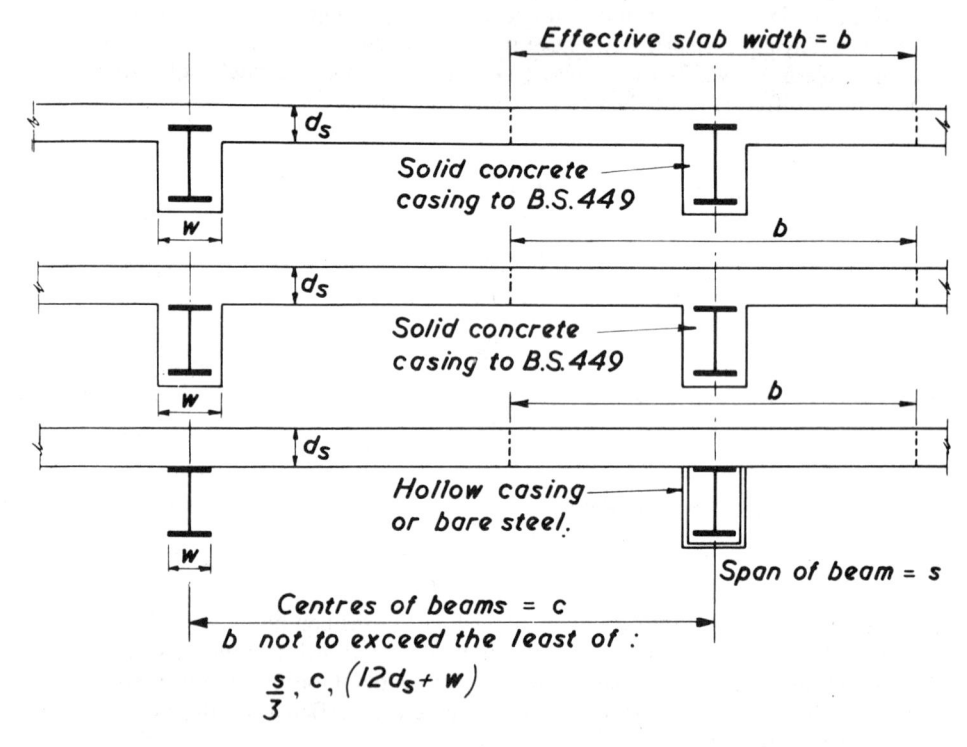

Fig. 3

    For L beams   : one sixth of the span of the beam
(slab on one side) : half the distance between centres of steel beams plus the width of the support
    Fig. 4   : four times the slab thickness plus the width of the support.

(*b*) *Elastic design using concrete cased beams*

Provided that the steel section has approximately equal flanges, and is fully encased in *in situ* concrete in accordance with the requirements for cased beams in Clause 21 of B.S. 449 and the top surface of the top flange is not less than 50 mm

DESIGN PROCEDURE

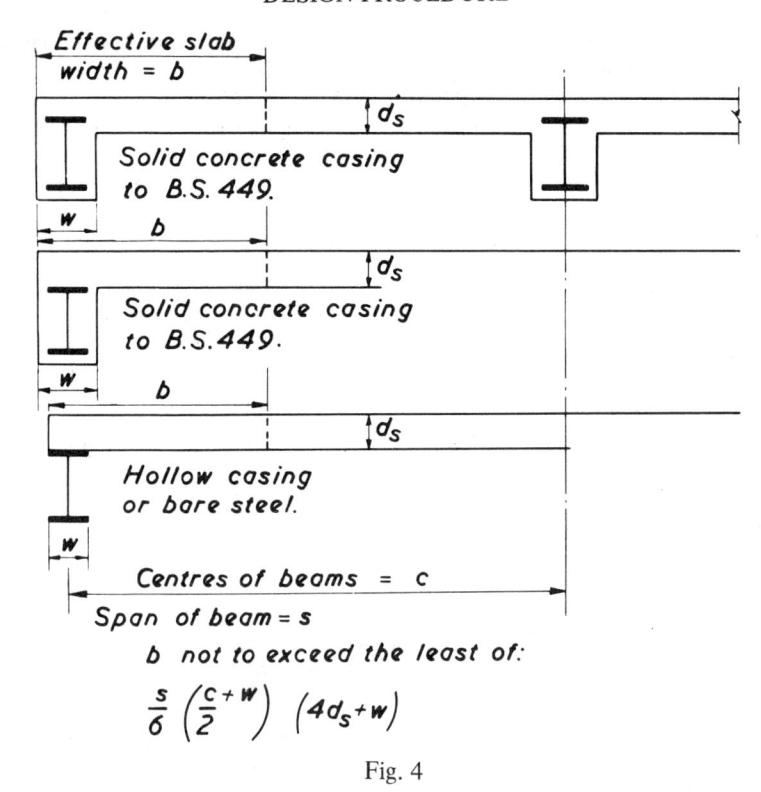

Fig. 4

Effective slab width = b

Solid concrete casing to B.S. 449.

Solid concrete casing to B.S. 449.

Hollow casing or bare steel.

Centres of beams = c
Span of beam = s
b not to exceed the least of:

$$\frac{s}{6} \quad \left(\frac{c+w}{2}\right) \quad (4d_s + w)$$

above the underside of the slab and also that the composite beam is not subjected to heavy concentrated loads, then the stirrups or binding specified are considered adequate to transfer the horizontal shear and no other form of connector is needed.

The detailed requirements for the concrete casing are:

minimum concrete strength, 21 N/mm² at 28 days.
minimum width of casing = b + 100 mm, where b = breadth of steel flange.
minimum cover to surfaces and edges of flanges = 50 mm.
stirrups to be at least 5 mm dia. at not more than 150 mm centres, and to pass through the centre of the cover to the edges and soffit of the lower flange (see Fig. 5).

(c) *Elastic design with hollow casing, or with solid concrete casing and haunches, or for beams having heavy concentrated loads*

In all these three cases of elastic design, shear connectors are required.

(d) *Load factor design*

As an alternative to elastic design, the composite section may be designed by the load factor method, using an overall factor of 1.75. When calculating the ultimate moment of resistance, the stresses to be taken are the specified yield strength of the steel and four ninths of the specified cube strength of the concrete; at 28 days if

ordinary Portland cement is used, or 7 days for concrete made with rapid-hardening Portland cement. For beams unpropped, the stresses in the steel section during the construction stage must meet the requirements of B.S. 449.

In all cases, the total elastic stress in the steel beam at working loads must not exceed nine tenths of the yield stress, not must the elastic stress in the concrete at working loads exceed one third of the specified cube strength.

(iii) Shear Connectors

(*a*) For all cases, other than that outlined in (*b*) above, shear connectors must be provided to transmit the horizontal shear, ignoring any bond between the steel beam and the concrete slab. The connectors must also prevent the slab from lifting.

In all cases, the shear connectors are to be designed by the load-factor method.

(*b*) The number of connectors provided must be sufficient to resist the maximum value of the total horizontal shear load to be transmitted at collapse, between points of maximum and zero moments, which is the total compressive force in the concrete.

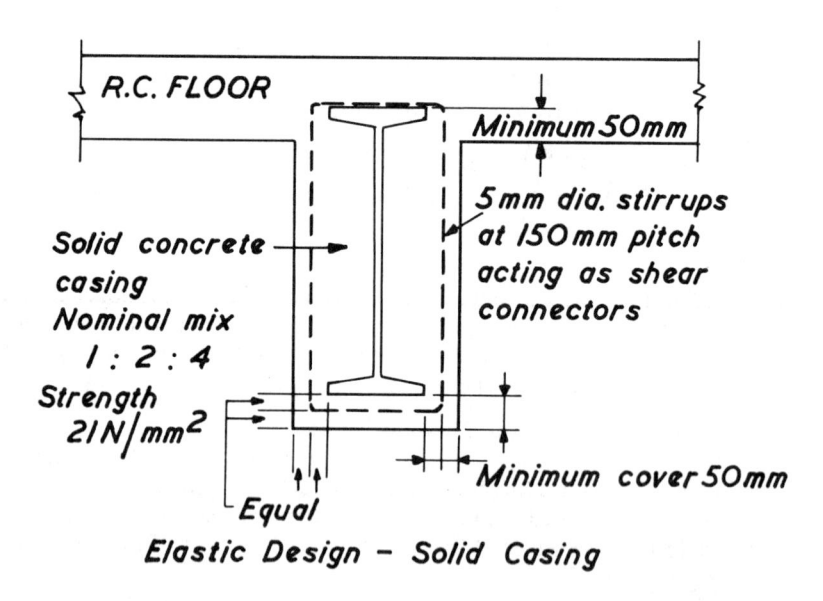

Fig. 5

(*c*) Types of and design loads for the most commonly used connectors are given in Fig. 6 and Table B, extracted from C.P. 117 and for other types of connector may be taken as 80 per cent of the lowest ultimate capacity as determined by tests in accordance with Clause 10 of the Code.

(*d*) If concrete haunches are used between the steel beam and the slab, with a slope steeper than 1 vertical to 3 horizontal, then tests must be made in accordance with Clause 10 of the Code to establish the design load for the shear connector. These test specimens must incorporate the proposed haunch and reinforcement.

DESIGN PROCEDURE

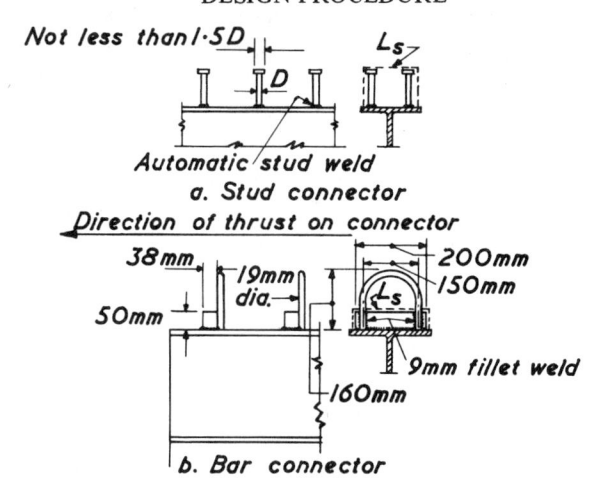

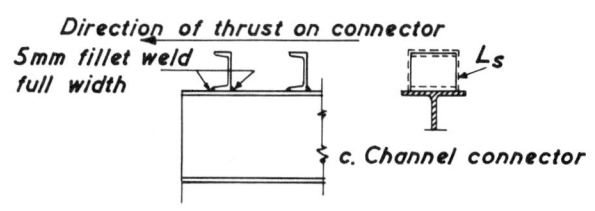

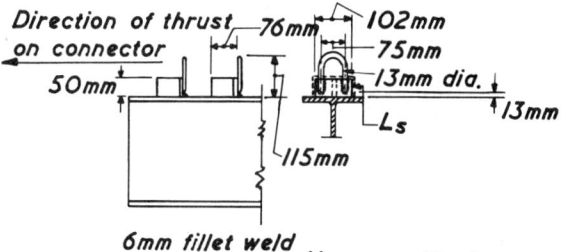

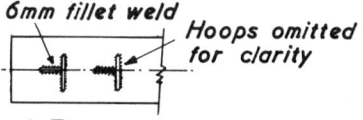

a. Stud connector

b. Bar connector

c. Channel connector

d. Tee connector

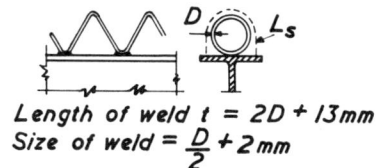

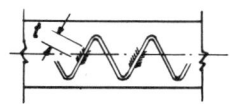

Length of weld $t = 2D + 13mm$
Size of weld $= \dfrac{D}{2} + 2mm$

e. Helical connector

Fig. 6

TABLE B
DESIGN VALUES FOR SHEAR CONNECTORS

Types of Connector	Connector Materials	Welds	Design values of Connectors for Concrete Strength (N/mm^2)		
			21	28	42
Headed studs Fig. 6 (c)	Min. yield stress 385 N/mm^2 Min. ult. tensile stress 494 N/mm^2	See Fig. 6 (a)	Load per stud, P_c, in kN		
Diam. (mm) overall (mm) height					
25 100			119	131	155
22 100			98	108	128
19 100			78	86	102
19 75			67	74	88
16 75			57	63	75
13 62			36	40	48
Bars with hoops Fig. 6 (b)	Grade 43	See Fig. 6 (b)	Load per bar, P_c, in kN		
50 x 38 x 200 mm bar			400	530	800
Channels, Fig. 6 (c)	Grade 43	See Fig. 6 (c)	Load per channel, P_c, in kN		
127 x 64 x 14.9 kg x 150 mm			228	252	300
102 x 51 x 10.4 kg x 150 mm			210	234	282
76 x 38 x 6.7 kg x 150 mm			198	222	270
Tees with hoops, Fig. 6 (d)	Grade 43	See Fig. 6 (d)	Load per connector P_c, kN		
102 x 76 x 12.7 mm T 50 mm high with 13 mm diam. bar loop			202	222	262
Helices Fig. 6 (e)	Grade 43	See Fig. 6 (e)	Load per pitch, P_c, kN		
Bar diam. mm Pitch circle diam. mm					
19 125			162	178	208
16 125			124	136	160
13 100			86	95	112
10 75			50	55	64

Note:
1. Connector values for bars and channels of less length than those quoted above are proportional to the length.
2. The values are not applicable where there is a concrete haunch between beam and slab with a slope steeper than 1 vertical and 3 horizontal.

DESIGN PROCEDURE

(e) Except in cases with heavy concentrated loads, for which the procedure laid down in Appendix C of the Code must be followed, the shear connectors may be spaced uniformly between each end of the beam and the point of maximum moment. Their spacing need not be in accordance with the shear diagram.

(f) The spacing of shear connectors must not exceed 600 mm nor four times the slab thickness.

The minimum longitudinal spacing of shear connectors will vary with the type of connector used, since it is governed by the load on one connector, the number of connectors in any one transverse line and the maximum horizontal shear force per mm run of beam, which is dealt with in the section which follows.

If a single row of studs is used they should be placed over the centre of the web of the steel section. The Code does not deal with the transverse spacing of studs, so that, if they are used in pairs, it is convenient from the fabricating angle to space them at the normal cross-centres for transverse flange holes for rolled sections. If the studs are regarded as rivets, then by analogy with the requirements for edge distance in Clause 51 of B.S. 449 there will be some limitation upon stud diameter in relation to flange width. These are listed in Table C, in which the flange widths given are the nominal widths using standard sizes of studs.

TABLE C

Nominal Flange Width mm	Standard Cross Centres mm	Stud Diameter mm
127	70	19
133	70	22
146	70	25
152	90	22
152–191	90	25
203–210	140	22
229	140	25

When three or more lines of studs are used, the maximum distance between the outermost studs will be governed by the permitted edge distances. Table D, showing minimum edge distances, has been compiled having regard to the recommendations of the makers.

TABLE D

Stud diameter mm	12	14	16	22	25
Edge distance mm	32	33	35	36	38

The minimum spacing of studs, governed by the welding gun, varies with the manufacturer, but ranges approximately from 45 mm for 12 mm studs to 56 mm for 25 mm studs.

(iv) Maximum Horizontal Shear Force

The shear force (N/mm run) of beam, Q,

$$= \frac{N_c \times \text{load in Newtons on one connector at ultimate load}}{\text{Longitudinal spacing of connectors in mm}}$$

This must not exceed either

(1) $C1 L_s \sqrt{u_w} + A_t . f_y . n$, or
(2) $C2 L_s \sqrt{u_w}$.

where $C1$ and $C2$ are constants based on units concerned.

Transverse steel reinforcement must be provided in the bottom of the slab or the haunch and this must not be less than $\dfrac{Q}{4f_y}$ mm²/mm run of beam.

The notation used in the expressions above is:

N_c = number of connectors at a cross-section.
u_w = specified cube strength of concrete in N/mm².
L_s = length in mm of the periphery of the connectors at a cross-section. This must not exceed the slab thickness for L beams nor twice the slab thickness for T beams. The thickness of the slab may be taken as the depth of the slab plus the haunch depth, provided that the haunch slope does not exceed 1 vertical to 3 horizontal.
A_t = area of transverse reinforcement mm²/mm run of beam in the bottom of the slab.
n = number of intersections of each lower transverse reinforcing bar and a shear surface, which equals 2 for T beams and 1 for L beams.
f_y = yield stress in N/mm² of the steel reinforcement.

(v) Tie Down of Slab

The overall height of a connector must not be less than 50 mm, and the minimum projection into the compression zone of the slab must not be less than 25 mm. The compression zone must be taken as that at the section of maximum bending moment calculated by the load-factor method. Studs must be headed, the diameter of the head being not less than $1\tfrac{1}{2}$ times the stud diameter.

(vi) Deflections

Should it be necessary to verify the deflection of the composite section, calculations should be made on the elastic basis, using a modular ratio of 15 for imposed loads and 30 for dead loads. The calculated deflection should not exceed that specified in B.S. 449.

Calculation of Section Properties

(a) Elastic Basis

The position of the neutral axis in this case is determined solely by the geometrical properties of the composite section and is independent of the stresses developed.

CALCULATION OF SECTION PROPERTIES 597

Two cases require consideration:

(1) Neutral axis within the depth of the concrete slab.
(2) Neutral axis below the slab and in the steel beam.

The depth to the neutral axis having been found by area moments about any convenient axis, the moment of inertia and section moduli are found in the usual manner.

$r = \dfrac{A_s}{b \cdot d_g}$ m = modular ratio

$d_e = d_g \left\{ \sqrt{m \cdot r (2 + m \cdot r)} - m \cdot r \right\}$

$I_g = \dfrac{b \cdot d_e^3}{3m} + I_s + A_s (d_g - d_e)^2$ in steel units.

Z_{st} = Section modulus for bottom flange of steel
  $= \dfrac{I_g}{d_b - d_e}$ In steel units.

Z_{cc} = Section modulus for top of slab
  $= \dfrac{m \cdot I_g}{d_e}$ In concrete units.

ELASTIC DESIGN — CASE I.
Neutral axis in slab.

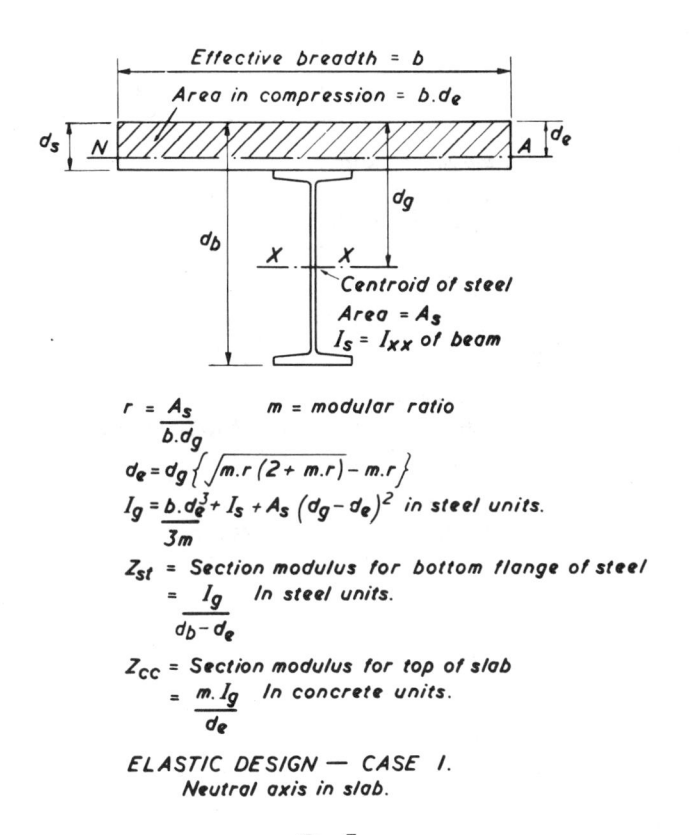

Fig. 7

In case 1, only that portion of the slab above the neutral axis is in compression and the formulae in Fig. 7 apply; they are the usual formulae for reinforced concrete beams, the steel section being the reinforcement.

In case 2, the whole of the slab and the portion of the steel beam above the neutral axis are in compression and the formulae in Fig. 8 should be used.

In both cases calculations are necessary for values of $m = 15$ and 30, the first for strength and the second for deflection. If it is required to find the stress in the top flange of the steel beam it can be done by proportion to the bottom flange stress in

the ratio $\dfrac{(d_e - d_s)}{\frac{1}{2} \text{steel beam depth}}$. This stress will be tension in Case 1 and compression

Figure 8 diagram: Composite section with effective breadth b, slab depth d_s, distances d_e, d_g, d_b, showing neutral axis N-A and centroid of steel.

Centroid of steel
Area = A_s
I_s = I_{xx} of beam.

m = modular ratio

$$d_e = \frac{\frac{b \cdot d_s^2}{2m} + A_s \cdot d_g}{\frac{b \cdot d_s}{m} + A_s}$$

$$I_g = \frac{b \cdot d_s^3}{12m} + \frac{b \cdot d_s}{m}\left(d_e - \frac{d_s}{2}\right)^2 + I_s + A_s\left(d_g - d_e\right)^2$$

in steel units.

Z_{st} = Section modulus for bottom flange of steel

$$\frac{I_g}{d_b - d_e} \quad \text{In steel units.}$$

Z_{cc} = Section modulus for top of slab

$$= \frac{m \cdot I_g}{d_e} \quad \text{In concrete units}$$

ELASTIC DESIGN - CASE 2.
Neutral axis below slab.

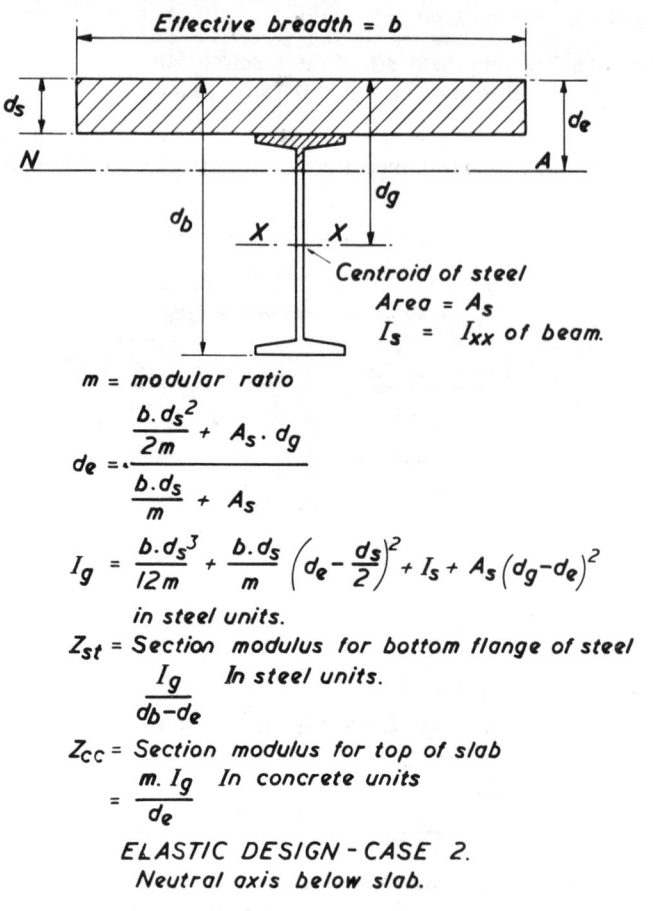

Fig. 8

in Case 2. The elastic properties are also required when load-factor design is used, in order to check that the elastic stresses at working loads do not exceed nine-tenths of the yield stress for the steel and one-third of the cube strength for the concrete.

(b) Load-factor Basis (Plastic Design)

The position of the plastic neutral axis is determined by the full plastic strength of the steel and concrete components of the composition section and is independent of the modular ratio.

The basic assumptions are that:

1. The whole of the steel beam is stressed to the yield point Y_s, whether in compression or tension, according to the position of the plastic neutral axis.
2. Concrete below the plastic neutral axis is unstressed.
3. Concrete above the neutral axis is stressed to its full compressive strength, taken as $\frac{4}{9}u_w$, where u_w = specified cube strength. This fraction is compounded

CALCULATION OF SECTION PROPERTIES

of two factors $\frac{2}{3} \times \frac{2}{3}$, the first being an allowance for the fact that the strength of concrete in a slab is assumed to be two-thirds of the cube strength, whilst the second is an increase in the load-factor on the concrete of 50 per cent to cover the greater variability of strength and lower ductility of concrete compared with steel.

As for elastic design, two cases require consideration:

1. Neutral axis within the depth of the concrete slab. This occurs when the fully plastic compression strength of the slab exceeds the fully plastic strength of the steel beam in tension.

2. Neutral axis below the slab and in the steel beam. This occurs when the fully plastic compression strength of the slab is less than that of the fully plastic steel beam. This case can be further subdivided according to whether the plastic neutral axis lies within the flange of the steel beam or within the web of the beam.

The depth to the plastic neutral axis having been found enables the full plastic moment of the composite section, M_r to be calculated. This must be equated to, or must exceed, the factored bending moment.

In Case 1, only that portion of the slab above the neutral axis is in compression and the ultimate stress conditions and formulae are given in Fig. 9. This occurs when αA_s is less than or equal to $b \cdot d_s$, the meaning of the symbols being as shown in the Figure.

$$\alpha = \frac{9 Y_s}{4 U_w}$$

Depth to plastic neutral axis
$$= d_n = \frac{\alpha \cdot A_s}{b}$$

Ultimate Moment of Resistance $= M_r = A_s \cdot Y_s \left(d_c + \frac{d_s - d_n}{2} \right)$

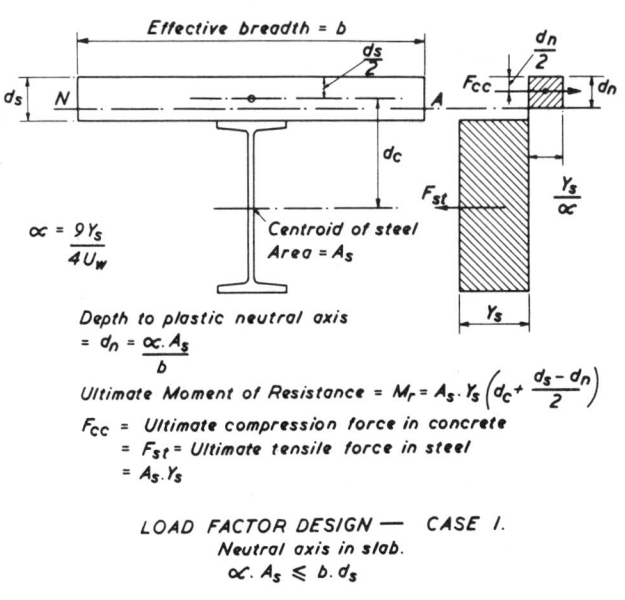

F_{cc} = Ultimate compression force in concrete
$\phantom{F_{cc}} = F_{st}$ = Ultimate tensile force in steel
$\phantom{F_{cc} = F_{st}} = A_s \cdot Y_s$

LOAD FACTOR DESIGN — CASE I.
Neutral axis in slab.
$\alpha \cdot A_s \leqslant b \cdot d_s$

Fig. 9

In Case 2a, in which the plastic neutral axis lies within the top flange of the steel beam, all material above the neutral axis is fully stressed in compression and the steel below the neutral axis is stressed to yield point in tension and the ultimate

COMPOSITE CONSTRUCTION

$\alpha = \dfrac{9Y_s}{4U_w}$

A_F = area of flange
    = $b_f \cdot t_f$

Depth to plastic neutral axis
= $d_n = d_s + \dfrac{\alpha A_s - b \cdot d_s}{2 b_f \cdot \alpha}$

Ultimate Moment of Resistance = $M_r = Y_s \left[A_s \cdot d_c - b_f \cdot d_n (d_n - d_s) \right]$

F_{cc} = Ultimate compression force in concrete
= $\dfrac{d_s \cdot b \cdot Y_s}{\alpha} = \dfrac{4}{9} U_w \cdot b \cdot d_s$

LOAD FACTOR DESIGN - CASE 2a.
Neutral axis in beam flange.
$b \cdot d_s < \alpha \cdot A_s < (b \cdot d_s + 2\alpha \cdot A_f)$

Fig. 10

$\alpha = \dfrac{9Y_s}{4U_w}$

A_F = area of flange
    = $b_f \cdot t_f$

Depth to plastic neutral axis
$d_n = d_s + t_f + \dfrac{\alpha (A_s - 2A_F) - b \cdot d_s}{2\alpha \cdot t_w}$

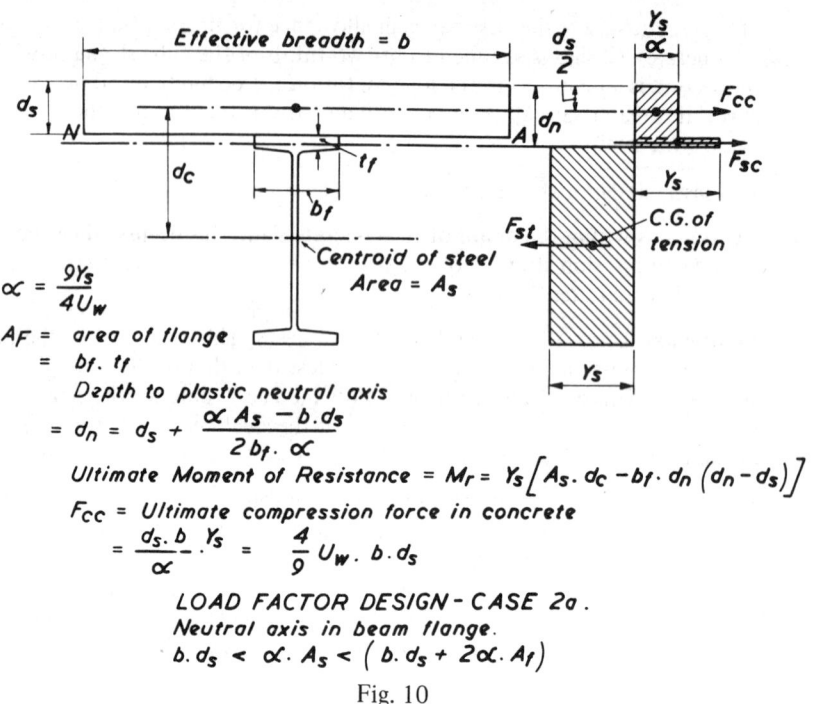

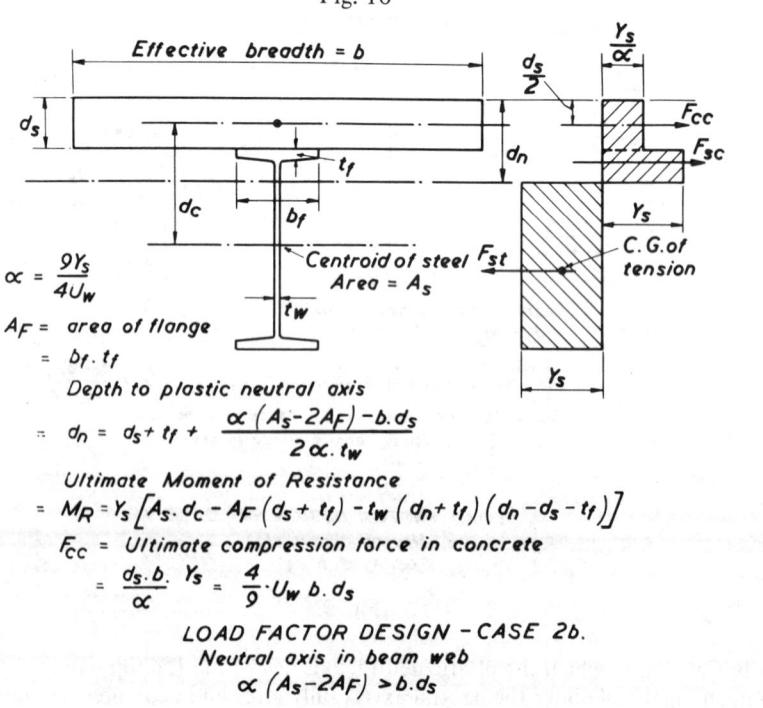

Ultimate Moment of Resistance
= $M_R = Y_s \left[A_s \cdot d_c - A_F (d_s + t_f) - t_w (d_n + t_f)(d_n - d_s - t_f) \right]$

F_{cc} = Ultimate compression force in concrete
= $\dfrac{d_s \cdot b \cdot Y_s}{\alpha} = \dfrac{4}{9} U_w b \cdot d_s$

LOAD FACTOR DESIGN - CASE 2b.
Neutral axis in beam web
$\alpha (A_s - 2A_F) > b \cdot d_s$

Fig. 11

DESIGN EXAMPLES

stress conditions and moment of resistance are as shown in Fig. 10. This case occurs when:

$$b \cdot d_s < \alpha A_s < (b \cdot d_s + 2\alpha \cdot A_f)$$

In Case 2b the plastic neutral axis lies within the web of the beam and the top flange is assumed to be a rectangle of area $b_f \times t_f$. As in Case 2a, all material above the neutral axis is fully stressed in compression, whilst that below is stressed to yield point. The ultimate conditions are those shown in Fig. 11. This case occurs when:

$$\alpha(A_s - 2A_f) > b \cdot d_s$$

Since the depth to the plastic neutral axis depends upon the strengths of the separate components, it thus follows that any change in the concrete or steel quality affects the position of the axis in each of the three cases above.

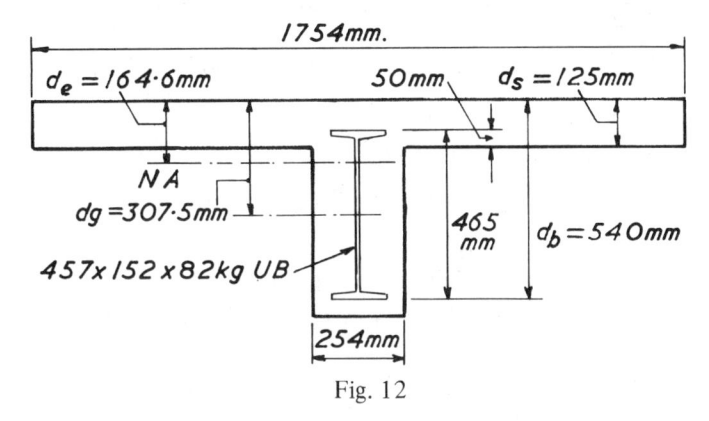

Fig. 12

Design Examples

(a) Elastic Design

Consider the composite section shown in Fig. 12, comprising a 457 mm × 152 mm × 82 kg U.B. spanning 7.315 m, spaced at 3.810 m centres, carrying a 125 mm slab. The top flange is embedded in the slab and the remainder of the beam is cased with solid concrete, reinforced as in Fig. 5. Shear connectors are not required.

The maximum effective width of slab in compression is given by twelve times the slab thickness plus the width of the casing and equals 1 754 mm.

The neutral axis in this case falls below the slab and the formulae given in Fig. 8 are applicable.

Depth from top of slab to elastic neutral axis

$$= d_e$$

$$= \frac{\dfrac{b \cdot d_s^2}{2m} + A_s \cdot d_g}{\dfrac{b \cdot d_s}{m} + A_s}$$

$$= \frac{\dfrac{1\,754 \times 125^2}{2 \times 15} + 104.4 \times 10^2 \times 307.5}{\dfrac{1\,754 \times 125}{15} + 104.4 \times 10^2} = \frac{4\,123\,841}{25\,057}$$

$$= 164.6 \text{ mm}$$

Moment of inertia about neutral axis

$$= I_g$$

$$= \frac{bd_s^3}{12m} + \frac{b \cdot d_s}{m}\left(d_e - \frac{d_s}{2}\right)^2 + I_s + A_s(d_g - d_e)^2$$

$$= \frac{1\,754 \times 125^3}{12 \times 15} + \frac{1\,754 \times 125}{15}(164.6 - 62.5)^2 +$$

$$+ 36\,160 \times 10^4 + 104.4 \times 10^2(307.5 - 164.6)^2$$

$$= 74\,619 \text{ cm}^4 \text{ in steel units.}$$

The tension section modulus in steel units

$$= Z_{st} = 74\,619 \times 10^4 \div (540 - 164.6)$$

$$= 1\,987.7 \text{ cm}^3.$$

The compression section modulus in concrete units

$$= Z_{cc} = 15 \times 74\,619 \times 10^4 \div 164.6$$

$$= 68\,000 \text{ cm}^3.$$

The corresponding Resistance Moments will therefore be:

		kNm
For Grade 43 steel	Stress 165 N/mm^2	328.0
For Grade 50 steel	Stress 230 N/mm^2	457.2
For 1:2:4 concrete	Stress 7 N/mm^2	476.0
For 1:1:2 concrete	Stress 10 N/mm^2	680.0

This indicates that a 1:2:4 mix of concrete for the slab is adequate, even when Grade 50 steel is used for the 457 x 152 mm U.B., so that if the beam is fully propped during erection the Resistance Moment of 457.2 kNm determines the load carrying capacity. The coexistent concrete stress will be

$$\frac{7 \times 457.2}{476.0} = 6.72 \text{ N/mm}^2$$

The maximum total load which can be carried is thus 500.0 kN if Grade 50 steel is used for the beam, the corresponding figure for a Grade 43 steel beam being 358.7 kN. It is quite clear that the shear strength of either beam is more than adequate.

Making allowances for finishes weighing 1.197 kN/m^2, 33.36 kN, slab at 2.993 kN/m^2, 83.40 kN and for beam and casing at 3.342 kN/m run, at 24.45 kN, totalling 141.2 kN, the balances available for superload are 358.8 kN and 217.5 kN respectively.

DESIGN EXAMPLES

The permissible superimposed loads are therefore:

	with Grade 50 steel	12.9 kN/m^2
	with Grade 43 steel	7.8 kN/m^2

The calculations which now follow demonstrate the reduction in carrying capacity if the beam is unpropped during the construction stage, when the steel beam alone carries the construction loads.

The construction loads are:

	slab	83.40 kN
	beam and casing	24.45 kN
	formwork	8.57 kN
		116.42 kN

The corresponding bending moment is 106.45 kNm and the bending stress on the steel beam, section modulus 1 555 cm^3, is thus 68.46 N/mm^2. At this stage the beam is laterally unsupported and with $l/r_y = 0.7 \times 7.315 \times 10^2/3.24 = 158.04$ and with $D/T = 24.6$, the safe working stress will be about 116.9 N/mm^2, so that lateral support is not necessary.

The steel stresses available for composite action are thus $230 - 68.46 = 161.54$ N/mm^2 for Grade 50 steel and $165 - 68.46 = 96.54$ N/mm^2 for Grade 43 steel. The corresponding bending moments are therefore $161.54 \times 1\ 987.7 \times 10^{-3} = 321$ kNm and $96.54 \times 1\ 987.7 \times 10^{-3} = 191.9$ kNm.

These are equivalent to total loads of 351.2 kN and 209.9 kN respectively and making the allowance for finishes of 33.4 kN, the balances available for superload are 317.8 kN and 176.5 kN.

The permissible superimposed loads are therefore:

	with Grade 50 steel	11.4 kN/m^2
	with Grade 43 steel	6.3 kN/m^2

The reduction in superload due to the extra stresses imposed by the lack of support during construction is thus 1.5 kN/m^2, that is, 12 per cent when high yield stress steel is used, and 19 per cent when mild steel is used.

As an indication of the manner in which the section moduli vary with variations in slab width and thickness, the relative figures for slabs of 75 mm, 100 mm, and 125 mm thicknesses in widths ranging from 0.6 m to 2.1 m used with a 457 x 191 x 82 kg U.B. are given in Fig. 13. They are based on a modular ratio of 15.

There is little point in using elastic design other than for cased sections with the advantage of lack of conventional shear connectors, and their use should be restricted to cases where solid concrete fire protection is required. As mentioned previously, shear connectors will be required if the beams are uncased or have hollow casing. Also the steel sections required by load-factor design will always be lighter or smaller than those required elastically. As the shear connectors, when used, are designed by load-factor methods, it is not proposed here to investigate their use with elastic designs. Their design is dealt with under the next heading and the same principles apply.

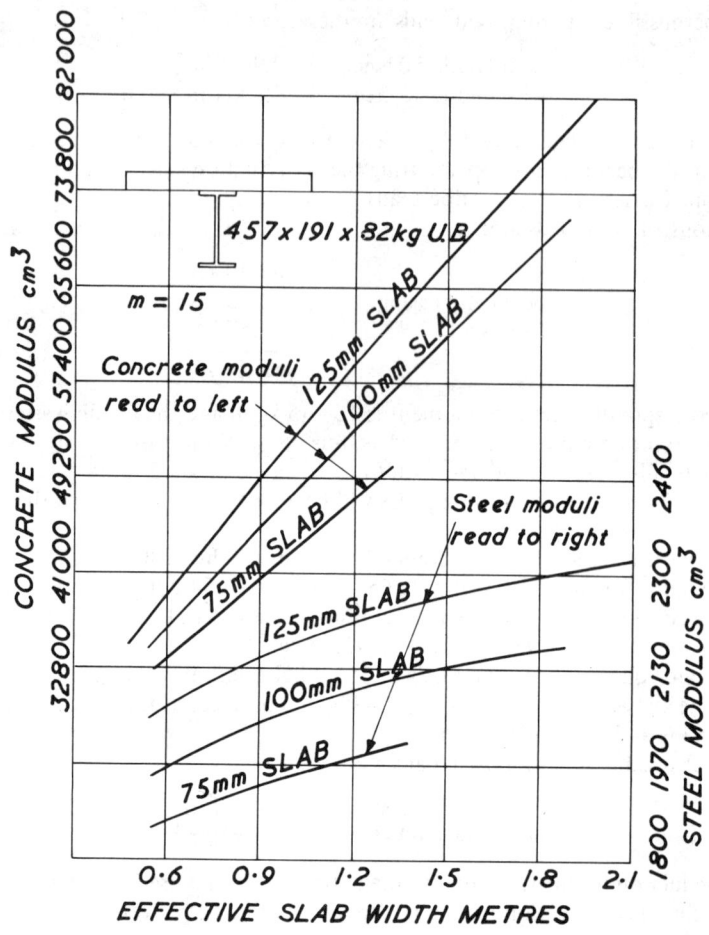

Fig. 13

(b) Load-Factor Design

Consider the composite section shown in Fig. 14 comprising a 457 x 191 x 82 kg U.B. acting in conjunction with a 125 mm slab of effective width 1 675 mm.

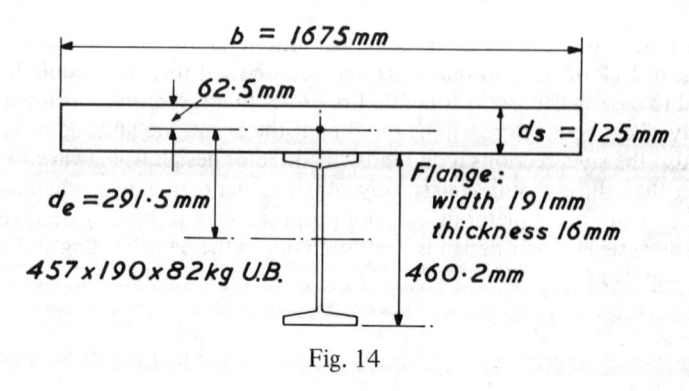

Fig. 14

DESIGN EXAMPLES

Using mild steel, yield point 245 N/mm² and 1:2:4 concrete with an ultimate cube strength of 21 N/mm², the following calculations are necessary:

$$\alpha = \frac{9}{4} \frac{Y_s}{u_w} = \frac{9}{4} \times \frac{245}{21} = 26.3$$

$\alpha A_s = 26.3 \times 104.4 \times 10^2 = 275\,000$ mm²
$A_f = 191 \times 16 = 3\,056$ mm² $ 2\alpha A_f = 2 \times 26.3 \times 3\,056 = 160\,400$ mm²
$bd_s = 1\,675 \times 125 = 210\,000$ mm²

For the plastic neutral axis to be within the slab, $\alpha A_s \leqslant bd_s$ (see Fig. 9); this is clearly not so, and therefore the neutral axis must be in the steel section.

For the plastic neutral axis to be in the top flange of the steel section (see Fig. 10):

$$bd_s < \alpha A_s < (bd_s + 2\alpha A_f)$$

i.e. $ 210\,000 < 275\,000 < (210\,000 + 160\,400)$

This condition is therefore satisfied and the section properties can be found by utilising the formulae in Fig. 10. Hence

d_n = depth to plastic neutral axis

$$= d_s + \frac{\alpha A_s - bd_s}{2b_f \alpha}$$

$$= 125 + \frac{275\,000 - 210\,000}{2 \times 191 \times 26.3} = 133.5 \text{ mm}.$$

The plastic moment of resistance, M_r

$= Y_s [A_s \cdot d_c - b_f d_n (d_n - d_s)]$
$= 245 [104.4 \times 10^2 \times 292.6 - 191 \times 133.5 (133.5 - 125)]$ N/mm
$= 245 [3\,060\,000 - 217\,000] \times 10^{-6}$ kNm
$= 696$ kNm.

The ultimate compression in the concrete, F_{cc}

$$= \frac{d_s b \cdot Y_s}{\alpha} = \frac{210\,000 \times 245}{26.3} \times 10^{-3}$$

or

$$= \frac{4}{9} u_w bd_s = \frac{4}{9} \times 21 \times 210\,000 \times 10^{-3}$$

$$= 1\,960 \text{ kN}.$$

If a 1:1:2 concrete is used, ultimate cube strength 31.5 N/mm², the position of the plastic neutral axis can be found in a similar manner.

$$\alpha = \frac{9}{4} \frac{Y_s}{u_w} = \frac{9}{4} \times \frac{245}{31.5} = 17.5$$

$\alpha A_s = 17.5 \times 104.4 \times 10^2 = 182\,500$ mm²
$bd_s = 1\,675 \times 125 = 210\,000$ mm²

606   COMPOSITE CONSTRUCTION

Since αA_s is less than bd_s it follows that the neutral axis falls within the slab and the formulae in Fig. 9 apply. Hence

$$d_n = \text{depth to plastic neutral axis}$$

$$= \frac{\alpha A_s}{b} = \frac{182\,500}{1\,675} = 109 \text{ mm}$$

The plastic moment of resistance, M_r

$$= A_s Y_s \left[d_c + \frac{(d_s - d_w)}{2} \right]$$

$$= 104.4 \times 10^2 \times 245(292.6 + 8) \times 10^{-6}$$

$$= 782 \text{ kNm.}$$

The ultimate compression force in the concrete slab is that carried by the concrete above the neutral axis and F_{cc} is therefore

$$\frac{4}{9} \times 31.5 \times 1\,675 \times 109 \times 10^{-3} = A_s Y_s = 104.4 \times 10^2 \times 245 \times 10^{-3} = 2\,560 \text{ kN}$$

These results are shown in Fig. 15.

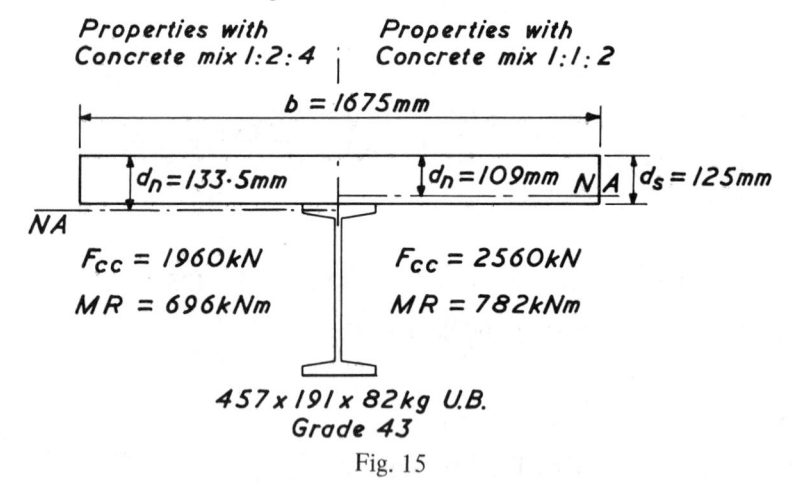

Properties with Concrete mix 1:2:4

$d_n = 133.5\text{mm}$
$F_{cc} = 1960\text{kN}$
$MR = 696\text{kNm}$

Properties with Concrete mix 1:1:2

$b = 1675\text{mm}$
$d_n = 109\text{mm}$ NA $d_s = 125\text{mm}$
$F_{cc} = 2560\text{kN}$
$MR = 782\text{kNm}$

457 x 191 x 82kg U.B.
Grade 43

Fig. 15

Similar calculations, using the same steel section but in Grade 50 material with the yield point of 355 N/mm² appropriate to the flange thickness of 16 mm, show that with both concrete mixes the plastic neutral axis lies within the top flange of the steel beam and the following figures result:

Concrete mix 1:2:4		Concrete mix 1:1:2
38.04	α	26.63
139.6 mm	d_n	133.3 mm
964.5 kNm	M_r	1 026.6 kNm
1 958.4 kN	F_{cc}	2 836.3 kN

These results are shown in Fig. 16.

DESIGN EXAMPLES

Properties with Concrete mix 1:2:4

$d_n = 139.6$ mm
$F_{cc} = 1958.4$ kN
$MR = 964.5$ kNm

Properties with Concrete mix 1:1:2

$d_n = 133.3$ mm
$F_{cc} = 2836.3$ kN
$MR = 1026.6$ kNm

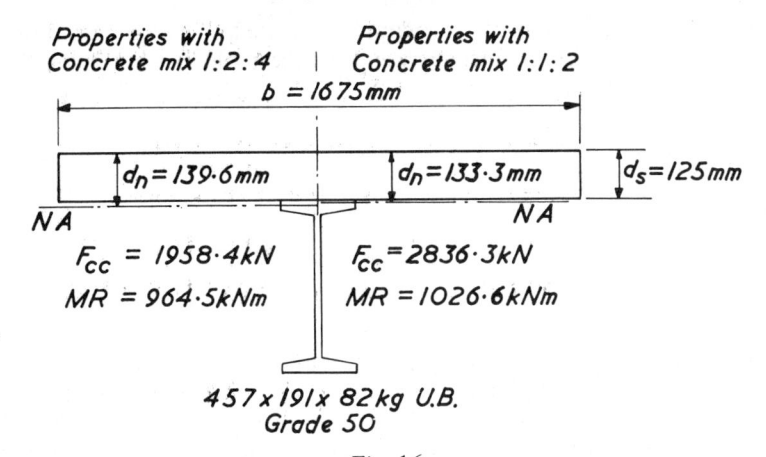

$d_s = 125$ mm

$b = 1675$ mm

457 × 191 × 82 kg U.B. Grade 50

Fig. 16

The Moment of Resistance so found must equal or exceed the factored bending moment on the section, that is, the statical bending moment multiplied by 1.75. The ultimate load conditions are not affected by the mode of construction.

The Code requires in addition that the elastic stresses at working loads shall not exceed $0.9 Y_s$ on the steel and $0.33 u_w$ on the concrete, so that the elastic properties of these sections, using $m = 15$, are also required. These are $Z_{st} = 2\,250$ cm^3 (steel units) and $Z_{cc} = 72\,400$ cm^3 (concrete units), from which safe working load bending moments can be obtained.

Comparative elastic and load-factor properties for the four sections are shown in Table E, bending moments and resistance moments being given in kNm.

The figures in italic are the controlling factors in elastic design and the degree of economy obtained by adopting load-factor design is shown by comparing them with the value of the ultimate moment of resistance divided by the load-factor which are given on the right of Table E.

TABLE E

Concrete Quality	Elastic Design — Working Load Moments in kNm					Load-factor Design			
	Grade 43		Grade 50		Ultimate Resistance Moment in kNm		RM ÷ 1.75 in kNm	
					Grade 43	Grade 50	Grade 43	Grade 50
	Concrete	Steel	Concrete	Steel				
1:2:4	498	*368*	498	*509*	696	965	398	552
1:1:2	746	*368*	746	*509*	782	1 027	447	586

If these sections are fully utilised under the ultimate load condition and fully propped during construction, it will be seen from the Table that in all cases the figure

$$\frac{\text{Ultimate Moment of Resistance}}{\text{Load-Factor}}$$

is less than

$$\frac{0.9 \times \text{yield stress in steel}}{\text{Working stress in steel}}$$

so that the steel stress at working loads will always be less than 0.9 × yield stress and need not be investigated. This will generally be the case for normal composite sections used in buildings. Regarding concrete stresses, it will be seen that only one case requires investigation, the 1:2:4 mix with Grade 50 steel, since it is the only case in which the reduced resistance moment exceeds the working bending moment based on concrete properties. Again the ratio of the figures is less than that of 0.444/0.333 and therefore the working stress will be within the prescribed limit. This will not always be the case, and the check is easily made from the elastic properties of the section, using the normal working stresses of 7 N/mm² for 1:2:4 concrete and 10.5 N/mm² for 1:1:2 concrete.

If, however, the sections are unpropped during construction, it is not possible to effect a direct comparison between stresses at working loads and ultimate loads as has been done in the previous paragraph and it is necessary to investigate the working load steel stresses as has been done for elastic design on page 603. The maximum total stress in the steel must not exceed 222 N/mm² for Grade 43 steel and 312 N/mm² for Grade 50 steel. Concrete stresses do not normally require investigation, since they only arise under the full composite action, but the values to be used are those given in the preceding paragraph.

(c) Shear Connectors

Shear connectors are to be designed by the load-factor method, irrespective of the method used for the design of the composite section. To demonstrate the calculations required, the shear connectors required for the first of the load-factor designs will be investigated, assuming the beam to span 9.15 m carrying a uniformly distributed load. The spacing of connectors must not be greater than four times the slab thickness, 500 mm, and they can be spaced equally between the sections of zero and maximum moment, 4.58 m.

The total ultimate compression in the concrete slab, F_{cc}, is 1 960 kN (see page 605) so that the shear load per mm run, Q, is 1 960/4 580 = 0.428 kN = 428 N.

It will be clear from Fig. 6, that the effective length of the periphery of the shear connectors is greater for studs in pairs than for two single studs so that studs in pairs will be adopted.

Using 75 mm × 19 mm headed studs in pairs, for which the load per pair is 2 × 67 = 134 kN at 300 mm centres the total shear resisted will be 15 × 134 = 2 000 kN, which is a little more than required.

Spacing the studs transversely at the usual gauge mark of 90 mm, it follows that the length around the periphery of a pair will be 270 mm (see Fig. 6) but as this exceeds twice the slab thickness, the figure of 250 mm must be used.

DESIGN EXAMPLES

The amount of transverse steel crossing the concrete table over the steel beam must be not less than $Q/4f_y$ per mm run of beam. In this example this will be

$$\frac{0.428 \times 10^3}{4 \times 245} = 0.436 \text{ mm}^2 \text{ per mm run, if mild steel bars are used, or a}$$

correspondingly smaller area if higher tensile material is employed.

C.P. 117 also states that the shear force per mm run should not exceed the lesser of the two following criteria (see page 596).

(1) $\qquad C1 L_s \sqrt{u_w} + A_t \cdot f_y \cdot n$

or

(2) $\qquad C2 L_s \sqrt{u_w}$

Let $C1$ equal 0.24 and $C2 = 0.64$.

Assuming that the minimum area of transverse reinforcement is provided, then

(1) $\quad C1 L_s \sqrt{u_w} + A_t \cdot f_y \cdot n = 0.24 \times 250 \times \sqrt{21} + 0.436 \times 245 \times 2$

$\qquad\qquad\qquad\qquad = 275 + 214$

$\qquad\qquad\qquad\qquad = 489$ N per mm run.

(2) $\qquad\qquad C2 L_s \sqrt{u_w} = 0.64 \times 250 \times \sqrt{21}$

$\qquad\qquad\qquad\qquad = 732$ N per mm run.

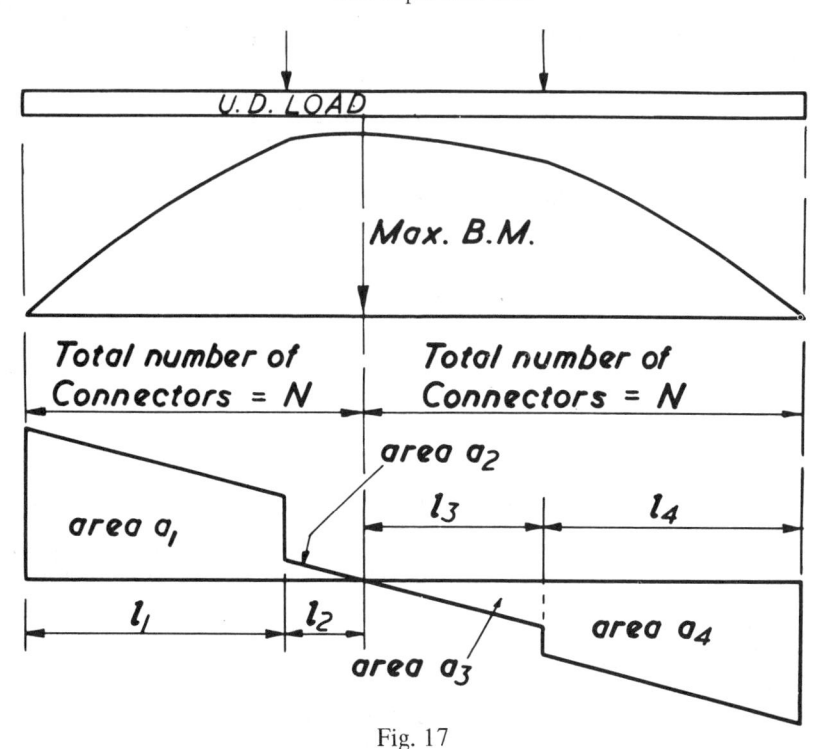

Fig. 17

Both criteria are thus satisfied but the area of the transverse reinforcement must be borne in mind when detailing the slab reinforcement. It can be provided by bars specially placed for the purpose or the whole or part can be provided by any of the main bottom reinforcement in the slab which is carried over the beam. Had the conditions been such that the minimum amount of transverse steel was insufficient to meet the requirements of criterion 1, then it must be increased accordingly.

The Code further states that such transverse reinforcement must be adequately anchored but gives no guidance as to the procedure to be adopted. It would seem logical, however, since it is ultimate load which is under consideration, to increase the elastic values by applying the load factor, giving bond stresses of 1.46 N/mm^2 for 1:2:4 concrete and 1.84 N/mm^2 for 1:1:2 concrete. With these values the required lengths of straight mild steel bars will be the diameter multiplied by 43 and 48 respectively, on each side of the beam centre line.

The sole remaining point to which attention should be directed is the method specified in the Code, Appendix C, dealing with the manner in which shear connectors should be spaced when concentrated loads occur resulting in large discontinuities in the shear diagram. The total number of shear connectors required on each side of the section of maximum moment is calculated in the usual manner from the ultimate compressive force in the concrete. This number should be distributed according to the areas of the shear diagram between the points of discontinuity but may be uniform between these points.

Referring to Fig. 17, if the total number of shear connectors required on each side is $N = n_1 + n_2 = n_3 + n_4$ the number of connectors required in each length is:

Length l_1, $\quad n_1 = N\left(\dfrac{a_1}{a_1 + a_2}\right)$

Length l_2, $\quad n_2 = N\left(\dfrac{a_2}{a_1 + a_2}\right)$

Length l_3, $\quad n_3 = N\left(\dfrac{a_3}{a_3 + a_4}\right)$

Length l_4, $\quad n_4 = N\left(\dfrac{a_4}{a_3 + a_4}\right)$

BIBLIOGRAPHY

1. VIEST, J. M., FOUNTAIN, R. S., and SINGLETON, R. C. *Composite Construction in Steel and Concrete.* McGraw Hill (1958).
2. SCOTT, W. BASIL. *Composite Construction in Steel Framed Buildings.* British Constructional Steelwork Association, Publication No. 25 (1965).
3. SATTLER, K. 'Composite Construction in Theory and Practice', *The Structural Engineer.* 1961, Vol. 39 (4) April and 1962, Vol. 40 (11) November. (A paper by one of the leading German engineers.)
4. WONG, F. K. C. 'The Horizontal Shear Resistance of Composite Beams', *The Structural Engineer.* 1963, Vol. 41 (8) August.
5. CHAPMAN, J. C. 'The Behaviour of Composite Beams in Steel and Concrete', *The Structural Engineer.* 1964, Vol. 42 (4) April.
6. CHAPMAN, J. C., and BALAKRISHNAN, S. 'Experiments on Composite Beams', *The Structural Engineer.* 1964, Vol. 42 (11) November.

BIBLIOGRAPHY

7. CREASY, L. R. 'Composite Construction', *The Structural Engineer.* 1964, Vol. 42 (12) December and 1965, Vol. 43 (5) May. (Contains much relative cost information.)
8. C.P. 117: Part 1: 1965. British Standard Code of Practice, Composite Construction in Structural Steel and Concrete. Part 1: Simply Supported Beams in Buildings. British Standards Institution. (Other parts in preparation will deal with Columns and Bridgework.)
9. PAGE, P. P., and GOLDREICH, J. D. 'Cut Composite Construction Costs', *Engineering News Record.* 1962, Vol. 168 (25) 21 June, p. 66.
10. PROCTOR, A. N. 'The Design of Composite Structural Members', *Concrete and Constructional Engineering.* 1962, Vol. 57 (8) August, pp. 292–302.
11. WILENKO, L. K. 'Building a Ten Storey Steel Frame Pre-stressed at Erection–The New Offices of the French Electricity Board in Paris', *Acier: Stahl: Steel.* 1962, Vol. 27 (11) November, pp. 479–486.
12. LANACH, W. J., and PARK, R. 'The Behaviour under Load of Six Castellated Composite T-Beams', *Civil Engineering.* 1964, Vol. 59 (692) March, pp. 339–343.
13. ANON. 'Weld Shear Connections', *Welding and Metal Fabrication.* 1964, Vol. 32 (3) March, pp. 94–98.
14. HUGHES, B. D. ' "Cyc-Arc" Stud Welded Concrete Anchors', *Civil Engineering.* 1964, Vol. 59 (695) June, pp. 723–727.

24. FOUNDATIONS

THE total load on the contact surface between the underside of the concrete block and the soil is that due to the imposed load upon the stanchion, plus the superimposed load at ground level on an area equal to that of the concrete block and plus also the weight of the concrete and the earth immediately above it. The pressure on the soil should not exceed the safe bearing capacity at this depth below the free surface.

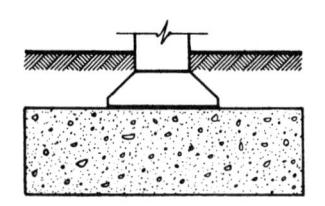

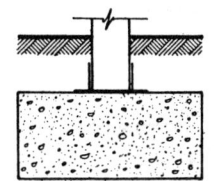

Fig. 1

In the remarks which follow:

P = vertical load on stanchion base.
W = weight of base, earth above it and superload if any.
R = total vertical load = $P + W$.
H = horizontal force at stanchion base.
M = moment at stanchion base.

Case 1

Central load:
Total load on soil = R

Soil pressure = $\dfrac{R}{BL}$.

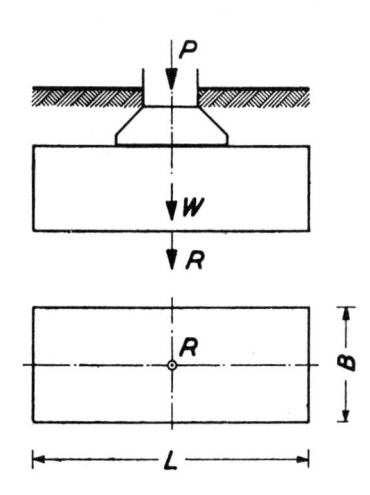

Fig. 2

FOUNDATIONS

Fig. 3

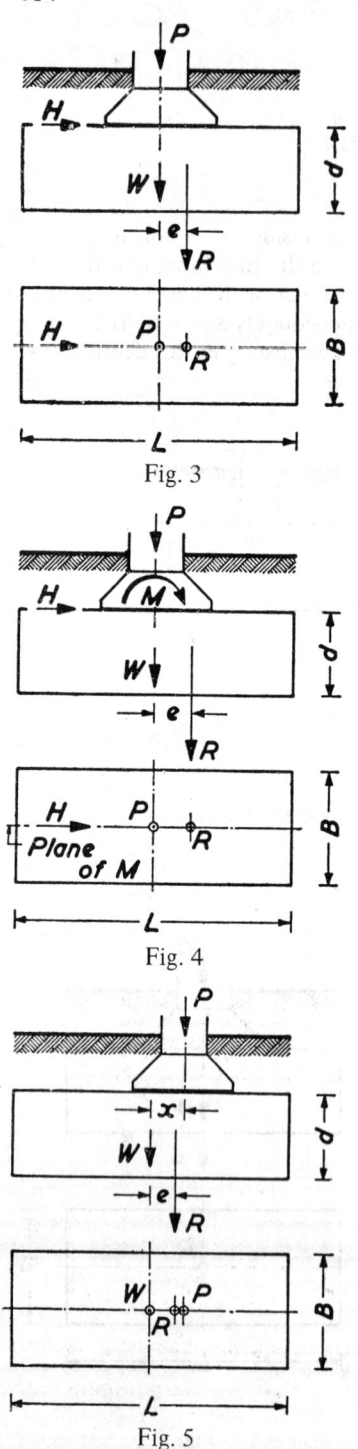

Case 2

Central load with horizontal force at base of stanchion:

Total load on soil = R

Eccentricity induced at underside of block due to horizontal force H

$$= \frac{Hd}{R} = e.$$

Fig. 4

Case 3

Central load with horizontal force and moment at stanchion base:

Total load on soil = R

Eccentricity induced at underside of block due to H and M.

$$= \frac{Hd + M}{R} = e$$

Fig. 5

Case 1a

Load eccentric on one axis:

Total load on soil = R

Eccentricity at underside of block

$$= \frac{Px}{R} = e$$

FOUNDATIONS

Case 2a

Load eccentric on one axis with horizontal force at stanchion base:
Total load on soil = R
Eccentricity at underside of block:

$$= \frac{Px + Hd}{R} = e$$

For load P to left of centre line:

$$e = \frac{-Px + Hd}{R}$$

Fig. 6

Case 3a

Load eccentric on one axis with horizontal load and moment at stanchion base:
base:
Total load on soil = R.
Eccentricity at underside of block

$$= \frac{Px + Hd + M}{R} = e$$

$$\left(\text{In general terms } e = \frac{\pm Px \pm Hd \pm M}{R}\right)$$

All the foregoing with the exception of Case 1 result in non-uniform loading on the soil and for the whole length L to be under pressure the eccentricity e must not exceed $L/6$.

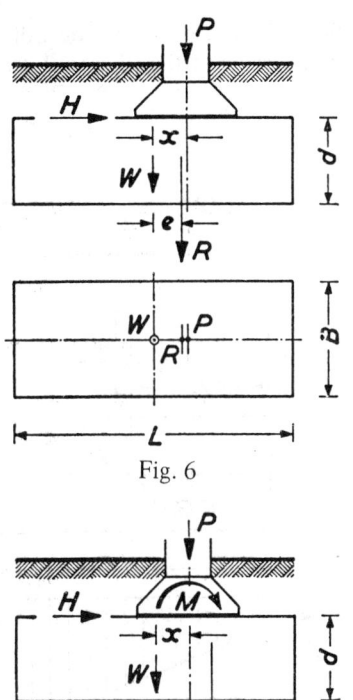

Fig. 7

With this proviso the stresses at the edges of the base are given by the expression:

$$p = \frac{R}{BL} \pm \frac{Re}{BL^2/6}$$

$$= \frac{R}{BL}\left(1 \pm \frac{6e}{L}\right)$$

The limiting values when $e = L/6$ are p_1 = zero and $p_2 = 2R/BL$ (see Fig. 8).

It is necessary to proceed on other lines should the value of e exceed $L/6$, since the preceding formula is based upon the ordinary bending theory which postulates the existence of tension. This cannot occur between the soil and the underside of the block. In such a case, let y be the distance from the line of action of R to the nearest edge of the base, i.e.

$$y = \frac{L}{2} - e$$

Since the stress distribution will be linear and the centroid of the diagram must be on the same vertical line as R it follows that the loaded length will be

$$3y = 3\left(\frac{L}{2} - e\right) \quad \text{(See Fig. 9.)}$$

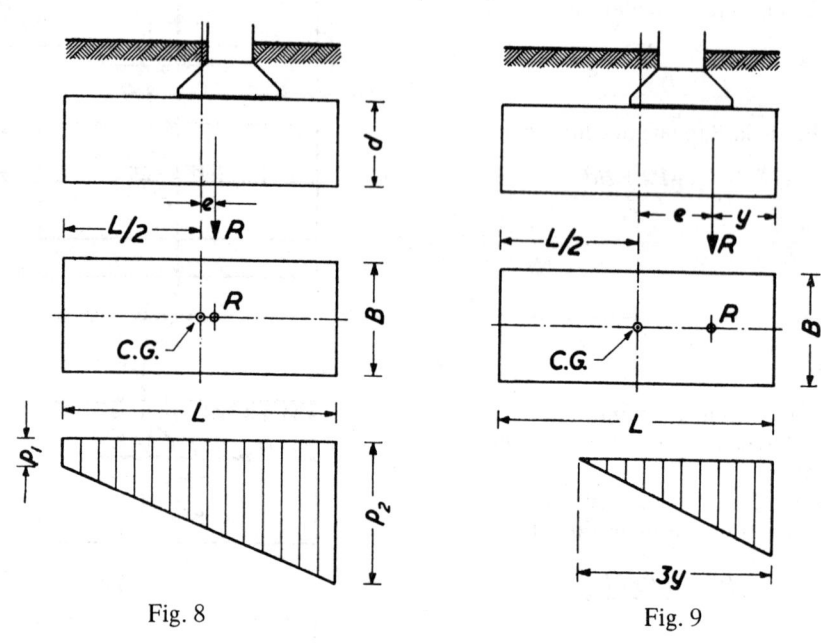

Fig. 8

Fig. 9

Consequently the maximum edge pressure

$$= p$$

$$= \frac{2R}{3 \cdot B \cdot y}$$

Loading Eccentric about both Axes

It will be seen by analogy with the foregoing that if e_L and e_B be the eccentricities of R with regard to L and B respectively then the stresses at the *four corners* of the base, using the ordinary bending theory, are given by the expression

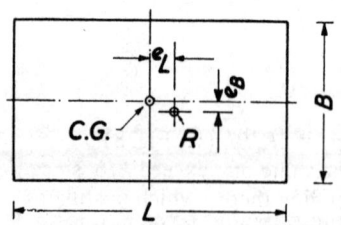

$$p = \frac{R}{BL} \pm \frac{Re_L}{BL^2/6} \pm \frac{Re_B}{B^2L/6}$$

$$= \frac{R}{BL}\left(1 \pm \frac{6e_L}{L} \pm \frac{6e_B}{B}\right).$$

Fig. 10

For the whole area of the base to be under pressure, i.e. for there to be no tension, R must act within the 'kern' of the section, shown hatched in Fig. 11. The limiting values of stresses for the configuration shown are minimum stress = zero and maximum stress = $2R/BL$.

SLAB BASES

Concrete bases, unreinforced, are normally not investigated for resistance to moment and shear, and the load from the stanchion base is assumed to disperse at an angle of 45° from which the thickness of the base is obtained as shown in Fig. 12.

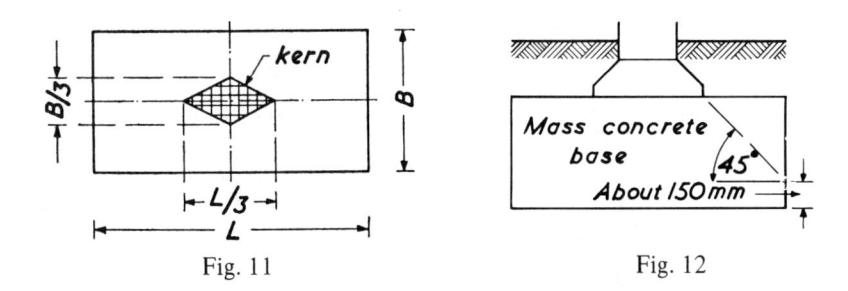

Fig. 11

Fig. 12

Safe bearing pressures between the stanchion base and concrete block should not exceed the values permitted by the local building regulations.

For full information on the design of reinforced concrete the reader is referred to any standard text-book on that subject, but adequate data for the design of simple square and rectangular bases carrying a single column is given on pages 625 to 635.

Slab Bases

Solid steel slab bases have to a great extent superseded grillages since the fabrication required is reduced to a minimum. If the load W on the steel slab is assumed uniformly distributed over an area $l \times d$ and the load under the slab also uniformly distributed over the concrete base, then

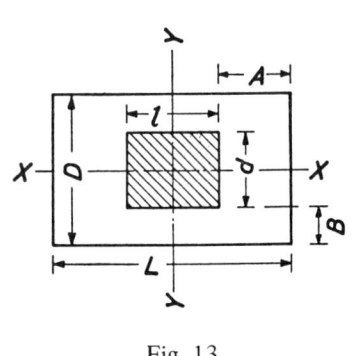

$$\text{B.M. in plane } XX = M_{XX} = \frac{W}{8}(L - l)$$

$$\text{and B.M. in plane } YY = M_{YY} = \frac{W}{8}(D - d)$$

Assuming Poisson's Ratio as $\frac{1}{4}$ it follows that the effective moments are:

$$\text{For breadth } D, \quad M_{XX} - \frac{M_Y}{4}$$

$$\text{and} \quad \text{for length } L, \quad M_{YY} - \frac{M_X}{4}$$

Fig. 13

It is, however, generally assumed that these calculated moments are excessive since they make no allowance for the 'dishing' action of the slab and maximum moments are assumed to occur at the edges of the upper rectangle $d \times l$.

It is then more convenient to express the load on the underside in terms of N/mm² — say w. Then if the overhang be A mm, the B.M. per mm run will be $wA^2/2$ Nmm and the moment in the direction at right angles will be $wB^2/2$ Nmm.

Allowance is normally made for the effect of two-dimensional stresses as outlined previously and therefore the effective moments for design purposes are:

$$\text{For breadth } D \ldots M = \frac{w}{2}\left(A^2 - \frac{B^2}{4}\right) \text{Nmm}$$

$$\text{for length } L \ldots M = \frac{w}{2}\left(B^2 - \frac{A^2}{4}\right) \text{Nmm}$$

If the base is square, i.e. $L = D$, and A is the *maximum* projection of the slab beyond the stanchion, the above reduces to

$$t = \sqrt{\left[\frac{3w}{p_{bct}}\left(A^2 - \frac{B^2}{4}\right)\right]},$$

wherein t = slab thickness in mm,

p_{bct} = permissible bending stress = 185 N/mm^2

This is the formula given in Clause 38.b of B.S. 449:Part 2:1969 which has remained unaltered since it first appeared in the 1937 edition. Current thought is, however, that a more correct value for Poisson's Ratio is 0.3, and this value should be used for bases where special calculations are made.

Alternatively, for a square base:

$$t = \frac{1}{2L}\sqrt{\left[W\left(A^2 - \frac{B^2}{4}\right)\right]},$$

where W is the total load and L is the length of one side of the base.

Where the slab is not uniformly loaded on its underside the formula given above does not apply and the example which follows illustrates the methods which can be used.

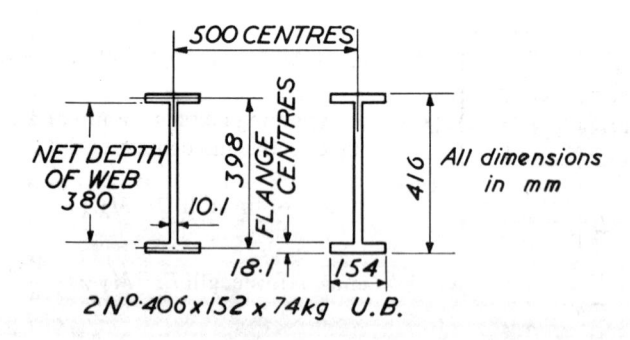

Fig. 14

Example 1. Eccentrically Loaded Foundation.

Figure 14 shows the foot of a heavy column for which a slab base and concrete foundation will be designed.

SLAB BASES

Data:

Direct central load = 1 520 kN
M_{XX} in plane XX = 145.8 kNm
M_{YY} in plane YY = 19.5 kNm

Permissible pressures:

on soil = 440 kN/m²
on concrete = 2 640 kN/m²

Assume that the concrete base is 3.0 m long and 1.5 m wide, as shown in Fig. 15.

Then the individual pressures on the soil are as follows:

Due to direct load = 1 520/4.5 = +337.8 kN/m²

due to M_{XX} $= \pm \dfrac{145.8 \times 6}{1.5 \times 3^2} = \pm\ 64.8$ kN/m²

due to M_{YY} $= \pm \dfrac{19.5 \times 6}{3 \times 1.5^2} = \pm\ 17.3$ kN/m²

When added algebraically, the resulting pressures are as shown in Fig. 15.
It will be noted that the whole base is in compression.

**Edge Stress Diagram
for concrete base**

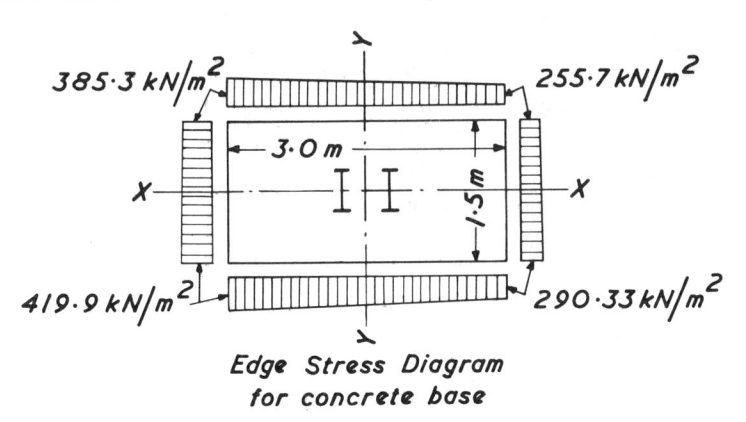

Fig. 15

Assume that the steel slab is 1.2 m long and 0.75 m wide. Then the individual pressures on the concrete, shown in Fig. 16, are as follows:

Due to direct load = 1 520/0.9 = + 1 688.9 kN/m²

due to M_{XX} $= \pm \dfrac{145.8 \times 6}{0.75 \times 1.2^2} = \pm\ 810$ kN/m²

due to M_{YY} $= \pm \dfrac{19.5 \times 6}{1.2 \times 0.75^2} = \pm\ 173.3$ kN/m²

When added algebraically, the resulting pressures are as shown in Fig. 16.

620   FOUNDATIONS

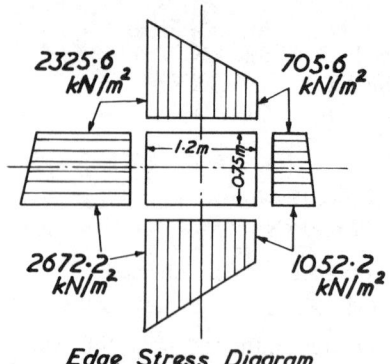

Edge Stress Diagram for slab base   Fig. 16

Consider the loads in the stanchions, as shown in Fig. 17.

The direct load in each stanchion = 760 kN.

This is divided proportionally to the area of the flanges,

$$154 \text{ mm} \times 18.1 \text{ mm} = 2\,787 \text{ mm}^2$$

and of the web $380 \text{ mm} \times 10.1 \text{ mm} = 3\,838 \text{ mm}^2$

to give loads of 225 kN in each flange and 310 kN in the web.

The loads in the webs at 500 mm centres due to M_{XX}

$$= \pm \frac{145.8 \times 10^3}{500} = \pm 291.6 \text{ kN}$$

The loads in the flanges at 398 mm centres due to M_{YY}

$$= \pm \frac{19.5 \times 10^3}{2 \times 398} = \pm 24.5 \text{ kN}$$

Now design for moments in the most heavily loaded quarter of the slab.

Check the moments about the lines XX, YY, AA (on the line of the stanchion web) and BB (at the centres of the stanchion flanges) shown in Fig. 18.

Loads on quarter slab

Upwards.   Due to direct load $= \frac{1\,520}{4} = $ = 380 kN

Due to M_{XX} $= \frac{810 \times 0.6 \times 0.375}{2} = $ 91.1 kN

Due to M_{YY} $= \frac{173.3 \times 0.6 \times 0.375}{2} = $ 19.5 kN

Downwards.   Flange load, due to direct load = 225 kN
                         due to M_{YY} = 24.5 kN
                                                 ─────────
                                                  249.5 kN

Web load, due to direct load $= \frac{310}{2} =$ 155 kN

due to M_{XX} $= \frac{291.6}{2} = $ 145.8 kN
                                       ─────────
                                        300.8 kN

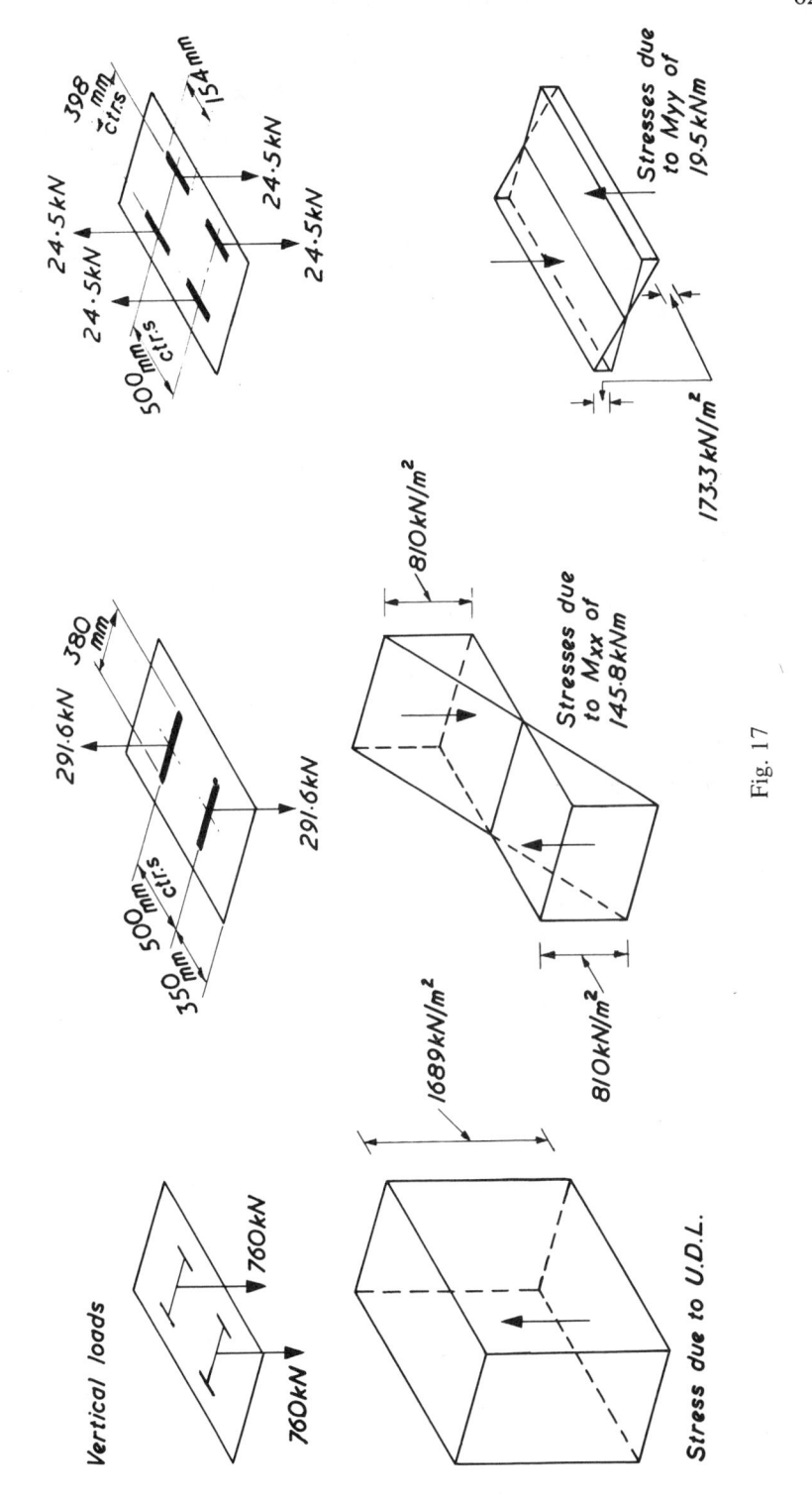

Fig. 17

622 FOUNDATIONS

Fig. 18

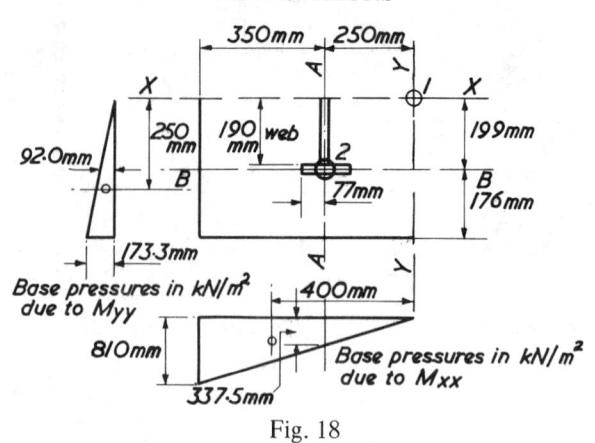

Moments about XX

Due to upward reactions from concrete slab:

$$\begin{aligned}
\text{Direct load} &= 380.0 \text{ kN} \times 190 \text{ mm} \times 10^{-3} = + 72.2 \\
M_{XX} &= 91.1 \text{ kN} \times 190 \text{ mm} \times 10^{-3} = + 17.3 \\
M_{YY} &= 19.5 \text{ kN} \times 250 \text{ mm} \times 10^{-3} = + 4.9
\end{aligned} \bigg\} +94.4$$

Due to downward loads from stanchion:

$$\begin{aligned}
\text{Flange load} &= 249.5 \text{ kN} \times 199 \text{ mm} \times 10^{-3} = - 49.7 \\
\text{Web load} &= 300.8 \text{ kN} \times 95 \text{ mm} \times 10^{-3} = - 28.6
\end{aligned} \bigg\} -78.3$$

$$+16.1 \text{ kNm}$$

Moments about YY

Due to upward loads:

$$\begin{aligned}
\text{Direct load} &= 380.0 \text{ kN} \times 300 \text{ mm} \times 10^{-3} = +114 \\
M_{XX} &= 91.1 \text{ kN} \times 400 \text{ mm} \times 10^{-3} = + 36.4 \\
M_{YY} &= 19.5 \text{ kN} \times 300 \text{ mm} \times 10^{-3} = + 5.9
\end{aligned} \bigg\} +156.3$$

Due to downward loads:

$$\begin{aligned}
\text{Flange load} &= 249.5 \text{ kN} \times 250 \text{ mm} \times 10^{-3} = - 62.4 \\
\text{Web load} &= 300.8 \text{ kN} \times 250 \text{ mm} \times 10^{-3} = - 75.2
\end{aligned} \bigg\} -137.6$$

$$+18.7 \text{ kNm}$$

Moments about AA (forces to left)

Due to upward loads:

Direct load $(1\,689 \times 0.35 \times 0.375)$ kN $\times 175$ mm $\times 10^{-3}$
$$= + 38.8$$

Pressure at AA due to M_{XX}

$$= \frac{810 \times 250 \times 10^{-3}}{600 \times 10^{-3}} = 337.5 \text{ kN/m}^2$$

$M_{XX}(337.5 \times 0.35 \times 0.375)$ kN $\times 175$ mm $\times 10^{-3} = + 7.8$

$\left(\dfrac{472.5}{2} \times 0.35 \times 0.375\right)$ kN $\times 233.3$ mm $\times 10^{-3} + 7.2$

$M_{YY}\left(\dfrac{173.3}{2} \times 0.35 \times 0.375\right)$ kN $\times 175$ mm $\times 10^{-3} = + 2.0$

$+55.8$

$+51.0$ kNm

Due to downward loads

Half flange load $= 124.8$ kN $\times 38.5 \times 10^{-3} \quad = \quad - 4.8$

SLAB BASES

Moments about BB (forces below line)

Due to upward loads:

$$\text{Direct load} = \frac{380 \times 0.176}{0.35} \text{kN} \times 88 \text{ mm} \times 10^{-3} = +16.8$$

$$M_{XX} = \frac{91.1 \times 0.176}{0.35} \text{kN} \times 88 \text{ mm} \times 10^{-3} = +4.0$$

Pressure at *BB* due to M_{YY}

$$= \frac{173.3 \times 199 \times 10^{-3}}{375 \times 10^{-3}} = 92.0 \text{ kN/m}^2$$

$M_{YY}(92.0 \times 0.6 \times 0.176)$ kN $\times$ 88 mm $\times 10^{-3}$ = + 0.9

$\left(\frac{81.3}{2} \times 0.6 \times 0.176\right)$ kN $\times$ 117 mm $\times 10^{-3}$ = + 0.5

$\}$ +22.2 kNm

Due to downward loads $\qquad$ = Nil

Taking Poisson's Ratio as 0.3, then

Moment at 1 for 375 mm width of slab = 18.7 − 0.3 × 16.1 = 13.9 kNm
for 600 mm width of slab = 16.1 − 0.3 × 18.7 = 10.5 kNm

Moment at 2 for 375 mm width of slab = 51.0 − 0.3 × 22.2 = 44.3 kNm
for 600 mm width of slab = 22.2 − 0.3 × 51.0 = 6.9 kNm

Hence design for 44.3 kNm on 375 mm width.
Therefore thickness of slab at 185 N/mm²

$$= \sqrt{\frac{6 \times 44.3 \times 10^6}{375 \times 185}} = 61.9 \text{ mm}$$

Use slab 1.2 m × 0.75 m × 65 mm thick.

Assuming the base to be of mass concrete, then with an angle of dispersion at 45°, the minimum thickness of concrete equals 1.5 − 0.6 = 0.9 m.

Use base 3 m × 1.5 m × 1.0 m thick.

Poisson's Ratio has been taken as 0.3 since this is the currently accepted value and special calculations have been made as the load is not uniformly distributed over the underside of the slab.

Inspection of the figures above will show that if 0.25 had been used, the increase in slab thickness would be about 1 mm so that the selected thickness would not be altered.

This slab base example has been dealt with at some length in order to demonstrate the basic principles which can be applied to design when the stanchion has more than one shaft and the load is not uniformly distributed.

Example of Design of Holding-down Bolts

Design of a base plate for a compound stanchion of the type used in a workshop building.

The centres of the two legs of the stanchion together with the worst conditions of loading are shown in Fig. 19.

624 FOUNDATIONS

In general, it will be found that a base plate of a length about 2½ to 3 times the centres of the stanchions will be suitable.

In this example, a length of base plate of 1.2 m will be used, with the holding-down bolts placed at 150 mm from the edge of the base plate.

The allowable pressure on the concrete below the base plate is 3 200 kN/m²

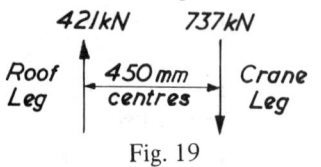

Figure 20 gives the dimensions of the base plate and the relative positions of stanchion legs and H.D. bolts.

It may be assumed that the B.M. caused by the stanchion loading is resisted by the two forces T and C, T being the tension in the H.D. bolts and C the compression between the base plates and the concrete base.

Fig. 19

The analogy of the resisting forces to those in a R.C. beam subjected to bending will be obvious.

The solution is approximate, because the forces on the base plate do not constitute a pure couple.

Adopting the R.C. beam analogy, the allowable maximum stresses in steel and concrete are as follows:

Assuming that 22 mm bolts will be used in tension, in accordance with Clause 50 of B.S. 449:Part 2:1969,

$$\text{Steel stress} = f_s = 130 + 25\% = 162.5 \text{ N/mm}^2$$

$$\text{Concrete stress} = f_c = 3\ 200 \text{ kN/m}^2 + 25\% = 4\ 000 \text{ kN/m}^2$$

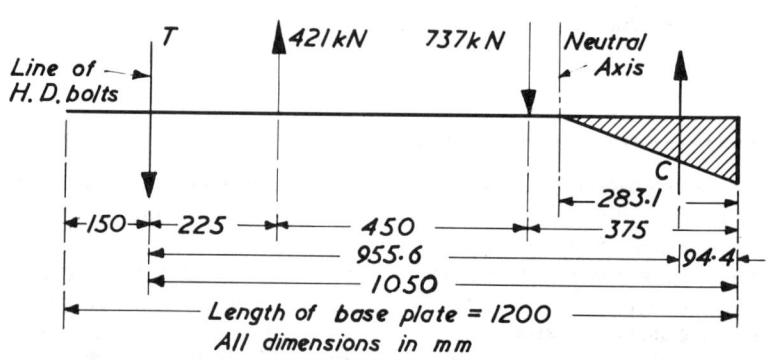

Fig. 20

The two allowable stresses may be increased by 25 per cent because the large proportion of the stanchion loads is due to wind forces.

$$f_s = 162.5 \text{ N/mm}^2$$
$$f_c = 4\ 000 \text{ kN/m}^2 = 4.0 \text{ N/mm}^2$$

Assuming that the two maximum stresses occur together and that the modular ratio, m, =15, then the depth to the neutral axis from the compression edge = $mf_c d/(mf_c + f_s)$ where d is the effective depth to the centres of the tension steel, 1 050 mm in the example.

$$\text{Depth to N.A.} = \frac{15 \times 4.0 \times 1\ 050}{15 \times 4.0 + 162.5} = 283.1 \text{ mm}$$

EXAMPLE OF DESIGN OF HOLDING-DOWN BOLTS

The compressive stress has a triangular distribution, as shown in Fig. 20 and the centre of compression is at 283.1/3 = 94.4 mm from the compression edge.

Taking moments about T, the line of the force in the H.D. bolts,

$$C = \frac{737 \times 675 - 421 \times 225}{955.6} = \frac{497\,475 - 94\,725}{955.6}$$

$$= 421.5 \text{ kN}$$

and

$$T = 421 + 421.5 - 737$$

$$= 105.5 \text{ kN}$$

The maximum allowable concrete pressure is 4 000 kN/m² and, if the width of the base plate is b mm, then

$$b = \frac{421.5 \times 2 \times 10^3}{283.1 \times 4\,000} = 744 \text{ mm}$$

$$= \text{say, } 775 \text{ mm}$$

The total tension to be supplied to the H.D. bolts = 105.5 kN.
Area of steel, A_s, required = $105.5 \times 10^3 / 162.5 = 649$ mm².
The area of the bottom of the thread of a 22 mm diameter bolt = 245 mm² and four such bolts would give a total area of 980 mm², which is satisfactory.

Design of Simple Reinforced Concrete Foundations

The design of reinforced concrete in buildings is governed by British Standard Code of Practice CP 114:Part 2:1969, 'The Structural Use of Reinforced Concrete in Buildings', and the remarks which follow are based on that document.

Concrete Stresses

Although the Code permits design by load factor methods, it is customary to adopt the elastic method of design for foundation blocks, using the prescribed modular ratio, $m = 15$. Table A gives permissible working stresses for the three most usual nominal mixes. These are basic working stresses for concrete made with Portland Cement to B.S. 12, 'Portland cement (ordinary and rapid hardening)', or with Portland-blast furnace cement to B.S. 146 and with aggregates complying with B.S. 882, 'Concrete aggregates from natural sources' or with B.S. 1047, 'Air-cooled blast furnace slag for concrete aggregate.'

TABLE A

Mix Proportions	Permissible Concrete Stresses (N/mm²)					
	Compressive		Shear	Bond	
	Direct	Due to Bending		Average	Local
1:1:2	7.6	10	0.9	1.00	1.50
1:1½:3	6.5	8.5	0.8	0.93	1.40
1:2:4	5.3	7	0.7	0.83	1.25

FOUNDATIONS

The stresses in Table A are based upon works test cube strengths at 28 days after pouring of 30.0 N/mm^2 for 1:1:2 mix, 25.5 N/mm^2 for 1:1½:3 mix, and 21.0 N/mm^2 for 1:2:4 mix. Modifications are permitted to the working stresses for (a) improved concrete strength, (b) age at loading and (c) wind forces. Modifications due to (a) are outside the scope of this section of the Manual but attention should be given to items (b) and (c).

Where it can be shown that a member will not receive its full design load within a period of 28 days after casting of the member, for example, foundations, the compressive stresses in Table A may be increased by multiplying by the factor given in Table B.

TABLE B

Age Factor for Permissible Compressive Stresses in Concrete

Minimum Age of Member when Full Design Load is Applied	Age Factor
Months	
1	1.0
2	1.10
3	1.16
6	1.20
12	1.24

Shear and bond stresses corresponding to the increased compressive stresses must be obtained from Table 10 of the Code and are related to an effective cube strength obtained by multiplying the specified cube strength appropriate to the mix by the age factor.

It would seem reasonable therefore to base designs for foundations to single-storey sheds on the specified stresses in Table A but to allow for, at least, an age factor of 3 months for foundations to multi-storey buildings. Design factors have therefore been given below for both these conditions.

TABLE C

| Nominal Mix | Permissible Concrete Stresses (N/mm^2) After Ageing for 3 months ||||||
| | Compression || Shear | Bond ||
	Direct	In Bending		Average	Local
1:1:2	8.8	11.6	0.9	1.0	1.5
1:1½:3	7.5	9.9	0.9	1.0	1.5
1:2:4	6.1	8.1	0.8	0.9	1.3

Table C gives particulars of permissible concrete stresses at age 3 months, being the basic stresses in Table A as corrected by Table B for compression and corrected as required by Table 10 of the Code for shear and bond stresses.

DESIGN FACTORS

Reinforcement Stresses

The permissible working stress for mild steel bars to B.S. 4449:1969, 'Hot rolled steel bars for the reinforcement of concrete and B.S. 4482:1969, Hard drawn mild steel wire for the reinforcement of concrete (metric units)' and all plain bars in tension is 140 N/mm² for effective diameters not exceeding 40 mm and 125 N/mm² above that size. The same tensile stresses may be used in shear reinforcement but this is seldom, if ever, used in block bases.

High bond bars, such as cold twisted steel bars complying with B.S. 4461:1969, 'Cold worked steel bars for the reinforcement of concrete' are used at stresses up to 55 per cent of the guaranteed yield point or proof stress, subject to the overriding values of 230 N/mm² for bars with an effective diameter not exceeding 20 mm and 210 N/mm² for larger bars.

Wind Loads

The permissible stresses in concrete and reinforcement may be increased by not more than 25 per cent, provided that the excess is solely due to stresses induced by wind loading, as is customary in steelwork design.

Design Factors

Design factors for rectangular slabs reinforced in tension only, as will normally be the case for foundation blocks, are given for ease of calculation in Table D, both for the basic stresses to be used with single storey sheds and for the stresses at 3 months after pouring, which are recommended for use with multi-storey buildings.

The symbols used are:

b = breadth of slab in bending (mm).

d_1 = effective depth to the tensile reinforcement (mm) (depth from top of concrete to centre of reinforcement).

d_n = neutral axis depth (mm).

    = depth of concrete in compression.

l_a = lever arm of the resistance moment (mm) (distance between centre of compression in concrete and centre of tensile steel).

M_r = moment of resistance of slab to bending (Nmm) when maximum concrete stress is used together with maximum tensile stress in steel).

    = total tension in steel, or total compression in concrete, multiplied by lever arm.

p = percentage tensile reinforcement required to develop simultaneously maximum concrete compressive stress and allowable tensile stress in steel, based on effective area of section $b \times d_1$.

The reinforced concrete code does not specifically deal with bases under steel slabs, but by analogy with Clause 340 of the Code the following requirements are obtained.

(a) The bending moment at any section of a base should be the moment of the forces over the entire area of the base to one side of the section. The critical section for bending in the concrete base should be taken at the edge of the slab base.

TABLE D

Nominal Mix	Factor	Design Factors Basic	Age 3 Months
1:1:2	Neutral axis depth, d_n (multiply by d_1)	0.517	0.554
	Percentage reinforcement (multiply by bd_1)	1.85	2.30
	Lever arm, l_a (multiply by d_1)	0.828	0.815
	Resistance moment, M_r (multiply by bd_1^2)	2.14	2.62
1:1½:3	Neutral axis depth, d_n (multiply by d_1)	0.477	0.515
	Percentage reinforcement (multiply by bd_1)	1.45	1.82
	Lever arm, l_a (multiply by d_1)	0.841	0.828
	Resistance moment, M_r (multiply by bd_1^2)	1.70	2.11
1:2:4	Neutral axis depth, d_n (multiply by d_1)	0.429	0.465
	Percentage reinforcement (multiply by bd_1)	1.07	1.34
	Lever arm, l_a (multiply by d_1)	0.857	0.845
	Resistance moment, M_r (multiply by bd_1^2)	1.29	1.59

(b) The reinforcement provided to resist the bending moments should be uniformly distributed across the full width of the section; except that for rectangular bases the reinforcement parallel to the short edge should be more closely spaced near the stanchion.

(c) The critical section for shear should be taken to be at a distance out from the edge of the steel slab equal to the effective depth of the concrete base.

(d) The critical section for local bond stress should be the same as that for bending moment.

Worked example: Square base

The following example will illustrate the calculations required. A stanchion in a multi-storey building carries a central load of 5 000 kN, sits on a square steel slab and is supported by a square reinforced concrete base of nominal mix 1:2:4, the soil pressure being 430 kN/m² (see Fig. 21). The further assumption made will be that the construction period for the building will exceed 3 months so that the design factors involved will be found at the bottom of the right-hand column in Table D.

The safe allowable compression on the concrete will be 5.3 x 1.16, say 6.1 kN/mm², so that the area of the steel slab base = 0.82 m². Use slab 1.0 m x 1.0 m = 1.0 m².

DESIGN FACTORS

Allow for weight of base at 200 kN, giving a total load on the soil of 5 200 kN. The area of concrete required at 430 kN/m² = 12.1 m². Use base 3.5 m square = 12.25 m².

The weight of the base and soil above it, if any, need not be taken into account when calculating bending moment and shear, since the base is directly supported by the soil under it. All subsequent calculations are therefore based upon the load of 5 000 kN, equivalent to a soil pressure of 408 kN/m², which has been rounded off to 410 kN/m².

The depth of the base will be decided either by the bending moment or the shear acting on it and it will generally be found that the latter is the controlling factor.

Dealing first with bending moment, the Code states that the critical section should be taken at the edge of the reinforced concrete column carried by the base, so with a steel slab, by analogy take the maximum bending moment as occurring at the edge of the slab base. The Code further states that the bending moment should be taken as the moment of forces over the entire area on one side of the section.

The load causing bending moment is thus that on the cantilever outstand of 1.25 m as shown in Fig. 21. This can be obtained by proportion, e.g.

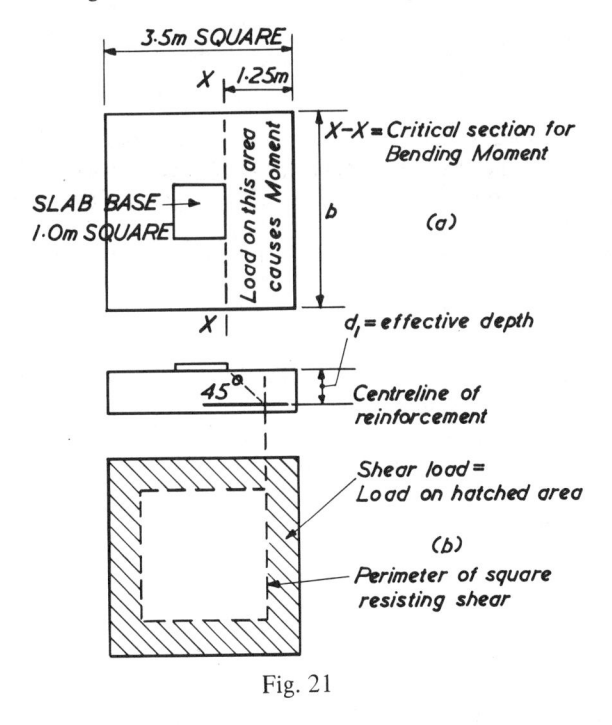

Fig. 21

$$\frac{1.25}{3.5} \times 5\ 000\ \text{kN} = 1\ 785\ \text{kN}$$

or by using the rounded off values for the soil pressure

$$= 3.5 \times 1.25 \times 410\ \text{kN}$$
$$= 1\ 793.8\ \text{kN}$$

FOUNDATIONS

Hence, moment at edge of slab base

$$= 1\ 793.8 \times 0.625 \text{ m}$$
$$= 1\ 121 \text{ kNm}$$

The resistance moment of a rectangular section reinforced in tension only with a concrete stress of 8.1 N/mm² at age 3 months, and a tensile steel stress of 140 N/mm², given in Table D

$$= 1.59\ bd_1^2 \text{ Nmm}$$

where b = effective breadth of slab (mm)
d_1 = effective depth of slab (mm)
= depth from top of concrete to centre-line of reinforcement.

The Code also states that the reinforcement should be distributed uniformly across the width of the base, thus clearly indicating that the full width of the base may be considered as resisting the moment.

Hence $\qquad b = 3.5 \text{ m} = 3\ 500 \text{ mm}$

Equating the resistance moment to the bending moment gives:

$$\text{Minimum effective depth, } d_1 = \sqrt{\frac{1\ 121 \times 10^6}{1.59 \times 3\ 500}}$$
$$= \sqrt{201\ 438} = 449 \text{ mm}$$

Dealing now with shear, the Code states that the critical sections for shear should be taken to be at a distance from the concrete column face (i.e. edge of steel slab in this example), equal to the effective depth of the base, which is equivalent to a 45° angle of dispersion.

The shear stress, q, in a reinforced concrete slab is obtained by dividing the shear load at the section by the product of the width acting in shear and the lever arm of the resistance moment, l_a. Reference to Table D shows that, for the concrete mix and age used, $l_a = 0.845 d_1$.

It can be argued that, as the whole width of the base is assumed to be effective in resisting the bending moment, it can also be effective in resisting shear and this will be true for a rectangular base of which the width does not exceed the width of the steel slab base plus twice the effective depth of the concrete block. A greater depth will, however, be given for a square base, if the critical sections for shear are taken as indicated in Fig. 21 (b), the load causing shear being that on the hatched area.

The depth required for shear may be found by trial and error or by solution of a quadratic equation, but as great refinement in calculation is unnecessary, many designers may choose the first method. Thus, on the trial and error basis, accepting provisionally the effective depth found necessary for moment, namely 449 mm, the side of the shear square will be 1 m + (2 × 449 mm) = 1.90 m, so that the hatched area will be $3.5^2 - 1.90^2 = 8.64$ m². The load causing shear will be

$$\frac{8.64 \times 5\ 000}{12.25} = 3\ 527 \text{ kN}$$

The safe shear at the permissible working stress of 0.8 N/mm²

$$= 4 \times 1.90 \times 10^3 \times 0.845 \times 449 \times 0.8 \times 10^{-3}$$
$$= 2\ 306.8 \text{ kN, which is insufficient}$$

DESIGN FACTORS

Repeating this calculation with an effective depth, d_1, of 560 mm, the side of the shear square will be 2.12 m resulting in a shear load of 3 165 kN. The safe shear will be

$$4 \times 2.12 \times 10^3 \times 0.845 \times 0.8 \times 10^{-3} = 3\ 210\ \text{kN}$$

which is sufficient.

If, however, the direct approach is preferred, the minimum depth required for shear can be found as follows: The bearing pressure on the soil being 410 kN/m², it follows from Fig. 21 that the force on the shaded area

$$= 410 \left[3.5^2 - \left(\frac{1\ 000 + 2d_1}{1\ 000} \right)^2 \right] \text{kN}$$

$$\text{The periphery of the shear square} = 4 \times \left(\frac{1\ 000 + 2d_1}{1\ 000} \right)$$

$$\text{Area in shear} = 4 \times \left(\frac{1\ 000 + 2d_1}{1\ 000} \right) \times 0.845 d_1$$

Hence

$$\text{Allowable shear load} = 0.8 \times 10^{-3} \times 4 \times \left(\frac{1\ 000 + 2d_1}{1\ 000} \right) \times 0.845 d_1 \text{ kN}$$

Equating actual load to allowable load and solving the resulting quadratic equation gives d_1 = 557.5 mm. Thus it is clear that the shear criterion governs the design, the effective depth from this viewpoint exceeding that required for bending moment. An overall depth of 720 mm will be used, which, when allowance has been made for two layers of reinforcement with adequate cover to the lower layer, will give an effective depth of 640 mm.

$$\text{The area of tensile steel required} = \frac{\text{Bending Moment}}{\text{Steel stress} \times \text{lever arm}}$$

$$= \frac{1\ 121 \times 10^6}{140 \times 0.845 \times 640} = 14\ 800\ \text{mm}^2$$

Allow 31 number 25 mm diameter bars in each direction = 15 220 mm² both ways.

Since both the effective depth and area of reinforcement provided are greater than the minima required, it follows that the actual stresses in compression, tension and shear will be less than the permissible stresses.

If it is necessary to calculate the actual stresses arising, for submission to an approving authority, the calculations must be based upon the actual sizes used. This is because the depth to the neutral axis depends solely upon the geometrical properties of the section and is independent of allowable stresses.

Calculations for this purpose now follow and the nomenclature used is set out below. The known properties are the width and effective depth of the concrete base and the area of reinforcement, whilst the unknown quantities are the depth to the neutral axis, lever arm of the resistance moment and the resulting stresses.

b = width of rectangular section = 3 500 mm
d_1 = effective depth of section = 640 mm
m = modular ratio = 15
A_{st} = cross-sectional area of reinforcement = 15 220 mm²

632   FOUNDATIONS

d_n = depth to neutral axis
p_{cb} = maximum compressive stress in concrete due to bending
p_{st} = tensile stress in reinforcement due to bending
} To be found

Making the standard elastic assumptions that, at any cross-section, plane sections remain plane, that all compression is taken by the concrete and that all tension is taken by the steel, referring to Fig. 22,

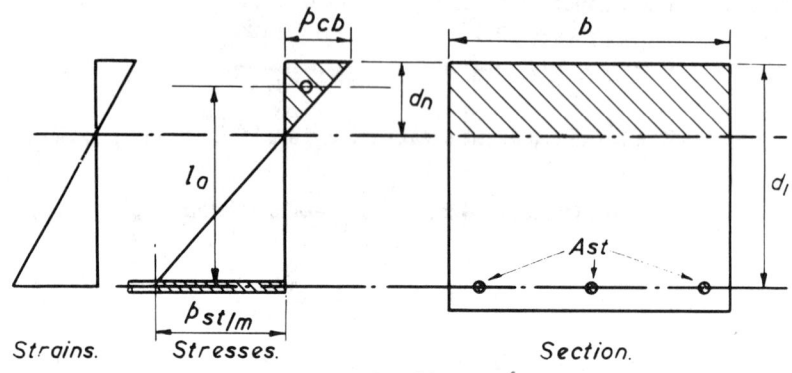

Fig. 22

Total compression in concrete = $\dfrac{b \cdot p_{cb} \cdot d_n}{2}$

Tensile stress in steel = $\dfrac{m(d_1 - d_n)}{d_n} \cdot p_{cb}$

Total tension in steel = $A_{st} \cdot p_{st}$

$= A_{st} \cdot \dfrac{m(d_1 - d_n)}{d_n} \cdot p_{cb}$

But the total compression must equal the total tension, hence, after removing the common factor p_{cb}

$$\dfrac{b \cdot d_n}{2} = A_{st} \cdot \dfrac{m(d_1 - d_n)}{d_n}$$

Simplifying and rearranging

$$\dfrac{b \cdot d_n^2}{2} - A_{st} \cdot m(d_1 - d_n) = 0$$

Inserting in this equation the only known quantities, b = 3 500 mm, d_1 = 640 mm, A_{st} = 15 220 mm^2 and m = 15, the equation reduces to:

$$35 d_n^2 + 4\,566 d_n - 2\,922\,240 = 0,$$

whence d_n = 231 mm

The lever arm $l_a = 640 - \dfrac{231}{3}$

$= 563$ mm

DESIGN FACTORS

Hence, tensile stress in steel due to bending $= \dfrac{1\,121 \times 10^6}{15\,220 \times 563}$

$= 130.8 \text{ N/mm}^2$

(Permissible, 140 N/mm^2)

and maximum compressive stress in concrete due to bending $= \dfrac{2 \times 1\,121 \times 10^6}{3\,500 \times 231 \times 563}$

$= 4.9 \text{ N/mm}^2$

(Permissible, 8.1 N/mm^2)

This calculation has been worked in detail to demonstrate the basic method of attack, but the same results can be obtained by the use of Table E which is based on the percentage ratio of A_{st} to bd_1 in rectangular beams.

The percentage reinforcement in the example is

$$\dfrac{100 \times 15\,220}{3\,500 \times 640} = 0.68$$

TABLE E

Percentage Reinforcement $\dfrac{100 A_{st}}{bd_1}$	Neutral Axis Depth d_n ($\times d_1$)	Lever Arm l_a ($\times d_1$)	Percentage Reinforcement $\dfrac{100 A_{st}}{bd_1}$	Neutral Axis Depth d_n ($\times d_1$)	Lever Arm l_a ($\times d_1$)
0·50	0·320	0·893	1·06	0·426	0·858
0·52	0·324	0·892	1·08	0·429	0·857
0·54	0·329	0·891	1·10	0·432	0·856
0·56	0·333	0·889	1·12	0·434	0·855
0·58	0·337	0·888	1·14	0·437	0·854
0·60	0·344	0·885	1·16	0·440	0·853
0·62	0·348	0·884	1·18	0·443	0·852
0·64	0·352	0·883	1·20	0·446	0·851
0·66	0·356	0·881	1·22	0·448	0·851
0·68	0·360	0·880	1·24	0·451	0·850
0·70	0·365	0·878	1·26	0·454	0·849
0·72	0·369	0·877	1·28	0·457	0·848
0·74	0·372	0·876	1·30	0·459	0·847
0·76	0·376	0·875	1·32	0·461	0·846
0·78	0·380	0·873	1·34	0·464	0·845
0·80	0·384	0·872	1·36	0·466	0·845
0·82	0·387	0·871	1·38	0·468	0·844
0·84	0·390	0·870	1·40	0·471	0·843
0·86	0·394	0·869	1·42	0·473	0·843
0·88	0·398	0·867	1·44	0·475	0·842
0·90	0·402	0·866	1·46	0·477	0·841
0·92	0·405	0·865	1·48	0·479	0·840
0·94	0·407	0·864	1·50	0·481	0·840
0·96	0·411	0·863	1·52	0·483	0·839
0·98	0·414	0·862	1·54	0·485	0·838
1·00	0·418	0·861	1·56	0·487	0·838
1·02	0·420	0·860	1·58	0·489	0·837
1·04	0·423	0·859	1·60	0·493	0·836

Rectangular beams reinforced in tension only, $m = 15$

whence from columns 2 and 3 in Table E, $d_n = 0.360 \times 640 = 231$ mm and $l_a = 0.880 \times 640 = 563$ mm.

It is not usually necessary to check either the average or local bond stress in block foundations, but attention is drawn to them so that the example may be complete.

The first is rarely critical and the average bond stress requirement will be met if the length of the bar measured from any section is not less than the

bar diameter × $\dfrac{\text{the tensile stress in the bar at the section}}{\text{four times the permissible average bond stress}}$.

The length determined may have deducted from it a length equal to the bond value of the hook at the end of the bar, which is 15 × bar diameter for the standard U hook and 8 × bar diameter for the standard square hook. Inspection shows that the reinforcement provided easily meets this requirement.

The local bond stress must be calculated for the section critical for bending moment, that is, at the edge of the steel slab. Its value is given by the equation.

$$\text{Local bond stress} = \frac{Q}{l_a \cdot o}$$

where Q = total shear across the section
l_a = arm of the resistance moment
o = sum of perimeters of the bars in tension.

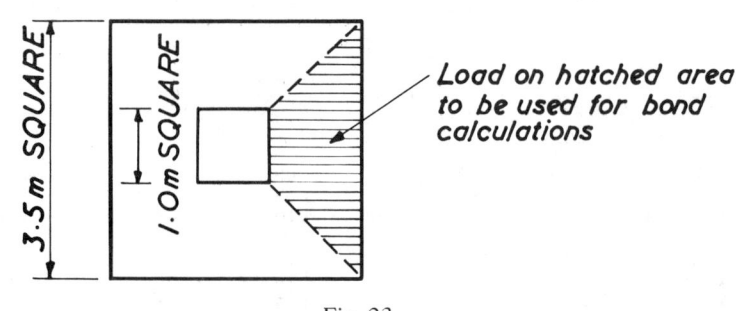

Fig. 23

Figure 23 shows the conditions; the total shear at the section is given by the load on the hatched area

$$= \frac{5\,000 \times 10^3}{4} \times \frac{3\,500^2 - 1000^2}{3\,500^2} = 1.148 \times 10^6 \text{ N}$$

$$\text{The local bond stress} = \frac{1.148 \times 10^6}{563 \times 31 \times (25 \times \pi)}$$

$$= 0.84 \text{ N/mm}^2$$

(Permissible, 1.3 N/mm²)

The maximum shear stress occurs at a distance of 640 mm out from the edge of the steel slab, so that the side of the shear square is 2.28 m and the load causing shear is

$$5\,000 \times 10^3 \times \frac{3\,500^2 - 2\,280^2}{3\,500^2} = 2.88 \times 10^6 \text{ N}$$

Hence $\quad$ shear stress $= \dfrac{2.88 \times 10^6}{4 \times 2\,280 \times 640}$

$$= 0.49 \text{ N/mm}^2$$

(Permissible 0.8 N/mm^2)

A conservative approach should always be used when calculating the shear resistance of foundation blocks where no shear reinforcement is provided, as has been done in this example.

Figure 24 gives details of the base as designed, except that 32 bars have been used for detailing symmetry. It must be pointed out that there are other methods of calculating moments and shears in foundation blocks in common use, but the method used in the example, based upon the requirements of C.P. 114, will be accepted by local authorities.

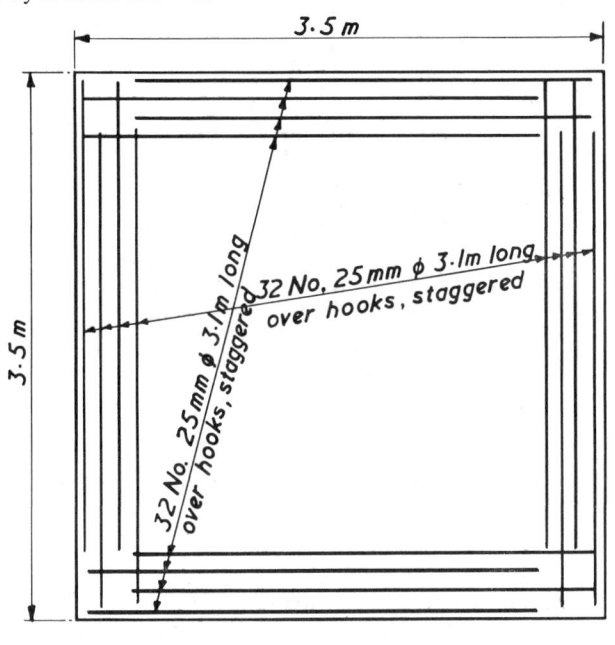

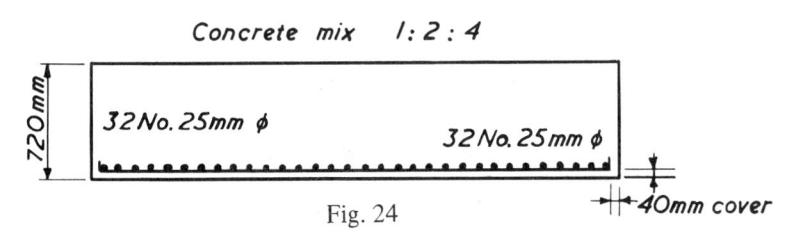

Fig. 24

Rectangular bases

Calculations should be made for both axes, as outlined for the square base, that is, moments taken about the edges of the steel slab, with appropriate critical shear sections. Calculations based on the width of the block will govern the design.

Pocket Bases

The attachment to concrete foundations of the bases of stanchions subjected to end moments can be made with anchor bolts as described on pages 623–625 or pages 526 and 527 where moments are large, but for the feet of single storey portal frames designed with fixed bases the use of the pocket type base shown in Fig. 25 is strongly recommended. It provides a very definite fixing and is economical both in cost, since it dispenses with a steel base and anchor bolts, and in site labour, as the templates necessary for accurate setting of foundation bolts are not required.

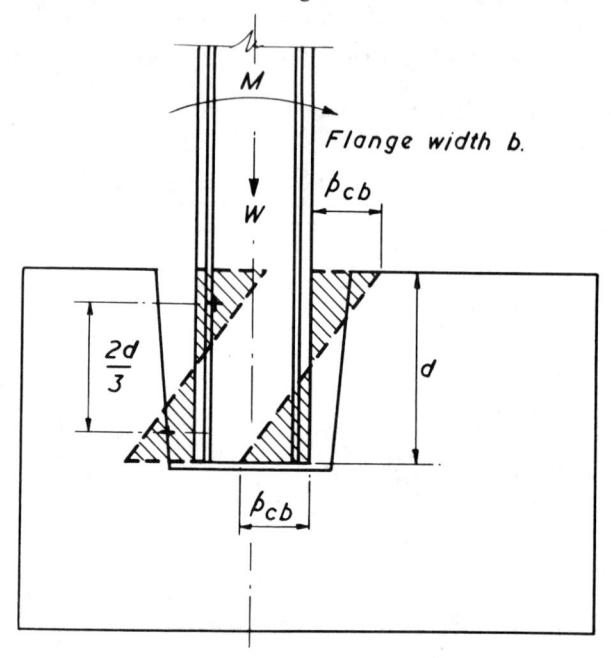

Cross section through Pocketed Base
Fig. 25

Assuming the triangular pressure distribution shown in Fig. 25 and taking pressure on both outer and inner faces of the stanchion flanges but neglecting the web thickness, then if

b = flange width (mm),
d = effective embedment (mm),
p_{cb} = permissible compressive stress on concrete

Total horizontal pressure above or below the neutral axis

$$= 2 \times \frac{p_{cb}}{2} \times \frac{bd}{2} = p_{cb} \times \frac{bd}{2}$$

Distance between centres of pressure = $\frac{2d}{3}$

Hence resistance moment = $p_{cb} \times \frac{bd}{2} \times \frac{2d}{3} = p_{cb} \times \frac{bd^2}{3} = 2 \times p_{cb} \times \frac{bd^2}{6}$

DESIGN FACTORS

Fig. 26 and Table F give the pocket dimensions for the range of universal beams and columns scheduled. The embedment has been calculated as that necessary to resist the full plastic moment, M_p of the section, divided by the load factor of 1.75, using a 1:2:4 concrete mix with maximum allowable compression on the concrete, $p_{cb} = 7$ N/mm². It will be found that the axial load is easily taken by the bond between the steel and the concrete.

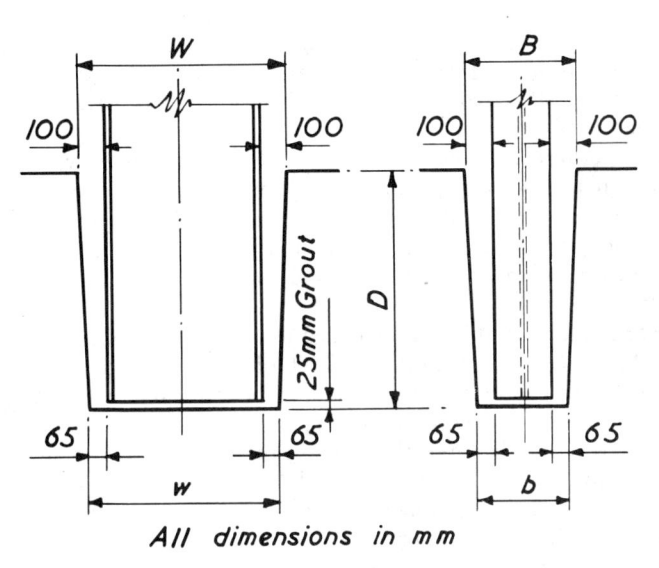

All dimensions in mm

Fig. 26

TABLE F

Section Millimetres	D mm	W mm	w mm	B mm	b mm
UB's					
254 x 102 x 22 kg	430	460	390	300	230
254 x 146 x 31 kg	430	460	390	350	280
305 x 165 x 40 kg	510	510	440	370	300
356 x 127 x 33 kg	510	560	490	330	260
356 x 171 x 45 kg	560	560	490	370	300
406 x 140 x 39 kg	590	610	540	340	270
406 x 178 x 54 kg	640	610	540	380	310
457 x 190 x 67 kg	710	660	590	390	320
533 x 210 x 82 kg	810	730	660	410	340
610 x 229 x 101 kg	920	810	740	430	360
UC's					
152 x 152 x 23 kg	330	350	280	350	280
203 x 203 x 46 kg	410	400	330	400	330
254 x 254 x 73 kg	510	460	390	460	390
305 x 305 x 97 kg	590	510	440	510	440

If it becomes necessary to increase the depth of pocket beyond 920 mm then it is recommended that breadths b and B should be a minimum of 460 mm and 530 mm respectively.

The concrete block thus falls within that shown in Case 3, Fig. 4 of this chapter, but it should be noted that, if the moment at the base is not reversible, economy in size of block and excavation will result if the stanchion is offset on the base by the amount e, thus giving a uniform soil pressure.

The concrete block may be unreinforced, in which case reference to Fig. 12 gives the manner in which the block thickness may be found. If a reinforced base is used, it should be designed on the lines indicated earlier in this section, the moments being taken at the stanchion flanges or flange tips as requisite, due allowance being made for non-uniform soil pressure if this exists. The critical sections for shear will be as before, at a distance equal to the effective depth of the base out from the flanges or flange tips.

The site procedure is simple but naturally it is necessary to provide the general contractor with a foundation plan, indicating base and pocket sizes, well in advance of the commencement of steel erection. After the pocketed bases have been cast by the general contractor, the steel erectors build up the bottom of the pockets to the correct level with steel packings grouted in place, which should be allowed to set before proceeding further.

Also prior to erection, it is recommended that the erectors scribe two setting out lines at right-angles on the concrete base, each offset a known distance from the required stanchion position to assist in landing the stanchions when being erected. A number of stanchions for, say, six frames, are landed in the pockets and wedged in place by timber wedges driven in at the top of the pocket. These wedges must not project too far into the pocket — say about $\frac{1}{4}$ depth, as they will interfere with the concrete filling when poured. Longitudinal members, bracing, etc., are next erected and the stanchions lined and plumbed.

The general contractor then fills the pockets with good quality concrete, using 10 mm aggregate up to the underside of the timber wedges. The stanchions remain undisturbed for 48 hours, after which the wedges are extracted and the filling of the pockets is completed.

25. STEEL PILING

Introduction

AMONG the tables in Chapter 40 giving the geometrical properties and other data relating to steel sections are a number appropriate to steel piling. Not only are there sections specifically designed for piling, but there are also sections, such as universal columns, which may augment the range of bearing piles, or tubes of large diameter, which may extend the range of hollow piling products. In addition, compound sections comprising sheet piling and universal beams, such as Unissen piles, may be fabricated when strength is required in excess of the capacity of the conventional piling sections.

It would be quite beyond the scope of this manual to give an exhaustive treatise on the uses of steel piling. In consequence, the general notes provided here are accompanied by a list of references from which much more information may be obtained.

Sheet Piles

Although bearing piles are isolated and their loads are normally applied vertically or along their longitudinal axes, sheet piling is used as a continuous wall to resist horizontal loads. Nevertheless, sheet piling can also carry significant vertical loading at little extra cost.

The basic sheet pile sections are either U-shaped, as in Larssen piling, or Z-shaped, as exemplified by the Frodingham sections. In recent years, the efficiency of these sections, i.e. their strength-to-weight ratio, has been increased by rolling piles of greater width with an increase in overall depth. The wider sections are slightly harder to drive, as resistance to penetration is partly dependent on width, as well as on interlock friction, so in some cases a heavier hammer may be needed. As, however, the increase in the efficiency of the section has brought about a reduction in the thickness of the sheet, the selection of a section to withstand heavy driving may become more critical. Nevertheless, where this is likely to occur the section may be rolled in high yield stress steel, instead of the traditional mild steel. There is a further advantage in using high yield stress steel in temporary works, as such piling can be re-used more often.

As far as retaining walls are concerned, there are a number of well-known international methods of design, which when compared sometimes give widely varying results. In some cases the methods are supported by useful experimental evidence and here attention is drawn to References 1 to 4. Where failures have occurred with anchored walls, they have been due to failure of the anchorage system or lack of penetration, rather than inadequate bending capacity in the sheet piling.

The design of cofferdams and strutted excavations is based on the empirical estimation of bracing loads, the problem being much more complex than that of anchored retaining walls and more subject to constructional procedures on the site.

The structures being temporary it sometimes happens that insufficient attention is given to safety precautions. The mechanisms of failure of cofferdams have been discussed by *S. Packshaw* in Reference 5.

It should be noted that further information on sheet piling may be obtained from the General Steels Division or the Regional Offices of the British Steel Corporation, in particular details of such accessories as junction and corner piles and the possible methods for capping or anchoring retaining walls.

Bearing Piles

There are three main types of steel bearing piles:—

1. H-piles
2. Pipe-piles
3. Box-piles

H-piles comprise the purpose-made universal bearing piles and also universal columns. For both types of section the breadths and depths are approximately equal. The universal bearing piles, however, have flanges and web of identical thickness, thus providing equal resistance to driving forces and any possible corrosion. H-piles are thus available in a very wide range of sizes and weights for the foundations of almost any type of structure.

The main advantages of H-piles are listed below.

1. Lengths up to 26 m are readily available from the mills.
2. Cut-offs and welded splices can be made quickly and economically.
3. By using welded splices lengths up to 60 m can be driven.
4. Accidental eccentric loading when driving has negligible effect on the pile.
5. Ability to withstand prolonged hard driving.
6. Ease of handling. The pile is light and can be slung from any point along its length.
7. Capacity to develop very high bearing values.
8. Small soil displacement.
9. Large friction area.

In foundations, when using mild steel, similar to Grade 43A to B.S. 4360, it has been American practice to limit the working stress to 12 000 lb/sq in (83 N/sq mm). Fortunately, all but the poorest soils will provide adequate lateral support to the embedded length of a pile shaft.

The recommended minimum spacing for bearing piles driven in groups is 1.1 m or three times the diagonal measurements of the pile, whichever is the greater. Where the piles derive their entire resistance from end bearing on a stratum of sound rock, the distance may be reduced to 750 mm or twice the width of the section.

Various authorities have undertaken research on pile caps and others publish standard designs, but it is recommended that References 6 and 7 be consulted.

Apart from certain characteristics associated with their shape, the remarks made with regard to H-piles apply generally to pipe and box-piles. Both may be provided with some kind of shoe when they become displacement piles and may produce ground heave. They are often driven open-ended, however, when heave is lessened and penetration is increased.

Pipes have a high radius of gyration and a constant section modulus in all

directions. Not being circular in section, box piles have small variations in these geometrical properties.

Both pipe and box piles have the advantage that they may be inspected after driving merely by lowering a light down inside them.

While such piles are frequently filled with concrete, this does not greatly increase the load bearing capacity and the usual reason for filling is to reduce internal corrosion although this is unlikely to be significant. Where the sections are so filled, however, a simple connection may be made to any pile cap merely by inserting a few short starter bars in the top of the pile. Where there is no concrete filling, the top of the pile should be closed with a flat steel plate welded in place.

While the characteristics of the three types of bearing piles are compared very thoroughly by *G. M. Cornfield* in Reference 8, it may generally be stated that the most economical for use in foundations are the H-piles, while where the piles project above bed level, as in marine structures, pipe and box piles are more advantageous.

Marine Structures

Although there has been a steady increase in the use of steel piles in dolphins, jetties and other marine structures, it is probable that the greatest advances in recent years have been made in the berthing installations for tankers and other bulk carriers, typical details of which are given in References 9 to 13.

The German Committee for Waterfront Structures have thoroughly investigated the problems associated with marine structures and their recommendations are contained in Reference 4.

Dolphins

The largest and most robust dolphins, described for example in Reference 14, usually comprise large hollow sections, constructed by driving a continuous wall of steel sheet piling, rather as in cofferdam construction, which are then filled with sand or other suitable material, topped with mass or reinforced concrete into which bollards and other ancillary equipment are embedded.

The more common dolphins, however, comprise either single piles or clusters of piles, linked or rigidly fastened together. Such dolphins may be designed to fulfil a number of requirements, but they are usually subjected to two types of force only:

1. A static horizontal pull from a moored vessel, the greatest stresses being imposed when the line is attached to the top of the dolphin.
2. Impact from a vessel, when the ship's energy must be counteracted by the internal work of the dolphin, which may be expressed mathematically as:

$$k \cdot \frac{m \cdot v^2}{2} = \frac{P \cdot d}{2} = A$$

where k = a reduction factor which varies with the size of ship, speed and angle of approach, position of dolphin, etc. It can be 0.5 for barges or as little as 0.2 for sea-going ships.

$\dfrac{m \cdot v^2}{2}$ = the energy of the ship.

P = the applied force.
d = the deflection of the dolphin at the level of the force.
A = the absorption capacity of the dolphin.

In this case, the worst conditions are at low water when the dolphin is less able to deflect under load.

The methods of design generally accepted in Europe have been described by G. B. *Godfrey* in Reference 15. They are based on the work of *Blum*, Reference 16, the validity of whose formulas has been checked by a number of full-scale tests, particularly in the Kiel Canal and in Hamburg, References 17 to 18, as a result of which certain modifications have been introduced.

While H-bearing piles are still favoured by American engineers for dolphins, pipe and box-piles are much more commonly used in Europe. Nevertheless, over 11 000 tons of H-piles were driven in the erection of the pier in Reference 9. However, the bending yield strength of a pipe is about one-third greater than the normal yield strength. In many European countries advantage is taken of this large shape factor in the design of dolphins and for pipe-piles the working stress is taken as the tensile yield stress of the material. Indeed, when designing dolphins in Amsterdam, *Risselada*, Reference 19, differentiated between the sections employed. While he designed in pipe up to the full tensile yield strength of the material, he was only prepared to design to 90 per cent of this value for square sections and 80 per cent for hexagonal sections.

Jetties

In their most simple form jetties consist of single fingers reaching out from the shore, but where the waters are shallow, they may comprise one or more heads running parallel to the shore joined together by a trunkway and connected with the land by a shore arm. The superstructure of the jetties, which may be built of steel, timber or precast concrete, or combinations of these materials, are usually designed in a manner similar to that employed for bridges.

In jetties all the three types of bearing piles are used, either as individual members or in trestles and bents.

In the Kwinana jetty, Western Australia, built for the British Petroleum Company, 936 Rendhex and Larssen box-piles, in lengths up to 30 m, were employed, 392 of which were raking, the superstructure being constructed from precast concrete elements.

As described by *Murray* and *Collett*, Reference 11, bents were used for the shore arm and trunkway. After being driven, the piles were filled with concrete with a special bottom opening skip which allowed the concrete to fall straight down the pile. The connexions between the piles and the transverse beams were effected by spun concrete muffs which were positioned and supported on temporary timber platforms while the annular spaces between the muffs and piles were filled with rich mix concrete, mild steel fillets having been welded to the pile heads to provide a key. Subsequently, the transverse and longitudinal beams and road trestles were placed in position and jointed.

Prior to driving, the piles were cleaned and given coats of Wailes Dove bitumastic primer and enamel. Further resistance to corrosion was given by a system of cathodic protection, designed by *Spencer*, as described in Reference 20.

Concrete-filled Rendhex piles were also used in the four tanker berths in the Aden Oil Harbour, described by *Palmer* and *Scrutton* in Reference 10 and

constructed for the same firm. In this case, however, the vertical piles were welded to a framework consisting of steel plate girders and broad flange beams. Raking piles were tied into the framework with a reinforced concrete cap. All the piles were sand-blasted and given three coats of Wailes Dove bitumastic paint, one in the shop and two on the site after driving. Subsequently, the piles were provided with cathodic protection designed by *Spencer*. Further details of cathodic protection in the Aden area are given in Reference 21.

Corrosion Resistance

Some methods of protecting marine structures against corrosion have just been mentioned. One of the most remarkable features of steel piles driven into undisturbed ground, however, is their almost complete immunity from corrosion. The standard work on this subject is the report by *Romanoff* in Reference 22. He examined piles which had been in service from 7 to 40 years and concluded that regardless of the soil characteristics and properties the type and amount of corrosion observed was not sufficient to affect the strength or useful life of piles as load-bearing structures.

REFERENCES

1. ROWE, P. W. 'A theoretical and experimental analysis of sheet-pile walls,' *Proceedings I.C.E.* (January 1955).
2. ROWE, P. W. 'Sheet-pile walls at failure,' *Proceedings I.C.E.* (May 1956).
3. ROWE, P. W. 'Sheet-pile walls in clay,' *Proceedings I.C.E.* (July 1957).
4. Empfehlungen des Arbeitsausschusses 'Ufereinfassungen' (EAU 1971) 4th Edition. Wilhelm Ernst & Sohn, Berlin 1971. (An English translation is available.)
5. PACKSHAW, S. 'Cofferdams,' *Proceedings I.C.E.* (February 1962).
6. ANON. 'Investigation of the strength of the connection between a concrete pile cap and the embedded end of a steel H-pile.' *State of Ohio Department of Highways Research Report No. 1* (December 1947).
7. ANON. 'Pile cap design for H-pile foundations,' United States Steel Corporation, Pittsburgh (July 1968).
8. CORNFIELD, G. M. 'Steel bearing piles,' CONSTRADO, London (1972).
9. McGOWAN, C. W. N., HARVEY, R. C. and LOWDON, J. W. 'Oil loading and cargo handling facilities at Mina al-Ahmadi, Persian Gulf,' *Proceedings I.C.E.* (June 1952).
10. PALMER, J. E. G. and SCRUTTON, H. 'The design and construction of Aden Oil Harbour,' *Proceedings I.C.E.* (July 1956).
11. MURRAY, P. and COLLETT, D. N. 'Kwinana jetty,' *Proceedings I.C.E.* (December 1956).
12. *20th. International Navigation Congress (Baltimore – U.S.A. 1961) Section II. Ocean Navigation – Subject 1.* General Secretariat P.I.A.N.C., Brussels.
13. *Conference on Tanker and Bulk Carrier Terminals*, I.C.E., 13th. November 1969.
14. RIJSSELBERGHE, L. van and DESCANS, L. 'Ducs d'Albe en palplanches métalliques,' *Annales des Travaux Publics de Belgique* (June 1957).
15. GODFREY, G. B. 'Steel dolphins,' CONSTRADO, London (1972).
16. BLUM, H. 'Wirtschaftliche Dalbenformen und deren Berechnung,' *Die Bautechnik* (May 1932).
17. MÜLLER, F. E. 'Stahlrammpfähle für Dalbenbau,' *Mitteilungen der Hannoverschen Versuchsanstalt für Grundbau und Wasserbau, Franzius-Institut der Technischen Hochschule, Hannover* (No. 5, 1954).
18. EBNER, H. and FORSTER, K. 'Die Hamburger Dalbenversuche,' *Hansa* (Nos. 16/17, 1957).
19. RISSELADA, Tj. J. 'Dolphins at the port of Amsterdam,' *Dock and Harbour Authority* (June–July 1954).
20. ANON. 'Cathodic protection installation for Kwinana Refinery jetty and water intake flume,' *Civil Engineering and Public Works Review* (July 1955).
21. WRIGHT, H. J. 'Cathodic protection,' *Proceedings I.C.E.* (April 1962).
22. ROMANOFF, M. 'Corrosion of steel pilings in soils,' National Bureau of Standards, Washington (September 1965).

26. GIRDERS

THE following types of girders are described and illustrated in this section:

1. Welded Compound Girders.
2. Welded Plate Girders.
3. Riveted Plate Girders.

Design of Welded Compound Girders

Where the B.M. to be resisted is greater than the resistance moment of a single universal beam, the bending capacity of the beam may be increased by the addition of flange plates, subject to checking the strength of the web in shear and buckling.

Clause 27a of B.S. 449 states that solid web girders should preferably be proportioned on the basis of the moment of inertia of the gross section, with the neutral axis taken at the centroid of the section, which will normally be the case with welded construction. This sub-clause also states that the effective sectional area for resisting shear shall be the product of the thickness of the web and the full depth of the rolled section.

Clause 27b of B.S. 449 states, *inter alia*, that each flange plate shall be extended beyond its theoretical cut-off point and the extension shall contain sufficient welds to develop in the plate the load calculated for the bending moment and girder section (taken to include the curtailed plate) at the theoretical cut-off point.

The theoretical points of cut-off can be found quite simply by calculation or by superimposing the B.M. diagram upon a diagram representing the variations in section modulus of the girder. The added plating is normally in one thickness on each flange, since each added plate requires two runs of fillet weld, the cost of which would normally outweigh any savings on material due to the curtailment of the outer plates as shown in Fig. 1. Reduction can, however, be made in the width or thickness of the flange plate where the B.M. permits, but this demands full-strength butt joints in the plates, which again, are likely to be more costly than using the thicker material throughout.

As an example, consider a beam to carry a uniformly distributed load of 1 000 kN over a span of 10.8 m, assuming full lateral support and using steel to Grade 43 B.S. 4360 Part 2:1969, for which the working stresses given in Table 2 of B.S. 449 Part 2:1969 are 165 N/mm^2 for flange plating up to and including 40 mm thick: 150 N/mm^2 for plating over 40 mm thick and 165 N/mm^2 for all universal beams, irrespective of flange thickness. If an allowance of 32 kN is made for the weight of the girder, the section modulus required at a working stress of 150 N/mm^2 will be 9 290 cm^3. If the overall depth of the girder is not to exceed 685 mm, then a suitable section would be a 610 x 229 x 140 kg U.B. with 330 mm x 30 mm plates on each flange, with section modulus of 9 421 cm^3 and overall depth of 677 mm.

646   GIRDERS

Fillet welds

reduced plate thickness

Fig. 1

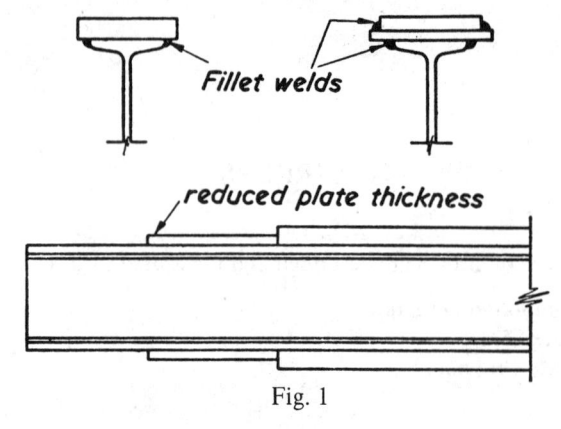

The theoretical cut-off points can be found graphically as shown in Fig. 2 or they may be calculated as follows:

The B.M. at any point distant x m from either end of the girder is

$$M = 516x - \left(\frac{1\,032}{10.8} \times \frac{x^2}{2}\right) \text{kNm}$$

$$= 516\left(x - \frac{x^2}{10.8}\right) \text{kNm}$$

GROSS MODULUS PROVIDED 9421 cm³
610×229×140 kg UB + 2/330×30 PLATES

MODULUS REQUIRED = 9288 cm³

LENGTH OF 330 mm PLT'S. = 9·0 m

1·32 m

THEORETICAL CUT-OFF POINT.

GROSS MODULUS OF 610×229×140 kg UB = 3620 cm³

150 mm, 180 mm, 150 mm Specified welding
162 mm, 168 mm, 162 mm Actual welding

SPAN 10·8 m
WELDED COMPOUND GIRDER.
Fig. 2

DESIGN OF WELDED COMPOUND GIRDERS

The theoretical cut-off points occur where M equals the resistance moment of the universal beam, of section modulus $3\,620\ \text{cm}^3$, so that with a working stress in bending of $165\ \text{N/mm}^2$, governed by the plate thickness of 30 mm, the section modulus

$$= \frac{516 \times 10^3}{165}\left(x - \frac{x^2}{10.8}\right) = 3\,620$$

or

$$\frac{516 \times 10^3 \times x^2}{165 \times 10.8} - \frac{516 \times 10^3 \times x}{165} + 3\,620 = 0$$

i.e.
$$x^2 - 10.8x + 12.50 = 0$$

whence
$$x = 1.32\ \text{m or } 9.48\ \text{m}$$

The vertical shear at these points is nearly 390 kN and it is from this value that the horizontal shear between beam and plate should be calculated. The horizontal shear/mm is given by

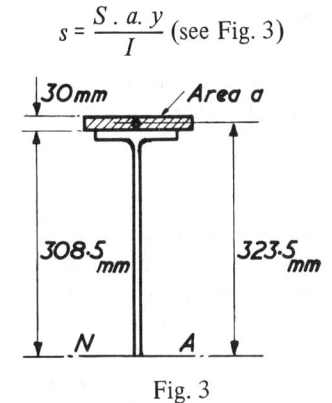

$$s = \frac{S \cdot a \cdot y}{I}\ \text{(see Fig. 3)}$$

Fig. 3

where s = horizontal shear in N/mm.
S = vertical shear at section in kN.
a = area of material to be connected.
y = distance from centroid of a to centroid of section.
and I = moment of inertia of compound section.

Hence
$$s = \frac{390 \times 10^3 \times (330 \times 30) \times 323.5}{318\,900 \times 10^4}$$

$$= 392\ \text{N/mm}$$

This is to be taken by two fillets and hence 4 mm continuous fillets would suffice for strength, but as the flange plates are 30 mm thick and the beam flange is 22.1 mm thick, it will be seen from Table C on page 886 that 6 mm is the minimum size weld allowable and this size will be adopted. The welds can be continuous throughout, as will be the case if automatic welding is used, or may be intermittent, if hand welding is employed. The ratio of effective length of intermittent weld to total length is found by dividing the actual shear/mm by the weld strength/mm, i.e.

$$\frac{196}{6 \times 0.7 \times 115} = 0.407$$

Welds 150 mm effective length and 200 mm spacing will give a ratio of 0.428 and can be adopted. The actual length of a weld is the effective length plus the weld size added to each free end and thus the welds are 162 mm long at 188 mm clear spacing, as shown in Fig. 2.

It is now necessary to check the web capacity of the beam. The end reaction is 516 kN, which is much less than the shear value of the web, 617 mm x 13.1 mm x 100 N/mm² = 808 kN, and if the end connection is made with riveted or welded end cleats or with a welded end plate, no further investigation need be made. If, however, the beam is supported on the bottom flange, i.e. by a seating cleat, then it is necessary to calculate web bearing and buckling capacities with reference to the length of stiff bearing.

Clause 28.a of B.S. 449 deals with web buckling on the following lines. The height of the column is to be the clear depth of web between root fillets, 543 mm and the slenderness ratio is specified as $\frac{d}{t}\sqrt{3}$, where

t = web thickness
d = clear depth of web

This equals $\frac{543 \times 1.732}{13.1} = 71.8$

The appropriate compressive stress given in Table 17A of B.S. 449 is 113 N/mm² so that the allowable buckling load per mm run of web = 13.1 x 113 = 1 480 N = 1.48 kN. The length of the column required is thus 516/1.48 = 349 mm, measured at the neutral axis of the section, as shown in Fig. 4, and allowing for 45° dispersion in accordance with Clause 28.a, the length of stiff bearing required is 40.5 mm.

The allowable bearing stress on steel to Grade 43 is given in Table 9 of B.S. 449 as 190 N/mm² and Clause 28.e states that the load can be dispersed at an angle of 30° to the plane of the flange.

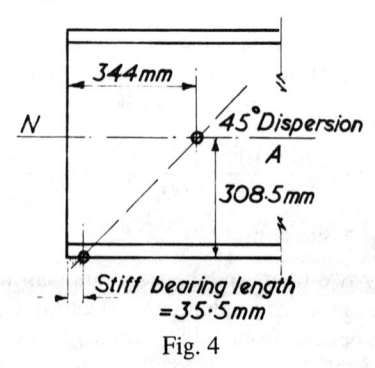

Fig. 4

The length of web required in bearing

$$= \frac{516 \times 10^3}{13.1 \times 190} = 207 \text{ mm}$$

DESIGN OF A WELDED PLATE GIRDER

measured at root fillet level, and with dispersion at 30° to the horizontal, the length of stiff bearing required under the flange

$$= 207 - 1.732 \times 37 = 207 - 64 = 143 \text{ mm}$$

Dispersion can continue through the seating at the same angle (see Fig. 5) to enable the stiffener outstand to be calculated.

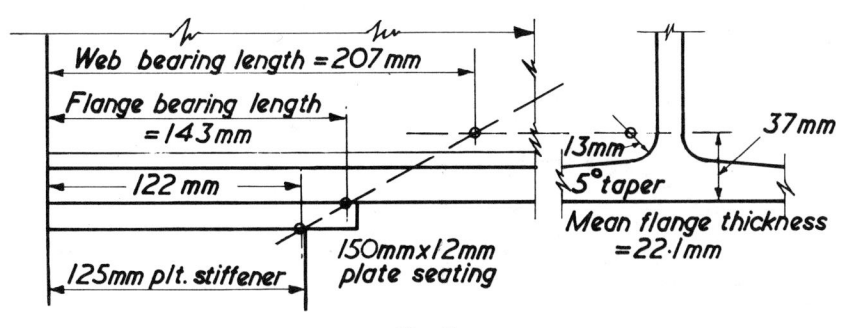

Fig. 5

It is now obvious that the web bearing criterion is more severe than web buckling and this governs the detail. If a welded tee bracket is used, then with a 150 mm x 12 mm plate seating, the stiffener plate should be 125 mm wide.

Design of a Welded Plate Girder

General Design Information

1. The girder designed in this example is 10.8 m span and is assumed to be simply supported at its ends.
2. The loading on the girder is shown in Fig. 6. In Case 1 loading (Fig. 6 (a)), both point loads are shown at their maximum magnitude. In Case 2 loading (Fig. 6 (e)), one of the loads is shown at its maximum magnitude and the other at its simultaneous minimum magnitude; this latter case gives the condition for maximum shear in the centre length of the girder.

The self-weight of the girder, which is first estimated, must eventually be checked when the girder is finally designed.

3. It is assumed that there are effective lateral restraints for the compression flange at positions A, B, C and D (Fig. 6 (a)).

Bending Moments and Shear Forces

The bending moments and shear forces acting on the girder are calculated by simple statics and are shown in Fig. 6 (b), (c), (f) and (g) for the alternative loading conditions.

Girder Cross-section

It is assumed that steel to B.S. 4360 Grade 43A will be used. The calculations are equally applicable to notch ductile steel to B.S. 4360 Grade 43C however.

650 GIRDERS

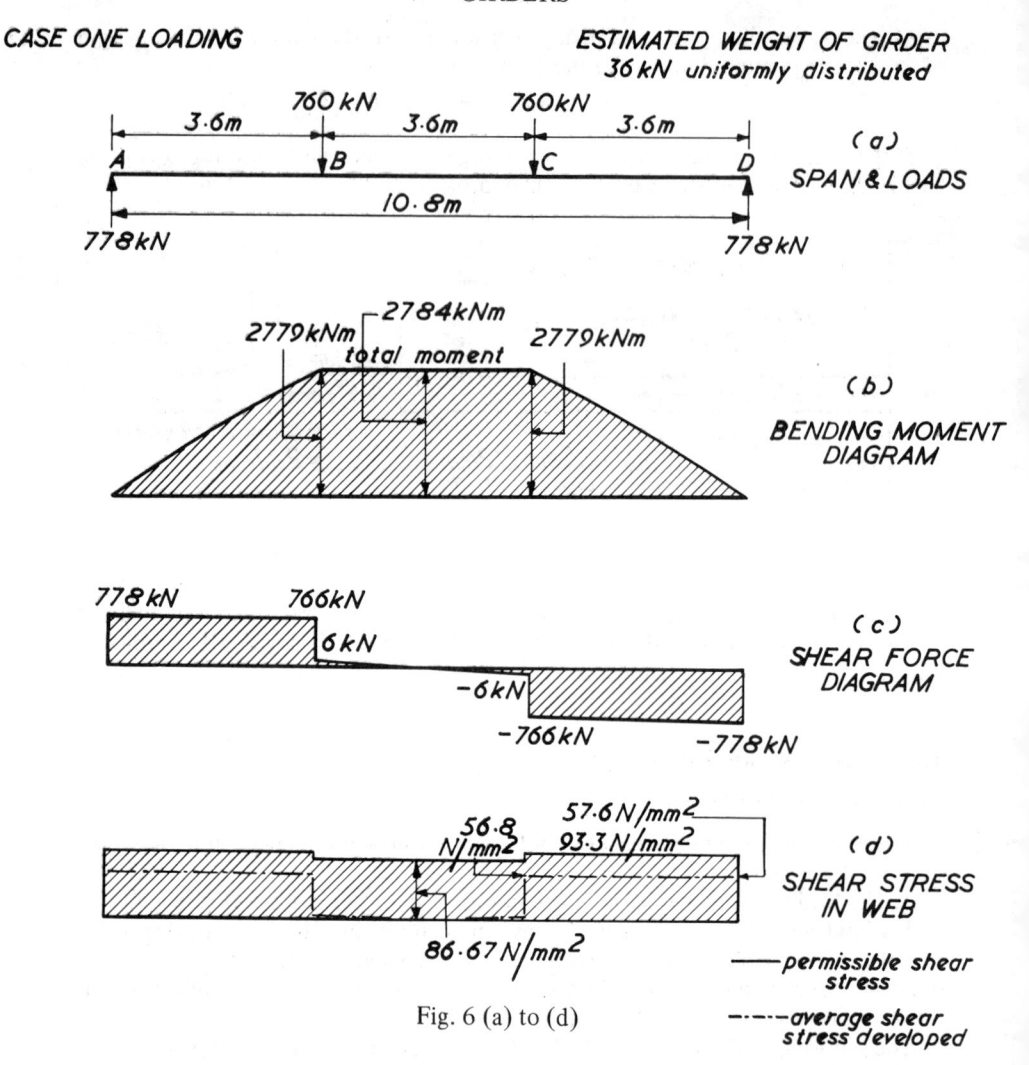

CASE ONE LOADING **ESTIMATED WEIGHT OF GIRDER**
36 kN uniformly distributed

(a) SPAN & LOADS

760 kN, 760 kN at B and C; 3.6 m spacing; A to D = 10.8 m; reactions 778 kN each.

(b) BENDING MOMENT DIAGRAM

2779 kNm, 2784 kNm total moment, 2779 kNm

(c) SHEAR FORCE DIAGRAM

778 kN, 766 kN, 6 kN, −6 kN, −766 kN, −778 kN

(d) SHEAR STRESS IN WEB

56.8 N/mm², 57.6 N/mm², 93.3 N/mm², 86.67 N/mm²

—— permissible shear stress
------ average shear stress developed

Fig. 6 (a) to (d)

A girder of I-shape will be used, as shown in Fig. 7.

It should be noted that this section is built up by welding directly together the flange and web plates, and there is no need for the flange angles used in riveted construction.

It is first decided that a depth of 1.35 m between flange plates and a width of flanges of 0.38 m is likely to prove a satisfactory basic dimension for the section. An approximate area of flange plate is obtained by dividing the maximum B.M. by the depth of the girder and using a permissible bending stress of 155 N/mm². An approximate required thickness of web is obtained based on the maximum shear force and a maximum permissible shear stress of 115 N/mm². The proposed section shown in Fig. 8 is determined in this manner and should next be checked by detailed calculations to ensure that the section is completely in accordance with the requirements of B.S. 449.

DESIGN OF A WELDED PLATE GIRDER

CASE TWO LOADING

ESTIMATED WEIGHT OF GIRDER
36kN uniformly distributed

(e) SPAN & LOADS

(f) BENDING MOMENT DIAGRAM

(g) SHEAR FORCE DIAGRAM

(h) SHEAR STRESS IN WEB

——— permissible shear stress
— · — average shear stress developed

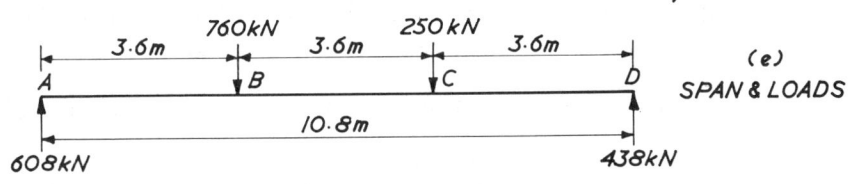

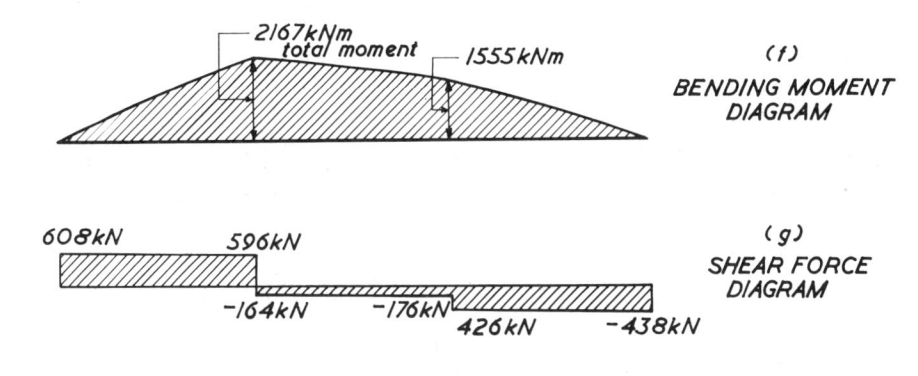

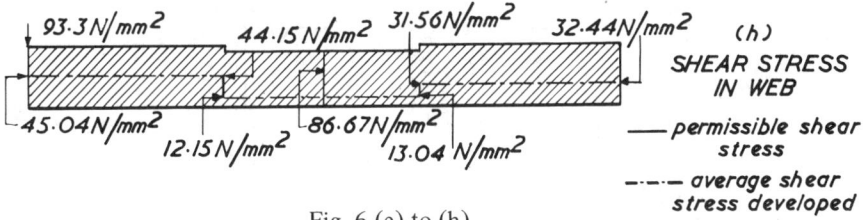

Fig. 6 (e) to (h)

Flanges

The flanges are designed in accordance with Clause 20 of B.S. 449.

In Table 2 of the Standard the allowable tensile stress p_{bt} in bending for a plate up to 40 mm thick is 155 N/mm².

The allowable compressive stress p_{bc} in a plate girder is the lesser of two values given in Table 2 and Table 8, the latter corresponding to the critical stress C_s obtained by following the detailed requirements of Clause 20.

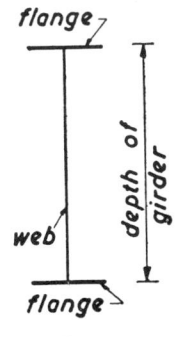

Fig. 7

As the flanges of the assumed section have equal moments of inertia about the $y - y$ axis of the girder, Case (I) applies.

Therefore $C_s = A$ N/mm².

The value of A depends on the two criteria l/r_y and D/T in Table 7.

GIRDERS

Now, l = the length between effective lateral restraint (see Clause 26.b)
  = 3.6 m = 3 600 mm.
r_y = radius or gyration about the $y - y$ axis of the gross cross-section of the whole girder at the point of maximum bending moment
  = 8.7 cm. (See below.)
D = overall depth of girder at the point of maximum bending moment
  = 1 410 mm.
T = effective thickness of compression flange
  = 30 mm.

Hence, $\quad\quad\quad\quad\quad l/r_y = 360/8.7 = 41.4$

and $\quad\quad\quad\quad\quad\quad D/T = 1\,410/30 = 47.0$

Properties of Cross-section

1 350 mm x 10 mm Web

380 mm x 30 mm Flanges

$I_x = 1 \times \dfrac{(135)^3}{12} + \dfrac{2 \times 38 \times (3)^3}{12}$

$\quad\quad + 2 \times (38 \times 3)(69)^2$

$\quad = 1\,290\,710$ cm^4

$I_y = 2 \times 3 \times \dfrac{(38)^3}{12} + \dfrac{135(1)^3}{12}$

$\quad = 27\,447$ cm^4

Area = 2 x 38 x 3 + 135 x 1
  = 363 cm^2

$r_x = \sqrt{\dfrac{1\,290\,710}{363}} = 59.6$ cm

$r_y = \sqrt{\dfrac{27\,447}{363}} = 8.7$ cm

a . y at joint of flange and web
  = 38 x 3 x 69 = 7 866 cm^3

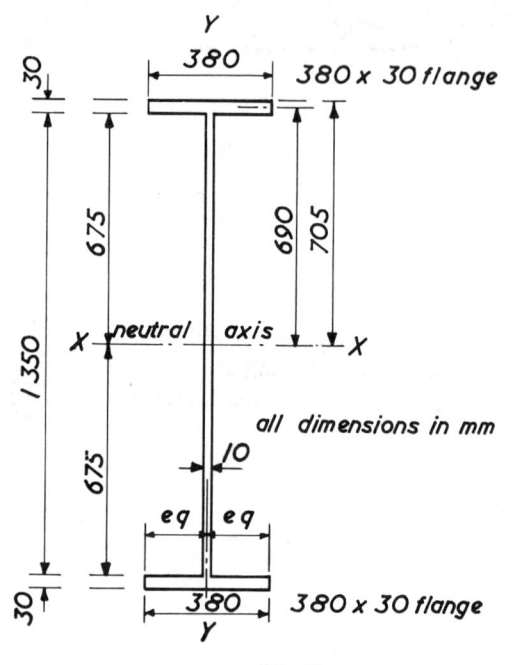

Fig. 8

On inspecting Table 7, it will be observed that the value of A exceeds 1 300. Therefore the corresponding value of p_{bc} in Table 8 is 165 N/mm^2.

As the value of p_{bc} given in Table 2 for a plate up to 40 mm thick is 155 N/mm^2, this value must be used in design.

From Fig. 6 (b) the maximum bending moment is 2 784 kNm.

Therefore the maximum tension or compression stress in the flanges due to bending,

$$f_{bt} = f_{bc} = M/Z = MD/2I$$

$$= \dfrac{2\,784 \times 10^6 \times 705}{1\,290\,710 \times 10^4}$$

$$= 152 \text{ N/mm}^2, \text{ which is satisfactory.}$$

DESIGN OF A WELDED PLATE GIRDER

It should be noted that if there had not been effective lateral restraints for the compression flange in this example, the permissible stress for compression would have been less than 155 N/mm². It would then have been economical to make the compression flange larger than the tension flange.

Web

The ratio d/t of the web is $1\,350/10 = 135$.

As this value exceeds the limiting value of 85 given in Clause 28.b of B.S. 449, vertical stiffeners must be provided at a distance apart not greater than $1\tfrac{1}{2}d = 2\,025$ mm.

The stiffener spacing assumed is shown in Fig. 9. In the outer-thirds of the length of the girder the spacing is $1\,200d/1\,350 = 0.889d$. In the middle third, the spacing is $1\,800d/1\,350 = 1.333d$.

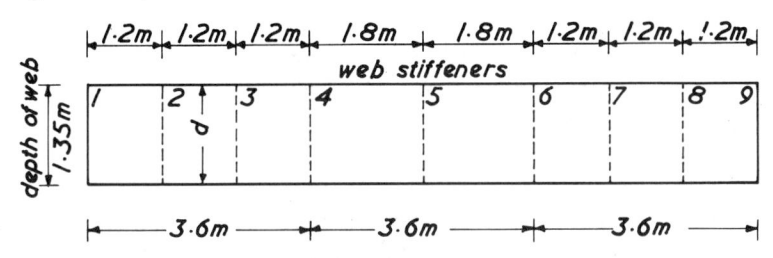

PROPOSED WEB STIFFENER SPACING

Fig. 9

Clause 23.b of B.S. 449 states that the average shear stress f_q'' on the gross section of a stiffened web shall not exceed the values of p_q' given in Table 12.

By interpolation, the value of p_q' in the outer-thirds of the girder is 93.3 N/mm² and in the middle third, 86.67. These values are inserted in Fig. 6 (d) and (h).

The maximum shear stress f_q'' induced in the girder by Case One Loading in Fig. 6 (a) occurs at the ends of the girder, where the maximum shear force is 778 kN.
Therefore

$$f_q'' = 778/d \times t = 778\,000/1\,350 \times 10$$

$$= 56.7 \text{ N/mm}^2$$

The variations in the value of f_q'' along the girder for Case One and Case Two Loading are shown in Figs. 6 (d) and (h) respectively.

It is interesting to consider in passing the actual stress conditions over the full depth of the girder. Figure 10 shows the actual shear per mm of girder, at either end, for the loading in Fig. 6 (a). At any point, the shear stress in the web is the shear per mm divided by t. For example, the maximum shear stress, at the neutral axis, is $614/10 = 61.4$ N/mm², compared with the average stress f_q'' already calculated of 57.6 N/mm².

Intermediate Stiffeners: Nos. 2, 3, 5, 7 and 8 in Fig. 9.

There is no applied end load on these stiffeners.

The stiffeners selected consist of flats arranged in pairs, that is to say, the flats are arranged on opposite sides of the web and directly opposite one another.

GIRDERS

```
         614 N/mm
        max. shear per
        mm of web
flange
                        480 N/mm
                        neutral
                        axis
flange  average shear
        per mm of
        web
        576 N/mm
```

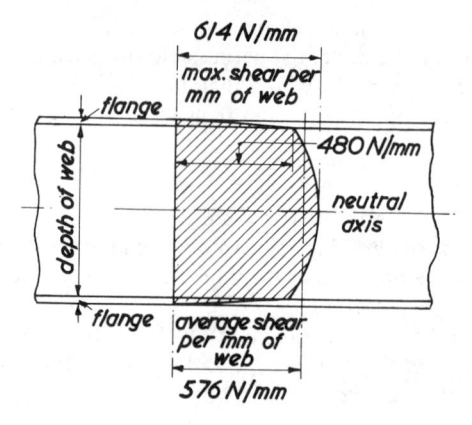

[as calculated on the approximate assumption that web resists total shear.]

Fig. 10

Clause 28.b (iii) of B.S. 449 states that unless the outer edge of each stiffener is continuously stiffened, the outstand of all flats shall be not more than $12t$, where t is the thickness of the flat.

Try a pair of flats, 100 mm x 10 mm.

I of complete stiffeners about the centre of the web $= \dfrac{10 \times (210)^3}{12 \times 10^4} = 772$ cm^4.

Clause 28.b (i) states that the stiffeners shall be so designed that I is not less than

$$1.5 \times \dfrac{d^3 \times t^3}{S^2}$$

where S = the maximum permitted clear distance between stiffeners for thickness t.
t = the minimum required thickness of web.

If $S = 1.5d = 1.5 \times 1\,350 = 2\,025$ mm and $t = 10$ mm.

$$\text{Minimum required } I = \dfrac{1.5 \times 1\,350^3 \times 10^3}{2\,025^2 \times 10^4}$$

$$= 90 \text{ cm}^4$$

Therefore, the section selected is more than adequate to meet this particular requirement in the Standard.

Stiffeners Under Loads: Nos. 4 and 6 in Fig. 9.

As shown in Fig. 6 (a), the load on each of these stiffeners is 760 kN, applied in the form of a line load across the top flange.

Try 180 mm x 15 mm flats, arranged in a pair located directly under the load.

In accordance with Clause 28.a (iii) of B.S. 449, the section for design as a column, shown in Fig. 11 (a), consists of the pair of stiffeners together with a

DESIGN OF A WELDED PLATE GIRDER 655

length of web on each side of the centre line of the stiffeners equal to 20 times the web thickness, i.e. 20 x 10 = 200 mm. The radius of gyration is taken about the axis parallel to the web of the girder, and the effective length of the column is assumed to be 0.7 of the actual length of the stiffener.

AREA USED FOR COLUMN LOADING
(a)

AREA USED FOR BEARING
(b)

STIFFENERS UNDER LOAD

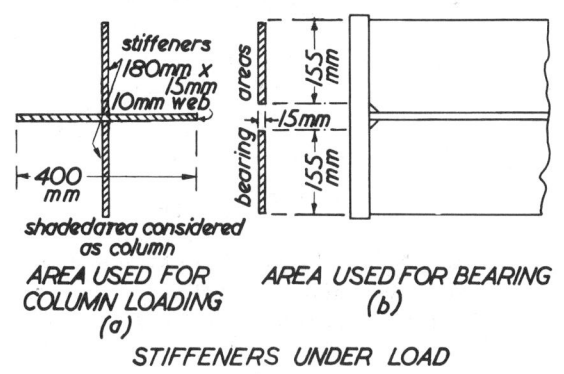

Fig. 11

$$I \text{ about centre line of web} = \frac{15 \times (370)^3}{12 \times 10^4} + \frac{(400 - 15) \times 10^3}{12 \times 10^4} = 6\,335 \text{ cm}^4$$

$$\text{Area} = 360 \times 15 \times 10^{-2} + 400 \times 10 \times 10^{-2} = 94 \text{ cm}^2$$

$$r \text{ about centre line of web} = \sqrt{\frac{6\,335}{94}} = 8.21$$

$$l/r = \frac{0.7 \times 135}{8.21} = 11.5$$

For this slenderness ratio, the allowable stress p_c interpolated from Table 17.a of B.S. 449 is 150 N/mm².

The actual stress f_c is $(760 \times 10^3)/94 \times 10^2) = 80.9$ N/mm².

The stiffeners must now be checked for bearing in accordance with Clause 28.a (iii), which also states that the outstanding legs of each pair of stiffeners shall be so proportioned that the bearing stress on that part of their area in contact with the flange clear of the welds does not exceed the bearing stress specified in Clause 22. The appropriate bearing stress p_b in Table 9 is 190 N/mm².

The stiffeners are notched 25 mm to clear the welds connecting the web to the flanges. Therefore, the area for bearing, shown in Fig. 11 (b), is 2 x 155 x 15 = 4 650 mm².

Hence, the actual bearing stress f_b = 760 000/4 650 = 163.3 N/mm².

The stiffeners selected are therefore satisfactory.

Stiffeners at Ends: Nos. 1 and 9 in Fig. 9.

The reaction at the end of the girder is 778 kN. In this example it is applied in the form of a line load across the bottom flange.

Try a 380 mm x 12 mm flat arranged across the end of the web and located directly above the applied line load.

GIRDERS

The section for design as a column consists of the stiffener together with a length of web extending 20 times the thickness of the web from the centre line of the stiffener, i.e. 20 x 10 mm = 200 mm, as shown in Fig. 12 (a).

$$I \text{ about centre line of web} = \frac{12 \times 380^3}{12 \times 10^4} + \frac{(200-6) \times 10^3}{12 \times 10^4} = 5\,489 \text{ cm}^4$$

$$\text{Area} = 380 \times 12 + (200-6) \times 10 = 65.0 \text{ cm}^2$$

$$r \text{ about centre line of web} = \sqrt{\frac{5\,489}{65}} = 9.2 \text{ cm}$$

$$\frac{l}{r} = \frac{0.7 \times 135}{9.2} = 10.3$$

The allowable stress p_c interpolated from Table 17.a of B.S. 449 is 151 N/mm². The actual stress f_c is 778 000/6 500 = 119.7 N/mm².

Now check for bearing at the bottom of the stiffener. The area available for bearing, as shown in Fig. 12 (b) is 380 mm x 12 mm = 45.6 cm².

STIFFENERS AT ENDS OF GIRDER

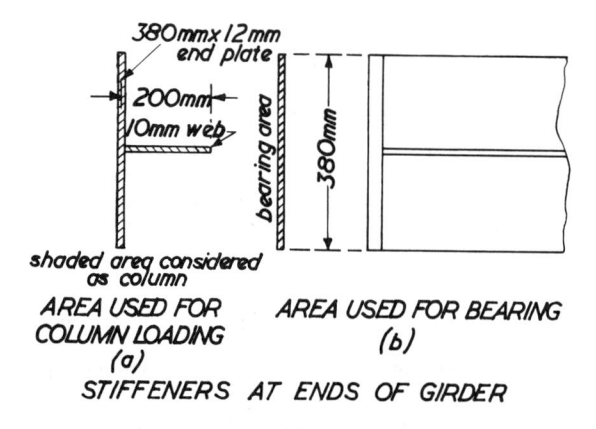

Fig. 12

The allowable bearing stress p_b is 190 N/mm².
The actual bearing stress f_b is 778 000/45.6 x 10² = 170.6 N/mm².
The stiffener selected is therefore satisfactory.

Welding: Web to Flange

The shear per linear mm between web and flange may be calculated from the formula:

$$\frac{S.a.y}{I} \text{ Newtons}$$

where S = the maximum shearing force in N.
      a = the area of either flange in sq. mm
         = 380 x 30 = 11 400 mm².
      y = the vertical distance between the neutral axis of the girder and the centroid of either flange

DESIGN OF A WELDED PLATE GIRDER

= 690 mm.

I = the moment of inertia of the complete cross-section of the girder in mm units

= 1 290 710 cm^4.

Consider the outer-thirds of the girder.
At the extreme ends of the girder, S = 778 kN.
Therefore, the shear

$$= \frac{778\,000 \times 11\,400 \times 690}{1\,290\,710 \times 10^4}$$

= 474 N/mm

Two fillet welds are used, one on each side of the web. Hence, the shear per mm on each weld = 474/2 = 237 N.

It will be seen in Table B on page 885 that 5 mm fillet welds would be sufficiently strong. However, the flanges are 30 mm thick, and as it will also be observed from Table C on page 886 that the minimum size of fillet for plates exceeding 16 mm, but not exceeding 30 mm thick, is 6 mm, this size must be adopted.

Therefore:
1. Two 6 mm fillet welds might be used continuous throughout the length of the girder. This would be satisfactory except that the amount of welding would be considerably greater than that required for strength in this example; or
2. Intermittent welds of the minimum size (here 6 mm) might be used to reduce the amount of welding.

The latter course will be adopted.

Now the appropriate part of Clause 54.c of B.S. 449 states that the distance along an edge of a part between effective lengths of consecutive intermittent fillet welds, whether the welds are in line or staggered on alternate sides of the edge, shall not exceed 16 times the thickness of the thinner part when in compression nor 24 times the thickness of the thinner part when in tension, and shall in no case exceed 300 mm.

Therefore, the maximum effective length of the space between intermittent fillet welds for a web 10 mm thick is 160 mm for the compression flange and 240 mm for the tension flange.

To meet the stress requirements, the minimum ratio of effective length of 6 mm intermittent weld to total length is:

$$\frac{\text{Actual shear/mm of joint}}{\text{Strength/mm of 6 mm weld}}$$

At the ends of the girder this ratio = 237/480 = 0.50.

Therefore, it is decided to use 160 mm effective lengths of 6 mm welds with 160 mm spaces for both the compression and the tension flanges over the outer 3.6 m lengths of the girder. Note that the spaces may need to be adjusted to suit the detail requirements of the structure, a point to be considered during the preparation of the detail drawings.

The maximum shear in the middle 3.6 m length of the girder, which occurs under Case Two Loading (see Fig. 6 (g)), is 176 kN. Therefore, the maximum shear

$$= \frac{176\,000 \times 11\,400 \times 690}{1\,290\,710 \times 10^4}$$

= 107 N/mm

while the ratio of actual shear to strength of 6 mm weld will be

$$\frac{107}{2} \times \frac{1}{480} = 0.11$$

Theoretically, a 20 mm length of intermittent weld with a 180 mm space would suffice. However, Clause 23 of B.S. 1856 states that the effective length of a fillet weld designed to transmit loading shall be not less than four times the size of the weld. Notwithstanding, on a structure of this size and type it is the usual practice to employ a minimum length of about 75 mm or 100 mm.

In this example, 75 mm effective lengths of weld and 150 mm spaces will be used for the middle 3.6 m length of girder. This provides more welding than required by calculation, but to reduce the amount of welding would be of doubtful economy.

The spaces in the intermittent welding of the web to the bottom tension flange could be increased to 225 mm, but detailing and marking-off will be simplified if the same spaces are used as for the top flange.

Welding – Intermediate Stiffeners: Nos. 2, 3, 5, 7 and 8 in Fig. 9.

There are no external loads applied to these stiffeners, their function being simply to prevent the web from buckling and hold the flanges square to the web. The welds provided in this example comply with the minimum sizes and spacing of welds laid down in Clause 54.c in B.S. 449.

4 mm fillet welds are used for connecting the stiffeners to the web, as shown in Fig. 13.

For the welds connecting the stiffeners to the flanges, 6 mm fillets are used, this being the minimum size in relation to the 30 mm flange thickness.

Welding – Load-bearing Stiffeners: Nos. 4 and 6 in Fig. 9.

Load-bearing stiffeners must be provided with sufficient welds to transmit the whole of the load to the web.

In this case the load per line of welds = 760/4 = 190 kN.

If continuous welds were used, the total length of each weld would be 1 300 mm (allowing for the 25 mm notches at each end) and the load per linear mm of weld would be 190 000/1 300 = 146 N.

The minimum size of fillet weld that could be used is 4 mm, the strength of which is 320 N/mm. It is economical, therefore, to use 4 mm intermittent welds. The lengths and spaces of welds are determined in the manner outlined for the web-to-flange welds, subject to the provisions of B.S. 449, Clause 54.c, and the arrangement adopted in this example for the stiffener-to-web welds is shown in Fig. 13.

It is assumed in this case that these stiffeners are fitted to bear tightly on the underside of the top flange plate and the welds between the ends of the stiffeners and the flange are not required to transmit load. They are made 6 mm fillets therefore, as for the intermediate stiffeners.

Welding – Load-bearing Stiffeners at Ends of Girder: Nos. 1 and 9 in Fig. 9.

Load per line of welds = 778/2 = 389 kN..

As these stiffeners are not notched at the ends, the total length of each continuous weld is 1 350 mm.

The load/mm of weld is therefore $\dfrac{389 \times 10^3 \text{ N}}{1\,350} = 288$ N.

4 mm fillet welds satisfy the requirements of both minimum size and strength (320 N/mm) and are adopted for the stiffener-to-web welds.

For the attachment of the ends of the stiffeners to the flanges, butt welds of single-bevel type provide a neat finish at the girder ends and are used as shown in Fig. 13. The bottom flange butt weld is required to transmit the full reaction of 778 kN, and must, therefore, be a properly sealed full-strength weld with a throat thickness of 12 mm. The weld is then capable of carrying the same load as the stiffener itself.

Notes on Foregoing Design of Plate Girder

The sizes of most of the fillet welds in this example are governed by the 'minimum size' requirements of B.S. 1856, Clause 14.a, Table 1. These 'minimum size' requirements must always be given attention as well as stress requirements.

The question whether, and to what extent, intermittent welds should be used involves consideration of the following points:

1. Intermittent fillet welds are not economical unless the 'minimum size' of weld is being used. A longer length of smaller fillet weld is usually more economical for the same strength. This is because the strength of a fillet weld increases only directly as the size but the weight of weld metal as the square of the size.
2. If automatic welding is to be used the welds should be continuous.
3. If the structure is exposed, the use of continuous welds may be considered preferable as conducive to greater ease of maintenance or longer life of the structure.
4. If severe dynamic loads act on the structure intermittent welds must not be used.
5. Intermittent butt welds may be used to resist shear forces only and, in general, are not to be recommended.

The girder as designed above is shown in Fig. 13.

For comparison, several alternative designs for the same conditions are shown in Fig. 14. In some of these the flange plates have been made in more than one length and reduced in thickness where the B.M. permits. This reduces the weight of parent metal but increases the amount of welding on account of the additional butt joints. Whether overall economy is effected depends on the conditions of each case. In some cases the web plate also can be made in more than one length and reduced in thickness where the shear permits. This again reduces the weight of parent metal but increases the amount of welding.

Riveted Plate Girders

In B.S. 449:1969 the principal Clauses relating to plate girders are Clauses 20, 23, 26, 27 and 28, the detailed requirements for sectional properties being given in Clause 27.

The web plate is the primary element in the girder and is normally the same thickness throughout, as, if altered, filler plates will be required under the flange angles for the length of web reduced in thickness. Subject to conditions relating to

660 GIRDERS

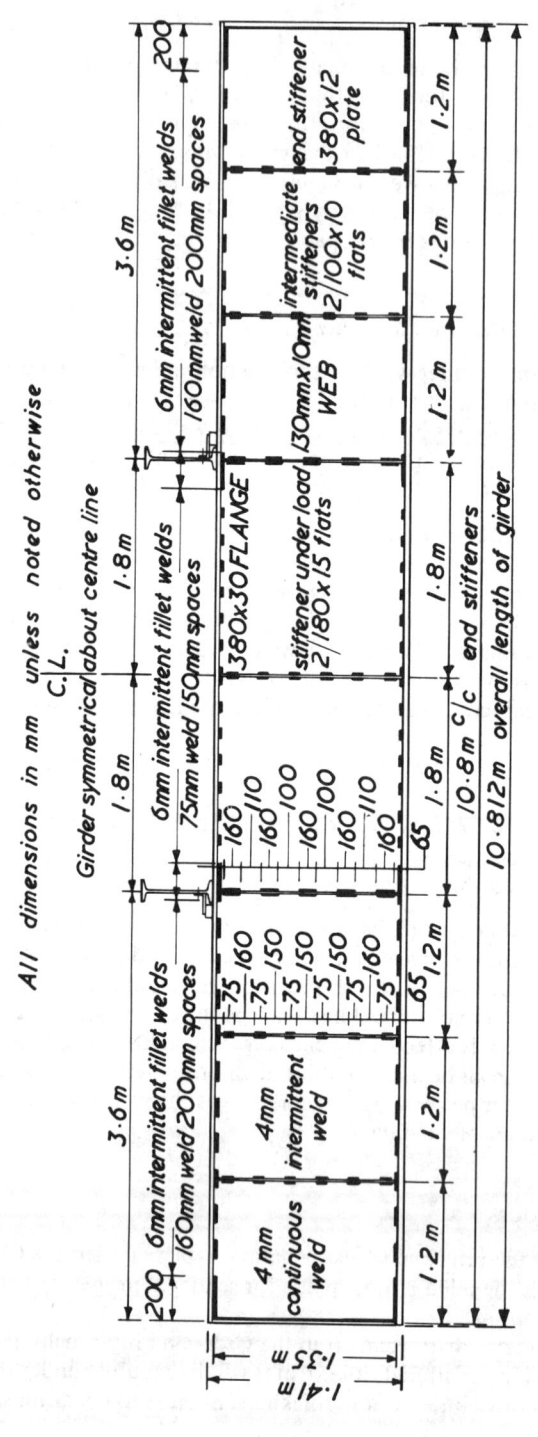

Fig. 13 (a)

Fig. 13 (b)

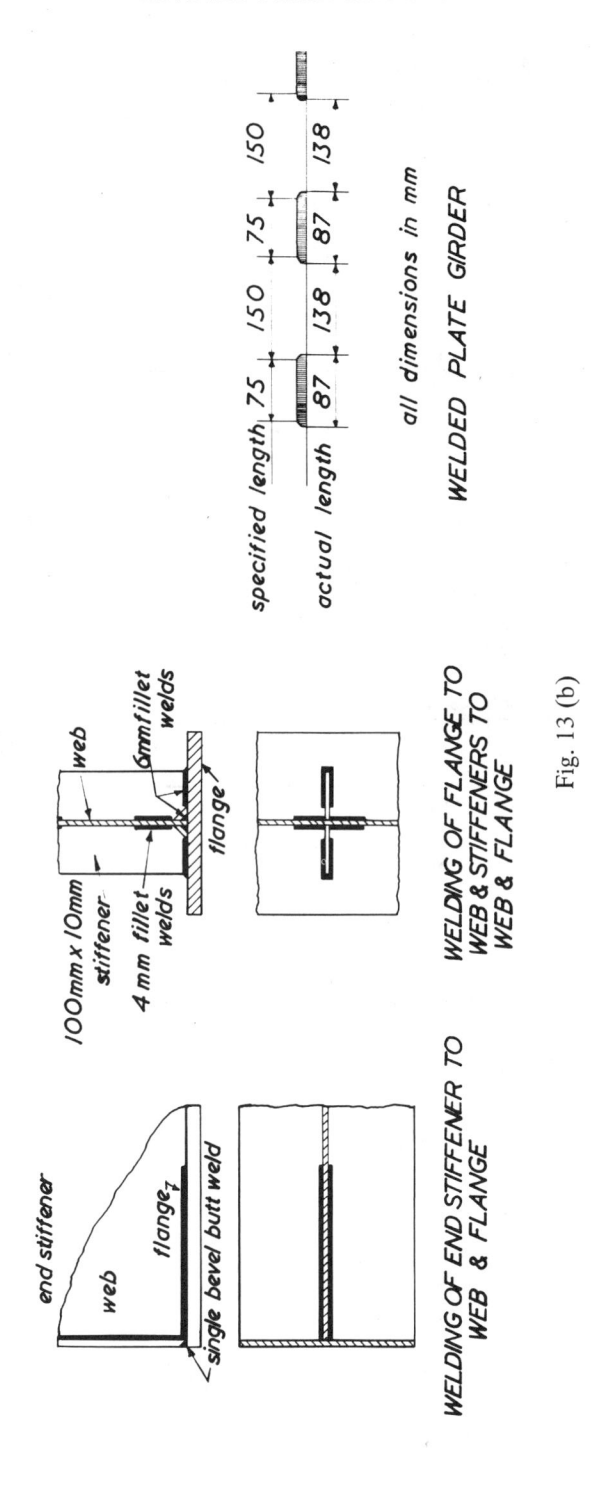

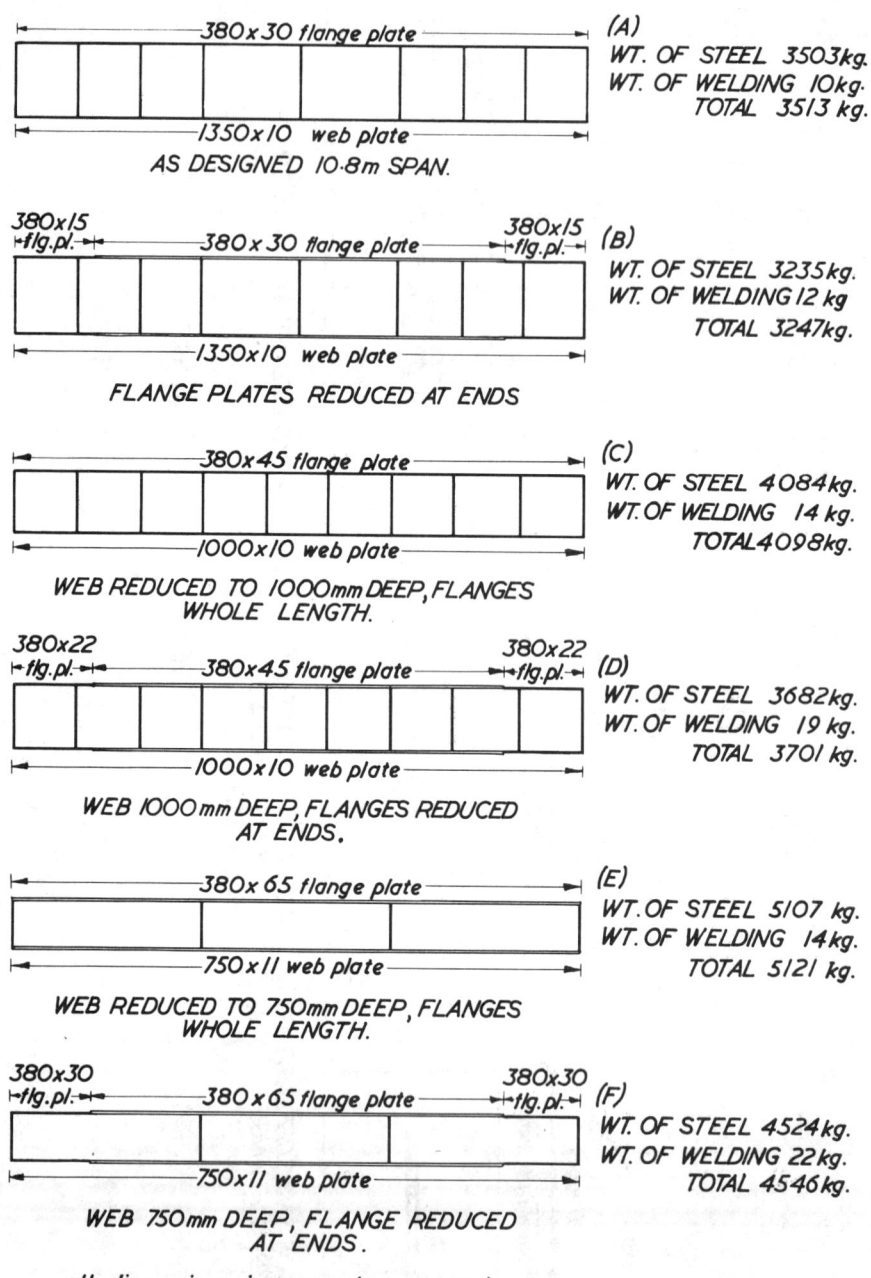

(A) AS DESIGNED 10·8m SPAN.
380 x 30 flange plate
1350 x 10 web plate
WT. OF STEEL 3503 kg.
WT. OF WELDING 10 kg.
TOTAL 3513 kg.

(B) FLANGE PLATES REDUCED AT ENDS
380 x 15 flg.pl.
380 x 30 flange plate
380 x 15 flg.pl.
1350 x 10 web plate
WT. OF STEEL 3235 kg.
WT. OF WELDING 12 kg.
TOTAL 3247 kg.

(C) WEB REDUCED TO 1000 mm DEEP, FLANGES WHOLE LENGTH.
380 x 45 flange plate
1000 x 10 web plate
WT. OF STEEL 4084 kg.
WT. OF WELDING 14 kg.
TOTAL 4098 kg.

(D) WEB 1000 mm DEEP, FLANGES REDUCED AT ENDS.
380 x 22 flg.pl.
380 x 45 flange plate
380 x 22 flg.pl.
1000 x 10 web plate
WT. OF STEEL 3682 kg.
WT. OF WELDING 19 kg.
TOTAL 3701 kg.

(E) WEB REDUCED TO 750 mm DEEP, FLANGES WHOLE LENGTH.
380 x 65 flange plate
750 x 11 web plate
WT. OF STEEL 5107 kg.
WT. OF WELDING 14 kg.
TOTAL 5121 kg.

(F) WEB 750 mm DEEP, FLANGE REDUCED AT ENDS.
380 x 30 flg.pl.
380 x 65 flange plate
380 x 30 flg.pl.
750 x 11 web plate
WT. OF STEEL 4524 kg.
WT. OF WELDING 22 kg.
TOTAL 4546 kg.

all dimensions in mm unless noted

COMPARISON OF ALTERNATIVE DESIGNS OF WELDED PLATE GIRDER

Fig. 14

transport and erection the web should be in as few pieces as possible. The maximum sizes of plates normally obtainable are 12 m in length, 3 m in width for all thicknesses from 16 mm to 40 mm, but there is in addition an overriding maximum area in one piece ranging from 33 sq. m for plates up to 20 mm thickness falling to 16 sq m at 40 mm thickness.

Flange angles and flange plates are riveted to the web plate. Angles are not usually readily obtainable in lengths greater than 18 m except by special arrangement. Flange plates up to 25 mm thickness can be obtained up to about 26–29 m in length in all suitable widths; the lengths fall to about half these values for plates 50 mm in thickness.

Flanges may be composed of angles only for light work, angles and flange plates for normal work and also angles, flange plates and flitch plates for heavy work as shown in Fig. 15.

The depth of plate girders, when not fixed by considerations of headroom, normally varies from about one-eighth of the span for short girders to about one-twelfth of the span for long girders. The girder of least total weight will not

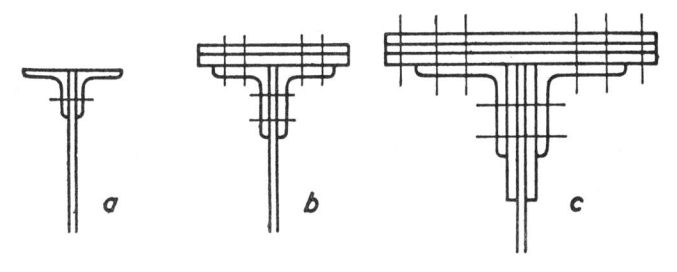

Fig. 15

necessarily be that with the least total cost since this depends upon elements other than weight such as transport and erection costs, so that it is generally economical to use a depth rather less than that giving minimum weight. It has been shown that the least weight depth is given approximately by

$$d = 1.1\sqrt{(M/f \cdot t)},$$

where M = maximum moment on girder,
$\quad f$ = allowable bending stress on flange
and $\quad t$ = web thickness.

This formula assumes that the bending resistance of the web is taken into account, that the flange plates are curtailed and that the weight of stiffeners is 60 per cent of the web weight. A reduction in the depth of 20 per cent from that calculated by the formula gives an increase in weight of about 2 per cent, and the normal depths in practice are usually considerably below those given by the formula. To apply the preceding principles a thickness of web must first be assumed and the appropriate depth found; the resulting proportions must then be checked against the specification requirements for the details of the design.

Moment of Resistance, Approximate Method

This should only be applied to girders with flanges of the type shown in Fig. 15 (b) by reason of the assumptions made which are set out in detail below. Consider

the girder shown in Fig. 16. The moment of inertia of the section equals

$$I = I_F + I_W,$$

where I_F = moment of inertia of two flanges about the neutral axis
and I_W = moment of inertia of web about the neutral axis,
let A_F = the net area of one flange (angles and flange plates), both flanges being identical,
A_W = the gross area of the web,
D = depth of girder overall,
d = depth back to back of angles
and h = effective depth or distance between centroids of flange areas.

Then $I_F = 2A_F(h/2)^2$, neglecting the moment of inertia of the flanges about their own centroids, and $I_W = td^3/12$. In the type of flange used, the centroid is not far

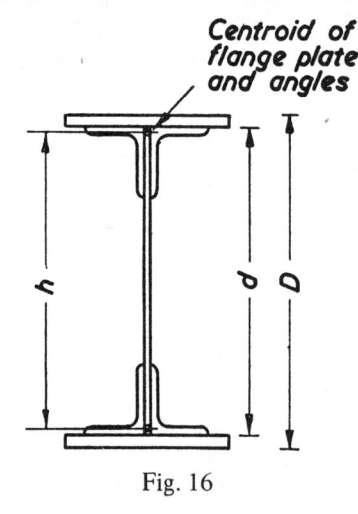

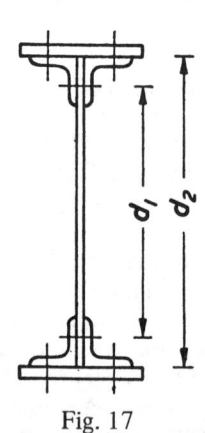

Fig. 16

Fig. 17

removed from the back of the angles, hence h is approximately equal to d. It is also assumed that the flange stress f is reached at the centroid of the flange and thus, by analogy with the basic flexure formula, $M = f \cdot I/y$, then

$$M = f \frac{[2 \cdot A_F(d/2)^2 + td^3/12]}{d/2}$$

$$= f \cdot d \left(A_F + \frac{A_W}{6} \right)$$

RIVETED PLATE GIRDERS

If it now be assumed that at the critical section there are holes in the web of diameter d at $4d$ pitch, then the net area of the web will be three-quarters of the gross area and $A_W/6$ will be reduced to $A_W/8$ and the formula becomes,

$$M = f \cdot d \left(A_F + \frac{A_W}{8} \right)$$

The quantity $A_W/8$ is often called the web equivalent and the total flange area is therefore

$$A_P + A_A + \frac{A_W}{8}$$

where A_P = net area of flange plates,
A_A = net area of flange angles,
and A_W = gross area of web.

Rivet pitch, Approximate Method

The rivets connecting the flange angles to the web plate must transfer from the web the horizontal shear necessary to induce the bending stresses in the flange angles and plates, and, in addition, if there is a vertical load on the flange plates must transfer this also.

Consider first the case of the flange with no vertical load. Then, for the rivets connecting the angles to the web, using the basic formula for intensity of horizontal shear stress at any point the pitch of rivets = $p_1 = r_1 d_1 / K_1 S$ (see Fig. 17); where

r_1 = safe load on rivet (double shear or bearing, whichever is least),
d_1 = depth between rivet lines,

$$K_1 = \frac{(A_P + A_A)}{(A_P + A_A + A_W/8)}$$

and S = vertical shear at section.

Similarly, if $K_2 = \dfrac{A_P}{(A_P + A_A + A_W/8)}$

r_2 = safe load on rivet in single shear
and d_2 = distance between flange plates,

then the pitch of flange-plate rivets,

$$p_2 = \frac{r_2 d_2}{K_2 S}$$

This is the staggered pitch, and the pitch on line will be twice this amount.

If there is a vertical load on the flange plate, such as a floor or a wall resulting in a shear load of w kN per mm run, this must, in the case of rivets connecting angles and web, be combined vectorially with the horizontal shear and

$$p_1 = \frac{r_1}{\sqrt{[(K_1 S)^2/d_1 + w^2]}}$$

Moment of Resistance of Girders with Sloping Flanges

If M is the total moment on the girder then it follows that the total flange force is M/d, where d = effective depth or distance back to back of angles. It therefore follows from what has been said earlier that the moment taken by the flange angles

and plates $= K_1 M$ whilst that taken by the web will be $M(1 - K_1)$. Thus for the girders shown in Fig. 18:

(a) $$C = \frac{K_1 M}{d} \quad \text{and} \quad T = \frac{K_1 M}{d \cos \alpha}$$

(b) and (c) $$C = \frac{K_1 M}{d \cos \beta} \quad \text{and} \quad T = \frac{K_1 M}{d \cos \alpha}$$

Inclined flanges affect the shear to be taken by the web. If the sections indicated on Fig. 18 are to the left of the point of maximum moment, assuming the girder to

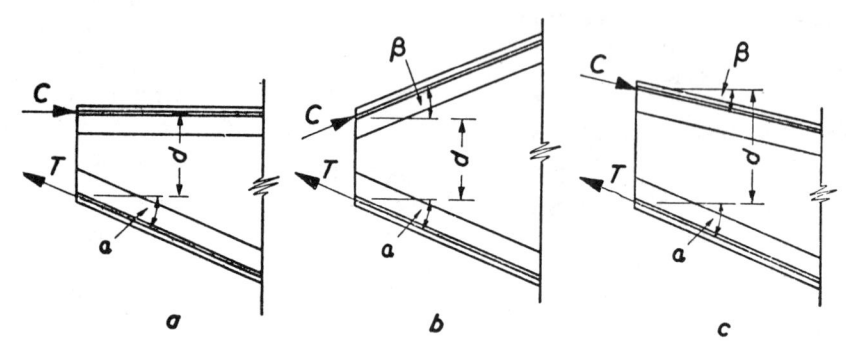

Fig. 18

be simply supported, S = vertical shear at section and S_W = shear to be taken by the web, then for the various cases shown:

(a) $$S_W = S - T \sin \alpha = S - \frac{K_1 M}{d} \tan \alpha$$

since the inclined tension absorbs shear.

(b) $$S_W = S - \frac{K_1 M}{d} (\tan \alpha + \tan \beta)$$

as both flanges absorb shear.

(c) $$S_W = S - \frac{K_1 M}{d} (\tan \alpha - \tan \beta)$$

In this case one flange absorbs shear and the other adds to the shear force.

It must be noted, however, that these computations refer only to the shear load in the web and not to the vertical reactions which support the girder.

The formulae previously given for rivet pitch require modification to allow for the increased flange loads and the fact that the pitch may be on a sloping line.

Design of a Riveted Plate Girder

The girder designed in this example is 10.0 m span, simply supported at the ends, and carries the loading shown in Fig. 19, which includes the weight of the girder, assumed as 60 kN, which must be checked after it has been designed. The figure also

DESIGN OF A RIVETED PLATE GIRDER

shows the shear force and bending moment diagrams. It is also assumed that effective lateral support is provided at each reaction and point load.

If it is assumed that Grade 43 steel plates not exceeding 40 mm thick will be used in the flanges, the maximum allowable bending stresses p_{bc} or p_{bt}, derived from

Fig. 19

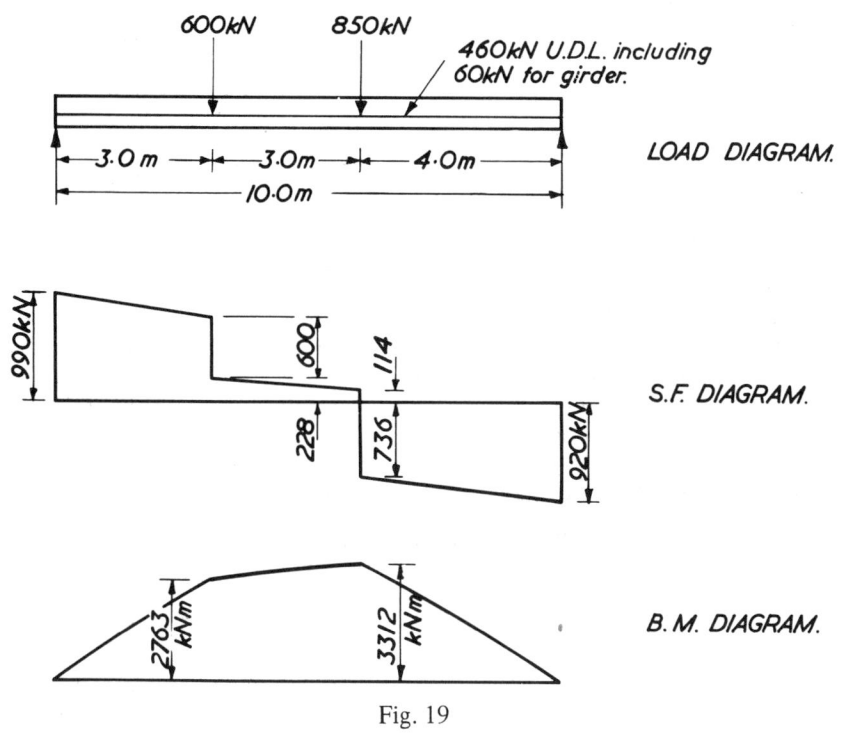

Table 2 of B.S. 449, is 155 N/mm². The corresponding allowable average shear stress p'_q in Table 11 is 100 N/mm². Assuming that a web thickness of 12 mm will be satisfactory, then the least weight depth, as given on page 663,

$$= 1.1 \sqrt{\frac{M}{f.t.}} = 1.1 \sqrt{\frac{3\,312 \times 10^6}{155 \times 12}}$$

$$= 1\,467 \text{ mm approximately}$$

This can be considerably reduced without sensible increase in weight, and a depth of web of 1 200 mm will be used, it being assumed that headroom considerations do not dictate a lesser depth. The depth d back-to-back of angles will be made 1 210 mm in order to clear the outside edges of the web plate.

Designing upon the Approximate Method

The flange area required

$$A = \frac{M}{f \times d} = \frac{3\,312 \times 10^6}{155 \times 1\,210 \times 10^2} = 176.6 \text{ cm}^2$$

This can be made up as follows:

$$\text{Web equivalent} = \frac{120 \times 1.2}{8} \qquad = 18.0 \text{ cm}^2$$

2 angles, 152 × 102 × 15.8   = 75.6

Plates, 400 × 30   = 120.0

     213.6 cm²

Less rivet holes, 2 × 24 × 43.8 + 24 × 43.6   = 31.5

     182.1 cm²

The slight excess area will compensate for the fact that in calculating the flange areas the depth has been taken as that over the angles, whereas the effective depth should be that between the centroids of the flanges.

It should be noted that B.S. 449 states that flange angles should form as large a part of the flange area as is practicable, and in the section given they form rather more than 39 per cent of the attached material. The specification also states that the number of flange plates shall be reduced to a practical minimum, and thus the 30 mm thickness will be made up from one 15 mm plate (net area 52.8 cm²) running the full length and one 15 mm plate which will be curtailed.

The net areas of the flange components are thus:

     Web equivalent   16.1 cm²
     Flange angles   60.4 cm²
     Inner plate   52.8 cm²
     Outer plate   52.8 cm²

These are used for the flange-plate curtailment diagram given in Fig. 20 giving theoretical points of cut-off of the outer plate at 2.61 and 2.83 m from the left and right ends respectively, the shears at these points being 870 and 790 kN.

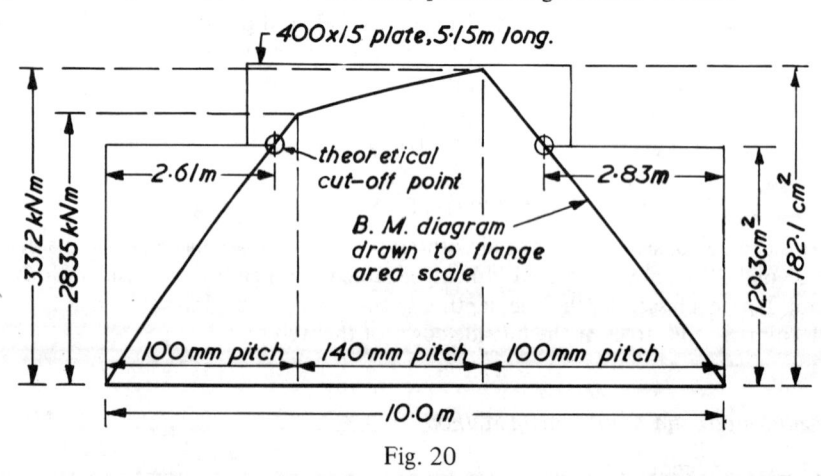

Fig. 20

The left-hand end of the girder is more highly stressed in shear and will be used to decide the pitch of rivets necessary as the pitch so obtained will be adequate for

DESIGN OF A RIVETED PLATE GIRDER

the other end. Dealing with the rivets connecting flange angles to web plate, the following figures result:

L.H. end.

Shear = 990 kN

Web equivalent	16.1	
Flange angles	60.4	$K_1 = \dfrac{113.2}{129.3} = 0.875$
One flange plate	52.8	
	$\overline{129.3}$ cm²	

The least strength of a 22 mm diameter Grade 43 steel power-driven shop rivet, taken as 24 mm diameter in accordance with Clause 17 (b) of B.S. 449 will be 90.7 kN in bearing on the web, at the enclosed bearing stress of 315 N/mm² permitted by Table 20. The horizontal shear load per mm, the distance between rivet lines being 1 100 mm,

$$= \frac{990 \times 0.875 \times 10^3}{1\ 100} = 787 \text{ N}$$

The vertical shear load per mm run due to the uniformly distributed load of 460 kN = 46 kN. The vertical effect of this vertical load is negligible, and we therefore have a resultant load of 787 N.

At point of cut-off of outer plate:

Shear = 870 kN

Web equivalent	16.1	
Flange angles	60.4	$K_1 = \dfrac{166}{182.1} = 0.915$
Two flange plates	105.6	
	$\overline{182.1}$	

$$\text{Horizontal shear load per mm} = \frac{870 \times 0.915 \times 10^3}{1\ 100} = 724 \text{ N}.$$

which is less than the corresponding figure for the left-hand end.

The maximum rivet pitch at the end of the girder is therefore 90 700/787 = 115 mm and a pitch of 100 mm will therefore be used for both end panels. The vertical shear in the centre panel has a maximum value of 252 kN only, but here the spacing of rivets is controlled by Clause 51c (ii) of B.S. 449 which limits the spacing to 12t or 200 mm, where t is the thickness of the thinnest element through which the rivets pass. The web thickness of 12 mm therefore dictates a maximum pitch of 140 mm in the centre panel.

Similarly, the calculations for the rivets connecting flange plates to flange angles follow, the safe load in the rivet being that in single shear, which is 49.8 kN, and the distance between shearing surfaces 1 210 mm under the action of horizontal shear only.

L.H. end. Shear 990 kN,

$$K_2 = \frac{56.3}{129.3} = 0.435$$

Horizontal shear load per mm $= \dfrac{990 \times 0.436 \times 10^3}{1\,210} = 356$ N

At point of cut-off of outer plate:
Shear = 870 kN,

$$K_2 = \frac{105.6}{182.1} = 0.579$$

Horizontal shear load per mm $= \dfrac{870 \times 0.579}{1\,210} = 416$ N

There are two rows of rivets connecting angles and flange plates and hence the straight line pitch

$$= \frac{2 \times 49.8 \times 10^3}{416} = 239 \text{ mm}$$

It is, however, not possible to use a 239 mm pitch, as Clause 51c (iii) states that the distance between centres of two consecutive rivets in a line adjacent and parallel to an edge of an outside plate shall not exceed $100 + 4t$. So that the 15 mm outside flange plate restricts the spacing to 160 mm for the length of cover plate. A pitch of 180 mm ($12t$ clause 51.c (ii)) would be permitted elsewhere. However, Clause 51.c (iv) states that when rivets are staggered at equal intervals and the gauge does not exceed 75 mm, the distances specified in (ii) and (iii) above, between centres of rivets may be increased by 50 per cent. For detailing purposes, however, a spacing of 150 mm will be used throughout.

The actual length required for the outer plates may now be calculated.

The B.M. at the theoretical cut-off points $= f \cdot a \cdot d = 155 \times 129.3 \times 10^2 \times 1\,210 \times 10^{-6} = 2\,425$ kNm.

The average stress in the outer tension plate at the theoretical cut-off points

$$= \frac{155 \times 129.3 \times 10^2 \times 1\,255}{182.1 \times 10^2 \times 1\,210}$$

$$= 114.2 \text{ N/mm}^2$$

Therefore. the actual load carried by the outer plate

$$= 114.2 \times 52.8 \times 10^{-1} = 603 \text{ kN}$$

The strength of one 22 mm diameter rivet in single shear is 49.8 kN. Hence, the number of rivets required beyond the theoretical cut-off points, to comply with Clause 27.b of B.S. 449, is $603/49.8 =$ say, 12 rivets. The arrangement of rivets is shown in Fig. 24, the actual length of the outer plate being 5.15 m.

A web thickness of 12 mm having been assumed, it is necessary to check that this complies with B.S. 449.

The ratio d/t of the web = $1\,007/12 = 83.8$, where $d =$ clear depth of web *between* flange angles and $t =$ the thickness of the web. Since this value is less than 85, the web does not require intermediate stiffeners. (See Clause 27.f of B.S. 449.)

DESIGN OF A RIVETED PLATE GIRDER

The allowable average shear stress p'_q in Table 11 for unstiffened webs not exceeding 40 mm thick is 100 N/mm².
The maximum shear force at the left-hand end of the girder is 990 kN.
The web is 1 200 mm deep.
Therefore, the minimum thickness required for the web

$$= \frac{990\,000}{100 \times 1\,200} = 8.25 \text{ mm}$$

Hence, the assumed thickness is satisfactory.

In order to comply with Clause 28.a (ii) of B.S. 449:1959, stiffeners must be provided at the points of concentrated load and at the points of support.

Clause 28.a (iii) states that load-bearing stiffeners shall be designed as struts, assuming the section to consist of the pair of stiffeners together with a length of web on each side of the centre line of the stiffeners equal, where available, to 20 times the web thickness. The radius of gyration shall be taken about the axis parallel to the web of the girder and the working stress shall be in accordance with the appropriate allowable value for a strut, assuming an effective length equal to 0.7 of the length of the stiffener.

The outstanding legs of each pair of stiffeners shall be so proportioned that the bearing stress on that part of their area in contact with the flange clear of the root of the flange angles does not exceed the bearing stress specified in Clause 22.

The stiffeners under the applied loads will be designed to suit the larger of the two loads, i.e. 850 kN. The allowable bearing stress p_b given in Table 9 is 190 N/mm². The minimum area required for bearing will be 850 000/190 = 44.7 cm². If it is assumed that the dimension of the out-standing leg of the stiffener is 127 mm, then the length of bearing between the flange angle and the stiffener is 116.3 mm, as shown in Fig. 21.

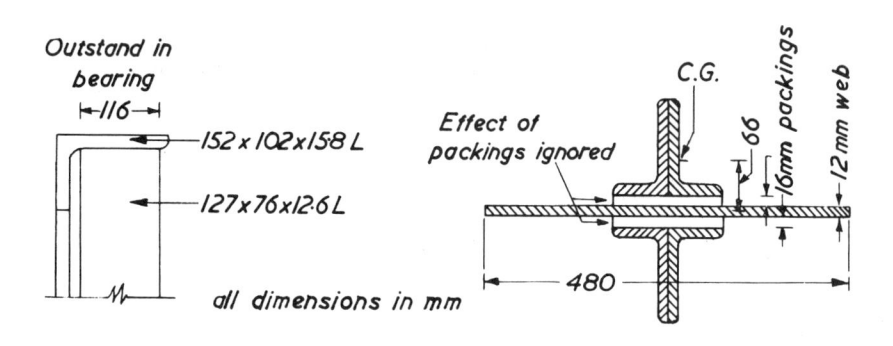

Fig. 21

Fig. 22

Assuming four angles, the minimum thickness

$$= \frac{4\,470}{4 \times 116.3} = 9.61 \text{ mm}$$

Use 127 x 76 x 12.6 mm angles, as shown in Fig. 22. The length of web included in the stiffener section is 2 x 20 x 12 = 480 mm, but the packings must be ignored.

Therefore, I about centre line of web

$$= \frac{480 \times 12^3}{12 \times 10^4} + 4(389 + 2\,409 \times 66^2 \times 10^{-4})$$

$$= 5\,760 \text{ cm}^4$$

$$\text{Area} = 480 \times 12 + 4 \times 2\,409$$

$$= 154 \text{ cm}^2$$

r about centre line of web

$$= \sqrt{(5\,760/154)} = 6.11 \text{ cm}$$

The effective length of the stiffener

$$= 0.7(1\,210 - 2 \times 16)$$

$$= 825 \text{ mm}$$

Designing as a strut,

$$l/r = 82.5/6.12 = 13.5$$

From Table 17.a of B.S. 449, the allowable stress p_c is 149 N/mm^2.
The actual stress $f_c = 850 \times 10^3 / 154 \times 10^2 = 55$ N/mm^2.
Hence, the section selected is satisfactory.

The detail of the stiffeners at the ends of the girder will be dependent upon the support detail, but assuming that 305 × 305 universal columns will be used, the arrangement shown in Fig. 23 will be investigated.

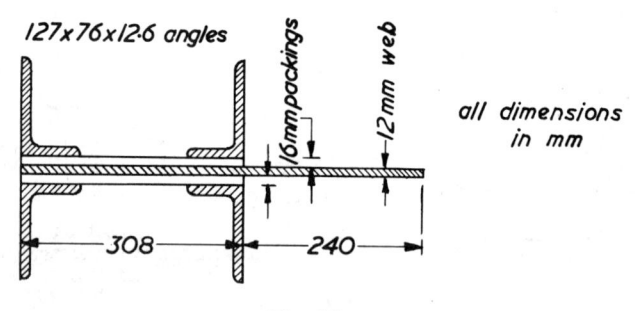

Fig. 23

The maximum reaction, at the left-hand end of the girder, is 990 kN. As the area required for bearing will be 990 000/190 = 52 cm^2, the minimum thickness of angle

$$\frac{5\,200}{4 \times 116.3} = 11.2 \text{ mm}$$

and, from this standpoint, 127 × 76 × 12.6 mm angles will be satisfactory.

Although it is reasonable to suppose that the end reaction is equally shared between both pairs of stiffeners, only the pair at the extreme end of the girder will be investigated. In their case, only 150 mm of web is available for inclusion in the section of the strut.

Thus, I about centre line of web

$$= \frac{150 \times 12^3}{12 \times 10^4} + 2(389 + 2\,409 \times 66^2 \times 10^{-4})$$

$$= 2\,874 \text{ cm}^4$$

$$\text{Area} = 150 \times 12 + 2 \times 24.09$$

$$= 66.18 \text{ cm}^2$$

r about centre line of web

$$= \sqrt{(2\,874/66.18)} = 6.58 \text{ cm}$$

As for the stiffeners under the loads,

$$l = 825 \text{ mm}$$

Hence,

$$l/r = 82.5/6.58 = 12.5$$

Interpolating from Table 17.a in B.S. 449,

the allowable stress $p_c = 150 \text{ N/mm}^2$

The actual stress $f_c = \dfrac{990\,000}{2 \times 66.18} = 74.7 \text{ N/mm}^2$

Therefore, the stiffeners selected are satisfactory.

In Clause 28.a (iii) it is stipulated that load-bearing stiffeners shall be provided with sufficient rivets or welds to transmit to the web the whole of the concentrated load. It is also stated that load-bearing stiffeners shall not be joggled but solidly packed throughout. Furthermore, in Clause 48.d, it is stated that the number of rivets carrying shear through packing shall be increased above the number required by normal calculations by $1\frac{1}{4}$ per cent for each 1 mm total thickness of packing, except that for packings having a thickness of 6 mm or less no increase need be made. For double shear connections packed on both sides the number of additional rivets required shall be determined from the thickness of the thicker packing. The additional rivets may be placed in an extension of the packing.

Considering the stiffeners in Fig. 23, the load on each pair is 990/2 = 495 kN. Assuming that 22 mm diameter rivets are used, the least strength of a rivet is 90.7 kN.

Then the number of rivets required by normal calculations is 495/90.7 = 6.

Both packings are 16 mm thick. Hence, the number of rivets must be increased by

$$\frac{1.25 \times 16}{1} = 20 \text{ per cent}$$

Therefore, the number of rivets required for each pair of stiffeners is 8.

The complete girder is shown in Fig. 24.

Designing upon the Exact Method

The design of the girder in this method is based upon the moment of inertia, and the following are the calculations where they differ from the preceding figures:

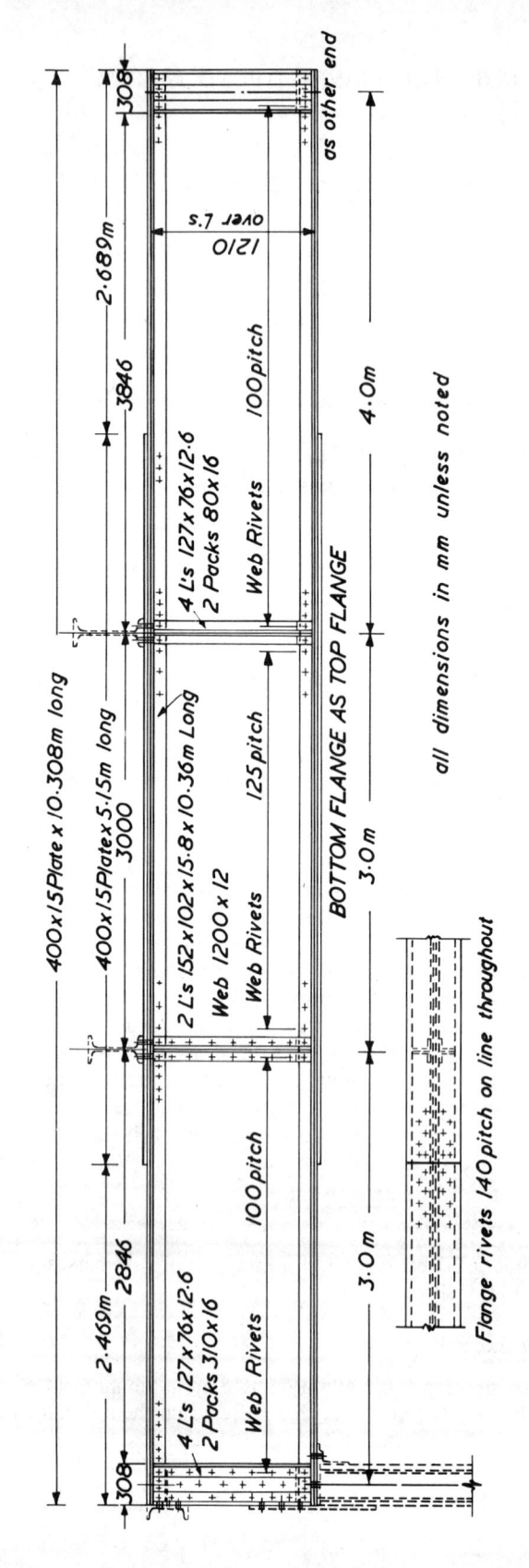

Fig. 24

DESIGN OF A RIVETED PLATE GIRDER

Moment of Inertia about centroid, see Fig. 25, at maximum section.

Web, 1 200 mm x 12 mm	= 172 800	p. 685
Flange angles, four 152 x 102 x 15.8 mm		
Self inertia, 4 x 309	= 1 236	p. 701
Ay^2, where y = 578.9 mm, = 4 x 37.78 x $(57.89)^2$	= 506 441	p. 701 and 693
Flange plates, two 400 x 30 mm		
Self inertia, 2 x 90	= 180	p. 692
Ay^2, where y = 620 mm = 2 x 40 x 3 x $(62.0)^2$	= 922 560	p. 693

$$\text{Gross } I = 1\ 603\ 217 \text{ cm}^4$$

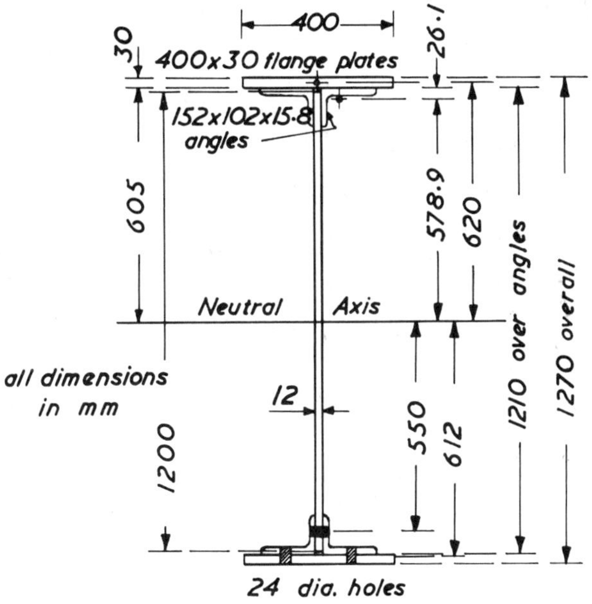

MAXIMUM CROSS SECTION
Fig. 25

Deduct rivet holes

  Ay^2 only, self inertia negligible

    y = 550 mm, 2 x 4.4 x 2.4 x $(55)^2$ = 63 888   p. 693

    y = 612 mm, 4 x 4.6 x 2.4 x $(61.2)^2$ = 165 398   p. 693

$$ 229\ 286$$

$$\text{Net } I = 1\ 373\ 931 \text{ cm}^4$$

In this example allowance has been made for 5 mm projection of flange angles, otherwise the tables in Chapter 27 could have been used throughout.

$$Z = \frac{1\,373\,931}{63.5} = 21\,637 \text{ cm}^3$$

$$Z \text{ required} = 3\,312 \times 10^6 / 155 \times 10^3 = 21\,368 \text{ cm}^3$$

Therefore the section is satisfactory.
    Moment of Inertia with 400 mm × 15 mm flange plates.

Web, as before	= 172 800
Flange angles, as before	= 1 236
	= 506 441

Flange plates, two 400 mm × 15 mm
    Self inertia, 2 × 11.2   = 22
    Ay^2, where $y = 61.25$ cm, = 2 × 40 × 1.6 × $(61.25)^2$ = 480 200

                                        Gross I = 1 160 699 cm⁴

Deduct rivet holes

$y = 550$ mm, as before   = 63 888
$y = 604.5$ mm, 4 × 3.1 × 2.4 × $(60.45)^2$ = 108 749

                                                172 637 cm⁴
                                 Net I = 988 062 cm⁴

$$Z = \frac{988\,062}{62.0} = 15\,936 \text{ cm}^3$$

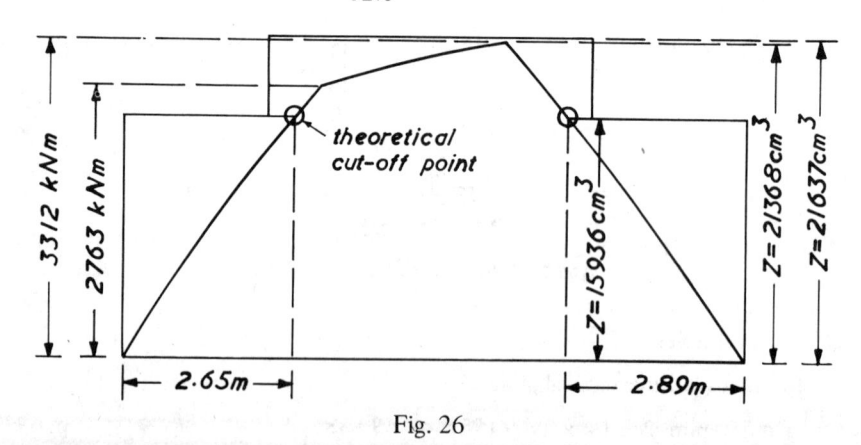

Fig. 26

This is used in the flange curtailment diagram given in Fig. 26. Note the cut-off points differ slightly from those found by the approximate method, the reason being that in the latter the effective depth is assumed to remain constant and equal to the depth over the angles, whereas by calculating on the moment of inertia allowance is made for the variation in effective depth. The difference is not, however, of any significance.

DESIGN OF A RIVETED PLATE GIRDER

Rivet pitches are calculated on the more exact formula, based upon the distribution of horizontal shear. The horizontal shear per lineal mm = $S \cdot a \cdot y/I$, where

S = shear at section,
a = area of section to be connected by rivets,
y = lever arm about NA of area a, and
I = moment of inertia of section about NA

Then, at the left-hand end, for the rivets connecting flange angle to web, we have, as shown in Fig. 27,

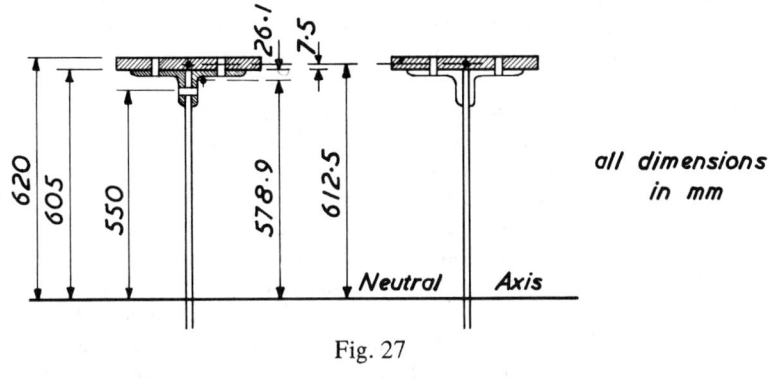

Fig. 27

Flange plate, $ay = 60 \times 61.25$ = 3 675
Flange angles, $ay = 2 \times 37.78 \times 57.89$ = 4 374

8 049

Less rivet holes

Flange, $ay = 2 \times 2.4 \times 3.1 \times 60.46 = 899$
Web, $ay = 2.4 \times 4.4 \times 55$ = 581

= 1 480

6 569

Hence, $$\text{shear/cm} = \frac{990 \times 6\ 569}{988\ 062} = 6.58 \text{ kN}$$

For the rivets between flange angles and flange plates,

$ay = 3\ 675 - (2 \times 2.4 \times 1.5 \times 61.25)$
= 3 234

Hence, $$\text{shear/cm} = \frac{990 \times 3\ 234}{988\ 062}$$
= 3.24 kN

Although both these values are less than those given by the approximate method, it is only possible to increase the rivet pitch between flange angles and web plate, as the other pitches are controlled by Clause 51.c. In the two end panels the

maximum pitch of the flange angle/web rivets could be 90.7/6.58 = 13.8 cm, so that a pitch of 130 mm could be used here, in lieu of 100 mm as required by the approximate method, but the spacing in the centre panel must remain unaltered at 140 mm.

Calculations at the point of cut-off of the outer plate are made in a similar manner.

We will now proceed to show that the flexural stress assumed is in accordance with Clause 20 of B.S. 449. The allowable compressive stress p_{bc} in a plate girder is the lesser of two values given in Table 2 and Table 8, the latter corresponding to the critical stress C_s obtained by following the detailed requirements of Clause 20.

As the flanges of the assumed section have equal moments of inertia about the $y - y$ axis of the girder, Case (I) applies.

Therefore $C_s = A$ N/mm^2.

The value of A depends on the two criteria l/r_y and D/T in Table 7.

Now l = the length between effective lateral restraint. (See Clause 26.b.)

r_y = radius of gyration about the $y - y$ axis of the gross section of the whole girder, at the point of maximum bending moment.

D = overall depth of girder at the point of maximum bending moment.

T = effective thickness of compression flange.

 = K_1 x mean thickness of the horizontal portion of the compression flange at the point of maximum bending moment. The coefficient K_1 makes allowance for reduction in thickness or breadth of flanges between points of effective lateral restraint and depends on N, the ratio of the total area of both flanges at the point of least bending moment to the corresponding area at the point of greatest bending moment between such points of restraint.

For plate girders with flange angles, the horizontal legs of the angles shall be included in the area of the horizontal portion.

The values of K_1 are given in Table 5 of B.S. 449.

The critical panel of the girder is the right-hand panel in which l = 4 000 mm.

r_y is derived from the data given in Chapters 27, 40 and Fig. 28.

Moment of inertia about the $y - y$ axis:

Web, 1 200 mm x 12 mm	= 17
Flange angles, four 152 x 102 x 15.8 mm	
Self inertia, 4 x 871	= 3 484
Ax^2, where x = 5.73 = 151.1 x 32.8	= 4 956
Flange plates, two 400 mm x 30 mm	
2 x 16 000	= 32 000
Gross I_y	= 40 457 cm^4

DESIGN OF A RIVETED PLATE GIRDER

$$\text{Area } A = 144 + 151.1 + 240 = 635.1 \text{ cm}^2$$

$$r_y = \sqrt{(I_y/A)} = \sqrt{(40\ 457/635.1)} = 8 \text{ cm}$$

Now

$$l = 400 \text{ cm}$$

Therefore

$$l/r_y = 400/8 = 50$$

As the flanges in the right-hand panel of the girder are curtailed, it is necessary to calculate the values of N and K_1 in Table 5 of B.S. 449.

The area of both flanges at the point of minimum bending moment

$$= 2(40 \times 1.5 + 30.5 \times 1.6) = 217.6 \text{ cm}^2$$

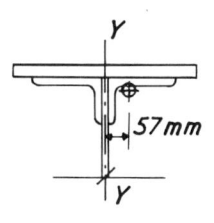

Fig. 28

The area of both flanges at the point of maximum bending moment

$$= 2(40 \times 3 + 30.5 \times 1.6) = 337.6 \text{ cm}^2$$

$$N = 217.6/337.6 = 0.64, \text{ whence } K_1 = 0.84$$

The mean thickness of the horizontal portion of the compression flange at the point of maximum bending moment

$$= \frac{40 \times 3.0 + 30.5 \times 1.6}{40} = 42.2 \text{ mm}$$

Therefore

$$T = 0.84 \times 42.2 = 35.4 \text{ mm}$$

Now

$$D = 1\ 270 \text{ mm}$$

Therefore

$$D/T = 1\ 270/35.4 = 35.9$$

Interpolating in Table 7 it will be found that the value of A exceeds 1 176. Hence, the allowable compression stress p_{bc} in bending given in Table 8 is 162 N/mm², which is more than the maximum permitted value of 155 N/mm² as given in Table 2 for plates not exceeding 40 mm in thickness.

It will be noted that in this case both the approximate and exact solutions give satisfactory designs complying with B.S. 449.

27. MOMENTS OF INERTIA OF PLATE GIRDERS

THE tables on pages 682 to 699 are designed to reduce the amount of work involved in calculating the moments of inertia of plate girders or other plated sections. Consider, for example, the plate girder shown in Fig. 1.

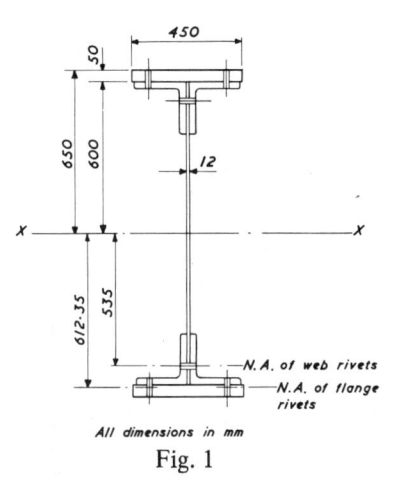

All dimensions in mm

Fig. 1

The web plate is 1.2 m deep and 12 mm thick, the flange plates are 450 mm wide and 50 mm deep, while the angles are 203 mm x 203 mm x 25.3 mm. The moment of inertia of this section, allowing for two rivet holes (22 mm diameter) in the web and two rivet holes in each flange, is calculated as follows:

Inertia about axis xx for Plate Girder in cm^4 units

Component		I_{xx}	A	Y	Y^2	Ay^2
Web Plate	(1)	172 800	–	–	–	–
Flange Plates	(2)	1 758 600	–	–	–	–
Flange Angles	(4)	1 144 556	–	–	–	–
Web Rivets	(2)	–	27.54	53.50	2 862.25	78 838
Flange Rivets	(4)	–	66.26	61.24	3 749.43	248 472
TOTAL		3 075 956				327 310

$$I_{xx} \text{ for girder} = (I_{xx} - Ay^2) = 2\ 748\ 646$$

Note

The values of the Inertia about axis xx of the full section for the web plate, flange plates and angles can be obtained direct from the tables. In the case of deductions for rivet holes, the self-inertia is so small this can be ignored and the deduction is therefore taken as equal to Ay^2.

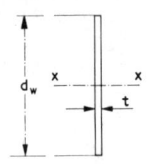

MOMENT OF INERTIA OF RECTANGULAR PLATES
about axis x–x

Depth d_w mm	\multicolumn{6}{c}{THICKNESS t MILLIMETRES}
	3	4	5	6	8	10
25	.391	.521	.651	.781	1.04	1.30
50	3.13	4.17	5.21	6.25	8.33	10.4
75	10.5	14.1	17.6	21.1	28.1	35.2
100	25.0	33.3	41.7	50.0	66.7	83.3
125	48.8	65.1	81.4	97.7	130	163
150	84.4	113	141	169	225	281
175	134	179	223	268	357	447
200	200	267	333	400	533	667
225	285	380	475	570	759	949
250	391	521	651	781	1042	1302
275	520	693	867	1040	1386	1733
300	675	900	1125	1350	1800	2250
325	858	1144	1430	1716	2289	2861
350	1072	1429	1786	2144	2858	3573
375	1318	1758	2197	2637	3516	4395
400	1600	2133	2667	3200	4267	5333
425	1919	2559	3199	3838	5118	6397
450	2278	3038	3797	4556	6075	7594
475	2679	3572	4465	5359	7145	8931
500	3125	4167	5208	6250	8333	10417
525	3618	4823	6029	7235	9647	12059
550	4159	5546	6932	8319	11092	13865
575	4753	6337	7921	9505	12674	15842
600	5400	7200	9000	10800	14400	18000
625	6104	8138	10173	12207	16276	20345
650	6866	9154	11443	13731	18308	22885
675	7689	10252	12814	15377	20503	25629
700	8575	11433	14292	17150	22867	28583
725	9527	12703	15878	19054	25405	31757
750	10547	14063	17578	21094	28125	35156
775	11637	15516	19395	23274	31032	38790
800	12800	17067	21333	25600	34133	42667
825	14038	18717	23396	28076	37434	46793
850	15353	20471	25589	30706	40942	51177
875	16748	22331	27913	33496	44661	55827
900	18225	24300	30375	36450	48600	60750

Moments of inertia are tabulated in cm^4.

INERTIA OF WEB PLATES

MOMENT OF INERTIA OF RECTANGULAR PLATES
about axis x—x

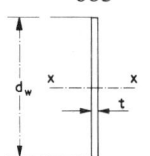

| THICKNESS t MILLIMETRES ||||||| Depth |
12	15	18	20	22	25	d_w mm
1.56	1.95	2.34	2.60	2.86	3.26	25
12.5	15.6	18.8	20.8	22.9	26.0	50
42.2	52.7	63.3	70.3	77.3	87.9	75
100	125	150	167	183	208	100
195	244	293	326	358	407	125
338	422	506	563	619	703	150
536	670	804	893	983	1117	175
800	1000	1200	1333	1467	1667	200
1139	1424	1709	1898	2088	2373	225
1563	1953	2344	2604	2865	3255	250
2080	2600	3120	3466	3813	4333	275
2700	3375	4050	4500	4950	5625	300
3433	4291	5149	5721	6293	7152	325
4288	5359	6431	7146	7860	8932	350
5273	6592	7910	8789	9668	10986	375
6400	8000	9600	10667	11733	13333	400
7677	9596	11515	12794	14074	15993	425
9113	11391	13669	15188	16706	18984	450
10717	13396	16076	17862	19648	22327	475
12500	15625	18750	20833	22917	26042	500
14470	18088	21705	24117	26529	30146	525
16638	20797	24956	27729	30502	34661	550
19011	23764	28516	31685	34853	39606	575
21600	27000	32400	36000	39600	45000	600
24414	30518	36621	40690	44759	50863	625
27463	34328	41194	45771	50348	57214	650
30755	38443	46132	51258	56384	64072	675
34300	42875	51450	57167	62883	71458	700
38108	47635	57162	63513	69864	79391	725
42188	52734	63281	70313	77344	87891	750
46548	58186	69823	77581	85339	96976	775
51200	64000	76800	85333	93867	106667	800
56152	70189	84227	93586	102945	116982	825
61413	76766	92119	102354	112590	127943	850
66992	83740	100488	111654	122819	139567	875
72900	91125	109350	121500	133650	151875	900

Moments of inertia are tabulated in cm^4.

MOMENTS OF INERTIA OF PLATE GIRDERS
MOMENT OF INERTIA
OF RECTANGULAR PLATES
about axis x—x

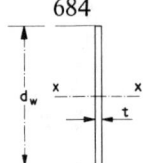

Depth d_w mm	THICKNESS t MILLIMETRES					
	3	4	5	6	8	10
1000	25000	33333	41667	50000	66667	83333
1100	33275	44367	55458	66550	88733	110917
1200	43200	57600	72000	86400	115200	144000
1300	54925	73233	91542	109850	146467	183083
1400	68600	91467	114333	137200	182933	228667
1500	84375	112500	140625	168750	225000	281250
1600	102400	136533	170667	204800	273067	341333
1700	122825	163767	204708	245650	327533	409417
1800	145800	194400	243000	291600	388800	486000
1900	171475	228633	285792	342950	457267	571583
2000	200000	266667	333333	400000	533333	666667
2100	231525	308700	385875	463050	617400	771750
2200	266200	354933	443667	532400	709867	887333
2300	304175	405567	506958	608350	811133	1013917
2400	345600	460800	576000	691200	921600	1152000
2500	390625	520833	651042	781250	1041667	1302083
2600	439400	585867	732333	878800	1171733	1464667
2700	492075	656100	820125	984150	1312200	1640250
2800	548800	731733	914667	1097600	1463467	1829333
2900	609725	812967	1016208	1219450	1625933	2032417
3000	675000	900000	1125000	1350000	1800000	2250000
3100	744775	993033	1241292	1489550	1986067	2482583
3200	819200	1092267	1365333	1638400	2184533	2730667
3300	898425	1197900	1497375	1796850	2395800	2994750
3400	982600	1310133	1637667	1965200	2620267	3275333
3500	1071875	1429167	1786458	2143750	2858333	3572917
3600	1166400	1555200	1944000	2332800	3110400	3888000
3700	1266325	1688433	2110542	2532650	3376867	4221083
3800	1371800	1829067	2286333	2743600	3658133	4572667
3900	1482975	1977300	2471625	2965950	3954600	4943250
4000	1600000	2133333	2666667	3200000	4266667	5333333
4100	1723025	2297367	2871708	3446050	4594733	5743417
4200	1852200	2469600	3087000	3704400	4939200	6174000
4300	1987675	2650233	3312792	3975350	5300467	6625583
4400	2129600	2839467	3549333	4259200	5678933	7098667
4500	2278125	3037500	3796875	4556250	6075000	7593750
4600	2433400	3244533	4055667	4866800	6489067	8111333
4700	2595575	3460767	4325958	5191150	6921533	8651917
4800	2764800	3686400	4608000	5529600	7372800	9216000
4900	2941225	3921633	4902042	5882450	7843267	9804083
5000	3125000	4166667	5208333	6250000	8333333	10416667

Moments of inertia are tabulated in cm^4.

INERTIA OF WEB PLATES

MOMENT OF INERTIA
OF RECTANGULAR PLATES
about axis x—x

\multicolumn{6}{c	}{THICKNESS t MILLIMETRES}	Depth
12	15	18	20	22	25	d_w mm
100000	125000	150000	166667	183333	208333	1000
133100	166375	199650	221833	244017	277292	1100
172800	216000	259200	288000	316800	360000	1200
219700	274625	329550	366167	402783	457708	1300
274400	343000	411600	457333	503067	571667	1400
337500	421875	506250	562500	618750	703125	1500
409600	512000	614400	682667	750933	853333	1600
491300	614125	736950	818833	900717	1023542	1700
583200	729000	874800	972000	1069200	1215000	1800
685900	857375	1028850	1143167	1257483	1428958	1900
800000	1000000	1200000	1333333	1466667	1666667	2000
926100	1157625	1389150	1543500	1697850	1929375	2100
1064800	1331000	1597200	1774667	1952133	2218333	2200
1216700	1520875	1825050	2027833	2230617	2534792	2300
1382400	1728000	2073600	2304000	2534400	2880000	2400
1562500	1953125	2343750	2604167	2864583	3255208	2500
1757600	2197000	2636400	2929333	3222267	3661667	2600
1968300	2460375	2952450	3280500	3608550	4100625	2700
2195200	2744000	3292800	3658667	4024533	4573333	2800
2438900	3048625	3658350	4064833	4471317	5081042	2900
2700000	3375000	4050000	4500000	4950000	5625000	3000
2979100	3723875	4468650	4965167	5461683	6206458	3100
3276800	4096000	4915200	5461333	6007467	6826667	3200
3593700	4492125	5390550	5989500	6588450	7486875	3300
3930400	4913000	5895600	6550667	7205733	8188333	3400
4287500	5359375	6431250	7145833	7860417	8932292	3500
4665600	5832000	6998400	7776000	8553600	9720000	3600
5065300	6331625	7597500	8442167	9286383	10552708	3700
5487200	6859000	8230800	9145333	10059867	11431667	3800
5931900	7414875	8897850	9886500	10875150	12358125	3900
6400000	8000000	9600000	10666667	11733333	13333333	4000
6892100	8615125	10338150	11486833	12635517	14358542	4100
7408800	9261000	11113200	12348000	13582800	15435000	4200
7950700	9938375	11926050	13251167	14576283	16563958	4300
8518400	10648000	12777600	14197333	15617067	17746667	4400
9112500	11390625	13668750	15187500	16706250	18984375	4500
9733600	12167000	14600400	16222667	17844933	20278333	4600
10382300	12977875	15573350	17303833	19034217	21629792	4700
11059200	13824000	16588800	18432000	20275200	23040000	4800
11764900	14706125	17647350	19608167	21568983	24510208	4900
12500000	15625000	18750000	20833333	22916667	26041667	5000

Moments of inertia are tabulated in cm^4.

MOMENT OF INERTIA OF TWO FLANGES
per millimetre of width

Distance d_w mm	\multicolumn{10}{c}{THICKNESS OF EACH FLANGE IN MILLIMETRES}
	10	12	15	18	20	22	25	28	30	32
1000	510.1	614.5	772.7	932.8	1041	1149	1314	1480	1592	1705
1100	616.1	742.0	932.5	1125	1255	1385	1582	1782	1916	2051
1200	732.1	881.4	1107	1335	1489	1643	1876	2112	2270	2429
1300	858.1	1033	1297	1564	1743	1923	2195	2469	2654	2839
1400	994.1	1196	1502	1810	2017	2224	2539	2855	3068	3282
1500	1140	1372	1721	2074	2311	2548	2907	3269	3512	3756
1600	1296	1559	1956	2356	2625	2894	3301	3711	3986	4262
1700	1462	1759	2206	2656	2959	3262	3720	4181	4490	4800
1800	1638	1970	2471	2975	3313	3652	4164	4679	5024	5371
1900	1824	2193	2750	3311	3687	4064	4632	5204	5588	5973
2000	2020	2429	3045	3665	4081	4498	5126	5758	6182	6607
2100	2226	2676	3355	4037	4495	4953	5645	6340	6806	7273
2200	2442	2936	3680	4428	4929	5431	6189	6950	7460	7971
2300	2668	3207	4019	4836	5383	5931	6757	7588	8144	8702
2400	2904	3491	4374	5262	5857	6453	7351	8254	8858	9464
2500	3150	3786	4744	5706	6351	6997	7970	8947	9602	10258
2600	3406	4094	5129	6169	6865	7563	8614	9669	10376	11084
2700	3672	4413	5528	6649	7399	8150	9282	10419	11180	11943
2800	3948	4744	5943	7147	7953	8760	9976	11197	12014	12833
2900	4234	5088	6373	7663	8527	9392	10695	12003	12878	13755
3000	4530	5443	6818	8198	9121	10046	11439	12837	13772	14709
3100	4836	5811	7277	8750	9735	10722	12207	13699	14696	15696
3200	5152	6190	7752	9320	10369	11420	13001	14588	15650	16714
3300	5478	6582	8242	9908	11023	12139	13820	15506	16634	17764
3400	5814	6985	8747	10515	11697	12881	14664	16452	17648	18846
3500	6160	7401	9266	11139	12391	13645	15532	17426	18692	19961
3600	6516	7828	9801	11781	13105	14431	16426	18428	19766	21107
3700	6882	8267	10351	12441	13839	15239	17345	19458	20870	22285
3800	7258	8719	10916	13120	14593	16069	18289	20515	22004	23495
3900	7644	9182	11495	13816	15367	16920	19257	21601	23168	24738
4000	8040	9658	12090	14530	16161	17794	20251	22715	24362	26012
4100	8446	10145	12700	15262	16975	18690	21270	23857	25586	27318
4200	8862	10645	13325	16012	17809	19608	22314	25027	26840	28656
4300	9288	11156	13964	16781	18663	20548	23382	26225	28124	30027
4400	9724	11679	14619	17567	19537	21510	24476	27450	29438	31429
4500	10170	12215	15289	18371	20431	22494	25595	28704	30782	32863
4600	10626	12762	15974	19193	21345	23499	26739	29986	32156	34329
4700	11092	13322	16673	20034	22279	24527	27907	31296	33560	35827
4800	11568	13893	17388	20892	23233	25577	29101	32634	34994	37358
4900	12054	14477	18118	21768	24207	26649	30320	34000	36458	38920
5000	12550	15072	18863	22662	25201	27743	31564	35393	37952	40514

Moments of inertia are tabulated in cm^4.

INERTIA OF FLANGE PLATES

MOMENT OF INERTIA OF TWO FLANGES
per millimetre of width

\multicolumn{9}{c	}{THICKNESS OF EACH FLANGE IN MILLIMETRES}	Distance
35	38	40	45	50	55	60	65	70	75	d_w mm
1875	2048	2164	2459	2758	3064	3374	3691	4013	4341	1000
2255	2461	2600	2951	3308	3671	4040	4416	4797	5184	1100
2670	2913	3076	3489	3908	4334	4766	5205	5651	6103	1200
3120	3402	3592	4072	4558	5052	5552	6060	6575	7097	1300
3604	3930	4148	4700	5258	5825	6398	6980	7569	8166	1400
4124	4495	4744	5372	6008	6652	7304	7965	8633	9309	1500
4679	5099	5380	6090	6808	7535	8270	9014	9767	10528	1600
5269	5740	6056	6853	7658	8473	9296	10129	10971	11822	1700
5893	6420	6772	7661	8558	9466	10382	11309	12245	13191	1800
6553	7137	7528	8513	9508	10513	11528	12554	13589	14634	1900
7248	7892	8324	9411	10508	11616	12734	13863	15003	16153	2000
7978	8686	9160	10354	11558	12774	14000	15238	16487	17747	2100
8742	9517	10036	11342	12658	13987	15326	16678	18041	19416	2200
9542	10387	10952	12374	13808	15254	16712	18183	19665	21159	2300
10377	11294	11908	13452	15008	16577	18158	19752	21359	22978	2400
11247	12240	12904	14575	16258	17955	19664	21387	23123	24872	2500
12151	13223	13940	15743	17558	19388	21230	23087	24957	26841	2600
13091	14245	15016	16955	18908	20875	22856	24852	26861	28884	2700
14066	15304	16132	18213	20308	22418	24542	26681	28835	31003	2800
15076	16401	17288	19516	21758	24016	26288	28576	30879	33197	2900
16120	17537	18484	20864	23258	25669	28094	30536	32993	35466	3000
17200	18710	19720	22256	24808	27376	29960	32561	35177	37809	3100
18315	19922	20996	23694	26408	29139	31886	34650	37431	40228	3200
19465	21171	22312	25177	28058	30957	33872	36805	39755	42722	3300
20649	22459	23668	26705	29758	32830	35918	39025	42149	45291	3400
21869	23784	25064	28277	31508	34757	38024	41310	44613	47934	3500
23124	25147	26500	29895	33308	36740	40190	43659	47147	50653	3600
24414	26549	27976	31558	35158	38778	42416	46074	49751	53447	3700
25738	27988	29492	33266	37058	40871	44702	48554	52425	56316	3800
27098	29466	31048	35018	39008	43018	47048	51099	55169	59259	3900
28493	30981	32644	36816	41008	45221	49454	53708	57983	62278	4000
29923	32535	34280	38659	43058	47479	51920	56383	60867	65372	4100
31387	34126	35956	40547	45158	49792	54446	59123	63821	68541	4200
32887	35756	37672	42479	47308	52159	57032	61928	66845	71784	4300
34422	37423	39428	44457	49508	54582	59678	64797	69939	75103	4400
35992	39128	41224	46480	51758	57060	62384	67732	73103	78497	4500
37596	40872	43060	48548	54058	59593	65150	70732	76337	81966	4600
39236	42653	44936	50660	56408	62180	67976	73797	79641	85509	4700
40911	44473	46852	52818	58808	64823	70862	76926	83015	89128	4800
42621	46330	48808	55021	61258	67521	73808	80121	86459	92822	4900
44365	48226	50804	57269	63758	70274	76814	83381	89973	96591	5000

Moments of inertia are tabulated in cm^4.

MOMENTS OF INERTIA OF PLATE GIRDERS

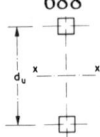

SECOND MOMENT OF A PAIR OF UNIT AREAS

about axis x—x

Distance d_u mm	0	5	10	15	20	25	30	35	40	45
500	1250	1275	1301	1326	1352	1378	1405	1431	1458	1485
550	1513	1540	1568	1596	1625	1653	1682	1711	1741	1770
600	1800	1830	1861	1891	1922	1953	1985	2016	2048	2080
650	2113	2145	2178	2211	2245	2278	2312	2346	2381	2415
700	2450	2485	2521	2556	2592	2628	2665	2701	2738	2775
750	2813	2850	2888	2926	2965	3003	3042	3081	3121	3160
800	3200	3240	3281	3321	3362	3403	3445	3486	3528	3570
850	3613	3655	3698	3741	3785	3828	3872	3916	3961	4005
900	4050	4095	4141	4186	4232	4278	4325	4371	4418	4465
950	4513	4560	4608	4656	4705	4753	4802	4851	4901	4950
1000	5000	5050	5101	5151	5202	5253	5305	5356	5408	5460
1050	5513	5565	5618	5671	5725	5778	5832	5886	5941	5995
1100	6050	6105	6161	6216	6272	6328	6385	6441	6498	6555
1150	6613	6670	6728	6786	6845	6903	6962	7021	7081	7140
1200	7200	7260	7321	7381	7442	7503	7565	7626	7688	7750
1250	7813	7875	7938	8001	8065	8128	8192	8256	8321	8385
1300	8450	8515	8581	8646	8712	8778	8845	8911	8978	9045
1350	9113	9180	9248	9316	9385	9453	9522	9591	9661	9730
1400	9800	9870	9941	10011	10082	10153	10225	10296	10368	10440
1450	10513	10585	10658	10731	10805	10878	10952	11026	11101	11175
1500	11250	11325	11401	11476	11552	11628	11705	11781	11858	11935
1550	12013	12090	12168	12246	12325	12403	12482	12561	12641	12720
1600	12800	12880	12961	13041	13122	13203	13285	13366	13448	13530
1650	13613	13695	13778	13861	13945	14028	14112	14196	14281	14365
1700	14450	14535	14621	14706	14792	14878	14965	15051	15138	15225
1750	15313	15400	15488	15576	15665	15753	15842	15931	16021	16110
1800	16200	16290	16381	16471	16562	16653	16745	16836	16928	17020
1850	17113	17205	17298	17391	17485	17578	17672	17766	17861	17955
1900	18050	18145	18241	18336	18432	18528	18625	18721	18818	18915
1950	19013	19110	19208	19306	19405	19503	19602	19701	19801	19900
2000	20000	20100	20201	20301	20402	20503	20605	20706	20808	20910
2050	21013	21115	21218	21321	21425	21528	21632	21736	21841	21945
2100	22050	22155	22261	22366	22472	22578	22685	22791	22898	23005
2150	23113	23220	23328	23436	23545	23653	23762	23871	23981	24090
2200	24200	24310	24421	24531	24642	24753	24865	24976	25088	25200
2250	25313	25425	25538	25651	25765	25878	25992	26106	26221	26335
2300	26450	26565	26681	26796	26912	27028	27145	27261	27378	27495
2350	27613	27730	27848	27966	28085	28203	28322	28441	28561	28680
2400	28800	28920	29041	29161	29282	29403	29525	29646	29768	29890
2450	30013	30135	30258	30381	30505	30628	30752	30876	31001	31125
2500	31250	31375	31501	31626	31752	31878	32005	32131	32258	32385
2550	32513	32640	32768	32896	33025	33153	33282	33411	33541	33670
2600	33800	33930	34061	34191	34322	34453	34585	34716	34848	34980
2650	35113	35245	35378	35511	35645	35778	35912	36046	36181	36315
2700	36450	36585	36721	36856	36992	37128	37265	37401	37538	37675

Second moments are tabulated in cm^4 and are for unit areas of 1 cm^2 each.

INERTIA OF UNIT AREAS

SECOND MOMENT OF A PAIR OF UNIT AREAS
about axis x—x

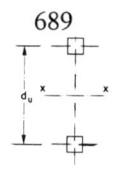

Distance d_u mm	0	5	10	15	20	25	30	35	40	45
2750	37813	37950	38088	38226	38365	38503	38642	38781	38921	39060
2800	39200	39340	39481	39621	39762	39903	40045	40186	40328	40470
2850	40613	40755	40898	41041	41185	41328	41472	41616	41761	41905
2900	42050	42195	42341	42486	42632	42778	42925	43071	43218	43365
2950	43513	43660	43808	43956	44105	44253	44402	44551	44701	44850
3000	45000	45150	45301	45451	45602	45753	45905	46056	46208	46360
3050	46513	46665	46818	46971	47125	47278	47432	47586	47741	47895
3100	48050	48205	48361	48516	48672	48828	48985	49141	49298	49455
3150	49613	49770	49928	50086	50245	50403	50562	50721	50881	51040
3200	51200	51360	51521	51681	51842	52003	52165	52326	52488	52650
3250	52813	52975	53138	53301	53465	53628	53792	53956	54121	54285
3300	54450	54615	54781	54946	55112	55278	55445	55611	55778	55945
3350	56113	56280	56448	56616	56785	56953	57122	57291	57461	57630
3400	57800	57970	58141	58311	58482	58653	58825	58996	59168	59340
3450	59513	59685	59858	60031	60205	60378	60552	60726	60901	61075
3500	61250	61425	61601	61776	61952	62128	62305	62481	62658	62835
3550	63013	63190	63368	63546	63725	63903	64082	64261	64441	64620
3600	64800	64980	65161	65341	65522	65703	65885	66066	66248	66430
3650	66613	66795	66978	67161	67345	67528	67712	67896	68081	68265
3700	68450	68635	68821	69006	69192	69378	69565	69751	69938	70125
3750	70313	70500	70688	70876	71065	71253	71442	71631	71821	72010
3800	72200	72390	72581	72771	72962	73153	73345	73536	73728	73920
3850	74113	74305	74498	74691	74885	75078	75272	75466	75661	75855
3900	76050	76245	76441	76636	76832	77028	77225	77421	77618	77815
3950	78013	78210	78408	78606	78805	79003	79202	79401	79601	79800
4000	80000	80200	80401	80601	80802	81003	81205	81406	81608	81810
4050	82013	82215	82418	82621	82825	83028	83232	83436	83641	83845
4100	84050	84255	84461	84666	84872	85078	85285	85491	85698	85905
4150	86113	86320	86528	86736	86945	87153	87362	87571	87781	87990
4200	88200	88410	88621	88831	89042	89253	89465	89676	89888	90100
4250	90313	90525	90738	90951	91165	91378	91592	91806	92021	92235
4300	92450	92665	92881	93096	93312	93528	93745	93961	94178	94395
4350	94613	94830	95048	95266	95485	95703	95922	96141	96361	96580
4400	96800	97020	97241	97461	97682	97903	98125	98346	98568	98790
4450	99013	99235	99458	99681	99905	100128	100352	100576	100801	101025
4500	101250	101475	101701	101926	102152	102378	102605	102831	103058	103285
4550	103513	103740	103968	104196	104425	104653	104882	105111	105341	105570
4600	105800	106030	106261	106491	106722	106953	107185	107416	107648	107880
4650	108113	108345	108578	108811	109045	109278	109512	109746	109981	110215
4700	110450	110685	110921	111156	111392	111628	111865	112101	112338	112575
4750	112813	113050	113288	113526	113765	114003	114242	114481	114721	114960
4800	115200	115440	115681	115921	116162	116403	116645	116886	117128	117370
4850	117613	117855	118098	118341	118585	118828	119072	119316	119561	119805
4900	120050	120295	120541	120786	121032	121278	121525	121771	122018	122265
4950	122513	122760	123008	123256	123505	123753	124002	124251	124501	124750

Second moments are tabulated in cm^4 and are for unit areas of 1 cm^2 each.

MOMENT OF INERTIA
OF FOUR EQUAL ANGLES
about axis x—x

Depth	ANGLES							
d_w		203 × 203				152 × 152			
mm	25.3	22.1	18.9	15.8	19.0	15.8	12.6	9.4
1000	764940	678609	589739	498289	455884	386738	313447	238826
1100	945066	837933	727782	614569	560468	475207	384935	293128
1200	1144556	1014341	880588	743252	675942	572865	463831	353042
1300	1363409	1207831	1048158	884338	802308	679714	550133	418566
1400	1601624	1418405	1230491	1037827	939564	795752	643842	489702
1500	1859202	1646062	1427588	1203720	1087710	920981	744958	566450
1600	2136142	1890803	1639448	1382016	1246748	1055399	853480	648808
1700	2432446	2152626	1866072	1572715	1416676	1199008	969410	736778
1800	2748112	2431533	2107459	1775817	1597675	1351806	1092747	830359
1900	3083140	2727523	2363609	1991323	1789205	1513794	1223490	929551
2000	3437532	3040596	2634523	2219231	1991805	1684973	1361640	1034355
2100	3811286	3370753	2920200	2459543	2205296	1865341	1507198	1144769
2200	4204403	3717992	3220640	2712259	2429678	2054899	1660162	1260795
2300	4616883	4082315	3535844	2977377	2664951	2253647	1820533	1382433
2400	5048726	4463722	3865812	3254899	2911114	2461585	1988310	1509681
2500	5499931	4862211	4210542	3544824	3168168	2678713	2163495	1642541
2600	5970499	5277523	4570036	3847152	3436113	2905031	2346087	1781012
2700	6460430	5710439	4944294	4161884	3714949	3140539	2536085	1925094
2800	6969723	6160178	5333315	4489018	4004675	3385237	2733491	2074787
2900	7498380	6627000	5737099	4828556	4305292	3639125	2938303	2230092
3000	8046399	7110906	6155647	5180497	4616800	3902203	3150522	2391008
3100	8613780	7611895	6588958	5544842	4939199	4174471	3370148	2557535
3200	9200525	8129966	7037032	5921590	5272488	4455929	3597181	2729673
3300	9806632	8665121	7499870	6310740	5616668	4746576	3831620	2907423
3400	10432102	9217360	7977472	6712295	5971739	5046414	4073467	3090784
3500	11076935	9786681	8469836	7126252	6337700	5355442	4322721	3279756
3600	11741130	10373086	8976964	7552613	6714553	5673659	4579381	3474339
3700	12424688	10976574	9498856	7991377	7102296	6001067	4843448	3674534
3800	13127609	11597145	10035511	8442544	7500929	6337664	5114922	3880339
3900	13849893	12234800	10586929	8906114	7910454	6683452	5393803	4091757
4000	14591539	12889537	11153111	9382088	8330869	7038429	5680091	4308785
4100	15352548	13561358	11734056	9870464	8762175	7402597	5973786	4531424
4200	16132920	14250262	12329764	10371244	9204372	7775954	6274888	4759675
4300	16932655	14956250	12940236	10884428	9657459	8158502	6583396	4993537
4400	17751752	15679320	13565472	11410014	10121438	8550239	6899312	5233010
4500	18590212	16419474	14205470	11948004	10596307	8951166	7222634	5478095
4600	19448035	17176711	14860232	12498397	11082066	9361283	7553363	5728791
4700	20325221	17951031	15529758	13061193	11578717	9780591	7891499	5985098
4800	21221769	18742434	16214047	13636393	12086258	10209088	8237042	6247016
4900	22137680	19550921	16913099	14223996	12604690	10646775	8589992	6514545
5000	23072954	20376491	17626915	14824002	13134012	11093652	8950349	6787686
Space 'S' in mm	15	15	15	15	12	12	12	12
I_y	32316	28143	24007	19904	10288	8529	6749	5015

Moments of Inertia I_x and I_y are tabulated in cm⁴.
Dimensions of angles are in mm.

INERTIA OF EQUAL ANGLES

MOMENT OF INERTIA
OF FOUR EQUAL ANGLES
about axis x–x

ANGLES									Depth
127 × 127			102 × 102			89 × 89		76 × 76		d_n
15.8	12.6	9.5	15.8	12.6	9.4	12.6	9.4	12.6	9.4	mm
325725	264888	203182	263075	213993	164182	187619	143892	160598	123230	1000
399420	324649	248888	321912	261720	200697	229215	175706	195987	150312	1100
480672	390521	299257	386704	314266	240889	274985	210704	234904	180086	1200
569479	462505	354289	457452	371631	284760	324927	248886	277348	212554	1300
665843	540600	413983	534155	433814	332309	379043	290251	323321	247715	1400
769763	624808	478340	616813	500817	383535	437332	334801	372822	285569	1500
881238	715128	547360	705426	572639	438439	499795	382535	425851	326117	1600
1000270	811560	621042	799995	649279	497021	566430	433453	482408	369358	1700
1126857	914103	699388	900519	730739	559281	637239	487555	542492	415293	1800
1261000	1022759	782396	1006999	817018	625219	712221	544841	606105	463920	1900
1402700	1137526	870067	1119433	908115	694834	791376	605311	673246	515242	2000
1551955	1258406	962400	1237823	1004032	768128	874705	668965	743915	569256	2100
1708766	1385397	1059396	1362168	1104767	845099	962206	735803	818112	625964	2200
1873133	1518500	1161055	1492469	1210321	925748	1053881	805826	895836	685365	2300
2045057	1657715	1267377	1628724	1320695	1010075	1149730	879032	977089	747459	2400
2224536	1803042	1378362	1770935	1435887	1098080	1249751	955422	1061870	812247	2500
2411571	1954481	1494009	1919102	1555898	1189762	1353946	1034997	1150179	879728	2600
2606162	2112032	1614319	2073223	1680729	1285123	1462314	1117755	1242016	949902	2700
2808309	2275695	1739292	2233300	1810378	1384161	1574855	1203697	1337380	1022769	2800
3018012	2445470	1868927	2399332	1944846	1486877	1691569	1292824	1436273	1098330	2900
3235271	2621356	2003225	2571320	2084133	1593271	1812457	1385134	1538694	1176585	3000
3460086	2803355	2142186	2749262	2228239	1703343	1937518	1480629	1644643	1257532	3100
3692457	2991465	2285810	2933160	2377164	1817092	2066752	1579307	1754120	1341173	3200
3932383	3185688	2434096	3123014	2530908	1934520	2200159	1681170	1867124	1427507	3300
4179866	3386022	2587045	3318822	2689471	2055625	2337740	1786216	1983657	1516535	3400
4434905	3592468	2744657	3520586	2852853	2180408	2479494	1894447	2103718	1608256	3500
4697500	3805026	2906932	3728305	3021054	2308869	2625421	2005862	2227307	1702670	3600
4967650	4023696	3073869	3941979	3194074	2441008	2775521	2120461	2354424	1799777	3700
5245357	4248478	3245470	4161609	3371913	2576825	2929795	2238243	2485068	1899578	3800
5530619	4479372	3421732	4387194	3554570	2716319	3088242	2359210	2619241	2002072	3900
5823438	4716378	3602658	4618734	3742047	2859492	3250862	2483361	2756942	2107260	4000
6123813	4959496	3788246	4856230	3934343	3006342	3417655	2610696	2898171	2215141	4100
6431743	5208726	3978497	5099681	4131458	3156870	3588622	2741215	3042928	2325715	4200
6747229	5464067	4173411	5349087	4333391	3311076	3763761	2874918	3191213	2438982	4300
7070272	5725521	4372988	5604448	4540144	3468960	3943074	3011805	3343025	2554943	4400
7400870	5993086	4577227	5865765	4751715	3630521	4126561	3151876	3498366	2673597	4500
7739024	6266763	4786129	6133036	4968106	3795760	4314220	3295131	3657235	2794944	4600
8084735	6546553	4999694	6406264	5189315	3964678	4506053	3441570	3819632	2918985	4700
8438001	6832454	5217921	6685446	5415344	4137273	4702059	3591194	3985557	3045719	4800
8798823	7124467	5440812	6970584	5646191	4313546	4902238	3744001	4155009	3175146	4900
9167201	7422592	5668365	7261677	5881857	4493496	5106591	3899992	4327990	3307267	5000
12	12	12	10	10	10	10	10	8	8	Space 'S' in mm
5077	4016	2995	2643	2078	1541	1430	1055	894	656	I_y

Moments of Inertia **Ix** and **Iy** are tabulated in cm⁴.
Dimensions of angles are in mm.

MOMENT OF INERTIA
OF FOUR UNEQUAL ANGLES
LONG LEGS BACK TO BACK
about axis x—x

Depth	ANGLES							
d_w		229 × 102			203 × 152			203 × 102	
mm	18.9	15.8	12.6	18.9	15.8	12.6	15.8	12.6
1000	418246	355254	287953	496224	419790	340293	342383	277742
1100	522140	443196	358970	614228	519305	420698	425399	344850
1200	637868	541125	438028	745052	629600	509790	517605	419365
1300	765429	649040	525125	888694	750678	607570	619000	501286
1400	904823	766943	620262	1045156	882537	714036	729586	590615
1500	1056051	894832	723438	1214437	1025179	829190	849362	687350
1600	1219111	1032708	834654	1396538	1178601	953030	978327	791493
1700	1394005	1180572	953909	1591457	1342806	1085558	1116483	903042
1800	1580732	1338422	1081204	1799196	1517792	1226772	1263829	1021998
1900	1779291	1506259	1216539	2019755	1703560	1376674	1420364	1148361
2000	1989684	1684082	1359913	2253132	1900110	1535263	1586090	1282131
2100	2211910	1871893	1511327	2499329	2107441	1702539	1761005	1423308
2200	2445970	2069691	1670780	2758346	2325554	1878501	1945110	1571892
2300	2691862	2277475	1838273	3030181	2554449	2063151	2138406	1727882
2400	2949588	2495246	2013806	3314836	2794126	2256488	2340891	1891279
2500	3219146	2723004	2197378	3612310	3044584	2458512	2552566	2062084
2600	3500538	2960750	2388990	3922604	3305824	2669223	2773432	2240295
2700	3793763	3208481	2588641	4245717	3577846	2888621	3003487	2425913
2800	4098821	3466200	2796332	4581649	3860649	3116706	3242732	2618938
2900	4415713	3733906	3012063	4930400	4154234	3353478	3491167	2819370
3000	4744437	4011599	3235833	5291971	4458601	3598937	3748792	3027208
3100	5084995	4299278	3467643	5666361	4773750	3853083	4015607	3242454
3200	5437385	4596944	3707492	6053570	5099680	4115916	4291612	3465106
3300	5801609	4904598	3955381	6453598	5436392	4387436	4576807	3695166
3400	6177666	5222238	4211310	6866446	5783886	4667644	4871192	3932632
3500	6565556	5549865	4475278	7292113	6142161	4956538	5174767	4177505
3600	6965279	5887479	4747286	7730600	6511218	5254119	5487532	4429785
3700	7376836	6235079	5027333	8181906	6891057	5560388	5809487	4689472
3800	7800225	6592667	5315420	8646031	7281678	5875343	6140632	4956565
3900	8235448	6960241	5611547	9122975	7683080	6198986	6480966	5231066
4000	8682504	7337803	5915713	9612739	8095264	6531315	6830491	5512974
4100	9141393	7725351	6227919	10115322	8518230	6872332	7189206	5802288
4200	9612115	8122886	6548164	10630724	8951978	7222035	7557110	6099009
4300	10094670	8530408	6876449	11158946	9396507	7580426	7934205	6403137
4400	10589058	8947917	7212774	11699986	9851818	7947503	8320489	6714672
4500	11095280	9375413	7557138	12253847	10317911	8323268	8715964	7033614
4600	11613334	9812896	7909542	12820526	10794785	8707720	9120628	7359963
4700	12143222	10260365	8269985	13400025	11282441	9100859	9534483	7693719
4800	12684943	10717822	8638468	13992343	11780879	9502684	9957527	8034881
4900	13238497	11185265	9014991	14597480	12290099	9913197	10389761	8383450
5000	13803884	11662695	9399553	15215437	12810100	10332397	10831186	8739427
Space 'S' in mm	15	15	15	15	15	15	15	15
Iy	3768	3068	2381	10736	8863	7007	3031	2360

Moments of Inertia Ix and Iy are tabulated in cm⁴.
Dimensions of angles are in mm.

MOMENT OF INERTIA
OF FOUR UNEQUAL ANGLES

INERTIA OF UNEQUAL ANGLES

LONG LEGS BACK TO BACK
about axis x—x

ANGLES											Depth
178 × 89		152 × 102			152 × 89		152 × 76		d_w
12.6	9.4	15.8	12.6	9.4	12.6	9.4	12.6	9.5	mm
249829	190882	307732	250347	192108	235073	179289	218832	168178	1000
309339	236192	379317	308405	236520	289890	220962	270182	207513	1100
375298	286399	458458	372575	285595	350508	267031	326997	251023	1200
447706	341500	545155	442856	339332	416926	317498	389277	298708	1300
526563	401498	639408	519249	397733	489145	372362	457022	350568	1400
611869	466391	741217	601755	460796	567164	431623	530233	406602	1500
703624	536179	850581	690372	528521	650984	495281	608909	466811	1600
801827	610863	967502	785101	600910	740605	563337	693050	531195	1700
906480	690443	1091979	885942	677961	836026	635789	782656	599753	1800
1017582	774918	1224012	992895	759675	937248	712639	877727	672486	1900
1135132	864288	1363600	1105960	846052	1044271	793886	978264	749394	2000
1259132	958555	1510745	1225137	937091	1157094	879530	1084266	830477	2100
1389580	1057717	1665445	1350426	1032793	1275717	969572	1195733	915734	2200
1526478	1161774	1827702	1481826	1133158	1400142	1064011	1312665	1005167	2300
1669824	1270727	1997514	1619339	1238186	1530367	1162846	1435063	1098774	2400
1819620	1384575	2174883	1762963	1347876	1666392	1266079	1562925	1196555	2500
1975864	1503319	2359807	1912700	1462230	1808218	1373710	1696253	1298512	2600
2138558	1626959	2552288	2068548	1581245	1955845	1485737	1835046	1404643	2700
2307700	1755494	2752324	2230508	1704924	2109272	1602162	1979304	1514949	2800
2483291	1888925	2959916	2398580	1833265	2268500	1722984	2129028	1629429	2900
2665331	2027251	3175065	2572765	1966269	2433528	1848203	2284216	1748085	3000
2853820	2170473	3397769	2753061	2103936	2604357	1977819	2444870	1870915	3100
3048759	2318591	3628029	2939469	2246266	2780987	2111832	2610989	1997920	3200
3250146	2471604	3865845	3131988	2393258	2963417	2250243	2782573	2129099	3300
3457982	2629512	4111217	3330620	2544913	3151648	2393051	2959623	2264454	3400
3672267	2792316	4364145	3535364	2701231	3345680	2540256	3142137	2403983	3500
3893001	2960016	4624629	3746219	2862212	3545512	2691858	3330117	2547687	3600
4120184	3132611	4892669	3963187	3027855	3751144	2847858	3523562	2695565	3700
4353816	3310102	5168265	4186266	3198161	3962578	3008254	3722472	2847618	3800
4593896	3492488	5451417	4415458	3373130	4179811	3173048	3926848	3003847	3900
4840426	3679770	5742125	4650761	3552761	4402846	3342239	4136688	3164249	4000
5093405	3871948	6040389	4892176	3737055	4631681	3515828	4351994	3328827	4100
5352833	4069021	6346209	5139703	3926012	4866317	3693813	4572765	3497579	4200
5618709	4270989	6659584	5393343	4119632	5106753	3876196	4799001	3670506	4300
5891035	4477853	6980516	5653094	4317915	5352990	4062976	5030703	3847608	4400
6169810	4689613	7309004	5918956	4520860	5605027	4254153	5267869	4028884	4500
6455033	4906268	7645047	6190931	4728468	5862865	4449727	5510501	4214336	4600
6746706	5127819	7988647	6469018	4940738	6126504	4649698	5758598	4403961	4700
7044827	5354265	8339802	6753217	5157672	6395943	4854067	6012161	4597762	4800
7349398	5585607	8698514	7043527	5379268	6671183	5062833	6271188	4795738	4900
7660417	5821845	9064781	7339950	5605527	6952224	5275996	6535681	4997888	5000
15	15	12	12	12	12	12	12	12	Space 'S' in mm
1664	1214	2791	2190	1621	1541	1125	1039	757	I_y

Moments of Inertia **Ix** and **Iy** are tabulated in cm⁴
Dimensions of angles are in mm.

MOMENT OF INERTIA OF FOUR UNEQUAL ANGLES
LONG LEGS BACK TO BACK
about axis x—x

Depth d_w mm	ANGLES									
	127 × 89			127 × 76		102 × 89			102 × 76	
	15.8	12.6	9.4	12.6	9.4	15.8	12.6	9.5	12.6	9.4
1000	266620	218473	166509	201880	154957	243141	199026	153074	182525	140011
1100	327687	268369	204420	248229	190426	297716	243577	187243	223554	171394
1200	395087	323428	246244	299398	229571	357834	292642	224865	268757	205961
1300	468822	383649	291979	355385	272395	423495	346221	265940	318132	243712
1400	548890	449032	341626	416191	318897	494698	404314	310469	371681	284646
1500	635292	519578	395185	481817	369076	571444	466921	358451	429404	328765
1600	728027	595286	452656	552261	422933	653733	534041	409887	491299	376068
1700	827097	676157	514039	627524	480469	741564	605675	464775	557368	426555
1800	932501	762190	579334	707606	541681	834938	681823	523117	627610	480226
1900	1044238	853386	648541	792507	606572	933855	762485	584913	702025	537081
2000	1162309	949744	721660	882227	675141	1038314	847661	650161	780613	597120
2100	1286714	1051264	798690	976766	747387	1148316	937350	718863	863375	660343
2200	1417453	1157947	879633	1076124	823312	1263861	1031553	791019	950310	726751
2300	1554526	1269793	964487	1180301	902914	1384948	1130270	866627	1041418	796342
2400	1697933	1386801	1053254	1289297	986194	1511578	1233501	945689	1136700	869117
2500	1847673	1508971	1145933	1403112	1073152	1643750	1341246	1028205	1236154	945076
2600	2003748	1636304	1242523	1521746	1163787	1781466	1453504	1114173	1339782	1024220
2700	2166156	1768799	1343025	1645199	1258101	1924724	1570276	1203595	1447583	1106547
2800	2334898	1906456	1447440	1773470	1356092	2073524	1691562	1296470	1559558	1192058
2900	2509974	2049176	1555766	1906561	1457761	2227867	1817362	1392799	1675705	1280754
3000	2691384	2197259	1668004	2044471	1563108	2387753	1947675	1492581	1796026	1372633
3100	2879128	2350404	1784154	2187199	1672133	2553182	2082503	1595816	1920520	1467697
3200	3073205	2508711	1904216	2334747	1784836	2724153	2221844	1702505	2049188	1565944
3300	3273617	2672181	2028190	2487114	1901216	2900667	2365699	1812646	2182028	1667376
3400	3480362	2840813	2156076	2644299	2021275	3082724	2514068	1926242	2319042	1771992
3500	3693441	3014608	2287874	2806304	2145011	3270323	2666950	2043290	2460229	1879791
3600	3912854	3193565	2423584	2973127	2272425	3463465	2824347	2163792	2605589	1990775
3700	4138601	3377685	2563206	3144769	2403517	3662149	2986257	2287747	2755123	2104943
3800	4370682	3566967	2706739	3321231	2538287	3866376	3152681	2415156	2908830	2222295
3900	4609097	3761411	2854185	3502511	2676734	4076146	3323619	2546017	3066710	2342831
4000	4853845	3961018	3005542	3688610	2818860	4291459	3499070	2680332	3228763	2466550
4100	5104927	4165787	3160812	3879529	2964663	4512314	3679036	2818101	3394989	2593454
4200	5362344	4375719	3319993	4075266	3114144	4738712	3863515	2959322	3565389	2723542
4300	5626094	4590813	3483087	4275822	3267303	4970652	4052508	3103998	3739962	2856814
4400	5896177	4811070	3650092	4481197	3424140	5208135	4246015	3252126	3918708	2993270
4500	6172595	5036489	3821009	4691391	3584654	5451161	4444035	3403708	4101628	3132910
4600	6455347	5267070	3995839	4906404	3748847	5699729	4646570	3558743	4288720	3275735
4700	6744432	5502814	4174580	5126236	3916717	5953841	4853618	3717231	4479986	3421743
4800	7039852	5743721	4357233	5350887	4088265	6213494	5065180	3879173	4675426	3570935
4900	7341605	5989790	4543798	5580357	4263491	6478691	5281256	4044568	4875038	3723311
5000	7649692	6241021	4734275	5814646	4442395	6749430	5501845	4213416	5078824	3878871
Space 'S' in mm	12	12	12	12	12	10	10	10	10	10
I_y	1934	1522	1113	1017	743	1834	1445	1069	952	697

Moments of Inertia I_x and I_y are tabulated in cm^4.
Dimensions of angles are in mm.

INERTIA OF UNEQUAL ANGLES

MOMENT OF INERTIA
OF FOUR UNEQUAL ANGLES
LONG LEGS BACK TO BACK
about axis x—x

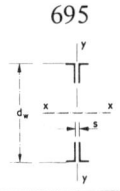

ANGLES										Depth
102 × 64		89 × 76		89 × 64		76 × 64			d_w
9.5	6.3	12.7	9.5	7.8	9.4	6.2	9.4	7.9	6.2	mm
128818	88041	172864	132718	111135	119859	81671	111989	95012	76353	1000
157836	107811	211331	162169	135759	146571	99817	136692	115941	93146	1100
189812	129591	253671	194578	162853	175976	119788	163859	138956	111610	1200
224745	153380	299884	229945	192417	208075	141584	193493	164057	131746	1300
262637	179178	349970	268269	224451	242867	165205	225591	191244	153553	1400
303486	206986	403928	309552	258954	280352	190650	260154	220517	177032	1500
347293	236804	461759	353792	295927	320531	217920	297183	251876	202183	1600
394057	268631	523463	400989	335370	363402	247015	336677	285322	229005	1700
443780	302468	589039	451145	377282	408968	277935	378637	320853	257498	1800
496460	338314	658489	504258	421665	457226	310680	423061	358471	287664	1900
552097	376170	731811	560329	468516	508178	345249	469951	398175	319500	2000
610693	416035	809006	619358	517838	561823	381643	519306	439965	353008	2100
672246	457910	890073	681344	569629	618162	419862	571127	483841	388188	2200
736757	501795	975013	746288	623890	677194	459906	625412	529803	425039	2300
804226	547689	1063826	814190	680621	738919	501774	682163	577851	463562	2400
874653	595592	1156512	885050	739821	803337	545468	741380	627985	503757	2500
948037	645505	1253071	958867	801491	870449	590986	803061	680205	545622	2600
1024379	697427	1353502	1035642	865631	940254	638328	867208	734512	589160	2700
1103679	751359	1457806	1115375	932240	1012752	687496	933820	790904	634369	2800
1185936	807301	1565983	1198066	1001319	1087944	738488	1002897	849383	681249	2900
1271151	865252	1678032	1283714	1072868	1165829	791306	1074439	909948	729801	3000
1359324	925212	1793954	1372320	1146887	1246408	845948	1148447	972599	780025	3100
1450455	987182	1913749	1463884	1223375	1329679	902414	1224920	1037335	831920	3200
1544543	1051162	2037417	1558406	1302333	1415644	960706	1303858	1104159	885487	3300
1641590	1117151	2164957	1655885	1383761	1504303	1020822	1385262	1173068	940725	3400
1741593	1185149	2296370	1756322	1467658	1595654	1082763	1469131	1244063	997635	3500
1844555	1255157	2431656	1859717	1554025	1689699	1146529	1555465	1317144	1056216	3600
1950474	1327175	2570815	1966069	1642862	1786438	1212120	1644264	1392312	1116469	3700
2059352	1401202	2713846	2075379	1734168	1885869	1279535	1735529	1469565	1178393	3800
2171186	1477239	2860750	2187647	1827944	1987994	1348775	1829259	1548905	1241989	3900
2285979	1555285	3011527	2302873	1924190	2092812	1419840	1925454	1630331	1307257	4000
2403729	1635341	3166177	2421057	2022906	2200324	1492730	2024114	1713842	1374196	4100
2524438	1717406	3324699	2542198	2124091	2310529	1567444	2125240	1799440	1442806	4200
2648103	1801481	3487094	2666297	2227746	2423427	1643984	2228830	1887124	1513088	4300
2774727	1887565	3653362	2793354	2333871	2539018	1722348	2334887	1976895	1585042	4400
2904308	1975659	3823502	2923368	2442465	2657303	1802537	2443408	2068751	1658667	4500
3036847	2065762	3997516	3056340	2553529	2778282	1884550	2554395	2162693	1733964	4600
3172344	2157875	4175402	3192270	2667063	2901953	1968389	2667846	2258722	1810932	4700
3310799	2251997	4357160	3331158	2783066	3028318	2054052	2783764	2356836	1889572	4800
3452211	2348129	4542792	3473003	2901539	3157376	2141540	2902146	2457037	1969883	4900
3596581	2446270	4732296	3617806	3022482	3289127	2230853	3022994	2559324	2051866	5000
10	10	10	10	10	10	10	8	8	8	Space 'S' in mm
434	279	949	696	568	425	273	401	331	258	I_y

Moments of Inertia **Ix** and **Iy** are tabulated in cm⁴.
Dimensions of angles are in mm.

MOMENT OF INERTIA
OF FOUR UNEQUAL ANGLES

SHORT LEGS BACK TO BACK
about axis x—x

MOMENTS OF INERTIA OF PLATE GIRDERS

BASED ON BS 449 1969

Depth	ANGLES							
d_w	229 × 102			203 × 152			203 × 102	
mm	18.9	15.8	12.6	18.9	15.8	12.6	15.8	12.6
1000	540312	458365	370987	548762	464012	375930	419728	340102
1100	659143	558899	452124	673238	568962	460707	512016	414670
1200	789807	669419	541300	810532	684694	554171	613494	496646
1300	932305	789927	638516	960646	811208	656323	724162	586028
1400	1086636	920421	743771	1123579	948504	767161	844020	682817
1500	1252799	1060902	857066	1299331	1096581	886687	973067	787013
1600	1430796	1211371	978401	1487903	1255440	1014899	1111305	898616
1700	1620626	1371826	1107775	1689294	1425080	1151799	1258733	1017626
1800	1822290	1542267	1245189	1903504	1605503	1297386	1415351	1144042
1900	2035786	1722696	1390642	2130534	1796707	1451659	1581158	1277866
2000	2261115	1913112	1544135	2370382	1998693	1614620	1756156	1419096
2100	2498278	2113514	1705668	2623051	2211460	1786268	1940343	1567733
2200	2747274	2323904	1875240	2888538	2435010	1966603	2133721	1723778
2300	3008103	2544280	2052852	3166845	2669341	2155624	2336289	1887229
2400	3280765	2774643	2238503	3457971	2914453	2353333	2548046	2058087
2500	3565260	3014993	2432194	3761916	3170348	2559729	2768993	2236351
2600	3861588	3265330	2633925	4078681	3437024	2774812	2999131	2422023
2700	4169750	3525654	2843695	4408265	3714482	2998582	3238458	2615102
2800	4489744	3795965	3061505	4750668	4002721	3231039	3486975	2815587
2900	4821572	4076262	3287354	5105891	4301743	3472183	3744683	3023479
3000	5165233	4366547	3521243	5473932	4611546	3722014	4011580	3238779
3100	5520727	4666818	3763172	5854794	4932130	3980532	4287667	3461485
3200	5888054	4977076	4013140	6248474	5263497	4247738	4572944	3691598
3300	6267215	5297321	4271148	6654974	5605645	4523630	4867411	3929117
3400	6658208	5627553	4537195	7074293	5958575	4808209	5171068	4174044
3500	7061035	5967772	4811282	7506431	6322286	5101475	5483915	4426378
3600	7475694	6317978	5093409	7951389	6696780	5403429	5805952	4686118
3700	7902187	6678171	5383575	8409166	7082055	5714069	6137179	4953265
3800	8340513	7048350	5681781	8879762	7478112	6033397	6477596	5227820
3900	8790672	7428515	5988026	9363178	7884950	6361411	6827203	5509781
4000	9252665	7818670	6302311	9859412	8302570	6698112	7186000	5799149
4100	9726490	8218810	6624636	10368467	8730972	7043501	7553986	6095924
4200	10212149	8628937	6955000	10890340	9170156	7397577	7931163	6400105
4300	10709640	9049051	7293404	11425033	9620121	7760339	8317530	6711694
4400	11218965	9479151	7639847	11972545	10080868	8131789	8713087	7030690
4500	11740123	9919239	7994330	12532876	10552397	8511926	9117833	7357092
4600	12273114	10369314	8356853	13106027	11034708	8900749	9531770	7690901
4700	12817939	10829375	8727415	13691997	11527800	9298260	9954896	8032117
4800	13374676	11299423	9106017	14290786	12031674	9704458	10387213	8380740
4900	13943087	11779458	9492658	14902394	12546330	10119343	10828719	8736770
5000	14523410	12269480	9887339	15526822	13071767	10542915	11279416	9100207
Space 'S' in mm	15	15	15	15	15	15	15	15
Iy	33307	27764	22053	23880	19818	15746	19797	15724

Moments of Inertia **Ix** and **Iy** are tabulated in cm⁴.
Dimensions of angles are in mm.

INERTIA OF UNEQUAL ANGLES
MOMENT OF INERTIA
OF FOUR UNEQUAL ANGLES

BASED ON BS 449 1969

SHORT LEGS BACK TO BACK
about axis x—x

ANGLES										Depth
178 × 89		152 × 102			152 × 89		152 × 76		d_w
12.6	9.4	15.8	12.6	9.4	12.6	9.4	12.6	9.5	mm
298475	227787	340673	277002	212431	266883	203388	255007	195799	1000
363670	277402	416072	338140	259187	325353	247823	310489	238283	1100
435315	331913	499026	405389	310606	389625	296655	371436	284942	1200
513408	391319	589537	478751	366687	459697	349885	437849	335776	1300
597950	455621	687603	558224	427431	535570	407511	509727	390785	1400
688942	524819	793226	643810	492838	617243	469535	587070	449968	1500
786382	598912	906404	735507	562907	704717	535956	669878	513326	1600
890271	677900	1027139	833316	637640	797991	606774	758151	580859	1700
1000609	761784	1155429	937237	717035	897066	681989	851890	652567	1800
1117396	850564	1291275	1047270	801093	1001942	761602	951093	728449	1900
1240632	944239	1434678	1163415	889813	1112618	845612	1055762	808506	2000
1370317	1042810	1585636	1285672	983197	1229095	934019	1165897	892738	2100
1506451	1146277	1744150	1414041	1081243	1351372	1026823	1281496	981144	2200
1649034	1254639	1910220	1548522	1183951	1479450	1124024	1402560	1073726	2300
1798066	1367896	2083847	1689114	1291323	1613329	1225623	1529090	1170482	2400
1953546	1486049	2265029	1835819	1403357	1753008	1331618	1661085	1271412	2500
2115476	1609098	2453767	1988635	1520054	1898488	1442011	1798545	1376518	2600
2283855	1737042	2650061	2147564	1641414	2049769	1556801	1941471	1485798	2700
2458683	1869882	2853911	2312604	1767436	2206850	1675989	2089861	1599253	2800
2639959	2007617	3065317	2483756	1898122	2369731	1799573	2243717	1716883	2900
2827685	2150248	3284279	2661020	2033470	2538414	1927555	2403038	1838687	3000
3021860	2297774	3510796	2844396	2173480	2712897	2059934	2567824	1964667	3100
3222483	2450196	3744870	3033884	2318154	2893180	2196710	2738076	2094820	3200
3429556	2607514	3986500	3229484	2467490	3079264	2337883	2913792	2229149	3300
3643077	2769727	4235686	3431196	2621489	3271149	2483454	3094974	2367653	3400
3863047	2936836	4492427	3639020	2780151	3468834	2633422	3281621	2510331	3500
4089467	3108840	4756725	3852955	2943475	3672320	2787787	3473733	2657184	3600
4322335	3285740	5028579	4073003	3111462	3881606	2946549	3671310	2808211	3700
4561652	3467535	5307988	4299162	3284112	4096693	3109708	3874353	2963414	3800
4807419	3654226	5594954	4531434	3461425	4317581	3277265	4082861	3122791	3900
5059634	3845812	5889475	4769817	3643400	4544269	3449218	4296834	3286343	4000
5318298	4042294	6191553	5014312	3830038	4776758	3625569	4516272	3454069	4100
5583411	4243672	6501186	5264920	4021339	5015048	3806317	4741175	3625971	4200
5854973	4449945	6818376	5521639	4217303	5259138	3991463	4971544	3802047	4300
6132984	4661114	7143121	5784470	4417929	5509028	4181005	5207378	3982298	4400
6417444	4877178	7475422	6053413	4623218	5764720	4374945	5448676	4166723	4500
6708353	5098138	7815280	6328468	4833170	6026211	4573282	5695441	4355323	4600
7005711	5323993	8162693	6609634	5047784	6293504	4776016	5947670	4548099	4700
7309518	5554744	8517662	6896913	5267062	6566597	4983147	6205365	4745048	4800
7619774	5790391	8880187	7190304	5491002	6845491	5194676	6468524	4946173	4900
7936479	6030933	9250268	7489806	5719604	7130185	5410602	6737149	5151472	5000
15	15	12	12	12	12	12	12	12	Space 'S' in mm
10726	8000	8444	6705	5018	6708	4984	6683	5005	I_y

Moments of Inertia **Ix** and **Iy** are tabulated in cm⁴.
Dimensions of angles are in mm.

MOMENT OF INERTIA
OF FOUR UNEQUAL ANGLES
SHORT LEGS BACK TO BACK
about axis x–x

Depth	ANGLES									
d_w	127 × 89			127 × 76		102 × 89			102 × 76	
mm	15.8	12.6	9.4	12.6	9.4	15.8	12.6	9.5	12.6	9.4
1000	287806	235746	179590	223514	171461	249419	204139	156984	192045	147272
1100	351270	287594	218977	272293	208779	304694	249260	191587	234127	179456
1200	421069	344605	262275	325891	249775	365512	298894	229643	280381	214824
1300	497201	406777	309486	384308	294449	431872	353042	271153	330810	253376
1400	579667	474112	360608	447544	342801	503775	411703	316116	385411	295112
1500	668467	546610	415642	515600	394830	581221	474879	364532	444185	340032
1600	763601	624270	474588	588474	450537	664209	542568	416402	507133	388136
1700	865069	707092	537447	666167	509923	752740	614771	471725	574254	439424
1800	972870	795077	604217	748679	572986	846814	691488	530501	645549	493896
1900	1087005	888225	674899	836010	639727	946430	772719	592731	721016	551552
2000	1207475	986535	749493	928159	710145	1051589	858463	658414	800657	612392
2100	1334278	1090007	827998	1025128	784242	1162291	948722	727550	884471	676416
2200	1467415	1198641	910416	1126916	862016	1278535	1043494	800139	972458	743625
2300	1606886	1312439	996746	1233523	943468	1400322	1142780	876182	1064619	814017
2400	1752690	1431398	1086988	1344949	1028599	1527652	1246579	955678	1160952	887593
2500	1904829	1555520	1181142	1461193	1117406	1660524	1354893	1038628	1261459	964354
2600	2063301	1684805	1279207	1582257	1209892	1798939	1467720	1125031	1366140	1044298
2700	2228108	1819251	1381185	1708139	1306056	1942897	1585061	1214887	1474993	1127427
2800	2399248	1958861	1487074	1838441	1405897	2092397	1706916	1308196	1588020	1213739
2900	2576722	2103632	1596876	1974362	1509417	2247440	1833285	1404959	1705220	1303236
3000	2760530	2253567	1710589	2114701	1616614	2408025	1964167	1505175	1826593	1395916
3100	2950671	2408663	1828214	2259860	1727489	2574154	2099564	1608845	1952139	1491781
3200	3147147	2568922	1949752	2409837	1842041	2745825	2239474	1715967	2081859	1590829
3300	3349956	2734344	2075201	2564633	1960272	2923038	2383898	1826543	2215752	1693062
3400	3559100	2904928	2204562	2724249	2082181	3105794	2532835	1940573	2353818	1798479
3500	3774577	3080674	2337835	2888683	2207767	3294093	2686287	2058056	2496057	1907080
3600	3996388	3261583	2475020	3057936	2337031	3487935	2844252	2178992	2642470	2018865
3700	4224533	3447654	2616117	3232008	2469973	3687319	3006731	2303381	2793056	2133833
3800	4459011	3638888	2761126	3410900	2606593	3892246	3173724	2431224	2947815	2251986
3900	4699824	3835284	2910047	3594610	2746891	4102716	3345231	2562520	3106747	2373323
4000	4946970	4036843	3062879	3783139	2890866	4318728	3521251	2697269	3269853	2497844
4100	5200451	4243564	3219624	3976487	3038519	4540283	3701786	2835471	3437132	2625549
4200	5460265	4455447	3380281	4174654	3189851	4767380	3886834	2977127	3608584	2756438
4300	5726413	4672493	3544849	4377640	3344860	5000020	4076396	3122237	3784209	2890511
4400	5998895	4894701	3713330	4585445	3503547	5238203	4270471	3270799	3964008	3027769
4500	6277711	5122072	3885722	4798069	3665911	5481929	4469061	3422815	4147979	3168210
4600	6562860	5354606	4062027	5015511	3831954	5731197	4672164	3578284	4336124	3311835
4700	6854344	5592301	4242243	5237773	4001674	5986008	4879781	3737207	4528443	3458644
4800	7152161	5835159	4426371	5464854	4175072	6246361	5091912	3899583	4724934	3608638
4900	7456312	6083180	4614412	5696754	4352149	6512257	5308557	4065412	4925599	3761815
5000	7766797	6336363	4806364	5933472	4532902	6783696	5529715	4234695	5130437	3918176
Space 'S' in mm	12	12	12	12	12	10	10	10	10	10
Iy	5016	4000	2960	3973	2962	2623	2078	1545	2060	1527

Moments of Inertia **Ix** and **Iy** are tabulated in cm⁴.
Dimensions of angles are in mm.

INERTIA OF UNEQUAL ANGLES   699

MOMENT OF INERTIA
OF FOUR UNEQUAL ANGLES
SHORT LEGS BACK TO BACK
about axis x−x

ANGLES										Depth
102 × 64		89 × 76			89 × 64			76 × 64			d_w
9.5	6.3	12.7	9.5	7.8	9.4	6.2	9.4	7.9	6.2	mm
138993	94937	177314	136115	113969	126088	85880	114860	97440	78295	1000
169128	115461	216270	165938	138903	153477	104484	139873	118630	95297	1100
202220	137996	259098	198719	166307	183561	124912	167351	141907	113971	1200
238269	162540	305799	234458	196181	216337	147165	197294	167270	134316	1300
277277	189093	356373	273155	228525	251807	171243	229703	194719	156333	1400
319242	217656	410820	314809	263338	289970	197146	264576	224255	180021	1500
364165	248229	469139	359421	300622	330826	224873	301915	255876	205381	1600
412046	280811	531331	406991	340374	374376	254425	341720	289583	232412	1700
462884	315402	597396	457518	382597	420619	285802	383989	325377	261115	1800
516680	352003	667334	511003	427289	469555	319004	428724	363256	291490	1900
573434	390614	741144	567446	474451	521185	354030	475924	403222	323536	2000
633146	431234	818827	626847	524082	575508	390882	525589	445274	357253	2100
695618	473864	900383	689205	576184	632524	429558	577720	489412	392642	2200
761442	518503	985812	754522	630755	692234	470059	632315	535636	429703	2300
830027	565152	1075113	822796	687795	754637	512384	689377	583946	468435	2400
901570	613810	1168287	894027	747306	819733	556535	748903	634342	508839	2500
976070	664478	1265334	968217	809286	887522	602510	810894	686825	550914	2600
1053528	717155	1366253	1045364	873736	958005	650310	875351	741393	594661	2700
1133944	771842	1471045	1125469	940655	1031182	699935	942273	798047	640079	2800
1217317	828538	1579710	1208531	1010044	1107051	751384	1011661	856788	687169	2900
1303649	887244	1692248	1294552	1081903	1185614	804659	1083513	917615	735930	3000
1392938	947959	1808658	1383530	1156232	1266870	859758	1157831	980528	786363	3100
1485184	1010684	1928942	1475466	1233030	1350820	916682	1234614	1045527	838468	3200
1580389	1075418	2053098	1570359	1312298	1437462	975430	1313863	1112612	892244	3300
1678551	1142162	2181126	1668210	1394036	1526798	1036004	1395576	1181783	947691	3400
1779671	1210915	2313028	1769020	1478243	1618828	1098402	1479755	1253040	1004810	3500
1883749	1281678	2448802	1872786	1564920	1713551	1162625	1566399	1326383	1063601	3600
1990784	1354451	2588449	1979511	1654067	1810967	1228673	1655509	1401813	1124063	3700
2100777	1429233	2731968	2089193	1745683	1911076	1296546	1747084	1479328	1186197	3800
2213728	1506024	2879361	2201833	1839770	2013879	1366243	1841124	1558930	1250002	3900
2329637	1584825	3030626	2317431	1936326	2119375	1437765	1937629	1640617	1315479	4000
2448503	1665636	3185764	2435986	2035351	2227564	1511112	2036599	1724391	1382627	4100
2570328	1748456	3344774	2557500	2136846	2338447	1586284	2138035	1810251	1451447	4200
2695110	1833285	3507658	2681971	2240811	2452023	1663280	2241936	1898197	1521939	4300
2822849	1920124	3674414	2809399	2347246	2568292	1742101	2348302	1988229	1594101	4400
2953547	2008973	3845043	2939786	2456150	2687255	1822748	2457134	2080347	1667936	4500
3087202	2099831	4019544	3073130	2567525	2808911	1905218	2568431	2174552	1743442	4600
3223815	2192699	4197918	3209432	2681368	2933260	1989514	2682193	2270842	1820620	4700
3363385	2287576	4380166	3348692	2797682	3060303	2075634	2798420	2369219	1899469	4800
3505914	2384463	4566285	3490909	2916465	3190039	2163579	2917112	2469681	1979989	4900
3651400	2483359	4756278	3636084	3037718	3322468	2253349	3038270	2572230	2062182	5000
10	10	10	10	10	10	10	8	8	8	Space 'S' in mm
1530	1006	1430	1056	866	1042	681	655	544	426	I_y

Moments of Inertia **Ix** and **Iy** are tabulated in cm⁴.
Dimensions of angles are in mm.

28. CONNECTIONS

Introduction

THE design of structural steelwork involves many problems, one of which is the design of connections.

It is essential that this stage of the work be given full consideration, for no structural frame will function correctly, unless its individual members are connected to each other in such a manner that all forces and moments can be transferred to each other as designed.

The methods used in forming connections are Riveting, Bolting, and Welding, each having its own advantages for specific cases and must be carefully considered in order to select the most suitable.

The following examples are given, to illustrate some of the various types of connections used in the construction of steel frames.

The notes and tables given on pages 703 to 708 inclusive for rivets and bolts, on pages 733 to 736 inclusive for high strength friction grip bolts and on pages 740 to 742 inclusive for welding are extracted from the Handbook on Structural Steelwork published jointly by: The British Constructional Steelwork Association and The Constructional Steel Research and Development Organisation, the allowable stresses, etc., being those given in B.S. 449 : Part 2 :1969.

Edge distance of holes

The minimum distances from the centre of any rivet or bolt hole to the edge of a plate shall be in accordance with Table 21 of B.S. 449, which is reproduced below.

Diameter of Hole	Distance to Sheared or Hand Flame Cut Edge	Distance to rolled, machine flame cut, sawn or planed edge
mm	mm	mm
39	68	62
36	62	56
33	56	50
30	50	44
26	42	36
24	38	32
22	34	30
20	30	28
18	28	26
16	26	24
14	24	22
12 or less	22	20

CONNECTIONS

The following tables give the safe load values in tension of black bolts between 16 mm and 48 mm diameter.

Bolts of grade designation 4.6 to B.S. 4190: 1967

Nominal diameter of bolt in mm	16	20	22	24	27	30	33	36	39	42	45	48
Tensile Stress Area mm²	157	245	303	353	459	561	694	817	976	1 120	1 300	1 470
Value in kilonewtons of one bolt at 130 N/mm²	20.4	31.9	39.4	45.9	59.7	72.9	90.2	106.2	126.9	145.6	169.0	191
Value in kilonewtons of one bolt at 130 + 25% N/mm²	25.5	39.8	49.2	57.4	74.6	91.2	112.8	132.8	158.6	182.0	211.3	238

Allowable stress obtained from table 20 in B.S. 449 Part 2: 1969.

Bolts of grade designation 8.8 to B.S. 3692: 1967

Nominal diameter of bolt in mm	16	20	22	24	27	30	33	36	39	42	45	48
Tensile Stress Area mm²	157	245	303	353	459	561	694	817	976	1 120	1 300	1 47
Value in kilonewtons of one bolt at 335 N/mm²	52.6	82.1	101.5	118.3	153.8	187.9	232.5	273.7	327.0	375.2	435.5	492.
Value in kilonewtons of one bolt at 335 + 25% N/mm²	65.7	102.6	126.9	147.8	192.2	234.9	296.6	342.1	408.7	490.0	544.4	615.

Allowable stress is $\frac{58.2}{22.6} \times 130$ from B.S. 449 Part 2: 1969

Safe load values for rivets and bolts

Working Stresses Safe load values are tabulated for rivets of material having a yield stress of 250 N/mm² and for bolts of strength grade designation 4.6 in accordance with the allowable stresses given in clause 50 of B.S. 449: 1969 as follows:—

Description	Allowable stresses in N/mm²	
	Single shear	Bearing (Double shear)
Power-driven shop rivets	110	315
Power-driven field rivets	100	290
Hand-driven rivets	90	265
Close tolerance turned bolts	95	300
Black bolts	80	200

Multiple Shear For rivets and bolts in double shear, the area to be assumed must be twice the area for single shear. Where the rivets or bolts are in single shear, the permissible bearing stress must be reduced by 20%.

Critical Values Bearing values printed in ordinary type are less than single shear. In these cases, the bearing values are the determining factors. Bearing values printed in prominent type are greater than single and less than double shear, so that in the case of:—

(a) single shear, the shearing value is the criterion.

(b) double shear, the bearing value is the criterion.

Bearing values printed in italic type are equal to or greater than double shear. In these cases, the shearing values are the criterion.

Other Grade Bolts For bolts of other strength grade designations, the allowable stresses shall be those for grade 4.6 bolts varied in the ratio of the specified stress under proof load in kgf/mm² to 22.6. In Table 13 of BS 3692, that stress for grade 8.8 bolts is given as 58.2 kgf/mm² so that the strength ratio multiplier is $58.2/22.6 = 2.575$.

When parts are connected together by bolts of a higher grade of material, the local bearing stress must not exceed 2.5 times the allowable stress in axial tension as given in Table 19 of BS 449: 1969 for the material of the connected part.

Tables on pages 704 to 710 are reproduced by permission of the British Constructional Steelwork Association Ltd., and the Constructional Steel Research and Development Organisation.

SHEARING AND BEARING VALUES

IN KILONEWTONS FOR POWER-DRIVEN SHOP RIVETS OF STEEL HAVING A YIELD STRESS OF 250 N/mm²

BASED ON BS 449 1969

Gross Dia. of Rivet after driving in mm	Area in cm²	Shearing Value @ 110 N/mm²		Simple Bearing Value @ 80% of 315 N/mm² and Enclosed Bearing Value @ 315 N/mm² (see footnote)											
		Single Shear	Double Shear	\multicolumn{12}{c}{Thickness in mm of plate passed through or of enclosed plate}
				5	6	7	8	9	10	12	15	18	20	22	25
10	0.79	8.64	17.3	12.6	15.1	17.6	20.2	22.7							
				15.8	18.9	22.0	25.2								
12	1.13	12.4	24.9	15.1	18.1	21.2	24.2	27.2	30.2	36.3					
				18.9	22.7	26.5	30.2	34.0							
14	1.54	16.9	33.9	17.6	21.2	24.7	28.2	31.8	35.3	42.3	52.9				
				22.0	26.5	30.9	35.3	39.7	44.1						
16	2.01	22.1	44.2	20.2	24.2	28.2	32.3	36.3	40.3	48.4	60.5	72.6			
				25.2	30.2	35.3	40.3	45.4	50.4	60.5					
18	2.54	28.0	56.0	22.7	27.2	31.8	36.3	40.8	45.4	54.4	68.0	81.6	90.7		
				28.3	34.0	39.7	45.4	51.0	56.7	68.0	85.0				
20	3.14	34.6	69.1	25.2	30.2	35.3	40.3	45.4	50.4	60.5	75.6	90.7	101		
				31.5	37.8	44.1	50.4	56.7	63.0	75.6	94.5	113			
22	3.80	41.8	83.6	27.7	33.3	38.8	44.4	49.9	55.4	66.5	83.2	99.8	111	122	
				34.6	41.6	48.5	55.4	62.4	69.3	83.2	104	125	139		
24	4.52	49.8	99.5	30.2	36.3	42.3	48.4	54.4	60.5	72.6	90.7	109	121	133	
				37.8	45.4	52.9	60.5	68.0	75.6	90.7	113	136	151		
27	5.73	63.0	126	34.0	40.8	47.6	54.4	61.2	68.0	81.6	102	122	136	150	170
				42.5	51.0	59.5	68.0	76.5	85.0	102	128	153	170		

Upper line Bearing Values for each diameter of rivet are Simple Bearing Values.
Lower line Bearing Values for each diameter of rivet are Enclosed Bearing Values.
For areas to be deducted from a bar for one hole, see table on page 709.
For explanation of table, see Notes.
1 kilonewton may be taken as 0.102 metric tonne (megagramme) force.

705

BASED ON
BS 449
1969

SHEARING AND BEARING VALUES

IN KILONEWTONS FOR POWER-DRIVEN FIELD RIVETS OF STEEL HAVING A YIELD STRESS OF 250 N/mm^2

Gross Dia. of Rivet after driving in mm	Area in cm^2	Shearing Value @ 100 N/mm^2 Single Shear	Shearing Value @ 100 N/mm^2 Double Shear	\multicolumn{11}{c}{Simple Bearing Value @ 80% of 290 N/mm^2 and Enclosed Bearing Value @ 290 N/mm^2 (see footnote) — Thickness in mm of plate passed through or of enclosed plate}
				5	6	7	8	9	10	12	15	18	20	22	25
10	0.79	7.85	15.7	11.6 / 14.5	13.9 / 17.4	16.2 / 20.3	18.6 / 23.2	20.9							
12	1.13	11.3	22.6	13.9 / 17.4	16.7 / 20.9	19.5 / 24.4	22.3 / 27.8	25.1 / 31.3	27.8	33.4					
14	1.54	15.4	30.8	16.2 / 20.3	19.5 / 24.4	22.7 / 28.4	26.0 / 32.5	29.2 / 36.5	32.5 / 40.6	39.0	48.7				
16	2.01	20.1	40.2	18.6 / 23.2	22.3 / 27.8	26.0 / 32.5	29.7 / 37.1	33.4 / 41.8	37.1 / 46.4	44.5 / 55.7	55.7	66.8			
18	2.54	25.4	50.9	20.9 / 26.1	25.1 / 31.3	29.2 / 36.5	33.4 / 41.8	37.6 / 47.0	41.8 / 52.2	50.1 / 62.6	62.6 / 78.3	75.2	83.5		
20	3.14	31.4	62.8	23.2 / 29.0	27.8 / 34.8	32.5 / 40.6	37.1 / 46.4	41.8 / 52.2	46.4 / 58.0	55.7 / 69.6	69.6 / 87.0	83.5 / 104	92.8		
22	3.80	38.0	76.0	25.5 / 31.9	30.6 / 38.3	35.7 / 44.7	40.8 / 51.0	45.9 / 57.4	51.0 / 63.8	61.2 / 76.6	76.6 / 95.7	91.9 / 115	102		
24	4.52	45.2	90.5	27.8 / 34.8	33.4 / 41.8	39.0 / 48.7	44.5 / 55.7	50.1 / 62.6	55.7 / 69.6	66.8 / 83.5	83.5 / 104	100 / 125	111 / 139	122	
27	5.73	57.3	115	31.3 / 39.1	37.6 / 47.0	43.8 / 54.8	50.1 / 62.6	56.4 / 70.5	62.6 / 78.3	75.2 / 94.0	94.0 / 117	113 / 141	125 / 157	138	157

Upper line Bearing Values for each diameter of rivet are Simple Bearing Values.
Lower line Bearing Values for each diameter of rivet are Enclosed Bearing Values.
For areas to be deducted from a bar for one hole, see table on page 709.
For explanation of table, see Notes.
1 kilonewton may be taken as 0.102 metric tonne (megagramme) force.

SHEARING AND BEARING VALUES

IN KILONEWTONS FOR HAND-DRIVEN RIVETS OF STEEL HAVING A YIELD STRESS OF 250 N/mm²

BASED ON BS 449 1969

Gross Dia. of Rivet after driving in mm	Area in cm²	Shearing Value @ 90 N/mm² Single Shear	Shearing Value @ 90 N/mm² Double Shear	5	6	7	8	9	10	12	15	18	20	22	25
10	0.79	7.07	14.1	10.6 / 13.3	12.7 / 15.9	14.8 / 18.5	17.0 / 21.2	19.1							
12	1.13	10.2	20.4	12.7 / 15.9	15.3 / 19.1	17.8 / 22.3	20.4 / 25.4	22.9 / 28.6	25.4	30.5					
14	1.54	13.9	27.7	14.8 / 18.5	17.8 / 22.3	20.8 / 26.0	23.7 / 29.7	26.7 / 33.4	29.7 / 37.1	35.6	44.5				
16	2.01	18.1	36.2	17.0 / 21.2	20.4 / 25.4	23.7 / 29.7	27.1 / 33.9	30.5 / 38.2	33.9 / 42.4	40.7 / 50.9	50.9	61.1			
18	2.54	22.9	45.8	19.1 / 23.8	22.9 / 28.6	26.7 / 33.4	30.5 / 38.2	34.3 / 42.9	38.2 / 47.7	45.8 / 57.2	57.2 / 71.5	68.7	76.3		
20	3.14	28.3	56.5	21.2 / 26.5	25.4 / 31.8	29.7 / 37.1	33.9 / 42.4	38.2 / 47.7	42.4 / 53.0	50.9 / 63.6	63.6 / 79.5	76.3 / 95.4	84.8		
22	3.80	34.2	68.4	23.3 / 29.1	28.0 / 35.0	32.6 / 40.8	37.3 / 46.6	42.0 / 52.5	46.6 / 58.3	56.0 / 70.0	70.0 / 87.5	84.0 / 105	93.3		
24	4.52	40.7	81.4	25.4 / 31.8	30.5 / 38.2	35.6 / 44.5	40.7 / 50.9	45.8 / 57.2	50.9 / 63.6	61.1 / 76.3	76.3 / 95.4	91.6 / 114	102 / 127	112	
27	5.73	51.5	103	28.6 / 35.8	34.3 / 42.9	40.1 / 50.1	45.8 / 57.2	51.5 / 64.4	57.2 / 71.5	68.7 / 85.9	85.9 / 107	103 / 129	114 / 143	126	143

Simple Bearing Value @ 80% of 265 N/mm² and Enclosed Bearing Value @ 265 N/mm² (SEE FOOTNOTE)
Thickness in mm of plate passed through or of enclosed plate

Upper line Bearing Values for each diameter of rivet are Simple Bearing Values.
Lower line Bearing Values for each diameter of rivet are Enclosed Bearing Values.
For areas to be deducted from a bar for one hole, see table on page 709.
For explanation of table, see Notes.
1 kilonewton may be taken as 0.102 metric tonne (megagramme) force.

SHEARING AND BEARING VALUES

IN KILONEWTONS FOR CLOSE TOLERANCE AND TURNED BOLTS OF STEEL OF STRENGTH GRADE DESIGNATION 4.6

BASED ON BS 449 1969

Dia. of Bolt Shank in mm	Area in cm²	Shearing Value @ 95 N/mm² Single Shear	Shearing Value @ 95 N/mm² Double Shear	5	6	7	8	9	10	12	15	18	20	22	25
10	0.79	7.46	14.9	12.0 / 15.0	14.4 / 18.0	16.8 / 21.0	19.2 / 24.0	21.6							
12	1.13	10.7	21.5	14.4 / 18.0	17.3 / 21.6	20.2 / 25.2	23.0 / 28.8	25.9	28.8						
14	1.54	14.6	29.2	16.8 / 21.0	20.2 / 25.2	23.5 / 29.4	26.9 / 33.6	30.2 / 37.8	33.6	40.3					
16	2.01	19.1	38.2	19.2 / 24.0	23.0 / 28.8	26.9 / 33.6	30.7 / 38.4	34.6 / 43.2	38.4 / 48.0	46.1	57.6				
18	2.54	24.2	48.3	21.6 / 27.0	25.9 / 32.4	30.2 / 37.8	34.6 / 43.2	38.9 / 48.6	43.2 / 54.0	51.8 / 64.8	64.8	77.8			
20	3.14	29.8	59.7	24.0 / 30.0	28.8 / 36.0	33.6 / 42.0	38.4 / 48.0	43.2 / 54.0	48.0 / 60.0	57.6 / 72.0	72.0 / 90.0	86.4	96.0		
22	3.80	36.1	72.2	26.4 / 33.0	31.7 / 39.6	37.0 / 46.2	42.2 / 52.8	47.5 / 59.4	52.8 / 66.0	63.4 / 79.2	79.2 / 99.0	95.0 / 119	106		
24	4.52	43.0	86.0	28.8 / 36.0	34.6 / 43.2	40.3 / 50.4	46.1 / 57.6	51.8 / 64.8	57.6 / 72.0	69.1 / 86.4	86.4 / 108	104 / 130	115		
27	5.73	54.4	109	32.4 / 40.5	38.9 / 48.6	45.4 / 56.7	51.8 / 64.8	58.3 / 72.9	64.8 / 81.0	77.8 / 97.2	97.2 / 122	117 / 146	130 / 162	143	

SIMPLE BEARING VALUE @ 80% OF 300 N/mm² AND ENCLOSED BEARING VALUE @ 300 N/mm² (SEE FOOTNOTE)

Thickness in mm of plate passed through or of enclosed plate

Upper line Bearing Values for each diameter of bolt are Simple Bearing Values.
Lower line Bearing Values for each diameter of bolt are Enclosed Bearing Values.
For areas to be deducted from a bar for one hole, see table on page 709.
For explanation of table, see Notes.
1 kilonewton may be taken as 0.102 metric tonne (megagramme) force.

SHEARING AND BEARING VALUES

IN KILONEWTONS FOR BLACK BOLTS OF STEEL OF STRENGTH GRADE DESIGNATION 4.6

BASED ON BS 449 1969

Dia. of Bolt Shank in mm	Area in cm²	Shearing Value @ 80 N/mm² Single Shear	Shearing Value @ 80 N/mm² Double Shear	\multicolumn{12}{c}{Simple Bearing Value @ 80% of 200 N/mm² and Enclosed Bearing Value @ 200 N/mm² (see footnote) — Thickness in mm of plate passed through or of enclosed plate}											
				5	6	7	8	9	10	12	15	18	20	22	25
10	0.79	6.28	12.6	**8.00** / 10.0	**9.60** / 12.0	**11.2** / 14.0	12.8 / 16.0	14.4 / 18.0	16.0						
12	1.13	9.05	18.1	**9.60** / 12.0	**11.5** / 14.4	**13.4** / 16.8	**15.4** / 19.2	**17.3** / 21.6	19.2 / 24.0	23.0	28.8				
14	1.54	12.3	24.6	**11.2** / 14.0	**13.4** / 16.8	**15.7** / 19.6	**17.9** / 22.4	**20.2** / 25.2	**22.4** / 28.0	26.9 / 33.6	33.6	40.3			
16	2.01	16.1	32.2	**12.8** / 16.0	**15.4** / 19.2	**17.9** / 22.4	**20.5** / 25.6	**23.0** / 28.8	**25.6** / 32.0	**30.7** / 38.4	38.4 / 48.0	46.1 / 57.6	51.2		
18	2.54	20.4	40.7	**14.4** / 18.0	**17.3** / 21.6	**20.2** / 25.2	**23.0** / 28.8	**25.9** / 32.4	**28.8** / 36.0	**34.6** / 43.2	43.2 / 54.0	51.8 / 64.8	57.6		
20	3.14	25.1	50.3	**16.0** / 20.0	**19.2** / 24.0	**22.4** / 28.0	**25.6** / 32.0	**28.8** / 36.0	**32.0** / 40.0	**38.4** / 48.0	**48.0** / 60.0	57.6 / 72.0	64.0 / 80.0	70.4	
22	3.80	30.4	60.8	17.6 / 22.0	21.1 / 26.4	24.6 / 30.8	28.2 / 35.2	**31.7** / 39.6	**35.2** / 44.0	**42.2** / 52.8	**52.8** / 66.0	63.4 / 79.2	70.4 / 88.0	77.4	
24	4.52	36.2	72.4	19.2 / 24.0	23.0 / 28.8	26.9 / 33.6	30.7 / 38.4	34.6 / 43.2	**38.4** / 48.0	**46.1** / 57.6	**57.6** / 72.0	**69.1** / 86.4	76.8 / 96.0	84.5 / 106	96.0
27	5.73	45.8	91.6	21.6 / 27.0	25.9 / 32.4	30.2 / 37.8	34.6 / 43.2	38.9 / 48.6	43.2 / 54.0	**51.8** / 64.8	**64.8** / 81.0	**77.8** / 97.2	86.4 / 108	95.0 / 119	108

Upper line Bearing Values for each diameter bolt are Simple Bearing Values.
Lower line Bearing Values for each diameter bolt are Enclosed Bearing Values.
For areas to be deducted from a bar for one hole, see table on page 709.
For explanation of table, see Notes.
1 kilonewton may be taken as 0.102 metric tonne (megagramme) force.

AREAS IN SQUARE CENTIMETRES TO BE DEDUCTED FOR ONE HOLE THROUGH A MEMBER

Dia. of Hole in mm	\multicolumn{14}{c}{THICKNESS OF MEMBER AT HOLE—IN MILLIMETRES}
	5	6	8	10	12	15	18	20	22	25	28	30	32	35	40
10	0.50	0.60	0.80	1.00	1.20	1.50	1.80	2.00	2.20	2.50	2.80	3.00	3.20	3.50	4.00
11	0.55	0.66	0.88	1.10	1.32	1.65	1.98	2.20	2.42	2.75	3.08	3.30	3.52	3.85	4.40
12	0.60	0.72	0.96	1.20	1.44	1.80	2.16	2.40	2.64	3.00	3.36	3.60	3.84	4.20	4.80
13	0.65	0.78	1.04	1.30	1.56	1.95	2.34	2.60	2.86	3.25	3.64	3.90	4.16	4.55	5.20
14	0.70	0.84	1.12	1.40	1.68	2.10	2.52	2.80	3.08	3.50	3.92	4.20	4.48	4.90	5.60
15	0.75	0.90	1.20	1.50	1.80	2.25	2.70	3.00	3.30	3.75	4.20	4.50	4.80	5.25	6.00
16	0.80	0.96	1.28	1.60	1.92	2.40	2.88	3.20	3.52	4.00	4.48	4.80	5.12	5.60	6.40
17	0.85	1.02	1.36	1.70	2.04	2.55	3.06	3.40	3.74	4.25	4.76	5.10	5.44	5.95	6.80
18	0.90	1.08	1.44	1.80	2.16	2.70	3.24	3.60	3.96	4.50	5.04	5.40	5.76	6.30	7.20
19	0.95	1.14	1.52	1.90	2.28	2.85	3.42	3.80	4.18	4.75	5.32	5.70	6.08	6.65	7.60
20	1.00	1.20	1.60	2.00	2.40	3.00	3.60	4.00	4.40	5.00	5.60	6.00	6.40	7.00	8.00
21	1.05	1.26	1.68	2.10	2.52	3.15	3.78	4.20	4.62	5.25	5.88	6.30	6.72	7.35	8.40
22	1.10	1.32	1.76	2.20	2.64	3.30	3.96	4.40	4.84	5.50	6.16	6.60	7.04	7.70	8.80
23	1.15	1.38	1.84	2.30	2.76	3.45	4.14	4.60	5.06	5.75	6.44	6.90	7.36	8.05	9.20
24	1.20	1.44	1.92	2.40	2.88	3.60	4.32	4.80	5.28	6.00	6.72	7.20	7.68	8.40	9.60
25	1.25	1.50	2.00	2.50	3.00	3.75	4.50	5.00	5.50	6.25	7.00	7.50	8.00	8.75	10.00
26	1.30	1.56	2.08	2.60	3.12	3.90	4.68	5.20	5.72	6.50	7.28	7.80	8.32	9.10	10.40
27	1.35	1.62	2.16	2.70	3.24	4.05	4.86	5.40	5.94	6.75	7.56	8.10	8.64	9.45	10.80
28	1.40	1.68	2.24	2.80	3.36	4.20	5.04	5.60	6.16	7.00	7.84	8.40	8.96	9.80	11.20
29	1.45	1.74	2.32	2.90	3.48	4.35	5.22	5.80	6.38	7.25	8.12	8.70	9.28	10.15	11.60

SPACING OF HOLES IN COLUMNS, BEAMS AND TEES

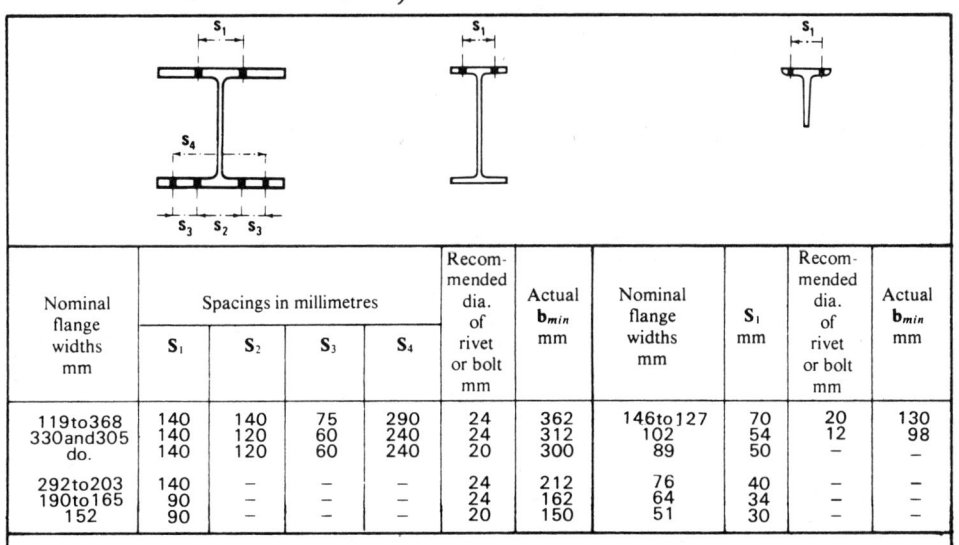

Nominal flange widths mm	Spacings in millimetres				Recommended dia. of rivet or bolt mm	Actual b_{min} mm	Nominal flange widths mm	S_1 mm	Recommended dia. of rivet or bolt mm	Actual b_{min} mm
	S_1	S_2	S_3	S_4						
119 to 368	140	140	75	290	24	362	146 to 127	70	20	130
330 and 305	140	120	60	240	24	312	102	54	12	98
do.	140	120	60	240	20	300	89	50	–	–
292 to 203	140	–	–	–	24	212	76	40	–	–
190 to 165	90	–	–	–	24	162	64	34	–	–
152	90	–	–	–	20	150	51	30	–	–

Note that the actual flange width for a universal section may be less than the nominal size and that the difference may be significant in determining the maximum diameter. The column headed b_{min} gives the actual minimum width of flange required to comply with Table 21 of BS 449:Part 2:1969.

The dimensions S_1 and S_2 have been selected for normal conditions but adjustments may be necessary for relatively large diameter fasteners or for particularly heavy weights of serial size.

SPACING OF HOLES IN CHANNELS

Nominal flange width mm	S_1 mm	Recommended dia. of rivet or bolt mm
102	55	24
89	55	20
76	45	20
64	35	16
51	30	10
38	22	–

SPACING OF HOLES IN ANGLES

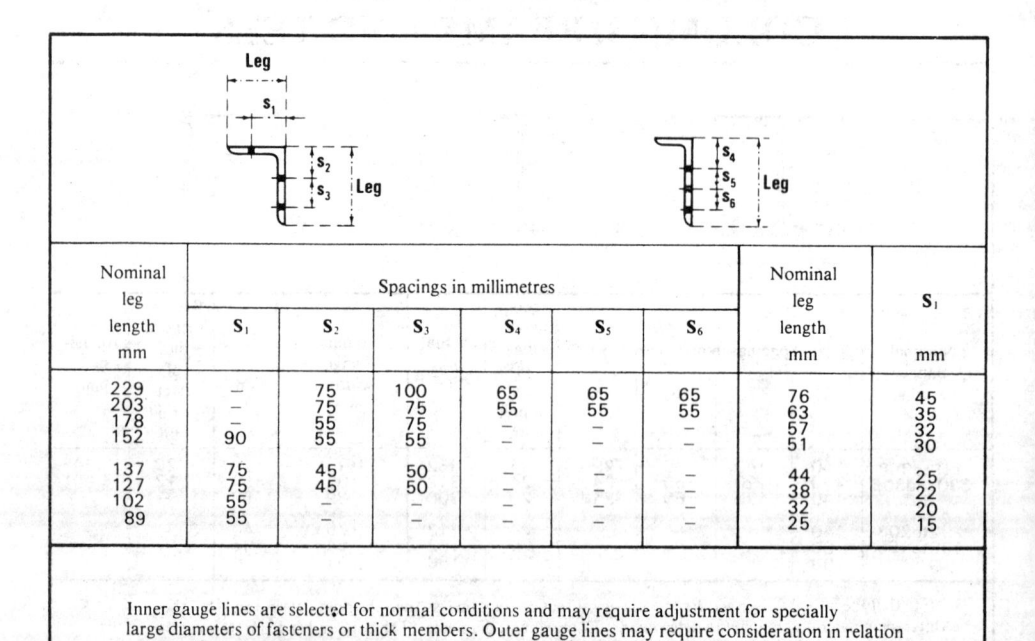

Nominal leg length mm	\multicolumn{6}{c}{Spacings in millimetres}	Nominal leg length mm	S_1 mm
	S_1	S_2	S_3	S_4	S_5	S_6		
229	–	75	100	65	65	65	76	45
203	–	75	75	55	55	55	63	35
178	–	55	75	–	–	–	57	32
152	90	55	55	–	–	–	51	30
137	75	45	50	–	–	–	44	25
127	75	45	50	–	–	–	38	22
102	55	–	–	–	–	–	32	20
89	55	–	–	–	–	–	25	15

Inner gauge lines are selected for normal conditions and may require adjustment for specially large diameters of fasteners or thick members. Outer gauge lines may require consideration in relation to a specified edge distance.

RIVETED CONNECTIONS

Riveted Connections

Although today most site connections are effected using black bolts or high strength friction grip bolts and shop fabrication is tending towards welding, riveted joints are still used for some types of connection. In view of this the following examples have been included.

Riveted Joints.

Tension Member
Using Grade 43 steel

Determine the number of 18 mm diameter shop driven rivets required to carry the load in the tension member shown in Fig. 1.

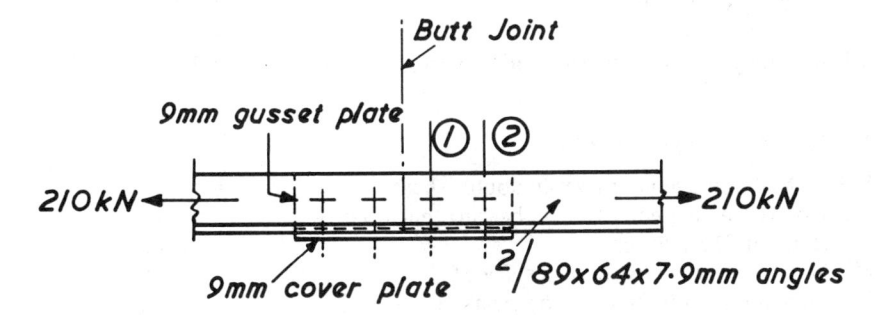

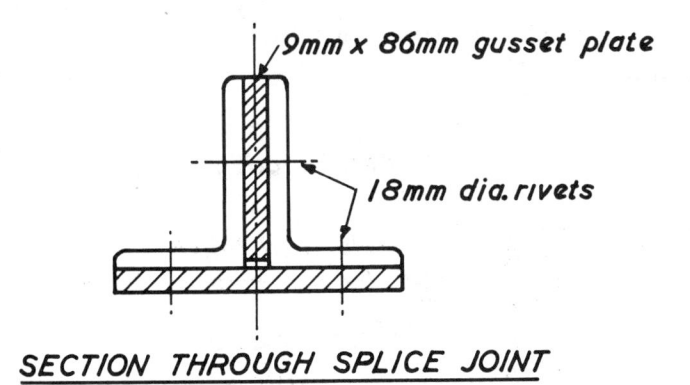

Fig. 1

Strength of Angles at 2–2

Area of 2/89 mm x 64 mm x 7.8 mm angles (2 x 1 137 mm²)	= 2 274 mm²
Less 4 – 20 mm diameter holes = 4(20 x 7.8)	= 632 mm²
∴ net area	= 1 642 mm²
∴ Maximum permissible tensile load in angles	= 1 642 x 155 N
	= 254.5 kN

Strength of gusset plates at section 1–1

Area of 9 mm thick gusset between angles	= 86 mm x 9 mm
	= 774 mm²
Less 1 – 20 mm diameter hole = 20 x 9	= 180 mm²
∴ net area	= 594 mm²
Area of 9 mm thick cover plate	= 137 mm x 9 mm
	= 1 233 mm²
Less 2 – 20 mm diameter holes = 2(20 x 9)	= 360 mm²
∴ net area	= 873 mm²
∴ Total net area of plates	= 594 + 873
	= 1 467 mm²
∴ Maximum permissible tensile load in plates.	= 1 467 x 155 N
	= 227.4 kN.

Strength of Rivets at 110 N/mm²

Value of 18 mm diameter rivet in double shear	= 69.1 kN
Value of 18 mm diameter rivet in bearing (enclosed bearing at 315 N/mm²)	= 56.7 kN
Maximum load taken by 2 rivets = 2 x 56.7	= 113.4 kN
Remainder of load to be taken by rivets in cover plate	= 210 – 113.4 kN
	= 96.6 kN
Value of 18 mm diameter rivet in single shear	= 34.6 kN
∴ No. required = $\dfrac{96.6}{34.6}$	= 2.79 kN.

Fig. 2

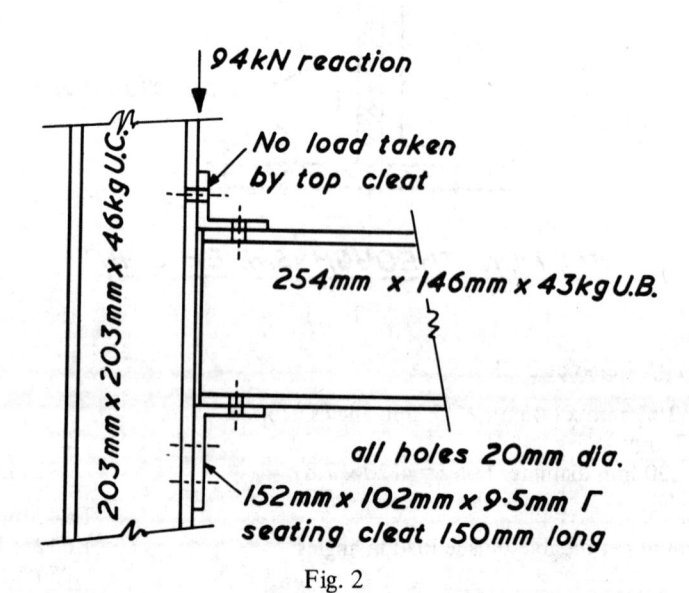

RIVETED CONNECTIONS 713

Use 4 No. 18 mm diameter rivets for practical reasons.

Beam End Connections. Simple Design Grade 43 steel.
Two Types (1) Seating cleat only.
(2) Web connection and seating cleat.
Type (1). Design a seating cleat for the loading shown in Fig. 2

Vertical load	=	94 kN
Value of 18 mm diameter rivet in single shear (20 mm finished diameter)	=	34.6 kN
Value of ditto in bearing (9.5 mm thick)	=	47.9 kN
∴ No. of rivets required $= \dfrac{94}{34.6}$	=	2.72.

Use 4 No. 18 mm diameter rivets.

Check web of 254 x 146 x 43 kg UB for buckling. See Fig. 3. See clause 28 (a) B.S. 449 : Part 2 : 1969.

The supporting bracket provides a stiff bearing of 9.5 mm which, with the dispersion of 45° to N.A. gives a total length B for buckling strength = (22.5 + 130) = 152.5 mm (scaled from diagram Fig. 3).

Slenderness ratio of web $= \dfrac{d}{t}\sqrt{3}$, where d = clear depth between roof fillets.
t = web thickness
$$= \dfrac{216}{7.3}\sqrt{3} = 51.2$$

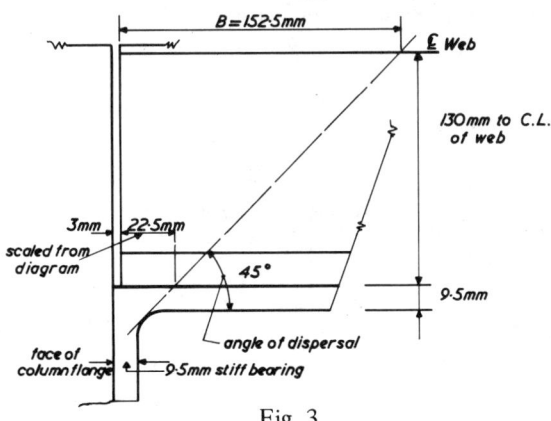

Fig. 3

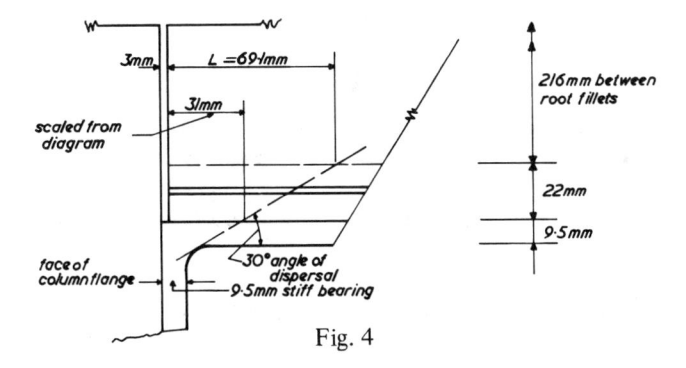

Fig. 4

714   CONNECTIONS

∴ Pc (Table 17(a) of B.S. 449) = 133 N/mm².
∴ permissible buckling load = $pc.t.B.$

$$= \frac{133 \times 7.3 \times 152.5}{10^3} = 148.06 \text{ kN}.$$

Check beam for bearing. Clause 27 (e) B.S. 449.
Length of bearing at junction of web and roof fillet

(see Fig. 4)

$$L = 31 \text{ mm} + \frac{22}{\tan 30°}$$

$$= 31 \text{ mm} + 38.1 \text{ mm}$$
$$L = 69.1 \text{ mm}$$

Allowable bearing stress = 190 N/mm²
∴ permissible bearing load = 190 × 69.1 × 7.3 N
= 95.8 kN

Check shear stress clause 23 (a) B.S. 449.
Allowable shear stress (Table 11)   = 100 N/mm²
Actual shear stress = $\dfrac{94 \times 10^3}{260 \times 7.3}$ = 49.5 N/mm².

These web capacities can be checked from the safe load tables.
∴ 254 mm × 146 mm × 43 kg UB is adequate for load.

Type (2).
Web connection with seating cleat.

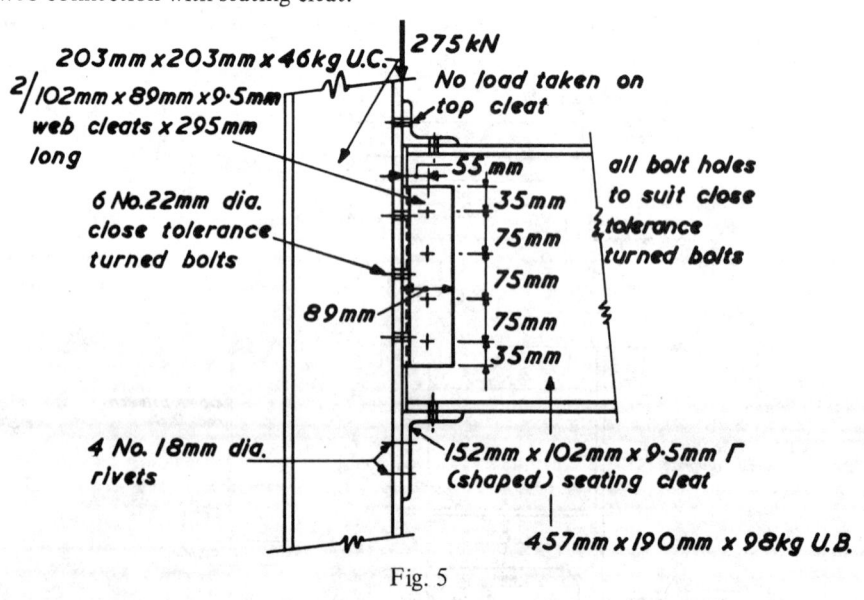

Fig. 5

vertical load = 275 kN.

RIVETED CONNECTIONS

Assuming size of seating cleat is limited to accommodate 4 No. 18 mm nominal (20 mm gross) diameter rivets only.

Load carried by seating cleat rivets $\qquad = 4 \times 34.6$
(taking critical value in single shear) $\qquad = 138.4$ kN

Remainder of load to be carried by web cleats $= 275 - 138.4 = 136.6$ kN
Using $2/102 \times 89 \times 9.5$ mm angle web cleats.
Check for eccentricity on web connection.

Vertical load per rivet $= \dfrac{136.6}{4} = 34.15$ kN.

Modulus of rivet group $= \dfrac{(2 \times 37.5^2) + (2 \times 112.5^2)}{112.5} = 250$ units.

$\therefore$ horizontal load per rivet $= \dfrac{136.6 \times 55}{250} = 30.05$ kN.

Vector sum $= \sqrt{34.15^2 + 30.05^2} = 45.49$ kN.
Value of 18 mm nominal diameter shop driven rivet $= 69.1$ kN
 (in double shear)
$\therefore$ 4 No. 18 mm nominal diameter shop driven rivets are adequate.

Connection of outstanding legs of web cleats to Column flanges. Using turned bolts in conjunction with rivets [see clause 48*b*, B.S. 449].
Value of 22 mm diameter close tolerance turned bolt in single shear $= 36.1$ kN.
$\therefore$ No. required $= 136.6/36.1 = 3.78$.

Use 6 – No. 22 mm diameter close tolerance turned bolts.

Check for buckling, bearing and shear as in previous example, or refer to web capacities given in Handbook on Structural Steelwork.

Eccentric Loading

When loads act at a distance from their support (i.e., eccentric about centre of gravity of rivet group), bending as well as shear stresses are set up, and these have to be dealt with in the following manner.

There are two general cases of moments on rivet groups:

> Case I. That in which the moment is applied in the plane of the connection and the centre of rotation is at the C.G. of the rivet group.
>
> Case II. That in which the moment is applied at right angles to the plane of connection. It will be assumed here that the centre of rotation is at the centre of the lowest rivet, although engineers are not unanimous on this point.

Consider the first case, an example of which is shown in Fig. 6.
Let $P =$ any vertical load,
 $e =$ the eccentricity of application of P about the C.G. of the rivet group,
 $n =$ the number of rivets in the group,
 $x, y =$ the co-ordinates of any rivet using the C.G. of the rivet group as origin,
  $z =$ the distance of any rivet from the C.G.
   $= \sqrt{(x^2 + y^2)}$,

CONNECTIONS

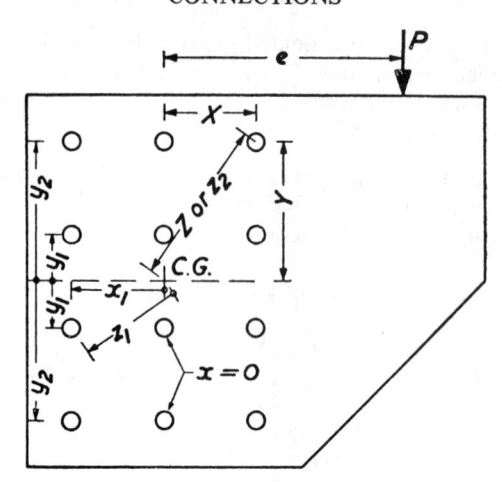

Fig. 6

X, Y = the co-ordinates of the rivet farthest from the C.G.
Z = the distance of this rivet from the C.G.
$= \sqrt{(X^2 + Y^2)}$

and F_b = the load on any rivet due to the moment $M = P \cdot e$
$= o \times z$, where o = load due to moment on a rivet at unit distance from the C.G. of the group.

Then the moment resisted by this rivet

$= $ load $\times$ lever arm to C.G.
$= o \times z^2 = o \times (x^2 + y^2)$.

Hence, the total moment resisted by the group

$$= o(\Sigma x^2 + \Sigma y^2) = P \cdot e$$

i.e.,
$$o = \frac{P \cdot e}{(\Sigma x^2 + \Sigma y^2)}.$$

Therefore the load due to the moment on the rivet receiving the maximum load

$= F_b = o \times Z$

$$= \frac{P \cdot e \times Z}{(\Sigma x^2 + \Sigma y^2)} = P \cdot e \times \frac{Z}{(\Sigma x^2 + \Sigma y^2)}$$

or
$$F = P \cdot e \div \frac{(\Sigma x^2 + \Sigma y^2)}{Z}.$$

Now consider the most heavily loaded rivet, as shown in Fig. 7.
Let F_s = the direct load on any rivet due to the load P
$= P/n$.

Then the resultant load F_R on the rivet is a vector quantity derived from the moment load F_b and the direct load F_s. The value of F_R can, of course, be found graphically and it is often convenient to do so.

RIVETED CONNECTIONS 717

Fig. 7

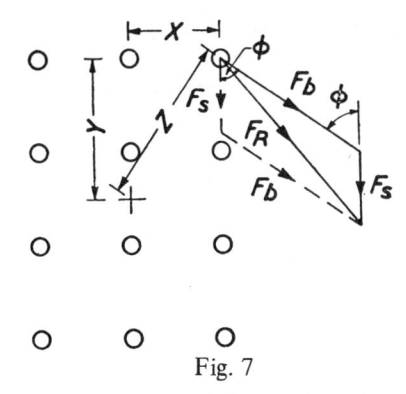

The horizontal component of $F_b = (Y/Z) \cdot F_b$, while the vertical component $= (X/Z) \cdot F_b$ (since the line of action of F_b is at right angles to the lever arm).

Hence
$$F_R = \sqrt{\left[\left(F_s + \frac{X}{Z} \cdot F_b\right)^2 + \left(\frac{Y}{Z} \cdot F_b\right)^2\right]}$$
$$= \sqrt{\left(F_s^2 + \frac{2X \cdot F_s \cdot F_b}{Z} + F_b^2\right)}$$
$$= \sqrt{(F_s^2 + F_b^2 + 2F_s \cdot F_b \cos \phi)}.$$

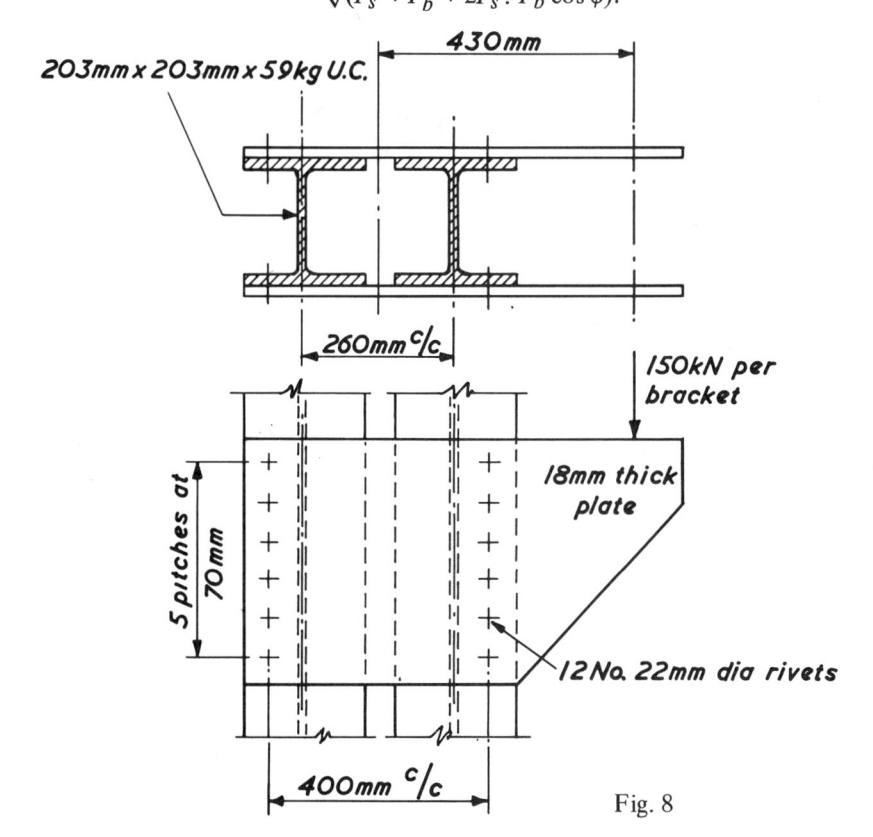

Fig. 8

718 CONNECTIONS

It is required to calculate the stress in the most highly stressed rivets, i.e., top and bottom right hand rivet in Fig. 8. For 205 mm flanges the spacing of the rivet holes is 140 mm so that the horizontal pitch for the two vertical rows of rivets is 260 + 140 = 400 mm.

Using 22 mm nominal diameter rivets at 70 mm vertical pitch the value $P \times e$ for each bracket is 150 × 430 = 64 500 kNmm.

The modulus of the rivet group, i.e., modulus $= \dfrac{(\Sigma x^2 + \Sigma y^2)}{z}$. See Fig. 9.

$$= \frac{651\,500}{265.75} = 2\,451.6 \text{ units.}$$

∴ $$F_b = \frac{P.e.}{\text{mod. of group}} = \frac{64\,500}{2\,451.6} = 26.3 \text{ kN}$$

now $$F_s = \frac{P}{n} = \frac{150}{12} = 12.5 \text{ kN.}$$

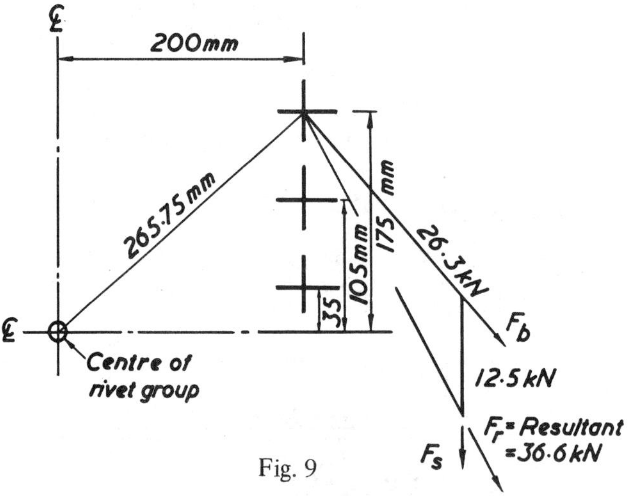

Fig. 9

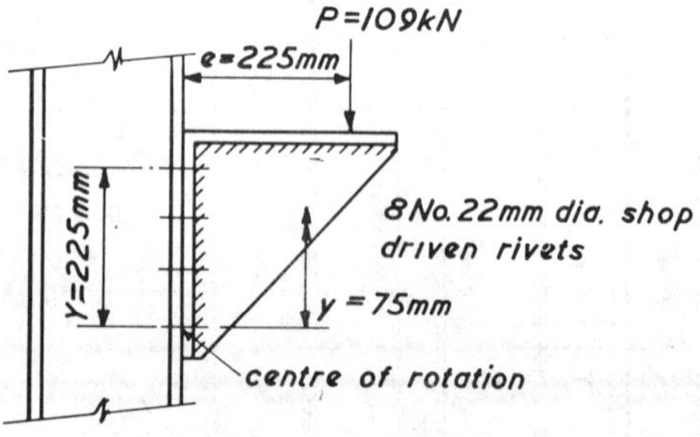

Fig. 10

If the values of F_b and F_s are drawn to scale as shown in Fig. 9 then the value of F_r may be found graphically to be approximately = 36.6 kN.

This is within the single shear value of a 22 mm rivet (24 mm diameter gross), i.e., 49.8 kN.

RIVETED CONNECTIONS

Figure 10 shows an example of the second case of a rivet group where the applied moment is at right angles to the plane of the connection. By analogy with the foregoing calculations it will be apparent that for one vertical row of rivets the tensile load on the topmost rivet due to the moment $P.e.$ will be

$$m = \frac{Pe \times Y}{\Sigma y^2}$$

Load P acting at a distance e, = 109 kN
Moment $P \times e$ = 109 × 225 = 24 525 kNmm
   = 78 750/225
Value of $\frac{\Sigma y^2}{Y}$ = 350 units
(for one row)

∴ maximum tension in top rivet $= \frac{24\ 525}{2 \times 350}$
(for two rows)

∴ F_t = 35.0 kN

direct force $F_s = \frac{P}{n}$

$= \frac{109}{8}$ = 13.6 kN

Maximum stresses

$$f_t \text{ tension per rivet} = \frac{35.0 \times 10^3}{452.4}$$

$$= 77.4 \text{ N/mm}^2$$

$$f_s \text{ shear per rivet} = \frac{109 \times 10^3}{8 \times 452.4}$$

$$= 30.1 \text{ N/mm}^2$$

The resultant stress f_r on the top rivet is a vector quantity derived from the stress f_b and the direct stress f_s (see Fig. 11) = 83.0 N/mm².

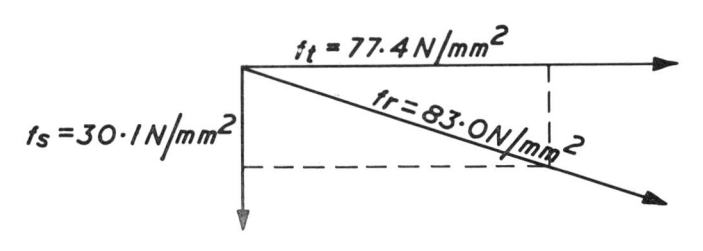

Fig. 11

Permissible stress B.S. 449 Table 20 (250 N/mm² yield stress)
   = 100 N/mm².
∴ Use 8 No. 22 mm diameter shop driven rivets.

Design of Gusset Plate

This particular part of a connection is often accepted as adequate. It is however possible in some types of connection such as a lattice girder, or a roof truss of large span, that the gusset plate could be overstressed due to the loads in the members which are connected to the gusset being out of balance. Moments are developed in the gusset plate, which must be analysed to check the stresses arising from this moment. The following example will illustrate this.

Eccentric Loading on Latticed Girder

It is required to design the connection for the two members of a lattice girder meeting as shown in Fig. 12.

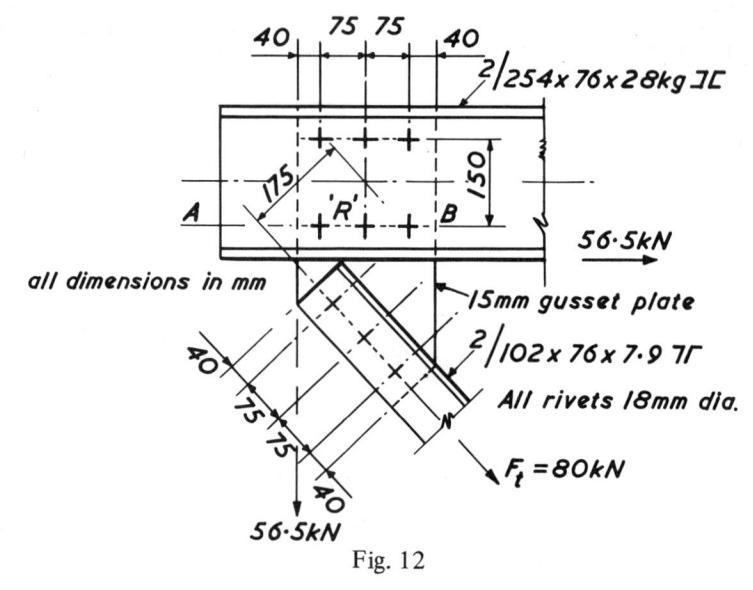

Fig. 12

Applied moment on connection due to 80 kN load acting at 175 mm from centre of rivet group

$$80 \times 175 \times 10^{-3} = 14 \text{ kNm}.$$

Assuming 3 m long top panel and depth of girder 3 m the relative stiffness are as follows:

I/L for compression boom $= \dfrac{6\,734}{300}$ $= 22.4467$

I/L for tension member $= \dfrac{276}{424.3}$ $= \dfrac{0.6505}{23.0972}$

Stiffness factor compression boom $= \dfrac{22.4467}{23.0972} = 0.9718$

Stiffness factor tension member $= \dfrac{0.6505}{23.0972} = 0.0282$

RIVETED CONNECTIONS   721

∴ moment taken by rivet group $= 14 \times 0.9718$
                                          $= 13.605$ kNm
  Value of rivet group (as example Fig. 8) $= (x^2 + y^2)$
                                          $= 530$ units.

  Load on rivet R due to moment applied $= \dfrac{13\,605}{530}$

                                          ∴ $F_b = 25.67$ kN

  direct load per rivet $F_s = \dfrac{80}{6}$ $= 13.33$ kN

  resultant value F_r on rivet R $= 25.67 + 13.33$
                                          $= 39.0$ kN

  Value of 18 mm nominal diameter rivet in D.S. $= 69.1$ kN
  (Shop driven 20 mm finished diameter.)

  Value of 18 mm diameter rivet in bearing $= 94.5$ kN
  ∴ Use 6 No. 18 mm nominal diameter rivets.

Strength of Gusset Plate

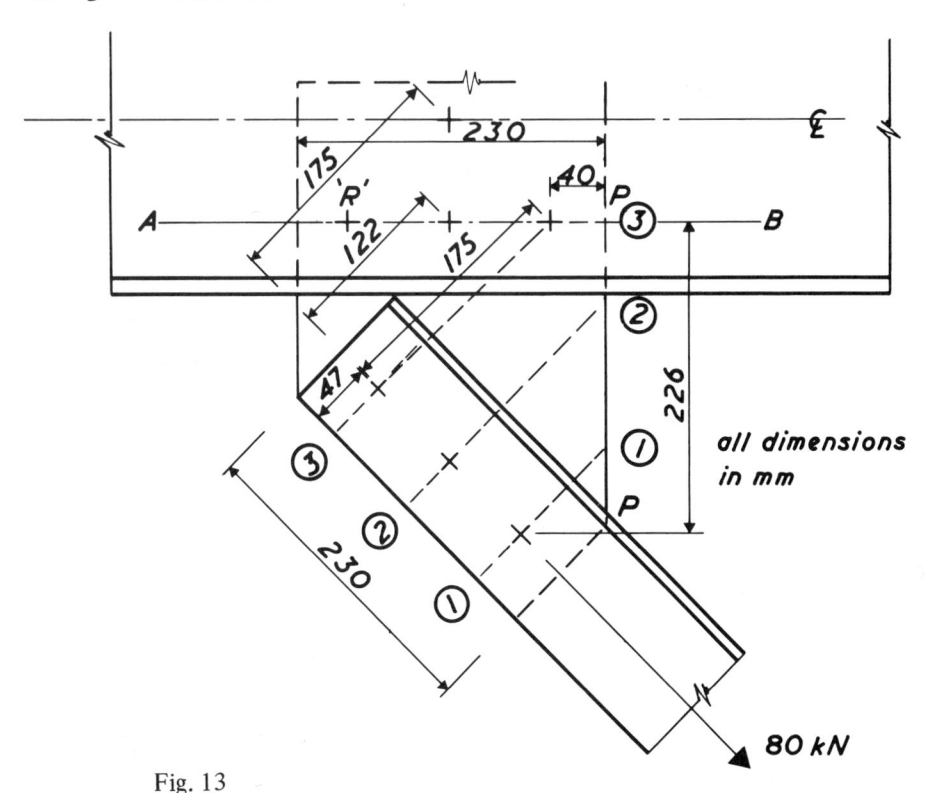

Fig. 13

Load at section 1–1 one rivet = 26.7 kN
Load at section 2–2 two rivets = 53.3 kN
Load at section 3–3 three rivets = 80.0 kN
Area at section 3–3 = (47 + 175 + 40)15
= 3 930 mm²

∴ Net area = 3 930 – 2(20 × 15) = 3 330 mm²

$$\left.\begin{array}{l}\text{permissible load at}\\ \text{section 3–3}\end{array}\right\} = 4.38 \times 9.5 = \frac{3\,330 \times 155}{10^3} = 516.15 \text{ kN}$$

Strength along line $A.B.$

$$\left.\begin{array}{l}\text{vertical component}\\ \text{horizontal component}\end{array}\right\} = \frac{80}{\sqrt{2}} = 56.5 \text{ kN}$$

$$\left.\begin{array}{l}\text{horizontal shear stress}\\ \text{vertical shear stress}\end{array}\right\} = \frac{\text{load}}{\text{net area}}$$

$$= \frac{56.5 \times 10^3}{[230 - (3 \times 20)]15} = 22.2 \text{ N/mm}^2$$

$B.M.$ about ℄ of $A.B.$ $= \frac{80 \times 122}{10^3}$ = 9.76 kNm.

Z $A.B.$ of plate

I_{xx} plate $= \frac{15 \times 230^3}{12 \times 10^3}$ = 1 521 cm⁴

I_{xx} holes (ignore centre hole and I_{cg} of holes) = $I_{cg} + AD^2$

$$= 2(15 \times 20) \times \frac{75^2}{10^4} = 337.5 \text{ cm}^4$$

∴ net I_{xx} at $A.B.$ = 1 521 – 337.5
= 1 183.5 cm⁴

∴ net Z_{xx} $= \frac{1\,183.5 \times 10^4}{115}$ = 102 913 mm³

stress fb_{ab} $= \frac{9.76 \times 10^6}{102\,913}$ = 94.8 N/mm²

Max. Tensile stress = 94.8 + 22.2 = 117 N/mm²
Max. Compressive stress = 94.8 – 22.2 = 72.6 N/mm².
 If the vertical edge of the gusset P–P is treated as a strut fixed at both ends (see Fig. 13), then the effective length = 0.7 × 226 = 158 mm.

r_{yy} for 15 mm thick plate = 4.33 mm.

∴ $\frac{l}{r_{yy}} = \frac{158}{4.33} = 36.5.$ ∴ P_c = 140 N/mm².

∴ allowable compressive and tensile bending stresses are 140 and 165 N/mm² respectively, and the gusset is adequate. It must be emphasized that usually no analysis is carried out for gusset plates of the size shown, and the example given is for demonstration only.

MOMENT CONNECTIONS

Wind Moment Connection

When designing large multi-storey steel framed buildings, the joints between the beams and columns have to be capable of resisting the moments due to wind forces. In most cases the applied moment is large, and consequently the usual type of connection is unable to resist the forces developed, i.e., force $F = M/D$ where M = bending moment, and D = depth of the beam. It is this force F in the flanges of the beam which requires a connection of the type illustrated in the following example.

Example: Split Beam Flange Connection

Vertical load = 120 kN
Moment due to Wind = 100 kNm
Design web cleats to take vertical load.
Value of 22 mm nominal diameter rivet in bearing = 68.7 kN
  (enclosed bearing 9.1 mm thick)

$\therefore$ No. required $= \dfrac{120}{68.7}$ = 1.74

Try 3 No. 22 mm nominal diameter rivets.

Check for eccentricity of reaction on rivet group.
(see Moment Fig. 14) = $120 \times 55/10^3$ = 6.6 kNm
Modulus value of rivet group $= \dfrac{75^2 + 75^2}{75}$ = 150 units
(assuming vertical spacing 75 mm)
$\therefore F_b$ (moment) = 6 600/150 = 44 kN
F_s (vert.) = 120/3 = 40 kN
Value F_r (found graphically) = 59.5 kN

Design of connection of outstanding legs to column flange
Value of 24 mm diameter bolt in single shear = 36.2 kN
$\therefore$ No. required = 120/36.2 = 3.3

Use 4 No. 24 mm diameter black bolts.

Design of Connection of 457 x 152 UB cutting to flange of Column.

Tension force $P = \dfrac{100 \times 10^3}{467.8}$ = 214 kN

Area at bottom of thread of 20 mm diameter bolt = 245 mm²
Safe load = $245 \times (130 + 25\%)/10^3$ = 39.8 kN

$\therefore$ No. required $= \dfrac{214}{39.8}$ = 5.37

Use 8 No. 20 mm diameter black bolts.

Design of bolts connecting horizontal leg to beam flange

Flange shear force $= \dfrac{100 \times 10^3}{457}$ = 219 kN

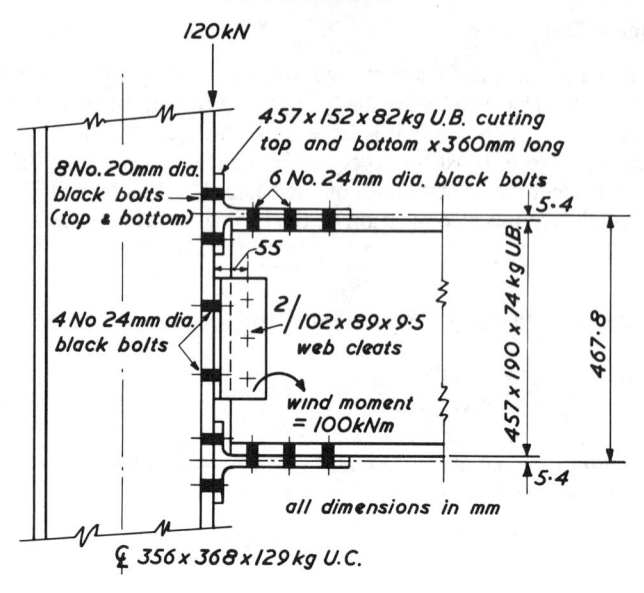

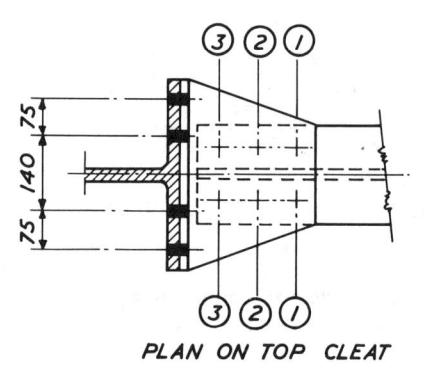

Fig. 14

Value of 24 mm diameter black bolt in single shear = 36.2 kN

plus 25% for wind $= \underline{ 9.1 \text{ kN}}$
$ 45.3 \text{ kN}$

No. required $ = \dfrac{219}{45.3} = 4.84$

∴ Use 6 No. 24 mm diameter black bolts in flange.

Strength of Stem

Load from 2 No. 24 mm diameter bolts at 1–1 = 2 × 36.5 = 73 kN
      4 No. 24 mm   2–2 = 4 × 36.5 = 146 kN
      6 No. 24 mm   3–3 = 6 × 36.5 = 219 kN

COLUMN BASES 725

Net area at section	1–1 = {227 – (2 x 27)}10.7	= 1 851 mm²
	2–2 = {282 – (2 x 27)}10.7	= 2 440 mm²
	3–3 = {338 – (2 x 27)}10.7	= 3 039 mm²
Permissible load at	1–1 = 1 851 x (130 + 25%)/10³	= 300.8 kN
	2–2 = 2 440 x (130 + 25%)/10³	= 396.5 kN
	3–3 = 3 039 x (130 + 25%)/10³	= 493.8 kN.

Thickness required at root of flange of joist cutting. See Fig. 15.

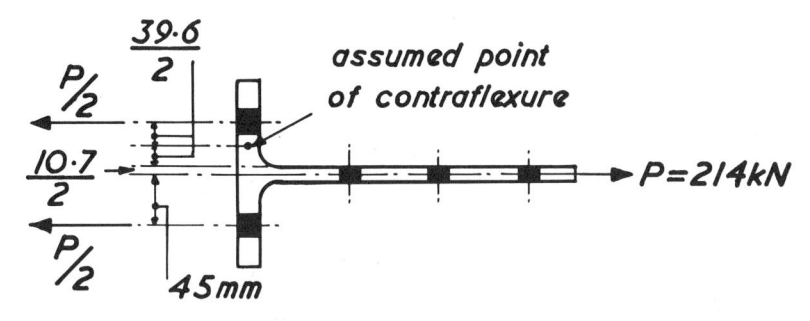

Fig. 15

BM on flange $= \dfrac{214}{2} \times \dfrac{39.6}{2} = 2\,124.0$ kNm

Modulus at root $= \dfrac{368 \times 18.9^2}{6} = 21\,909$ mm³

$\therefore F_b \qquad = \dfrac{2\,124.0 \times 10^3}{21\,909} = 96.9$ N/mm²

Permissible stress = 165 + 25%   = 206.25 N/mm²

∴ 457 x 152 x 82 kg UB cutting is adequate.

Check BM on Column flanges.

It may be necessary to place stiffeners between column flanges to ensure that the loads from the top and bottom flange cuttings are dispersed.

Another method of dealing with this joint is to use a welded connection of the type illustrated in the example on page 752.

Column Bases

The type of column base required for large multi-storey columns, with heavy loads of say 400 kN and upwards, is a slab. These slabs vary in thickness from about 18 mm to say 100 mm thick according to the requirements, the thickness being calculated in accordance with clause 38 (b) B.S. 449. They are machined on the top face to give perfect bearing contact with the bottom of the column, which is also machined, and fastened to the column by angle cleats sufficient to hold the base in position.

726 CONNECTIONS

Gusseted, or built up bases, are required only when bending moments are applied to the base of the column as illustrated in the following example.

Design of Stanchion Base carrying vertical load plus Wind Moment

W = Vertical load on stanchion base.

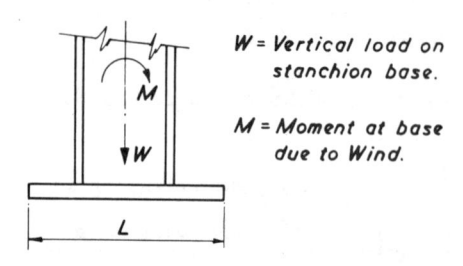

M = Moment at base due to Wind.

Fig. 16

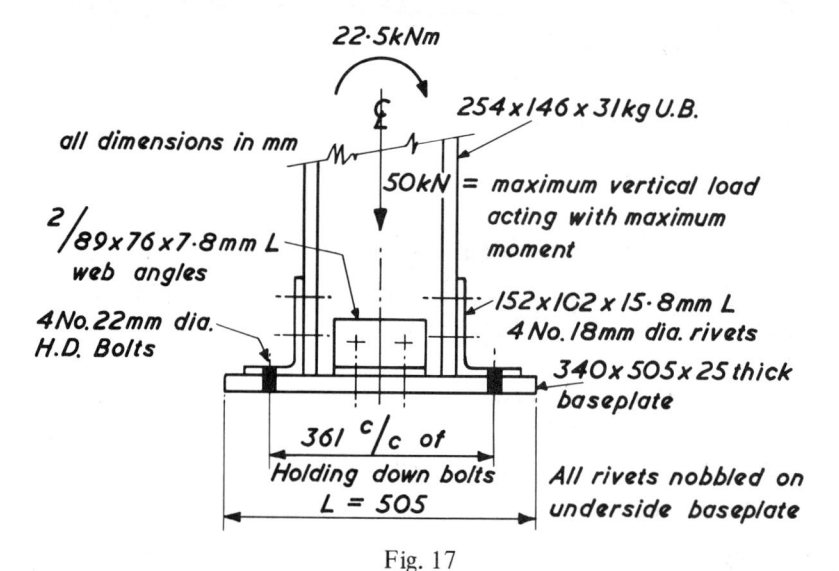

22.5kNm

254 x 146 x 31kg U.B.

all dimensions in mm

50kN = maximum vertical load acting with maximum moment

2/89 x 76 x 7.8mm L web angles

4 No. 22mm dia. H.D. Bolts

152 x 102 x 15.8mm L
4 No. 18mm dia. rivets

340 x 505 x 25 thick baseplate

361 c/c of Holding down bolts
L = 505

All rivets nobbled on underside baseplate

Fig. 17

This type of base is common to single storey shed buildings, i.e., dead and live loads are small in comparison to the overturning moments due to wind pressure on the side of the building.

Example

W = vertical load on Stanchion Base of which 15 kN = dead load only.

Ends of stanchion and gussets are not faced for bearing. (Clause 38 (*a*) B.S. 449.)

Length of base plate parallel to stanchion web = 251 + 204 + 50 = 505 mm (assuming two 152 x 102 x 15.8 mm base cleats).

Length parallel to stanchion flanges (assumed) = 300 mm.

$$\text{Position of resultant upward thrust} = \frac{M}{W} = \frac{22\,500}{50}$$

$$= 450 \text{ mm.}$$

COLUMN BASES

Where the resultant thrust falls within the middle third, i.e., $M/W \not> L/6$ there will be no tension on base plate. If, as in this example M/W exceeds $L/6$ it is not possible for tension to develop between baseplate and the concrete, we must draw an analogy to a reinforced concrete section which will be approximate.

Using the reinforced concrete beam analogy, assuming the two stresses occur simultaneously, and the modular ratio $m = 15$

then depth to N.A. $= \dfrac{m \cdot f_c \cdot d}{m \cdot f_c + f_s}$

where f_s (Grade 4.6 bolts, tension) = 130 + 25% = 162.5 N/mm²

Concrete stress, f_c at 4 000 kN/m² + 25% = 5 000 kN/m²

  = 5.0 N/mm²

∴ depth d to N.A. $= \dfrac{15 \times 5.0 \times 433}{(15 \times 5.0) + 162.5}$ = 136.7 mm.

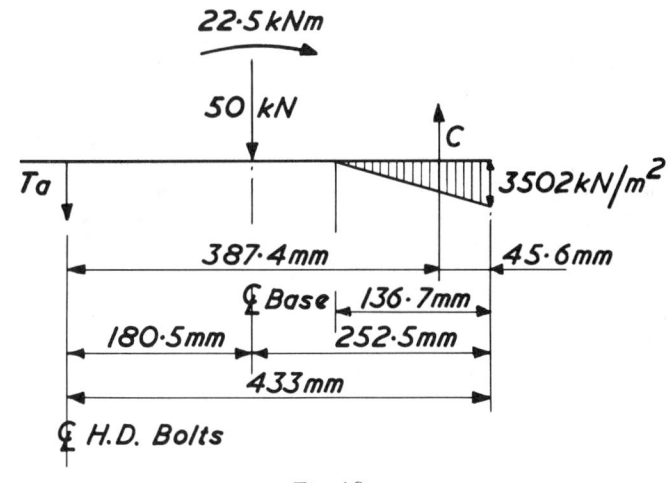

Fig. 18

centre of compression for a triangular distribution

$$= 136.7/3 = 45.6 \text{ mm}$$

take moments about Ta, ℄ of H.D. bolts.

$$C = \frac{22.5 \times 10^3 + 50 \times 180.5}{387.4} = 81.38 \text{ kN}$$

$$Ta = 81.38 - 50 = 31.38 \text{ kN.}$$

breadth of base plate $\quad b = \dfrac{81.38 \times 2 \times 10^6}{136.7 \times 5\,000} = 238 \text{ mm}$

(i.e., upward force = downward force)
increase breadth by 40% (Clause 10 (b) B.S. 449).

$$= 238 \times 1.4 = 333 \text{ mm.}$$

Holding Down Bolts

Minimum vertical load = 15 kN

$$C = \frac{22.5 \times 1\,000 + 15 \times 180.5}{387.4} = 65.1 \text{ kN}$$

$$Ta = 65.1 - 15.0 = 50.1 \text{ kN}$$

which is greater than with maximum vertical load of 50 kN. (Clause 10 (*b*), B.S. 449.) This value of T_a must be increased by the factor of 1.4.
∴ Maximum tensional load in H.D. Bolts.

$$= 50.1 \times 1.4 = 71.4 \text{ kN}.$$

Permissible value in tension for

22 mm diameter bolts = (130 + 25%) x 303/10³ = 49.2 kN
for 2 bolts   = 2 x 49.2   = 98.4 kN

∴ Use 2 No. 22 mm diameter H.D. Bolts each side.

The above 4-bolts will be adequate to resist the horizontal shear caused by the Wind loading, in this case 10 kN acting at eaves level 2.25 m above base level.

Flange rivets: As end of stanchion will be cold sawn, and underside of base cleats are not machined for bearing on baseplate, the flange rivets will be designed to carry maximum vertical load together with maximum vertical load due to wind moment, the rivets taking all the load.

$$\therefore \text{ vertical shear} = \frac{50}{2} + \frac{22.5 \times 10^3}{251} = 114.6 \text{ kN}$$

Value of 18 mm nominal diameter shop rivet in single shear
(Min. plate thickness 8 mm)
$$= 34.6 + 25\% = 43.25 \text{ kN}$$
No. required $= 114.6/43.25 = 2.65$

∴ Use 4 No. 18 mm nominal diameter rivets in each flange.

Base Angle Thicknesses

Assume base angle acts independently of base plate and for an angle of dispersion from H.D. bolts of 120° see Fig. 19.

Maximum B.M. in angle = tension load in one H.D. Bolt x lever arm, force in bolt not increased by factor of 1.4.

$$= \frac{50.1}{2}(55 - t) \text{ kNm.}$$

Effective length $le = (55 - t) \times 2\sqrt{3}$ mm

Resistance moment of (102 leg) angle $= (55 - t) \times 2\sqrt{3} \times \frac{t^2}{6} \times P_{bc}$ Nmm.

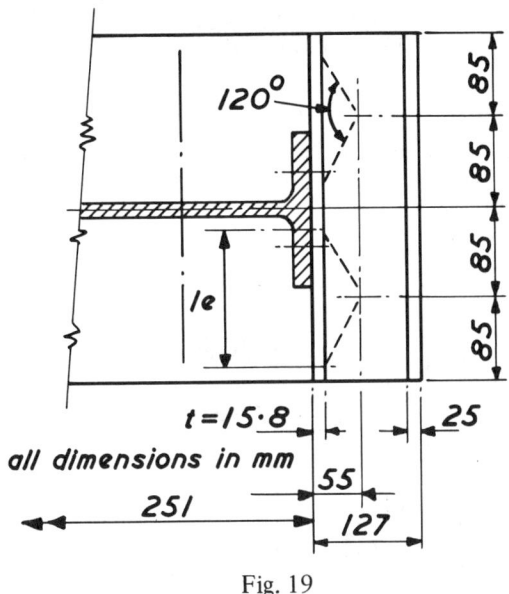

COLUMN BASES

Fig. 19

Where P_{bc} = 165 N/mm² + 25% due to wind
    = 206.25 N/mm²
B.M. = Moment of resistance.

$$\therefore T_a(55 - t) = (55 - t) \times 2\sqrt{3} \times \frac{t^2}{6} \times 206.25$$

Simplifying gives $t^2 = \dfrac{\sqrt{3} \times T_a}{N \times 206.25} = \dfrac{1.732 T_a}{206.25 N}$

Where N = Number of H.D. Bolts and T_a = actual load tension

$$\therefore \quad t = 0.0916 \sqrt{\frac{50\,100}{2}} \qquad = 14.5 \text{ mm}$$

Use angle 152 x 102 x 15.8 mm.

Length $le = 2\sqrt{3}(55 - t)$ = 2 x 1.732 x 39.2
                                 = 140.3 mm
∴ Minimum length of base angle   = 2 x 140.3 = 280.6 mm.
∴ Use 340 mm breadth of base plate (see page 727).

Base Plate Thickness

Pressure distribution resulting from maximum vertical load, and maximum wind acting together.

Maximum pressure

load on bolts C = 81.38 kN

730   CONNECTIONS

average pressure $= \dfrac{81.38 \times 10^6}{136.7 \times 340} = 1\,751$ kN/m²

∴ maximum pressure
at base edge $= 1\,751 \times 2 = 3\,502$ kN/m².

Consider base as a beam supported at the centre of the base angle vertical legs and loaded with the pressure diagram shown. Fig. 20. This is an approximate method, but is much easier to use.

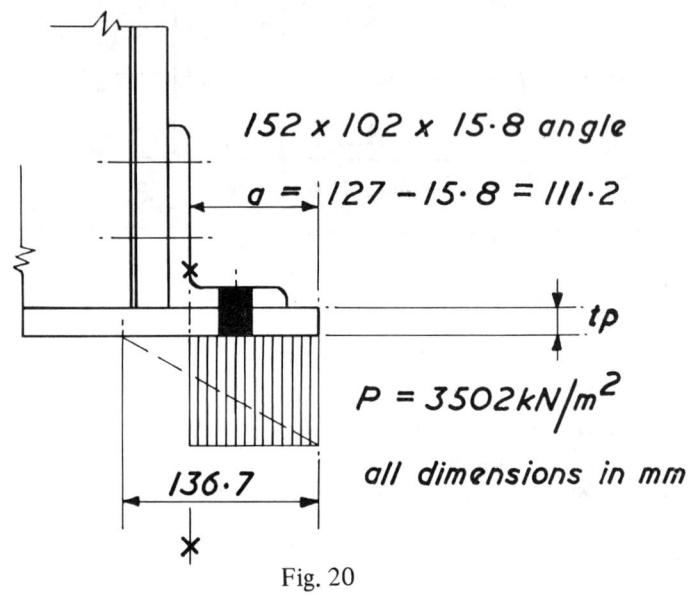

Fig. 20

Length of overhang of baseplate beyond angle thickness
$$a = 127 - 15.8 = 111.2 \text{ mm}$$

Assume maximum pressure P
as constant over this length $= 3\,502$ kN/m²

B.M. at outside face of
vertical leg of angle point XX $= \dfrac{P \times a^2}{10^6 \times 2}$ kNmm for 1 mm breadth of plate

Moment of resistance of plate
at XX for 1 mm breadth $= \dfrac{P_{bc} \times tp^2}{6}$

∴ $\dfrac{(P_{bc} + 25\%) \times tp^2}{6} = \dfrac{P \times a^2}{10^6 \times 2}$

∴ $tp^2 = \dfrac{P \times a^2 \times 6}{206.25 \times 10^6 \times 2} = \dfrac{Pa^2}{68.75 \times 10^6}$

$tp = \dfrac{a}{8\,291.6}\sqrt{P} = \dfrac{111.2}{8\,291.6}\sqrt{3\,502\,000}$

$= 25.1$ mm

∴ thickness of base required $= 26$ mm.

COLUMN SPLICES   731

Using the more accurate method, that is the base is considered as a beam, supported at the centre of the thickness of the vertical legs of the base angles and loaded with the triangle of pressure shown in Fig. 20,

the thickness required = 20.8 mm.

In view of the small amount of difference, and considering the time taken in this latter method, it may be considered reasonable to adopt the quicker method as illustrated.
Add web cleats to base to assist in the distribution of the load to the base plate.

Column Splice

The design of multi-storey columns is to some degree affected by questions of transport of columns to the site and also regulated by the length in which the sections are normally available.

Because of this it is sometimes desirable to design the length of column through two or more storeys using a size of universal column which could be spliced at the appropriate level to a length of lighter section. Clause 32 (*b*), B.S. 449 sets the requirements of such spliced joints and the following example illustrates the design of a typical splice for the stanchion of a multi-storey building.

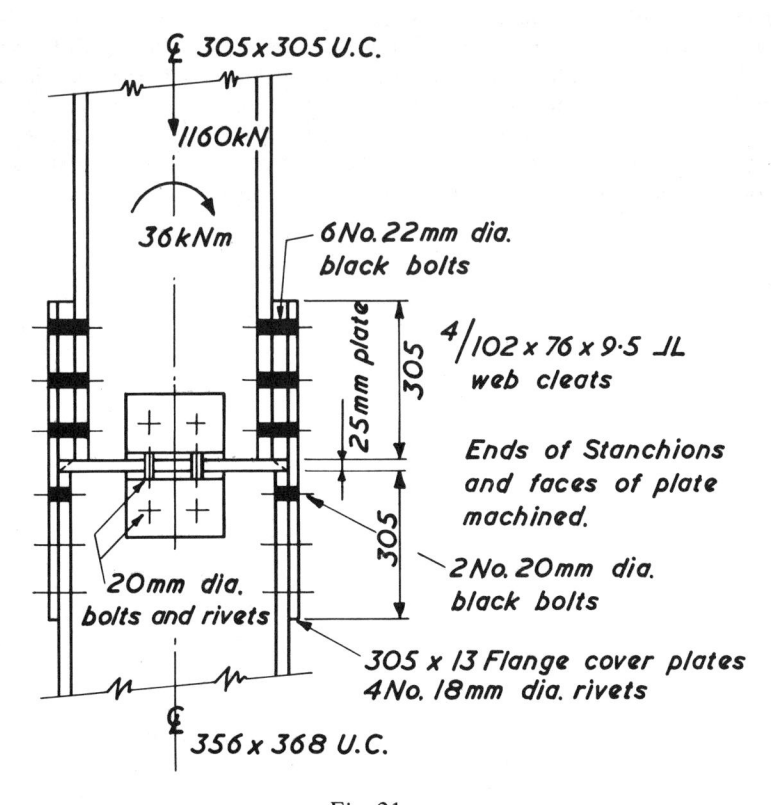

Fig. 21

Design of Stanchion Splice

Vertical load $P \begin{pmatrix} \text{dead load + live load} \\ \text{560 kN + 600 kN} \end{pmatrix}$ = 1 160 kN

Wind Moment M = 36 kNm
Shear (wind loading) = 20 kN

Ends machined for bearing, check for tension in column flanges due to wind moment.

$$\text{Vertical load in flanges due to dead load only} = \frac{560}{2} = 280 \text{ kN}.$$

Compression and tension forces in flanges due to wind moment

$$= \frac{36 \times 10^3}{308} = 117 \text{ kN} \quad \text{(upper length)}$$

i.e., no tension present in flanges.

Maximum compression force $= \frac{1\,160}{2} + 117 = 697$ kN.

As no tension stresses are developed, it is only necessary that the splice plate be sufficient to hold the connected members in place.

It is usual to make the length of the flange cover plates each side of the splice, equal to the flange width or 230 mm whichever is the greater, and the thickness, half the thickness of the upper column flange or 10 mm whichever is the greater. Web cleats and a division plate must be provided, when the depth of the section changes, to give a load dispersal of 45°. The width of the flange cover plates is made a nominal size of 300 mm or to suit upper column size.

Notes on High Strength Friction Grip Bolts

Dimensions and Properties The bolts, nuts and washers must comply with BS 4395: 1969, 'High strength friction grip bolts and associated nuts and washers for structural engineering, Part 1. General grade'.

Application This must conform to BS 4604: 1970, 'The use of high strength friction grip bolts in structural steelwork, Part 1. General grade'.

Length of Bolts The length of the bolt should be calculated by adding to the grip the allowance given in Table 1 below to allow for the thickness of one nut and of one flat washer and for sufficient protrusion of the bolt end. Where taper washers are used instead of flat washers an additional allowance of up to 2 mm for each taper washer may be necessary when calculating the length of the bolt.

Table 1. Bolt Length Allowances BS 4604: Part 1: 1970

Nominal size and thread diameter	M12	M16	M20	M22	M24	M27	M30	M36
Allowance to be added to the grip in mm	22	26	30	34	36	39	42	48

Surfaces in Contact At the time of assembly, surfaces in contact must be free of paint or any other applied finish, oil, dirt, loose rust, loose scale, burrs and other defects which would prevent solid seating of the parts or would interfere with the development of friction between them.

If any other surface condition, including a machined surface, is desired, it will be the responsibility of the Engineer to determine the slip factor to be used in the particular case.

Holes in Members All holes must be drilled and burrs must be removed. Where the number of plies in the grip does not exceed three, the diameters of the holes must be 2 mm larger than those of the bolts for bolt diameters up to 24 mm, and 3 mm larger than those of the bolts for diameters larger than 24 mm. Where the number of plies in the grip exceeds three, the nominal diameters of the holes in the two outer plies must be as above and the diameters of the holes in the inner plies must be not more than 3 mm larger than those of the bolts.

Where high strength friction grip bolts are used, the deduction in cross sectional area of connected tension members must be in accordance with BS 449: 1969 except that, in calculating the area to be

CONNECTIONS

deducted, the actual diameter of the hole must be used. No deduction should be made in the case of compression members.

The distance from the centre of any hole to the edge of a member and the distance between the centres of holes must be in accordance with BS 449: 1969.

Minimum Ply Thickness

In connections using these bolts, no outer ply must be smaller in thickness than half the diameter of the bolt or 10 mm whichever is less.

Wherever possible, this condition for minimum thickness should be observed for inner plies.

Design

(*a*) *Shear Connections*. In connections subject only to shear in the plane of the friction faces, the number of friction grip bolts and their disposition must be such that the resulting load at any bolt position does not exceed the value:—

$$\frac{\text{Slip factor}}{\text{Load factor}} \times \text{number of effective interfaces} \times \text{proof load of one bolt}$$

in which

Slip factor is the ratio of the load per effective interface required to produce slip in a pure shear joint to the nominal shank tension (i.e proof load) induced in the bolt or bolts.

Load factor is the numerical value by which the load which would cause slip in a joint is divided to give the permissible working load on the joint.

Effective interface is a common contact surface between two load-transmitting plies, excluding packing pieces, through which the bolt passes.

Proof load is the appropriate load given in Table 2 below.

The load factor may be taken as 1.4 for structures and materials covered by BS 449: 1969. Where the effect of wind forces on the structure has to be taken into consideration, this load factor may be reduced to 1.2 provided the connections are adequate when wind forces are neglected. No additional factor is required to take account of fatigue conditions.

In all cases where surfaces in contact comply with the conditions set out above, the slip factor may be taken as 0.45.

Shear values of bolts per interface have been tabulated in Parts II and III, using the above numerical factors as the basis.

(b) *Connections subject to external tension only in the direction of the bolt axes.* In these cases, the maximum permissible external tension on any bolt must not exceed 0.6 of the proof load of the bolts used, as given in Table 2. However, where fatigue conditions are involved, the maximum permissible external tension on any bolt must be limited to 0.5 of the proof load.

(c) *Connections subject to external tension in addition to shear.* An externally applied tension in the direction of the bolt axis reduces the effective clamping action of a bolt which has been tightened to induce shank tension. To allow for this effect, the permissible resulting load at any bolt position, as calculated from the expression in (a) above, must be reduced by substituting for the proof load of the bolt an effective clamping force obtained by subtracting 1.7 times the applied external tensile load from the proof load.

Under this rule, the effective clamping action of a bolt is considered to cease when the externally applied tension reaches 0.6 of its proof load, which is the maximum permissible value—see (b) above.

Table 2. Proof Loads of Bolts (minimum shank tensions)
BS 4604: Part 1: 1970

Nominal size and thread diameter	Proof load (minimum shank tension) kN	Nominal size and thread diameter	Proof load (minimum shank tension) kN
M12	49.4	M24	207
M16	92.1	M27	234
M20	144	M30	286
M22	177	M36	418

NOTE 1. The proof loads in the Table are those specified in Table 4 of BS 4395, 'High strength friction grip bolts and associated nuts and washers for structural engineering. Metric series', Part 1, 'General grade'.

NOTE 2. The torque necessary to induce a specified tension is determined by actual site conditions and equipment.

NOTE 3. For calibration purposes, the minimum shank tensions are to be increased by 10 per cent (see clause 4.3 of BS 4604).

HIGH STRENGTH FRICTION GRIP BOLTS

TO BS 4395 : PART 1 : 1969

GENERAL GRADE

SHEAR VALUES OF BOLTS PER INTERFACE

Diameter of Bolt Shank in mm	Shear Value without wind kilonewtons	Shear Value including wind kilonewtons
12	15.9	18.5
16	29.6	34.5
20	46.3	54.0
22	56.9	66.4
24	66.5	77.6
27	75.2	87.8
30	91.9	107
36	134	157

1 kilonewton may be taken as 0.102 metric tonne (megagramme) force.

HIGH STRENGTH BOLTED CONNECTIONS

Connections using High Strength Friction Grip Bolts

Example. Using previous examples, page 718 Fig. 10, determine the number of H.S.F.G. bolts required, as compared to 8 No. 22 mm nominal diameter rivets (shop driven).

B.M. as previous example = 24 525 kNmm

Value of $\dfrac{\Sigma y^2}{Y} = \dfrac{78\,750}{225}$ = 350 units

∴ Maximum tension in top bolt (for two rows) $= \dfrac{24\,525}{2 \times 350}$

∴ F_t = 35.0 kN

direct force $F_s = \dfrac{109}{8}$ = 13.6 kN

∴ resultant force found graphically = 37.42 kN.

Permissible load on 22 mm diameter H.S.F.G. Bolt in combined shear and tension (see notes).

$$= \dfrac{0.45}{1.4} [1 \times (177 - 1.7 \times 35.0)] = 37.8 \text{ kN}.$$

Therefore 8 No. 22 mm diameter H.S.F.G. Bolts would be required which is the same number as rivets required in previous example, page 726.

It is not advantageous to use H.S.F.G. Bolts for connections involving tension as the shear value is lessened by the tension force exerted in the bolt, thus defeating the purpose of these bolts.

Beam Splices

Continuous beam spliced at point of contraflexure.

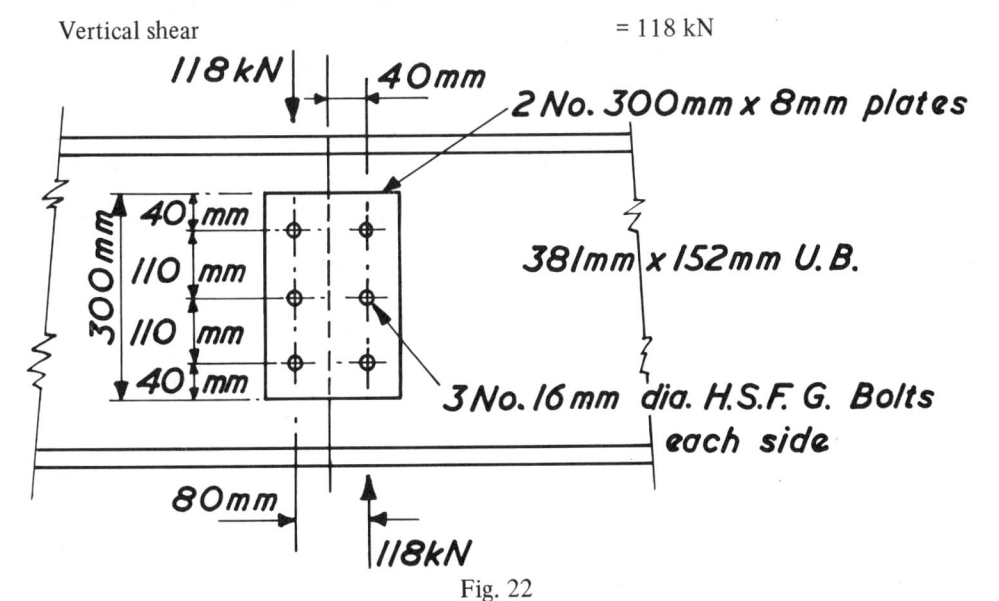

Vertical shear = 118 kN

Fig. 22

Area required for plates $= \dfrac{118 \times 10^3}{115}$ $= 1\,026.1 \text{ mm}^2$

Net thickness of plates required $= \dfrac{1\,026.1}{300 - (3 \times 18)}$

$= \dfrac{1\,026.1}{246}$ $= 4.17 \text{ mm}$

∴ Two 300 mm × 8 mm plates are adequate

Vertical force per bolt $= \dfrac{118}{3}$ $= 39.33 \text{ kN}$

Moment due to force horizontal force acting on outermost bolt $= 118 \times 80$ $= 9\,440 \text{ kNmm}$

$= \dfrac{9\,440}{\text{modulus of bolt group (1 line)}}$

$= 9\,440/220$ $= 42.91 \text{ kN}$

∴ resultant force F_r $= \sqrt{39.33^2 + 42.91^2}$ $= 58.2 \text{ kN}$
permissible value of 16 mm diameter H.S.F.G. Bolt $= 29.6 \text{ kN/interface}$

∴ Value of 16 mm diameter H.S.F.G. Bolt = 29.6 × 2 = 59.2 kN

∴ Use 3 No. 16 mm H.S.F.G. Bolts each side of splice.

Design of Connection for Continuous Beams

i.e., Moment at Support.

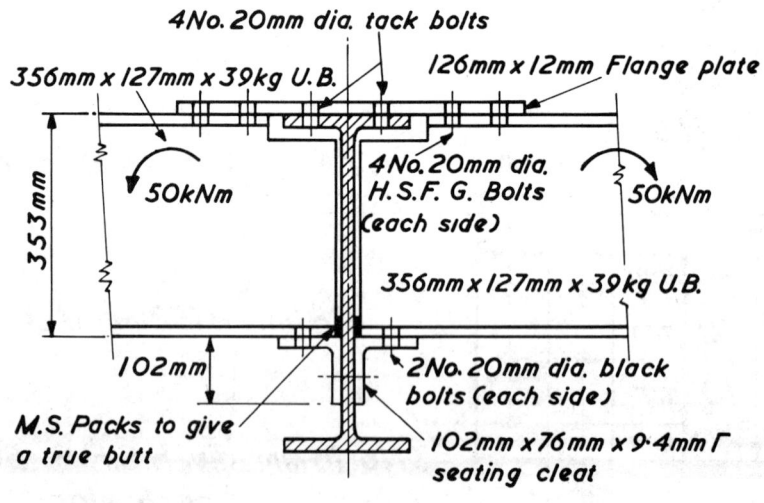

Fig. 23

Support Moment $= 50 \text{ kNm}$

Shear force on bolts due to moment $= \dfrac{50 \times 10^3}{353}$ $= 141.6 \text{ kN}$

WELDING

Permissible load on one 20 mm diameter
H.S.F.G. Bolt (see tables)   = 46.3 kN

$$\therefore \text{ No. required} = \frac{141.6}{46.3} = 3.06$$

Use 4 No. 20 mm diameter H.S.F.G. Bolts each side.

Tensile force in top flange   = 141.6 kN

$$\text{Area required in flange plate} = \frac{141.6 \times 10^3}{155} = 913.5 \text{ mm}^2$$

$$\text{thickness required} = \frac{913.5}{[126 - (2 \times 22)]} = 11.1 \text{ mm}$$

$\therefore$ 126 mm x 12 mm cover plate is adequate.

The compressive forces due to the moment in the beam are dealt with by inserting mild steel packs as necessary, thus transferring all forces into the web of the main cross beam.

Compressive force   = 141.6 kN
possible area of contact = width of flange x thickness of flange
            = 126 x 10.7 = 1 348.2 mm^2
$\therefore$ permissible force   = 1 348.2 x 190 N
            = 256.2 kN.

therefore web of main beam will not be overstressed. The rivets in the vertical leg of the seating cleat should be checked for carrying the beam reaction load.

Welding

The tendency today to use welding for all manner of connections has greatly increased.

The ideal situation is to have shop welded connections in fabrication, and bolted site connections. There is no doubt also that site welding of joints for rigid frame construction (see later Rigid Frame Joints notes) can lead to a more competitive design, as it approaches to a homogeneous structure, as in reinforced concrete construction.

It is now possible to carry out on the site ultrasonic testing of welds. B.S. 3923 "Methods for Ultrasonic Examination of Welds."

Notes on Welding

Electrodes BS 449 refers to BS 639, 'Covered electrodes for the manual metal-arc welding of mild steel and medium-tensile steel' and states that when electrodes complying with Sections 1 and 2 of BS 639 are used for the welding of grade 43 steel, or with Sections 1 and 4 of BS 639 are used for the welding of grade 50 steel, or with Sections 1 and 4 are used for the welding of grade 55 steel (see clause 4.1.1 of BS 639) and the yield stress of an all-weld tensile test specimen is not less than 430 N/mm^2 when tested in accordance with Appendix D of BS 639, the following shall apply:—

Butt Welds (i) *Butt welds.* Butt weld shall be treated as parent metal with a throat thickness (or a reduced throat thickness as specified in clause 54 for certain butt welds) and the stresses shall not exceed those allowed in BS 449 for the parent metal.

Fillet Welds (ii) *Fillet welds.* The allowable stresses in fillet welds, based on a thickness equal to the throat thickness, shall be 115 N/mm^2 for grade 43 steel or 160 N/mm^2 for grade 50 steel or 195 N/mm^2 for grade 55 steel.

Mixed Grades (iii) When electrodes appropriate to a lower grade of steel are used for welding together parts of material of a higher grade of steel, the allowable stresses for the lower grade of steel shall apply.

Combined Stresses (iv) When a weld is subject to a combination of stresses, the stresses shall be combined as required by subclauses 14.c and d, the value of the equivalent stress f_e being not greater than that permitted for the parent metal.

WELDING

STRENGTH OF FILLET WELDS

FOR GRADE 43 STEEL

PERMISSIBLE LOADS IN KILONEWTONS PER mm RUN
WITH ELECTRODES TO BS 639, SECTIONS 1 AND 2.

Leg length in mm	Throat thickness in mm	Load at 115 N/mm²	Leg length in mm	Throat thickness in mm	Load at 115 N/mm²
3	2.1	0.24	12	8.4	0.97
4	2.8	0.32	15	10.5	1.21
5	3.5	0.40	18	12.6	1.45
6	4.2	0.48	20	14.0	1.61
8	5.6	0.64	22	15.4	1.77
10	7.0	0.80	25	17.5	2.01

STRENGTH OF FULL PENETRATION BUTT WELDS

FOR GRADE 43 STEEL

PERMISSIBLE LOADS IN KILONEWTONS PER mm RUN
WITH ELECTRODES TO BS 639, SECTIONS 1 and 2

Thickness in mm	Shear at 100 N/mm²	Tension or Compression at 155 N/mm²	Thickness in mm	Shear at 100 N/mm²	Tension or Compression at 155 N/mm²
6	0.60	0.93	22	2.20	3.41
8	0.80	1.24	25	2.50	3.88
10	1.00	1.55	28	2.80	4.34
12	1.20	1.86	30	3.00	4.65
15	1.50	2.33	35	3.50	5.43
18	1.80	2.79	40	4.00	6.20
20	2.00	3.10			

STRENGTH OF FILLET WELDS
FOR GRADE 50 STEEL
PERMISSIBLE LOADS IN KILONEWTONS PER mm RUN
WITH ELECTRODES TO BS 639, SECTIONS 1 and 4

Leg length in mm	Throat thickness in mm	Load at 160 N/mm²	Leg length in mm	Throat thickness in mm	Load at 160 N/mm²
3	2.1	0.34	12	8.4	1.34
4	2.8	0.45	15	10.5	1.68
5	3.5	0.56	18	12.6	2.02
6	4.2	0.67	20	14.0	2.24
8	5.6	0.90	22	15.4	2.46
10	7.0	1.12	25	17.5	2.80

STRENGTH OF FULL PENETRATION BUTT WELDS
FOR GRADE 50 STEEL
PERMISSIBLE LOADS IN KILONEWTONS PER mm RUN
WITH ELECTRODES TO BS 639, SECTIONS 1 and 4

Thickness in mm	Shear at 140 N/mm²	Tension or Compression at 215 N/mm²	Thickness in mm	Shear at 140 N/mm²	Tension or Compression at 215 N/mm²
6	0.84	1.29	22	3.08	4.73
8	1.12	1.72	25	3.50	5.38
10	1.40	2.15	28	3.92	6.02
12	1.68	2.58	30	4.20	6.45
15	2.10	3.23	35	4.90	7.53
18	2.52	3.87	40	5.60	8.60
20	2.80	4.30			

WELD GROUPS

WELDS NOT IN THE PLANE OF THE FORCE

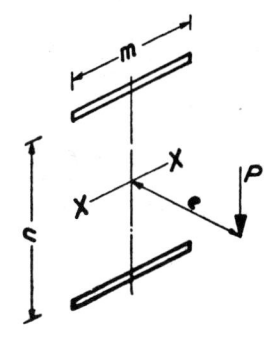

$$F_s = \frac{P}{2m}$$

$$F_b = \frac{P \cdot e \cdot n}{2I_{XX}} = \frac{P \cdot e}{Z_{XX}}$$

$$F_R = \sqrt{(F_s^2 + F_b^2)}$$

VALUES OF I_{XX} (in cm⁴) FOR 1 mm THROAT THICKNESS

		\multicolumn{10}{c}{Values of m in mm}
		50	75	100	125	150	175	200	225	250	275	300
Values of n in mm	50	6.3	9.4	12.5	15.6	18.8	21.9	25.0	28.1	31.3	34.4	37.5
	75	14.1	21.1	28.1	35.2	42.2	49.2	56.3	63.3	70.3	77.3	84.4
	100	25.0	37.5	50.0	62.5	75.0	87.5	100.0	112.5	125.0	137.5	150.0
	125	39.1	58.6	78.1	97.7	117.2	136.7	156.3	175.8	195.3	214.8	234.4
	150	56.3	84.4	112.5	140.6	168.8	196.9	225.0	253.1	281.3	309.4	337.5
	175	76.6	114.8	153.1	191.4	229.7	268.0	306.3	344.5	382.8	421.1	459.4
	200	100.0	150.0	200.0	250.0	300.0	350.0	400.0	450.0	500.0	550.0	600.0
	225	126.6	189.8	253.1	316.4	379.7	443.0	506.3	569.5	632.8	696.1	759.4
	250	156.3	234.4	312.5	390.6	468.8	546.9	625.0	703.1	781.3	859.4	937.5
	275	189.1	283.6	378.1	472.7	567.2	661.7	756.3	850.8	945.3	1 039.8	1 134.4
	300	225.0	337.5	450.0	562.5	675.0	787.5	900.0	1 012.5	1 125.0	1 237.5	1 350.0
	325	264.1	396.1	528.1	660.2	792.2	924.2	1 056.3	1 188.3	1 320.3	1 452.3	1 584.4
	350	306.3	459.4	612.5	765.6	918.8	1 071.9	1 225.0	1 378.1	1 531.3	1 684.4	1 837.5
	375	351.6	527.3	703.1	878.9	1 054.7	1 230.5	1 406.3	1 582.0	1 757.8	1 933.6	2 109.4
	400	400.0	600.0	800.0	1 000.0	1 200.0	1 400.0	1 600.0	1 800.0	2 000.0	2 200.0	2 400.0
	425	451.6	677.3	903.1	1 128.9	1 354.7	1 580.5	1 806.3	2 032.0	2 257.8	2 483.6	2 709.4
	450	506.3	759.4	1 012.5	1 265.6	1 518.8	1 771.9	2 025.0	2 278.1	2 531.3	2 784.4	3 037.5
	475	564.1	846.1	1 128.1	1 410.2	1 692.2	1 974.2	2 256.3	2 538.3	2 820.3	3 102.3	3 384.4
	500	625.0	937.5	1 250.0	1 562.5	1 875.0	2 187.5	2 500.0	2 812.5	3 125.0	3 437.5	3 750.0
	525	689.1	1 033.6	1 378.1	1 722.7	2 067.2	2 411.7	2 756.3	3 100.8	3 445.3	3 789.8	4 134.4
	550	756.3	1 134.4	1 512.5	1 890.6	2 268.8	2 646.9	3 025.0	3 403.1	3 781.3	4 159.4	4 537.5
	575	826.6	1 239.8	1 653.1	2 066.4	2 479.7	2 893.0	3 306.3	3 719.5	4 132.8	4 546.1	4 959.4
	600	900.0	1 350.0	1 800.0	2 250.0	2 700.0	3 150.0	3 600.0	4 050.0	4 500.0	4 950.0	5 400.0

WELDS NOT IN THE PLANE OF THE FORCE

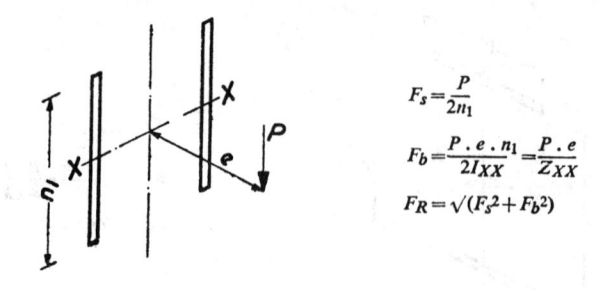

$$F_s = \frac{P}{2n_1}$$

$$F_b = \frac{P \cdot e \cdot n_1}{2I_{XX}} = \frac{P \cdot e}{Z_{XX}}$$

$$F_R = \sqrt{(F_s^2 + F_b^2)}$$

VALUES OF I_{XX} (in cm^4)
FOR 1 mm THROAT THICKNESS

n	I_{XX}
50	2.1
75	7.0
100	16.7
125	32.6
150	56.3
175	89.3
200	133.3
225	189.8
250	260.4
275	346.6
300	450.0
325	572.1
350	714.6
375	878.9
400	1 066.7
425	1 279.4
450	1 518.8
475	1 786.2
500	2 083.3
525	2 411.7
550	2 772.9
575	3 168.5
600	3 600.0

WELD GROUPS

WELDS IN THE PLANE OF THE FORCE

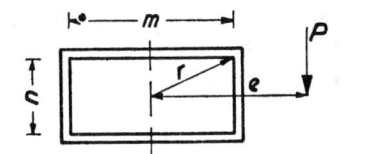

$$F_s = \frac{P}{2(m+n)}$$

$$F_b = \frac{P \cdot e \cdot r}{I_p}$$

VALUES OF I_p (in cm^4) FOR 1 mm THROAT THICKNESS

		\multicolumn{10}{c	}{Values of m in mm}
		50	75	100	125	150	175	200	225	250	275	300
Values of n in mm	50	16.7	32.6	56.3	89.3	133.3	189.8	260.4	346.6	450.0	572.1	714.6
	75	32.6	56.3	89.3	133.3	189.8	260.4	346.6	450.0	572.1	714.6	878.9
	100	56.3	89.3	133.3	189.8	260.4	346.6	450.0	572.1	714.6	878.9	1 066.7
	125	89.3	133.3	189.8	260.4	346.6	450.0	572.1	714.6	878.9	1 066.7	1 279.4
	150	133.3	189.8	260.4	346.6	450.0	572.1	714.6	878.9	1 066.7	1 279.4	1 518.8
	175	189.8	260.4	346.6	450.0	572.1	714.6	878.9	1 066.7	1 279.4	1 518.8	1 786.2
	200	260.4	346.6	450.0	572.1	714.6	878.9	1 066.7	1 279.4	1 518.8	1 786.2	2 083.3
	225	346.6	450.0	572.1	714.6	878.9	1 066.7	1 279.4	1 518.8	1 786.2	2 083.3	2 411.7
	250	450.0	572.1	714.6	878.9	1 066.7	1 279.4	1 518.8	1 786.2	2 083.3	2 411.7	2 772.9
	275	572.1	714.6	878.9	1 066.7	1 279.4	1 518.8	1 786.2	2 083.3	2 411.7	2 772.9	3 168.5
	300	714.6	878.9	1 066.7	1 279.4	1 518.8	1 786.2	2 083.3	2 411.7	2 772.9	3 168.5	3 600.0
	325	878.9	1 066.7	1 279.4	1 518.8	1 786.2	2 083.3	2 411.7	2 772.9	3 168.5	3 600.0	4 069.0
	350	1 066.7	1 279.4	1 518.8	1 786.2	2 083.3	2 411.7	2 772.9	3 168.5	3 600.0	4 069.0	4 577.1
	375	1 279.4	1 518.8	1 786.2	2 083.3	2 411.7	2 772.9	3 168.5	3 600.0	4 069.0	4 577.1	5 125.8
	400	1 518.8	1 786.2	2 083.3	2 411.7	2 772.9	3 168.5	3 600.0	4 069.0	4 577.1	5 125.8	5 716.7
	425	1 786.2	2 083.3	2 411.7	2 772.9	3 168.5	3 600.0	4 069.0	4 577.1	5 125.8	5 716.7	6 351.3
	450	2 083.3	2 411.7	2 772.9	3 168.5	3 600.0	4 069.0	4 577.1	5 125.8	5 716.7	6 351.3	7 031.3
	475	2 411.7	2 772.9	3 168.5	3 600.0	4 069.0	4 577.1	5 125.8	5 716.7	6 351.3	7 031.3	7 758.1
	500	2 772.9	3 168.5	3 600.0	4 069.0	4 577.1	5 125.8	5 716.7	6 351.3	7 031.3	7 758.1	8 533.3
	525	3 168.5	3 600.0	4 069.0	4 577.1	5 125.8	5 716.7	6 351.3	7 031.3	7 758.1	8 533.3	9 358.6
	550	3 600.0	4 069.0	4 577.1	5 125.8	5 716.7	6 351.3	7 031.3	7 758.1	8 533.3	9 358.6	10 235.4
	575	4 069.0	4 577.1	5 125.8	5 716.7	6 351.3	7 031.3	7 758.1	8 533.3	9 358.6	10 235.4	11 165.4
	600	4 577.1	5 125.8	5 716.7	6 351.3	7 031.3	7 758.1	8 533.3	9 358.6	10 235.4	11 165.4	12 150.0

CONNECTIONS
WELDS IN THE PLANE OF THE FORCE

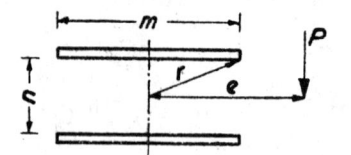

$$F_s = \frac{P}{2m}$$

$$F_b = \frac{P \cdot e \cdot r}{I_p}$$

VALUES OF I_p (in cm^4) FOR 1 mm THROAT THICKNESS

		\multicolumn{10}{c	}{Values of m in mm}
		50	75	100	125	150	175	200	225	250	275	300
Values of n in mm	50	8.3	16.4	29.2	48.2	75.0	111.2	158.3	218.0	291.7	381.0	487.5
	75	16.1	28.1	44.8	67.7	98.4	138.5	189.6	253.1	330.7	424.0	534.4
	100	27.1	44.5	66.7	95.1	131.3	176.8	233.3	302.3	385.4	484.1	600.0
	125	41.1	65.6	94.8	130.2	173.4	226.0	289.6	365.6	455.7	561.5	684.4
	150	58.3	91.4	129.2	173.2	225.0	286.2	358.3	443.0	541.7	656.0	787.5
	175	78.6	121.9	169.8	224.0	285.9	357.3	439.6	534.4	643.2	767.7	909.4
	200	102.1	157.0	216.7	282.6	356.3	439.3	533.3	639.8	760.4	896.6	1 050.0
	225	128.6	196.9	269.8	349.0	435.9	532.3	639.6	759.4	893.2	1 042.7	1 209.4
	250	158.3	241.4	329.2	423.2	525.0	636.2	758.3	893.0	1 041.7	1 206.0	1 387.5
	275	191.1	290.6	394.8	505.2	623.4	751.0	889.6	1 040.6	1 205.7	1 386.5	1 584.4
	300	227.1	344.5	466.7	595.1	731.3	876.8	1 033.3	1 202.3	1 385.4	1 584.1	1 800.0
	325	266.1	403.1	544.8	692.7	848.4	1 013.5	1 189.6	1 378.1	1 580.7	1 799.0	2 034.4
	350	308.3	466.4	629.2	798.2	975.0	1 161.2	1 358.3	1 568.0	1 791.7	2 031.0	2 287.5
	375	353.6	534.4	719.8	911.5	1 110.9	1 319.8	1 539.6	1 771.9	2 018.2	2 280.2	2 559.4
	400	402.1	607.0	816.7	1 032.6	1 256.5	1 489.3	1 733.3	1 989.8	2 260.4	2 546.6	2 850.0
	425	453.6	684.4	919.8	1 161.5	1 410.9	1 669.8	1 939.6	2 221.9	2 518.2	2 830.2	3 159.4
	450	508.3	766.4	1 029.2	1 298.2	1 575.0	1 861.2	2 158.3	2 468.0	2 791.7	3 131.0	3 487.5
	475	566.1	853.1	1 144.8	1 442.7	1 748.4	2 063.5	2 389.6	2 728.1	3 080.7	3 449.0	3 834.4
	500	627.1	944.5	1 266.7	1 595.1	1 931.3	2 276.8	2 633.3	3 002.3	3 385.4	3 784.1	4 200.0
	525	691.1	1 040.6	1 394.8	1 755.2	2 123.4	2 501.0	2 889.6	3 290.6	3 705.7	4 136.5	4 584.4
	550	758.3	1 141.4	1 529.2	1 923.2	2 325.0	2 736.2	3 158.3	3 593.0	4 041.7	4 506.0	4 987.5
	575	828.6	1 246.9	1 669.8	2 099.0	2 535.9	2 982.3	3 439.6	3 909.4	4 393.2	4 892.7	5 409.4
	600	902.1	1 357.0	1 816.7	2 282.6	2 756.5	3 239.3	3 733.3	4 239.8	4 760.4	5 296.6	5 850.0

WELD GROUPS

WELDS IN THE PLANE OF THE FORCE

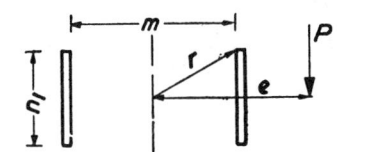

$$F_s = \frac{P}{2n_1}$$

$$F_b = \frac{P \cdot e \cdot r}{I_p}$$

VALUES OF I_p (in cm^4) FOR 1 mm THROAT THICKNESS

		\multicolumn{10}{c}{Values of m in mm}
		50	75	100	125	150	175	200	225	250	275	300
Values of n in mm	50	8.3	16.1	27.11	41.1	58.3	78.6	102.1	128.6	158.3	191.1	227.1
	75	16.4	28.1	44.5	65.6	91.4	121.9	157.0	196.9	241.4	290.6	344.5
	100	29.2	44.8	66.7	94.8	129.2	169.8	216.7	269.8	329.2	394.8	466.7
	125	48.2	67.7	95.1	130.2	173.2	224.0	282.6	349.0	423.2	505.2	595.1
	150	75.0	98.4	131.3	173.4	225.0	285.9	356.3	435.9	525.0	623.4	731.3
	175	111.0	138.5	176.8	226.0	286.2	357.3	439.3	532.3	636.2	751.0	876.8
	200	158.3	189.6	233.3	289.6	358.3	439.6	533.3	639.6	758.3	889.6	1 033.3
	225	218.0	253.1	302.3	365.6	443.0	534.4	639.8	759.4	893.0	1 040.6	1 202.3
	250	291.7	330.7	385.4	455.7	541.7	643.2	760.4	893.2	1 041.7	1 205.7	1 385.4
	275	381.0	424.0	484.1	561.5	656.0	767.7	896.6	1 042.7	1 206.0	1 386.5	1 584.1
	300	487.5	534.4	600.0	684.4	787.5	909.4	1 050.0	1 209.4	1 387.5	1 584.4	1 800.0
	325	612.8	663.5	734.6	826.0	937.8	1 069.8	1 222.1	1 394.8	1 587.8	1 801.0	2 034.6
	350	758.3	813.0	889.6	988.0	1 108.3	1 250.5	1 414.6	1 600.5	1 808.3	2 038.0	2 289.6
	375	925.8	984.4	1 066.4	1 171.9	1 300.8	1 453.1	1 628.9	1 828.1	2 050.8	2 296.9	2 566.4
	400	1 116.7	1 179.2	1 266.7	1 379.2	1 516.7	1 679.2	1 866.7	2 079.2	2 316.7	2 579.2	2 866.7
	425	1 332.6	1 399.0	1 491.9	1 611.5	1 757.6	1 930.2	2 129.4	2 355.2	2 607.6	2 886.5	3 191.9
	450	1 575.0	1 645.3	1 743.8	1 870.3	2 025.0	2 207.8	2 418.8	2 657.8	2 925.0	3 220.3	3 543.8
	475	1 845.6	1 919.8	2 023.7	2 157.3	2 320.6	2 513.5	2 736.2	2 988.5	3 270.6	3 582.3	3 923.7
	500	2 145.8	2 224.0	2 333.3	2 474.0	2 645.8	2 849.0	3 083.3	3 349.0	3 645.8	3 974.0	4 333.3
	525	2 477.3	2 559.4	2 674.2	2 821.9	3 002.3	3 215.6	3 461.7	3 740.6	4 052.3	4 396.9	4 774.2
	550	2 841.7	2 927.6	3 047.9	3 202.6	3 391.7	3 615.1	3 872.9	4 165.1	4 491.7	4 852.6	5 247.9
	575	3 240.4	3 330.2	3 456.0	3 617.7	3 815.4	4 049.0	4 318.5	4 624.0	4 965.4	5 342.7	5 756.0
	600	3 675.0	3 768.8	3 900.0	4 068.8	4 275.0	4 518.8	4 800.0	5 118.8	5 475.0	5 868.8	6 300.0

CONNECTIONS
WELDS IN THE PLANE OF THE FORCE

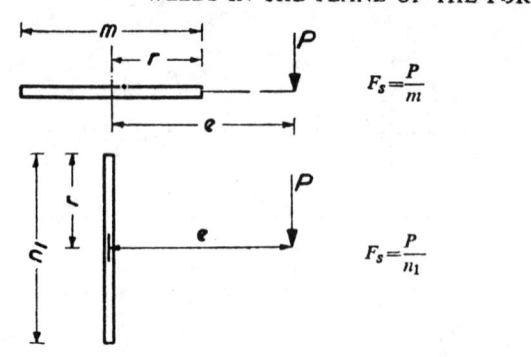

$$F_s = \frac{P}{m}$$

$$F_b = \frac{P \cdot e \cdot r}{I_p}$$

$$F_s = \frac{P}{n_1}$$

VALUES OF I_p (in cm^4)
FOR 1 mm THROAT THICKNESS

m or n_1	I_p
50	1.0
75	3.5
100	8.3
125	16.3
150	28.1
175	44.7
200	66.7
225	94.9
250	130.2
275	173.3
300	225.0
325	286.1
350	357.3
375	439.5
400	533.3
425	639.7
450	759.4
475	893.1
500	1 041.7
525	1 205.9
550	1 386.5
575	1 584.2
600	1 800.0

WELDED CONNECTIONS

Design of Welded Connections

Example. Single Angle. Tension Load = 110 kN.

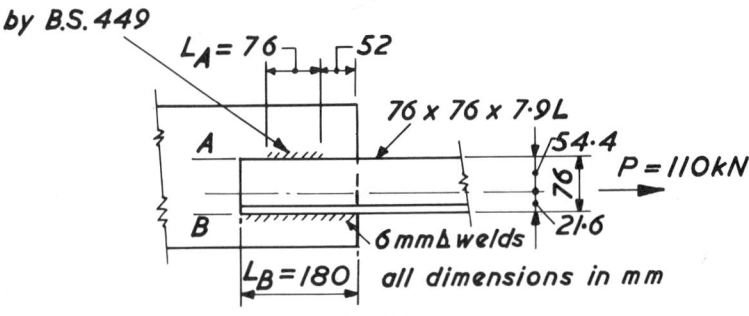

Fig. 24

Design end connection using balanced welds, to keep distortion to a minimum.
Use 6 mm fillet welds.
Strength of welds = 0.48 kN/mm run.

Side B. Taking moments about A.

$L_B \times f \times 76 = P \times 54.4$ where L_B = length of weld in mm.

f = strength of weld

$$\therefore L_B = \frac{110 \times 54.4}{0.48 \times 76} = 164.6 \text{ mm}$$

make 168 mm, plus one end return of 2 × 6 mm.

∴ Total length = 180 mm

Side A. Taking moments about B.

$L_A \times f \times 76 = P \times 21.6$

$$\therefore L_A = \frac{110 \times 21.6}{0.48 \times 76} = 65 \text{ mm}$$

Make total length = 76 mm (see B.S. 449 Clause 54 (f)).

Example. Double angle. Tension Load = 600 kN.

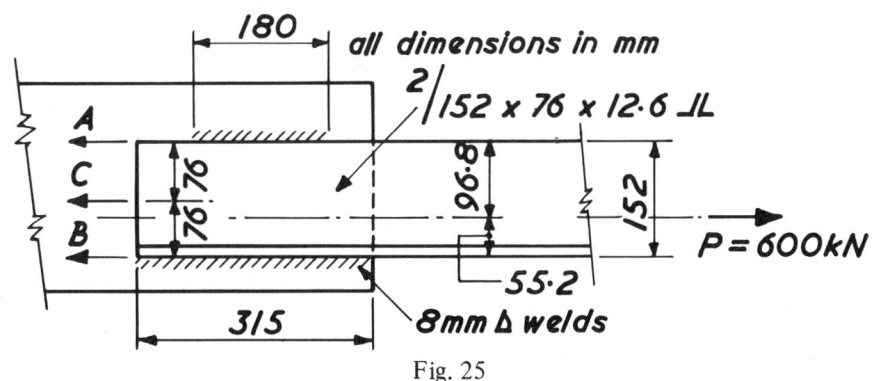

Fig. 25

Using 8 mm fillet welds.
strength of weld = 0.64 kN/mm run.

Side A. Taking moments about B.

length required $= \dfrac{600 \times 55.2}{2 \times 152 \times 0.64} = 170.2$ mm.

Use 180 mm run.

Side B. Taking moments about A.

length required $= \dfrac{600 \times 96.8}{2 \times 152 \times 0.64} = 300$ mm.

Use 315 mm run.

If the length of side weld is restricted, then the end of the angle can be welded. This will reduce the side welds as follows:

Side A

length required $= \dfrac{600 \times 55.2}{2 \times 152 \times 0.64} - \dfrac{152 \times 0.64 \times 76}{0.64 \times 152} = 94.2$ mm.

Use 100 mm run.

Side B

length required $= \dfrac{600 \times 96.8}{2 \times 152 \times 0.64} - 76 = 225.5$ mm.

Use 240 mm run.

In practice it is generally considered adequate to carry the weld all round the angles, but to decrease the lengths of welds to suit.

Moment Connection with Welded End Plate

In order to avoid excessive local flange bending in the column due to the externally applied end moment on the beam it is desirable that the cross-centres of the bolts be kept to a minimum, i.e., as close as possible to the column root. Therefore the end plate width, assuming 150 mm centres of bolts, will be in the order of 225 mm.

It will be appreciated that almost invariably web stiffeners will be required in the column opposite the beam flange position, under heavy loading, i.e., A and B (Fig. 26). The end plate will be fully welded to the beam: the weld around the top flange and locally in the web resisting the force due to the end moment and the remaining web welds resisting the vertical end shear.

In fillet welds, failures generally occur in the tension flange of the beams and it is significant that the initial crack invariably starts near the root of the section.

It is important, therefore, to design the tension flange welds to develop the full strength of the flange and also to continue these welds down the web to avoid any discontinuity adjacent to the root. (Reference by L. G. Johnson: Tests on Weld Connections, etc., B.W.J. Jan. 1959 (6).)

WELDED CONNECTIONS 751

In order to determine the size of the weld, end plate thickness and bolt diameter required to resist the tensile forces created by the applied moment it will be assumed that rotation takes place about a point adjacent underside of bottom flange. (See Fig. 26.)

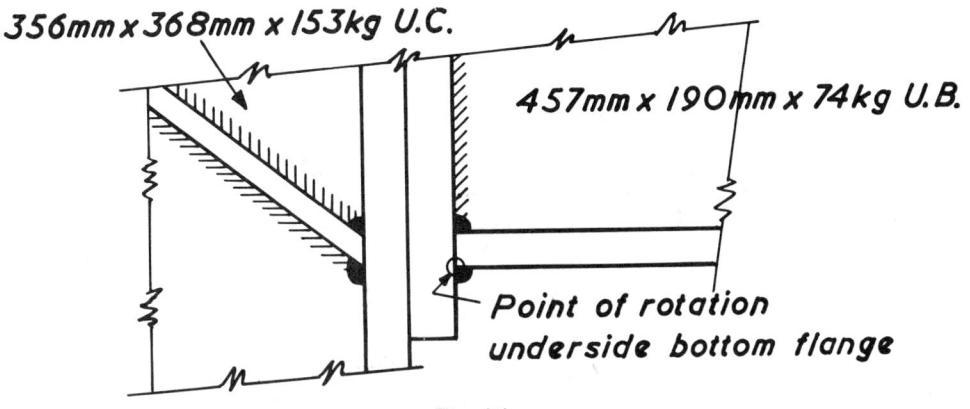

Fig. 26

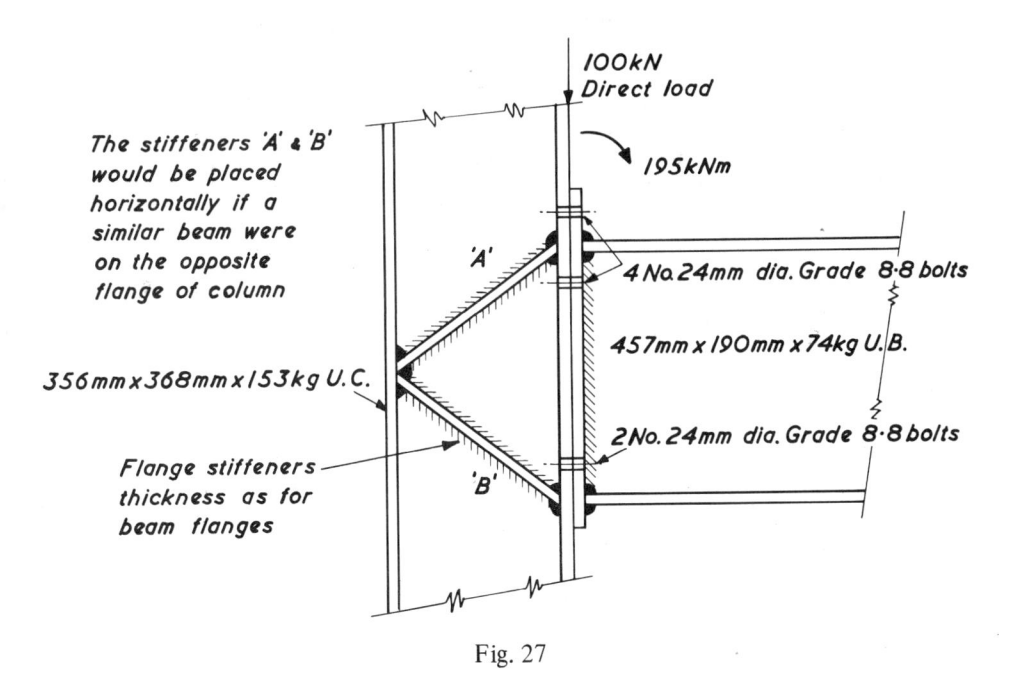

Fig. 27

In this particular example since the end moment does not change sign the forces on the bottom flange are compressive and only a nominal connection will be necessary.

If end moments from wind forces have to be considered, the foregoing statement will not necessarily hold, as it may be possible to have tensile forces applied to the

lower flange connection. Should this occur, the following treatment should be applied to both top and bottom flanges bearing in mind any increase in permissible stress.

The size of weld required to connect the top flange to the end plate is now calculated. It is assumed that the force in the top flange is applied to the centre of gravity of the weld group (Fig. 28) acting about the centre of rotation (lever arm) adjacent to the bottom flange (Fig. 26).

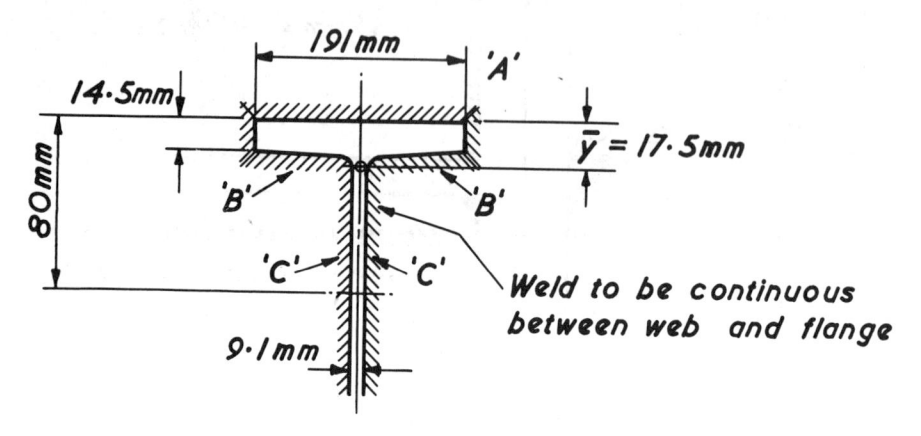

Fig. 28

To calculate centre of gravity of top flange weld.
Assuming weld to be unity.

Take moments about top flange.
Weld A 191   x 1 x 0   =   0
Weld B 2 x   90   x 1 x 14.5   = 2 610
Weld C 2 x 59.5 x 1 x (20.5 + 29.75) = 5 986
                                         ———
                                         8 596

length of weld group = 191 + (2 x 90) + (2 x 59.5) = 490 mm.
   Position of C. of G. of weld group from top flange

$$\bar{y} = \frac{8\,596}{490} = 17.5 \text{ mm}$$

∴ Lever arm d = 457 mm (beam depth) − 17.5 ($\bar{y}$)
         = 439.5 mm.
Force applied to C. of G. of weld group = M/d

$$= \frac{195 \times 10^3}{439.5} = 443.7 \text{ kN}$$

To determine Weld size around top flange
length of weld in top flange group = 490 mm

WELDED CONNECTIONS

load per mm run of weld = $\dfrac{443.7}{490}$ = 0.91 kN.

Use 12 mm Fillet Weld (0.97 kN/mm provided).

To determine weld around remainder of section. Let weld on remainder of end plate support the 100 kN vertical reaction

$\therefore$ size of weld = $\dfrac{100}{\text{length of weld}}$

= $\dfrac{100}{(2 \times 377 \text{ mm}) + (2 \times 190 \text{ mm})} = \dfrac{100}{1\,134} = 0.09$ kN/mm

$\therefore$ Use 5 mm fillet weld.

To determine the size and number of bolts in the Upper Group.

This group is assumed to carry the tension from the beam end moment. Assume the force applied to the bolts to be equal to that applied to the weld group, but acting at a point at the centre of the flange thickness.

Assume No. of bolts required = 4

$\therefore$ tensile load per bolt = $\dfrac{443.7}{4} = 110.9$ kN

Therefore use 4 No. 24 mm diameter Grade 8.8 bolts (118.2 kN).

To determine number of bolts to carry 100 kN Vertical Load

Assuming a bolt size of 24 mm diameter Grade 8.8

(Since bearing value of column flange is criterion, value of 79.5 kN to be used.)

No. of bolts = $\dfrac{100}{79.5}$ = 1.26 bolts

Therefore use 2 No. 24 mm diameter Grade 8.8 bolts in bottom group.

To determine thickness of end plate

To calculate the thickness of the end plate it is necessary to know the position of the top moment resisting group of bolts in the end plate.

To obtain a minimum end plate thickness the bolt group should be as compact as possible.

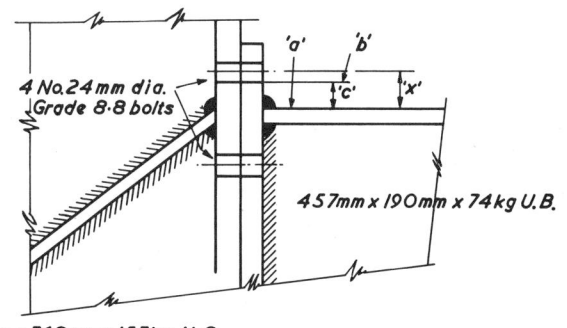

Fig. 29

Assume the end plate deforms in double curvature, then the effective lever arm is taken as the distance from the back of the weld (i.e., flange face a) to the edge of the bolt hole (b) with a point of contraflexure occurring midway between these points, i.e. $c/2$.

In this example let the distance above the flange to the centre line of the hole = 50 mm.

$$\therefore c = 50 - \frac{26}{2} \text{ (hole diameter)} = 50 - 13$$
$$= 37 \text{ mm}.$$

For end plate thickness

Assume the force applied to the bolt group is divided equally between the bolts above and below the flange.

$$\text{Elastic moment induced in plate} = \frac{c}{2} \times \frac{\text{flange force}}{2}$$
$$= \frac{37}{2} \times \frac{440.6}{2}$$
$$= 4\,076 \text{ kNmm}.$$

Design end plate plastically.

$$\text{Modulus required } Z_p = \frac{4\,076 \times 10^3 \times 1.75}{245 \times 10^3} = 29.1 \text{ cm}^3$$

$$t^2 = \frac{4Z_p}{b} \text{ where } t = \text{thickness of plate}$$
$$b = \text{breadth of plate}$$

$$= \frac{4 \times 29.1 \times 10^3}{225}$$

$$\therefore \quad t^2 = 517.6$$
$$\therefore \quad t = 22.8 \text{ mm say 23 mm}.$$

Calculate the size of welds for the joist cutting used as a bracket in Fig. 30.

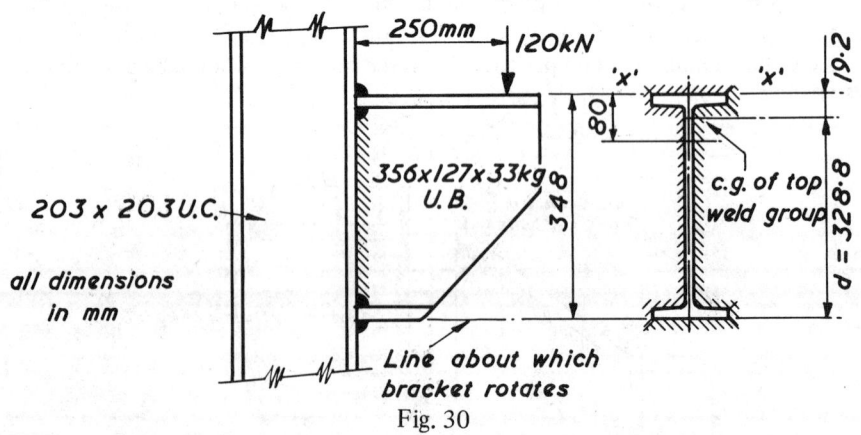

Fig. 30

Calculate:
Centre of gravity of top flange weld.

WELDED BRACKETS

Assume end moment is resisted by the top flange weld rotating about the bottom flange.

The vertical shear may be assumed to be resisted by the remaining 6 mm fillet weld which extends along each side of the web and encompasses the bottom flange.

Moment = 120 kN x 250 mm = 30 kNm.

To calculate C. of G. of top flange weld. Take moment about $X-X$ top of top flange.

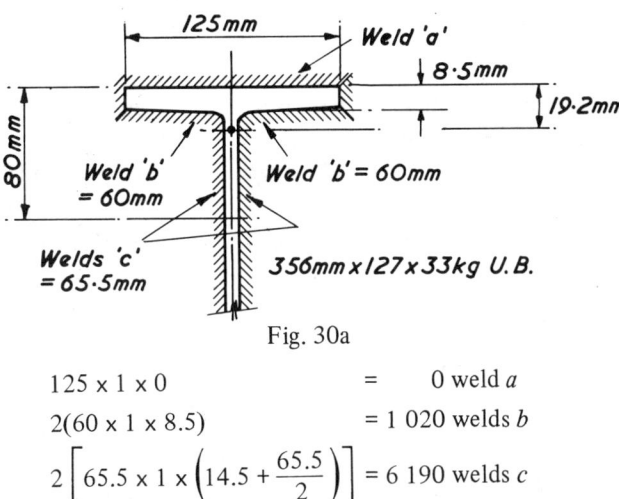

Fig. 30a

$$125 \times 1 \times 0 = 0 \text{ weld } a$$
$$2(60 \times 1 \times 8.5) = 1\,020 \text{ welds } b$$
$$2\left[65.5 \times 1 \times \left(14.5 + \frac{65.5}{2}\right)\right] = 6\,190 \text{ welds } c$$
$$A\bar{y} = 7\,210$$

Where $A = \Sigma$ weld areas.

$\bar{y}$ = distance from top flange to C. of G.

$$[(125 \times 1) + (2 \times 60 \times 1) + (2 \times 65.5 \times 1)]\bar{y} = 7\,210$$
$$[125 + 120 + 131]\bar{y} = 7\,210$$
$$376\bar{y} = 7\,210$$
$$\therefore \bar{y} = \frac{7\,210}{376} = 19.2 \text{ mm}.$$

Position of C. of G. of top flange weld from bottom flange point of rotation

$$d = 348 - 19.2$$
$$= 328.8 \text{ mm}$$

Force in top flange weld group $= \dfrac{M}{d} = \dfrac{30 \times 10^3}{328.8}$

$$= 91.2 \text{ kN}$$

$\therefore$ load per mm run of weld $= \dfrac{91.2}{\text{length of top flange weld}}$

length of top flange and local weld = 37 mm

$\therefore$ load per mm run of weld $= \dfrac{91.2}{376} = 0.243 \text{ kN}$

756 CONNECTIONS

Value of 6 mm F.W. is 0.48 kN/mm run.

Therefore 6 mm F. Weld is adequate.

Since remainder of weld is assumed to resist the vertical shear from the 120 kN end reaction the load per mm run of weld

$$= \frac{\text{shear force}}{\text{length of weld}}$$

$$= \frac{120}{(125 + 120 + 507)} = \frac{120}{752}$$

$$= 0.16 \text{ kN/mm run}$$

since value of 5 mm F.W. is 0.40 kN/mm, this is more than adequate.

In practice one would probably stipulate a 6 mm F.W. continuous round the whole connection, there is very little difference between a 5 mm and a 6 mm F.W. as regards cost.

Example. Calculate the size of welds required for the bracket shown in Fig. 31. The force P on each plate = 120 kN.

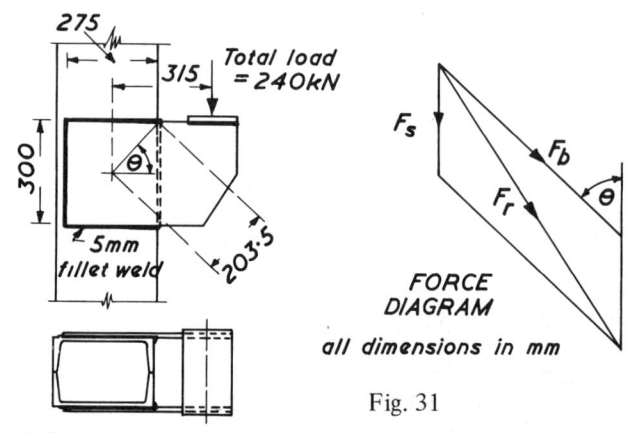

Fig. 31

Considering Table page 745 and assuming a unit throat thickness, the loads are calculated as follows:

$$F_s = \frac{P}{2(m+n)} = \frac{120 \times 10^3}{2(275 + 300)}$$

$$= 104.3 \text{ N/mm run}$$

$$F_b = \frac{P \cdot e \cdot r}{I_p} = \frac{120 \times 10^3 \times 315 \times 203.5}{3\,168.5 \times 10^4}$$

$$= 242.7 \text{ N/mm run}$$

$$F_R = \sqrt{(F_s^2 + F_b^2 + 2F_s \cdot F_b \cdot \cos\theta)}.$$

Now $\cos\theta = \dfrac{137.5}{203.5} = 0.675$

Hence, $F_R = \sqrt{[104.3^2 + 242.7^2 + (2 \times 104.3 \times 242.7 \times 0.675)]}$

$$= 322.5 \text{ N/mm run}.$$

Use 5 mm fillet welds throughout.

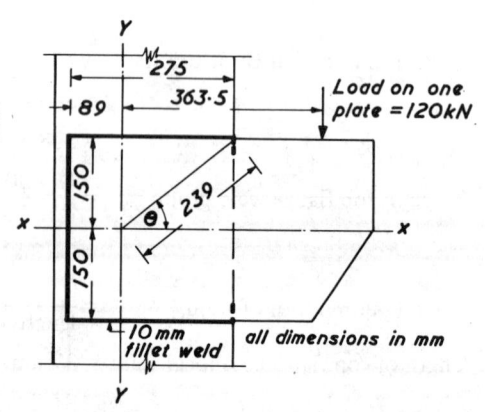

Fig. 32

WELDED SEATING CLEATS

Example. Calculate the size of welds required if the brackets in Example 2 are welded on three sides only as shown in Fig. 32.

First find the position of the centroid of the welds by taking moments about the left-hand edge of the plate. Assume welds of unit throat thickness,

$$\bar{x} = \frac{(2 \times 275 \times 137.5) + (300 \times 0)}{2 \times 275 + 300} = 89 \text{ mm}$$

then

$$e = 315 + 137.5 - 89 = 363.5 \text{ mm}$$

$$I_{xx} = \left[(2 \times 275 \times 150^2) + \frac{300^3}{12} \right] 10^{-4}$$

$$= 1\,462.5 \text{ cm}^4$$

$$I_{yy} = \left[\left(2 \times \frac{275^3}{12} \right) + (2 \times 275 \times 48.5^2) + (300 \times 89^2) \right] 10^{-4}$$

$$= 713.6 \text{ cm}^4$$

$$I_p = I_{xx} + I_{yy}$$

$$= 2\,176.1 \text{ cm}^4.$$

$$\text{Moment} = 120 \times 363.5 \times 10^{-3}$$
$$= 43.62 \text{ kNm}$$

$$\text{Maximum } r = \sqrt{150^2 + 186^2}$$
$$= 239 \text{ mm}$$

Now $\cos \theta = \frac{186}{239} = 0.778$.

$$F_s = \frac{120 \times 10^3}{(2 \times 275) + 300} = 218.2 \text{ N/mm run}$$

$$F_b = \frac{43.62 \times 10^6 \times 239}{2\,176 \times 10^4} = 479.1 \text{ N/mm run}$$

$$F_R = \sqrt{[218.2^2 + 479.1^2 + (2 \times 218.2 \times 479.1 \times 0.778)]}$$

$$= 663.2 \text{ N/mm run}.$$

Use 10 mm fillet weld (0.80 kN/mm run).

Example. Calculate the size of welds required for the seating cleat shown in Fig. 33.

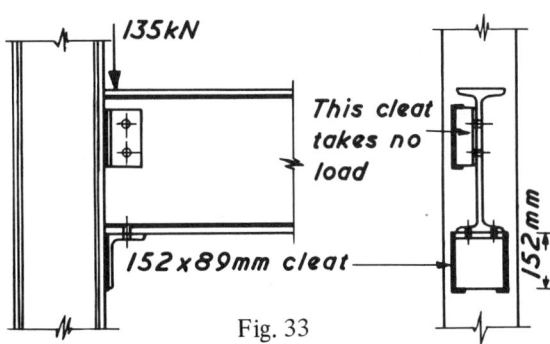

Fig. 33

If the cleat were stiffened and the eccentricity of the beam reaction increased, the welds would have to be calculated as in Example (page 749), but in this case there is direct load only.

The reaction = 135 kN.

Effective length of fillet welds connecting cleat to stanchion = say, 290 mm.

$$\text{Load per mm run} = \frac{135}{290}$$

$$= 0.47 \text{ kN}.$$

Use 6 mm fillet weld (0.48 kN/mm run).

Example. Design a gusseted base for the 457 x 190 x 67 kg U.B. shown in Fig. 34, which is subjected to an axial load of 800 kN and a B.M. of 152 kNm.

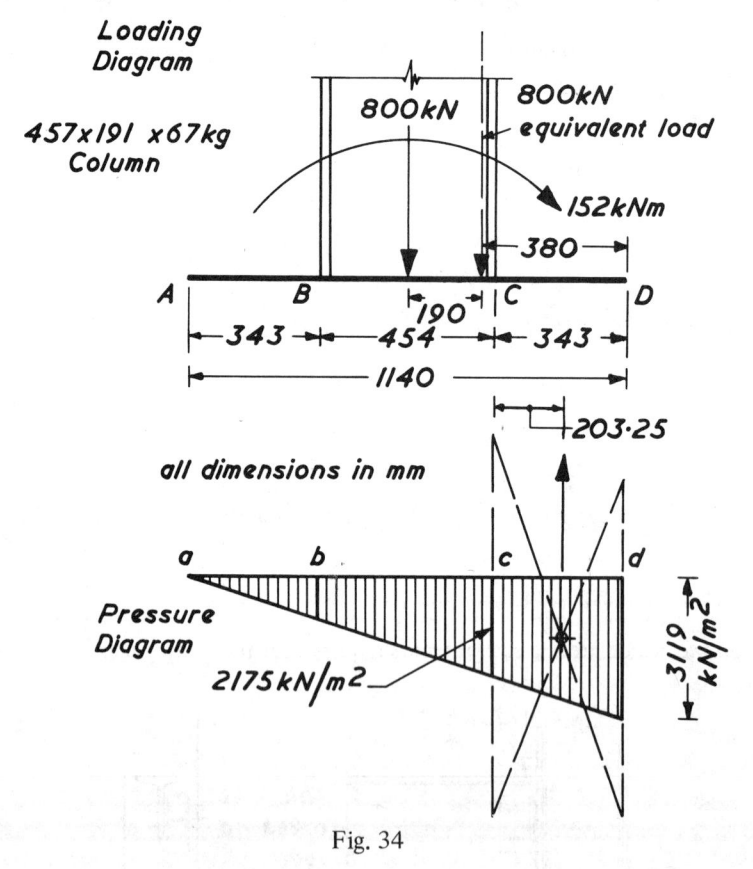

Fig. 34

The loading system can be replaced by a single equivalent load of 800 kN with an eccentricity about the vertical axis of the column of $152 \times 10^3/800 = 190$ mm, as shown in the figure.

If the base is made 1 140 mm long, then the equivalent load is placed 380 mm, or one-third of the base length, from the end *D* and the pressure diagram will be

WELDED BASES

triangular as shown, varying from zero at a to a maximum at d, the value being $2P/A$, where P is the axial load and A the area of the base.

A convenient base width is 450 mm. Checking the maximum pressure on the concrete foundation,

$$f = \frac{2P}{A} = \frac{2 \times 800}{1\,140 \times 454 \times 10^{-6}}$$

$$= 3\,118.9 \text{ kN/m}^2.$$

The base will be of the form shown in Fig. 35.

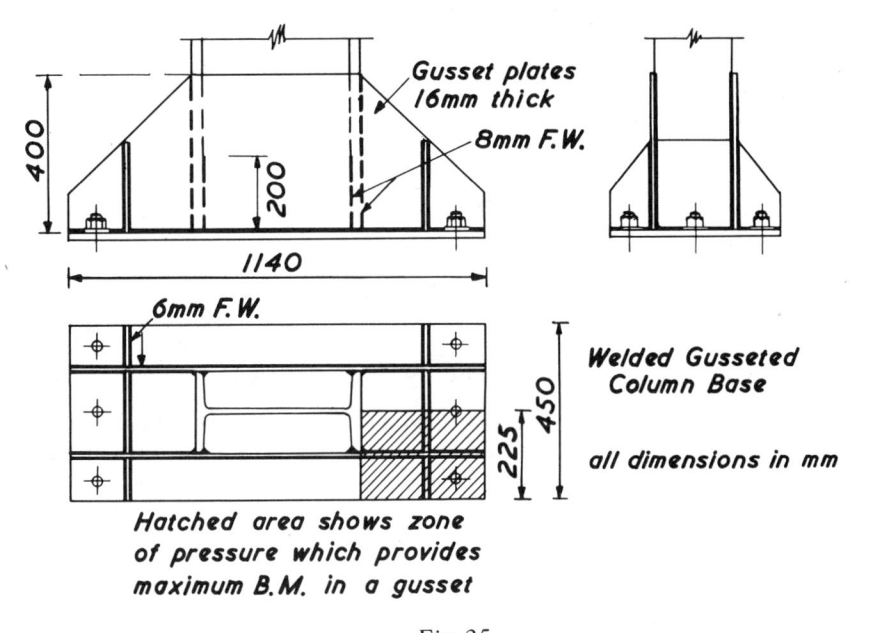

Fig. 35

Welded Gusseted Column Base

all dimensions in mm

Hatched area shows zone of pressure which provides maximum B.M. in a gusset

Gusset Plates

Assume that the two gusset plates are 400 mm high and 16 mm thick.

The critical points for B.M. in a gusset are at the faces of the column flanges. In this case the maximum B.M. is at C in Fig. 34. The load consists of the upward pressure on the base plate in the zone hatched in Fig. 35. The intensity of pressure is shown in the pressure diagram in Fig. 34 and the lever arm, which may be obtained by calculation or graphically as shown, is 203.25 mm. Therefore,

$$M_{\max} = \text{load} \times \text{lever arm}$$

$$= \left(\frac{3\,119 + 2\,175}{2} \times \frac{343 \times 225}{10^6} \right) \times \frac{203.25}{10^3}$$

$$= 209.4 \times 0.20325$$

$$= 42.55 \text{ kNm.}$$

At *C* the section modulus

$$Z = \frac{bd^2}{6} = \frac{16 \times 400^2}{6 \times 10^3}$$

$$= 426.67 \text{ cm}^3$$

Hence, the maximum stress

$$f = \frac{M}{Z} = \frac{42.55 \times 10^6}{426.67 \times 10^3}$$

$$= 99.73 \text{ N/mm}^2.$$

As the gusset plates are unstiffened along the compression edge, this stress is not unduly low and it would not be prudent to reduce the height of the gusset plates.

The maximum shear stress

$$f_q = \frac{3}{2} \times \frac{209.4 \times 10^3}{16 \times 400}$$

$$= 49.07 \text{ N/mm}^2$$

This is satisfactory.

Welds

In the absence of any specific guidance in B.S. 449 : Part 2 : 1969, the following design is based on Clause 50.c of B.S. 449 : 1948, which stated:

> "In riveted and welded construction for stanchions with gusseted bases, the gusset plates, angle cleats, stiffeners, fastenings, etc., in combination with the bearing area of the shaft, all fabricated flush for bearing, shall be sufficient to take the loads, bending moments and reactions to the base plate without exceeding specified stresses, but not less than 60 per cent of the axial load shall be calculated as taken by the fastenings.
>
> In the case of bending moments sufficient rivets or welding shall be provided to transmit the full bending moment to the base.
>
> Where the end of the stanchion shaft and the gusset plates are not faced for complete bearing, the fastenings connecting them to the base plate shall be sufficient to transmit all the forces to which the base is subjected."

The forces are assumed to be imposed by the column through its flanges. Then the force at *C*, due to the direct load, assuming flush bearing

$$F_s = \frac{800}{2} \times 60\% = 240 \text{ kN},$$

while the force at *C* due to B.M.

$$F_b = \frac{\text{B.M.}}{\text{Column depth}} = \frac{152 \times 10^3}{454}$$

$$= 334.8 \text{ kN}.$$

Therefore the total load in the flange at *C* = 240 + 334.8 = 574.8 kN, and the force applied to one gusset plate = 574.8/2 = 287.4 kN.

WELDED ROOF TRUSSES

The maximum length of weld which can be laid on the inside face of the flange is about 200 mm. Allowing 200 mm inside and 400 mm outside and deducting, say, 50 mm for end craters, the length available for welding = 550 mm.

Hence, the force per mm run

$$= \frac{287.4}{550} = 0.522 \text{ kN}$$

∴ Use 8 mm fillet welds (0.64 kN/mm run).

When there is tension in the base plate the welds connecting the gusset plates to the base plates must be capable of carrying the same load as the holding-down bolts. In this case no tension exists and the whole of the remaining welds can be of nominal size sufficient to resist handling and erection stresses, 6 mm fillet welds being suitable.

Typical Roof-truss Connections

Figure 36 illustrates some typical node connections for welded roof trusses.

The connection between the rafter and the tie at the shoe consists of a single 'V' butt joint, while at the apex joints are butt or fillet welded. Everywhere else lap joints are used. Lap joints are usually welded all round to resist corrosion. Appendix B in B.S. 1856 states that the minimum amount of lap shall be four times the thickness of the thinner part, while Table 1 indicates that the minimum leg length of single-run fillet weld shall not be less than 5 mm.

Having decided the amount of overlap to be used it is simple to calculate the length of weld available all round the joint. Then using the allowable stress of 115 N/mm² for all fillet welds in Grade 43 steel and factor of 0.7 for the throat thickness, the leg length of the weld is calculated from the following formula:

$$\text{Leg length} = \frac{\text{Force in member}}{\text{Length of weld} \times 115 \times 0.7}.$$

Considering the central member shown in Fig. 36 (c) and assuming that the force in the member is 15 kN and the length of weld all round is 250 mm, then the leg length of the weld required

$$= \frac{15 \times 10^3}{250 \times 115 \times 0.7} = 0.75 \text{ mm}.$$

This calculated size is much less than the 5 mm which is the minimum allowable leg length.

If 325 mm of 6 mm butt weld are used for the connection in Fig. 36 (a), then the maximum permissible force along the line of weld is

$$\frac{325 \times 6 \times 115}{10^3} = 224.25 \text{ kN}.$$

This is a much greater force than that which might be found in the tie or rafter of a truss.

From the foregoing it will be understood why the welds for such roof trusses are not usually calculated. The shoe members are butt jointed and special arrangements made for site joints, but otherwise 5 mm lap joints are frequently sufficient.

762   CONNECTIONS

The common assumptions made in the determination of forces in lattice frames are that the external loads are applied at the node points or joints and that all the members of the frame meeting at a point are represented by lines which meet at that point. Logically, it therefore follows that the frame should be detailed in

WELDED ROOF TRUSSES

a

S.V.B.W.

Alternative shoes

stiffener

b

S.V.B.W.
F.W.
site joint
F.W.
flat
angles

Typical apex joints for small or large spans

c

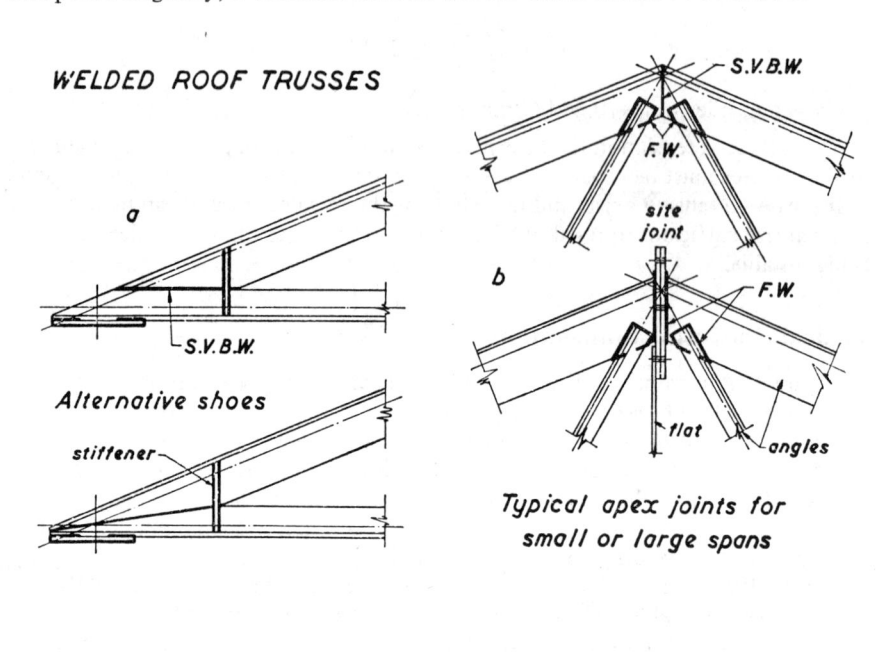

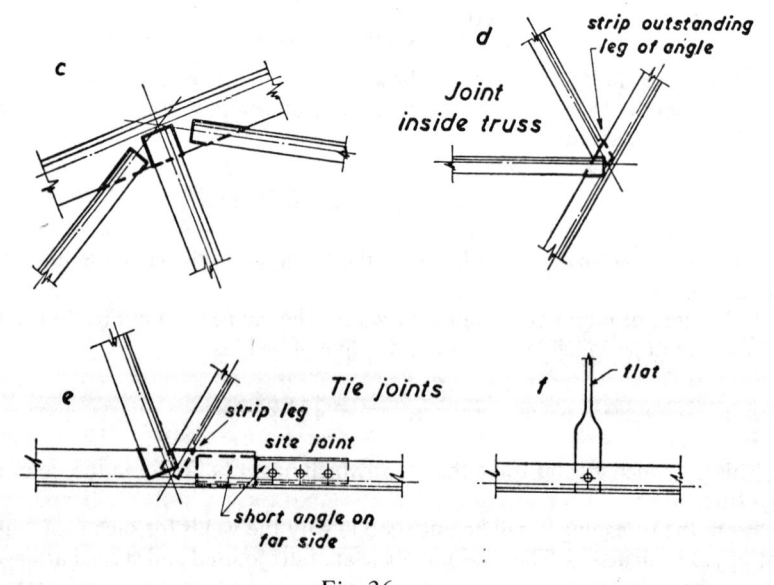

d
Joint inside truss
strip outstanding
leg of angle

e
Tie joints
strip leg
site joint
short angle on far side

f
flat

Fig. 36

ROOF TRUSSES

accordance with the design assumptions. This means that the members should be set out so that the lines of their C.G.s or gravity axes meet at points.

Two common cases of the correct procedure are illustrated in Fig. 37.

If the gravity axes do not intersect, then in addition to the direct forces in the members there will be a B.M. in the joint. See Fig. 38 (a).

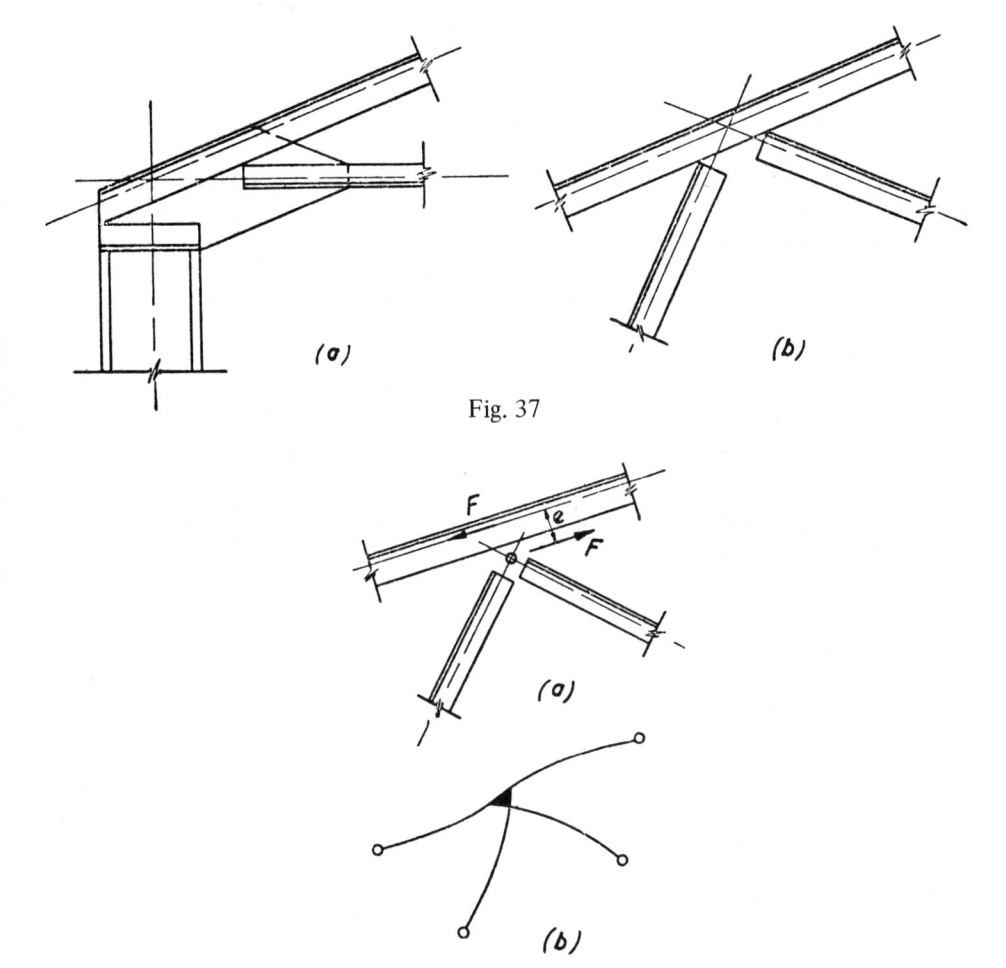

Fig. 37

Fig. 38

This B.M. will be distributed into the members meeting at the joint in proportion to their stiffness. See Fig. 38 (b).

The types of joint shown in Fig. 37 are more economical than that shown in Fig. 38 from the standpoint of the members, but they do involve somewhat larger gusset plates.

Figure 39 illustrates a method of reducing the size of the gusset without losing the advantage of having the gravity axes meet at a point. The cropping of the angles will, of course, add to the cost.

Fig. 39

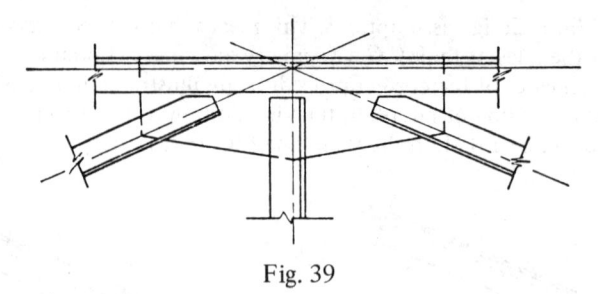

While it is a comparatively simple matter to arrange that the members meeting at a joint carry direct forces only, it is not so simple to arrange that the gusset plates themselves carry direct forces only.

In order that there shall be no B.M. in the gusset plate, the resultant force at any section must lie on the centre line of the plate at that section. It will be found that

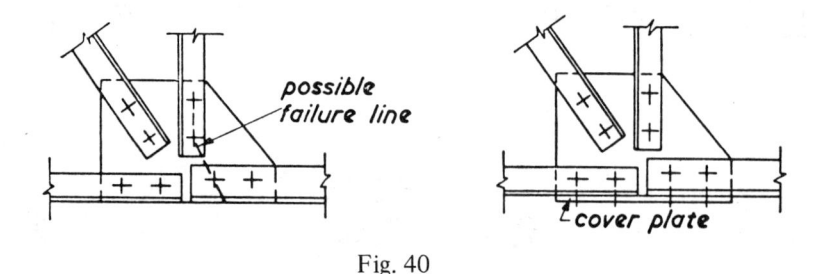

Fig. 40

this can rarely be done except on the simplest of connections, and consequently the gusset plates should be investigated from the point of view of B.M. as well as direct load. *See example page 726.*

Where members which are substantially in the same straight line are discontinuous at a joint, the gusset plate should not be used as both gusset and cover plate unless

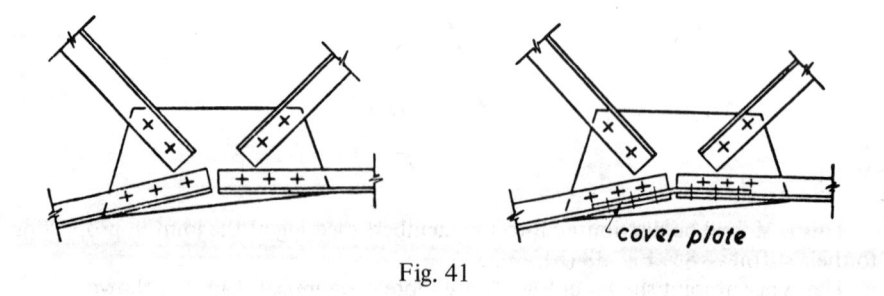

Fig. 41

the calculated stresses show that it is capable of doing so. The use of cover plates in such cases is, in general, more economical than the use of a gusset in the dual capacity. Figures 40 and 41 illustrate the two most common instances of this type of connection.

RIGID-FRAME JOINTS

Although the general principles on which the design of rigid joints is based are fairly well established, the details offer considerable scope for ingenuity. While these joints may be riveted, bolted or welded, the theory to be developed in the following pages will, for simplicity, be largely associated with welding.

A variety of sections may be used in rigid frames, but the most common are Universal Beams and Columns or built-up plate sections.

Knees for Rectangular Frames

Consider a simple knee for a rectangular frame, as shown in Fig. 42. It does not matter, in principle, whether the cross-beam butts against the column or rests on top of it, but it is assumed that there are two sets of stiffeners, so that the web of the joint has a rectangular frame around it.

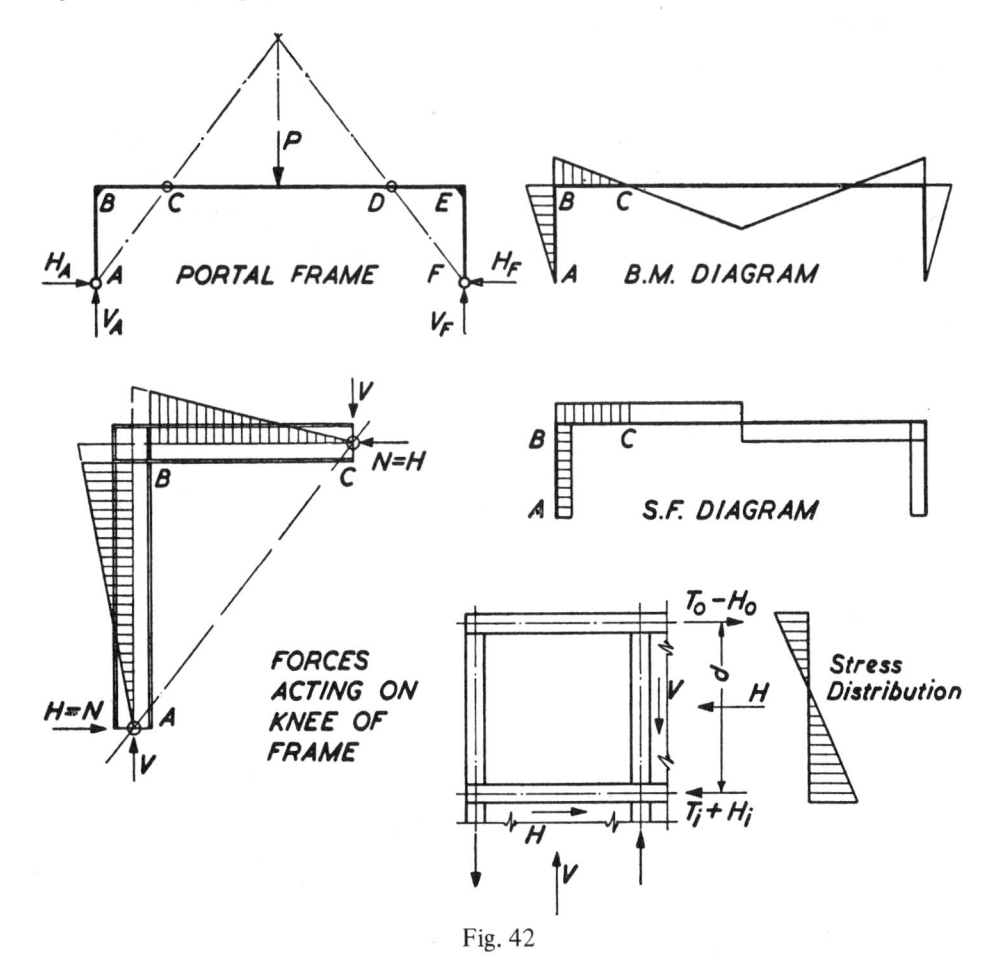

Fig. 42

At any section of a member, the stresses may be found from the normal expression of stress derived in the chapter on Bending and Axial Stresses:

$$f = \frac{P}{A} \pm \frac{M_{xx} y}{I_{xx}},$$

where f = the stress in any fibre,
  P = the longitudinal thrust (i.e., N or H in this case),
  A = the cross-sectional area of the member,
  M_{xx} = the B.M. at the section,
  y = the distance from the neutral axis to the fibre being considered
and I_{xx} = the moment of inertia of the member.

Whilst the maximum B.M. in the frame occurs at the intersection of the neutral axes of the girder and column, the B.M. taken for design purposes can be that at the limits of the knee, i.e., in line with the inside flange of the girder or column.

The shear stress may be found in the strictly accurate manner, giving the distribution shown in Fig. 43 or, in accordance with B.S. 449, by dividing the shear force by the gross web area. Considering the forces applied to the joint, if

  f_o and f_i = the average bending stresses in the outside and inside flanges respectively;

  A_o and A_i = the cross-sectional areas of the outside and inside flanges respectively;

  T_o and T_i = the forces in the outside and inside flanges respectively; while

  H_o and H_i = the components of the horizontal or normal thrust H (or N) in the outside and inside flanges respectively; then

$$T_o = A_o f_o \quad \text{and} \quad T_i = A_i f_i$$

$$H_o = \frac{A_o}{A} H \quad \text{and} \quad H_i = \frac{A_i}{A} H.$$

Consequently, the girder will impose a tensile force of $T_o - H_o$ in the outside flange and a compressive force of $T_i + H_i$ in the inside flange at the boundaries of the knee.

The foregoing analysis is theoretical and can safely be simplified by assuming that the flanges of the girder take the whole of the B.M. and transmit the whole of the thrusts, while the web transmits only the shear.

Then, $\qquad T_o = T_i = \dfrac{M}{d}$ (where d is the depth of the girder)

$$H_o = \frac{A_o}{A_o + A_i} \cdot H$$

$$H_i = \frac{A_i}{A_o + A_i} \cdot H.$$

In rolled sections, where the flanges are equal in size, $H_o = H_i$.

As for the knee itself, experimental evidence shows that there is no tensile stress at the extreme corner as the load takes a direct path across the web. It is possible to assume, therefore, that the tensile forces in the outer flanges vary uniformly from a

RIGID FRAMES

maximum at points in line with the inside flanges of the frame, to zero at the outside corner, as shown in Fig. 43.

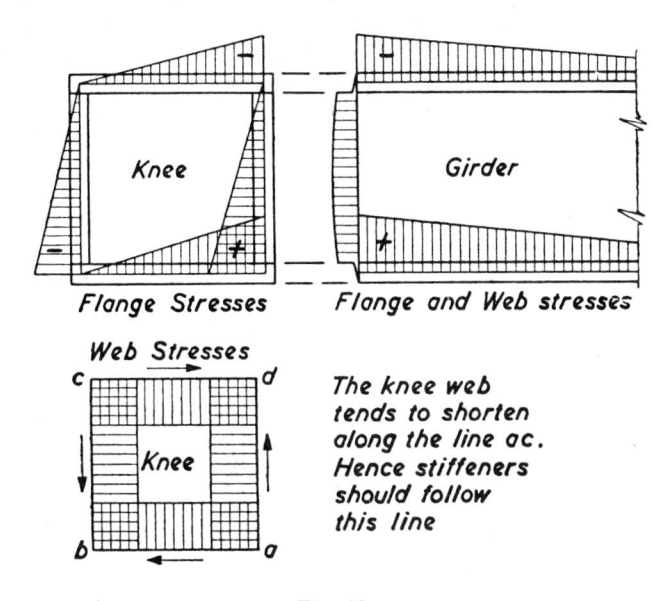

Fig. 43

Each of the flange loads is transmitted into the knee web plate within the lengths of its sides, and this plate is the only means by which the B.M. is transferred from the girder to the column. Consequently, there are heavy shear forces in the knee.

Considering the shear, if

L = the length of the side of the web plate being considered,

t = the web thickness

and T = the total thrust in the flange,

then, the shear per unit length of plate = $\dfrac{T}{L}$,

while the shear stress = $\dfrac{T}{L \times t}$.

In welded knees, the load per unit length of fillet weld (one each side of the web plate) is $T/2L$.

In the top edge of the web plate the thrust T is equal to $T_o - H_o$, while in the bottom edge it is equal to $(T_i + H_i) - H$, as shown in Fig. 42. But $(T_i + H_i) - H = (T_i + H_i) - (H_i + H_o) = T_i - H_o = T_o - H_o$. Therefore the shear forces in the top and bottom edges of the web plate are equal. Similarly, those in the outside and inside vertical edges are equal.

If the forces in the web tend to cause overstressing, the web may be increased in thickness or provided with suitable stiffeners. The normal procedure for simple knees is to use diagonal stiffeners.

American Research on Rectangular Knees

Much of the experimental work carried out on the knees of portal frames has been done in America, particularly by the American Bureau of Standards and at Lehigh University. One of the more easily adaptable groups of formulae for rectangular web plates, attributed to Osgood, was published in Research Paper, R.P. 1130 (reference 4). The theory and example which follow are based on extracts from this paper.

Consider a flat rectangular plate of uniform thickness t, loaded by forces and couples, as shown in Fig. 44. For equilibrium,

$$M_x - a(2F_{xy} + F_y) = M_y - b(2F_{yx} + F_x).$$

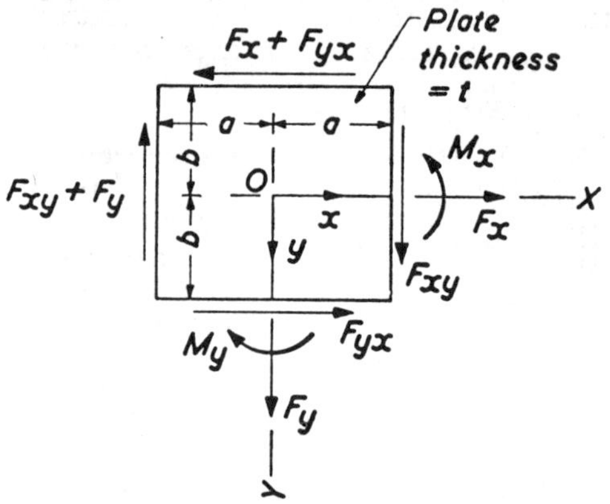

Fig. 44

It is required to determine the stress conditions in the plate, assuming that the normal stresses f_x and f_y along the boundaries $x = a$ and $y = b$ respectively are uniformly varying and along the boundaries $x = -a$ and $y = -b$ are everywhere zero.

Such a condition may be derived from the Airy Stress Function:

$$\phi = b_2 xy + \tfrac{1}{6}(3c_3 + d_4 y)(a + x)y^2 + \tfrac{1}{6}(3b_3 + b_4 x)(b + y)x^2.$$

It can be shown that:

$$b_2 = \frac{1}{4abt}\left[M_x - a(2F_{xy} + F_y) - \frac{3}{2}(M_x + M_y)\right]$$

$$b_3 = \frac{F_y}{4abt}, \quad c_3 = \frac{F_x}{4abt}$$

$$b_4 = \frac{3M_y}{4a^3bt}, \quad d_4 = \frac{3M_x}{4ab^3t}, \text{ from which it may be derived that:}$$

$$f_x = \frac{1}{4abt}\left(F_x + \frac{3M_x}{b^2}y\right)(a + x)$$

$$f_y = \frac{1}{4abt}\left(F_y + \frac{3M_y}{a^2}x\right)(b + y)$$

$$v_{xy} = -\frac{1}{4abt}\left[M_x - a(2F_{xy} + F_y) + F_y \cdot x + F_x \cdot y - \frac{3M_y}{2}\left(1 - \frac{x^2}{a^2}\right) - \frac{3M_x}{2}\left(1 - \frac{y^2}{b^2}\right)\right]$$

where v_{xy} = the shear stress

$$F_x = -(1-k-j)H - \left(\frac{m}{1-k+j} - \frac{n}{1+k-j}\right)\frac{M}{b}$$

$$F_y = -(1-2p)V$$

$$F_{xy} = V - pV - \frac{r}{p}(M_O - Hb)$$

$$F_{yx} = H - jH - \frac{nM}{(1+k-j)b}$$

$$M_x = -\left(1 - \frac{m}{1-k+j} - \frac{n}{1+k-j}\right)M$$

$$M_y = -(1-2r)(M_O - Hb).$$

These formulae have been quoted in their original form, but the stresses f_x and f_y will be negative if compressive and positive if tensile, the signs being opposite to those given elsewhere in this section.

In the foregoing formulae p and r are the proportions of V and M respectively, which are taken by each flange of the column at the edge $y = b$. If a flange is not wholly continuous at the knee, as in riveted construction, it will transmit stress only partially across the discontinuous section. Consequently, k and j are the proportions of H, and m and n are the proportions of M, which are taken by the top and bottom flanges of the beam portion of the knee at the edge $x = a$. Had there been no discontinuity, k would equal j and m would equal n. It is further assumed that the flanges carry no transverse shear.

For a welded frame,

$$M_O = V(a + A) = H(b + B), \qquad M = M_O - Va$$

$$M_x = -M(1 - 2n), \qquad M_y = -(M_O - Hb)(1 - 2p)$$

$$F_x = -(1-2j)H, \qquad F_y = -(1-2p)V$$

$$F_{xy} = V - pV - \frac{r}{a}(M_O - Hb)$$

$$F_{yx} = H - jH - \frac{nM}{b}$$

$$j = \frac{\text{Area of one flange of beam}}{\text{Total sectional area of beam}},$$

$$n = \frac{\text{Moment of inertia } I \text{ of one flange of beam}}{\text{Total } I \text{ for the beam}}, \text{ and}$$

p and r = the corresponding quantities for the column.

Principal Stresses and Greatest Shear Stresses

Although not required by B.S. 449 : Part 2 : 1969, the principal stresses in the knee web of a frame can be computed from the usual formula:

$$f = \frac{f_x + f_y}{2} \pm \sqrt{\left[\left(\frac{f_x - f_y}{2}\right)^2 + v_{xy}^2\right]}.$$

Normally the greatest stress occurs at the inside corner of the knee where $x = +a$, $y = +b$.

The greatest shear stress in the web occurs at the point:

$$x = -\frac{b_3}{b_4}, \quad y = \frac{c_3}{d_4}, \quad \text{i.e.,} \quad x = -\frac{F_y \cdot a^2}{3M_y}, \quad y = -\frac{F_x \cdot b^2}{3M_x},$$

the maximum stress being computed from the formula:

$$v_{max} = \sqrt{\left[\left(\frac{f_x - f_y}{2}\right)^2 + v_{xy}^2\right]}.$$

It will be seen that it is necessary to compute two sets of coefficients, first to obtain the principal stresses, and then to obtain the greatest shear stress. Now the point where the greatest shear stress occurs is very near the centre of the web. Noting this fact, a Canadian engineer, Prof. D. T. Wright, has produced a formula for the maximum shear stress which gives results within about 2 per cent of the exact figures derived from Osgood's formula, viz.:

$$v = \frac{M_c}{4abt}\left(1 + \frac{a^2 t}{3Z_a} + \frac{b^2 t}{3Z_b}\right),$$

where $M_c = M_O - H \cdot b - V \cdot a$ = the moment of the inside corner of the frame.

Z_a = the section modulus of the knee along a horizontal axis, the section including the vertical flanges as well as the web plate, and Z_b = the corresponding section modulus along a vertical axis.

It is often quicker to use this formula than that evolved by Osgood.

The various formulae will be demonstrated by an example.

Example. Figure 46 shows a square knee for a rectangular portal frame in which the flanges are 300 mm x 25 mm in cross-section, and the web plate is 1.0 m x 1.0 m x 12 mm thick.

Fig. 46

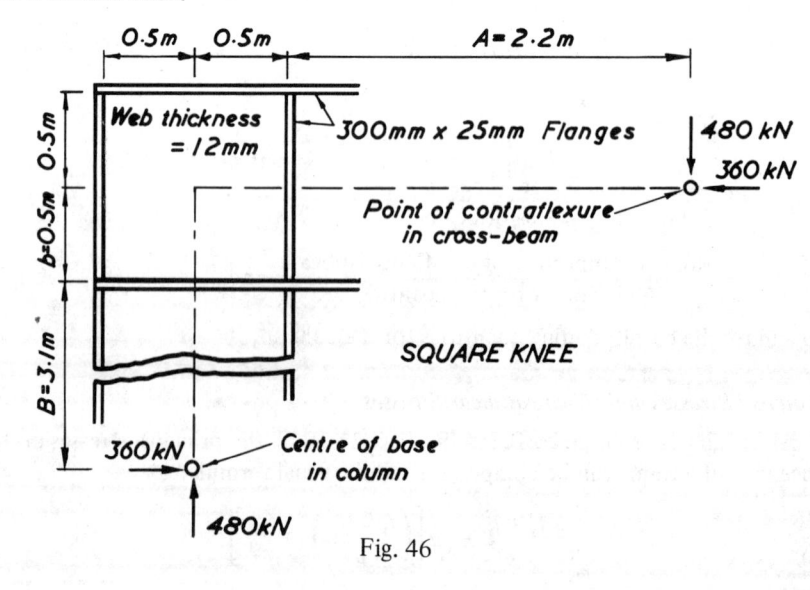

RIGID FRAMES

Using Osgood's formulae,

$$a = b = 500 \text{ mm} \qquad t = 12 \text{ mm}$$
$$A = 2.2 \text{ mm} \qquad B = 3.1 \text{ m}$$
$$V = 480 \text{ kN} \qquad H = 360 \text{ kN}.$$

The proportion of the reaction V taken by each flange of the column,

$$p = \frac{\text{Area of one flange}}{\text{Area of whole section}} = \frac{A_f}{2A_f + A_w}$$

(where f and w refer to flange and web respectively)

$$= \frac{(300 \times 25)}{(2 \times 300 \times 25) + (1\,000 \times 12)} = 0.2778.$$

The proportion of the moment M taken by each flange

$$r = \frac{I_f}{2I_f + I_w} = \frac{300 \times 25 \times 512.5^2/10^4}{2(300 \times 25 \times 512.5^2/10^4) + (12 \times 1\,000^3/12 \times 10^4)}$$

$$= \frac{196\,992.1875}{493\,984.3750} = 0.3988.$$

The proportion of the thrust H taken by each flange of the beam,

$$k = j = 0.2778 \quad \text{(as above for } p\text{)}.$$

The proportion of the moment M taken by each flange,

$$m = n = 0.3988 \quad \text{(as above for } r\text{)}$$

The moment about the centre of gravity of the knee web,

$$M_O = V(a + A) = H(b + B)$$
$$= 480(0.5 + 2.2)$$
$$= 1\,296 \text{ kNm}.$$

The moment at the junction with the beam (where $x = a$),

$$M = M_O - Va = 1\,296 - (480 \times 0.5)$$
$$= 1\,056 \text{ kNm}.$$

$$F_x = -(1 - 2j)H = -160.000 \text{ kN}$$
$$F_y = -(1 - 2p)V = -213.333 \text{ kN}.$$

$$F_{xy} = V - pV - \frac{r}{a}(M_O - Hb)$$

$$= 480 - (0.2778 \times 480) - \frac{0.3988(1\,296 - 360 \times 0.5)}{0.5}$$

$$= -543.415 \text{ kN}$$

$$F_{yx} = H - jH - \frac{nM}{b}$$

$$= 360 - (0.2778 \times 360) - \frac{0.3988 \times 1\,056}{0.5}$$

$$= -582.228 \text{ kN}$$

$$M_x = -M(1 - 2m)$$

$$= -1\,056(1 - 2 \times 0.3988)$$

$$= -213.772 \text{ kNm}$$

$$M_y = -(M_O - Hb)(1 - 2r)$$

$$= -(1\,296 - 360 \times 0.5)(1 - 2 \times 0.3988)$$

$$= -225.918 \text{ kNm.}$$

Checking for equilibrium,

$$M_x - a(2F_{xy} + F_y) = M_y - b(2F_{yx} + F_x)$$

$$-213.772 - 0.5(-1\,086.830 - 213.333) = -225.918 - 0.5(-1\,164.456 - 160.000)$$

or

$$436.310 = 436.310$$

which satisfies the expression $\Sigma M = 0$.

Stresses

The greatest compressive stresses occur at the inside corner where $x = a$, $y = b$. Considering the stresses at this point,

$$f_x = \frac{1}{4abt}\left(F_x + \frac{3M_x \cdot y}{b^2}\right)(a + x)$$

$$= \frac{1}{4 \times 0.5 \times 0.5 \times 12 \times 10^{-3} \times 10^3}\left(-160.0 + \frac{-3 \times 213.772 \times 0.5}{0.5^2}\right)(0.5 + 0.5)$$

$$= -120.219 \text{ N/mm}^2$$

$$f_y = \frac{1}{4abt}\left(F_y + \frac{3M_y \cdot x}{a^2}\right)(b + y)$$

$$= \frac{1}{4 \times 0.5 \times 0.5 \times 12 \times 10^{-3} \times 10^3}\left(-213.333 + \frac{-3 \times 225.918 \times 0.5}{0.5^2}\right)(0.5 + 0.5)$$

$$= -130.737 \text{ N/mm}^2$$

$$v_{xy} = -\frac{1}{4abt}[M_x - a(2F_{xy} + F_y) + F_y \cdot x + F_x \cdot y + 0 + 0]$$

$$= -\frac{1}{4 \times 0.5 \times 0.5 \times 12 \times 10^{-3} \times 10^3}[-213.772 - 0.5(-1\,086.830 - 213.333)$$

$$-(213.333 \times 0.5) - (160.00 \times 0.5)]$$

$$= -20.804 \text{ N/mm}^2.$$

The shear stress has been computed as it is needed for the calculation of the principal stresses, but this stress is not the greatest shear stress in the knee. This will be computed later. The maximum principal stress, in compression, is:

$$f_{max} = \frac{f_x + f_y}{2} - \sqrt{\left[\left(\frac{f_x - f_y}{2}\right)^2 + v_{xy}^2\right]}$$

$$= \frac{-120.219 - 130.737}{2} - \sqrt{\left[\left(\frac{-120.219 + 130.737}{2}\right)^2 + 20.804^2\right]}$$

$$= -125.478 - 21.458$$

$$= -146.936 \text{ N/mm}^2$$

The formula evolved by Prof. Wright will be used to find the greatest shear stress:

$$v = \frac{M_c}{4abt}\left(1 + \frac{a^2 t}{3Z_a} + \frac{b^2 t}{3Z_b}\right).$$

Now $\quad M_c = M_O - H \cdot b - V \cdot a$

$$= 1\,296 - (360 \times 0.5) - (480 \times 0.5)$$

$$= 876 \text{ kNm}$$

$$Z_a = Z_b = \text{Total } I/a = 493\,984.375/50 \text{ cm}^3.$$

Hence,

$$v = \left[\frac{876}{4 \times 0.5 \times 0.5 \times 12 \times 10^{-3}}\left(1 + \frac{2 \times 0.5^2 \times 12 \times 10^{-3} \times 50}{3 \times 493\,984.375 \times 10^{-6}}\right)\right] 10^{-3}$$

$$= 87.778 \text{ N/mm}^2.$$

Knees for Rigid Frames with Pitched Roofs

It is quite common to haunch the knees of frames with pitched roofs, as shown in Fig. 47, or by curving the inner flange, a method of treatment to be described later.

There is ample experimental evidence to show that the neutral axis of stress in haunched knees moves towards the inside of the frame, as shown in Fig. 47 (a). Consequently the approximate force in the inside flange may be found by assuming that the neutral axis occurs at the third point along the diagonal from the inside flange to the outside corner of the frame. Alternatively the force in the flange may be resolved from the triangle of forces shown in Fig. 47 (a), the force being known that in the inside flange of the rafter. In the type of joint shown in Fig. 47 (b) it is often assumed that the whole of the inside flange forces are resolved into the bracket flange, but, as an alternative, it may be assumed that, say, one-half of the force continues along the flange of the main member, as shown in the diagram.

The stresses in knees of the type shown in Fig. 47 (a), (c) and (d) can be found by using Vierendeel's Tapered Beam formulae or Olander's formulae which are described later. See Figs. 60 and 62. Considering the knee in Fig. 48, all the sections between *AA* and *BB*, except in the hatched areas can be analysed by the above formulae. However, if the knee is not over-stressed elsewhere, it is very unlikely to be so in the hatched zones.

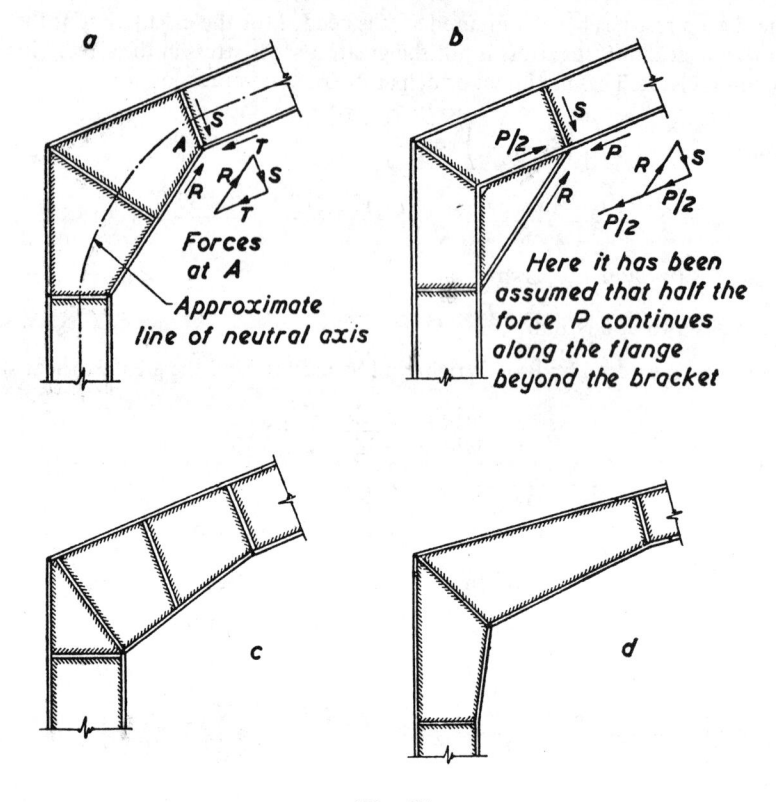

Fig. 47

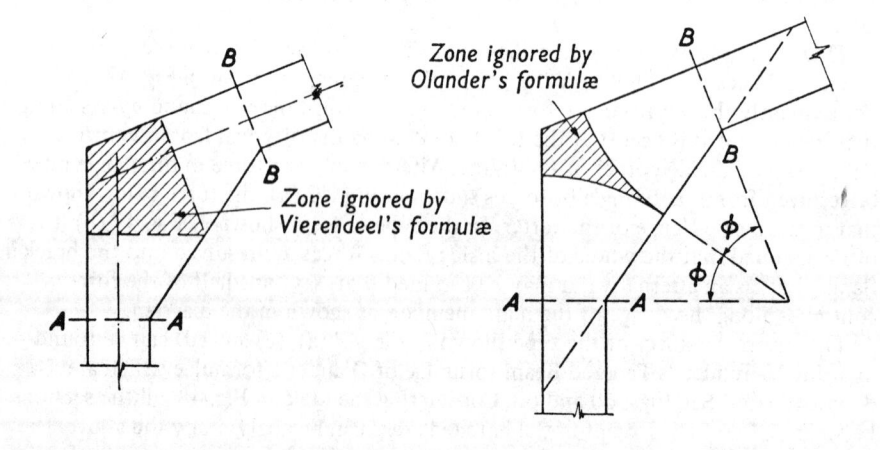

Fig. 48

RIGID FRAME JOINTS

The forces in the stiffeners at the limits of the knee are found by a resolution of forces, as at point A in Fig. 47 (a) where S is the appropriate stiffener force, and the stiffeners and welds are designed accordingly. The remaining stiffeners inside the knee can be of nominal size, their primary function being to prevent local buckling of the web and lateral failure of the inner flange.

Ridges in Pitched Roofs

Ridges in pitched roofs are designed in precisely the same manner as obtuse-angled knees. Normally they present less difficulty than knee joints as the angle between the rafters is very obtuse and the forces are much less than in knees. If joist sections are employed it may be unnecessary to add brackets, while in lightly loaded structures it is sufficient to butt weld the ends of the rafters. Some typical joints are shown in Fig. 49.

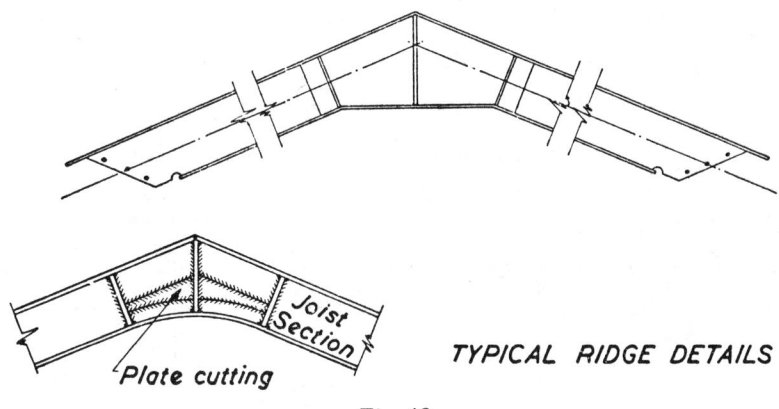

Fig. 49

Knees with Curved Flanges

It is probable that joints with curved flanges will always be the subject of some controversy but here it is proposed to describe and illustrate the articles published or research work carried out both in this country and abroad.

It is of some interest to consider a formula, known as the Winkler-Résal formula, which can be used for an *initially* curved bar with parallel flanges, such as that shown in Fig. 50.

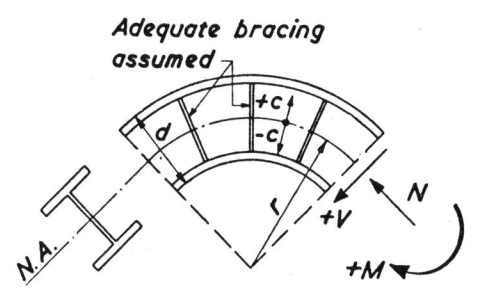

Fig. 50

776   CONNECTIONS

The formula for the stress in any fibre of the bar is

$$f = \frac{N}{A} - \frac{M}{r.A} - \frac{M.c}{U} \times \frac{r}{r+c},$$

where N = the normal thrust,
  A = the cross-sectional area of the bar,
  M = the applied B.M.
  r = the initial radius of curvature of the bar taken to the neutral axis N.A. of the section,
  c = the distance from the N.A. to the fibre being considered, being positive when measured away from the centre of curvature and negative when measured towards it
and U = a figure analogous to the moment of inertia I and which may be replaced by I when the value of r is greater than twice the depth d of the bar.

When r is less than twice d and the section is composed of rectangles,

$$U = r^2(2.30258 r \Sigma b . \log w_1/w_2 - A),$$

where the symbols have the significance shown in Fig. 51.

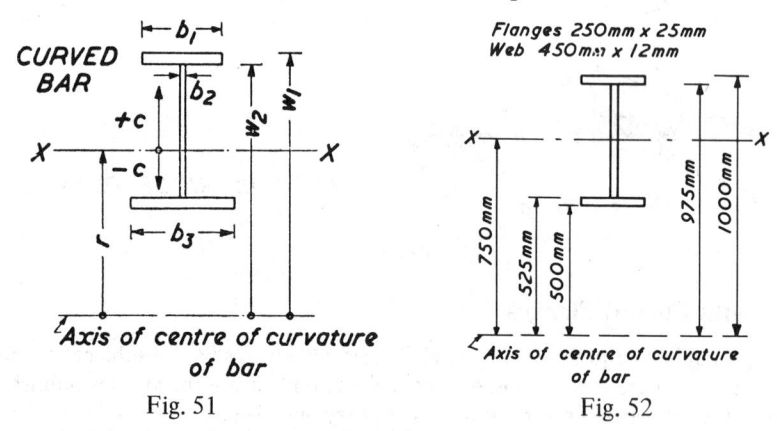

Fig. 51

Fig. 52

Suppose the member is of the section shown in Fig. 52, the properties being as follows:

  Flanges = 250 mm x 25 mm   Web = 450 mm x 12 mm
  Area   = 17 900 mm² r = 750 mm (r/d = 1.5).

Then, $U = r^2(2.30258 r \Sigma b . \log w_1/w_2 - A)$.
Now,
        250 log 1 000/975 = 2.74875
         12 log 975/525 = 3.22614
        250 log 525/500 = 5.29733
                          ─────────
        $\Sigma b . \log w_1/w_2$ = 11.27222

RIGID FRAME JOINTS

Therefore,

$$U = 750^2(2.30258 \times 750 \times 11.27222 - 17\,900) \times 10^{-4}$$
$$= 88\,109.37 \text{ cm}^4.$$

It is interesting to note that the moment of inertia $I = 250 \times 500^3/12 \times 10^4 - 238 \times 450^3/12 \times 10^4 = 79\,685.42$ cm^4.

It will be observed that the general form of the stress equation resembles the normal stress formula, i.e., $N/A \pm M/Z$.

The Winkler-Résal formula should be used for curved members when r/d is less than 2.5. When this ratio exceeds 2.5, the normal formula can be used with safety.

Example. Considering the member shown in Fig. 52, and assuming that it is subjected to a B.M. of 475 kNm,

$$f = \frac{N}{A} - \frac{M}{r.A} - \frac{M.c}{U} \times \frac{r}{r+c} = \frac{N}{A} - \frac{M}{r.A} - \frac{M.r}{U} \times \frac{c}{r+c}$$

$$= 0 - \frac{475 \times 10^6}{750 \times 17\,900} - \frac{475 \times 10^6 \times 750}{88\,109 \times 10^4} \times \frac{c}{750+c}$$

$$= -35.382 - 404.327 \times \frac{c}{750+c} \text{ (N/mm}^2\text{)}.$$

If values are plotted for various depths c from the neutral axis, the stress diagram shown in Fig. 53 is obtained.

It should be noted that the Winkler-Résal formula makes suitable allowance for the shift of the neutral plane of bending from the N.A. of the section towards the inner flange.

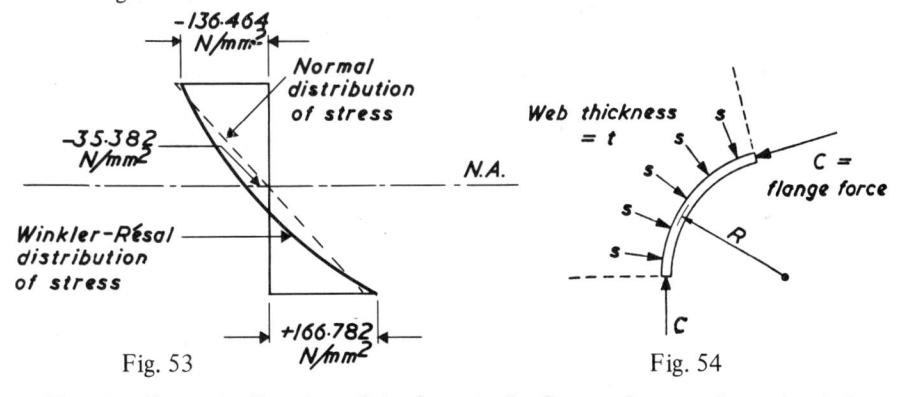

Fig. 53

Fig. 54

Now the change in direction of the force in the flange of a curved member induces radial stresses in the web which can be calculated from the following formula, due to Professor Campus of Liège:

$$s = C/Rt, \quad \text{(see Fig. 54)}$$

where s = the unit radial stress,
  C = the total flange force,
  R = the radius of curvature of the flange being considered
and t = web thickness.

Professor Magnel (reference 6) states that the radial stress should be added to that due to the shear across the web.

The radial force is applied at the junction of web and flange and causes cross-bending, the edges of the flanges moving away from the centre of curvature when the flange is compressed and towards it when it is in tension, as shown in Fig. 54.

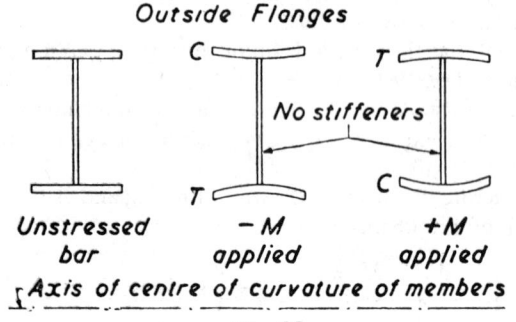

Fig. 55

Dr. Hans Bleich (reference 7) investigated the effects of this phenomenon and produced two coefficients ν and μ (nu and mu), the first being associated with the longitudinal stresses in the flanges and the second the transverse stresses.

If f is the mean stress derived from the Winkler-Résal formula, then max. $f = f/\nu$ and $f' = \mu$ max. f, where ν and μ have the following values with respect to the expression b^2/Rt, the symbols for which are shown in Fig. 56:

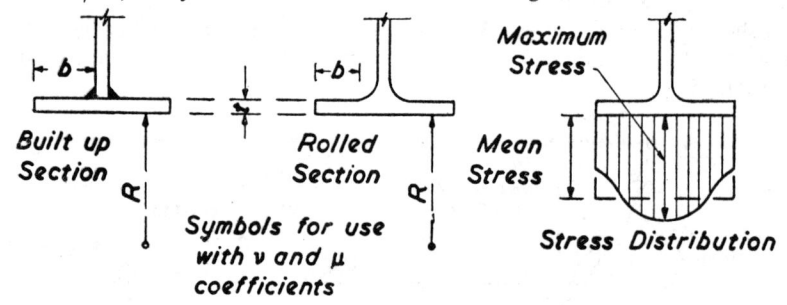

Fig. 56

Bleich's Coefficients

$\dfrac{b^2}{Rt}$	0	0.1	0.2	0.3	0.4	0.5	0.6	0.7	0.8	0.9
ν	1.000	0.994	0.977	0.950	0.915	0.878	0.838	0.800	0.762	0.726
μ	0	0.297	0.580	0.836	1.056	1.238	1.382	1.495	1.577	1.636
$\dfrac{b^2}{Rt}$	1.0	1.1	1.2	1.3	1.4	1.5	2.0	3.0	4.0	5.0
ν	0.693	0.663	0.636	0.611	0.589	0.569	0.495	0.414	0.367	0.334
μ	1.677	1.703	1.721	1.728	1.732	1.732	1.707	1.671	1.680	1.700

RIGID FRAME JOINTS

If the radial forces tend to lead to overstressing, the flanges must be braced either by entire web stiffeners or by small gussets, the spacing of which is a matter for engineering judgment as no rules, mathematical or empirical, have been derived.

Some practical examples of curved knees closely resembling curved bars are shown in Figs. 57 and 58, the latter being described in reference 8, as well as in other papers.

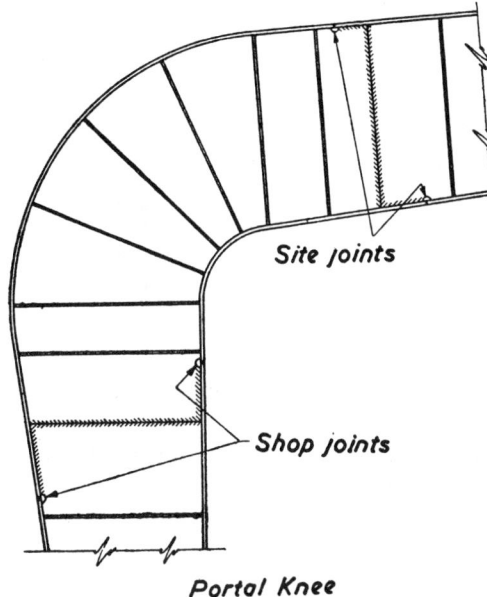

Portal Knee for Bus Garage
Fig. 57

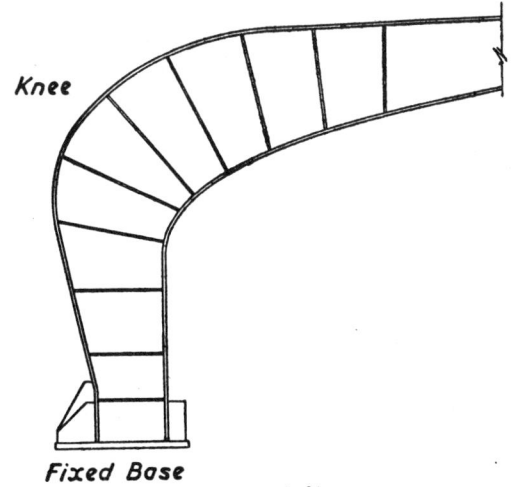

Portal Knee for Bridge
Fig. 58

780 CONNECTIONS

Generally knees are not shaped like curved bars, the majority being of the type shown in Figs. 59 (a) and (b) where the outside flange is straight. The rapid change of section at the knee, and the curvature of the centre-line affect both the magnitude and the distribution of the fibre stresses.

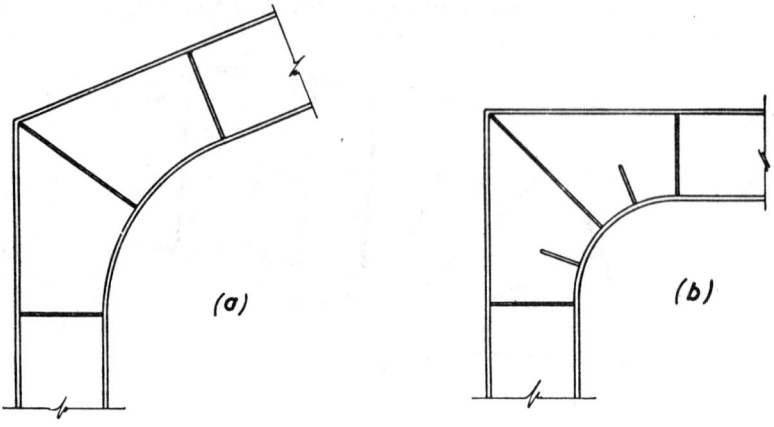

Fig. 59

Professor Vierendeel devised Tapered Beam formulae for the knee shown in Fig. 60, the formulae being as follows:

On any section AA,

$$f_o = \frac{P}{A} - \frac{Ma_o}{I}$$

$$f_i = \frac{P}{A} + \frac{Ma_i}{I}$$

$$v = \frac{1}{bd}(V + f_i F_i \sin \phi),$$

VIERENDEEL'S TAPERED BEAM SYMBOLS

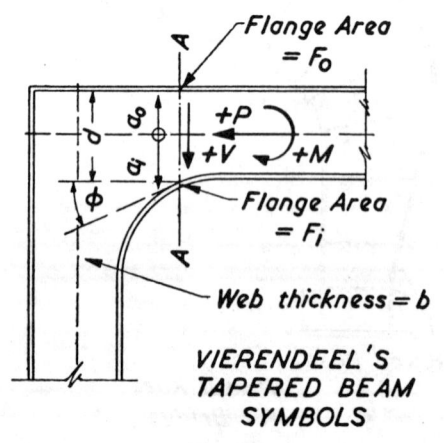

Fig. 60

where f_o and f_i = the mean stresses in the outside and inside flanges, respectively,

a_o and a_i = the distances of the centroids of the outside and inside flanges, respectively, from the axis shown,

v = the shear stress

M, V and P = the bending moment, shear force and thrust at the section AA,

b = the web thickness,

d = the web depth,

$A = bd + F_o + F_i \cos \phi$

and $I = bd^3/12 + F_o a_o^2 + F_i \cos \phi \, a_i^2$.

These formulae are logical and give reasonable results.

When the mean flange stresses have been calculated, the maximum stresses may be computed using Dr. H. H. Bleich's coefficients and suitable stiffeners added where necessary.

It should be noted that the Tapered Beam formulae can be applied to knees of the type shown in Fig. 47.

In the U.S.A. a number of investigators have produced formulae for curved knees, using circular sections as shown in Fig. 61.

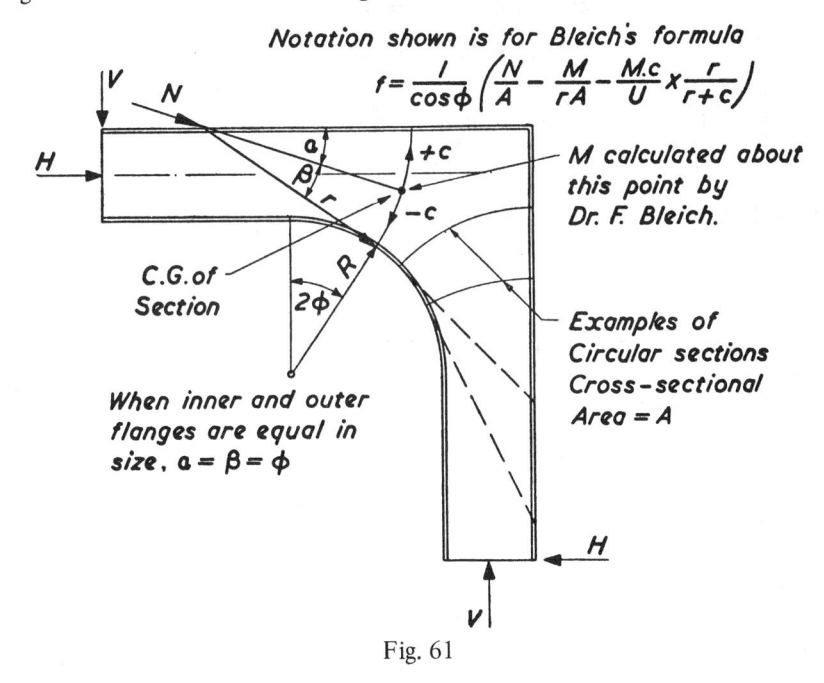

Fig. 61

Osgood's formulae (reference 5) are mathematically exact but are lengthy and cumbersome for use in design and are not quoted here.

Dr. Friedrich Bleich (reference 9) adapted the Winkler-Résal formula, adjusting it by multiplying the whole expression by $1/\cos \phi$. It will be noted from Fig. 61 that r and c are measured in a different way, but the principles of calculation are identical with those for Example 1.

The amount of work involved in using Bleich's method is less than with Osgood's method but greater than that with a recent method (reference 10) due to Harvey C. Olander, which gives reasonable results and can be recommended for use in design.

Olander states that the method is simply to take circular sections that cut the extreme fibres at right angles, such as section AB, Fig. 62, develop the section as shown and obtain the cross-sectional area A and moment of inertia I of the developed section. Next, resolve all forces to the right of section AB into the values P_O and M_O about the point O, the centre of the arc. P_O passes through the centre of gravity of section AB, and M_O is the moment of the forces about O. Then, with these values, the stresses are calculated as for an ordinary beam, except that the

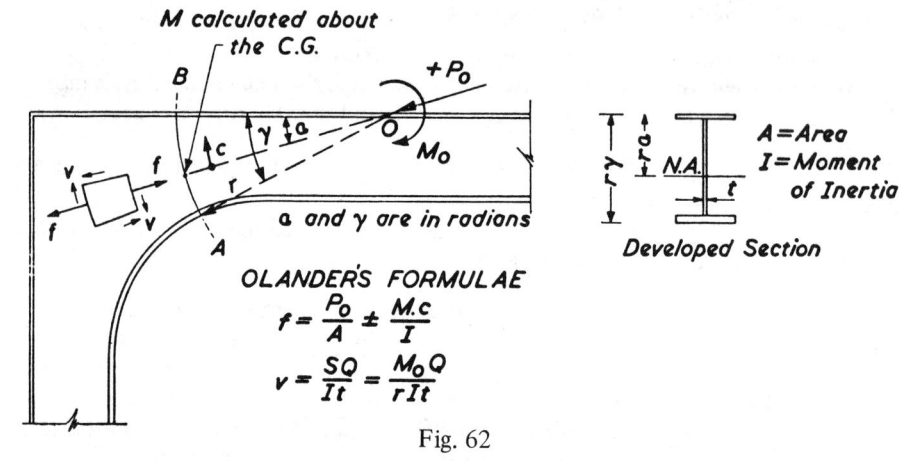

Fig. 62

OLANDER'S FORMULAE
$$f = \frac{P_O}{A} \pm \frac{M.c}{I}$$
$$v = \frac{SQ}{It} = \frac{M_O Q}{rIt}$$

A = Area
I = Moment of Inertia

Developed Section

shear is determined from M_O. The total shear on section $AB = S = M_O/r$. Then the unit shear along the section,

$$v = \frac{SQ}{It} = \frac{M_O Q}{rIt}$$

where Q is the statical moment of the area of the section about the point being considered. (Cf. British notation, $v = S \cdot a \cdot y/It$.)

The stresses normal to the section,

$$f = \frac{P_O}{A} \pm \frac{M \cdot c}{I}$$

where M is the B.M. at the C.G. of the section AB.

An example will help to explain the method.

Example. Figure 62 shows a curved knee joining a 305 mm × 305 mm × 198 kg UC to a 610 mm × 305 mm × 238 kg UB. The properties of the two sections are as follows:

Section (mm)	Flange Thickness (mm)	Web Thickness (mm)	Area A (cm²)	Z_{xx} (cm³)
305 × 305	31.4	19.2	252	2 991
610 × 305	31.4	18.6	304	6 549

RIGID FRAME JOINTS

The cross-beam is chosen to resist the B.M. in the centre of the beam and will be understressed at its junction with the knee. If the radius of the inner flange is 1.2 m, then the radius will be almost twice the depth of the larger section or four times that of the smaller section.

It is convenient to continue the flange breadth of the narrower section around the knee, i.e., to use a 312 mm x 32 mm plate. The web of the knee will be 20 mm thick, i.e., a 'preferred' thickness close to those of the two sections.

The knee can be divided into any number of circular sections. As, however, the greatest stress is situated just inside the knees, these areas should be investigated. Consequently, sections are usually chosen at 15° or 18° intervals along the inside of the knee.

If 18° intervals are used in this case, there will be six cross-sections to consider, as shown in Fig. 63.

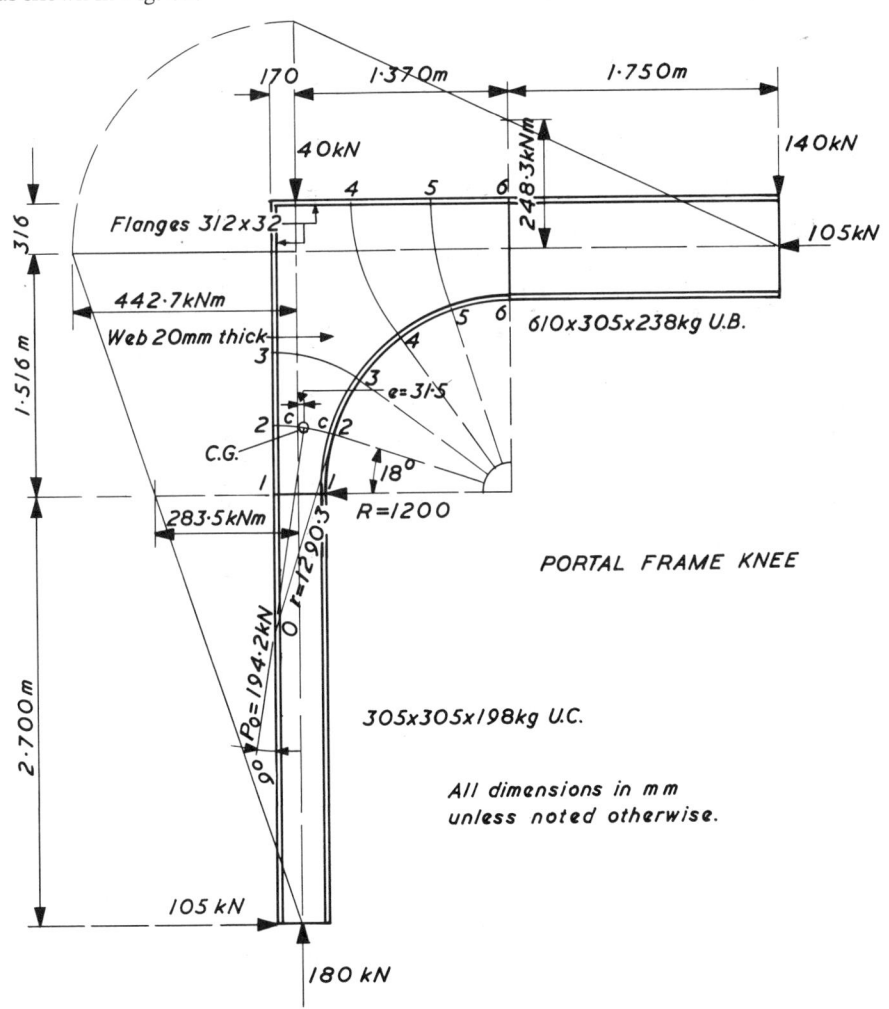

Fig. 63

It is advisable to check that sections 1–1 and 6–6 are in order first, before embarking upon the knee proper.

Considering section 1–1,

$$P = 180 \text{ kN}, \quad A = 252 \text{ cm}^3,$$
$$M_{xx} = 283.5 \text{ kNm}, \quad Z_{xx} = 2\,991 \text{ cm}^3$$

Then
$$f = \frac{P}{A} \pm \frac{M_{xx}}{Z_{xx}}$$
$$= \frac{180 \times 10^3}{252 \times 10^2} \pm \frac{283.5 \times 10^6}{2\,991 \times 10^3}$$
$$= +101.93 \text{ or } -87.64 \text{ N/mm}^2.$$

Using the normal B.S. 449 procedure for joists, the shear stress

$$v = \frac{\text{load}}{\text{gross web area}} = \frac{105 \times 10^3}{340 \times 19.2}$$
$$= 16.08 \text{ N/mm}^2.$$

Considering section 6–6, the properties of the rolled section should again be taken, as those for the adjoining knee section are greater in magnitude.

$$P = 105 \text{ kN} \quad A = 304 \text{ cm}^2$$
$$M_{xx} = 248.3 \text{ kNm} \quad Z_{xx} = 6\,549 \text{ cm}^3$$

Then
$$f = \frac{P}{A} \pm \frac{M_{xx}}{Z_{xx}}$$
$$= \frac{105 \times 10^3}{304 \times 10^2} \pm \frac{248.3 \times 10^6}{6\,549 \times 10^3}$$
$$= +41.37 \text{ or } -34.46 \text{ N/mm}^2$$

and
$$v = \frac{140 \times 10^3}{633 \times 18.6}$$
$$= 11.89 \text{ N/mm}^2.$$

Consequently, the stresses at sections 1–1 and 6–6 are acceptable.

Although Olander's method is quicker than some other methods of analysis, a certain amount of time is taken in calculating the geometrical properties of the circular sections. It is desirable to devise some system of tabulation for the calculations.

Figure 64 shows the symbols which have to be calculated for each section.

$$r = \frac{h + R(1 - \cos 2\phi)}{\sin 2\phi}$$
$$a = R \cdot \sin 2\phi$$
$$(a + b) = r \cdot \cos 2\phi$$
$$d = r \cdot \cos \phi$$
$$c = \phi \cdot r \quad (\phi \text{ in radians})$$
$$(g + e) = r \cdot \sin \phi,$$

where h, R, ϕ and g are known quantities.

RIGID FRAME JOINTS

Fig. 64

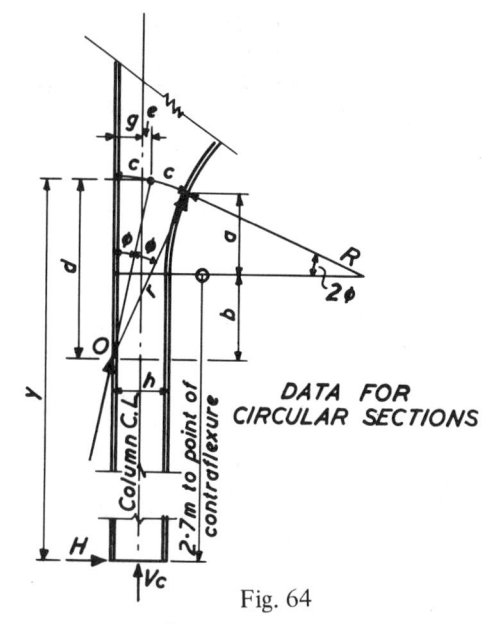

For values of 2ϕ not exceeding 45° the radius r is measured along the column, but between 45° and 90° it is measured along the cross-beam. (The angle subtended by the inner flange should always be bisected in this way, even if the angle is acute as in the case of a frame with a pitched roof.)

The area and moment of inertia of the 'I'-sections traced by each circular section are found in the normal way, the total depth being the arc length.

For sections 2–2 and 3–3,

$$P_O = V_c \cos \phi + H \sin \phi$$
$$M = H . y - V_c . e.$$

For sections 5–5 and 4–4,

$$P_O = H \cos \phi + V_b \sin \phi$$
$$M = V_b . x - H . e$$

where y is the vertical distance between the bottom of the column and the C.G. of the section, x is the horizontal distance from the end of the beam to the C.G. of the section and V_c and V_b refer to the column and beam respectively.

Proceeding, the relevant properties of, and the stresses in, the four sections inside the knee are as follows:

Section	P_O	A	M	c	I_{xx}	$\dfrac{P_O}{A} \pm \dfrac{M_c}{I_{xx}}$
	kN	cm²	kNm	mm	cm⁴	N/mm²
2–2	194.2	268.0	321.7	202.7	76 390.5	+92.59 or −78.10
3–3	203.6	308.6	348.7	304.2	192 935.5	+61.59 or −48.39
4–4	143.1	371.2	358.6	460.8	500 399.5	+36.87 or −29.16
5–5	125.6	327.5	304.9	351.6	268 629.5	+43.70 or −36.03

786   CONNECTIONS

Figure 65: Diagram showing final design and flange stresses. Tensile stresses at top: 29.16, 36.03, 34.46 (points 4, 5, 6), with 3.45 at far right. Stress assumed to be zero at top left. Left side tensile stresses: 48.39 (3), 78.10 (2), 87.64 (1), with 101.93 adjacent. Bottom: 7.14. Compressive stresses on right: 41.37 (6), 43.70 (5), 36.87 (4), 61.59 (3), 92.59 (2), 101.93 (1), with 41.37 at top of curve. All stresses are in N/mm².

FINAL DESIGN AND FLANGE STRESSES

Fig. 65

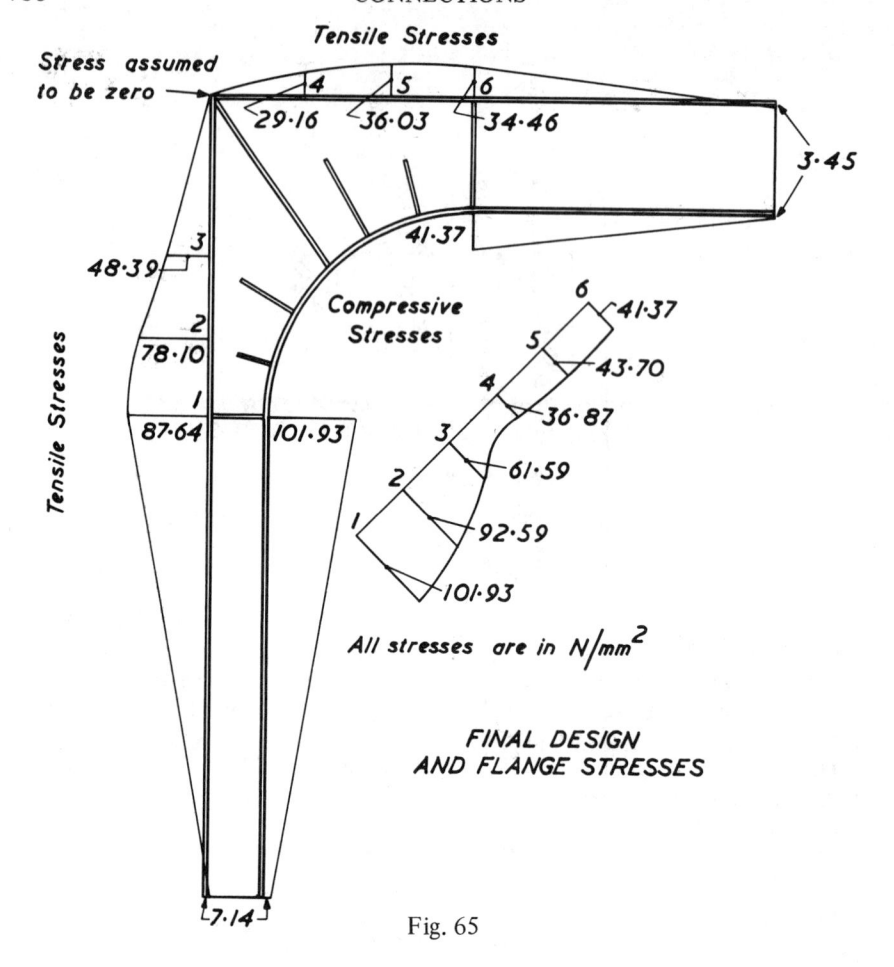

The complete flange stresses have been plotted in Fig. 65.

The stresses given for the inner flange presuppose that it is properly stiffened to resist cross-bending. If unstiffened, the stresses can be calculated by Dr. H. H. Bleich's formulae. (See p. 781.)

Considering section 2–2,

$$\max. f = \frac{f}{v}.$$

The value of b^2/Rt for the flange

$$= \left(\frac{312 - 20}{2}\right)^2 \times \frac{1}{1\,200 \times 32} = 0.56$$

Hence $v = 0.86$ (by interpolation) and $\max. f = \frac{f}{v}$.

$$= \frac{92.59}{0.86} = 107.67 \text{ N/mm}^2$$

RIGID FRAME JOINTS

Now the cross-bending stress

$$f' = \mu \max. f$$

$$\mu = 1.32 \text{ (by interpolation)}.$$

Therefore $\quad f' = 1.32 \times 107.67$

$$= 142.1 \text{ N/mm}^2.$$

As this does not exceed the permissible stress, stiffeners are theoretically not required. However it is considered good engineering practice to nominally stiffen the knee. Where stiffeners are needed, the shear stresses can be calculated by using Campus's formula (see page 777) or by Olander's method (see page 782).

Valley Joints

The principles involved in the design of the Y-shaped valley joints in multi-bay construction are the same as those for knee joints. It is quite reasonable to design the inner flange of each rafter section as though the joint were a knee, the other rafter being ignored, as shown in Fig. 67. The detailing should be as simple as possible. Two examples showing haunched and curved inner flanges are shown in Figs. 68 and 69 (references 11 and 12 respectively).

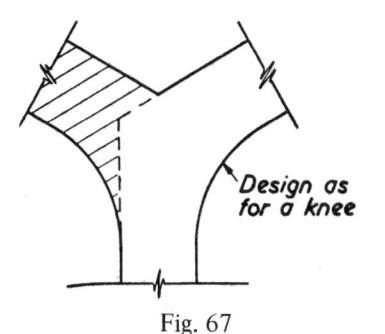

To reduce the cost of heating the building or for aesthetic reasons roofs are often made of low pitch, and practical difficulties can arise in accommodating the valley gutter. The problem may be overcome as shown in Fig. 69, the pitch being fairly steep at the feet of the rafters and low over the central portion of the roof.

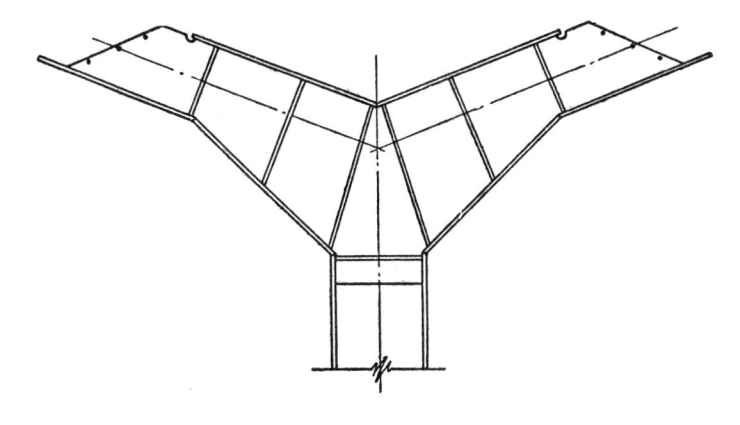

Fig. 67

Fig. 68

788   CONNECTIONS

Fig. 69

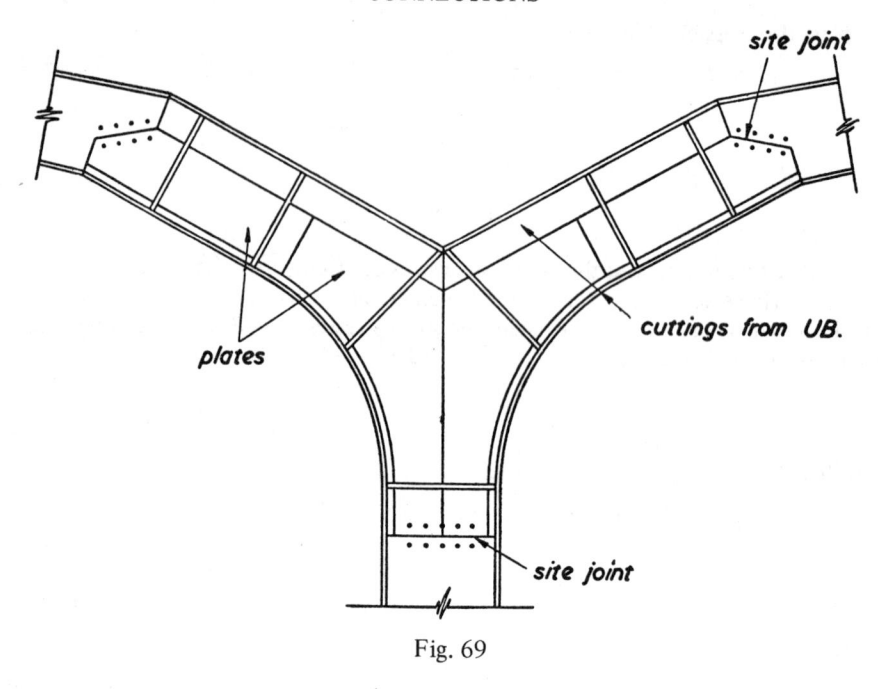

Splice Connections

Splice connections in rigid frames should be arranged at or near the dead-load points of contraflexure. Figures 49, 57, 68 and 69 all incorporate splice connections. Some typical joints for a column and cross-beam are shown in Fig. 70, but it should be noted that it is not essential that joints should occur at stiffeners. Further details may be found in reference 13.

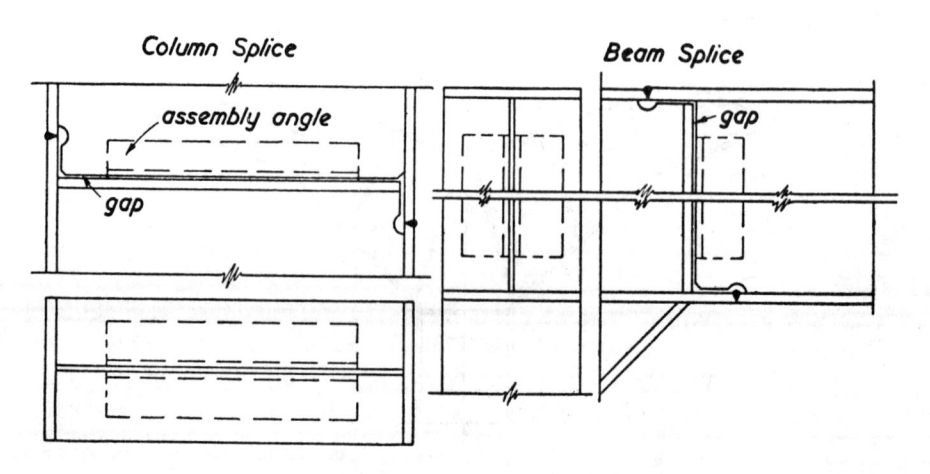

Fig. 70

Rigid Joints in Multi-storey Buildings

Although some of the calculations involved in designing a joint in a multi-storey building may be lengthy, the underlying principles on which the work is based are comparatively elementary.

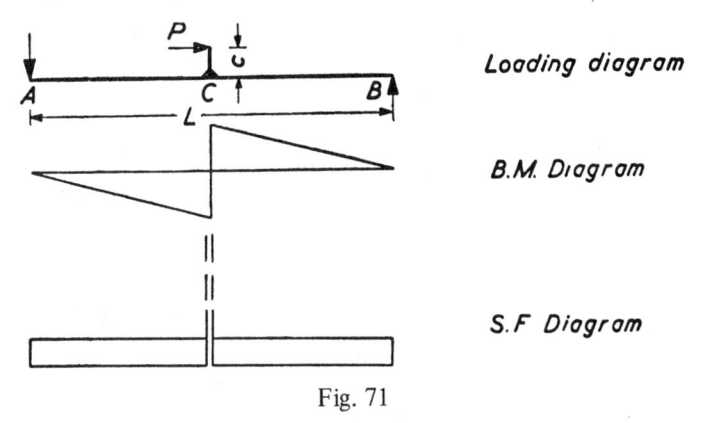

Fig. 71

Consider the simply supported prismatic beam shown in Fig. 71. When the load P is applied, a moment M is induced at C which is distributed between the portions of the beam AC and CB as shown. The reactions at the supports are:

$$R_A = \frac{-M}{L} \quad \text{and} \quad R_B = \frac{+M}{L}.$$

Now the shear in AC and CB may be obtained by calculating the slope of the B.M. diagram. Hence, the value of the shear = M/L, which is, of course, equal to the reactions at the supports A or B.

At C the moment diagram undergoes an abrupt vertical change of moment, equal to M, and the corresponding shear is infinite. This results from the assumption that the moment is applied at a point, as drawn in Fig. 71.

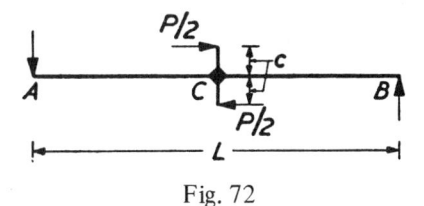

Fig. 72

Much the same state would result if the moment M were applied through a couple as shown in Fig. 72.

If, however, the moment M were to be applied through a couple formed by two equal and opposite forces of value P as shown in Fig. 73, then the change of moment at C would be less abrupt and the S.F. between the two loads could be calculated.

Thus,
$$V = \frac{M_{CA} + M_{CB}}{c} = \frac{P(L-c)}{L}.$$

790 CONNECTIONS

As before, the shear on either side of the applied moment would be of constant value, being equal to cP/L.

Now it is proposed to give a practical example to demonstrate how such joints can be treated.

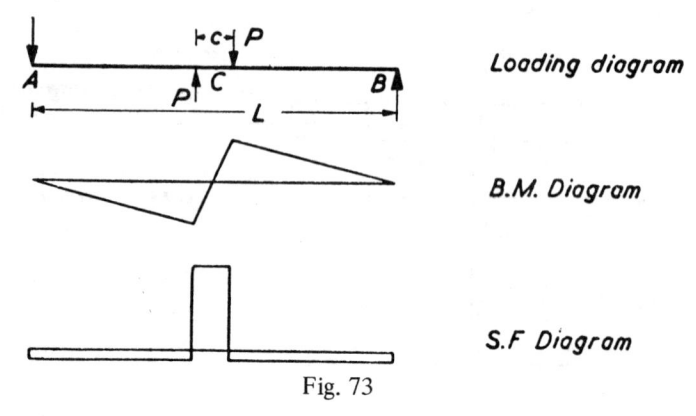

Fig. 73

Example. Figure 74 shows a portion of a rigid-frame building, consisting of a section or girder *BD* and sections of two columns *AB* and *BC*; *A*, *C* and *D* being the points of contraflexure in the B.M. diagram for the worst conditions of loading on this portion of the frame.

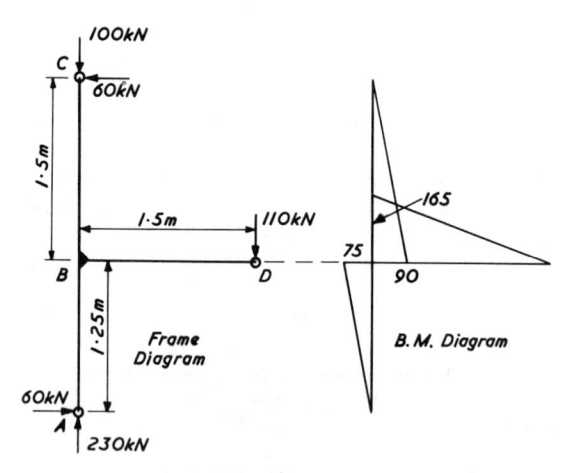

Fig. 74

The maximum B.M.s at the junction of the neutral axes of the girder and columns are as follows:

$$M_{BC} = +90 \text{ kNm}$$
$$M_{BA} = -75 \text{ kNm}$$
$$M_{BD} = -165 \text{ kNm}.$$

It is not proposed to design the columns or cross-beam. It will be assumed that the sections shown in Fig. 75 are adequate.

MULTI-STOREY FRAMES

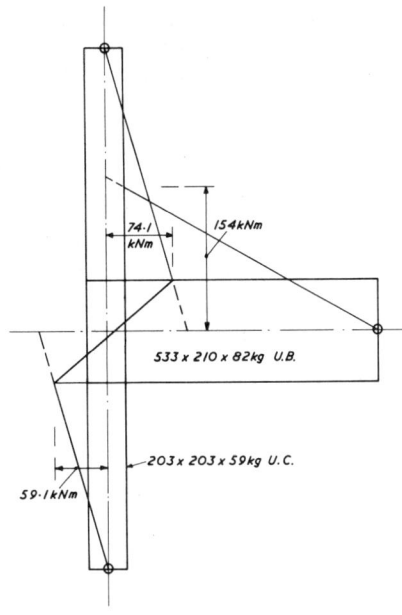

Fig. 75

Considering the joint at B, the shear across B is calculated from the slope of the B.M. diagram.

Hence
$$V = \frac{(74.1 + 59.1)10^3}{533}$$

$$= 249.9 \text{ kN}.$$

Now the web thickness of a 203 x 203 x 59 kg U.C. is 9.3 mm and the gross area of the web = 210 mm x 9.3 mm = 1 953 mm^2.

Therefore the shear stress in the web,

$$v = \frac{249.9 \times 10^3}{1\ 953} = 128.0 \text{ N/mm}^2.$$

This stress exceeds the maximum permissible shear stress of 115 N/mm^2. A number of expedients may be adopted to increase the resistance to shear.

One of the most common methods is to weld two rectangular plates to the web of the column. The clear web depth of a 203 x 203 x 59 kg U.C. is 161 mm. If two plates, 160 mm wide and 5 mm thick, were welded, one each side of the web, as shown in Fig. 76, then the shear stress would be,

$$v = \frac{249.9 \times 10^3}{210 \times 19.3} = 61.7 \text{ N/mm}^2$$

which would be acceptable.

It will be noticed that two sets of stiffeners have been introduced at the level of the flanges of the girder. These should be of approximately the same section as the girder flanges and, in this case, are made 15 mm thick.

CONNECTIONS

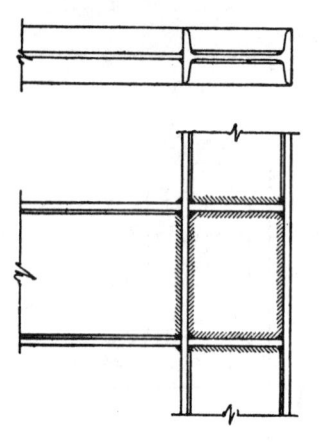

Fig. 76

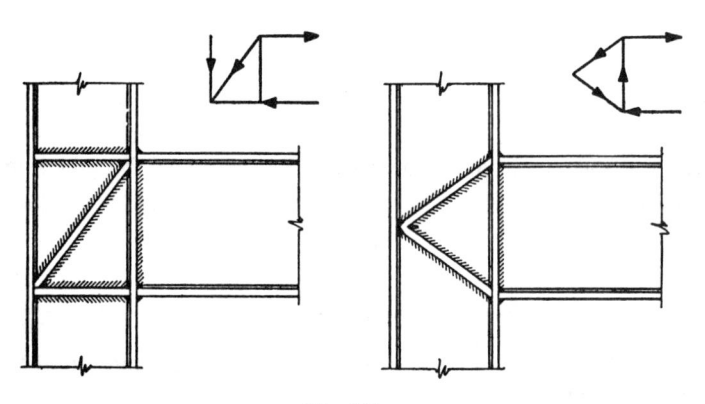

Fig. 77

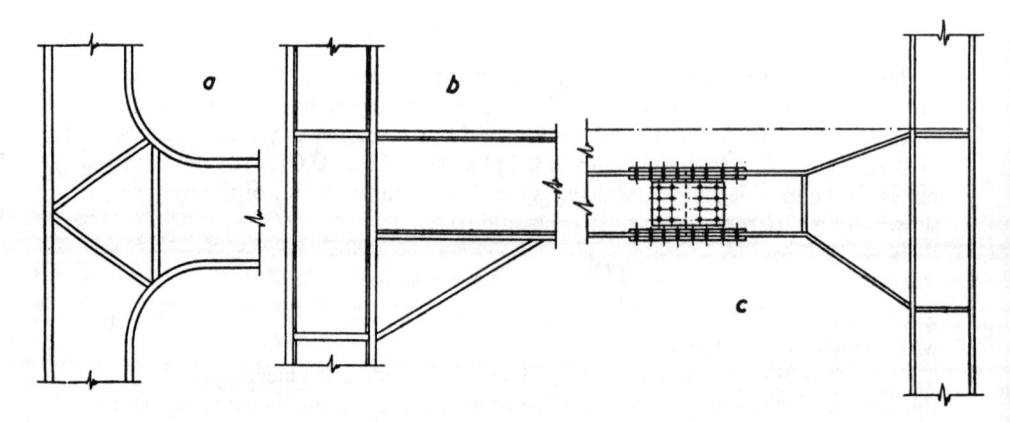

Fig. 78

Instead of thickening up the web some kind of triangulated system of stiffeners may be used, such as those shown in Fig. 77. In such cases the diagonal or oblique stiffeners may supplement the strength of the web, or the web may be entirely ignored, when the stiffeners form a truss framework and are designed accordingly.

Another method to avoid overstressing the web of the column is to deepen the end of the girder or to provide brackets at the junction of the girder and column, the object being to reduce the steep slope of the B.M. diagram, i.e., to reduce the shear force in the joint. Various expedients are shown in Fig. 78.

Open-frame or Vierendeel Girders

The most obvious characteristic of open-frame girders, some examples of which are shown in Fig. 79, is the complete absence of diagonal members in the panels, the girders depending on the rigidity of the joints for their stability.

The open-frame girder is not commonly used in Great Britain, but many examples exist in Europe, particularly in Belgium, where they are associated with the pioneer work and development of the late Professor Vierendeel after whom they are usually named.

Vierendeel Girders

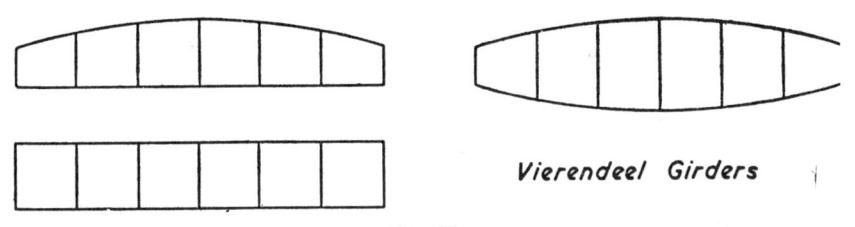

Fig. 79

Although there are many Vierendeel bridges in Belgium, only one such bridge, for pedestrians, has been built in Great Britain. Examples in structures taking static loads have usually resulted from a demand for a free unobstructed space where the use of diagonals has been precluded. Consequently, Vierendeel girders can be used for clerestory lighting in churches and other structures or for spanning any gap where a plate girder or truss would be used but for the fact that the web would provide an obstruction.

Many foreign authors have devoted whole books to the analysis of Vierendeel girders, but, unfortunately, the literature relating to the design of joints is comparatively scanty. Here, it is proposed to deal with girders with parallel top and bottom booms. Provided that the booms are parallel, analysis by Slope Deflection is possible but lengthy, but if the top and bottom booms are of identical section in each panel the girders may be analysed quickly and accurately by Naylor's application of Moment Distribution, as demonstrated in Example 15 in the section on that method of analysis. See page 267.

The analysis and design of a joint for a multi-storey building frame has just been given. Now a single-bay, multi-storey building of the type described in Examples 9 and 11 in the section on Moment Distribution is really a Vierendeel girder erected vertically, so that the principles underlying the design have already been described. However, it is usual for the axial forces in the members of Vierendeel girders to be very great compared with those in the analogous members in building frames.

CONNECTIONS

Most of the loads applied to Vierendeel girders are applied at the panel points but sometimes the booms take comparatively light loads, usually uniformly distributed, when a design by Moment Distribution will incorporate the devices associated with inter-panel loading (c.f. Example 16 in Moment Distribution). Inter-panel loading can also be treated by Slope Deflection.

Formulae for Joints

Vierendeel evolved a number of formulae for different types of joint. For the 'T'-joint shown in Fig. 80, the formulae resemble the Tapered Beam formulae given earlier in this section, viz.:

$$f_1 = \frac{P}{A} - \frac{Ma_1}{I}$$

$$f_2 = \frac{P}{A} + \frac{Ma_2}{I}$$

$v = \dfrac{1}{bd}[V - f_1 F_1 \sin \phi_1 + f_2 F_2 \sin \phi_2]$. (using appropriate signs for f_1 and f_2)

$A = bd + F_1 \cos \phi_1 + F_2 \cos \phi_2$.

$I = \dfrac{bd^3}{12} + F_1 \cos \phi_1 a_1^2 + F_2 \cos \phi_2 a_2^2$.

These formulae may be used for a joint in a multi-storey building where the radius of the top flange is small so that the flange does not project above floor level. Normally in Vierendeel girders $\phi_1 = \phi_2$.

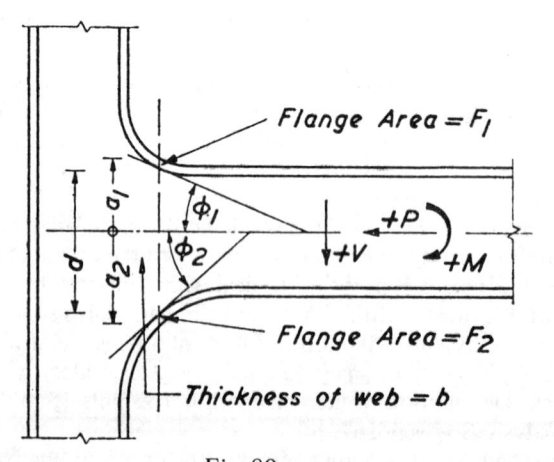

Fig. 80

This type of joint is not used in Belgium for bridges or other structures taking dynamic loading. It is the invariable practice to employ the type shown in Fig. 81, where the posts are planted on the booms and the radii of curvature may be as much as one-third of the panel length.

VIERENDEEL GIRDERS

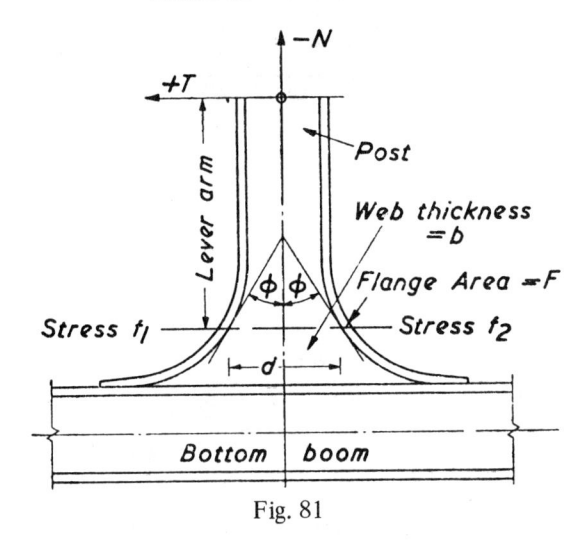

Fig. 81

In this case the post is symmetrical about the vertical axis, and the flange stresses *due to the moment only* are:

$$f_1 = -f_2 = \frac{M}{d(F \cos \phi + bd/6)}.$$

M equals T times the lever arm about the section considered.

At the junction of post and boom, as $\cos \phi = 0$, the moment is taken by the web only.

The stresses due to $-N = -N/A$. The values of v and A are found as before.

The B.M. diagram for a boom is drawn as shown in Fig. 82. The lines between the tangent points of the post are parabolas.

Professor Magnel of Ghent (reference 6) quoted the following formulae for the section shown in Fig. 83:

$$T = P_3 + P_1 \sin \phi - P_2 \sin \theta$$
$$T \cdot OX - M = P_3 \cdot OX$$
$$N = P_1 \cos \phi + P_2 \cos \theta,$$

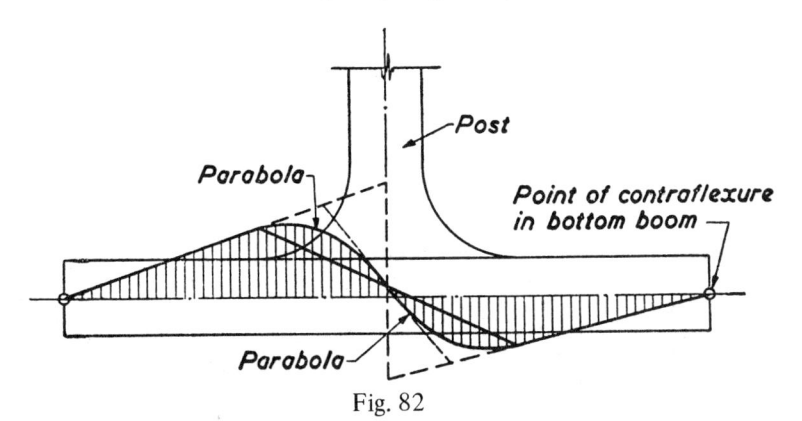

Fig. 82

from which

$$P_1 = \frac{N \cdot \sin\theta + \dfrac{M}{OX}\cos\theta}{\sin(\phi+\theta)}$$

$$P_2 = \frac{N \cdot \sin\phi - \dfrac{M}{OX}\cos\phi}{\sin(\phi+\theta)}$$

$$P_3 = T - \frac{M}{OX}.$$

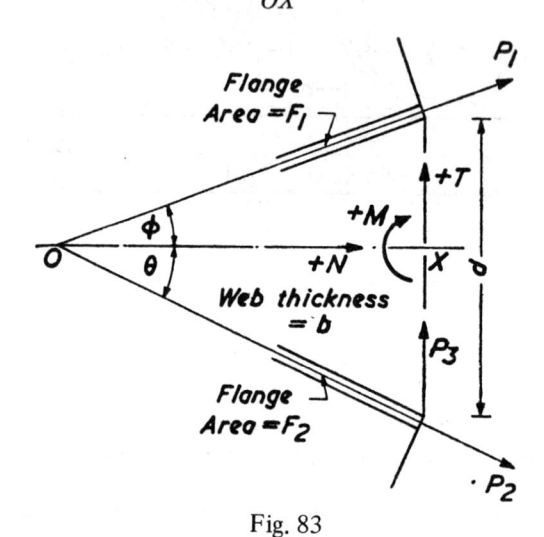

Fig. 83

The member shown has straight sloping sides, but the same formulae could be used for curved flanges although Magnel stated that the formulae should not be used when $(\phi + \theta)$ exceeds 90°. However, he quoted experiments carried out by Professor Campus in which he proved that in joints such as that shown in Fig. 81 the critical sections were those in which $(\phi + \theta)$ varied from 0 to 90°.

The stresses derived from Magnel's formulae are as follows:

$$f_1 = \frac{P_1}{F_1}; \quad f_2 = \frac{P_2}{F_2}; \quad v = \frac{P_3}{bd}.$$

BIBLIOGRAPHY

1. HENDRY, A. W. 'An Investigation of the Stress Distribution in Steel Portal Frame Knees,' *The Structural Engineer* (March–April 1947 and December 1947).
2. HENDRY, A. W. 'An Investigation of the Strength of Welded Portal Frame Connections,' *The Structural Engineer* (October 1950 and September 1951).
3. HENDRY, A. W. 'An Investigation of Certain Welded Portal Frames in Relation to the Plastic Method of Design', *The Structural Engineer* (December 1950 and September 1951).

4. STANG, A., GREENSPAN, M., AND OSGOOD, W. R. 'Strength of a Riveted Steel Rigid Frame having Straight Flanges', R.P. 1130, U.S. National Bureau of Standards, *Journal of Research*, Vol. XXI (1938).
5. STANG, A., GREENSPAN, M., AND OSGOOD, W. R. 'Strength of a Riveted Steel Rigid Frame having a Curved Inner Flange', R.P. 1161, U.S. National Bureau of Standards, *Journal of Research*, Vol. XXI (1938).
6. MAGNEL, G. *Stabilité des Constructions,* Vol. I.
7. BLEICH, H. H. 'Spannungsverteilung in den Gurtungen Gekrümmter Stäbe mit T- und I-förmigen Querschnitt', *Der Stahlbau* (January 1933).
8. KURSBATT, I. 'The Kidlington Bridges', *Journal of the Institution of Civil Engineers* (January 1940).
9. BLEICH, F. 'Design of Rigid Frame Knees', *American Institute of Steel Construction* (August 1952).
10. OLANDER, H. C. 'A method of calculating stresses in Rigid Frame Corners', *Journal of the American Society of Civil Engineers* (August 1953).
11. PHILLIPS, J. T., AND DANIELS, F. T. 'Welded Steel Framed Buildings', *The Structural Engineer* (March 1949).
12. ANON. 'New Tube Factory for Liverpool', *Welding and Metal Fabrication* (January 1951).
13. ANON. 'Welded Portal Frame Retort House', *Welding and Metal Fabrication* (April 1953).
14. BEEDLE, LYNN S. *Plastic Design of Steel Frames,* John Wiley, 1958.

29. SURFACE PREPARATION OF STRUCTURAL STEELWORK

IN recent years, advances in design techniques and the development of higher strength steels of welding quality have established a trend towards more efficient lighter weight steel construction and placed greater emphasis on the protection of structures which will be exposed in service to a corrosive environment. This trend has been paralleled by the development of new methods of protective treatment and in particular, new techniques for the surface preparation of steelwork.

The essential element in a first-class anti-corrosion treatment is the removal of the mill scale which forms on the surface of the steel plates and sections during the hot rolling process, before the protective coating is applied. The traditional method of wire brushing, sometimes combined with extensive weathering at site, is still widely used for preparation of the surface and may be adequate in those cases where the steelwork will be exposed in an environment which is at worst only marginally corrosive. However, this method does not remove tightly adhering mill scale, an essential prerequisite for the successful use of modern high grade protective systems. For such systems, more consistent and rapid means of surface preparation are required.

Methods used to prepare the surface of structural steelwork in ascending order of their effectiveness are:

    Flame Cleaning.
    Pickling.
    Blast Cleaning.

(1) Flame Cleaning

In this process, an oxy-acetylene flame is passed over the surface of the steelwork rapidly heating it locally to between $95°$ and $150°C$ when the differential expansion rates of the scale and steel results in the loosening of mill scale and at the same time facilitates the removal of the rust. The surface is cleaned down by wire brushing and immediately given a priming coat of paint, preferably before the steel has cooled down. Not all the tightly adherent scale is removed by this method of treatment. With plates and sections less than 6 mm thick, the heat of the flame may result in buckling and distortion. Where steelwork has to be painted in the open, at site for example, flame cleaning is often the most satisfactory method of preparing the surface.

(2) Pickling

This term is used to describe the process in which steelwork is immersed in a tank filled with cold dilute hydrochloric or hot dilute sulphuric acid for as long as is necessary for the iron oxides to be removed by chemical action, this period varying

from a few hours to a few minutes, depending on the type and concentration of acid used. Individual components up to 12 m in length may be treated. Inhibitors are usually added to the acid to slow down the attack on the base steel after the scale has been removed.

This process is the established means of preparing steelwork for hot dip galvanising, but has not been widely adopted for the preparation of structural steelwork for painting, although it gives results comparable with blast cleaning. However, the Footner pickling process in which steel is immersed in hot dilute sulphuric acid then rinsed in hot water and finally dipped in hot dilute phosphoric acid is extensively used for the surface preparation of steel plates for the roofs, sides and bottom plates of oil storage tanks. The priming paint is applied over the thin phosphate film formed on the surface of the steelwork, preferably while the metal is still warm.

(3) Blast Cleaning

The most effective means of removing mill scale is by blast cleaning, a term used to describe the projection at high velocities of a hard abrasive material on to the surface of the steel. There are two basic processes in use, one in which the abrasive is propelled by compressed air, usually termed 'hand' or 'manual' blasting, and the other in which the abrasive is thrown from the rim of large diameter impeller wheels rotating at high speed. The latter method of cleaning is known as 'airless' or 'mechanised' blasting.

(a) Manual Blasting

With manually operated equipment, the abrasive is ejected by compressed air at pressures of 420 to 560 kN/m^2, through chilled iron or ceramic nozzles usually 6 mm to 10 mm diameter. Cleaning rates are of the order of 3.7 to 11.2 m^2 per hour per operator, depending on the shape and size of the steelwork and the degree of cleaning required. Although an effective means of removing mill scale, manual blasting is slow and a relatively expensive process, but the equipment is cheap and portable and is widely used where the volume of cleaning is small or the steelwork is of intricate shape and therefore unsuitable for mechanised blasting. This process is also used for preparing steelwork on site for maintenance painting. Where blasting is carried out on site or in fabricators' works on an occasional basis only, it is usual to provide temporary enclosures and to use an expendable abrasive, but for a more regular demand, steel-plated rooms large enough to accommodate finished fabrications are provided complete with abrasive recovery and ventilating systems. This type of fixed installation is gradually being replaced by the mechanical blasting plant described later.

A refinement of the manual type of equipment is an enclosed system in which the nozzle is surrounded by a suction hood which draws the abrasive back into the collecting chamber after impingement on the surface. The abrasive is then cleaned and returned to the hopper feeding the nozzle. A range of machines is available, from small portable units to large machines designed for specific applications. In general the method is slower than open blasting but the advantage of this type of enclosed system is that there is no necessity to transfer the steelwork to separate shops or temporary enclosures and it has particular value where localised areas only are required to be cleaned.

(b) Airless or Mechanical Blasting

For cleaning structural steelwork on a large scale, manual blasting equipment is gradually being replaced by airless or mechanised plants. These permanent installations are normally located in the steel fabricators' works and can be divided into two distinctive types, (i) Standard machines sub-divided into equipment designed to clean steel plates up to 4.0 m wide and section machines which can deal with beams up to 920 mm deep as received from the rolling mills, (ii) Custom built machines which are designed for the treatment of finished fabrications ranging from simple beamwork to the large and complex fabrications required for major bridges.

Although these plants differ in such details as size of blasting chamber, the number and location of impellers and method of work handling handling, their operating principles are similar. These are illustrated in Fig. 1 which relates to equipment designed for cleaning finished fabrications up to 5 m deep by 1.8 m wide by 36 m long by 50 000 kg in weight. The steelwork is traversed through the chamber on specially designed bogies at a pre-selected speed which can be infinitely varied between 0 and 3 m per minute. A faster speed is available for moving the bogies when they are unloaded.

The impellers are mounted on a heavy frame and so positioned as to give complete coverage of steelwork passing through the plant including stiffeners, etc., at right angles to the line of travel. Fabrications up to 1.8 m deep, depending on shape, can be cleaned in one pass and deeper fabrications are dealt with by raising the bank of the impellers and reversing the steelwork through the plant.

After striking the steel surface the abrasive falls through an open grid floor and via belt and bucket conveyors is returned to the hoppers feeding the impellers, passing on route through desiltering equipment which removes fines. The interior of the plant is under suction and all the dust formed during the blasting operations is drawn through ducting to dust collecting equipment.

In typical plate cleaning machines for the descaling of plates prior to fabrication the impellers are positioned to give complete coverage of both sides of plates as they are passed through the plant on power-driven rollers. Mechanised handling equipment is usually incorporated on the input and output side of the machine. Small sections such as light angles can also be blast cleaned in this type of plant.

Type of Abrasive

In mechanised impeller plants, chilled iron angular grit or round steel shot are the two types of abrasive in general use, the choice being normally determined by the primary function of the plant.

In all purpose plants designed to prepare the surface of finished fabrications to receive metal or paint primers, chilled iron angular grit is generally used. This abrasive produces the type of surface profile and depth between peak and valley necessary to ensure that metal coatings are satisfactorily keyed to the base metal. Typical of the abrasives used is a mixture of G47 grit and S340 chilled iron shot. As a result of the immediate breaking down of the round iron shot on impact and the continual removal of fines by the exhaust and desiltering equipment, the abrasive in circulation is a graded mixture containing a wide range of angular particle sizes.

Chilled iron angular grit, though considerably cheaper per tonne in first cost, is less economical than round steel shot owing to higher consumption per unit of area

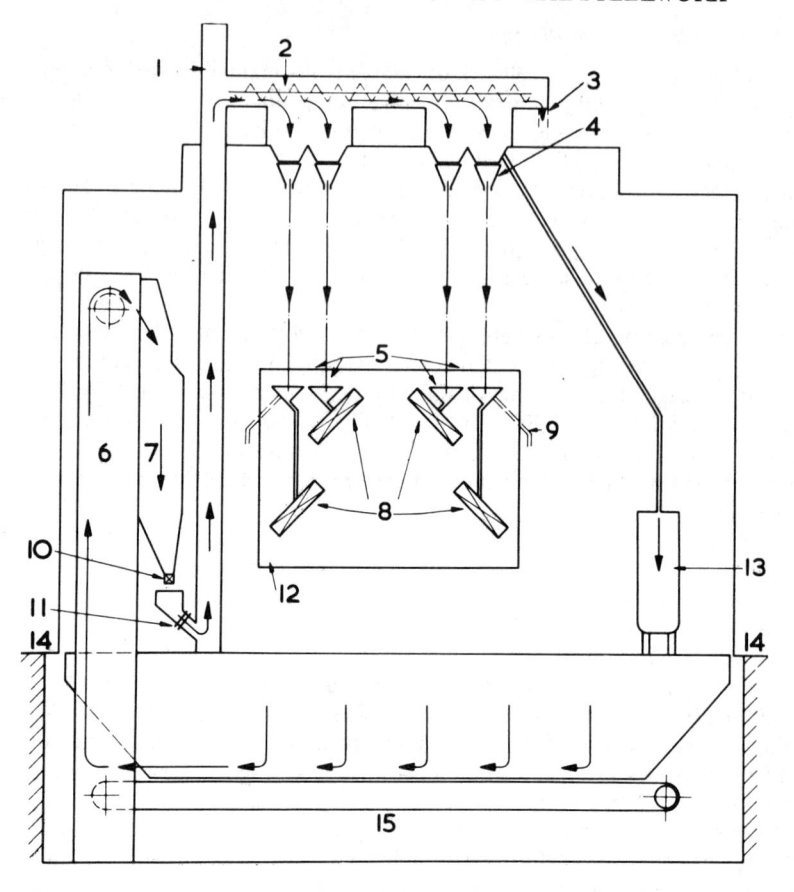

Fig. 1

1. Elevator
2. Spiral screw feed
3. Overflow 1 cwt/min
4. Adjustable chutes
5. Tundishes
6. Elevator
7. Desilter
8. Impellors
9. Overflow
10. Solenoid feed valve
11. Adjustable volume control slide
12. Rise and fall panel
13. Pressure vessel for hand blasting
14. Ground level
15. Conveyor

blasted and higher plant maintenance costs as a result of greater wear and tear on the impellers and other items of equipment.

For these reasons, round steel shot is generally used in pre-fabrication blast cleaning plants. A typical grade of abrasive is 330 gauge round annealed steel shot and this produces a surface which is suitable for pre-fabrication primers. The steelwork is normally coated with a very thin film of the primer, 12 to 25 microns in thickness, immediately after blasting to prevent re-rusting and to ensure protection of the surface during the fabrication processes.

In manually-operated compressed-air equipment, depending on whether the surface is being prepared for metal or paint primers, either chilled iron grit or round steel shot is used where it is possible to recover and recirculate the abrasive. Where conditions of operation are such that the abrasive must be regarded as expendable, cheaper materials are used such as crushed slag or non-metallic silica-free compounds.

Quality of Surface

When assessing the quality of surface preparation produced by a blast cleaning process, there are two criteria: (a) surface cleanliness and (b) surface roughness.

(a) Surface Cleanliness

For any particular plant, whether it be a mechanised or manual type, the degree of surface cleanliness achieved is primarily determined by the blasting rate. For mechanised plants, this is a function of the linear speed at which the steelwork is traversed past the impellers. In practice, speeds generally range between 1 m and 2.4 m per minute, depending in the case of finished fabrications on the contour of the surface to be cleaned as well as the degree of cleaning required.

With manual equipment, for a given quality of surface, the cleaning rate depends on the size of nozzle used and in practice ranges between 0.05 and 0.20 m^2 per minute.

CP 2008 : 1966, 'Protection of Iron and Steel Structures from Corrosion', recommends that two standards of cleanliness should be adopted for structural steel, First Quality (white metal) and the other Second Quality. These are defined in B.S. 4232 1967 and generally as follows:

First Quality (*White Metal Finish*) is a surface with a grey white metallic cover, roughened to form a suitable anchor pattern for coatings. The entire surface should show evidence of blast cleaning and should be clean base metal. No part of the surface being inspected should contain or be discoloured by mill scale, rust, rust stain, residues of paint or other coating or any other form of contamination.

Second Quality Finish – the entire surface should show evidence of blast cleaning and 90 per cent of the surface should be clean base metal. In addition, no single square of the surface of side 25 mm should contain more than 20 per cent of its area discoloured by discontinuous areas of mill scale, rust, rust stain, residues of paint or other coating or any other form of contamination.

It is possible to over-specify the grade of surface cleanliness required since there is a considerable difference between the cost of achieving the above qualities of finish.

A white metal finish is essential as a preparation for metal spraying and for some of the modern and sophisticated protective paint systems. However, for a conventional paint system it is generally accepted that the lower standard of finish is satisfactory, a lower standard still corresponding to a higher blasting rate might be accepted where the service conditions are only marginally corrosive.

(b) Surface Roughness

This is best defined as the maximum amplitude between adjacent peak and valley. For new steel, the maximum amplitude is a function of the type and grade

of abrasive used. Where a paint system is subsequently to be applied, the smoothest possible surface, consistent with complete removal of mill scale, should be produced by using a fine grade of abrasive, but economic and practical considerations determine the minimum grades which can be successfully used. The grades referred to previously can be expected to produce a maximum surface amplitude of the order of 75 to 100 microns in mechanised impeller plants and a somewhat rougher surface in manual equipment.

It is necessary to consider the roughness of a blast-cleaned surface relative to the subsequent treatment. When cleaned, the steel is in a vulnerable condition and not only must it be primed as soon as practicable (within 4 hours is generally specified), but also sufficient paint must be applied before it leaves the shops for the site or for storage in the open, to ensure that the film thickness over the peaks is adequate to afford full protection until the final coats of paint are applied. In practice, this means a minimum of two coats with most paints.

Modern priming paints for application to blast-cleaned steelwork are formulated so that they dry rapidly; in the case of pre-fabrication primers, within seconds of application. This reduces any tendency for the paint to drain away from the peaks and thus helps to ensure that the steel is fully protected during fabrication. Primers which have a high percentage of zinc in the dry film can afford protection under cover in the shops for up to 6 months, and the best will give a substantial degree of protection fully exposed out of doors.

Application of Metal Coatings

The optimum anti-corrosion treatment adopted after the removal of mill scale and rust depends on the service conditions. Such factors as the type of environment to which the steelwork will be exposed, access for maintenance painting and its expected life, will determine the relative economics of the various treatments available. Three or four coats of paint is the most common treatment for steelwork which has been blast cleaned but there are applications where the adoption of hot dip galvanising or metal sprayed coatings has special merit and is economically attractive.

The principal fields of application for metal coatings (other than sheeting) so far have been steelwork for bridges, transmission towers, railway electrification overhead structures and light lattice construction. Power Station Switch House steelwork is a new example where access for maintenance painting presents a special problem and zinc coatings have been adopted.

Zinc or aluminium are the two metals involved in sprayed coatings which are generally specified for priming steelwork, with zinc the pre-eminent material in this field.

Metal Spraying

As previously referred to, the first requirement is to blast clean the steel using an abrasive which will remove mill scale and provide the necessary mechanical key or anchor pattern as it is termed, for the metal.

Metal spraying equipment consists essentially of a gun which feeds the metal in the form of wire or powder through an oxy-fuel gas flame where it is fused and atomised and then projected on to the steel surface by a stream of compressed air.

In the case of the wire process, a compressed air driven rotor feeds 3 or 5 mm diameter wire from a coil, through the oxy-fuel gas flame; in the powder process the powder is conveyed by compressed air from a storage hopper along the supply tubing and through the central aperture of the nozzle into the melting zone. In both cases, compressed air issuing from an annular ring of apertures in the nozzle projects the molten, atomised zinc particles on to the surface of the steel.

Although the zinc coatings produced by the two methods differ slightly in appearance there is no significant difference in their performance and the choice of process is based on practical and economic considerations. 5 mm wire guns spray zinc at the rate of 27/30 kg per hour and the larger type of powder gun has a throughput rate of the order of 36/40 kg of zinc per hour. Losses occur in the fusing and atomising processes and the deposition factor, i.e., weight of metal sprayed to weight deposited depends to a large extent on the size of the areas being treated. For structural steelwork a deposition factor of 50/60 per cent is about average.

Overall spraying rates between 5.6 and 11.2 m^2 per hour per operator for a 75 microns coating are usual, depending on the type and shape of the fabrication. Coating thickness of 75 to 100 microns are generally specified but where necessary, considerably heavier coatings can be applied.

Although zinc is the most widely used metal for the protection of structural steelwork, aluminium is also used, particularly where sulphur compounds are expected to be present in the environment. Recently introduced zinc/aluminium alloys are said to give better results than either metal alone for some applications, but these can only be applied by the powder process.

Hot Dip Galvanising

Hot dip galvanising is one of the oldest and still one of the most effective methods of protecting steel against corrosion.

Essentially the process consists of the removal of rust and mill scale by pickling in hydrochloric acid or sulphuric acid, followed by fluxing with ammonium chloride and them immersion in a bath of virtually pure molten zinc.

As a refinement of this basic process, water wash and pre-fluxing tanks containing zinc ammonium chloride are sometimes introduced immediately after the pickling process. The fluxed work is then passed through a drying oven prior to immersion in the galvanising bath. During dipping, the zinc reacts with the steel forming a zinc iron alloy layer and during withdrawal from the bath, an additional thin layer of pure zinc is added.

The weight of coating deposited on the article is influenced by the bath temperature, time of immersion, and withdrawal speed as well as the shape and size of the article.

In normal practice, these factors are more or less constant for any given item of steelwork and the coating weight cannot be varied significantly. It is generally between 610 and 765 g/m^2, 75 to 100 microns thickness for most classes of structural steelwork.

This is satisfactory for most purposes but if in special circumstances a heavier coating is required, this can be produced by grit blasting the steel prior to pickling and galvanising. The roughened surface results in a greater build-up of the alloy layer and a coating weight of the order of 916 to 1 220 g/m^2 can be obtained. The

use of steel with a high silicon content is an alternative method of obtaining heavier coatings which has been adopted in Sweden but not so far in the U.K.

Galvanised coatings vary in appearance from dull grey to bright metallic depending chiefly on the composition and thickness of the steel. Heavy items of structural steelwork cool slowly after being withdrawn from the bath and the alloy layer continues to grow, thus reducing the thickness of the pure zinc layer. In extreme cases, there is virtually no pure zinc present and the article has a dull grey appearance. These grey coats are in no way inferior in respect of their anti-corrosion properties, although they do tend to acquire a 'patina' superficially resembling rust which although disfiguring, does not indicate a failure of the coating.

The period over which protection is afforded by zinc coatings is a linear function of the thickness of the coating and can be regarded as independent of the method by which the zinc is applied provided the coating is uniform. In a non-industrial atmosphere the corrosion rate of zinc is about 1/15 to 1/20 of that of steel and a life of 20/25 years may be expected from coatings of 75 to 100 microns thickness. In industrial environments the corrosion rate generally varies between 1/5 to 1/10 to that of steel depending on the sulphur content of the atmosphere and except in extreme conditions, lives of 5–8 years may be anticipated.

The relative economics of hot dip galvanising and zinc spraying depend on the nature of the work and in particular, the ratio of surface area to weight. The higher this ratio is, the more economically attractive becomes the galvanising process and conversely zinc spraying shows to a special advantage when dealing with articles of low surface area to weight ratio.

In general, for most classes of structural steelwork which can be accommodated by existing plant, the galvanising process should prove to be the lower cost process. Steelwork up to 10.5 m long by 1 m by 1.2 m represents the approximate limits of size which can be galvanised, but where the cross-section is small, lengths above 10.5 m can be dealt with by a double dipping technique. The Hot Dip Galvanisers Association publish full details of the plant installed by their members and the maximum size of pieces which can be handled.

Within these size limits, zinc spraying may prove to be more suitable for light and bulky fabrications because of distortion problems in galvanising and where selective application of a metal coating is required, e.g., exterior surface only of box members.

There are virtually no size limits as far as metal spraying is concerned and the largest components for bridges and other structures can be treated.

Summary

Thorough surface preparation is the essential foundation for a high grade anti-corrosion treatment for steel structures which will be exposed to a corrosive environment in service. Blast cleaning, pickling and to a lesser extent flame cleaning, are effective methods of removing mill scale and rust and the installation in fabricators' works of high production mechanised impeller blast cleaning plants for the treatment of finished fabrications, or alternatively, plates and sections prior to fabrication, has resulted in a marked increase in cleaning rates and a significant reduction in costs, compared with manual methods of blasting using portable equipment.

SUMMARY

By varying the cleaning rates, different standards of surface cleanliness can be achieved, from a white metal finish which is essential for metal sprayed coatings and the more sophisticated paints, to the lower standards suitable for conventional protective paint systems. The development of more efficient methods of blast cleaning, combined with improvements in the performance and efficiency of metal spraying and galvanising equipment, has led to a position where the constructional steelwork industry can offer a range of effective anti-corrosion treatments for its products to suit any environment to which they may be exposed.

30. DESIGN OF MULTI-STOREY STANCHIONS

Multi-Storey Steel Frameworks

BEFORE considering the detailed design of multi-storey stanchions it is necessary to consider the design of multi-storey steel frameworks as a whole. Some attention will be given to this in the following paragraphs.

Stability

All multi-storey frameworks must be capable of resisting the effects of horizontal forces both laterally and longitudinally. There are several alternative ways of doing this, of which it is frequently considered that the most economical is to introduce a series of vertical shear walls or bracing systems. In these systems (i) the floors act as deep horizontal girders to transmit the horizontal forces to the vertical stiffening elements and these (ii) serve to hold the structure in position and transmit the horizontal forces to the foundations.

An alternative method of providing the necessary stability is the use of rigid joints in the framework so that the latter is made capable of transferring the horizontal forces to the foundations without undue sway.

Emphasis will be based in this chapter on the design of multi-storey stanchions where the stability is provided by shear walls or other similar construction and for the design of multi-storey stanchions subject to sway conditions, reference should be made to suitable text books or other publications of a similar nature.

Design

At present the accepted design standard for multi-storey steel frameworks is B.S. 449: Part 2: 1970 'The Use of Structural Steel in Buildings'. This standard is at present under review and it is possible that when the revised version is issued rules for guidance in the design of multi-storey structures will be modified. At present, however, the standard permits the use of one of three design methods, namely, 'simple', 'semi-rigid' and 'rigid'. Each of these methods is considered in more detail below.

(a) Simple Design

Where beams are supported by seating cleats, web cleats or simple end plates designed to carry the beam reactions only. The only bending moments which need to be allowed for in the design of the stanchions are those induced by the eccentricities of the beam reactions; B.S. 449 gives certain rules for ascertaining these eccentricities and for applying the bending moments induced thereby.

The detailed calculations which are given in this chapter are based on this method of design.

(b) Semi-rigid Design

As the types of connections usually adopted between beams and columns inevitably provide a degree of restraint, this will tend to reduce the bending moments in the beams. The Steel Structures Research Committee which investigated this problem some years ago produced a report which permitted the transference of some moments between the beams and columns. The method suggested was not universally adopted, however, a simplified rule is written into B.S. 449 whereby advantage can be taken of the semi-rigid nature of the connections. References 1 and 2 give details of the application of semi-rigid design methods.

(c) Rigid Design

Where beams are rigidly connected to the stanchions by means of welding or high strength friction grip bolts the connections can be made capable of developing the full strength of the members, the analysis of the framework thereby becomes extremely complex. The standard permits the use of two design techniques, namely 'elastic' and 'plastic'.

(i) Elastic

For this purpose B.S. 449 states 'The design shall be carried out in accordance with accurate methods of elastic analysis and to the limiting stresses permitted in this British Standard'.

A number of multi-storey frameworks have been designed and constructed on this basis and details are given in various publications. The development of the use of the electronic computer for the solution of engineering problems has made the method of analysis much less tedious, whilst at the same time providing a degree of accuracy of information which enhances the value of the method of design. An example of this treatment is given in reference 3.

(ii) Plastic

Again B.S. 449 states 'Alternatively it shall be based on the principles of plastic design so as to provide an adequate load factor, and with the deflections under working loads not in excess of the limits implied in this British Standard.

A Joint Committee of the Institution of Structural Engineers and the Institute of Welding has examined the problems of the design of rigid frame structures and have made recommendations of a design method which allows for a degree of plasticity in the beams whilst the stanchions are designed elastically. Reference should be made to the second report of the Joint Committee for details of the design method proposed. (See reference 4.)

Alternative plastic design methods will be found in reference 5.

Multi-storey Steel Stanchions

The general description 'multi-storey' applies equally to stanchions two or three storeys high and to stanchions 28 to 30 storeys high. This wide variation in height and correspondingly in loads to be carried, had led to the demand from the

designer of a wide range of structural sections from the smaller hollow steel section to the larger universal beam and column section.

Since the range of sections available is at present under review internationally, reference should be made to the current Safe Load Tables for details of the maximum axial loads which the various sections are capable of carrying on a range of heights. These Safe Load Tables also include figures for the different grades of steel commonly used for structural purposes, namely grades 43, 50 and 55 in B.S. 4360.

This chapter will be confined to the consideration of universal column sections as multi-storey stanchions, since these are the sections most frequently used for this purpose. Examples of calculations to B.S. 449: Part 2: (1970) will be given but it is first desirable to examine the various factors affecting the design method used.

Column Behaviour

A 'short' column fails by permanent deformation of the material at the yield stress.

A 'long' column becomes unstable at the 'Euler' load, and fails by buckling. The Euler load is given by the expression $\pi^2 EA/(l/r)^2$, where E is Young's Modulus, A is the cross-sectional area, l is the length, and r is the radius of gyration, a geometrical property of the column section. The term (l/r) is referred to as the 'slenderness ratio', and is a critical factor in column design.

Practical columns are neither 'short' nor 'long' but fall somewhere between the two extremes, failing by a combination of yielding and buckling. Also, they suffer from certain imperfections such as (i) departure from ideal straightness, known as 'initial curvature', (ii) variation of yield point across the section and (iii) residual stresses due to the method of manufacture.

The formula given in B.S. 449 for calculating the safe loads on practical columns is a modification of the Perry-Robertson formula, which was evolved by Professor Perry on the assumption that all practical imperfections can be represented by a hypothetical initial curvature of the column. The formula gives the intensity of end loading which, at an extreme fibre at the mid-height of the column, would cause the combined axial and bending stresses to reach the yield point. This intensity of end loading, when divided by a suitable 'load factor' is referred to as the 'permissible axial stress'.

The practical application of the formula was made possible by Professor Robertson, who carried out a series of closely controlled experiments which enabled him to recommend suitable values for both the load factor, and for the expression in the formula representing the initial curvature. With the passage of time, other experimenters have suggested modifications to these values, all tending to greater economy. A more detailed account of the historical background, and of the work leading up to the most recent changes will be found in Reference 6.

The latest revision to B.S. 449 (Appendix B) states:

'The average stress on the gross sectional area of a strut or other compression member in steel with a specified minimum yield stress shall not exceed the value of p_c obtained by the formula:

$$K_2 p_c = \frac{Y_s + (\eta + 1)C_0}{2} - \sqrt{\left\{\left(\frac{Y_s + (\eta + 1)C_0}{2}\right)^2 - Y_s C_0\right.}$$

DESIGN OF MULTI-STOREY STANCHIONS

where p_c = the permissible average stress, N/mm².
K_2 = load factor or coefficient, taken as 1.7 for the purposes of this standard.
Y_s = minimum yield stress, N/mm².
C_0 = Euler critical stress = $\dfrac{\pi^2 E}{(l/r)^2} = \dfrac{\pi^2 210\,000}{(l/r)^2}$ N/mm².
$\eta = 0.3(l/100r)^2$.
l/r = slenderness ratio = $\dfrac{\text{effective length}}{\text{radius of gyration}}$.

Values derived from this formula are given in Table 17a (Grade 43 steel), Table 17b (Grade 50 steel) and Table 17c (Grade 55 steel) of B.S. 449 Part 2 which are reproduced on pages 816 to 818 inc.

Effective Length Factors

The length, l, to be used in the column formula is the 'effective length', which is the actual length measured centre to centre of floor beams, multiplied by an effective length factor. The value of this factor depends upon whether the ends of the column are (i) held in position, and (ii) restrained in direction, i.e., fixed, partially fixed or pinned. Effective length factors for a variety of end conditions are given in Clause 31 and Appendix D of B.S. 449.

Maximum Slenderness Ratio

Clause 33 of B.S. 449 states that for any member carrying loads resulting from dead weights, with or without imposed loads, the maximum slenderness ratio shall not exceed 180. This limitation clearly applies to multi-storey stanchions.

Bending and Axial Compression

Clause 14.a of B.S. 449 states that:
'Members subject to both axial compression and bending stresses shall be so proportioned that the quantity

$$\frac{f_c}{p_c} + \frac{f_{bc}}{p_{bc}}$$

does not exceed unity at any point, where

f_c = the calculated average axial compressive stress.
p_c = the allowable compressive stress in axially loaded struts (see Table 17).
f_{bc} = the resultant compressive stress due to bending about both rectangular axes.
p_{bc} = the appropriate allowable compressive stress for members subject to bending (see Clause 19).

If unity represents the carrying capacity of the member as an axially loaded stanchion, the first factor in the expression represents the fraction of this capacity which is actually being utilised.

FIRE PROTECTION AND STRUCTURAL CASING

Similarly, if unity represents the carrying capacity of the member as a beam subject to bending only, the second factor indicates the fraction of this capacity which is actually being utilised.

Clearly, if the member carries axial loads and bending moments simultaneously, the sum of these two factors should not exceed unity.

Fire Protection and Structural Casing

In multi-storey buildings it is often necessary to provide a degree of fire resistance to the steelwork to satisfy bye-law requirements.

A number of proprietary systems of hollow lightweight casings are available, and details are given in Reference 7.

Where solid reinforced concrete casings are used for fire protection, the casing and the steel core may be assumed to act compositely in carrying the load, provided that the casing complies with the requirements of Clause 30.b of B.S. 449, illustrated diagrammatically in Fig. 1.

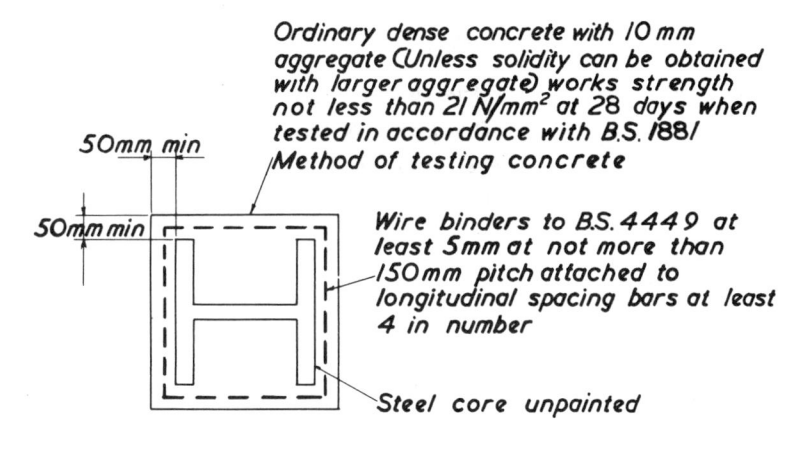

Fig. 1

It is normal practice for the structural casing of the stanchions to be carried out storey by storey, in conjunction with the placing of the floors. Where this is not done, the construction loads on the uncased stanchions should be limited to safe values.

Design Examples

Typical details of part of a multi-storey steel framework are given in Fig. 2. The building is stiffened laterally and longitudinally by vertical shéar walls or bracing systems and by the floors. The design of the stiffening elements is considered to be outside the scope of this chapter.

814 DESIGN OF MULTI-STOREY STANCHIONS

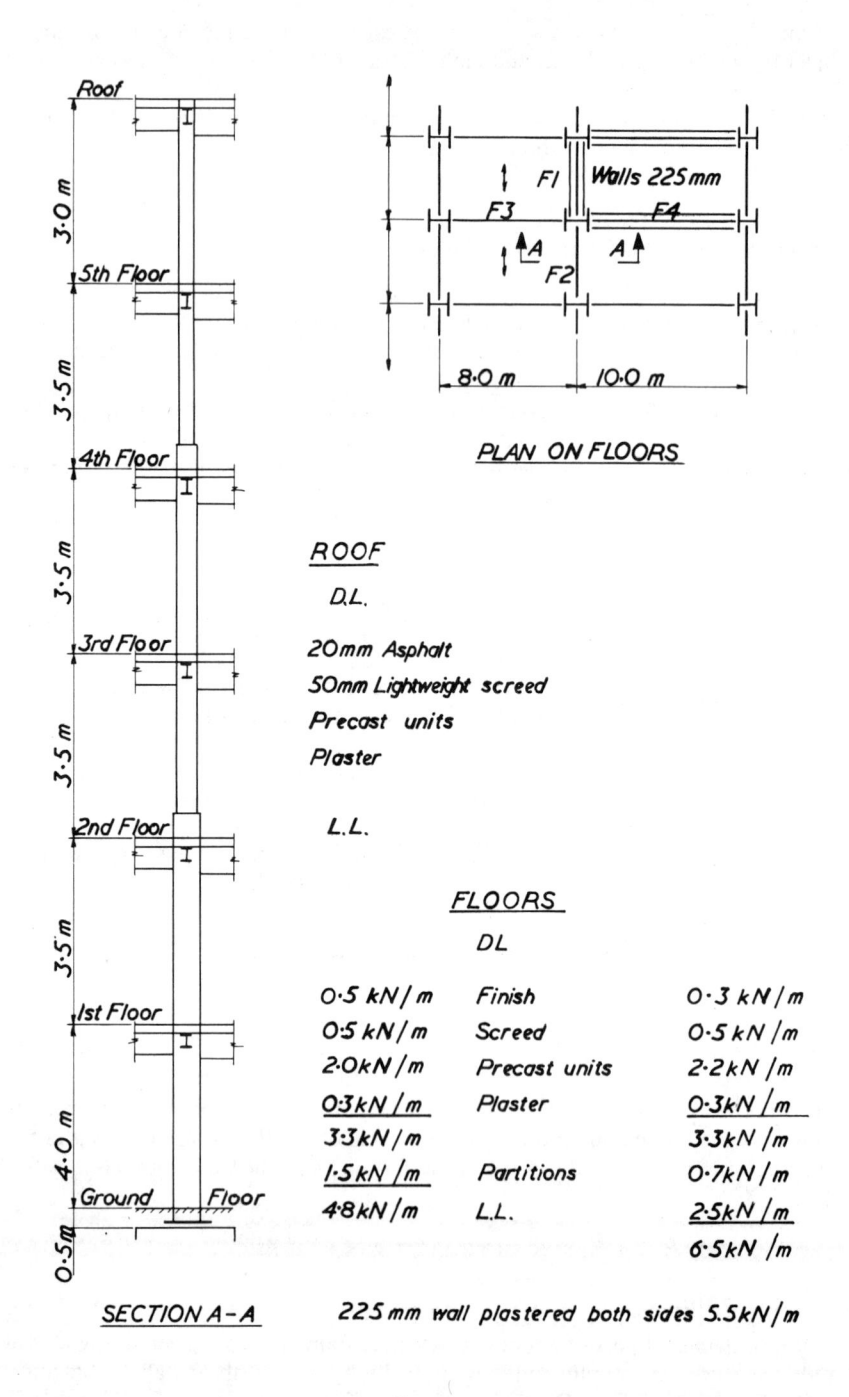

PLAN ON FLOORS

ROOF

D.L.

20mm Asphalt
50mm Lightweight screed
Precast units
Plaster

L.L.

FLOORS

DL

0·5 kN/m	Finish	0·3 kN/m
0·5 kN/m	Screed	0·5 kN/m
2·0 kN/m	Precast units	2·2 kN/m
0·3 kN/m	Plaster	0·3 kN/m
3·3 kN/m		3·3 kN/m
1·5 kN/m	Partitions	0·7 kN/m
4·8 kN/m	L.L.	2·5 kN/m
		6·5 kN/m

SECTION A-A   225mm wall plastered both sides 5.5 kN/m

Fig. 2

TABLE 1

ECCENTRICITY OF BEAM REACTIONS RELATIVE TO COLUMN AXES WHEN UNSTIFFENED BRACKETS USED AND REACTION ACTS AT 100 MM FROM FACE OF SECTION

Serial Size	Wt/m	Ecc. xx	Ecc. yy	Serial Size	Wt/m	Ecc. xx	Ecc. yy
mm mm	kg	mm	mm	mm mm	kg	mm	mm
356 × 406	634	337	124	305 × 305	137	260	107
	551	328	121		118	257	106
	467	318	118		97	254	105
	393	310	115	254 × 254	167	245	110
	340	303	113		132	238	108
	287	297	111		107	233	107
	235	291	109		89	230	105
Column core	477	314	124		73	227	104
356 × 368	202	287	108	203 × 203	86	211	107
	177	284	107		71	208	105
	153	281	106		60	205	105
	129	278	105		52	203	104
305 × 305	283	283	113		46	202	104
	240	276	112	152 × 152	37	181	104
	198	270	110		30	179	103
	158	264	108		23	176	103

Comparative designs will be given for the stanchion shown on Section A.A. as follows:

Example 1A   Lightweight cased—Grade 43
Example 1B   Concrete encased —Grade 43
Example 2A   Lightweight cased—Grade 50
Example 2B   Concrete encased —Grade 50
Example 3A   Lightweight cased—Grade 55
Example 3B   Concrete cased   —Grade 55

For all examples the following notes should be considered.

Dead Loads

The make-up of the floor and roof loading are given in Fig. 2 and can be considered as representative for this type of structure.

Imposed Loads

Imposed loads for the roof and floors are also given in Fig. 2, and are based on the requirements of CP 3 Chapter V: Part 1:1967 'Loading' applied to office buildings.

TABLE 17a. ALLOWABLE STRESS p_o ON GROSS SECTION FOR AXIAL COMPRESSION

l/r	\multicolumn{10}{c}{p_o (N/mm²) for grade 43 steel}
	0	1	2	3	4	5	6	7	8	9
0	155	155	154	154	153	153	153	152	152	151
10	151	151	150	150	149	149	148	148	148	147
20	147	146	146	146	145	145	144	144	144	143
30	143	142	142	142	141	141	141	140	140	139
40	139	138	138	137	137	136	136	136	135	134
50	133	133	132	131	130	130	129	128	127	126
60	126	125	124	123	122	121	120	119	118	117
70	115	114	113	112	111	110	108	107	106	105
80	104	102	101	100	99	97	96	95	94	92
90	91	90	89	87	86	85	84	83	81	80
100	79	78	77	76	75	74	73	72	71	70
110	69	68	67	66	65	64	63	62	61	61
120	60	59	58	57	56	56	55	54	53	53
130	52	51	51	50	49	49	48	48	47	46
140	46	45	45	44	43	43	42	42	41	41
150	40	40	39	39	38	38	38	37	37	36
160	36	35	35	35	34	34	33	33	33	32
170	32	32	31	31	31	30	30	30	29	29
180	29	28	28	28	28	27	27	27	26	26
190	26	26	25	25	25	25	24	24	24	24
200	24	23	23	23	23	22	22	22	22	22
210	21	21	21	21	21	20	20	20	20	20
220	20	19	19	19	19	19	19	18	18	18
230	18	18	18	18	17	17	17	17	17	17
240	17	16	16	16	16	16	16	16	16	15
250	15									
300	11									
350	8									

Intermediate values may be obtained by linear interpolation.

NOTE. For material over 40 mm thick, other than rolled I-beams or channels, and for Universal columns of thicknesses exceeding 40 mm, the limiting stress is 140 N/mm².

TABLE 17b. ALLOWABLE STRESS p_c ON GROSS SECTION FOR AXIAL COMPRESSION

l/r	\multicolumn{10}{c}{p_c (N/mm²) for grade 50 steel}
	0	1	2	3	4	5	6	7	8	9
0	215	214	214	213	213	212	212	211	211	210
10	210	209	209	208	208	207	207	206	206	205
20	205	204	204	203	203	202	202	201	201	200
30	200	199	199	198	197	197	196	196	195	194
40	193	193	192	191	190	189	188	187	186	185
50	184	183	181	180	179	177	176	174	173	171
60	169	168	166	164	162	160	158	156	154	152
70	150	148	146	144	142	140	138	135	133	131
80	129	127	125	123	121	119	117	115	113	111
90	109	107	106	104	102	100	99	97	95	94
100	92	91	89	88	86	85	84	82	81	80
110	78	77	76	75	74	72	71	70	69	68
120	67	66	65	64	63	62	61	60	60	59
130	58	57	56	55	55	54	53	52	52	51
140	50	50	49	48	48	47	47	46	45	45
150	44	44	43	43	42	42	41	41	40	40
160	39	39	38	38	37	37	36	36	36	35
170	35	34	34	34	33	33	33	32	32	31
180	31	31	30	30	30	30	29	29	29	28
190	28	28	27	27	27	27	26	26	26	26
200	25	25	25	25	24	24	24	24	23	23
210	23	23	23	22	22	22	22	22	21	21
220	21	21	21	20	20	20	20	20	20	19
230	19	19	19	19	19	18	18	18	18	18
240	18	18	17	17	17	17	17	17	17	16
250	16									
300	11									
350	8									

Intermediate values may be obtained by linear interpolation.

NOTE. For material over 65 mm thick, the allowable stress p_c on gross section for axial compression shall be calculated in accordance with the procedure in Appendix B taking Y_s equal to the value of the yield stress agreed with the manufacturer, with a maximum value of 350 N/mm².

TABLE 17c. ALLOWABLE STRESS p_c ON GROSS SECTION FOR AXIAL COMPRESSION

l/r	\multicolumn{10}{c}{p_c (N/mm²) for grade 55 steel}
	0	1	2	3	4	5	6	7	8	9
0	265	264	264	263	262	262	261	260	260	259
10	258	258	257	256	256	255	254	254	253	252
20	252	251	250	250	249	248	248	247	246	246
30	245	244	244	243	242	241	240	239	239	238
40	236	235	234	233	232	230	229	227	226	224
50	222	220	219	217	214	212	210	208	205	203
60	200	197	195	192	189	186	183	180	178	175
70	172	169	166	163	160	157	154	151	148	146
80	143	140	138	135	133	130	128	125	123	121
90	118	116	114	112	110	108	106	104	102	100
100	99	97	95	93	92	90	89	87	86	84
110	83	82	80	79	78	76	75	74	73	72
120	71	69	68	67	66	65	64	63	62	62
130	61	60	59	58	57	56	56	55	54	53
140	53	52	51	50	50	49	49	48	47	47
150	46	45	45	44	44	43	43	42	42	41
160	41	40	40	39	39	38	38	37	37	37
170	36	36	35	35	34	34	34	33	33	33
180	32	32	32	31	31	31	30	30	30	29
190	29	29	28	28	28	28	27	27	27	27
200	26	26	26	25	25	25	25	25	24	24
210	24	24	23	23	23	23	23	22	22	22
220	22	22	21	21	21	21	21	20	20	20
230	20	20	20	19	19	19	19	19	19	18
240	18	18	18	18	18	18	17	17	17	17
250	17									
300	12									
350	9									

Intermediate values may be obtained by linear interpolation.

NOTE. For material over 40 mm thick, other than rolled I-beams or channels, and for Universal columns of thicknesses exceeding 40 mm, the limiting stress is 245 N/mm².

PERMISSIBLE AXIAL STRESSES

For certain occupancies, including offices, Clause 5(a) of this Code permits reductions to be made in the total imposed floor loads carried by stanchions and their foundations, as follows:

Number of Floors including the Roof Carried by Member under Consideration	Per Cent Reduction of Imposed Load on all Floors above the Member under Consideration
1	0
2	10
3	20
4	30
5 to 10	40
over 10	50

In this example the reductions have been calculated as follows:

Area supported at any one level = 9 x 3.5 = 31.5 m².

Imposed load at roof level = 31.5 x 1.5 = 47 kN.

Imposed load at any one floor level = 31.5 x 2.5 = 79 kN.

Storey	Number of Floors	% Reduction	Reduction in Imposed Load (kN)
Roof – 5th	1	0	0
5th – 4th	2	10	13
4th – 3rd	3	20	41
3rd – 2nd	4	30	85
2nd – 1st	5	40	145
1st – G	6	40	177

Effective Length Factors

From an examination of Clause 31 and Appendix D of B.S. 449, it appears that an effective length coefficient of 0.7 is appropriate for both axes for the two upper lengths and 0.85 for the lower length. As the effective lengths about both axes are similar in this example, it is only necessary to calculate the slenderness ratio about the y–y axis, since for a Universal Column the radius of gyration is least about this axis. This is shown in the calculations, and for convenience the actual length is measured base to floor or floor to floor. In every case the slenderness ratio is much less than the limiting value of 180.

Permissible Axial Stresses

The permissible axial stresses for steel stanchions are given in Tables 17a, 17b and 17c of B.S. 449 and for convenience these tables have already been reproduced.

The limitations of stress mentioned at the foot of the various tables refer to heavier sections than will be required for this example.

For uncased stanchions the slenderness ratio is calculated using the radius of gyration of the steel section and the permissible stress is considered as acting on the area of steel section. For stanchions encased in concrete, which complies with the requirements given in B.S. 449, the radius of gyration about the weaker axis of the section may be calculated having regard to the assistance provided by the concrete and the area of concrete encasement within certain limits, may be used to assist in carrying the axial load.

The detailed calculations show the method of utilising these concessions, but attention is drawn to the limitations imposed in B.S. 449, which are not applicable in this example.

These are as follows:

1. In no case shall the actual load on a cased strut exceed twice that which would be permitted on the uncased section.
2. The slenderness ratio of the uncased section on the full length, centre to centre of connection must not exceed 250.
3. The amount of concrete cover used in computing the allowable axial load on the cased stanchion must not exceed 75 mm.
4. The clauses permitting calculations for cased stanchions do not apply when the overall dimensions of the steel member exceed 1 000 mm x 500 mm nor do they apply to hollow box sections.

Eccentricity of Beam Reactions

Clause 34 of B.S. 449 specifies that the eccentricity shall be taken as 100 mm from the face of the section, or at the centre of the bearing, whichever dimension gives the greater eccentricity. Since in this example it has been assumed that the beams are supported by unstiffened angle cleats the former requirement is applicable. Table 1 on page 815 gives the eccentricities of beam reactions relative to the two axes for the universal column sections when the reactions are applied at 100 mm from the face of the section.

Distribution of Bending Moments Produced by Eccentricity of Beam Reactions

According to Clause 34.b of B.S. 449 the effect of the moments produced at any floor need only be considered at that floor. The clause also states that the moments may be divided equally between the stanchion lengths above and below that floor, provided the I/L value of one length does not exceed 1.5 x the I/L value of the other length. Where this is not the case the moments must be proportioned in accordance with the I/L values.

In this example each stanchion section extends over two storeys so that it is only necessary to check the lower storey length of each section. The design moments from the floor at mid-height of each section will be divided equally between the upper and lower storey lengths since the I/L ratio is almost the same for both.

PERMISSIBLE BENDING STRESSES

The upper storey length of each section, being over-designed for axial loads, will amply cater for the additional moments induced by the change of section at the upper splice.

Permissible Bending Stresses

The permissible compressive bending stresses for the different grades of steel are given in Tables 2 and 3a, 3b and 3c of B.S. 449, and depend upon the D/T and l/r_y ratios. In this case l is the effective length of the compression flange in bending as defined in Clause 26.

Comparison of the beam effective length factors in Clause 26, with the stanchion effective length factors in Clause 31, suggests that when establishing the permissible compressive bending stress for multi-storey stanchions it is reasonable to use the same l/r_y ratio that is used for finding the permissible axial compressive stress.

The maximum permissible bending stresses given in Table 2 are 165, 230 and 280 N/mm² for grades 43, 50 and 55 respectively but these are to be reduced when the material thickness exceeds specified values. The sections used in these examples, however, do not exceed these limits and therefore the maximum values stated can be used. Again, it will be seen from Tables 3a, 3b and 3c that provided the l/r_y ratios do not exceed 90, 80 or 75 respectively no reduction is required for the $l/r_y - D/T$ ratios and again the sections used in these examples do not necessitate a reduction on account of the flange instability.

In considering concrete encased stanchions B.S. 449 permits the r_y value of the cased section to be used in determining the l/r_y ratio provided that the permissible bending stress for the cased section does not exceed $1\frac{1}{2}$ times that which would be permitted for the uncased section. For cased columns used in the context shown by these examples, this is never critical and may be ignored.

The actual bending stresses are calculated on the steel section alone and the concrete encasement is only allowed to assist in carrying the axial load, this is obviously conservative but is a requirement of the present B.S. 449.

Combined Axial Load and Bending

Reference has been made earlier to Clause 14.a of B.S. 449 where it is stated that the sum of the ratios of

$$\frac{\text{actual axial stress}}{\text{permissible axial stress}} + \frac{\text{actual bending stress}}{\text{permissible bending stress}}$$

must not exceed unity. When dealing with concrete encased stanchions the first ratio is usually taken as actual axial load divided by actual permissible load, since this is in effect identical with the former requirement.

Calculations

Calculations in a tabular form are given on the following pages for the six alternative forms of construction referred to earlier, but to assist in understanding these expanded calculations will be given for one shaft in each of the six cases.

Storey Height m	Loading details	Item of load	Loads kN	Total on Storey kN	Total on Shaft kN	Live load reduction kN	Design Load kN
Roof							
3.0 5th		1
2
3
4
O.W.+C.	75
5
93
5
6	184	184	–	184
3.5 4th		1
2
3
4
O.W.+C.	99
34
210
5
7	355	539	13	526
3.5 3rd		1
2
3
4
O.W.+C.	99
34
210
5
9	357	896	41	855
3.5 2nd		1
2
3
4
O.W.+C.	99
34
210
5
9	357	1253	85	1168
3.5 1st		1
2
3
4
O.W.+C.	99
34
210
5
13	361	1614	145	1469
4.0 Grd.		1
2
3
4
O.W.+C.	99
34
210
5
15	363	1977	177	1800
0.5 Base							

Section and Properties	Permissible Stresses N/mm²	Design Calculations
Grade 43 steel. Concrete encased		
4th - Roof Steel size = 152 x 152 x 37 kg UC Cased size = 262 x 255 mm Area of steel = 47.4 cm² Area of conc. = 666 cm² $r_{min.}$ uncased = 3.87 cm $r_{min.}$ cased = 0.2 x 255 = 5.1 cm Z_{xx} for steel = 274 cm³ Z_{yy} for steel = 92 cm³ Ecc xx axis = 181 mm Ecc yy axis = 104 mm	L = 3.5 coef. = 0.7 l = 2.45 l/r = 48 p_c = 135 p_{bc} = 165 Axial stress on concrete = 0.19 x 165 = 31.4 $= \frac{135}{31.4} = 4.3$	Axial load = 526 kN BM_{xx} = (210 − 99) x 181 = 20 200 kNmm BM_{yy} = (34 − 5) x 104 = 3 020 kNmm $f_{bc} = \frac{20200 \times 10^3}{2 \times 274 \times 10^3} + \frac{3020 \times 10^3}{2 \times 92 \times 10^3}$ = 37 + 16 = 53 N/mm² Safe axial load on cased shaft Steel = 47.4 x 10² x 135 x 10⁻³ = 640 kN Concrete = 664 x 10² x 4.3 x 10⁻³ = 286 kN $\overline{926\ kN}$ Ratio = $\frac{\text{Act. axial load}}{\text{Safe axial load}} + \frac{f_{bc}}{p_{bc}}$ = $\frac{526}{926} + \frac{53}{165}$ = 0.57 + 0.32 = 0.89 Section proves O.K.
2nd − 4th Steel size = 203 x 203 x 60 kg UC Cased size = 310 x 305 mm Area of steel = 75.8 cm² Area of conc. = 945 cm² $r_{min.}$ uncased = 5.19 cm $r_{min.}$ cased = 0.2 x 305 = 6.1 cm Z_{xx} for steel = 581 cm³ Z_{yy} for steel = 199 cm³ Ecc xx axis = 205 mm Ecc yy axis = 105 mm	L = 3.5 coef. = 0.7 l = 2.45 l/r = 40 p_c = 139 p_{bc} = 165 Axial stress on concrete = 0.19 x 165 = 31.4 $\frac{139}{31.4} = 4.43$	Axial load = 1168 kN BM_{xx} = (210 − 99) x 205 = 22 800 kNmm BM_{yy} = (34 − 5) x 105 = 3 050 kNmm $f_{bc} = \frac{22800 \times 10^3}{2 \times 581 \times 10^3} + \frac{3050 \times 10^3}{2 \times 199 \times 10^3}$ = 20 + 8 = 28 N/mm² Safe axial load on cased shaft Steel = 75.8 x 10² x 139 x 10⁻³ = 1060 kN Concrete = 940 x 10² x 4.43 x 10⁻³ = 420 kN $\overline{1480\ kN}$ Ratio = $\frac{1168}{1480} + \frac{28}{165}$ = 0.79 + 0.17 = 0.96 Section proves O.K.
B − 2nd Steel size = 254 x 254 x 89 kg UC Cased size = 360 x 360 mm Area of steel = 114 cm² Area of conc. = 1 300 cm² $r_{min.}$ uncased = 6.52 cm $r_{min.}$ cased = 0.2 x 356 = 7.12 cm Z_{xx} for steel = 1099 cm³ Z_{yy} for steel = 379 cm³ Ecc xx axis = 230 mm Ecc yy axis = 105 mm	L = 4.5 coef. = 0.85 l = 3.82 l/r = 54 p_c = 130 p_{bc} = 165 Axial stress on concrete = 0.19 x 165 = 31.4 $\frac{130}{31.4} = 4.15$	Axial load = 1800 kN BM_{xx} = (210 − 99) x 230 = 25 600 kNmm BM_{yy} = (34 − 5) x 105 = 3 050 kNmm $f_{bc} = \frac{25600 \times 10^3}{2 \times 1099 \times 10^3} + \frac{3050 \times 10^3}{2 \times 379 \times 10^3}$ = 12 + 4 = 16 N/mm² Safe axial load on cased shaft Steel = 114 x 10² x 130 x 10⁻³ = 1480 kN Concrete = 1300 x 10² x 4.15 x 10⁻³ = 640 kN $\overline{2120\ kN}$ Ratio = $\frac{1800}{2120} + \frac{16}{165}$ = 0.85 + 0.10 = 0.95 Section proves O.K.

Storey Height m	Loading details	Item of load	Loads kN	Total on Storey kN	Total on Shaft kN	Live load reduction kN	Design Load kN
Roof							
3·0 5th		1 2 3 4 o.w.+c.	75 5 93 5 4	182	182	—	182
3·5 4th		1 2 3 4 o.w.+c.	99 34 210 5 5	353	535	13	522
3·5 3rd		1 2 3 4 o.w.+c.	99 34 210 5 6	354	889	41	848
3·5 2nd		1 2 3 4 o.w.+c.	99 34 210 5 6	354	1243	85	1158
3·5 1st		1 2 3 4 o.w.+c.	99 34 210 5 10	358	1601	145	1456
4·0 Grd.		1 2 3 4 o.w.+c.	99 34 210 5 12	360	1961	177	1784
Base 0·5							

Section and Properties	Permissible Stresses N/mm²	Design Calculations
Grade 43 steel. Lightweight casing		
<u>4th- Roof</u> Steel size = 203 x 203 x 46 kg UC Cased size = 260 x 260 mm Area of steel = 58·8 cm² r_{min} uncased = 5·11 cm Z_{xx} for steel = 449 cm³ Z_{yy} for steel = 151 cm³ Ecc xx axis = 202 mm Ecc yy axis = 104 mm	L = 3·5 coef. = 0·7 l = 2·45 l/r = 48 p_c = 135 p_{bc} = 165	Axial load = 522 kN BM_{xx} = (210 − 99) x 202 = 22 420 kNmm BM_{yy} = (34 − 5) x 104 = 3 020 kNmm Axial stress f_c = $\frac{522 \times 10^3}{58·8 \times 10^2}$ = 89 N/mm² Bending stress f_{bc} = $\frac{22\,420 \times 10^3}{2 \times 449 \times 10^3} + \frac{3\,020 \times 10^3}{2 \times 151 \times 10^3}$ = 25 + 10 = 35 N/mm² Ratio $\frac{f_c}{p_c} + \frac{f_{bc}}{p_{bc}}$ = $\frac{89}{136} + \frac{35}{165}$ = 0·66 + 0·21 = 0·87 Section proves O.K.
<u>2nd- 4th</u> Steel size = 254 x 254 x 73 kg UC Cased size = 310 x 310 mm Area of steel = 92·9 cm² r_{min} uncased = 6·46 cm Z_{xx} for steel = 894 cm³ Z_{yy} for steel = 305 cm³ Ecc xx axis = 227 mm Ecc yy axis = 104 mm	L = 3·5 coef. = 0·7 l = 2·45 l/r = 38 p_c = 140 p_{bc} = 165	Axial load = 1 158 kN BM_{xx} = (210−99) x 227 = 25 200 kNmm BM_{yy} = (34 − 5) x 104 = 3 020 kNmm Axial stress f_c = $\frac{1\,158 \times 10^3}{92·9 \times 10^2}$ = 124 N/mm² Bending stress f_{bc} = $\frac{25\,200 \times 10^3}{2 \times 894 \times 10^3} - \frac{3\,020 \times 10^3}{2 \times 305 \times 10^3}$ = 14 + 5 = 19 N/mm² Ratio $\frac{f_c}{p_c} + \frac{f_{bc}}{p_{bc}}$ = $\frac{124}{140} + \frac{19}{165}$ = 0·89 + 0·11 = 1·0 Section proves O.K.
<u>B - 2nd</u> Steel size = 305 x 305 x 118 kg UC Cased size = 360 x 360 mm Area of steel = 149·8 cm² r_{min} uncased = 7·75 cm Z_{xx} for steel = 1755 cm³ Z_{yy} for steel = 587 cm³ Ecc xx axis = 258 mm Ecc yy axis = 106 mm	L = 4·5 coef. = 0·85 l = 3·82 l/r = 49 p_c = 135 p_{bc} = 165	Axial load = 1 784 kN BM_{xx} = (210 − 99) x 258 = 28 640 kNmm BM_{yy} = (34 − 5) x 106 = 3 070 kNmm Axial stress f_c = $\frac{1\,784 \times 10^3}{149·8 \times 10^2}$ = 119 N/mm² Bending stress f_{bc} = $\frac{28\,640 \times 10^3}{2 \times 1755 \times 10^3} + \frac{3\,070 \times 10^3}{2 \times 587 \times 10^3}$ = 8 + 3 = 11 N/mm² Ratio $\frac{f_c}{p_c} + \frac{f_{bc}}{p_{bc}}$ = $\frac{119}{135} + \frac{11}{165}$ = 0·88 + 0·07 = 0·95 Section proves O.K.

Storey Height m	Loading details	Item of load	Loads kN	Total on Storey kN	Total on Shaft kN	Live load reduction kN	Design Load kN
Roof							
3.0 (5th)		1	75	182	182	–	182
		2	5				
		3	93				
		4	5				
		o.w.+c.	4				
3.5 (4th)		1	99	352	534	13	521
		2	34				
		3	210				
		4	5				
		o.w.+c.	4				
3.5 (3rd)		1	99	353	887	41	846
		2	34				
		3	210				
		4	5				
		o.w.+c.	5				
3.5 (2nd)		1	99	353	1 240	85	1 155
		2	34				
		3	210				
		4	5				
		o.w.+c.	5				
3.5 (1st)		1	99	356	1 596	145	1 451
		2	34				
		3	210				
		4	5				
		o.w.+c.	8				
4.0 (Grd.)		1	99	358	1 954	177	1 777
		2	34				
		3	210				
		4	5				
		o.w.+c.	10				
Base 0.5							

Section and Properties	Permissible Stresses N/mm^2	Design Calculations
Grade 50 steel. Lightweight casing		
4th - Roof Steel size = 152 x 152 x 37 kg UC Cased size = 210 x 210 mm Area of steel = 47.4 cm^2 r_{min} uncased = 3.87 cm Z_{xx} for steel = 274 cm^3 Z_{yy} for steel = 92 cm^3 Ecc xx axis = 181 mm Ecc yy axis = 104 mm	L = 3.5 coef. = 0.7 l = 2.45 l/r = 63 p_c = 164 p_{bc} = 230	Axial load = 521 kN BM_{xx} = (210 − 99) x 181 = 20 200 kNmm BM_{yy} = (34 − 5) x 104 = 3 020 kNmm Axial stress $f_c = \dfrac{521 \times 10^3}{474 \times 10^2}$ = 110 N/mm^2 Bending stress $f_{bc} = \dfrac{20\,200 \times 10^3}{2 \times 274 \times 10^3} + \dfrac{3020 \times 10^3}{2 \times 92 \times 10^3}$ = 37 + 16 = 53 N/mm^2 Ratio $\dfrac{f_c}{p_c} + \dfrac{f_{bc}}{p_{bc}} = \dfrac{110}{164} + \dfrac{53}{230}$ = 0.67 + 0.23 = 0.90 Section proves O.K.
2nd - 4th Steel size = 203 x 203 x 60 kg UC Cased size = 260 x 260 mm Area of steel = 75.8 cm^2 r_{min} uncased = 5.19 cm Z_{xx} for steel = 581 cm^3 Z_{yy} for steel = 199 cm^3 Ecc xx axis = 205 mm Ecc yy axis = 105 mm	L = 3.5 coef. = 0.7 l = 2.45 l/r = 47 p_c = 187 p_{bc} = 230	Axial load = 1 155 kN BM_{xx} = (210 − 99) x 205 = 22 800 kNmm BM_{yy} = (34 − 5) x 105 = 3 050 kNmm Axial stress $f_c = \dfrac{1\,155 \times 10^3}{75.8 \times 10^2}$ = 153 N/mm^2 Bending stress $f_{bc} = \dfrac{22\,800 \times 10^3}{2 \times 581 \times 10^3} + \dfrac{3050 \times 10^3}{2 \times 199 \times 10^3}$ = 20 + 8 = 28 N/mm^2 Ratio $\dfrac{f_c}{p_c} + \dfrac{f_{bc}}{p_{bc}} = \dfrac{153}{187} + \dfrac{28}{230}$ = 0.82 + 0.12 = 0.94 Section proves O.K.
B - 2nd Steel size = 254 x 254 x 89 kg UC Cased size = 310 x 310 mm Area of steel = 114 cm^2 r_{min} uncased = 6.52 cm Z_{xx} for steel = 1099 cm^3 Z_{yy} for steel = 379 cm^3 Ecc xx axis = 230 mm Ecc yy axis = 105 mm	L = 4.5 coef. = 0.85 l = 3.82 l/r = 59 p_c = 171 p_{bc} = 230	Axial load = 1 777 kN BM_{xx} = (210 − 99) x 230 = 25 600 kNmm BM_{yy} = (34 − 5) x 105 = 3 050 kNmm Axial stress $f_c = \dfrac{1\,777 \times 10^3}{114 \times 10^2}$ = 156 N/mm^2 Bending stress $f_{bc} = \dfrac{25\,600 \times 10^3}{2 \times 1099 \times 10^3} + \dfrac{3050 \times 10^3}{2 \times 379 \times 10^3}$ = 12 + 4 = 16 N/mm^2 Ratio $\dfrac{f_c}{p_c} + \dfrac{f_{bc}}{p_{bc}} = \dfrac{156}{171} + \dfrac{16}{230}$ = 0.91 + 0.07 = 0.98 Section proves O.K.

Storey Height m	Loading details	Item of load	Loads kN	Total on Storey kN	Total on Shaft kN	Live load reduction kN	Design Load kN
Roof							
3·0 5th	(diagram: 1—2—3—4)	1 2 3 4 O.W.+ C.	75 5 93 5 5	183	183	–	183
3·5 4th	(diagram: 1—2—3—4)	1 2 3 4 O.W.+ C.	99 34 210 5 5	353	536	13	523
3·5 3rd	(diagram: 1—2—3—4)	1 2 3 4 O.W.+ C.	99 34 210 5 7	355	891	41	850
3·5 2nd	(diagram: 1—2—3—4)	1 2 3 4 O.W.+ C.	99 34 210 5 7	355	1246	85	1161
3·5 1st	(diagram: 1—2—3—4)	1 2 3 4 O.W.+ C.	99 34 210 5 12	360	1606	145	1461
4·0 Grd	(diagram: 1—2—3—4)	1 2 3 4 O.W.+ C.	99 34 210 5 14	362	1968	177	1791
Base 0·5							

Section and Properties	Permissible Stresses N/mm^2	Design Calculations
Grade 50 steel. Concrete encased		
4th – Roof Steel size = 152 x 152 x 30 kg UC Cased size = 260 x 255 mm Area of steel = 38·2 cm^2 Area of conc. = 664 cm^2 r_{min} uncased = 3·82 cm r_{min} cased = 0·2 x 253 = 5·06 cm Z_{xx} for steel = 221 cm^3 Z_{yy} for steel = 73 cm^3 Ecc xx axis = 179 mm Ecc yy axis = 104 mm	L = 3·5 coef. = 0·7 l = 2·45 l/r = 48 p_c = 186 p_{bc} = 230 Axial stress on concrete = 0·19 x 230 = 43·9 = $\frac{186}{43·9}$ = 4·25	Axial load = 523 kN BM_{xx} = (210 − 99) x 179 = 19 900 kNmm BM_{yy} = (34 − 5) x 104 = 3 020 kNmm $f_{bc} = \frac{19\,900 \times 10^3}{2 \times 221 \times 10^3} + \frac{3\,020 \times 10^3}{2 \times 73 \times 10^3}$ = 45 + 21 = 66 N/mm^2 Safe axial load on cased shaft Steel = 38·2 x 10^2 x 186 x 10^{-3} = 710 kN Concrete: 664 x 10^2 x 4·25 x 10^{-3} = $\underline{282\ kN}$ 992 kN Ratio: $\frac{Act.\ axial\ load}{Safe\ axial\ load} + \frac{f_{bc}}{p_{bc}}$ = $\frac{523}{992} + \frac{66}{230}$ = 0·53 + 0·29 = 0·82 Section proves O.K.
2nd – 4th Steel size = 203 x 203 x 46 kg UC Cased size = 305 x 305 mm Area of steel = 58·8 cm^2 Area of conc. = 930 cm^2 r_{min} uncased = 5·11 cm r_{min} cased = 0·2 x 303 = 6·06 cm Z_{xx} for steel = 449 cm^3 Z_{yy} for steel = 152 cm^3 Ecc xx axis = 202 mm Ecc yy axis = 104 mm	L = 3·5 coef. = 0·7 l = 2·45 l/r = 40 p_c = 193 p_{bc} = 230 Axial stress on concrete = $\frac{193}{43·9}$ = 4·4	Axial load = 1161 kN BM_{xx} = (210 − 99) x 202 = 22 400 kNmm BM_{yy} = (34 − 5) x 104 = 3 020 kNmm $f_{bc} = \frac{22\,400 \times 10^3}{2 \times 449 \times 10^3} + \frac{3\,020 \times 10^3}{2 \times 152 \times 10^3}$ = 25 + 10 = 35 N/mm^2 Safe axial load on cased shaft Steel = 58·8 x 10^2 x 193 x 10^{-3} = 1130 kN Concrete: 930 x 10^2 x 4·4 x 10^{-3} = $\underline{410\ kN}$ 1540 kN Ratio = $\frac{1161}{1540} + \frac{35}{230}$ = 0·76 + 0·15 = 0·91 Section proves O.K.
B – 2nd Steel size = 254 x 254 x 73 kg UC Cased size = 355 x 355 mm Area of steel = 92·9 cm^2 Area of conc. = 1260 cm^2 r_{min} uncased = 6·46 cm r_{min} cased = 0·2 x 354 = 7·08 cm Z_{xx} for steel = 894 cm^3 Z_{yy} for steel = 305 cm^3 Ecc xx axis = 227 mm Ecc yy axis = 104 mm	L = 4·5 coef. = 0·85 l = 3·82 l/r = 54 p_c = 179 p_{bc} = 230 Axial stress on concrete = $\frac{179}{43·9}$ = 4·1	Axial load = 1791 kN BM_{xx} = (210 − 99) x 227 = 25 200 kNmm BM_{yy} = (34 − 5) x 104 = 3 020 kNmm $f_{bc} = \frac{25\,200 \times 10^3}{2 \times 894 \times 10^3} + \frac{3\,020 \times 10^3}{2 \times 227 \times 10^3}$ = 14 + 5 = 19 N/mm^2 Safe axial load on cased shaft Steel = 92·9 x 10^2 x 179 x 10^{-3} = 1660 kN Concrete = 1260 x 10^2 x 4·1 x 10^{-3} = $\underline{518\ kN}$ 2178 kN Ratio = $\frac{1791}{2178} + \frac{19}{230}$ = 0·83 + 0·08 = 0·91 Section proves O.K.

Storey Height m	Loading details	Item of load	Loads kN	Total on Storey kN	Total on Shaft kN	Live load reduction kN	Design Load kN
Roof							
3·0 5th		1 2 3 4 o.w.+c.	75 5 93 5 4	182	182	—	182
3·5 4th		1 2 3 4 o.w.+c.	99 34 210 5 4	352	534	13	521
3·5 3rd		1 2 3 4 o.w.+c.	99 34 210 5 5	353	887	41	846
3·5 2nd		1 2 3 4 o.w.+c.	99 34 210 5 5	353	1240	85	1155
3·5 1st		1 2 3 4 o.w.+c.	99 34 210 5 7	355	1595	145	1450
4·0 Grd.		1 2 3 4 o.w.+c.	99 34 210 5 9	357	1952	177	1775
Base 0·5							

Section and Properties	Permissible Stresses N/mm^2	Design Calculations
Grade 55 steel. Lightweight casing		
4th - Roof Steel size = 152 x 152 x 30 kg UC Cased size = 210 x 210 mm Area of steel = 38·2 cm^2 r_{min} uncased = 3·82 cm Z_{xx} for steel = 221 cm^3 Z_{yy} for steel = 73 cm^3 Ecc xx axis = 179 mm Ecc yy axis = 104 mm	L = 3·5 coef. = 0·7 l = 2·45 l/r = 64 p_c = 189 p_{bc} = 280	Axial load = 521 kN BM_{xx} = (210 − 99) x 179 = 19 900 kNmm BM_{yy} = (34 − 5) x 104 = 3 020 kNmm Axial stress f_c = $\frac{521 \times 10^3}{38 \cdot 2 \times 10^2}$ = 136 N/mm^2 Bending stress f_{bc} = $\frac{19\,900 \times 10^3}{2 \times 221 \times 10^3} + \frac{3\,020 \times 10^3}{2 \times 73 \times 10^3}$ = 45 + 21 = 66 N/mm^2 Ratio $\frac{f_c}{p_c} + \frac{f_{bc}}{p_{bc}}$ = $\frac{136}{189} + \frac{66}{280}$ = 0·72 + 0·24 = 0·96 Section proves O.K.
2nd - 4th Steel size = 203 x 203 x 46 kg UC Cased size = 255 x 255 mm Area of steel = 58·8 cm^2 r_{min} uncased = 5·11 cm Z_{xx} for steel = 449 cm^3 Z_{yy} for steel = 152 cm^3 Ecc xx axis = 202 mm Ecc yy axis = 104 mm	L = 3·5 coef. = 0·7 l = 2·45 l/r = 50 p_c = 222 p_{bc} = 280	Axial load = 1155 kN BM_{xx} = (210 − 99) x 202 = 22 400 kNmm BM_{yy} = (34 − 5) x 104 = 3 020 kNmm Axial stress f_c = $\frac{1155 \times 10^3}{58 \cdot 8 \times 10^2}$ = 197 N/mm^2 Bending stress f_{bc} = $\frac{22\,400 \times 10^3}{2 \times 449 \times 10^3} + \frac{3\,020 \times 10^3}{2 \times 152 \times 10^3}$ = 25 + 10 = 35 N/mm^2 Ratio $\frac{f_c}{p_c} + \frac{f_{bc}}{p_{bc}}$ = $\frac{197}{222} + \frac{35}{280}$ = 0·88 + 0·12 = 1·0 Section proves O.K.
B - 2nd Steel size = 254 x 254 x 73 kg UC Cased size = 305 x 305 mm Area of steel = 92·9 cm^2 r_{min} uncased = 6·46 cm Z_{xx} for steel = 894 cm^3 Z_{yy} for steel = 305 cm^3 Ecc xx axis = 227 mm Ecc yy axis = 104 mm	L = 4·5 coef. = 0·85 l = 3·82 l/r = 59 p_c = 203 p_{bc} = 280	Axial load = 1775 kN BM_{xx} = (210 − 99) x 227 = 25 200 kNmm BM_{yy} = (34 − 5) x 104 = 3 020 kNmm Axial stress f_c = $\frac{1775 \times 10^3}{92 \cdot 9 \times 10^2}$ = 190 N/mm^2 Bending stress f_{bc} = $\frac{25\,200 \times 10^3}{2 \times 894 \times 10^3} + \frac{3\,020 \times 10^3}{2 \times 305 \times 10^3}$ = 14 + 5 = 19 N/mm^2 Ratio $\frac{f_c}{p_c} + \frac{f_{bc}}{p_{bc}}$ = $\frac{190}{203} + \frac{19}{280}$ = 0·93 + 0·07 = 1·0 Section proves O.K.

Storey Height m	Loading details	Item of load	Loads kN	Total on Storey kN	Total on Shaft kN	Live load reduction kN	Design Load kN
Roof							
3.0　　5th		1 2 3 4 O.W.+C.	75 5 93 5 5	183	183	—	183
3.5　　4th		1 2 3 4 O.W.+C.	99 34 210 5 5	353	536	13	523
3.5　　3rd		1 2 3 4 O.W.+C.	99 34 210 5 7	355	891	41	850
3.5　　2nd		1 2 3 4 O.W.+C.	99 34 210 5 7	355	1246	85	1161
3.5　　1st		1 2 3 4 O.W.+C.	99 34 210 5 12	360	1606	145	1461
4.0　　Grd.		1 2 3 4 O.W.+C.	99 34 210 5 14	362	1968	177	1791
Base 0.5							

Section and Properties	Permissible Stresses N/mm²	Design Calculations
Grade 55 steel. Concrete encased		
4th – Roof Steel size = 152 x 152 x 23 kg UC Cased size = 255 x 255 mm Area of steel = 29·8 cm² Area of conc. = 650 cm² $r_{min.}$ uncased = 3·68 cm $r_{min.}$ cased = 0·2 x 252 = 5·04 cm Z_{xx} for steel = 166 cm³ Z_{yy} for steel = 53 cm³ Ecc xx axis = 176 mm Ecc yy axis = 103 mm	L = 3·5 coef. = 0·7 l = 2·45 l/r = 49 p_c = 224 p_{bc} = 280 Axial stress on concrete = 0·19 x 280 = 53·2 = $\frac{224}{53·2}$ = 4·2	Axial load = 523 kN BM_{xx} = (210 − 99) x 176 = 19 600 kNmm BM_{yy} = (34 − 5) x 103 = 2 990 kNmm $f_{bc} = \frac{19\,600 \times 10^3}{2 \times 166 \times 10^3} + \frac{2\,990 \times 10^3}{2 \times 53 \times 10^3}$ = 59 + 28 = 87 N/mm² Safe axial load on cased shaft Steel = 29·8 x 10² x 224 x 10⁻³ = 670 kN Concrete = 650 x 10² x 4·2 x 10⁻³ = 274 kN 944 kN Ratio = $\frac{Act.\ axial\ load}{Safe\ axial\ load} + \frac{f_{bc}}{p_{bc}}$ = $\frac{523}{944} + \frac{87}{280}$ 0·56 + 0·31 = 0·87 Section proves O.K.
2nd – 4th Steel size = 203 x 203 x 46 kg UC Cased size = 305 x 305 mm Area of steel = 58·8 cm² Area of conc. = 930 cm² $r_{min.}$ uncased = 5·11 cm $r_{min.}$ cased = 0·2 x 303 = 6·06 cm Z_{xx} for steel = 449 cm³ Z_{yy} for steel = 152 cm³ Ecc xx axis = 202 mm Ecc yy axis = 104 mm	L = 3·5 coef. = 0·7 l = 2·45 l/r = 40 p_c = 236 p_{bc} = 280 Axial stress on concrete = $\frac{236}{53·2}$ = 4·4	Axial load = 1 161 kN BM_{xx} = (210 − 99) x 202 = 22 400 kNmm BM_{yy} = (34 − 5) x 104 = 3 020 kNmm $f_{bc} = \frac{22\,400 \times 10^3}{2 \times 449 \times 10^3} + \frac{3\,020 \times 10^3}{2 \times 152 \times 10^3}$ = 25 + 10 = 35 N/mm² Safe axial load on cased shaft Steel = 58·8 x 10² x 236 x 10⁻³ = 1390 kN Concrete = 930 x 10² x 4·4 x 10⁻³ = 410 kN 1800 kN Ratio = $\frac{1\,161}{1\,800} + \frac{35}{280}$ 0·65 + 0·09 = 0·74 Section proves O.K.
B – 2nd Steel size = 254 x 254 x 73 kg UC Cased size = 355 x 355 mm Area of steel = 92·9 cm² Area of conc. = 1 260 cm² $r_{min.}$ uncased = 6·46 cm $r_{min.}$ cased = 0·2 x 354 = 7·08 cm Z_{xx} for steel = 894 cm³ Z_{yy} for steel = 305 cm³ Ecc xx axis = 227 mm Ecc yy axis = 104 mm	L = 4·5 coef. = 0·85 l = 3·82 l/r = 54 p_c = 214 p_{bc} = 280 Axial stress on concrete = $\frac{214}{53·2}$ = 4·0	Axial load = 1 791 kN BM_{xx} = (210 − 99) x 227 = 25 200 kNmm BM_{yy} = (34 − 5) x 104 = 3 020 kNmm $f_{bc} = \frac{25\,200 \times 10^3}{2 \times 894 \times 10^3} + \frac{3\,020 \times 10^3}{2 \times 305 \times 10^3}$ = 14 + 5 = 19 N/mm² Safe axial load on cased shaft Steel = 92·9 x 10² x 214 x 10⁻³ = 1990 kN Concrete = 1260 x 10² x 4 x 10⁻³ = 504 kN 2494 kN Ratio = $\frac{1\,791}{2\,494} + \frac{19}{280}$ 0·72 + 0·07 = 0·79 Section proves O.K.

834 DESIGN OF MULTI-STOREY STANCHIONS

Example 1A. Grade 43 Steel—Lightweight Cased

Examine shaft between 2nd and 4th floors.

The critical loading conditions will be those at the 3rd floor level. These are shown in sketch below, Fig. 3.

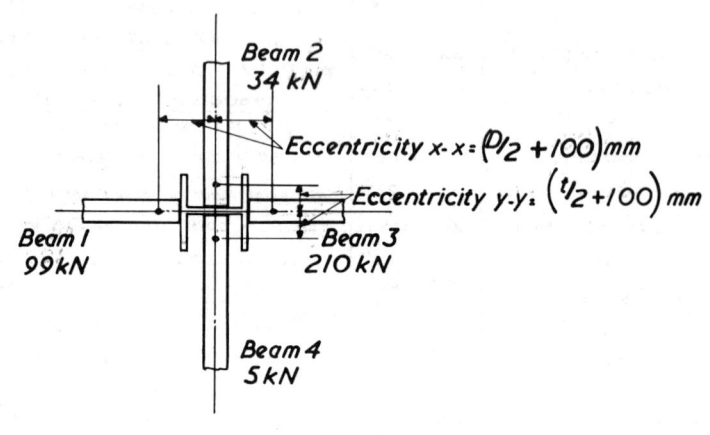

Fig. 3

Beam 2: 34 kN
Beam 1: 99 kN
Beam 3: 210 kN
Beam 4: 5 kN

Eccentricity x-x = $(D/2 + 100)$ mm
Eccentricity y-y = $(t/2 + 100)$ mm

The axial load on the shaft under consideration will be the load from above plus the sum of the loads at the level shown including an allowance for the weight of this stack of stanchion and its casing, less the permitted reduction in live load, thus:—

Load from above (see tabulation)	= 889 kN
Load at 3rd level	= 348 kN
Weight of stanchion and casing	= 6 kN
	1 243 kN
Deduct for live load reduction	85 kN
Nett axial load	= 1 158 kN

The effective length of the stack under consideration is the actual length floor to floor multiplied by the coefficient. In this case since the web beams are load carrying this will be taken as 0.7.

Then:—

Actual length floor to floor = 3.5 m

Coefficient = 0.7

Effective length = 0.7 × 3.5 = 2.45 m.

Try section 254 × 254 × 73 kg U.C.

Actual size = 254 × 254 mm.

Area A = 92.9 cm² : Web thickness t = 8.6 mm
Radius of gyration r_{xx} = 11.1 cm : r_{yy} = 6.46 cm
Modulus of section Z_{xx} = 894 cm³ : Z_{yy} = 305 cm³

$$l/r_{yy} = \frac{2.45 \times 10^3}{6.46 \times 10} = 38 \;:\; \text{Permissible axial stress} \quad p_c = 140 \text{ N/mm}^2$$
$$\text{Permissible bending stress } p_{bc} = 165 \text{ N/mm}^2$$

Eccentricities:—
$$xx = \left(\frac{D}{2} + 100\right) = \left(\frac{254}{2} + 100\right) = 227 \text{ mm}$$

$$yy = \left(\frac{t}{2} + 100\right) = \left(\frac{8.6}{2} + 100\right) = 104 \text{ mm}$$

$M_{xx} = (210 - 99) \times 227 = 25\,200$ kNmm $\quad M_{yy} = (34 - 5) \times 104 = 3\,020$ kNmm

Bending moments divided equally into stack above and below

Actual axial stress $\quad f_c = \dfrac{W}{A} = \dfrac{1\,158 \times 10^3}{92.9 \times 10^2} = 124$ N/mm^2

Actual bending stress $\quad f_{bc} = \dfrac{M_{xx}}{2 \times Z_{xx}} + \dfrac{M_{yy}}{2 \times Z_{yy}}$

$$= \frac{25\,200 \times 10^3}{2 \times 894 \times 10^3} + \frac{3\,020 \times 10^3}{2 \times 305 \times 10^3}$$

$$= 14 + 5 = 19 \text{ N/mm}^2.$$

Ratio of $\dfrac{f_c}{p_c} + \dfrac{f_{bc}}{p_{bc}}$ not to exceed unity.

Then $\quad \dfrac{f_c}{p_c} + \dfrac{f_{bc}}{p_{bc}} = \dfrac{124}{140} + \dfrac{19}{165} = 0.89 + 0.11 = 1.00.$

∴ Section selected is satisfactory.

Example 1B. Grade 43 Steel Encased in Concrete to requirements of B.S. 449

Examine shaft between 2nd and 4th floors

Loading conditions at the 3rd floor level as sketch: Fig. 4.

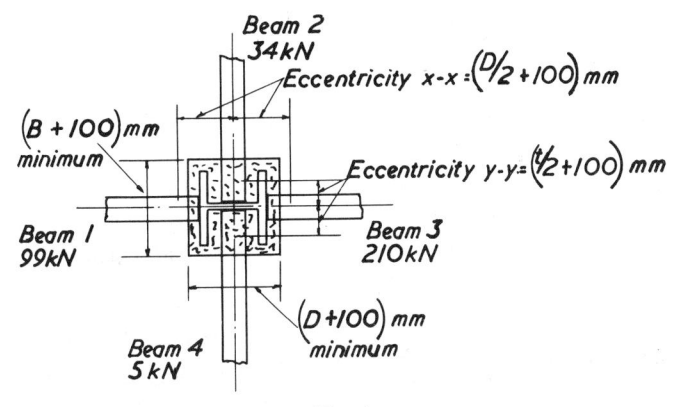

Fig. 4

DESIGN OF MULTI-STOREY STANCHIONS

Axial load on shaft under consideration

    Load from above (see tabulation)   = 896 kN
    Load at 3rd level   = 348 kN
    Weight of stanchion and casing   =   9 kN
                              Total   = 1 253 kN
    Deduct for live load reduction   =   85 kN
                 Nett axial load   = 1 168 kN

  Actual length floor to floor = 3.5 m
  Coefficient   = 0.7
  Effective length = 0.7 × 3.5 = 2.45 m

Try section 203 × 203 × 60 kg U.C.

  Actual size = 210 × 205 mm : Z_{xx} = 581 cm³ : Z_{yy} = 199 cm³
  Cased size = 310 × 305 mm say.
  Area of steel = 75.8 cm² : Web thickness = 9.3 mm
  Area of concrete casing = 310 × 305 = 94 500 mm²
  Radius of gyration r_{xx} = 8.96 cm : r_{yy} = 5.19 cm
  Effective minimum radius of gyration of cased section

$$= 0.2(B + 100) = 0.2(205 + 100) = 61 \text{ mm}$$
$$= 6.1 \text{ cm}$$

$l/r_{min} = \dfrac{2.45 \times 10^3}{6.1 \times 10} = 40$: Permissible axial stress p_c = 139 N/mm²
                                               Permissible bending stress p_{bc} = 165 N/mm²

These stresses are on the steel shaft only but the concrete casing can be used to assist in carrying the axial load.

Permissible stress on concrete casing (ignoring any casing in excess of 75 mm beyond overall dimensions of steel)

$$= \dfrac{p_c}{0.19 \times p_{bc}} = \dfrac{139}{0.19 \times 165} = 4.43 \text{ N/mm}^2.$$

The treatment for bending is similar to uncased shafts.
Then

Eccentricity $xx = \left(\dfrac{D}{2} + 100\right) = \left(\dfrac{210}{2} + 100\right) = 205$ mm

Eccentricity $yy = \left(\dfrac{t}{2} + 100\right) = \left(\dfrac{9.3}{2} + 100\right) = 105$ mm

$$M_{xx} = (210 - 99) \times 205 = 111 \times 205 = 22\,800 \text{ kNmm}$$
$$M_{yy} = (34 - 5) \times 105 = 29 \times 105 = 3\,050 \text{ kNmm}$$

Actual bending stress $f_{bc} = \dfrac{22\,800 \times 10^3}{2 \times 581 \times 10^3} + \dfrac{3\,050 \times 10^3}{2 \times 199 \times 10^3}$

$$= 20 + 8 = 28 \text{ N/mm}^2.$$

DESIGN EXAMPLES

The axial load is treated by determining the safe axial load on the steel plus the concrete thus

Safe load on steel $= 75.8 \times 10^2 \times 139 \times 10^{-3} = 1\,060$ kN
Safe load on concrete $= 94\,500 \times 4.43 \times 10^{-3} = \underline{420 \text{ kN}}$

$\phantom{Safe load on concrete = 94 500 \times 4.43 \times 10^{-3}}$ Total safe load $= 1\,480$ kN

Then ratio of $\dfrac{\text{Actual axial load}}{\text{Safe axial load}} + \dfrac{f_{bc}}{p_{bc}}$ must not exceed unity.

$$\frac{1\,168}{1\,480} + \frac{28}{165} = 0.79 + 0.17 = 0.96.$$

∴ Section selected is satisfactory.

Example 2A. Grade 50 Steel—Lightweight Cased

Examine shaft between base and 2nd floor.

The critical loading conditions will be those at the 1st floor level. These are as shown in sketch, Fig. 3 previously given.

The axial load on the shaft under consideration will be the load from above plus the sum of the loads at the level shown including an allowance for the weight of this stack of stanchion and its casing, less the permitted reduction in live load, thus:—

Load from above (see tabulation)	$= 1\,596$ kN
Load at 1st level	$=348$ kN
Weight of stanchion and casing	$=10$ kN
Total	$= 1\,954$ kN
Deduct for live load reduction	177 kN
Nett axial load	$= 1\,777$ kN

The effective length of the stack under consideration is the actual length base to 1st floor multiplied by the coefficient. In this case since the web beams are load carrying this will be taken as 0.85.
Then:—

Actual length base to 1st floor $= 4.5$ m
Coefficient $= 0.85$
Effective length $= 0.85 \times 4.5 = 3.82$ m

Try section 254 × 254 × 89 kg U.C.

Actual size $= 260 \times 260$ mm
Area $A = 114$ cm² : Web thickness $t = 10.5$ mm
Radius of gyration $r_{xx} = 11.2$ cm : $r_{yy} = 6.52$ cm
Modulus of section $Z_{xx} = 1\,099$ cm³ : $Z_{yy} = 379$ cm³

$l/r_{yy} = \dfrac{3.82 \times 10^3}{6.52 \times 10} = 59$: Permissible axial stress $p_c = 171$ N/mm²
$\phantom{l/r_{yy} = \dfrac{3.82 \times 10^3}{6.52 \times 10} = 59 :\ }$ Permissible bending stress $p_{bc} = 230$ N/mm²

Eccentricities:— $xx = \left(\dfrac{D}{2} + 100\right) = \left(\dfrac{260}{2} + 100\right) = 230$ mm

$yy = \left(\dfrac{t}{2} + 100\right) = \left(\dfrac{10.5}{2} + 100\right) = 105$ mm

$M_{xx} = (210 - 99) \times 230 = 25\,600$ kNmm : $M_{yy} = (34 - 5) \times 105 = 3\,050$ kNmm

Bending moments divided equally into stack above and below.

Actual axial stress $\quad f_c = \dfrac{W}{A} = \dfrac{1\,777 \times 10^3}{114 \times 10^2} = 156$ N/mm^2

Actual bending stress $\quad f_{bc} = \dfrac{M_{xx}}{2 \times Z_{xx}} + \dfrac{M_{yy}}{2 \times Z_{yy}}$

$\qquad\qquad\qquad\qquad\quad = \dfrac{25\,600 \times 10^3}{2 \times 1\,099 \times 10^3} + \dfrac{3\,050 \times 10^3}{2 \times 379 \times 10^3}$

$\qquad\qquad\qquad\qquad\quad = 12 + 4 = 16$ N/mm^2.

Ratio of $\dfrac{f_c}{p_c} + \dfrac{f_{bc}}{p_{bc}}$ not to exceed unity.

Then $\qquad \dfrac{f_c}{p_c} + \dfrac{f_{bc}}{p_{bc}} = \dfrac{156}{171} + \dfrac{16}{230} = 0.91 + 0.07 = 0.98$

∴ Section selected is satisfactory.

Example 2B. Grade 50 Steel Encased in Concrete to requirements of B.S. 449

Examine shaft between base and 2nd floor.

Loading conditions at the 1st floor level as sketch: Fig. 4 previously given.

Axial load on shaft under consideration
    Load from above (see tabulation)   = 1 606 kN
    Load at 1st level   =   348 kN
    Weight of stanchion and casing   =   14 kN
              Total   = 1 968 kN
    Deduct for live load reduction   177 kN
           Nett axial load   = 1 791 kN

Actual length base to 1st floor = 4.5 m
Coefficient   = 0.85
Effective length = 0.85 × 4.5 = 3.82 m

Try section 254 × 254 × 73 kg U.C.

Actual size = 254 × 254 mm : $Z_{xx} = 894$ cm^3 : $Z_{yy} = 305$ cm^3
Cased size = 355 × 355 mm say

DESIGN EXAMPLES

Area of steel = 92.9 cm² : Web thickness = 8.6 mm
Area of concrete casing = 355 x 355 = 126 000 mm²
Radius of gyration r_{xx} = 11.1 cm : r_{yy} = 6.46 cm
Effective minimum radius of gyration of cased section

$$= 0.2(B + 100) = 0.2(254 + 100) = 70.8 \text{ mm}$$
$$= 7.08 \text{ cm}.$$

$l/r_{min} = \dfrac{3.82 \times 10^3}{7.08 \times 10} = 54$: Permissible axial stress $\quad p_c = 179$ N/mm²
$\qquad\qquad\qquad\qquad\qquad\quad$ Permissible bending stress $p_{bc} = 230$ N/mm²

These stresses are on the steel shaft only but the concrete casing can be used to assist in carrying the axial load.

Permissible stress on concrete casing (ignoring any casing in excess of 75 mm beyond overall dimensions of steel)

$$= \frac{p_c}{0.19 \times p_{bc}} = \frac{179}{0.19 \times 230} = 4.1 \text{ N/mm}^2.$$

The treatment for bending is similar to uncased shafts.
Then:—

Eccentricity $\quad xx = \left(\dfrac{D}{2} + 100\right) = \left(\dfrac{254}{2} + 100\right) = 227$ mm

Eccentricity $\quad yy = \left(\dfrac{t}{2} + 100\right) = \left(\dfrac{8.6}{2} + 100\right) = 104$ mm

$$M_{xx} = (210 - 99) \times 227 = 111 \times 227 = 25\,200 \text{ kNmm}$$
$$M_{yy} = (34 - 5) \times 104 = 29 \times 104 = 3\,020 \text{ kNmm}.$$

Actual bending stress $\quad f_{bc} = \dfrac{25\,200 \times 10^3}{2 \times 894 \times 10^3} + \dfrac{3\,020 \times 10^3}{2 \times 305 \times 10^3}$

$$= 14 + 5 = 19 \text{ N/mm}^2$$

The axial load is treated by determining the safe axial load on the steel plus the concrete thus

Safe load on steel $\quad = 92.9 \times 10^2 \times 179 \times 10^{-3} = 1\,660$ kN
Safe load on concrete $= 126\,000 \times 4.1 \times 10^{-3} \quad = \underline{518 \text{ kN}}$
$\qquad\qquad\qquad\qquad\qquad$ Total safe load $= \underline{2\,178 \text{ kN}}$

Then ratio of $\dfrac{\text{Actual axial load}}{\text{Safe axial load}} + \dfrac{f_{bc}}{p_{bc}}$ must not exceed unity.

$$\dfrac{1\,791}{2\,178} + \dfrac{19}{230} = 0.83 + 0.08 = 0.91.$$

∴ Section selected is satisfactory.

Example 3A. Grade 55 Steel—Lightweight Cased

Examine shaft between 4th floor and roof.

The critical loading conditions will be those at the 5th floor level. These are as shown in sketch, Fig. 3 previously given.

The axial load on the shaft under consideration will be the load from above plus the sum of the loads at the level shown including an allowance for the weight of this stack of stanchion and its casing, less the permitted reduction in live load, thus:—

Load from above (see tabulation)	= 182 kN
Load at 5th level	= 348 kN
Weight of stanchion and casing	= 4 kN
Total	= 534 kN
Deduct for live load reduction	13 kN
Nett axial load	= 521 kN

The effective length of the stack under consideration is the actual length 4th to 5th floor multiplied by the coefficient. In this case since the web beams are load carrying this will be taken as 0.7.

Then:—

Actual length	4th to 5th = 3.5 m
Coefficient	= 0.7
Effective length	= 0.7 x 3.5 = 2.45 m

Try section 152 x 152 x 30 kg U.C.

Actual size = 158 x 153 mm
Area A = 38.2 cm^2 : Web thickness t = 6.6 mm
Radius of gyration r_{xx} = 6.75 cm : r_{yy} = 3.82 cm
Modulus of section Z_{xx} = 221 cm^3 : Z_{yy} = 73 cm^3

$l/r_{yy} = \dfrac{2.45 \times 10^3}{3.82 \times 10} = 64$: Permissible axial stress p_c = 189 N/mm^2
Permissible bending stress p_{bc} = 280 N/mm^2

Eccentricities:— $xx = \left(\dfrac{D}{2} + 100\right) = \left(\dfrac{158}{2} + 100\right) = 179$ mm

$yy = \left(\dfrac{t}{2} + 100\right) = \left(\dfrac{6.6}{2} + 100\right) = 104$ mm

M_{xx} = (210 – 99) x 179 = 19 900 kNmm : M_{yy} = (34 – 5) x 104 = 3 020 kNmm.

Bending moments divided equally into stack above and below.

Actual axial stress $\quad f_c = \dfrac{W}{A} = \dfrac{521 \times 10^3}{38.2 \times 10^2} = 136$ N/mm^2

Actual bending stress $\quad f_{bc} = \dfrac{M_{xx}}{2 \times Z_{xx}} + \dfrac{M_{yy}}{2 \times Z_{yy}}$

$= \dfrac{19\,900 \times 10^3}{2 \times 221 \times 10^3} + \dfrac{3\,020 \times 10^3}{2 \times 73 \times 10^3}$

$= 45 + 21 = 66$ N/mm^2.

DESIGN EXAMPLES

Ratio of $\dfrac{f_c}{p_c} + \dfrac{f_{bc}}{p_{bc}}$ not to exceed unity.

Then $\quad \dfrac{f_c}{p_c} + \dfrac{f_{bc}}{p_{bc}} = \dfrac{136}{189} + \dfrac{66}{280} = 0.72 + 0.24 = 0.96$

∴ Section selected is satisfactory.

Example 3B. Grade 55 Steel Encased in Concrete to requirements of B.S. 449

Examine shaft between 4th floor and roof.

Loading conditions at the 5th floor level as sketch: Fig. 4 previously given.

Axial load on shaft under consideration
```
   Load from above (see tabulation)    = 183 kN
   Load at 5th level                   = 348 kN
   Weight of stanchion and casing      =   5 kN
                             Total     = 536 kN
   Deduct for live load reduction         13 kN
                    Nett axial load    = 523 kN
```

Actual length 4th to 5th = 3.5 m
Coefficient = 0.7
Effective length = 0.7 x 3.5 = 2.45 m.

Try section 152 x 152 x 23 kg U.C.

Actual size = 152 x 152 mm : Z_{xx} = 166 cm^3 : Z_{yy} = 53 cm^3
Cased size = 255 x 255 mm say.
Area of steel = 29.8 cm^2 : Web thickness = 6.1 mm
Area of concrete casing = 255 x 255 = 65 000 mm^2
Radius of gyration r_{xx} = 6.51 cm : r_{yy} = 3.68 cm
Effective minimum radius of gyration of cased section

$$= 0.2(B + 100) = 0.2(152 + 100) = 5.04 \text{ mm}$$
$$= 5.04 \text{ cm.}$$

$l/r_{min} = \dfrac{2.45 \times 10^3}{5.04 \times 10} = 49$: Permissible axial stress $\quad p_c$ = 224 N/mm^2
Permissible bending stress p_{bc} = 280 N/mm^2

These stresses are on the steel shaft only, but the concrete casing can be used to assist in carrying the axial load.

Permissible stress on concrete casing (ignoring any casing in excess of 75 mm beyond overall dimensions of steel)

$$= \dfrac{p_c}{0.19 \times p_{bc}} = \dfrac{224}{0.19 \times 280} = 4.2 \text{ N/mm}^2$$

The treatment for bending is similar to uncased shafts. Then:—

Eccentricity $xx = \left(\dfrac{D}{2} + 100\right) = \left(\dfrac{152}{2} + 100\right) = 176$ mm

Eccentricity $yy = \left(\dfrac{t}{2} + 100\right) = \left(\dfrac{6.1}{2} + 100\right) = 103$ mm

$$M_{xx} = (210 - 99) \times 176 = 111 \times 176 = 19\,600 \text{ kNmm}$$

$$M_{yy} = (34 - 5) \times 103 = 29 \times 103 = 2\,990 \text{ kNmm}$$

Actual bending stress $f_{bc} = \dfrac{19\,600 \times 10^3}{2 \times 166 \times 10^3} + \dfrac{2\,990 \times 10^3}{2 \times 53 \times 10^3}$

$$= 59 + 28 = 87 \text{ N/mm}^2.$$

The axial load is treated by determining the safe axial load on the steel plus the concrete thus

Safe load on steel $= 29.8 \times 10^2 \times 224 \times 10^{-3} = 670$ kN
Safe load on concrete $= 65\,000 \times 4.2 \times 10^{-3} = \underline{274 \text{ kN}}$
Total safe load $= \underline{944 \text{ kN}}$

Then ratio of $\dfrac{\text{Actual axial load}}{\text{Safe axial load}} + \dfrac{f_{bc}}{p_{bc}}$ must not exceed unity

$$\dfrac{523}{944} + \dfrac{87}{280} = 0.56 + 0.31 = 0.87$$

∴ Section selected is satisfactory.

Connection Details (See Fig. 5)

The figure indicates one method of construction only, namely shop riveted and site bolted construction. The details given can readily be adapted to suit fully bolted, or shop welded and site bolted construction.

Beam to Stanchion Connections

The design of these connections is outside the scope of this chapter, but it is perhaps of interest to note that there is a tendency on the part of designers to discard the traditional top and bottom cleats in favour of end plates welded to the beams or bolted double cleated ends.

Splices

The ends of each section of the stanchion will be machined over the whole area. The calculations show that the bending stresses are small compared with the direct compressive stresses, so that no tension develops in the stanchion. In these circumstances the only requirement of Clause 32.b of B.S. 449 is, that the splice

DESIGN EXAMPLES

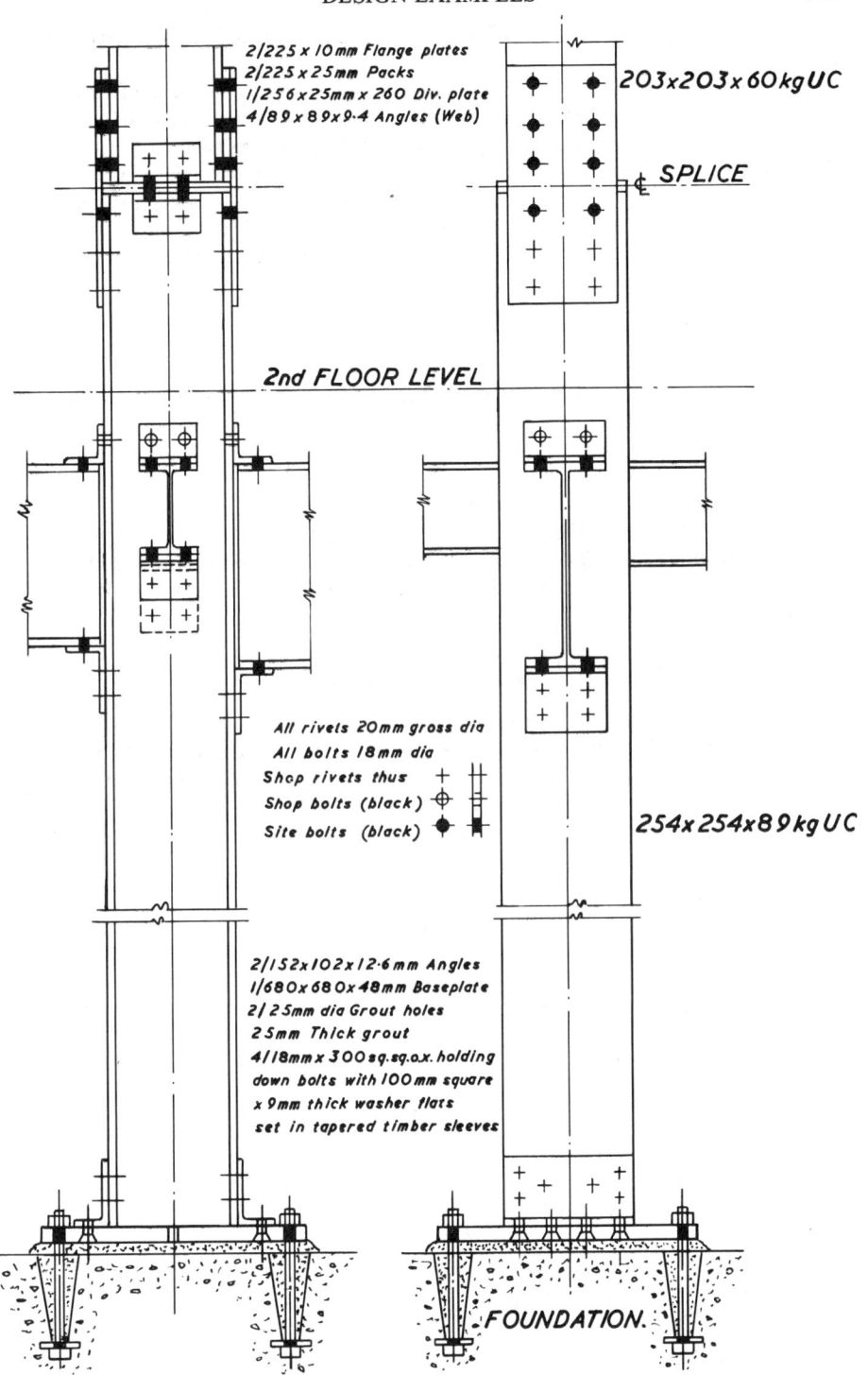

Fig. 5

plates and connections shall be sufficient to hold the connected members accurately in place.

In the absence of guidance from the Standard the splice shown in the figure has been proportioned in accordance with the following empirical rules (a) the projection of the flange plates beyond the end of the members is equal to the upper flange width or 225 mm, whichever is greater, (b) the thickness of the flange plate is half the thickness of the upper flange or 8 mm, whichever is greater, (c) nominal web plates are provided when the serial size of the members is the same above and below the splice, (d) web cleats and a division plate are provided where the serial size is different, to give a load dispersal of 45°.

Slab Bases

The end of the stanchion will be machined and fastenings provided sufficient to hold the slab in position. Slabs 50 mm or less in thickness will be flattened. Slabs over 50 mm thick will be machined on the upper surface only. Grout holes will be provided where necessary.

It will be assumed that the concrete foundations are of Ordinary quality 1:2:4 mix with a permissible bearing pressure of 4 000 kN/m².

General information relating to slab bases is given in Clauses 67 and 76a of B.S. 449 and Clause 38b gives the following formula for calculating the thickness of slab required:

$$t = \sqrt{\left\{\frac{3w}{p_{bct}}\left(A^2 - \frac{B^2}{4}\right)\right\}}$$

where t = the slab thickness in millimetres.
A = the greater projection of the plate beyond the stanchion in millimetres.
B = the lesser projection of the plate beyond the stanchion in millimetres.
w = the pressure or loading on the underside of the base in N/mm².
p_{bct} = the permissible bending stress in the steel specified as 185 N/mm² for all steels.

Example: Base to stanchion as example 1B.
Shaft 254 x 254 x 89 kg U.C. Total axial load 1 800 kN..

$$\text{Area of base required} = \frac{1\,800}{4\,000} = 0.45 \text{ m}^2$$

Therefore base slab 0.68 x 0.68 m is satisfactory.

$$\text{Greater projection} = \frac{0.68 \times 10^3 - 256}{2} = 212 \text{ mm}$$

$$\text{Lesser projection} = \frac{0.68 \times 10^3 - 260}{2} = 210 \text{ mm}$$

$$\text{Base pressure} = \frac{1\,800 \times 10^3}{0.68 \times 0.68 \times 10^6} = 3.9 \text{ N/mm}^2$$

$$\text{Then } t = \sqrt{\frac{3 \times 3.9}{185}\left(212^2 - \frac{210^2}{4}\right)} = 47 \text{ mm}.$$

Use flattened plate 48 mm thick.

DESIGN EXAMPLES

Comments on the Results of the Design Examples

Eccentricity of Beam Reactions

In these examples the beams are all positioned on the stanchion centre lines. Where in practice some of the beams may be offset from the centre lines they will of course produce bending moments about both axes of the stanchion simultaneously, and these bending moments must be provided for.

Joints

Joints have been provided at 2 storey intervals, as giving a convenient and economical arrangement but joints at three or even four storey intervals are often adopted. From the point of view of erecting the steelwork the 2 storey interval has obvious advantages since the length of the members do not become such as to be difficult in handling.

Sections

It is frequently considered that the minimum size of universal column sections suitable for multi-storey work is 203 x 203 mm. Smaller sections almost inevitably require beams connecting to the web to be notched and limit the space available for the end connections. Despite these factors the use of 152 x 152 universal column sections, where possible, have been recommended as it is considered that these will give overall economy.

Steel Quality

For comparison purposes examples have been provided in Grades 43, 50 and 55 steels. It should, however, be appreciated that Grade 55 is not as yet readily available in small quantities. The user who can justify an order of 50 tons or more will not experience great difficulty in obtaining supplies. Orders for smaller amounts depend on the prospects for further demands or the ingot stock position.

Economics

The economic factors governing the adoption of one or other type of construction cannot be discussed in detail here, since these can change rapidly. However, it is interesting to note that using the uncased stanchion in grade 43 steel as unity there is a saving of approximately $33\frac{1}{3}\%$ in the weight of the steel core when this is made of grade 43 steel concrete encased or when it is made of grade 50 steel uncased. To offset against this saving it must be remembered that the concrete encasure could be more expensive than the lightweight casing and alternatively the cost of the raw material in grade 50 is more than that in grade 43. There are, however, other factors which require to be taken into consideration such as the reduction in size of the member which cannot be discussed fully in this chapter.

Apart from very large tall buildings where there could be obvious economies in using grade 55 steel, one of the main advantages can be gained by using a common section throughout a project with the different loads being covered by the selective

use of the three grades of steel. Such a procedure could simplify the architectural requirements and when pre-formed lightweight fire protection is used effect a considerable saving due to the limited number of moulds required in casting this fire protection.

Further information on the economy of stanchion design can be found in Reference 8.

BIBLIOGRAPHY

1. B.C.S.A. *Notes on the Modified Method of Semi-Rigid Design,* 1960.
2. ALLWOOD, B. O., HEATON, H. and NELSON, K. *Steel Frames for Multi-Storey Buildings.* B.C.S.A. Publication No. 16, 1961.
3. BATES, W. *Modern Design of Steel Frames for Multi-Storey Buildings.* B.C.S.A. Publication No. 20, 1963.
4. 'Joint Committee Second Report on Fully Rigid Multi-Storey Welded Steel Frames'. The Institution of Structural Engineers and the Institute of Welding, 1971.
5. BURNETT, N., JOHNSON, L. G., MORRIS, L. J., RANDALL, A. L. and THOMPSON, C. P. *Plastic Design,* B.C.S.A. Publication No. 28, 1965.
6. GODFREY, G. B. 'The Allowable Stresses in Axially Loaded Steel Struts'. The Structural Engineer, Vol. XL No. 3, March, 1962.
7. B.C.S.A. *Modern Fire Protection for Structural Steelwork.* Publication No. FP2, 1963.
8. B.C.S.A. *The Economics of Structural Members in High Strength Steel to B.S. 968 :1962.* Publication No. H.S.1, 1962.

31. WIND ON MULTI-STOREY BUILDINGS

THE wind loads required to be used by practically all Local Authorities in Great Britain are those set out in the British Standard Code of Practice, C.P.3: Code of Basic Data for the Design of Buildings, Chapter V, Loading: Part II: 1970 Wind Loads.

This 1970 edition of C.P.3, Chapter V, presents wind loading in a far more detailed manner than any of the former editions, being based on a careful study of wind data which have only become available over the last few years.

The wind speed has been investigated in many places in the British Isles and by statistical analysis of the records of this investigation values of the basic wind speed have been evaluated for all parts. These are indicated in the Code and range from 38 m/sec in the London area to 56 m/sec in the North of Scotland. The values represent the three second gust speed at ten metres above ground in an open situation which is not likely to be exceeded on the average more than once in fifty years.

The basic wind speed is adjusted to give a design wind speed by the application of three factors, namely,

S1 topographic factor,
S2 ground roughness, building size and height factor,
S3 building life factor.

A full explanation of these factors is given in the Code and need not be repeated here. It is sufficient to appreciate that the design wind speed is governed by all the relevant factors which could affect the free flow and is variable for a particular structure dependent upon the height of the elements above the ground.

From the design wind speed the dynamic pressure, q, is found by applying the formula

$$q = kV_s^2$$

where k is a constant taken as 0.613 in SI units, V_s is the design wind speed.

The application of the dynamic pressure for any particular example is governed by the size and shape of the building and whether or not the cladding or the structure is being examined; for the purpose of this Chapter the latter only will be considered.

The total horizontal force, F, to be considered is found from the expression

$$F = C_f q A_e$$

where,	C_f = force coefficient obtained from the Code

q = dynamic pressure of wind

A_e = area of exposed face.

Since q varies with height above ground, the total area A_e is usually sub-divided to suit each change in the value of q.

In certain circumstances, apart from the direct pressure on the exposed face given by the foregoing, frictional drag must be taken into consideration. Examination of the Code will determine if this is necessary in a particular case.

The 1952 edition of Chapter V permitted the designer to omit direct calculation for the effect of wind forces provided that certain requirements as to the building dimensions and the relationship between these were satisfied. The present edition omits this relaxation but does not preclude the use of shear walls, core structures etc., in assisting the resistance to wind loads. For the purpose of this Chapter, however, such aids have been ignored and it has been assumed that the whole of the wind forces are to be resisted by the steel framework.

Clause 13 of B.S. 449: Part 2: 1969, permits certain stresses to be exceeded by 25% in cases where such increases in stress are solely due to wind forces, provided that the steel section is not less than that needed if the wind stresses were neglected. The higher working stress is allowed because of the transient nature of the load and also because the steel structure is sufficiently elastic to allow it to absorb such design loads without permanent deflection. The increase is not permissible on foundations and provision for the over-turning effect must be made at normal stresses.

Simple Design

Prior to the introduction of the rigid frame it was customary to design the frames as portals or cantilevers to resist the lateral wind loads, with connections designed to transfer wind moments from beams to stanchions. Such methods are still adequate for relatively low buildings provided that the frame has been designed on the basis of simple design as set out in B.S. 449. There are several approximate methods by which the forces may be calculated and the three methods in most common use are briefly dealt with in the notes which follow.

The basic assumptions are three in number:

(a) Wind loads are applied at floor levels.
(b) The total horizontal shear at any level is resisted by the columns at points of of contraflexure immediately below that level.
(c) There is a point of contraflexure in each column between floor levels and at mid-height.

These assumptions are illustrated in Fig. 1 (a), which for the sake of simplicity is shown as a single bay frame.

Referring to Fig. 1 (b), the upper portion of the diagram shows the forces on the portion of the frame above the points of contraflexure A and D, while the lower portion shows the forces induced in the portion $ABCDEF$ of the frame. The vertical forces are found by taking moments about the appropriate points of contraflexure. The shear force in the beam BE is the difference between the vertical forces in the columns, i.e.

$$\frac{W_1 H_1}{L} + \frac{(W_1 + W_2)H_2}{2L} - \frac{W_1 H_1}{2L} = \frac{W_1 H_1 + (W_1 + W_2)H_2}{2L}.$$

The resulting moments in the frame are shown in Fig. 1 (c).

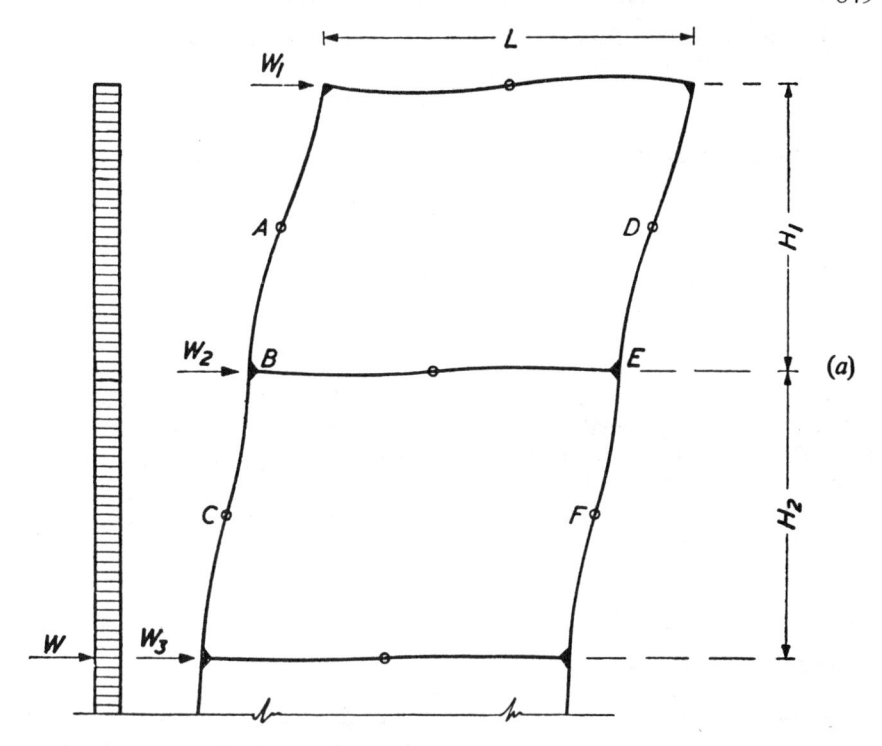

Columns assumed to be of equal strength

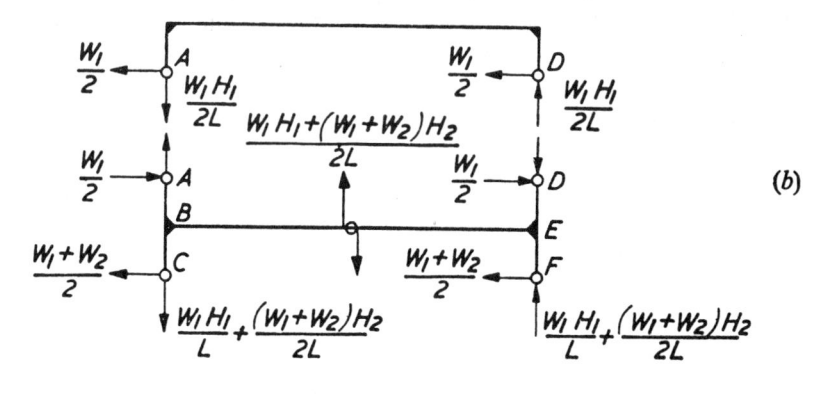

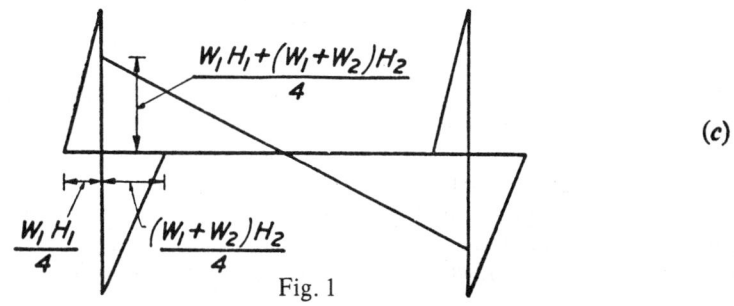

Fig. 1

The moment in the upper portion of the column at B due to the force at A

$$= \frac{W_1}{2} \times \frac{H_1}{2} = \frac{W_1 H_1}{4}$$

while that in the lower portion of the column at B due to the force at C is

$$\frac{W_1 + W_2}{2} \times \frac{H_2}{2} = \frac{(W_1 + W_2)H_2}{4},$$

The moment in the beam at B = the shear force in the beam × $L/2$

$$= \frac{W_1 H_1 + (W_1 + W_2)H_2}{2L} \times \frac{L}{2} = \frac{W_1 H_1 + (W_1 + W_2)H_2}{4},$$

from which it will be seen that the moment in the beam at B is equal to the sum of the moments in the column, both upper and lower, at B. This is true for every joint in the frame and irrespective of the method used, i.e. the algebraic sum of the moments at any joint must be zero.

The three methods previously referred to differ only in the distribution of vertical and horizontal loads.

Method 1. Continuous Portal

The assumptions made are:

(a) there is a point of contraflexure at mid-height of each column;
(b) the axial *stress* in the columns is proportional to the distances of the columns from the centre of gravity of the columns as a whole;
(c) the total horizontal shear is divided between columns in proportions to their stiffness in the direction of the wind forces.

Method 2. Cantilever

(a)
(b) } As Method 1.
(c) There is a point of contraflexure at mid-span of each beam.

Method 3. Portal

(a) As Method 1.
(b) As (c), Method 2.
(c) Each bay acts as a simple portal and the total horizontal load is divided between the bays in proportion to the spans of the bays. With equal bays this results in no vertical load in the internal columns.

An example will serve to demonstrate the differences in the results obtained by the three methods. This is based on a ten storey building 21 m in width, in three equal bays of 7 m, and 45 m long, in ten equal bays of 4.5 m. The storey heights are 3.5 m each which gives a total height of 35 m.

To illustrate the application of C.P. 3: Chapter V: Part 2: 1970, in determining the wind loads on a structure of this nature a building located in the vicinity of London will be considered, where a basic wind speed of 40 m/sec is appropriate.

SIMPLE DESIGN

With factors for S1 and S3 of unity, the design wind speed will be

$$V_s = 40 \times 1 \times S2 \times 1 = 40 S2 \text{ m/sec}$$

The value of S2 depends on ground roughness, building size, class and height above the ground. Taking ground roughness Category (3) and building size Class B, the value of S2 can be obtained from Table 3. Though variations for the factor are given for certain values of the height, for convenience in the particular example the total height of the building will be divided into four parts, giving the following:—

Ground to 3rd floor — S2 = 0.74; V_s = 40 x 0.74 = 29.6 m/sec
3rd to 6th floor — S2 = 0.9 ; V_s = 40 x 0.9 = 36.0 m/sec
6th to 8th floor — S2 = 0.97; V_s = 40 x 0.97 = 38.8 m/sec
8th to roof — S2 = 1.01; V_s = 40 x 1.01 = 40.4 m/sec

The dynamic pressure of wind, q, can be calculated from the formula $q = 0.613 V_s^2$ or read from Table 4, approximating as necessary.

The force coefficient C_f for the example is obtained from Table 10 and can be taken as 1.05. (The tables referred to are from C.P.3 Chapter V.)
Then

Ground to 3rd floor q = 540 N/m² F = 1.05 x 540 x A_e = 594 A_e
3rd to 6th floor q = 794 N/m² F = 1.05 x 794 x A_e = 834 A_e
6th to 8th floor q = 930 N/m² F = 1.05 x 930 x A_e = 976 A_e
8th to roof q = 1010 N/m² F = 1.05 x 1 010 x A_e = 1 060 A_e

These loads are multiplied by the storey height and bay widths to give a total load on each storey as follows

Roof to 9th $W1 = 3.5 \times 4.5 \times 1\,060 \times 10^{-3} = 16.8$ kN
9th to 8th $W2 = 3.5 \times 4.5 \times 1\,060 \times 10^{-3} = 16.8$ kN
8th to 7th $W3 = 3.5 \times 4.5 \times 976 \times 10^{-3} = 15.4$ kN
7th to 6th $W4 = 3.5 \times 4.5 \times 976 \times 10^{-3} = 15.4$ kN
6th to 5th $W5 = 3.5 \times 4.5 \times 834 \times 10^{-3} = 13.2$ kN
5th to 4th $W6 = 3.5 \times 4.5 \times 834 \times 10^{-3} = 13.2$ kN
4th to 3rd $W7 = 3.5 \times 4.5 \times 834 \times 10^{-3} = 13.2$ kN
3rd to 2nd $W8 = 3.5 \times 4.5 \times 594 \times 10^{-3} = 9.4$ kN
2nd to 1st $W9 = 3.5 \times 4.5 \times 594 \times 10^{-3} = 9.4$ kN
1st to Grd $W10 = 3.5 \times 4.5 \times 594 \times 10^{-3} = 9.4$ kN

This shows reactions at the various levels as follows

Reaction at roof = 8.4 kN
,, ,, 9th floor = 16.8 kN
,, ,, 8th floor = 16.1 kN
,, ,, 7th floor = 15.4 kN
,, ,, 6th floor = 14.3 kN
,, ,, 5th floor = 13.2 kN
,, ,, 4th floor = 13.2 kN
,, ,, 3rd floor = 11.3 kN
,, ,, 2nd floor = 9.4 kN
,, ,, 1st floor = 9.4 kN
,, ,, Ground = 4.7 kN

852   WIND ON MULTI-STOREY BUILDINGS

The above loads are all indicated in Fig. 2 and form the basis from which wind loads and moments are to be calculated.

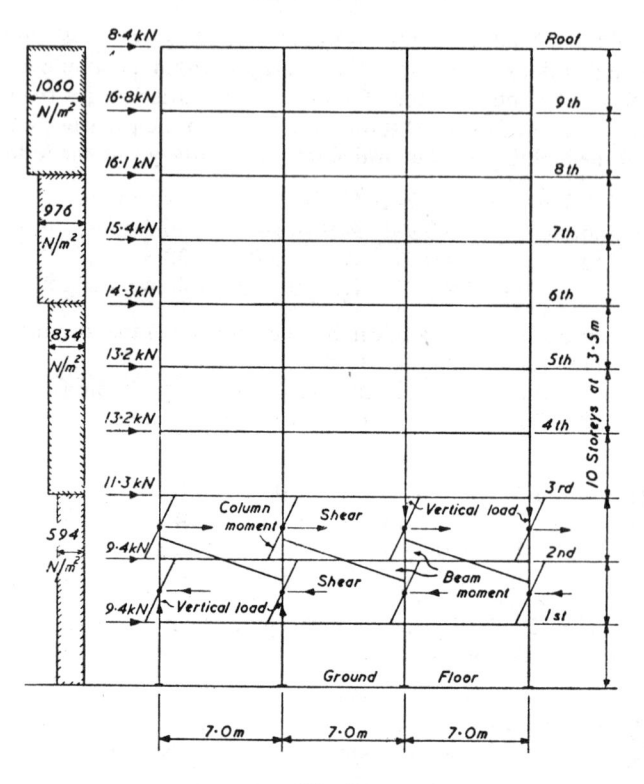

Fig. 2

Comparative results have been calculated and are shown in Table (A) for beams and columns at second floor level.

Method 1 fails if there are more than four equal bays since then there is no point of contraflexure in the outer beams.

Method 3 is undoubtedly the simplest from the design and detail angle.

It must be stressed that these methods are only applicable to multi-storey buildings designed on the basis of simple design to B.S. 449 and they must *not* be applied to a frame designed on any other basis.

The beams are designed as simply supported under the vertical loading only, at normal stresses, and checked with the increased stresses applied to the slightly higher maximum bending moment obtained when the fixing moment diagram is imposed. The moment connections at their ends are designed in the usual way to resist moment and shear at the increased stresses and for vertical loads only at the normal stresses.

A completely worked example of an eight-storey building, using the Portal method, is given in Brochure No. 16, 'Steel Frames for Multi-Storey Buildings', published by the British Constructional Steelwork Association.

RIGID FRAME DESIGN

TABLE A

Comparison of results for approximate methods

		Method 1	Method 2	Method 3
External column				
Above 2nd floor				
Shear	kN	27.2	16.3	18.1
Vertical load	kN	18.8	18.8	20.9
Bending moment	kNm	48.0	28.6	31.8
Below 2nd floor				
Shear	kN	29.5	17.7	19.7
Vertical load	kN	23.6	23.6	26.3
Bending moment	kNm	52.0	31.0	34.6
Internal column				
Above 2nd floor				
Shear	kN	27.2	38.0	36.2
Vertical load	kN	6.3	6.3	nil
Bending moment	kNm	48.0	66.5	63.6
Below 2nd floor				
Shear	kN	29.5	41.2	39.4
Vertical load	kN	7.9	7.9	nil
Bending moment	kNm	52.0	72.0	69.2
Beams at 2nd floor				
External bay				
Moment at outer end	kNm	100	60	66.4
Moment at inner end	kNm	20	60	66.4
Internal bay				
Moment at both ends	kNm	80	80	66.4

Post-war experience and the introduction of much higher buildings led to a re-appraisal of the methods of dealing with wind loads in order to reduce costs, since the fabrication and erection costs of riveted or bolted moment connections became prohibitively high when moments became large. This led to the use of rigid frames with welded or high-strength friction grip bolted joints, with the frame taking all the loads and moments. The current trend in Great Britain, however, is for maximum utilisation of all structural components of the building and the transfer of wind loads to points where they can be dealt with in a less costly manner than by dealing with them at their points of application. Architectural and planning requirements, particularly the demand for large unobstructed floor areas in office buildings frequently dictate the manner by which wind loads are to be handled. Both systems, rigid frames taking all loads or simple design frames with wind loads carried by shear walls or braced frames at suitable intervals, have their merits and combinations of the two can be very effective.

Rigid Frame Design

The calculations for a multi-storey rigid frame are lengthy and complex and can only be carried out economically by using an electronic computer. In every case it is necessary to select provisional sizes for the beams and columns before the analysis of displacements, loads and moments can be carried out by the computer. There is

no direct approach to this part of the design, and hence accuracy in provisional sizes depends upon the experience of the designer.

Clause 9.*b*.3 of B.S. 449 states that:

'Fully Rigid Design

This method, as compared with the method for simple and semi-rigid design, will give the greatest rigidity and economy in weight of steel used when applied in appropriate cases. For this purpose the design shall be carried out in accordance with accurate methods of elastic analysis and to the limiting stresses permitted in this British Standard. . . .'

Analysis by the slope-deflection method (see Chapter 13) is accurate when correctly applied. It is usual to design the frame in the first instance for vertical loads only at normal stresses and then to make a separate calculation for wind loads, the results of which are applied, at the increases stresses, to the sections obtained for vertical loads. For information generally on rigid frame design by computer, the reader is referred to Chapter 38.

It is not possible to give a complete example of a frame designed by this method, owing to limitations of space, but an excellent exposition of the method is given in Brochure No. 20, 'Modern Design of Steel Frames for Multi-storey Buildings', published by the British Constructional Steelwork Association. This Brochure repays detailed study as it gives a complete and exhaustive analysis by the slope-deflection method of an office building 43.5 m height, including the effect of wind loading, to the former edition of C.P.3. Chapter V.

It must be emphasised that, as analysis of the rigid frame has been based on fully fixed joints throughout, it must be detailed accordingly. Joints can be made with high-strength friction-grip bolts but the use of these necessitates additional jointing material such as end plates to beams and splice plates to columns, which can add appreciably to the weight of the main material. Erection, however, can be carried out in the conventional manner. Alternatively, site welding can be used and the beam to column connection made by using either fillet welds of the requisite strength round the beam profile or by butt-welding the beam to the column. Both methods necessitate some temporary connection between beam and column prior to welding.

Fillet welds do not require any end preparation to the beam, so that shop fabrication is simple. The welds must, however, be designed to the stresses laid down in B.S. 449, namely 115 N/mm^2 for steel to Grade 43 and 160 N/mm^2 for steel to Grade 50. It is also necessary to use overhead site welding for the undersides of beam flanges.

Butt welds on the other hand require costly end preparation to the beams but use less weld metal since the welds can be stressed to the design values for the parent metal on a thickness equal to the throat thickness or a reduced throat thickness if an incomplete penetration butt weld is used (see B.S. 449, Clauses 53.*a* (i), 53.*d* and 54.*b*). It therefore follows that the butt welds need not be designed if they are of the full thickness of the beam flange.

Braced Frames

A very economical solution, if architectural and planning requirements permit, is given by the use of braced frames at selected positions, such as gable walls or walls enclosing lift and staircase wells as is shown in Fig. 3. Such braced frames can of

course be used with the main building frame designed on any basis, e.g. simple, semi-rigid or fully rigid, but necessitate some form of horizontal girder to transfer the wind loads from the walls of the building to the braced frames.

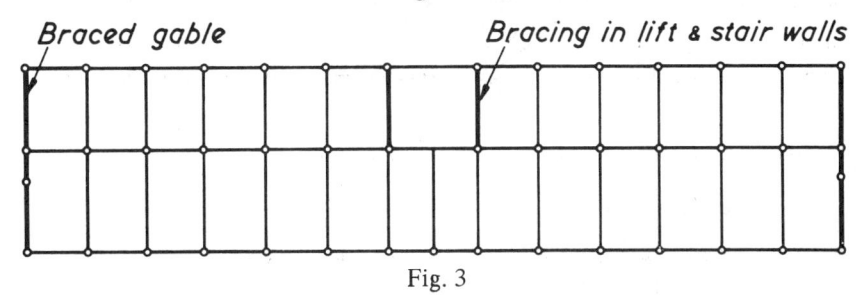

Braced gable   *Bracing in lift & stair walls*

Fig. 3

The horizontal wind girders can conveniently be made of steel, where a false ceiling is incorporated for purposes of fire resistance, thus giving a void in which the diagonals can be placed, as shown in Fig. 4.

On the other hand, in most districts, fire-resistance requirements necessitate the use of some kind of reinforced concrete floor, which can usefully fulfil the additional function of transferring the wind loads to the selected points. There is no doubt about the capability of a solid concrete floor to perform this duty, since even in narrow slab type buildings of about 15 m width and with frames spaced about 60 m apart, the concrete stresses when the floor acts as a horizontal girder, both compression and shear, will be virtually negligible and the edge reinforcement required to take the tension will be of the order of 6.5 cm^2. If the depth of the horizontal girder is restricted to half the building width, the distance between frames can be of the order of 45 m with approximately the same results.

Many types of precast concrete floors can also be used in this way.

The braced frames may extend the whole depth of the gable wall, or may be in pairs to give a clear corridor space, thereby allowing an unobstructed window at the end of the corridor, as shown in Fig. 5. The design of the frames as vertical cantilevers is conventional, the main booms being a pair of stanchions, with floor beams acting as struts with diagonals as requisite. Allowance must be made for reversal of loads and the bracing may be of N, K or X type or of X type with diagonals acting in tension only. Alternatively, braced portals can be used and these may be required to meet fenestration demands. There is thus considerable scope for ingenuity in arrangement of bracing and consultation with the architect is essential at the early planning stage in order to arrive at an economical and satisfactory solution.

As the design of the braced frames is on the normal statical basis it is not proposed to include a detailed example and the reader is referred to the BCSA Brochure No. 16 previously mentioned; this gives complete information on the design and detailing of frames for an office building about 70 m high by 14 m wide.

Typical solutions taken from practice are given in Figs. 6 to 11 which follow.

Figure 6 indicates schematically the solution adopted for the Seagram Building in New York. This building is about 160 m high and the tower block shown in plan is about 42 m wide and 33 m deep. The wind loads are taken by braced frames up to the 29th floor, placed in the walls between the lifts and staircases and between the lifts and toilet areas. Above the 29th floor the main frame absorbs the wind load, whilst below the 17th floor the braced frame acts in conjunction with the reinforced concrete walls in which it is embedded.

[Figure 4: Elevation and Plan showing Horizontal wind girders, Braced gables, Gable bracing]

Fig. 4

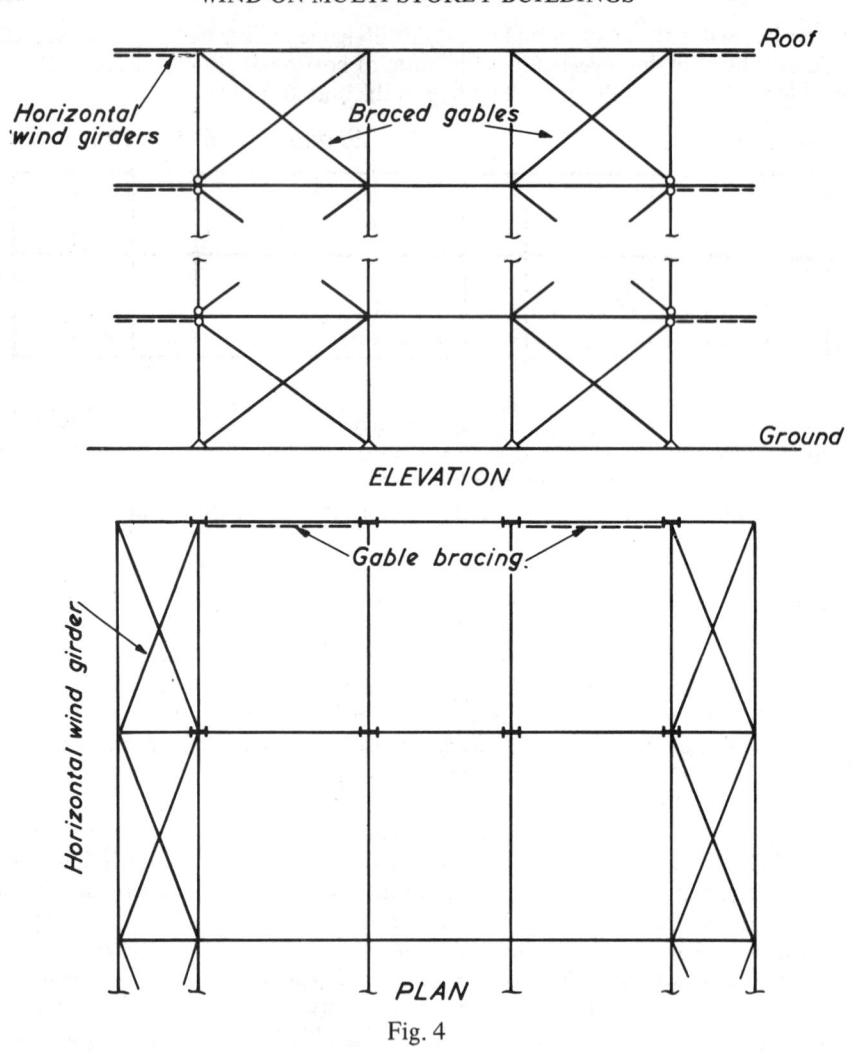

The method used in the building of the Statistical Service Headquarters of Western Germany is shown in Fig. 7. This fourteen-storey building is about 60 m long by 19 m wide and the transverse braced frames are in pairs in the gables and in the walls adjacent to the main entrance halls. The gable walls enclosing the bracing are solid and the windows in the ends are restricted to the centre portions of the walls. In order to equalise foundation loads as far as possible the transverse frames are on one side of the building for the first four storeys and on the other side for the remainder.

Also in Germany is the Thyssen Hochhaus at Düsseldorf and the scheme used here is a good example of the method. The building, some 64 m in height, consists of three blocks about 6.4 m wide staggered in plan as shown in Fig. 8. The wind load is taken entirely by two braced vertical cantilevers with main booms at about 14 m centres and an architectural feature is that the bracing is exposed in the gaps of about 2 m between pairs of blocks. Each braced frame extends across two of the

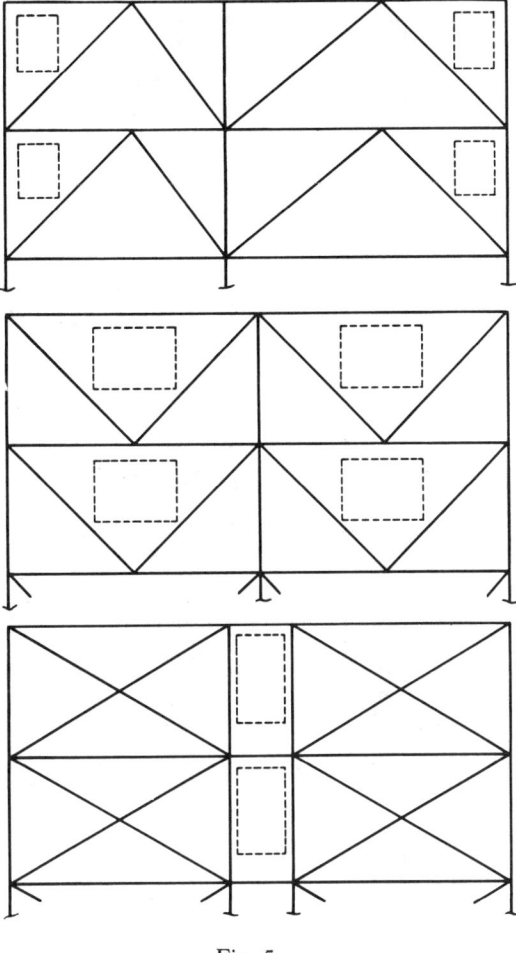

Fig. 5

blocks only, leaving the third entirely unobstructed; in the outer blocks the frames are in the gable walls, whilst in the centre block they are concealed within the staircase walls.

An example from Great Britain is the Shell Tower. This tower block is about 53 m by 24 m in plan, with a projecting wing about 17 m by 4.5 m in plan. There were severe restrictions on the position of diagonal wind bracing.

Up to the second floor level the steelwork is constructed as two-storey three-bay portal frames, interconnected by main beams with rigid joints to provide for wind forces in conjunction with the vertical bracing adjacent to the lift shafts. The steelwork above the second floor level is designed to act as a rigid frame in conjunction with braced frames and typical cross-sections are shown in Fig. 9. The main portal frames up to second floor level were site welded but above this the moment joints in the rigid frame portion were made with special grade high-strength friction-grip bolts in haunched beams.

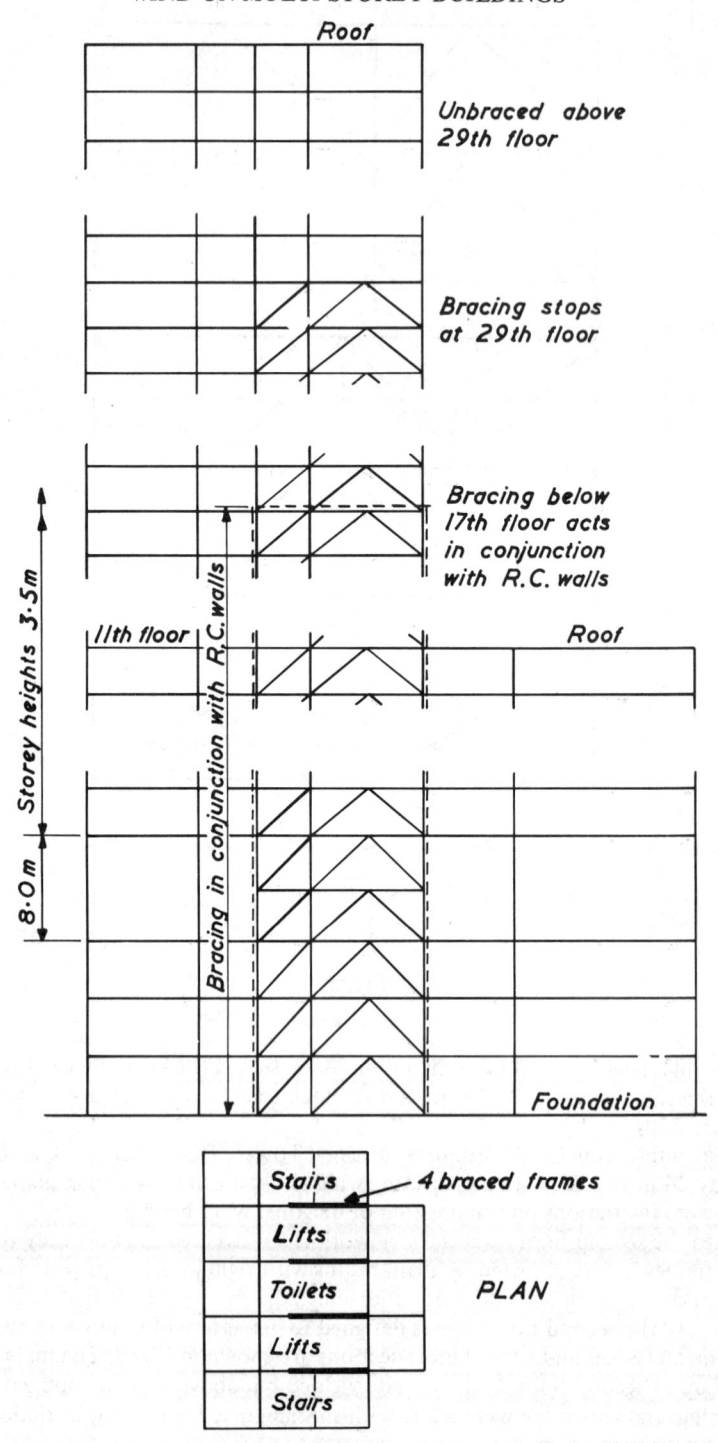

Fig. 6

Mixed or Composite Construction

The walls around staircase wells and lift lobbies, which isolate means of access from the body of the building, are always required to be of fire-resisting construction, partly to form a safe means of escape in case of fire and partly to permit safe access by personnel of the fire brigade when a fire actually occurs. These walls may, on occasion, be constructed in reinforced concrete and in such cases it would appear logical to investigate the possibility of the tower so formed acting additionally as a vertical cantilever to resist wind loads, so taking the place of braced frames.

From this consideration has arisen what can be termed 'Core' or 'Mixed' construction whereby all lateral loads are resisted by a reinforced concrete core surrounding the service and access areas of the building, leaving the steel frame to transmit vertical loads only.

The elements of such a scheme are shown in Fig. 10. The floor beams span between external columns and the concrete core, to which they transfer all lateral loads and part of the floor load. The stanchions carry vertical loads only and thus the steelwork scheme becomes very simple. There is a number of buildings in Europe utilising this principle which has also been used in a building for the Cooperative Insurance Society at Manchester.

An interesting and ingenious use of the composite principle is shown in the Albany block of flats in Bournemouth, 18 storeys high. A typical floor plan is given in Fig. 11, from which it will be seen that the building is in the form of two 'Ys' with the stems joined together to form a spine. The steel frame, which takes no wind loads, is

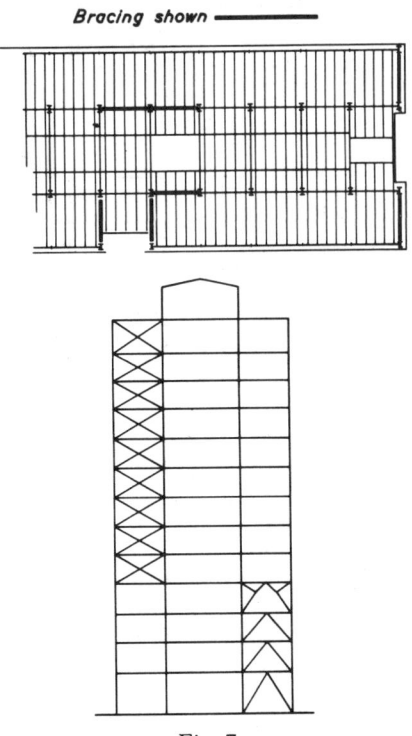

Fig. 7

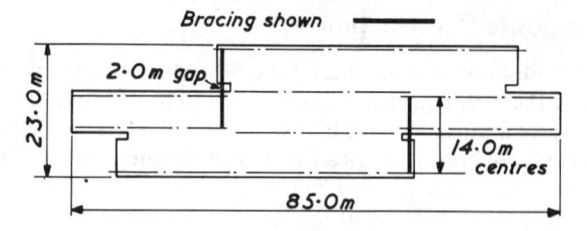

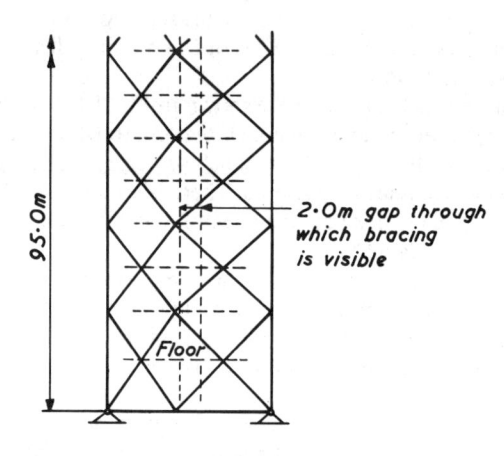

Fig. 8

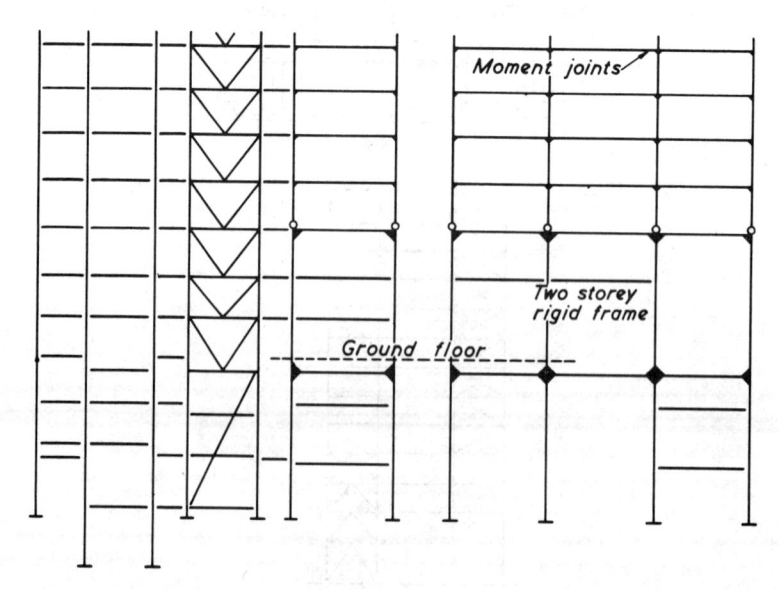

Fig. 9

Fig. 10

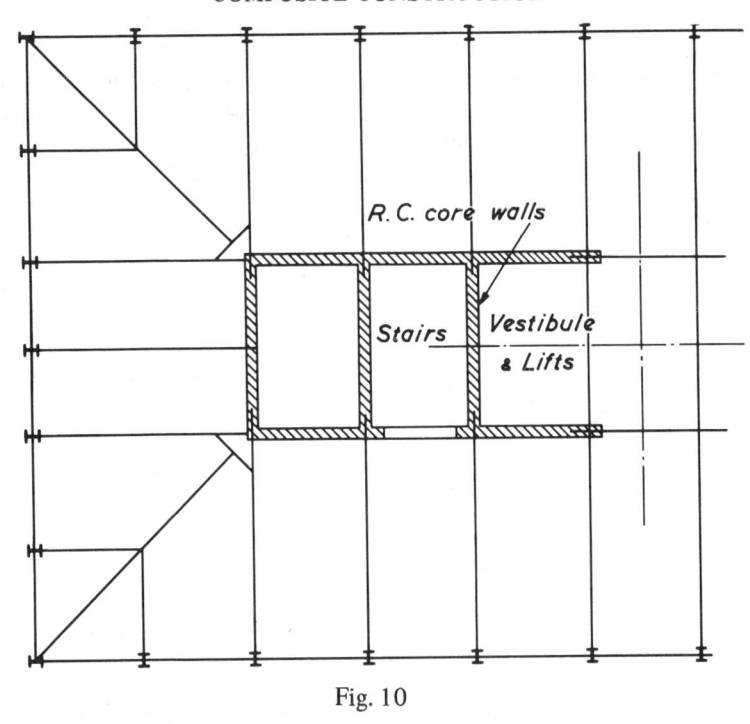

stiffened by the two triangular cores, which accommodate service and access facilities, with their reinforced concrete walls and by other reinforced concrete walls which have been located to accord with the planning of the flats. Steel beams in the planes of stiffening walls are double channels, the walls being continuous through them.

The floors act as horizontal wind girders to transmit wind loads to the stiffening walls and are of reinforced concrete ribbed form, using woodwool units left permanently in place to form the ribs, which together with the 63 mm thick topping are cast *in situ*. The lightweight floor with high insulating and acoustic properties has been found adequate as a lateral girder.

Other Methods

There are occasions when the architectural features permit of wind bracing being placed in external walls thereby allowing the internal steel framing to be designed for vertical loads only. A classic and unusual example of this type of construction is provided by the framing of the IBM building in Pittsburgh, U.S.A. This thirteen-storey building has a diamond pattern structural steel grid in the external walls to transmit wind loads to the foundations in addition to providing support for the floor beams which frame into it.

References

A list of articles dealing with the treatment of wind loads in completed steel framed buildings, supplied by the BCSA, will be found at the end of this Chapter.

Side Sway

Two questions are often asked but seldom answered satisfactorily: they are, what is the deflection or lateral displacement at the top of the building under the action of wind and what limit, if any, should be placed upon it.

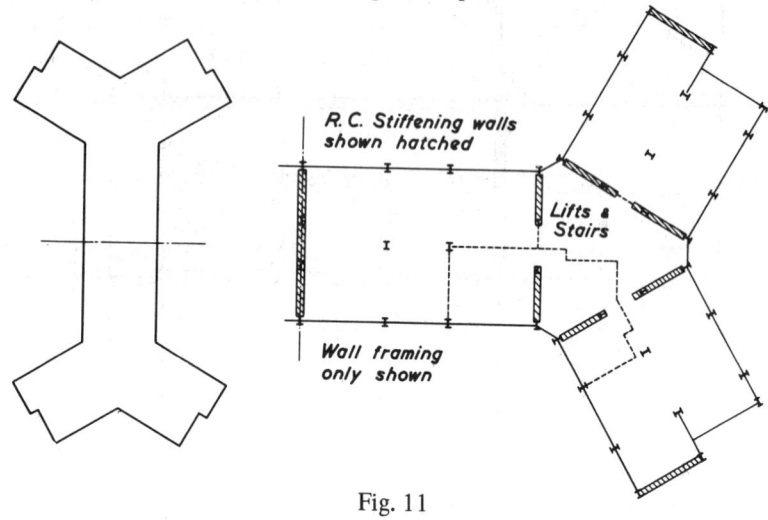

Fig. 11

Theoretical calculations may be made for the first of these but no guidance is given by B.S. 449 on the second point, as clearly the deflection limits for simply supported beams in Clause 15 and for single storey stanchions in Clause 31.b, are not necessarily applicable.

The first problem facing the designer is to decide whether the steel frame is, or is not, prevented from sidesway at loads less than would cause failure of the individual members of the frame. Much has been written on this point and some more or less approximate methods have been propounded, but in the main, this is still a matter of engineering judgement.

In the opinion of the authors, the frame may be considered as prevented from sidesway, and hence from failure by frame instability, if it has been designed by any elastic method and lateral stability provided either by shear walls, diagonal bracing, braced towers, braced portals, core-type construction or any combination of these. The same can apply if there are shear walls, such as gables, which have not been included in the wind resistance calculations.

Where curtain walling is used and the building has no positive lateral bracing system, then sidesway should be considered in tall buildings. It is not possible to define the word 'tall', since it depends on overall building stiffness and all that can be said is that the problem should not arise in a building with height/depth ratio of about 8-10 less, if the frame has the conventional storey heights and plan grid. Large beam spans accentuate the trouble.

It is quite common to make the theoretical deflection calculations on the bare steel frame, using the wind loads used in the stress calculations but the results obtained in this way are usually greatly in excess of the real answer, for a variety of reasons, all of which tend to reduce the calculated figure. The safety of the structure having been secured by any of the means so far discussed, it is, in the author's

opinion, essential to adopt a more realistic approach when considering deflection and particularly so, should displacement be a criterion, leading to the need for producing an estimate of what is likely to happen under the worst conditions and not a purely hypothetical figure known to be too large.

Whilst accepting that from considerations of safety and strength, it is prudent to design the frame on the assumption that it carries the whole of the wind load, the same assumption need not necessarily be made when calculating sidesway movements under wind, when there are other stiffening media which can come into operation. The building acts as a vertical cantilever and any sidesway of the frame must necessarily be accompanied by movement of the building as a whole which will bring into action any non-structural components which may feature in the layout, such as cross walls, staircase walls, etc. If such exist, then some attempt should be made to assess the relief which they can give to the steel frame: appropriate stresses are given for reinforced concrete in C.P. 114: The structural use of Reinforced Concrete in Buildings, and for brickwork in C.P. 111: Structural Recommendations for Load-bearing Walls. The total wind load on the building should then be apportioned between the frame and the other components and the frame deflection calculated on the reduced load. This applies irrespective of the manner in which the frame has been designed, or the type of bracing used.

There will, of course, be many cases, such as office blocks, where cross walls do not exist and the frame alone must take the total load.

It is interesting to note that blocks of flats eleven storeys in height have been built solely of brickwork, the cross walls between flats being amply strong enough to resist the wind loads when considered as vertical cantilevers.

When dealing with calculations for the frame, allowance can be made for the stiffening effect of solid casing where used, both for columns and beams.

If the frame has been designed on the rigid, elastic basis the deflections can be obtained as outlined in Chapter 13. If, however, the frame has been designed using one of the approximate methods given earlier in this chapter, then an assessment of sideways deflection by similar simple calculations is not possible. An approximate answer can be found by treating the frame, storey by storey, as fixed ended portals, using the moments given by the selected method, on the lines of the second example in Chapter 13. Alternatively, the deflection may be calculated using the moment of inertia of stanchions only, calculated about the central axis of the frame. If, however, the deflection is to be a criterion, then the frame should not be designed upon any approximate method, but should be designed on a more accurate basis.

The deflections of braced frames can be calculated arithmetically, or graphically by means of a Williot-Mohr diagram, as outlined in Chapter 11. If, however, the depth of members is large compared with their length, with consequent increased gusset plate sizes, it is recommended that they be analysed making allowance for joint stiffness and this is best done by computer. Should the bracing be embedded in a wall, then allowance should be made for the stiffening effect and the loads apportioned accordingly. It will be clear that in order to keep the deflection as low as possible the depth of the braced frame should be the largest which can conveniently be accommodated in the structural layout. It will be noted from Fig. 8 that a satisfactory solution has been obtained with frames of height/depth ratio nearly 7 and the same applies to Fig. 7. With regard to core construction, using reinforced concrete as the constructional medium, it is not usual to calculate deflections and it is assumed that sufficient stiffness will be assured if the height/depth

ratio of the vertical cantilever does not exceed 10 if a 1:2:4 mix is used or 9 if a 1:1:2 mix is used.

There now remains the vexed question as to what limit, if any, should be placed upon the lateral deflection. It must be remembered that deflection as such, is not dangerous and the purpose of a limit is not to lay down a safety requirement, since this is controlled by stresses, but to give a limit which will avoid damage to non-structural components and avoid objectionable vibration. The maximum deflection of a simply supported beam, given in Clause 15 of B.S. 449 due to imposed loads, is given as 1/360 of the span which is equivalent to 1/180 of the length of a cantilever beam; whilst Clause 31.b states that in single storey buildings, the deflections at the tops of stanchions due to lateral forces must not exceed 1/325 of the stanchion height, except that greater deflections may be permitted provided that they do not impair the strength of the structure or lead to damage to finishings. No guidance is given on the question under review, nor so far as can be ascertained in 1965, is any limit given in any European or American regulations.

It is thus a question of opinion. A figure of 1/500 of the height has been mentioned but in the opinion of the authors this is too severe a restriction. Bearing in mind the transitory nature of wind gusts, the wind loads and their applications specified in C.P. 3: Chapter V, and the fact that calculated deflections will not be realised in practice, it seems to the authors that the maximum calculated deflections at the top of the building could well be satisfactory if not more than 1/300 of the building height.

BIBLIOGRAPHY

List of Articles dealing with steel-framed buildings having reference to the method of dealing with wind loads.

A. Simple and Rigid Frame Design

1. ANON. 'Huge Frame Nears Completion', *Engineering News Record,* 1962, Vol. 168 (4). Jan. 25, p. 49.
2. DE UGARTE, JUAN M. 'The S.E.A.T. Multi-Storey Building for Car Showrooms, Barcelona', *Acier-Stahl-Steel,* 1962, Vol. 27 (3). March, pp. 116–118.
3. ANON. 'New York Double Header', *Engineering News Record,* 1962, Vol. 168 (21). May 24, p. 30. (Pan-Am Building.)
4. ANON. 'A Giant Rises from the Tracks', *Engineering News Record,* 1962, Vol. 168 (25). June 2, pp. 52–60. (Pan-Am Building.)
5. WIENHOLD, WOLFRAM A. 'The Pan-Am Building a Bold Achievement in Steelwork', *Acier-Stahl-Steel,* 1963, Vol. 28 (5). May, pp. 208–210.
6. ANON. 'Two Giants Spearhead Boom in Hotel and Motel Construction', *Engineering News Record,* 1962, Vol. 169 (13). Sept. 27, pp. 38–41. (New York Hilton Hotel.)
7. ANON. 'Blue-Tinted Hotel Tower Pierces New York Skyline', *Engineering News Record,* 1963, Vol. 171 (1). July 1, p. 31. (Hilton Hotel.)
8. ANON. 'Column-Free Floors Make Room for Research', *Engineering News Record,* 1963, Vol. 170 (23). June 6, pp. 40–41. (Research Building, California University.)
9. ANON. 'Boston's Pru Tower Puts New Life Into Back Bay', *Engineering News Record,* 1964, Vol. 172 (8). Feb. 20, pp. 38–39.
10. BLODGETT, OMER W. 'The Efficient Use of Steel in Multi-Storey Structures', *Welding Journal* (U.S.A.), 1962, Vol 41 (3). March. pp. 209–218.
11. ANON. 'The C.I.L. House, Montreal', *Acier-Stahl-Steel,* 1962, Vol. 27 (3), March, pp. 116–118.
12. ROGERS, ERNESTO. 'Departmental Store, Rome', *Architectural Design,* 1962, Vol. 32 (6). June, pp. 286–289.
13. ANON. 'Welded Goal-Post Bents Speed Erection of Apartment House', *Engineering News Record,* 1962, Vol. 168 (26). June 28, p. 44. (Single-storey frames in 10 storey structure.)
14. GROTENHUIS, EUGENE N. 'Building Economies Realised Through', *Civil Engineering* (U.S.A.), 1962, Vol. 32 (8), Aug., pp. 49–50.
15. COVRE, G. 'The New Steel-Framed Rinascente Building in Rome', *Acier-Stahl-Steel,* 1963, Vol. 38 (1). Jan., pp. 1–5.
16. ANON. 'United Engineering Centre, New York', *Acier-Stahl-Steel,* 1963, Vol. 28 (1). Jan., pp. 15–18.
17. McHALFIE CLARK. 'New Headquarters Building for ISCOR, Pretoria', *Acier-Stahl-Steel,* 1963, Vol. 28 (2). Feb., pp. 51–56.
18. ANON. 'Tailored Sections Frame Welded Tower', *Engineering News Record,* 1963, Vol. 170 (20). May 16, pp. 36–27. (Sunset-Vine Tower, Los Angeles.)
19. ANON. 'Sunset-Vine Tower, Los Angeles', *Acier-Stahl-Steel,* 1965, Vol. 30 (3). March, pp. 143–144.
20. JOLLIOT, ROBERT. 'Design of a 27-Storey Building under Construction at Nancy', *Acier-Stahl-Steel,* 1963, Vol. 28 (6). June, pp. 291–296.
21. TREBOUET, PAUL. 'Functional New Headquarters Building for the S.A.F. Esso Standard, Courbevoie'. *Acier-Stahl-Steel,* 1963, Vol. 28 (40). Oct., pp. 437–445.
22. ANON. 'Multi-Storey Frame in High-Yield-Stress Steel', *Engineer,* 1963, Vol. 217 (5635). Jan., 24, pp. 173–174. (Littlewoods Mail Order Stores, Liverpool.)
23. ZACSEK, STEPHANE and TAMIGNIAUX, RENE. 'New Grouped Sales and Administrative Departments in Brussels for Anc. Etabl. d'Iteren Freres', *Acier-Stahl-Steel,* 1964, Vol 29 (1). Jan., pp. 1–10.
24. ANON. 'Offices, Detroit', *Architect and Building News,* 1964, Vol. 226 (13). Sept. 23, pp. 579–582.
25. ANON. 'Building Has Removable 28-Storey Section', *Engineering News Record,* 1965, Vol. 174 (3). Jan. 21, pp. 34–35.

B. **Diagonal Bracing, Braced Towers and Core Construction**

1. FINZI, L. and NOVA, E. 'The E.N.I. Administration Block at the Rome E.U.R.', *Acier-Stahl-Steel*, 1962, Vol. 27 (1). Jan., pp. 5–12.
2. ANON. 'Headquarters Building for the Rhone Poulenc Company, Paris', *Acier-Stahl-Steel*, 1962, Vol. 27 (2). Feb., pp. 71–75.
3. MEASOR, E. O. and WILLIAMS, G. M. J. 'Features in the Design and Construction of the Shell Centre, London', *Proceedings, Institution of Civil Engineers*, 1962, Vol. 21. March, pp. 475–502.
4. ANON. 'Engineering Features of the Shell Centre', *Surveyor*, 1962, Vol. 121 (3646). April 21, pp. 547–549.
5. ANON. 'The Shell Centre', *Civil Engineering and Public Works Review*, 1962, Vol. 57 (671). June, pp. 749–751.
6. ANON. 'The Building of the Shell Centre', *Architect and Building News*, 1962, Vol. 222 (5). Aug. 1, pp. 161–164.
7. 'Features in the Design and Construction of the Shell Centre', *Proceedings, Institution of Civil Engineers*, 1963, Vol. 24. March, pp. 409–424. (Discussion on item B3.)
8. VERSWIJVEREN, A. 'The New K.N.H.S. Centre Laboratory at Ijmuiden', *Acier-Stahl-Steel*, 1962, Vol. 27 (3). March, pp. 111–115.
9. ANON. 'Shear Studs Stiffen 42-Storey Bank Tower', *Engineering News Record*, 1962, Vol. 168 (9). March 8, pp. 28–29.
10. DEFAY, A. 'Steel Framed Building at Brussels', *Acier-Stahl-Steel*, 1962, Vol. 27 (5). May, pp. 202–206.
11. CAIN, GEORGE B. 'Composite Building Construction comes of Age', *Civil Engineering* (U.S.A.), 1962, Vol. 32 (6). June, pp. 50–53.
12. ANON. 'Royal Hilton Hotel, Teheran', *Acier-Stahl-Steel*, 1962, Vol. 27 (9). Sept., p. 400.
13. ANON. 'Royal Teheran Hilton Hotel', *Architect and Building News*, 1963, Vol. 223 (13). March 27, pp. 464–466.
14. ANON. 'The Offices of the Nillmij Life Assurance Company at The Hague', *Acier-Stahl-Steel*, 1962, Vol. 27 (10). Oct., pp. 403–408.
15. JEAN, POL and PILLARO, JACQUES. 'Jewish Refugee Centre in Paris', *Acier-Stahl-Steel*, 1962, Vol. 27 (10). Oct., pp. 416–418.
16. ANON. 'Prefabricated Office Block', *Engineer*, 1962, Vol. 214 (5574). Nov. 23, pp. 911–912. (Design Centre for Krupps.)
17. ANON. 'Composite Design Lightens Tower', *Engineering News Record*, 1962, Vol. 169 (24). Dec. 13, p. 38. (Wenner-Gren Centre, Stockholm.)
18. ANON. 'C.I.S. Offices, Manchester', *Architect and Building News*, 1963, Vol. 223 (3). Jan. 16, pp. 85–94.
19. ANON. 'Manchester, New Offices for the C.I.S. and C.W.S.', *Architects Journal*, 1963, Vol. 137 (7). Feb. 13, p. 341.
20. ANON. 'C.I.S. Building, Manchester', *Builder*, 1963, Vol. 121 (6251). March 8, pp. 489–493.
21. ANON. 'Connection Details Simplify Hotel's Frame', *Engineering News Record*, 1963, Vol. 170 (10). March 7, pp. 44–45.
22. EIERMANN, EGON. 'Steel Company Offices, Offenburg', *Architectural Design*, 1963, Vol. 33 (6). June, pp. 270–273.
23. ANON. 'Skyscraper is a Synthetics Showcase', *Engineering News Record*, 1963, Vol. 170 (25). June 20, p. 149. (33-storey building at Leverkusen.)
24. LEABU, VICTOR F. 'Detroit's Automated Post Office', *Civil Engineering* (U.S.A.), 1963, Vol. 33 (3). March, pp. 34–38.
25. LAVEND'HOMME, R. 'New Building "Delta-Hainault" at Mons', *Acier-Stahl-Steel*, 1963, Vol. 28 (10). Oct., pp. 413–419.
26. GOFFAUX, R. and HEYWANG, C. 'The Westbury Hotel at Brussels', *Acier-Stahl-Steel*, 1964, Vol. 29 (2). Feb., pp. 75–78.
27. PASCAUD, S. 'The New Faculty of Law at Paris', *Acier-Stahl-Steel*, 1964, Vol 29 (6). June, pp. 291–300.
28. ANON. 'Stronger Steels Cut Column Costs', *Engineering News Record*, 1964, Vol. 173 (12). Sept. 17, p. 171. (22-storey building at Tampa, Florida.)
29. ANON. 'Steel Pipe Columns Support 20-storey Walls', *Engineering News Record*, 1964, Vol. 173 (15). Oct. 8, pp. 66–68. (20-storey I.B.M. building at Seattle.)
30. ANON. 'Office Block, Belgium', *Architect and Building News*, 1965, Vol. 227 (3). Jan. 20, pp. 121–122.

31. CULOT-FONTAINE, M. 'The United States Gypsum Building, Chicago', *Acier-Stahl-Steel*, 1965, Vol. 30 (1). Jan., pp. 14–16.
32. ANON. 'West's Tallest Skyscraper opens in Downtown, Dallas', *Engineering News Record*, 1965, Vol. 174 (5). Feb. 4, p. 29. (52-storey building.)

C. Miscellaneous Types, not Covered Above

1. ANON. 'Pittsburgh Office Building Features Special Steel Design', *Civil Engineering* (U.S.A.), 1962, Vol. 32 (6). June, p. 96. (13-storey I.B.M. building with "diamond" pattern wall framing.)
2. ANON. 'Building's Lattice Wall Carries Beams, Resists Wind', *Engineering News Record*, 1962, Vol. 169 (10). Sept. 6, pp. 34–41. (I.B.M. building, as Item C1.)
3. GRINDROD, J. 'Super-Strength Steel in Pittsburgh Building', *Builder*, 1963, Vol. 121 (6245). Jan. 25, p. 200. (I.B.M. building.)
4. ANON. 'Twin Towers go to 110 storeys', *Engineering News Record*, 1964, Vol. 172 (4). Jan. 23, pp. 33–34. (World Trade Centre, New York.)
5. ANON. 'How Columns will be Designed for 110-storey Buildings', *Engineering News Record*, 1964, Vol. 172 (14). April 2, pp. 48–49. (World Trade Centre, New York.)
6. GOETA, F. 'Steelwork for the second S.N.A.M. Office Block at S. Donato Milanese', *Acier-Stahl-Steel*, 1964, Vol. 29 (12). Dec., pp. 566–570.
7. ANON. 'Pyramid Will Rise 100-Storeys', *Engineering News Record*, 1965, Vol. 174 (13). April 1, p. 14. (John Hancock Centre, Chicago.)

32. FLOORS

THIS section deals with floors that are generally of some form of reinforced concrete construction. Many criteria may be used for classifying the various types of floors and the following table gives a convenient pattern as, in addition to solid *in situ* concrete floors, there are types corresponding to almost every combination of these characteristics.

TABLE 1

CLASSIFICATION OF FLOORS

Method of Construction	Initial Strength	Materials Used Apart from in situ Concrete
Mainly *in situ* Partially prefabricated Completely prefabricated	None, requiring complete support until concrete hardens. Adequate to carry construction loads (*a*) with or (*b*) without intermediate supports Fully self-supporting at all stages after erection	Clay or concrete blocks or tiles Precast (and possibly prestressed) concrete beams of channel, hollow rectangle, joist, rib or tee profile Precast (and possibly prestressed) hollow or solid planks and panels. Sheet steel Wood-wool formers

Most of the floors embodying preformed units are proprietary products and have been patented. Particulars of manufacturers and other information may be obtained from the Association of Constructional Floor Specialists, 13 Goodwin's Court, St. Martin's Lane, London, W.C.2. Much information on a wide variety of such floors is contained in a booklet issued jointly by the British Constructional Steelwork Association and the British Steel Makers and entitled Publication No. M2, 1964: (Revised 1965) 'Prefabricated Floors for Use in Steel Framed Buildings'.

The design of reinforced concrete slabs is adequately treated in most textbooks on reinforced concrete. Other information may be obtained from the relevant British Standard Codes of Practice.

In many cases, floors are designed to be simply supported over single spans and this treatment can be economic for various reasons including convenience for initial planning and possible future alterations also suitability for mass production. Other types of floors, probably mainly those with *in situ* ribs, are more readily adapted to take advantage of the reduced bending moments associated with

continuity. In such cases, the bending moment must allow for possible variations in loading on different spans. Table 2 gives the bending moment values for nominally uniformly loaded slabs continuous over three or more approximately equal spans, that is, not differing by more than 15 per cent of the longer span.

TABLE 2

BENDING MOMENTS IN UNIFORMLY LOADED CONTINUOUS SLABS

	Near Middle of End Span	At Support Next to End Support	At Middle of Interior Span	At Other Interior Supports
Moment due to dead load	$\dfrac{W_d \cdot l}{12}$	$-\dfrac{W_d \cdot l}{10}$	$+\dfrac{W_d \cdot l}{24}$	$-\dfrac{W_d \cdot l}{12}$
Moment due to imposed load	$+\dfrac{W_s \cdot l}{10}$	$-\dfrac{W_s \cdot l}{9}$	$+\dfrac{W_s \cdot l}{12}$	$-\dfrac{W_s \cdot l}{9}$

Where W_d = total dead load per span
W_s = total imposed load per span

When proprietary floors are used, advice should be obtained from the makers regarding any reinforcement that may be necessary. The main reinforcement is often fully supplied within precast concrete units but many types allow for reinforcing bars within *in situ* concrete ribs. Continuity or anti-cracking reinforcement is often necessary or desirable over supports.

The selection of the most appropriate floor for any application requires consideration of many factors in addition to initial cost and ability to carry the floor loading. Modern buildings involve many services and several types of floors incorporate ducts for such services: in other types services may be laid in a concrete screed. Either deliberately or incidentally the floor serves also as a fire barrier between two storeys. Many types of floors have considerable inherent fire resistance, and in all cases the resistance can be augmented by screeding on the top or by applying an incombustible material such as asbestos or plaster to the soffit. Suspended ceilings can be provided to enhance fire resistance and to modify acoustic properties. They can serve also to conceal services suspended below the floor.

THE BUILDING REGULATIONS, 1965, which came into operation on 1st February, 1966, include a list of thicknesses of materials in concrete floors for various periods of fire resistance. A metric version of this list is given below as Table 3.

The floor can also contribute substantially to the stability of the structure by acting as a horizontal girder carrying lateral forces to selected strong points such as braced frames or walls. This is amplified later.

Reference to tables of floor sizes and capacities will show that in some systems the strength varies with the depth while in other systems a constant depth is retained and the reinforcement is varied. Nevertheless the weights for different types of floors of various depths do lie within fairly clearly defined boundaries as shown by the lines in Fig. 1.

TABLE 3

EXTRACTED FROM "THE BUILDING REGULATIONS"

PART VIII: CONCRETE FLOORS

Construction and materials	Minimum thickness of solid substance including screed in millimetres	Ceiling finish (in millimetres) for a fire resistance of:				
		4 hours	2 hours	1½ hours	1 hour	½ hour
Solid flat slab or filler joist floor. Units of channel or T section	90	25 mm V or 25 mm A	10 mm V or 13 mm A	10 mm V or 13 mm A	7 mm V or 7 mm A	nil
	100	19 mm V or 19 mm A	7 mm V	7 mm V	nil	nil
	125	10 mm V or 13 mm A	nil	nil	nil	nil
	150	nil	nil	nil	nil	nil
Solid flat slab or filler joist floor with 25 mm wood-wool slab ceiling base	90	13 mm G	nil	13 mm G	nil	nil
	100		nil	nil	nil	nil
	125	nil	nil	nil	nil	nil
	150		nil	nil	nil	nil
Units of inverted U section with minimum thickness at crown	63	nil	nil	nil	nil	nil
	75		nil	nil	nil	nil
	100					nil
	150					nil
Hollow block construction or units of box or I section	63	nil	nil	nil	nil	nil
	75					nil
	90					nil
	125					nil
Cellular steel with concrete topping	63	13 mm V suspended on metal lathing or 13 mm A (direct)	13 mm G suspended on metal lathing	13 mm G suspended on metal lathing	13 mm G suspended on metal lathing	nil

"V"—vermiculite-gypsum plaster   "A"—sprayed asbestos in accordance with B.S. 3590 : 1963   "G"—gypsum plaster

871

FLOORS

The solid concrete slab obviously has a constant weight of about 2 kg/m² of floor area per mm of thickness. For all other types of floor, the minimum weight per m² seems to be about 0.8 kg/mm of thickness. For floors using wood-wool or similar formers the upper limit of weight seems to be about 1.2 kg/mm of thickness and for other types using precast units of almost any shape, the upper limit seems to be

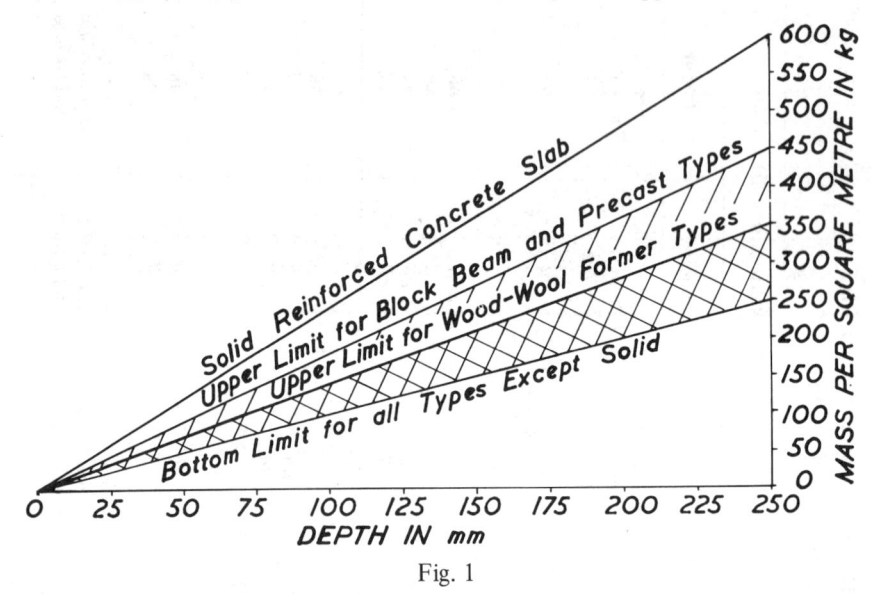

Fig. 1

about 1.5 kg/mm of thickness. These figures allow generally for all the materials of the floor including load bearing structural concrete toppings but they do not allow for finishing screeds or plastering applied to the floor. Whilst a lighter floor obviously means less load to be carried by superstructure and foundations a heavier floor may have counterbalancing qualities for other reasons, such as fire resistance or reduced noise transmission.

The wide variations in carrying capacities for floors of similar depths or weights indicate that any attempt at such correlation could hardly produce useful results. The prefabricated floors booklet previously mentioned gives limiting spans for various loadings for each individual floor listed.

General descriptions of some of the types follow. The sequence is the more or less alphabetical arrangement of materials used as given in Table 1 and there is no attempt to arrange types in order of merit or importance nor to illustrate individual makers' products.

Clay or concrete blocks or tiles are generally hollow with intermediate webs in the larger sizes. Standard sizes for some such clay blocks are specified in B.S. 1190. For an *in situ* hollow block floor the tiles may be laid end to end in rows when the reinforcement runs in one direction as shown in Fig. 2 or the reinforcement may be run in two directions, at right angles to each other, when the tiles are isolated and provision is made to prevent concrete entering the open ends. Hollow tile floors are designed as T-beams, the breadth of the ribs being regulated by the amount of reinforcement to be housed and the depth by the span and load to be carried. Hollow tile *in situ* floors are usually constructed to be continuous over several spans.

These are various methods of determining the bending moment on such slabs and one set of values has been quoted earlier.

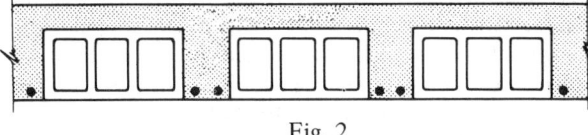

Fig. 2

Floors of this type require shuttering and external supports until the concrete has hardened sufficiently. The shuttering may be conventional formwork supported on struts but telescopic centerings are available which are designed for frequent re-use to avoid obstruction to the floor under.

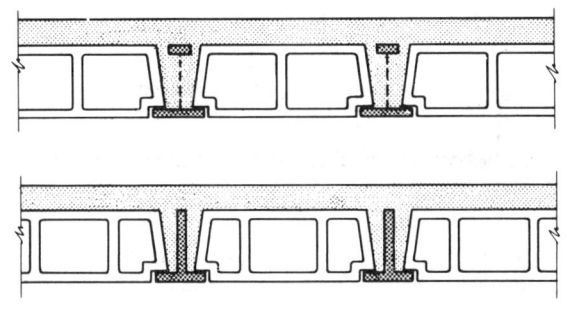

Fig. 3

Hollow concrete or clay blocks are used in another type of floor illustrated generally by Fig. 3. In this, the cross-hatched members are precast beams which are usually designed to span between the main supports and be capable of supporting the constructional loads unaided until the added concrete has hardened. The hollow tiles may be of various widths, they serve as shuttering between the precast beams and a wide range of capacities can be achieved by suitable combinations of reinforcement within the precast members and thickness of *in situ* topping.

The precast beams in these floors may be of various sections, the most usual being tees, joists and steel assemblies having the bottom member enclosed within a concrete flange. The strength of this type of floor can be varied also by using two beams side by side at each rib (Fig. 4) and again by using the beams side by side without spacing tiles (Fig. 5).

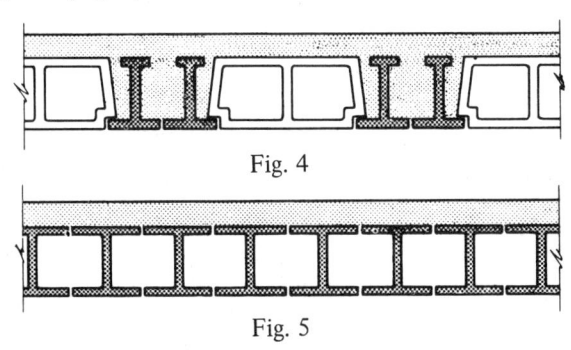

Fig. 4

Fig. 5

The next type of floor is essentially precast and possibly prestressed units. It comprises beams of channel, Fig. 6, or hollow rectangular form, Fig. 7, which are designed to carry the construction loads unaided and frequently also to be able to support the total load without *in situ* structural concrete topping although topping can easily be placed either to augment the capacity or simply to provide a more continuous surface and to accommodate service ducting. Many of these floors use units having sloping sides or with a ledge at the bottom edge to facilitate the grouting that is often used to bind the components into a continuous floor.

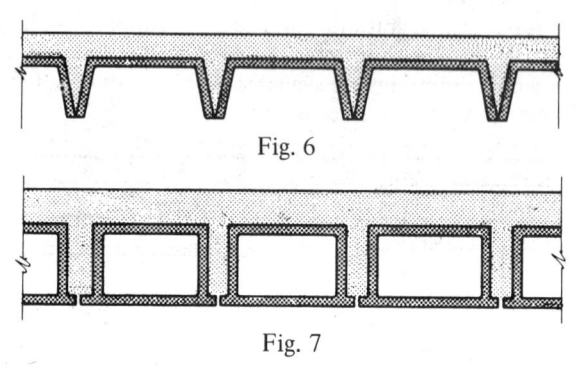

Fig. 6

Fig. 7

Some manufacturers make these hollow rectangular floors in multi-unit widths so that each component comprises a substantial panel of flooring, Fig. 8.

Fig. 8

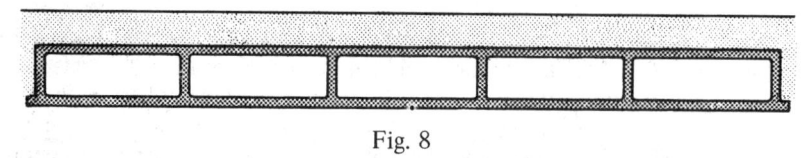

Figure 9 shows another type of floor of substantial breadth. The units are precast double tees, having prestressed tendons.

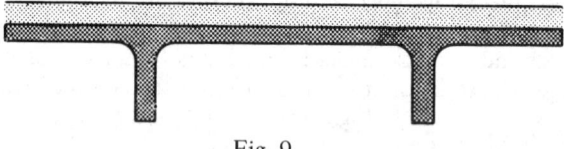

Fig. 9

A similar finished profile is shown by the rib type floors of Fig. 10 which utilise sheet steel formers which may be used as permanent shuttering and reinforcement, or may be removed for use elsewhere. The steel troughs may be supported on the lower flanges of precast tee beams which may incorporate a latticed steel rib or the

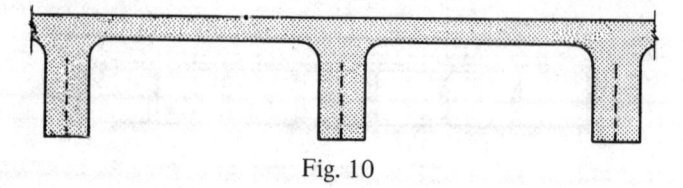

Fig. 10

troughs may be designed to span between main supports possibly with only a few intermediate props. The strength of such floors can be widely varied by altering the amounts of the reinforcement and *in situ* concrete.

An almost axiomatic feature of a plank floor is that it is intended to act compositely with a sufficient thickness of added *in situ* concrete to carry the working load. The plank will often incorporate the main reinforcement for the full load. Whilst a plank suggests a solid rectangle as shown in Fig. 11, yet the base

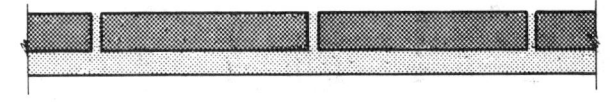

Fig. 11

elements of such floors may also be as shown in Fig. 7. In a variant of this type of floor shown in Fig. 12, precast concrete shutter panels span between secondary steel or concrete beams which may be spaced up to about 1.5 m apart. These secondary beams are normally designed to act compositely with the concrete.

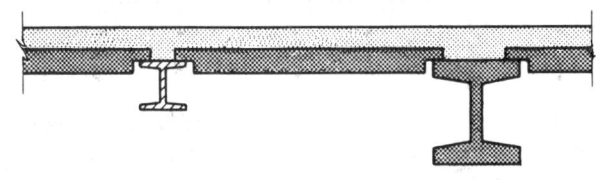

Fig. 12

Floors using sheet steel formers have been mentioned above. Other types are shown in Fig. 13. The sheets may vary in thickness or in shape and these variations together with different thicknesses of *in situ* structural concrete allow a wide range of carrying capacities. The steel sheet formers may sometimes require propping until the concrete has hardened but additional supports are not always necessary. The

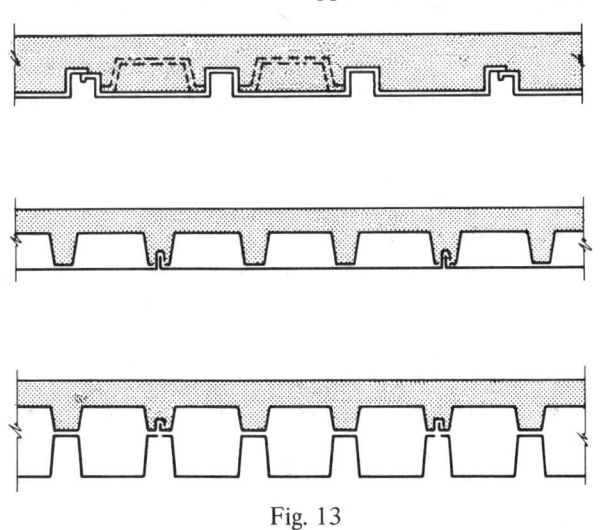

Fig. 13

manufacturers make a feature of the facility with which electrical services, for example, can be passed along the cavities. In one type, the cavity element which is shown dotted, can be omitted or incorporated where it is wanted. In the other sketches, the cavities are an automatic result of the pattern. The wide variations in cavities made it impossible to show these floors adequately in Fig. 1.

The last type of floor mentioned in Table 1 is simply described as woodwool formers. Some patterns are shown in Fig. 14.

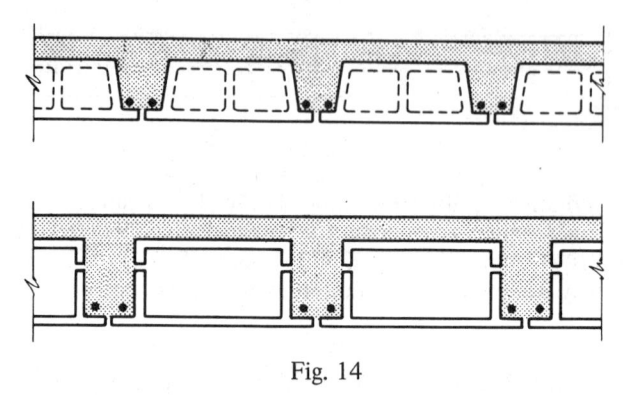

Fig. 14

The formers may be solid or hollow as indicated by the dotted lines. Alternatively two components may be superimposed to form a unit. These formers are usually made in convenient lengths for delivery and site handling so that temporary support is usual until the *in situ* concrete has hardened. The reinforcement is introduced into the ribs as required, the design being generally in accordance with that for normal reinforced rib floors.

The foregoing has considered these floors almost without regard to how they are integrated into the building. The easiest way to support such floors on a beam is obviously to sit them on the top flange. Where this is not practicable, they may be supported on ledge angles attached to the web of the beam or they may sit on concrete haunches which should be adequately secured. This may necessitate specially shaped ends to the floor units and many manufacturers have appropriate end details as standard patterns. Other standard variations are often available for special loadings such as partitions, for bridles at openings in floors or for special width units at boundaries.

Whilst the floors are designed primarily to resist vertical loading, yet many types are able to carry substantial additional horizontal forces from wind loading without exceeding the appropriate stresses. A series of tests* was carried out on full-size floors in three square panels giving an overall size of about 10 m by $3\frac{1}{3}$ m and made of hollow rectangular precast floor units placed side by side and having the joints grouted in normal fashion. Lightly reinforced 75 mm thick *in situ* concrete edge trimmers were provided to finish the floor. Loading was applied horizontally by hydraulic jacks at the third-points of the 10 m length and substantial loads were carried on both the unloaded floors and on the floors with weights added to simulate the live load. The tests were made with the floor units parallel with the long sides and also with the floor units parallel with the short sides.

* Reported in *Building with Steel*, Vol. 2, No. 8.

FLOORS

The design of beams taking account of composite action with the floor slab is considered in Chapter 23 and little need be included here. The typical sketches show that some types of floor are more readily suitable than others for this application. In some cases, it may be desirable to have a more or less substantial area of solid concrete adjacent to the steel beam and some types are more readily adaptable to this.

33. FLOOR PLATES

THE various types of rolled steel floor plates with non-slip raised patterns have now been reduced and the British Steel Corporation roll only Durbar plate which is illustrated in Fig. 1.

Fig. 1

These plates are normally supplied in standard widths, lengths and thicknesses but the rolling mills are prepared to discuss variations from these standards if a sufficiently large order of a particular size is required.

Table 1 gives the metric thicknesses of the plain plate together with the weight in kg/m. It should be noted that the pattern projects approximately 1.5 mm above the plain plate.

TABLE 1

Thickness on plain mm	Approximate weight kg/m^2
3.0	28.7
4.5	40.5
6.0	52.2
8.0	67.9
10.0	83.6
12.5	103.1

The recommended standard sizes of plates are:—

      600 mm wide in lengths of 2 000, 2 500 and 3 000 mm
    1 000 mm ,, ,, ,, ,, 2 000 mm
    1 250 mm ,, ,, ,, ,, 2 500 mm
    1 500 mm ,, ,, ,, ,, 3 000 mm
    1 750 mm ,, ,, ,, ,, 4 000 and 6 000 mm
    2 000 mm ,, ,, ,, ,, 6 000 mm

FLOOR PLATES

These must be cut to the required finished size and it should be noted that the detail of the pattern adopted is such that plates can be used in either direction without spoiling the appearance of a floor area. This was not possible with the old diamond pattern plate which had to be cut to match in one direction only thus frequently incurring considerable waste.

It is usual to consider floor plates as supported on all four edges even though two edges may only be supported by stiffeners or joint covers. If the plates are securely bolted or welded to the supporting system they may be considered as encastré which increases the load carrying capacity slightly but reduces the deflection considerably.

Safe loads have been calculated for rectangular plates either simply supported or encastré on all four edges using Pounder's formulae and these are tabulated in Tables 2 and 3. The formulae used are as follows:—

Plates simply supported on all four edges

1. $$p = \frac{4ft^2}{3kB^2 \left[1 + \frac{14}{75}(1-k) + \frac{20}{57}(1-k)^2\right]}$$

2. $$d = \frac{m^2 - 1}{m^2} \times \frac{5kpB^4}{32Et^3} \left[1 + \frac{37}{175}(1-k) + \frac{79}{201}(1-k)^2\right]$$

Plates encastré on all four edges

3. $$p = \frac{2ft^2}{kB^2 \left[1 + \frac{11}{35}(1-k) + \frac{79}{141}(1-k)^2\right]}$$

4. $$d = \frac{m^2 - 1}{m^2} \times \frac{kpB^4}{32Et^3} \left[1 + \frac{47}{210}(1-k) + \frac{200}{517}(1-k)^2\right]$$

where

L = length of plate in mm,
B = breadth of plate in mm,
t = thickness of plate in mm,

$$k = \left(\frac{L^4}{L^4 + B^4}\right),$$

f = allowable skin stress = 165 N/mm² for Grade 43 Steel,
p = pressure on plate in N/mm²,
E = Young's modulus of section = 2.1 × 10⁵ N/mm²,

$\frac{1}{m}$ = Poisson's ratio (m is taken at 3.0 in this exercise).

d = maximum deflection in mm.

FLOOR PLATES

TABLE 2

Safe uniformly distributed loads in kN/m² on steel floor plates stressed to 165 N/mm²

Simply supported on all four edges

Thickness in mm	Length L in mm							Breadth B in mm
	600	1 000	1 250	1 500	1 750	2 000	
3.0	9.31	6.06	5.73	5.61	5.56	5.54	600
		3.35	2.57	2.28	2.15	2.08	1 000
			2.15	1.71	1.52	1.42	1 250
				1.49	1.22	1.09	1 500
4.5	20.96	13.62	12.90	12.63	12.51	12.46	600
		7.54	5.79	5.13	4.83	4.68	1 000
			4.83	3.85	3.41	3.19	1 250
				3.35	2.75	2.45	1 500
					2.46	2.07	1 750
6.0	37.26	24.22	22.93	22.45	22.25	22.14	600
		13.41	10.30	9.12	8.58	8.31	1 000
			8.58	6.84	6.06	5.67	1 250
				5.96	4.89	4.35	1 500
					4.38	3.67	1 750
						3.35	2 000
8.0	66.23	43.06	40.77	39.84	39.55	39.37	600
		23.84	18.31	16.21	15.26	14.78	1 000
			15.26	12.17	10.78	10.08	1 250
				10.60	8.69	7.73	1 500
					7.79	6.53	1 750
						5.96	2 000
10.0	103.5	67.30	63.67	62.25	61.79	61.51	600
		37.26	28.61	25.32	23.84	23.09	1 000
			23.84	19.01	16.84	15.74	1 250
				16.56	13.58	12.09	1 500
					12.15	10.20	1 750
						9.31	2 000
12.5	161.7	105.1	99.49	97.26	96.55	96.11	600
		58.21	44.71	39.57	37.25	36.08	1 000
			37.26	29.70	26.32	24.60	1 250
				25.87	21.22	18.88	1 500
					19.01	15.94	1 750
						14.55	2 000

Note: Loads to right of zig-zag line cause deflection greater than $B/100$.

FLOOR PLATES

TABLE 3

Safe uniformly distributed loads in kN/m² on steel plates stressed to 165 N/mm²

Encastré on all four edges

Thickness in mm	\multicolumn{6}{c}{Length L in mm}	Breadth B in mm
	600	1 000	1 250	1 500	1 750	2 000	
3.0	12.71	8.93 4.58	8.54 3.68 2.93	8.39 3.33 2.42 2.04	8.33 3.17 2.20 1.72	8.30 3.09 2.08 1.57	600 1 000 1 250 1 500
4.5	28.62	20.10 10.30	19.22 8.27 6.59	18.88 7.50 5.45 4.58	18.74 7.14 4.95 3.88 3.36	18.66 6.96 4.69 3.53 2.90	600 1 000 1 250 1 500 1 750
6.0	50.88	35.72 18.32	34.16 14.71 11.72	33.57 13.33 9.70 8.14	33.31 12.70 8.80 6.90 5.98	33.18 12.37 8.33 6.27 5.16 4.58	600 1 000 1 250 1 500 1 750 2 000
8.0	90.45	63.51 32.56	60.73 26.15 20.84	59.51 23.70 17.24 14.47	59.22 22.57 15.65 12.26 10.63	58.99 21.99 14.82 11.15 9.17 8.14	600 1 000 1 250 1 500 1 750 2 000
10.0	141.3	99.23 50.88	94.89 40.85 32.56	92.98 37.03 26.95 22.61	92.52 35.27 24.45 19.15 16.61	92.17 34.36 23.15 17.43 14.33 12.72	600 1 000 1 250 1 500 1 750 2 000
12.5	220.8	155.1 79.50	148.3 63.83 50.88	145.7 57.86 42.10 35.33	144.6 55.10 38.20 29.94 25.96	144.0 53.69 36.17 27.23 22.39 19.87	600 1 000 1 250 1 500 1 750 2 000

Note: Deflection caused by all above loads is less than $B/100$.

34. WELDING PRACTICE

British Standards

THE following British Standards are associated with welding:

B.S. 4360 : Part 2. *Weldable structural steels.*
B.S. 639. *Covered electrodes for the manual metal-arc welding of mild steel and medium-tensile steel.*
B.S. 1719. *Classification, coding and marking of covered electrodes for metal-arc welding.*
B.S. 1856. *General requirements for the metal-arc welding of mild steel.*
B.S. 2642. *General requirements for the arc welding of carbon manganese steels.*
B.S. 449. *The use of structural steel in building.*
B.S. 499. *Welding terms and symbols.*

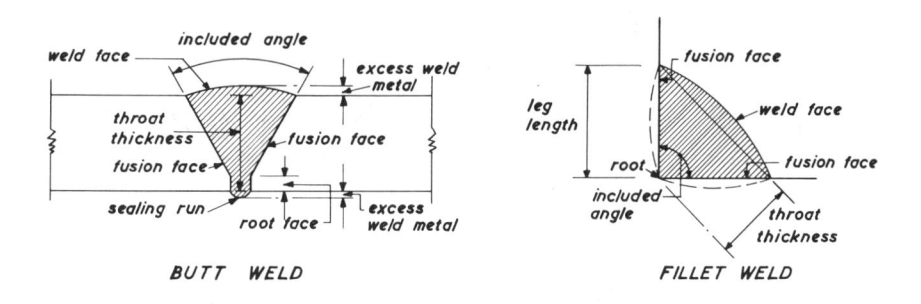

Fig. 1

Types of Weld

Fillet and butt welds are normally used in structural steelwork, the appropriate nomenclature and details of typical welds being shown in Fig. 1.

Butt Welds

Notes on Butt Welds are given in Appendix A of B.S. 1856.

The size of a butt weld is specified by its throat thickness, which is taken as the thickness of the thinner plate when the plate thicknesses vary and as the plate thickness when the plates are equal in thickness. For full stress butt welds must be reinforced with excess weld metal and sealed as shown in Fig. 1 although the additional thickness of weld involved is ignored in calculations. Where a sealing run cannot be employed, say in a single 'V' butt weld where the back of the plates is inaccessible, the effective throat thickness is taken as five-eighths of the thinner plate thickness.

The permissible stresses in mild-steel butt welds, calculated on an area equal to the length of the weld times the throat thickness, must not exceed the permissible tensile, compressive and shear stresses of the parent metal. Consequently there is normally no need to make calculations for complete penetration welds.

Fillet Welds

The size of a normal penetration fillet weld shall be taken as the minimum leg length of a convex or flat fillet weld or 1.41 times the effective throat thickness of a concave fillet weld, as shown in Fig. 2.

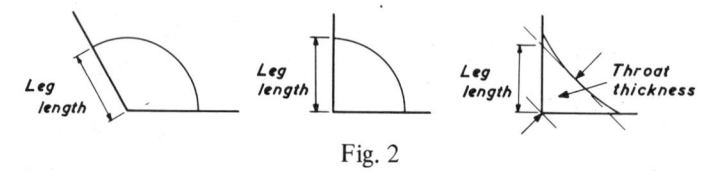

Fig. 2

A deep penetration weld is one in which the depth of penetration beyond the root is 2.4 mm or over. Consequently, the size of a deep penetration weld is taken as the minimum leg length plus 2.4 mm in the case of a convex or flat fillet weld and as 1.41 times the effective throat thickness plus 2.4 mm in the case of a concave fillet weld.

Notes on fillet welds are given in Appendix B of B.S. 1856. The following are of interest here:

Angle between fusion faces. Fillet welds connecting parts, the fusion faces of which form an angle of more than 120° or less than 60°, should not be relied upon to transmit calculated loads at the full working stresses unless permitted to do so by the standard for the particular application.

It is appreciated that full penetration to the root of a 60° fillet weld may not be attained, but this point is allowed for in determining the effective throat thickness (see Table A).

End returns. Fillet welds terminating at the ends or sides of parts, or members, should be returned continuously around the corners for a distance not less than twice the size of the weld. This provision should apply particularly to

TYPES OF WELD

TABLE A. – EFFECTIVE THROAT THICKNESS OF FLAT OR CONVEX FILLET WELDS

Angle between fusion faces	60–90°	91–100°	101–106°	107–113°	114–120°
Factor by which fillet size is multiplied to give throat thickness	0.70	0.65	0.60	0.55	0.50

side and end fillet welds connecting brackets, seats and similar connections at the tension side of such connections.

Size at toe of rolled section. Where a fillet weld is applied to a rounded toe of a rolled section, the specified size of the weld should generally not exceed 75 per cent of the thickness of the section at the toe.

The stress in a fillet weld must not exceed the permissible shear stress in the parent metal, e.g., for Grade 43 steel the maximum permissible stress is 115 N/mm². The following table gives the strengths of 60–90° angle fillet welds, i.e., based on a throat thickness of 0.7 times the leg length. For different angles between the fusion faces the values of the strength per mm run must be varied directly with the factor given in Table A.

TABLE B. – STRENGTH OF FILLET WELDS IN kN/mm RUN FOR 60–90° ANGLES BETWEEN FUSION FACES

Leg Length in mm	Throat Thickness in mm	Safe Loads on Fillet Welds in kN/mm run — Grade 43, 115 N/mm²	Grade 50, 160 N/mm²	Grade 55, 195 N/mm²
5	3.5	0.40	0.56	0.68
6	4.2	0.48	0.67	0.82
8	5.6	0.64	0.90	1.09
10	7.0	0.80	1.12	1.37
12	8.4	0.97	1.34	1.64
15	10.5	1.21	1.68	2.05
18	12.6	1.45	2.02	2.46
20	14.0	1.61	2.24	2.73
22	15.4	1.77	2.46	3.00
25	17.5	2.01	2.80	3.41

Clause 14a of B.S. 1856 (see also Appendix D, B.S. 1856) states:

Minimum size of first run. On parts 9.5 mm and over in thickness, the minimum size of a single run fillet weld or the first run in a multi-run fillet weld made by a manual process using electrodes of Class 2 or 3 to B.S. 1719, Part 1, shall be as specified in Table C to avoid the risk of cracking without preheating.

Where the thicker part is more than 50.8 mm thick, special precautions shall be taken to ensure weld soundness.

TABLE C
MINIMUM SIZE OF SINGLE RUN FILLET WELDS FOR MANUAL WELDING USING ELECTRODES OF CLASS 2 OR 3 TO B.S. 1719, PART 1

Thickness of Thicker Part		Minimum size of Single Run Fillet Welds
Over	Up to and Including	
mm	mm	mm
9	16	5
16	30	6
30 and over	–	8

Symbols

The type and size of weld is indicated on drawings in one of the following ways:

(*a*) By the use of the symbols in B.S. 499 : Part 2 : 1965.

(*b*) By the use of the following letter symbols with the appropriate dimension:

Fillet Weld	F.W.
Single V Butt Weld	S.V.B.W.
Double V Butt Weld	D.V.B.W.
Single U Butt Weld	S.U.B.W.
Double U Butt Weld	D.U.B.W.
Single J Butt Weld	S.J.B.W.
Double J Butt Weld	D.J.B.W.
Single Bevel Butt Weld	S.B.B.W.
Double Bevel Butt Weld	D.B.B.W.

(*c*) By a complete description, e.g., 10 mm fillet weld.

Deep-penetration welds must be separately and clearly defined.

35. STEEL SHEET ROOFING AND CLADDING

General

PROFILED steel sheet is commonly used in building as a cladding for roofs and sidewalls, its main properties being strength, light weight, durability, ease and speed of erection and low cost. Cladding sheets are now available in long lengths, thus obviating end laps.

Types of Surface Protection

Cladding sheets are formed from cold-reduced steel sheet or coil of a quality described in B.S. 1449. Appropriate surface treatments are as follows:

Galvanised Coating

The standard protective surface is a tight-bonded uniform coating of 98.5 per cent zinc applied by the continuous hot-dip process. According to the conditions under which the coating is applied, the resulting surface is either a matt grey or a crystalline spangle.

TABLE 1

TABLE OF ZINC COATING WEIGHTS

TYPES OF MATERIALS AND WEIGHTS OF COATING
(ABSTRACTED FROM B.S. 3083:1959)

Type	Weight of Zinc Coating (including both sides)	
	Min.	Max.
	g/m^2	g/m^2
125	381	455
150	458	548
180	550	608
200	610	762

Note. The weight of zinc on one side of a sheet is half that quoted above. The first-listed coating type is the most readily available commercial quality.

When exposed to the atmosphere, the zinc layer corrodes at a relatively slow rate until it is all consumed. This rate of corrosion depends on the nature of the atmosphere, being quickest in polluted industrial atmospheres and in coastal areas subjected to winds off the sea. Elsewhere the zinc is a long-lasting, reliable protection. The heavier the coating, the longer the life of the galvanised sheet.

Perforated Galvanised Sheet

In certain industries such as laundries, etc., where steam and humidity levels are high, direct ventilation of the cladding sheet is desirable. Corrugated galvanised sheets are available with specially cut crest perforations which allow air penetration but exclude rain.

Aluminium Coating

Similar in nature to the standard galvanised sheet is a sheet coated with a thin layer of tightly-bonded aluminium of about the same thickness (0.025 mm) as the zinc on a galvanised sheet. The aluminium coating is considered to have certain advantages such as greater corrosion resistance in industrial atmospheres, brighter appearance and greater solar heat reflection.

Plastic Coating

Protection of exceptional durability is obtained when a tightly-bonded layer of plastic is applied to either a plain or a galvanised steel sheet. The plastic, on one or both sides of the sheet as required, is applied either as a laminate or as a roll-coated plastisol. A plastic surface permits colour and attractive surface textures to be featured on steel sheet claddings. Sheets coloured differently on opposite sides are obtainable. The bond between plastic and steel is such that in places where accidental blemishes occur, corrosion is confined to that area and does not spread beneath the plastic. Profile forming of sheets does not affect either the plastic or the bond between steel and plastic. Plastic coated sheet needs no maintenance and is impervious to most chemicals, polluted atmospheres, frost, hot sun, and prolonged dampness. It does not support fire.

Vitreous Enamel Coating

Vitreous enamel is a glass-like porcelain which, when fused to steel sheet, forms a tough, permanent, colour-fast protection. As a cladding, vitreous enamelled sheet is most frequently used as a flat curtain walling or exterior panelling material. Profiled roof tiles and cladding sheets are, however, available to order. The durability and longevity of vitreous enamelled steel is its main feature, maintenance not being required. The material is impervious to chemicals, polluted atmospheres, frost, sun, dampness and abrasion, and will not support fire.

Pre-painted Sheet

One of the difficulties of painting cladding sheets on site is to ensure that the sheets are initially clean and free from dampness and that paint is applied in a uniform coating. Such difficulties can be avoided if the paint is applied under controlled factory conditions before delivery of the sheet. Pre-painted steel cladding sheet is now available. Such features as a uniform thickness of the correct quality paint properly bonded to the sheet can thus be assured.

Built-up Weatherproofing

Composite steel claddings comprising a trough sectioned profile with layers of weatherproofing, vapour seals and tough outside surface layers built up on the top

corrugated.

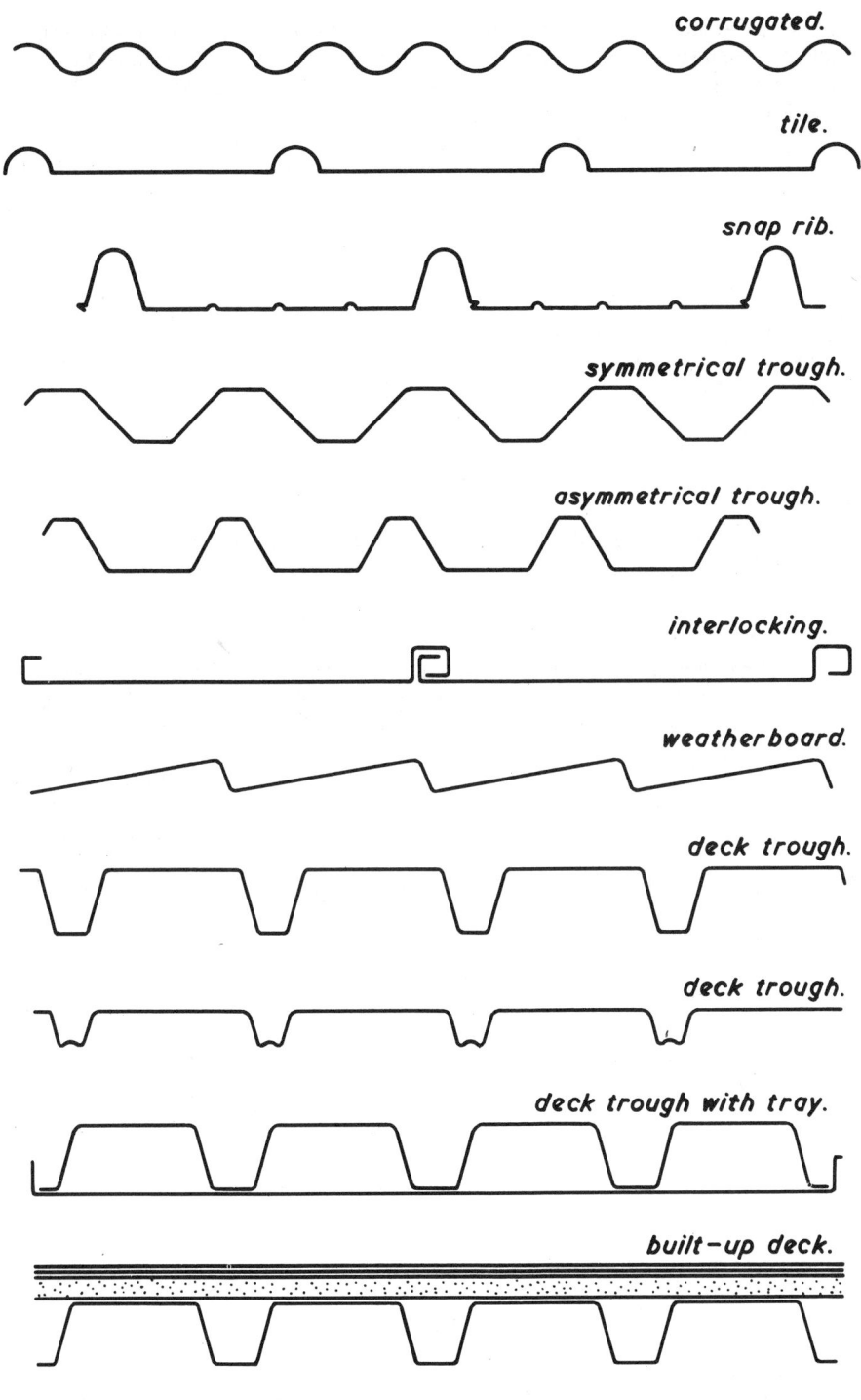

tile.

snap rib.

symmetrical trough.

asymmetrical trough.

interlocking.

weatherboard.

deck trough.

deck trough.

deck trough with tray.

built-up deck.

PROFILE SHAPES.   Fig. 1

surface of the sheet (see Fig. 1) are in widespread use, particularly for industrial buildings. Such systems are exceptionally durable and are strong enough to form horizontal roof decks able to carry pedestrian and other traffic. Thermal insulation is normally incorporated within the built-up layers. Since the vapour seals and thermal insulation are attached to the exterior surface of the cladding sheet, condensation problems are by-passed, for the steel remains on the warm, interior side of the deck.

Bitumen-Asbestos Felted Sheet

A common protective for cladding sheets on industrial buildings is a layer of blended bitumen and asbestos felt. The felt is tightly bonded to both sides of a corrugated or trough profiled sheet and forms a durable weather resistant protection with good fire-resisting properties. The felt can be coloured.

Asbestos Covered Sheet

Whilst steel claddings as a whole have a Class A rating for resistance to spread of flame, as specified in B.S. 476, asbestos covered steel sheets which offer exceptional resistance to fire are available. The profile is the standard 76.2 mm corrugated and the steel sheet is sandwiched between pressed-on layers of asbestos.

Types of Profile

A detailed description of all the cladding profiles manufactured in the United Kingdom, including details of manufacturers, coatings and design spans may be obtained from Strip Mills Division of the British Steel Corporation.

Profile development is increasing rapidly, spurred in part by developments in modular construction and by the greater range of coatings now available.

For details of and design data for all profiles other than the corrugated profiles described in the following pages, it is necessary to consult manufacturers' literature.

Specification for Galvanised Corrugated Cladding Sheets

Shape

A symmetrical, sinusoidal waveform, both edges of a sheet turning downwards from the crest of the outermost corrugation.

Dimensions and Coverage

TABLE 2

Pitch of Corrugation*	Depth of Corrugation*	Number of Corrugations Per Sheet	Overall Width of Sheet after Corrugating*	Coverage		
				1 Corrugation Side Lap	1½ Corrugation Side Lap	2 Corrugation Side Lap
mm	mm		mm	mm	mm	mm
76.2	19	8 10 12	660 812 965	610 762 914	572 724 876	533 686 838
127	32	5 6	711 838	635 762		

* Tolerances should be as given in Clause 12 of B.S. 3083.

TYPES OF PROFILE

Note 1. Sheet width tolerances are given in B.S. 3083.

Note 2. The most commonly used widths of the 76.2 mm profile are those with 8 and 10 corrugations. The wider sheets are more economical to use since less material is wasted in sidelaps; furthermore, the total number of fastenings will be less since the number used per sheet is normally the same for all widths.

Length

TABLE 3

Pitch	Length Range*	Usual Increments
mm	mm	mm
76.2	1 219 to 7 620	150
127	1 219 to 3 048	150

* Depending on gauge.

Certain economies can be effected with longer sheets, notably the reduced number of fixings required, the elimination of sheet wastage at endlaps, and the possibility of using flatter roof pitches where the sheet is continuous between ridge and eaves.

Strength and Stiffness

The resistance of a cladding sheet to longitudinal bending is proportional to the depth of corrugation and the gauge thickness. The strength of a sheet, as indicated by the yield stress in the outside fibres of the steel, is proportional to the section modulus of the profile. The stiffness of the sheet, as indicated by its deflection under load, is proportional to the moment of inertia of the profile section. Properties are as follows:—

TABLE 4

Pitch of Corrugation*	Gauge† B.S. 3083	Approximate Moment of Inertia per Metre Width	Approximate Modulus of Section per Metre Width
mm		cm^4	cm^3
76.2	14	10.65	11.29
	15	9.56	10.21
	16	8.46	9.13
	17	7.64	8.06
	18	6.69	7.01
	19	6.00	6.45
	20	5.32	5.37
	21	4.78	5.00
	22	4.32	4.46
	23	3.82	4.03
	24	3.41	3.60
	25	3.00	3.17
	26	2.73	2.85

* This dimension is designated as nominal to cover the slight differences between the results obtained by the various types plant used by manufacturers.
† Gauge to B.S. 3083 should not be confused with Birmingham Gauge.

STEEL SHEET ROOFING AND CLADDING

TABLE 4 (continued)

Pitch of Corrugation*	Gauge† B.S. 3083	Approximate Moment of Inertia per Metre Width	Approximate Modulus of Section per Metre Width
mm		cm^4	cm^3
127	14	30.44	19.08
	15	27.03	17.00
	16	23.89	15.05
	17	21.57	13.55
	18	18.98	11.93
	19	16.93	10.64
	20	15.01	9.46
	21	13.51	8.49
	22	12.01	7.58
	23	10.79	6.77
	24	9.82	6.07
	25	8.87	5.59
	26	8.05	5.05

* This dimension is designated as nominal to cover the slight differences between the results obtained by the various types of plant used by manufacturers.
† Gauge to B.S. 3083 should not be confused with Birmingham Gauge.

TABLE 5

MAXIMUM UNSUPPORTED SPANS OF CLADDING SHEETS ON PITCHED ROOFS

Gauge to B.S. 3083	Total Imposed Load (kN/m^2) normal to Roof Slope									
	0.50 kN/m^2		0.75 kN/m^2		1.00 kN/m^2		1.25 kN/m^2		1.50 kN/m^2	
	Corrugation		Corrugation		Corrugation		Corrugation		Corrugation	
	mm 76.2	mm 127	mm 76.2	mm 127	mm 76.2	mm 127	mm 76.2	mm 127	mm 76.2	mm 127
14	3.37 m (3.00 m)	4.72 m (4.20 m)	3.00 m (2.67 m)	4.27 m (3.80 m)	2.77 m (2.45 m)	3.95 m (3.50 m)	2.62 m (2.33 m)	3.75 m (3.33 m)	2.47 m (2.20 m)	3.50 m (3.11 m)
16	3.15 m (2.80 m)	4.50 m (4.00 m)	2.85 m (2.53 m)	4.05 m (3.60 m)	2.62 m (2.33 m)	3.75 m (3.33 m)	2.47 m (2.20 m)	3.30 m (2.93 m)	2.32 m (2.00 m)	3.15 m (2.80 m)
18	2.92 m (2.60 m)	4.20 m (3.73 m)	2.62 m (2.33 m)	3.67 m (3.25 m)	2.40 m (2.10 m)	3.22 m (2.85 m)	2.22 m (2.00 m)	3.00 m (2.67 m)	2.17 m (1.90 m)	2.77 m (2.45 m)
20	2.77 m (2.45 m)	3.82 m (3.40 m)	2.47 m (2.20 m)	3.22 m (2.85 m)	2.17 m (1.95 m)	2.85 m (2.50 m)	1.95 m (1.70 m)	2.50 m (2.20 m)	1.80 m (1.60 m)	2.40 m (2.10 m)
22	2.62 m (2.33 m)	3.37 m (3.00 m)	2.22 m (1.95 m)	2.92 m (2.60 m)	1.95 m (1.70 m)	2.50 m (2.20 m)	1.80 m (1.60 m)	2.32 m (2.00 m)	1.65 m (1.45 m)	2.10 m (1.85 m)
24	2.40 m (2.10 m)	3.15 m (2.80 m)	2.02 m (1.80 m)	2.62 m (2.33 m)	1.72 m (1.50 m)	2.25 m (2.00 m)	1.57 m (1.40 m)	2.02 m (1.80 m)	1.42 m (1.25 m)	1.87 m (1.65 m)
26	2.17 m (1.95 m)	2.85 m (2.50 m)	1.80 m (1.60 m)	2.40 m (2.10 m)	1.57 m (1.40 m)	2.22 m (1.80 m)	1.42 m (1.25 m)	1.87 m (1.65 m)	1.27 m (1.10 m)	1.72 m (1.50 m)

TYPES OF PROFILE

Notes:
1. The quoted maximum unsupported spans include an allowance for the dead weight of the cladding as laid.
2. Limiting fibre stresses: 115 N/mm² — 14 gauge; 93 N/mm² — 26 gauge; intermediate values by interpolation. ($M = WL/10$.)
3. Limiting deflection taken as Span ÷ 100. (Deflection = $3WL^3/384EI$.)
4. The spans quoted above do not apply to curved roofs, whose spans would normally be greater.
5. Where the sheets are used over single spans the figures in brackets should be adopted.

Curved Sheeting

Corrugated sheets wholly or partially curved along their length are available. Curvature radius limitations, which depend upon gauge, are as follows:—

TABLE 6

MINIMUM RADIUS OF CURVATURE

Gauge*	Minimum Radius of Curvature
	mm
16G	914
17G	914
18G	762
19G	762
20G	610
21G	457
22G	457
23G and over	305

* The gauge numbers specified above are in accordance with the accepted practice of the trade and are related to the number of sheets in 1 016 kg in accordance with Appendix B, B.S. 3083.

Pitch of Roofs

In order to ensure weather-tightness, the following minimum roof pitches for various types of steel roofing are recommended:—

TABLE 7

Roof Type	Pitch
1. Built-up felted roofs.	1° (1 in 60)
2. Long, corrugated sheets with no end laps between ridge and eaves.	5°
3. Long sheets in other profiles such as trough, tiles, etc., with no end laps between ridge and eaves.	5°
4. All unprotected, profiled sheets where end laps are necessary.	15°

The above recommendations are based upon accepted practice in the building industry. The modern trend in roofing design is towards flatter pitches, and it should be possible to use a pitch less than the standard minimum of 15° for standard roofs, although special sealing arrangements for end-laps would be necessary since, at low pitches, end-laps are vulnerable to rain penetration.

Sidelaps

Typical side-lap arrangements are shown in Fig. 2. A one-corrugation overlap is suitable for sheltered side-wall claddings, but not generally for roofing. A 1½ corrugation overlap is suitable for most roofs subject to moderate exposure: the 1½ effect is obtained by inverting alternate sheets. In cases where buildings are very exposed,

10/76·2mm corrugated steel sheeting.

6/127mm corrugated steel sheeting.

TYPICAL SIDELAP ARRANGEMENTS.

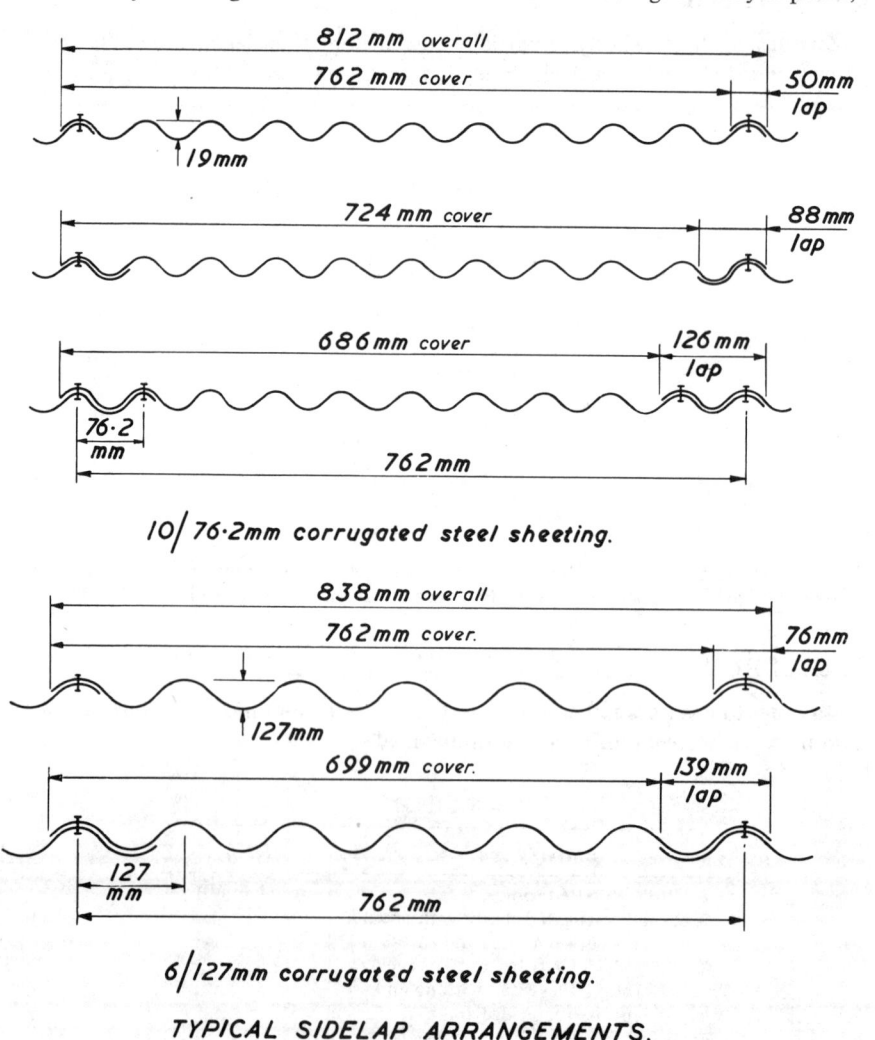

Fig. 2

a 2 corrugation overlap is recommended. The edge of a top overlapping sheet should always turn downwards to ensure maximum resistance to rain penetration.

In the case of the 127 mm corrugated profile and the various trough profiles 25 mm or more deep, a one corrugation side-lap is normally sufficient for roofs.

When cladding a building, it is advisable to lay sheets such that exposed side-laps are downwind from the prevailing wind.

Endlaps

Normal practice is as follows:

Side-wall Sheeting: 100 mm minimum when the lap is over a support purlin, otherwise 150 mm.

Roof Sheeting: a minimum of 150 mm for slopes of 20° and over. For flatter slopes, either a 225 mm lap or 150 mm with mastic sealing, especially where exposed to driving rain.

Gauge

Steel sheets are normally sold by the tonne, sheet thickness being defined by the number of sheets of a particular size in a parcel of a certain weight, as listed in B.S. 3083.

Gauges commonly used for moderate-sized buildings are 20–26. 26 Gauge galvanised corrugated sheets are acceptable in B.S. 2053, 'General Purpose Farm Buildings of Framed Construction', as a material suitable for cladding farm buildings.

The life of a galvanised sheet is proportional to the thickness of the zinc coating and not to the thickness of the steel. Once the zinc has been consumed, a 22-gauge sheet (say) will last only marginally longer than a 26-gauge sheet. The surface area covered, however, by a tonne of 26-gauge sheets is considerably greater than by a tonne of 22-gauge sheets.

Weight

TABLE 8

Gauge	Approx. Weight per m^2 of Galvanised Corrugated Sheet	Approx. Weight per m^2 of Cladding as Fixed
	N	N
16	160	180
17	135	165
18	121	145
19	106	130
20	95	115
21	86	105
22	77	95
23	70	85
24	64	75
25	55	65
26	48	60

Notes:
1. The weights per m² include side and end laps.
2. Where long sheets with no end laps are used, deduct about $3\frac{1}{2}$ per cent from the weight per m².
3. Add 5 N/m² for hook bolts, seam bolts and washers.
4. The table above is sufficiently accurate for design purposes, using either 76.2 mm or 127 mm profiles, but should not be used for estimating.

Layout of Sheeting

Accessories

Ridgecaps, corner and eaves closures, aprons, louvres and flashings are usually made from flat, galvanised sheet and bent to whatever angle is required. Such pieces are normally cut from standard width sheets of 300 mm, 375 mm or 450 mm in the same gauge as the cladding sheet. Zinc sheet (to B.S. 849) flashings of 21 S.W. gauge minimum may be used, or alternatively lead sheet flashings (to B.S. 1178) of Code No. 4 minimum weight. End-laps of 75 mm are usually sufficient, although the open lap end should face downwind from the prevailing wind.

General advice concerning the installation of all types of accessory is contained in C.P. 143, Part 2. Specifications for steel gutters and other fittings are laid down in B.S. 1091, which includes sizes, shapes and fixing methods.

Specifications regarding fixing accessories are laid down in B.S. 1494.

Numerous fastening methods are now available for attaching cladding sheets to purlins and for securing side-laps. A selection of fastening systems is shown in Fig. 4. 8 mm diameter hook bolts and 6 mm diameter galvanised bolts are still commonly used for fastening, respectively, sheets to purlins and side-laps. Other methods, however, such as self-tapping screws, rivetting, drive screws, stud welding, bullet studs and curtain wall studs are often used and are gaining favour.

Also gaining favour are plastic cappings to protect projecting bolt heads, and plastic sealing washers instead of the traditional felt type. Steel washers, usually diamond shaped, are used in conjunction with sealing washers.

Bolt holes are usually drilled through the crest of a corrugation, but in some instances, such as with bullet studs or curtain wall studs, trough fastening is used.

Loading

B.S. Code of Functional Requirements of Buildings, C.P. 3, Chapter V, 'Loading', contains basic loading data and gives guidance on loadings for various degrees of building exposure, building height and roof slope. It is recommended that the Code be studied carefully before the design of any building is carried out.

In general, all roof sheeting should be designed to resist the probable snow, wind, maintenance traffic and other anticipated loads. However, special provisions such as walkways may avoid the necessity of designing for loads occurring during maintenance.

Normally side wall cladding has only to resist moderate wind pressure or wind suction. Chapter V indicates, however, that wind pressures on the walls of tall buildings may be very high and indeed exceed the combined snow and wind loads on roofs.

897

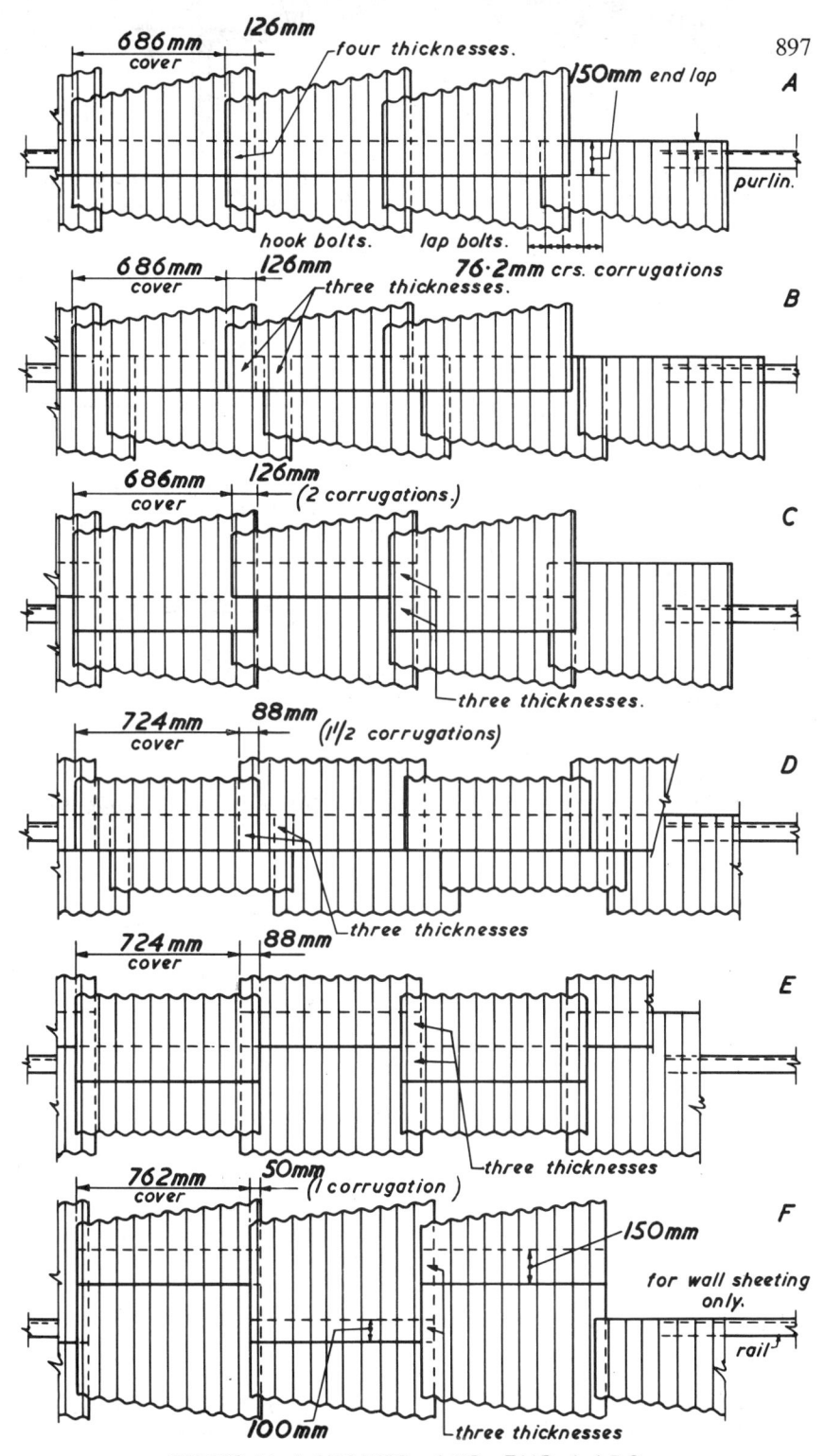

TYPICAL LAYOUTS AND END LAPS
OF 76·2mm CORRUGATED STEEL SHEETING.

Fig. 3

898

Seam Fastener.
6mm dia. galvanised steel bolt.
Placed at 450 centres for sidelaps.
plastic or felt washer.

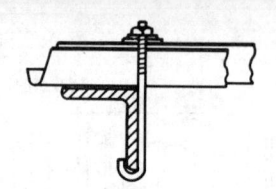

Purlin Fastener.
8mm dia. galvanised hook bolt.
Placed at 375 centres max.

(dimensions in mm)

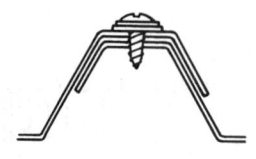

Seam Fastener.
Self tapping screw.

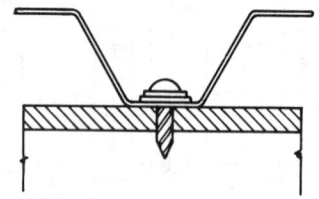

Purlin Fastener.
Drive screw.

Seam Fastener.
Blind rivet.

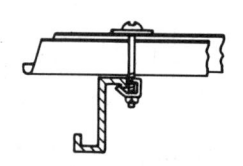

Purlin Fastener.
Galvanised bolt, movable hook.

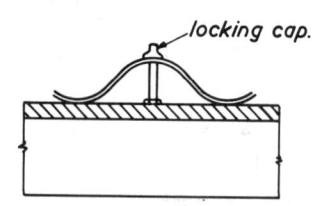

Purlin Fastener.
Stud welded bolt.
locking cap.

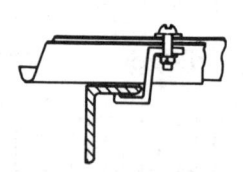

Purlin Fastener.
Sheetclip.

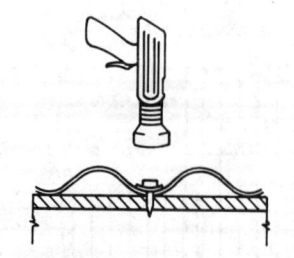

Purlin Fastener.
Pistol fired stud.

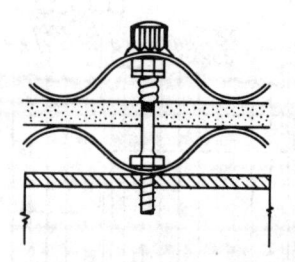

Purlin and insulation Fastener.
Curtain wall stud.

SELECTION OF FASTENING METHODS.

Fig 4

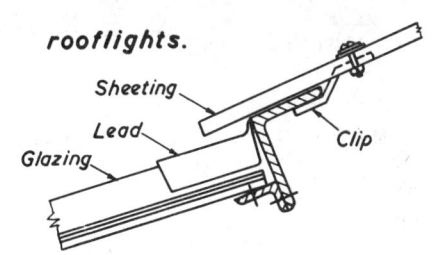

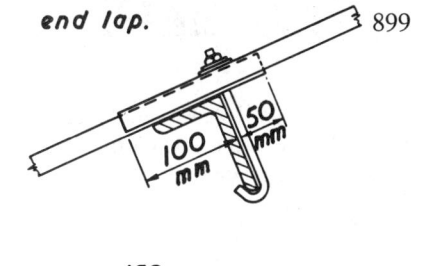

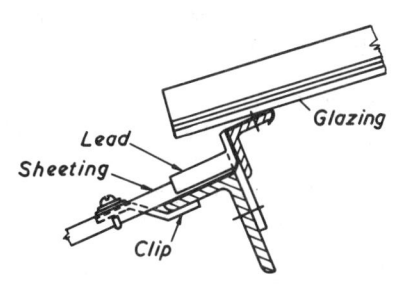

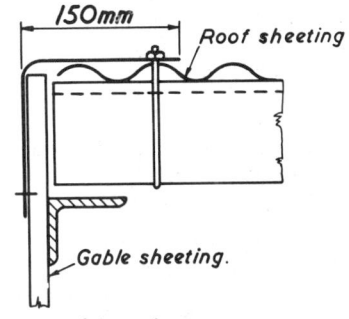

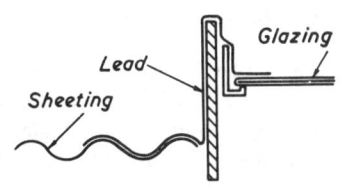

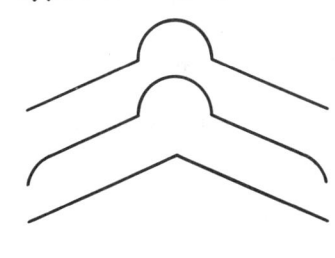

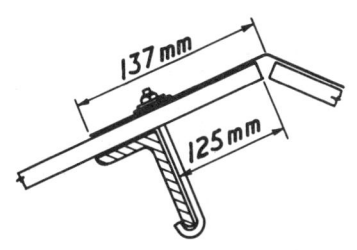

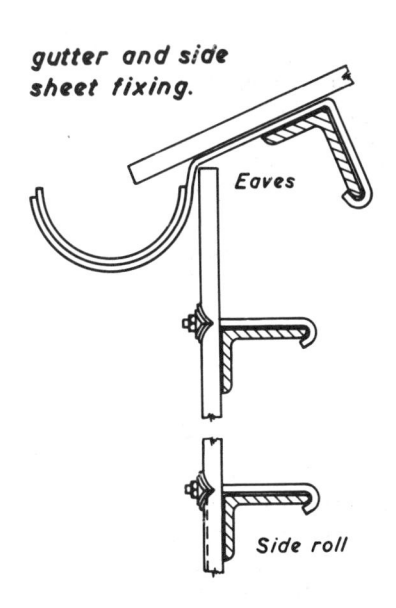

CLADDING ACCESSORIES.

Fig. 5

Design

C.P. 143, Part 2 gives guidance on the design and detailing of standard 76.2 mm (B.S. 3083) cladding. The Code, moreover, states that any profile is permissible which provides adequate strength for the particular loading conditions. It should be borne in mind that the Code of Practice applies to steel claddings generally. Where specific standards are available they should be used, e.g., B.S. 1754, Steel barns with curved roofs, and B.S. 2053, General purpose farm buildings of framed construction.

Good design should not only ensure a sound weathertight and economical roof but should also take into account the prevention of corrosion (see section on 'Maintenance').

C.P. 143, Part 2 includes recommendations regarding end and side-laps for sheets and methods of fixing. Typical sheeting layouts are shown in Fig. 3. Method A is customarily adopted.

Where long sheets are employed it is possible to take advantage of continuity of span. Table 5 refers to continuous spans, which should, in the case of simply supported sheets, be reduced by 10 per cent.

The use of long sheets (spanning two or more purlins) results not only in economies in material and elimination of end laps but speeds erection and simplifies the layout of sheeting. Sheets are normally supplied in 150 mm increments of length up to 7.620 m. Where transport can be arranged, however, longer sheets are obtainable.

Standard 76.2 mm corrugated sheets may be designed either from first principles or by reference to Table 5 which gives purlin spacings for 0.25 kN/m^2 increments of loading between 0.5 and 1.5 kN per m^2. Approximate section properties of sheeting are shown in Table 4. For estimating and design purposes the approximate weights per square metre of 76.2 mm and 127 mm corrugated sheet are listed in Table 8.

C.P. 143, Part 2 indicates the maximum permissible stresses for the usual gauges of B.S. 3083 sheeting (see footnotes to Table 5).

The maximum permissible deflection for B.S. 3083 sheeting is given in the Code of Practice as the span ÷ 100. This is for profiled cladding sheet. Where built-up weather-proofing layers and/or thermal insulation are to be attached to the profiled sheet, a deflection limit of span ÷ 240 is considered adequate.

Where special profiles are to be designed, it is advisable to use the stress and deflection limits quoted above.

Thermal Insulation

The cost of heating buildings is high, and the Ministry of Power has long fostered the idea of economy through adequate thermal insulation.

A great variety of materials, ranging from the traditional fibreboard to modern cellular plastics, can be used for insulating a steel cladding sheet which of itself has little insulation value. The Ministry of Power's Thermal Insulation (Industrial Buildings) Act of 1957 lays down minimum U value requirements for the roofs of industrial buildings. As a result, a minimum U value of 1.7 W/m^2/h/°C has become a standard insulation requirement for commercial buildings where people are working in a heated environment. U values lower than this can readily be obtained, however.

Insulation material can be applied in one of three ways: (i) on the exterior surface of the cladding sheet; this is the method commonly used in built-up cladding

systems and has certain basic advantages, such as weather protection for the sheet, easy fastening and the avoidance of condensation problems. (ii) Sandwiched between cladding sheet and support purlins. (iii) Attached to the inside of the support purlins: of particular advantage where unbroken internal wall or ceiling surfaces are desired.

Careful consideration must be given to the problem of cold bridging, which limits the efficiency of an insulation system, and to condensation which is best avoided through careful design, the ventilation of potentially cold surfaces, and the use of proper vapour barriers. Such barriers inhibit the movement of moist air through insulation materials, most of which are porous, and therefore restrict the flow of moist air against cold surfaces.

Thermal insulation, however, is a specialised subject and guidance either from experts or from the extensive technical literature on the subject should be sought.

Examples of typical U values of finished steel claddings:

TABLE 9

Component Materials	U Value $W/m^2/h/°C$
Corrugated steel sheet; 25 mm thick mineral wool pad; 6 mm asbestos board backing	0.85
Corrugated steel sheet; 25 mm expanded polystyrene bonded on	1.00
Corrugated steel sheet; 150 mm thick compressed straw slab below purlins	1.10
Built-up steel deck comprising; troughed steel sheet; 25 mm asbestos fibre board; rubberised/bituminous weatherproof felt.	1.10

Maintenance and Sheet Protection

The life of a galvanised sheet is dependent on the nature of the atmosphere to which it is exposed. It is recognised that galvanised cladding sheets subjected to urban or marine atmospheres deteriorate more quickly than those inland rural areas and require, therefore, paint treatment at regular intervals.

Other types of steel claddings, such as those with plastic or vitreous enamel coatings or those with built-up weatherproof felting, need no such maintenance.

Galvanised sheets will last indefinitely if properly maintained.

New galvanised sheets may be satisfactorily painted straight away. Dirt and any residual grease should first be removed with white spirit, naphtha, etc., after which calcium plumbate primer should be applied. Either a further coat of primer and a finish coat of alkyd gloss, can then be applied, or simply one or two finishing coats of alkyd gloss. A total paint thickness of about 25 mm is desirable.

In areas of exceptionally corrosive atmospheres where paint manufacturers may advise the use of special paints such as chlorinated rubber, epoxy, polyurethane, etc., plumbate primers cannot be used. A special 'etch' primer such as a phosphoric acid wash is then necessary so as to key the galvanised surface, after which normal paints can be used. But the etch primer must be thoroughly washed off first.

Galvanised sheets can alternatively be erected and allowed to weather until chemically inactive, then painted. No special primers are then necessary although

plumbate primers are often used since they are hardly different in price from standard chromate or red lead primers. It is essential, however, that before painting, the sheets should be quite dry and completely free from dirt, oil, scale and other grime.

The majority of paint failures are attributable to inadequate preparation and such carelessness as painting damp sheets.

Careful cleaning and preparation particularly applies to old sheets where the zinc coating has been consumed and rust has appeared. In areas of rust, cleaning (by sand blasting, mechanical wire brushing, power hosing, etc.) down to the bare metal is essential. Then a primer and two topcoats can be applied.

Paint manufacturers should generally be consulted, especially when unusually corrosive atmospheres are likely to be encountered.

British Standards

B.S. 476:Part 1:1953 — *Fire tests on building materials and structures.*
B.S. 476:Part 2:1955 — *Flammability test for thin flexible materials* (withdrawn).
B.S. 476:Part 3:1958 — *External fire exposure roof tests.*
B.S. 476:Part 4:1970 — *Non-combustibility test for materials.*
B.S. 476:Part 5:1968 — *Ignitability test for materials.*
B.S. 476:Part 6:1968 — *Fire propagation test for materials.*
B.S. 849:1939 — *Plain sheet zinc roofing.*
B.S. 1091:1963 — *Presses steel gutters, rainwater pipes, fittings and accessories.*
B.S. 1178:1969 — *Milled lead sheet and strip for building purposes.*
B.S. 1449:Part 1B:1962 — *Carbon steel sheet and coil, rolled by the continuous process.*
B.S. 1494:1964 — *Part 1 Fixing accessories for building purposes.*
B.S. 1754:1961 — *Steel barns with curved roofs.*
B.S. 2053:1965 — *General purpose farm buildings of framed construction.*
B.S. 2989:1967 — *Hot-dip galvanised plain steel sheet and coil.*
B.S. 3083:1959 — *Hot-dipped galvanised corrugated steel sheets for general purposes.*

C.P. 3, Chapter V:1952 — *Code of functional requirements of buildings: Loading.*
C.P. 3, Chapter V:Part 1:1967 — *Dead and imposed loads.*
C.P. 3, Chapter V:Part 2:1970 — *Wind Loads.*
C.P. 143:Part 2:1961 — *Sheet roof and wall coverings. Galvanised corrugated steel.*

36. STEEL WINDOWS AND PATENT GLAZING

Steel Windows

THREE different types of steel windows commonly used are:

(1) Metal Casement Windows and Casement Doors.
(2) Steel Windows for Industrial Buildings.
(3) Agricultural Windows.

The windows described in the first classification are of the domestic variety, and, although they are used in the administrative blocks of industrial buildings, they are normally set in brickwork or masonry. Consequently, they are not considered here. The windows in the second class are those which are used in the production blocks, boiler-houses, etc., where they are frequently attached directly to steel frames.

Although the methods shown are by no means exhaustive, the diagrams in Fig. 1 demonstrate some of the methods of fixing the windows and show how the lead flashings may be arranged.

Patent Glazing

General

The following information on Patent Glazing must be considered as of general use and more specific information on individual systems must be obtained from the manufacturer concerned.

While the sheets of glass used for patent glazing are frequently 600 mm wide, the spacing of the glazing bars may be varied according to design conditions and as glass can be supplied in other widths it is a matter for the designer to adopt the most suitable dimensions for any particular application.

Minimum Slopes

Orthodox patent glazing can be fixed with complete success at any slope between the vertical and the recommended minimum pitch of 20°. The lower limit is dictated by two separate factors which are common to glass disposed at shallow pitches:

(*a*) the tendency of condensation to collect more readily into large drops and thus fall directly from the glass.
(*b*) in exposed positions rainwater sluggishly drained away on the outside surface tends to be blown under the top flashings.

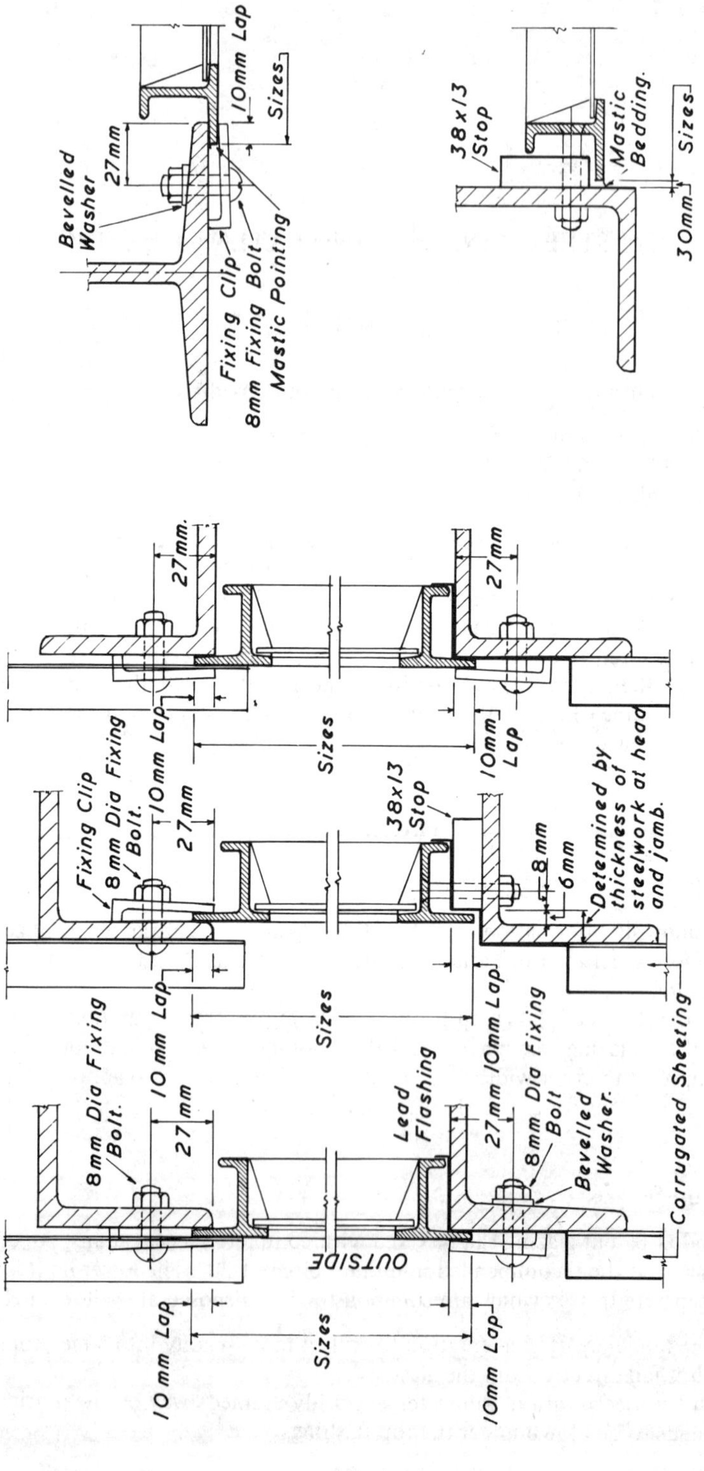
FIXING TO STEELWORK.
Bevelled Washer
27mm
10mm Lap
Sizes
Fixing Clip
8mm Fixing Bolt
Mastic Pointing
38x13 Stop
Mastic Bedding.
30mm
OUTSIDE.
10 mm Lap
8mm Dia Fixing Bolt.
27 mm
Sizes
OUTSIDE
Lead Flashing
27 mm
10mm Lap
8mm Dia Fixing Bolt
Bevelled Washer.
Corrugated Sheeting
Fixing Clip
8 mm Dia Fixing Bolt.
10mm Lap
27 mm
38x13 Stop
8 mm
6 mm
Determined by thickness of steelwork at head and jamb.
27mm.
27mm
10mm Lap

Fig. 1

Shallower slopes of patent glazing are successfully installed under certain conditions using special sealing techniques, but the advice of a patent glazier should be sought before a pitch lower than 20° is decided upon.

Mid-Roof and North-Light Glazing

Methods of fixing mid-roof and north-light glazing are shown in Figs. 2 and 3.

Mid-Roof Glazing

Fig. 2

North Light Glazing

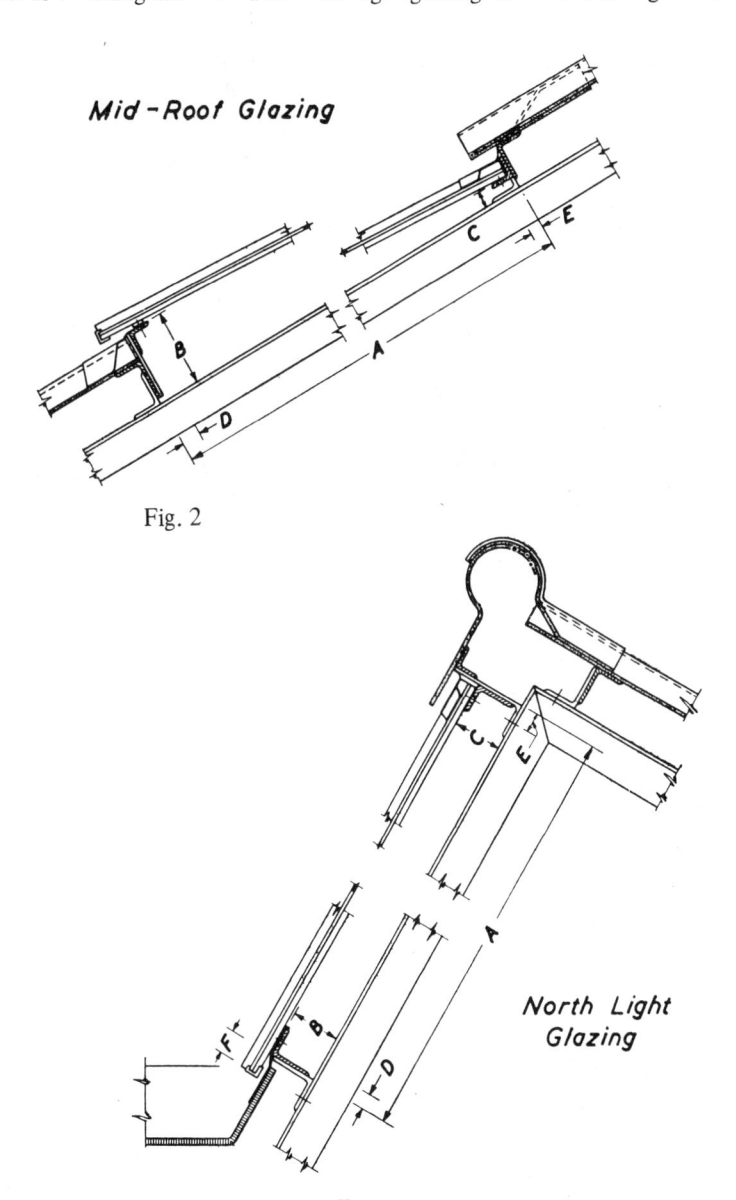

Fig. 3

STEEL WINDOWS AND PATENT GLAZING

The dimensions to specify are as follows:

	Mid-roof Glazing	North-light Glazing
Pitch	Roof pitch on rafter back.	Roof pitch on rafter back.
A	Back to back of purlin cleats.	Back to back of purlin cleats.
B	Height—lower purlin above rafter back.	Height—lower purlin above rafter back.
C	Height—upper purlin above rafter back.	Height—upper purlin above rafter back.
D	Back of purlin cleat to lower fixing hole.	Back of purlin cleat to lower fixing hole.
E	Back of purlin cleat to upper fixing hole.	Back of purlin cleat to upper fixing hole.
F		Overhang required.
Run	Total run of glazing between end rails.	Total run of glazing between end rails.
Height	Height of glazing above ground level.	Height of glazing above ground level.

Ends of patent glazing runs may be finished in various ways, a typical example being shown in Fig. 4. The spacing for the fixing bolts should not exceed 400 mm.

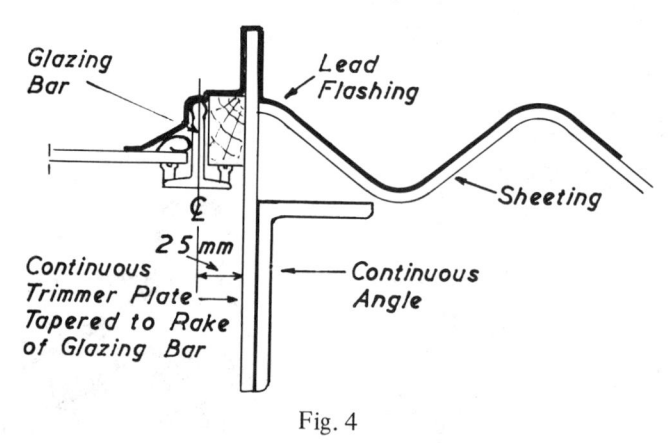

Fig. 4

Vertical Patent Glazing

Recent years have seen an increasing emphasis on glazed wall construction, and vertical glazing has achieved popularity in meeting this demand. Typical methods of fixing vertical glazing are shown in Fig. 5.

Patent Double Glazing

The modern need for fuel economy has increased the demand for double glazing owing to its greater thermal insulating qualities and a number of proprietary systems are available. The use of patent double glazing reduces heat losses by approximately 50 per cent as compared with single glazing.

Condensation is also eliminated or reduced, thereby offering advantages where falling droplets could be a nuisance or could cause corrosion or spoilation of stock or equipment.

907

TYPICAL DETAILS FOR VERTICAL GLAZING

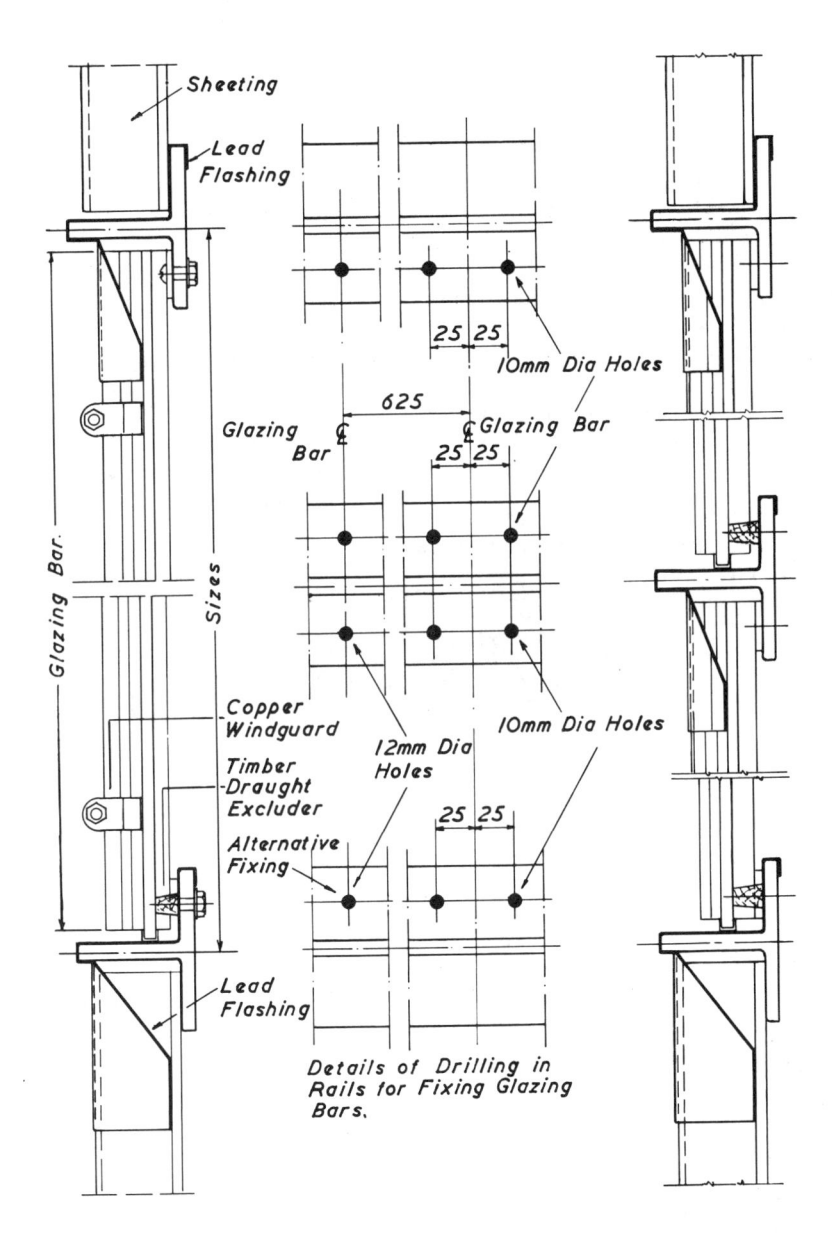

Fig. 5

Weight per Square Metre

Patent glazing is usually reckoned in weight at 30 kg/m² for the lead-clothed bar system. This figure is inclusive of the weight of of 6 mm thick glass and envisages the bars being spaced at approximately 600 mm centres.

Ventilation

Ventilation, either controllable or permanent, is often required in patent glazing, and, in the former type, the 'top hung' method has become general. Most makers are prepared to supply ventilators either in separate units or continuous ranges. They can be constructed in steel, galvanised or lead covered.

Since these ventilators are situated in remote positions, mechanical opening gear is frequently necessary to operate them and a variety of methods, manually, electrically or hydraulically controlled, is available.

Details of construction both for the ventilators and the operating gear differ considerably, and early consultation with the makers is recommended to ensure that satisfactory preparations are made for fixing them.

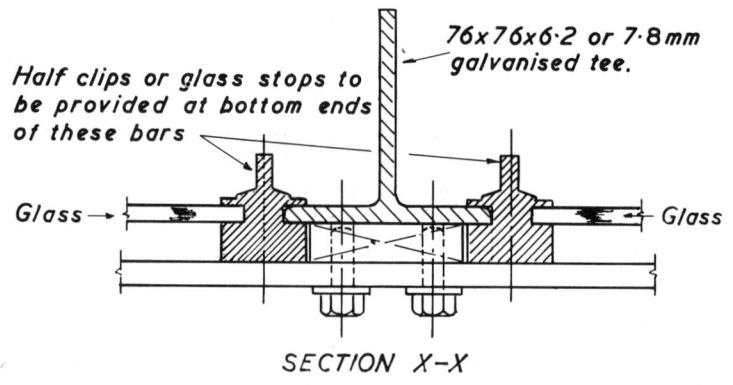

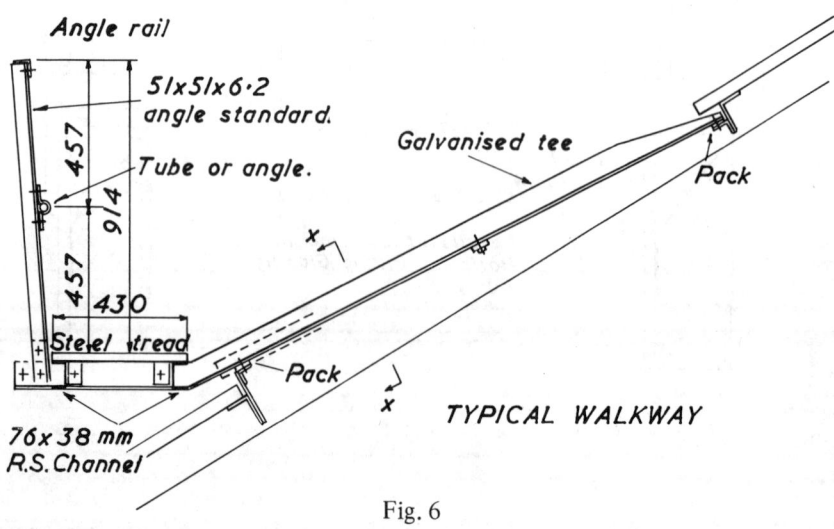

Fig. 6

Walkways

The Code of Practice, C.P. 145, for patent glazing, recommends that walkways should not be attached to ordinary glazing bars, but difficulties often arise in attaching them to, or through, other forms of roof covering. An illustration is given in Fig. 6 which shows a method by which both these objections are avoided.

This consists of introducing 'T'-bearers fixed direct to the purlins between pairs of closely spaced glazing bars, so that additional loads are not imposed on the bars. The bearers extend beyond the lower edge of the glazing and are cranked to form a horizontal cantilever support for the walkway. To simplify the design of purlins and to minimise the number of bearers required, the latter are intended to be fixed immediately over the roof trusses.

The major dimensions have been interpreted in accordance with the Factories Act and Building Regulations. It should, however, be borne in mind that such walkways are intended as working platforms only, and not as gangways for the passage or stacking of materials which would, under the Building Regulations, necessitate heavier and wider construction.

Glass

The varieties of glass which are used in patent glazing are:

(*a*) 6 mm wired cast (hexagon mesh);
(*b*) 6 mm Georgian wired cast (square mesh);
(*c*) rough cast;
(*d*) clear plate;
(*e*) heat-absorbing glass;
(*f*) the heavier types of sheet glass;
(*g*) 6 mm diamond wired;
(*h*) toughened cast, plate and sheet;
(*j*) toughened heat absorbing;
(*k*) toughened coloured cast and plate;
(*l*) laminated.

It may perhaps be mentioned here that wire-reinforced glass has achieved considerable popularity because of its ability to delay the spread of fire and the additional safety factor which it confers. This is not, as is sometimes supposed, due to any greater resistance to breakage but rather to the effect of the wire mesh in restraining the glass from falling after it has once been broken. Glazing bars are usually spaced about 600 mm from centre to centre and, although wider spacing is occasionally employed, it should only be specified with due caution and recognition given to the increased difficulties of obtaining and handling the wider squares of glass involved. Glazing bars should not normally exceed 3 m in length. Here, again, convenience of handling the glass is one of the criteria, but it is also important to remember that the cost of glass replacements in these sizes is proportionately high.

Lead Flashings

Lead flashings are a necessary adjunct to patent glazing and generally 20 kg/m^2 lead flashings are used, but in exposed situations the weight should be increased to 24 kg/m^2.

37. WALLS

General

THIS chapter describes the basic features of external brick walls, curtain walling and precast concrete facing panels. In addition, the basic rules for the determination of wall thicknesses are set out for single-storey buildings and typical fixing details for both curtain walling and precast concrete facing panels are given. Mention is also made of other aspects of wall construction which are of interest to the steel designer such as clearances, dimensional tolerances, fire resistance and jointing.

Design of External Brick Walls for Single Storey Buildings

Introduction

The necessary thickness of brick walls can be determined by one of two methods:

(a) The thickness of the wall can be related to the height and length of the wall.
(b) The thickness can be determined in relation to the load to be carried by the wall in conjunction with certain specified permissible stresses.

Method (a) is dealt with in the relevant clauses of the Ministry of Housing and Local Government, Series IV, 'Model Byelaws', 'Buildings' (1953).
Method (b) is dealt with in C.P. 111:1970, 'Structural Recommendations for Loadbearing Walls', Part 2. Metric Units.

Design

The following example is based upon two typical methods of construction for single-storey sheds (for details of the buildings see Figs. 1–3 incl.):

(1) Portal frame construction where the side wall brickwork is non-loadbearing, but the gable end wall supports the purlins.
(2) Roof truss construction where the side wall brickwork supports the roof truss and the gable end wall the purlins.

The design is carried out in both cases by Method (a). However, where large point loads can occur, as in the case of roof truss construction, the stability of the supporting wall should be checked against Method (b).

The relevant points relating to the design of such single-storey buildings are summarised below.

The height of the wall shall be measured from the base to the highest part of the wall, or in the case of a storey comprising of a gable, to half the gable height. Hence the mean height of the gable is:

$$6.0 \text{ m} + \frac{4.5 \text{ m}}{2} = 8.25 \text{ m}.$$

WALLS

BRICKWORK SIZES FOR ROOF TRUSS CONSTRUCTION

Fig. 1

Fig. 2

BRICKWORK SIZES FOR PORTAL FRAME BUILDING

25mm CLEARANCE BETWEEN FACE OF WALL AND PORTAL FRAME

PADSTONE

225, 1125, $L_1/4$, $L_1 = 4.5m$, $L_1/2$

$L_2/6 = 1406.3$ min

MEAN HEIGHT OF GABLE WALL ÷ 14 = 675

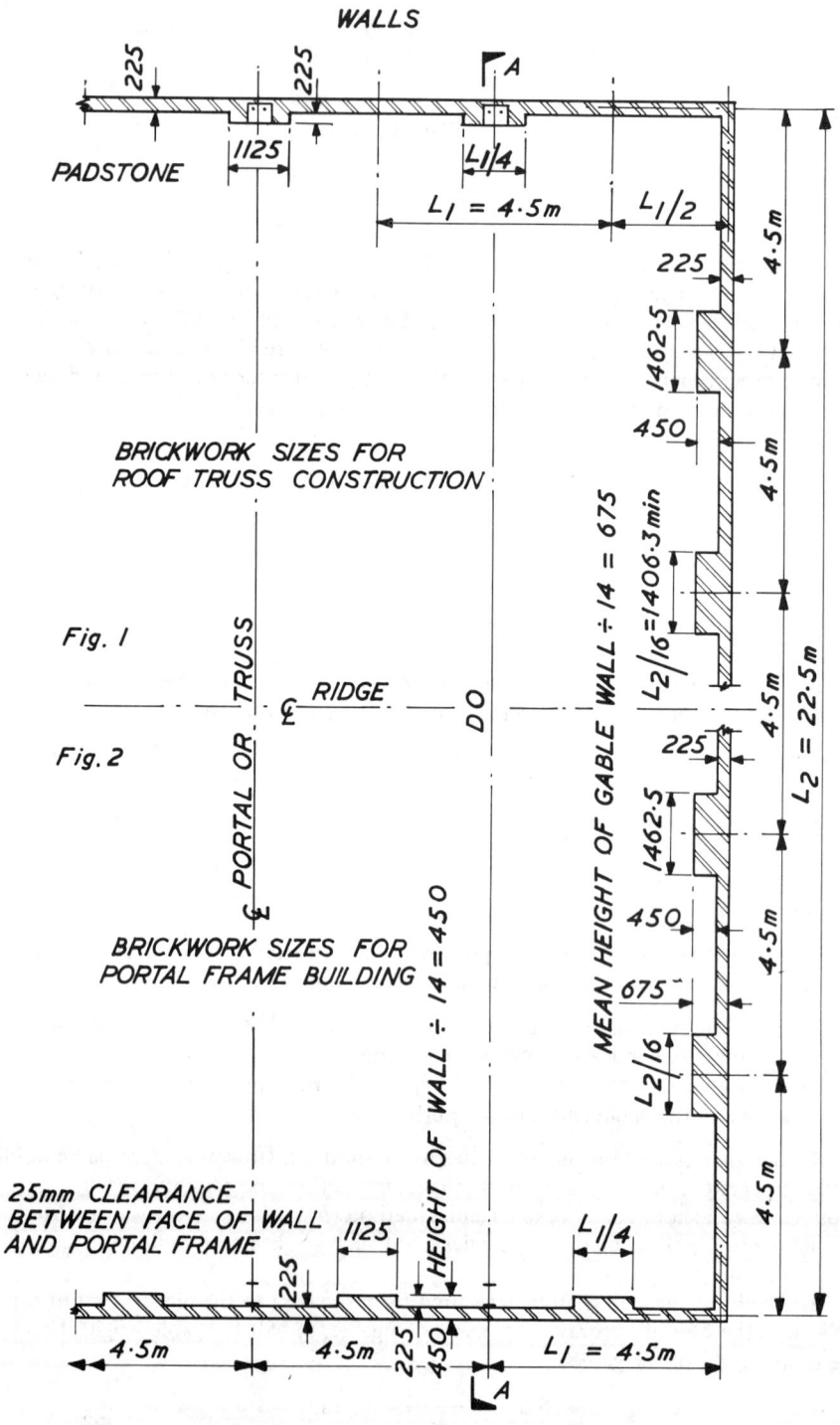

€ PORTAL OR TRUSS

€ RIDGE

DO

HEIGHT OF WALL ÷ 14 = 450

$L_2/6$

$L_2 = 22.5m$

4.5m (repeated), 225, 1462.5, 450, 675

all dimensions in mm unless noted

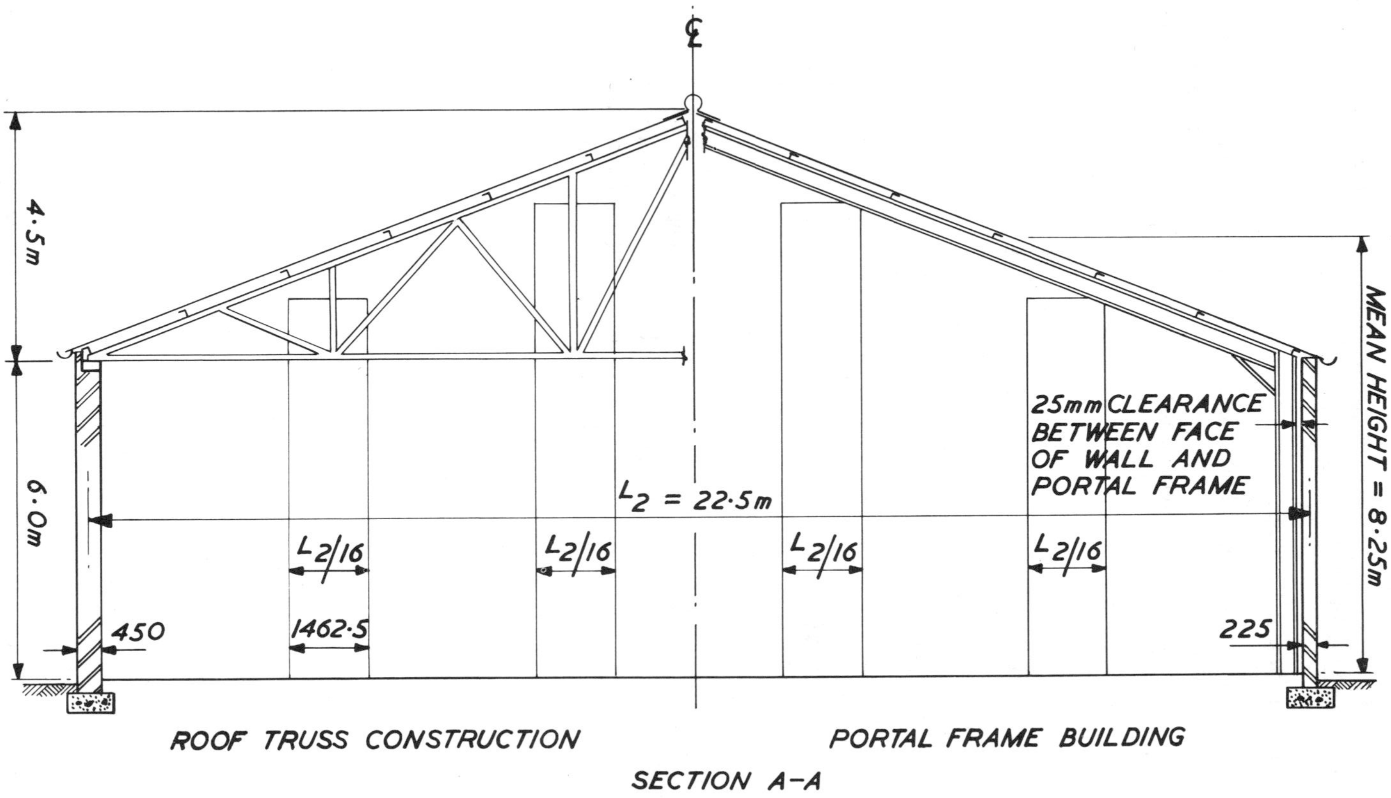

MEAN HEIGHT = 8·25m

25mm CLEARANCE BETWEEN FACE OF WALL AND PORTAL FRAME

225

$L_2/16$

$L_2/16$

$L_2 = 22·5m$

$L_2/16$

$L_2/16$

1462·5

450

4·5m   6·0m

ROOF TRUSS CONSTRUCTION   PORTAL FRAME BUILDING

SECTION A-A

Fig. 3

all dimensions in mm unless noted

914 WALLS

All measurements for the length of walls shall be made from the centre of the return walls, piers or buttresses.

Definition of Pier or Buttress

Piers and buttresses shall have dimensions (1) not less in thickness (to include the thickness of the wall) of 3 × wall thickness, (2) not less in breadth than ½ × wall thickness, (3) can be of height extending upwards from the base to within a distance from the top of the wall equal to 3 × wall thickness. (Fig. 4.)

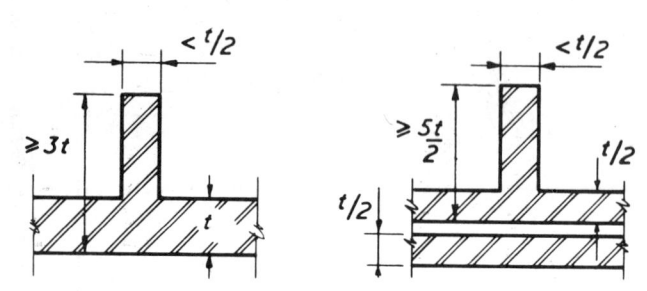

MINIMUM DIMENSIONS FOR PIERS AND BUTTRESSES.

Fig. 4

Numerical Example. The thickness of any wall of a building of the warehouse class shall not be less than 1/14 of the height (H). However, for buildings of only one storey, the thickness may be taken as 225 mm with suitable reinforcing walls of thickness $H/14$. The reinforcing walls shall have an aggregate length of 1/4 of the total wall length. For practical purposes, all wall thickness measurements are taken to the nearest half brick size.

For the side wall in the example:

$$\frac{\text{Height}}{14} = \frac{6.0 \text{ m}}{14} = 450 \text{ mm}$$

$$\frac{\text{Length}}{4} = \frac{4.5 \text{ m}}{4} = 1\ 125 \text{ mm}$$

Hence, 450 mm walls are placed at 4.5 m centres; are of length 1 125 mm and have 225 mm walls between them, as shown in Figs. 1 and 2.

For the gable end walls:

$$\frac{\text{Height}}{14} = \frac{8.25 \text{ m}}{14} = 675 \text{ mm}$$

$$\frac{\text{Length}}{4} = \frac{22.5 \text{ m}}{4} = 5\ 625 \text{ mm spread over 4 reinforcing walls.}$$

Therefore, the length of the reinforcing wall = 1 462.5 mm is 675 mm thick, spaced at 4.5 m centres with 225 mm walls between. These reinforcing walls fall inside the category of a pier and can be stopped short in height a distance of 3 × wall thickness from the top of the building, i.e., 3 × 225 mm = 675 mm.

Cavity Walls

Cavity walls have a thickness equivalent to the sum of the two leaves as shown in Fig. 4. The cavity itself shall not be less than 50 mm and no more than 75 mm. Ties should be placed at 900 mm horizontal and 450 mm vertical centres.

Curtain Walling

Curtain walling in its broadest sense, is any form of walling which is non-load-bearing, and recent years have seen considerable developments in this field. There has been an increasing tendency to replace brick and stone walls by thin sheet materials in large panels or by lightweight frameworks with various forms of infilling. The main function of any form of walling is protection against the weather. In addition to providing this, curtain walling also affords rapid erection, fixing from the inside of the building, a saving in foundation loads, increased floor areas and good insulating properties.

Requirements

Ideally, curtain walling should be durable, weatherproof, simple to attach to steelwork or to a floor, attractive in appearance, self-cleaning, easily removable, easily maintained and have sufficient strength to withstand wind pressure.

There are many types on the market which possess some or all these requirements. The panel frames are made of steel, aluminium, or timber, but the infilling includes materials such as thin stone, asbestos cement, porcelain enamelled steel, aluminium sheet, stove enamelled glass, timber, plastic sheet, P.V.C. coated steel and various forms of sandwich construction.

Jointing

Joints must be completely waterproof and yet allow for thermal and structural movement which can be as much as 25 mm in 45 m: allowance should also be made for the building tolerances. Obviously the larger the panels, the less the number of joints required, but fabrication and erection problems then occur. Broadly speaking, there are three different ways of sealing joints:

1. Rubber or plastic strips.
2. Clamping or interlocking connections.
3. Plastic jointing compounds.

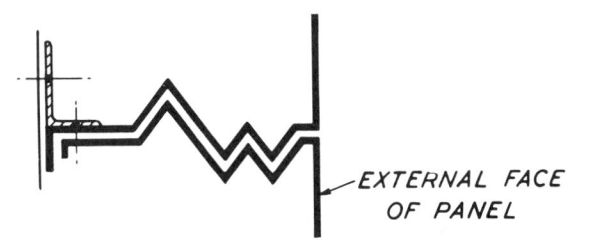

Fig. 5

A good example of simple interlocking design is shown in Fig. 5.

All manufacturers have devised their own method of jointing, but these are too numerous to list.

Insulation

The 'semi-curtain wall' which is frequently employed, embodies a separation between the outer skin and the thermal insulation behind it. The insulating backing is generally built up from hollow blocks, clinker blocks or cellular concrete. This type of wall must also be provided with a ventilated air cavity and a vapour barrier to prevent internal condensation. With the 'integral curtain wall' on the other hand, all these items are incorporated in a single prefabricated composite unit.

Fixing to the Framework

The fastening devices must secure the panels rigidly, even if the structure undergoes deformation due to temperature variations. Moreover, they should be corrosion resistant and as far as possible fireproof. Scaffolding should be kept to a minimum and hence internal fastenings are to be preferred. As there are tolerances in the manufacture of the cladding units and in the structure as a whole, fixing devices must be capable of masking and compensating for dimensional discrepancies.

Fire Resistance

Fire protection requirements vary considerably with the location, size and use of the building, but most regulations call for a minimum rating of 1 hour for the external walls of buildings. Most curtain wall systems have a low fire resistance and this has resulted in the widespread adoption of a back-up wall behind the cladding Fig. 6. The point to remember in design is to prevent flames which are escaping out of a window from spreading to the floor above.

Precast Concrete Panels

Introduction

Concrete facing panels can be an attractive and economical means of cladding many different types of multi-storey structure. They have been widely specified in recent years in the construction of blocks of flats, offices and school buildings. Although they form an integral part of a number of proprietary building systems, large quantities are also purpose made. Concrete facing slabs are also used as infilling panels below windows and as an external cover to brickwork, lightweight blocks and *in situ* concrete: such applications, however, are not widespread.

Surface Finish

As both concrete and aggregate can be obtained in different colours, many varieties of finish are possible. Variations in texture depend on the size of aggregate

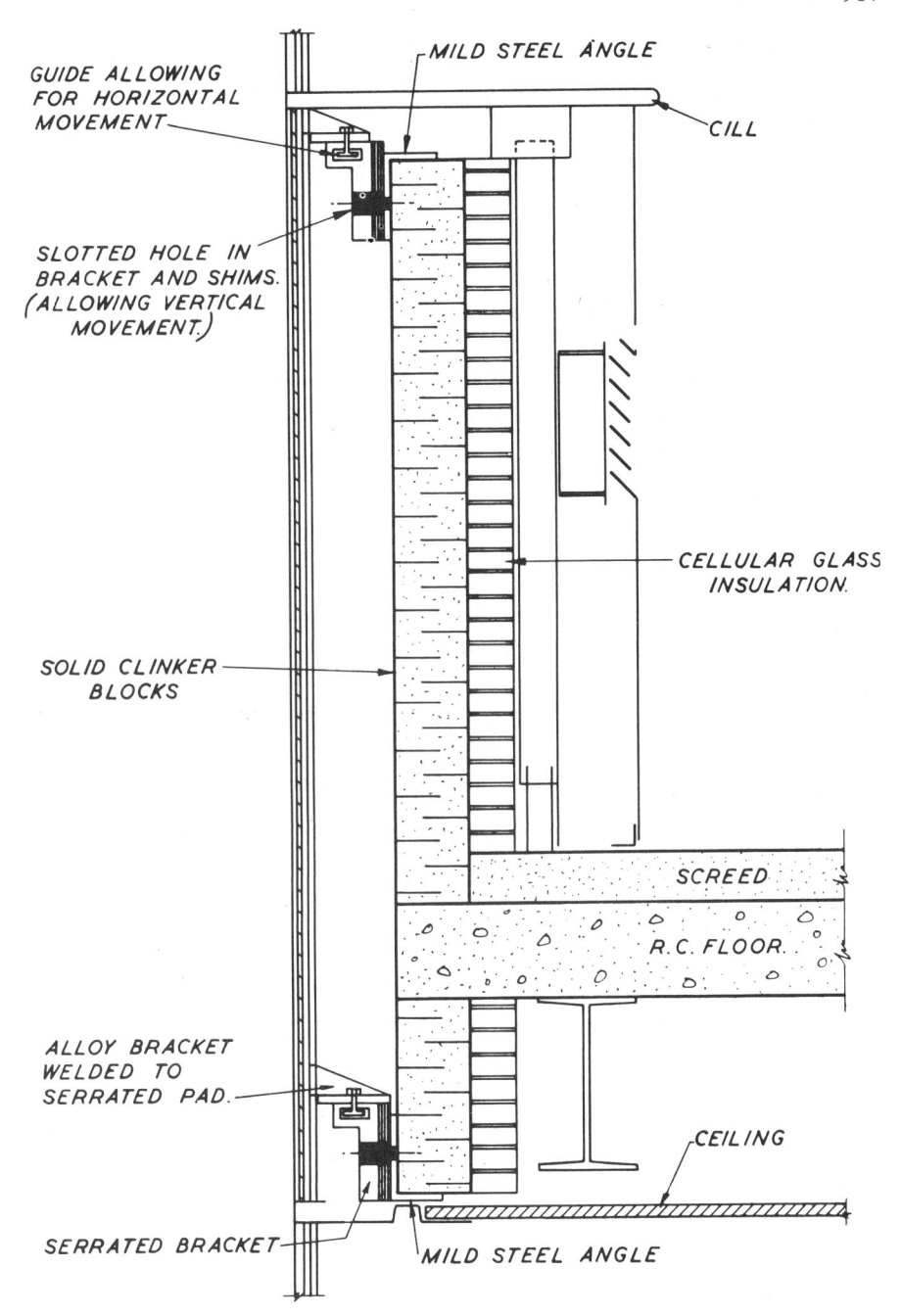

TYPICAL CURTAIN WALLING INSTALLATION.

Fig. 6

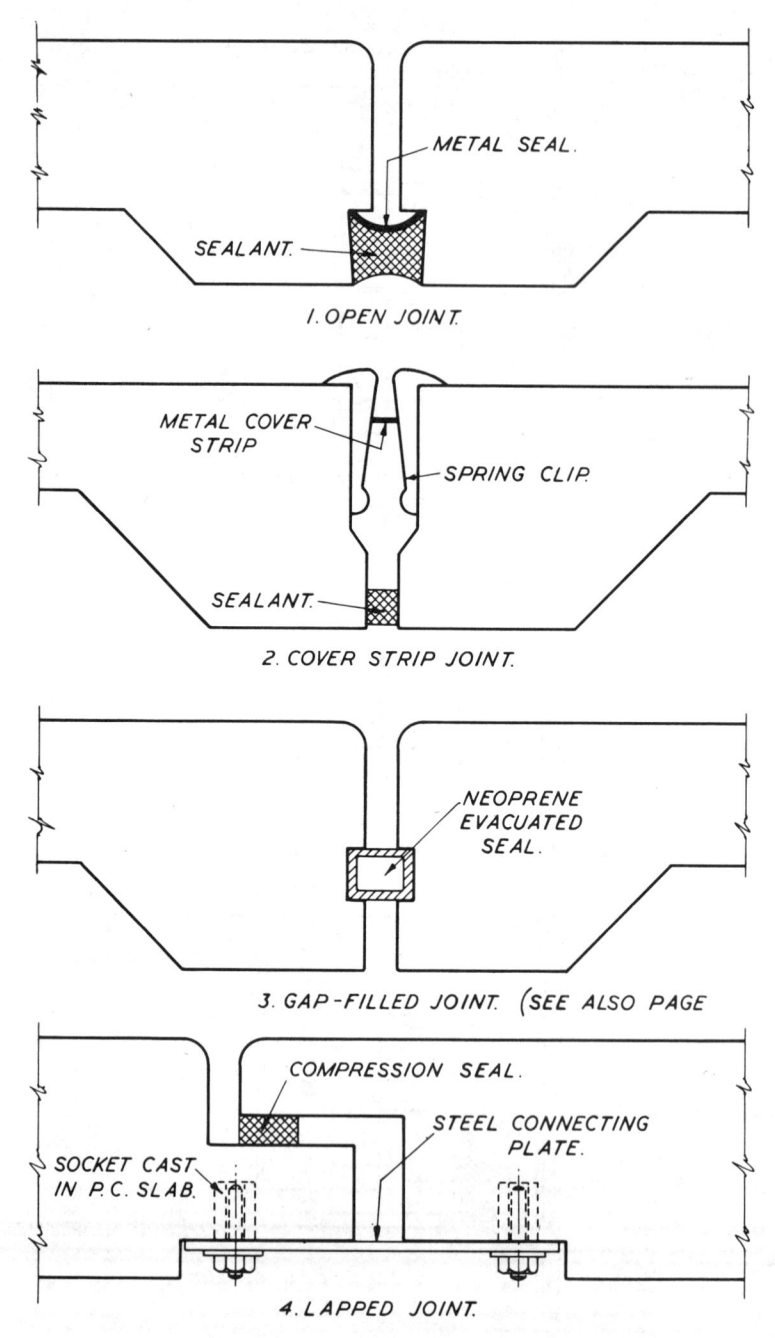

TYPICAL VERTICAL JOINTS
FOR PRE-CAST CONCRETE PANELS.

Fig. 7

used (which can range from 6 mm to 50 mm) and on the final treatment given to the face of the slabs. For example, by grinding with carborundum, a smooth face is obtained and a medium texture is possible if, before hardening, the surface of the concrete is sprayed with water. Various patterns may be obtained by profiling the slabs and/or applying a special treatment to the joints.

Maximum Sizes and Tolerances

Panels should not weigh more than 55 kg unless mechanical lifting aids are available. When framed slabs are used, the thickness of the material between the stiffening ribs should not be less than 50 mm. Unless special precautions are taken, inaccuracies are likely to occur in length, out of squareness, lack of straightness, twisting and flatness. For slabs of reasonable proportions (up to 3 m in length) a minimum tolerance of 5 mm should be allowed in each of these cases. Reinforcing bars or wire mesh is generally used in the slab as a safety precaution against fracture.

Jointing

Where large precast panels are used, there are four types of vertical joint in common use:

1. Open drained joints.
2. Joints sealed over by cover strips.
3. Gap filled joints.
4. Lapped joints.

A great deal of importance should be attached to the choice of a suitable joint. Not only has it to resist rain and wind, but it must provide a degree of insulation equivalent to the panels themselves. Examples of such vertical joints are shown in Fig. 7.

Horizontal rebated joints are generally adopted for both light and heavy forms of construction. Free drainage must always be maintained and special attention is necessary to avoid capillary action.

Fixing

There are numerous methods of fixing precast concrete panels to the framework. The fixing device must allow for a certain amount of movement, without impairing the efficiency of the joint. Two good examples are shown in Fig. 8.

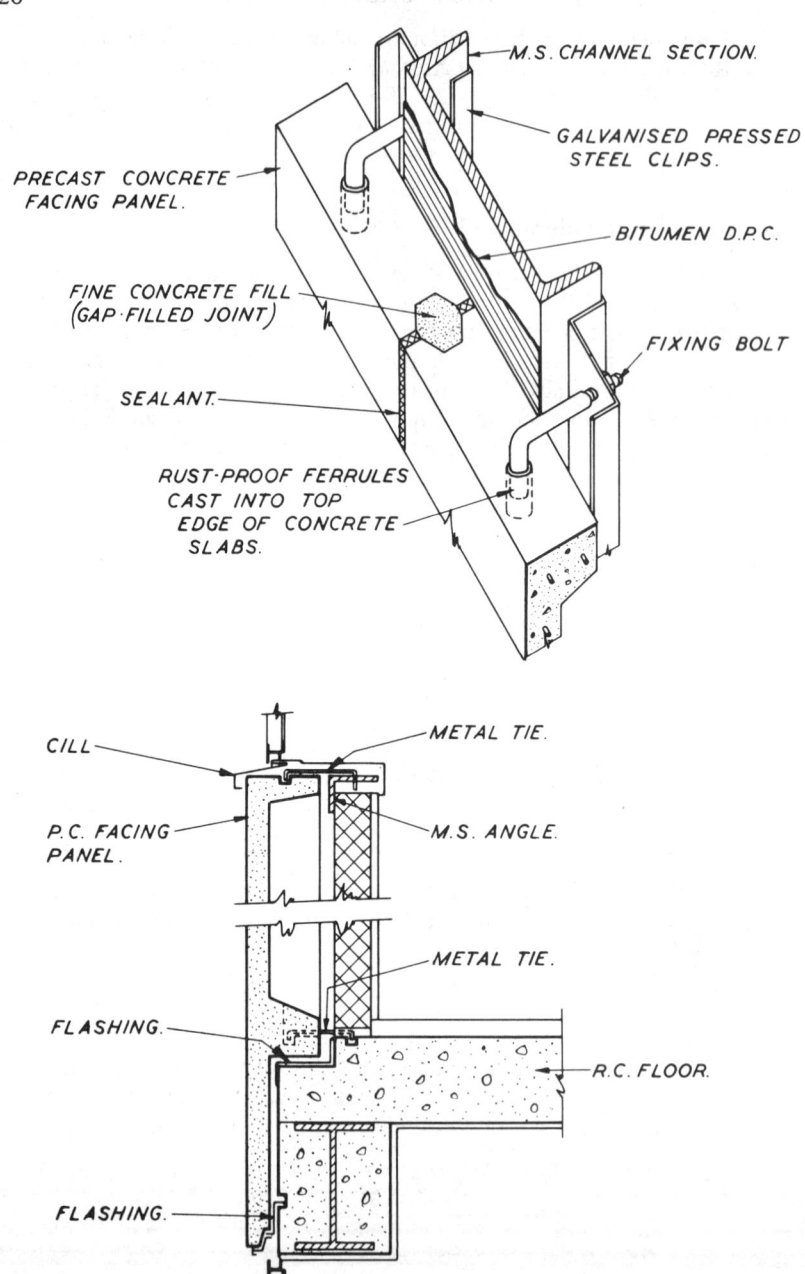

TYPICAL DETAILS FOR FIXING PRE-CAST
CONCRETE PANELS TO FRAMEWORK.

Fig. 8

38. USE OF COMPUTERS IN STRUCTURAL DESIGN

Introduction

SINCE this chapter was first introduced into the Manual in 1966, the use of computers in structural design has developed to a stage when few engineers have not had occasion to use one or other of the many services available. The general information which follows may, however, enable a better appreciation to be made of the advantages to be gained by the proper use of the merits of computers.

Mechanical methods of solving arithmetical problems are not new. Even the use of the fingers for counting can be considered a mechanical method. The development of the use of electronics, however, is of comparatively recent date and has progressed rapidly as the advantages of speedy calculations became apparent.

In the design of structural steelwork, the most common tool in the past has, of course, been the slide rule. If at the outset the computer is considered as an electronic slide rule, much more powerful, speedy and accurate, its advantages will immediately become apparent.

There are two main forms of electronic computer with different basic characteristics:

(a) Analogue computer.
(b) Digital computer.

Whilst in both cases the source of power is electricity, the analogue computer is used to solve one problem at a time based on the problem being simulated by varying voltages, whereas the digital computer works by a series of electrical impulses of constant magnitude.

Several kinds of digital computer are made, some of these being suitable for commercial purposes and others for scientific and technical problems and this chapter will be concerned with the latter type of machine.

Digital Computers

Once it is appreciated that numbers can be represented by a sequence of digits, the simple mechanics of an electrical device to represent such numbers can easily be understood.

All numbers can be shown in binary form consisting of ones and zeros. The binary table can be written as follows:

Number	Binary Equivalent
1	0000001
2	0000010
3	0000011
4	0000100
5	0000101
6	0000110

7	0000111
8	0001000
9	0001001
10	0001010
11	0001011
12	0001100
13	0001101
14	0001110
15	0001111
16	0010000
..	....
32	0100000
..	....
64	1000000
..	....
etc.	etc.

Hence the number 57 for instance can be represented in binary by the figure 0111001 made up of:

32	0100000
16	0010000
8	0001000
1	0000001
——	——
57	0111001

Once it has been appreciated that the figure one or the zero can be represented by an electrical switch, i.e. the switch 'on' meaning figure 1 and the switch 'off' meaning zero, the recording of a binary number electronically by a series of switches becomes very simple.

The arithmetical rules of adding, subtracting, multiplying and dividing can be applied to binary numbers as well as to ordinary numbers, particularly when it is realised that these arithmetical functions are all basically concerned with addition and subtraction. Thus multiplication can be replaced by the successive additions of the same number to itself, and division by a similar succession of subtractions.

Regarding the number 57 used in the previous example; the multiplication of this by 3 in binary digits can be represented by the following sum:

$$
\begin{aligned}
&0111001\\
&+0111001\\
&+0111001\\
\hline
&=10101011
\end{aligned}
$$

which is, of course, the binary equivalent of 171.

Considering again the electrical switches which are either 'on' or 'off', the simple multiplication just shown can be represented electronically by the flow of current instead of physically by writing, printing, etc.

In order that simple calculations of the type just illustrated can be carried out, it is obvious that there must be some method of instructing the machine. This is

known as programming and consists of writing down a sequence of arithmetical operations with the necessary instructions to carry out the operations in the correct order.

Some means must be devised for conveying these instructions and the data with respect to the particular problem to the computer, and this can be done by preparing either punched paper tape, punched cards, or a magnetic tape, whichever is appropriate for the particular computer being used. In a similar manner the results of the instructions when they have been carried out by the machine must be recorded or conveyed to the user.

The basic equipment of a computer, or 'hardware' as it is normally called, consists of an input device, a storage unit, an arithmetical unit and an output device, the diagrammatic layout being shown in the sketch, Fig. 1.

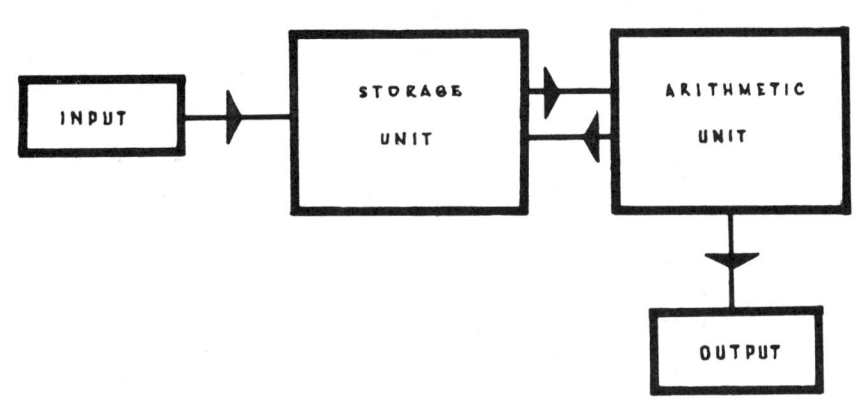

Fig. 1

The function of the input and output devices has already been referred to, the former reads the instructions from the 'software', i.e. the punched tape or other medium, by means of an electronic eye, and the latter performs a similar function in reverse, i.e. produces punched tape, cards, etc., giving the answers.

The second and perhaps the most important part of the 'hardware' is the storage unit. This is designed to accept and retain binary numbers representing instructions or figures in specific locations from which they can be withdrawn as required. The storage unit of a computer is governed by its size, and this, in turn, governs the magnitude of the operation which can be successfully and rapidly carried out.

The final item in the hardware is the arithmetical unit in which the actual calculations are performed. This arithmetical unit has facilities for withdrawing from the storage unit the numbers it requires, performing arithmetical operations on these numbers and accumulating the results until the completion of the operations, when the product can be transferred back into the storage unit or delivered via the output unit as an answer depending upon the instruction given.

Uses of a Computer

There are three main ways in which the digital computer can be employed with advantage as a mechanical calculator viz.:

(a) Solution of a series of simple arithmetical expressions.
(b) Solution of repetitive mathematical arguments.
(c) Solution of a multiplicity of simultaneous equations.

In considering these, two points should be borne in mind:

(i) Few people have constant access to a computer without the necessity of reserving time or without involving travel.
(ii) Being a costly precision instrument the use of a computer is expensive.

Because of these factors the carrying out of simple calculations at random is generally uneconomic. Nevertheless, it is useful to examine all three uses outlined above.

(a) *Solution of an Arithmetical Expression*

In structural design, an appreciable amount of time is often spent in working out by a series of simple calculations the loads to be carried by the different components of a structure and in selecting suitable sizes for these. For example, the following type of calculation is frequently met with in beam design:

Beam 21.   Span 8.0 m.

Loading

Dead load from floor = 8 x 4   x 3.75 kN/m^2 = 120 kN
Live load from floor = 8 x 4   x 5.0 kN/m^2 = 160 kN
Wall load   = 8 x 3.5 x 4.5 kN/m^2 = 126 kN
Weight of beam, etc. = 8   x 5.0 kN/m   = 40 kN

Total = 446 kN

B.M. at $\dfrac{wl}{8} = \dfrac{446 \times 10^3 \text{ N} \times 8 \times 10^3 \text{ mm}}{8} = 446 \times 10^6$ Nmm

Z required in Grade 43 steel = $\dfrac{446 \times 10^6 \text{ Nmm}}{165 \text{ N/mm}^2} = 2.7 \times 10^6$ mm^3

= 2 700 cm^3

Use 533 x 210 x 122 kg UB

These calculations are usually carried out by a slide rule or desk calculator but there is no difficulty in programming a computer to perform them both quickly and accurately. The only problem, apart from the two points already made, is in presenting the necessary data. By the time that this has been recorded for input into the computer the answer can frequently have been found with sufficient accuracy for all practical purposes.

With the development of desk-top computers it has become possible to carry out this type of calculation by machine much more economically. Reference will be made to such machines later in the chapter.

One further observation can be made regarding this type of calculation, namely that should there be a number of governing factors controlling the selection of the size of the beam — such as limiting depth, deflection, etc. — the computer is able to assess these factors and make the final selection.

(b) *Solution of Mathematical Arguments*

In engineering, it is often necessary to calculate a series of values for a particular function; this occurs frequently in the design of steel structures. For instance, the eccentric loads which can safely be carried by a rivet or weld group form an excellent example.

Provided that the problem can be expressed as a logical mathematical function, the computer can be instructed to work out a whole series of values very swiftly, and by the correct directives can also be made to print out the results in tabular form so that they can be issued direct to the user.

(c) *Solution of Simultaneous Equations*

Many structural problems such as the analysis of rigid frame structures require the solution of numerous simultaneous equations for an accurate answer. Though the evaluating of these can be done by hand, the more complex the problem, the more laborious this becomes.

Because of this time-consuming labour, approximate methods of analysis such as moment distribution were developed, though with many problems even these prove to be both lengthy and tedious.

Now by the use of matrix algebra the solution of simultaneous equations becomes simply a successive series of arithmetical operations — eminently suitable for a digital computer — and hence accurate analysis becomes as easy to apply as the approximate methods.

Computer Programmes

Reference has already been made to the necessity of computer programming. Too much consideration cannot be given to this aspect of the subject since the machine can only perform those operations which it is instructed to do and cannot itself correct errors or otherwise develop routines *without a programme.*

Numerous programmes suitable for use on one or other computer have already been written on various aspects of structural engineering both for analysis and for design. These programmes are owned or administered by organisations which can be divided into a number of groups, viz.:

(a) Universities, Colleges and other Public Bodies.
(b) Computer Services.
(c) Private Companies.
(d) Government sponsored bodies such as the National Computing Centre, Genesys Centre, etc.

and each of these function in a different manner.

(a) *Programmes owned by University Departments, etc.*

These generally comprise programmes written in the course of studies carried out by research students and members of the staff. These cover a wide range of subjects of which some have little interest for the designer. They are, however, usually made available to interested parties without charge other than for computer time.

In this connection it can be noted that the original programme for the analysis of rigid frames was written as a subject of research though this has been developed since by many people in slightly different ways.

(b) *Computer Services*

Most of the larger manufacturers of computers maintain some form of computer service as a means of encouraging the use of their products. These services are usually organised from some central establishment where suitable equipment is available for both the preparation of the input tapes or cards, the computer itself and finally the print-out of the results.

These computer services maintain a library of standard programmes either written for them or prepared by their own staff and since the actual work in running these is done by non-technical operatives a very detailed specification as to the manner of presenting the data and interpreting the answers is usually available for each.

Full details of the available programmes and other facilities can be obtained on application to the makers concerned.

(c) *Private Programmes*

A number of private companies and consulting engineers who have found the use of computers advantageous have had personnel trained to produce computer programmes to suit their needs. These are not usually made available to others though the engineers concerned are frequently prepared to discuss them with other interested parties.

The most efficient use of the computer is achieved by the preparation of programmes designed to suit particular problems.

(d) *Government sponsored organisations*

The National Computing Centre was established to serve as a central organisation for the dissimilation of information on both computers and programmes whilst the Genesys Centre is concerned with the development of a general system of programmes for various applications of computers to civil and structural engineering.

More details of these two centres are given later in this chapter.

Computers for Structural Design

As already mentioned, not all computers are considered suitable for technical calculations, some being mainly for commercial application. However, there are several machines which can be used with advantage for this purpose.

The cost of using computers is frequently based on an hourly rate for the time which the machine itself is engaged on a problem. This rate can vary from £10.00 per hour for the smaller machines to as much as £250.00 per hour for the larger, and the fee generally includes for ancillary work away from the machine which usually takes far longer. It does not, however, include for the cost of preparing programmes.

Apart from the Universities and Colleges, a number of private companies and national bodies who own computers are willing to rent time on their machines and hence there should be facilities available within easy reach of most design offices.

Preparation of Programme

The preparation of a computer programme for a particular item of structural analysis involves three main stages:

(a) Engineering the problem.
(b) Reducing it to a logical mathematical exposition.
(c) Translating this to machine instructions.

From this it would appear that the ideal team would consist of an Engineer, a Mathematician and a Programmer. This is not the case, however, since the numerical strength of the team multiplies the possibility of errors and also makes it more difficult to trace these.

It has already been proved that the ideal way to produce a programme is for the engineer himself to perform all three of the operations, and since he must already have a knowledge of mathematics it then only becomes necessary for him to learn programming.

The most economic way in which to programme problems is in machine code. This necessitates a knowledge of the particular computer which is to be used since the machine code varies with different makes and sizes of machine and therefore has limited use unless it is known for certain that such a machine will always be available.

Again, to learn a machine code requires far more study of a computer than most engineers have the time for, and as a result, simplified programme languages have been devised which can be more quickly assimilated and applied.

These simplified languages were originally written for particular machines and were not interchangeable. This again presents problems if different machines are available and so in many instances the manufacturers have themselves produced special programmes which enable their machines to translate other people's simplified coding and act upon it.

Simplified codes are frequently referred to as Autocode and there are Autocode languages for many different makes of machine. These vary in many respects and so it becomes desirable to adopt a particular Autocode which is suitable to the machine most likely to be available.

One of the larger firms producing computers has produced a language which they call 'Fortran' (or Formulae Translation) and again this is written so that with minor variations it can be used for any of their machines provided the necessary translation is first fed into the computer.

It is appreciated that considerable advantage can ensue from the adoption of a universal simplified language and attempts to achieve this have been made, this particular language being known as 'Algol', unfortunately with little success to date.

In order fully to understand the special codings referred to, it is necessary to study these in some detail. Fortunately courses are run at frequent intervals by the computer services, and others, which occupy very little time and sufficient knowledge can be acquired to write a simple programme without errors.

Earlier it was stated that the preparation of a computer programme required three stages and so far only the last stage has been considered. The first and second stages, however, do not generally require any knowledge of the computer but mainly a clear understanding of the engineering of the problem and the ability to reduce this into a logical sequence of mathematical steps.

Existing Computer Programmes

As previously stated, programmes exist and are available for use on a variety of aspects of structural engineering, including:

(a) Analysis of rigid plane frames.
(b) Influence lines for continuous bridges.
(c) Suspension cables.
(d) Floor grillages.
   etc., etc.

Of these, the programme for the analysis of rigid frames is probably the most widely known and most of the computer services offer it in one form or another.

This programme is based on the slope-deflection method of analysis, since the facility of solving simultaneous equations by means of matrices and using a computer has removed the time-consuming drudgery from the use of this old-established method.

In order to understand and accept the accuracy of the computer programme it is useful to examine briefly its derivation.

Analysis of Rigid Frames by Computer

Figure 2 shows an inclined member 1.2 subjected to three unit displacements:

(a) Unit axial displacement;
(b) Unit transverse displacement;
(c) Unit rotation;

and the forces and moments necessary to maintain the member in its displaced condition.

It will be appreciated that a proper sign convention must be adopted at the outset. This is indicated on the diagram and is:

Horizontal displacement to the right . . . . positive.
Vertical displacement upwards . . . . positive.
Counterclockwise rotation . . . . positive.

The same sign convention is also used for forces and moments.

The member 1.2 is inclined at an angle α to the horizontal and, by geometry, the displacement shown can be resolved into the axes X and Y using the properties of the angle. This is illustrated and it can be seen that the unit displacement of end 1 in both the X and Y axes due to the axial and lateral displacement given can be written.

Unit displacement $X = \cos \alpha \times$ axial displacement
$\qquad - \sin \alpha \times$ lateral displacement.
Unit displacement $Y = \sin \alpha \times$ axial displacement
$\qquad + \cos \alpha \times$ lateral displacement.

The reason for the introduction of the negative sign can be followed by referring to the figures including the modified slope-deflection diagrams. Hence the adjusted displacements are:

(a1) Unit displacement X,
(b1) Unit displacement Y,
(c1) Unit rotation θ,

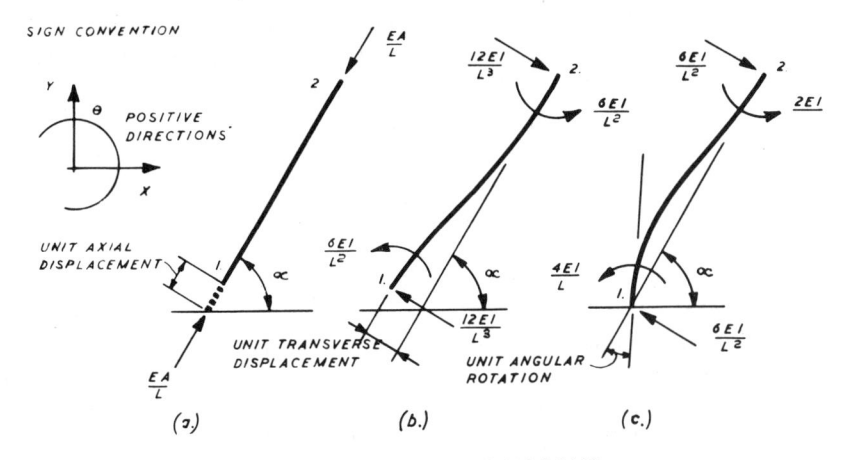

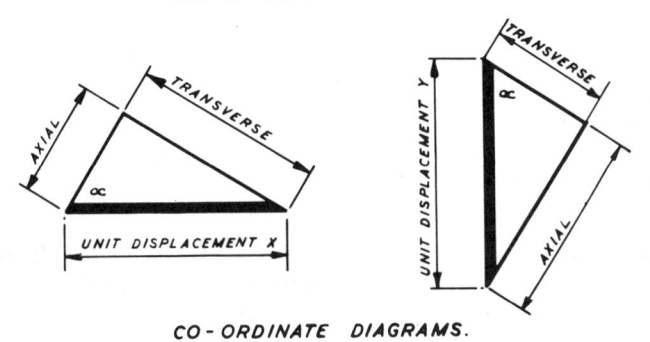

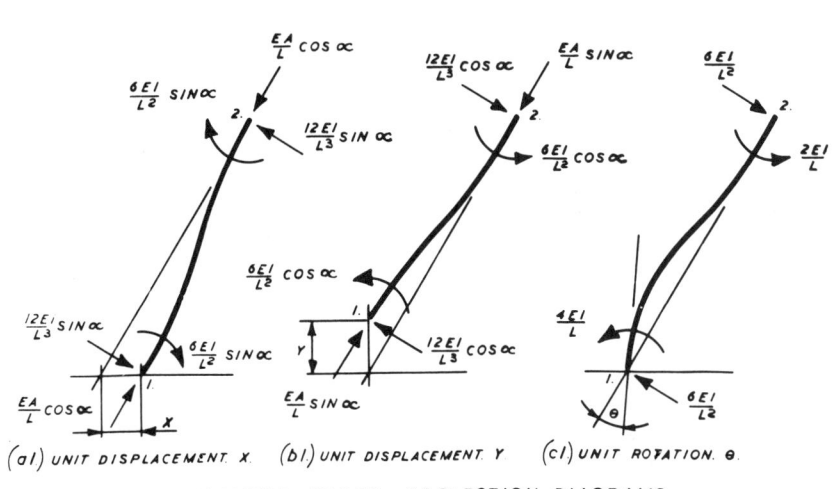

Fig. 2

together with the forces and moments which produce these. The diagram (c1) is, of course, a repeat of the original rotation diagram since rotation alone does not cause any translation of ends 1 and 2.

In the diagrams the forces and moments are shown as products of the properties of the member, i.e.:

E = Young's modulus of elasticity,
A = Area of cross-section,
I = Moment of inertia of section in plane of bending,
L = Length of member,

and from this it can be seen that the theory can apply to a structure built up of any shape of member, and is not limited to any particular structural shapes or indeed to any particular material. It is essential, however, to use members of constant section between joints and the theory given does not apply to sections having a variable moment of inertia.

The following symbols are used:

Fx = Force in X direction.
Fy = Force in Y direction.
M = Bending moment.
x = Displacement in X direction.
y = Displacement in Y direction.
θ = Angular rotation.

In order to distinguish between ends 1 and 2 these are indicated as suffixes. For each member, six equations can be written, three for end 1 and three for end 2. These equations contain six unknowns, i.e. the displacement in the X and Y axis and the rotations at each end.

The equations are as follows:

Slope-deflection Equations for Members 1.2
For end 1.

$$Fx_1 = \left(\frac{EA}{L}\cos^2\alpha + \frac{12EI}{L^3}\sin^2\alpha\right)x_1 + \left(\frac{EA}{L} - \frac{12EI}{L^3}\right)(\sin\alpha\cos\alpha)y_1$$

$$+ \left(-\frac{6EI}{L^2}\sin\alpha\right)\theta_1 + \left(-\frac{EA}{L}\cos^2\alpha - \frac{12EI}{L^3}\sin^2\alpha\right)x_2$$

$$+ \left(-\frac{EA}{L} + \frac{12EI}{L^3}\right)(\sin\alpha\cos\alpha)y_2 + \left(-\frac{6EI}{L^2}\sin\alpha\right)\theta_2$$

$$Fy_1 = \left(\frac{EA}{L} - \frac{12EI}{L^3}\right)(\sin\alpha\cos\alpha)x_1 + \left(\frac{EA}{L}\sin^2\alpha + \frac{12EI}{L^3}\cos^2\alpha\right)y_1$$

$$+ \left(\frac{6EI}{L^2}\cos\alpha\right)\theta_1 + \left(-\frac{EA}{L} + \frac{12EI}{L^3}\right)(\sin\alpha\cos\alpha)x_2$$

$$+ \left(-\frac{EA}{L}\sin^2\alpha - \frac{12EI}{L^3}\cos^2\alpha\right)y_2 + \left(\frac{6EI}{L^2}\cos\alpha\right)\theta_2$$

$$M_1 = \left(-\frac{6EI}{L^2}\sin\alpha\right)x_1 + \left(\frac{6EI}{L^2}\cos\alpha\right)y_1 + \left(\frac{4EI}{L}\right)\theta_1$$

$$+ \left(\frac{6EI}{L^2}\sin\alpha\right)x_2 + \left(-\frac{6EI}{L^2}\cos\alpha\right)y_2 + \left(\frac{2EI}{L}\right)\theta_2$$

For end 2.

$$Fx_2 = \left(-\frac{EA}{L}\cos^2\alpha - \frac{12EI}{L^3}\sin^2\alpha\right)x_1 + \left(-\frac{EA}{L} + \frac{12EI}{L^3}\right)(\sin\alpha\cos\alpha)y_1$$
$$+ \left(\frac{6EI}{L^2}\sin\alpha\right)\theta_1 + \left(\frac{EA}{L}\cos^2\alpha + \frac{12EI}{L^3}\sin^2\alpha\right)x_2$$
$$+ \left(\frac{EA}{L} - \frac{12EI}{L^3}\right)(\sin\alpha\cos\alpha)y_2 + \left(\frac{6EI}{L^2}\sin\alpha\right)\theta_2$$

$$Fy_2 = \left(-\frac{EA}{L} + \frac{12EI}{L^3}\right)(\sin\alpha\cos\alpha)x_1 + \left(-\frac{EA}{L}\sin^2\alpha - \frac{12EI}{L^3}\cos^2\alpha\right)y_1$$
$$+ \left(-\frac{6EI}{L^2}\cos\alpha\right)\theta_1 + \left(\frac{EA}{L} - \frac{12EI}{L^3}\right)(\sin\alpha\cos\alpha)x_2$$
$$+ \left(\frac{EA}{L}\sin^2\alpha + \frac{12EI}{L^3}\cos^2\alpha\right)y_2 + \left(-\frac{6EI}{L^2}\cos\alpha\right)\theta_2$$

$$M_2 = \left(-\frac{6EI}{L^2}\sin\alpha\right)x_1 + \left(\frac{6EI}{L^2}\cos\alpha\right)y_1 + \left(\frac{2EI}{L}\right)\theta_1$$
$$+ \left(\frac{6EI}{L^2}\sin\alpha\right)x_2 + \left(-\frac{6EI}{L^2}\cos\alpha\right)y_2 + \left(\frac{4EI}{L}\right)\theta_2$$

Certain constants can be identified from these equations and labelled K1, etc.:

Constants for Stiffness Matrices

$$\frac{EA}{L}\cos^2\alpha + \frac{12EI}{L^3}\sin^2\alpha = K_1$$

$$\left(\frac{EA}{L} - \frac{12EI}{L^3}\right)(\sin\alpha\cos\alpha) = K_2$$

$$\frac{EA}{L}\sin^2\alpha + \frac{12EI}{L^3}\cos^2\alpha = K_3$$

$$\frac{6EI}{L^2}\sin\alpha = K_4$$

$$\frac{6EI}{L^2}\cos\alpha = K_5$$

$$\frac{2EI}{L} = K_6$$

These constants are related to the properties of the members of the frame and the slope of these members relative to the horizontal. Hence for a given frame they can be calculated without regard to the externally applied loading, and the six equations for the member 1.2 rewritten in a simpler form, thus:

Modified Equations

$$Fx_1 = K_1(x_1 - x_2) + K_2(y_1 - y_2) - K_4(\theta_1 + \theta_2)$$
$$Fy_1 = K_2(x_1 - x_2) + K_3(y_1 - y_2) + K_5(\theta_1 + \theta_2)$$
$$M_1 = K_4(x_2 - x_1) + K_5(y_1 - y_2) + K_6(2\theta_1 + \theta_2)$$
$$Fx_2 = K_1(x_2 - x_1) + K_2(y_2 - y_1) - K_4(\theta_1 + \theta_2)$$
$$Fy_2 = K_2(x_2 - x_1) + K_3(y_2 - y_1) - K_5(\theta_1 + \theta_2)$$
$$M_2 = K_4(x_2 - x_1) + K_5(y_1 - y_2) + K_6(\theta_1 + 2\theta_2)$$

From the member details provided by the designer, the constants K1, etc., are calculated by the computer and stored for use later as required. These constants form the stiffness matrices of the members and are used for the distribution of the applied loads and moments.

It is now necessary to consider the joints in the frame and the conditions under which these will remain in equilibrium.

Figure 3 shows a joint 1 in part of a rigid frame.

The three members meeting at this joint are:

$$1.2$$
$$1.3$$
$$1.4$$

and for each member, equations can be obtained as already shown for Fx, Fy and M. Since, however, the remote ends of the members will form other joints in the

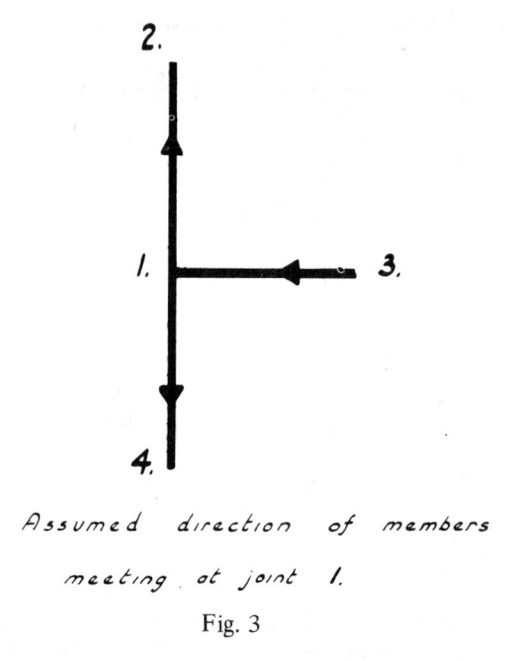

Assumed direction of members meeting at joint 1.

Fig. 3

frame, it is necessary to give an indication as to the interpretation of ends 1 and 2 in the stiffness matrices. An arrow is, therefore, given on each member as shown.

Hence for the member between:

> joints 1 and 2, end 1 occurs at joint 1 and end 2 at joint 2.
> joints 1 and 3, end 1 occurs at joint 3 and end 2 at joint 1.
> joints 1 and 4, end 1 occurs at joint 1 and end 2 at joint 4.

The equations for all members meeting at joint 1 must be assembled using the right stiffness matrices having regard to the directions of the members, and these must be equated to the external loads and moments. Then the equations are solved by means of matrix algebra to give first the displacements of all joints in the frame and secondly the distribution of the forces and moments.

The information given by the computer from this programme consists of:

(a) Displacements and rotations of all joints;
(b) Shears, thrusts and moments on all members;

and this is sufficient to provide a complete picture of the behaviour of the frame under the given loading.

Desk-Top Computers

There are a number of desk-top computers now available, which can bring the merits of simple computing within easy reach of the designer. They are generally advanced electric calculating machines and can be used as such, but in addition include memory registers which can retain either instructions or numerical data to be called into use when required by means of a simple programme.

The extent of the memory store and thereby the scope when used as a computer, depends on the price of the machine, but a number costing around £3,000 have a sufficient capacity to deal with many routine design problems in structural engineering.

The programming technique for this kind of computer is relatively simple and can be quickly mastered by the designer. The suppliers frequently organise courses to which personnel can be sent to obtain programming instruction. Usually the arrangement of these courses is an 'after-sales' service provided free of charge by the suppliers.

Typical items met in structural steelwork design which can be programmed on desk-top computers are:—

(a) Properties of structural shapes.
(b) Bending moment, shear force and deflection calculations.
(c) Safe loads on varying spans.
(d) Calculations for rolling loads
    etc., etc.

Since programmes can normally be recorded on magnetic tape, cards or in some other manner, the use of these computers for routine calculations where there are a limited number of cases becomes a viable proposition, since it is only necessary to feed in the correct programme — a second or so — and the machine is ready.

National Computing Centre

The N.C.C. was established by the Government to promote the more effective use of computers in every field of national or commercial activity.

This is covered by the terms of reference of the Centre which include the following:—

(a) By the provision of services, assistance, advice, and information to users or manufacturers of computers.
(b) For the dissemination of information about computers, their use, programming and operation.
(c) To acquire computer programmes and specifications of computer programmes and to provide information when required to potential users.

Membership of the N.C.C. is available to computer manufacturers and to any company or organisation requiring access to the available information. Full details can be obtained from the Registered Office, Quay House, Quay Street, Manchester M3 3HU.

Genesys Centre

The Genesys Centre was established by the Ministry of the Environment following exhaustive investigations into the viability of developing a library of reliable tested programmes for all aspects of Civil and Structural Engineering.

The system consists of a master programme called the Genesys system which selects sub-systems for particular requirements on receiving commands from the user together with his numerical data.

The Genesys sub-systems already in an advanced stage of preparation or being planned include the following:—

(a) Frame Analysis — a general analysis sub-system.
(b) Reinforced Concrete Buildings — a complete design and detailing package.
(c) Bridge Design — calculations for bridges, designed as a series of continuous beams.
(d) Slab Bridges — finite element analysis of bridge structures.
(e) Highways — a complete design of highways.
(f) Steel Buildings — a design and detail package for structural steel buildings.

All the Genesys programmes are written in a language known as Gentran which is a variation of Fortran IV and it is intended that users can write their own particular variations in the same language so that they can be coupled to the Genesys system.

Membership of the Genesys Centre can be secured for a nominal charge and apart from being kept informed of developments, etc., the member is entitled to use the Genesys system at a cheaper rate than the non-member.

As and when the various sub-systems have been thoroughly tested and approved it is intended that they will be released to computer bureaux throughout the country so that access to the system should be readily available to all designers.

Full details of the Genesys system can be obtained from Genesys Centre, University of Loughborough, England.

Computer Services and Programmes

A comprehensive list of computer programmes, services, etc., is maintained by the Institution of Structural Engineers, 11 Upper Belgrave Street, London, S.W.1, and copies of many of the programmes themselves are held in the Institution library. Full details of these can be obtained from the Secretary.

39. FIRE RESISTING CONSTRUCTION

List of Contents

(1) Introduction.
(2) A brief history of fire protection and development of relevant Building Regulations, Standards and Constructional By-laws.
(3) Review of current legislation.
(4) Some methods of fire protection.
(5) Possible future developments.
(6) Bibliography.

(1) Introduction

In Great Britain certain structural elements of permanent buildings are required by Building Regulations and By-laws to possess fire resistance dependent upon purpose or occupancy group, height, capacity and floor area of building and its separation from others.

In general, the beams and columns supporting floors and walls of multi-storey buildings require fire protection. In the case of single storey buildings the roof framing and supporting structure may be left unprotected, but irrespective of the number of storeys however, where the external wall is near a boundary, it may be required to possess fire resistance from both sides so that it acts as a barrier to a fire within, thus reducing the hazard to neighbouring property. According to The Building Regulations 1972 and the Building Standards (Scotland) (Consolidation) Regulations 1971, such a wall requires fire resistance from both sides if it is on or within 1 metre of the boundary. A similar concept is embodied in the London Building (Constructional) By-laws 1972.

The requirements of fire resistance are set out in:—

(a) The Building Regulations 1972 (1).*
(b) The Building Standards (Scotland) (Consolidation) Regulations 1971 (2).
(c) The London Building (Constructional) By-laws 1972 (3).
(d) The Greater London Council's Code of Practice for buildings requiring approval under Section 20 of the London Building Acts (Amendment) Act 1939 (for certain large buildings in the Inner London area) (4).

All of these regulations and By-laws contain requirements for the adequate fire resistance of different elements of structure. In addition, those relating to Scotland and other regulations applicable in Inner London, contain provisions for means of escape. In Scotland and for 'Section 20' buildings in the Inner London area, the regulations also provide for assistance to the Fire Service. All these measures contribute to the protection of the contents although this is mainly an insurance requirement.

To provide the necessary fire resistance in any structural element two basic factors must be established:—

*The references are listed in Section 6 of this Chapter.

1. The fire resistance period required.
2. The efficacy of the fire protection.

The purpose or occupancy groups in the regulations are intended to group buildings according to their hazard and fire load of building and contents. For example, an office may have an average fire load of only 20 kg/m^2 (this is the equivalent weight of timber per unit floor area with the same calorific value as the contents) requiring less than 1 hour fire resistance depending on height, whereas a warehouse may have a fire load of 250 kg/m^2 requiring a fire resistance of 4 hours to resist collapse. Here, fire resistance is expressed as the period for which the structural element resists collapse and, where appropriate, resists the passage of flame and excessive heat transfer to the unexposed face when the element is subjected to the standard fire test conditions specified in B.S. 476(5). Under The Building Regulations 1972, the grades of fire resistance adopted are ½, 1, 1½, 2 and 4 hours; in The Building Standards (Scotland) (Consolidation) Regulations 1971, 3 hours is also quoted, whereas periods of up to 6 hours are included in the 'Section 20' Code of Practice applicable to Inner London.

It is recommended that before the type and amount of structural fire protection is specified, the Local Authority responsible for the enforcement of the regulations is consulted. This will ensure a correct interpretation insofar as the fire grading period and the admissability of the method of fire protection proposed.

(2) A Brief History of Fire Protection and Development of Relevant Building Regulations and By-laws

Earliest records indicate that a fire fighting service existed in Rome and show that in 300 B.C. bands of slaves were employed as fire fighters in that city. It is also known that the Romans had fire fighters known as 'vigiles' in Chester during their occupation but with the withdrawal of the Roman troops they passed out of history. As the people of Britain gradually became urbanised and began living in permanent settlements fire became a serious risk and a law was passed and first enforced at Oxford in 872 A.D. that at the ringing of an evening bell all house fires were to be extinguished; and this was continued by William following the invasion of 1066.

From then onward there are records of fires in many ecclesiastical buildings but apparently little effort was made to fight them, for often they were believed to be an Act of God.

During the 12th Century there were severe fires at Winchester, Worcester, Bath, Lincoln, Chichester, Rochester, Peterborough, York, Nottingham, Glastonbury and Carlisle.

It was in the reign of Richard I in 1189 that the Mayor of London, Henry Fitz-Alwyn, issued his 'assize' or requirement stipulating that houses in London were to be constructed of stone and that they were to be covered with slate or burnt tile; thus dangerous thatch was banned. It was therefore at this time that building regulations concerning fire came into existence. These requirements also provided for right of light, thickness of party walls and construction of cess pits and were enforced by the sherriffs. Crude means of fire fighting were also provided such as barrels of water and large iron crooks with wooden handles – some of these latter implements still survive.

In 1212 a disaster occurred in London which, if records are correct, caused the greatest death toll in any British fire — three thousand fatalities. Then in 1666 the Great Fire of London occurred when four fifths of the city was destroyed. In this conflagration 13,200 houses, 84 churches, 44 Livery companies and most of the Public Buildings were burnt out. Strangely, only six deaths by fire were recorded but many died of disease and exposure in the shanty towns that flourished thereafter. Fortunately the City Council set about rebuilding with vigour and it was this fire that engendered the Metropolitan Building Acts in 1667.

These Acts lasted for over 100 years and were extended to St. Pancras, Paddington, Westminster, Marylebone and Chelsea. They provided for the employment of Statutory Surveyors until 1844 when District Surveyors were appointed who are still effective. They now enforce mainly the London Building Acts and Constructional By-laws within the area of the City of London and the twelve Inner London Boroughs, these being the City of Westminster, and the Boroughs of Camden, Islington, Hackney, Tower Hamlets, Greenwich, Lewisham, Southwark, Lambeth, Wandsworth, Hammersmith and the Royal Borough of Kensington and Chelsea.

In 1845 the first Public Health Act was passed largely to assist in fighting the squalor in Liverpool and eventually these ideals spread to the whole country. Housing was the main target and apart from the risk of fire it sought to put right poor sanitation, rising damp, instability of structure and lack of ventilation and light. In general, however, these Acts were less stringent than those applying to the Metropolitan area.

Until 1909, however, the regulations were applicable only to load bearing masonry structures; nevertheless in 1796 an iron framed mill was built in Shrewsbury; in 1899 a steel framed building was erected in West Hartlepool followed by the Ritz Hotel and Selfridges in London, both steel framed, built in 1904 and 1906 respectively. The London Building Act issued in 1909 dealt with framed buildings and prescribed details of fire protection for structural members, from this time regulations have been issued under subsequent London Building Acts.

As far as Scotland is concerned the evolution of fire requirements has been a process of some complexity. Until 1892 building control was exercised by the Dean of Guild Courts in burghs on an arbitrary basis but in the Burgh Police Act of that year a detailed set of building rules was laid down for these Courts to apply. The Public Health Act of 1897 subsequently conferred powers on the County Councils to make By-laws to regulate building in the landward areas. No standards were, however, set within that Act.

After the First World War various bodies proposed standardisation and model By-laws were drafted. During this time the larger Local Authorities were promoting their private legislation which in turn meant that building requirements were scattered through many By-laws, rules and statutes, local and general.

In 1957 the Committee of Building Legislation in Scotland, with C. W. Graham Guest, Q.C. as Chairman, presented their report which culminated in the Building (Scotland) Act, 1959, which gave powers to make national building regulations. In 1963 the Building Standards (Scotland) Regulations were issued which have now been consolidated in the Building Standards (Scotland) (Consolidation) Regulations 1971.

These cover, as far as fire is concerned, both the Structural Fire Precautions

(Part D) and Means of Escape from Fire and Assistance to Fire Service (Part E) together with their relating Schedules 5 and 6, the first half of Schedule 9, and Part D in Schedule 10.

Distinct from Government control, the Fire Offices' Committee (F.O.C.) formed in 1868, was originally sponsored by Fire Insurance Concerns, who were active in fire fighting and prevention as an extension of their underwriting work. This committee issued fire regulations for buildings as early as 1896 and in 1908 opened a testing laboratory in Manchester, in 1935 their testing was transferred to a new site at Boreham Wood which has now become the Fire Research Station.

In 1897 another body known as the British Fire Protection Committee (B.F.P.C.) was formed following a severe fire in Cripplegate in 1897, the B.F.P.C. established their own testing station and issued test reports and other recommendations known as Red Books. With the death of their Chairman in 1920 the activities of this committee regressed although some of their publications were later issued by the former Department of Scientific and Industrial Research (D.S.I.R.).

The growing importance of loss from fire led to the convening of a Royal Commission in 1921, which reported in 1923, recommending the formation of a Central Advisory Board to deal with all the aspects of fire hazard.

At the instigation of the Royal Institute of British Architects (R.I.B.A.), the British Engineering Standards Association (now the B.S.I.) formed a committee whose work resulted in the issue of B.S. 476 in 1932 entitled 'British Standard Definitions for Fire Resistance, Incombustibility and Non-Inflammability of Building Materials and Structures'. This was issued to standardise the material in the B.F.P.C. Red Books and the standard testing procedure. British Standard 476 was re-issued in 1953 as 'Fire Tests on Building Materials and Structures' with subsequent revisions, and is now issued in seven parts. Although Part 2 has been withdrawn, other parts are in the course of preparation.

In 1937 the model By-laws issued by the Ministry of Health contained fire resistance requirements for adoption by Local Authorities, these and the 1938 L.C.C. By-laws both made reference to the 1932 B.S. 476.

A joint committee on fire grading was formed in 1938 with representatives from both the D.S.I.R. and F.O.C., thus forming the body known as the Joint Fire Research Organisation (J.F.R.O.). The work of this committee was interrupted by the outbreak of war, and their report was finally published in 1946 as Post War Building Study No. 20 with the title 'Fire Grading of Buildings — Part 1, General Principles and Structural Precautions'.

This comprehensive report contained fire grading periods for buildings based upon various classifications of occupancy and size. Information on the necessary thicknesses of different protective claddings for structural steelwork to withstand fire periods from $\frac{1}{2}$ to 6 hours were also given in this publication. This information, amended to provide a maximum fire rating of 4 hours, is now to be found in the various tables attached to the current Regulations and By-laws. The exception is a six hour requirement by the G.L.C. for basements with abnormal fire load or with access difficulties. More recent events include the metrication of the regulations and revisions to take account of research and up-to-date knowledge insofar as it affects the metricated figures.

As these regulations are subject to constant amendment, care should be taken that a design allows for the latest provisions.

(3) Review of Current Legislation

The principal sets of documents which cover the fire protection aspects of the Building Regulations in Great Britain and which are of concern to the steel designer are as follows:—

(i) The Building Regulations 1972 (England and Wales except the G.L.C. Inner London Area previously noted) — Part E, Structural Fire Precautions — deal only with structural precautions necessary for personal safety and do not include for safeguarding property. Unfortunately, until recently the enabling powers have not permitted means of escape to be included* (as they are in both the Scottish Regulations and the London Building Acts), but there are a number of regulations which, whilst not directly relating to means of escape, set standards of fire resistance for elements of structure and provision of protective shafts which include all enclosed staircases and surface linings. The principal aim is to limit the spread of fire within buildings and from one to another. The risk depends largely on the use to which the building is put and the ability of the structural elements to adequately withstand the effects of fire.

The whole of the fire regulations are grouped together under Part E for ease of reference. Because of the concepts embodied in the Regulations which are distinct from those previously contained in the model By-laws the Minister has retained the right to grant relaxation or dispensation, and by this means is able to gain experience in the working of these particular regulations and ascertain whether any modifications are required. An example of this is the position regarding multi-storey steel framed car parks which may now be constructed without any fire protection within certain parameters, which are set out in Circular 17/68(6) issued by the Ministry of Housing and Local Government (now the Department of the Environment).

Part E of the Regulations commences with Notes on interpretation which are reproduced here† as follows:

Interpretation of Part E

E1.—(1) In this Part and in Schedules 8 and 9—

"basement storey" means a storey which is below the ground storey; or, if there is no ground storey, means a storey the floor of which is situated at such a level or levels that some point on its perimeter is more than 1.2 m below the level of the finished surface of the ground adjoining the building in the vicinity of that point;

"compartment" means any part of a building which is separated from all other parts by one or more compartment walls or compartment floors or by both such walls and floors; and for the purposes of this Part, if any part of the top storey of a building is within a compartment, the compartment shall also include any roof space above such part of the top storey;

"compartment wall" and "compartment floor" mean respectively a wall and a floor which complies with regulation E9 and which is provided as such for the

---

*Section 11 of the Fire Precautions Act 1971 gives the Secretary of State for the Environment power to impose requirements as to the provision of means of escape in case of fire, and work is proceeding on the preparation of the Regulations.

†Where quotation of the Regulations is verbatim this is indicated by a double rule on the right-hand margin.

purposes of regulation E4 or to divide a building into compartments for any purpose in connection with regulations E5 or E7;

"door" includes any shutter, cover or other form of protection to an opening in any wall or floor of a building, or in the structure surrounding a protected shaft, whether the door is constructed of one or more leaves;

"element of structure" means—

- (a) any member forming part of the structural frame of a building or any other beam or column (not being a member forming part of a roof structure only);
- (b) a floor, including a compartment floor, other than the lowest floor of a building;
- (c) an external wall;
- (d) a separating wall;
- (e) a compartment wall;
- (f) structure enclosing a protected shaft;
- (g) a load-bearing wall or load-bearing part of a wall; and
- (h) a gallery;

"externally non-combustible" means externally faced with, or otherwise externally consisting of, non-combustible material;

"fire resistance" has the meaning ascribed to that expression in regulation E6(1);

"fire stop" means a barrier or seal which would prevent or retard the passage of smoke or flame within a cavity or around a pipe or duct where it passes through a wall or floor or at a junction between elements of structure; and "fire-stopped" shall be construed accordingly;

'"ground storey" means a storey the floor of which is situated at such a level or levels that any given point on its perimeter is at or about, or not more than 1.2 m below, the level of the finished surface of the ground adjoining the building in the vicinity of that point; or, if there are two or more such storeys, means the higher or highest of these;".

"height of a building" has the meaning ascribed to it in regulation E3;

"open carport" means a carport of not more than one storey which is open on two or more of its sides; and for the purpose of this definition a side which includes or consists of a door shall not for that reason be regarded as an open side;

"permitted limit of unprotected areas" means the maximum aggregate area of unprotected areas in any side or external wall of a building or compartment, which complies with the requirements of Schedule 9 for such building or compartment;

"protected shaft" means a stairway, lift, escalator, chute, duct or other shaft which enables persons, things or air to pass between different compartments, and which complies with the requirements of regulation E10;

"the relevant boundary", in relation to any side of a building or compartment (including any external wall or part of an external wall), means (unless otherwise specified) that side, unless there is adjacent to that side land belonging to such building or compartment (such land being deemed to include any abutting portion of any street, canal or river up to the centre line thereof) in which case the relevant boundary means that part of the boundary of such land which is either parallel to, or at an angle of not more than 80 degrees with, that wall or side;

"separating wall" means a wall or a part of a wall which is common to two adjoining buildings;

"unprotected area", in relation to an external wall or side of a building, means—
  (a) a window, door or other opening;
  (b) any part of the external wall which has fire resistance less than that specified by this Part for that wall; and
  (c) any part of the external wall which has combustible material more than 1 mm thick attached or applied to its external face, whether for cladding or any other purpose.

(2) Any reference in this Part to a roof or part of a roof of a specified designation shall be construed as meaning a roof or part of a roof so constructed as to be capable of satisfying the relevant test criteria specified in respect of that designation of roof in B.S. 476: Part 3: 1958:

Provided that any roof or part of a roof shall be deemed to be of such a designation if—
  (a) *it conforms with one of the specifications set out against the designation in Schedule 10; or*
  (b) *a similar part made to the same specification as that roof is proved to satisfy the relevant test criteria.*

(3) Any reference in this Part to a building shall, in any case where two or more houses adjoin, be construed as a reference to one of those houses.

(4) If any part of a building other than a single storey building—
  (a) **consists of a ground storey only;**
  (b) **has a roof to which there is only such access as may be necessary for the purposes of maintenance or repair; and**
  (c) **is completely separated from all other parts of the building by a compartment wall or compartment walls in the same continuous vertical plane, that part may be treated, for the purposes of this Part, as a part of a single storey building.**

The Regulations permit the single storey portion with one or more basement storeys of a multi-storey building to be considered separately as a ground storey thus not being subject to E5(1) or E5(2)(a)(i), (ii) and (iii) (see E5(2)(b), (c) and (c)(i)), provided it is completely compartmented from the remainder of the building.

It has been argued that the single storey portion might constitute a fire risk to the upper storey of the multi-storey building by way of fire spread through the roof, if the fire resistance of the roof or the openings in the wall above were not controlled; but this problem will be no greater than at present permitted for buildings or compartments of varying heights in other circumstances.

Further notes on definitions are as follows:—

"Basement Storey" and "Ground Storey" are included in order that the fire resistance requirement of E5 may be appropriately applied.

"Compartment walls" and "Compartment floors" are those elements used to divide a building into compartments and include certain walls and floors referred to in regulation E4(2) and E4(3). Any compartment extending to the upper storey of a building also includes the roof space immediately above.

"Elements of structure" are the parts of a building which are subject to the fire resisting requirements of regulation E5. It will be noted that the roof and staircase are not structural elements unless they are structural members without which the building would be in danger of collapse.

"Relevant boundary." This term is not to be confused with that of a boundary of a site as defined in Part A, but is the relevant portion of a boundary in relation to any side of a building (or compartment) and could be the external wall of the building. Any boundary which is at an angle of more than 80° to the side of the building is excluded, and where the building or site abuts on a street or watercourse, the relevant boundary can be considered as the centre line thereof.

"Separating wall" describes a wall or part of a wall which is common to two adjoining buildings. This is therefore broadly speaking what used to be called a party wall under the By-laws, but this term had legal connections with ownership, etc., other than just those of fire resistance. Because houses, whether in pairs or terraces are considered as separate buildings, the walls dividing them are separating walls.

If two adjoining buildings are erected at one and the same time, the wall between might be considered as a separating wall, particularly if the buildings are in separate ownerships; but from the regulation point of view, the complex could be considered as one building in which case presumably the wall would then be a compartment wall (assuming the building to be of an overall size requiring compartmentation). If, however, the buildings were erected at separate times each with its own [external] wall abutting one another, these walls too, may then be considered as a separating wall for the purpose of the regulations.

"Unprotected area." This can be either a door, a window, or other opening, or a wall which does not have the degree of fire resistance as specified in E5 for that wall, or a wall which has the degree of fire resistance required but is clad externally with combustible material more than 1 mm thick and in this case only 50 per cent of the area is taken into account. The expression is used in regulation E7(i) and Schedule 9 and the percentage of unprotected area in any side of a building will determine the distance to the relevant boundary.

There follows Regulation E2 which is reproduced as follows:—

Designation of purpose groups

E2. For the purposes of this Part every building or compartment shall be regarded according to its use or intended use as falling within one of the purpose groups set out in the Table to this regulation and, where a building is divided into compartments used or intended to be used for different purposes, the purpose group of each compartment shall be determined separately:

Provided that where the whole or part of a building or compartment (as the case may be) is used or intended to be used for more than one purpose, only the main purpose of use of that building or compartment shall be taken into account in determining into which purpose group it falls.

The Scottish regulations are given an even more detailed split and may give guidance in cases of doubt. Personal hazard is taken into consideration in all groups and its degree of importance is relevant to the fire load and varies from group to group.

Purpose group II takes into account the types of accommodation which have a sleeping risk attached to youth or infirmity. Groups I to III comprise all habitable accommodation.

TABLE TO REGULATION E2
(Designation of purpose groups)

Purpose group (1)	Descriptive title (2)	Purposes for which building or compartment is intended to be used (3)
I	Small residential	Private dwellinghouse (not including a flat or maisonette)*
II	Institutional	Hospital, home, school or other similar establishment used as living accommodation for, or for treatment, care or maintenance of, persons suffering from disabilities due to illness or old age or other physical or mental disability or under the age of 5 years, where such persons sleep in the premises.
III	Other residential	Accommodation for residential purposes other than any premises comprised in groups I and II.
IV	Office	Office, or premises used for office purposes, meaning thereby the purpose of administration, clerical work (including writing, book-keeping, sorting papers, filing, typing, duplicating, machine-calculating, drawing and the editorial preparation of matter for publication), handling money and telephone and telegraph operating; or as premises occupied with an office for the purposes of the activities there carried on.
V	Shop	Shop, or shop premises, meaning thereby premises not being a shop but used for the carrying on there of retail trade or business (including the sale to members of the public of food or drink for immediate consumption, retail sales by auction, the business of lending books or periodicals for the purpose of gain, and the business of a barber or hairdresser), and premises to which members of the public are invited to resort for the purpose of delivering there goods for repair or other treatment or of themselves carrying out repairs to, or other treatment of, goods.
VI	Factory	Factory within the meaning ascribed to that word by section 175 of the Factories Act 1961 (a) (but not including slaughterhouses and other premises referred to in paragraphs (d) and (e) of subsection (1) of that section).
VII	Other place of assembly	Place, whether public or private, used for the attendance of persons for or in connection with their social, recreational, educational, business or other activities, and not comprised within groups I to VI.
VIII	Storage and general	Place for storage, deposit or parking of goods and materials (including vehicles), and any other premises not comprised in groups I to VII.*

(a) 1961 c. 34.

Note: By regulation E20 certain small garages and open carports are treated as being of purpose group I.

When a building is divided into separate compartments each used for a different purpose, the purpose group of each must be considered separately, but where a building or compartment is intended to be used for more than one purpose, it is the main purpose group only that can be considered. An uncompartmented building or a compartment may only belong to one group and any subsidiary uses are therefore discounted.

The amendment insertion of the footnote to the Table draws attention to the classification of certain small garages and carports as purpose group I buildings. (*See* E20.)

Regulation E3 sets forth the rules for measuring height, area and cubic capacity of a building or compartment as follows:—

Rules for measurement

E3. In this Part—
(*a*) the height of a building, or (where relevant) of part of a building as described in regulation E5(3)(*b*), means the height of such building or part, measured from the mean level of the ground adjoining the outside of the external walls of the building to the level of half the vertical height of the roof of the building or part, or to the top of the walls or of the parapet (if any), whichever is the higher;
(*b*) the area of—
  (i) any storey of a building or compartment shall be taken to be the total area in that storey bounded by the finished inner surfaces of the enclosing walls or, on any side where there is no enclosing wall, by the outermost edge of the floor on that side;
  (ii) any room or garage shall be taken to be the total area of its floor bounded by the inner finished surfaces of the walls forming the room or garage;
  (iii) any part of a roof shall be taken to be the actual visible area of such part measured on a plane parallel to the pitch of the roof;
(*c*) the cubic capacity of a building or compartment shall be ascertained by measuring the volume of space contained within—
  (i) the finished inner surfaces of the enclosing walls or, on any side where there is no enclosing wall, a plane extending vertically above the outermost edge of the floor on that side; and
  (ii) the upper surface of its lowest floor; and
  (iii) in the case of a building or a compartment which extends to a roof, the under surface of the roof or, in the case of any other compartment, the under surface of the ceiling of the highest storey within that compartment;
including the space occupied by any other walls, or any shafts, ducts, or structure within the space to be so measured.

Further Regulations of consequence to the steel designer are reproduced here as follows:—

Provision of compartment walls and compartment floors

E4.—(1) Any building of a purpose group specified in column (1) of the Table to this regulation and which has—

(a) any storey the floor area of which exceeds that specified as relevant to a building of that purpose group and height in column (3) of the Table; or

(b) a cubic capacity which exceeds that specified as so relevant in column (4) of the Table,

shall be so divided into compartments by means of compartment walls or compartment floors or both that—

(i) no such compartment has any storey the floor area of which exceeds the area specified as relevant to the building in column (3) of the Table; and

(ii) no such compartment has a cubic capacity which exceeds that specified as so relevant in column (4) of the Table:

Provided that if any building of purpose group V is fitted throughout with an automatic sprinkler system which complies with the relevant recommendations of CP 402.201:1952, this paragraph shall have effect in relation to that building as if the limits of dimensions specified in columns (3) and (4) of the Table to this regulation were doubled.

(2) In any building which exceeds 28 m in height, any floor which separates one storey from another storey, other than a floor which is—

(a) within a maisonette; or
(b) above the ground storey but at a height not exceeding 9 m above the adjoining ground,

shall be constructed as a compartment floor.

(3) The following walls and floors shall be constructed as compartment walls or compartment floors—

(a) any floor in a building of purpose group II;
(b) any wall or floor separating a flat or maisonette from any other part of the same building;
(c) any wall or floor separating part of a building from any other part of the same building which is used or intended to be used mainly for a purpose falling within a different purpose group in the Table to regulation E2; and
(d) any floor immediately over a basement storey if such storey—
  (i) forms part of a building of purpose group I which has three or more storeys or a building or compartment of purpose group III or V; and
  (ii) has an area exceeding 100 m^2.

With reference to regulation E4(1) it is of interest that the spaces which are separated either horizontally or vertically are called "compartments", and the walls and floors which separate them are termed "compartment walls" or "compartment floors". All buildings, with certain exceptions which exceed a given overall height, area or cubic capacity, are required to be "compartmented" so as to reduce the building to these maximum sized units which will be better able to contain a fire and so stop the spread throughout the building. The exceptions to this requirement for compartmentation are single storey buildings (other than in purpose groups II and III) and buildings other than single storey in groups I, IV and VII. The Table forming part of this regulation sets forth the maximum sizes of such compartments for the various purpose groups. However, it may be desirable to compartment any building in order to reduce the period of fire resistance required.

FIRE RESISTING CONSTRUCTION

TABLE TO REGULATION E4

(Dimensions of buildings and compartments)

Purpose group	Height of building	Limits of dimensions	
		Floor area of storey in building or compartment (in m^2)	Cubic capacity of building or compartment (in m^3)
(1)	(2)	(3)	(4)

Part 1—Buildings other than single storey buildings

(1)	(2)	(3)	(4)
II (Institutional)	Any height	2 000	No limit
III (Other residential)	Not exceeding 28 m	3 000	8 500
,, ,, ,,	Exceeding 28 m	2 000	5 500
V (Shop)	Any height	2 000	7 000
VI (Factory)	Not exceeding 28 m	No limit	28 000
,, ,,	Exeeding 28 m	2 000	5 500
VIII (Storage and general)	Not exceeding 28 m	No limit	21 000
,, ,, ,, ,,	Exceeding 28 m	1 000	No limit

Part 2—Single storey buildings

II (Institutional)	Any height	3 000	No limit
III (Other residential)	Any height	3 000	No limit

Sprinklers or drenchers are not considered when arriving at the maximum sizes of compartments (other than those now permitted for Shops). According to the BRAC report, this has not been allowed because it was not possible to specify all devices for the control of fire. However, sprinklers have proved most effective in containing a fire and it is hoped that further consideration will be given to this provision. Some effective means would need to be found to ensure adequate water supply and even more important periodical maintenance inspection, this latter responsibility would presumably be undertaken by the Insurance Company or local fire authority. If a sprinkler system is installed in accordance with CP 402.201 (1952), it ought to be feasible to increase the overall size of the compartments, especially in groups V, VI and VIII, which form the Shops, Factories and Storage groups and are likely to be large buildings.

Where sprinklers are installed throughout in accordance with the Code of Practice, the permitted floor area and cubic capacity for shops having a 2 hour standard of fire resistance in accordance with Table A Part 1 to regulation E5, may be doubled.

The credit that can be afforded to sprinklers is difficult to adduce, but doubling the size of the compartment is considered reasonable.

Factories and Storage buildings are still omitted presumably largely because the

fire risk can vary so considerably and each case must still be dealt with by Ministry relaxation according to the particular circumstances.

In a high building of over 28 m, any floor over 9 m must be a compartment floor unless it is the intermediate floor of a maisonette.

All floors in an Institutional building (group II) must be formed as compartment floors and each flat or maisonette must have compartment floors and compartment walls to separate them from another, or from any other part of the same building.

When part of a building is separated from another part of the same building and each part is used for a different purpose, any wall or floor between each part must be a compartment floor or a compartment wall, but there is no mandatory requirement to provide such a wall or floor. When a compartment wall or floor are provided they should comply with the requirements of the purpose group having the greater fire risk. Similarly, where there is no wall, the greater fire resistance requirements will apply to the whole building.

The floor over a basement storey which exceeds 100 m^2 in area must be a compartment floor if the basement forms part of a building or compartment which is used for either residential or shop purposes.

It is noted that no account appears to be taken of the actual fire load within a compartment. This does seem to indicate that when a building is used for another purpose within the same group, there is no additional requirement, although the fire load within the same fire group could vary considerably particularly within Factories and Storage buildings where some processes or goods to be stored have a very low fire risk whilst others have a much greater fire risk. The requirements may, therefore, be onerous in some respects but not onerous enough in others.

Fire resistance

E5.—(1) Subject to any express provision to the contrary, every element of structure shall be so constructed as to have fire resistance for not less than the relevant period specifed in Table A to this regulation, having regard to the purpose group of the building of which it forms part and the dimensions specified in that Table.

   (2) (*a*) In addition to any relevant requirement under paragraph (1)—
   (i) any external wall shall have fire resistance of not less than half an hour;
   (ii) any separating wall shall have fire resistance of not less than 1 hour.
   (iii) any compartment wall or compartment floor which separates a part of a building falling within purpose group II or III from any other part of the same building falling within a different purpose group from purpose group II or III shall have fire resistance of not less than 1 hour.

  (*b*) Nothing in paragraph (1) or in sub-paragraph (*a*) of this paragraph shall apply to any part of an external wall which is non-loadbearing and may, in accordance with regulation E7, be an unprotected area.

  (*c*) In the case of a single storey building or a building consisting of a ground storey and one or more basement storeys, nothing in paragraph (1) or in sub-paragraph (*a*) of this paragraph shall apply to any element of structure which forms part of the ground storey and consists of—
   (i) a structural frame or a beam or column: Provided that any beam or column (whether or not it forms part of a structural frame) which is

within or forms part of a wall, and any column which gives support to a wall or gallery, shall have fire resistance of not less than the minimum period, if any, required by these regulations for that wall or that gallery;
  (ii) an internal load-bearing wall or a load-bearing part of a wall, unless that wall or part is, or forms part of, a compartment wall or a separating wall, or forms part of the structure enclosing a protected shaft or supports a gallery; or
  (iii) part of an external wall which does not support a gallery and which may, in accordance with regulation E7, be an unprotected area.

(3) (*a*) In this regulation and in Table A thereto (subject to the provisions of sub-paragraph (*b*) of this paragraph and any other express provision to the contrary) any reference to a building of which an element of structure forms part means the building or (if a building is divided into compartments) any compartment of the building of which the element forms part.

(*b*) In this regulation and in Table A thereto, any reference to height means the height of a building, not of any compartment in the building, but if any part of the building is completely separated throughout its height both above and below ground from all other parts by a compartment wall or compartment walls in the same continuous vertical plane, any reference to height in relation to that part means the height solely of that part.

(*c*) If any element of structure forms part of more than one building or compartment and the requirements of fire resistance specified in Table A in respect of one building or compartment differ from those specified in respect of any other building or compartment of which the element forms part, such element shall be so constructed as to comply with the greater or greatest of the requirements specified.

(4) Any element of structure shall have fire resistance of not less than the minimum period required by these regulations for any element which it carries.

(5) Any compartment wall separating a flat or maisonette from any other part of the same building shall not be required to have fire resistance exceeding 1 hour unless—
  (i) the wall is a load-bearing wall or a wall forming part of a protected shaft; or
  (ii) the part of the building from which the wall separates the flat or maisonette is of a different purpose group and the minimum period of fire resistance required by the provisions of this regulation for any element of structure in that part is $1\frac{1}{2}$ hours or more.

(6) In the application of this regulation to floors, no account shall be taken of any fire resistance attributable to any suspended ceiling other than a suspended ceiling constructed as described in Table B.

TABLE A TO REGULATION E5
(Minimum periods of fire resistance)

In this Table—.

"cubic capacity" means the cubic capacity of the building or, if the building is divided into compartments, the compartment of which the element of structure forms part;

"floor area" means the floor area of each storey in the building or, if the building is divided into compartments, of each storey in the compartment of which the element of structure forms part;

"height" has the meaning assigned to that expression by regulation E5(3)(*b*).

Part 1—*Buildings other than single storey buildings* (see next page)

REVIEW OF CURRENT LEGISLATION

Part 1—Buildings other than single storey buildings

Purpose group	Maximum dimensions			Minimum period of fire resistance (in hours) for elements of structure(*) forming part of—		
	Height (in m)	Floor area (in m²)	Cubic capacity (in m³)	ground storey or upper storey	basement storey	
(1)	(2)	(3)	(4)	(5)	(6)	
I (Small residential) House having not more than three storeys	No limit	No limit	No limit	½	1(*a*)	x
House having four storeys	No limit	250	No limit	1(*b*)	1	
House having any number of storeys	No limit	No limit	No limit	1	1½	
II (Institutional)	28 over 28	2 000 2 000	No limit No limit	1 1½	1½ 2	
III (Other residential) Building or part(†) having not more than two storeys	No limit	500	No limit	½	1	x
Building or part(†) having three storeys	No limit	250	No limit	1(*b*)	1	
Building having any number of storeys	28	3 000	8 500	1	1½	
Building having any number of storeys	No limit	2 000	5 500	1½	2	
IV (Office)	7.5 7.5 15 28 No limit	250 500 No limit 5 000 No limit	No limit No limit 3 500 14 000 No limit	0 ½ 1(*b*) 1 1½	1(*c*) 1 1 1½ 2	x
V (Shop)	7.5 7.5 15 28 No limit	150 500 No limit 1 000 2 000	No limit No limit 3 500 7 000 7 000	0 ½ 1(*b*) 1 2	1(*c*) 1 1 2 4	x y
VI (Factory)	7.5 7.5 15 28 28 over 28	250 No limit No limit No limit No limit 2 000	No limit 1 700 4 250 8 500 28 000 5 500	0 ½ 1(*b*) 1 2 2	1(*c*) 1 1 2 4 4	x

FIRE RESISTING CONSTRUCTION

Part 1—Buildings other than single storey buildings (contd.)

Purpose group	Maximum dimensions			Minimum period of fire resistance (in hours) for elements of structure(*) forming part of—		
	Height (in m)	Floor area (in m^2)	Cubic capacity (in m^3)	ground storey or upper storey	basement storey	
(1)	(2)	(3)	(4)	(5)	(6)	
VII (Assembly)	7.5	250	No limit	0	1(c)	x
	7.5	500	No limit	$\frac{1}{2}$	1	
	15	No limit	3 500	1(b)	1	
	28	5 000	14 000	1	$1\frac{1}{2}$	
	No limit	No limit	No limit	$1\frac{1}{2}$	2	
VIII (Storage and general)	7.5	150	No limit	0	1(c)	x
	7.5	300	No limit	$\frac{1}{2}$	1	
	15	No limit	1 700	1(b)	1	
	15	No limit	3 500	1	2	
	28	No limit	7 000	2	4	
	28	No limit	21 000	4	4	
	over 28	1 000	No limit	4	4	

Notes to Part 1

For the purpose of regulation E5(1), the period of fire resistance to be taken as being relevant to an element of structure is the period included in column (5) or (6), whichever is appropriate, in the line of entries which specifies dimensions with all of which there is conformity or, if there are two or more such lines, in the topmost of those lines.

(*) A floor which is immediately over a basement storey shall be deemed to be an element of structure forming part of a basement storey.
(†) The expression "part" means a part which is separated as described in regulation E5(3)(b).
(a) The period is half an hour for elements forming part of a basement storey which has an area not exceeding 50 m^2.
(b) This period is reduced to half an hour in respect of a floor which is not a compartment floor, except as to the beams which support the floor or any part of the floor which contributes to the structural support of the building as a whole.
(c) No fire resistance is required if the elements form part of a basement storey which has an area not exceeding 50 m^2.
x The items thus marked are applicable only to buildings, not to compartments, except in relation to purpose group III; see also regulations E7(2)(a) proviso (i) and E8(7)(a).
y If the building is fitted throughout with an automatic sprinkler system which complies with the relevant recommendations of CP 402.201: 1952, any maximum limits specified in columns (3) and (4) shall be doubled.

TABLE A TO REGULATION E5—*continued*

(Minimum periods of fire resistance)

Part 2—*Single storey buildings* (see next page)

Part 2—Single storey buildings

Purpose group	Maximum floor area (in m²)	Minimum period of fire resistance (in hours) for elements of structure *	
(1)	(2)	(3)	
I (Small residential)	No limit	½	z
II (Institutional)	3 000	½	z
III (Other residential)	3 000	½	z
IV (Office)	3 000 No limit	½ 1	z
V (Shop)	2 000 3 000 No limit	½ 1 2	z
VI (Factory)	2 000 3 000 No limit	½ 1 2	z
VII (Assembly)	3 000 No limit	½ 1	z
VIII (Storage and general)	500 1 000 3 000 No limit	½ 1 2 4	z

Notes to Part 2

For the purpose of regulation E5(1), the period of fire resistance to be taken as being relevant to an element of structure is the period included in column (3) in the line of entries which specifies the floor area with which there is conformity or, if there are two or more such lines, in the topmost of those lines.

z   See regulations E7(2)(*a*) proviso (i) and E8(7)(*a*).

* *Note:* Structural steel frames are generally immune from this requirement. See regulation E5(2)(*c*).

Further to regulation E5 it should be noted that every element of structure must be so constructed as to have the periods of fire resistance set forth in the Table which forms part of this regulation, bearing in mind the particular size and purpose group of the building or compartment.

If a building is compartmentated, it is the size and purpose group of each compartment and not that of the building generally, which will decide the fire resistance required. The overall size of the building or compartment affects the severity of the fire, hence the fire resistance is increased for the elements as the size of the structure increases. It should be noted that if an element of structure is part of more than one building or compartment for which different fire resistances are required, the higher fire resistance will apply for that element.

FIRE RESISTING CONSTRUCTION
TABLE B TO REGULATION E5
(Suspended ceilings)

Height of building (1)	Type of floor (2)	Required fire resistance of floor (3)	Description of suspended ceiling (4)
Less than 15 m	Non-compartment	1 hour or less	Surface of ceiling exposed within the cavity not lower than Class 1 (as to surface spread of flame).
	Compartment	Less than 1 hour	
	Compartment	1 hour	Surface of ceiling exposed within the cavity not lower than Class O (as to surface spread of flame); supports and fixings for the ceiling non-combustible.
15 m or more	Any	1 hour or less	Surface of ceiling exposed within the cavity not lower than Class O (as to surface spread of flame) and jointless; supports and fixings for the ceiling non-combustible.
Any	Any	More than 1 hour	Ceiling of non-combustible construction and jointless; supports and fixings for the ceiling non-combustible.

Note: References to classes are to classes as specified in regulation E14.

Regulation E5(2)(*c*) has the effect of removing the requirement for fire resistance for structural frames, columns, beams and walls which perform no other function than that of supporting the roof of a single storey building. This is particularly significant and not necessarily apparent if Part 2 of Table A is studied without reference to the text.

Compartment walls or compartment floors constructed in compliance with regulation E4(3)(*c*) and separating accommodation falling within purpose groups II or III from that in another purpose group, should have a minimum fire resistance of one hour. The risk of fire spreading unnoticed is highest at night, and where there is a sleeping risk over a non-sleeping risk is particularly pertinent, as any outbreak of fire in a lower compartment may well become established before being discovered, hence the requirement for a higher minimum standard of fire resistance in this situation.

The height referred to in Table A is that of the building and not of any compartment within that building, unless a compartment wall completely splits the building vertically from top to bottom in which case the heights of each section are considered. The floor area is presumed to be that of each storey of a building or compartment, whilst the cubic capacity is that of the building or compartment.

Although no compartment walls are required in single storey buildings, used for one purpose only (unless to reduce the requirement for fire resistance), E4(3)(*c*)

requires a compartment wall to separate the parts of a building of different purpose groups (unless the higher fire resistance is taken).

In addition to the requirements laid down in these Tables any external wall must have a fire resistance of not less than half an hour (unless it is non-load-bearing and may, in accordance with regulation E7, be considered as an unprotected area), and any separating wall not less than one hour.

Any element of structure carrying another element must have at least the fire resistance of the element it carries.

Compartment walls surrounding flats or maisonettes need not have a fire resistance exceeding one hour except in the special circumstances described in Regulation E5.

No account can be taken of any suspended ceiling when considering the fire resistance of a compartment floor, unless it be of a type referred to in Table B. Jointed ceilings have apparently been omitted on the grounds that panels are often removed or adapted for services and therefore the fire resistance would seriously be impaired thereby.

Tests of fire resistance

E6.–(1) For the purposes of regulation E5, requirements as to fire resistance shall be construed as meaning that an element of structure shall be capable of resisting the action of fire for the specified period under the conditions of test appropriate to such element in accordance with B.S. 476: Part 1: 1953, subject to such modifications or applications of such conditions of test as are prescribed in this regulation.

(2) Any compartment floor shall, if the underside of such floor is exposed to test by fire, have fire resistance for not less than the minimum period required by the provisions of regulation E5 for elements of structure forming part of the compartment immediately below such floor.

(3) Any structure (other than an external wall) enclosing a protected shaft shall, if each side of the wall is separately exposed to test by fire, have fire resistance for not less than the minimum period required by the provisions of regulation E5.

(4) Any compartment wall or separating wall shall, if each side of the wall is separately exposed to test by fire, have fire resistance for not less than the minimum period required by regulation E5.

(5) Any part of an external wall which constitutes, or is situated less than 1 m from any point on, the relevant boundary shall, if each side of the wall is separately exposed to test by fire, have fire resistance for not less than the minimum period required by regulation E5.

(6) Any part of an external wall which is situated 1 m or more from the relevant boundary and which is required by the provisions of these regulations to have fire resistance, shall, if the inside of the wall is exposed to test by fire, have fire resistance for not less than the minimum period required by regulation E5:

Provided that, for the purposes of this paragraph, the wall shall be capable of satisfying the requirements of Clause 11C of Section 3 of B.S. 476: Part 1: 1953 as to insulation for not less than 15 minutes.

(7) In any building of purpose group 1 which has two storeys the floor of the upper storey shall, if the underside of such floor is exposed to test by fire in

accordance with B.S. 476: Part 1: 1953, be capable of satisfying the requirements of that test as to freedom from collapse for a period of not less than half an hour and as to insulation and resistance to passage of flame for not less than 15 minutes.

(8) *Any element of structure shall be deemed to have the requisite fire resistance if—*
   (a) *it is constructed in accordance with one of the specifications given in Schedule 8, and the notional period of fire resistance given in that Schedule as being appropriate to that type of construction and other relevant factors is not less than the requisite fire resistance: or*
   (b) *a similar part made to the same specification as that element is proved to have the requisite fire resistance under the conditions of test prescribed in the foregoing paragraphs of this regulation.*

Elements of structure must be capable of withstanding or resisting the action of fire for a specified period under certain conditions of test. The tests are in accordance with B.S. 476: Part 1: 1953 and the element is deemed to have the requisite fire resistance if a similar part when tested satisfies these test requirements as to:—

   (i)   Collapse: The element must not collapse at any stage of the test.
   (ii) Passage of flame: The element must not develop cracks through which flame or hot gases can pass.
   (iii) Insulation: The element must have sufficient resistance to the passage of heat that the temperature of the unexposed face does not rise by more than a prescribed amount.

B.S. 476 allows for separating elements to be tested from one side only or either side separately depending on the functions of the element and this regulation sets out these requirements.

A compartment floor when tested on the underside must have the requisite minimum fire resistance for the structural elements in the compartment below the floor.

A compartment wall, separating wall, protecting structure (enclosing a protected shaft), or an external wall situated on or within 1 m of the relevant boundary must stand up to tests on each side, whilst an external wall situated 1 m or more from the relevant boundary is tested on the inside only. Generally the fire resistance is that period of time in hours and minutes for which the element is capable of withstanding collapse, the passage of flame and the passage of heat (insulation). However, in two cases, external walls 1 m or more from the boundary and the upper floors of two storey houses, the passage of heat requirement is reduced to 15 minutes, and in the latter case the passage of flame requirement is also only 15 minutes. In the second case this relaxation is known as a "modified half hour requirement" and is noted as such in Schedule 8, Part VII.

Schedule 8 indicates the type of specifications which are deemed-to-satisfy the requirements of E6(7), these are not exhaustive and are only given for guidance. Other specifications may be applied providing they can be proved by tests to comply with the fire resistance requirements of E6(7).

External walls

E7.—(1) Subject to the provisions of regulations E18 and E19 concerning small garages and open carports, any side of a building shall comply with any relevant requirements relating to the permitted limits of unprotected areas specified in

Schedule 9 unless the building is so situated that such side might in accordance with Schedule 9 consist entirely of any unprotected area.

(2) (a) Any external wall which constitutes, or is situated within a distance of 1 m from any point on, the relevant boundary or is a wall of a building which exceeds 15 m in height shall—

(i) be constructed wholly of non-combustible materials apart from any external cladding which complies with paragraph (3) of this regulation or any internal lining which complies with regulation E15; and

(ii) be so constructed as to attain any fire resistance required by this Part without assistance from any combustible material permitted by this sub-paragraph:

Provided that the requirements of this sub-paragraph shall not apply to—

(i) an external wall of a building which is within the limits of size indicated by the letter "x" in Part 1 of Table A to regulation E5 or of a building which is not divided into compartments and is within the limits of size indicated by the letter "z" in Part 2 of the table if, in either case, that building does not exceed 15 m in height;

(ii) an external wall of a building or part of purpose group III which consists of flats or maisonettes if that building has not more than three storeys or that part is separated as described in regulation E5(3)(b) and has not more than three storeys;

(iii) an external wall of a part of a building if that wall is situated 1 m or more from the relevant boundary and that part is separated as described in regulation E5(3)(b) and does not exceed 15 m in height.

(b) Any beam or column forming part of, and any structure carrying, an external wall which is required to be constructed of non-combustible materials shall comply with the provisions of sub-paragraph (a) as to non-combustibility.

(3) (a) Any cladding on any external wall, if such cladding is situated less than 1 m from any point on the relevant boundary, shall have a surface complying with the requirements for Class O specified in regulation E15(1)(e); and

(b) Any cladding on any external wall situated 1 m or more from the relevant boundary shall, if the building is more than 15 m in height, have a surface complying with the requirements specified for Class O in regulation E15(1)(e), except that any part of such cladding below a height of 15 m from the ground may consist of timber of not less than 9 mm finished thickness or of a material having a surface which, when tested in accordance with B.S. 476: Part 6: 1968, has an index of performance (I) not exceeding 20.

(4) For the purposes of this regulation—

(a) any part of a roof shall be deemed to be part of an external wall or side of a building if it is pitched at an angle of 70 degrees or more to the horizontal and adjoins a space within the building to which persons have access not limited to the purposes of maintenance or repair;

(b) any reference to Schedule 9 shall be construed as referring to the provisions of Part I of that Schedule, together with (at the option of the person intending to erect the building) either the provisions of Part II or those of

Part III or, if the building is one to which Part IV applies, those of that Part or of Part II or III.

(5) If—

(a) any building is to be erected on land occupied with any other building, or two or more detached buildings are to be erected on land in common occupation; and

(b) either of those buildings is of purpose group I or III (other than a detached building which consists only of a garage or of an open carport or of both and complies with regulation E18 or E19 as the case may be),

in the application of the provisions of this regulation to any external wall of any building to be so erected which faces an external wall of such other building—

(i) the relevant boundary shall be a notional boundary passing between those buildings and such boundary must be capable of being situated in such a position as to enable the external walls of those buildings to comply with the requirements of this regulation; and

(ii) if such other building is an existing building, it shall be deemed to be a building to be erected on the site which it occupies, being of the same purpose group and having the same unprotected areas and fire resistance as the existing building.

As has already been stated an external wall on or within 1 m of the boundary must be fire resistant from both sides, whilst one situated 1 m or more from the boundary is tested from the inside only.

Regulation E7 does not control the space separation between buildings on the same site and in the same ownership (unless one of them is a residential building). Admittedly such a control might be unduly onerous on an industrial site but it is felt that some "modified" control might be available to restrict the erection of buildings so close together that they form a fire hazard one to another. It would also assist in fire fighting if adequate access were available on all sides of each building.

There are certain small unprotected areas listed in Schedule 9, Part 1 of which no account need be taken and this means that a wall on the boundary may include these certain small openings, otherwise it needs to be imperforate and to have the required fire resistance. In addition to the requirement of fire resistance the regulations require external walls to be non-combustible if they are on or within 1 m of the boundary, or form part of a building more than 15 m high. In Table A, Part 1 and Part 2 to regulation E5 exceptions to the rule are noted for small buildings falling within the lines marked "x" or "z" respectively, but it should be noted that buildings falling within purpose group II of Part I are not included. Similarly, parts of a building which do not exceed 15 m in height, providing they are completely separated from the main building, which may exceed 15 m high by a compartment wall, and they are 1 m or more from the boundary, need not comply with the requirement for non-combustibility. Where external walls are required to be non-combustible they may have combustible inner linings, providing these comply with E15 regarding surface spread of flame, or external cladding so long as this complies with E7(3)(b).

In applying this regulation, any beams, columns or other structure associated with an external wall must comply with the same requirements as to fire resistance and non-combustibility as the external wall.

Any part of a steeply pitched roof such as a Mansard and which contains

accommodation to which persons have access, is considered as an external wall if pitched at an angle of 70° or more to the horizontal.

Where no boundaries exist between buildings and either of them is a residential building (groups I and III) (other than small garages), a notional boundary is assumed which is plotted in such a position that the external walls of both buildings will comply with the regulations, but if one is an existing building, the notional boundary is placed so that the existing building complies, before considering the new building. This rule also applies to estates which are developed by local authorities and the ownership of the land and buildings remains under their jurisdiction.

The Regulations also contain other sections which are not perhaps so relevant to the steel designer. They are as follows:

    E8 Separating walls
    E9 Special requirements as to compartment walls and compartment floors
    E10 Protected shafts
    E11 Fire-resisting doors
    E12 Exceptions permitting use of certain doors in lift shafts
    E13 Stairways
    E14 Fire-stopping
    E15 Restriction of spread of flame over surfaces of walls and ceilings
    E16 Exceptions permitting ceilings to consist of plastics materials
    E17 Roofs
    E18 Small garages
    E19 Small open carports
    E20 Purpose group of small garages and open carports

With reference to the Deemed to Satisfy clauses; Schedule 8 — Part V, Tables A and B provide specifications for thickness of protection for fire resistance for structural steel.

It is interesting to note that the specifications provide for the protection of steel of minimum weights. This gives an opening for further research whereby the thickness of material required would vary with the section size. Some of the specifications shown are obsolete and the tables are incomplete because some new materials and methods are not included.

However, regulation E6(8)(*b*) permits the use of material other than that indicated in Schedule 8 provided it is proved to have the requisite fire resistance under conditions of the prescribed test. Pending a complete revision of the Deemed to Satisfy tables this allows the use of recently developed materials and methods or revised specifications.

Administration and enforcement of the preceding regulations is the responsibility of Local Authorities in England and Wales.

(ii) The Building Standards (Scotland) (Consolidation) Regulations 1971 — these regulations originally introduced in 1964 brought a new approach to fire protection and replaced a series of statutes and local By-laws which had tended to make a rational interpretation somewhat complex. They are most exhaustive in detail

SCHEDULE 8 NOTIONAL PERIODS OF FIRE RESISTANCE Part V : STRUCTURAL STEEL

A. *Encased steel stanchions (Mass per metre not less than 45 kilogrammes)*

Construction and materials	Minimum thickness (in millimetres) of protection for a fire resistance of:					
	4 hours	2 hours	1½ hours	1 hour	½ hour
(A) Solid Protection* (unplastered)					
1. Concrete not leaner than 1 : 2 : 4 mix with natural aggregate:					
(a) concrete not assumed to be loadbearing, reinforced†	50	25	25	25	25
(b) concrete assumed to be loadbearing, reinforced in accordance with B.S. 449: Part 2: 1969	75	50	50	50	50
2. Solid bricks of clay, composition or sand-lime	75	50	50	50	50
3. Solid blocks of foamed slag or pumice concrete reinforced† in every horizontal joint	62	50	50	50	50
4. Sprayed asbestos — 140 to 240 kilogrammes per cubic metre	44	19	15	10	10
5. Sprayed vermiculite-cement		38	32	19	12.5
(B) Hollow Protection‡					
1. Solid bricks of clay, composition or sand-lime reinforced in every horizontal joint, unplastered	115	50	50	50	50
2. Solid blocks of foamed slag or pumice concrete reinforced§ in every horizontal joint, unplastered	75	50	50	50	50
3. Metal lath with gypsum or cement-lime plaster of thickness of		§38	25	19	12.5
4. (a) Metal lath with vermiculite-gypsum or perlite-gypsum plaster of thickness of	§50	19	16	13	13
(b) metal lath spaced 25 mm from flanges with vermiculite-gypsum or perlite-gypsum plaster of thickness of	44	19	12.5	12.5	12.5
5. Gypsum plasterboard with 1.6 mm wire binding at 100 mm pitch:					
(a) 9.5 mm plasterboard with gypsum plaster of thickness of				12.5	12.5
(b) 19 mm plasterboard with gypsum plaster of thickness of		12.5	10	7	7
6. Gypsum plasterboard with 1.6 mm wire binding at 100 mm pitch:					
(a) 9.5 mm plasterboard with vermiculite-gypsum plaster of thickness of		16	12.5	10	7
(b) 19 mm plasterboard with vermiculite-gypsum plaster of thickness of		10	10	7	7
7. Metal lath with sprayed asbestos of thickness of	§32	19	15	10	10
8. Vermiculite-cement slabs of 4 : 1 mix reinforced with wire mesh and finished with plaster skim. Slabs of thickness of	44				
9. Asbestos insulating boards of density 510 to 880 kilogrammes per cubic metre (screwed to 25 mm thick asbestos battens for ½ hour and 1 hour periods)	63	25	25	25	25
		25	19	12	9

* Solid protection means a casing which is bedded close to the steel without intervening cavities and with all joints in that casing made full and solid.

† Reinforcement shall consist of steel binding wire not less than 2.3 mm in thickness, or a steel mesh weighing not less than 0.48 Kg/m². In concrete protection the spacing of that reinforcement shall not exceed 150 mm in any direction.

‡ Hollow protection means that there is a void between the protective material and the steel. All hollow protection to columns shall be effectively sealed at each floor level.

§ Light mesh reinforcement required 12.5 to 19 mm below surface unless special corner beads are used.

SCHEDULE 8 NOTIONAL PERIODS OF FIRE RESISTANCE Part V : STRUCTURAL STEEL (continued)

B. Encased steel beams (Mass per metre not less than 30 kilogrammes)

Construction and materials	4 hours	2 hours	1½ hours	1 hour	½ hour
(A) Solid Protection* (unplastered)					
1. Concrete not leaner than 1 : 2 : 4 mix with natural aggregates:					
(a) concrete not assumed to be loadbearing, reinforced†	63	25	25	25	25
(b) concrete assumed to be loadbearing, reinforced in accordance with B.S. 449: Part 2: 1969					
2. Sprayed asbestos — 140 to 240 kilogrammes per cubic metre	75	50	50	50	50
3. Sprayed vermiculite-cement	44	19	15	10	10
		38	32	19	12.5
(B) Hollow Protection‡					
1. Metal lathing:					
(a) with cement-lime plaster of thickness of		38	25	19	12.5
(b) with gypsum plaster of thickness of		22	19	16	12.5
(c) with vermiculite-gypsum or perlite-gypsum plaster of thickness of	32	12.5	12.5	12.5	12.5
2. Gypsum plasterboard with 1.6 mm wire binding at 100 mm pitch:					
(a) 9.5 mm plasterboard with gypsum plaster of thickness of				12.5	12.5
(b) 19 mm plasterboard with gypsum plaster of thickness of		12.5	10	7	7
3. Plasterboard with 1.6 mm wire binding at 100 mm pitch:					
(a) 9.5 mm plasterboard nailed to wooden cradles finished with gypsum plaster of thickness of					12.5
(b) 9.5 mm plasterboard with vermiculite-gypsum plaster of thickness of		16	12.5	10	7
(c) 19 mm plasterboard with vermiculite-gypsum plaster of thickness of		10	10	7	7
(d) 19 mm plasterboard with gypsum plaster of thickness of	§32	12.5			
4. Metal lathing with sprayed asbestos 140 to 240 kilogrammes per cubic metre of thickness of	44	19	15	10	10
5. Asbestos insulating boards of density 510 to 880 kilogrammes per cubic metre (screwed to 25.4 mm thick asbestos battens for ½ hour and 1 hour periods)		25	19	12	9
6. Vermiculite cement slabs of 4 : 1 mix reinforced with wire mesh and finished with plaster skim. Slabs of thickness of	63	25	25	25	25
7. Gypsum-sand plaster 12.5 mm thick applied to heavy duty (Type I as designated in B.S. 1105: 1963) woodwool slabs of thickness of		50	38	38	38

* Solid protection means a casing which is bedded close to the steel without intervening cavities and with all joints in that casing made full and solid.

† Reinforcement shall consist of steel binding wire not less than 2.3 mm in thickness, or a steel mesh weighing not less than 0.48 Kg/m². In concrete protection, the spacing of that reinforcement shall not exceed 150 mm in any direction.

‡ Hollow protection means that there is a void between the protective material and the steel. All hollow protection to columns shall be effectively sealed at each floor level.

§ Light mesh reinforcement required 12.5 to 19 mm below surface unless special corner beads are used.

content and represent a serious attempt by legislators to deal with the problems involved, employing the result of recent research and development. A notable feature of the regulations is the provision for means of escape and assistance to the fire service which establishes and combines all requirements concerning exits, stairways, enclosures, lighting of exits, fire mains and fire lifts.

(iii) The London Building (Constructional) By-laws 1972 contain fire resisting requirements in Part XI which is sub-headed 'Fire resistance of elements of construction and separations between buildings'.

The purpose of this part of these By-laws is to minimise the risk of the spread of fire between adjoining buildings by a stable and durable form of construction to prevent the untimely collapse of buildings in the event of fire and to minimise the risk of the spread of fire between specified parts of buildings.

Besides the above, details are provided in these By-laws for the construction of party walls, parapets and party floors. Clause 11.05 summarises the parameters of fire resistance requirements and Table 11 is the classification showing fire resistance periods, both these being reproduced below, as follows:—

11.05

Fire resistance period

(1) Each element of construction in a building, part of a building or division other than in a single-storey building or a division of a single-storey building not exceeding 7079.2 m^3 in cubical extent or 7.500 m in height, except separations between different tenancies and differing users, shall be capable of resisting the action of fire for a period not less than that specified in Table 11 according to the use and size of the building or part of a building.

(2) In a building of Class Nos. III or IV the elements of construction of the part used for office purposes, or for dwelling purposes or other purposes within the meaning of Class No. II (as the case may be), shall be capable of resisting the action of fire for the same period as that required if the whole building were used for purposes in Class No. II.

(3) Where, in any building, the level of the surface of any floor is more than 12.800 m above the level of the footway immediately in front of the centre of the face of the building, or if there is no footway, above the level of the ground before excavation, the elements of construction of that building shall be capable of resisting the action of fire for a period of not less than one hour, and shall be of non-combustible construction.

(4) Where, in the same building or part of a building, a different period of resistance to the action of fire is required by this By-law according to whether regard is had to the cubical extent, or to the level of the surface of any floor of the building or part or division of the building, the longer period shall be taken.

(5) Each element of construction of any basement storey (which for the purpose of this paragraph of this By-law

TABLE 11
CLASSIFICATION TABLE SHOWING FIRE-RESISTANCE PERIODS (IN HOURS)

Class No.	Use of building, or division, or part of a building	Not Exceeding 710 m³	Exceeding 710 m³ Not exceeding 1420 m³	Exceeding 1420 m³ Not exceeding 2130 m³	Exceeding 2130 m³ Not exceeding 3550 m³	Exceeding 3550 m³
I	As a warehouse or for trade or manufacture—the building or division so used	½	½	½	1	2*
II	For offices and/or for dwelling purposes, or for a purpose not included in this table, except as a public building—the building or part so used	Nil	Nil	½	½	1
III	Partly for offices or for other purposes under Class II, except for dwelling purposes and partly as a warehouse or for trade or manufacture—the part used as a warehouse or for trade or manufacture	½	½	1	1	2*
IV	Partly for dwelling purposes and partly as a warehouse or for trade or manufacture—the part used as a warehouse or for trade or manufacture	½	1	1	2	2*
V	For housing high-voltage power type electrical transformers and/or high-voltage power type switchgear or for a purpose involving a similar risk—the building or part so used	2	2	2	2	2*
		Not exceeding 142 m³	142 m³ to 1420 m³			
VI	For housing or displaying a petrol driven vehicle—the part so used	½	1	1	2	2*

* Limit of cubical extent of building or division if used as a warehouse or for purposes of trade or manufacture—7 079.20 m³

shall be deemed to include the floor over the basement storey), except in any building in Classes Nos. III and IV not exceeding 1420 m^3 in extent, shall be capable of resisting the action of fire for a period of not less than twice that required for the elements of construction of the building or part or division, of the building in which that basement storey is situated:

Provided that in no such case need that period exceed two hours.

Floors within separate dwellings (6) Notwithstanding the foregoing provisions of this By-law, where, in any building or part of a building, separate dwellings are to be formed, within any such dwelling comprising not more than two storeys the floors between those storeys need not be capable of resisting the action of fire for a period longer than half-an-hour.

Landings and stairs (7) Where in any building or part of a building the elements of construction are required to be capable of resisting the action of fire for a period longer than one hour, except where stairs or landings form the separation between different tenancies or different uses, any stairs (including landings and supports) need not be constructed to resist the action of fire for a period longer than one hour:

Provided that in any case where a staircase (including its landings and supports) is constructed of non-combustible material, it shall be deemed to comply with the provisions of this part of these By-laws if the whole of that staircase (including its landings and supports) is enclosed by partitions or walls capable of resisting the action of fire for a period not less than that prescribed by this part of these By-laws for the building or part of the building within which it is situated.

(8) Where a building, or part of a building, is provided with only one staircase, and the stairs, landings and the supports thereof are required by the provisions of this part of the By-laws to be capable of resisting the action of fire, those landings, stairs and supports shall be constructed of non-combustible material:

Provided that this paragraph of this By-law shall not apply to the stairs, landings and the supports thereof within the building or part of the building used as a separate dwelling.

The By-laws also provide for separation between tenancies and different users, openings and doors, borrowed lights, openings in floors, insulation of pipes and ventilating ducts, joints in separations and supports. By-laws 11.12 and 11.13 relate to existing buildings and fire resistance of materials respectively. Schedules VI and VII to the By-laws set out in detail various methods of providing the required standard of fire resistance and they are shortly to be metrically rounded and brought up to date in accordance with the latest advice from the Fire Research Station.

Public Buildings are not included in the Classification table as these must be

constructed in such a manner as may be approved by the District Surveyor.

Schedule IV Table D of the By-laws sets out various specifications "Deemed to Satisfy the By-laws by providing the required Standard of fire resistance.

In the Inner London area there are also a number of Acts (some of which apply to the country as a whole), Regulations and Codes of Practice that relate to fire which may be relevant to the steel designer. The Acts and Regulations are those concerned with office, shop, railway and factory premises and with places of public entertainment and steel rolling shutters. There are also Codes of Practice concerning means of escape and for buildings requiring approval under Section 20 of the London Building Acts (Amendment) Act 1939 namely

(a) all buildings with a storey or part of a storey at a greater height than
  (i) 30.480 m or
  (ii) 24.384 m if the area of the building exceeds 929.030 m^2.
(b) a warehouse building or a building used for trade or manufacture exceeding 7 079.210 m^3.

It is Part V (clause 5.08) of this code which contains the six hour fire rating requirement, and there are many other detailed requirements concerning multi-storey car parks and other special fire risks, partitions, wall and ceiling linings, the siting of buildings and the maximum size of compartments, etc.

It is important to remember that in the G.L.C. Inner London area previously defined the enforcement of legislation is undertaken by the District Surveyors and it is from them that advice should be obtained in the first instance regarding specific projects.

(4) Some Methods of Fire Protection

Fire protective claddings for structural steelwork can be classified broadly under two headings, which are:—

(a) methods executed by general contractors as builders work within a main building contract, or
(b) methods executed by nominated or specified sub-contractors.

Generally the traditional forms of solid encasements which are slow, heavy and laborious are carried out by the general contractors, whereas the lightweight casing or pre-formed methods, besides the sprayed and painted systems, are applied by approved applicators being somewhat more specialised operations and thus becoming nominated or specified.

The thickness of traditional claddings specified in By-laws have, over the years, been progressively reduced as tests to B.S. 476 indicate that the required protection can be obtained with lesser thicknesses.

The use of solid concrete casing to steel beams and stanchions usually involved *in-situ* concrete work poured into specially constructed formwork and is normally programmed by the general contractor to coincide with the pouring of *in-situ* concrete floors; By-laws and regulations specify the amount of wire fabric necessary to maintain the concrete cladding in position without spalling if exposed to fire, but irrespective of mandatory requirements it is obviously good practice to provide some light steel mesh wrapping and B.S. 4483 1969(7) includes meshes D49 at 0.770 kg/m^2

and D31 at 0.492 kg/m². Additional reinforcement is required if structural assistance is being considered.

The concurrent use of concrete casings as fire protection and assisting in carrying load has received more attention following the revision of B.S. 449(8) in 1959, in that in the revised standard, designs are permitted to include the effect of the load-carrying capacity of the concrete casing as distinct from merely taking advantage only of the stiffening effect as was previously allowed.

A series of tests carried out jointly at the Fire Research and Building Research Stations have indicated that cased stanchions, when furnace tested with the design load which takes the strength of the concrete into account, have fire resistance periods of 3.33 hours and 4.20 hours respectively; a lightweight concrete casing with expanded clay aggregate has even greater periods.

From these tests it was also apparent that the load factors for cased stanchions at normal temperatures were very high, irrespective of the type of concrete used.

The use of brickwork for fire protection is now infrequent although still desirable in some situations.

Lightweight concrete blocks prove to be a very satisfactory and economical method of fire protection when used in the right circumstances, this is particularly so when blockwork is being used as an inner skin on an external cavity wall in situations where steel columns occur on the perimeter of the building. It is then only necessary to extend the blockwork around the column to provide for fire protection and the cost is only for blockwork to the sides of the column.

Low density asbestos insulating board can provide a very satisfactory means of fire protection, and the manufacturers of this material will supply specifications and fixing details to comply with the various fire ratings.

Within the scope of work executed by sub-contractors there are now available both gravel concrete and lightweight concrete pre-cast units for encasing steelwork. The use of these represent a significant saving of time on site as no formwork is required but because of the many joints, they may need plastering on any contract where this finish is desirable, and in some cases to upgrade the fire rating where required to be over two hours.

Using pre-cast lightweight concrete interlocking blocks as beam casings, it is possible to support the edges of floors within the web. This can provide the advantage of saving overall height in a building as well as reducing the cost of fire protecting the sides of the beams. It is also possible to pre-cast insulating concrete to steelwork and erect the members complete with casing except for the ends where site connections are to be formed. The advantages mentioned for pre-cast casing will also obtain in this case, but the erection of the members will obviously be affected by their increased weight. Also the programming of steel delivery to allow time for casting and hardening of the casing is entailed. In precasting insulating concrete around steel sections it is usual to provide a casing of sheet steel — this can provide a variety of shapes and finishes, for example, galvanised or p.v.c.

It is significant that the tabulated data on conventional periods of fire resistance for various structural elements specified in the Schedules of the Building Regulations differ between concrete casing used structurally and that used merely to obtain fire protection.

The limitations involved in the use of the traditional methods of cladding has resulted in the development and adoption of some lightweight techniques. Most of these methods are executed as sub-contracts and can be classified as *in-situ*, generally

METHODS OF FIRE PROTECTION

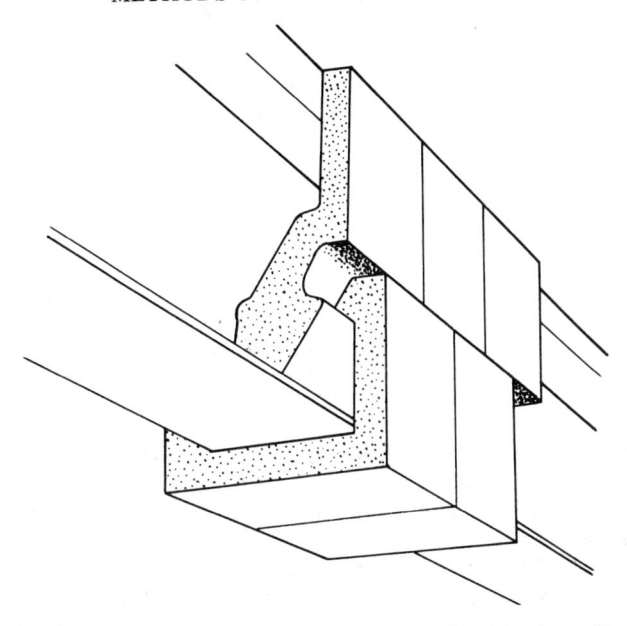

Fig. 1. Precast lightweight concrete interlocking blocks fixed to beam.

following the profile of the steel, or preformed, which are normally of square or rectangular section. Amongst the former are the Intumescent materials (paint and mastic).

The Paint has restricted application because it provides only a ½ hour rating and is only suitable for internal use and in environments which are not chemically corrosive or of high humidity. It provides a decorative finish, within a thickness of 2.54 mm, takes up no floor area and does not increase the dead load, and these advantages can offset the high cost. The finish is similar to a gloss paint but with a reduced sheen and a slightly dimpled or 'orange peel' surface.

The Intumescent mastic currently available provides for fire ratings of up to two hours, but because of the fairly high cost of executing this work, its use is mostly confined to the plant and process industries. It is particularly suitable for external use and can be executed whilst plant is operational.

Perhaps the most economical *in-situ* method is sprayed asbestos. This material has long been recognised for its fire resistance properties and these characteristics are exploited in spraying whereby the controlled moisturised stream of asbestos fibre forms a homogeneous coating over the steel surface to be protected. Thus it can be applied to any structure regardless of shape or size. In addition there are surface treatments available which prove visually more acceptable than the normal pressed finish, but in any case this method is not recommended if robustness is a prime requirement. In specifying this method of fire protection consideration should be taken of the Asbestos Regulations 1969(9) which require operatives to wear protective clothing and provides for cleanliness of plant and equipment, storage, distribution and prohibits the employment of young persons.

Further *in-situ* sprayed methods include specially prepared vermiculite and cement, also cement and sand with the inclusion of foaming agents.

Other methods of fire protecting steel are the dry preformed systems entailing

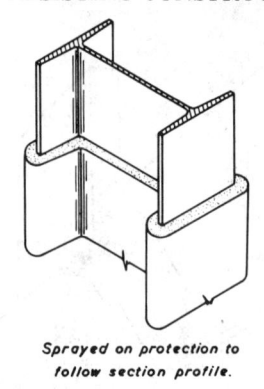

Sprayed on protection to follow section profile.

Fig. 2

the use of boards, blocks, slabs or pads of material. Special fixing clips, blocks, battens or other framing are some of the means employed to apply the dry materials, and it is possible to mount certain types directly on to the steel surface using a special adhesive.

The majority of boarding uses vermiculite aggregate bound with plaster or other binder, or alternatively ordinary plaster board can be used with a gypsum finish or gypsum and vermiculite (Carlite) being governed by the thickness the fire rating requires. Likewise expanded metal lathing can be wrapped around the steel to form a key for Carlite plaster. This is particularly useful where a complex steel framing is to be protected.

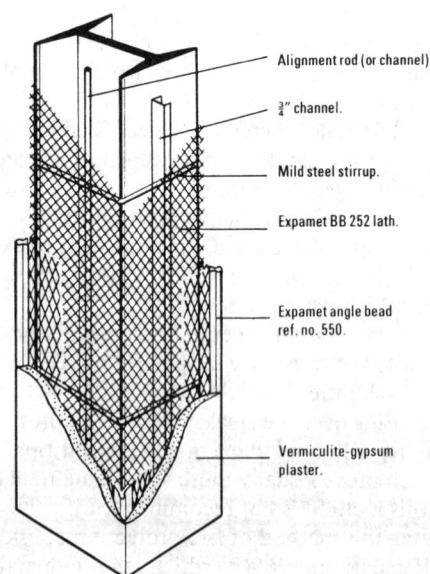

Fig. 3. Expamet lath and plaster encasement to a column.

Suspended ceilings which are frequently used in conjunction with structural steel frames offer several attractive features which include the ability to accommodate service engineering equipment whilst providing an aesthetically

acceptable soffit which, if required, can be profiled as an architectural feature. The choice of suspended ceilings is governed by Building Regulations; depending mainly on the height of building, type of floor and fire resistance of floor there are requirements for:—

(*a*) surface spread of flame of the unexposed surface of the ceiling;
(*b*) non-combustible supports and fixings;
(*c*) jointless as opposed to panel type ceilings.

If a building is more than 15 m high or if the fire rating is more than one hour then the ceiling must be jointless. In Inner London whenever the suspended ceiling is accepted as giving fire resisting protection to structural steel work (such a ceiling not being accepted as protection to columns) the ceiling must be jointless and imperforate, and the ceiling void must be used for no purpose other than for electric wiring in steel conduit and metal pipes other than gas pipes. In the Scottish Regulations, Schedule 5 details the design and construction of suspended ceilings contributing to fire resistance.

Several types of suspended ceiling can also provide many other desirable characteristics, such as sound absorption and insulation.

The fire protective elements are manufactured from materials which are good insulators and which do not deteriorate upon exposure to high temperatures. Some of the more common materials used are the vermiculite, perlite, asbestos, gypsum plaster and foamed slag. Most of these are used singly or combined with others to form proprietary materials.

Where traditional cladding is used this must normally be constructed as part of the structural building work and often becomes the sole work in progress for a long period during the contract. When modern protection is used, however, the fixing takes place during the finishing operations on the building. The effect of this is to allow the main contractor greater flexibility in planning and also, as the fire protection is no longer likely to be affected by adverse weather conditions, will invariably result in a reduced contract period which, with consequent reductions in the contractor's preliminaries, reduces overall building costs. The use of pre-cast flooring and roofing is frequently associated with modern fire protection of the structural steelwork and this also tends to reduce site erection time to a minimum; when pre-cast floors are used they can be seated directly on the top flanges of supporting beams which must then be protected on the remaining three sides or a suspended ceiling provided, or fixed by means of the pre-cast blocks previously described. A means of reducing headroom is for the floor units to be carried on shelf angles set so that the top flange of the steel beam is usually at least 25 mm below the top of the structural floor.

The outstanding feature of most forms of modern fire protection is extreme lightness in weight — many being less than 25 kg/m^2 for 50 mm thick cover. This factor means that the cladding weight can be reduced by as much as 10 per cent of the total load, the effect of this is to reduce steel section sizes and ease foundation problems, especially in areas when settlement due to dead load is a design criterion. Again, these factors reduce building costs.

The final advantage offered by lightweight fire protection is the ease with which it can be modified to suit changes of building layout or service alterations. In some instances it may be necessary to modify the degree of protection offered and this too can readily be accommodated by the flexibility available.

In order to make the most of the above advantages it is usually necessary to detail structural steelwork with reference to the form of protection envisaged — the small thickness of protection required precludes the use of cleats, splice plates or bolt heads which, as they cannot be accommodated within the cladding, would result in unsightly connections; for this reason it may in some cases be desirable to use welded end plate connections for beam to column joints. If seating cleats are required for the purpose of erection these may be simply bolted on and subsequently removed. Similarly, stanchion splices should be located and detailed with care to achieve the desired clean lines; for universal sections of the same serial size this may be achieved by the use of internal splice plates.

The question of paint protection for structural steelwork to be encased with lightweight material must be decided upon taking all factors into consideration; obviously all loose mill scale which in itself can be a source of corrosion, must be removed prior to cladding but in buildings where no aggressive features are present it is unlikely that the steelwork will deteriorate if left unpainted. For some forms of cladding which are applied to the steel surface, however, a special pre-treatment is often applied and the presence of a conventional paint film may even be disadvantageous in such circumstances.

Thickness of different types of modern protection are contained in the schedules to the Building Regulations and Constructional By-laws quoted earlier in this chapter. General constructional details of many proprietary types of cladding are given in a booklet (10) published by the British Constructional Steelwork Association, a new edition of which is being prepared by 'CONSTRADO'.*

(5) Possible Future Developments

There seems to be no doubt that in the future there will be a continued emphasis on fire protection as far as building is concerned. With capital intensive plant, for example, computer installations, destruction by fire or damage by water could become critical and designers will more and more recognise that fire protection should not be dealt with in isolation from the rest of the building, but must be considered and co-ordinated from the time the design is first conceived. This will ensure that the maximum economy and integration will be provided initially, not only in regard to the requirements of the various Building Regulations but also for fire prevention and fire fighting. It is also anticipated that the Regulations will be based exclusively on scientific research, and therefore greater flexibility will be allowed and further relaxations permitted when research indicates that the present requirements are too stringent.

It may be optimistic to consider that some completely new fire resisting material is going to be developed that is comparable or lower in cost to materials already in use, but there certainly will be new methods of application of existing material. Research will pursue the development of admixes for adhesion, aeration or foaming which will enable *in-situ* mixes to be applied more quickly and cheaply. Also, the expected rationalisation of structural sections due to metrication should reduce the number of sizes of sections rolled thus making pre-formed boarded methods more viable. Again, there may be a move away from some of the heavier methods such as concrete to the lightweight pre-formed or *in-situ* systems.

* Constructional Steel Research & Development Organisation. Albany House, Petty France, London, S.W.1.

A further development now actively being pursued is the use of water-filled hollow section steel structures; this started on the basis of columns only but now includes the complete structural frame. Some testing has been carried out both here and overseas and the results are sufficiently encouraging to warrant further development of this system. In the meantime there are a number of buildings using this system erected outside the U.K. The problems of freezing and of corrosion seem to have been overcome by suitable additives to the water.

Intumescent material will also be the subject of further development. With paint the aim must be at least for a one hour rating with an ultimate aim of two hours coupled with a complete range of colours and also a higher resistance to moisture. For external use and for ratings exceeding two hours, it is anticipated the Intumescent mastic should become more popular dependent upon the ability to provide an acceptable finish at a more reasonable cost than at present.

With the use of weathering steel, a new design concept has arisen whereby structural elements of the external walling are placed outside the façade and the exposed steel frame becomes an architectural feature. This presents an entirely new fire situation, where conventional fire protection is not acceptable for aesthetic reasons. However, where elements are not subject to severe radiation or flame impingement, as with most applications, there is no need for fire protection. Further research is necessary before legislation can be framed to provide for these conditions and at present each design is examined by the Joint Fire Research Organisation and recommendations formulated.

Given time, design rules will be produced which will enable the designer to compute the position of the exposed columns to eliminate the need for fire protection. The important parameters are the sizes of the steel sections and their distances from the façade, details of window openings and the proposed occupancy or fire load.

In a situation where it is not possible to eliminate fire protection completely, suitable heat shields can be provided. These may consist of a sheet steel fabrication fixed to the fire exposed face of the column or beam at fire risk with provision for an insulating void and overlap.

Again, research has shown (11) that it is safe, for certain values of fire load and ventilation opening, to employ unprotected steelwork internally in offices and similar buildings. Present legislation prohibits this, however.

All these findings stem from the ability to calculate severity of a fire — that is, its maximum temperature and duration — from a knowledge of the nature, amount and arrangement of the fuel, and also the size and shape of the room and window(s) and thermal insulation of walls and ceiling. Knowing the severity of the real fire it is possible to correlate this with a certain period of the standard fire test. (12)

The B.S. 476 fire resistance test applies to isolated elements, usually tested in a simply supported state. To examine the fire behaviour of full size assemblies of beams and columns, a new experimental station has been built for the European Convention for Structural Steelwork at Maiziérs-Les-Metz, in France. This enables realistic conditions of jointing, continuity and structural restraint to be examined under the ISO time temperature curve (this is virtually identical to that of B.S. 476). The test results, now becoming available, will provide much needed information on the behaviour of an assembled structure which is not necessarily the same as that of its component parts.

The administration of the Building Regulations 1972, the G.L.C. Constructional By-laws and the Building Standards (Scotland) (Consolidation) Regulations 1971 could be improved by uniformity so that one set would apply nationally.

The history and reasons for the differing regulations has already been traced but it should now be recognised that rationalisation is necessary.

This could be achieved when Local Government is reorganised into larger administrative areas and could take the form of a set of National Regulations. The use of these should show considerable saving in design costs and simplify enforcement.

(6) Bibliography

(1) The Building Regulations 1972. H.M.S.O.
(2) The Building Standards (Scotland) (Consolidation) Regulations 1971. H.M.S.O.
(3) London Building (Constructional) By-laws 1972. Greater London Council.
(4) Code of Practice for buildings of excess height and/or additional cubical extent requiring approval under Section 20 of the London Building Acts (Amendment) Act 1939. Greater London Council.
(5) B.S. 476 Part 1:1953 with Amendment No. 1, 26th January 1970. British Standards Institution.
(6) Joint Circular from the Ministry of Housing and Local Government and Welsh Office (17/68 and 11/68 respectively) 'The Building Regulations 1965 Multi-Storey Car Parks'. H.M.S.O.
(7) B.S. 4483: 1969 Steel fabric for the reinforcement of concrete (metric units). British Standards Institution.
(8) B.S. 449: 1959 The use of Structural Steel in Building. British Standards Institution.
(9) The Asbestos Regulations 1969. H.M.S.O.
(10) Modern Fire Protection for Structural Steelwork FP3. British Constructional Steelwork Association Ltd.
(11) The temperature attained by steel in building fires. Fire Research Technical paper No. 15 1966. H.M.S.O.
(12) Behaviour of Structural Steel in Fire. JFRO Symposium No. 2. H.M.S.O.

Other References

(a) W. H. CUTMORE, *Shaw's Commentary on The Building Regulations.* Shaw & Sons Ltd. The publishers gratefully acknowledge permission to reproduce sections of parts of the commentary.
(b) London Building Acts 1930–39. Greater London Council.
(c) Explanatory Memorandum to the Building Standards (Scotland) Regulations 1971 – Parts D and E. H.M.S.O.
(d) Report of the Committee on Building Legislation in Scotland. Cmnd. 269. H.M.S.O.
(e) Report of the Departmental Committee on the Fire Service. Cmnd. 7371 H.M.S.O.
(f) Glossary of terms associated with fire B.S. 4422 Part 1. British Standards Institution.
(g) Notes on some methods of fire protection of Structural Steelwork currently in use. British Steel Corporation.
(h) CARPENTER, JOHN and WHITTINGTON, RICHARD (compiled by), *Liber Albus,* translated by H. T. Riley. Richard Griffin and Company.
(i) G. V. BLACKSTONE, *A History of the British Fire Service.* Routledge and Kegan Paul.

BIBLIOGRAPHY

(j)   Building with Steel issue No. 3 May 1970. Fire Protection.
         The Spread of Fire in Buildings.
         Structural Steel and Fire.
         The Economics of Fire Protection.
         Car Parks and Fire.
         Fire and Steel Components.
            British Steel Corporation.

(k)   Multi-storey Car Parks — Relaxations of the Building Regulations 1965 to permit the use of Exposed Steel Framework. The British Constructional Steelwork Association Ltd.

(l)   Conference on 'Steel in Architecture' November 1969, organised by and papers published by the British Constructional Steelwork Association Ltd.

40. MISCELLANEOUS TABLES

CONVERSION TABLES
MOMENTS OF INERTIA
INCHES^4 UNITS TO CENTIMETRES^4 UNITS

in.^{4}	0	·1	·2	·3	·4	·5	·6	·7	·8	·9
	cm.^{4}	cm.^{4}	cm.^{4}	cm.^{4}	cm.^{4}	cm.^{4}	cm.^{4}	cm.^{4}	cm.^{4}	cm.^{4}
1	41·62314256	45·785	49·948	54·110	58·272	62·435	66·597	70·759	74·922	79·084
2	83·24628	87·409	91·571	95·733	99·895	104·058	108·220	112·382	116·545	120·707
3	124·86942	129·032	133·194	137·356	141·519	145·681	149·843	154·006	158·168	162·330
4	166·49257	170·655	174·817	178·979	183·142	187·304	191·466	195·629	199·791	203·953
5	208·11571	212·278	216·440	220·603	224·765	228·927	233·089	237·252	241·414	245·576
6	249·73886	253·901	258·063	262·226	266·388	270·550	274·713	278·875	283·037	287·200
7	291·36200	295·524	299·687	303·849	308·011	312·173	316·336	320·498	324·660	328·883
8	332·98514	337·147	341·310	345·472	349·634	353·797	357·959	362·121	366·284	370·446
9	374·60829	378·771	382·933	387·095	391·257	395·420	399·582	403·744	407·907	412·069
10	416·23143									

CENTIMETRES^4 UNITS TO INCHES^4 UNITS

cm.^{4}	0	·1	·2	·3	·4	·5	·6	·7	·8	·9
	in.^{4}	in.^{4}	in.^{4}	in.^{4}	in.^{4}	in.^{4}	in.^{4}	in.^{4}	in.^{4}	in.^{4}
1	0·02402510	0·02643	0·02883	0·03123	0·03363	0·03604	0·03844	0·04084	0·04324	0·04565
2	0·04805019	0·05045	0·05285	0·05526	0·05766	0·06006	0·06246	0·06487	0·06727	0·06967
3	0·07207529	0·07448	0·07688	0·07928	0·08168	0·08409	0·08649	0·08889	0·09129	0·09370
4	0·09610038	0·09850	0·10091	0·10331	0·10571	0·10811	0·11052	0·11292	0·11532	0·11772
5	0·12012548	0·12253	0·12493	0·12733	0·12974	0·13214	0·13454	0·13694	0·13935	0·14175
6	0·14415058	0·14655	0·14895	0·15136	0·15376	0·15616	0·15856	0·16097	0·16337	0·16577
7	0·16817567	0·17058	0·17298	0·17538	0·17779	0·18019	0·18259	0·18499	0·18740	0·18980
8	0·19220077	0·19460	0·19701	0·19941	0·20181	0·20421	0·20662	0·20902	0·21142	0·21382
9	0·21622586	0·21863	0·22103	0·22343	0·22584	0·22824	0·23064	0·23304	0·23545	0·23785
10	0·24025096									

MODULI OF SECTION
INCHES^3 UNITS TO CENTIMETRES^3 UNITS

in.^{3}	0	·1	·2	·3	·4	·5	·6	·7	·8	·9
	cm.^{3}	cm.^{3}	cm.^{3}	cm.^{3}	cm.^{3}	cm.^{3}	cm.^{3}	cm.^{3}	cm.^{3}	cm.^{3}
1	16·387064	18·026	19·664	21·303	22·942	24·580	26·219	27·858	29·497	31·135
2	32·77413	34·413	36·051	37·690	39·329	40·967	42·606	44·245	45·884	47·522
3	49·16119	50·800	52·438	54·077	55·716	57·355	58·993	60·632	62·271	63·909
4	65·54826	67·187	68·826	70·464	72·103	73·742	75·380	77·019	78·658	80·297
5	81·93532	83·574	85·213	86·851	88·490	90·129	91·767	93·406	95·045	96·684
6	98·32238	99·961	101·600	103·238	104·877	106·516	108·154	109·793	111·432	113·071
7	114·70945	116·348	117·987	119·625	121·624	122·903	124·542	126·180	127·819	129·458
8	131·09651	132·735	134·374	136·012	137·651	139·290	140·929	142·567	144·206	145·845
9	147·48358	149·122	150·761	152·400	154·038	155·677	157·316	158·954	160·593	162·232
10	163·87064									

CENTIMETRES^3 UNITS TO INCHES^3 UNITS

cm.^{3}	0	·1	·2	·3	·4	·5	·6	·7	·8	·9
	in.^{3}	in.^{3}	in.^{3}	in.^{3}	in.^{3}	in.^{3}	in.^{3}	in.^{3}	in.^{3}	in.^{3}
1	0·061024	0·06713	0·07323	0·07933	0·08543	0·09154	0·09764	0·10374	0·10984	0·11594
2	0·122047	0·12815	0·13425	0·14035	0·14646	0·15256	0·15866	0·16476	0·17087	0·17697
3	0·183071	0·18917	0·19527	0·20138	0·20748	0·21358	0·21968	0·22579	0·23189	0·23799
4	0·244095	0·25020	0·25630	0·26240	0·26850	0·27461	0·28071	0·28681	0·29291	0·29902
5	0·305119	0·31122	0·31732	0·32342	0·32953	0·33563	0·34173	0·34783	0·35394	0·36004
6	0·366142	0·37224	0·37835	0·38445	0·39055	0·39665	0·40276	0·40886	0·41496	0·42106
7	0·427166	0·43327	0·43937	0·44547	0·45158	0·45768	0·46378	0·46988	0·47599	0·48209
8	0·488190	0·49429	0·50040	0·50650	0·51260	0·51870	0·52480	0·53091	0·53701	0·54311
9	0·549213	0·55532	0·56142	0·56752	0·57362	0·57973	0·58583	0·59193	0·59803	0·60414
10	0·610237									

Based on 1 inch = 25·4 millimetres

MISCELLANEOUS TABLES

CONVERSION TABLES
Extract from B.S. 350 : 1944

WEIGHTS PER UNIT LENGTH

POUNDS PER FOOT TO KILOGRAMS PER METRE

Based on 1 inch = 25·4 millimetres ; 1 pound = 0·45359243 kilograms

Pounds per ft.	0	1	2	3	4	5	6	7	8	9
	kg./m.	kg./m.	kg./m.	kg./m.	kg./m.	kg./m.	kg./m.	kg./m.	kg./m.	kg./m.
—	—	1·4882	2·9763	4·4645	5·9527	7·4408	8·9290	10·417	11·905	13·393
10	14·882	16·370	17·858	19·346	20·834	22·322	23·811	25·299	26·787	28·275
20	29·763	31·251	32·740	34·228	35·716	37·204	38·692	40·180	41·669	43·157
30	44·645	46·133	47·621	49·109	50·598	52·086	53·574	55·062	56·550	58·038
40	59·527	61·015	62·503	63·991	65·479	66·967	68·456	69·944	71·432	72·920
50	74·408	75·896	77·385	78·873	80·361	81·849	83·337	84·825	86·314	87·802
60	89·290	90·778	92·266	93·754	95·242	96·731	98·219	99·707	101·195	102·683
70	104·171	105·660	107·148	108·636	110·124	111·612	113·100	114·589	116·077	117·565
80	119·053	120·541	122·029	123·518	125·006	126·494	127·982	129·470	130·958	132·447
90	133·935	135·423	136·911	138·399	139·887	141·376	142·864	144·352	145·840	147·328
100	148·816									

KILOGRAMS PER METRE TO POUNDS PER FOOT

Based on 1 inch = 25·4 millimetres ; 1 pound = 0·45359243 kilograms

Kg. per m.	0	1	2	3	4	5	6	7	8	9
	lb./ft.	lb./ft.	lb./ft.	lb./ft.	lb./ft.	lb./ft.	lb./ft.	lb./ft.	lb./ft.	lb./ft.
—	—	0·67197	1·3439	2·0159	2·6879	3·3598	4·0318	4·7038	5·3758	6·0477
10	6·7197	7·3917	8·0636	8·7356	9·4076	10·0795	10·7515	11·4235	12·0954	12·7674
20	13·4394	14·1113	14·7833	15·4553	16·1273	16·7992	17·4712	18·1432	18·8151	19·4871
30	20·1591	20·8310	21·5030	22·1750	22·8469	23·5189	24·1909	24·8629	25·5348	26·2068
40	26·8788	27·5507	28·2227	28·8947	29·5666	30·2386	30·9106	31·5825	32·2545	32·9265
50	33·5984	34·2704	34·9424	35·6144	36·2863	36·9583	37·6303	38·3022	38·9742	39·6462
60	40·3181	40·9901	41·6621	42·3340	43·0060	43·6780	44·3500	45·0219	45·6939	46·3659
70	47·0378	47·7098	48·3818	49·0537	49·7257	50·398	51·070	51·742	52·414	53·086
80	53·758	54·429	55·101	55·773	56·445	57·117	57·789	58·461	59·133	59·805
90	60·477	61·149	61·821	62·493	63·165	63·837	64·509	65·181	65·853	66·525
100	67·197	67·869	68·541	69·213	69·885	70·557	71·229	71·901	72·573	73·245
110	73·917	74·589	75·261	75·932	76·604	77·276	77·948	78·620	79·292	79·964
120	80·636	81·308	81·980	82·652	83·324	83·996	84·668	85·340	86·012	86·684
130	87·356	88·028	88·700	89·372	90·044	90·716	91·388	92·060	92·732	93·404
140	94·076	94·748	95·420	96·092	96·764	97·436	98·107	98·779	99·451	100·123

MISCELLANEOUS TABLES

CONVERSION TABLES
Extract from B.S. 350: 1944

WEIGHTS
POUNDS TO KILOGRAMS
Based on 1 pound = 0·45359243 kilograms

lb.	0	1	2	3	4	5	6	7	8	9
	kg.	kg.	kg.	kg.	kg.	kg.	kg.	kg.	kg.	kg.
—	—	0·45359	0·90718	1·36078	1·81437	2·26796	2·72155	3·17515	3·62874	4·08233
10	4·53592	4·98952	5·4431	5·8967	6·3503	6·8039	7·2575	7·7111	8·1647	8·6183
20	9·0718	9·5254	9·9790	10·4326	10·8862	11·3398	11·7934	12·2470	12·7006	13·1542
30	13·6078	14·0614	14·5150	14·9686	15·4221	15·8757	16·3293	16·7829	17·2365	17·6901
40	18·1437	18·5973	19·0509	19·5045	19·9581	20·4117	20·8653	21·3188	21·7724	22·2260
50	22·6796	23·1332	23·5868	24·0404	24·4940	24·9476	25·4012	25·8548	26·3084	26·7620
60	27·2155	27·6691	28·1227	28·5763	29·0299	29·4835	29·9371	30·3907	30·8443	31·2979
70	31·7515	32·2051	32·6587	33·1122	33·5658	34·0194	34·4730	34·9266	35·3802	35·8338
80	36·2874	36·7410	37·1946	37·6482	38·1018	38·5554	39·0089	39·4625	39·9161	40·3697
90	40·8233	41·2769	41·7305	42·1841	42·6377	43·0913	43·5449	43·9985	44·4521	44·9057
100	45·3592	45·8128	46·2664	46·7200	47·1736	47·6272	48·0808	48·5344	48·9880	49·4416
10	49·8952	50·349	50·802	51·256	51·710	52·163	52·617	53·070	53·524	53·977
20	54·431	54·885	55·338	55·792	56·245	56·699	57·153	57·606	58·060	58·513
30	58·967	59·421	59·874	60·328	60·781	61·235	61·689	62·142	62·596	63·049
40	63·503	63·957	64·410	64·864	65·317	65·771	66·224	66·678	67·132	67·585
50	68·039	68·492	68·946	69·400	69·853	70·307	70·760	71·214	71·668	72·121
60	72·575	73·028	73·482	73·936	74·389	74·843	75·296	75·750	76·204	76·657
70	77·111	77·564	78·018	78·471	78·925	79·379	79·832	80·286	80·739	81·193
80	81·647	82·100	82·554	83·007	83·461	83·915	84·368	84·822	85·275	85·729
90	86·183	86·636	87·090	87·543	87·997	88·451	88·904	89·358	89·811	90·265
200	90·718	91·172	91·626	92·079	92·533	92·986	93·440	93·894	94·347	94·801
10	95·254	95·708	96·162	96·615	97·069	97·522	97·976	98·430	98·883	99·337
20	99·790	100·244	100·698	101·151	101·605	102·058	102·512	102·965	103·419	103·873
30	104·326	104·780	105·233	105·687	106·141	106·594	107·048	107·501	107·955	108·409
40	108·862	109·316	109·769	110·223	110·677	111·130	111·584	112·037	112·491	112·945
50	113·398	113·852	114·305	114·759	115·212	115·666	116·120	116·573	117·027	117·480
60	117·934	118·388	118·841	119·295	119·748	120·202	120·656	121·109	121·563	122·016
70	122·470	122·924	123·377	123·831	124·284	124·738	125·192	125·645	126·099	126·552
80	127·006	127·459	127·913	128·367	128·820	129·274	129·727	130·181	130·635	131·088
90	131·542	131·995	132·449	132·903	133·356	133·810	134·263	134·717	135·171	135·624
300	136·078	136·531	136·985	137·439	137·892	138·346	138·799	139·253	139·706	140·160
10	140·614	141·067	141·521	141·974	142·428	142·882	143·335	143·789	144·242	144·696
20	145·150	145·603	146·057	146·510	146·964	147·418	147·871	148·325	148·778	149·232
30	149·686	150·139	150·593	151·046	151·500	151·953	152·407	152·861	153·314	153·768
40	154·221	154·675	155·129	155·582	156·036	156·489	156·943	157·397	157·850	158·304
50	158·757	159·211	159·665	160·118	160·572	161·025	161·479	161·932	162·386	162·840
60	163·293	163·747	164·200	164·654	165·108	165·561	166·015	166·468	166·922	167·376
70	167·829	168·283	168·736	169·190	169·644	170·097	170·551	171·004	171·458	171·912
80	172·365	172·819	173·272	173·726	174·179	174·633	175·087	175·540	175·994	176·447
90	176·901	177·355	177·808	178·262	178·715	179·169	179·623	180·076	180·530	180·983
400	181·437	181·891	182·344	182·798	183·251	183·705	184·159	184·612	185·066	185·519
10	185·973	186·426	186·880	187·334	187·787	188·241	188·694	189·148	189·602	190·055
20	190·509	190·962	191·416	191·870	192·323	192·777	193·230	193·684	194·138	194·591
30	195·045	195·498	195·952	196·406	196·859	197·313	197·766	198·220	198·673	199·127
40	199·581	200·034	200·488	200·941	201·395	201·849	202·302	202·756	203·209	203·663
50	204·117	204·570	205·024	205·477	205·931	206·385	206·838	207·292	207·745	208·199
60	208·653	209·106	209·560	210·013	210·467	210·920	211·374	211·828	212·281	212·735
70	213·188	213·642	214·096	214·549	215·003	215·456	215·910	216·364	216·817	217·271
80	217·724	218·178	218·632	219·085	219·539	219·992	220·446	220·900	221·353	221·807
90	222·260	222·714	223·167	223·621	224·075	224·528	224·982	225·435	225·889	226·343

MISCELLANEOUS TABLES

CONVERSION TABLES
Extract from B.S. 350: 1944

WEIGHTS

POUNDS TO KILOGRAMS (continued)
Based on 1 pound = 0·45359243 kilograms

lb.	0	1	2	3	4	5	6	7	8	9
	kg.	kg.	kg.	kg.	kg.	kg.	kg.	kg.	kg.	kg.
500	226·796	227·250	227·703	228·157	228·611	229·064	229·518	229·971	230·425	230·879
10	231·332	231·786	232·239	232·693	233·147	233·600	234·054	234·507	234·961	235·414
20	235·868	236·322	236·775	237·229	237·682	238·136	238·590	239·043	239·497	239·950
30	240·404	240·858	241·311	241·765	242·218	242·672	243·126	243·579	244·033	244·486
40	244·940	245·394	245·847	246·301	246·754	247·208	247·661	248·115	248·569	249·022
50	249·476	249·929	250·383	250·837	251·290	251·744	252·197	252·651	253·105	253·558
60	254·012	254·465	254·919	255·373	255·826	256·280	256·733	257·187	257·640	258·094
70	258·548	259·001	259·455	259·908	260·362	260·816	261·269	261·723	262·176	262·630
80	263·084	263·537	263·991	264·444	264·898	265·352	265·805	266·259	266·712	267·166
90	267·620	268·073	268·527	268·980	269·434	269·887	270·341	270·795	271·248	271·702
600	272·155	272·609	273·063	273·516	273·970	274·423	274·877	275·331	275·784	276·238
10	276·691	277·145	277·599	278·052	278·506	278·959	279·413	279·867	280·320	280·774
20	281·227	281·681	282·134	282·588	283·042	283·495	283·949	284·402	284·856	285·310
30	285·763	286·217	286·670	287·124	287·578	288·031	288·485	288·938	289·392	289·846
40	290·299	290·753	291·206	291·660	292·114	292·567	293·021	293·474	293·928	294·381
50	294·835	295·289	295·742	296·196	296·649	297·103	297·557	298·010	298·464	298·917
60	299·371	299·825	300·278	300·732	301·185	301·639	302·093	302·546	303·000	303·453
70	303·907	304·361	304·814	305·268	305·721	306·175	306·628	307·082	307·536	307·989
80	308·443	308·896	309·350	309·804	310·257	310·711	311·164	311·618	312·072	312·525
90	312·979	313·432	313·886	314·340	314·793	315·247	315·700	316·154	316·608	317·061
700	317·515	317·968	318·422	318·875	319·329	319·783	320·236	320·690	321·143	321·597
10	322·051	322·504	322·958	323·411	323·865	324·319	324·772	325·226	325·679	326·133
20	326·587	327·040	327·494	327·947	328·401	328·855	329·308	329·762	330·215	330·669
30	331·122	331·576	332·030	322·483	332·937	333·390	333·844	334·298	334·751	335·205
40	335·658	336·112	336·566	337·019	337·473	337·926	338·380	338·834	339·287	339·741
50	340·194	340·648	341·101	341·555	342·009	342·462	342·916	343·369	343·823	344·277
60	344·730	345·184	345·637	346·091	346·545	346·998	347·452	347·905	348·359	348·813
70	349·266	349·720	350·173	350·627	351·081	351·534	351·988	352·441	352·895	353·349
80	353·802	354·256	354·709	355·163	355·616	356·070	356·524	356·977	357·431	357·884
90	358·338	358·792	359·245	359·699	360·152	360·606	361·060	361·513	361·967	362·420
800	362·874	363·328	363·781	364·235	364·688	365·142	365·595	366·049	366·503	366·956
10	367·410	367·863	368·317	368·771	369·224	369·678	370·131	370·585	371·039	371·492
20	371·946	372·399	372·853	373·307	373·760	374·214	374·667	375·121	375·575	376·028
30	276·482	376·935	377·389	377·842	378·296	378·750	379·203	379·657	380·110	380·564
40	381·018	381·471	381·925	382·378	382·832	383·286	383·739	384·193	384·646	385·100
50	385·554	386·007	386·461	386·914	387·368	387·822	388·275	388·729	389·182	389·636
60	390·089	390·543	390·997	391·450	391·904	392·357	392·811	393·265	393·718	394·172
70	394·625	395·079	395·533	395·986	396·440	396·893	397·347	397·801	398·254	398·708
80	399·161	399·615	400·069	400·522	400·976	401·429	401·883	402·336	402·790	403·244
90	403·697	404·151	404·604	405·058	405·512	405·965	406·419	406·872	407·326	407·780
900	408·233	408·687	409·140	409·594	410·048	410·501	410·955	411·408	411·862	412·316
10	412·769	413·223	413·676	414·130	414·583	415·037	415·491	415·944	416·398	416·851
20	417·305	417·759	418·212	418·666	419·119	419·573	420·027	420·481	420·934	421·387
30	421·841	422·295	422·748	423·202	423·655	424·109	424·563	425·016	425·470	425·923
40	426·377	426·830	427·284	427·738	428·191	428·645	429·098	429·552	430·006	430·459
50	430·913	431·366	431·820	432·274	432·727	433·181	433·634	434·088	434·542	434·995
60	435·449	435·902	436·356	436·810	437·263	437·717	438·170	438·624	439·077	439·531
70	439·985	440·438	440·892	441·345	441·799	442·253	442·706	443·160	443·613	444·067
80	444·521	444·974	445·428	445·881	446·335	446·789	447·242	447·696	448·149	448·603
90	449·057	449·510	449·964	450·417	450·871	451·324	451·778	452·232	452·685	453·139
1000	453·592									

MISCELLANEOUS TABLES

CONVERSION TABLES
Extract from B.S. 350 : 1944

KILOGRAMS TO POUNDS

Based on 1 pound = 0·45359243 kilograms

Kg.	0	1	2	3	4	5	6	7	8	9
	lb.	lb.	lb.	lb.	lb.	lb.	lb.	lb.	lb.	lb.
—	—	2·20462	4·40924	6·6139	8·8185	11·0231	13·2277	15·4324	17·6370	19·8416
10	22·0462	24·2508	26·4555	28·6601	30·8647	33·0693	35·2740	37·4786	39·6832	41·8878
20	44·0924	46·2971	48·5017	50·706	52·911	55·116	57·320	59·525	61·729	63·934
30	66·139	68·343	70·548	72·753	74·957	77·162	79·366	81·571	83·776	85·980
40	88·185	90·390	92·594	94·799	97·003	99·208	101·413	103·617	105·822	108·026
50	110·231	112·436	114·640	116·845	119·050	121·254	123·459	125·663	127·868	130·073
60	132·277	134·482	136·687	138·891	141·096	143·300	145·505	147·710	149·914	152·119
70	154·324	156·528	158·733	160·937	163·142	165·347	167·551	169·756	171·961	174·165
80	176·370	178·574	180·779	182·984	185·188	187·393	189·598	191·802	194·007	196·211
90	198·416	200·621	202·825	205·030	207·235	209·439	211·644	213·848	216·053	218·258
100	220·462	222·667	224·871	227·076	229·281	231·485	233·690	235·895	238·099	240·304
10	242·508	244·713	246·918	249·122	251·327	253·532	255·736	257·941	260·145	262·350
20	264·555	266·759	268·964	271·169	273·373	275·578	277·782	279·987	282·192	284·396
30	286·601	288·806	291·010	293·215	295·419	297·624	299·829	302·033	304·238	306·442
40	308·647	310·852	313·056	315·261	317·466	319·670	321·875	324·079	326·284	328·489
50	330·693	332·898	335·103	337·307	339·512	341·716	343·921	346·126	348·330	350·535
60	352·740	354·944	357·149	359·353	361·558	363·763	365·967	368·172	370·377	372·581
70	374·786	376·990	379·195	381·400	383·604	385·809	388·014	390·218	392·423	394·627
80	396·832	399·037	401·241	403·446	405·651	407·855	410·060	412·264	414·469	416·674
90	418·878	421·083	423·288	425·492	427·697	429·901	432·106	434·311	436·515	438·720
200	440·924	443·129	445·334	447·538	449·743	451·948	454·152	456·357	458·561	460·766
10	462·971	465·175	467·380	469·585	471·789	473·994	476·198	478·403	480·608	482·812
20	485·017	487·222	489·426	491·631	493·835	496·040	498·245	500·45	502·65	504·86
30	507·06	509·27	511·47	513·68	515·88	518·09	520·29	522·50	524·70	526·90
40	529·11	531·31	533·52	535·72	537·93	540·13	542·34	544·54	546·75	548·95
50	551·16	553·36	555·56	557·77	559·97	562·18	564·38	566·59	568·79	571·00
60	573·20	575·41	577·61	579·82	582·02	584·22	586·43	588·63	590·84	593·04
70	595·25	597·45	599·66	601·86	604·07	606·27	608·48	610·68	612·89	615·09
80	617·29	619·50	621·70	623·91	626·11	628·32	630·52	632·73	634·93	637·14
90	639·34	641·55	643·75	645·95	648·16	650·36	652·57	654·77	656·98	659·18
300	661·39	663·59	665·80	668·00	670·21	672·41	674·61	676·82	679·02	681·23
10	683·43	685·64	687·84	690·05	692·25	694·46	696·66	698·87	701·07	703·27
20	705·48	707·68	709·89	712·09	714·30	716·50	718·71	720·91	723·12	725·32
30	727·53	729·73	731·93	734·14	736·34	738·55	740·75	742·96	745·16	747·37
40	749·57	751·78	753·98	756·19	758·39	760·59	762·80	765·00	767·21	769·41
50	771·62	773·82	776·03	778·23	780·44	782·64	784·85	787·05	789·25	791·46
60	793·66	795·87	798·07	800·28	802·48	804·69	806·89	809·10	811·30	813·51
70	815·71	817·91	820·12	822·32	824·53	826·73	828·94	831·14	833·35	835·55
80	837·76	839·96	842·17	844·37	846·57	848·78	850·98	853·19	855·39	857·60
90	859·80	862·01	864·21	866·42	868·62	870·83	873·03	875·24	877·44	879·64
400	881·85	884·05	886·26	888·46	890·67	892·87	895·08	897·28	899·49	901·69
10	903·90	906·10	908·30	910·51	912·71	914·92	917·12	919·33	921·53	923·74
20	925·94	928·15	930·35	932·56	934·76	936·96	939·17	941·37	943·58	945·78
30	947·99	950·19	952·40	954·60	956·81	959·01	961·22	963·42	965·62	967·83
40	970·03	972·24	974·44	976·65	978·85	981·06	983·26	985·47	987·67	989·88
50	992·08	994·28	996·49	998·69	1000·90	1003·10	1005·31	1007·51	1009·72	1011·92
60	1014·13	1016·33	1018·54	1020·74	1022·94	1025·15	1027·35	1029·56	1031·76	1033·97
70	1036·17	1038·38	1040·58	1042·79	1044·99	1047·20	1049·40	1051·60	1053·81	1056·01
80	1058·22	1060·42	1062·63	1064·83	1067·04	1069·24	1071·45	1073·65	1075·86	1078·06
90	1080·26	1082·47	1084·67	1086·88	1089·08	1091·29	1093·49	1095·70	1097·90	1100·11

MISCELLANEOUS TABLES

CONVERSION TABLES
Extract from B.S. 350: 1944

KILOGRAMS TO POUNDS (continued)

Based on 1 pound = 0·45359243 kilograms

Kg.	0	1	2	3	4	5	6	7	8	9
	lb.	lb.	lb.	lb.	lb.	lb.	lb.	lb.	lb.	lb.
500	1102·31	1104·52	1106·72	1108·93	1111·13	1113·33	1115·54	1117·74	1119·95	1122·15
10	1124·36	1126·56	1128·77	1130·97	1133·18	1135·38	1137·59	1139·79	1141·99	1144·20
20	1146·40	1148·61	1150·81	1153·02	1155·22	1157·43	1159·63	1161·84	1164·04	1166·25
30	1168·45	1170·65	1172·86	1175·06	1177·27	1179·47	1181·68	1183·88	1186·09	1188·29
40	1190·50	1192·70	1194·91	1197·11	1199·31	1201·52	1203·72	1205·93	1208·13	1210·34
50	1212·54	1214·75	1216·95	1219·16	1221·36	1223·57	1225·77	1227·97	1230·18	1232·38
60	1234·59	1236·79	1239·00	1241·20	1243·41	1245·61	1247·82	1250·02	1252·23	1254·43
70	1256·63	1258·84	1261·04	1263·25	1265·45	1267·66	1269·86	1272·07	1274·27	1276·48
80	1278·68	1280·89	1283·09	1285·29	1287·50	1289·70	1291·91	1294·11	1296·32	1298·52
90	1300·73	1302·93	1305·14	1307·34	1309·55	1311·75	1313·95	1316·16	1318·36	1320·57
600	1322·77	1324·98	1327·18	1329·39	1331·59	1333·80	1336·00	1338·21	1340·41	1342·62
10	1344·82	1347·02	1349·23	1351·43	1353·64	1355·84	1358·05	1360·25	1362·46	1364·66
20	1366·87	1369·07	1371·28	1373·48	1375·68	1377·89	1380·09	1382·30	1384·50	1386·71
30	1388·91	1391·12	1393·32	1395·53	1397·73	1399·94	1402·14	1404·34	1406·55	1408·75
40	1410·96	1413·16	1415·37	1417·57	1419·78	1421·98	1424·19	1426·39	1428·60	1430·80
50	1433·00	1435·21	1437·41	1439·62	1441·82	1444·03	1446·23	1448·44	1450·64	1452·85
60	1455·05	1457·26	1459·46	1461·66	1463·87	1466·07	1468·28	1470·48	1472·69	1474·89
70	1477·10	1479·30	1481·51	1483·71	1485·92	1488·12	1490·32	1492·53	1494·73	1496·94
80	1499·14	1501·35	1503·55	1505·76	1507·96	1510·17	1512·37	1514·58	1516·78	1518·98
90	1521·19	1523·39	1525·60	1527·80	1530·01	1532·21	1534·42	1536·62	1538·83	1541·03
700	1543·24	1545·44	1547·64	1549·85	1552·05	1554·26	1556·46	1558·67	1560·87	1563·08
10	1565·28	1567·49	1569·69	1571·90	1574·10	1576·30	1578·51	1580·71	1582·92	1585·12
20	1587·33	1589·53	1591·74	1593·94	1596·15	1598·35	1600·56	1602·76	1604·97	1607·17
30	1609·37	1611·58	1613·78	1615·99	1618·19	1620·40	1622·60	1624·81	1627·01	1629·22
40	1631·42	1633·63	1635·83	1638·03	1640·24	1642·44	1644·65	1646·85	1649·06	1651·26
50	1653·47	1655·67	1657·88	1660·08	1662·29	1664·49	1666·69	1668·90	1671·10	1673·31
60	1675·51	1677·72	1679·92	1682·13	1684·33	1686·54	1688·74	1690·95	1693·15	1695·35
70	1697·56	1699·76	1701·97	1704·17	1706·38	1708·58	1710·79	1712·99	1715·20	1717·40
80	1719·61	1721·81	1724·01	1726·22	1728·42	1730·63	1732·83	1735·04	1737·24	1739·45
90	1741·65	1743·86	1746·06	1748·27	1750·47	1752·67	1754·88	1757·08	1759·29	1761·49
800	1763·70	1765·90	1768·11	1770·31	1772·52	1774·72	1776·93	1779·13	1781·33	1783·54
10	1785·74	1787·95	1790·15	1792·36	1794·56	1796·77	1798·97	1801·18	1803·38	1805·59
20	1807·79	1809·99	1812·20	1814·40	1816·61	1818·81	1821·02	1823·22	1825·43	1827·63
30	1829·84	1832·04	1834·25	1836·45	1838·65	1840·86	1843·06	1845·27	1847·47	1849·68
40	1851·88	1854·09	1856·29	1858·50	1860·70	1862·91	1865·11	1867·32	1869·52	1871·72
50	1873·93	1876·13	1878·34	1880·54	1882·75	1884·95	1887·16	1889·36	1891·57	1893·77
60	1895·98	1898·18	1900·38	1902·59	1904·79	1907·00	1909·20	1911·41	1913·61	1915·82
70	1918·02	1920·23	1922·43	1924·64	1926·84	1929·04	1931·25	1933·45	1935·66	1937·86
80	1940·07	1942·27	1944·48	1946·68	1948·89	1951·09	1953·30	1955·50	1957·70	1959·91
90	1962·11	1964·32	1966·52	1968·73	1970·93	1973·14	1975·34	1977·55	1979·75	1981·96
900	1984·16	1986·36	1988·57	1990·77	1992·98	1995·18	1997·39	1999·59	2001·80	2004·00
10	2006·21	2008·41	2010·62	2012·82	2015·02	2017·23	2019·43	2021·64	2023·84	2026·05
20	2028·25	2030·46	2032·66	2034·87	2037·07	2039·28	2041·48	2043·68	2045·89	2048·09
30	2050·30	2052·50	2054·71	2056·91	2059·12	2061·32	2063·53	2065·73	2067·94	2070·14
40	2072·35	2074·55	2076·75	2078·96	2081·16	2083·37	2085·57	2087·78	2089·98	2092·19
50	2094·39	2096·60	2098·80	2101·01	2103·21	2105·41	2107·62	2109·82	2112·03	2114·23
60	2116·44	2118·64	2120·85	2123·05	2125·26	2127·46	2129·67	2131·87	2134·07	2136·28
70	2138·48	2140·69	2142·89	2145·10	2147·30	2149·51	2151·71	2153·92	2156·12	2158·33
80	2160·53	2162·73	2164·94	2167·14	2169·35	2171·55	2173·76	2175·96	2178·17	2180·37
90	2182·58	2184·78	2186·99	2189·19	2191·39	2193·60	2195·80	2198·01	2200·21	2202·42
1000	2204·62									

MISCELLANEOUS TABLES

CONVERSION TABLES
Extract from B.S. 350 : 1944

FEET AND INCHES TO METRES
Based on 1 inch = 25·4 millimetres

Uncontracted values

Feet	Inches 0	1	2	3	4	5	Differences for sixteenths of an inch	
	m.	m.	m.	m.	m.	m.		m.
0	—	0·0254	0·0508	0·0762	0·1016	0·1270		
1	0·3048	0·3302	0·3556	0·3810	0·4064	0·4318		
2	0·6096	0·6350	0·6604	0·6858	0·7112	0·7366	1	0·0016
3	0·9144	0·9398	0·9652	0·9906	1·0160	1·0414		
4	1·2192	1·2446	1·2700	1·2954	1·3208	1·3462		
5	1·5240	1·5494	1·5748	1·6002	1·6256	1·6510	2	0·0032
6	1·8288	1·8542	1·8796	1·9050	1·9304	1·9558		
7	2·1336	2·1590	2·1844	2·2098	2·2352	2·2606		
8	2·4384	2·4638	2·4892	2·5146	2·5400	2·5654	3	0·0048
9	2·7432	2·7686	2·7940	2·8194	2·8448	2·8702		
10	3·0480	3·0734	3·0988	3·1242	3·1496	3·1750		
11	3·3528	3·3782	3·4036	3·4290	3·4544	3·4798	4	0·0064
12	3·6576	3·6830	3·7084	3·7338	3·7592	3·7846		
13	3·9624	3·9878	4·0132	4·0386	4·0640	4·0894		
14	4·2672	4·2926	4·3180	4·3434	4·3688	4·3942	5	0·0079
15	4·5720	4·5974	4·6228	4·6482	4·6736	4·6990		
16	4·8768	4·9022	4·9276	4·9530	4·9784	5·0038		
17	5·1816	5·2070	5·2324	5·2578	5·2832	5·3086	6	0·0095
18	5·4864	5·5118	5·5372	5·5626	5·5880	5·6134		
19	5·7912	5·8166	5·8420	5·8674	5·8928	5·9182		
20	6·0960	6·1214	6·1468	6·1722	6·1976	6·2230	7	0·0111
21	6·4008	6·4262	6·4516	6·4770	6·5024	6·5278		
22	6·7056	6·7310	6·7564	6·7818	6·8072	6·8326		
23	7·0104	7·0358	7·0612	7·0866	7·1120	7·1374	8	0·0127
24	7·3152	7·3406	7·3660	7·3914	7·4168	7·4422		
25	7·6200	7·6454	7·6708	7·6962	7·7216	7·7470		
26	7·9248	7·9502	7·9756	8·0010	8·0264	8·0518	9	0·0143
27	8·2296	8·2550	8·2804	8·3058	8·3312	8·3566		
28	8·5344	8·5598	8·5852	8·6106	8·6360	8·6614		
29	8·8392	8·8646	8·8900	8·9154	8·9408	8·9662	10	0·0159
30	9·1440	9·1694	9·1948	9·2202	9·2456	9·2710		
31	9·4488	9·4742	9·4996	9·5250	9·5504	9·5758		
32	9·7536	9·7790	9·8044	9·8298	9·8552	9·8806	11	0·0175
33	10·0584	10·0838	10·1092	10·1346	10·1600	10·1854		
34	10·3632	10·3886	10·4140	10·4394	10·4648	10·4902		
35	10·6680	10·6934	10·7188	10·7442	10·7696	10·7950	12	0·0190
36	10·9728	10·9982	11·0236	11·0490	11·0744	11·0998		
37	11·2776	11·3030	11·3284	11·3538	11·3792	11·4046		
38	11·5824	11·6078	11·6332	11·6586	11·6840	11·7094	13	0·0206
39	11·8872	11·9126	11·9380	11·9634	11·9888	12·0142		
40	12·1920	12·2174	12·2428	12·2682	12·2936	12·3190		
41	12·4968	12·5222	12·5476	12·5730	12·5984	12·6238	14	0·0222
42	12·8016	12·8270	12·8524	12·8778	12·9032	12·9286		
43	13·1064	13·1318	13·1572	13·1826	13·2080	13·2334		
44	13·4112	13·4366	13·4620	13·4874	13·5128	13·5382	15	0·0238
45	13·7160	13·7414	13·7668	13·7922	13·8176	13·8430		
46	14·0208	14·0462	14·0716	14·0970	14·1224	14·1478		
47	14·3256	14·3510	14·3764	14·4018	14·4272	14·4526		
48	14·6304	14·6558	14·6812	14·7066	14·7320	14·7574		
49	14·9352	14·9606	14·9860	15·0114	15·0368	15·0622		

MISCELLANEOUS TABLES

CONVERSION TABLES
Extract from B.S. 350: 1944

FEET AND INCHES TO METRES

Based on 1 inch = 25·4 millimetres

Uncontracted values.

Feet	Inches						Differences for sixteenths of an inch	
	6	7	8	9	10	11		
	m.	m.	m.	m.	m.	m.		m.
0	0·1524	0·1778	0·2032	0·2286	0·2540	0·2794		
1	0·4572	0·4826	0·5080	0·5334	0·5588	0·5842		
2	0·7620	0·7874	0·8128	0·8382	0·8636	0·8890	1	0·0016
3	1·0668	1·0922	1·1176	1·1430	1·1684	1·1938		
4	1·3716	1·3970	1·4224	1·4478	1·4732	1·4986		
5	1·6764	1·7018	2·7272	1·7526	1·7780	1·8034	2	0·0032
6	1·9812	2·0066	2·0320	2·0574	2·0828	2·1082		
7	2·2860	2·3114	2·3368	2·3622	2·3876	2·4130		
8	2·5908	2·6162	2·6416	2·6670	2·6924	2·7178	3	0·0048
9	2·8956	2·9210	2·9464	2·9718	2·9972	3·0226		
10	3·2004	3·2258	3·2512	3·2766	3·3020	3·3274		
11	3·5052	3·5306	3·5560	3·5814	3·6068	3·6322	4	0·0064
12	3·8100	3·8354	3·8608	3·8862	3·9116	3·9370		
13	4·1148	4;1402	4·1656	4·1910	4·2164	4·2418		
14	4·4196	4·4450	4·4704	4·4958	4·5212	4·5466	5	0·0079
15	4·7244	4·7498	4·7752	4·8006	4·8260	4·8514		
16	5·0292	5·0546	5·0800	5·1054	5·1308	5·1562		
17	5·3340	5·3594	5·3848	5·4102	5·4356	5·4610	6	0·0095
18	5·6388	5·6642	5·6896	5·7150	5·7404	5·7658		
19	5·9436	5·9690	5·9944	6·0198	6·0452	6·0706		
20	6·2484	6·2738	6·2992	6·3246	6·3500	6·3754	7	0·0111
21	6·5532	6·5786	6·6040	6·6294	6·6548	6·6802		
22	6·8580	6·8834	6·9088	6·9342	6·9596	6·9850		
23	7·1628	7·1882	7·2136	7·2390	7·2644	7·2898	8	0·0127
24	7·4676	7·4930	7·5184	7·5438	7·5692	7·5946		
25	7·7724	7·7978	7·8232	7·8486	7·8740	7·8994		
26	8·0772	8·1026	8·1280	8·1534	8·1788	8·2042	9	0·0143
27	8·3820	8·4074	8·4328	8·4582	8·4836	8·5090		
28	8·6868	8·7122	8·7376	8·7630	8·7884	8·8138		
29	8·9916	9·0170	9·0424	9·0678	9·0932	9·1186	10	0·0159
30	9·2964	9·3218	9·3472	9·3726	9·3980	9·4234		
31	9·6012	9·6266	9·6520	9·6774	9·7028	9·7282		
32	9·9060	9·9314	9·9568	9·9822	10·0076	10·0330	11	0·0175
33	10·2108	10·2362	10·2616	10·2870	10·3124	10·3378		
34	10·5156	10·5410	10·5664	10·5918	10·6172	10·6426		
35	10·8204	10·8458	10·8712	10·8966	10·9220	10·9474	12	0·0190
36	11·1252	11·1506	11·1760	11·2014	11·2268	11·2522		
37	11·4300	11·4554	11·4808	11·5062	11·5316	11·5570		
38	11·7348	11·7602	11·7856	11·8110	11·8364	11·8618	13	0·0206
39	12·0396	12·0650	12·0904	12·1158	12·1412	12·1666		
40	12·3444	12·3698	12·3952	12·4206	12·4460	12·4714		
41	12·6492	12·6746	12·7000	12·7254	12·7508	12·7762	14	0·0222
42	12·9540	12·9794	13·0048	13·0302	13·0556	13·0810		
43	13·2588	13·2842	13·3096	13·3350	13·3604	13·3858		
44	13·5636	13·5890	13·6144	13·6398	13·6652	13·6906	15	0·0238
45	13·8684	13·8938	13·9192	13·9446	13·9700	13·9954		
46	14·1732	14·1986	14·2240	14·2494	14·2748	14·3002		
47	14·4780	14·5034	14·5288	14·5542	14·5796	14·6050		
48	14·7828	14·8082	14·8336	14·8590	14·8844	14·9098		
49	15·0876	15·1130	15·1384	15·1638	15·1892	15·2146		

MISCELLANEOUS TABLES

CONVERSION TABLES
Extract from B.S. 350 : 1944

FEET AND INCHES TO METRES

Based on 1 inch = 25·4 millimetres

Uncontracted values

Feet	\multicolumn{6}{c}{Inches}	\multicolumn{2}{c}{Differences for sixteenths of an inch}
	0	1	2	3	4	5		
	m.	m.	m.	m.	m.	m.		m.
50	15·2400	15·2654	15·2908	15·3162	15·3416	15·3670		
51	15·5448	15·5702	15·5956	15·6210	15·6464	15·6718		
52	15·8496	15·8750	15·9004	15·9258	15·9512	15·9766	1	0·0016
53	16·1544	16·1798	16·2052	16·2306	16·2560	16·2814		
54	16·4592	16·4846	16·5100	16·5354	16·5608	16·5862		
55	16·7640	16·7894	16·8148	16·8402	16·8656	16·8910	2	0·0032
56	17·0688	17·0942	17·1196	17·1450	17·1704	17·1958		
57	17·3736	17·3990	17·4244	17·4498	17·4752	17·5006		
58	17·6784	17·7038	17·7292	17·7546	17·7800	17·8054	3	0·0048
59	17·9832	18·0086	18·0340	18·0594	18·0848	18·1102		
60	18·2880	18·3134	18·3388	18·3642	18·3896	18·4150		
61	18·5928	18·6182	18·6436	18·6690	18·6944	18·7198	4	0·0064
62	18·8976	18·9230	18·9484	18·9738	18·9992	19·0246		
63	19·2024	19·2278	19·2532	19·2786	19·3040	19·3294		
64	19·5072	19·5326	19·5580	19·5834	19·6088	19·6342	5	0·0079
65	19·8120	19·8374	19·8628	19·8882	19·9136	19·9390		
66	20·1168	20·1422	20·1676	20·1930	20·2184	20·2438		
67	20·4216	20·4470	20·4724	20·4978	20·5232	20·5486	6	0·0095
68	20·7264	20·7518	20·7772	20·8026	20·8280	20·8534		
69	21·0312	21·0566	21·0820	21·1074	21·1328	21·1582		
70	21·3360	21·3614	21·3868	21·4122	21·4376	21·4630	7	0·0111
71	21·6408	21·6662	21·6916	21·7170	21·7424	21·7678		
72	21·9456	21·9710	21·9964	22·0218	22·0472	22·0726		
73	22·2504	22·2758	22·3012	22·3266	22·3520	22·3774	8	0·0127
74	22·5552	22·5806	22·6060	22·6314	22·6568	22·6822		
75	22·8600	22·8854	22·9108	22·9362	22·9616	22·9870		
76	23·1648	23·1902	23·2156	23·2410	23·2664	23·2918	9	0·0143
77	23·4696	23·4950	23·5204	23·5458	23·5712	23·5966		
78	23·7744	23·7998	23·8252	23·8506	23·8760	23·9014		
79	24·0792	24·1046	24·1300	24·1554	24·1808	24·2062	10	0·0159
80	24·3840	24·4094	24·4348	24·4602	24·4856	24·5110		
81	24·6888	24·7142	24·7396	24·7650	24·7904	24·8158		
82	24·9936	25·0190	25·0444	25·0698	25·0952	25·1206	11	0·0175
83	25·2984	25·3238	25·3492	25·3746	25·4000	25·4254		
84	25·6032	25·6286	25·6540	25·6794	25·7048	25·7302		
85	25·9080	25·9334	25·9588	25·9842	26·0096	26·0350	12	0·0190
86	26·2128	26·2382	26·2636	26·2890	26·3144	26·3398		
87	26·5176	26·5430	26·5684	26·5938	26·6192	26·6446		
88	26·8224	26·8478	26·8732	26·8986	26·9240	26·9494	13	0·0206
89	27·1272	27·1526	27·1780	27·2034	27·2288	27·2542		
90	27·4320	27·4574	27·4828	27·5082	27·5336	27·5590		
91	27·7368	27·7622	27·7876	27·8130	27·8384	27·8638	14	0·0222
92	28·0416	28·0670	28·0924	28·1178	28·1432	28·1686		
93	28·3464	28·3718	28·3972	28·4226	28·4480	28·4734		
94	28·6512	28·6766	28·7020	28·7274	28·7528	28·7782	15	0·0238
95	28·9560	28·9814	29·0068	29·0322	29·0576	29·0830		
96	29·2608	29·2862	29·3116	29·3370	29·3624	29·3878		
97	29·5656	29·5910	29·6164	29·6418	29·6672	29·6926		
98	29·8704	29·8958	29·9212	29·9466	29·9720	29·9974		
99	30·1752	30·2006	30·2260	30·2514	30·2768	30·3022		
100	30·4800							

MISCELLANEOUS TABLES

CONVERSION TABLES
Extract from B.S. 350 : 1944

FEET AND INCHES TO METRES

Based on 1 inch = 25·4 millimetres

Uncontracted values

Feet	\multicolumn{6}{c	}{Inches}	\multicolumn{2}{c}{Differences for sixteenths of an inch}
	6	7	8	9	10	11		
	m.	m.	m.	m.	m.	m.		m.
50	15·3924	15·4178	15·4432	15·4686	15·4940	15·5194		
51	15·6972	15·7226	15·7480	15·7734	15·7988	15·8242		
52	16·0020	16·0274	16·0528	16·0782	16·1036	16·1290	1	0·0016
53	16·3068	16·3322	16·3576	16·3830	16·4084	16·4338		
54	16·6116	16·6370	16·6624	16·6878	16·7132	16·7386		
55	16·9164	16·9418	16·9672	16·9926	17·0180	17·0434	2	0·0032
56	17·2212	17·2466	17·2720	17·2974	17·3228	17·3482		
57	17·5260	17·5514	17·5768	17·6022	17·6276	17·6530		
58	17·8304	17·8562	17·8816	17·9070	17·9324	17·9578	3	0·0048
59	18·1356	18·1610	18·1864	18·2118	18·2372	18·2626		
60	18·4404	18·4658	18·4912	18·5166	18·5420	18·5674		
61	18·7452	18·7706	18·7960	18·8214	18·8468	18·8722	4	0·0064
62	19·0500	19·0754	19·1008	19·1262	19·1516	19·1770		
63	19·3548	19·3802	19·4056	19·4310	19·4564	19·4818		
64	19·6596	19·6850	19·7104	19·7358	19·7612	19·7866	5	0·0079
65	19·9644	19·9898	20·0152	20·0406	20·0660	20·0914		
66	20·2692	20·2946	20·3200	20·3454	20·3708	20·3962		
67	20·5740	20·5994	20·6248	20·6502	20·6756	20·7010	6	0·0095
68	20·8788	20·9042	20·9296	20·9550	20·9804	21·0058		
69	21·1836	21·2090	21·2344	21·2598	21·2852	21·3106		
70	21·4884	21·5138	21·5392	21·5646	21·5900	21·6154	7	0·0111
71	21·7932	21·8186	21·8440	21·8694	21·8948	21·9202		
72	22·0980	22·1234	22·1488	22·1742	22·1996	22·2250		
73	22·4028	22·4282	22·4536	22·4790	22·5044	22·5298	8	0·0127
74	22·7076	22·7330	22·7584	22·7838	22·8092	22·8346		
75	23·0124	23·0378	23·0632	23·0886	23·1140	23·1394		
76	23·3172	23·3426	23·3680	23·3934	23·4188	23·4442	9	0·0143
77	23·6220	23·6474	23·6728	23·6982	23·7236	23·7490		
78	23·9268	23·9522	23·9776	24·0030	24·0284	24·0538		
79	24·2316	24·2570	24·2824	24·3078	24·3332	24·3586	10	0·0159
80	24·5364	24·5618	24·5872	24·6126	24·6380	24·6634		
81	24·8412	24·8666	24·8920	24·9174	24·9428	24·9682		
82	25·1460	25·1714	25·1968	25·2222	25·2476	25·2730	11	0·0175
83	25·4508	25·4762	25·5016	25·5270	25·5524	25·5778		
84	25·7556	25·7810	25·8064	25·8318	25·8572	25·8826		
85	26·0604	26·0858	26·1112	26·1366	26·1620	26·1874	12	0·0190
86	26·3652	26·3906	26·4160	26·4414	26·4668	26·4922		
87	26·6700	26·6954	26·7208	26·7462	26·7716	26·7970		
88	26·9748	27·0002	27·0256	27·0510	27·0764	27·1018	13	0·0206
89	27·2796	27·3050	27·3304	27·3558	27·3812	27·4066		
90	27·5844	27·6098	27·6352	27·6606	27·6860	27·7114		
91	27·8892	27·9146	27·9400	27·9654	27·9908	28·0162	14	0·0222
92	28·1940	28·2194	28·2448	28·2702	28·2956	28·3210		
93	28·4988	28·5242	28·5496	28·5750	28·6004	28·6258		
94	28·8036	28·8290	28·8544	28·8798	28·9052	28·9306	15	0·0238
95	29·1084	29·1338	29·1592	29·1846	29·2100	29·2354		
96	29·4132	29·4386	29·4640	29·4894	29·5148	29·5402		
97	29·7180	29·7434	29·7688	29·7942	29·8196	29·8450		
98	30·0228	30·0482	30·0736	30·0990	30·1244	30·1498		
99	30·3276	30·3530	30·3784	30·4038	30·4292	30·4546		

MISCELLANEOUS TABLES

CONVERSION TABLES
Extract from B.S. 350: 1944

METRES TO FEET

Based on 1 inch = 25·4 millimetres

m.	0	1	2	3	4	5	6	7	8	9
	ft.	ft.	ft.	ft.	ft.	ft.	ft.	ft.	ft.	ft.
—	—	3·28084	6·5617	9·8425	13·1234	16·4042	19·6850	22·9659	26·2467	29·5276
10	32·8084	36·0892	39·3701	42·6509	45·9317	49·2126	52·493	55·774	59·055	62·336
20	65·617	68·898	72·178	75·459	78·740	82·021	85·302	88·583	91·863	95·144
30	98·425	101·706	104·987	108·268	111·549	114·829	118·110	121·391	124·672	127·953
40	131·234	134·514	137·795	141·076	144·357	147·638	150·919	154·199	157·480	160·761
50	164·042	167·323	170·604	173·884	177·165	180·446	183·727	187·008	190·289	193·570
60	196·850	200·131	203·412	206·693	209·974	213·255	216·535	219·816	223·097	226·378
70	229·659	232·940	236·220	239·501	242·782	246·063	249·344	252·625	255·905	259·186
80	262·467	265·748	269·029	272·310	275·590	278·871	282·152	285·433	288·714	291·995
90	295·276	298·556	301·837	305·118	308·399	311·680	314·961	318·241	321·522	324·803
100	328·084	331·365	334·646	337·926	341·207	344·488	347·769	351·050	354·331	357·611
10	360·892	364·173	367·454	370·735	374·016	377·296	380·577	383·858	387·139	390·420
20	393·701	396·982	400·262	403·543	406·824	410·105	413·386	416·667	419·947	423·228
30	426·509	429·790	433·071	436·352	439·632	442·913	446·194	449·475	452·756	456·037
40	459·317	462·598	465·879	469·160	472·441	475·722	479·002	482·283	485·564	488·845
50	492·126	495·407	498·688	501·97	505·25	508·53	511·81	515·09	518·37	521·65
60	524·93	528·22	531·50	534·78	538·06	541·34	544·62	547·90	551·18	554·46
70	557·74	561·02	564·30	567·59	570·87	574·15	577·43	580·71	583·99	587·27
80	590·55	593·83	597·11	600·39	603·67	606·96	610·24	613·52	616·80	620·08
90	623·36	626·64	629·92	633·20	636·48	639·76	643·04	646·33	649·61	652·89
200	656·17	659·45	662·73	666·01	669·29	672·57	675·85	679·13	682·41	685·70
10	688·98	692·26	695·54	698·82	702·10	705·38	708·66	711·94	715·22	718·50
20	721·78	725·07	728·35	731·63	734·91	738·19	741·47	744·75	748·03	751·31
30	754·59	757·87	761·15	764·44	767·72	771·00	774·28	777·56	780·84	784·12
40	787·40	790·68	793·96	797·24	800·52	803·81	807·09	810·37	813·65	816·93
50	820·21	823·49	826·77	830·05	833·33	836·61	839·89	843·18	846·46	849·74
60	853·02	856·30	859·58	862·86	866·14	869·42	872·70	875·98	879·26	882·55
70	885·83	889·11	892·39	895·67	898·95	902·23	905·51	908·79	912·07	915·35
80	918·63	921·92	925·20	928·48	931·76	935·04	938·32	941·60	944·88	948·16
90	951·44	954·72	958·00	961·29	964·57	967·85	971·13	974·41	977·69	980·97
300	984·25	987·53	990·81	994·09	997·38	1000·66	1003·94	1007·22	1010·50	1013·78
10	1017·06	1020·34	1023·62	1026·90	1030·18	1033·46	1036·75	1040·03	1043·31	1046·59
20	1049·87	1053·15	1056·43	1059·71	1062·99	1066·27	1069·55	1072·83	1076·12	1079·40
30	1082·68	1085·96	1089·24	1092·52	1095·80	1099·08	1102·36	1105·64	1108·92	1112·20
40	1115·49	1118·77	1122·05	1125·33	1128·61	1131·89	1135·17	1138·45	1141·73	1145·01
50	1148·29	1151·57	1154·86	1158·14	1161·42	1164·70	1167·98	1171·26	1174·54	1177·82
60	1181·10	1184·38	1187·66	1190·94	1194·23	1197·51	1200·79	1204·07	1207·35	1210·63
70	1213·91	1217·19	1220·47	1223·75	1227·03	1230·31	1233·60	1236·88	1240·16	1243·44
80	1246·72	1250·00	1253·28	1256·56	1259·84	1263·12	1266·40	1269·68	1272·97	1276·25
90	1279·53	1282·81	1286·09	1289·37	1292·65	1295·93	1299·21	1302·49	1305·77	1309·05
400	1312·34	1315·62	1318·90	1322·18	1325·46	1328·74	1332·02	1335·30	1338·58	1341·86
10	1345·14	1348·42	1351·71	1354·99	1358·27	1361·55	1364·83	1368·11	1371·39	1374·67
20	1377·95	1381·23	1384·51	1387·79	1391·08	1394·36	1397·64	1400·92	1404·20	1407·48
30	1410·76	1414·04	1417·32	1420·60	1423·88	1427·16	1430·45	1433·73	1437·01	1440·29
40	1443·57	1446·85	1450·13	1453·41	1456·69	1459·97	1463·25	1466·54	1469·82	1473·10
50	1476·38	1479·66	1482·94	1486·22	1489·50	1492·78	1496·06	1499·34	1502·62	1505·91
60	1509·19	1512·47	1515·75	1519·03	1522·31	1525·59	1528·87	1532·15	1535·43	1538·71
70	1541·99	1545·28	1548·56	1551·84	1555·12	1558·40	1561·68	1564·96	1568·24	1571·52
80	1574·80	1578·08	1581·36	1584·65	1587·93	1591·21	1594·49	1597·77	1601·05	1604·33
90	1607·61	1610·89	1614·17	1617·45	1620·73	1624·02	1627·30	1630·58	1633·86	1637·14

MISCELLANEOUS TABLES

CONVERSION TABLES

Extract from B.S. 350 : 1944

METRES TO FEET

Based on 1 inch = 25·4 millimetres

m.	0	1	2	3	4	5	6	7	8	9
	ft.	ft.	ft.	ft.	ft.	ft.	ft.	ft.	ft.	ft.
500	1640·42	1643·70	1646·98	1650·26	1653·54	1656·82	1660·10	1663·39	1666·67	1669·95
10	1673·23	1676·51	1679·79	1683·07	1686·35	1689·63	1692·91	1696·19	1699·48	1702·76
20	1706·04	1709·32	1712·60	1715·88	1719·16	1722·44	1725·72	1729·00	1732·28	1735·56
30	1738·84	1742·13	1745·41	1748·69	1751·97	1755·25	1758·53	1761·81	1765·09	1768·37
40	1771·65	1774·93	1778·22	1781·50	1784·78	1788·06	1791·34	1794·62	1797·90	1801·18
50	1804·46	1807·74	1811·02	1814·30	1817·59	1820·87	1824·15	1827·43	1830·71	1833·99
60	1837·27	1840·55	1843·83	1847·11	1850·39	1853·67	1856·96	1860·24	1863·52	1866·80
70	1870·08	1873·36	1876·64	1879·92	1883·20	1886·48	1889·76	1893·04	1896·33	1899·61
80	1902·89	1906·17	1909·45	1912·73	1916·01	1919·29	1922·57	1925·85	1929·13	1932·41
90	1935·70	1938·98	1942·26	1945·54	1948·82	1952·10	1955·38	1958·66	1961·94	1965·22
600	1968·50	1971·78	1975·07	1978·35	1981·63	1984·91	1988·19	1991·47	1994·75	1998·03
10	2001·31	2004·59	2007·87	2011·15	2014·44	2017·72	2021·00	2024·28	2027·56	2030·84
20	2034·12	2037·40	2040·68	2043·96	2047·24	2050·52	2053·81	2057·09	2060·37	2063·65
30	2066·93	2070·21	2073·49	2076·77	2080·05	2083·33	2086·61	2089·90	2093·18	2096·46
40	2099·74	2103·02	2106·30	2109·58	2112·86	2116·14	2119·42	2122·70	2125·98	2129·27
50	2132·55	2135·83	2139·11	2142·39	2145·67	2148·95	2152·23	2155·51	2158·79	2162·07
60	2165·35	2168·63	2171·92	2175·20	2178·48	2181·76	2185·04	2188·32	2191·60	2194·88
70	2198·16	2201·44	2204·72	2208·01	2211·29	2214·57	2217·85	2221·13	2224·41	2227·69
80	2230·97	2234·25	2237·53	2240·81	2244·09	2247·38	2250·66	2253·94	2257·22	2260·50
90	2263·78	2267·06	2270·34	2273·62	2276·90	2280·18	2283·46	2286·75	2290·03	2293·31
700	2296·59	2299·87	2303·15	2306·43	2309·71	2312·99	2316·27	2319·55	2322·83	2326·12
10	2329·40	2332·68	2335·96	2339·24	2342·52	2345·80	2349·08	2352·36	2355·64	2358·92
20	2362·20	2365·49	2368·77	2372·05	2375·33	2378·61	2381·89	2385·17	2388·45	2391·73
30	2395·01	2398·29	2401·57	2404·86	2408·14	2411·42	2414·70	2417·98	2421·26	2424·54
40	2427·82	2431·10	2434·38	2437·66	2440·94	2444·23	2447·51	2450·79	2454·07	2457·35
50	2460·63	2463·91	2467·19	2470·47	2473·75	2477·03	2480·31	2483·60	2486·88	2490·16
60	2493·44	2496·72	2500·00	2503·28	2506·56	2509·84	2513·12	2516·40	2519·69	2522·97
70	2526·25	2529·53	2532·81	2536·09	2539·37	2542·65	2545·93	2549·21	2552·49	2555·77
80	2559·06	2562·34	2565·62	2568·90	2572·18	2575·46	2578·74	2582·02	2585·30	2588·58
90	2591·86	2595·14	2598·43	2601·71	2604·99	2608·27	2611·55	2614·83	2618·11	2621·39
800	2624·67	2627·95	2631·23	2634·51	2637·80	2641·08	2644·36	2647·64	2650·92	2654·20
10	2657·48	2660·76	2664·04	2667·32	2670·60	2673·88	2677·17	2680·45	2683·73	2687·01
20	2690·29	2693·57	2696·85	2700·13	2703·41	2706·69	2709·97	2713·25	2716·54	2719·82
30	2723·10	2726·38	2729·66	2732·94	2736·22	2739·50	2742·78	2746·06	2749·34	2752·62
40	2755·91	2759·19	2762·47	2765·75	2769·03	2772·31	2775·59	2778·87	2782·15	2785·43
50	2788·71	2791·99	2795·28	2798·56	2801·84	2805·12	2808·40	2811·68	2814·96	2818·24
60	2821·52	2824·80	2828·08	2831·36	2834·65	2837·93	2841·21	2844·49	2847·77	2851·05
70	2854·33	2857·61	2860·89	2864·17	2867·45	2870·73	2874·02	2877·30	2880·58	2883·86
80	2887·14	2890·42	2893·70	2896·98	2900·26	2903·54	2906·82	2910·10	2913·39	2916·67
90	2919·95	2923·23	2926·51	2929·79	2933·07	2936·35	2939·63	2942·91	2946·19	2949·48
900	2952·76	2956·04	2959·32	2962·60	2965·88	2969·16	2972·44	2975·72	2979·00	2982·28
10	2985·56	2988·85	2992·13	2995·41	2998·69	3001·97	3005·25	3008·53	3011·81	3015·09
20	3018·37	3021·65	3024·93	3028·22	3031·50	3034·78	3038·06	3041·34	3044·62	3047·90
30	3051·18	3054·46	3057·74	3061·02	3064·30	3067·59	3070·87	3074·15	3077·43	3080·71
40	3083·99	3087·27	3090·55	3093·83	3097·11	3100·39	3103·67	3106·96	3110·24	3113·52
50	3116·80	3120·08	3123·36	3126·64	3129·92	3133·20	3136·48	3139·76	3143·04	3146·33
60	3149·61	3152·89	3156·17	3159·45	3162·73	3166·01	3169·29	3172·57	3175·85	3179·13
70	3182·41	3185·70	3188·98	3192·26	3195·54	3198·82	3202·10	3205·38	3208·66	3211·94
80	3215·22	3218·50	3221·78	3225·07	3228·35	3231·63	3234·91	3238·19	3241·47	3244·75
90	3248·03	3251·31	3254·59	3257·87	3261·15	3264·44	3267·72	3271·00	3274·28	3277·56
1000	3280·84									

MISCELLANEOUS TABLES

CONVERSION TABLES
Extract from B.S. 350 : 1944

SQUARE FEET TO SQUARE METRES
Based on 1 inch = 25·4 millimetres

Sq. ft.	0	1	2	3	4	5	6	7	8	9
	m.²	m.²	m.²	m.²	m.²	m.²	m.²	m.²	m.²	m.²
—	—	0·09290	0·18581	0·27871	0·37161	0·46452	0·55742	0·65032	0·74322	0·83613
10	0·92903	1·02193	1·11484	1·20774	1·30064	1·39355	1·48645	1·57935	1·67225	1·76516
20	1·85806	1·95096	2·04387	2·13677	2·22967	2·32258	2·41548	2·50838	2·60129	2·69419
30	2·78709	2·87999	2·97290	3·06580	3·15870	3·25161	3·34451	3·43741	3·53032	3·62322
40	3·71612	3·80902	3·90193	3·99483	4·08773	4·18064	4·27354	4·36644	4·45935	4·55225
50	4·64515	4·73806	4·83096	4·92386	5·0168	5·1097	5·2026	5·2955	5·3884	5·4813
60	5·5742	5·6671	5·7600	5·8529	5·9458	6·0387	6·1316	6·2245	6·3174	6·4103
70	6·5032	6·5961	6·6890	6·7819	6·8748	6·9677	7·0606	7·1535	7·2464	7·3393
80	7·4322	7·5251	7·6180	7·7110	7·8039	7·8968	7·9897	8·0826	8·1755	8·2684
90	8·3613	8·4542	8·5471	8·6400	8·7329	8·8258	8·9187	9·0116	9·1045	9·1974
100	9·2903	9·3832	9·4761	9·5690	9·6619	9·7548	9·8477	9·9406	10·0335	10·1264
10	10·2193	10·3122	10·4051	10·4980	10·5909	10·6838	10·7768	10·8697	10·9626	11·0555
20	11·1484	11·2413	11·3342	11·4271	11·5200	11·6129	11·7058	11·7987	11·8916	11·9845
30	12·0774	12·1703	12·2632	12·3561	12·4490	12·5419	12·6348	12·7277	12·8206	12·9135
40	13·0064	13·0993	13·1922	13·2851	13·3780	13·4709	13·5638	13·6567	13·7496	13·8426
50	13·9355	14·0284	14·1213	14·2142	14·3071	14·4000	14·4929	14·5858	14·6787	14·7716
60	14·8645	14·9574	15·0503	15·1432	15·2361	15·3290	15·4219	15·5148	15·6077	15·7006
70	15·7935	15·8864	15·9793	16·0722	16·1651	16·2580	16·3509	16·4438	16·5367	16·6296
80	16·7225	16·8155	16·9084	17·0013	17·0942	17·1871	17·2800	17·3729	17·4658	17·5587
90	17·6516	17·7445	17·8374	17·9303	18·0232	18·1161	18·2090	18·3019	18·3948	18·4877
200	18·5806	18·6735	18·7664	18·8593	18·9522	19·0451	19·1380	19·2309	19·3238	19·4167
10	19·5096	19·6025	19·6954	19·7883	19·8813	19·9742	20·0671	20·1600	20·2529	20·3458
20	20·4387	20·5316	20·6245	20·7174	20·8103	20·9032	20·9961	21·0890	21·1819	21·2748
30	21·3677	21·4606	21·5535	21·6464	21·7393	21·8322	21·9251	22·0180	22·1109	22·2038
40	22·2967	22·3896	22·4825	22·5754	22·6683	22·7612	22·8541	22·9471	23·0400	23·1329
50	23·2258	23·3187	23·4116	23·5045	23·5974	23·6903	23·7832	23·8761	23·9690	24·0619
60	24·1548	24·2477	24·3406	24·4335	24·5264	24·6193	24·7122	24·8051	24·8980	24·9909
70	25·0838	25·1767	25·2696	25·3625	25·4554	25·5483	25·6412	25·7341	25·8270	25·9199
80	26·0129	26·1058	26·1987	26·2916	26·3845	26·4774	26·5703	26·6632	26·7561	26·8490
90	26·9419	27·0348	27·1277	27·2206	27·3135	27·4064	27·4993	27·5922	27·6851	27·7780
300	27·8709	27·9638	28·0567	28·1496	28·2425	28·3354	28·4283	28·5212	28·6141	28·7070
10	28·7999	28·8928	28·9857	29·0787	29·1716	29·2645	29·3574	29·4503	29·5432	29·6361
20	29·7290	29·8219	29·9148	30·0077	30·1006	30·1935	30·2864	30·3793	30·4722	30·5651
30	30·6580	30·7509	30·8438	30·9367	31·0296	31·1225	31·2154	31·3083	31·4012	31·4941
40	31·5870	31·6799	31·7728	31·8657	31·9586	32·0515	32·1445	32·2374	32·3303	32·4232
50	32·5161	32·6090	32·7019	32·7948	32·8877	32·9806	33·0735	33·1664	33·2593	33·3522
60	33·4451	33·5380	33·6309	33·7238	33·8167	33·9096	34·0025	34·0954	34·1883	34·2812
70	34·3741	34·4670	34·5599	34·6528	34·7457	34·8386	34·9315	35·0244	35·1173	35·2103
80	35·3032	35·3961	35·4890	35·5819	35·6748	35·7677	35·8606	35·9535	36·0464	36·1393
90	36·2322	36·3251	36·4180	36·5109	36·6038	36·6967	36·7896	36·8825	36·9754	37·0683
400	37·1612	37·2541	37·3470	37·4399	37·5328	37·6257	37·7186	37·8115	37·9044	37·9973
10	38·0902	38·1831	38·2761	38·3690	38·4619	38·5548	38·6477	38·7406	38·8335	38·9264
20	39·0193	39·1122	39·2051	39·2980	39·3909	39·4838	39·5767	39·6696	39·7625	39·8554
30	39·9483	40·0412	40·1341	40·2270	40·3199	40·4128	40·5057	40·5986	40·6915	40·7844
40	40·8773	40·9702	41·0631	41·1560	41·2489	41·3419	41·4348	41·5277	41·6206	41·7135
50	41·8064	41·8993	41·9922	42·0851	42·1780	42·2709	42·3638	42·4567	42·5496	42·6425
60	42·7354	42·8283	42·9212	43·0141	43·1070	43·1999	43·2928	43·3857	43·4786	43·5715
70	43·6644	43·7573	43·8502	43·9431	44·0360	44·1289	44·2218	44·3148	44·4077	44·5006
80	44·5935	44·6864	44·7793	44·8722	44·9651	45·0580	45·1509	45·2438	45·3367	45·4296
90	45·5225	45·6154	45·7083	45·8012	45·8941	45·9870	46·0799	46·1728	46·2657	46·3586

MISCELLANEOUS TABLES

CONVERSION TABLES
Extract from B.S. 350 : 1944

SQUARE FEET TO SQUARE METRES (continued)

Based on 1 inch = 25·4 millimetres

Sq. ft.	0	1	2	3	4	5	6	7	8	9
	m.²	m.²	m.²	m.²	m.²	m.²	m.²	m.²	m.²	m.²
500	46·4515	46·5444	46·6373	46·7302	46·8231	46·9160	47·0089	47·1018	47·1947	47·2876
10	47·3806	47·4735	47·5664	47·6593	47·7522	47·8451	47·9380	48·0309	48·1238	48·2167
20	48·3096	48·4025	48·4954	48·5883	48·6812	48·7741	48·8670	48·9599	49·0528	49·1457
30	49·2386	49·3315	49·4244	49·5173	49·6102	49·7031	49·7960	49·8889	49·9818	50·075
40	50·168	50·261	50·353	50·446	50·539	50·632	50·725	50·818	50·911	51·004
50	51·097	51·190	51·282	51·375	51·468	51·561	51·654	51·747	51·840	51·933
60	52·026	52·119	52·212	52·304	52·397	52·490	52·583	52·676	52·769	52·862
70	52·955	53·048	53·141	53·233	53·326	53·419	53·512	53·605	53·698	53·791
80	53·884	53·977	54·070	54·162	54·255	54·348	54·441	54·534	54·627	54·720
90	54·813	54·906	54·999	55·092	55·184	55·277	55·370	55·463	55·556	55·649
600	55·742	55·835	55·928	56·021	56·113	56·206	56·299	56·392	56·485	56·578
10	56·671	56·764	56·857	56·950	57·042	57·135	57·228	57·321	57·414	57·507
20	57·600	57·693	57·786	57·879	57·971	58·064	58·157	58·250	58·343	58·436
30	58·529	58·622	58·715	58·808	58·901	58·993	59·086	59·179	59·272	59·365
40	59·458	59·551	59·644	59·737	59·830	59·922	60·015	60·108	60·201	60·294
50	60·387	60·480	60·573	60·666	60·759	60·851	60·944	61·037	61·130	61·223
60	61·316	61·409	61·502	61·595	61·688	61·781	61·873	61·966	62·059	62·152
70	62·245	62·338	62·431	62·524	62·617	62·710	62·802	62·895	62·988	63·081
80	63·174	63·267	63·360	63·453	63·546	63·639	63·731	63·824	63·917	64·010
90	64·103	64·196	64·289	64·382	64·475	64·568	64·661	64·753	64·846	64·939
700	65·032	65·125	65·218	65·311	65·404	65·497	65·590	65·682	65·775	65·868
10	65·961	66·054	66·147	66·240	66·333	66·426	66·519	66·611	66·704	66·797
20	66·890	66·983	67·076	67·169	67·262	67·355	67·448	67·541	67·633	67·726
30	67·819	67·912	68·005	68·098	68·191	68·284	68·377	68·470	68·562	68·655
40	68·748	68·841	68·934	69·027	69·120	69·213	69·306	69·399	69·491	69·584
50	69·677	69·770	69·863	69·956	70·049	70·142	70·235	70·328	70·421	70·513
60	70·606	70·699	70·792	70·885	70·978	71·071	71·164	71·257	71·350	71·442
70	71·535	71·628	71·721	71·814	71·907	72·000	72·093	72·186	72·279	72·371
80	72·464	72·557	72·650	72·743	72·836	72·929	73·022	73·115	73·208	73·300
90	73·393	73·486	73·579	73·672	73·765	73·858	73·951	74·044	74·137	74·230
800	74·322	74·415	74·508	74·601	74·694	74·787	74·880	74·973	75·066	75·159
10	75·251	75·344	75·437	75·530	75·623	75·716	75·809	75·902	75·995	76·088
20	76·180	76·273	76·366	76·459	76·552	76·645	76·738	76·831	76·924	77·017
30	77·110	77·202	77·295	77·388	77·481	77·574	77·667	77·760	77·853	77·946
40	78·039	78·131	78·224	78·317	78·410	78·503	78·596	78·689	78·782	78·875
50	78·968	79·060	79·153	79·246	79·339	79·432	79·525	79·618	79·711	79·804
60	79·897	79·990	80·082	80·175	80·268	80·361	80·454	80·547	80·640	80·733
70	80·826	80·919	81·011	81·104	81·197	81·290	81·383	81·476	81·569	81·662
80	81·755	81·848	81·940	82·033	82·126	82·219	82·312	82·405	82·498	82·591
90	82·684	82·777	82·870	82·962	83·055	83·148	83·241	83·334	83·427	83·520
900	83·613	83·706	83·799	83·891	83·984	84·077	84·170	84·263	84·356	84·449
10	84·542	84·635	84·728	84·820	84·913	85·006	85·099	85·192	85·285	85·378
20	85·471	85·564	85·657	85·750	85·842	85·935	86·028	86·121	86·214	86·307
30	86·400	86·493	86·586	86·679	86·771	86·864	86·957	87·050	87·143	87·236
40	87·329	87·422	87·515	87·608	87·700	87·793	87·886	87·979	88·072	88·165
50	88·258	88·351	88·444	88·537	88·630	88·722	88·815	88·908	89·001	89·094
60	89·187	89·280	89·373	89·466	89·559	89·651	89·744	89·837	89·930	90·023
70	90·116	90·209	90·302	90·395	90·488	90·580	90·673	90·766	90·859	90·952
80	91·045	91·138	91·231	91·324	91·417	91·509	91·602	91·695	91·788	91·881
90	91·974	92·067	92·160	92·253	92·346	92·439	92·531	92·624	92·717	92·810
1000	92·903									

MISCELLANEOUS TABLES

CONVERSION TABLES
Extract from B.S. 350 : 1944

SQUARE METRES TO SQUARE FEET

Based on 1 inch = 25·4 millimetres

m².	0	1	2	3	4	5	6	7	8	9
	sq. ft.	sq. ft.	sq. ft.	sq. ft.	sq. ft.	sq. ft.	sq. ft.	sq. ft.	sq. ft.	sq. ft.
	—	—	10·7639	21·5278	32·2917	43·0556	53·820	64·583	75·347	86·111	96·875
10	107·639	118·403	129·167	139·931	150·695	161·459	172·223	182·986	193·750	204·514
20	215·278	226·042	236·806	247·570	258·334	269·098	279·862	290·626	301·389	312·153
30	322·917	333·681	344·445	355·209	365·973	376·737	387·501	398·265	409·029	419·792
40	430·556	441·320	452·084	462·848	473·612	484·376	495·140	505·90	516·67	527·43
50	538·20	548·96	559·72	570·49	581·25	592·01	602·78	613·54	624·31	635·07
60	645·83	656·60	667·36	678·13	688·89	699·65	710·42	721·18	731·95	742·71
70	753·47	764·24	775·00	785·77	796·53	807·29	818·06	828·82	839·58	850·35
80	861·11	871·88	882·64	893·40	904·17	914·93	925·70	936·46	947·22	957·99
90	968·75	979·52	990·28	1001·04	1011·81	1022·57	1033·34	1044·10	1054·86	1065·63
100	1076·39	1087·15	1097·92	1108·68	1119·45	1130·21	1140·97	1151·74	1162·50	1173·27
10	1184·03	1194·79	1205·56	1216·32	1227·09	1237·85	1248·61	1259·38	1270·14	1280·91
20	1291·67	1302·43	1313·20	1323·96	1334·72	1345·49	1356·25	1367·02	1377·78	1388·54
30	1399·31	1410·07	1420·84	1431·60	1442·36	1453·13	1463·89	1474·66	1485·42	1496·18
40	1506·95	1517·71	1528·48	1539·24	1550·00	1560·77	1571·53	1582·29	1593·06	1603·82
50	1614·59	1625·35	1636·11	1646·88	1657·64	1668·41	1679·17	1689·93	1700·70	1711·46
60	1722·23	1732·99	1743·75	1754·52	1765·28	1776·04	1786·81	1797·57	1808·34	1819·10
70	1829·86	1840·63	1851·39	1862·16	1872·92	1883·68	1894·45	1905·21	1915·98	1926·74
80	1937·50	1948·27	1959·03	1969·80	1980·56	1991·32	2002·09	2012·85	2023·61	2034·38
90	2045·14	2055·91	2066·67	2077·43	2088·20	2098·96	2109·73	2120·49	2131·25	2142·02
200	2152·78	2163·55	2174·31	2185·07	2195·84	2206·60	2217·37	2228·13	2238·89	2249·66
10	2260·42	2271·18	2281·95	2292·71	2303·48	2314·24	2325·00	2335·77	2346·53	2357·30
20	2368·06	2378·82	2389·59	2400·35	2411·12	2421·88	2432·64	2443·41	2454·17	2464·94
30	2475·70	2486·46	2497·23	2507·99	2518·75	2529·52	2540·28	2551·05	2561·81	2572·57
40	2583·34	2594·10	2604·87	2615·63	2626·39	2637·16	2647·92	2658·69	2669·45	2680·21
50	2690·98	2701·74	2712·51	2723·27	2734·03	2744·80	2755·56	2766·32	2777·09	2787·85
60	2798·62	2809·38	2820·14	2830·91	2841·67	2852·44	2863·20	2873·96	2884·73	2895·49
70	2906·26	2917·02	2927·78	2938·55	2949·31	2960·07	2970·84	2981·60	2992·37	3003·13
80	3013·89	3024·66	3035·42	3046·19	3056·95	3067·71	3078·48	3089·24	3100·01	3110·77
90	3121·53	3132·30	3143·06	3153·83	3164·59	3175·35	3186·12	3196·88	3207·64	3218·41
300	3229·17	3239·94	3250·70	3261·46	3272·23	3282·99	3293·76	3304·52	3315·28	3326·05
10	3336·81	3347·58	3358·34	3369·10	3379·87	3390·63	3401·40	3412·16	3422·92	3433·69
20	3444·45	3455·21	3465·98	3476·74	3487·51	3498·27	3509·03	3519·80	3530·56	3541·33
30	3552·09	3562·85	3573·62	3584·38	3595·15	3605·91	3616·67	3627·44	3638·20	3648·97
40	3659·73	3670·49	3681·26	3692·02	3702·78	3713·55	3724·31	3735·08	3745·84	3756·60
50	3767·37	3778·13	3788·90	3799·66	3810·42	3821·19	3831·95	3842·72	3853·48	3864·24
60	3875·01	3885·77	3896·53	3907·30	3918·06	3928·83	3939·59	3950·35	3961·12	3971·88
70	3982·65	3993·41	4004·17	4014·94	4025·70	4036·47	4047·23	4057·99	4068·76	4079·52
80	4090·29	4101·05	4111·81	4122·58	4133·34	4144·10	4154·87	4165·63	4176·40	4187·16
90	4197·92	4208·69	4219·45	4230·22	4240·98	4251·74	4262·51	4273·27	4284·04	4294·80
400	4305·56	4316·33	4327·09	4337·86	4348·62	4359·38	4370·15	4380·91	4391·67	4402·44
10	4413·20	4423·97	4434·73	4445·49	4456·26	4467·02	4477·79	4488·55	4499·31	4510·08
20	4520·84	4531·61	4542·37	4553·13	4563·90	4574·66	4585·43	4596·19	4606·95	4617·72
30	4628·48	4639·24	4650·01	4660·77	4671·54	4682·30	4693·06	4703·83	4714·59	4725·36
40	4736·12	4746·88	4757·65	4768·41	4779·18	4789·94	4800·70	4811·47	4822·23	4833·00
50	4843·76	4854·52	4865·29	4876·05	4886·81	4897·58	4908·34	4919·11	4929·87	4940·63
60	4951·40	4962·16	4972·93	4983·69	4994·45	5005·2	5016·0	5026·7	5037·5	5048·3
70	5059·0	5069·8	5080·6	5091·3	5102·1	5112·9	5123·6	5134·4	5145·1	5155·9
80	5166·7	5177·4	5188·2	5199·0	5209·7	5220·5	5231·3	5242·0	5252·8	5263·6
90	5274·3	5285·1	5295·8	5306·6	5317·4	5328·1	5338·9	5349·7	5360·4	5371·2

MISCELLANEOUS TABLES

CONVERSION TABLES
Extract from B.S. 350 : 1944

SQUARE METRES TO SQUARE FEET (continued)

Based on 1 inch = 25·4 millimetres

m².	0	1	2	3	4	5	6	7	8	9
	sq. ft.	sq. ft.	sq. ft.	sq. ft.	sq. ft.	sq. ft.	sq. ft.	sq. ft.	sq. ft.	sq. ft.
500	5382·0	5392·7	5403·5	5414·2	5425·0	5435·8	5446·5	5457·3	5468·1	5478·8
10	5489·6	5500·4	5511·1	5521·9	5532·6	5543·4	5554·2	5564·9	5575·7	5586·5
20	5597·2	5608·0	5618·8	5629·5	5640·3	5651·1	5661·8	5672·6	5683·3	5694·1
30	5704·9	5715·6	5726·4	5737·2	5747·9	5758·7	5769·5	5780·2	5791·0	5801·7
40	5812·5	5823·3	5834·0	5844·8	5855·6	5866·3	5877·1	5887·9	5898·6	5909·4
50	5920·1	5930·9	5941·7	5952·4	5963·2	5974·0	5984·7	5995·5	6006·3	6017·0
60	6027·8	6038·6	6049·3	6060·1	6070·8	6081·6	6092·4	6103·1	6113·9	6124·7
70	6135·4	6146·2	6157·0	6167·7	6178·5	6189·2	6200·0	6210·8	6221·5	6232·3
80	6243·1	6253·8	6264·6	6275·4	6286·1	6296·9	6307·7	6318·4	6329·2	6339·9
90	6350·7	6361·5	6372·2	6383·0	6393·8	6404·5	6415·3	6426·1	6436·8	6447·6
600	6458·3	6469·1	6479·9	6490·6	6501·4	6512·2	6522·9	6533·7	6544·5	6555·2
10	6566·0	6576·7	6587·5	6598·3	6609·0	6619·8	6630·6	6641·3	6652·1	6662·9
20	6673·6	6684·4	6695·2	6705·9	6716·7	6727·4	6738·2	6749·0	6759·7	6770·5
30	6781·3	6792·0	6802·8	6813·6	6824·3	6835·1	6845·8	6856·6	6867·4	6878·1
40	6888·9	6899·7	6910·4	6921·2	6932·0	6942·7	6953·5	6964·2	6975·0	6985·8
50	6996·5	7007·3	7018·1	7028·8	7039·6	7050·4	7061·1	7071·9	7082·7	7093·4
60	7104·2	7114·9	7125·7	7136·5	7147·2	7158·0	7168·8	7179·5	7190·3	7201·1
70	7211·8	7222·6	7233·3	7244·1	7254·9	7265·6	7276·4	7287·2	7297·9	7308·7
80	7319·5	7330·2	7341·0	7351·7	7362·5	7373·3	7384·0	7394·8	7405·6	7416·3
90	7427·1	7437·9	7448·6	7459·4	7470·2	7480·9	7491·7	7502·4	7513·2	7524·0
700	7534·7	7545·5	7556·3	7567·0	7577·8	7588·6	7599·3	7610·1	7620·8	7631·6
10	7642·4	7653·1	7663·9	7674·7	7685·4	7696·2	7707·0	7717·7	7728·5	7739·3
20	7750·0	7760·8	7771·5	7782·3	7793·1	7803·8	7814·6	7825·4	7836·1	7846·9
30	7857·7	7868·4	7879·2	7889·9	7900·7	7911·5	7922·2	7933·0	7943·8	7954·5
40	7965·3	7976·1	7986·8	7997·6	8008·3	8019·1	8029·9	8040·6	8051·4	8062·2
50	8072·9	8083·7	8094·5	8105·2	8116·0	8126·8	8137·5	8148·3	8159·0	8169·8
60	8180·6	8191·3	8202·1	8212·9	8223·6	8234·4	8245·2	8255·9	8266·7	8277·4
70	8288·2	8299·0	8309·7	8320·5	8331·3	8342·0	8352·8	8363·6	8374·3	8385·1
80	8395·8	8406·6	8417·4	8428·1	8438·9	8449·7	8460·4	8471·2	8482·0	8492·7
90	8503·5	8514·3	8525·0	8535·8	8546·5	8557·3	8568·1	8578·8	8589·6	8600·4
800	8611·1	8621·9	8632·7	8643·4	8654·2	8664·9	8675·7	8686·5	8697·2	8708·0
10	8718·8	8729·5	8740·3	8751·1	8761·8	8772·6	8783·3	8794·1	8804·9	8815·6
20	8826·4	8837·2	8847·9	8858·7	8869·5	8880·2	8891·0	8901·8	8912·5	8923·3
30	8934·0	8944·8	8955·6	8966·3	8977·1	8987·9	8998·6	9009·4	9020·2	9030·9
40	9041·7	9052·4	9063·2	9074·0	9084·7	9095·5	9106·3	9117·0	9127·8	9138·6
50	9149·3	9160·1	9170·9	9181·6	9192·4	9203·1	9213·9	9224·7	9235·4	9246·2
60	9257·0	9267·7	9278·5	9289·3	9300·0	9310·8	9321·5	9332·3	9343·1	9353·8
70	9364·6	9375·4	9386·1	9396·9	9407·7	9418·4	9429·2	9439·9	9450·7	9461·5
80	9472·2	9483·0	9493·8	9504·5	9515·3	9526·1	9536·8	9547·6	9558·4	9569·1
90	9579·9	9590·6	9601·4	9612·2	9622·9	9633·7	9644·5	9655·2	9666·0	9676·8
900	9687·5	9698·3	9709·0	9719·8	9730·6	9741·3	9752·1	9762·9	9773·6	9784·4
10	9795·2	9805·9	9816·7	9827·4	9838·2	9849·0	9859·7	9870·5	9881·3	9892·0
20	9902·8	9913·6	9924·3	9935·1	9945·9	9956·6	9967·4	9978·1	9988·9	9999·7
30	10010·4	10021·2	10032·0	10042·7	10053·5	10064·3	10075·0	10085·8	10096·5	10107·3
40	10118·1	10128·8	10139·6	10150·4	10161·1	10171·9	10182·7	10193·4	10204·2	10214·9
50	10225·7	10236·5	10247·2	10258·0	10268·8	10279·5	10290·3	10301·1	10311·8	10322·6
60	10333·4	10344·1	10354·9	10365·6	10376·4	10387·2	10397·9	10408·7	10419·5	10430·2
70	10441·0	10451·8	10462·5	10473·3	10484·0	10494·8	10505·6	10516·3	10527·1	10537·9
80	10548·6	10559·4	10570·2	10580·9	10591·7	10602·5	10613·2	10624·0	10634·7	10645·5
90	10656·3	10667·0	10677·8	10688·6	10699·3	10710·1	10720·9	10731·6	10742·4	10753·1
1000	10763·9									

MISCELLANEOUS TABLES

PRESSURE, STRESS

Pressure, Stress: UK tons-force per square inch to meganewtons per square metre (newtons per square millimetre)
Basis: 1 UK ton = 2240 lb; 1 lbf = 0.453 592 37 kgf; 1 kgf = 9.806 65 N; 1 in = 0.0254 m.

UK tonf/in²	0	0·1	0·2	0·3	0·4	0·5	0·6	0·7	0·8	0·9
				meganewtons per square metre						
0	0·00000	1·54443	3·08885	4·63328	6·17770	7·72213	9·26655	10·8110	12·3554	13·8998
1	15·4443	16·9887	18·5331	20·0775	21·6220	23·1664	24·7108	26·2552	27·7997	29·3441
2	30·8885	32·4329	33·9774	35·5218	37·0662	38·6106	40·1551	41·6995	43·2439	44·7883
3	46·3328	47·8772	49·4216	50·9660	52·5105	54·0549	55·5993	57·1437	58·6882	60·2326
4	61·7770	63·3215	64·8659	66·4103	67·9547	69·4992	71·0436	72·5880	74·1324	75·6769
5	77·2213	78·7657	80·3101	81·8546	83·3990	84·9434	86·4878	88·0323	89·5767	91·1211
6	92·6655	94·2100	95·7544	97·2988	98·8432	100·388	101·932	103·477	105·021	106·565
7	108·110	109·654	111·199	112·743	114·287	115·832	117·376	118·921	120·465	122·010
8	123·554	125·098	126·643	128·187	129·732	131·276	132·821	134·365	135·909	137·454
9	138·998	140·543	142·087	143·632	145·176	146·720	148·265	149·809	151·354	152·898
10	154·443	155·987	157·531	159·076	160·620	162·165	163·709	165·254	166·798	168·342
11	169·887	171·431	172·976	174·520	176·065	177·609	179·153	180·698	182·242	183·787
12	185·331	186·876	188·420	189·964	191·509	193·053	194·598	196·142	197·686	199·231
13	200·775	202·320	203·864	205·409	206·953	208·497	210·042	211·586	213·131	214·675
14	216·220	217·764	219·308	220·853	222·397	223·942	225·486	227·031	228·575	230·119
15	231·664	233·208	234·753	236·297	237·842	239·386	240·930	242·475	244·019	245·564
16	247·108	248·635	250·197	251·741	253·286	254·830	256·375	257·919	259·464	261·008
17	262·552	264·097	265·641	267·186	268·730	270·274	271·819	273·363	274·908	276·452
18	277·997	279·541	281·085	282·630	284·174	285·719	287·263	288·808	290·352	291·896
19	293·441	294·985	296·530	298·074	299·619	301·163	302·707	304·252	305·796	307·341
20	308·885	310·430	311·974	313·518	315·063	316·607	318·152	319·696	321·241	322·785
21	324·329	325·874	327·418	328·963	330·507	332·052	333·596	335·140	336·685	338·229
22	339·774	341·318	342·862	344·407	345·951	347·496	349·040	350·585	352·129	353·673
23	355·218	356·762	358·307	359·851	361·396	362·940	364·484	366·029	367·573	369·118
24	370·662	372·207	373·751	375·295	376·840	378·384	379·929	381·473	383·018	384·562

Supplement No. 1 (1967) to
B.S. 350 : Part 2 : 1962

PRESSURE, STRESS

Pressure, Stress: UK tons-force per square inch to meganewtons per square metre
Basis: 1 UKton = 2240 lb; 1 lbf = 0.453 592 37 kgf; 1 kgf = 9.806 65 N; 1 in = 0.0254 m.

UK tonf/in²	0	0·1	0·2	0·3	0·4	0·5	0·6	0·7	0·8	0·9
				meganewtons per square metre						
25	386·106	387·651	389·195	390·740	392·284	393·829	395·373	396·917	398·462	400·006
26	401·551	403·095	404·640	406·184	407·728	409·273	410·817	412·362	413·906	415·450
27	416·995	418·539	420·084	421·628	423·173	424·717	426·261	427·806	429·350	430·895
28	432·439	433·984	435·528	437·072	438·617	440·161	441·706	443·250	444·795	446·339
29	447·883	449·428	450·972	452·517	454·061	455·606	457·150	458·694	460·239	461·783
30	463·328	464·872	466·417	467·961	469·505	471·050	472·594	474·139	475·683	477·228
31	478·772	480·316	481·861	483·405	484·950	486·494	488·039	489·583	491·127	492·672
32	494·216	495·761	497·305	498·849	500·394	501·938	503·483	505·027	506·572	508·116
33	509·660	511·205	512·749	514·294	515·838	517·383	518·927	520·471	522·016	523·560
34	525·105	526·649	528·194	529·738	531·282	532·827	534·371	535·916	537·460	539·005
35	540·549	542·093	543·638	545·182	546·727	548·271	549·816	551·360	552·904	554·449
36	555·993	557·538	559·082	560·627	562·171	563·715	565·260	566·804	568·349	569·893
37	571·437	572·982	574·526	576·071	577·615	579·160	580·704	582·248	583·793	585·337
38	586·882	588·426	589·971	591·515	593·059	594·604	596·148	597·693	599·237	600·782
39	602·326	603·870	605·415	606·959	608·504	610·048	611·593	613·137	614·681	616·226
40	617·770	619·315	620·859	622·404	623·948	625·492	627·037	628·581	630·126	631·670
41	633·215	634·759	636·303	637·848	639·392	640·937	642·481	644·025	645·570	647·114
42	648·659	650·203	651·748	653·292	654·836	656·381	657·925	659·470	661·014	662·559
43	664·103	665·647	667·192	668·736	670·281	671·825	673·370	674·914	676·458	678·003
44	679·547	681·092	682·636	684·181	685·725	687·269	688·814	690·358	691·903	693·447
45	694·992	696·536	698·080	699·625	701·169	702·714	704·258	705·803	707·347	708·891
46	710·436	711·980	713·525	715·069	716·613	718·158	719·702	721·247	722·791	724·336
47	725·880	727·424	728·969	730·513	732·058	733·602	735·147	736·691	738·235	739·780
48	741·324	742·869	744·413	745·958	747·502	749·046	750·591	752·135	753·680	755·224
49	756·769	758·313	759·857	761·402	762·946	764·491	766·035	767·580	769·124	770·668

Pressure, stress:
UK tons-force per square inch to meganewtons per square metre

UK tonf/in²	0	0·1	0·2	0·3	0·4	0·5	0·6	0·7	0·8	0·9
				meganewtons per square metre						
50	772·213	773·757	775·302	776·846	778·391	779·935	781·479	783·024	784·568	786·113
51	787·657	789·201	790·746	792·290	793·835	795·379	796·924	798·468	800·012	801·557
52	803·101	804·646	806·190	807·735	809·279	810·823	812·368	813·912	815·457	817·001
53	818·546	820·090	821·634	823·179	824·723	826·268	827·812	829·357	830·901	832·445
54	833·990	835·534	837·079	838·623	840·168	841·712	843·256	844·801	846·345	847·890
55	849·434	850·979	852·523	854·067	855·612	857·156	858·701	860·245	861·790	863·334
56	864·878	866·423	867·967	869·512	871·056	872·600	874·145	875·689	877·234	878·778
57	880·323	881·867	883·411	884·956	886·500	888·045	889·589	891·134	892·678	894·222
58	895·767	897·311	898·856	900·400	901·945	903·489	905·033	906·578	908·122	909·667
59	911·211	912·756	914·300	915·844	917·389	918·933	920·478	922·022	923·567	925·111
60	926·655	928·200	929·744	931·289	932·833	934·378	935·922	937·466	939·011	940·555
61	942·100	943·644	945·188	946·733	948·277	949·822	951·366	952·911	954·455	955·999
62	957·544	959·088	960·633	962·177	963·722	965·266	966·810	968·355	969·899	971·444
63	972·988	974·533	976·077	977·621	979·166	980·710	982·255	983·799	985·344	986·888
64	988·432	989·977	991·521	993·066	994·610	996·155	997·699	999·243	1000·79	1002·33
65	1003·88	1005·42	1006·97	1008·51	1010·05	1011·60	1013·14	1014·69	1016·23	1017·78
66	1019·32	1020·87	1022·41	1023·95	1025·50	1027·04	1028·59	1030·13	1031·68	1033·22
67	1034·77	1036·31	1037·85	1039·40	1040·94	1042·49	1044·03	1045·58	1047·12	1048·67
68	1050·21	1051·75	1053·30	1054·84	1056·39	1057·93	1059·48	1061·02	1062·56	1064·11
69	1065·65	1067·20	1068·74	1070·29	1071·83	1073·38	1074·92	1076·46	1078·01	1079·55
70	1081·10	1082·64	1084·19	1085·73	1087·28	1088·82	1090·36	1091·91	1093·45	1095·00
71	1096·54	1098·09	1099·63	1101·18	1102·72	1104·26	1105·81	1107·35	1108·90	1110·44
72	1111·99	1113·53	1115·08	1116·62	1118·16	1119·71	1121·25	1122·80	1124·34	1125·89
73	1127·43	1128·98	1130·52	1132·06	1133·61	1135·15	1136·70	1138·24	1139·79	1141·33
74	1142·87	1144·42	1145·96	1147·51	1149·05	1150·60	1152·14	1153·69	1155·23	1156·77

Pressure, stress:
UK tons-force per square inch to meganewtons per square metre

UK tonf/in²	0	0·1	0·2	0·3	0·4	0·5	0·6	0·7	0·8	0·9
				meganewtons per square metre						
75	1158·32	1159·86	1161·41	1162·95	1164·50	1166·04	1167·59	1169·13	1170·67	1172·22
76	1173·76	1175·31	1176·85	1178·40	1179·94	1181·49	1183·03	1184·57	1186·12	1187·66
77	1189·21	1190·75	1192·30	1193·84	1195·39	1196·93	1198·47	1200·02	1201·56	1203·11
78	1204·65	1206·20	1207·74	1209·29	1210·83	1212·37	1213·92	1215·46	1217·01	1218·55
79	1220·10	1221·64	1223·19	1224·73	1226·27	1227·82	1229·36	1230·91	1232·45	1234·00
80	1235·54	1237·08	1238·63	1240·17	1241·72	1243·26	1244·81	1246·35	1247·90	1249·44
81	1250·98	1252·53	1254·07	1255·62	1257·16	1258·71	1260·25	1261·80	1263·34	1264·88
82	1266·43	1267·97	1269·52	1271·06	1272·61	1274·15	1275·70	1277·24	1278·78	1280·33
83	1281·87	1283·42	1284·96	1286·51	1288·05	1289·60	1291·14	1292·68	1294·23	1295·77
84	1297·32	1298·86	1300·41	1301·95	1303·50	1305·04	1306·58	1308·13	1309·67	1311·22
85	1312·76	1314·31	1315·85	1317·40	1318·94	1320·48	1322·03	1323·57	1325·12	1326·66
86	1328·21	1329·75	1331·29	1332·84	1334·38	1335·93	1337·47	1339·02	1340·56	1342·11
87	1343·65	1345·19	1346·74	1348·28	1349·83	1351·37	1352·92	1354·46	1356·01	1357·55
88	1359·09	1360·64	1362·18	1363·73	1365·27	1366·82	1368·36	1369·91	1371·45	1372·99
89	1374·54	1376·08	1377·63	1379·17	1380·72	1382·26	1383·81	1385·35	1386·89	1388·44
90	1389·98	1391·53	1393·07	1394·62	1396·16	1397·71	1399·25	1400·79	1402·34	1403·88
91	1405·43	1406·97	1408·52	1410·06	1411·61	1413·15	1414·69	1416·24	1417·78	1419·33
92	1420·87	1422·42	1423·96	1425·50	1427·05	1428·59	1430·14	1431·68	1433·23	1434·77
93	1436·32	1437·86	1439·40	1440·95	1442·49	1444·04	1445·58	1447·13	1448·67	1450·22
94	1451·76	1453·30	1454·85	1456·39	1457·94	1459·48	1461·03	1462·57	1464·12	1465·66
95	1467·20	1468·75	1470·29	1471·84	1473·38	1474·93	1476·47	1478·02	1479·56	1481·10
96	1482·65	1484·19	1485·74	1487·28	1488·83	1490·37	1491·92	1493·46	1495·00	1496·55
97	1498·09	1499·64	1501·18	1502·73	1504·27	1505·81	1507·36	1508·90	1510·45	1511·99
98	1513·54	1515·08	1516·63	1518·17	1519·71	1521·26	1522·80	1524·35	1525·89	1527·44
99	1528·98	1530·53	1532·07	1533·61	1535·16	1536·70	1538·25	1539·79	1541·34	1542·88
100	1544·43	—								—

MISCELLANEOUS TABLES

PRESSURE, STRESS

Pressure, Stress: Meganewtons per square metre (newtons per square millimetre) to UK tons-force per square inch
Basis: 1 UKton = 2240 lb; 1 lbf = 0.453 592 37 kgf; 1 kgf = 9.806 65 N; 1 in = 0.0254 m.

MN/m²	0	1	2	3	4	5	6	7	8	9
				UK tons-force per square inch						
0	0·00000	0·06475	0·12950	0·19425	0·25900	0·32374	0·38849	0·45324	0·51799	0·58274
10	0·64749	0·71224	0·77699	0·84174	0·90649	0·97123	1·03598	1·10073	1·16548	1·23023
20	1·29498	1·35973	1·42448	1·48923	1·55398	1·61872	1·68347	1·74822	1·81297	1·87772
30	1·94247	2·00722	2·07197	2·13672	2·20147	2·26621	2·33096	2·39571	2·46046	2·52521
40	2·58996	2·65471	2·71946	2·78421	2·84896	2·91370	2·97845	3·04320	3·10795	3·17270
50	3·23745	3·30220	3·36695	3·43170	3·49645	3·56119	3·62594	3·69069	3·75544	3·82019
60	3·88494	3·94969	4·01444	4·07919	4·14394	4·20868	4·27343	4·33818	4·40293	4·46768
70	4·53243	4·59718	4·66193	4·72668	4·79143	4·85617	4·92092	4·98567	5·05042	5·11517
80	5·17992	5·24467	5·30942	5·37417	5·43892	5·50366	5·56841	5·63316	5·69791	5·76266
90	5·82741	5·89216	5·95691	6·02166	6·08641	6·15115	6·21590	6·28065	6·34540	6·41015
100	6·47490	6·53965	6·60440	6·66915	6·73389	6·79864	6·86339	6·92814	6·99289	7·05764
110	7·12239	7·18714	7·25189	7·31664	7·38138	7·44613	7·51088	7·57563	7·64038	7·70513
120	7·76988	7·83463	7·89938	7·96413	8·02887	8·09362	8·15837	8·22312	8·28787	8·35262
130	8·41737	8·48212	8·54687	8·61162	8·67636	8·74111	8·80586	8·87061	8·93536	9·00011
140	9·06486	9·12961	9·19436	9·25911	9·32385	9·38860	9·45335	9·51810	9·58285	9·64760
150	9·71253	9·77710	9·84185	9·90660	9·97134	10·0361	10·1008	10·1656	10·2303	10·2951
160	10·3598	10·4246	10·4893	10·5541	10·6188	10·6836	10·7483	10·8131	10·8778	10·9426
170	11·0073	11·0721	11·1368	11·2016	11·2663	11·3311	11·3958	11·4606	11·5253	11·5901
180	11·6548	11·7196	11·7843	11·8491	11·9138	11·9786	12·0433	12·1081	12·1728	12·2376
190	12·3023	12·3671	12·4318	12·4966	12·5613	12·6261	12·6908	12·7556	12·8203	12·8850
200	12·9498	13·0145	13·0793	13·1440	13·2088	13·2735	13·3383	13·4030	13·4678	13·5325
210	13·5973	13·6620	13·7268	13·7915	13·8563	13·9210	13·9858	14·0505	14·1153	14·1800
220	14·2448	14·3095	14·3743	14·4390	14·5038	14·5685	14·6333	14·6980	14·7628	14·8275
230	14·8923	14·9570	15·0218	15·0865	15·1513	15·2160	15·2808	15·3455	15·4103	15·4750
240	15·5398	15·6045	15·6693	15·7340	15·7988	15·8635	15·9283	15·9930	16·0577	16·1225

PRESSURE, STRESS

Pressure, Stress: Meganewtons per square metre to UK tons-force per square inch
Basis: 1 UKton = 2240 lb; 1 lbf = 0.453 592 37 kgf; 1 kgf = 9.806 65 N; 1 in = 0.0254 m.

MN/m²	0	1	2	3	4	5	6	7	8	9
					UK tons-force per square inch					
250	16.1872	16.2520	16.3167	16.3815	16.4462	16.5110	16.5757	16.6405	16.7052	16.7700
260	16.8347	16.8995	16.9642	17.0290	17.0937	17.1585	17.2232	17.2880	17.3527	17.4175
270	17.4822	17.5470	17.6117	17.6765	17.7412	17.8060	17.8707	17.9355	18.0002	18.0650
280	18.1297	18.1945	18.2592	18.3240	18.3887	18.4535	18.5182	18.5830	18.6477	18.7125
290	18.7772	18.8420	18.9067	18.9715	19.0362	19.1010	19.1657	19.2305	19.2952	19.3599
300	19.4247	19.4894	19.5542	19.6189	19.6837	19.7484	19.8132	19.8779	19.9427	20.0074
310	20.0722	20.1369	20.2017	20.2664	20.3312	20.3959	20.4607	20.5254	20.5902	20.6549
320	20.7197	20.7844	20.8492	20.9139	20.9787	21.0434	21.1082	21.1729	21.2377	21.3024
330	21.3672	21.4319	21.4967	21.5614	21.6262	21.6909	21.7557	21.8204	21.8852	21.9499
340	22.0147	22.0794	22.1442	22.2089	22.2737	22.3384	22.4032	22.4679	22.5326	22.5974
350	22.6621	22.7269	22.7916	22.8564	22.9211	22.9859	23.0506	23.1154	23.1801	23.2449
360	23.3096	23.3744	23.4391	23.5039	23.5686	23.6334	23.6981	23.7629	23.8276	23.8924
370	23.9571	24.0219	24.0866	24.1514	24.2161	24.2809	24.3456	24.4104	24.4751	24.5399
380	24.6046	24.6694	24.7341	24.7989	24.8636	24.9284	24.9931	25.0579	25.1226	25.1874
390	25.2521	25.3169	25.3816	25.4464	25.5111	25.5759	25.6406	25.7053	25.7701	25.8348
400	25.8996	25.9643	26.0291	26.0938	26.1586	26.2233	26.2881	26.3528	26.4176	26.4823
410	26.5471	26.6118	26.6766	26.7413	26.8061	26.8708	26.9356	27.0003	27.0651	27.1298
420	27.1946	27.2593	27.3241	27.3888	27.4536	27.5183	27.5831	27.6478	27.7126	27.7773
430	27.8421	27.9068	27.9716	28.0363	28.1011	28.1658	28.2306	28.2953	28.3601	28.4248
440	28.4896	28.5543	28.6191	28.6838	28.7486	28.8133	28.8780	28.9428	29.0075	29.0723
450	29.1370	29.2018	29.2665	29.3313	29.3960	29.4608	29.5255	29.5903	29.6550	29.7198
460	29.7845	29.8493	29.9140	29.9788	30.0435	30.1083	30.1730	30.2378	30.3025	30.3673
470	30.4320	30.4968	30.5615	30.6263	30.6910	30.7558	30.8205	30.8853	30.9500	31.0148
480	31.0795	31.1443	31.2090	31.2738	31.3385	31.4033	31.4680	31.5328	31.5975	31.6623
490	31.7270	31.7918	31.8565	31.9213	31.9860	32.0508	32.1155	32.1802	32.2450	32.3097

Pressure, stress:
Meganewtons per square metre to UK tons-force per square inch

MN/m²	0	1	2	3	4	5	6	7	8	9
				UK tons-force per square inch						
500	32·3745	32·4392	32·5040	32·5687	32·6335	32·6982	32·7630	32·8277	32·8925	32·9572
510	33·0220	33·0867	33·1515	33·2162	33·2810	33·3457	33·4105	33·4752	33·5400	33·6047
520	33·6695	33·7342	33·7990	33·8637	33·9285	33·9932	34·0580	34·1227	34·1875	34·2522
530	34·3170	34·3817	34·4465	34·5112	34·5760	34·6407	34·7055	34·7702	34·8350	34·8997
540	34·9645	35·0292	35·0940	35·1587	35·2235	35·2882	35·3529	35·4177	35·4824	35·5472
550	35·6119	35·6767	35·7414	35·8062	35·8709	35·9357	36·0004	36·0652	36·1299	36·1947
560	36·2594	36·3242	36·3889	36·4537	36·5184	36·5832	36·6479	36·7127	36·7774	36·8422
570	36·9069	36·9717	37·0364	37·1012	37·1659	37·2307	37·2954	37·3602	37·4249	37·4897
580	37·5544	37·6192	37·6839	37·7487	37·8134	37·8782	37·9429	38·0077	38·0724	38·1372
590	38·2019	38·2667	38·3314	38·3962	38·4609	38·5256	38·5904	38·6551	38·7199	38·7846
600	38·8494	38·9141	38·9789	39·0436	39·1084	39·1731	39·2379	39·3026	39·3674	39·4321
610	39·4969	39·5616	39·6264	39·6911	39·7559	39·8206	39·8854	39·9501	40·0149	40·0796
620	40·1444	40·2091	40·2739	40·3386	40·4034	40·4681	40·5329	40·5976	40·6624	40·7271
630	40·7919	40·8566	40·9214	40·9861	41·0509	41·1156	41·1804	41·2451	41·3099	41·3746
640	41·4394	41·5041	41·5689	41·6336	41·6983	41·7631	41·8278	41·8926	41·9573	42·0221
650	42·0868	42·1516	42·2163	42·2811	42·3458	42·4106	42·4753	42·5401	42·6048	42·6696
660	42·7343	42·7991	42·8638	42·9286	42·9933	43·0581	43·1228	43·1876	43·2523	43·3171
670	43·3818	43·4466	43·5113	43·5761	43·6408	43·7056	43·7703	43·8351	43·8998	43·9646
680	44·0293	44·0941	44·1588	44·2236	44·2883	44·3531	44·4178	44·4826	44·5473	44·6121
690	44·6768	44·7416	44·8063	44·8711	44·9358	45·0005	45·0653	45·1300	45·1948	45·2595
700	45·3243	45·3890	45·4538	45·5185	45·5833	45·6480	45·7128	45·7775	45·8423	45·9070
710	45·9718	46·0365	46·1013	46·1660	46·2308	46·2955	46·3603	46·4250	46·4898	46·5545
720	46·6193	46·6840	46·7488	46·8135	46·8783	46·9430	47·0078	47·0725	47·1373	47·2020
730	47·2668	47·3315	47·3963	47·4610	47·5258	47·5905	47·6553	47·7200	47·7848	47·8495
740	47·9143	47·9790	48·0438	48·1085	48·1732	48·2380	48·3027	48·3675	48·4322	48·4970

Pressure, stress:
Meganewtons per square metre to UK tons-force per square inch

MN/m²	0	1	2	3	4	5	6	7	8	9
					UK tons-force per square inch					
750	48·5617	48·6265	48·6912	48·7560	48·8207	48·8855	48·9502	49·0150	49·0797	49·1445
760	49·2092	49·2740	49·3387	49·4035	49·4682	49·5330	49·5977	49·6625	49·7272	49·7920
770	49·8567	49·9215	49·9862	50·0510	50·1157	50·1805	50·2452	50·3100	50·3747	50·4395
780	50·5042	50·5690	50·6337	50·6985	50·7632	50·8280	50·8927	50·9575	51·0222	51·0870
790	51·1517	51·2165	51·2812	51·3459	51·4107	51·4754	51·5402	51·6049	51·6697	51·7344
800	51·7992	51·8639	51·9287	51·9934	52·0582	52·1229	52·1877	52·2524	52·3172	52·3819
810	52·4467	52·5114	52·5762	52·6409	52·7057	52·7704	52·8352	52·8999	52·9647	53·0294
820	53·0942	53·1589	53·2237	53·2884	53·3532	53·4179	53·4827	53·5474	53·6122	53·6769
830	53·7417	53·8064	53·8712	53·9359	54·0007	54·0654	54·1302	54·1949	54·2597	54·3244
840	54·3892	54·4539	54·5186	54·5834	54·6481	54·7129	54·7776	54·8424	54·9071	54·9719
850	55·0366	55·1014	55·1661	55·2309	55·2956	55·3604	55·4251	55·4899	55·5546	55·6194
860	55·6841	55·7489	55·8136	55·8784	55·9431	56·0079	56·0726	56·1374	56·2021	56·2669
870	56·3316	56·3964	56·4611	56·5259	56·5906	56·6554	56·7201	56·7849	56·8496	56·9144
880	56·9791	57·0439	57·1086	57·1734	57·2381	57·3029	57·3676	57·4324	57·4971	57·5619
890	57·6266	57·6914	57·7561	57·8208	57·8856	57·9503	58·0151	58·0798	58·1446	58·2093
900	58·2741	58·3388	58·4036	58·4683	58·5331	58·5978	58·6626	58·7273	58·7921	58·8568
910	58·9216	58·9863	59·0511	59·1158	59·1806	59·2453	59·3101	59·3748	59·4396	59·5043
920	59·5691	59·6338	59·6986	59·7633	59·8281	59·8928	59·9576	60·0223	60·0871	60·1518
930	60·2166	60·2813	60·3461	60·4108	60·4756	60·5403	60·6051	60·6698	60·7346	60·7993
940	60·8641	60·9288	60·9935	61·0583	61·1230	61·1878	61·2525	61·3173	61·3820	61·4468
950	61·5115	61·5763	61·6410	61·7058	61·7705	61·8353	61·9000	61·9648	62·0295	62·0943
960	62·1590	62·2238	62·2885	62·3533	62·4180	62·4828	62·5475	62·6123	62·6770	62·7418
970	62·8065	62·8713	62·9360	63·0008	63·0655	63·1303	63·1950	63·2598	63·3245	63·3893
980	63·4540	63·5188	63·5835	63·6483	63·7130	63·7778	63·8425	63·9073	63·9720	64·0368
990	64·1015	64·1662	64·2310	64·2957	64·3605	64·4252	64·4900	64·5547	64·6195	64·6842
1000	64·7490									

THE GREEK ALPHABET

Name	Capital Letter	Small Letter	English Equivalent	Name	Capital Letter	Small Letter	English Equivalent
Alpha	A	α	a	Nu	N	ν	n
Beta	B	β	b	Xi	Ξ	ξ	x
Gamma	Γ	γ	g	Omicron	O	o	short o
Delta	Δ	δ	d	Pi	Π	π	p
Epsilon	E	ϵ	short e	Rho	P	ρ	rh
Zeta	Z	ζ	z	Sigma	Σ	σ	s
Eta	H	η	long e	Tau	T	τ	t
Theta	Θ	θ	th	Upsilon	Y	υ	u
Iota	I	ι	i	Phi	Φ	ϕ	ph
Kappa	K	κ	k	Chi	X	χ	ch
Lambda	Λ	λ	l	Psi	Ψ	ψ	ps
Mu	M	μ	m	Omega	Ω	ω	long o

GEOMETRICAL PROPERTIES OF PLANE SECTIONS

Section	Area	Position of Centroid	Moments of Inertia	Section Moduli
TRIANGLE	$A = \dfrac{bh}{2}$	$e_x = \dfrac{h}{3}$	$I_{XX} = bh^3/36$ $I_{YY} = hb^3/48$ $I_{aa} = bh^3/4$ $I_{bb} = bh^3/12$	Z_{XX} base $= bh^2/12$ apex $= bh^2/24$ $Z_{YY} = bh^2/24$
RECTANGLE	$A = bd$	$e_x = \dfrac{h}{2}$	$I_{XX} = bd^3/12$ $I_{YY} = db^3/12$ $I_{bb} = bd^3/3$	$Z_{XX} = bd^2/6$ $Z_{YY} = db^2/6$
RECTANGLE axis on diagonal	$A = bd$	$e_x = \dfrac{bd}{\sqrt{b^2+d^2}}$	$I_{XX} = \dfrac{b^3 d^3}{6(b^2+d^2)}$	$Z_{XX} = \dfrac{b^2 d^2}{6\sqrt{b^2+d^2}}$
RECTANGLE axis through C.G.	$A = bd$	$e_x = \dfrac{b\sin\theta + d\cos\theta}{2}$	$I_{XX} = \dfrac{bd(b^2\sin^2\theta + d^2\cos^2\theta)}{12}$	$Z_{XX} = \dfrac{bd(b^2\sin^2\theta + d^2\cos^2\theta)}{6(b\sin\theta + d\cos\theta)}$
SQUARE	$A = s^2$	$e_x = \dfrac{s}{2}$ $e_v = \dfrac{s}{\sqrt{2}}$	$I_{XX} = I_{YY} = s^4/12$ $I_{bb} = s^4/3$ $I_{VV} = s^4/12$	$Z_{XX} = Z_{YY} = \dfrac{s^3}{6}$ $Z_{VV} = \dfrac{s^3}{6\sqrt{2}}$
TRAPEZIUM	$A = \dfrac{d(a+b)}{2}$	$e_{x_1} = \dfrac{d(2a+b)}{3(a+b)}$	$I_{XX} = \dfrac{d^3(a^2+4ab+b^2)}{36(a+b)}$ $I_{YY} = \dfrac{d(a^3+a^2b+ab^2+b^3)}{48}$	$Z_{XX} = \dfrac{I_{XX}}{d-e_x}$ (two values) $Z_{YY} = \dfrac{2 I_{YY}}{b}$
DIAMOND	$A = \dfrac{bd}{2}$	$e_x = \dfrac{d}{2}$	$I_{XX} = \dfrac{bd^3}{48}$ $I_{YY} = \dfrac{db^3}{48}$	$Z_{XX} = \dfrac{bd^2}{24}$ $Z_{YY} = \dfrac{db^2}{24}$
HEXAGON	$A = 0.866 d^2$	$e_x = 0.866 s$ $= d/2$	$I_{XX} = I_{YY} = I_{VV}$ $= 0.0601 d^4$	$Z_{XX} = 0.1203 d^3$ $Z_{YY} = Z_{VV}$ $= 0.1042 d^3$

GEOMETRICAL PROPERTIES OF PLANE SECTIONS

Section	Area	Position of Centroid	Moments of Inertia	Section Moduli
OCTAGON	$A = 0.8284 d^2$ $s = 0.4142 d$	$e_x = \dfrac{d}{2}$ $e_v = 0.541 d$	$I_{XX} = I_{YY} = I_{VV}$ $= 0.0547 d^4$	$Z_{XX} = Z_{YY}$ $= 0.1095 d^3$ $Z_{VV} = 0.1011 d^3$
POLYGON (n sides, Regular figure)	$A = \dfrac{n s^2 \cot\theta}{4}$ $A = n r^2 \tan\theta$ $A = \dfrac{n R^2 \sin 2\theta}{2}$	$e = r$ or R depending on the axis and value of n	$I_1 = I_2$ $= \dfrac{A(6R^2 - s^2)}{24}$ $= \dfrac{A(12 r^2 + s^2)}{48}$	$Z = \dfrac{I}{e}$
CIRCLE	$A = \pi r^2$ $A = 0.7854 d^2$	$e = r = \dfrac{d}{2}$	$I = \dfrac{\pi d^4}{64}$ $I = 0.7854 r^4$	$Z = \dfrac{\pi d^3}{32}$ $Z = 0.7854 r^3$
SEMI-CIRCLE	$A = 1.5708 r^2$	$e_x = 0.424 r$	$I_{XX} = 0.1098 r^4$ $I_{YY} = 0.3927 r^4$	Z_{XX} base $= 0.2587 r^3$ crown $= 0.1907 r^3$ $Z_{YY} = 0.3927 r^3$
SEGMENT	$A =$ $\dfrac{r^2}{2}\left(\dfrac{\pi\theta°}{180°} - \sin\theta\right)$	$e_0 = \dfrac{c^3}{12 A}$ $e_x = e_0 - r\cos\dfrac{\theta}{2}$	$I_{XX} = \dfrac{r^4}{16}\left(\dfrac{\pi\theta°}{90°} - \sin 2\theta\right)$ $- \dfrac{20 r^4(1 - \cos\theta)^3}{\pi\theta° - 180°\sin\theta}$ $I_{YY} = \dfrac{r^4}{48}\left(\dfrac{\pi\theta°}{30°} - 8\sin\theta + \sin 2\theta\right)$	Z_{XX} base $= I_{XX}/e_x$ crown $= \dfrac{I_{XX}}{b - e_x}$ $Z_{YY} = \dfrac{2 I_{YY}}{c}$
SECTOR	$A = \dfrac{\theta°}{360°}\pi r^2$	$e_x = \dfrac{2}{3} r \dfrac{c}{a}$ $e_x = \dfrac{r^2 c}{3 A}$	$I_{XX} = I_0 - \dfrac{360°}{8°\pi}\sin^2\theta \cdot \dfrac{4 r^4}{2 \cdot 3}$ $I_{YY} = \dfrac{r^4}{8}\left(\dfrac{\pi\theta°}{180°} - \sin\theta\right)$ $I_0 = \dfrac{r^4}{8}\left(\dfrac{\pi\theta°}{180°} + \sin\theta\right)$	Z_{XX} centre $= I_{XX}/e_x$ crown $= \dfrac{I_{XX}}{r - e_x}$ $Z_{YY} = \dfrac{2 I_{YY}}{c}$
QUADRANT	$A = \dfrac{\pi r^2}{4}$	$e_x = 0.424 r$ $e_v = 0.6 r$ $e_u = 0.707 r$	$I_{XX} = I_{YY} = 0.0549 r^4$ $I_{bb} = 0.1963 r^4$ $I_{UU} = 0.0714 r^4$ $I_{VV} = 0.0384 r^4$	Minimum Values $Z_{XX} = Z_{YY}$ $= 0.0953 r^3$ $Z_{UU} = 0.1009 r^3$ $Z_{VV} = 0.064 r^3$
COMPLEMENT	$A = 0.2146 r^2$	$e_x = 0.777 r$ $e_v = 1.098 r$ $e_u = 0.707 r$ $e_a = 0.316 r$ $e_b = 0.391 r$	$I_{XX} = I_{YY} = 0.0076 r^4$ $I_{UU} = 0.012 r^4$ $I_{VV} = 0.0031 r^4$	Minimum Values $Z_{XX} = Z_{YY}$ $= 0.0097 r^3$ $Z_{UU} = 0.017 r^3$ $Z_{VV} = 0.0079 r^3$

GEOMETRICAL PROPERTIES OF PLANE SECTIONS

Section	Area	Position of Centroid	Moments of Inertia	Section Moduli
ELLIPSE	$A = \pi ab$	$e_x = a$ $e_y = b$	$I_{XX} = 0.7854 ba^3$ $I_{YY} = 0.7854 ab^3$	$Z_{XX} = 0.7854 ba^2$ $Z_{YY} = 0.7854 ab^2$
SEMI-ELLIPSE	$A = \dfrac{\pi ab}{2}$	$e_x = 0.424 a$ $e_y = b$	$I_{XX} = 0.1098 ba^3$ $I_{YY} = 0.3927 ab^3$ $I_{base} = 0.3927 ba^3$	Z_{XX} - base $= 0.2587 ba^2$ Z_{XX} - crown $= 0.1907 ba^2$ $Z_{YY} = 0.3927 ab^2$
¼ ELLIPSE	$A = 0.7854 ab$	$e_x = 0.424 a$ $e_y = 0.424 b$	$I_{XX} = 0.0549 ba^3$ $I_{YY} = 0.0549 ab^3$ $I_{b_1 a_1} = 0.1963 ba^3$ $I_{b_1 c_1} = 0.1963 ab^3$	Z_{XX} - base $= 0.1293 ba^2$ Z_{XX} - crown $= 0.0953 ba^2$ Z_{YY} - base $= 0.1293 ab^2$ Z_{YY} - crown $= 0.0953 ab^2$
COMPLEMENT	$A = 0.2146 ab$	$e_x = 0.777 a$ $e_y = 0.777 b$	$I_{XX} = 0.0076 ba^3$ $I_{YY} = 0.0076 ab^3$	Z_{XX} - base $= 0.0338 ba^2$ Z_{XX} - apex $= 0.0097 ba^2$ Z_{YY} - base $= 0.0338 ab^2$ Z_{YY} - apex $= 0.0097 ab^2$
FULL PARABOLA	$A = \dfrac{4ab}{3}$	$e_x = \dfrac{2a}{5}$ $e_y = b$	$I_{XX} = 0.0914 ba^3$ $I_{YY} = 0.2666 ab^3$ $I_{base} = 0.3048 ba^3$	Z_{XX} - base $= 0.2286 ba^2$ Z_{XX} - crown $= 0.1524 ba^2$ $Z_{YY} = 0.2666 ab^2$
SEMI-PARABOLA	$A = \dfrac{2ab}{3}$	$e_x = \dfrac{2a}{5}$ $e_y = \dfrac{3b}{8}$	$I_{XX} = 0.0457 ba^3$ $I_{YY} = 0.0396 ab^3$ $I_{b_1 a_1} = 0.1524 ba^3$ $I_{b_1 c_1} = 0.1333 ab^3$	Z_{XX} - base $= 0.1143 ba^2$ Z_{XX} - crown $= 0.076 ba^2$ Z_{YY} - base $= 0.1055 ab^2$ Z_{YY} - crown $= 0.0633 ab^2$
COMPLEMENT	$A = \dfrac{ab}{3}$	$e_x = \dfrac{7a}{10}$ $e_y = \dfrac{3b}{4}$	$I_{XX} = 0.0176 ba^3$ $I_{YY} = 0.0125 ab^3$ $I_{a_1 b_1} = 0.181 ba^3$ $I_{b_1 c_1} = 0.2 ab^3$	Z_{XX} - base $= 0.0587 ba^2$ Z_{XX} - apex $= 0.0252 ba^2$ Z_{YY} - base $= 0.05 ab^2$ Z_{YY} - apex $= 0.0167 ab^2$
FILLET	$A = \dfrac{s^2}{6}$	$e_u = e_v = \dfrac{4s}{5}$	$I_{UU} = I_{VV} = 0.00524 s^4$ $I_{ab} = 0.1119 a^4$	$Z_{UU} = Z_{VV}$ base $= 0.0262 a^3$ apex $= 0.0066 a^3$

STEEL PILING

FRODINGHAM SHEET PILING
DIMENSIONS AND PROPERTIES

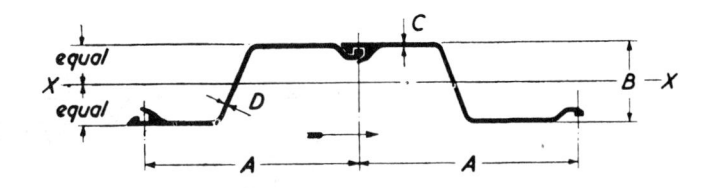

1. Profile of Nos. 1A, 1B, 1BXN, 2N, 3N & 4N sections

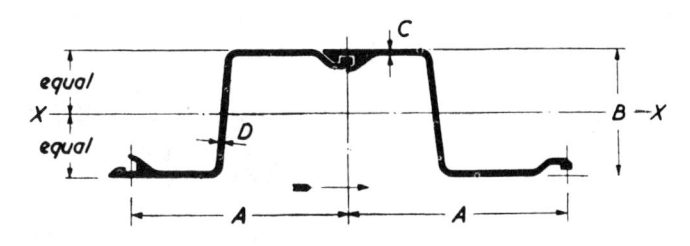

2. Profile of No. 5 section

Section No.	Size in mm A	B	C	D	Weight per lin m in kg	Weight per sq m of wall in kg	Section modulus cm³/m of wall
1A	400	146	6.9	6.9	35.64	89.1	563
1B	400	133	9.5	9.5	42.13	105.3	562
1BXN	476	143	12.7	12.7	62.1	130.4	688
2N	483	235	9.7	8.4	54.21	112.3	1 150
3N	483	283	11.7	8.9	66.15	137.1	1 688
4N	483	330	14.0	10.4	82.45	170.8	2 414
5	425	311	17.0	11.9	100.76	236.9	3 168

1002 STEEL PILING

FRODINGHAM HIGH MODULUS SHEET PILING
DIMENSIONS AND PROPERTIES

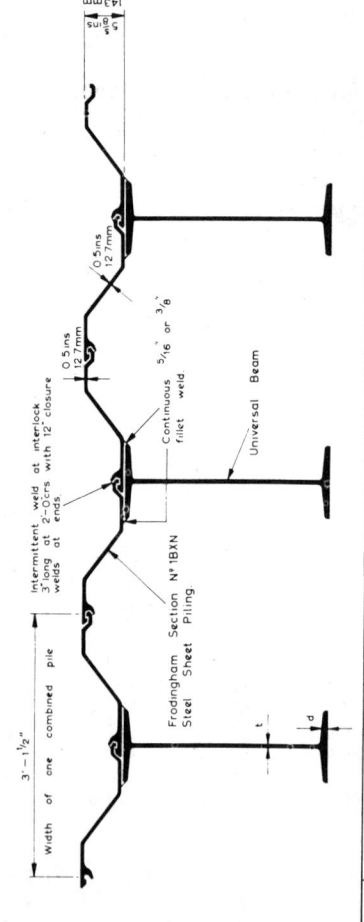

Section No.	Universal Beam Size British Standard 4	Dimension d in	Dimension d mm	Dimension t in	Dimension t mm	Weight of one Combined Pile lb per lin ft	Weight of one Combined Pile kg per lin m	Weight per Unit Area of Wall lb per sq ft	Weight per Unit Area of Wall kg per sq m	Minimum Section Modulus about Neutral Axis ins³ per ft	Minimum Section Modulus about Neutral Axis cm³ per m
	in and lb per foot										
10X	24 × 9 × 94	0.87	22.1	0.52	13.2	178.4	265.5	57.08	278.7	95	5110
11X	27 × 10 × 102	0.83	22.1	0.52	13.2	186.4	277.4	59.64	291.2	110	5910
12X	27 × 10 × 114	0.93	23.6	0.57	14.5	198.4	295.2	63.48	310.0	125	6720
13X	30 × 10½ × 116	0.85	21.6	0.56	14.2	200.4	298.2	64.12	313.1	136	7310
14X	30 × 10½ × 132	1.00	25.4	0.62	15.7	216.4	322.0	69.24	338.1	155	8330
15X	36 × 12 × 150	0.94	23.9	0.63	16.0	234.4	348.8	75.00	366.2	200	10800
16X	36 × 12 × 170	1.10	27.9	0.68	17.3	254.4	378.6	81.40	397.5	230	12400
17X	36 × 12 × 194	1.26	32.0	0.77	19.6	278.4	414.3	89.08	435.0	260	14000

STEEL PILING

FRODINGHAM DOUBLE BOX PILES
DIMENSIONS AND PROPERTIES

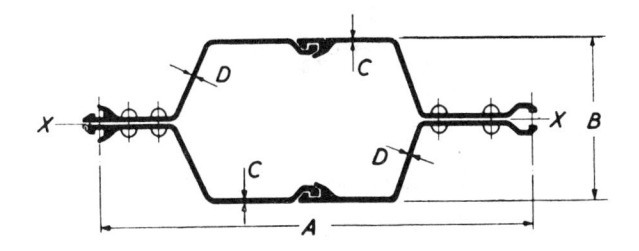

1. Profile of Nos. 1A, 1B, 1BXN, 2N, 3N & 4N sections

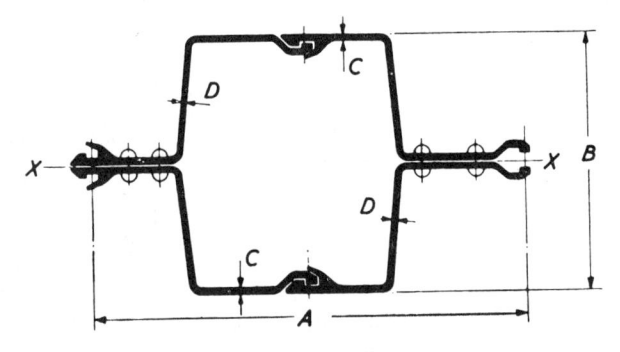

2. Profile of No. 5 section

Section No.	Dimensions in mm A	Dimensions in mm B	Cross Section sq cm Steel only	Cross Section sq cm Whole Pile	Weight kg per lin m	Moment of Inertia about XX cm^4	Section Modulus about XX cm^3	Radius of Gyration about XX cm
1A	800	292	182	1258	145.1	16275	1115	9.4
1B	800	267	215	1174	171.0	15565	1165	8.5
1BXN	953	286	317	1523	252.4	25515	1785	9.0
2N	965	470	277	2406	219.3	64305	2735	15.2
3N	965	565	338	2884	268.6	113415	4015	18.3
4N	965	660	421	3387	333.7	191660	5805	21.3
5	850	622	514	2903	407.0	208430	6700	20.1

STEEL PILING

FRODINGHAM PLATED BOX PILES
DIMENSIONS AND PROPERTIES

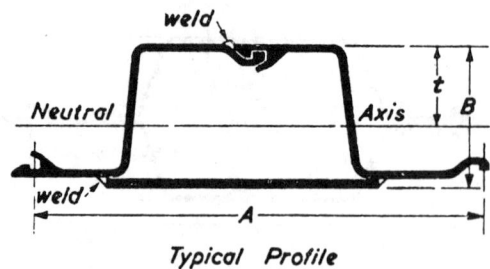

Typical Profile

Section No.	Dimensions in mm A	Dimensions in mm B	Dimensions in mm t	Size of Standard Plate mm	Cross Section sq cm Steel only	Cross Section sq cm Whole Pile	*Weight kg per lin m	Moment of Inertia about Neutral Axis cm^4	Minimum Section Modulus about Neutral Axis cm^3	Radius of Gyration about Neutral Axis cm
1A	800	156	100	508 x 10	139	677	109.2	5200	520	6.1
1B	800	144	93	559 x 11	170	652	132.9	5035	540	5.5
1BXN	953	157	102	711 x 14	259	858	203.9	8490	830	5.7
2N	965	248	164	660 x 13	222	1290	174.2	21060	1280	9.7
3N	965	297	195	660 x 14	263	1548	206.3	36375	1870	11.8
4N	965	346	223	660 x 16	315	1813	247.1	59475	2705	13.7
5	850	330	202	533 x 19	359	1555	281.2	61805	3055	13.1

* Weights quoted are for piles with full length plate.

STEEL PILING

FRODINGHAM BOX PILES
DIMENSIONS AND PROPERTIES

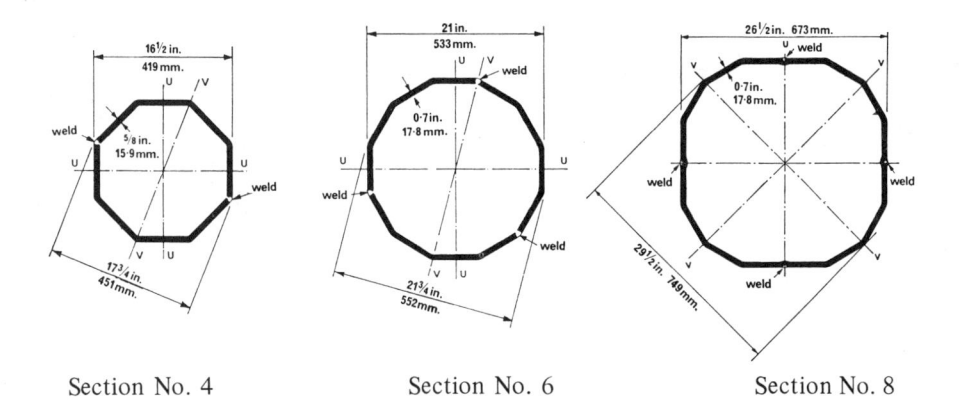

Section No. 4   Section No. 6   Section No. 8

Section No.	Weight including welds kg/m	Cross-sectional area sq cm	Overall area of section sq cm	Outside perimeter mm	Moment of inertia – any axis cm⁴	Radius of gyration any axis mm	Section modulus maximum (axis UU) cm³	Section modulus minimum (axis VV) cm³
4	167.6	213.7	1460	1390	45620	146	2183	2025
6	233.4	297.9	2560	1715	101345	185	3802	3670
8	313	398	3970	2273	237650	245	7063	6342

FRODINGHAM STRAIGHT WEB PILING
DIMENSIONS AND PROPERTIES

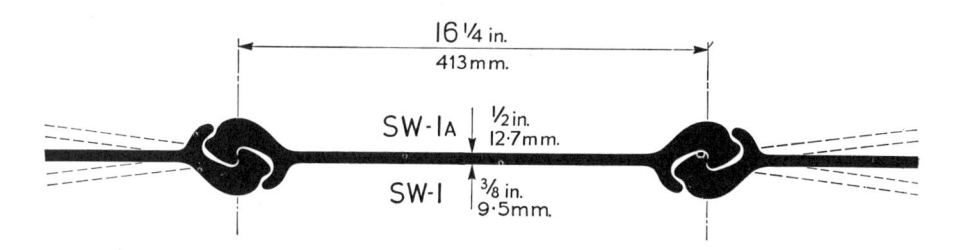

Section No.	Weight Single pile kg/m	Weight Per unit area kg/sq m	Minimum ultimate strength of interlock tonnes/m	Minimum section modulus of single pile cm³
SW1	55.32	134.0	285	27.9
SW1A	63.77	154.5	285	27.9

STEEL PILING

LARSSEN SHEET PILING

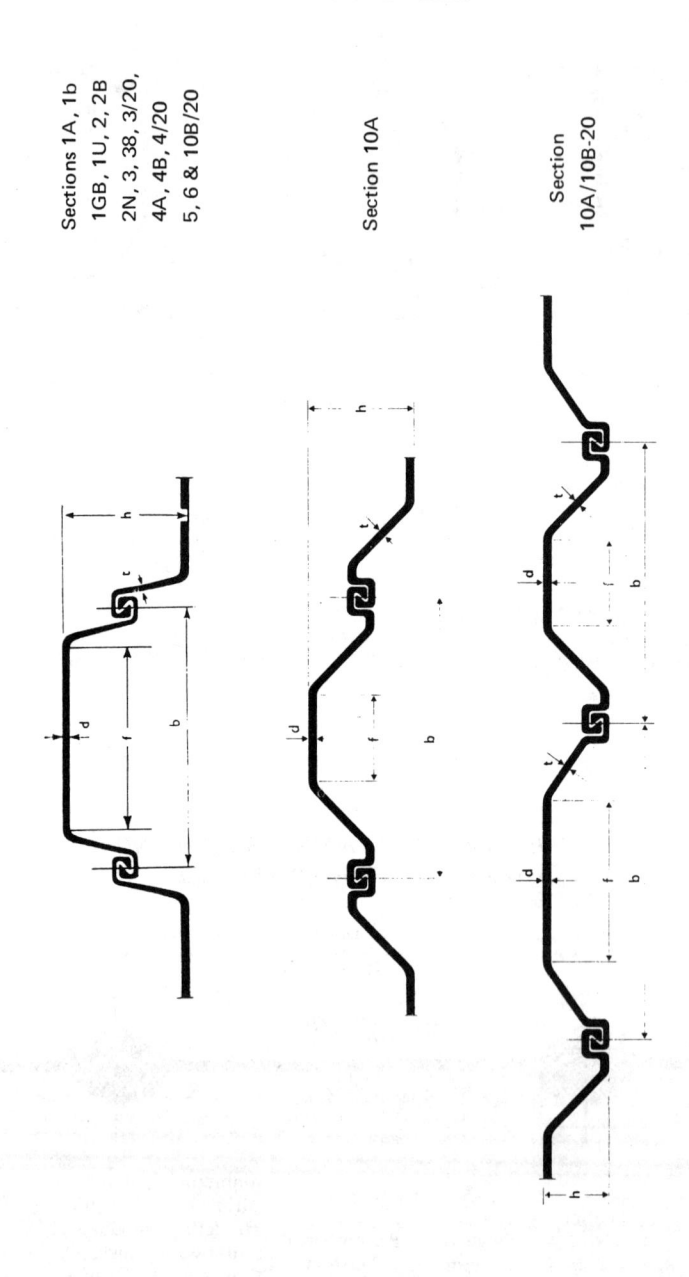

Sections 1A, 1b
1GB, 1U, 2, 2B
2N, 3, 38, 3/20,
4A, 4B, 4/20
5, 6 & 10B/20

Section 10A

Section 10A/10B-20

STEEL PILING

DIMENSIONS AND PROPERTIES

Section	b mm	h mm	d mm	t mm (nominal)	f Flat of Web mm	Sectional Area cm²/m of wall	Weight per linear metre Kg	Weight per m² of wall Kg	Combined Moment of Inertia cm⁴/m	Section Modulus cm³/m
1A	400	130	7.2	5.8	302	107	33.6	84	2496	384
1B	400	178	7.1	6.4	305	114	35.6	89	4998	562
1GB	400	130	8.1	5.8	302	115	36.2	90	2729	419
1U	400	130	9.4	9.4	302	135	42.4	106	3184	489
2	400	200	10.2	7.8	270	156	48.8	122	8494	850
2B	400	270	8.6	7.1	248	149	46.7	117	13663	1013
2N	400	270	9.4	7.1	248	156	48.8	122	14855	1101
3	400	247	14.0	8.9	248	198	62.0	155	16839	1360
3B	400	298	13.5	8.9	235	198	62.1	155	23910	1602
3/20	508	343	11.7	8.4	330	175	69.6	137	28554	1665
4A	400	381	15.7	9.4	219	236	74.0	185	45160	2371
4B	420	343	15.5	10.9	257	256	84.5	201	39165	2285
4/20 {	508	381	14.3	9.4	321	207	82.5	162	43167	2266
	508	381	15.7	9.4	321	218	86.8	171	45924	2414
5	420	343	22.1	11.9	257	303	100.0	238	50777	2962
6 {	420	440	22.0	14.0	248	370	122.0	290	92298	4200
	420	440	25.4	14.0	251	398	131.0	312	101689	4618
	420	440	28.6	14.0	251	421	138.7	330	109968	5000
10A	450	171	12.7	12.7	130	176	62.2	138	4166	486
10B/20	508	171	12.7	12.7	273	167	66.4	131	6054	706
10A-10B/20	450/ 508	108	12.7/ 12.7	12.7/ 12.7	130/ 273	171	62.2/ 66.4	134	2250	356

1008 STEEL PILING

FRODINGHAM HIGH MODULUS SHEET PILING DIMENSIONS AND PROPERTIES

Universal Beam		Crs. of U.B.'s mm	Weight		Y mm	Mom. of Inertia XX		Sectn. Modulus XX		Rad. of Gyration XX mm
Serial Size mm	Weight Kg/m		Kg/m	Kg/m²		cm⁴/m wall	cm⁴/single pile	cm³/m wall	cm³/single pile	
533 × 210	101	690	165.5	240.0	384.8	156141	107737	4059	2801	226.1
533 × 330	167	810	231.0	285.3	349.8	201060	162859	5749	4657	235.2
533 × 330	211	813	275.6	339.0	343.4	247148	200931	7197	5851	239.0
610 × 229	113	708	177.4	250.7	425.2	207792	147117	4887	3460	255.0
610 × 305	149	784	213.1	271.8	406.4	242651	190238	5971	4681	264.4
686 × 254	152	734	216.1	294.5	455.4	316460	232282	6950	5101	290.3
762 × 267	173	746	236.9	317.6	493.0	412292	307570	8363	6239	319.0
762 × 267	196	748	260.7	348.8	487.4	464832	347694	9537	7134	323.3
838 × 292	194	772	257.7	334.0	533.1	525497	405684	9858	7610	351.3
915 × 305	223	784	287.5	367.0	563.9	672330	527107	11923	9348	379.0
915 × 305	253	785	317.3	404.2	558.8	757222	594419	13551	10637	383.3
915 × 305	289	787	353.0	448.4	553.5	850280	669170	15362	12090	385.3
915 × 419	387	900	451.2	501.4	530.4	987327	888595	18614	16753	392.7

SHEET PILING

LARSSEN SHEET PILING DIMENSIONS AND PROPERTIES FOR BOX PILING

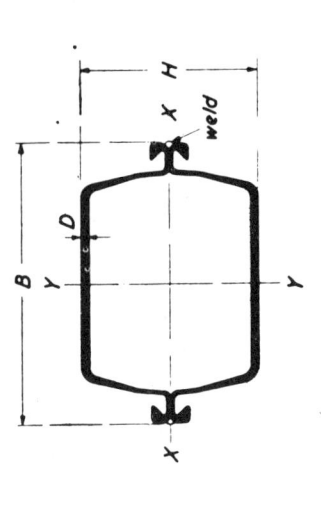

Section	B mm	H mm	d mm	Weight per metre Kg	Sectional Area cm² Steel only	Sectional Area cm² Whole Piles	Least Rad. of Gyr. cm	Approx. Perimeter cm	Moment of Inertia cm⁴ About XX	Moment of Inertia cm⁴ About YY	Section Modulus cm³ About XX	Section Modulus cm³ About YY
BP 1A	432	167	7.2	67.3	86	597	6.36	112	3492	18142	420	840
BP 1B	432	214	7.1	71.3	91	774	7.59	122	6048	19246	565	891
BP 1GB	432	165	8.1	72.3	92	619	6.40	122	3775	19614	457	980
BP 1U	432	165	9.4	84.8	108	619	6.40	122	4400	23620	533	1087
BP 2	438	240	10.2	97.6	124	858	9.07	124	10198	25777	850	1177
BP 2B	436	314	8.6	93.4	119	1077	10.57	140	15430	26156	982	1201
BP 2N	436	314	9.5	97.6	124	1084	11.61	142	16791	26672	1068	1224
BP 3	438	289	14.2	124.0	158	1019	11.05	132	19251	29761	1332	1358
BP 3B	438	343	13.5	124.1	158	1161	12.75	142	26152	30218	1526	1380
BP 3/20	544	390	11.7	139.1	177	1581	12.78	165	38789	56807	1991	2089
BP 4A	436	429	15.7	148.0	189	1413	13.77	160	47529	35833	2217	1645
BP 4B	467	394	15.5	168.9	215	1432	14.58	155	44682	47500	2270	2078
BP 4/20	543	427	14.3	165.0	210	1855	16.46	178	56828	64316	2661	2366
	543	427	15.7	173.6	222	1855	16.51	178	60474	67209	2832	2900
BP 5	467	394	22.1	200.0	255	1432	14.22	155	58501	51471	2973	2252
BP 6	464	502	22.1	243.6	311	1794	14.76	208	105731	67563	4215	2912
	464	502	25.4	262.0	335	1794	14.51	208	116174	70356	4631	3037
	464	502	28.6	277.4	354	1794	14.27	208	125893	71958	5019	3104
BP 10B/20	541	215	12.7	132.7	170	923	7.87	132	10547	49906	980	1844

RENDHEX FOUNDATION COLUMNS DIMENSIONS AND PROPERTIES

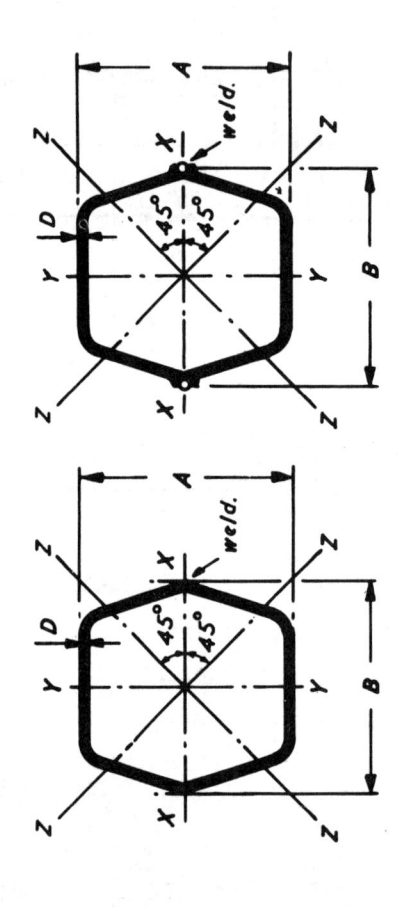

Rendhex Column No.	Size-mm A	Size-mm B	Size-mm t	Weight of complete Col. (incl. Weld) kg/m	Sectional Area (excl. welds) sq cm Steel only	Sectional Area (excl. welds) sq cm Whole column	Approx. Perimeter cm	Moment of Inertia-cm⁴ About XX	Moment of Inertia-cm⁴ About YY	Moment of Inertia-cm⁴ About ZZ	Radius of Gyration-cm About XX	Radius of Gyration-cm About YY	Radius of Gyration-cm About ZZ	Section Modulus-cm³ About XX	Section Modulus-cm³ About YY	Section Modulus-cm³ About ZZ
No. 3	330.20	327.03	12.70	105.71	134.25	867.5	112	19488	15445	17466	12.00	10.69	11.38	1180.00	995.00	942.00
No. 4	406.40	414.34	15.88	171.44	217.42	1421.6	142	50110	43142	46743	15.11	14.05	14.63	2466.10	2082.60	1934.50
No. 6	508.00	508.00	17.78	240.33	305.25	2163.3	178	110622	90728	101170	19.02	17.22	18.19	4355.70	3571.96	3395.40

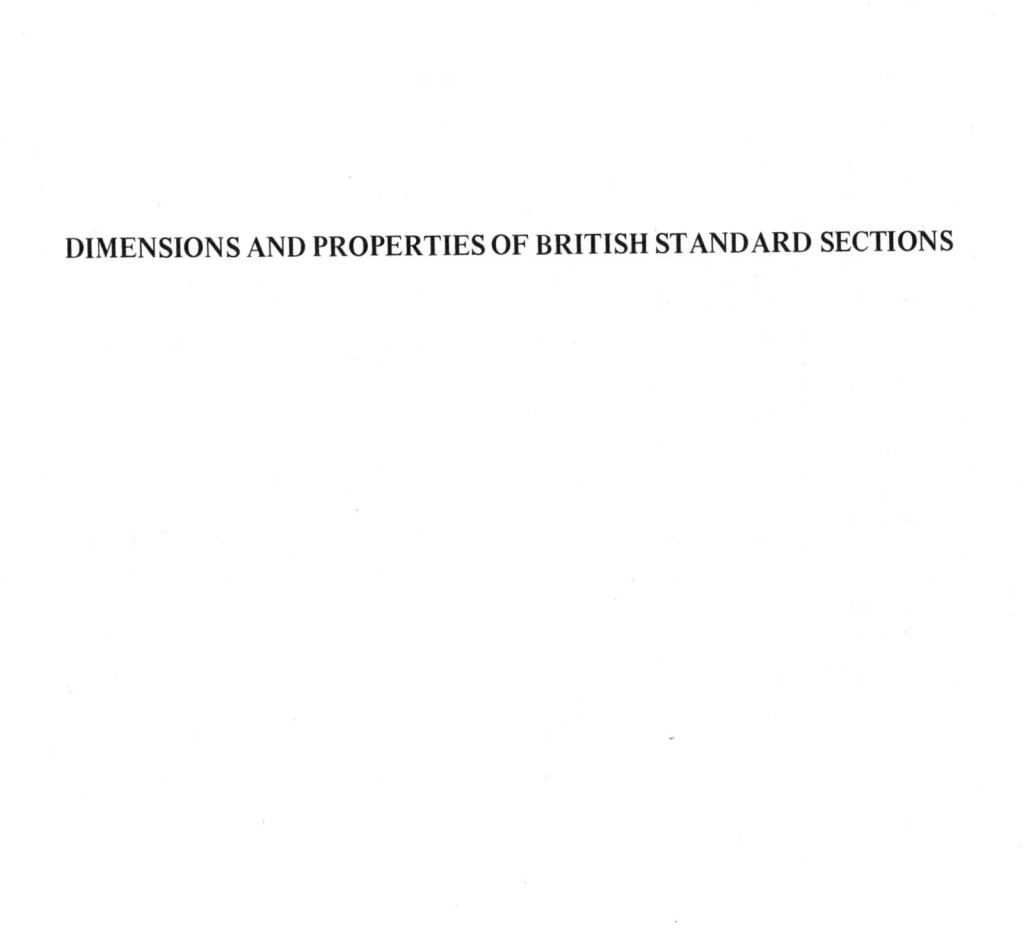

DIMENSIONS AND PROPERTIES OF BRITISH STANDARD SECTIONS

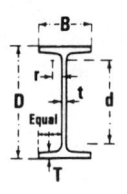

UNIVERSAL BEAMS

DIMENSIONS AND PROPERTIES

Serial Size	Mass per metre	Depth of Section D	Width of Section B	Thickness Web t	Thickness Flange T	Root Radius r	Depth between Fillets d	Area of Section
mm	kg	mm	mm	mm	mm	mm	mm	cm²
914 × 419	388	920.5	420.5	21.5	36.6	24.1	791.5	493.9
	343	911.4	418.5	19.4	32.0	24.1	791.5	436.9
914 × 305	289	926.6	307.8	19.6	32.0	19.1	819.2	368.5
	253	918.5	305.5	17.3	27.9	19.1	819.2	322.5
	224	910.3	304.1	15.9	23.9	19.1	819.2	284.9
	201	903.0	303.4	15.2	20.2	19.1	819.2	256.1
838 × 292	226	850.9	293.8	16.1	26.8	17.8	756.4	288.4
	194	840.7	292.4	14.7	21.7	17.8	756.4	246.9
	176	834.9	291.6	14.0	18.8	17.8	756.4	223.8
762 × 267	197	769.6	268.0	15.6	25.4	16.5	681.2	250.5
	173	762.0	266.7	14.3	21.6	16.5	681.2	220.2
	147	753.9	265.3	12.9	17.5	16.5	681.2	187.8
686 × 254	170	692.9	255.8	14.5	23.7	15.2	610.6	216.3
	152	687.6	254.5	13.2	21.0	15.2	610.6	193.6
	140	683.5	253.7	12.4	19.0	15.2	610.6	178.4
	125	677.9	253.0	11.7	16.2	15.2	610.6	159.4
610 × 305	238	633.0	311.5	18.6	31.4	16.5	531.6	303.5
	179	617.5	307.0	14.1	23.6	16.5	531.6	227.7
	149	609.6	304.8	11.9	19.7	16.5	531.6	189.9
610 × 229	140	617.0	230.1	13.1	22.1	12.7	543.1	178.2
	125	611.9	229.0	11.9	19.6	12.7	543.1	159.4
	113	607.3	228.2	11.2	17.3	12.7	543.1	144.3
	101	602.2	227.6	10.6	14.8	12.7	543.1	129.0
610 × 178	91	602.5	178.4	10.6	15.0	12.7	547.1	115.9
	82	598.2	177.8	10.1	12.8	12.7	547.1	104.4
533 × 330	212	545.1	333.6	16.7	27.8	16.5	450.1	269.6
	189	539.5	331.7	14.9	25.0	16.5	450.1	241.2
	167	533.4	330.2	13.4	22.0	16.5	450.1	212.7
533 × 210	122	544.6	211.9	12.8	21.3	12.7	472.7	155.6
	109	539.5	210.7	11.6	18.8	12.7	472.7	138.4
	101	536.7	210.1	10.9	17.4	12.7	472.7	129.1
	92	533.1	209.3	10.2	15.6	12.7	472.7	117.6
	82	528.3	208.7	9.6	13.2	12.7	472.7	104.3
533 × 165	73	528.8	165.6	9.3	13.5	12.7	476.5	93.0
	66	524.8	165.1	8.8	11.5	12.7	476.5	83.6
457 × 191	98	467.4	192.8	11.4	19.6	10.2	404.4	125.2
	89	463.6	192.0	10.6	17.7	10.2	404.4	113.8
	82	460.2	191.3	9.9	16.0	10.2	404.4	104.4
	74	457.2	190.5	9.1	14.5	10.2	404.4	94.9
	67	453.6	189.9	8.5	12.7	10.2	404.4	85.4

Note: These tables are based on Universal Beams with tapered flanges.
Universal Beams with parallel flanges have properties at least equal to the values given.
Both Taper and Parallel Flange Beams comply with the requirements of the British Standard 4: Part 1:1971 and are interchangeable.

PROPERTIES OF STEEL SECTIONS

UNIVERSAL BEAMS

DIMENSIONS AND PROPERTIES

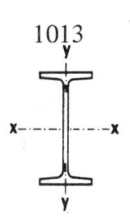

1013

Serial Size	Moment of Inertia Axis x–x Gross	Moment of Inertia Axis x–x Net	Moment of Inertia Axis y–y	Radius of Gyration Axis x–x	Radius of Gyration Axis y–y	Elastic Modulus Axis x–x	Elastic Modulus Axis y–y	Ratio $\dfrac{D}{T}$
mm	cm⁴	cm⁴	cm⁴	cm	cm	cm³	cm³	
914 × 419	717325	639177	42481	38.1	9.27	15586	2021	25.2
	623866	555835	36251	37.8	9.11	13691	1733	28.5
914 × 305	503781	469903	14793	37.0	6.34	10874	961.3	29.0
	435796	406504	12512	36.8	6.23	9490	819.2	32.9
	375111	350209	10425	36.3	6.05	8241	685.6	38.1
	324715	303783	8632	35.6	5.81	7192	569.1	44.7
838 × 292	339130	315153	10661	34.3	6.08	7971	725.9	31.8
	278833	259625	8384	33.6	5.83	6633	573.6	38.7
	245412	228867	7111	33.1	5.64	5879	487.6	44.4
762 × 267	239464	221138	7699	30.9	5.54	6223	574.6	30.3
	204747	189341	6376	30.5	5.38	5374	478.1	35.3
	168535	156213	5002	30.0	5.16	4471	377.1	43.1
686 × 254	169843	156106	6225	28.0	5.36	4902	486.8	29.2
	150015	137965	5391	27.8	5.28	4364	423.7	32.7
	135972	125156	4789	27.6	5.18	3979	377.5	36.0
	117700	108580	3992	27.2	5.00	3472	315.5	41.8
610 × 305	207252	192203	14973	26.1	7.02	6549	961.3	20.2
	151312	140269	10571	25.8	6.81	4901	688.6	26.2
	124341	115233	8471	25.6	6.68	4079	555.9	30.9
610 × 229	111673	101699	4253	25.0	4.88	3620	369.6	27.9
	98408	89675	3676	24.8	4.80	3217	321.1	31.2
	87260	79645	3184	24.6	4.70	2874	279.1	35.1
	75549	69132	2658	24.2	4.54	2509	233.6	40.7
610 × 178	63970	57238	1427	23.5	3.51	2124	160.0	40.2
	55779	50076	1203	23.1	3.39	1865	135.3	46.7
533 × 330	141682	121777	16064	22.9	7.72	5199	963.2	19.6
	125618	107882	14093	22.8	7.64	4657	849.6	21.6
	109109	93647	12057	22.6	7.53	4091	730.3	24.2
533 × 210	76078	68719	3208	22.1	4.54	2794	302.8	25.6
	66610	60218	2755	21.9	4.46	2469	261.5	28.7
	61530	55671	2512	21.8	4.41	2293	239.2	30.8
	55225	50040	2212	21.7	4.34	2072	211.3	34.2
	47363	43062	1826	21.3	4.18	1793	175.0	40.0
533 × 165	40414	35752	1027	20.8	3.32	1528	124.1	39.2
	35083	31144	863	20.5	3.21	1337	104.5	45.6
457 × 191	45653	40469	2216	19.1	4.21	1954	229.9	23.8
	40956	36313	1960	19.0	4.15	1767	204.2	26.2
	37039	32869	1746	18.8	4.09	1610	182.6	28.8
	33324	29570	1547	18.7	4.04	1458	162.4	31.5
	29337	26072	1328	18.5	3.95	1293	139.9	35.7

Note: One hole is deducted from each flange under 300 mm wide (serial size) and two holes from each flange 300 mm and over (serial size), in calculating the Net Moment of Inertia about x–x.

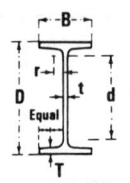

UNIVERSAL BEAMS

DIMENSIONS AND PROPERTIES

Serial Size	Mass per metre	Depth of Section D	Width of Section B	Thickness Web t	Thickness Flange T	Root Radius r	Depth between Fillets d	Area of Section
mm	kg	mm	mm	mm	mm	mm	mm	cm²
457 × 152	82	465.1	153.5	10.7	18.9	10.2	404.4	104.4
	74	461.3	152.7	9.9	17.0	10.2	404.4	94.9
	67	457.2	151.9	9.1	15.0	10.2	404.4	85.3
	60	454.7	152.9	8.0	13.3	10.2	407.7	75.9
	52	449.8	152.4	7.6	10.9	10.2	407.7	66.5
406 × 178	74	412.8	179.7	9.7	16.0	10.2	357.4	94.9
	67	409.4	178.8	8.8	14.3	10.2	357.4	85.4
	60	406.4	177.8	7.8	12.8	10.2	357.4	76.1
	54	402.6	177.6	7.6	10.9	10.2	357.4	68.3
406 × 152	74	416.3	153.7	10.1	18.1	10.2	357.4	94.8
	67	412.2	152.9	9.3	16.0	10.2	357.4	85.3
	60	407.9	152.2	8.6	13.9	10.2	357.4	75.8
406 × 140	46	402.3	142.4	6.9	11.2	10.2	357.4	58.9
	39	397.3	141.8	6.3	3.6	10.2	357.4	49.3
381 × 152	67	388.6	154.3	9.7	16.3	10.2	333.2	85.4
	60	384.8	153.4	8.7	14.4	10.2	333.2	75.9
	52	381.0	152.4	7.8	12.4	10.2	333.2	66.4
356 × 171	67	364.0	173.2	9.1	15.7	10.2	309.1	85.3
	57	358.6	172.1	8.0	13.0	10.2	309.1	72.1
	51	355.6	171.5	7.3	11.5	10.2	309.1	64.5
	45	352.0	171.0	6.9	9.7	10.2	309.1	56.9
356 × 127	39	352.8	126.0	6.5	10.7	10.2	309.1	49.3
	33	348.5	125.4	5.9	8.5	10.2	309.1	41.7
305 × 165	54	310.9	166.8	7.7	13.7	8.9	262.6	68.3
	46	307.1	165.7	6.7	11.8	8.9	262.6	58.8
	40	303.8	165.1	6.1	10.2	8.9	262.6	51.4
305 × 127	48	310.4	125.2	8.9	14.0	8.9	262.6	60.8
	42	306.6	124.3	8.0	12.1	8.9	262.6	53.1
	37	303.8	123.5	7.2	10.7	8.9	262.6	47.4
305 × 102	33	312.7	102.4	6.6	10.8	7.6	275.3	41.8
	28	308.9	101.9	6.1	8.9	7.6	275.3	36.3
	25	304.8	101.6	5.8	6.8	7.6	275.3	31.4
254 × 146	43	259.6	147.3	7.3	12.7	7.6	216.2	55.0
	37	256.0	146.4	6.4	10.9	7.6	216.2	47.4
	31	251.5	146.1	6.1	8.6	7.6	216.2	39.9
254 × 102	28	260.4	102.1	6.4	10.0	7.6	224.5	36.2
	25	257.0	101.9	6.1	8.4	7.6	224.5	32.1
	22	254.0	101.6	5.8	6.8	7.6	224.5	28.4
203 × 133	30	206.8	133.8	6.3	9.6	7.6	169.9	38.0
	25	203.2	133.4	5.8	7.8	7.6	169.9	32.3

Note: These tables are based on Universal Beams with tapered flanges.
Universal Beams with parallel flanges have properties at least equal to the values given.
Both Taper and Parallel Flange Beams comply with the requirements of the British Standard 4: Part 1:1971 and are interchangeable.

PROPERTIES OF STEEL SECTIONS

UNIVERSAL BEAMS

DIMENSIONS AND PROPERTIES

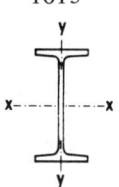

Serial Size	Moment of Inertia Axis x–x Gross	Moment of Inertia Axis x–x Net	Moment of Inertia Axis y–y	Radius of Gyration Axis x–x	Radius of Gyration Axis y–y	Elastic Modulus Axis x–x	Elastic Modulus Axis y–y	Ratio $\dfrac{D}{T}$
mm	cm⁴	cm⁴	cm⁴	cm	cm	cm³	cm³	
457 × 152	36160	32058	1093	18.6	3.24	1555	142.5	24.6
	32380	28731	963	18.5	3.18	1404	126.1	27.1
	28522	25342	829	18.3	3.12	1248	109.1	30.5
	25464	22613	794	18.3	3.23	1120	104.0	34.2
	21345	19034	645	17.9	3.11	949.0	84.61	41.3
406 × 178	27279	23981	1448	17.0	3.91	1322	161.2	25.8
	24279	21357	1269	16.9	3.85	1186	141.9	28.6
	21520	18928	1108	16.8	3.82	1059	124.7	31.8
	18576	16389	922	16.5	3.67	922.8	103.8	36.9
406 × 152	26938	23811	1047	16.9	3.32	1294	136.2	23.0
	23798	21069	908	16.7	3.26	1155	118.8	25.8
	20619	18283	768	16.5	3.18	1011	100.9	29.3
406 × 140	15603	13699	500	16.3	2.92	775.6	70.26	35.9
	12408	10963	373	15.9	2.75	624.7	52.61	46.2
381 × 152	21276	18817	947	15.8	3.33	1095	122.7	23.8
	18632	16489	814	15.7	3.27	968.4	106.2	26.7
	16046	14226	685	15.5	3.21	842.3	89.96	30.7
356 × 171	19483	17002	1278	15.1	3.87	1071	147.6	23.2
	16038	14018	1026	14.9	3.77	894.3	119.2	27.6
	14118	12349	885	14.8	3.71	794.0	103.3	30.9
	12052	10578	730	14.6	3.58	684.7	85.39	36.3
356 × 127	10054	8688	333	14.3	2.60	570.0	52.87	33.0
	8167	7099	257	14.0	2.48	468.7	40.99	41.0
305 × 165	11686	10119	988	13.1	3.80	751.8	118.5	22.7
	9924	8596	825	13.0	3.74	646.4	99.54	26.0
	8500	7368	691	12.9	3.67	559.6	83.71	29.8
305 × 127	9485	8137	438	12.5	2.68	611.1	69.94	22.2
	8124	6978	367	12.4	2.63	530.0	58.99	25.3
	7143	6142	316	12.3	2.58	470.3	51.11	28.4
305 × 102	6482	5792	189	12.5	2.13	414.6	37.00	29.0
	5415	4855	153	12.2	2.05	350.7	30.01	34.7
	4381	3959	116	11.8	1.92	287.5	22.85	44.8
254 × 146	6546	5683	633	10.9	3.39	504.3	85.97	20.4
	5544	4814	528	10.8	3.34	433.1	72.11	23.5
	4427	3859	406	10.5	3.19	352.1	55.53	29.2
254 × 102	4004	3565	174	10.5	2.19	307.6	34.13	26.0
	3404	3041	144	10.3	2.11	264.9	28.23	30.6
	2863	2572	116	10.0	2.02	225.4	22.84	37.4
203 × 133	2880	2469	354	8.71	3.05	278.5	52.85	21.5
	2348	2020	280	8.53	2.94	231.1	41.92	26.1

Note: One hole is deducted from each flange under 300 mm wide (serial size) and two holes from each flange 300 mm and over (serial size), in calculating the Net Moment of Inertia about x–x.

PROPERTIES OF STEEL SECTIONS

UNIVERSAL BEAMS

PLASTIC MODULI—MAJOR AXIS

Serial Size	Mass per metre	Plastic Modulus Axis x—x	Reduced Values of Plastic Modulus under Axial Load		
			Lower Values of n	Change formula at n =	Higher Values of n
mm	kg	cm³	cm³		cm³
914 × 419	388	17628	$17628 - 28366n^2$	0.368	$1453(1-n)(14.65+n)$
	343	15445	$15445 - 24603n^2$	0.375	$1176(1-n)(15.93+n)$
914 × 305	289	12566	$12566 - 17318n^2$	0.459	$1088(1-n)(14.70+n)$
	253	10930	$10930 - 15026n^2$	0.463	$845.2(1-n)(16.52+n)$
	224	9505	$9505 - 12764n^2$	0.481	$661.2(1-n)(18.61+n)$
	201	8345	$8345 - 10787n^2$	0.512	$526.6(1-n)(20.96+n)$
838 × 292	226	9144	$9144 - 12919n^2$	0.443	$731.9(1-n)(15.77+n)$
	194	7635	$7635 - 10363n^2$	0.473	$548.4(1-n)(17.92+n)$
	176	6795	$6795 - 8947n^2$	0.497	$453.5(1-n)(19.60+n)$
762 × 267	197	7156	$7156 - 10057n^2$	0.446	$600.8(1-n)(15.05+n)$
	173	6186	$6186 - 8478n^2$	0.465	$473.6(1-n)(16.71+n)$
	147	5163	$5163 - 6836n^2$	0.492	$342.9(1-n)(19.65+n)$
686 × 254	170	5616	$5616 - 8069n^2$	0.432	$455.6(1-n)(15.45+n)$
	152	4989	$4989 - 7100n^2$	0.440	$360.1(1-n)(17.48+n)$
	140	4552	$4552 - 6417n^2$	0.446	$334.8(1-n)(17.21+n)$
	125	3987	$3987 - 5430n^2$	0.471	$270.0(1-n)(19.01+n)$
610 × 305	238	7447	$7447 - 12384n^2$	0.349	$745.2(1-n)(11.89+n)$
	179	5512	$5512 - 9190n^2$	0.352	$433.3(1-n)(15.22+n)$
	149	4562	$4562 - 7573n^2$	0.357	$299.7(1-n)(18.31+n)$
610 × 229	140	4141	$4141 - 6063n^2$	0.420	$350.6(1-n)(14.68+n)$
	125	3672	$3672 - 5338n^2$	0.426	$290.6(1-n)(15.78+n)$
	113	3283	$3283 - 4648n^2$	0.444	$225.3(1-n)(18.45+n)$
	101	2877	$2877 - 3927n^2$	0.471	$176.8(1-n)(20.97+n)$
610 × 178	91	2484	$2484 - 3166n^2$	0.520	$205.1(1-n)(16.02+n)$
	82	2194	$2194 - 2699n^2$	0.556	$138.3(1-n)(21.58+n)$
533 × 330	212	5849	$5849 - 10877n^2$	0.302	$556.1(1-n)(12.21+n)$
	189	5212	$5212 - 9760n^2$	0.301	$449.2(1-n)(13.48+n)$
	167	4560	$4560 - 8440n^2$	0.308	$347.3(1-n)(15.33+n)$
533 × 210	122	3198	$3198 - 4731n^2$	0.413	$280.1(1-n)(14.13+n)$
	109	2820	$2820 - 4131n^2$	0.421	$222.4(1-n)(15.79+n)$
	101	2616	$2616 - 3825n^2$	0.422	$205.7(1-n)(15.85+n)$
	92	2362	$2362 - 3390n^2$	0.436	$160.0(1-n)(18.59+n)$
	82	2051	$2051 - 2832n^2$	0.463	$122.8(1-n)(21.43+n)$
533 × 165	73	1776	$1776 - 2327n^2$	0.498	$143.7(1-n)(16.12+n)$
	66	1562	$1562 - 1984n^2$	0.524	$118.6(1-n)(17.49+n)$
457 × 191	98	2229	$2229 - 3436n^2$	0.389	$208.9(1-n)(13.00+n)$
	89	2012	$2012 - 3055n^2$	0.399	$166.7(1-n)(14.83+n)$
	82	1830	$1830 - 2754n^2$	0.405	$144.7(1-n)(15.61+n)$
	74	1654	$1654 - 2474n^2$	0.410	$119.1(1-n)(17.21+n)$
	67	1469	$1469 - 2143n^2$	0.426	$95.25(1-n)(19.32+n)$

Let p = mean axial stress
Ys = yield stress
then n = p/Ys

UNIVERSAL BEAMS

PLASTIC MODULI—MINOR AXIS

Serial Size	Mass per metre	Plastic Modulus Axis x—x	Reduced Values of Plastic Modulus under Axial Load		
			Lower Values of **n**	Change formula at **n** =	Higher Values of **n**
mm	kg	cm³	cm³		cm³
914 × 419	388	3206	$3206 - 662.5n^2$	0.401	$7744(1-n)(0.267+n)$
	343	2756	$2756 - 523.7n^2$	0.405	$6823(1-n)(0.253+n)$
914 × 305	289	1552	$1552 - 366.3n^2$	0.493	$4978(1-n)(0.087+n)$
	253	1322	$1322 - 283.0n^2$	0.493	$4320(1-n)(0.079+n)$
	224	1112	$1112 - 222.9n^2$	0.508	$3895(1-n)(0.042+n)$
	201	932.2	$932.2 - 181.6n^2$	0.536	$3667(1-n)(n-0.019)$
838 × 292	226	1166	$1166 - 244.4n^2$	0.475	$3588(1-n)(0.115+n)$
	194	929.4	$929.4 - 181.2n^2$	0.501	$3182(1-n)(0.056+n)$
	176	796.6	$796.6 - 150.0n^2$	0.522	$2982(1-n)(0.008+n)$
762 × 267	197	924.8	$924.8 - 203.9n^2$	0.479	$2863(1-n)(0.110+n)$
	173	773.4	$773.4 - 159.1n^2$	0.495	$2566(1-n)(0.072+n)$
	147	615.2	$615.2 - 117.0n^2$	0.518	$2260(1-n)(0.018+n)$
686 × 254	170	780.8	$780.8 - 168.9n^2$	0.464	$2302(1-n)(0.139+n)$
	152	680.5	$680.5 - 136.3n^2$	0.469	$2065(1-n)(0.124+n)$
	140	608.2	$608.2 - 116.4n^2$	0.475	$1901(1-n)(0.108+n)$
	125	512.5	$512.5 - 93.71n^2$	0.498	$1753(1-n)(0.058+n)$
610 × 305	238	1522	$1522 - 363.9n^2$	0.388	$3454(1-n)(0.306+n)$
	179	1092	$1092 - 209.8n^2$	0.382	$2522(1-n)(0.299+n)$
	149	884.1	$884.1 - 147.8n^2$	0.382	$2076(1-n)(0.290+n)$
610 × 229	140	591.0	$591.0 - 128.7n^2$	0.454	$1677(1-n)(0.163+n)$
	125	514.2	$514.2 - 103.8n^2$	0.457	$1495(1-n)(0.150+n)$
	113	448.7	$448.7 - 85.72n^2$	0.471	$1385(1-n)(0.116+n)$
	101	378.6	$378.6 - 69.12n^2$	0.495	$1281(1-n)(0.064+n)$
610 × 178	91	256.2	$256.2 - 55.70n^2$	0.551	$1051(1-n)(n-0.044)$
	82	217.5	$217.5 - 45.56n^2$	0.579	$1014(1-n)(n-0.105)$
533 × 330	212	1518	$1518 - 333.2n^2$	0.338	$3034(1-n)(0.399+n)$
	189	1340	$1340 - 269.6n^2$	0.333	$2678(1-n)(0.401+n)$
	167	1156	$1156 - 212.0n^2$	0.336	$2352(1-n)(0.389+n)$
533 × 210	122	484.0	$484.0 - 111.2n^2$	0.448	$1334(1-n)(0.179+n)$
	109	418.5	$418.5 - 88.81n^2$	0.452	$1187(1-n)(0.164+n)$
	101	383.4	$383.4 - 77.69n^2$	0.453	$1099(1-n)(0.158+n)$
	92	339.6	$339.6 - 64.87n^2$	0.462	$1015(1-n)(0.134+n)$
	82	283.5	$283.5 - 51.47n^2$	0.486	$930.8(1-n)(0.081+n)$
533 × 165	73	197.3	$197.3 - 40.93n^2$	0.529	$744.4(1-n)(0.001+n)$
	66	166.9	$166.9 - 33.27n^2$	0.553	$701.1(1-n)(n-0.053)$
457 × 191	98	365.8	$365.8 - 83.80n^2$	0.426	$938.9(1-n)(0.225+n)$
	89	325.4	$325.4 - 69.86n^2$	0.432	$861.2(1-n)(0.207+n)$
	82	291.5	$291.5 - 59.25n^2$	0.436	$789.9(1-n)(0.193+n)$
	74	259.6	$259.6 - 49.24n^2$	0.439	$716.5(1-n)(0.183+n)$
	67	224.7	$224.7 - 40.15n^2$	0.452	$654.2(1-n)(0.152+n)$

UNIVERSAL BEAMS

PLASTIC MODULI—MAJOR AXIS

Serial Size	Mass per metre	Plastic Modulus Axis x–x	Reduced Values of Plastic Modulus under Axial Load		
			Lower Values of **n**	Change formula at **n** =	Higher Values of **n**
mm	kg	cm³	cm³		cm³
457 × 152	82	1797	$1797 - 2545n^2$	0.438	$174.5(1-n)(12.91+n)$
	74	1620	$1620 - 2274n^2$	0.446	$146.3(1-n)(13.96+n)$
	67	1439	$1439 - 2000n^2$	0.453	$128.0(1-n)(14.24+n)$
	60	1284	$1284 - 1802n^2$	0.449	$99.17(1-n)(16.41+n)$
	52	1094	$1094 - 1455n^2$	0.490	$67.64(1-n)(21.11+n)$
406 × 178	74	1502	$1502 - 2319n^2$	0.390	$122.5(1-n)(14.98+n)$
	67	1343	$1343 - 2072n^2$	0.392	$102.0(1-n)(16.13+n)$
	60	1195	$1195 - 1856n^2$	0.389	$82.74(1-n)(17.69+n)$
	54	1046	$1046 - 1536n^2$	0.423	$66.91(1-n)(19.55+n)$
406 × 152	74	1486	$1486 - 2226n^2$	0.403	$151.0(1-n)(12.07+n)$
	67	1323	$1323 - 1958n^2$	0.412	$125.0(1-n)(13.07+n)$
	60	1158	$1158 - 1672n^2$	0.429	$99.66(1-n)(14.52+n)$
406 × 140	46	886.3	$886.3 - 1255n^2$	0.444	$63.01(1-n)(17.79+n)$
	39	718.7	$718.7 - 964.8n^2$	0.482	$47.94(1-n)(19.43+n)$
381 × 152	67	1254	$1254 - 1880n^2$	0.404	$117.3(1-n)(13.15+n)$
	60	1106	$1106 - 1656n^2$	0.406	$98.55(1-n)(13.82+n)$
	52	959.0	$959.0 - 1413n^2$	0.418	$71.08(1-n)(16.80+n)$
356 × 171	67	1210	$1210 - 2000n^2$	0.355	$104.4(1-n)(13.88+n)$
	57	1007	$1007 - 1624n^2$	0.370	$73.11(1-n)(16.68+n)$
	51	892.9	$892.9 - 1424n^2$	0.376	$61.31(1-n)(17.70+n)$
	45	771.7	$771.7 - 1172n^2$	0.404	$45.34(1-n)(21.08+n)$
356 × 127	39	651.8	$651.8 - 935.0n^2$	0.434	$51.70(1-n)(15.82+n)$
	33	537.9	$537.9 - 738.3n^2$	0.465	$37.78(1-n)(18.25+n)$
305 × 165	54	843.4	$843.4 - 1515n^2$	0.318	$72.92(1-n)(13.56+n)$
	46	721.3	$721.3 - 1292n^2$	0.321	$54.56(1-n)(15.56+n)$
	40	623.1	$623.1 - 1084n^2$	0.335	$40.91(1-n)(18.10+n)$
305 × 127	48	704.9	$704.9 - 1037n^2$	0.413	$73.97(1-n)(11.75+n)$
	42	609.2	$609.2 - 881.4n^2$	0.426	$54.65(1-n)(13.90+n)$
	37	539.3	$539.3 - 780.0n^2$	0.427	$47.17(1-n)(14.26+n)$
305 × 102	33	479.6	$479.6 - 660.3n^2$	0.459	$43.31(1-n)(14.07+n)$
	28	406.9	$406.9 - 539.5n^2$	0.489	$32.50(1-n)(16.24+n)$
	25	337.5	$337.5 - 424.1n^2$	0.532	$28.77(1-n)(15.62+n)$
254 × 146	43	567.4	$567.4 - 1038n^2$	0.309	$53.24(1-n)(12.42+n)$
	37	484.5	$484.5 - 877.7n^2$	0.315	$38.90(1-n)(14.60+n)$
	31	394.8	$394.8 - 653.9n^2$	0.357	$27.99(1-n)(16.94+n)$
254 × 102	28	353.1	$353.1 - 511.2n^2$	0.427	$30.32(1-n)(14.53+n)$
	25	305.3	$305.3 - 423.6n^2$	0.455	$25.38(1-n)(15.28+n)$
	22	261.5	$261.5 - 347.7n^2$	0.485	$22.53(1-n)(15.01+n)$
203 × 133	30	312.6	$312.6 - 571.6n^2$	0.310	$27.53(1-n)(13.26+n)$
	25	259.1	$259.1 - 448.7n^2$	0.334	$20.90(1-n)(14.69+n)$

Let p = mean axial stress
Ys = yield stress
then n = p/Ys

PROPERTIES OF STEEL SECTIONS

UNIVERSAL BEAMS

PLASTIC MODULI—MINOR AXIS

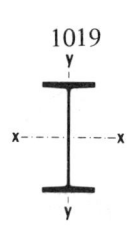

Serial Size	Mass per metre	Plastic Modulus Axis x—x	Reduced Values of Plastic Modulus under Axial Load		
			Lower Values of n	Change formula at $n =$	Higher Values of n
mm	kg	cm³	cm³		cm³
457 × 152	82	229.2	$229.2 - 58.56n^2$	0.477	$682.0(1-n)(0.128+n)$
	74	203.0	$203.0 - 48.80n^2$	0.481	$621.8(1-n)(0.113+n)$
	67	176.1	$176.1 - 39.81n^2$	0.488	$558.9(1-n)(0.094+n)$
	60	162.9	$162.9 - 31.70n^2$	0.479	$515.3(1-n)(0.101+n)$
	52	133.2	$133.2 - 24.58n^2$	0.514	$484.6(1-n)(0.024+n)$
406 × 178	74	256.6	$256.6 - 54.50n^2$	0.422	$659.7(1-n)(0.226+n)$
	67	226.2	$226.2 - 44.53n^2$	0.422	$588.1(1-n)(0.220+n)$
	60	198.8	$198.8 - 35.62n^2$	0.417	$515.0(1-n)(0.224+n)$
	54	167.2	$167.2 - 28.99n^2$	0.448	$482.2(1-n)(0.158+n)$
406 × 152	74	217.5	$217.5 - 54.02n^2$	0.443	$581.7(1-n)(0.196+n)$
	67	190.1	$190.1 - 44.18n^2$	0.449	$525.1(1-n)(0.177+n)$
	60	162.1	$162.1 - 35.25n^2$	0.463	$474.4(1-n)(0.144+n)$
406 × 140	46	113.1	$113.1 - 21.53n^2$	0.472	$349.3(1-n)(0.115+n)$
	39	85.84	$85.84 - 15.30n^2$	0.508	$306.3(1-n)(0.036+n)$
381 × 152	67	196.1	$196.1 - 46.94n^2$	0.441	$524.8(1-n)(0.196+n)$
	60	169.8	$169.8 - 37.45n^2$	0.441	$460.9(1-n)(0.190+n)$
	52	144.1	$144.1 - 28.93n^2$	0.448	$406.2(1-n)(0.169+n)$
356 × 171	67	233.7	$233.7 - 50.01n^2$	0.388	$540.7(1-n)(0.295+n)$
	57	189.5	$189.5 - 36.22n^2$	0.398	$459.2(1-n)(0.267+n)$
	51	164.8	$164.8 - 29.24n^2$	0.403	$408.9(1-n)(0.253+n)$
	45	137.4	$137.4 - 22.97n^2$	0.427	$371.6(1-n)(0.199+n)$
356 × 127	39	85.07	$85.07 - 17.23n^2$	0.465	$254.3(1-n)(0.133+n)$
	33	66.63	$66.63 - 12.50n^2$	0.493	$222.7(1-n)(0.070+n)$
305 × 165	54	186.9	$186.9 - 37.52n^2$	0.351	$391.6(1-n)(0.366+n)$
	46	157.4	$157.4 - 28.18n^2$	0.350	$333.7(1-n)(0.360+n)$
	40	133.0	$133.0 - 21.77n^2$	0.360	$293.7(1-n)(0.333+n)$
305 × 127	48	112.4	$112.4 - 29.73n^2$	0.455	$307.7(1-n)(0.179+n)$
	42	94.91	$94.91 - 23.00n^2$	0.462	$271.3(1-n)(0.155+n)$
	37	82.33	$82.33 - 18.49n^2$	0.462	$238.7(1-n)(0.149+n)$
305 × 102	33	59.10	$59.10 - 13.94n^2$	0.494	$190.6(1-n)(0.084+n)$
	28	48.18	$48.18 - 10.65n^2$	0.519	$173.1(1-n)(0.025+n)$
	25	37.24	$37.24 - 8.070n^2$	0.564	$161.2(1-n)(n-0.071)$
254 × 146	43	135.4	$135.4 - 29.18n^2$	0.344	$276.3(1-n)(0.384+n)$
	37	113.9	$113.9 - 21.94n^2$	0.346	$236.6(1-n)(0.373+n)$
	31	88.78	$88.78 - 15.86n^2$	0.384	$208.1(1-n)(0.290+n)$
254 × 102	28	54.09	$54.09 - 12.56n^2$	0.461	$155.4(1-n)(0.153+n)$
	25	45.07	$45.07 - 10.05n^2$	0.488	$143.4(1-n)(0.093+n)$
	22	36.80	$36.80 - 7.939n^2$	0.519	$132.6(1-n)(0.025+n)$
203 × 133	30	83.70	$83.70 - 17.41n^2$	0.343	$171.1(1-n)(0.383+n)$
	25	67.04	$67.04 - 12.81n^2$	0.365	$147.4(1-n)(0.333+n)$

UNIVERSAL BEAMS IN TORSION

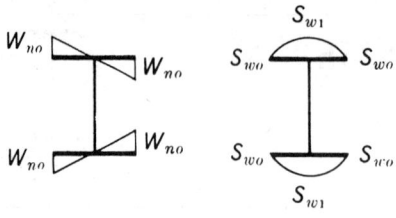

TORSIONAL PROPERTIES

Section mm	Mass per metre kg	K cm^4	a cm	I_W cm^6
914 x 419	388	1824.84	343.349	82977969.09
	343	1263.23	379.242	70077928.62
914 x 305	289	967.26	281.659	29597383.67
	253	655.09	313.328	24806649.98
	224	441.56	346.760	20479437.96
	201	308.19	376.123	16817076.68
838 x 292	226	538.06	295.333	18101756.75
	194	322.87	336.009	14060468.96
	176	234.16	362.062	11840034.67
762 x 267	197	422.90	255.645	10660517.46
	173	280.56	284.155	8737947.04
	147	169.83	321.735	6780657.77
686 x 254	170	320.85	237.321	6970209.91
	152	230.37	259.591	5987762.07
	140	177.95	277.545	5287275.59
	125	123.25	303.183	4369875.19
610 x 305	238	825.37	206.290	13547840.00
	179	360.85	258.763	9319524.14
	149	214.07	298.758	7369985.92
610 x 229	140	227.64	206.988	3761808.39
	125	162.48	226.812	3223986.55
	113	118.35	246.387	2771142.89
	101	81.86	269.494	2293186.99
610 x 178	91	69.03	215.035	1231174.80
	82	49.01	233.464	1030334.51
533 x 330	212	610.36	213.646	10745833.67
	189	445.84	232.867	9325315.88
	167	308.89	257.247	7884382.71
533 x 210	122	187.57	174.202	2195536.46
	109	132.00	191.498	1867098.75
	101	106.99	202.590	1693775.00
	92	80.42	218.496	1480846.03
	82	54.44	240.150	1210953.10
533 x 165	73	45.51	197.092	681922.44
	66	31.95	214.775	568409.55
457 x 191	98	126.47	150.915	1111052.77
	89	95.00	163.071	974377.60
	82	72.88	175.063	861570.06
	74	54.95	189.121	758099.81
	67	39.36	206.226	645744.47

PROPERTIES OF STEEL SECTIONS

UNIVERSAL BEAMS IN TORSION

TORSIONAL PROPERTIES		STATICAL MOMENTS	
W_{no}	S_{w1}	Q_f	Q_w
cm²	cm⁴	cm³	cm³
929.214	35728.52	3270.374	8813.983
919.941	30800.90	2852.625	7722.583
688.322	16949.82	2091.961	6282.918
680.106	14512.23	1821.822	5464.924
673.906	12232.12	1554.062	4752.599
669.537	10254.13	1313.461	4172.407
605.203	11909.85	1556.062	4571.861
598.612	9501.54	1257.843	3817.339
595.026	8154.41	1087.989	3397.636
498.619	8485.35	1210.820	3578.209
493.668	7106.42	1026.729	3093.218
488.340	5676.05	832.385	2581.419
427.942	6477.95	969.569	2808.048
424.035	5666.25	858.518	2494.566
421.561	5074.04	774.483	2275.800
418.612	4284.55	659.749	1993.639
468.504	11445.10	1396.071	3723.290
455.834	8265.16	1041.384	2755.831
449.515	6742.71	864.502	2281.182
342.242	4361.48	723.376	2070.365
339.051	3805.88	638.694	1836.093
336.618	3326.96	563.079	1641.414
334.274	2811.83	479.728	1438.491
261.991	1753.90	379.631	1242.159
260.208	1477.72	323.451	1096.774
431.343	10004.03	1151.428	2924.521
426.694	8853.93	1031.422	2606.217
422.185	7657.19	901.917	2279.892
277.169	3132.58	563.247	1599.172
274.270	2715.40	494.706	1409.879
272.709	2491.74	457.282	1307.824
270.790	2213.31	409.933	1180.857
268.742	1851.86	346.177	1025.569
213.351	1193.61	280.611	887.976
211.868	1001.76	238.297	780.858
215.882	2035.50	401.251	1114.577
214.001	1813.07	361.038	1005.887
212.475	1626.17	326.453	915.202
210.846	1453.82	294.747	827.091
209.328	1262.04	258.024	734.282

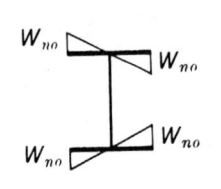

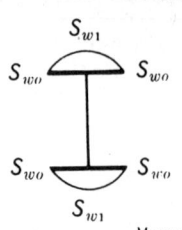

PROPERTIES OF STEEL SECTIONS

UNIVERSAL BEAMS IN TORSION

TORSIONAL PROPERTIES

Section mm	Mass per metre kg	K cm^4	a cm	I_w cm^6
457 x 152	82	92.64	123.404	544173.70
	74	69.24	133.354	474949.85
	67	49.56	145.608	405281.57
	60	33.63	172.705	386898.65
	52	21.26	194.604	310544.72
406 x 178	74	66.15	149.454	569932.83
	67	48.56	162.597	495229.08
	60	35.14	177.948	429248.64
	54	24.22	194.543	353623.87
406 x 152	74	79.39	116.407	414950.92
	67	57.39	126.905	356519.53
	60	39.47	139.946	298166.73
406 x 140	46	20.12	157.002	191314.64
	39	11.26	180.072	140807.32
381 x 152	67	60.33	118.757	328208.42
	60	42.56	130.443	279303.38
	52	28.65	145.131	232769.67
356 x 171	67	58.36	131.235	387680.48
	57	35.07	150.495	306371.06
	51	25.10	164.534	262102.26
	45	16.81	181.606	213903.88
356 x 127	39	15.53	127.568	97458.79
	33	9.13	145.206	74255.82
305 x 165	54	36.49	124.504	218180.70
	46	23.73	140.155	179785.88
	40	15.80	156.317	148949.13
305 x 127	48	32.66	87.377	96178.12
	42	21.87	97.051	79469.68
	37	15.63	106.035	67802.26
305 x 102	33	12.31	95.336	43147.21
	28	7.74	107.313	34386.42
	25	4.72	118.962	25768.12
254 x 146	43	25.44	99.165	96483.83
	37	16.49	111.626	79252.39
	31	9.38	128.536	59774.41
254 x 102	28	9.79	85.033	27298.16
	25	6.56	93.750	22232.11
	22	4.38	102.373	17717.88
203 x 133	30	10.86	90.570	34361.98
	25	6.58	102.547	26675.94

PROPERTIES OF STEEL SECTIONS

UNIVERSAL BEAMS IN TORSION

TORSIONAL PROPERTIES		STATICAL MOMENTS	
w_{no}	S_{w1}	Q_f	Q_w
cm²	cm⁴	cm³	cm³
171.183	1241.14	305.103	898.634
169.578	1099.89	273.604	809.988
167.988	954.66	240.290	719.504
168.659	857.83	217.382	642.162
167.235	694.29	177.870	547.193
178.216	1276.79	272.609	750.760
176.622	1128.94	243.763	671.695
174.943	997.46	218.162	597.311
173.920	839.46	184.592	523.049
153.047	1062.19	261.802	743.158
151.486	928.28	231.584	661.730
149.960	791.44	199.802	579.236
139.221	553.78	151.679	443.153
137.749	421.66	117.290	359.343
143.644	900.79	222.255	627.203
142.039	781.55	195.619	552.969
140.419	665.86	169.245	479.521
150.832	1025.35	227.508	605.181
148.711	833.76	187.960	503.748
147.487	727.38	165.624	446.473
146.356	608.73	139.806	385.855
107.760	362.07	112.175	325.915
106.562	284.20	89.566	268.973
123.888	708.38	164.058	421.716
122.344	598.72	140.660	360.663
121.193	508.23	120.614	311.550
92.787	406.53	122.700	352.459
91.499	343.72	105.609	304.624
90.525	298.99	93.210	269.634
77.259	212.92	79.541	239.787
76.390	172.43	65.368	203.429
75.684	131.35	50.442	168.727
90.913	425.16	110.931	283.694
89.683	358.39	94.893	242.234
88.661	279.57	74.597	197.378
63.905	163.25	61.378	176.559
63.326	134.75	51.153	152.655
62.781	108.95	41.831	130.775
65.952	211.82	61.318	156.296
65.134	169.87	49.701	129.553

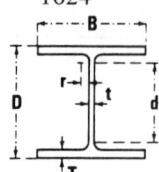

UNIVERSAL COLUMNS
Parallel Flanges
DIMENSIONS AND PROPERTIES

Serial Size	Mass per metre	Depth of Section D	Width of Section B	Thickness Web t	Thickness Flange T	Root Radius r	Depth between Fillets d	Area of Section
mm	kg	mm	mm	mm	mm	mm	mm	cm²
356 × 406	634	474.7	424.1	47.6	77.0	15.2	290.1	808.1
	551	455.7	418.5	42.0	67.5	15.2	290.1	701.8
	467	436.6	412.4	35.9	58.0	15.2	290.1	595.5
	393	419.1	407.0	30.6	49.2	15.2	290.1	500.9
	340	406.4	403.0	26.5	42.9	15.2	290.1	432.7
	287	393.7	399.0	22.6	36.5	15.2	290.1	366.0
	235	381.0	395.0	18.5	30.2	15.2	290.1	299.8
Column Core	477	427.0	424.4	48.0	53.2	15.2	290.1	607.2
356 × 368	202	374.7	374.4	16.8	27.0	15.2	290.1	257.9
	177	368.3	372.1	14.5	23.8	15.2	290.1	225.7
	153	362.0	370.2	12.6	20.7	15.2	290.1	195.2
	129	355.6	368.3	10.7	17.5	15.2	290.1	164.9
305 × 305	283	365.3	321.8	26.9	44.1	15.2	246.6	360.4
	240	352.6	317.9	23.0	37.7	15.2	246.6	305.6
	198	339.9	314.1	19.2	31.4	15.2	246.6	252.3
	158	327.2	310.6	15.7	25.0	15.2	246.6	201.2
	137	320.5	308.7	13.8	21.7	15.2	246.6	174.6
	118	314.5	306.8	11.9	18.7	15.2	246.6	149.8
	97	307.8	304.8	9.9	15.4	15.2	246.6	123.3
254 × 254	167	289.1	264.5	19.2	31.7	12.7	200.2	212.4
	132	276.4	261.0	15.6	25.1	12.7	200.2	167.7
	107	266.7	258.3	13.0	20.5	12.7	200.2	136.6
	89	260.4	255.9	10.5	17.3	12.7	200.2	114.0
	73	254.0	254.0	8.6	14.2	12.7	200.2	92.9
203 × 203	86	222.3	208.8	13.0	20.5	10.2	160.8	110.1
	71	215.9	206.2	10.3	17.3	10.2	160.8	91.1
	60	209.6	205.2	9.3	14.2	10.2	160.8	75.8
	52	206.2	203.9	8.0	12.5	10.2	160.8	66.4
	46	203.2	203.2	7.3	11.0	10.2	160.8	58.8
152 × 152	37	161.8	154.4	8.1	11.5	7.6	123.4	47.4
	30	157.5	152.9	6.6	9.4	7.6	123.4	38.2
	23	152.4	152.4	6.1	6.8	7.6	123.4	29.8

PROPERTIES OF STEEL SECTIONS 1025

UNIVERSAL COLUMNS
Parallel Flanges
DIMENSIONS AND PROPERTIES

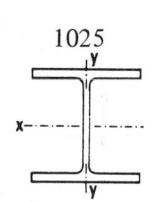

Serial Size	Moment of Inertia Axis x–x Gross	Moment of Inertia Axis x–x Net	Moment of Inertia Axis y–y	Radius of Gyration Axis x–x	Radius of Gyration Axis y–y	Elastic Modulus Axis x–x	Elastic Modulus Axis y–y	Ratio $\dfrac{D}{T}$
mm	cm⁴	cm⁴	cm⁴	cm	cm	cm³	cm³	
356 × 406	275140	243076	98211	18.5	11.0	11592	4632	6.2
	227023	200312	82665	18.0	10.9	9964	3951	6.8
	183118	161331	67905	17.5	10.7	8388	3293	7.5
	146765	129159	55410	17.1	10.5	7004	2723	8.5
	122474	107667	46816	16.8	10.4	6027	2324	9.5
	99994	87843	38714	16.5	10.3	5080	1940	10.8
	79110	69424	31008	16.2	10.2	4153	1570	12.6
Column Core	172391	152936	68057	16.8	10.6	8075	3207	8.0
356 × 368	66307	57806	23632	16.0	9.57	3540	1262	13.9
	57153	49798	20470	15.9	9.52	3104	1100	15.5
	48525	42250	17470	15.8	9.46	2681	943.8	17.5
	40246	35040	14555	15.6	9.39	2264	790.4	20.3
305 × 305	78777	72827	24545	14.8	8.25	4314	1525	8.3
	64177	59295	20239	14.5	8.14	3641	1273	9.4
	50832	46935	16230	14.2	8.02	2991	1034	10.8
	38740	35766	12524	13.9	7.89	2368	806.3	13.1
	32838	30314	10672	13.7	7.82	2049	691.4	14.8
	27601	25472	9006	13.6	7.75	1755	587.0	16.8
	22202	20488	7268	13.4	7.68	1442	476.9	20.0
254 × 254	29914	27171	9796	11.9	6.79	2070	740.6	9.1
	22416	20350	7444	11.6	6.66	1622	570.4	11.0
	17510	15890	5901	11.3	6.57	1313	456.9	13.0
	14307	12976	4849	11.2	6.52	1099	378.9	15.1
	11360	10297	3873	11.1	6.46	894.5	305.0	17.9
203 × 203	9462	8374	3119	9.27	5.32	851.5	298.7	10.8
	7647	6758	2536	9.16	5.28	708.4	246.0	12.5
	6088	5383	2041	8.96	5.19	581.1	199.0	14.8
	5263	4653	1770	8.90	5.16	510.4	173.6	16.5
	4564	4035	1539	8.81	5.11	449.2	151.5	18.5
152 × 152	2218	1932	709	6.84	3.87	274.2	91.78	14.1
	1742	1515	558	6.75	3.82	221.2	73.06	16.8
	1263	1104	403	6.51	3.68	165.7	52.95	22.4

Note: One hole is deducted from each flange under 300 mm wide (serial size) and two holes from each flange 300 mm and over (serial size), in calculating the Net Moment of Inertia about x–x.

PROPERTIES OF STEEL SECTIONS
UNIVERSAL COLUMNS
Parallel Flanges
PLASTIC MODULI—MAJOR AXIS

Serial Size	Mass per metre	Plastic Modulus Axis x—x	Reduced Values of Plastic Modulus under Axial Load		
			Lower Values of **n**	Change formula at n =	Higher Values of **n**
mm	kg	cm³	cm³		cm³
356 × 406	634	14247	$14247 - 34295n^2$	0.189	$3850(1-n)(3.981+n)$
	551	12078	$12078 - 29319n^2$	0.192	$2946(1-n)(4.428+n)$
	467	10009	$10009 - 24694n^2$	0.193	$2151(1-n)(5.042+n)$
	393	8229	$8229 - 20495n^2$	0.196	$1542(1-n)(5.806+n)$
	340	6994	$6994 - 17659n^2$	0.196	$1164(1-n)(6.553+n)$
	287	5818	$5818 - 14816n^2$	0.198	$840.4(1-n)(7.573+n)$
	235	4689	$4689 - 12147n^2$	0.197	$571.6(1-n)(8.991+n)$
Column Core	477	9700	$9700 - 19203n^2$	0.253	$2174(1-n)(4.963+n)$
356 × 368	202	3976	$3976 - 9899n^2$	0.209	$444.6(1-n)(9.869+n)$
	177	3457	$3457 - 8785n^2$	0.206	$342.5(1-n)(11.14+n)$
	153	2964	$2964 - 7561n^2$	0.207	$257.9(1-n)(12.70+n)$
	129	2482	$2482 - 6355n^2$	0.208	$184.4(1-n)(14.91+n)$
305 × 305	283	5101	$5101 - 12071n^2$	0.206	$1012(1-n)(5.506+n)$
	240	4245	$4245 - 10154n^2$	0.208	$736.3(1-n)(6.319+n)$
	198	3436	$3436 - 8290n^2$	0.210	$507.9(1-n)(7.443+n)$
	158	2680	$2680 - 6448n^2$	0.215	$329.0(1-n)(9.006+n)$
	137	2298	$2298 - 5523n^2$	0.218	$248.7(1-n)(10.25+n)$
	118	1953	$1953 - 4714n^2$	0.219	$185.5(1-n)(11.70+n)$
	97	1589	$1589 - 3837n^2$	0.222	$125.5(1-n)(14.12+n)$
254 × 254	167	2417	$2417 - 5873n^2$	0.203	$427.0(1-n)(6.189+n)$
	132	1861	$1861 - 4507n^2$	0.210	$270.9(1-n)(7.554+n)$
	107	1485	$1485 - 3591n^2$	0.214	$180.4(1-n)(9.103+n)$
	89	1228	$1228 - 3092n^2$	0.207	$128.7(1-n)(10.53+n)$
	73	988.5	$988.5 - 2507n^2$	0.208	$86.22(1-n)(12.68+n)$
203 × 203	86	978.8	$978.8 - 2330n^2$	0.214	$145.1(1-n)(7.432+n)$
	71	802.4	$802.4 - 2013n^2$	0.204	$100.7(1-n)(8.763+n)$
	60	652.0	$652.0 - 1546n^2$	0.222	$70.26(1-n)(10.31+n)$
	52	568.1	$568.1 - 1380n^2$	0.218	$54.26(1-n)(11.63+n)$
	46	497.4	$497.4 - 1186n^2$	0.224	$43.12(1-n)(12.86+n)$
152 × 152	37	310.1	$310.1 - 693.2n^2$	0.236	$36.73(1-n)(9.438+n)$
	30	247.1	$247.1 - 554.0n^2$	0.239	$23.72(1-n)(11.70+n)$
	23	184.3	$184.3 - 363.5n^2$	0.283	$14.67(1-n)(14.47+n)$

Let p = mean axial stress
Ys = yield stress
then n = p/Ys

PROPERTIES OF STEEL SECTIONS

UNIVERSAL COLUMNS
Parallel Flanges
PLASTIC MODULI—MINOR AXIS

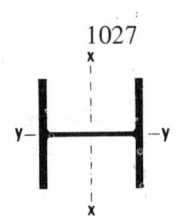

Serial Size mm	Mass per metre kg	Plastic Modulus Axis x—x cm³	Reduced Values of Plastic Modulus under Axial Load		
			Lower Values of n cm³	Change formula at n =	Higher Values of n cm³
356 × 406	634	7114	$7114 - 3439n^2$	0.280	$10520(1-n)(0.624+n)$
	551	6058	$6058 - 2702n^2$	0.273	$9046(1-n)(0.618+n)$
	467	5038	$5038 - 2031n^2$	0.263	$7571(1-n)(0.615+n)$
	393	4157	$4157 - 1496n^2$	0.256	$6303(1-n)(0.610+n)$
	340	3541	$3541 - 1152n^2$	0.249	$5387(1-n)(0.609+n)$
	287	2952	$2952 - 850.5n^2$	0.243	$4519(1-n)(0.605+n)$
	235	2384	$2384 - 589.8n^2$	0.235	$3656(1-n)(0.606+n)$
Column Core	477	4979	$4979 - 2159n^2$	0.338	$8585(1-n)(0.495+n)$
356 × 368	202	1917	$1917 - 443.8n^2$	0.244	$3022(1-n)(0.584+n)$
	177	1668	$1668 - 345.9n^2$	0.237	$2615(1-n)(0.589+n)$
	153	1430	$1430 - 263.2n^2$	0.234	$2249(1-n)(0.587+n)$
	129	1196	$1196 - 191.2n^2$	0.231	$1889(1-n)(0.585+n)$
305 × 305	283	2337	$2337 - 888.9n^2$	0.273	$3626(1-n)(0.588+n)$
	240	1947	$1947 - 662.4n^2$	0.265	$3042(1-n)(0.585+n)$
	198	1576	$1576 - 468.3n^2$	0.259	$2485(1-n)(0.580+n)$
	158	1228	$1228 - 309.4n^2$	0.255	$1965(1-n)(0.570+n)$
	137	1052	$1052 - 237.8n^2$	0.253	$1697(1-n)(0.564+n)$
	118	891.7	$891.7 - 178.4n^2$	0.250	$1446(1-n)(0.562+n)$
	97	723.5	$723.5 - 123.4n^2$	0.247	$1182(1-n)(0.557+n)$
254 × 254	167	1132	$1132 - 390.0n^2$	0.261	$1749(1-n)(0.594+n)$
	132	869.9	$869.9 - 254.4n^2$	0.257	$1370(1-n)(0.581+n)$
	107	695.5	$695.5 - 175.0n^2$	0.254	$1110(1-n)(0.573+n)$
	89	575.4	$575.4 - 124.7n^2$	0.240	$904.9(1-n)(0.586+n)$
	73	462.4	$462.4 - 84.89n^2$	0.235	$730.3(1-n)(0.584+n)$
203 × 203	86	455.9	$455.9 - 136.3n^2$	0.263	$724.0(1-n)(0.574+n)$
	71	374.2	$374.2 - 96.02n^2$	0.244	$582.9(1-n)(0.592+n)$
	60	302.8	$302.8 - 68.60n^2$	0.257	$492.8(1-n)(0.558+n)$
	52	263.7	$263.7 - 53.53n^2$	0.248	$425.5(1-n)(0.566+n)$
	46	230.0	$230.0 - 42.59n^2$	0.252	$377.7(1-n)(0.553+n)$
152 × 152	37	140.1	$140.1 - 34.70n^2$	0.277	$236.1(1-n)(0.528+n)$
	30	111.2	$111.2 - 23.22n^2$	0.272	$189.2(1-n)(0.523+n)$
	23	80.87	$80.87 - 14.55n^2$	0.312	$154.7(1-n)(0.434+n)$

Note: For explanation of tables, see notes commencing page 102.

PROPERTIES OF STEEL SECTIONS

UNIVERSAL BEAMS IN TORSION

TORSIONAL PROPERTIES

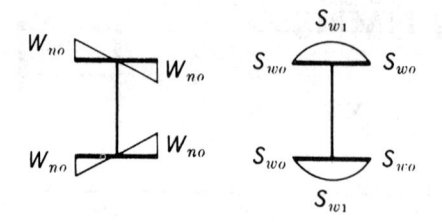

Section mm	Mass per metre kg	K cm^4	a cm	I_W cm^6
356 × 406	634	13746.48	85.578	38831514.04
	551	9239.92	93.471	31137966.72
	467	5816.77	104.153	24338169.39
	393	3550.91	117.630	18951254.00
	340	2337.66	130.972	15466825.17
	287	1443.71	148.906	12347224.89
	235	812.33	174.501	9541082.03
427 × 424	477	5698.47	104.003	23774743.22
356 × 368	202	560.38	181.755	7140335.89
	177	382.88	202.780	6072642.56
	153	250.65	229.394	5087384.79
	129	153.26	265.280	4160228.85
305 × 305	283	2034.82	89.800	6329087.34
	240	1273.25	101.047	5014483.34
	198	734.05	116.769	3860520.55
	158	378.73	139.865	2857654.20
	137	250.25	157.098	2382222.04
	118	160.40	178.419	1969499.60
	97	91.07	210.338	1554128.65
254 × 254	167	624.65	82.050	1622036.20
	132	314.12	98.471	1174836.13
	107	173.44	115.607	894102.38
	89	103.76	133.735	715787.95
	73	57.32	158.716	556900.23
203 × 203	86	138.06	77.199	317358.98
	71	81.47	89.184	249947.71
	60	46.59	104.122	194808.06
	52	31.98	116.022	166068.97
	46	22.24	128.715	142121.64
152 × 152	37	19.46	72.996	40003.92
	30	10.49	87.011	30620.85
	23	4.87	106.713	21373.32

UNIVERSAL COLUMNS IN TORSION

TORSIONAL PROPERTIES		STATICAL MOMENTS	
W_{no}	S_{w_1}	Q_f	Q_w
cm²	cm⁴	cm³	cm³
421.602	34432.51	2890.992	7123.608
406.081	28681.44	2474.000	6039.074
390.346	23335.42	2074.084	5004.652
376.380	18853.23	1721.243	4114.480
366.225	15818.59	1474.592	3497.167
356.312	12982.95	1235.530	2909.051
346.413	10321.64	1004.058	2344.562
396.646	22374.65	1877.978	4849.933
325.397	8223.40	847.060	1988.249
320.456	7102.59	741.609	1728.701
315.877	6037.05	637.961	1481.817
311.328	5009.36	536.115	1240.853
258.386	9166.47	1050.752	2550.492
250.178	7504.24	882.734	2122.609
242.194	5970.12	720.662	1718.190
234.618	4563.23	564.535	1339.755
230.631	3870.38	485.698	1149.195
226.870	3253.33	414.353	976.260
222.851	2613.82	338.608	794.566
170.201	3568.15	504.219	1208.721
163.936	2684.23	390.592	930.306
158.980	2107.08	313.722	742.620
155.463	1725.45	262.399	614.109
152.290	1370.61	212.312	494.269
105.321	1128.52	204.701	489.386
102.339	915.10	170.639	401.216
100.207	728.43	137.564	326.015
98.745	630.26	120.755	284.053
97.639	545.51	105.476	248.707
58.015	258.29	64.222	155.042
56.598	202.72	51.616	123.547
55.461	144.38	37.222	92.142

PROPERTIES OF STEEL SECTIONS

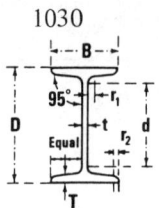

JOISTS

DIMENSIONS AND PROPERTIES

Nominal Size	Mass per metre	Depth of Section D	Width of Section B	Thickness Web t	Thickness Flange T	Radius Root r_1	Radius Toe r_2	Depth between Fillets d	Area of Section
mm	kg	mm	mm	mm	mm	mm	mm	mm	cm²
203 × 102	25.33	203.2	101.6	5.8	10.4	9.4	3.2	161.0	32.3
178 × 102	21.54	177.8	101.6	5.3	9.0	9.4	3.2	138.2	27.4
152 × 89	17.09	152.4	88.9	4.9	8.3	7.9	2.4	117.9	21.8
127 × 76	13.36	127.0	76.2	4.5	7.6	7.9	2.4	94.2	17.0
102 × 64	9.65	101.6	63.5	4.1	6.6	6.9	2.4	73.2	12.3
76 × 51	6.67	76.2	50.8	3.8	5.6	6.9	2.4	50.3	8.49

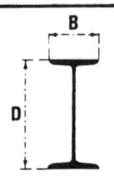

JOISTS

PLASTIC MODULI—MAJOR AXIS

Nominal Size	Mass per metre	Plastic Modulus Axis x—x	Reduced Values of Plastic Modulus under Axial Load Lower Values of n	Change formula at n =	Higher Values of n
mm	kg	cm³	cm³		cm³
203 × 102	25.33	256.3	$256.3 - 448.7n^2$	0.324	$27.23(1-n)(11.04+n)$
178 × 102	21.54	193.0	$193.0 - 355.2n^2$	0.305	$19.78(1-n)(11.33+n)$
152 × 89	17.09	131.0	$131.0 - 241.7n^2$	0.302	$14.08(1-n)(10.78+n)$
127 × 76	13.36	85.23	$85.23 - 161.0n^2$	0.292	$10.02(1-n)(9.786+n)$
102 × 64	9.65	48.98	$48.98 - 92.11n^2$	0.292	$6.148(1-n)(9.155+n)$
76 × 51	6.67	25.07	$25.07 - 47.43n^2$	0.284	$3.752(1-n)(7.624+n)$

Let p = mean axial stress
Ys = yield stress
then n = p/Ys

JOISTS

DIMENSIONS AND PROPERTIES

Nominal Size	Moment of Inertia Axis x–x Gross	Moment of Inertia Axis x–x Net	Axis y–y	Radius of Gyration Axis x–x	Radius of Gyration Axis y–y	Elastic Modulus Axis x–x	Elastic Modulus Axis y–y	Ratio $\dfrac{D}{T}$
mm	cm⁴	cm⁴	cm⁴	cm	cm	cm³	cm³	
203 × 102	2294	2023	162.6	8.43	2.25	225.8	32.02	19.5
178 × 102	1519	1340	139.2	7.44	2.25	170.9	27.41	19.8
152 × 89	881.1	762.1	85.98	6.36	1.99	115.6	19.34	18.4
127 × 76	475.9	399.8	50.18	5.29	1.72	74.94	13.17	16.7
102 × 64	217.6	181.9	25.30	4.21	1.43	42.84	7.97	15.4
76 × 51	82.58	68.85	11.11	3.12	1.14	21.67	4.37	13.6

Note: One hole is deducted from each flange in calculating the Net Moment of Inertia about x–x.

JOISTS

PLASTIC MODULI—MINOR AXIS

Nominal Size	Mass per metre	Plastic Modulus Axis y–y	Reduced Values of Plastic Modulus under Axial Load Lower Values of n	Change formula at n =	Higher Values of n
mm	kg	cm³	cm³		cm³
203 × 102	25.33	51.79	$51.79 - 12.81 n^2$	0.365	$109.6(1-n)(0.355+n)$
178 × 102	21.54	44.48	$44.48 - 10.59 n^2$	0.343	$89.21(1-n)(0.395+n)$
152 × 89	17.09	31.29	$31.29 - 7.772 n^2$	0.343	$62.26(1-n)(0.400+n)$
127 × 76	13.36	21.29	$21.29 - 5.704 n^2$	0.336	$41.04(1-n)(0.422+n)$
102 × 64	9.65	12.91	$12.91 - 3.717 n^2$	0.339	$24.75(1-n)(0.424+n)$
76 × 51	6.67	7.14	$7.142 - 2.366 n^2$	0.341	$13.36(1-n)(0.439+n)$

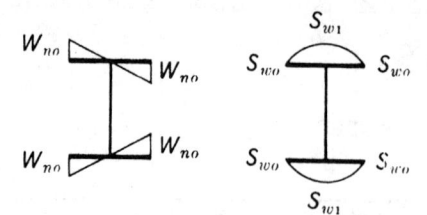

JOISTS IN TORSION

TORSIONAL PROPERTIES

Section mm	Mass per metre kg	K cm^4	a cm	I_w cm^6
203 × 102	25.33	10.649	60.672	15119.765
178 × 102	21.54	7.381	59.009	9913.856
152 × 89	17.09	4.818	49.023	4466.378
127 × 76	13.36	3.275	37.629	1788.625
102 × 64	9.65	1.825	28.479	570.853
76 × 51	6.67	1.011	18.831	138.292

JOISTS IN TORSION

TORSIONAL PROPERTIES		STATICAL MOMENTS	
W_{no}	S_{w1}	Q_f	Q_w
cm²	cm⁴	cm³	cm³
48.981	128.929	48.988	128.129
42.864	98.450	37.716	96.510
32.036	58.776	25.586	65.479
22.747	32.909	16.750	42.616
15.081	15.810	9.606	24.488
8.961	**6.417**	4.884	12.533

CHANNELS

DIMENSIONS AND PROPERTIES

Nominal Size mm	Mass per metre in kg	Depth of Section D mm	Width of Section B mm	Thickness Web t mm	Thickness Flange T mm	Radius Root r_1 mm	Radius Toe r_2 mm	Depth between Fillets d mm	Ratio $\dfrac{D}{T}$	Area of Section cm²
432 × 102	65.54	431.8	101.6	12.2	16.8	15.2	4.8	362.5	25.7	83.49
381 × 102	55.10	381.0	101.6	10.4	16.3	15.2	4.8	312.4	23.4	70.19
305 × 102	46.18	304.8	101.6	10.2	14.8	15.2	4.8	239.3	20.6	58.83
305 × 89	41.69	304.8	88.9	10.2	13.7	13.7	3.2	245.4	22.2	53.11
254 × 89	35.74	254.0	88.9	9.1	13.6	13.7	3.2	194.8	18.7	45.52
254 × 76	28.29	254.0	76.2	8.1	10.9	12.2	3.2	203.7	23.3	36.03
229 × 89	32.76	228.6	88.9	8.6	13.3	13.7	3.2	169.9	17.2	41.73
229 × 76	26.06	228.6	76.2	7.6	11.2	12.2	3.2	178.1	20.4	33.20
203 × 89	29.78	203.2	88.9	8.1	12.9	13.7	3.2	145.3	15.8	37.94
203 × 76	23.82	203.2	76.2	7.1	11.2	12.2	3.2	152.4	18.1	30.34
178 × 89	26.81	177.8	88.9	7.6	12.3	13.7	3.2	120.9	14.5	34.15
178 × 76	20.84	177.8	76.2	6.6	10.3	12.2	3.2	128.8	17.3	26.54
152 × 89	23.84	152.4	88.9	7.1	11.6	13.7	3.2	97.0	13.1	30.36
152 × 76	17.88	152.4	76.2	6.4	9.0	12.2	2.4	105.9	16.9	22.77
127 × 64	14.90	127.0	63.5	6.4	9.2	10.7	2.4	84.1	13.8	18.98
102 × 51	10.42	101.6	50.8	6.1	7.6	9.1	2.4	65.8	13.4	13.28
76 × 38	6.70	76.2	38.1	5.1	6.8	7.6	2.4	45.7	11.2	8.53

PROPERTIES OF STEEL SECTIONS

CHANNELS

DIMENSIONS AND PROPERTIES

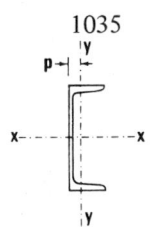

Nominal Size mm	Dimension p cm	Moment of Inertia Axis x–x Gross cm⁴	Moment of Inertia Axis x–x Net cm⁴	Axis y–y cm⁴	Radius of Gyration Axis x–x cm	Radius of Gyration Axis y–y cm	Elastic Modulus Axis x–x cm³	Elastic Modulus Axis y–y cm³
432 × **102**	2.32	21399	17602	628.6	16.0	2.74	991.1	80.15
381 × **102**	2.52	14894	12060	579.8	14.6	2.87	781.8	75.87
305 × **102**	2.66	8214	6587	499.5	11.8	2.91	539.0	66.60
305 × **89**	2.18	7061	5824	325.4	11.5	2.48	463.3	48.49
254 × **89**	2.42	4448	3612	302.4	9.88	2.58	350.2	46.71
254 × **76**	1.86	3367	2673	162.6	9.67	2.12	265.1	28.22
229 × **89**	2.53	3387	2733	285.0	9.01	2.61	296.4	44.82
229 × **76**	2.00	2610	2040	158.7	8.87	2.19	228.3	28.22
203 × **89**	2.65	2491	1996	264.4	8.10	2.64	245.2	42.34
203 × **76**	2.13	1950	1506	151.4	8.02	2.23	192.0	27.59
178 × **89**	2.76	1753	1397	241.0	7.16	2.66	197.2	39.29
178 × **76**	2.20	1337	1028	134.0	7.10	2.25	150.4	24.73
152 × **89**	2.86	1166	923.7	215.1	6.20	2.66	153.0	35.70
152 × **76**	2.21	851.6	654.3	113.8	6.12	2.24	111.8	21.05
127 × **64**	1.94	482.6	367.5	67.24	5.04	1.88	75.99	15.25
102 × **51**	1.51	207.7	167.9	29.10	3.96	1.48	40.89	8.16
76 × **38**	1.19	74.14	54.52	10.66	2.95	1.12	19.46	4.07

One hole is deducted from each flange in calculating the Net Moment of Inertia about **x–x**.

PROPERTIES OF STEEL SECTIONS

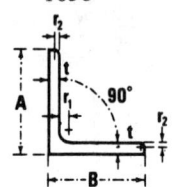

EQUAL ANGLES

DIMENSIONS AND PROPERTIES

Nominal Size	Leg Lengths A×B	Actual Thickness	Mass per metre	Radii Root r_1	Radii Toe r_2	Area of Section	Centre of Gravity Cx	Centre of Gravity Cy
mm	mm	mm	kg	mm	mm	cm²	cm	cm
203 × 203	203.2 × 203.2	25.3	76.00	15.2	4.8	96.81	5.99	5.99
		23.7	71.51	15.2	4.8	91.09	5.93	5.93
		22.1	67.05	15.2	4.8	85.42	5.87	5.87
		20.5	62.56	15.2	4.8	79.69	5.81	5.81
		18.9	57.95	15.2	4.8	73.82	5.75	5.75
		17.3	53.30	15.2	4.8	67.89	5.69	5.69
		15.8	48.68	15.2	4.8	62.02	5.63	5.63
152 × 152	152.4 × 152.4	22.1	49.32	12.2	4.8	62.83	4.60	4.60
		20.5	46.03	12.2	4.8	58.63	4.54	4.54
		19.0	42.75	12.2	4.8	54.45	4.49	4.49
		17.3	39.32	12.2	4.8	50.09	4.42	4.42
		15.8	36.07	12.2	4.8	45.95	4.37	4.37
		14.2	32.62	12.2	4.8	41.55	4.31	4.31
		12.6	29.07	12.2	4.8	37.03	4.24	4.24
		11.0	25.60	12.2	4.8	32.61	4.18	4.18
		9.4	22.02	12.2	4.8	28.06	4.11	4.11
127 × 127	127.0 × 127.0	19.0	35.16	10.7	4.8	44.80	3.85	3.85
		17.4	32.47	10.7	4.8	41.37	3.79	3.79
		15.8	29.66	10.7	4.8	37.78	3.73	3.73
		14.2	26.80	10.7	4.8	34.14	3.67	3.67
		12.6	23.99	10.7	4.8	30.56	3.61	3.61
		11.0	21.14	10.7	4.8	26.93	3.55	3.55
		9.5	18.30	10.7	4.8	23.31	3.49	3.49
102 × 102	101.6 × 101.6	19.0	27.57	9.1	4.8	35.12	3.22	3.22
		17.4	25.48	9.1	4.8	32.45	3.16	3.16
		15.8	23.37	9.1	4.8	29.78	3.10	3.10
		14.2	21.17	9.1	4.8	26.96	3.04	3.04
		12.6	18.91	9.1	4.8	24.09	2.98	2.98
		11.0	16.69	9.1	4.8	21.27	2.92	2.92
		9.4	14.44	9.1	4.8	18.39	2.86	2.86
		7.8	12.06	9.1	4.8	15.37	2.79	2.79

Some of the thicknesses given in this table are obtained by raising the rolls (Practice in this respect is not uniform throughout the industry). In such cases the legs will be slightly longer and the backs of the toes will be slightly rounded.

PROPERTIES OF STEEL SECTIONS

EQUAL ANGLES

DIMENSIONS AND PROPERTIES

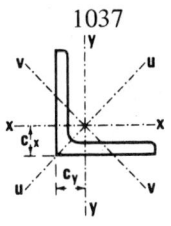

Moment of Inertia				Radius of Gyration				Elastic Modulus	
Axis x–x	Axis y–y	Axis u–u Max.	Axis v–v Min.	Axis x–x	Axis y–y	Axis u–u Max.	Axis v–v Min.	Axis x–x	Axis y–y
cm⁴	cm⁴	cm⁴	cm⁴	cm	cm	cm	cm	cm³	cm³
3686	3686	5845	1527	6.17	6.17	7.77	3.97	257	257
3491	3491	5540	1442	6.19	6.19	7.80	3.98	243	243
3294	3294	5232	1357	6.21	6.21	7.83	3.99	228	228
3094	3094	4916	1271	6.23	6.23	7.85	3.99	213	213
2885	2885	4587	1183	6.25	6.25	7.88	4.00	198	198
2671	2671	4248	1093	6.27	6.27	7.91	4.01	183	183
2455	2455	3907	1004	6.29	6.29	7.94	4.02	167	167
1321	1321	2089	553	4.58	4.58	5.77	2.97	124	124
1243	1243	1968	517	4.60	4.60	5.79	2.97	116	116
1164	1164	1846	482	4.62	4.62	5.82	2.98	108	108
1080	1080	1714	446	4.64	4.64	5.85	2.98	99.8	99.8
999	999	1587	411	4.66	4.66	5.88	2.99	91.9	91.9
911	911	1448	374	4.68	4.68	5.90	3.00	83.3	83.3
819	819	1303	335	4.70	4.70	5.93	3.01	74.5	74.5
727	727	1156	297	4.72	4.72	5.96	3.02	65.7	65.7
631	631	1003	258	4.74	4.74	5.98	3.03	56.7	56.7
651	651	1028	273	3.81	3.81	4.79	2.47	73.5	73.5
607	607	961	253	3.83	3.83	4.82	2.47	68.1	68.1
560	560	888	232	3.85	3.85	4.85	2.48	62.4	62.4
511	511	811	211	3.87	3.87	4.87	2.48	56.6	56.6
462	462	734	190	3.89	3.89	4.90	2.49	50.8	50.8
411	411	654	169	3.91	3.91	4.93	2.50	44.9	44.9
359	359	571	147	3.93	3.93	4.95	2.51	39.0	39.0
317	317	497	136	3.00	3.00	3.76	1.97	45.6	45.6
296	296	466	126	3.02	3.02	3.79	1.97	42.3	42.3
275	275	434	116	3.04	3.04	3.82	1.97	38.9	38.9
252	252	399	105	3.06	3.06	3.84	1.97	35.4	35.4
228	228	361	94.3	3.07	3.07	3.87	1.98	31.7	31.7
203	203	323	83.8	3.09	3.09	3.90	1.99	28.1	28.1
178	178	283	73.1	3.11	3.11	3.92	1.99	24.4	24.4
150	150	239	61.7	3.13	3.13	3.95	2.00	20.4	20.4

Finished sections in which the angle between the legs is not less than 89° and not more than 91° shall be deemed to comply with the requirements of the standard. Angles may be ordered by width of flanges and thickness, or by width of flanges and mass per metre, but not by both thickness and mass per metre.

PROPERTIES OF STEEL SECTIONS

EQUAL ANGLES

DIMENSIONS AND PROPERTIES

Nominal Size	Leg Lengths A×B	Actual Thickness	Mass per metre	Radii Root r_1	Radii Toe r_2	Area of Section	Centre of Gravity Cx	Centre of Gravity Cy
mm	mm	mm	kg	mm	mm	cm^2	cm	cm
89 × 89	88.9 × 88.9	15.8	20.10	8.4	4.8	25.61	2.78	2.78
		14.2	18.31	8.4	4.8	23.32	2.72	2.72
		12.6	16.38	8.4	4.8	20.87	2.66	2.66
		11.0	14.44	8.4	4.8	18.40	2.60	2.60
		9.4	12.50	8.4	4.8	15.92	2.54	2.54
		7.9	10.58	8.4	4.8	13.47	2.48	2.48
		6.3	8.49	8.4	4.8	10.81	2.41	2.41
76 × 76	76.2 × 76.2	14.3	15.50	7.6	4.8	19.74	2.41	2.41
		12.6	13.85	7.6	4.8	17.64	2.35	2.35
		11.0	12.20	7.6	4.8	15.55	2.29	2.29
		9.4	10.57	7.6	4.8	13.47	2.23	2.23
		7.8	8.93	7.6	4.8	11.37	2.16	2.16
		6.2	7.16	7.6	4.8	9.12	2.10	2.10
64 × 64	63.5 × 63.5	12.5	11.31	6.9	2.4	14.41	2.03	2.03
		11.0	10.12	6.9	2.4	12.89	1.98	1.98
		9.4	8.78	6.9	2.4	11.18	1.92	1.92
		7.9	7.45	6.9	2.4	9.48	1.86	1.86
		6.2	5.96	6.9	2.4	7.59	1.80	1.80
57 × 57	57.2 × 57.2	9.3	7.74	6.6	2.4	9.86	1.76	1.76
		7.8	6.55	6.6	2.4	8.35	1.70	1.70
		6.2	5.35	6.6	2.4	6.82	1.64	1.64
		4.6	4.01	6.6	2.4	5.11	1.57	1.57
51 × 51	50.8 × 50.8	9.4	6.85	6.1	2.4	8.72	1.60	1.60
		7.8	5.80	6.1	2.4	7.39	1.54	1.54
		6.3	4.77	6.1	2.4	6.08	1.49	1.49
		4.6	3.58	6.1	2.4	4.56	1.42	1.42

Some of the thicknesses given in this table are obtained by raising the rolls (Practice in this respect is not uniform throughout the industry). In such cases the legs will be slightly longer and the backs of the toes will be slightly rounded.

PROPERTIES OF STEEL SECTIONS
EQUAL ANGLES
DIMENSIONS AND PROPERTIES

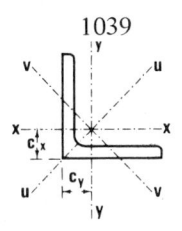

Moment of Inertia				Radius of Gyration				Elastic Modulus	
Axis x–x	Axis y–y	Axis u–u Max.	Axis v–v Min.	Axis x–x	Axis y–y	Axis u–u Max.	Axis v–v Min.	Axis x–x	Axis y–y
cm⁴	cm⁴	cm⁴	cm⁴	cm	cm	cm	cm	cm³	cm³
178	178	280	75.7	2.63	2.63	3.30	1.72	29.1	29.1
164	164	259	69.1	2.65	2.65	3.33	1.72	26.6	26.6
149	149	235	62.0	2.67	2.67	3.36	1.72	23.9	23.9
133	133	211	55.0	2.69	2.69	3.38	1.73	21.1	21.1
116	116	185	47.9	2.70	2.70	3.41	1.74	18.3	18.3
99.8	99.8	159	41.0	2.72	2.72	3.43	1.74	15.6	15.6
81.0	81.0	129	33.3	2.74	2.74	3.45	1.75	12.5	12.5
99.6	99.6	157	42.7	2.25	2.25	2.82	1.47	19.1	19.1
90.4	90.4	143	38.2	2.26	2.26	2.84	1.47	17.1	17.1
80.9	80.9	128	33.8	2.28	2.28	2.87	1.47	15.2	15.2
71.1	71.1	113	29.5	2.30	2.30	2.89	1.48	13.2	13.2
60.9	60.9	96.8	25.1	2.31	2.31	2.92	1.49	11.2	11.2
49.6	49.6	78.8	20.3	2.33	2.33	2.94	1.49	8.97	8.97
50.4	50.4	78.9	21.8	1.87	1.87	2.34	1.23	11.7	11.7
45.8	45.8	72.1	19.5	1.89	1.89	2.37	1.23	10.5	10.5
40.5	40.5	64.0	17.0	1.90	1.90	2.39	1.23	9.15	9.15
35.0	35.0	55.5	14.6	1.92	1.92	2.42	1.24	7.80	7.80
28.6	28.6	45.4	11.8	1.94	1.94	2.45	1.25	6.28	6.28
28.6	28.6	45.0	12.1	1.70	1.70	2.14	1.11	7.22	7.22
24.7	24.7	39.1	10.3	1.72	1.72	2.16	1.11	6.15	6.15
20.6	20.6	32.6	8.53	1.74	1.74	2.19	1.12	5.05	5.05
15.8	15.8	25.0	6.51	1.76	1.76	2.21	1.13	3.80	3.80
19.6	19.6	30.8	8.42	1.50	1.50	1.88	.98	5.64	5.64
17.0	17.0	26.8	7.17	1.52	1.52	1.91	.98	4.81	4.81
14.3	14.3	22.7	5.95	1.53	1.53	1.93	.99	3.98	3.98
11.0	11.0	17.4	4.54	1.55	1.55	1.95	1.00	3.00	3.00

Finished sections in which the angle between the legs is not less than 89° and not more than 91° shall be deemed to comply with the requirements of the standard. Angles may be ordered by width of flanges and thickness, or by width of flanges and mass per metre, but not by both thickness and mass per metre.

PROPERTIES OF STEEL SECTIONS

EQUAL ANGLES

DIMENSIONS AND PROPERTIES

Nominal Size	Leg Lengths A × B	Actual Thickness	Mass per metre	Radii Root r_1	Radii Toe r_2	Area of Section	Centre of Gravity Cx	Centre of Gravity Cy
mm	mm	mm	kg	mm	mm	cm²	cm	cm
45 × 45	44.5 × 44.5	7.9	5.06	5.8	2.4	6.45	1.39	1.39
		6.1	4.02	5.8	2.4	5.12	1.32	1.32
		4.7	3.13	5.8	2.4	3.99	1.26	1.26
38 × 38	38.1 × 38.1	7.8	4.24	5.3	2.4	5.40	1.23	1.23
		6.3	3.50	5.3	2.4	4.46	1.17	1.17
		4.7	2.68	5.3	2.4	3.41	1.11	1.11
32 × 32	31.8 × 31.8	6.2	2.83	5.1	2.4	3.61	1.01	1.01
		4.6	2.16	5.1	2.4	2.75	0.95	0.95
		3.1	1.49	5.1	2.4	1.90	0.88	0.88
25 × 25	25.4 × 25.4	6.4	2.23	4.6	2.4	2.84	0.85	0.85
		4.7	1.72	4.6	2.4	2.19	0.79	0.79
		3.1	1.19	4.6	2.4	1.52	0.73	0.73

Some of the thicknesses given in this table are obtained by raising the rolls (Practice in this respect is not uniform throughout the industry). In such cases the legs will be slightly longer and the backs of the toes will be slightly rounded.

PROPERTIES OF STEEL SECTIONS

EQUAL ANGLES

DIMENSIONS AND PROPERTIES

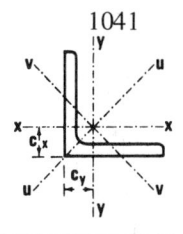

Moment of Inertia				Radius of Gyration				Elastic Modulus	
Axis x–x	Axis y–y	Axis u–u Max.	Axis v–v Min.	Axis x–x	Axis y–y	Axis u–u Max.	Axis v–v Min.	Axis x–x	Axis y–y
cm⁴	cm⁴	cm⁴	cm⁴	cm	cm	cm	cm	cm³	cm³
11.1	11.1	17.5	4.75	1.31	1.31	1.65	.86	3.64	3.64
9.09	9.09	14.4	3.80	1.33	1.33	1.68	.86	2.91	2.91
7.24	7.24	11.5	3.00	1.35	1.35	1.70	.87	2.28	2.28
6.69	6.69	10.5	2.92	1.11	1.11	1.39	.73	2.59	2.59
5.67	5.67	8.94	2.41	1.13	1.13	1.42	.73	2.15	2.15
4.47	4.47	7.08	1.86	1.14	1.14	1.44	.74	1.66	1.66
3.10	3.10	4.87	1.34	.93	.93	1.16	.61	1.43	1.43
2.45	2.45	3.87	1.03	.94	.94	1.19	.61	1.10	1.10
1.74	1.74	2.75	.72	.96	.96	1.20	.62	.76	.76
1.50	1.50	2.33	.68	.73	.73	.90	.49	.89	.89
1.20	1.20	1.89	.51	.74	.74	.93	.48	.69	.69
.86	.86	1.37	.36	.75	.75	.95	.49	.48	.48

Finished sections in which the angle between the legs is not less than 89° and not more than 91° shall be deemed to comply with the requirements of the standard. Angles may be ordered by width of flanges and thickness, or by width of flanges and mass per metre, but not by both thickness and mass per metre.

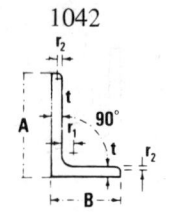

UNEQUAL ANGLES

DIMENSIONS AND PROPERTIES

Nominal Size	Leg Lengths A×B	Actual Thickness	Mass per metre	Radii Root r_1	Radii Toe r_2	Area of Section	Centre of Gravity Cx	Centre of Gravity Cy
mm	mm	mm	kg	mm	mm	cm^2	cm	cm
229 × 102	228.6 × 101.6	22.1	53.77	13.0	4.8	68.49	8.73	2.41
		20.6	50.21	13.0	4.8	63.97	8.67	2.35
		18.9	46.45	13.0	4.8	59.17	8.60	2.29
		17.4	42.87	13.0	4.8	54.61	8.54	2.23
		15.8	39.20	13.0	4.8	49.93	8.47	2.17
		14.2	35.43	13.0	4.8	45.13	8.40	2.10
		12.6	31.56	13.0	4.8	40.20	8.33	2.04
203 × 152	203.2 × 152.4	22.1	58.09	13.7	4.8	74.00	6.59	4.07
		20.5	54.22	13.7	4.8	69.07	6.53	4.01
		18.9	50.32	13.7	4.8	64.10	6.47	3.95
		17.3	46.30	13.7	4.8	58.99	6.41	3.89
		15.8	42.32	13.7	4.8	53.91	6.35	3.83
		14.2	38.29	13.7	4.8	48.78	6.29	3.77
		12.6	34.10	13.7	4.8	43.44	6.22	3.70
203 × 102	203.2 × 101.6	19.0	42.75	12.2	4.8	54.45	7.46	2.41
		17.3	39.32	12.2	4.8	50.09	7.40	2.35
		15.8	36.07	12.2	4.8	45.95	7.33	2.29
		14.2	32.62	12.2	4.8	41.55	7.27	2.23
		12.6	29.07	12.2	4.8	37.03	7.20	2.16
178 × 89	177.8 × 88.9	15.8	31.30	11.2	4.8	39.87	6.49	2.08
		14.2	28.28	11.2	4.8	36.02	6.43	2.01
		12.6	25.31	11.2	4.8	32.24	6.36	1.95
		11.1	22.36	11.2	4.8	28.48	6.29	1.89
		9.4	19.22	11.2	4.8	24.48	6.22	1.83
152 × 102	152.4 × 101.6	19.0	35.16	10.7	4.8	44.80	5.25	2.73
		17.4	32.47	10.7	4.8	41.37	5.19	2.67
		15.8	29.66	10.7	4.8	37.78	5.13	2.61
		14.2	26.80	10.7	4.8	34.14	5.07	2.55
		12.6	23.99	10.7	4.8	30.56	5.00	2.48
		11.0	21.14	10.7	4.8	26.93	4.94	2.42
		9.5	18.30	10.7	4.8	23.31	4.88	2.36

Some of the thicknesses given in this table are obtained by raising the rolls (Practice in this respect is not uniform throughout the industry). In such cases the legs will be slightly longer and the backs of the toes will be slightly rounded.

PROPERTIES OF STEEL SECTIONS

UNEQUAL ANGLES

DIMENSIONS AND PROPERTIES

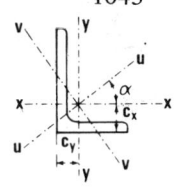

Moment of Inertia				Radius of Gyration				Angle	Elastic Modulus	
Axis x–x	Axis y–y	Axis u–u Max.	Axis v–v Min.	Axis x–x	Axis y–y	Axis u–u Max.	Axis v–v Min.	Axis x–x to Axis u–u	Axis x–x	Axis y–y
cm⁴	cm⁴	cm⁴	cm⁴	cm	cm	cm	cm	tan α	cm³	cm³
3606	447	3747	306	7.26	2.56	7.40	2.11	.207	255	57.8
3388	423	3523	287	7.28	2.57	7.42	2.12	.209	239	54.1
3154	396	3283	267	7.30	2.59	7.45	2.13	.211	221	50.3
2929	369	3051	248	7.32	2.60	7.47	2.13	.213	205	46.6
2695	342	2808	229	7.35	2.62	7.50	2.14	.214	187	42.8
2451	313	2556	208	7.37	2.63	7.53	2.15	.216	170	38.9
2197	283	2292	187	7.39	2.65	7.55	2.16	.218	151	34.8
2992	1439	3648	783	6.36	4.41	7.02	3.25	.545	218	129
2811	1355	3432	734	6.38	4.43	7.05	3.26	.547	204	121
2625	1268	3209	684	6.40	4.45	7.08	3.27	.548	190	112
2432	1177	2976	633	6.42	4.47	7.10	3.28	.550	175	104
2237	1085	2740	582	6.44	4.49	7.13	3.29	.551	160	95.1
2037	990	2497	530	6.46	4.51	7.15	3.30	.553	145	86.3
1826	890	2240	476	6.48	4.53	7.18	3.31	.554	129	77.1
2277	386	2409	253	6.47	2.66	6.65	2.15	.256	177	49.7
2109	359	2234	234	6.49	2.68	6.68	2.16	.259	163	46.0
1947	333	2064	216	6.51	2.69	6.70	2.17	.260	150	42.4
1773	305	1881	197	6.53	2.71	6.73	2.18	.262	136	38.5
1591	276	1689	177	6.55	2.73	6.75	2.19	.264	121	34.5
1280	217	1355	142	5.67	2.33	5.83	1.89	.257	113	31.9
1165	199	1235	129	5.69	2.35	5.86	1.89	.260	103	28.9
1051	181	1115	117	5.71	2.37	5.88	1.90	.262	92.0	26.0
935	162	993	104	5.73	2.38	5.90	1.91	.264	81.4	23.1
810	141	861	90.3	5.75	2.40	5.93	1.92	.265	70.1	20.0
1015	358	1161	212	4.76	2.83	5.09	2.17	.427	102	48.2
945	335	1083	196	4.78	2.84	5.12	2.18	.430	94.1	44.7
871	309	1000	180	4.80	2.86	5.14	2.19	.432	86.1	41.0
794	283	913	164	4.82	2.88	5.17	2.19	.435	78.0	37.2
716	257	825	148	4.84	2.90	5.20	2.20	.437	70.0	33.4
637	229	734	132	4.86	2.92	5.22	2.21	.439	61.8	29.6
555	201	641	115	4.88	2.93	5.24	2.22	.441	53.6	25.7

Finished sections in which the angle between the legs is not less than 89° and not more than 91° shall be deemed to comply with the requirements of the standard. Angles may be ordered by width of flanges and thickness, or by width of flanges and mass per metre, but not by both thickness and mass per metre.

PROPERTIES OF STEEL SECTIONS

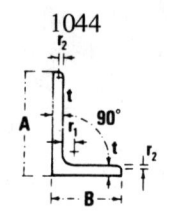

UNEQUAL ANGLES

DIMENSIONS AND PROPERTIES

Nominal Size mm	Leg Lengths A×B mm	Actual Thickness mm	Mass per metre kg	Radii Root r_1 mm	Radii Toe r_2 mm	Area of Section cm²	Centre of Gravity Cx cm	Centre of Gravity Cy cm
152 × 89	152.4 × 88.9	15.7	27.99	10.4	4.8	35.66	5.37	2.22
		14.2	25.46	10.4	4.8	32.43	5.31	2.16
		12.6	22.77	10.4	4.8	29.00	5.25	2.10
		11.1	20.12	10.4	4.8	25.63	5.19	2.04
		9.4	17.26	10.4	4.8	21.99	5.12	1.97
		7.8	14.44	10.4	4.8	18.40	5.04	1.91
152 × 76	152.4 × 76.2	15.8	26.52	9.9	4.8	33.78	5.65	1.86
		14.2	23.99	9.9	4.8	30.56	5.59	1.80
		12.6	21.45	9.9	4.8	27.33	5.52	1.74
		11.0	18.92	9.9	4.8	24.10	5.46	1.68
		9.5	16.39	9.9	4.8	20.87	5.39	1.62
		7.8	13.69	9.9	4.8	17.44	5.32	1.55
127 × 89	127.0 × 88.9	15.8	24.86	9.7	4.8	31.67	4.29	2.40
		14.2	22.64	9.7	4.8	28.84	4.24	2.34
		12.6	20.26	9.7	4.8	25.81	4.17	2.28
		11.1	17.89	9.7	4.8	22.79	4.11	2.22
		9.4	15.35	9.7	4.8	19.56	4.04	2.16
		7.9	12.94	9.7	4.8	16.48	3.98	2.10
127 × 76	127.0 × 76.2	14.2	21.17	9.1	4.8	26.96	4.47	1.95
		12.6	18.91	9.1	4.8	24.09	4.41	1.89
		11.0	16.69	9.1	4.8	21.27	4.35	1.83
		9.4	14.44	9.1	4.8	18.39	4.28	1.77
		7.8	12.06	9.1	4.8	15.37	4.21	1.70
102 × 89	101.6 × 88.9	15.8	21.75	8.9	4.8	27.71	3.27	2.64
		14.2	19.67	8.9	4.8	25.06	3.21	2.58
		12.6	17.72	8.9	4.8	22.57	3.15	2.52
		11.0	15.62	8.9	4.8	19.90	3.09	2.46
		9.5	13.55	8.9	4.8	17.27	3.03	2.40
		7.8	11.31	8.9	4.8	14.41	2.96	2.33

Some of the thicknesses given in this table are obtained by raising the rolls (Practice in this respect is not uniform throughout the industry). In such cases the legs will be slightly longer and the backs of the toes will be slightly rounded.

PROPERTIES OF STEEL SECTIONS

UNEQUAL ANGLES

DIMENSIONS AND PROPERTIES

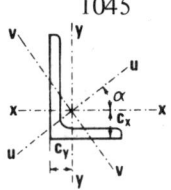

Moment of Inertia				Radius of Gyration				Angle	Elastic Modulus	
Axis x–x	Axis y–y	Axis u–u Max.	Axis v–v Min.	Axis x–x	Axis y–y	Axis u–u Max.	Axis v–v Min.	Axis x–x to Axis u–u	Axis x–x	Axis y–y
cm⁴	cm⁴	cm⁴	cm⁴	cm	cm	cm	cm	tan α	cm³	cm³
828	208	907	129	4.82	2.42	5.04	1.90	.336	83.9	31.2
759	192	833	118	4.84	2.43	5.07	1.91	.338	76.5	28.5
685	174	752	107	4.86	2.45	5.09	1.92	.341	68.5	25.6
610	156	671	95.2	4.88	2.47	5.12	1.93	.343	60.7	22.7
528	136	581	82.6	4.90	2.48	5.14	1.94	.345	52.1	19.6
445	115	490	69.9	4.92	2.50	5.16	1.95	.347	43.7	16.5
786	132	830	87.2	4.82	1.98	4.96	1.61	.253	81.9	22.9
717	121	759	79.3	4.84	1.99	4.98	1.61	.256	74.3	20.8
647	110	685	71.5	4.87	2.01	5.01	1.62	.259	66.6	18.7
575	98.5	610	63.7	4.89	2.02	5.03	1.63	.261	58.8	16.6
503	86.7	534	55.7	4.91	2.04	5.06	1.63	.263	51.0	14.4
424	73.6	450	47.2	4.93	2.05	5.08	1.64	.265	42.7	12.1
496	198	580	114	3.96	2.50	4.28	1.90	.470	59.0	30.5
456	183	534	104	3.97	2.52	4.30	1.90	.473	53.8	27.9
412	166	484	93.9	3.99	2.54	4.33	1.91	.476	48.3	25.1
367	149	432	83.6	4.01	2.55	4.35	1.92	.479	42.7	22.3
318	129	375	72.5	4.03	2.57	4.38	1.93	.481	36.8	19.2
271	110	319	61.7	4.05	2.59	4.40	1.94	.483	31.0	16.2
430	116	475	71.5	4.00	2.07	4.20	1.63	.351	52.3	20.4
389	105	429	64.4	4.02	2.09	4.22	1.63	.355	46.9	18.3
346	94.2	383	57.3	4.04	2.10	4.25	1.64	.358	41.5	16.3
302	82.8	335	50.0	4.06	2.12	4.27	1.65	.360	35.9	14.1
255	70.2	283	42.3	4.07	2.14	4.29	1.66	.362	30.1	11.9
262	186	357	91.5	3.08	2.59	3.59	1.82	.743	38.1	29.7
240	170	328	83.0	3.10	2.61	3.62	1.82	.746	34.6	27.0
219	156	299	75.0	3.11	2.63	3.64	1.82	.748	31.2	24.4
195	139	268	66.6	3.13	2.64	3.67	1.83	.750	27.6	21.6
171	122	235	58.2	3.15	2.66	3.69	1.84	.752	24.0	18.8
145	103	199	49.1	3.17	2.68	3.72	1.85	.753	20.1	15.8

Finished sections in which the angle between the legs is not less than 89° and not more than 91° shall be deemed to comply with the requirements of the standard. Angles may be ordered by width of flanges and thickness, or by width of flanges and mass per metre, but not by both thickness and mass per metre.

PROPERTIES OF STEEL SECTIONS

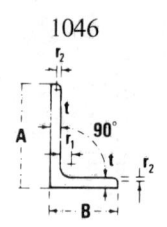

UNEQUAL ANGLES

DIMENSIONS AND PROPERTIES

Nominal Size	Leg Lengths A×B	Actual Thickness	Mass per metre	Radii Root r_1	Radii Toe r_2	Area of Section	Centre of Gravity Cx	Centre of Gravity Cy
mm	mm	mm	kg	mm	mm	cm^2	cm	cm
102 × 76	101.6 × 76.2	14.2	18.31	8.4	4.8	23.32	3.40	2.14
		12.6	16.38	8.4	4.8	20.87	3.34	2.08
		11.0	14.44	8.4	4.8	18.40	3.28	2.02
		9.4	12.50	8.4	4.8	15.92	3.22	1.96
		7.9	10.58	8.4	4.8	13.47	3.16	1.90
102 × 64	101.6 × 63.5	11.0	13.40	8.1	4.8	17.07	3.51	1.62
		9.5	11.61	8.1	4.8	14.79	3.45	1.56
		7.8	9.69	8.1	4.8	12.35	3.38	1.49
		6.3	7.89	8.1	4.8	10.05	3.31	1.43
89 × 76	88.9 × 76.2	14.2	16.83	8.1	4.8	21.44	2.89	2.26
		12.7	15.20	8.1	4.8	19.36	2.84	2.21
		11.0	13.40	8.1	4.8	17.07	2.77	2.14
		9.5	11.61	8.1	4.8	14.79	2.71	2.08
		7.8	9.69	8.1	4.8	12.35	2.65	2.02
		6.3	7.89	8.1	4.8	10.05	2.58	1.96
89 × 64	88.9 × 63.5	11.0	12.20	7.6	4.8	15.55	2.97	1.71
		9.4	10.57	7.6	4.8	13.47	2.91	1.65
		7.8	8.93	7.6	4.8	11.37	2.85	1.59
		6.2	7.16	7.6	4.8	9.12	2.78	1.53
76 × 64	76.2 × 63.5	11.0	11.17	7.4	4.8	14.23	2.46	1.83
		9.4	9.68	7.4	4.8	12.33	2.40	1.77
		7.9	8.19	7.4	4.8	10.43	2.34	1.71
		6.2	6.56	7.4	4.8	8.36	2.27	1.64
76 × 51	76.2 × 50.8	11.0	10.12	6.9	2.4	12.89	2.68	1.42
		9.4	8.78	6.9	2.4	11.18	2.62	1.36
		7.9	7.45	6.9	2.4	9.48	2.56	1.30
		6.2	5.96	6.9	2.4	7.59	2.49	1.24
		4.7	4.62	6.9	2.4	5.88	2.43	1.18

Some of the thicknesses given in this table are obtained by raising the rolls (Practice in this respect is not uniform throughout the industry). In such cases the legs will be slightly longer and the backs of the toes will be slightly rounded.

PROPERTIES OF STEEL SECTIONS

UNEQUAL ANGLES

DIMENSIONS AND PROPERTIES

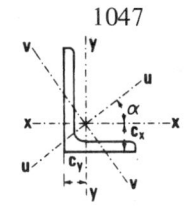

Moment of Inertia				Radius of Gyration				Angle	Elastic Modulus	
Axis x–x	Axis y–y	Axis u–u Max.	Axis v–v Min.	Axis x–x	Axis y–y	Axis u–u Max.	Axis v–v Min.	Axis x–x to Axis u–u	Axis x–x	Axis y–y
cm⁴	cm⁴	cm⁴	cm⁴	cm	cm	cm	cm	tan α	cm³	cm³
228	109	277	60.4	3.13	2.16	3.45	1.61	.537	33.8	19.9
207	98.8	251	54.3	3.15	2.18	3.47	1.61	.540	30.3	17.8
185	88.5	225	48.2	3.17	2.19	3.50	1.62	.544	26.8	15.8
162	77.8	197	42.0	3.19	2.21	3.52	1.62	.547	23.3	13.7
138	66.8	169	35.9	3.20	2.23	3.54	1.63	.549	19.7	11.7
174	51.9	194	31.4	3.19	1.74	3.37	1.36	.380	26.1	11.0
152	45.8	171	27.4	3.21	1.76	3.40	1.36	.383	22.7	9.56
129	38.9	145	23.2	3.23	1.78	3.42	1.37	.386	19.0	8.02
106	32.2	119	19.1	3.25	1.79	3.44	1.38	.388	15.4	6.54
155	104	207	52.4	2.69	2.20	3.11	1.56	.710	25.9	19.4
142	95.4	190	47.5	2.71	2.22	3.13	1.57	.713	23.4	17.6
127	85.4	170	42.0	2.73	2.24	3.16	1.57	.715	20.7	15.6
111	75.1	150	36.7	2.74	2.25	3.18	1.58	.718	18.0	13.6
94.2	63.7	127	30.9	2.76	2.27	3.21	1.58	.720	15.1	11.4
77.5	52.5	104	25.5	2.78	2.29	3.23	1.59	.721	12.3	9.27
118	49.8	140	28.2	2.76	1.79	3.00	1.35	.489	20.0	10.7
104	43.9	123	24.6	2.78	1.80	3.02	1.35	.493	17.4	9.34
88.8	37.7	106	21.0	2.79	1.82	3.05	1.36	.496	14.7	7.92
72.1	30.7	85.8	17.0	2.81	1.83	3.07	1.37	.498	11.8	6.37
76.6	47.8	99.9	24.4	2.32	1.83	2.65	1.31	.669	14.8	10.6
67.3	42.1	88.2	21.3	2.34	1.85	2.67	1.31	.673	12.9	9.20
57.8	36.2	75.9	18.1	2.35	1.86	2.70	1.32	.676	10.9	7.81
46.9	29.5	61.7	14.7	2.37	1.88	2.72	1.33	.678	8.77	6.27
71.7	25.1	81.7	15.1	2.36	1.40	2.52	1.08	.420	14.5	6.86
63.2	22.3	72.3	13.2	2.38	1.41	2.54	1.09	.426	12.6	5.99
54.5	19.3	62.5	11.3	2.40	1.43	2.57	1.09	.431	10.8	5.12
44.4	15.9	51.1	9.20	2.42	1.45	2.59	1.10	.436	8.66	4.13
34.9	12.6	40.2	7.26	2.44	1.46	2.62	1.11	.438	6.72	3.22

Finished sections in which the angle between the legs is not less than 89° and not more than 91° shall be deemed to comply with the requirements of the standard. Angles may be ordered by width of flanges and thickness, or by width of flanges and mass per metre, but not by both thickness and mass per metre.

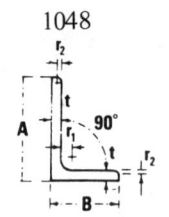

UNEQUAL ANGLES

DIMENSIONS AND PROPERTIES

Nominal Size mm	Leg Lengths A×B mm	Actual Thickness mm	Mass per metre kg	Radii Root r_1 mm	Radii Toe r_2 mm	Area of Section cm^2	Centre of Gravity Cx cm	Centre of Gravity Cy cm
64 × 51	63.5 × 50.8	9.3	7.74	6.6	2.4	9.86	2.09	1.46
		7.8	6.55	6.6	2.4	8.35	2.03	1.40
		6.2	5.35	6.6	2.4	6.82	1.97	1.35
		4.6	4.01	6.6	2.4	5.11	1.90	1.28
64 × 38	63.5 × 38.1	7.8	5.80	6.1	2.4	7.39	2.26	1.00
		6.3	4.77	6.1	2.4	6.08	2.20	0.94
		4.6	3.58	6.1	2.4	4.56	2.13	0.88
51 × 38	50.8 × 38.1	7.9	5.06	5.8	2.4	6.45	1.73	1.10
		6.1	4.02	5.8	2.4	5.12	1.66	1.03
		4.7	3.13	5.8	2.4	3.99	1.60	0.98

Some of the thicknesses given in this table are obtained by raising the rolls (Practice in this respect is not uniform throughout the industry). In such cases the legs will be slightly longer and the backs of the toes will be slightly rounded.

PROPERTIES OF STEEL SECTIONS

UNEQUAL ANGLES

DIMENSIONS AND PROPERTIES

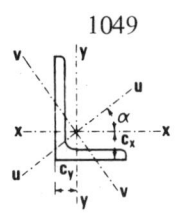

Moment of Inertia				Radius of Gyration				Angle	Elastic Modulus	
Axis x–x	Axis y–y	Axis u–u Max.	Axis v–v Min.	Axis x–x	Axis y–y	Axis u–u Max.	Axis v–v Min.	Axis x–x to Axis u–u	Axis x–x	Axis y–y
cm⁴	cm⁴	cm⁴	cm⁴	cm	cm	cm	cm	tan α	cm³	cm³
37.2	20.9	47.0	11.1	1.94	1.46	2.18	1.06	.613	8.73	5.78
32.1	18.1	40.7	9.50	1.96	1.47	2.21	1.07	.618	7.44	4.93
26.7	15.1	34.0	7.86	1.98	1.49	2.23	1.07	.622	6.10	4.06
20.4	11.6	26.0	6.01	2.00	1.51	2.26	1.08	.625	4.59	3.06
15.5	7.37	18.8	4.15	1.55	1.07	1.71	.80	.532	4.64	2.72
12.7	6.04	15.4	3.32	1.57	1.09	1.73	.81	.540	3.70	2.17
10.1	4.83	12.3	2.63	1.59	1.10	1.75	.81	.544	2.89	1.70
29.2	7.79	32.1	4.87	1.99	1.03	2.08	.81	.347	7.13	2.77
24.4	6.59	27.0	4.05	2.00	1.04	2.11	.82	.353	5.89	2.30
18.7	5.10	20.7	3.10	2.03	1.06	2.13	.82	.358	4.43	1.74

Finished sections in which the angle between the legs is not less than 89° and not more than 91° shall be deemed to comply with the requirements of the standard. Angles may be ordered by width of flanges and thickness, or by width of flanges and mass per metre, but not by both thickness and mass per metre.

PROPERTIES OF STEEL SECTIONS

EQUAL ANGLES B.S. 4848:Part 4:1972

Dimensions and Properties of Equal and Unequal Angles in B.S. Range to be introduced in January 1973. B.S. 4848:Part 4:1972.

Equal Angles

DESIGNATION			DIMENSIONS			Mass/unit length	Area of section	Distance of centre of gravity	MOMENT OF INERTIA				RADIUS OF GYRATION				Elastic modulus	
Size	Thickness mm		Leg Length A	Thickness t	Radius Root r₁				About X-X, Y-Y	About U-U	About V-V		About X-X, Y-Y	About U-U	About V-V		About X-X, Y-Y	
Standard	Standard Min.	Max.	mm	mm	mm	kg/m	cm²	c cm	cm⁴	cm⁴	cm⁴		cm	cm	cm		cm³	
25 × 25	3	3	6	25	3	3·5	1·11	1·42	0·72	0·80	1·26	0·33		0·75	0·94	0·48		0·45	
25 × 25	4	3	6	25	4	3·5	1·45	1·85	0·76	1·01	1·60	0·43		0·74	0·93	0·48		0·58	
25 × 25	5	3	6	25	5	3·5	1·77	2·26	0·80	1·20	1·89	0·52		0·73	0·91	0·48		0·71	
30 × 30	3	3	6	30	3	5	1·36	1·74	0·84	1·40	2·23	0·58		0·90	1·13	0·58		0·65	
30 × 30	4	3	6	30	4	5	1·78	2·27	0·88	1·80	2·85	0·75		0·89	1·12	0·58		0·85	
30 × 30	5	3	6	30	5	5	2·18	2·78	0·92	2·16	3·41	0·92		0·88	1·11	0·57		1·04	
40 × 40	4	3	9	40	4	6	2·42	3·08	1·12	4·47	7·09	1·85		1·21	1·52	0·78		1·55	
40 × 40	5	3	9	40	5	6	2·97	3·79	1·16	5·43	8·60	2·26		1·20	1·51	0·77		1·91	
40 × 40	6	3	9	40	6	6	3·52	4·48	1·20	6·31	9·98	2·65		1·19	1·49	0·77		2·26	
45 × 45	4	3	9	45	4	7	2·74	3·49	1·23	6·43	10·2	2·67		1·36	1·71	0·87		1·97	
45 × 45	5	3	9	45	5	7	3·38	4·30	1·28	7·84	12·4	3·25		1·35	1·70	0·87		2·43	
45 × 45	6	3	9	45	6	7	4·00	5·09	1·32	9·16	14·5	3·82		1·34	1·69	0·87		2·88	
50 × 50	5	4	9	50	5	7	3·77	4·80	1·40	11·0	17·4	4·54		1·51	1·90	0·97		3·05	
50 × 50	6	4	9	50	6	7	4·47	5·69	1·45	12·8	20·4	5·33		1·50	1·89	0·97		3·61	
50 × 50	8	4	9	50	8	7	5·82	7·41	1·52	16·3	25·7	6·87		1·48	1·86	0·96		4·68	
60 × 60	5	4·5	11	60	5	8	4·57	5·82	1·64	19·4	30·7	8·02		1·82	2·30	1·17		4·45	
60 × 60	6	4·5	11	60	6	8	5·42	6·91	1·69	22·8	36·2	9·43		1·82	2·29	1·17		5·29	
60 × 60	8	4·5	11	60	8	8	7·09	9·03	1·77	29·2	46·2	12·1		1·80	2·26	1·16		6·89	
60 × 60	10	4·5	11	60	10	8	8·69	11·1	1·85	34·9	55·1	14·8		1·78	2·23	1·16		8·41	
70 × 70	6	5	12	70	6	9	6·38	8·13	1·93	36·9	53·5	15·2		2·13	2·68	1·37		7·27	
70 × 70	8	5	12	70	8	9	8·36	10·6	2·01	47·5	75·3	19·7		2·11	2·66	1·36		9·52	
70 × 70	10	5	12	70	10	9	10·3	13·1	2·09	57·2	90·5	23·9		2·09	2·63	1·35		11·7	
80 × 80	6	5	16	80	6	10	7·34	9·35	2·17	55·8	88·5	23·1		2·44	3·08	1·57		9·57	
80 × 80	8	5	16	80	8	10	9·63	12·3	2·26	72·2	115	29·8		2·43	3·06	1·56		12·6	
80 × 80	10	5	16	80	10	10	11·9	15·1	2·34	87·5	139	36·3		2·41	3·03	1·55		15·4	

PROPERTIES OF STEEL SECTIONS

EQUAL ANGLES B.S. 4848:Part4:1972

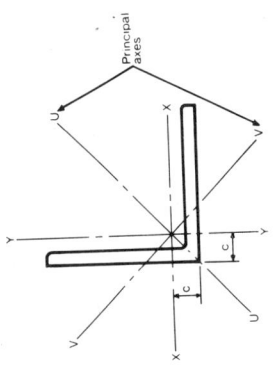

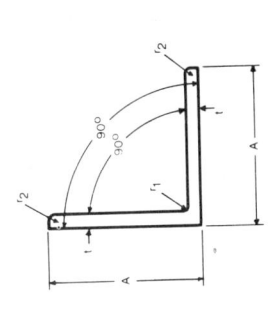

DESIGNATION		DIMENSIONS					Area of section	Distance of centre of gravity	MOMENT OF INERTIA			RADIUS OF GYRATION				Elastic modulus
Size	Thickness	Leg Length A	Thick-ness t	Radius Root r₁	Mass/unit length				About X-X, Y-Y	About U-U	About V-V	About X-X, Y-Y	About U-U	About V-V		About X-X, Y-Y
	Standard Min. Max.	mm	mm	mm	kg/m	cm²	cm	cm⁴	cm⁴	cm⁴	cm	cm	cm		cm³
Standard																
90 × 90	6 6 16	90	6	11	8·30	10·6	2·41	80·3	127	33·3	2·76	3·47	1·78		12·2
90 × 90	8 6 16	90	8	11	10·9	13·9	2·50	104	166	43·1	2·74	3·45	1·76		16·1
90 × 90	10 6 16	90	10	11	13·4	17·1	2·58	127	201	52·8	2·72	3·42	1·76		19·8
90 × 90	12 6 16	90	12	11	15·9	20·3	2·66	148	234	62·0	2·70	3·40	1·75		23·3
100 × 100	8 6·5 19	100	8	12	12·2	15·5	2·74	145	230	59·8	3·06	3·85	1·96		19·9
100 × 100	12 6·5 19	100	12	12	17·8	22·7	2·90	207	328	85·7	3·02	3·80	1·94		29·1
100 × 100	15 6·5 19	100	15	12	21·9	27·9	3·02	249	393	104	2·98	3·75	1·93		35·6
120 × 120	8 8 19	120	8	13	14·7	18·7	3·23	255	405	105	3·69	4·65	2·37		29·1
120 × 120	10 8 19	120	10	13	18·2	23·2	3·31	313	497	129	3·67	4·63	2·36		36·0
120 × 120	12 8 19	120	12	13	21·6	27·5	3·40	368	584	151	3·65	4·60	2·35		42·7
120 × 120	15 8 19	120	15	13	26·6	33·9	3·51	445	705	185	3·62	4·56	2·33		52·4
150 × 150	10 10 25	150	10	16	23·0	29·3	4·03	624	991	258	4·62	5·82	2·97		56·9
150 × 150	12 10 25	150	12	16	27·3	34·8	4·12	737	1170	303	4·60	5·80	2·95		67·7
150 × 150	15 10 25	150	15	16	33·8	43·0	4·25	898	1430	370	4·57	5·76	2·93		83·5
150 × 150	18 10 25	150	18	16	10·1	51·0	4·37	1050	1670	435	4·54	5·71	2·92		98·7
200 × 200	16 12 28	200	16	18	48·5	61·8	5·52	2340	3720	959	6·16	7·76	3·94		162
200 × 200	18 12 28	200	18	18	54·2	69·1	5·60	2600	4130	1070	6·13	7·73	3·93		181
200 × 200	20 12 28	200	20	18	59·9	76·3	5·68	2850	4530	1170	6·11	7·70	3·92		199
200 × 200	24 12 28	200	24	18	71·1	90·6	5·84	3330	5280	1380	6·06	7·64	3·90		235

Note 1. Some of the thicknesses given in the tables are obtained by raising the rolls. (Practice in this respect is not uniform throughout the Division). In such cases the flanges will be slightly longer and the back of the toes will be slightly rounded.

Note 2. Angles should be ordered by flange length and thickness.

Note 3. Finished sections in which the angle between the flanges is not less than 89° and not more than 91° will be accepted as conforming to the standard.

PROPERTIES OF STEEL SECTIONS
UNEQUAL ANGLES B.S. 4848: Part 4: 1972

Dimensions and Properties of Equal and Unequal Angles in B.S. Range to be introduced in January 1973. B.S. 4848:Part 4:1972.

Unequal Angles

Size	Thickness Standard (mm)	Thickness Min. (mm)	Thickness Max. (mm)	Leg Lengths A (mm)	Leg Lengths B (mm)	Thickness t (mm)	Radius Root r (mm)	Mass/unit length (kg/m)	Area of section (cm²)	Distance c_x (cm)	Distance c_y (cm)	I About X-X (cm⁴)	I About Y-Y (cm⁴)	I About U-U (cm⁴)	I About V-V (cm⁴)	r About X-X (cm)	r About Y-Y (cm)	r About U-U (cm)	r About V-V (cm)	Angle tan α	Z About X-X (cm³)	Z About Y-Y (cm³)
40 x 25	4	3	6	40	25	4	4	1.91	2.46	1.36	0.62	3.89	1.16	4.35	0.70	1.26	0.69	1.33	0.53	0.380	1.47	0.62
60 x 30	5	3	8	60	30	5	6	3.37	4.29	2.15	0.68	15.6	2.60	16.5	1.69	1.90	0.78	1.96	0.63	0.256	4.04	1.12
60 x 30	6	3	8	60	30	6	6	3.99	5.08	2.20	0.72	18.2	3.02	19.2	1.99	1.89	0.77	1.95	0.63	0.252	4.78	1.32
65 x 50	5	4	10	65	50	5	6	4.35	5.54	1.99	1.25	23.2	11.9	28.8	6.32	2.05	1.47	2.28	1.07	0.577	5.14	3.19
65 x 50	6	4	10	65	50	6	6	5.16	6.58	2.04	1.29	27.2	14.0	33.8	7.43	2.03	1.46	2.27	1.06	0.575	6.10	3.77
65 x 50	8	4	10	65	50	8	6	6.75	8.60	2.11	1.37	34.8	17.7	43.0	9.57	2.01	1.44	2.23	1.05	0.569	7.93	4.89
75 x 50	6	5	10	75	50	6	7	5.65	7.19	2.44	1.21	40.5	14.4	46.6	8.36	2.37	1.42	2.55	1.08	0.435	8.01	3.81
75 x 50	8	5	10	75	50	8	7	7.39	9.41	2.52	1.29	52.0	18.4	59.6	10.8	2.35	1.40	2.52	1.07	0.430	10.1	4.95
80 x 60	6	5	11	80	60	6	8	6.37	8.11	2.47	1.48	51.4	24.8	62.8	13.4	2.52	1.75	2.78	1.29	0.547	9.29	5.49
80 x 60	7	5	11	80	60	7	8	7.36	9.38	2.51	1.52	59.0	28.4	72.0	15.4	2.51	1.74	2.77	1.28	0.546	10.7	6.34
80 x 60	8	5	11	80	60	8	8	8.34	10.6	2.55	1.56	66.3	31.8	80.8	17.3	2.50	1.73	2.76	1.28	0.544	12.2	7.16
100 x 65	7	6	14	100	65	7	10	8.77	11.2	3.23	1.51	113	37.6	128	22.0	3.17	1.83	3.39	1.40	0.415	16.6	7.53
100 x 65	8	6	14	100	65	8	10	9.94	12.7	3.27	1.55	127	42.2	144	24.8	3.16	1.83	3.37	1.40	0.414	18.9	8.54
100 x 65	10	6	14	100	65	10	10	12.3	15.6	3.36	1.63	154	51.0	175	30.1	3.14	1.81	3.35	1.39	0.410	23.2	10.5
100 x 75	8	6	14	100	75	8	10	10.6	13.5	3.10	1.87	133	64.1	163	34.6	3.14	2.18	3.47	1.60	0.547	19.3	11.4
100 x 75	10	6	14	100	75	10	10	13.0	16.6	3.19	1.95	162	77.6	197	42.1	3.12	2.16	3.45	1.59	0.544	23.8	14.0
100 x 75	12	6	14	100	75	12	10	15.4	19.7	3.27	2.03	189	90.2	230	49.5	3.10	2.14	3.42	1.59	0.540	28.0	16.5
125 x 75	8	7	16	125	75	8	11	12.2	15.5	4.14	1.68	247	67.6	274	41.1	4.00	2.09	4.20	1.63	0.359	29.6	11.6
125 x 75	10	7	16	125	75	10	11	15.0	19.1	4.23	1.76	302	82.1	334	50.0	3.97	2.07	4.18	1.62	0.356	36.5	14.3
125 x 75	12	7	16	125	75	12	11	17.8	22.7	4.31	1.84	354	95.5	391	58.8	3.95	2.05	4.15	1.61	0.353	43.2	16.9

PROPERTIES OF STEEL SECTIONS

1053

UNEQUAL ANGLES B.S. 4848:Part 4:1972

DESIGNATION			DIMENSIONS				Mass/unit length	Area of section	Distance of centre of gravity		MOMENT OF INERTIA					RADIUS OF GYRATION				Angle	Elastic modulus	
Size	Thickness mm			Leg Lengths		Thickness t	Radius Root r_1				About X-X	About Y-Y	About U-U	About V-V		About X-X	About Y-Y	About U-U	About V-V	$\tan \alpha$	About X-X	About Y-Y
	Standard	Min.	Max.	A	B					$-x$	$-y$												
	mm	mm	mm	mm	mm	mm	mm	kg/m	cm²	cm	cm	cm⁴	cm⁴	cm⁴	cm⁴	cm	cm	cm	cm		cm³	cm³
150 × 75	10	8	19	150	75	10	11	17.0	21.6	5.32	1.61	501	85.8	532	55.3	4.81	1.99	4.96	1.60	0.261	51.8	14.6
150 × 75	12	8	19	150	75	12	11	20.2	25.7	5.41	1.69	589	99.9	624	64.9	4.79	1.97	4.93	1.59	0.259	61.4	17.2
150 × 75	15	8	19	150	75	15	11	24.8	31.6	5.53	1.81	713	120	754	78.8	4.75	1.94	4.88	1.58	0.254	75.3	21.0
150 × 90	10	8	19	150	90	10	12	18.2	23.2	5.00	2.04	533	146	591	88.3	4.80	2.51	5.05	1.95	0.360	53.3	21.0
150 × 90	12	8	19	150	90	12	12	21.6	27.5	5.08	2.12	627	171	694	104	4.77	2.49	5.02	1.94	0.358	63.3	24.8
150 × 90	15	8	19	150	90	15	12	26.6	33.9	5.21	2.23	761	205	841	126	4.74	2.46	4.98	1.93	0.354	77.7	30.4
200 × 100	10	10	23	200	100	10	15	23.0	29.2	6.93	2.01	1220	210	1290	135	6.46	2.68	6.65	2.15	0.263	93.2	26.3
200 × 100	12	10	23	200	100	12	15	27.3	34.8	7.03	2.10	1440	247	1530	159	6.43	2.67	6.63	2.14	0.262	111.0	31.3
200 × 100	15	10	23	200	100	15	15	33.7	43.0	7.16	2.22	1758	299	1863	194	6.40	2.64	6.58	2.13	0.259	137.0	38.4
200 × 150	12	11	23	200	150	12	15	32.0	40.8	6.08	3.61	1652	803	2024	431	6.36	4.44	7.04	3.25	0.552	119.0	70.5
200 × 150	15	11	23	200	150	15	15	39.6	50.5	6.21	3.73	2022	979	2475	527	6.33	4.40	7.00	3.23	0.550	147.0	86.9
200 × 150	18	11	23	200	150	18	15	47.1	60.0	6.33	3.85	2376	1146	2902	620	6.29	4.37	6.95	3.21	0.548	174.0	103

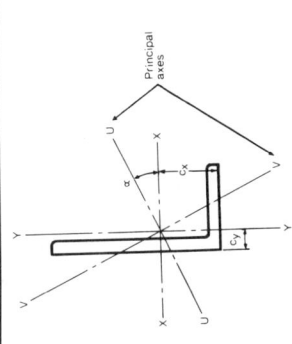

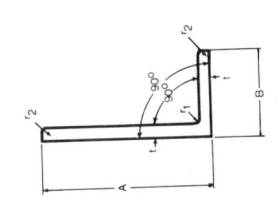

Note 1. Some of the thicknesses given in the tables are obtained by raising the rolls. (Practice in this respect is not uniform throughout the Division). In such cases the flanges will be slightly longer and the back of the toes will be slightly rounded.

Note 2. Angles should be ordered by flange length and thickness.

Note 3. Finished sections in which the angle between the flanges is not less than 89° and not more than 91° will be accepted as conforming to the standard.

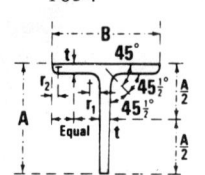

T—BARS

DIMENSIONS AND PROPERTIES

Designation			Width of Section B	Depth of Section A	Thick-ness t	Radius Root r_1	Radius Toe r_2	Area of Section
Nominal Size		Mass per metre						
mm		kg	mm	mm	mm	mm	mm	cm²
152 × 152		36	152.4	152.4	15.9	12.2	8.6	45.97
152 × 152		29	152.4	152.4	12.7	12.2	8.6	37.23
152 × 102		30	152.4	101.6	15.9	10.7	7.4	37.94
152 × 102		24	152.4	101.6	12.7	10.7	7.4	30.78
152 × 76		22	152.4	76.2	12.7	9.9	6.9	27.55
152 × 76		16	152.4	76.2	9.5	9.9	6.9	21.02
127 × 102		22	127.0	101.6	12.7	9.9	6.9	27.55
127 × 102		16	127.0	101.6	9.5	9.9	6.9	20.96
127 × 76		19	127.0	76.2	12.7	9.1	6.4	24.32
127 × 76		15	127.0	76.2	9.5	9.1	6.4	18.58
102 × 102		19	101.6	101.6	12.7	9.1	6.4	24.25
102 × 102		15	101.6	101.6	9.5	9.1	6.4	18.51
102 × 76		16	101.6	76.2	12.7	8.4	5.8	21.02
102 × 76		13	101.6	76.2	9.5	8.4	5.8	16.13
76 × 76		11	76.2	76.2	9.5	7.6	5.3	13.67
64 × 64		9	63.5	63.5	9.5	6.9	4.8	11.22
64 × 64		6	63.5	63.5	6.4	6.9	4.8	7.74
51 × 51		5	50.8	50.8	6.4	6.1	4.3	6.06
38 × 38		4	38.1	38.1	6.4	5.3	3.8	4.45

PROPERTIES OF STEEL SECTIONS 1055

T–BARS

DIMENSIONS AND PROPERTIES

Centre of Gravity c_x	Moment of Inertia Axis x–x	Moment of Inertia Axis y–y	Radius of Gyration Axis x–x	Radius of Gyration Axis y–y	Elastic Modulus Axis x–x	Elastic Modulus Axis y–y
cm	cm^4	cm^4	cm	cm	cm^3	cm^3
4.29	970.2	452.4	4.57	3.12	88.49	59.32
4.14	792.5	356.3	4.62	3.10	71.45	46.70
2.59	304.7	454.9	2.84	3.45	40.31	59.65
2.46	252.7	359.6	2.87	3.53	32.77	47.19
1.73	109.5	360.9	1.98	3.61	18.68	47.36
1.60	85.74	266.4	2.03	3.56	14.26	34.90
2.67	240.2	209.0	2.95	2.77	32.12	32.94
2.54	186.1	154.0	2.97	2.72	24.42	24.25
1.88	104.5	209.8	2.08	2.95	18.19	32.94
1.75	82.00	154.8	2.18	2.90	13.93	24.42
2.95	224.8	107.8	3.05	2.11	31.14	21.30
2.79	174.4	79.08	3.07	2.06	23.76	15.57
2.08	98.65	108.2	2.16	2.26	17.70	21.30
1.96	77.42	79.50	2.18	2.21	13.60	15.73
2.21	71.18	33.71	2.29	1.57	13.11	8.85
1.90	39.96	19.56	1.88	1.32	9.01	6.23
1.78	28.30	12.49	1.90	1.27	6.23	3.93
1.47	14.15	6.66	1.52	1.04	3.93	2.62
1.17	5.83	2.91	1.12	0.79	2.13	1.47

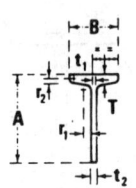

LONG STALK T–BARS

DIMENSIONS AND PROPERTIES

Designation Nominal Size mm	Mass per metre kg	Width of Section B mm	Depth of Section A mm	Thickness T mm	Thickness t_1 mm	Thickness t_2 mm	Root Radius r_1 mm	Toe Radius r_2 mm	Area of Section cm²
127 × 254	35.42	127.0	254.0	18.3	9.4	8.9	13.5	6.6	45.35
102 × 203	25.02	101.6	203.2	16.3	8.4	7.9	12.2	7.6	31.93
89 × 178	20.42	88.9	177.8	15.2	7.9	7.4	11.2	6.4	26.06
76 × 152	16.30	76.2	152.4	14.2	7.4	6.9	10.2	6.4	20.90
64 × 127	12.62	63.5	127.0	13.4	6.9	6.4	8.9	5.1	16.13
44 × 114	7.44	44.5	114.3	9.5	5.1	5.1	7.6	3.8	9.48
25 × 76	4.65	25.4	76.2	6.4	4.4	4.4	5.1	3.8	4.64

LONG STALK T–BARS

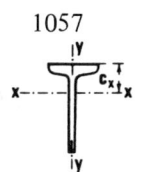

DIMENSIONS AND PROPERTIES

Centre of Gravity c_x cm	Moment of Inertia Axis x–x cm⁴	Moment of Inertia Axis y–y cm⁴	Radius of Gyration Axis x–x cm	Radius of Gyration Axis y–y cm	Elastic Modulus Axis x–x cm³	Elastic Modulus Axis y–y cm³
6.93	2811	273.0	7.85	2.46	153	42.9
5.84	1289	124.9	6.38	1.98	89.0	24.6
5.18	804.9	79.49	5.56	1.75	63.7	17.9
4.44	468.2	46.61	4.72	1.50	43.4	12.3
3.76	248.5	25.80	3.94	1.27	27.9	8.19
3.66	126.1	7.08	3.63	0.86	16.2	3.11
2.82	27.89	0.83	2.44	0.43	5.74	0.66

PROPERTIES OF STEEL SECTIONS

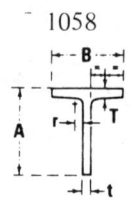

STRUCTURAL TEES

Cut from Universal Beams

DIMENSIONS AND PROPERTIES

Serial Size	Mass per metre	Width of Section B	Depth of Section A	Thickness Web t	Thickness Flange T	Root Radius r	Slope inside Flange	Area of Section
mm	kg	mm	mm	mm	mm	mm	per cent	cm²
305 × 457	127	305.5	459.2	17.3	27.9	19.1	5	161.2
305 × 457	112	304.1	455.2	15.9	23.9	19.1	5	142.5
305 × 457	101	303.4	451.5	15.2	20.2	19.1	5	128.0
292 × 419	113	293.8	425.5	16.1	26.8	17.8	5	144.2
292 × 419	97	292.4	420.4	14.7	21.7	17.8	5	123.4
292 × 419	88	291.6	417.4	14.0	18.8	17.8	5	111.9
267 × 381	99	268.0	384.8	15.6	25.4	16.5	5	125.3
267 × 381	87	266.7	381.0	14.3	21.6	16.5	5	110.1
267 × 381	74	265.3	376.9	12.9	17.5	16.5	5	93.9
254 × 343	85	255.8	346.5	14.5	23.7	15.2	5	108.2
254 × 343	76	254.5	343.8	13.2	21.0	15.2	5	96.8
254 × 343	70	253.7	341.8	12.4	19.0	15.2	5	89.2
254 × 343	63	253.0	339.0	11.7	16.2	15.2	5	79.7
305 × 305	119	311.5	316.5	18.6	31.4	16.5	5	151.8
305 × 305	90	307.0	308.7	14.1	23.6	16.5	5	113.8
305 × 305	75	304.8	304.8	11.9	19.7	16.5	5	94.9
229 × 305	70	230.1	308.5	13.1	22.1	12.7	5	89.1
229 × 305	63	229.0	305.9	11.9	19.6	12.7	5	79.7
229 × 305	57	228.2	303.7	11.2	17.3	12.7	5	72.2
229 × 305	51	227.6	301.1	10.6	14.8	12.7	5	64.5
178 × 305	46	178.4	301.2	10.6	15.0	12.7	0	57.9
178 × 305	41	177.8	299.1	10.1	12.8	12.7	0	52.2
330 × 267	106	333.6	272.5	16.7	27.8	16.5	5	134.8
330 × 267	95	331.7	269.7	14.9	25.0	16.5	5	120.6
330 × 267	84	330.2	266.7	13.4	22.0	16.5	5	106.3
210 × 267	61	211.9	272.3	12.8	21.3	12.7	5	77.8
210 × 267	55	210.7	269.7	11.6	18.8	12.7	5	69.2
210 × 267	51	210.1	268.4	10.9	17.4	12.7	5	64.6
210 × 267	46	209.3	266.6	10.2	15.6	12.7	5	58.8
210 × 267	41	208.7	264.2	9.6	13.2	12.7	5	52.1
165 × 267	37	165.6	264.4	9.3	13.5	12.7	0	46.5
165 × 267	33	165.1	262.4	8.8	11.5	12.7	0	41.8
191 × 229	49	192.8	233.7	11.4	19.6	10.2	5	62.6
191 × 229	45	192.0	231.8	10.6	17.7	10.2	5	56.9
191 × 229	41	191.3	230.1	9.9	16.0	10.2	5	52.2
191 × 229	37	190.5	228.6	9.1	14.5	10.2	5	47.4
191 × 229	34	189.9	226.8	8.5	12.7	10.2	5	42.7

These tables are based on Structural Tees cut from Universal Beams having the flange slope shown in the table. Structural Tees cut from Universal Beams with parallel flanges have properties approximately equal to the values given for tapered flange sections. A taper of 5% corresponds to a slope of 2° 52′.

PROPERTIES OF STEEL SECTIONS

STRUCTURAL TEES

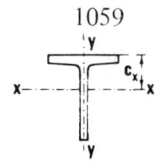

Cut from Universal Beams

DIMENSIONS AND PROPERTIES

Gravity Centre Distance C_x	Moment of Inertia Axis $x-x$	Moment of Inertia Axis $y-y$	Radius of Gyration Axis $x-x$	Radius of Gyration Axis $y-y$	Elastic Modulus Axis $x-x$ C_x	Elastic Modulus Axis $x-x$ E_x	Elastic Modulus Axis $y-y$	Cut from Universal Beam
cm	cm^4	cm^4	cm	cm	cm^3	cm^3	cm^3	mm × mm @ kg/m
12.03	32664	6256	14.2	6.23	2716	963.7	409.6	914 × 305 @ 253
12.16	29001	5212	14.3	6.05	2386	869.3	342.8	914 × 305 @ 224
12.56	26399	4316	14.4	5.81	2101	810.2	284.5	914 × 305 @ 201
10.84	24636	5331	13.1	6.08	2272	777.2	362.9	838 × 292 @ 226
11.11	21354	4192	13.2	5.83	1922	690.4	286.8	838 × 292 @ 194
11.39	19560	3555	13.2	5.64	1718	644.3	243.8	838 × 292 @ 176
9.91	17512	3850	11.8	5.54	1766	613.0	287.3	762 × 267 @ 197
10.01	15477	3188	11.9	5.38	1547	550.9	239.1	762 × 267 @ 173
10.20	13308	2501	11.9	5.16	1304	484.1	188.6	762 × 267 @ 147
8.69	12025	3113	10.5	5.36	1384	463.2	243.4	686 × 254 @ 170
8.61	10726	2695	10.5	5.28	1246	416.2	211.8	686 × 254 @ 152
8.66	9926	2395	10.5	5.18	1146	389.1	188.7	686 × 254 @ 140
8.88	8984	1996	10.6	5.00	1011	359.2	157.7	686 × 254 @ 125
7.12	12283	7487	9.00	7.02	1726	500.7	480.7	610 × 305 @ 238
6.66	8939	5285	8.86	6.81	1341	369.2	344.3	610 × 305 @ 179
6.45	7355	4236	8.80	6.68	1140	306.1	277.9	610 × 305 @ 149
7.62	7739	2126	9.32	4.88	1016	333.1	184.8	610 × 229 @ 140
7.56	6904	1838	9.31	4.80	913.7	299.7	160.5	610 × 229 @ 125
7.62	6288	1592	9.34	4.70	825.6	276.4	139.5	610 × 229 @ 113
7.82	5702	1329	9.40	4.54	729.6	255.7	116.8	610 × 229 @ 101
8.68	5351	713.5	9.61	3.51	616.3	249.6	80.0	610 × 178 @ 91
8.90	4848	601.3	9.64	3.39	544.7	230.8	67.6	610 × 178 @ 82
5.56	7381	8032	7.40	7.72	1329	340.2	481.6	533 × 330 @ 212
5.36	6484	7046	7.33	7.64	1209	300.0	424.8	533 × 330 @ 189
5.23	5678	6029	7.31	7.53	1085	264.9	365.2	533 × 330 @ 167
6.68	5178	1604	8.16	4.54	775.1	252.0	151.4	533 × 210 @ 122
6.61	4588	1377	8.14	4.46	694.5	225.3	130.7	533 × 210 @ 109
6.58	4277	1256	8.14	4.41	649.9	211.2	119.6	533 × 210 @ 101
6.58	3900	1106	8.14	4.34	593.0	194.2	105.7	533 × 210 @ 92
6.75	3511	912.8	8.21	4.18	520.3	178.5	87.5	533 × 210 @ 82
7.35	3258	513.6	8.37	3.32	443.0	170.7	62.0	533 × 165 @ 73
7.55	2949	431.5	8.40	3.21	390.6	157.8	52.3	533 × 165 @ 66
5.56	2976	1108	6.90	4.21	535.4	167.1	114.9	457 × 191 @ 98
5.50	2698	980.1	6.89	4.15	490.5	152.7	102.1	457 × 191 @ 89
5.49	2479	873.1	6.89	4.09	451.9	141.5	91.3	457 × 191 @ 82
5.43	2244	773.6	6.88	4.04	413.4	128.7	81.2	457 × 191 @ 74
5.48	2034	664.2	6.90	3.95	371.5	118.2	70.0	457 × 191 @ 67

PROPERTIES OF STEEL SECTIONS

STRUCTURAL TEES

Cut from Universal Beams

DIMENSIONS AND PROPERTIES

Serial Size mm	Mass per metre kg	Width of Section B mm	Depth of Section A mm	Thickness Web t mm	Thickness Flange T mm	Root Radius r mm	Slope inside Flange per cent	Area of Section cm²
152 × 229	41	153.5	232.5	10.7	18.9	10.2	5	52.2
152 × 229	37	152.7	230.6	9.9	17.0	10.2	5	47.4
152 × 229	34	151.9	228.6	9.1	15.0	10.2	5	42.7
152 × 229	30	152.9	227.3	8.0	13.3	10.2	0	38.0
152 × 229	26	152.4	224.9	7.6	10.9	10.2	0	33.2
178 × 203	37	179.7	206.4	9.7	16.0	10.2	5	47.4
178 × 203	34	178.8	204.7	8.8	14.3	10.2	5	42.7
178 × 203	30	177.8	203.2	7.8	12.8	10.2	5	38.0
178 × 203	27	177.6	201.3	7.6	10.9	10.2	5	34.2
152 × 203	37	153.7	208.2	10.1	18.1	10.2	5	47.4
152 × 203	34	152.9	206.1	9.3	16.0	10.2	5	42.7
152 × 203	30	152.2	204.0	8.6	13.9	10.2	5	37.9
140 × 203	23	142.4	201.2	6.9	11.2	10.2	5	29.4
140 × 203	20	141.8	198.6	6.3	8.6	10.2	5	24.7
152 × 191	34	154.3	194.3	9.7	16.3	10.2	5	42.7
152 × 191	30	153.4	192.4	8.7	14.4	10.2	5	38.0
152 × 191	26	152.4	190.5	7.8	12.4	10.2	5	33.2
171 × 178	34	173.2	182.0	9.1	15.7	10.2	5	42.7
171 × 178	29	172.1	179.3	8.0	13.0	10.2	5	36.0
171 × 178	26	171.5	177.8	7.3	11.5	10.2	5	32.2
171 × 178	23	171.0	176.0	6.9	9.7	10.2	5	28.4
127 × 178	20	126.0	176.4	6.5	10.7	10.2	5	24.7
127 × 178	17	125.4	174.2	5.9	8.5	10.2	5	20.9
165 × 152	27	166.8	155.4	7.7	13.7	8.9	5	34.2
165 × 152	23	165.7	153.5	6.7	11.8	8.9	5	29.4
165 × 152	20	165.1	151.9	6.1	10.2	8.9	5	25.7
127 × 152	24	125.2	155.2	8.9	14.0	8.9	5	30.4
127 × 152	21	124.3	153.3	8.0	12.1	8.9	5	26.6
127 × 152	19	123.5	151.9	7.2	10.7	8.9	5	23.7
102 × 152	17	102.4	156.3	6.6	10.8	7.6	2	20.9
102 × 152	14	101.9	154.4	6.1	8.9	7.6	2	18.1
102 × 152	13	101.6	152.4	5.8	6.8	7.6	2	15.7
146 × 127	22	147.3	129.8	7.3	12.7	7.6	5	27.5
146 × 127	19	146.4	128.0	6.4	10.9	7.6	5	23.7
146 × 127	16	146.1	125.7	6.1	8.6	7.6	5	20.0
102 × 127	14	102.1	130.2	6.4	10.0	7.6	2	18.1
102 × 127	13	101.9	128.5	6.1	8.4	7.6	2	16.1
102 × 127	11	101.6	127.0	5.8	6.8	7.6	2	14.2
133 × 102	15	133.8	103.4	6.3	9.6	7.6	5	19.0
133 × 102	13	133.4	101.6	5.8	7.8	7.6	5	16.1

These tables are based on Structural Tees cut from Universal Beams having the flange slope shown in the table. Structural Tees cut from Universal Beams with parallel flanges have properties approximately equal to the values given for tapered flange sections.
A taper of 5% corresponds to a slope of 2°52'.
A taper of 2% corresponds to a slope of 1°9'.

PROPERTIES OF STEEL SECTIONS

STRUCTURAL TEES

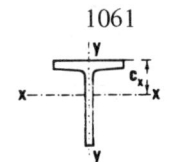

Cut from Universal Beams

DIMENSIONS AND PROPERTIES

Gravity Centre Distance C_x	Moment of Inertia Axis x–x	Moment of Inertia Axis y–y	Radius of Gyration Axis x–x	Radius of Gyration Axis y–y	Elastic Modulus Axis x–x C_x	Elastic Modulus Axis x–x E_x	Axis y–y	Cut from Universal Beam
cm	cm^4	cm^4	cm	cm	cm^3	cm^3	cm^3	mm × mm @ kg/m
6.03	2606	546.7	7.07	3.24	431.8	151.3	71.2	457 × 152 @ 82
5.99	2362	481.3	7.06	3.18	394.3	138.4	63.0	457 × 152 @ 74
5.99	2126	414.4	7.06	3.12	354.7	126.0	54.6	457 × 152 @ 67
5.82	1870	397.2	7.02	3.23	321.4	110.6	52.0	457 × 152 @ 60
6.03	1667	322.4	7.08	3.11	276.3	101.3	42.3	457 × 152 @ 52
4.81	1756	724.0	6.08	3.91	365.2	110.9	80.6	406 × 178 @ 74
4.74	1572	634.3	6.07	3.85	331.7	99.9	71.0	406 × 178 @ 67
4.62	1382	554.2	6.03	3.82	299.2	88.0	62.3	406 × 178 @ 60
4.82	1280	460.9	6.12	3.67	265.6	83.6	51.9	406 × 178 @ 54
5.14	1823	523.3	6.20	3.32	354.3	116.3	68.1	406 × 152 @ 74
5.11	1638	454.2	6.20	3.26	320.8	105.6	59.4	406 × 152 @ 67
5.12	1462	384.0	6.21	3.18	285.5	95.7	50.5	406 × 152 @ 60
5.06	1129	250.1	6.19	2.92	223.1	75.0	35.1	406 × 140 @ 46
5.29	966.4	186.5	6.26	2.75	182.8	66.3	26.3	406 × 140 @ 39
4.75	1427	473.4	5.78	3.33	300.8	97.2	61.4	381 × 152 @ 67
4.67	1261	407.0	5.76	3.27	269.8	86.6	53.1	381 × 152 @ 60
4.61	1097	342.7	5.75	3.21	238.1	76.0	45.0	381 × 152 @ 52
4.01	1157	639.2	5.21	3.87	288.3	81.6	73.8	356 × 171 @ 67
3.96	977.9	513.0	5.21	3.77	247.2	70.0	59.6	356 × 171 @ 57
3.93	876.6	442.7	5.21	3.71	222.9	63.3	51.6	356 × 171 @ 51
4.03	790.3	365.1	5.27	3.58	196.0	58.2	42.7	356 × 171 @ 45
4.42	718.5	166.5	5.40	2.60	162.6	54.4	26.4	356 × 127 @ 39
4.54	617.2	128.5	5.44	2.48	136.0	47.9	20.5	356 × 127 @ 33
3.20	635.8	494.1	4.31	3.80	198.9	51.5	59.3	305 × 165 @ 54
3.09	540.3	412.4	4.29	3.74	174.7	44.1	49.8	305 × 165 @ 46
3.07	475.4	345.5	4.30	3.67	154.7	39.2	41.9	305 × 165 @ 40
3.92	653.1	219.0	4.64	2.68	166.7	56.3	35.0	305 × 127 @ 48
3.86	567.2	183.3	4.62	2.63	147.1	49.4	29.5	305 × 127 @ 42
3.81	503.8	157.9	4.61	2.58	132.2	44.3	25.6	305 × 127 @ 37
4.15	486.5	94.68	4.83	2.13	117.3	42.4	18.5	305 × 102 @ 33
4.23	426.5	76.41	4.85	2.05	100.8	38.0	15.0	305 × 102 @ 28
4.48	375.4	58.05	4.89	1.92	83.8	34.9	11.4	305 × 102 @ 25
2.67	348.8	316.6	3.56	3.39	130.5	33.8	43.0	254 × 146 @ 43
2.58	296.4	263.8	3.54	3.34	114.8	29.0	36.1	254 × 146 @ 37
2.69	262.7	202.8	3.63	3.19	97.7	26.6	27.8	254 × 146 @ 31
3.26	278.7	87.12	3.93	2.19	85.6	28.6	17.1	254 × 102 @ 28
3.36	252.5	71.89	3.96	2.11	75.2	26.6	14.1	254 × 102 @ 25
3.49	227.1	58.00	4.00	2.02	65.1	24.7	11.4	254 × 102 @ 22
2.10	152.4	176.8	2.83	3.05	72.5	18.5	26.4	203 × 133 @ 30
2.13	133.5	139.8	2.88	2.94	62.7	16.6	21.0	203 × 133 @ 25

PROPERTIES OF STEEL SECTIONS

STRUCTURAL TEES

Cut from Universal Columns

DIMENSIONS AND PROPERTIES

Serial Size	Mass per metre	Width of Section B	Depth of Section A	Thickness Web t	Thickness Flange T	Root Radius r	Slope inside Flange	Area of Section
mm	kg	mm	mm	mm	mm	mm	per cent	cm²
406 × 178	118	395.0	190.5	18.5	30.2	15.2	0	149.9
368 × 178	101	374.4	187.3	16.8	27.0	15.2	0	129.0
368 × 178	89	372.1	184.2	14.5	23.8	15.2	0	112.9
368 × 178	77	370.2	181.0	12.6	20.7	15.2	0	97.6
368 × 178	65	368.3	177.8	10.7	17.5	15.2	0	82.5
305 × 152	79	310.6	163.6	15.7	25.0	15.2	0	100.6
305 × 152	69	308.7	160.3	13.8	21.7	15.2	0	87.3
305 × 152	59	306.8	157.2	11.9	18.7	15.2	0	74.9
305 × 152	49	304.8	153.9	9.9	15.4	15.2	0	61.6
254 × 127	66	261.0	138.2	15.6	25.1	12.7	0	83.9
254 × 127	54	258.3	133.4	13.0	20.5	12.7	0	68.3
254 × 127	45	255.9	130.2	10.5	17.3	12.7	0	57.0
254 × 127	37	254.0	127.0	8.6	14.2	12.7	0	46.4
203 × 102	43	208.8	111.1	13.0	20.5	10.2	0	55.0
203 × 102	36	206.2	108.0	10.3	17.3	10.2	0	45.5
203 × 102	30	205.2	104.8	9.3	14.2	10.2	0	37.9
203 × 102	26	203.9	103.1	8.0	12.5	10.2	0	33.2
203 × 102	23	203.2	101.6	7.3	11.0	10.2	0	29.4
152 × 76	19	154.4	80.9	8.1	11.5	7.6	0	23.7
152 × 76	15	152.9	78.7	6.6	9.4	7.6	0	19.1
152 × 76	12	152.4	76.2	6.1	6.8	7.6	0	14.9

PROPERTIES OF STEEL SECTIONS
STRUCTURAL TEES
Cut from Universal Columns

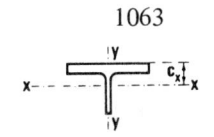

DIMENSIONS AND PROPERTIES

Gravity Centre Distance C_x	Moment of Inertia Axis x–x	Moment of Inertia Axis y–y	Radius of Gyration Axis x–x	Radius of Gyration Axis y–y	Elastic Modulus Axis x–x C_x	Elastic Modulus Axis x–x E_x	Elastic Modulus Axis y–y	Cut from Universal Column
cm	cm^4	cm^4	cm	cm	cm^3	cm^3	cm^3	mm × mm @ kg/m
3.41	2886	15504	4.39	10.2	846.3	184.5	785.1	356 × 406 @ 235
3.32	2500	11816	4.40	9.57	754.1	162.1	631.2	356 × 368 @ 202
3.10	2099	10235	4.31	9.52	677.4	137.0	550.1	356 × 368 @ 177
2.92	1765	8735	4.25	9.46	605.6	116.3	471.9	356 × 368 @ 153
2.73	1451	7278	4.20	9.39	531.1	96.4	395.2	356 × 368 @ 129
3.04	1529	6262	3.90	7.89	502.9	114.9	403.2	305 × 305 @ 158
2.86	1291	5336	3.85	7.82	450.9	98.1	345.7	305 × 305 @ 137
2.69	1075	4503	3.79	7.75	399.9	82.5	293.5	305 × 305 @ 118
2.50	858.1	3634	3.73	7.68	343.1	66.6	238.5	305 × 305 @ 97
2.72	886.7	3722	3.25	6.66	325.6	79.9	285.2	254 × 254 @ 132
2.47	683.0	2951	3.16	6.57	277.0	62.8	228.5	254 × 254 @ 107
2.24	534.8	2424	3.06	6.52	238.7	49.6	189.5	254 × 254 @ 89
2.06	418.7	1936	3.00	6.46	203.7	39.3	152.5	254 × 254 @ 73
2.22	379.8	1560	2.63	5.32	171.0	42.7	149.4	203 × 203 @ 86
1.98	288.1	1268	2.52	5.28	145.3	32.7	123.0	203 × 203 @ 71
1.88	241.2	1021	2.52	5.19	128.3	28.1	99.5	203 × 203 @ 60
1.76	203.1	885.0	2.47	5.16	115.2	23.8	86.8	203 × 203 @ 52
1.71	179.6	769.4	2.47	5.11	105.3	21.2	75.7	203 × 203 @ 46
1.55	94.67	354.3	2.00	3.87	61.2	14.5	45.9	152 × 152 @ 37
1.41	72.79	279.2	1.95	3.82	51.5	11.3	36.5	152 × 152 @ 30
1.43	61.11	201.7	2.03	3.68	42.7	9.88	26.5	152 × 152 @ 23

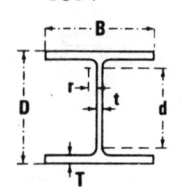

UNIVERSAL BEARING PILES
Parallel Flanges
DIMENSIONS AND PROPERTIES

Serial Size	Mass per metre	Depth of Section D	Width of Section B	Thickness Web t	Thickness Flange T	Root Radius r	Depth between Fillets d	Area of Section
mm	kg	mm	mm	mm	mm	mm	mm	cm²
356 × 368	174	361.5	378.1	20.4	20.4	15.2	290.1	222.2
	152	356.4	375.5	17.9	17.9	15.2	290.1	193.6
	133	351.9	373.3	15.6	15.6	15.2	290.1	169.0
	109	346.4	370.5	12.9	12.9	15.2	290.1	138.4
305 × 305	110	307.9	310.3	15.4	15.4	15.2	246.6	140.4
	79	299.2	306.0	11.1	11.1	15.2	246.6	100.4
254 × 254	85	254.3	259.7	14.3	14.3	12.7	200.2	108.1
	63	246.9	256.0	10.6	10.6	12.7	200.2	79.7
203 × 203	54	203.9	207.2	11.3	11.3	10.2	160.8	68.4

PROPERTIES OF STEEL SECTIONS 1065

UNIVERSAL BEARING PILES
Parallel Flanges
DIMENSIONS AND PROPERTIES

Serial Size mm	Moment of Inertia Axis x–x Gross cm⁴	Moment of Inertia Axis x–x Net cm⁴	Moment of Inertia Axis y–y cm⁴	Radius of Gyration Axis x–x cm	Radius of Gyration Axis y–y cm	Elastic Modulus Axis x–x cm³	Elastic Modulus Axis y–y cm³	Ratio $\dfrac{D}{T}$
356 × 368	51134	44954	18444	15.2	9.11	2829	975.7	17.7
	43916	38578	15799	15.1	9.03	2464	841.5	19.9
	37840	33248	13576	15.0	8.96	2150	727.4	22.6
	30515	26784	10901	14.8	8.87	1762	588.5	26.9
305 × 305	23580	21865	7689	13.0	7.40	1532	495.6	20.0
	16400	15201	5292	12.8	7.26	1096	345.9	27.0
254 × 254	12264	11192	4188	10.7	6.22	964.5	322.6	17.8
	8775	8005	2971	10.5	6.11	710.9	232.2	23.3
203 × 203	4987	4441	1683	8.54	4.96	489.2	162.4	18.0

Note: One hole is deducted from each flange under 300 mm wide (serial size) and two holes from each flange 300 mm and over (serial size), in calculating the Net Moment of Inertia about x–x.

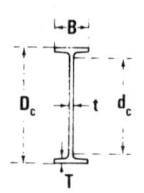

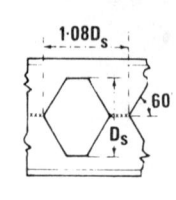

CASTELLATED UNIVERSAL BEAMS

DIMENSIONS AND PROPERTIES

Serial Size Original	Serial Size Castellated	Mass per metre	Depth of Section Dc	Width of Section B	Thickness Web t	Thickness Flange T	Depth between Fillets dc	Area of Section Gross	Area of Section Net
mm	mm	kg	mm	mm	mm	mm	mm	cm²	cm²
914 × 419	1371 × 419	388	1377.5	420.5	21.5	36.6	1248.5	592.2	395.7
		343	1368.4	418.5	19.4	32.0	1248.5	525.6	348.3
914 × 305	1371 × 305	289	1383.6	307.8	19.6	32.0	1276.2	458.0	278.9
		253	1375.5	305.5	17.3	27.9	1276.2	401.5	243.4
		224	1367.3	304.1	15.9	23.9	1276.2	357.6	212.3
		201	1360.0	303.4	15.2	20.2	1276.2	325.6	186.6
838 × 292	1257 × 292	226	1269.9	293.8	16.1	26.8	1175.4	355.9	221.0
		194	1259.7	292.4	14.7	21.7	1175.4	308.5	185.3
		176	1253.9	291.6	14.0	18.8	1175.4	282.5	165.2
762 × 267	1143 × 267	197	1150.6	268.0	15.6	25.4	1062.2	310.0	191.1
		173	1143.0	266.7	14.3	21.6	1062.2	274.7	165.7
		147	1134.9	265.3	12.9	17.5	1062.2	237.0	138.7
686 × 254	1029 × 254	170	1035.9	255.8	14.5	23.7	953.6	266.1	166.6
		152	1030.6	254.5	13.2	21.0	953.6	238.9	148.3
		140	1026.5	253.7	12.4	19.0	953.6	220.9	135.9
		125	1020.9	253.0	11.7	16.2	953.6	199.5	119.3
610 × 305	915 × 305	238	938.0	311.5	18.6	31.4	836.6	360.3	246.8
		179	922.5	307.0	14.1	23.6	836.6	270.7	184.7
		149	914.6	304.8	11.9	19.7	836.6	226.2	153.6
610 × 229	915 × 229	140	922.0	230.1	13.1	22.1	848.1	218.2	138.3
		125	916.9	229.0	11.9	19.6	848.1	195.7	123.1
		113	912.3	228.2	11.2	17.3	848.1	178.5	110.1
		101	907.2	227.6	10.6	14.8	848.1	161.4	96.7
610 × 178	915 × 178	91	907.5	178.4	10.6	15.0	852.1	148.2	83.5
		82	903.2	177.8	10.1	12.8	852.1	135.2	73.6
533 × 330	800 × 330	212	811.6	333.6	16.7	27.8	716.6	314.1	225.1
		189	806.0	331.7	14.9	25.0	716.6	280.9	201.5
		167	799.9	330.2	13.4	22.0	716.6	248.4	177.0
533 × 210	800 × 210	122	811.1	211.9	12.8	21.3	739.2	189.8	121.5
		109	806.0	210.7	11.6	18.8	739.2	169.4	107.5
		101	803.2	210.1	10.9	17.4	739.2	158.2	100.1
		92	799.6	209.3	10.2	15.6	739.2	144.8	90.4
		82	794.8	208.7	9.6	13.2	739.2	129.9	78.7
533 × 165	800 × 165	73	795.3	165.6	9.3	13.5	743.0	117.8	68.3
		66	791.3	165.1	8.8	11.5	743.0	107.0	60.1
457 × 191	686 × 191	98	695.9	192.8	11.4	19.6	632.9	151.2	99.1
		89	692.1	192.0	10.6	17.7	632.9	138.0	89.6
		82	688.7	191.3	9.9	16.0	632.9	127.1	81.8
		74	685.7	190.5	9.1	14.5	632.9	115.7	74.1
		67	682.1	189.9	8.5	12.7	632.9	104.8	65.9

The overall depth, **Dc**, of the castellated section = $D + D_s/2$
where **D** = actual depth of original section
and **Ds** = serial depth of original section

PROPERTIES OF STEEL SECTIONS

CASTELLATED UNIVERSAL BEAMS

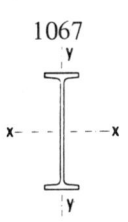

DIMENSIONS AND PROPERTIES

Serial Size	Moment of Inertia (Net) Axis x–x	Moment of Inertia (Net) Axis y–y	Design Radius of Gyration Axis x–x	Design Radius of Gyration Axis y–y	Elastic Modulus (Net) Axis x–x	Elastic Modulus (Net) Axis y–y	Pitch of Standard Castellation 1.08 Ds	Ratio Dc / T
mm	cm^4	cm^4	cm	cm	cm^3	cm^3	mm	
1371 × 419	1661103	42443	59.95	9.42	24118	2019	987.1	37.6
	1449837	36223	59.62	9.25	21190	1731	987.1	42.8
1371 × 305	1161303	14765	58.76	6.48	16787	959	987.1	43.2
	1007335	12493	58.54	6.38	14647	818	987.1	49.3
	869734	10409	58.06	6.20	12722	685	987.1	57.2
	755158	8619	57.38	5.97	11105	568	987.1	67.3
1257 × 292	779766	10647	54.26	6.21	12281	725	905.0	47.4
	643992	8373	53.57	5.97	10225	573	905.0	58.1
	568302	7101	53.07	5.79	9065	487	905.0	66.7
1143 × 267	552705	7687	49.08	5.67	9607	574	823.0	45.3
	474232	6366	48.65	5.51	8298	477	823.0	52.9
	391781	4995	48.11	5.30	6904	377	823.0	64.9
1029 × 254	391973	6216	44.38	5.47	7568	486	740.9	43.7
	347016	5384	44.19	5.39	6734	423	740.9	49.1
	315378	4784	43.95	5.30	6145	377	740.9	54.0
	273808	3987	43.51	5.13	5364	315	740.9	63.0
915 × 305	474179	14957	40.72	7.12	10110	960	658.8	29.9
	349028	10563	40.37	6.91	7567	688	658.8	39.1
	287953	8467	40.18	6.77	6297	556	658.8	46.4
915 × 229	257736	4247	39.58	4.98	5591	369	658.8	41.7
	227784	3672	39.39	4.90	4969	321	658.8	46.8
	202408	3181	39.13	4.80	4437	279	658.8	52.7
	175762	2655	38.73	4.65	3875	233	658.8	61.3
915 × 178	149144	1424	38.02	3.62	3287	160	658.8	60.5
	130248	1200	37.65	3.51	2884	135	658.8	70.6
800 × 330	326980	16054	35.70	7.80	8058	962	575.6	29.2
	290902	14085	35.60	7.72	7218	849	575.6	32.2
	253598	12052	35.43	7.61	6341	730	575.6	36.4
800 × 210	174817	3203	34.83	4.62	4311	302	575.6	38.1
	153530	2751	34.65	4.55	3810	261	575.6	42.9
	142133	2509	34.54	4.50	3539	239	575.6	46.2
	127785	2209	34.38	4.43	3196	211	575.6	51.3
	109944	1824	34.02	4.28	2767	175	575.6	60.2
800 × 165	93996	1025	33.53	3.42	2364	124	575.6	58.9
	81825	861	33.18	3.31	2068	104	575.6	68.8
686 × 191	104993	2213	30.01	4.28	3017	230	493.6	35.5
	94405	1958	29.88	4.22	2728	204	493.6	39.1
	85606	1744	29.74	4.16	2486	182	493.6	43.0
	77175	1546	29.64	4.11	2251	162	493.6	47.3
	68119	1327	29.44	4.02	1997	140	493.6	53.7

Design Radius of Gyration is the average between the values for the gross and net section. These tables are based on Universal Beams with tapered flanges.

PROPERTIES OF STEEL SECTIONS

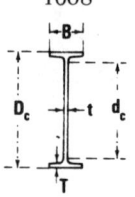

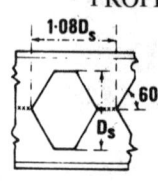

CASTELLATED UNIVERSAL BEAMS

DIMENSIONS AND PROPERTIES

Serial Size Original	Serial Size Castellated	Mass per metre	Depth of Section Dc	Width of Section B	Thickness Web t	Thickness Flange T	Depth between Fillets dc	Area of Section Gross	Area of Section Net
mm	mm	kg	mm	mm	mm	mm	mm	cm²	cm²
457 × 152	686 × 152	82	693.6	153.5	10.7	18.9	632.9	128.8	79.9
		74	689.8	152.7	9.9	17.0	632.9	117.5	72.3
		67	685.7	151.9	9.1	15.0	632.9	106.1	64.5
		60	683.2	152.9	8.0	13.3	636.2	94.2	57.7
		52	678.3	152.4	7.6	10.9	636.2	83.9	49.1
406 × 178	609 × 178	74	615.8	179.7	9.7	16.0	560.4	114.6	75.2
		67	612.4	178.8	8.8	14.3	560.4	103.3	67.5
		60	609.4	177.8	7.8	12.8	560.4	91.9	60.3
		54	605.6	177.6	7.6	10.9	560.4	83.8	52.9
406 × 152	609 × 152	74	619.3	153.7	10.1	18.1	560.4	115.3	74.3
		67	615.2	152.9	9.3	16.0	560.4	104.2	66.5
		60	610.9	152.2	8.6	13.9	560.4	93.3	58.4
406 × 140	609 × 140	46	605.3	142.4	6.9	11.2	560.4	72.9	44.9
		39	600.3	141.8	6.3	8.6	560.4	62.1	36.5
381 × 152	572 × 152	67	579.1	154.3	9.7	16.3	523.7	103.9	66.9
		60	575.3	153.4	8.7	14.4	523.7	92.5	59.4
		52	571.5	152.4	7.8	12.4	523.7	81.3	51.5
356 × 171	534 × 171	67	542.0	173.2	9.1	15.7	487.1	101.5	69.1
		57	536.6	172.1	8.0	13.0	487.1	86.3	57.8
		51	533.6	171.5	7.3	11.5	487.1	77.5	51.5
		45	530.0	171.0	6.9	9.7	487.1	69.2	44.6
356 × 127	534 × 127	39	530.8	126.0	6.5	10.7	487.1	60.9	37.7
		33	526.5	125.4	5.9	8.5	487.1	52.2	31.2
305 × 165	458 × 165	54	463.4	166.8	7.7	13.7	415.1	80.1	56.6
		46	459.6	165.7	6.7	11.8	415.1	69.1	48.6
		40	456.3	165.1	6.1	10.2	415.1	60.7	42.1
305 × 127	458 × 127	48	462.9	125.2	8.9	14.0	415.1	74.3	47.2
		42	459.1	124.3	8.0	12.1	415.1	65.3	40.9
		37	456.3	123.5	7.2	10.7	415.1	58.4	36.4
305 × 102	458 × 102	33	465.2	102.4	6.6	10.8	427.8	51.8	31.7
		28	461.4	101.9	6.1	8.9	427.8	45.6	27.0
		25	457.3	101.6	5.8	6.8	427.8	40.2	22.5
254 × 146	381 × 146	43	386.6	147.3	7.3	12.7	343.2	64.3	45.8
		37	383.0	146.4	6.4	10.9	343.2	55.5	39.3
		31	378.5	146.1	6.1	8.6	343.2	47.7	32.2
254 × 102	381 × 102	28	387.4	102.1	6.4	10.0	351.5	44.3	28.1
		25	384.0	101.9	6.1	8.4	351.5	39.9	24.4
		22	381.0	101.6	5.8	6.8	351.5	35.8	21.0
203 × 133	305 × 133	30	308.3	133.8	6.3	9.6	271.4	44.4	31.6
		25	304.7	133.4	5.8	7.8	271.4	38.2	26.4

The overall depth, Dc, of the castellated section = D + Ds/2
where D = actual depth of original section
and Ds = serial depth of original section.

PROPERTIES OF STEEL SECTIONS

CASTELLATED UNIVERSAL BEAMS

DIMENSIONS AND PROPERTIES

Serial Size	Moment of Inertia (Net) Axis x–x	Moment of Inertia (Net) Axis y–y	Design Radius of Gyration Axis x–x	Design Radius of Gyration Axis y–y	Elastic Modulus (Net) Axis x–x	Elastic Modulus (Net) Axis y–y	Pitch of Standard Castellation 1.08 Ds	Ratio Dc / T
mm	cm⁴	cm⁴	cm	cm	cm³	cm³	mm	
686 × 152	83405	1091	29.51	3.31	2405	142	493.6	36.7
	74893	961	29.36	3.26	2171	126	493.6	40.6
	66207	827	29.17	3.19	1931	109	493.6	45.7
	59154	794	29.20	3.31	1732	104	493.6	51.4
	49742	644	28.81	3.20	1467	85	493.6	62.2
609 × 178	62799	1446	26.65	3.97	2040	161	438.5	38.5
	56053	1268	26.55	3.92	1831	142	438.5	42.8
	49804	1108	26.51	3.88	1635	125	438.5	47.6
	43142	921	26.17	3.75	1425	104	438.5	55.6
609 × 152	61953	1045	26.54	3.38	2001	136	438.5	34.2
	54919	907	26.38	3.32	1785	119	438.5	38.5
	47753	767	26.17	3.25	1563	101	438.5	43.9
609 × 140	36292	500	25.96	2.98	1199	70	438.5	54.0
	29003	373	25.53	2.82	966	53	438.5	69.8
572 × 152	49011	945	24.87	3.39	1693	123	411.5	35.5
	43080	813	24.75	3.33	1498	106	411.5	40.0
	37194	685	24.63	3.28	1302	90	411.5	46.1
534 × 171	44793	1277	23.62	3.92	1653	147	384.5	34.5
	37049	1025	23.42	3.83	1381	119	384.5	41.3
	32718	885	23.30	3.76	1226	103	384.5	46.4
	28024	730	23.06	3.65	1057	85	384.5	54.6
534 × 127	23424	333	22.76	2.65	883	53	384.5	49.6
	19108	257	22.47	2.54	726	41	384.5	61.9
458 × 165	26927	988	20.38	3.85	1162	118	329.4	33.8
	22959	824	20.29	3.79	999	100	329.4	38.9
	19730	691	20.16	3.71	865	84	329.4	44.7
458 × 127	21926	437	19.77	2.74	947	70	329.4	33.1
	18847	366	19.64	2.68	821	59	329.4	37.9
	16633	315	19.55	2.63	729	51	329.4	42.6
458 × 102	14857	189	19.73	2.18	639	37	329.4	43.1
	12467	153	19.48	2.11	540	30	329.4	51.8
	10151	116	19.08	1.98	444	23	329.4	67.3
381 × 146	15099	633	16.99	3.43	781	86	274.3	30.4
	12844	527	16.90	3.37	671	72	274.3	35.1
	10322	405	16.60	3.23	545	55	274.3	44.0
381 × 102	9183	174	16.58	2.24	474	34	274.3	38.7
	7849	144	16.34	2.16	409	28	274.3	45.7
	6637	116	16.09	2.07	348	23	274.3	56.0
305 × 133	6646	353	13.58	3.09	431	53	219.2	32.1
	5455	279	13.39	2.98	358	42	219.2	39.1

Design Radius of Gyration is the average between the values for the gross and net section.

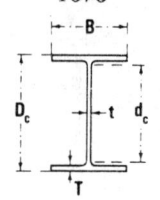

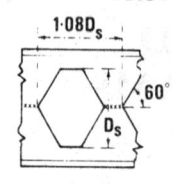

CASTELLATED UNIVERSAL COLUMNS

DIMENSIONS AND PROPERTIES

Serial Size Original (mm)	Serial Size Castellated (mm)	Mass per metre (kg)	Depth of Section Dc (mm)	Width of Section B (mm)	Thickness Web t (mm)	Thickness Flange T (mm)	Depth between Fillets dc (mm)	Area of Section Gross (cm²)	Area of Section Net (cm²)
356 × 406	546 × 406	634	665.2	424.1	47.6	77.0	480.6	892.8	723.3
		551	646.2	418.5	42.0	67.5	480.6	776.6	627.1
		467	627.1	412.4	35.9	58.0	480.6	659.4	531.6
		393	609.6	407.0	30.6	49.2	480.6	555.3	446.4
		340	596.9	403.0	26.5	42.9	480.6	479.8	385.5
		287	584.2	399.0	22.6	36.5	480.6	406.2	325.8
		235	571.5	395.0	18.5	30.2	480.6	332.7	266.9
Column Core	559 × 424	477	630.0	424.4	48.0	53.2	493.1	704.6	509.8
356 × 368	534 × 368	202	552.7	374.4	16.8	27.0	468.1	287.8	228.0
		177	546.3	372.1	14.5	23.8	468.1	251.5	199.9
		153	540.0	370.2	12.6	20.7	468.1	217.6	172.8
		129	533.6	368.3	10.7	17.5	468.1	184.0	145.9
305 × 305	458 × 305	283	517.8	321.8	26.9	44.1	399.1	401.4	319.4
		240	505.1	317.9	23.0	37.7	399.1	340.7	270.6
		198	492.4	314.1	19.2	31.4	399.1	281.6	223.1
		158	479.7	310.6	15.7	25.0	399.1	225.2	177.3
		137	473.0	308.7	13.8	21.7	399.1	195.7	153.6
		118	467.0	306.8	11.9	18.7	399.1	167.9	131.6
		97	460.3	304.8	9.9	15.4	399.1	138.4	108.2
254 × 254	381 × 254	167	416.1	264.5	19.2	31.7	327.2	236.8	188.0
		132	403.4	261.0	15.6	25.1	327.2	187.5	147.9
		107	393.7	258.3	13.0	20.5	327.2	153.2	120.1
		89	387.4	255.9	10.5	17.3	327.2	127.3	100.6
		73	381.0	254.0	8.6	14.2	327.2	103.8	82.0
203 × 203	305 × 203	86	323.8	208.8	13.0	20.5	262.3	123.3	96.9
		71	317.4	206.2	10.3	17.3	262.3	101.5	80.6
		60	311.1	205.2	9.3	14.2	262.3	85.3	66.4
		52	307.7	203.9	8.0	12.5	262.3	74.6	58.3
		46	304.7	203.2	7.3	11.0	262.3	66.3	51.4
152 × 152	228 × 152	37	237.8	154.4	8.1	11.5	199.4	53.6	41.2
		30	233.5	152.9	6.6	9.4	199.4	43.3	33.2
		23	228.4	152.4	6.1	6.8	199.4	34.4	25.2

The overall depth, D_c, of the castellated section = $D + D_s/2$
where D = actual depth of original section
and D_s = serial depth of original section, except for the 356 × 406 Series, and the Column Core Section.

PROPERTIES OF STEEL SECTIONS

CASTELLATED UNIVERSAL COLUMNS

DIMENSIONS AND PROPERTIES

Serial Size mm	Moment of Inertia (Net) Axis x–x cm⁴	Moment of Inertia (Net) Axis y–y cm⁴	Design Radius of Gyration Axis x–x cm	Design Radius of Gyration Axis y–y cm	Elastic Modulus (Net) Axis x–x cm³	Elastic Modulus (Net) Axis y–y cm³	Pitch of Standard Castellation 1.08 D_s mm	Ratio $\dfrac{D_c}{T}$
546 × 406	577088	98051	27.03	11.07	17351	4624	411.5	8.6
	483788	82555	26.57	10.90	14973	3945	411.5	9.6
	396642	67836	26.13	10.72	12650	3290	411.5	10.8
	322847	55368	25.72	10.56	10592	2721	411.5	12.4
	272525	46788	25.43	10.45	9131	2322	411.5	13.9
	225110	38697	25.13	10.33	7707	1940	411.5	16.0
	180239	30999	24.85	10.22	6308	1570	411.5	18.9
559 × 424	408431	67869	26.58	10.69	12966	3198	438.5	11.8
534 × 368	151992	23625	24.64	9.62	5500	1262	384.5	20.5
	131805	20466	24.52	9.57	4825	1100	384.5	23.0
	112595	17467	24.37	9.51	4170	944	384.5	26.1
	93964	14554	24.23	9.44	3522	790	384.5	30.5
458 × 305	171956	24520	22.14	8.29	6642	1524	329.4	11.7
	141928	20224	21.85	8.18	5620	1272	329.4	13.4
	113936	16221	21.56	8.06	4628	1033	329.4	15.7
	88053	12519	21.24	7.93	3671	806	329.4	19.2
	75184	10669	21.08	7.86	3179	691	329.4	21.8
	63624	9004	20.94	7.80	2725	587	329.4	25.0
	51555	7267	20.78	7.72	2240	477	329.4	29.9
381 × 254	66885	9789	18.00	6.83	3215	740	274.3	13.1
	50944	7440	17.69	6.70	2526	570	274.3	16.1
	40329	5899	17.45	6.61	2049	457	274.3	19.2
	33246	4847	17.34	6.56	1716	379	274.3	22.4
	26631	3872	17.20	6.49	1398	305	274.3	26.8
305 × 203	21439	3118	14.17	5.35	1324	299	219.2	15.8
	17509	2535	14.07	5.30	1103	246	219.2	18.3
	14092	2041	13.86	5.22	906	199	219.2	21.9
	12253	1770	13.80	5.19	796	174	219.2	24.6
	10683	1539	13.71	5.14	701	151	219.2	27.7
228 × 152	5052	708	10.50	3.89	425	92	164.2	20.7
	4003	558	10.41	3.85	343	73	164.2	24.8
	2937	403	10.16	3.71	257	53	164.2	33.6

Design Radius of Gyration is the average between the values for the gross and net section

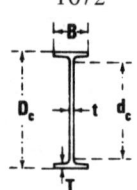

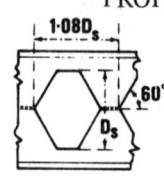

CASTELLATED JOISTS

DIMENSIONS AND PROPERTIES

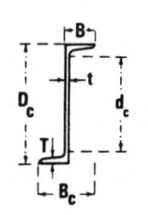

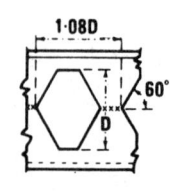

Nominal Size Original	Nominal Size Castellated	Mass per metre	Depth of Section D_c	Width of Section B	Thickness Web t	Thickness Flange T	Depth between Fillets d_c	Area of Section Gross	Area of Section Net
mm	mm	kg	mm	mm	mm	mm	mm	cm²	cm²
203 × 102	305 × 102	25.33	304.7	101.6	5.8	10.4	262.5	38.2	26.4
178 × 102	267 × 102	21.54	266.8	101.6	5.3	9.0	227.2	32.2	22.7
152 × 89	228 × 89	17.09	228.4	88.9	4.9	8.3	193.9	25.5	18.0
127 × 76	191 × 76	13.36	190.5	76.2	4.5	7.6	157.7	19.9	14.2
102 × 64	153 × 64	9.65	152.6	63.5	4.1	6.6	124.2	14.4	10.2
76 × 51	114 × 51	6.67	114.2	50.8	3.8	5.6	88.3	9.9	7.1

CASTELLATED ZED BEAMS

DIMENSIONS AND PROPERTIES

Original Channel Section D×B	Castellated Zed Section $D_c × B_c$	Mass per metre	Thickness Web	Thickness Flange	Depth between Fillets d_c	Area of Section Gross	Area of Section Net	Angle α (Net Section) tan α	Ratio $\dfrac{D_c}{T}$
mm	mm	kg	mm	mm	mm	cm²	cm²		
432 × 102	647.7 × 191.0	65.54	12.2	16.8	578.4	109.85	57.17	0.0908	38.6
381 × 102	571.5 × 192.8	55.10	10.4	16.3	502.9	89.98	50.36	0.1133	35.1
305 × 102	457.2 × 193.0	46.18	10.2	14.8	391.7	74.43	43.34	0.1483	30.9
305 × 89	457.2 × 167.6	41.69	10.2	13.7	397.8	68.72	37.63	0.1208	33.4
254 × 89	381.0 × 168.7	35.74	9.1	13.6	321.8	57.02	33.91	0.1603	28.0
254 × 76	381.0 × 144.3	28.29	8.1	10.9	330.7	46.28	25.71	0.1249	35.0
229 × 89	342.9 × 169.2	32.76	8.6	13.3	284.2	51.52	31.86	0.1861	25.8
229 × 76	342.9 × 144.8	26.06	7.6	11.2	292.4	41.86	24.49	0.1484	30.6
203 × 89	304.8 × 169.7	29.78	8.1	12.9	246.9	46.14	29.68	0.2181	23.6
203 × 76	304.8 × 145.3	23.82	7.1	11.2	254.0	37.54	23.12	0.1772	27.2
178 × 89	266.7 × 170.2	26.81	7.6	12.3	209.8	40.89	27.38	0.2593	21.7
178 × 76	266.7 × 145.8	20.84	6.6	10.3	217.7	32.41	20.67	0.2088	25.9
152 × 89	228.6 × 170.7	23.84	7.1	11.6	173.2	35.77	24.95	0.3151	19.7
152 × 76	228.6 × 146.0	17.88	6.4	9.0	182.1	27.69	17.93	0.2459	25.4
127 × 64	190.5 × 120.6	14.90	6.4	9.2	147.6	23.08	14.95	0.2542	20.7
102 × 51	152.4 × 95.5	10.42	6.1	7.6	116.6	16.38	10.18	0.2454	20.1
76 × 38	114.3 × 71.1	6.70	5.1	6.8	83.8	10.48	6.60	0.2530	16.8

The overall depth, D_c, of the castellated section = $3D/2$
where D = depth of original section.

PROPERTIES OF STEEL SECTIONS 1073

CASTELLATED JOISTS

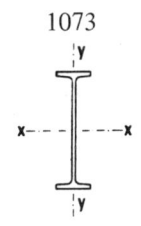

DIMENSIONS AND PROPERTIES

Nominal Size	Moment of Inertia (Net) Axis x—x	Moment of Inertia (Net) Axis y—y	Design Radius of Gyration Axis x—x	Design Radius of Gyration Axis y—y	Elastic Modulus (Net) Axis x—x	Elastic Modulus (Net) Axis y—y	Pitch of Standard Castellation 1.08 Ds	Ratio $\frac{Dc}{T}$
mm	cm⁴	cm⁴	cm	cm	cm³	cm³	mm	
305 × 102	5372	162.5	13.29	2.27	352.6	31.98	219.2	29.3
267 × 102	3562	139.1	11.70	2.28	267.0	27.39	192.2	29.6
228 × 89	2065	85.91	10.00	2.01	180.8	19.33	164.2	27.5
191 × 76	1122	50.13	8.33	1.74	117.7	13.16	137.2	25.1
153 × 64	515.6	25.27	6.65	1.45	67.57	7.96	110.2	23.1
114 × 51	196.3	11.09	4.94	1.16	34.38	4.37	82.1	20.4

CASTELLATED ZED BEAMS

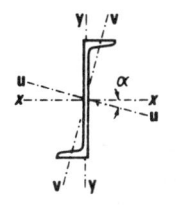

DIMENSIONS AND PROPERTIES

Axis x—x	Axis y—y	Axis u—u Max.	Axis v—v Min.	Axis x—x	Axis y—y	Axis u—u Max.	Axis v—v Min.	Axis x—x	Axis y—y	Size of Castellated Section
cm⁴	cm⁴	cm⁴	cm⁴	cm	cm	cm	cm	cm³	cm³	mm
50022	868.6	50430	459.8	26.3	3.36	26.4	2.51	1545	90.96	647.7 × 191.0
34830	858.1	35271	416.7	23.6	3.61	23.8	2.59	1219	89.01	571.5 × 192.8
19259	770.0	19675	354.5	19.1	3.72	19.3	2.59	842.5	79.79	457.2 × 193.0
16538	472.1	16776	234.1	18.8	3.09	18.9	2.23	723.4	56.33	457.2 × 167.6
10453	476.5	10716	213.3	15.9	3.32	16.1	2.29	548.7	56.50	381.0 × 168.7
7879	238.0	8000	117.0	15.7	2.66	15.8	1.91	413.6	32.99	381.0 × 144.3
7980	468.6	8249	199.3	14.5	3.43	14.7	2.30	465.4	55.39	342.9 × 169.2
6120	245.1	6252	112.7	14.3	2.79	14.5	1.95	357.0	33.85	342.9 × 144.8
5885	454.4	6156	183.2	13.0	3.53	13.2	2.30	386.2	53.55	304.8 × 169.7
4587	247.1	4727	106.4	12.9	2.92	13.0	1.97	301.0	34.01	304.8 × 145.3
4157	433.6	4425	165.1	11.4	3.62	11.8	2.30	311.7	50.95	266.7 × 170.2
3151	226.8	3284	93.44	11.3	2.98	11.6	1.97	236.3	31.11	266.7 × 145.8
2777	406.1	3038	144.7	9.84	3.70	10.3	2.27	242.9	47.58	228.6 × 170.7
2009	195.3	2126	78.58	9.75	2.98	10.0	1.94	175.8	26.75	228.6 × 146.0
1147	117.0	1218	45.90	8.07	2.53	8.30	1.63	120.4	19.41	190.5 × 120.6
494.4	48.40	523.0	19.83	6.38	1.95	6.54	1.29	64.88	10.14	152.4 × 95.5
177.9	18.14	188.8	7.22	4.76	1.49	4.89	.969	31.13	5.10	114.3 × 71.1

Design Radius of Gyration is the average between the values for the gross and net section.

PROPERTIES OF STEEL SECTIONS

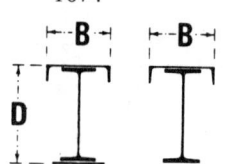

GANTRY GIRDERS

COMPOSITION AND DIMENSIONS

Size D×B mm	Composed of - Universal Beam	Composed of - Top Flange Channel	Composed of - Bottom Flange Plate	Mass per metre in kg	Extreme Fibre Distances n_c cm	Extreme Fibre Distances n_t cm	Ratio D/T Top Flange
954 × 432	914 × 305 @ 289	432 × 102 @ 66	350 × 15	396.0	44.96	50.42	26.3
939 × 432	289	432 × 102 @ 66	---	354.8	39.19	54.69	25.9
931 × 432	253	432 × 102 @ 66	---	318.7	37.93	55.14	28.1
923 × 432	224	432 × 102 @ 66	---	289.2	36.67	55.58	30.6
878 × 432	838 × 292 @ 226	432 × 102 @ 66	350 × 15	333.2	40.97	46.84	27.9
863 × 432	226	432 × 102 @ 66	---	292.0	34.46	51.85	27.4
853 × 432	194	432 × 102 @ 66	---	259.3	32.91	52.38	30.6
782 × 432	762 × 267 @ 197	432 × 102 @ 66	---	262.2	30.36	47.82	27.0
774 × 432	173	432 × 102 @ 66	---	238.4	29.15	48.27	29.3
772 × 381	173	381 × 102 @ 55	---	228.0	30.29	46.95	29.1
705 × 432	686 × 254 @ 170	432 × 102 @ 66	---	235.4	26.52	43.99	26.1
700 × 432	152	432 × 102 @ 66	---	217.5	25.57	44.41	27.7
694 × 381	140	381 × 102 @ 55	---	195.2	25.98	43.41	29.1
660 × 432	610 × 305 @ 238	432 × 102 @ 66	350 × 15	345.0	30.94	35.08	18.4
645 × 432	238	432 × 102 @ 66	---	303.8	26.28	38.24	18.0
630 × 432	179	432 × 102 @ 66	---	244.3	24.10	38.87	21.1
622 × 432	149	432 × 102 @ 66	---	214.6	22.73	39.45	23.1
627 × 381	610 × 229 @ 140	381 × 102 @ 55	---	195.0	23.59	39.15	25.7
622 × 381	125	381 × 102 @ 55	---	180.2	22.73	39.50	27.2
618 × 305	113	305 × 89 @ 42	---	155.0	23.53	38.22	25.8
613 × 381	610 × 178 @ 91	381 × 102 @ 55	---	146.1	20.36	40.93	34.1
608 × 305	82	305 × 89 @ 42	---	123.7	21.24	39.60	33.3
555 × 381	533 × 210 @ 122	381 × 102 @ 55	---	177.3	20.27	35.23	24.3
550 × 305	109	305 × 89 @ 42	---	150.4	20.84	34.13	23.0
543 × 305	92	305 × 89 @ 42	---	134.0	19.74	34.59	25.2
539 × 305	533 × 165 @ 73	305 × 89 @ 42	---	114.7	18.27	35.63	29.6
535 × 305	66	305 × 89 @ 42	---	107.3	17.51	35.99	31.4
478 × 305	457 × 191 @ 98	305 × 89 @ 42	---	139.9	17.77	29.99	20.6
472 × 254	89	254 × 76 @ 28	---	117.6	18.67	28.50	21.4
465 × 254	74	254 × 76 @ 28	---	102.8	17.67	28.86	23.8
473 × 254	457 × 152 @ 82	254 × 76 @ 28	---	110.2	18.37	28.95	23.5
465 × 229	67	229 × 76 @ 26	---	93.0	17.56	28.92	25.5
423 × 305	406 × 178 @ 74	305 × 89 @ 42	---	116.2	14.67	27.63	21.1
418 × 254	67	254 × 76 @ 28	---	95.3	15.52	26.23	22.3
411 × 254	54	254 × 76 @ 28	---	81.9	14.35	26.72	25.3
424 × 254	406 × 152 @ 74	254 × 76 @ 28	---	102.7	16.18	26.26	21.6
416 × 229	60	229 × 76 @ 26	---	85.6	15.32	26.23	23.8

NOTE: The above properties are based on compound girders of welded construction.

PROPERTIES OF STEEL SECTIONS

GANTRY GIRDERS

PROPERTIES

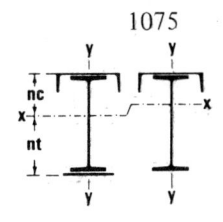

Area in cm²	Moment of Inertia in cm⁴ Axis x—x	Axis y—y Top Flange only	Axis y—y Complete Section	Radius of Gyration Axis y—y cm	Elastic Modulus in cm³ Axis x—x n_t	n_f	Axis y—y Top Flange	Horizontal Shear Coefficients Top Flange	Bottom Flange	Size D×B mm
504.5	788234	28770	41552	9.08	17531	15634	1333	0.0045	0.0033	954×432
452.0	643676	28770	36192	8.95	16423	11770	1333	0.0048	---	939×432
406.0	569695	27637	33911	9.14	15021	10331	1280	0.0052	---	931×432
368.4	503131	26597	31823	9.29	13721	9052	1232	0.0057	---	923×432
424.4	578285	26716	37420	9.39	14116	12345	1237	0.0056	0.0042	878×432
371.9	450992	26716	32060	9.28	13087	8698	1237	0.0060	---	863×432
330.3	384020	25581	29783	9.50	11669	7331	1185	0.0067	---	853×432
334.0	327603	25238	29098	9.33	10792	6850	1169	0.0071	---	782×432
303.7	288265	24578	27775	9.56	9890	5972	1138	0.0078	---	774×432
290.4	276716	18073	21269	8.56	9136	5894	949	0.0070	---	772×381
299.8	238272	24504	27624	9.60	8984	5417	1135	0.0085	---	705×432
277.1	215264	24088	26789	9.83	8418	4847	1116	0.0090	---	700×432
248.6	190409	17283	19683	8.90	7328	4387	907	0.0087	---	694×381
439.5	339296	28871	41731	9.74	10968	9671	1337	0.0070	0.0053	660×432
387.0	269004	28871	36372	9.69	10236	7034	1337	0.0074	---	645×432
311.2	206109	26678	31969	10.14	8550	5303	1236	0.0088	---	630×432
273.4	175036	25631	29870	10.45	7702	4436	1187	0.0097	---	622×432
248.4	155702	17015	19147	8.78	6600	3977	893	0.0095	---	627×381
229.6	140305	16728	18570	8.99	6172	3552	878	0.0101	---	622×381
197.4	120699	8650	10245	7.20	5130	3158	568	0.0094	---	618×305
186.1	100423	15604	16321	9.37	4933	2453	819	0.0125	---	613×381
157.5	85203	7660	8264	7.24	4012	2151	503	0.0119	---	608×305
225.8	108741	16493	18,101	8.95	5366	3086	866	0.0115	---	555×381
191.5	92516	8435	9815	7.16	4440	2710	553	0.0107	---	550×305
170.7	79333	8165	9272	7.37	4018	2294	536	0.0118	---	543×305
146.2	62348	7573	8088	7.44	3412	1750	497	0.0137	---	539×305
136.7	55835	7491	7924	7.61	3188	1552	492	0.0146	---	535×305
178.3	64373	8167	9277	7.21	3622	2147	536	0.0129	---	478×305
149.8	54524	4345	5327	5.96	2921	1913	342	0.0111	---	472×254
130.9	45912	4139	4914	6.13	2599	1591	326	0.0124	---	465×254
140.4	49533	3911	4460	5.64	2697	1711	308	0.0120	---	473×254
118.5	39855	3023	3439	5.39	2269	1378	264	0.0130	---	465×229
148.0	40525	7783	8509	7.58	2763	1467	511	0.0164	---	423×305
121.4	34001	4000	4635	6.18	2191	1296	315	0.0145	---	418×254
104.4	27330	3827	4288	6.41	1904	1023	301	0.0165	---	411×254
130.9	37305	3888	4413	5.81	2305	1421	306	0.0138	---	424×254
109.0	29253	2993	3378	5.57	1909	1115	262	0.0151	---	416×229

BRITISH STANDARDS

(*Obtainable from BSI Sales Department at Newton House, 101 Pentonville Road, London, N1 9ND; Telephone 01-837 8801; Telex 23218*)

    4 *Structural Steel Sections:*
          Part 1: Hot-rolled sections.
          Part 2: Hot-rolled hollow sections.
153 *Steel girder bridges:*
          Parts 1 and 2: 1. Materials and Workmanship.
                      2. Weighing, shipping and erection.
          Part 3A: Loads
          Parts 3B and 4: 3B. Stresses.
                      4. Design and construction.
275 *Dimensions of rivets.*
350 *Conversion factors and tables:*
          Part 1. Basis of tables, conversion factors.
          Part 2. Detailed conversion tables.
          Supplement No. 1. Additional table for S.I. Conversions.
449 *The use of structural steel in building:*
          Part 1. Imperial units.
          Part 2. Metric units.
499 *Welding terms and symbols:*
          Part 1. Welding, brazing and thermal cutting glossary.
          Part 2. Symbols for welding.
          Part 3. Terminology of and abbreviations for fusion weld
                imperfections as revealed by radiography.
499C *Chart of British Standard welding symbols (based on 449: Part 2).*
639 *Covered electrodes for the manual metal-arc welding of mild steel and medium tensile steel.*
648 *Schedule of weights of building materials.*
877 *Foamed or expanded blast furnace slag lightweight aggregate for concrete.*
916 *Black bolts, screws and nuts.*
938 *General requirements for the metal-arc welding of structural steel tubes to B.S. 1775.*
990 *Steel windows generally for domestic and similar buildings.*
1083 *Precision hexagon bolts, screws and nuts (B.S.W. and B.S.F. threads).*
1091 *Pressed steel gutters, rainwater pipes, fittings and accessories.*
1449 *Steel plate, sheet and strip.*
1719 *Classification, coding and marking of covered electrodes for metal-arc welding.*
1775 *Steel tubes for mechanical, structural and general engineering purposes.*
1787 *Steel windows for industrial buildings.*
1856 *General requirements for the metal-arc welding of mild steel.*
2503 *Steel windows for agricultural use.*
2642 *General requirements for the arc welding of carbon manganese steels.*
2994 *Cold rolled steel sections.*

BRITISH STANDARDS

- 3139 *High strength friction grip bolts for structural engineering.*
- 3294 *The use of high strength friction grip bolts in structural steelwork.*
- 4360 *Weldable structural steels.*
- 4395 *High strength friction grip bolts and associated nuts and washers for structural engineering. Metric series.*
- 4449 *Hot rolled steel bars for the reinforcement of concrete.*
- 4461 *Cold worked steel bars for the reinforcement of concrete.*
- 4482 *Hard drawn mild steel wire for the reinforcement of concrete.*
- 4483 *Steel fabric for the reinforcement of concrete.*
- 4486 *Cold worked high tensile alloy steel bars for prestressed concrete.*
- CP3 *Chapter V. Loading.*
- CP111 *Structural recommendations for loadbearing walls:*
    Part 1. Imperial units.
    Part 2. Metric units.
- CP114 *Structural use of reinforced concrete in buildings.*
- CP114 Part 2. *Structural use of reinforced concrete in buildings. Metric units.*
- CP2008 *Protection of iron and steel structures from corrosion.*

It is important that users of British Standards should ascertain that they are in possession of the latest amendments or editions.

INDEX

A

ABRASIVES, 801-4
ACOUSTIC GAUGE, THE MAIHAK, xi
AIRLESS BLAST CLEANING, 801
ALBANY FLATS, Bournemouth, 859-61
ALGOL language for computers, 927
ALUMINIUM COATING, 888
ANGLES
   Backmarks in, 710
   Properties of
      equal, 1036-41, 1050-1
      unequal, 1042-9, 1052-3
   Struts, design of, 487-96
ANHEUSER, 114
ARCHES OR PORTALS
   Comparison of types, 432
   Hingeless, formulae for, 305-8, 313-7, 323-6
   Multiple bay, hinged at feet
      formulae for, 332-44
   Three-hinged, 132-4, 432
   Two-hinged, 125-8, 430-2
      formulae for, 309-12, 313-7, 327-9, 330-1
AREA-MOMENT METHOD, 85-102, 195-224
   Application in two stages, 197-200
   Closed frames, 223-4
   Industrial building frame, 210-6
   Portal frames, 200-10
   Reciprocal theorem, 195-7
   Symmetrical multi-storey frames, 216-22
   Vertical loading, 222
AREAS
   Sections (see Section Tables)
   Simpson's rule, 21
ASBESTOS COVERED SHEET, 890
ASYMMETRICAL PORTAL-FRAMES, 284-8
AUTOCODE for computers, 927
AXIAL STRESSES
   (see Bending and Axial Stresses)

B

BAKER, Sir John, x
BASES
   (see Stanchions)
BATHO, Dr C., xi
BEAMS
   (see also Girders)
   British Standard, properties of, 1012-23

BEAMS *(cont.)*
   Castellated, 1066-9
   Compound (see Girders)
   Continuous
      analysis by slope deflection, 277-8
      charts for B.M. and reactions, 57-9
      Clapeyron's theorem of three moments, 51-6
      definition, 51
      deflection of, 97-100
      influence lines, 176-94
      moment distribution, 227-34
      plastic design, 542-4
      slope deflection analysis, 277-83
      spliced at point of contraflexure, 737-9
      worked examples, 52-6
   Deflection, 15-16
   Fixed, built-in or encastré, 39-49
      charts for loading, moment, shear, deflection, 43-9
      definition, 39
      influence lines, 172, 175-6
      principle of reciprocal moments, 42
      shear forces in, 41-2
      supports at different levels, 41-2
      supports at same level, 39-41
      worked examples, 40-2
   Moment of resistance, 1
   Redundant, in plastic design, 536-40
   Simply supported, 17-38
      charts for loading, moment, shear, deflection, 31-8
      worked examples, 22-6
   Theory, 1-3
   Torsion (see Torsion, Beams in)
   Universal, 1012-23
      T-bars cut from, 1058-61
   Z-polygon, 3-5

BENDING
   Circular, 1
   Principal axes, 8-11
   Unsymmetrical, 3

BENDING AND AXIAL STRESSES, 5-7
   Circle of inertia, 13
   Compression, 5
   General expression for stress, 7-8
   Neutral axis, location of, 11-15
   Principal axes, 8-11
   Unsymmetrical bending formulae, 3

BENDING AND AXIAL STRESSES (cont.)
    worked examples, 2, 9, 11
    Z-polygons, 3–5
BENDING MOMENTS (B.M.)
    Definition, 17
    Diagrams, 22–38
    Parallel boom girders, 123
    Relationship with loading and S.F., 3–7, 17–20
BETHLEHEM STEEL CO., 114
BITUMEN-ASBESTOS FELTED SHEET, 890
BLAST CLEANING, 800–4
BLUM, 642
BOLTS
    (see Rivets and Bolts; Connections)
BOLTS, HOLDING-DOWN, 623–5
BORNSCHEUER, 114
BOX PILING
    Dimensions and properties, 1003–5, 1009
BRACED VAULT CONSTRUCTION, 472–7
BRACING
    (see Engineering Workshop Design; Knee-braced Frames; Single-Storey Sheds)
BRICK WALLS, 911–6
BRITISH STANDARDS, 1077–8
BUCKLING OF WEBS, 645, 648–9
BUILT-IN BEAMS
    (see Beams, Fixed, Built-in or Encastré)
BUILT-UP WEATHERPROOFING, 888–90

C

CAMPUS, Professor, formula, 777
CANTILEVERS AND SUSPENDED SPANS
    Charts for loading, moment, shear, deflection, 29–30, 62–6
    Deflection, 100–2
CANTILEVERS, PROPPED, 67–79
    Charts for loading, moment, shear, deflection, 72–9
    Examples of solution
      by assimilation with encastré beams, 70–1
      by deflection formulae, 69–70
      by Theorem of Three Moments, 67–9
    Formulae for B.M. and S.F., 72–9
    Worked examples, 68–71
CASSIE, W. Fisher, 114
CASTELLATED SECTIONS,
    Properties of, 1066–73
    Space structures, use in, 446–50
CENTROID, POSITION OF, 998–1000
CHANNELS, BRITISH STANDARD
    Properties of, 1034–5

CHEQUERED PLATES, 879–82
CIRCLE OF INERTIA, 13–15
CIRCULAR BENDING
    (see Bending)
CLADDING
    (see Roofing and Cladding)
CLAPEYRON
    (see Theorem of Three Moments)
CLOSED FRAMES, 223–4
COLUMN BASES, 725–31
COLUMN SPLICE, 731
COLUMNS
    (see Stanchions)
COMMON PLANE SECTIONS
    Areas, position of centroid, moments of inertia, section moduli, 998–1000
COMPOSITE CONSTRUCTION, 585–611
    Bibliography, 610–1
    C.P. 117, 586
    economy by use of, 586–8
    Multi-storey buildings, 859–61
    Section properties calculation, 596–601
    Shear connectors, 592–5, 601–10
COMPOUND BEAMS
    (see Girders, Compound)
COMPUTERS, 921–34
    Analogue and digital, 921–3
    Desk-top computers, 933
    Genesys Centre, 934
    National Computing Centre, 933
    Programmes, 925–6
      existing and available for structural engineering, 927–8, 934
      preparation of, 926–7
      Rigid frame analysis, 928–33
      Vierendeel girders, 419–20
CONCRETE
    (see also Reinforced)
    Floors, 871
    Pre-cast panels, 916–20
CONNECTIONS, 701–797
    (see also Joints in Rigid Frame Structures)
    Beam end, 712
    Beam splices, 737–9
    Bibliography, 796–7
    Column bases, 725–31
    Column splice, 731
    Eccentric loading, 715–22
    Gusset plate design, 720
    High strength friction grip bolts, 733–9, 857
    Multi-storey stanchions, 842–4
    Plastic design, 579–80
    Rivet groups, 715–22
    Riveted, 711–5
    Rivets and bolts, tables, 704–10
    Roof truss, 761–4
    Splices, 731–2, 737–9, 788

INDEX

CONNECTIONS (cont.)
 Tension member, 711
 in Vierendeel girders, 415 et seq.
 Welded, 739–64
 Wind moment, 723–31
CONTINUOUS BEAMS
 (see Beams)
CONVERSION TABLES
 Feet and inches to metres, 979–82
 $Ft^2 - m^2$, 985–6
 Kg – pounds, 977–8
 Kg per m – pounds per ft, 974
 Metres – ft, 983–4
 $M^2 - ft^2$, 987–8
 $MN/m^2 - tonf/in^2$, 993–6
 Moments of inertia, 973
 Pounds per ft – kg per m, 974
 Pounds – kg, 975–6
 $Tonf/in^2 - MN/m^2$, 989–92
COOPERATIVE INSURANCE SOCIETY
  BUILDING, MANCHESTER, 859
CORROSION RESISTANCE
 Steel piles, 642, 643
CORRUGATED CLADDING SHEETS,
  889–96
CRANES
 Gantry girders, 497, 506–12
  dimensions and properties, 1074–5
CROSS, Professor Hardy, 195
CURTAILMENT OF FLANGE
  PLATES, 103–6
CURTAIN WALLING, 915–6

D

DEFLECTION OF BEAMS AND
  GIRDERS
 Area-moment method (Mohr's theorem),
  85–91
 Built-in beams (see Fixed beams, below)
 Cantilevers,
  charts, 29–30
 Compound girders with curtailed flange
  plates, 103–6
  formulae, 104
 Continuous beams,
  worked example, 97–100
 Fixed beams, 92–7
  charts, 43–9
  worked examples, 93
 Gantry girders
  worked example, 511–2
 Mathematical (slope-deflection) method,
  83–4
 Propped cantilevers, 100–2
  charts, 72–9
  worked examples, 68–71
 Simple cantilevered beams, 91–2
  charts, 31–8
  worked example, 83–4

DEFLECTION OF BEAMS AND
  GIRDERS (cont.)
 Unsymmetrical sections, 15–6
DEFLECTION OF FRAMED
  STRUCTURES, 135–54
 Due to temperature changes, 137
 Formula for deflection, 137
 Graphical method, 142–54
 Mathematical method, 135–42
 Principle of work, 135
 Williot and Williot-Mohr diagrams, 142–54
 Worked examples, 138–42
DEFLECTION OF RIGID FRAMES,
  289–97
 Area-moment method, 85–92, 289
 Eaves joints of portal frames, 203–4
 Non-symmetrical frames, 297
 Slope deflection method, 276–88
DOLPHINS, 641–2
DOMES, 477–9

E

EAVES BRACING, 522
ENCASTRÉ BEAMS
 (see Beams, Fixed, Built-in or Encastré)
ENGINEERING WORKSHOP DESIGN,
  497–529
 Bases of stanchions, 524–7
 Bracings, 521–4, 529
 Foundation bolts, 526
 Gable members, 528–9
 Gantry girder, 497, 506–12, 528
 Glazing purlin, 502
 Lacings, 518, 521
 Purlins, 501–2
 Roof truss, 502–5
 Side and gable rails, 512–4
 Stanchions
  corner, 528–9
  main, 499, 514–21
  side, 516–8, 519–21
 Valley beams, 506
 Vertical bracing, 523–4
 Wind loading, 498–501
EULER, 811
 Load on columns, 811

F

FIRE PROTECTION AND
  STRUCTURAL CASING, 917–8
FIRE-RESISTING CONSTRUCTION.
  935–71
 Bibliography, 970–1
 Building legislation, current, 939–63
  history and development, 936–8
 Building Regulations 1972, Part E, 939–57
  compartments, 944–7
  external walls, 954–7
  interpretation, 939–42
  measurement, 944

FIRE-RESISTING CONSTRUCTION
(*cont.*)
   minimum thickness of materials, 958–9
   purpose groups, 942–4
   resistance periods, 947–53
   resistance tests, 953–4
   Building Standards (Scotland) (Consolidation) Regulations 1971, 957–60
   Future developments, 968–70
   London Building (Constructional) Amending By-laws (No. 1) 1964, 960–3
   Methods of fire protection, 963–8
FIXED BEAMS
   (see Beams, Fixed, Built-in or Encastré)
FLAME CLEANING, 799
FLANGE PLATES
   (see also Girders)
   Curtailment, cut-off points
     affecting deflection, 103–6
     welded compound girders, 645–7
   Moments of inertia, tables of, 686–7
FLANGE STABILITY
   Plastic design, 549–53, 580
FLOOR PLATES, 879–82
   Durbar, 879
   Safe loads, 880–2
   Weights and dimensions, 879
FLOORS, 869–77
   Concrete, 871
FORCES IN PLANE FRAMES, 119–34
   Arch, three-hinged, 132
   B.M. and S.F. diagrams, solution by, 123
   Equilibrium polygon solution method, 124–5
   Graphical solution, 119–20
   Joint resolution method of solution, 122–3
   Knee-braced frames, 125–32
   Method of sections or moments, 121–2
   Resultant force resolution method, 124
   Sections or moments method of solution, 121–2
   Three-hinged arch, 132
   Trigonometrical solution, 122–3
FORTRAN
   (formulae translation for computers), 927
FOUNDATION BOLTS, 526
FOUNDATIONS, 613–38
   Bearing piles, 640–1
   Eccentric loading, 614–7
   Holding-down bolts, 623–5
   Pocket bases, 636–8
   Reinforced concrete, 625–35
   'Rendhex' steel box columns, 1010
   Slab bases, 617–23
FRAMES
   (see Deflection of Framed Structures; Forces in Plane Frames; Influence Lines; Knee-Braced Frames; Rigid Frames and Sheds)

   With non-prismatic columns, 263–66

G

GALVANISING, 805–6, 887
GANTRY GIRDERS
   Dimensions and properties, 1074–5
GEOLOGICAL MUSEUM, THE NEW, xi
GEOMETRICAL PROPERTIES OF SECTIONS
   (see Section Tables)
   Areas of common plane sections, 998–1000
   Centroid positions in common plane sections, 998–1000
   common plane sections, 998–1000
GERBER, 61
GERMAN COMMITTEE FOR WATER-FRONT STRUCTURES, 641
GIRDERS, 645–79
   Castellated beams, 1066–9
   Compound, deflection of, 103–6
   Compound, welded, 645–9
   Gantry, dimensions and properties, 1074–5
   Hog-backed, influence line example, 169–70
   Lattice (see Lattice Girders)
   Plate, moments of inertia of, 681–99
     calculation method, 681
   Plate, riveted, 659–79
     design, 666–79
     flange, curtailment, 668–70, 676
     flange types, 663
     flanges, sloping, 665–6
     resistance moment, 663–6
     rivets, 669–70, 673, 677
     stiffeners, 671–4
     web design, 670–1
   Plate, welded, 649–60
     alternative designs, 659
     flanges, 651–3
     selection of section, 649–53
     stiffeners, 653–6, 658–9
     web design, 653
     welding calculations, 656–9
   Vierendeel, 413–28
GLAZING, PATENT, 903–8
   Mid-roof, 905–6
   North-light, 905–6
GREENE, C.E., 85
GREEK ALPHABET, 997
GRIDS, 442–50

H

HEYMAN, J., 531
HIGH STRENGTH FRICTION GRIP BOLTS, 733–9, 860
HOG-BACKED GIRDER
   Influence lines example, 169–70
HOLDING-DOWN BOLTS, 623–5
HOT DIP GALVANISING, 805–6

I.B.M. BUILDING, Pittsburgh, USA, 861
INDETERMINATE STRUCTURES
  Moment distribution, 225–7
INDUSTRIAL BUILDING FRAME, 210–6
INERTIA
  (see Moments of Inertia; Geometrical Properties of Sections)
INFLUENCE LINES, 155–94
  Beams
    built-in, 172–6
    continuous, 176–94
      charts, 180–90
      notes on charts, 179
    simply supported, 155–64
      point loads, 155–9
      uniformly distributed loads, 159–64
  Definition, 155
  Framed structures, 164–72
    lattice girder with K-bracing, 172–4
    Pratt or N-truss examples, 166–8

J

JETTIES, 642
JOINTS IN RIGID FRAMES, 765–97
  Bibliography, 796–7
  Bleich, F., 781
  Bleich, H. H., 778, 781, 786
  Campus, Professor, 777
  Circular sections, 783–5
  Flange stresses, 786
  Formulae
    Bleich, F., 781
    Bleich, H. H., 778, 781, 786
    Campus, 777
    Magnel, 778
    Olander, 782, 784–5
    Osgood, 768–9, 781
    tapered beam, 780–1
    Vierendeel, 780–1
    Winkler-Résal, 775–7
    Wright, D. T., 770
  Knees
    curved flanges, 773, 775–87
    haunches, 773
    pitched roofs, 775
  Magnel, Professor, 778, 795–6
  Multi-storey buildings, 789–93
  Olander, H. C., 782, 784–5
  Osgood, W. R., 768–9, 781
  Portal, 779
  Portal frame, 783
    rectangular, American research on, 768–73
    frames for, 765–7
    with pitched roof, 773
  Résal, 775–7
  Ridge joints, 775
  Sections, circular, 783–5

JOINTS IN RIGID FRAMES (cont.)
  Splice connections, 788
  Stiffeners, 775
  Stress function, Airy, 768
  Stresses
    flange, 767, 786
    principal and greatest shear, 769–70
    shear, 772–3
    tensile, 786
    web, 767
  Valley joints, 787–8
  Vierendeel, Professor, 780, 793
  Vierendeel girders, 793–6
    formulae, 780–1
  Winkler, 775–7
  Worked examples, 770–3, 782–7
  Wright, D. T., 770
JOISTS
  Dimensions and properties, 1030–3
  Castellated, 1072–3

K

KLEINLOGEL, Professor
  Extracts from "Rahmenformeln", 299, 305–44
KNEE-BRACED FRAMES, 125–32
  Avoiding moments on foundations, 429–30

L

LATTICE FRAMES, 762
LATTICE GIRDERS, 433–40
  Influence lines, K-braced, 173–4
  Ridges of multi-span roofs, 433–5
  Space structures, in, 446–50
  Stress coefficients, 435–40
LATYMER COURT, xi
LOADS
  Axial, in plastic design, 547–8
  Euler, 811
  Factors, in plastic design, 535–6
  Interpanel, on frames and Vierendeel girders, 267–9
  Repeated, in plastic design, 578–9

M

MAIHAK ACOUSTIC GAUGE, xi
MANEY, Professor, 195
MARINE STRUCTURES, 641–3
MECHANICAL BLAST CLEANING, 801
METAL COATINGS AND SPRAYING, 804–5
MISCELLANEOUS TABLES, 973–1075

INDEX

MODULUS OF SECTION
(see Section Modulus)
MOHR
  Area-moment method, 85, 195
  Williot-Mohr diagrams, 148–54
MOMENT DISTRIBUTION, 225–75
  Cantilever ends, 232
  Continuous beams, 227–34
  Fixed-end moments (F.E.Ms)
    charts for various conditions, 271–5
  Frames
    interpanel loading, 268
    non-prismatic columns, with, 263–6
    multi-storey, 255–6
    pitched roofs, 244–55
    portal, 234–43, 259–61
    symmetrical, 256–9
    unsymmetrical vertical loads, 270
  Loading, interpanel, 267–8
  Naylor's method, 256–9, 268
  Portal frames, 234–43, 259–61
  Principles, 225–7
  Side loading on symmetrical frames, 256–9
  Side-sway
    definition, 234
    relative movement of joints, 241–2
    treatment of, 236, 239, 241
  Sign convention, 225
  Stiffness factor, definitions, 226, 236, 255–6
  Unsymmetrical vertical loads, 270
  Vierendeel girders, 267
MOMENTS OF INERTIA
  Circle of inertia, 13–15
  Common plane sections, 998–1000
  Conversion tables, 973
  Flanges, 686–7
  Plate girders, 681–99
    calculation method, 681
  Tables
    box piling, 1003–5, 1009
    equal angles, 1036–41, 1050–1
    four equal angles (in plate girders), 690–1
    four unequal angles (in plate girders), 692–9
    foundation columns, 1010
    plane sections, 998–1000
    plate girders, 682–5
    rectangular plates, 682–5
    unequal angles, 1042–9, 1052–3
    unit areas
      second moment of a pair, 688–9
    universal beams, 1012–5
    universal columns, 1024–5
MÜLLER-BRESLAU, H.F.B., 85
MULTI-STOREY RIGID FRAMES, 255–6, 262
  Stanchions, 809–46
  Wind moment connection, 726–31

N

NAYLOR, N., 256, 268
NEUTRAL AXIS, LOCATION OF, 11–15
NORTH-LIGHT ROOF
  Design, 452–8
N-TRUSS, 170–2
  Influence lines examples, 166–8

O

OVERHEAD CRANES
  Gantry girders, 497, 506–12
    dimensions and properties, 1074–5

P

PARABOLA
  Properties of, 95
PATENT GLAZING, 903–8
PERFORATED GALVANISED SHEET, 888
PERRY-ROBERTSON FORMULA, 811–2
PICKLING, 799–800
PILES, STEEL, 639–43
  Aden Oil Harbour, 642
  Bearing, 640–1
    minimum spacing of, 640
  Corrosion resistance, 642, 643
  Kwinana jetty, 642
  Marine structures, 641–3
  Sheet, 639–40
  Universal bearing, dimensions and properties, 1001–10, 1064–5
PITCHED ROOF RIGID FRAMES, 244–55
  Plastic design, 565–76
PLANE FRAMES
  (see Forces in)
PLANE SECTIONS
  Geometrical properties, 998–1000
PLASTIC COATING, 888
PLASTIC THEORY AND DESIGN, 531–83
  Axial load, 547–8
  Basis of theory, 533–4
  Bibliography, 531
  Column design, 577–8
  Connections, 579–80
  Continuous beams, 542–4
  Deflections, 576–7
  Equilibrium, 538, 540–1, 558, 570
  Fixed base rectangular frame, 560–5
  Flange stability, 549–53
  Frame analysis, 553–7
  Full plastic moments, 545–6
  Load factors, 535–6

PLASTIC THEORY AND DESIGN *(cont.)*
  Plastic moduli of sections
    joists, 1030–1
    universal beams, 1016–9
    universal columns, 1026–7
  Portal frames, 553–7
    pitched roof, 565–76
  Principles, 532
  Redundant beams, 536–40
  Repeated loading, 578–9
  Shear, 548–9
  Single span, slenderness ratio, 578
  Universal beams and columns, tables, 581
  Upper and lower bounds, 540–2
  Virtual work, 557–60
PLATES
  (see Floor plates)
PLATE GIRDERS
  (see Girders and Moments of Inertia)
POCKET BASES, 636–8
PORTAL ARCH, 132–4
PORTAL FRAMES
  Area-moment method of design, 200–10
  Asymmetrical, slope-deflection design, 284–8
  Deflection of eaves joints, 203–4
  Moment distribution, 234–43, 259–61
  Pitched roof, plastic design, 565–76
  Plastic design analysis, 553–7
  Single-bay, 259–61
  Single storey sheds, 433–5
  Snow load, 203
  Symmetrical, slope deflection design, 283–4
  Wind loading, 205–10
PRAGUE THEATRE
  Space structure, 480–1
PRATT TRUSS
  Influence line examples, 166–8
PRE-PAINTED SHEET, 888
PRINCIPAL AXES, 3
  (see also Geometrical Properties of Sections)
  Bending about, 8–11
PRODUCT OF INERTIA
  (see Geometrical Properties of Sections)
PROPERTIES OF SECTIONS
  (see Section Tables)
PROPPED CANTILEVERS
  (see Cantilevers, Propped)
PURE TENSION OR COMPRESSION, 5

R

RADIUS OF GYRATION
  Plane sections, geometrical properties of, 998–1000

RADIUS OF GYRATION *(cont.)*
  Steel sections
    angles
      equal, 1036–41, 1050–1
      unequal, 1042–9, 1052–3
    castellated universal beams, 1066–9
    channels, 1034–5
    piling, 1001–10, 1064–5
    T-bars and structural tees, 1054–63
RECIPROCAL THEOREM, 195–7
REINFORCED CONCRETE
  Encasement of stanchions, 822–42
  Floor slabs in composite construction, 588–9
  Foundations, 625–38
RIGID FRAMES
  (see also Deflection of Rigid Frames; Joints in Rigid Frame Structures; Moment Distribution; Plastic Theory; Semi-Graphical Integration; Slope Deflection)
  Analysis by computer, 928–33
  Joints, 765–97
  Kleinlogel formulae explained, 299–301
    arrangement, 302
    checking of calculations, 303–4
    sign conventions, 302–3
  Knees
    for rectangular frames, 765–7
    for rigid frames with pitched roofs, 773–5
    with curved flanges, 775–87
  Plastic design, 553–7
  Slopes and deflections, 289–97
  Symmetrical loads conversion coefficients, 3
RIGID FRAMES: CHARTS
  Single-bay rectangular
    fixed base, 345–51
    hinged base, 380–3
  Single-bay ridged (pitch 1 in 5)
    fixed base, 352–65
    hinged base, 384–90
  Single-bay ridged (pitch 1 in 25)
    fixed base, 366–79
    hinged base, 391–7
  Twin-bay ridged (pitch 1 in 25)
    hinged base, 398–411
RIGID FRAMES: KLEINLOGEL FORMULAE
  Rectangular portal frame
    hingeless, 305–8
    two-hinged, 309–12
  Ridged frame with vertical legs
    hingeless, 313–7
    two-hinged, 318–22
  Skew-cornered frame
    hingeless, 323–6
    two-hinged, 327–9
  Triangular frame with hinged feet, 330–1

INDEX

RIGID FRAMES: KLEINOGEL FORMULAE (cont.)
  Twin-ridged frame with hinged feet, 332–44
RISSELADA, 642
RIVETS AND BOLTS
  (see also Connections)
  Pitch in plate girders, 665
  Shearing and bearing values, spacing, 702–10
  Tables, 702–10
ROBERTSON, Andrew, x
ROOF TRUSSES
  Engineering workshop, design, 502–5
  Welded connections, 761–4
ROOFING AND CLADDING, 887–902
  British Standards and advisory literature, 902
  Corrugated sheets, 889–94
  Endlaps, 895
  Fastenings and fittings, 897–9
  Galvanised coating, 887, 901
  Maintenance, 901–2
  Pitch to ensure weather-tightness, 893–4
  Protection, 887–90
  Sidelaps, 894–5
  Strength, 891
  Supports, spacing of, 892
  Thermal insulation, 900–1
  Weights, 895
ROOFS, PITCHED
  Frame moment distribution, 244–55
ROUGHNESS OF STEEL SURFACE, 803–4

S

SEAGRAM BUILDING, New York, 855
SECTION MODULUS (Z)
  Conversion tables, 973
  Geometrical sections, 998–1000
  Steel sections (see Section Tables)
SECTION TABLES
  Angles
    backmarks in, 710
    equal, 1036–41, 1050–1,
    unequal, 1042–9, 1052–3
  Broad-flange beams (see Universal Columns)
  Castellated sections, 1066–73
  Channels, 1034–5
  Common plane sections, 998–1000
  Gantry girders, 1074–5
  Joists, 1030–3
  Sheet and box piling, 1001–10,
  T's, 1054–7
    cut from Universal beams, 1058–61
    cut from Universal columns, 1062–3
  Universal
    beams, 1012–23

SECTION TABLES (cont.)
  bearing piles, 1064–5
  columns, 1024–9
SEMI-GRAPHICAL INTEGRATION, 195
SHEAR CONNECTORS
  Composite construction, in, 592–5, 601–10
SHEAR FORCE (S.F.)
  Definition, 17
  Diagrams, 22–38
  Plastic design, 548–9
  Relationship with loading and B.M., 3–7, 17–20
SHEAR WALLS
  Stabilising multi-storey frameworks, 809
SHEDS, SINGLE-STOREY
  (see Single-Storey Sheds)
SHEET PILING, 639–40
  Dimensions and properties, 1001–2, 1005–9
SHELL TOWER, London, 857
SIDE LOADING ON SYMMETRICAL FRAMES, 256–9
SIDE-SWAY
  (see Moment Distribution)
SIMPLY SUPPORTED BEAMS, 17–38
  Charts for B.M., S.F. and deflection, 31–8
SIMPSON'S RULE
  Application finding areas and centres of gravity, 21–2
  Definition, 21
SINGLE-STOREY, SHEDS, 429–40
  Bracing details, 433
  Charts, 437–40
  Flat-roofed, 435
  Framing, 429–40
  Gable frames, 431
  Multi-span ridged, 433–5
  Portal frame construction, 432
  Single-span, 429–32
    knee-braced trusses, 430
    stanchions with fixed bases, 429–30
  Stress coefficients
    lattice girders, 435–40
    roof trusses, 429–30
    shallow pitch, 431
SLAB BASES, 617–23
SLENDERNESS RATIO
  Angle struts, 487
  Columns in plastic design, 578
  Definition, 812
SLOPE
  (see also Deflection of Rigid Frames)
  Rigid frames, 289–97
SLOPE-DEFLECTION METHOD OF ANALYSIS, 276–88
  Continuous beams, 277–83

INDEX 1087

SLOPE-DEFLECTION METHOD OF ANALYSIS (cont.)
  Deflection, of framed structures, 135–42
  Fixed-end moments (F.E.Ms), 277
  Frames
    asymmetrical portal, 284–8
    symmetrical portal, 283–4
  Sign convention, 276
  Standard formulae, 277
SNOW LOAD, 203
SPACE FRAMES, 441–85
  Bibliography, 483
  Braced vault construction, 472–7
  Castellated beams, 447–9
  Domes, 477–9
  Grids, 442–50
  Lattice girders, 447–9, 457, 462
  Materials of construction, 480–3
  Resoluble into a series of plane frames, 450–71
  Suspension and tension supported, 480–1
  Transmission towers, 451–2
SPENCER, 642
SPRAYING, METAL, 804–5
STANCHIONS (COLUMNS)
  Allowable stresses, 816–8
  Bases, 524–5
  Behaviour of short and long, 811–2
  Bending and axial compression, 812–3
  Design examples, 813–46
  Design of multi-storey, 809–46
  Effective length factors, 812
  Engineering workshops, 499
  Euler load, 811
  Fire protection and structural casing, 813
  Lacings, 518, 521
  Multi-storey, design, 809–46
  Perry-Robertson formula, 811
  Pocket bases, 636–8
  'Rendhex' foundation columns, 642, 1010
  Slab bases, 617–23
  Slenderness ratio (L/r), 578, 812
  Splice, 732
  Stresses, 816–8
  Universal columns, 1024–9
STEEL PILING
  (see Piles)
STEEL STRUCTURES RESEARCH COMMITTEE, x–xii
STEEL WINDOWS
  (see Windows)
STRESSES
  (see also Bending and Axial Stresses; Joints in Rigid Frame Structures; Stanchions; Welding Practice)
  Bending and axial, 5–7
  General expression for, 7–8
  Pure tension or compression, 5
  Rivets and bolts, 702

STRUCTURAL ANALYSIS, METHODS OF, 195–288
  (see also Moment Distribution; Slope Deflection)
STRUCTURAL CASING, 813
  Stanchion design examples, 813–46
STRUTS
  (see also Stanchions)
  Design of angle, 487–96
SUM CURVE, EXAMPLES, 26–8
SURFACE PREPARATION OF STRUCTURAL STEELWORK, 799–807
  Blast cleaning, 800–4
  Preparatory treatments, 799–803
  Protection of steel sheet, 887–90
  Roughness, 803–4
  Quality, 803–4
SUSPENDED SPANS, 61–6
SUSPENSION STRUCTURES, 480–2
SYMMETRICAL MULTI-STOREY FRAMES, 216–22
  Portal frames, 283–4

T

TEES
  Properties of bars, 1054–7
    cut from Universal beams, 1058–61
    cut from Universal columns, 1062–3
TENSION SUPPORTED STRUCTURES, 480–1
THEOREM OF THREE MOMENTS
  Continuous beams, 51–6
  Propped cantilevers, 67–9
THREE-HINGED ARCH, 132–4
  Lattice or braced, 134
  Methods of solution, 132–3
THYSSEN HOCHHAUS, Düsseldorf, 856
TORSION, BEAMS IN, 107–18
  Bibliography, 116–7
  Bredt's formulae, 111–2
  Comparison of sections, 115
  Non-uniform, 116
  Stokes's Law, 111–2
  St-Vénant's analysis, 107–8, 114
  Timoshenko, 113
  Uniform, 108–116
  Warping, 114–5
TORSIONAL PROPERTIES
  Joists, 1032–3
  Universal beams, 1020–3
  Universal columns, 1028–9
TRANSMISSION TOWER
  As space frame, 451–2
TRUSSES
  Node connections, 762–3
  Roof construction, 913
  Welded roof, 761–4

INDEX

U

UNIVERSAL BEAMS, COLUMNS, BEARING PILES, 1012–29, 1064–5
   T-bars cut from, 1058–63
UNSYMMETRICAL BENDING, 3
UNSYMMETRICAL SECTIONS
   Deflection, 15–16
UNSYMMETRICAL VERTICAL LOADS ON RIGID FRAMES, 270

V

VAULTS
   Braced construction, 472–7
VIERENDEEL GIRDERS, 413–28
   Computer analysis, 419–20
   Moment distribution, 267
   Statically determinated analysis, 416
VIRTUAL WORK, 557–60
VITREOUS ENAMEL COATING, 888

W

WALKWAYS, 909
WALLS, 911–20
   Bricks for single storey buildings, 911
   Cavity walls, 915
   CP 111: 1970, 911
   Curtain walling, 915
   Design, 911
   Insulation, 916
   Jointing, 915, 919
   Model Byelaws (1953), 911
   Pier or buttress, definition, 914
   Precast concrete panels, 916
WARPING, 114–5
WARREN GIRDER, 139–42
   Influence lines examples, 165
WEB PLATES
   (see Girders)
WELDING, 739–64
   Base for stanchion, 760–1
   Brackets, 754–7
   Butt welds, 740, 741–2, 883, 884
   Compound girder calculations, 645–9
   Connections, 749–54
   Electrodes, 740
   Fillet welds, 740, 741–2, 883, 884
   Nomenclature, 883
   Plate girder calculations, 645–60
   Practice, 883–6
   Roof trusses, 761–4
   Seating cleats, 757–8
   Stresses, 769–70
   Symbols, 886
   Types of weld, 884–5

WELDING (*cont.*)
   Weld groups
      in plane of force, 745–8
      not in plane of force, 743–4
WILLIOT AND WILLIOT-MOHR DIAGRAMS, 144–54, 863
WILSON, Professor, 195
WIND MOMENT ON CONNECTIONS, 723–31
WIND ON MULTI-STOREY BUILDINGS, 847–67
   Albany Flats, Bournemouth, 859–61
   Bibliography, 865–7
   Braced frames, 854–8
   Brickwork, 863
   Cantilever method of design, 850
   Casing, solid, stiffening effect, 863
   Continuous portal method of design, 850
   Cooperative Insurance Society building, Manchester, 861
   CP3, 847
   Curtain walling, 862
   Deflection, 863
      lateral, 863
   Design wind speed, factors, 847
   I.B.M. building, Pittsburgh, 861
   Methods of design, 852–63
   Mixed or composite construction, 859–61
   Moment connection, 723–31
   Portal method of design, 850
   Precast concrete floors, 855
   Rigid frame design, 853–4
   Rigid, fully, design, 854
   Seagram building, New York, 855
   Shell Tower, London, 857
   Side sway, 862–4
   Thyssen Hochhaus, Düsseldorf, 856
   Total horizontal force, 847
WIND ON SINGLE STOREY SHEDS
   Engineering workshop, 498–501
   Knee-braced frames, 125–32
   Multi-storey rigid frame connections, 726–31
   Portal frames, 205–10
WINDOWS, STEEL, 903–9
   Condensation, 906
   Fixing to steelwork, 904
   Glass, varieties of, 909
   Lead flashings, 903, 909
   Mid-roof and north-light glazing, 905–6
   Minimum slopes, 903–5
   Patent glazing, 903–8
      double, 906
      vertical, 906–7
   Rainwater, 903
   Ventilation, 908
   Walkways, 909
   Weight per m^2, 908

WINKLER-RÉSAL FORMULA, 775–7
WORK
    Principal of, 135
    Virtual, 557–60
WORKSHOPS
    (see Engineering Workshop Design)

Z

Z-BEAMS, CASTELLATED, 1072–3
Z-POLYGONS, 3–5
ZINC COATING AND SPRAYING, 805–6